# Engineering Mathematics

**K.A. Stroud**

*Formerly Principal Lecturer*
*Department of Mathematics*
*Coventry University*

with

**Dexter J. Booth**

*Formerly Principal Lecturer*
*School of Computing and Engineering*
*University of Huddersfield*

**SEVENTH EDITION**



Industrial Press, Inc.
New York

Library of Congress Cataloging-in-Publication Data

Stroud, K. A.
Engineering mathematics / K.A. Stroud ; with additions by Dexter J. Booth—7th ed.
p. cm.
Includes index.
ISBN 978-0-8311-3470-9
1. Engineering mathematics—Programmed instruction. I. Booth, Dexter J. II. Title.

pending from Library of Congress

Published in North America under license from Palgrave Publishers Ltd, Houndmills, Basingstoke, Hants RG21 6XS, United Kingdom



**Industrial Press, Inc.**
989 Avenue of the Americas
New York, NY 10018



10 9 8 7 6 5 4 3 2

# Summary of contents

## Part 1 Foundation topics



## Part II

# Contents

## Programme 28 Data handling and statistics 1045



## Programme 29 1076

# Preface to the first edition

The purpose of this book is to provide a complete year's course in mathematics for those studying in the engineering, technical and scientific fields. The material has been specially written for courses leading to

(i) Part I of B.Sc. Engineering Degrees,
(ii) Higher National Diploma and Higher National Certificate in technological subjects, and for other courses of a comparable level. While formal proofs are included where necessary to promote understanding, the emphasis throughout is on providing the student with sound mathematical skills and with a working knowledge and appreciation of the basic concepts involved. The programmed structure ensures that the book is highly suited for general class use and for individual self-study, and also provides a ready means for remedial work or subsequent revision.

The book is the outcome of some eight years' work undertaken in the development of programmed learning techniques in the Department of Mathematics at the Lanchester College of Technology, Coventry. For the past four years, the whole of the mathematics of the first year of various Engineering Degree courses has been presented in programmed form, in conjunction with seminar and tutorial periods. The results obtained have proved to be highly satisfactory, and further extension and development of these learning techniques are being pursued.

Each programme has been extensively validated before being produced in its final form and has consistently reached a success level above 80/80, i.e. at least 80% of the students have obtained at least 80% of the possible marks in carefully structured criteria tests. In a research programme, carried out against control groups receiving the normal lectures, students working from programmes have attained significantly higher mean scores than those in the control groups and the spread of marks has been considerably reduced. The general pattern has also been reflected in the results of the sessional examinations.

The advantages of working at one's own rate, the intensity of the student involvement and the immediate assessment of responses, are well known to those already acquainted with programmed learning activities. Programmed learning in the first year of a student's course at a college or university provides the additional advantage of bridging the gap between the rather highly organised aspect of school life and the freer environment and greater personal responsibility for his own progress which faces every student on entry to the realms of higher education.

Acknowledgement and thanks are due to all those who have assisted in any way in the development of the work, including whose who have been actively engaged in validation processes. I especially wish to record my sincere thanks for the continued encouragement and support which I received from my present head of Department at the College, Mr. J.E. Sellars, M.Sc., A.F.R.Ae.S., F.I.M.A., and also from Mr. R. Wooldridge, M.C., B.Sc., F.I.M.A., formerly Head of Department, now Principal of Derby College of Technology. Acknowledgement is also made of the many sources, too numerous to list, from which the selected examples quoted in the programmes have been gleaned over the years. Their inclusion contributes in no small way to the success of the work.

K.A. Stroud

# Preface to the second edition

The continued success of *Engineering Mathematics* since its first publication has been reflected in the number of courses for which it has been adopted as the official class text and also in the correspondence from numerous individuals who have welcomed the self-instructional aspects of the work.

Over the years, however, syllabuses of existing courses have undergone some modification and new courses have been established. As a result, suggestions have been received from time to time requesting the inclusion of further programme topics in addition to those already provided as core material for relevant undergraduate and comparable courses. Unlimited expansion of the book to accommodate all the topics requested is hardly feasible on account of the physical size of the book and the commercial aspects of production. However, in the light of these representations and as a result of further research undertaken by the author and the publishers, it has now been found possible to provide a new edition of *Engineering Mathematics* incorporating three of the topics for which there is clearly a wide demand.

The additional programmes cover the following topics:

(a) Matrices: definitions; types of matrices; operations; transpose; inverse; solution of linear equations; eigenvalues and eigenvectors.
(b) Curves and curve fitting: standard curves; asymptotes; systematic curve sketching; curve recognition; curve fitting; method of least squares.
(c) Statistics: discrete and continuous data; grouped data; frequency and relative frequency; histograms; central tendency – mean, mode and median; coding; frequency polygons and frequency curves; dispersion – range, variance and standard deviation; normal distribution and standardised normal curve.

The three new programmes follow the structure of the previous material and each is provided with numerous worked examples and exercises. As before, each programme concludes with a short Test Exercise for self-assessment and set of Further Problems provides valuable extra practice. A complete set of answers is available at the end of the book.

Advantage has also been taken during the revision of the book to amend a small number of minor points in other parts of the text and it is anticipated that, in its new updated form, the book will have an even greater appeal and continue to provide a worthwhile service.

K.A. Stroud

# Preface to the third edition

Following the publication of the enlarged second edition of *Engineering Mathematics*, which included a programme on the introduction to Statistics, requests were again received for an associated programme on Probability. This has now been incorporated as Programme XXVIII of the current third edition of the book.

The additional programme follows the established pattern and structure of the previous sections of the text, including the customary worked examples through which the student is guided with progressive responsibility and concluding with the Text Exercise and set of Further Problems for essential practice. Answers to all problems are provided. The opportunity has also been taken to make one or two minor modifications to the remainder of the text.

*Engineering Mathematics*, originally devised as a first year mathematics course for engineering and science degree undergraduates and students of comparable courses, is widely sought both for general class use and for individual study. A companion volume and sequel, *Further Engineering Mathematics*, dealing with core material of a typical second/third year course, is also now available through the normal channels. The two texts together provide a comprehensive and integrated course of study and have been well received as such.

My thanks are due, once again, to the publishers for their ready cooperation and helpful advice in the preparation of the material for publication.

K.A.S.

# Preface to the fourth edition

Since the publication of the third edition of *Engineering Mathematics*, considerable changes in the syllabus and options for A-level qualifications in Mathematics have been introduced nationally, as a result of which numbers of students with various levels of mathematical background have been enrolling for undergraduate courses in engineering and science. In view of the widespread nature of the situation, requests have been received from several universities for the inclusion in the new edition of *Engineering Mathematics* of what amounts to a bridging course of material in relevant topics to ensure a solid foundation on which the main undergraduate course can be established.

Accordingly, the fourth edition now includes ten new programmes ranging from Number Systems and algebraic processes to an introduction to the Calculus. These Foundation Topics constitute Part I of the current book and precede the well-established section of the text now labelled as Part II.

For students already well versed in the contents of the Part I programmes the Test Exercises and Further Problems afford valuable revision and should not be ignored.

With the issue of the new edition, the publishers have undertaken the task of changing the format of the pages and of resetting the whole of the text to provide a more open presentation and improved learning potential for the student.

Once again, I am indebted to the publishers for their continued support and close cooperation in the preparation of the text in its new form and particularly to all those directly involved in the editorial, production and marketing processes both at home and overseas.

K.A.S.

# Preface to the seventh edition

*Engineering Mathematics* by Ken Stroud has been a favoured textbook of science and engineering students for almost 43 years and to have been asked to contribute to the fifth edition of this remarkable book gave me great pleasure and no little trepidation. And now, twelve years later, we have the seventh edition.

A clear requirement of any additions or changes to such a well established textbook is the retention of the very essence of the book that has contributed to so many students' mathematical abilities over the years. In line with this I have taken great care to preserve the time-tested Stroud format and close attention to technique development throughout the book which have made *Engineering Mathematics* the tremendous success it is. The largest part of my work for previous editions was to re-structure, re-organize and expand the Foundation section as well as the addition of a Programme on Laplace Transforms. In this edition the major change is a new Programme dealing with sequences and functions of a discrete variable. Discrete mathematics is playing an ever-increasing role in scientific endeavour and this Programme is included as an introduction to the important topic of *Linear Difference Equations*. One immediate consequence of this introduction was the need to split the old, over-large Functions programme into two parts. The first part deals with algebraic functions in general up to composition and the second part deals with the trigonometric, exponential and logarithmic functions. In addition various house-keeping tasks have been completed such as updating the Graphs Programme to cater for the latest edition of the Excel™ spreadsheet.

The Personal Tutor, which can be accessed at www.palgrave.com/stroud, provides the answers and working for a large number of questions selected from the book for which, in the text, no worked solutions are given. Using the Personal Tutor the reader is closely guided through the solutions to these questions, so confirming and adding to skills in mathematical techniques and knowledge of mathematical ideas. On the same site there is a collection of problems set within engineering and scientific contexts. Lecturers will also find Powerpoint™ lecture slides which include all the figures and graphs from the book.

The work involved in creating a new edition of an established textbook is always a cooperative, team effort and this book is no exception. I was fortunate enough to be able to meet Ken Stroud in the months before he died and to be able to discuss with him ideas for the fifth edition. He was very enthusiastic about taking the book forward with new technology and both his eagerness and his concerns were taken into account in the development of the Personal Tutor. The enormous task which Ken undertook in writing the original book and three subsequent editions cannot be underestimated. Ken's achievement was an extraordinary one and it has been a great privilege for us all to be able to work on such a book. I should like to thank the Stroud family for their support in my work for this new edition and the editorial team for their close attention to detail, their appropriate comments on the text and the assiduous checking of everything that I have written. As with any team the role of the leader is paramount and I should particularly like to thank my Editor Helen Bugler whose continued care over this project inspires us all.

*Huddersfield*
*January 2013*

Dexter J. Booth

# How to use this book

This book contains forty-two lessons called *Programmes*. Each Programme has been written in such a way as to make learning more effective and more interesting. It is like having a personal tutor because you proceed at your own rate of learning and any difficulties you may have are cleared before you have the chance to practise incorrect ideas or techniques.

You will find that each Programme is divided into numbered sections called *frames*. When you start a Programme, begin at Frame 1. Read each frame carefully and carry out any instructions or exercise that you are asked to do. In almost every frame, you are required to make a response of some kind, testing your understanding of the information in the frame, and you can immediately compare your answer with the correct answer given in the next frame. *To obtain the greatest benefit, you are strongly advised to cover up the following frame until you have made your response.* When a series of dots occurs, you are expected to supply the missing word, phrase, number or mathematical expression. At every stage you will be guided along the right path. There is no need to hurry: read the frames carefully and follow the directions exactly. In this way, you must learn.

Each Programme opens with a list of **Learning outcomes** which specify exactly what you will learn by studying the contents of the Programme. The Programme ends with a matching checklist of **Can You?** questions that enables you to rate your success in having achieved the **Learning outcomes**. If you feel sufficiently confident then tackle the short **Test exercise** which follows. This is set directly on what you have learned in the Programme: the questions are straightforward and contain no tricks. To provide you with the necessary practice, a set of **Further problems** is also included: do as many of these problems as you can. Remember, that in mathematics, as in many other situations, practice makes perfect - or more nearly so.

Of the forty-two Programmes, the first thirteen are at Foundation level. Some of these will undoubtedly contain material with which you are already familiar. However, read the Programme's **Learning outcomes** and if you feel confident about them try the **Quiz** that immediately follows - you will soon find out if you need a refresher course. Indeed, even if you feel you have done some of the topics before, it would still be worthwhile to work steadily through the Programme: it will serve as useful revision and fill any gaps in your knowledge that you may have.

When you have come to the end of a Foundation level Programme and have rated your success in achieving the **Learning outcomes** using the **Can You?** checklist, go back to the beginning of the Programme and try the **Quiz** before you complete the Programme with the **Test exercise** and the **Further problems**. This way you will get even more practice.

## The Personal Tutor

Alongside this book is an interactive Personal Tutor program that contains a bank of questions for you to answer using your computer. In response to comments from our many readers the Personal Tutor has been re-designed and updated from the previous edition. It is extremely intuitive and easy to navigate and offers **Hints** whenever you need them. If you are having difficulties you can click on either **Check your answers** for a particular **Step** or **Show full working**. There are no scores for the questions and you are guided every inch of the way without having to worry about getting any answers wrong. Using the Personal Tutor will give you more practice and increase your confidence in your learning of the mathematics. As with the exercises in the book, take your time, make mistakes, correct them using all the assistance available to you and you will surely learn.

The bank consists of odd numbered questions from the **Quizzes**, **Test exercises** and **Further problems**. A PERSONAL TUTOR symbol next to an exercise in the book indicates that it is also on the program.

The Personal Tutor is available online through the companion website and also to download for offline use.

## The companion website – www.palgrave.com/stroud

You are recommended to visit the book's website at www.palgrave.com/stroud. There you will find a growing resource to accompany the text including mathematical questions set within engineering and scientific contexts. From here you can also access the Personal Tutor online or download it to your computer so you can work offline. There is an email address for you to communicate your comments on the book, critical or otherwise. This book is for you and so the more that we know about your wishes and desires the more likely they are to be accommodated in future editions.

# Useful background information

## Symbols used in the text

| Symbol | Meaning |
|---|---|
| $=$ | is equal to |
| $\approx$ | is approximately equal to |
| $>$ | is greater than |
| $\geq$ | is greater than or equal to |
| $n!$ | factorial $n = 1 \times 2 \times 3 \times \ldots \times n$ |
| $\lvert k \rvert$ | modulus of $k$, i.e. size of $k$ irrespective of sign |
| $\sum$ | summation |
| $\rightarrow$ | tends to |
| $\neq$ | is not equal to |
| $\equiv$ | is identical to |
| $<$ | is less than |
| $\leq$ | is less than or equal to |
| $\infty$ | infinity |
| $\underset{n \to \infty}{Lim}$ | limiting value as $n \rightarrow \infty$ |

## Useful mathematical information

### 1 Algebraic identities

$$(a+b)^2 = a^2 + 2ab + b^2 \qquad (a+b)^3 = a^3 + 3a^2b + 3ab^2 + b^3$$

$$(a-b)^2 = a^2 - 2ab + b^2 \qquad (a-b)^3 = a^3 - 3a^2b + 3ab^2 - b^3$$

$$(a+b)^4 = a^4 + 4a^3b + 6a^2b^2 + 4ab^3 + b^4$$

$$(a-b)^4 = a^4 - 4a^3b + 6a^2b^2 - 4ab^3 + b^4$$

$$a^2 - b^2 = (a-b)(a+b) \qquad a^3 - b^3 = (a-b)(a^2 + ab + b^2)$$

$$a^3 + b^3 = (a+b)(a^2 - ab + b^2)$$

### 2 Trigonometrical identities

(a) $\sin^2\theta + \cos^2\theta = 1;\ \sec^2\theta = 1 + \tan^2\theta;\ {}^2\theta = 1 + \cot^2\theta$

(b) $\sin(A+B) = \sin A \cos B + \cos A \sin B$

$$\sin(A-B) = \sin A \cos B - \cos A \sin B$$

$$\cos(A+B) = \cos A \cos B - \sin A \sin B$$

$$\cos(A-B) = \cos A \cos B + \sin A \sin B$$

$$\tan(A+B) = \frac{\tan A + \tan B}{1 - \tan A \tan B}$$

$$\tan(A-B) = \frac{\tan A - \tan B}{1 + \tan A \tan B}$$

(c) Let $A = B = \theta \quad \therefore \quad \sin 2\theta = 2\sin\theta\cos\theta$

$$\cos 2\theta = \cos^2\theta - \sin^2\theta = 1 - 2\sin^2\theta = 2\cos^2\theta - 1$$

$$\tan 2\theta = \frac{2\tan\theta}{1 - \tan^2\theta}$$

(d) Let $\theta = \frac{\phi}{2}$ $\therefore$ $\sin\phi = 2\sin\frac{\phi}{2}\cos\frac{\phi}{2}$

$$\cos\phi = \cos^2\frac{\phi}{2} - \sin^2\frac{\phi}{2} = 1 - 2\sin^2\frac{\phi}{2} = 2\cos^2\frac{\phi}{2} - 1$$

$$\tan\phi = \frac{2\tan\frac{\phi}{2}}{1 - 2\tan^2\frac{\phi}{2}}$$

(e) $\sin C + \sin D = 2\sin\frac{C+D}{2}\cos\frac{C-D}{2}$

$\sin C - \sin D = 2\cos\frac{C+D}{2}\sin\frac{C-D}{2}$

$\cos C + \cos D = 2\cos\frac{C+D}{2}\cos\frac{C-D}{2}$

$\cos D - \cos C = 2\sin\frac{C+D}{2}\sin\frac{C-D}{2}$

(f) $2\sin A\cos B = \sin(A+B) + \sin(A-B)$

$2\cos A\sin B = \sin(A+B) - \sin(A-B)$

$2\cos A\cos B = \cos(A+B) + \cos(A-B)$

$2\sin A\sin B = \cos(A-B) - \cos(A+B)$

(g) Negative angles: $\sin(-\theta) = -\sin\theta$

$\cos(-\theta) = \cos\theta$

$\tan(-\theta) = -\tan\theta$

(h) Angles having the same trigonometrical ratios:

(i) Same sine: $\theta$ and $(180^\circ - \theta)$

(ii) Same cosine: $\theta$ and $(360^\circ - \theta)$, i.e. $(-\theta)$

(iii) Same tangent: $\theta$ and $(180^\circ + \theta)$

(i) $a\sin\theta + b\cos\theta = A\sin(\theta + \alpha)$

$a\sin\theta - b\cos\theta = A\sin(\theta - \alpha)$

$a\cos\theta + b\sin\theta = A\cos(\theta - \alpha)$

$a\cos\theta - b\sin\theta = A\cos(\theta + \alpha)$

where $\begin{cases} A = \sqrt{a^2 + b^2} \\ \alpha = \tan^{-1}\frac{b}{a} \quad (0^\circ < \alpha < 90^\circ) \end{cases}$

## 3 Standard curves

(a) *Straight line*

Slope, $m = \frac{dy}{dx} = \frac{y_2 - y_1}{x_2 - x_1}$

Angle between two lines, $\tan\theta = \frac{m_2 - m_1}{1 + m_1 m_2}$

For parallel lines, $m_2 = m_1$

For perpendicular lines, $m_1 m_2 = -1$

Equation of a straight line (slope $= m$)

(i) Intercept $c$ on real $y$-axis: $y = mx + c$

(ii) Passing through $(x_1, y_1)$: $y - y_1 = m(x - x_1)$

(iii) Joining $(x_1, y_1)$ and $(x_2, y_2)$: $\frac{y - y_1}{y_2 - y_1} = \frac{x - x_1}{x_2 - x_1}$

(b) *Circle*

Centre at origin, radius $r$: $x^2 + y^2 = r^2$

Centre $(h, k)$, radius $r$: $(x - h)^2 + (y - k)^2 = r^2$

General equation: $x^2 + y^2 + 2gx + 2fy + c = 0$

with centre $(-g, -f)$: radius $= \sqrt{g^2 + f^2 - c}$

Parametric equations: $x = r\cos\theta,\ y = r\sin\theta$

(c) *Parabola*

Vertex at origin, focus $(a, 0)$: $y^2 = 4ax$

Parametric equations : $x = at^2,\ y = 2at$

(d) *Ellipse*

Centre at origin, foci $\left(\pm\sqrt{a^2 + b^2}, 0\right)$: $\dfrac{x^2}{a^2} + \dfrac{y^2}{b^2} = 1$

where $a$ = semi-major axis, $b$ = semi-minor axis

Parametric equations: $x = a\cos\theta,\ y = b\sin\theta$

(e) *Hyperbola*

Centre at origin, foci $\left(\pm\sqrt{a^2 + b^2}, 0\right)$: $\dfrac{x^2}{a^2} - \dfrac{y^2}{b^2} = 1$

Parametric equations: $x = a\sec\theta,\ y = b\tan\theta$

Rectangular hyperbola:

Centre at origin, vertex $\pm\left(\dfrac{a}{\sqrt{2}}, \dfrac{a}{\sqrt{2}}\right)$: $xy = \dfrac{a^2}{2} = c^2$

i.e. $xy = c^2$ where $c = \dfrac{a}{\sqrt{2}}$

Parametric equations: $x = ct,\ y = c/t$

## 4 Laws of mathematics

(a) *Associative laws* – for addition and multiplication

$a + (b + c) = (a + b) + c$

$a(bc) = (ab)c$

(b) *Commutative laws* – for addition and multiplication

$a + b = b + a$

$ab = ba$

(c) *Distributive laws* – for multiplication and division

$a(b + c) = ab + ac$

$\dfrac{b + c}{a} = \dfrac{b}{a} + \dfrac{c}{a}$ (provided $a \neq 0$)

# PART I
# Foundation topics

# Arithmetic

## Learning outcomes

When you have completed this Programme you will be able to:

- ☐ Carry out the basic rules of arithmetic with integers
- ☐ Check the result of a calculation making use of rounding
- ☐ Write a whole as a product of prime numbers
- ☐ Find the highest common factor and lowest common multiple of two whole numbers
- ☐ Manipulate fractions, ratios and percentages
- ☐ Manipulate decimal numbers
- ☐ Manipulate powers
- ☐ Use standard or preferred standard form and complete a calculation to the required level of accuracy
- ☐ Understand the construction of various number systems and convert from one number system to another.

If you already feel confident about these why not try the quiz over the page? You can check your answers at the end of the book.

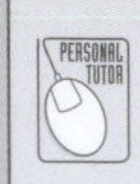

Questions marked with this icon can be found at www.palgrave.com/stroud. There you can go through the question step by step and follow online hints, with full working provided.

# Quiz F.1

**Frames**

**1** Place the appropriate symbol < or > between each of the following pairs of numbers:
(a) $-3 \quad -2$ (b) $8 \quad -13$ (c) $-25 \quad 0$ — Frames 1 to 4

**2** Find the value of each of the following:
(a) $13 + 9 \div 3 - 2 \times 5$ (b) $(13 + 9) \div (3 - 2) \times 5$ — Frames 5 to 12

**3** Round each number to the nearest 10, 100 and 1000:
(a) 1354 (b) 2501 (c) $-2452$ (d) $-23\,625$ — Frames 13 to 15

**4** Write each of the following as a product of prime factors:
(a) 170 (b) 455 (c) 9075 (d) 1140 — Frames 19 to 22

**5** Find the HCF and the LCM of each pair of numbers:
(a) 84, 88 (b) 105, 66 — Frames 23 to 24

**6** Reduce each of the following fractions to their lowest terms:
(a) $\dfrac{12}{18}$ (b) $\dfrac{144}{21}$ (c) $-\dfrac{49}{14}$ (d) $\dfrac{64}{4}$ — Frames 28 to 36

**7** Evaluate the following:
(a) $\dfrac{3}{7} \times \dfrac{2}{3}$ (b) $\dfrac{11}{30} \div \dfrac{5}{6}$ (c) $\dfrac{3}{7} + \dfrac{4}{13}$ (d) $\dfrac{5}{16} - \dfrac{4}{3}$ — Frames 37 to 46

**8** Write the following proportions as ratios:
(a) $\dfrac{1}{2}$ of A, $\dfrac{1}{5}$ of B and $\dfrac{3}{10}$ of C
(b) $\dfrac{1}{4}$ of P, $\dfrac{1}{3}$ of Q, $\dfrac{1}{5}$ of R and the remainder S — Frames 47 to 48

**9** Complete the following:
(a) $\dfrac{4}{5} = \quad \%$ (b) $48\%$ of $50 =$
(c) $\dfrac{9}{14} = \quad \%$ (d) $15\%$ of $25 =$ — Frames 49 to 52

**10** Round each of the following decimal numbers, first to 3 significant figures and then to 2 decimal places:
(a) 21·355 (b) 0·02456
(c) 0·3105 (d) 5134·555 — Frames 56 to 65

**11** Convert each of the following to decimal form to 3 decimal places:
(a) $\dfrac{4}{15}$ (b) $-\dfrac{7}{13}$ (c) $\dfrac{9}{5}$ (d) $-\dfrac{28}{13}$ — Frames 66 to 67

**12** Convert each of the following to fractional form in lowest terms:
(a) 0·8 (b) 2·8 (c) $3{\cdot}\dot{3}\dot{2}$ (d) $-5{\cdot}5$ — Frames 68 to 73

**13** Write each of the following in abbreviated form:
(a) 1·010101... (b) 9·2456456456... — Frames 70 to 71

**14** Write each of the following as a number raised to a power:
(a) $3^6 \times 3^3$ (b) $4^3 \div 2^5$ (c) $(9^2)^3$ (d) $(7^0)^{-8}$ — Frames 78 to 89

**Frames**

**15** Find the value of each of the following to 3 dp:
(a) $15^{\frac{1}{3}}$ (b) $\sqrt[5]{5}$ (c) $(-27)^{\frac{1}{3}}$ (d) $(-9)^{\frac{1}{2}}$ — Frames 90 to 94

**16** Write each of the following as a single decimal number:
(a) $3{\cdot}2044 \times 10^3$ (b) $16{\cdot}1105 \div 10^{-2}$ — Frames 95 to 98

**17** Write each of the following in standard form:
(a) 134·65 (b) 0·002401 — Frames 99 to 101

**18** Write each of the following in preferred standard form:
(a) $16{\cdot}1105 \div 10^{-2}$ (b) 9·3304 — Frames 102 to 104

**19** In each of the following the numbers have been obtained by measurement. Evaluate each calculation to the appropriate level of accuracy:
(a) $11{\cdot}4 \times 0{\cdot}0013 \div 5{\cdot}44 \times 8{\cdot}810$
(b) $\dfrac{1{\cdot}01 \div 0{\cdot}00335}{9{\cdot}12 \times 6{\cdot}342}$ — Frames 105 to 108

**20** Express the following numbers in denary form:
(a) $1011{\cdot}01_2$ (b) $456{\cdot}721_8$
(c) $123{\cdot}\Lambda29_{12}$ (d) $CA1{\cdot}B22_{16}$ — Frames 112 to 126

**21** Convert $15{\cdot}605_{10}$ to the equivalent octal, binary, duodecimal and hexadecimal forms. — Frames 127 to 149

# Types of number

1

## The natural numbers

The first numbers we ever meet are the *whole numbers*. These, together with *zero*, are called the *natural numbers*, and are written down using *numerals*.

## Numerals and place value

The *natural numbers* are written using the ten numerals 0, 1, . . ., 9 where the position of a numeral dictates the value that it represents. For example:

246 stands for 2 hundreds and 4 tens and 6 units. That is $200 + 40 + 6$

Here the numerals 2, 4 and 6 are called the hundreds, tens and unit *coefficients* respectively. This is the place value principle.

## Points on a line and order

The natural numbers can be represented by equally spaced points on a straight line where the first natural number is zero 0.

The natural numbers are ordered – they progress from small to large. As we move along the line from left to right the numbers increase as indicated by the arrow at the end of the line. On the line, numbers to the left of a given number are *less than* $(<)$ the given number and numbers to the right are *greater than* $(>)$ the given number. For example, $8 > 5$ because 8 is represented by a point on the line to the right of 5. Similarly, $3 < 6$ because 3 is to the left of 6.

*Now move on to the next frame*

2

## The integers

If the straight line displaying the natural numbers is extended to the left we can plot equally spaced points to the left of zero.

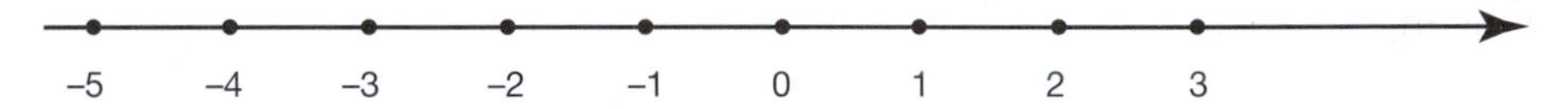


These points represent *negative* numbers which are written as the natural number preceded by a minus sign, for example $-4$. These positive and negative whole numbers and zero are collectively called the *integers*. The notion of order still applies. For example, $-5 < 3$ and $-2 > -4$ because the point on the line representing $-5$ is to the *left* of the point representing 3. Similarly, $-2$ is to the *right* of $-4$.

The numbers $-10$, 4, 0, $-13$ are of a type called . . . . . . . . . . .

*You can check your answer in the next frame*

3

Integers

They are integers. The natural numbers are all positive. Now try this:

Place the appropriate symbol $<$ or $>$ between each of the following pairs of numbers:

(a) $-3 \quad -6$
(b) $\ \ 2 \quad -4$
(c) $-7 \quad 12$

*Complete these and check your results in the next frame*

4

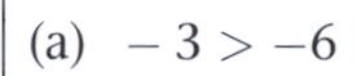

(a) $-3 > -6$
(b) $\ \ 2 > -4$
(c) $-7 < 12$

The reasons being:

(a) $-3 > -6$ because $-3$ is represented on the line to the *right* of $-6$
(b) $\ \ 2 > -4$ because 2 is represented on the line to the *right* of $-4$
(c) $-7 < 12$ because $-7$ is represented on the line to the *left* of 12

*Now move on to the next frame*

5

## Brackets

Brackets should be used around negative numbers to separate the minus sign attached to the number from the arithmetic operation symbol. For example, $5 - -3$ should be written $5 - (-3)$ and $7 \times -2$ should be written $7 \times (-2)$. *Never write two arithmetic operation symbols together without using brackets.*

## Addition and subtraction

Adding two numbers gives their *sum* and subtracting two numbers gives their *difference*. For example, $6 + 2 = 8$. Adding moves to the right of the first number and subtracting moves to the left of the first number, so that $6 - 2 = 4$ and $4 - 6 = -2$:

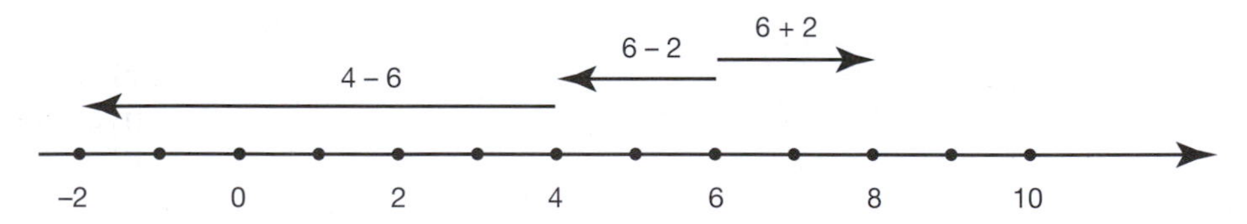


Adding a negative number is the same as subtracting its positive counterpart. For example $7 + (-2) = 7 - 2$. The result is 5. Subtracting a negative number is the same as adding its positive counterpart. For example $7 - (-2) = 7 + 2 = 9$.

So what is the value of:

(a) $8 + (-3)$
(b) $9 - (-6)$
(c) $(-4) + (-8)$
(d) $(-14) - (-7)$?

*When you have finished these check your results with the next frame*

**6**

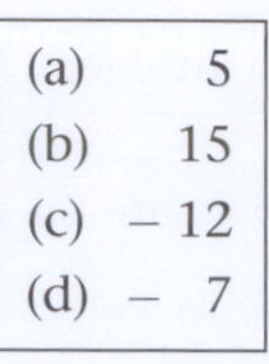

| | |
|---|---|
| (a) | 5 |
| (b) | 15 |
| (c) | – 12 |
| (d) | – 7 |

*Move now to Frame 7*

**7**

## Multiplication and division

Multiplying two numbers gives their *product* and dividing two numbers gives their *quotient*. Multiplying and dividing two positive or two negative numbers gives a positive number. For example:

$12 \times 2 = 24$ and $(-12) \times (-2) = 24$

$12 \div 2 = 6$ and $(-12) \div (-2) = 6$

Multiplying or dividing a positive number by a negative number gives a negative number. For example:

$12 \times (-2) = -24$, $(-12) \div 2 = -6$ and $8 \div (-4) = -2$

So what is the value of:

(a) $(-5) \times 3$
(b) $12 \div (-6)$
(c) $(-2) \times (-8)$
(d) $(-14) \div (-7)$?

*When you have finished these check your results with the next frame*

**8**

| | |
|---|---|
| (a) | – 15 |
| (b) | – 2 |
| (c) | 16 |
| (d) | 2 |

*Move on to Frame 9*

**9**

## Brackets and precedence rules

Brackets and the precedence rules are used to remove ambiguity in a calculation. For example, $14 - 3 \times 4$ could be either:

$11 \times 4 = 44$ or $14 - 12 = 2$

depending on which operation is performed first.

To remove the ambiguity we rely on the precedence rules:

In any calculation involving all four arithmetic operations we proceed as follows:

(a) Working from the left evaluate divisions and multiplications as they are encountered;

this leaves a calculation involving just addition and subtraction.

(b) Working from the left evaluate additions and subtractions as they are encountered.

▶

For example, to evaluate:

$4 + 5 \times 6 \div 2 - 12 \div 4 \times 2 - 1$

a first sweep from left to right produces:

$4 + 30 \div 2 - 3 \times 2 - 1$

a second sweep from left to right produces:

$4 + 15 - 6 - 1$

and a final sweep produces:

$19 - 7 = 12$

If the calculation contains brackets then these are evaluated first, so that:

$$\begin{aligned}(4 + 5 \times 6) \div 2 - 12 \div 4 \times 2 - 1 &= 34 \div 2 - 6 - 1 \\ &= 17 - 7 \\ &= 10\end{aligned}$$

This means that:

$$\begin{aligned}14 - 3 \times 4 &= 14 - 12 \\ &= 2\end{aligned}$$

because, reading from the left we multiply before we subtract. Brackets must be used to produce the alternative result:

$$\begin{aligned}(14 - 3) \times 4 &= 11 \times 4 \\ &= 44\end{aligned}$$

because the precedence rules state that brackets are evaluated first.

So that

$34 + 10 \div (2 - 3) \times 5 =$ ...........

*Result in the next frame*

---

**10**

$\boxed{-16}$

Because

$$\begin{aligned}34 + 10 \div (2 - 3) \times 5 &= 34 + 10 \div (-1) \times 5 && \text{we evaluate the bracket first} \\ &= 34 + (-10) \times 5 && \text{by dividing} \\ &= 34 + (-50) && \text{by multiplying} \\ &= 34 - 50 && \text{finally we subtract} \\ &= -16\end{aligned}$$

Notice that when brackets are used we can omit the multiplication signs and replace the division sign by a line, so that:

$5 \times (6 - 4)$ becomes $5(6 - 4)$

and

$(25 - 10) \div 5$ becomes $(25 - 10)/5$ or $\dfrac{25 - 10}{5}$

▶

When evaluating expressions containing *nested* brackets the innermost brackets are evaluated first. For example:

| | |
|---|---|
| $3(4-2[5-1]) = 3(4-2\times 4)$ | evaluating the innermost bracket [...] first |
| $= 3(4-8)$ | multiplication before subtraction inside the (...) bracket |
| $= 3(-4)$ | subtraction completes the evaluation of the (...) bracket |
| $= -12$ | multiplication completes the calculation |

so that

$5-\{8+7[4-1]-9/3\} = \ldots\ldots\ldots\ldots$

*Work this out, the result is in the following frame*

**11**

$\boxed{-21}$

Because

$$\begin{aligned}5-\{8+7[4-1]-9/3\} &= 5-\{8+7\times 3-9\div 3\}\\ &= 5-\{8+21-3\}\\ &= 5-\{29-3\}\\ &= 5-26\\ &= -21\end{aligned}$$

*Now move to Frame 12*

**12**

## Basic laws of arithmetic

All the work that you have done so far has been done under the assumption that you know the rules that govern the use of arithmetic operations as, indeed, you no doubt do. However, there is a difference between knowing the rules innately and being consciously aware of them, so here they are. The four basic arithmetic operations are:

addition and subtraction

multiplication and division

where each pair may be regarded as consisting of 'opposites' – in each pair one operation is the reverse operation of the other.

### 1 Commutativity

Two integers can be added or multiplied in either order without affecting the result. For example:

$5+8=8+5=13$ and $5\times 8=8\times 5=40$

***We say that addition and multiplication are commutative operations***

The order in which two integers are subtracted or divided *does* affect the result. For example:

$4-2\neq 2-4$ because $4-2=2$ and $2-4=-2$

Notice that $\neq$ means *is not equal to*. Also

$4\div 2\neq 2\div 4$

***We say that subtraction and division are not commutative operations***

▶

### 2 Associativity

The way in which three or more integers are associated under addition or multiplication does not affect the result. For example:

$$3 + (4 + 5) = (3 + 4) + 5 = 3 + 4 + 5 = 12$$

and

$$3 \times (4 \times 5) = (3 \times 4) \times 5 = 3 \times 4 \times 5 = 60$$

***We say that addition and multiplication are associative operations***

The way in which three or more integers are associated under subtraction or division does affect the result. For example:

$3 - (4 - 5) \neq (3 - 4) - 5$ because
$3 - (4 - 5) = 3 - (-1) = 3 + 1 = 4$ and $(3 - 4) - 5 = (-1) - 5 = -6$

Also

$24 \div (4 \div 2) \neq (24 \div 4) \div 2$ because
$24 \div (4 \div 2) = 24 \div 2 = 12$ and $(24 \div 4) \div 2 = 6 \div 2 = 3$

***We say that subtraction and division are not associative operations***

### 3 Distributivity

Consider the equations:

$$3 \times (4 + 5) = 3 \times 9 = 27$$

and

$$(3 \times 4) + (3 \times 5) = 12 + 15 = 27$$

From this we can deduce that:

$$3 \times (4 + 5) = (3 \times 4) + (3 \times 5)$$

We say that *multiplication distributes itself over addition from the left*. Multiplication is also distributive over addition from the right. For example:

$$(3 + 4) \times 5 = (3 \times 5) + (4 \times 5) = 35$$

The same can be said of multiplication and subtraction: *multiplication is distributive over subtraction from both the left and the right*. For example:

$3 \times (4 - 5) = (3 \times 4) - (3 \times 5) = -3$ and $(3 - 4) \times 5 = (3 \times 5) - (4 \times 5) = -5$

Division is distributed over addition and subtraction from the right but not from the left. For example:

$(60 + 15) \div 5 = (60 \div 5) + (15 \div 5)$ because

$(60 + 15) \div 5 = 75 \div 5 = 15$ and $(60 \div 5) + (15 \div 5) = 12 + 3 = 15$

However, $60 \div (15 + 5) \neq (60 \div 15) + (60 \div 5)$ because

$60 \div (15 + 5) = 60 \div 20 = 3$ and $(60 \div 15) + (60 \div 5) = 4 + 12 = 16$

Also:

$(20 - 10) \div 5 = (20 \div 5) - (10 \div 5)$ because

$(20 - 10) \div 5 = 10 \div 5 = 2$ and $(20 \div 5) - (10 \div 5) = 4 - 2 = 2$

but $20 \div (10 - 5) \neq (20 \div 10) - (20 \div 5)$ because

$20 \div (10 - 5) = 20 \div 5 = 4$ and $(20 \div 10) - (20 \div 5) = 2 - 4 = -2$

**13**

## Estimating

Arithmetic calculations are easily performed using a calculator. However, by pressing a wrong key, wrong answers can just as easily be produced. Every calculation made using a calculator should at least be checked for the reasonableness of the final result and this can be done by *estimating* the result using *rounding*. For example, using a calculator the sum $39 + 53$ is incorrectly found to be 62 if $39 + 23$ is entered by mistake. If, now, 39 is rounded up to 40, and 53 is rounded down to 50 the reasonableness of the calculator result can be simply checked by adding 40 to 50 to give 90. This indicates that the answer 62 is wrong and that the calculation should be done again. The correct answer 92 is then seen to be close to the approximation of 90.

## Rounding

An integer can be rounded to the nearest 10 as follows:

If the number is less than halfway to the next multiple of 10 then the number is rounded *down* to the previous multiple of 10. For example, 53 is rounded down to 50.

If the number is more than halfway to the next multiple of 10 then the number is rounded *up* to the next multiple of 10. For example, 39 is rounded up to 40.

If the number is exactly halfway to the next multiple of 10 then the number is rounded *up*. For example, 35 is rounded up to 40.

This principle also applies when rounding to the nearest 100, 1000, 10 000 or more. For example, 349 rounds up to 350 to the nearest 10 but rounds down to 300 to the nearest 100, and 2501 rounds up to 3000 to the nearest 1000.

Try rounding each of the following to the nearest 10, 100 and 1000 respectively:

(a) 1846
(b) $-638$
(c) 445

*Finish all three and check your results with the next frame*

**14**

| |
|---|
| (a) 1850, 1800, 2000 |
| (b) $-640$, $-600$, $-1000$ |
| (c) 450, 400, 0 |

Because

(a) 1846 is nearer to 1850 than to 1840, nearer to 1800 than to 1900 and nearer to 2000 than to 1000.

(b) $-638$ is nearer to $-640$ than to $-630$, nearer to $-600$ than to $-700$ and nearer to $-1000$ than to 0. The negative sign does not introduce any complications.

(c) 445 rounds to 450 because it is halfway to the next multiple of 10, 445 is nearer to 400 than to 500 and nearer to 0 than 1000.

How about estimating each of the following using rounding to the nearest 10:

(a) $18 \times 21 - 19 \div 11$
(b) $99 \div 101 - 49 \times 8$

*Check your results in Frame 15*

15

(a) 398
(b) $-499$

Because

(a) $18 \times 21 - 19 \div 11$ rounds to $20 \times 20 - 20 \div 10 = 398$
(b) $99 \div 101 - 49 \times 8$ rounds to $100 \div 100 - 50 \times 10 = -499$

***At this point let us pause and summarize the main facts so far***

## Review summary

16

1 The integers consist of the positive and negative whole numbers and zero.
2 The integers are ordered so that they range from large negative to small negative through zero to small positive and then large positive. They are written using the ten numerals 0 to 9 according to the principle of place value where the place of a numeral in a number dictates the value it represents.
3 The integers can be represented by equally spaced points on a line.
4 The four arithmetic operations of addition, subtraction, multiplication and division obey specific precedence rules that govern the order in which they are to be executed:

*In any calculation involving all four arithmetic operations we proceed as follows:*
(a) *working from the left evaluate divisions and multiplications as they are encountered.*
*This leaves an expression involving just addition and subtraction:*
(b) *working from the left evaluate additions and subtractions as they are encountered.*

5 Multiplying or dividing two positive numbers or two negative numbers produces a positive number. Multiplying or dividing a positive number and a negative number produces a negative number.
6 Brackets are used to group numbers and operations together. In any arithmetic expression, the contents of brackets are evaluated first.
7 Integers can be rounded to the nearest 10, 100 etc. and the rounded values used as estimates for the result of a calculation.

## Review exercise

17

1 Place the appropriate symbol $<$ or $>$ between each of the following pairs of numbers:
(a) $-1 \quad -6$
(b) $5 \quad -29$
(c) $-14 \quad 7$

▶

2 Find the value of each of the following:

(a) $16 - 12 \times 4 + 8 \div 2$

(b) $(16 - 12) \times (4 + 8) \div 2$

(c) $9 - 3(17 + 5[5 - 7])$

(d) $8(3[2 + 4] - 2[5 + 7])$

3 Show that:

(a) $6 - (3 - 2) \neq (6 - 3) - 2$

(b) $100 \div (10 \div 5) \neq (100 \div 10) \div 5$

(c) $24 \div (2 + 6) \neq (24 \div 2) + (24 \div 6)$

(d) $24 \div (2 - 6) \neq (24 \div 2) - (24 \div 6)$

4 Round each number to the nearest 10, 100 and 1000:

(a) 2562

(b) 1500

(c) $-3451$

(d) $-14\,525$

---

**18**

1 (a) $-1 > -6$ because $-1$ is represented on the line to the right of $-6$

(b) $5 > -29$ because 5 is represented on the line to the right of $-29$

(c) $-14 < 7$ because $-14$ is represented on the line to the left of 7

2 (a) $16 - 12 \times 4 + 8 \div 2 = 16 - 48 + 4 = 16 - 44 = -28$
divide and multiply before adding and subtracting

(b) $(16 - 12) \times (4 + 8) \div 2 = (4) \times (12) \div 2 = 4 \times 12 \div 2 = 4 \times 6 = 24$
brackets are evaluated first

(c)
$$\begin{aligned} 9 - 3(17 + 5[5 - 7]) &= 9 - 3(17 + 5[-2]) \\ &= 9 - 3(17 - 10) \\ &= 9 - 3(7) \\ &= 9 - 21 = -12 \end{aligned}$$

(d)
$$\begin{aligned} 8(3[2 + 4] - 2[5 + 7]) &= 8(3 \times 6 - 2 \times 12) \\ &= 8(18 - 24) \\ &= 8(-6) = -48 \end{aligned}$$

3 (a) Left-hand side (LHS) $= 6 - (3 - 2) = 6 - (1) = 5$
Right-hand side (RHS) $= (6 - 3) - 2 = (3) - 2 = 1 \neq$ LHS

(b) Left-hand side (LHS) $= 100 \div (10 \div 5) = 100 \div 2 = 50$
Right-hand side (RHS) $= (100 \div 10) \div 5 = 10 \div 5 = 2 \neq$ LHS

(c) Left-hand side (LHS) $= 24 \div (2 + 6) = 24 \div 8 = 3$
Right-hand side (RHS) $= (24 \div 2) + (24 \div 6) = 12 + 4 = 16 \neq$ LHS

(d) Left-hand side (LHS) $= 24 \div (2 - 6) = 24 \div (-4) = -6$
Right-hand side (RHS) $= (24 \div 2) - (24 \div 6) = 12 - 4 = 8 \neq$ LHS

4 (a) 2560, 2600, 3000

(b) 1500, 1500, 2000

(c) $-3450$, $-3500$, $-3000$

(d) $-14\,530$, $-14\,500$, $-15\,000$

*So now on to Frame 19*

# Factors and prime numbers

## Factors

19

Any pair of whole numbers are called *factors* of their product. For example, the numbers 3 and 6 are factors of 18 because $3 \times 6 = 18$. These are not the only factors of 18. The complete collection of factors of 18 is 1, 2, 3, 6, 9, 18 because

$$\begin{aligned} 18 &= 1 \times 18 \\ &= 2 \times 9 \\ &= 3 \times 6 \end{aligned}$$

So the factors of:

(a) 12

(b) 25

(c) 17 are . . . . . . . . . . .

*The results are in the next frame*

20

(a) 1, 2, 3, 4, 6, 12
(b) 1, 5, 25
(c) 1, 17

Because

(a) $12 = 1 \times 12 = 2 \times 6 = 3 \times 4$
(b) $25 = 1 \times 25 = 5 \times 5$
(c) $17 = 1 \times 17$

*Now move to the next frame*

## Prime numbers

21

If a whole number has only two factors which are itself and the number 1, the number is called a *prime number*. The first six prime numbers are 2, 3, 5, 7, 11 and 13. The number 1 is *not* a prime number for the reason given below.

## Prime factorization

Every whole number can be written as a product involving only prime factors. For example, the number 126 has the factors 1, 2, 3, 6, 7, 9, 14, 18, 21, 42, 63 and 126, of which 2, 3 and 7 are prime numbers and 126 can be written as:

$126 = 2 \times 3 \times 3 \times 7$

To obtain this *prime factorization* the number is divided by successively increasing prime numbers thus:

| | |
|---|---|
| 2 | 126 |
| 3 | 63 |
| 3 | 21 |
| 7 | 7 |
| | 1 |

so that $126 = 2 \times 3 \times 3 \times 7$

Notice that a prime factor may occur more than once in a prime factorization. ▶

## Fundamental theorem of arithmetic

The fundamental theorem of arithmetic states that the prime factorization of a natural number is unique. For example, we can write the prime factorization of 126 differently by rearranging the prime factors but it will always contain one 2, two 3s and one 7. This explains why 1 is not a prime number because if it were we could write:

$$\begin{aligned}126 &= 2 \times 3 \times 3 \times 7\\ &= 1 \times 2 \times 3 \times 3 \times 7\\ &= 1 \times 1 \times 2 \times 3 \times 3 \times 7\\ &= \ldots\end{aligned}$$

and the prime factorization would no longer be unique.

Now find the prime factorization of:

(a) 84 (b) 512

*Work these two out and check the working in Frame 22*

**22**

(a) $84 = 2 \times 2 \times 3 \times 7$
(b) $512 = 2 \times 2 \times 2 \times 2 \times 2 \times 2 \times 2 \times 2 \times 2$

Because

(a)

| 2 | 84 |
|---|---|
| 2 | 42 |
| 3 | 21 |
| 7 | 7 |
| | 1 |

so that $84 = 2 \times 2 \times 3 \times 7$

(b) The only prime factor of 512 is 2 which occurs 9 times. The prime factorization is:

$$512 = 2 \times 2 \times 2 \times 2 \times 2 \times 2 \times 2 \times 2 \times 2$$

*Move to Frame 23*

**23**

## Highest common factor (HCF)

The *highest common factor* (HCF) of two whole numbers is the largest factor that they have in common. For example, the prime factorizations of 144 and 66 are:

$$\begin{aligned}144 &= 2 \times 2 \times 2 \times 2 \times 3 \times 3\\ 66 &= 2 \qquad\qquad\quad \times 3 \qquad \times 11\end{aligned}$$

Only the 2 and the 3 are common to both factorizations and so the highest factor that these two numbers have in common (HCF) is $2 \times 3 = 6$.

## Lowest common multiple (LCM)

The smallest whole number that each one of a pair of whole numbers divides into a whole number of times is called their *lowest common multiple* (LCM). This is also found from the prime factorization of each of the two numbers. For example:

$$\begin{aligned}144 &= 2 \times 2 \times 2 \times 2 \times 3 \times 3\\ 66 &= 2 \qquad\qquad\quad \times 3 \qquad \times 11\\ \text{LCM} &= 2 \times 2 \times 2 \times 2 \times 3 \times 3 \times 11 = 1584\end{aligned}$$

The HCF and LCM of 84 and 512 are ...........

24

HCF: 4
LCM: 10 752

Because
84 and 512 have the prime factorizations:

$84 = 2 \times 2 \qquad\qquad\qquad\qquad \times 3 \times 7$

$512 = 2 \times 2 \times 2 \times 2 \times 2 \times 2 \times 2 \times 2 \times 2 \qquad HCF = 2 \times 2 = 4$

$LCM = 2 \times 2 \times 2 \times 2 \times 2 \times 2 \times 2 \times 2 \times 2 \times 3 \times 7 = 10\,752$

***At this point let us pause and summarize the main facts on factors and prime numbers***

## Review summary

25

1 A pair of whole numbers are called factors of their product.
2 If a whole number only has one and itself as factors it is called a prime number.
3 Every whole number can be written as a product of its prime factors, some of which may be repeated.
4 The highest common factor (HCF) is the highest factor that two whole numbers have in common.
5 The lowest common multiple (LCM) is the lowest whole number that two whole numbers will divide into a whole number of times.

## Review exercise

26

1 Write each of the following as a product of prime factors:
(a) 429 (b) 1820 (c) 2992 (d) 3185
2 Find the HCF and the LCM of each pair of numbers:
(a) 63, 42 (b) 92, 34

27

1 (a)

| | |
|---|---|
| 3 | 429 |
| 11 | 143 |
| 13 | 13 |
| | 1 |

$429 = 3 \times 11 \times 13$

(b)

| | |
|---|---|
| 2 | 1820 |
| 2 | 910 |
| 5 | 455 |
| 7 | 91 |
| 13 | 13 |
| | 1 |

$1820 = 2 \times 2 \times 5 \times 7 \times 13$

▶

(c)

$$\begin{array}{r|r} 2 & 2992 \\ \hline 2 & 1496 \\ \hline 2 & 748 \\ \hline 2 & 374 \\ \hline 11 & 187 \\ \hline 17 & 17 \\ \hline & 1 \end{array} \qquad 2992 = 2 \times 2 \times 2 \times 2 \times 11 \times 17$$

(d)

$$\begin{array}{r|r} 5 & 3185 \\ \hline 7 & 637 \\ \hline 7 & 91 \\ \hline 13 & 13 \\ \hline & 1 \end{array} \qquad 3185 = 5 \times 7 \times 7 \times 13$$

2 (a) The prime factorizations of 63 and 42 are:

$$\begin{aligned} 63 &= \phantom{2 \times{}} 3 \times 3 \times 7 \\ 42 &= 2 \times 3 \phantom{{}\times 3} \times 7 \qquad \text{HCF } 3 \times 7 = 21 \\ \text{LCM} &= 2 \times 3 \times 3 \times 7 = 126 \end{aligned}$$

(b) The prime factorizations of 34 and 92 are:

$$\begin{aligned} 34 &= 2 \phantom{{}\times 2} \times 17 \\ 92 &= 2 \times 2 \phantom{{}\times 17} \times 23 \qquad \text{HCF } 2 \\ \text{LCM} &= 2 \times 2 \times 17 \times 23 = 1564 \end{aligned}$$

*Now on to the next topic*

---

# Fractions, ratios and percentages

**28**

## Division of integers

A fraction is a number which is represented by one integer – the *numerator* – divided by another integer – the *denominator* (or the *divisor*). For example, $\frac{3}{5}$ is a fraction with numerator 3 and denominator 5. Because fractions are written as one integer divided by another – a *ratio* – they are called *rational* numbers. Fractions are either *proper*, *improper* or *mixed*:

- in a proper fraction the numerator is less than the denominator, for example, $\dfrac{4}{7}$
- in an improper fraction the numerator is greater than the denominator, for example $\dfrac{12}{5}$
- a mixed fraction is in the form of an integer and a fraction, for example $6\frac{2}{3}$

So that $-\dfrac{8}{11}$ is a ........... fraction?

*The answer is in the next frame*

29

Proper

Fractions can be either positive or negative.

*Now to the next frame*

## Multiplying fractions

30

Two fractions are multiplied by multiplying their respective numerators and denominators independently. For example:

$$\frac{2}{3} \times \frac{5}{7} = \frac{2 \times 5}{3 \times 7} = \frac{10}{21}$$

Try this one for yourself.

$$\frac{5}{9} \times \frac{2}{7} = \ldots\ldots\ldots\ldots$$

31

$$\frac{10}{63}$$

Because

$$\frac{5}{9} \times \frac{2}{7} = \frac{5 \times 2}{9 \times 7} = \frac{10}{63}$$

*Correct? Then on to the next frame*

## Of

32

The word 'of' when interposed between two fractions means multiply. For example:

Half of half a cake is one-quarter of a cake. That is

$$\frac{1}{2} \textit{ of } \frac{1}{2} = \frac{1}{2} \times \frac{1}{2} = \frac{1 \times 1}{2 \times 2} = \frac{1}{4}$$

So that, for example:

$$\frac{1}{3} \textit{ of } \frac{2}{5} = \frac{1}{3} \times \frac{2}{5} = \frac{1 \times 2}{3 \times 5} = \frac{2}{15}$$

So that $\frac{3}{8} \textit{ of } \frac{5}{7} = \ldots\ldots\ldots\ldots$

33

$$\frac{15}{56}$$

Because

$$\frac{3}{8} \textit{ of } \frac{5}{7} = \frac{3}{8} \times \frac{5}{7} = \frac{3 \times 5}{8 \times 7} = \frac{15}{56}$$

*On now to the next frame*

**34**

## Equivalent fractions

Multiplying the numerator and denominator by the same number is equivalent to multiplying the fraction by unity, that is by 1:

$$\frac{4\times 3}{5\times 3}=\frac{4}{5}\times\frac{3}{3}=\frac{4}{5}\times 1=\frac{4}{5}$$

Now, $\frac{4\times 3}{5\times 3}=\frac{12}{15}$ so that the fraction $\frac{4}{5}$ and the fraction $\frac{12}{15}$ *both represent the same number* and for this reason we call $\frac{4}{5}$ and $\frac{12}{15}$ *equivalent fractions.*

A second fraction, equivalent to a first fraction, can be found by multiplying the numerator and the denominator of the first fraction by the same number.

So that if we multiply the numerator and denominator of the fraction $\frac{7}{5}$ by 4 we obtain the equivalent fraction ...........

*Check your result in Frame 35*

**35**

$$\boxed{\frac{28}{20}}$$

Because

$$\frac{7\times 4}{5\times 4}=\frac{28}{20}$$

We can reverse this process and find the equivalent fraction that has the smallest numerator by *cancelling out* common factors. This is known as reducing the fraction to its *lowest terms*. For example:

$\frac{16}{96}$ can be reduced to its lowest terms as follows:

$$\frac{16}{96}=\frac{4\times 4}{24\times 4}=\frac{4\times \cancel{4}}{24\times \cancel{4}}=\frac{4}{24}$$

by cancelling out the 4 in the numerator and the denominator

The fraction $\frac{4}{24}$ can also be reduced:

$$\frac{4}{24}=\frac{4}{6\times 4}=\frac{\cancel{4}}{6\times \cancel{4}}=\frac{1}{6}$$

Because $\frac{1}{6}$ cannot be reduced further we see that $\frac{16}{96}$ reduced to its lowest terms is $\frac{1}{6}$.

How about this one? The fraction $\frac{84}{108}$ reduced to its lowest terms is ...........

*Check with the next frame*

**36**

$$\boxed{\frac{7}{9}}$$

Because

$$\frac{84}{108}=\frac{7\times 3\times 4}{9\times 3\times 4}=\frac{7\times \cancel{3}\times \cancel{4}}{9\times \cancel{3}\times \cancel{4}}=\frac{7}{9}$$

*Now move on to the next frame*

## Dividing fractions

37

The expression $6 \div 3$ means the number of 3's in 6, which is 2. Similarly, the expression $1 \div \frac{1}{4}$ means the number of $\frac{1}{4}$'s in 1, which is, of course, 4. That is:

$1 \div \frac{1}{4} = 4 = 1 \times \frac{4}{1}$. Notice how the numerator and the denominator of the divisor are switched and the division replaced by multiplication.

Two fractions are divided by switching the numerator and the denominator of the divisor and multiplying the result. For example:

$$\frac{2}{3} \div \frac{5}{7} = \frac{2}{3} \times \frac{7}{5} = \frac{14}{15}$$

So that $\frac{7}{13} \div \frac{3}{4} =$ ...........

---

38

$$\boxed{\frac{28}{39}}$$

Because

$$\frac{7}{13} \div \frac{3}{4} = \frac{7}{13} \times \frac{4}{3} = \frac{28}{39}$$

In particular:

$$1 \div \frac{3}{5} = 1 \times \frac{5}{3} = \frac{5}{3}$$

The fraction $\frac{5}{3}$ is called the *reciprocal* of $\frac{3}{5}$

So that the reciprocal of $\frac{17}{4}$ is ...........

---

39

$$\boxed{\frac{4}{17}}$$

Because

$$1 \div \frac{17}{4} = 1 \times \frac{4}{17} = \frac{4}{17}$$

And the reciprocal of $-5$ is ...........

---

40

$$\boxed{-\frac{1}{5}}$$

Because

$$1 \div (-5) = 1 \div \left(-\frac{5}{1}\right) = 1 \times \left(-\frac{1}{5}\right) = -\frac{1}{5}$$

*Move on to the next frame*

**41**

## Adding and subtracting fractions

Two fractions can only be added or subtracted immediately if they both possess the same denominator, in which case we add or subtract the numerators and divide by the common denominator. For example:

$$\frac{2}{7}+\frac{3}{7}=\frac{2+3}{7}=\frac{5}{7}$$

If they do not have the same denominator they must be rewritten in equivalent form so that they do have the same denominator – called the *common denominator*. For example:

$$\frac{2}{3}+\frac{1}{5}=\frac{10}{15}+\frac{3}{15}=\frac{10+3}{15}=\frac{13}{15}$$

The common denominator of the equivalent fractions is the LCM of the two original denominators. That is:

$$\frac{2}{3}+\frac{1}{5}=\frac{2\times 5}{3\times 5}+\frac{1\times 3}{5\times 3}=\frac{10}{15}+\frac{3}{15} \text{ where 15 is the LCM of 3 and 5}$$

So that $\frac{5}{9}+\frac{1}{6}=$ . . . . . . . . . . .

*The result is in Frame 42*

**42**

$$\boxed{\frac{13}{18}}$$

Because

The LCM of 9 and 6 is 18 so that $\frac{5}{9}+\frac{1}{6}=\frac{5\times 2}{9\times 2}+\frac{1\times 3}{6\times 3}=\frac{10}{18}+\frac{3}{18}$

$$=\frac{10+3}{18}=\frac{13}{18}$$

*There's another one to try in the next frame*

**43**

Now try $\frac{11}{15}-\frac{2}{3}=$ . . . . . . . . . . .

**44**

$$\boxed{\frac{1}{15}}$$

Because

$$\frac{11}{15}-\frac{2}{3}=\frac{11}{15}-\frac{2\times 5}{3\times 5}=\frac{11}{15}-\frac{10}{15}$$

$$=\frac{11-10}{15}=\frac{1}{15} \quad \text{(15 is the LCM of 3 and 15)}$$

*Correct? Then on to Frame 45*

## Fractions on a calculator

45

The calculator we shall use is the Casio *fx-85GT PLUS* and on this calculator the fraction key is denoted by the symbol:

*on* the key and the symbol for a mixed fraction:

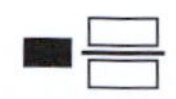

*above* the key – accessed by using in combination with the SHIFT key. The REPLAY key is used to move the flashing cursor around the screen to enable functions to be entered and manipulated with the results given in fractional form. For example, to evaluate $\frac{2}{3} \times 2\frac{3}{4}$ using this calculator [*note*: your calculator may not have the identical display in what follows, indeed the symbol $a^b\!/_c$ is often used for the fraction key]:

Press the fraction key
  the fraction symbol is then displayed on the screen
Enter the number 2
  the numerator in the display
Press the down arrow on the REPLAY key
  the flashing cursor moves to the denominator
Enter the number 3
  the denominator in the display
Press the right arrow on the REPLAY key
  the flashing cursor moves to the right of the display
Press the × key
Press the SHIFT key and then the fraction key
  the mixed fraction symbol is then displayed on the screen
Enter the number 2
Press the right arrow on the REPLAY key
Enter the number 3
Press the down arrow on the REPLAY key
Enter the number 4

The display then should look just like:

$\frac{2}{3} \times 2\frac{3}{4}$

Press the = key and the result appears in the bottom right-hand corner of the screen:

$$\frac{11}{6}$$

Now press the SHIFT key and then the S⇔D key to change the display to the mixed fraction:

$1\frac{5}{6}$

That is:

$$\begin{aligned}\tfrac{2}{3} \times 2\tfrac{3}{4} &= \frac{2}{3} \times \frac{11}{4}\\ &= \frac{11}{6}\\ &= 1\tfrac{5}{6}\end{aligned}$$

▶

Now use your calculator to evaluate each of the following:

(a) $\frac{5}{7} + 3\frac{2}{3}$

(b) $\dfrac{8}{3} - \dfrac{5}{11}$

(c) $\dfrac{13}{5} \times \dfrac{4}{7} - \dfrac{2}{9}$

(d) $4\frac{1}{11} \div \left(-\dfrac{3}{5}\right) + \dfrac{1}{8}$

*Check your answers in Frame 46*

**46**

(a) $\frac{5}{7} + 3\frac{2}{3} = \frac{92}{21} = 4\frac{8}{21}$

(b) $\dfrac{8}{3} - \dfrac{5}{11} = \dfrac{73}{33} = 2\frac{7}{33}$

(c) $\dfrac{13}{5} \times \dfrac{4}{7} - \dfrac{2}{9} = \dfrac{398}{315} = 1\frac{83}{315}$

(d) $4\frac{1}{11} \div \left(-\dfrac{3}{5}\right) + \dfrac{1}{8} = -\dfrac{589}{88} = -6\frac{61}{88}$

*On now to the next frame*

**47**

## Ratios

If a whole number is separated into a number of fractional parts where each fraction has the same denominator, the numerators of the fractions form a *ratio*. For example, if a quantity of brine in a tank contains $\dfrac{1}{3}$ salt and $\dfrac{2}{3}$ water, the salt and water are said to be in the ratio 'one-to-two' – written 1 : 2.

What ratio do the components A, B and C form if a compound contains $\dfrac{3}{4}$ of A, $\dfrac{1}{6}$ of B and $\dfrac{1}{12}$ of C?

*Take care here and check your results with Frame 48*

**48**

9 : 2 : 1

Because the LCM of the denominators 4, 6 and 12 is 12, then:

$\dfrac{3}{4}$ of A is $\dfrac{9}{12}$ of A, $\dfrac{1}{6}$ of B is $\dfrac{2}{12}$ of B and the remaining $\dfrac{1}{12}$ is of C. This ensures that the components are in the ratio of their numerators. That is: 9 : 2 : 1

Notice that the sum of the numbers in the ratio is equal to the common denominator.

*On now to the next frame*

**49**

## Percentages

A percentage is a fraction whose denominator is equal to 100. For example, if 5 out of 100 people are left-handed then the fraction of left-handers is $\dfrac{5}{100}$ which is written as 5%, that is 5 *per cent* (%).

So if 13 out of 100 cars on an assembly line are red, the percentage of red cars on the line is ...........

50

13%

Because

The fraction of cars that are red is $\frac{13}{100}$ which is written as 13%.

Try this. What is the percentage of defective resistors in a batch of 25 if 12 of them are defective?

51

48%

Because

The fraction of defective resistors is $\frac{12}{25} = \frac{12 \times 4}{25 \times 4} = \frac{48}{100}$ which is written as 48%.

Notice that this is the same as:

$$\left(\frac{12}{25} \times 100\right)\% = \left(\frac{12}{25} \times 25 \times 4\right)\% = (12 \times 4)\% = 48\%$$

*A fraction can be converted to a percentage by multiplying the fraction by 100.*

To find the percentage part of a quantity we multiply the quantity by the percentage written as a fraction. For example, 24% of 75 is:

$$24\% \textit{ of } 75 = \frac{24}{100} \textit{ of } 75 = \frac{24}{100} \times 75 = \frac{6 \times 4}{25 \times 4} \times 25 \times 3 = \frac{6 \times \not{4}}{\not{25} \times \not{4}} \times \not{25} \times 3$$
$$= 6 \times 3 = 18$$

So that 8% of 25 is ...........

*Work it through and check your results with the next frame*

52

2

Because

$$\frac{8}{100} \times 25 = \frac{2 \times 4}{25 \times 4} \times 25 = \frac{2 \times \not{4}}{\not{25} \times \not{4}} \times \not{25} = 2$$

***At this point let us pause and summarize the main facts on fractions, ratios and percentages***

## Review summary

53

1 A fraction is a number represented as one integer (the numerator) divided by another integer (the denominator or divisor).

2 The same number can be represented by different but equivalent fractions.

3 A fraction with no common factors other than unity in its numerator and denominator is said to be in its lowest terms.

4 Two fractions are multiplied by multiplying the numerators and denominators independently.

5 Two fractions can only be added or subtracted immediately when their denominators are equal.

6 A ratio consists of the numerators of fractions with identical denominators.

7 The numerator of a fraction whose denominator is 100 is called a percentage.

## Review exercise

**54**

1 Reduce each of the following fractions to their lowest terms:

(a) $\frac{24}{30}$ (b) $\frac{72}{15}$ (c) $-\frac{52}{65}$ (d) $\frac{32}{8}$

2 Evaluate the following:

(a) $\frac{5}{9} \times \frac{2}{5}$ (b) $\frac{13}{25} \div \frac{2}{15}$ (c) $\frac{5}{9} + \frac{3}{14}$ (d) $\frac{3}{8} - \frac{2}{5}$

(e) $\frac{12}{7} \times \left(-\frac{3}{5}\right)$ (f) $\left(-\frac{3}{4}\right) \div \left(-\frac{12}{7}\right)$ (g) $\frac{19}{2} + \frac{7}{4}$ (h) $\frac{1}{4} - \frac{3}{8}$

3 Write the following proportions as ratios:

(a) $\frac{1}{2}$ of A, $\frac{2}{5}$ of B and $\frac{1}{10}$ of C

(b) $\frac{1}{3}$ of P, $\frac{1}{5}$ of Q, $\frac{1}{4}$ of R and the remainder S

4 Complete the following:

(a) $\frac{2}{5} =$ % (b) 58% of 25 =

(c) $\frac{7}{12} =$ % (d) 17% of 50 =

**55**

1 (a) $\frac{24}{30} = \frac{2 \times 2 \times 2 \times 3}{2 \times 3 \times 5} = \frac{2 \times 2}{5} = \frac{4}{5}$

(b) $\frac{72}{15} = \frac{2 \times 2 \times 2 \times 3 \times 3}{3 \times 5} = \frac{2 \times 2 \times 2 \times 3}{5} = \frac{24}{5}$

(c) $-\frac{52}{65} = -\frac{2 \times 2 \times 13}{5 \times 13} = -\frac{2 \times 2}{5} = -\frac{4}{5}$

(d) $\frac{32}{8} = \frac{2 \times 2 \times 2 \times 2 \times 2}{2 \times 2 \times 2} = 4$

2 (a) $\frac{5}{9} \times \frac{2}{5} = \frac{5 \times 2}{9 \times 5} = \frac{2}{9}$

(b) $\frac{13}{25} \div \frac{2}{15} = \frac{13}{25} \times \frac{15}{2} = \frac{13 \times 15}{25 \times 2} = \frac{13 \times 3 \times 5}{5 \times 5 \times 2} = \frac{39}{10}$

(c) $\frac{5}{9} + \frac{3}{14} = \frac{5 \times 14}{9 \times 14} + \frac{3 \times 9}{14 \times 9} = \frac{70}{126} + \frac{27}{126} = \frac{70 + 27}{126} = \frac{97}{126}$

(d) $\frac{3}{8} - \frac{2}{5} = \frac{3 \times 5}{8 \times 5} - \frac{2 \times 8}{5 \times 8} = \frac{15}{40} - \frac{16}{40} = \frac{15 - 16}{40} = -\frac{1}{40}$

(e) $\frac{12}{7} \times \left(-\frac{3}{5}\right) = \frac{12 \times (-3)}{7 \times 5} = \frac{-36}{35} = -\frac{36}{35}$

(f) $\left(-\frac{3}{4}\right) \div \left(-\frac{12}{7}\right) = \left(-\frac{3}{4}\right) \times \left(-\frac{7}{12}\right) = \frac{(-3) \times (-7)}{4 \times 12} = \frac{3 \times 7}{4 \times 3 \times 4} = \frac{7}{16}$

(g) $\frac{19}{2} + \frac{7}{4} = \frac{38}{4} + \frac{7}{4} = \frac{45}{4}$

(h) $\frac{1}{4} - \frac{3}{8} = \frac{2}{8} - \frac{3}{8} = -\frac{1}{8}$

▶

3 (a) $\frac{1}{2}, \frac{2}{5}, \frac{1}{10} = \frac{5}{10}, \frac{4}{10}, \frac{1}{10}$ so ratio is 5 : 4 : 1

(b) $\frac{1}{3}, \frac{1}{5}, \frac{1}{4} = \frac{20}{60}, \frac{12}{60}, \frac{15}{60}$ and $\frac{20}{60} + \frac{12}{60} + \frac{15}{60} = \frac{47}{60}$

so the fraction of S is $\frac{13}{60}$

so P, Q , R and S are in the ratio 20 : 12 : 15 : 13

4 (a) $\frac{2}{5} = \frac{2 \times 20}{5 \times 20} = \frac{40}{100}$ that is 40% or $\frac{2}{5} = \frac{2}{5} \times 100\% = 40\%$

(b) $\frac{58}{100} \times 25 = \frac{58}{4} = \frac{29}{2} = 14\frac{1}{2}$

(c) $\frac{7}{12} = \frac{7}{12} \times 100\% = \frac{700}{12}\% = \frac{58 \times 12 + 4}{12}\% = 58\frac{4}{12}\% = 58\frac{1}{3}\%$

(d) $\frac{17}{100} \times 50 = \frac{17}{2} = 8\frac{1}{2}$

*Now let's look at decimal numbers*

# Decimal numbers

## Division of integers

**56**

If one integer is divided by a second integer that is not one of the first integer's factors the result will not be another integer. Instead, the result will lie between two integers. For example, using a calculator it is seen that:

$25 \div 8 = 3{\cdot}125$

which is a number greater than 3 but less than 4. As with integers, the position of a numeral within the number indicates its value. Here the number 3·125 represents

3 units + 1 tenth + 2 hundredths + 5 thousandths.

That is $3 + \frac{1}{10} + \frac{2}{100} + \frac{5}{1000}$

where the decimal point shows the separation of the units from the tenths. Numbers written in this format are called *decimal numbers*.

*On to the next frame*

## Rounding

**57**

All the operations of arithmetic that we have used with the integers apply to decimal numbers. However, when performing calculations involving decimal numbers it is common for the end result to be a number with a large quantity of numerals after the decimal point. For example:

$15{\cdot}11 \div 8{\cdot}92 = 1{\cdot}6939461883\ldots$

To make such numbers more manageable or more reasonable as the result of a calculation, they can be rounded either to a specified number of *significant figures* or to a specified number of *decimal places*.

*Now to the next frame*

**58**

## Significant figures

Significant figures are counted from the first non-zero numeral encountered starting from the left of the number. When the required number of significant figures has been counted off, the remaining numerals are deleted with the following proviso:

If the first of a group of numerals to be deleted is a 5 or more, the last significant numeral is increased by 1. For example:

9·4534 to two significant figures is 9·5, to three significant figures is 9·45, and 0·001354 to two significant figures is 0·0014

Try this one for yourself. To four significant figures the number 18·7249 is ...........

*Check your result with the next frame*

**59**

18·72

Because

The first numeral deleted is a 4 which is less than 5.

There is one further proviso. If the only numeral to be dropped is a 5 then the last numeral retained is rounded up. So that 12·235 to four significant figures (abbreviated to *sig fig*) is 12·24 and 3·465 to three sig fig is 3·47.

So 8·1265 to four sig fig is ...........

*Check with the next frame*

**60**

8·127

Because

The only numeral deleted is a 5 and the last numeral is rounded up.

*Now on to the next frame*

**61**

## Decimal places

Decimal places are counted to the right of the decimal point and the same rules as for significant figures apply for rounding to a specified number of decimal places (abbreviated to *dp*). For example:

123·4467 to one decimal place is 123·4 and to two dp is 123·45

So, 47·0235 to three dp is ...........

**62**

47·024

Because

The only numeral dropped is a 5 and the last numeral retained is rounded up.

*Now move on to the next frame*

63

## Trailing zeros

Sometimes zeros must be inserted within a number to satisfy a condition for a specified number of either significant figures or decimal places. For example:

12 645 to two significant figures is 13 000, and 13·1 to three decimal places is 13·100.

These zeros are referred to as *trailing zeros*.

So that 1515 to two sig fig is . . . . . . . . . . .

64

1500

And 25·13 to four dp is . . . . . . . . . . .

65

25·1300

*On to the next frame*

66

## Fractions as decimals

Because a fraction is one integer divided by another it can be represented in decimal form simply by executing the division. For example:

$$\frac{7}{4} = 7 \div 4 = 1{\cdot}75$$

So that the decimal form of $\frac{3}{8}$ is . . . . . . . . . . .

67

0·375

Because

$$\frac{3}{8} = 3 \div 8 = 0{\cdot}375$$

*Now move on to the next frame*

68

## Decimals as fractions

A decimal can be represented as a fraction. For example:

$$1{\cdot}224 = \frac{1224}{1000} \text{ which in lowest terms is } \frac{153}{125}$$

So that 0·52 as a fraction in lowest terms is . . . . . . . . . . .

69

$\frac{13}{25}$

Because

$$0{\cdot}52 = \frac{52}{100} = \frac{13}{25}$$

*Now move on to the next frame*

## 70 Unending decimals

Converting a fraction into its decimal form by performing the division always results in an infinite string of numerals after the decimal point. This string of numerals may contain an infinite sequence of zeros or it may contain an infinitely repeated pattern of numerals. A repeated pattern of numerals can be written in an abbreviated format. For example:

$$\frac{1}{3} = 1 \div 3 = 0{\cdot}3333\ldots$$

Here the pattern after the decimal point is of an infinite number of 3's. We abbreviate this by placing a dot over the first 3 to indicate the repetition, thus:

$0{\cdot}3333\ldots = 0{\cdot}\dot{3}$ (described as zero point 3 recurring)

For other fractions the repetition may consist of a sequence of numerals, in which case a dot is placed over the first and last numeral in the sequence. For example:

$$\frac{1}{7} = 0{\cdot}142857142857142857\ldots = 0{\cdot}\dot{1}4285\dot{7}$$

So that we write $\frac{2}{11} = 0{\cdot}181818\ldots$ as ...........

## 71

$\boxed{0{\cdot}\dot{1}\dot{8}}$

Sometimes the repeating pattern is formed by an infinite sequence of zeros, in which case we simply omit them. For example:

$\frac{1}{5} = 0{\cdot}20000\ldots$ is written as $0{\cdot}2$

*Next frame*

## 72 Unending decimals as fractions

Any decimal that displays an unending repeating pattern can be converted to its fractional form. For example:

To convert $0{\cdot}181818\ldots = 0{\cdot}\dot{1}\dot{8}$ to its fractional form we note that because there are two repeating numerals we multiply by 100 to give:

$$100 \times 0{\cdot}\dot{1}\dot{8} = 18{\cdot}\dot{1}\dot{8}$$

Subtracting $0{\cdot}\dot{1}\dot{8}$ from both sides of this equation gives:

$$100 \times 0{\cdot}\dot{1}\dot{8} - 0{\cdot}\dot{1}\dot{8} = 18{\cdot}\dot{1}\dot{8} - 0{\cdot}\dot{1}\dot{8}$$

That is:

$$99 \times 0{\cdot}\dot{1}\dot{8} = 18{\cdot}0$$

This means that:

$$0{\cdot}\dot{1}\dot{8} = \frac{18}{99} = \frac{2}{11}$$

▶

Similarly, the fractional form of $2{\cdot}0\dot{3}1\dot{5}$ is found as follows:

$2{\cdot}0\dot{3}1\dot{5} = 2{\cdot}0 + 0{\cdot}0\dot{3}1\dot{5}$ and, because there are three repeating numerals:

$$1000 \times 0{\cdot}0\dot{3}1\dot{5} = 31{\cdot}5\dot{3}1\dot{5}$$

Subtracting $0{\cdot}0\dot{3}1\dot{5}$ from both sides of this equation gives:

$$1000 \times 0{\cdot}0\dot{3}1\dot{5} - 0{\cdot}0\dot{3}1\dot{5} = 31{\cdot}5\dot{3}1\dot{5} - 0{\cdot}0\dot{3}1\dot{5} = 31{\cdot}5$$

That is:

$$999 \times 0{\cdot}0\dot{3}1\dot{5} = 31{\cdot}5 \text{ so that } 0{\cdot}0\dot{3}1\dot{5} = \frac{31{\cdot}5}{999} = \frac{315}{9990}$$

This means that:

$$2{\cdot}0\dot{3}1\dot{5} = 2{\cdot}0 + 0{\cdot}0\dot{3}1\dot{5} = 2 + \frac{315}{9990} = 2\tfrac{35}{1110} = 2\tfrac{7}{222}$$

What are the fractional forms of $0{\cdot}\dot{2}\dot{1}$ and $3{\cdot}2\dot{1}$?

*The answers are in the next frame*

---

**73**

$\frac{7}{33}$ and $3\frac{19}{90}$

Because

$$100 \times 0{\cdot}\dot{2}\dot{1} = 21{\cdot}\dot{2}\dot{1} \text{ so that } 99 \times 0{\cdot}\dot{2}\dot{1} = 21 \text{ giving}$$

$$0{\cdot}\dot{2}\dot{1} = \frac{21}{99} = \frac{7}{33}$$

and

$$3{\cdot}2\dot{1} = 3{\cdot}2 + 0{\cdot}0\dot{1} \text{ and } 10 \times 0{\cdot}0\dot{1} = 0{\cdot}1\dot{1} \text{ so that } 9 \times 0{\cdot}0\dot{1} = 0{\cdot}1 \text{ giving}$$

$$0{\cdot}0\dot{1} = \frac{0{\cdot}1}{9} = \frac{1}{90}, \text{ hence}$$

$$3{\cdot}2\dot{1} = \frac{32}{10} + \frac{1}{90} = \frac{289}{90} = 3\tfrac{19}{90}$$

---

**74**

## Rational, irrational and real numbers

A number that can be expressed as a fraction is called a *rational* number. An *irrational* number is one that *cannot* be expressed as a fraction and has a decimal form consisting of an infinite string of numerals that does not display a repeating pattern. As a consequence it is not possible either to write down the complete decimal form or to devise an abbreviated decimal format. Instead, we can only round them to a specified number of significant figures or decimal places. Alternatively, we may have a numeral representation for them, such as $\sqrt{2}$, $e$ or $\pi$. The complete collection of rational and irrational numbers is called the collection of *real* numbers.

***At this point let us pause and summarize the main facts so far on decimal numbers***

## Review summary

**75**

1 Every fraction can be written as a decimal number by performing the division.

2 The decimal number obtained will consist of an infinitely repeating pattern of numerals to the right of one of its digits.

3 Other decimals, with an infinite, non-repeating sequence of numerals after the decimal point are the irrational numbers.

4 A decimal number can be rounded to a specified number of significant figures (sig fig) by counting from the first non-zero numeral on the left.

5 A decimal number can be rounded to a specified number of decimal places (dp) by counting from the decimal point.

## Review exercise

**76**

1 Round each of the following decimal numbers, first to 3 significant figures and then to 2 decimal places:

(a) 12·455 (b) 0·01356 (c) 0·1005 (d) 1344·555

2 Write each of the following in abbreviated form:

(a) 12·110110110... (b) 0·123123123...

(c) $-3{\cdot}11111\ldots$ (d) $-9360{\cdot}936093609360\ldots$

3 Convert each of the following to decimal form to 3 decimal places:

(a) $\frac{3}{16}$ (b) $-\frac{5}{9}$ (c) $\frac{7}{6}$ (d) $-\frac{24}{11}$

4 Convert each of the following to fractional form in lowest terms:

(a) 0·6 (b) $1{\cdot}\dot{4}$ (c) $1{\cdot}\dot{2}\dot{4}$ (d) $-7{\cdot}3$

**77**

1 (a) 12·5, 12·46 (b) 0·0136, 0·01 (c) 0·101, 0·10 (d) 1340, 1344·56

2 (a) $12{\cdot}\dot{1}1\dot{0}$ (b) $0{\cdot}\dot{1}2\dot{3}$ (c) $-3{\cdot}\dot{1}$ (d) $-9360{\cdot}\dot{9}36\dot{0}$

3 (a) $\frac{3}{16} = 0{\cdot}1875 = 0{\cdot}188$ to 3 dp

(b) $-\frac{5}{9} = -0{\cdot}555\ldots = -0{\cdot}556$ to 3 dp

(c) $\frac{7}{6} = 1{\cdot}1666\ldots = 1{\cdot}167$ to 3 dp

(d) $-\frac{24}{11} = -2{\cdot}1818\ldots = -2{\cdot}182$ to 3 dp

4 (a) $0{\cdot}6 = \frac{6}{10} = \frac{3}{5}$

(b) $1{\cdot}\dot{4} = 1 + \frac{4}{9} = \frac{13}{9}$

(c) $1{\cdot}\dot{2}\dot{4} = 1 + \frac{24}{99} = \frac{123}{99} = \frac{41}{33}$

(d) $-7{\cdot}3 = -\frac{73}{10}$

*Now move on to the next topic*

# Powers

## Raising a number to a power

78

The arithmetic operation of raising a number to a *power* is devised from repetitive multiplication. For example:

$10 \times 10 \times 10 \times 10 = 10^4$ that is, 4 number 10s multiplied together

The power is also called an *index* and the number to be raised to the power is called the *base*. Here the number 4 is the power (index) and 10 is the base.

So $5 \times 5 \times 5 \times 5 \times 5 \times 5 =$ ........... (in the form of 5 raised to a power)

*Compare your answer with the next frame*

---

79

$5^6$

Because the number 5 (the base) is multiplied by itself 6 times (the power or index).

*Now to the next frame*

---

## The laws of powers

80

The laws of powers are contained within the following set of rules:

- **Power unity**

  Any number raised to the power 1 equals itself.

  $3^1 = 3$

So $99^1 =$ ...........

*On to the next frame*

---

81

99

Because any number raised to the power 1 equals itself.

- **Multiplication of numbers and the addition of powers**

  If two numbers are each written as a given base raised to some power then the *product of the two numbers* is equal to the same base raised to the *sum of the powers*. For example, $16 = 2^4$ and $8 = 2^3$ so:

$$\begin{aligned} 16 \times 8 &= 2^4 \times 2^3 \\ &= (2 \times 2 \times 2 \times 2) \times (2 \times 2 \times 2) \\ &= 2 \times 2 \times 2 \times 2 \times 2 \times 2 \times 2 \\ &= 2^7 \\ &= 2^{4+3} \\ &= 128 \end{aligned}$$

  Multiplication requires powers to be added.

So $8^3 \times 8^5 =$ ........... (in the form of 8 raised to a power)

*Next frame*

**82**

$8^8$

Because multiplication requires powers to be added.

Notice that we cannot combine different powers with different bases. For example:

$2^2 \times 4^3$ cannot be written as $8^5$

but we can combine different bases to the same power. For example:

$3^4 \times 5^4$ can be written as $15^4$ because

$$\begin{aligned} 3^4 \times 5^4 &= (3 \times 3 \times 3 \times 3) \times (5 \times 5 \times 5 \times 5) \\ &= 15 \times 15 \times 15 \times 15 \\ &= 15^4 \\ &= (3 \times 5)^4 \end{aligned}$$

So that $2^3 \times 4^3$ can be written as ...........

(in the form of a number raised to a power)

**83**

$8^3$

*Next frame*

**84**

- **Division of numbers and the subtraction of powers**

  If two numbers are each written as a given base raised to some power then the *quotient of the two numbers* is equal to the same base raised to the *difference of the powers*. For example:

$$\begin{aligned} 15\,625 \div 25 &= 5^6 \div 5^2 \\ &= (5 \times 5 \times 5 \times 5 \times 5 \times 5) \div (5 \times 5) \\ &= \frac{5 \times 5 \times 5 \times 5 \times 5 \times 5}{5 \times 5} \\ &= 5 \times 5 \times 5 \times 5 \\ &= 5^4 \\ &= 5^{6-2} \\ &= 625 \end{aligned}$$

  Division requires powers to be subtracted.

So $12^7 \div 12^3 =$ ........... (in the form of 12 raised to a power)

*Check your result in the next frame*

**85**

$12^4$

Because division requires the powers to be subtracted.

- **Power zero**

  Any number raised to the power 0 equals unity. For example:

$$\begin{aligned} 1 &= 3^1 \div 3^1 \\ &= 3^{1-1} \\ &= 3^0 \end{aligned}$$

So $193^0 =$ ...........

86

$\boxed{1}$

Because any number raised to the power 0 equals unity.

- **Negative powers**

A number raised to a negative power denotes the reciprocal. For example:

$$\begin{aligned} 6^{-2} &= 6^{0-2} \\ &= 6^0 \div 6^2 && \text{subtraction of powers means division} \\ &= 1 \div 6^2 && \text{because } 6^0 = 1 \\ &= \frac{1}{6^2} \end{aligned}$$

Also $6^{-1} = \dfrac{1}{6}$

A negative power denotes the reciprocal.

So $3^{-5} = \dots\dots\dots$

87

$\boxed{\dfrac{1}{3^5}}$

Because

$$3^{-5} = 3^{0-5} = 3^0 \div 3^5 = \frac{1}{3^5}$$

A negative power denotes the reciprocal.

*Now to the next frame*

88

- **Multiplication of powers**

If a number is written as a given base raised to some power then that number *raised to a further power* is equal to the base raised to the *product of the powers*. For example:

$$\begin{aligned} (25)^3 &= \left(5^2\right)^3 \\ &= 5^2 \times 5^2 \times 5^2 \\ &= 5 \times 5 \times 5 \times 5 \times 5 \times 5 \\ &= 5^6 \\ &= 5^{2\times 3} \\ &= 15\,625 && \text{Notice that } \left(5^2\right)^3 \neq 5^{2^3} \text{ because } 5^{2^3} = 5^8 = 390\,625. \end{aligned}$$

Raising to a power requires powers to be multiplied.

So $\left(4^2\right)^4 = \dots\dots\dots$ (in the form of 4 raised to a power)

89

$\boxed{4^8}$

Because raising to a power requires powers to be multiplied.

*Now to the next frame*

**90**

## Powers on a calculator

Powers on a calculator can be evaluated by using the $x^{\blacksquare}$ key (again, on other calculators this is often annotated as $x^y$). For example, enter the number 4, press the $x^{\blacksquare}$ key, enter the number 3 and press $=$. The result is 64 which is $4^3$.

Try this one for yourself. To two decimal places, the value of $1{\cdot}3^{3{\cdot}4}$ is ...........

*The result is in the following frame*

**91**

2·44

Because

Enter the number 1·3
Press the $x^{\blacksquare}$ key
Enter the number 3·4
Press the $=$ key

The number displayed is 2·44 to 2 dp.

Now try this one using the calculator:

$8^{\frac{1}{3}} =$ ........... The 1/3 is a problem, use the fraction key 

*Check your answer in the next frame*

**92**

2

Because

Enter the number 8
Press the $x^{\blacksquare}$ key
Enter the number 1
Press the fraction key
Enter the number 3
Press $=$ the number 2 is displayed.

*Now move on to the next frame*

**93**

## Fractional powers and roots

We have just seen that $8^{\frac{1}{3}} = 2$. We call $8^{\frac{1}{3}}$ the *third root* or, alternatively, the *cube root* of 8 because:

$\left(8^{\frac{1}{3}}\right)^3 = 8$ *the number 8 is the result of raising the 3rd root of 8 to the power 3*

Roots are denoted by such fractional powers. For example, the 5th root of 6 is given as $6^{\frac{1}{5}}$ because:

$$\left(6^{\frac{1}{5}}\right)^5 = 6$$

and by using a calculator $6^{\frac{1}{5}}$ can be seen to be equal to 1·431 to 3 dp. Odd roots are unique in the real number system but even roots are not. For example, there are two 2nd roots – *square roots* – of 4, namely:

$4^{\frac{1}{2}} = 2$ and $4^{\frac{1}{2}} = -2$ because $2 \times 2 = 4$ and $(-2) \times (-2) = 4$

Similarly:

$81^{\frac{1}{4}} = \pm 3$

▶

Odd roots of negative numbers are themselves negative. For example:

$(-32)^{\frac{1}{5}} = -2$ because $\left[(-32)^{\frac{1}{5}}\right]^5 = (-2)^5 = -32$

Even roots of negative numbers, however, pose a problem. For example, because

$\left[(-1)^{\frac{1}{2}}\right]^2 = (-1)^1 = -1$

we conclude that the square root of $-1$ is $(-1)^{\frac{1}{2}}$. Unfortunately, we cannot write this as a decimal number – we cannot find its decimal value because there is no decimal number which when multiplied by itself gives $-1$. We decide to accept the fact that, for now, we cannot find the even roots of a negative number. We shall return to this problem in a later programme when we introduce complex numbers.

## Surds

An alternative notation for the square root of 4 is the surd notation $\sqrt{4}$ and, by convention, this is always taken to mean the positive square root. This notation can also be extended to other roots, for example, $\sqrt[7]{9}$ is an alternative notation for $9^{\frac{1}{7}}$.

Use your calculator to find the value of each of the following roots to 3 dp:

(a) $16^{\frac{1}{7}}$ (b) $\sqrt{8}$ (c) $19^{\frac{1}{4}}$ (d) $\sqrt{-4}$

*Answers in the next frame*

---

94

| | |
|---|---|
| (a) 1·486 | use the fraction key |
| (b) 2·828 | the positive value only |
| (c) $\pm 2{\cdot}088$ | there are two values for even roots |
| (d) We cannot find the square root of a negative number | |

*On now to Frame 95*

---

95

## Multiplication and division by integer powers of 10

If a decimal number is multiplied by 10 raised to an integer power, the decimal point moves the integer number of places to the right if the integer is positive and to the left if the integer is negative. For example:

$1{\cdot}2345 \times 10^3 = 1234{\cdot}5$ (3 places to the right) and

$1{\cdot}2345 \times 10^{-2} = 0{\cdot}012345$ (2 places to the left).

Notice that, for example:

$1{\cdot}2345 \div 10^3 = 1{\cdot}2345 \times 10^{-3}$ and

$1{\cdot}2345 \div 10^{-2} = 1{\cdot}2345 \times 10^2$

So now try these:

(a) $0{\cdot}012045 \times 10^4$

(b) $13{\cdot}5074 \times 10^{-3}$

(c) $144{\cdot}032 \div 10^5$

(d) $0{\cdot}012045 \div 10^{-2}$

*Work all four out and then check your results with the next frame*

**96**

(a) 120·45
(b) 0·0135074
(c) 0·00144032
(d) 1·2045

Because

(a) multiplying by $10^4$ moves the decimal point 4 places to the right
(b) multiplying by $10^{-3}$ moves the decimal point 3 places to the left
(c) $144{\cdot}032 \div 10^5 = 144{\cdot}032 \times 10^{-5}$ move the decimal point 5 places left
(d) $0{\cdot}012045 \div 10^{-2} = 0{\cdot}012045 \times 10^2$ move the decimal point 2 places right

*Now move on to the next frame*

**97**

## Precedence rules

With the introduction of the arithmetic operation of raising to a power we need to amend our earlier precedence rules – *evaluating powers is performed before dividing and multiplying*. For example:

$$\begin{aligned}5(3 \times 4^2 \div 6 - 7) &= 5(3 \times 16 \div 6 - 7)\\ &= 5(48 \div 6 - 7)\\ &= 5(8 - 7)\\ &= 5\end{aligned}$$

So that $14 \div (125 \div 5^3 \times 4 + 3) = \ldots\ldots\ldots\ldots$

*Check your result in the next frame*

**98**

2

Because

$$\begin{aligned}14 \div (125 \div 5^3 \times 4 + 3) &= 14 \div (125 \div 125 \times 4 + 3)\\ &= 14 \div (4 + 3)\\ &= 2\end{aligned}$$

**99**

## Standard form

Any decimal number can be written as a decimal number greater than or equal to 1 and less than 10 (called the *mantissa*) multiplied by the number 10 raised to an appropriate power (the power being called the *exponent*). For example:

$$\begin{aligned}57{\cdot}3 &= 5{\cdot}73 \times 10^1\\ 423{\cdot}8 &= 4{\cdot}238 \times 10^2\\ 6042{\cdot}3 &= 6{\cdot}0423 \times 10^3\end{aligned}$$

and

$$\begin{aligned}0{\cdot}267 &= 2{\cdot}67 \div 10 = 2{\cdot}67 \times 10^{-1}\\ 0{\cdot}000485 &= 4{\cdot}85 \div 10^4 = 4{\cdot}85 \times 10^{-4} \text{ etc.}\end{aligned}$$

So, written in standard form:

(a) 52 674 = ...........
(b) 0·00723 = ...........
(c) 0·0582 = ...........
(d) 1 523 800 = ...........

100

| (a) $5{\cdot}2674 \times 10^4$ | (c) $5{\cdot}82 \times 10^{-2}$ |
|---|---|
| (b) $7{\cdot}23 \times 10^{-3}$ | (d) $1{\cdot}5238 \times 10^6$ |

## Working in standard form

Numbers written in standard form can be multiplied or divided by multiplying or dividing the respective mantissas and adding or subtracting the respective exponents. For example:

$$\begin{aligned} 0{\cdot}84 \times 23\,000 &= (8{\cdot}4 \times 10^{-1}) \times (2{\cdot}3 \times 10^4) \\ &= (8{\cdot}4 \times 2{\cdot}3) \times 10^{-1} \times 10^4 \\ &= 19{\cdot}32 \times 10^3 \\ &= 1{\cdot}932 \times 10^4 \end{aligned}$$

Another example:

$$\begin{aligned} 175{\cdot}4 \div 6340 &= (1{\cdot}754 \times 10^2) \div (6{\cdot}34 \times 10^3) \\ &= (1{\cdot}754 \div 6{\cdot}34) \times 10^2 \div 10^3 \\ &= 0{\cdot}2767 \times 10^{-1} \\ &= 2{\cdot}767 \times 10^{-2} \text{ to 4 sig fig} \end{aligned}$$

Where the result obtained is not in standard form, the mantissa is written in standard number form and the necessary adjustment made to the exponent.

In the same way, then, giving the results in standard form to 4 dp:

(a) $472{\cdot}3 \times 0{\cdot}000564 = \ldots\ldots\ldots\ldots$
(b) $752\,000 \div 0{\cdot}862 = \ldots\ldots\ldots\ldots$

101

| (a) $2{\cdot}6638 \times 10^{-1}$ |
|---|
| (b) $8{\cdot}7239 \times 10^5$ |

Because

(a) $$\begin{aligned} 472{\cdot}3 \times 0{\cdot}000564 &= (4{\cdot}723 \times 10^2) \times (5{\cdot}64 \times 10^{-4}) \\ &= (4{\cdot}723 \times 5{\cdot}64) \times 10^2 \times 10^{-4} \\ &= 26{\cdot}638 \times 10^{-2} = 2{\cdot}6638 \times 10^{-1} \end{aligned}$$

(b) $$\begin{aligned} 752\,000 \div 0{\cdot}862 &= (7{\cdot}52 \times 10^5) \div (8{\cdot}62 \times 10^{-1}) \\ &= (7{\cdot}52 \div 8{\cdot}62) \times 10^5 \times 10^1 \\ &= 0{\cdot}87239 \times 10^6 = 8{\cdot}7239 \times 10^5 \end{aligned}$$

For *addition and subtraction in standard form* the approach is slightly different.

**Example 1**

$4{\cdot}72 \times 10^3 + 3{\cdot}648 \times 10^4$

Before these can be added, the powers of 10 must be made the same:

$$\begin{aligned} 4{\cdot}72 \times 10^3 + 3{\cdot}648 \times 10^4 &= 4{\cdot}72 \times 10^3 + 36{\cdot}48 \times 10^3 \\ &= (4{\cdot}72 + 36{\cdot}48) \times 10^3 \\ &= 41{\cdot}2 \times 10^3 = 4{\cdot}12 \times 10^4 \text{ in standard form} \end{aligned}$$

Similarly in the next example.

▶

**Example 2**

$13{\cdot}26 \times 10^{-3} - 1{\cdot}13 \times 10^{-2}$

Here again, the powers of 10 must be equalized:

$$13{\cdot}26 \times 10^{-3} - 1{\cdot}13 \times 10^{-2} = 1{\cdot}326 \times 10^{-2} - 1{\cdot}13 \times 10^{-2}$$
$$= (1{\cdot}326 - 1{\cdot}13) \times 10^{-2}$$
$$= 0{\cdot}196 \times 10^{-2} = 1{\cdot}96 \times 10^{-3} \text{ in standard form}$$

## Using a calculator

Numbers given in standard form can be manipulated on the Casio calculator by making use of the Mode key.

Press SHIFT followed by MODE
Select option 7 : Sci — This is the standard form display
Enter 5 — This is the number of significant figures

Now enter the number 12345 and press the = key to give the display:

$1{\cdot}2345 \times 10^4$

Numbers can be entered in standard form as well.

Press AC to clear the screen
Enter the number 1·234
Press ×
Press SHIFT followed by the log key (this accesses $10^{\blacksquare}$)
Enter the number 3
Press =

The number displayed is then $1{\cdot}234 \times 10^3$. We can also manipulate numbers in standard form on the calculator. For example:

Enter $1{\cdot}234 \times 10^3 + 2{\cdot}6 \times 10^2$ and press =

This results in the display:

$1{\cdot}4940 \times 10^3$

Therefore, working in standard form:

(a) $43{\cdot}6 \times 10^2 + 8{\cdot}12 \times 10^3$ = ...........
(b) $7{\cdot}84 \times 10^5 - 12{\cdot}36 \times 10^3$ = ...........
(c) $4{\cdot}25 \times 10^{-3} + 1{\cdot}74 \times 10^{-2}$ = ...........

---

**102**

(a) $1{\cdot}248 \times 10^4$
(b) $7{\cdot}7164 \times 10^5$
(c) $2{\cdot}165 \times 10^{-2}$

## Preferred standard form

In the SI system of units, it is recommended that when a number is written in standard form, the power of 10 should be restricted to powers of $10^3$, i.e. $10^3$, $10^6$, $10^{-3}$, $10^{-6}$, etc. Therefore in this *preferred standard form* up to three figures may appear in front of the decimal point.

In practice it is best to write the number first in standard form and to adjust the power of 10 to express this in preferred standard form.

**Example 1**

$5{\cdot}2746 \times 10^4$ in standard form
$= 5{\cdot}2746 \times 10 \times 10^3$
$= 52{\cdot}746 \times 10^3$ in preferred standard form

**Example 2**

$3{\cdot}472 \times 10^8$ in standard form
$= 3{\cdot}472 \times 10^2 \times 10^6$
$= 347{\cdot}2 \times 10^6$ in preferred standard form

**Example 3**

$3{\cdot}684 \times 10^{-2}$ in standard form
$= 3{\cdot}684 \times 10 \times 10^{-3}$
$= 36{\cdot}84 \times 10^{-3}$ in preferred standard form

If you are using the Casio calculator then all this can be achieved by using the ENG key whose action is to alter the display of a number in standard form to preferred standard form. For example,

Enter the number 123456 and press the = key
Press the ENG key and the display changes to $123{\cdot}456 \times 10^3$

So, rewriting the following in preferred standard form, we have

(a) $8{\cdot}236 \times 10^7 =$ ...........
(b) $1{\cdot}624 \times 10^{-4} =$ ...........
(c) $4{\cdot}827 \times 10^4 =$ ...........
(d) $6{\cdot}243 \times 10^5 =$ ...........
(e) $3{\cdot}274 \times 10^{-2} =$ ...........
(f) $5{\cdot}362 \times 10^{-7} =$ ...........

---

**103**

(a) $82{\cdot}36 \times 10^6$
(b) $162{\cdot}4 \times 10^{-6}$
(c) $48{\cdot}27 \times 10^3$
(d) $624{\cdot}3 \times 10^3$
(e) $32{\cdot}74 \times 10^{-3}$
(f) $536{\cdot}2 \times 10^{-9}$

One final exercise on this piece of work:

**Example 4**

The product of $(4{\cdot}72 \times 10^2)$ and $(8{\cdot}36 \times 10^5)$

(a) in standard form = ...........
(b) in preferred standard form = ...........

---

**104**

(a) $3{\cdot}9459 \times 10^8$
(b) $394{\cdot}59 \times 10^6$

Because

(a) $(4{\cdot}72 \times 10^2) \times (8{\cdot}36 \times 10^5) = (4{\cdot}72 \times 8{\cdot}36) \times 10^2 \times 10^5$
$= 39{\cdot}459 \times 10^7$
$= 3{\cdot}9459 \times 10^8$ in standard form

(b) $(4{\cdot}72 \times 10^2) \times (8{\cdot}36 \times 10^5) = 3{\cdot}9459 \times 10^2 \times 10^6$
$= 394{\cdot}59 \times 10^6$ in preferred standard form

*Now move on to the next frame*

## 105 Checking calculations

When performing a calculation involving decimal numbers it is always a good idea to check that your result is reasonable and that an arithmetic blunder or an error in using the calculator has not been made. This can be done using standard form. For example:

$$\begin{aligned}59{\cdot}2347 \times 289{\cdot}053 &= 5{\cdot}92347 \times 10^1 \times 2{\cdot}89053 \times 10^2\\ &= 5{\cdot}92347 \times 2{\cdot}89053 \times 10^3\end{aligned}$$

This product can then be estimated for reasonableness as:

$6 \times 3 \times 1000 = 18\,000$ (see Frames 22–24)

The answer using the calculator is 17 121·968 to three decimal places, which is 17 000 when rounded to the nearest 1000. This compares favourably with the estimated 18 000, indicating that the result obtained could be reasonably expected.

So, the estimated value of $800{\cdot}120 \times 0{\cdot}007953$ is ...........

*Check with the next frame*

## 106

6·4

Because

$$\begin{aligned}800{\cdot}120 \times 0{\cdot}007953 &= 8{\cdot}00120 \times 10^2 \times 7{\cdot}953 \times 10^{-3}\\ &= 8{\cdot}00120 \times 7{\cdot}9533 \times 10^{-1}\end{aligned}$$

This product can then be estimated for reasonableness as:

$8 \times 8 \div 10 = 6{\cdot}4$

The exact answer is 6·36 to two decimal places.

*Now move on to the next frame*

## 107 Accuracy

Many calculations are made using numbers that have been obtained from measurements. Such numbers are only accurate to a given number of significant figures but using a calculator can produce a result that contains as many figures as its display will permit. Because any calculation involving measured values will not be accurate to *more significant figures than the least number of significant figures in any measurement*, we can justifiably round the result down to a more manageable number of significant figures. For example:

The base length and height of a rectangle are measured as 114·8 mm and 18 mm respectively. The area of the rectangle is given as the product of these lengths. Using a calculator this product is 2066·4 mm$^2$. Because one of the lengths is only measured to 2 significant figures, the result cannot be accurate to more than 2 significant figures. It should therefore be read as 2100 mm$^2$.

Assuming the following contains numbers obtained by measurement, use a calculator to find the value to the correct level of accuracy:

$19{\cdot}1 \times 0{\cdot}0053 \div 13{\cdot}345$

108

0·0076

Because

The calculator gives the result as 0·00758561 but because 0·0053 is only accurate to 2 significant figures the result cannot be accurate to more than 2 significant figures, namely 0·0076.

***At this point let us pause and summarize the main facts so far on powers***

## Review summary

109

1 Powers are devised from repetitive multiplication of a given number.
2 Negative powers denote reciprocals and any number raised to the power 0 is unity.
3 Multiplication of a decimal number by 10 raised to an integer power moves the decimal point to the right if the power is positive and to the left if the power is negative.
4 A decimal number written in standard form is in the form of a mantissa (a number between 1 and 10 but excluding 10) multiplied by 10 raised to an integer power, the power being called the exponent.
5 Writing decimal numbers in standard form permits an estimation of the reasonableness of a calculation.
6 In preferred standard form the powers of 10 in the exponent are restricted to multiples of 3.
7 If numbers used in a calculation are obtained from measurement, the result of the calculation is a number accurate to no more than the least number of significant figures in any measurement.

## Review exercise

110

1 Write each of the following as a number raised to a power:
(a) $5^8 \times 5^2$ (b) $6^4 \div 6^6$ (c) $(7^4)^3$ (d) $(19^{-8})^0$
2 Find the value of each of the following to 3 dp:
(a) $16^{\frac{1}{4}}$ (b) $\sqrt[3]{3}$ (c) $(-8)^{\frac{1}{5}}$ (d) $(-7)^{\frac{1}{4}}$
3 Write each of the following as a single decimal number:
(a) $1{\cdot}0521 \times 10^3$ (b) $123{\cdot}456 \times 10^{-2}$
(c) $0{\cdot}0135 \div 10^{-3}$ (d) $165{\cdot}21 \div 10^4$
4 Write each of the following in standard form:
(a) 125·87 (b) 0·0101 (c) 1·345 (d) 10·13
5 Write each of the following in preferred standard form:
(a) $1{\cdot}3204 \times 10^5$ (b) 0·0101 (c) 1·345 (d) $9{\cdot}5032 \times 10^{-8}$

▶

6 In each of the following the numbers have been obtained by measurement. Evaluate each calculation to the appropriate level of accuracy:

(a) $13{\cdot}6 \div 0{\cdot}012 \times 7{\cdot}63 - 9015$ (b) $\dfrac{0{\cdot}003 \times 194}{13{\cdot}6}$

(c) $19{\cdot}3 \times 1{\cdot}04^{2{\cdot}00}$ (d) $\dfrac{18 \times 2{\cdot}1 - 3{\cdot}6 \times 0{\cdot}54}{8{\cdot}6 \times 2{\cdot}9 + 5{\cdot}7 \times 9{\cdot}2}$

---

**111**

1 (a) $5^8 \times 5^2 = 5^{8+2} = 5^{10}$ (b) $6^4 \div 6^6 = 6^{4-6} = 6^{-2}$ (c) $(7^4)^3 = 7^{4\times3} = 7^{12}$

(d) $(19^{-8})^0 = 1$ as any number raised to the power 0 equals unity

2 (a) $16^{\frac{1}{4}} = \pm 2{\cdot}000$ (b) $\sqrt[3]{3} = 1{\cdot}442$ (c) $5(-8)^{\frac{1}{5}} = -1{\cdot}516$

(d) $(-7)^{\frac{1}{4}}$ You cannot find the even root of a negative number

3 (a) $1{\cdot}0521 \times 10^3 = 1052{\cdot}1$ (b) $123{\cdot}456 \times 10^{-2} = 1{\cdot}23456$

(c) $0{\cdot}0135 \div 10^{-3} = 0{\cdot}0135 \times 10^3 = 13{\cdot}5$

(d) $165{\cdot}21 \div 10^4 = 165{\cdot}21 \times 10^{-4} = 0{\cdot}016521$

4 (a) $125{\cdot}87 = 1{\cdot}2587 \times 10^2$ (b) $0{\cdot}0101 = 1{\cdot}01 \times 10^{-2}$

(c) $1{\cdot}345 = 1{\cdot}345 \times 10^0$ (d) $10{\cdot}13 = 1{\cdot}013 \times 10^1 = 1{\cdot}013 \times 10$

5 (a) $1{\cdot}3204 \times 10^5 = 132{\cdot}04 \times 10^3$ (b) $0{\cdot}0101 = 10{\cdot}1 \times 10^{-3}$

(c) $1{\cdot}345 = 1{\cdot}345 \times 10^0$ (d) $9{\cdot}5032 \times 10^{-8} = 95{\cdot}032 \times 10^{-9}$

6 (a) $13{\cdot}6 \div 0{\cdot}012 \times 7{\cdot}63 - 9015 = -367{\cdot}\dot{6} = -370$ to 2 sig fig

(b) $\dfrac{0{\cdot}003 \times 194}{13{\cdot}6} = 0{\cdot}042794\ldots = 0{\cdot}04$ to 1 sig fig

(c) $19{\cdot}3 \times 1{\cdot}04^{2{\cdot}00} = 19{\cdot}3 \times 1{\cdot}0816 = 20{\cdot}87488 = 20{\cdot}9$ to 3 sig fig

(d) $\dfrac{18 \times 2{\cdot}1 - 3{\cdot}6 \times 0{\cdot}54}{8{\cdot}6 \times 2{\cdot}9 + 5{\cdot}7 \times 9{\cdot}2} = \dfrac{35{\cdot}856}{77{\cdot}38} = 0{\cdot}46337554\ldots = 0{\cdot}46$ to 2 sig fig

*On now to the next topic*

---

## Number systems

**112**

### 1 Denary (or decimal) system

This is our basic system in which quantities large or small can be represented by use of the symbols 0, 1, 2, 3, 4, 5, 6, 7, 8, 9 together with appropriate place values according to their positions.

| For example | 2 | 7 | 6 | 5 | · | 3 | 2 | $1_{10}$ |
|---|---|---|---|---|---|---|---|---|
| has place values | $10^3$ | $10^2$ | $10^1$ | $10^0$ | | $10^{-1}$ | $10^{-2}$ | $10^{-3}$ |
| | 1000 | 100 | 10 | 1 | | $\frac{1}{10}$ | $\frac{1}{100}$ | $\frac{1}{1000}$ |

In this case, the place values are powers of 10, which gives the name *denary* (or *decimal*) to the system. The denary system is said to have a *base* of 10. You are, of course, perfectly familiar with this system of numbers, but it is included here as it leads on to other systems which have the same type of structure but which use different place values.

*So let us move on to the next system*

113

**2 Binary system**

This is widely used in all forms of switching applications. The only symbols used are 0 and 1 and the place values are powers of 2, i.e. the system has a base of 2.

| For example | 1 | 0 | 1 | 1 | · | 1 | 0 | $1_2$ |
|---|---|---|---|---|---|---|---|---|
| has place values | $2^3$ | $2^2$ | $2^1$ | $2^0$ | | $2^{-1}$ | $2^{-2}$ | $2^{-3}$ |
| i.e. | 8 | 4 | 2 | 1 | | $\frac{1}{2}$ | $\frac{1}{4}$ | $\frac{1}{8}$ |

So: 1 0 1 1 · 1 0 1 in the binary system

$= 1\times 8 \quad 0\times 4 \quad 1\times 2 \quad 1\times 1 \quad 1\times\frac{1}{2} \quad 0\times\frac{1}{4} \quad 1\times\frac{1}{8}$

$= 8 + 0 + 2 + 1 + \frac{1}{2} + 0 + \frac{1}{8}$ in denary

$= 11\frac{5}{8} = 11{\cdot}625$ in the denary system.

Therefore $1011{\cdot}101_2 = 11{\cdot}625_{10}$

The small subscripts 2 and 10 indicate the bases of the two systems. In the same way, the denary equivalent of $1\ 1\ 0\ 1\cdot 0\ 1\ 1_2$ is:

........... to 3dp.

114

$13{\cdot}375_{10}$

Because

$$\begin{array}{l} 1 \quad 1 \quad 0 \quad 1 \cdot 0 \quad 1 \quad 1_2 \\ = 8 + 4 + 0 + 1 + 0 + \frac{1}{4} + \frac{1}{8} \end{array}$$

$= 13\frac{3}{8} = 13{\cdot}375_{10}$

**3 Octal system (base 8)**

This system uses the symbols

0, 1, 2, 3, 4, 5, 6, 7

with place values that are powers of 8.

| For example | 3 | 5 | 7 | · | 3 | 2 | 1 | in the octal system |
|---|---|---|---|---|---|---|---|---|
| has place values | $8^2$ | $8^1$ | $8^0$ | | $8^{-1}$ | $8^{-2}$ | $8^{-3}$ | |
| i.e. | 64 | 8 | 1 | | $\frac{1}{8}$ | $\frac{1}{64}$ | $\frac{1}{512}$ | |

So 3 5 7 · 3 2 $1_8$

$= 3\times 64 \quad 5\times 8 \quad 7\times 1 \quad 3\times\frac{1}{8} \quad 2\times\frac{1}{64} \quad 1\times\frac{1}{512}$

$= 192 + 40 + 7 + \frac{3}{8} + \frac{1}{32} + \frac{1}{512}$

$= 239\frac{209}{512} = 239{\cdot}4082_{10}$

That is

$357{\cdot}321_8 = 239{\cdot}408_{10}$ to 3 dp

As you see, the method is very much as before: the only change is in the base of the place values.

In the same way then, $263{\cdot}452_8$ expressed in denary form is

........... to 3 dp.

**115**

$179{\cdot}582_{10}$

Because

$$\begin{aligned}&2\,6\,3\cdot 4\,5\,2_8\\&=2\times 8^2+6\times 8^1+3\times 8^0+4\times 8^{-1}+5\times 8^{-2}+2\times 8^{-3}\\&=2\times 64+6\times 8+3\times 1+4\times \tfrac{1}{8}+5\times \tfrac{1}{64}+2\times \tfrac{1}{512}\\&=128+48+3+\tfrac{1}{2}+\tfrac{5}{64}+\tfrac{1}{256}\\&=179\tfrac{149}{256}\\&=179{\cdot}582_{10}\text{ to 3 dp}\end{aligned}$$

Now we come to the duodecimal system, which has a base of 12.

*So move on to the next frame*

**116**

**4 Duodecimal system (base 12)**

With a base of 12, the units column needs to accommodate symbols up to 11 before any carryover to the second column occurs. Unfortunately, our denary symbols go up to only 9, so we have to invent two extra symbols to represent the values 10 and 11. Several suggestions for these have been voiced in the past, but we will adopt the symbols X and Λ for 10 and 11 respectively. The first of these calls to mind the Roman numeral for 10 and the Λ symbol may be regarded as the two strokes of 11 tilted together 1 1 to join at the top.

The duodecimal system, therefore, uses the symbols

0, 1, 2, 3, 4, 5, 6, 7, 8, 9, X, Λ

and has place values that are powers of 12.

| | | | | | | | |
|---|---|---|---|---|---|---|---|
| For example | 2 | X | 5 | · | 1 | 3 | $6_{12}$ |
| has place values | $12^2$ | $12^1$ | $12^0$ | | $12^{-1}$ | $12^{-2}$ | $12^{-3}$ |
| i.e. | 144 | 12 | 1 | | $\frac{1}{12}$ | $\frac{1}{144}$ | $\frac{1}{1728}$ |

So 2 X 5 · 1 3 $6_{12}$

$$=2\times 144+10\times 12+5\times 1+1\times \tfrac{1}{12}+3\times \tfrac{1}{144}+6\times \tfrac{1}{1728}$$

$= \ldots\ldots\ldots\ldots_{10}$ to 3 dp

*Finish it off*

**117**

$413{\cdot}108_{10}$

Because

$$\begin{aligned}&2\,X\,5\cdot 1\,3\,6_{12}\\&=288+120+5+\tfrac{1}{12}+\tfrac{1}{48}+\tfrac{1}{288}\\&=413\tfrac{31}{288}\end{aligned}$$

Therefore $2X5{\cdot}136_{12}=413{\cdot}108_{10}$ to 3 dp.

**5 Hexadecimal system (base 16)**

This system has computer applications. The symbols here need to go up to an equivalent denary value of 15, so, after 9, letters of the alphabet are used as follows:

0, 1, 2, 3, 4, 5, 6, 7, 8, 9, A, B, C, D, E, F

The place values in this system are powers of 16.

| | | | | | | | |
|---|---|---|---|---|---|---|---|
| For example | 2 | A | 7 | · | 3 | E | $2_{16}$ |
| has place values | $16^2$ | $16^1$ | $16^0$ | | $16^{-1}$ | $16^{-2}$ | $16^{-3}$ |
| i.e. | 256 | 16 | 1 | | $\frac{1}{16}$ | $\frac{1}{256}$ | $\frac{1}{4096}$ |

Therefore $2\,A\,7 \cdot 3\,E\,2_{16} =$ ........... to 3 dp.

---

**118**

$679{\cdot}243_{10}$

Here it is: $2\;A\;7 \cdot 3\;E\;2_{16}$

$$= 2 \times 256 + 10 \times 16 + 7 \times 1 + 3 \times \frac{1}{16} + 14 \times \frac{1}{256} + 2 \times \frac{1}{4096}$$

$$= 679\tfrac{497}{2048} = 679{\cdot}243_{10} \text{ to 3 dp.}$$

And now, two more by way of practice.

Express the following in denary form:

(a) $3\,\Lambda\,4 \cdot 2\,6\,5_{12}$
(b) $3\,C\,4 \cdot 2\,1\,F_{16}$

*Finish both of them and check the working with the next frame*

---

**119**

Here they are:

(a)

| | | | | | | | |
|---|---|---|---|---|---|---|---|
| | 3 | Λ | 4 | · | 2 | 6 | $5_{12}$ |
| Place values | 144 | 12 | 1 | | $\frac{1}{12}$ | $\frac{1}{144}$ | $\frac{1}{1728}$ |

So $3\,\Lambda\,4 \cdot 2\,6\,5_{12}$

$$= 3 \times 144 + 11 \times 12 + 4 \times 1 + 2 \times \frac{1}{12} + 6 \times \frac{1}{144} + 5 \times \frac{1}{1728}$$

$$= 432 + 132 + 4 + \frac{1}{6} + \frac{1}{24} + \frac{5}{1728}$$

$$= 568\tfrac{365}{1728} = 568{\cdot}211_{10}$$

Therefore $3\,\Lambda\,4 \cdot 265_{12} = 568{\cdot}211_{10}$ to 3 dp.

(b)

| | | | | | | | |
|---|---|---|---|---|---|---|---|
| | 3 | C | 4 | · | 2 | 1 | $F_{16}$ |
| Place values | 256 | 16 | 1 | | $\frac{1}{16}$ | $\frac{1}{256}$ | $\frac{1}{4096}$ |

So $3\,C\,4 \cdot 2\,1\,F_{16}$

$$= 3 \times 256 + 12 \times 16 + 4 \times 1 + 2 \times \frac{1}{16} + 1 \times \frac{1}{256} + 15 \times \frac{1}{4096}$$

$$= 768 + 192 + 4 + \frac{1}{8} + \frac{1}{256} + \frac{15}{4096}$$

$$= 964\tfrac{543}{4096} = 964{\cdot}133_{10}$$

Therefore $3\,C\,4 \cdot 2\,1\,F_{16} = 964{\cdot}133_{10}$ to 3 dp.

**120**

## An alternative method

So far, we have changed numbers in various bases into the equivalent denary numbers from first principles. Another way of arriving at the same results is by using the fact that two adjacent columns differ in place values by a factor which is the base of the particular system. An example will show the method.

Express the octal $357{\cdot}121_8$ in denary form.

First of all, we will attend to the whole-number part $357_8$. Starting at the left-hand end, multiply the first column by the base 8 and add the result to the entry in the next column (making 29).

$$\begin{array}{ccccc} 3 & & 5 & & 7 \\ \times\ 8 & \rightarrow & 24 & \rightarrow & 232 \\ \hline 24 & & 29 & & 239 \\ & & \times\ 8 & & \\ \hline & & 232 & & \end{array}$$

Now repeat the process. Multiply the second column total by 8 and add the result to the next column. This gives 239 in the units column.

So $357_8 = 239_{10}$

*Now we do much the same with the decimal part of the octal number*

**121**

The decimal part is $0 \cdot 1\,2\,1_8$

$$\begin{array}{cccccc} 0 & \cdot\ 1 & & 2 & & 1 \\ & \times\ 8 & \rightarrow & 8 & \rightarrow & 80 \\ \hline & 8 & & 10 & & 81 \\ & & & \times\ 8 & & \\ \hline & & & 80 & & \end{array}$$

Starting from the left-hand column immediately following the decimal point, multiply by 8 and add the result to the next column. Repeat the process, finally getting a total of 81 in the end column.

But the place value of this column is ...........

**122**

$8^{-3}$

The denary value of $0{\cdot}121_8$ is $81 \times 8^{-3}$ i.e. $81 \times \dfrac{1}{8^3} = \dfrac{81}{512} = 0{\cdot}1582_{10}$

Collecting the two partial results together, $357{\cdot}121_8 = 239{\cdot}1582_{10}$ to 4 dp.

In fact, we can set this out across the page to save space, thus:

$$\begin{array}{ccccccccccc} 3 & & 5 & & 7 & \cdot & 1 & & 2 & & 1 \\ \times\ 8 & \rightarrow & 24 & \rightarrow & 232 & & \times\ 8 & \rightarrow & 8 & \rightarrow & 80 \\ \hline 24 & & 29 & & 239 & & 8 & & 10 & & 81 \\ & & \times\ 8 & & & & & & \times\ 8 & & \\ \hline & & 232 & & & & & & 80 & & \end{array}$$

$81 \times \dfrac{1}{8^3} = \dfrac{81}{512} = 0{\cdot}1582_{10}$ Therefore $357{\cdot}121_8 = 239{\cdot}158_{10}$

▶

Now you can set this one out in similar manner.

Express the duodecimal $245{\cdot}136_{12}$ in denary form.

Space out the duodecimal digits to give yourself room for the working:

$2\ \ 4\ \ 5\ \cdot\ 1\ \ 3\ \ 6_{12}$

Then off you go.

$245{\cdot}136_{12} =$ ........... to 4 dp.

---

**123**

$341{\cdot}1076_{10}$

Here is the working as a check:

| 2 | 4 | 5 · | 1 | 3 | $6_{12}$ |
|---|---|---|---|---|---|
| × 12 | 24 | 336 | × 12 | 12 | 180 |
| 24 | 28 | 341 | 12 | 15 | 186 |
| | × 12 | | | × 12 | |
| | 336 | | | 180 | |

Place value of last column is $12^{-3}$, therefore

$$0{\cdot}136_{12} = 186 \times 12^{-3} = \frac{186}{1728} = 0{\cdot}1076_{10}$$

So $245{\cdot}136_{12} = 341{\cdot}1076_{10}$ to 4 dp.

*On to the next*

---

**124**

Now for an easy one. Find the denary equivalent of the binary number $1\,1\,0\,1\,1\cdot1\,0\,1\,1_2$.

Setting it out in the same way, the result is ........... to 4 dp.

---

**125**

$27{\cdot}6875_{10}$

| 1 | 1 | 0 | 1 | 1 · | 1 | 0 | 1 | 1 |
|---|---|---|---|---|---|---|---|---|
| ×2 | 2 | 6 | 12 | 26 | ×2 | 2 | 4 | 10 |
| 2 | 3 | 6 | 13 | 27 | 2 | 2 | 5 | 11 |
| | ×2 | ×2 | ×2 | | | ×2 | ×2 | |
| | 6 | 12 | 26 | | | 4 | 10 | |

$$11 \times 2^{-4} = \frac{11}{16} = 0{\cdot}6875_{10}$$

Therefore $1\,1\,0\,1\,1\cdot1\,0\,1\,1_2 = 27{\cdot}6875_{10}$ to 4 dp.

And now a hexadecimal. Express $4\,C\,5\cdot2\,B\,8_{16}$ in denary form. Remember that C = 12 and B = 11. There are no snags.

$4\,C\,5\cdot2\,B\,8_{16} =$ ........... to 4 dp.

**126**

$1221{\cdot}1699_{10}$

Here it is:

| 4 | C | 5 · | 2 | B | 8 |
|---|---|---|---|---|---|
| × 16 → | 64 | → 1216 | × 16 → | 32 | → 688 |
| 64 | 76 | 1221 | 32 | 43 | 696 |
| | × 16 | | | × 16 | |
| | 456 | | | 258 | |
| | 76 | | | 43 | |
| | 1216 | | | 688 | |

Place value $16^{-3}$

So $\dfrac{696}{4096} = 0{\cdot}16992_{10}$

Therefore $4\,C\,5 \cdot 2\,B\,8_{16} = 1221{\cdot}1699_{10}$ to 4 dp.

They are all done in the same way.

By now then, we can change any binary, octal, duodecimal or hexadecimal number – including a decimal part – into the equivalent denary number. So here is a brief review summary followed by a short review exercise for valuable practice. Complete the set and check your results with Frame 129.

## Review summary

**127**

1 *Denary (or decimal) system*: Base 10. Place values powers of 10. Symbols 0, 1, 2, 3, 4, 5, 6, 7, 8, 9.

2 *Binary system*: Base 2. Place values powers of 2. Symbols 0, 1.

3 *Octal system*: Base 8. Place values powers of 8. Symbols 0, 1, 2, 3, 4, 5, 6, 7.

4 *Duodecimal system*: Base 12. Place values powers of 12. Symbols 0, 1, 2, 3, 4, 5, 6, 7, 8, 9, X, Λ

5 *Hexadecimal system*: Base 16. Place values powers of 16. Symbols 0, 1, 2, 3, 4, 5, 6, 7, 8, 9, A, B, C, D, E, F.

*Now move on to the next frame*

## Review exercise

**128**

Express each of the following in denary form to 3 dp.

(a) $1\,1\,0\,0\,1 \cdot 1\,1_2$

(b) $7\,7\,6 \cdot 1\,4\,3_8$

(c) $4\,X\,9 \cdot 2\,\Lambda\,5_{12}$

(d) $6\,F\,8 \cdot 3\,D\,5_{16}$

**129**

| |
|---|
| (a) $25{\cdot}750_{10}$ |
| (b) $510{\cdot}193_{10}$ |
| (c) $705{\cdot}246_{10}$ |
| (d) $1784{\cdot}240_{10}$ |

Just in case you have made a slip anywhere, here is the working.

(a)
$$\begin{array}{ccccccc}
1 & 1 & 0 & 0 & 1 & \cdot\ 1 & 1\\
\underline{\times 2} & \underline{2} & \underline{6} & \underline{12} & \underline{24} & \underline{\times 2} & \underline{2}\\
\underline{2} & 3 & 6 & 12 & \underline{25_{10}} & \underline{2} & \underline{3}\times 2^{-2} = \frac{3}{4} = 0{\cdot}75_{10}\\
 & \underline{\times 2} & \underline{\times 2} & \underline{\times 2} & & & \\
 & \underline{6} & \underline{12} & \underline{24} & & &
\end{array}$$

$$11001{\cdot}11_2 = 25{\cdot}750_{10}$$

(b)
$$\begin{array}{cccccc}
7 & 7 & 6 & \cdot\ 1 & 4 & 3\\
\underline{\times 8} & \underline{56} & \underline{504} & \underline{\times 8} & \underline{8} & \underline{96}\\
\underline{56} & 63 & \underline{510_{10}} & \underline{8} & 12 & \underline{99}\times 8^{-3} = \frac{99}{512}\\
 & \underline{\times 8} & & & \underline{\times 8} & = 0{\cdot}19336_{10}\\
 & \underline{504} & & & \underline{96} &
\end{array}$$

$$776{\cdot}143_8 = 510{\cdot}193_{10}$$

(c)
$$\begin{array}{cccccc}
4 & X & 9 & \cdot\ 2 & \Lambda & 5\\
\underline{\times 12} & \underline{48} & \underline{696} & \underline{\times 12} & \underline{24} & \underline{420}\\
\underline{48} & 58 & \underline{705_{10}} & \underline{24} & 35 & \underline{425}\times 12^{-3} = \frac{425}{1728}\\
 & \underline{\times 12} & & & \underline{\times 12} & = 0{\cdot}2459_{10}\\
 & \underline{696} & & & \underline{420} &
\end{array}$$

$$4X9{\cdot}2\Lambda5_{12} = 705{\cdot}246_{10}$$

(d)
$$\begin{array}{cccccc}
6 & F & 8 & \cdot\ 3 & D & 5\\
\underline{\times 16} & \underline{96} & \underline{1776} & \underline{\times 16} & \underline{48} & \underline{976}\\
\underline{96} & 111 & \underline{1784_{10}} & \underline{48} & 61 & \underline{981}\times 16^{-3} = \frac{981}{4096}\\
 & \underline{\times 16} & & & \underline{\times 16} & = 0{\cdot}2395_{10}\\
 & 666 & & & 366 & \\
 & \underline{111} & & & \underline{61} & \\
 & \underline{1776} & & & \underline{976} &
\end{array}$$

$$6F8{\cdot}3D5_{16} = 1784{\cdot}240_{10}$$

In all the previous examples, we have changed binary, octal, duodecimal and hexadecimal numbers into their equivalent denary forms. The reverse process is also often required, so we will now see what is involved.

*So on then to the next frame*

# Change of base from denary to a new base

**130**

### Binary form

The simplest way to do this is by repeated division by 2 (the new base), noting the remainder at each stage. Continue dividing until a final zero quotient is obtained.

For example, to change $245_{10}$ to binary:

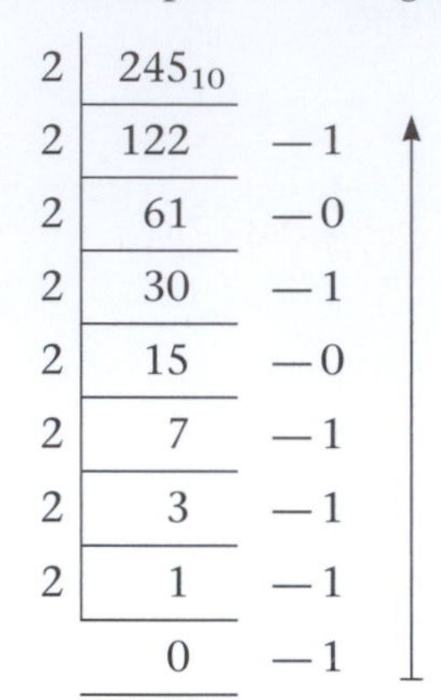


Now write all the remainders in the reverse order, i.e. from bottom to top.

Then $245_{10} = 11110101_2$

### Octal form

The method here is exactly the same except that we divide repeatedly by 8 (the new base). So, without more ado, changing $524_{10}$ to octal gives ............

**131**

$1014_8$

For:

| | | |
|---|---|---|
| 8 | $524_{10}$ | |
| 8 | 65 | — 4 |
| 8 | 8 | — 1 |
| 8 | 1 | — 0 |
| | 0 | — 1 |

As before, write the remainders in order, i.e. from bottom to top.

$\therefore\ 524_{10} = 1014_8$

### Duodecimal form

Method as before, but this time we divide repeatedly by 12.

So $897_{10} =$ ............

**132**

$629_{12}$

Because

| | | |
|---|---|---|
| 12 | $897_{10}$ | |
| 12 | 74 | — 9 |
| 12 | 6 | — 2 |
| | 0 | — 6 |

$\therefore\ 897_{10} = 629_{12}$

*Now move to the next frame*

133

The method we have been using is quick and easy enough when the denary number to be changed is a whole number. When it contains a decimal part, we must look further.

**A denary decimal in octal form**

To change $0{\cdot}526_{10}$ to octal form, we multiply the decimal repeatedly by the new base, in this case 8, but on the second and subsequent multiplication, we do not multiply the whole-number part of the previous product.

$$\begin{array}{r@{\,\cdot\,}l}
0 & 526_{10} \\
 & \quad 8 \\ \hline
4 & 208 \\
 & \quad 8 \\ \hline
1 & 664 \\
 & \quad 8 \\ \hline
5 & 312 \\
 & \quad 8 \\ \hline
2 & 496 \\ \hline
\end{array}$$

Now multiply again by 8, but treat only the decimal part

and so on

Finally, we write the whole-number numerals downwards to form the required octal decimal.

Be careful *not* to include the zero unit digit in the original denary decimal. In fact, it may be safer simply to write the decimal as $\cdot 526_{10}$ in the working.

So $$0{\cdot}526_{10} = 0{\cdot}4152_8$$

Converting a denary decimal into any new base is done in the same way. If we express $0{\cdot}306_{10}$ as a duodecimal, we get ...........

*Set it out in the same way: there are no snags*

134

$0{\cdot}3809_{12}$

$$\begin{array}{r@{\,\cdot\,}l}
 & 306_{10} \\
 & \quad 12 \\ \hline
3 & 672 \\
 & \quad 12 \\ \hline
8 & 064 \\
 & \quad 12 \\ \hline
0 & 768 \\
 & \quad 12 \\ \hline
9 & 216 \\
 & \quad 12 \\ \hline
2 & 592 \\ \hline
\end{array}$$

There is no carryover into the units column, so enter a zero.

etc.

$$\therefore\ 0{\cdot}306_{10} = 0{\cdot}3809_{12}$$

*Now let us go one stage further – so on to the next frame*

**135**

If the denary number consists of both a whole-number and a decimal part, the two parts are converted separately and united in the final result. The example will show the method.

Express $492{\cdot}731_{10}$ in octal form:

$$\begin{array}{r|l l}
8 & 492_{10} & \\
8 & 61 & -4 \\
8 & 7 & -5 \\
 & 0 & -7
\end{array}
\qquad
\begin{array}{r l}
 & \cdot\ 731_{10} \\
 & \quad\ 8 \\
\hline
5 & \cdot\ 848 \\
 & \quad\ 8 \\
\hline
6 & \cdot\ 784 \\
 & \quad\ 8 \\
\hline
6 & \cdot\ 272 \\
 & \quad\ 8 \\
\hline
2 & \cdot\ 176
\end{array}$$

Then $492{\cdot}731_{10} = 754{\cdot}5662_8$

In similar manner, $384{\cdot}426_{10}$ expressed in duodecimals becomes ...........

*Set the working out in the same way*

**136**

$280{\cdot}5142_{12}$

Because we have:

$$\begin{array}{r|l l}
12 & 384_{10} & \\
12 & 32 & -0 \\
12 & 2 & -8 \\
 & 0 & -2
\end{array}
\qquad
\begin{array}{r l}
 & \cdot\ 426_{10} \\
 & \quad\ 12 \\
\hline
5 & \cdot\ 112 \\
 & \quad\ 12 \\
\hline
1 & \cdot\ 344 \\
 & \quad\ 12 \\
\hline
4 & \cdot\ 128 \\
 & \quad\ 12 \\
\hline
1 & \cdot\ 536 \\
 & \quad\ 12 \\
\hline
6 & \cdot\ 432
\end{array}$$

$\therefore\ 384{\cdot}426_{10} = 280{\cdot}5142_{12}$

That is straightforward enough, so let us now move on to see a very helpful use of octals in the next frame.

▶

## Use of octals as an intermediate step

137

This gives us an easy way of converting denary numbers into binary and hexadecimal forms. As an example, note the following.

Express the denary number $348{\cdot}654_{10}$ in octal, binary and hexadecimal forms.

(a) First we change $348{\cdot}654_{10}$ into octal form by the usual method.

This gives $348{\cdot}654_{10} =$ ...........

138

$534{\cdot}517_8$

(b) Now we take this octal form and write the binary equivalent of each digit in groups of three binary digits, thus:

101 011 100 · 101 001 111

Closing the groups up we have the binary equivalent of $534{\cdot}517_8$

i.e. $$348{\cdot}654_{10} = 534{\cdot}517_8 = 101011100{\cdot}101001111_2$$

(c) Then, starting from the decimal point and working in each direction, regroup the same binary digits in groups of four. This gives ...........

139

0001 0101 1100 · 1010 0111 1000

completing the group at either end by addition of extra zeros, as necessary.

(d) Now write the hexadecimal equivalent of each group of four binary digits, so that we have ............

140

1 5 (12) · (10) 7 8

Replacing (12) and (10) with the corresponding hexadecimal symbols, C and A, this gives $1\,5\,\mathrm{C}\cdot\mathrm{A}\,7\,8_{16}$

So, collecting the partial results together:

$$348{\cdot}654_{10} = 534{\cdot}517_8 = 101011100{\cdot}101001111_2 = 15\mathrm{C}{\cdot}\mathrm{A}78_{16}$$

*Next frame*

141

We have worked through the previous example in some detail. In practice, the method is more concise. Here is another example.

Change the denary number $428{\cdot}371_{10}$ into its octal, binary and hexadecimal forms.

(a) First of all, the octal equivalent of $428{\cdot}371_{10}$ is ............

142

$654{\cdot}276_8$

(b) The binary equivalent of each octal digit in groups of three is ...........

**143**

$$110\ 101\ 100\cdot 010\ 111\ 110_2$$

(c) Closing these up and rearranging in groups of four in each direction from the decimal point, we have ............

**144**

$$0001\ 1010\ 1100\cdot 0101\ 1111_2$$

(d) The hexadecimal equivalent of each set of four binary digits then gives .........

**145**

$$1AC{\cdot}5F_{16}$$

$$\begin{aligned}\text{So } 428{\cdot}371_{10} &= 654{\cdot}276_8\\ &= 110101100{\cdot}010111110_2\\ &= 1AC{\cdot}5F_{16}\end{aligned}$$

This is important, so let us do one more.

Convert $163{\cdot}245_{10}$ into octal, binary and hexadecimal forms.

*You can do this one entirely on your own. Work right through it and then check with the results in the next frame*

**146**

$$\begin{aligned}163{\cdot}245_{10} &= 243{\cdot}175_8\\ &= 010100011{\cdot}001111101_2\\ &= 1010\ 0011\cdot 0011\ 1110\ 1000_2\\ &= A3{\cdot}3E8_{16}\end{aligned}$$

And that is it.

*On to the next frame*

## Reverse method

**147** Of course, the method we have been using can be used in reverse, i.e. starting with a hexadecimal number, we can change it into groups of four binary digits, regroup these into groups of three digits from the decimal point, and convert these into the equivalent octal digits. Finally, the octal number can be converted into denary form by the usual method.

Here is one you can do with no trouble.

Express the hexadecimal number $4B2{\cdot}1A6_{16}$ in equivalent binary, octal and denary forms.

(a) Rewrite $4B2{\cdot}1A6_{16}$ in groups of four binary digits.
(b) Regroup into groups of three binary digits from the decimal point.
(c) Express the octal equivalent of each group of three binary digits.
(d) Finally convert the octal number into its denary equivalent.

*Work right through it and then check with the solution in the next frame*

148

$4B2{\cdot}1A6_{16} =$

(a) $0100\ 1011\ 0010 \cdot 0001\ 1010\ 0110_2$

(b) $010\ 010\ 110\ 010 \cdot 000\ 110\ 100\ 110_2$

(c) $2\ 2\ 6\ 2 \cdot 0\ 6\ 4\ 6_8$

(d) $1202{\cdot}103_{10}$

Now one more for good measure.

Express $2E3{\cdot}4D_{16}$ in binary, octal and denary forms.

*Check results with the next frame*

149

$$\begin{aligned} 2E3{\cdot}4D_{16} &= 0010\ 1110\ 0011 \cdot 0100\ 1101_2 \\ &= 001\ 011\ 100\ 011 \cdot 010\ 011\ 010_2 \\ &= 1\ 3\ 4\ 3 \cdot 2\ 3\ 2_8 \\ &= 739{\cdot}301_{10} \end{aligned}$$

***Let us pause and summarize the main facts on number systems***

## Review summary

150

| | | | |
|---|---|---|---|
| 1 | *Denary (or decimal) system* | Base 10<br>Numerals | Place values – powers of 10<br>0, 1, 2, 3, 4, 5, 6, 7, 8, 9 |
| 2 | *Binary system* | Base 2<br>Numerals | Place values – powers of 2<br>0, 1 |
| 3 | *Octal system* | Base 8<br>Numerals | Place values – powers of 8<br>0, 1, 2, 3, 4, 5, 6, 7 |
| 4 | *Duodecimal system* | Base 12<br>Numerals | Place values – powers of 12<br>0, 1, 2, 3, 4, 5, 6, 7, 8, 9, X, Λ<br>(10, 11) |
| 5 | *Hexadecimal system* | Base 16<br>Numerals | Place values – powers of 16<br>0, 1, 2, 3, 4, 5, 6, 7, 8, 9, A, B, C, D, E, F<br>(10,11,12,13,14,15) |

## Review exercise

151

1 Express the following numbers in denary form:
(a) $1110{\cdot}11_2$ (b) $507{\cdot}632_8$ (c) $345{\cdot}2\Lambda7_{12}$ (d) $2B4{\cdot}CA3_{16}$

2 Express the denary number $427{\cdot}362_{10}$ as a duodecimal number.

3 Convert $139{\cdot}825_{10}$ to the equivalent octal, binary and hexadecimal forms.

**152**

1 (a) $14{\cdot}75_{10}$ (b) $327{\cdot}801_{10}$ (c) $485{\cdot}247_{10}$ (d) $692{\cdot}790_{10}$ to 3 dp

2 $2\Lambda 7{\cdot}442_{12}$

3 $213{\cdot}646_8$, 10 001 011 · 110 100 $110_2$ and $8B{\cdot}D3_{16}$

**153**

You have now come to the end of this Programme. A list of **Can You?** questions follows for you to gauge your understanding of the material in the Programme. You will notice that these questions match the **Learning outcomes** listed at the beginning of the Programme so go back and try the **Quiz** that follows them. After that try the **Test exercise**. *Work through these at your own pace, there is no need to hurry.* A set of **Further problems** provides additional valuable practice.

## Can You?

**154**

### Checklist F.1

Check this list before and after you try the end of Programme test.

**On a scale of 1 to 5 how confident are you that you can:** Frames

- Carry out the basic rules of arithmetic with integers? 1 to 12
  *Yes* ☐ ☐ ☐ ☐ ☐ *No*
- Check the result of a calculation making use of rounding? 13 to 15
  *Yes* ☐ ☐ ☐ ☐ ☐ *No*
- Write a whole number as a product of prime numbers? 19 to 22
  *Yes* ☐ ☐ ☐ ☐ ☐ *No*
- Find the highest common factor and lowest common multiple of two whole numbers? 23 to 24
  *Yes* ☐ ☐ ☐ ☐ ☐ *No*
- Manipulate fractions, ratios and percentages? 28 to 52
  *Yes* ☐ ☐ ☐ ☐ ☐ *No*
- Manipulate decimal numbers? 56 to 74
  *Yes* ☐ ☐ ☐ ☐ ☐ *No*
- Manipulate powers? 78 to 98
  *Yes* ☐ ☐ ☐ ☐ ☐ *No*
- Use standard or preferred standard form and complete a calculation to the required level of accuracy? 99 to 108
  *Yes* ☐ ☐ ☐ ☐ ☐ *No*
- Understand the construction of various number systems and convert from one number system to another? 112 to 149
  *Yes* ☐ ☐ ☐ ☐ ☐ *No*

##  Test exercise F.1

155

Questions marked with [Personal Tutor icon] can be found online at **www.palgrave.com/stroud**.

There you can go through the question step by step and follow online hints, with full working provided.

Frames

1 Place the appropriate symbol < or > between each of the following pairs of numbers:

(a) $-12 \quad -15$ (b) $9 \quad -17$ (c) $-11 \quad 10$ — Frames 1 to 4

2 Find the value of each of the following:

(a) $24 - 3 \times 4 + 28 \div 14$ (b) $(24 - 3) \times (4 + 28) \div 14$ — Frames 5 to 12

3 Write each of the following as a product of prime factors:

(a) 156 (b) 546 (c) 1445 (d) 1485 — Frames 19 to 22

4 Round each number to the nearest 10, 100 and 1000:

(a) 5045 (b) 1100 (c) $-1552$ (d) $-4995$ — Frames 13 to 15

5 Find (i) the HCF and (ii) the LCM of:

(a) 1274 and 195 (b) 64 and 18 — Frames 23 to 24

6 Reduce each of the following fractions to their lowest terms:

(a) $\frac{8}{14}$ (b) $\frac{162}{36}$ (c) $-\frac{279}{27}$ (d) $-\frac{81}{3}$ — Frames 28 to 36

7 Evaluate each of the following, giving your answer as a fraction:

(a) $\frac{1}{3} + \frac{3}{5}$ (b) $\frac{2}{7} - \frac{1}{9}$ (c) $\frac{8}{3} \times \frac{6}{5}$ (d) $\frac{4}{5}$ *of* $\frac{2}{15}$

(e) $\frac{9}{2} \div \frac{3}{2}$ (f) $\frac{6}{7} - \frac{4}{5} \times \frac{3}{2} \div \frac{7}{5} + \frac{9}{4}$ — Frames 37 to 46

8 In each of the following the proportions of a compound are given. Find the ratios of the components in each case:

(a) $\frac{3}{4}$ of A and $\frac{1}{4}$ of B

(b) $\frac{2}{3}$ of P, $\frac{1}{15}$ of Q and the remainder of R

(c) $\frac{1}{5}$ of R, $\frac{3}{5}$ of S, $\frac{1}{6}$ of T and the remainder of U — Frames 47 to 48

9 What is:

(a) $\frac{3}{5}$ as a percentage?

(b) 16% as a fraction in its lowest terms?

(c) 17·5% of £12·50? — Frames 49 to 52

10 Evaluate each of the following (i) to 4 sig fig and (ii) to 3 dp:

(a) $13{\cdot}6 \times 25{\cdot}8 \div 4{\cdot}2$

(b) $13{\cdot}6 \div 4{\cdot}2 \times 25{\cdot}8$

(c) $9{\cdot}1(17{\cdot}43 + 7{\cdot}2(8{\cdot}6 - 4{\cdot}1^2 \times 3{\cdot}1))$

(d) $-8{\cdot}4((6{\cdot}3 \times 9{\cdot}1 + 2{\cdot}2^{1{\cdot}3}) - (4{\cdot}1^{-3{\cdot}1} \div 3{\cdot}3^3 - 5{\cdot}4))$ — Frames 56 to 65

▶

Frames

**11** Convert each of the following to decimal form to 3 decimal places:

(a) $\dfrac{3}{17}$ (b) $-\dfrac{2}{15}$ (c) $\dfrac{17}{3}$ (d) $-\dfrac{24}{11}$ — 66 to 67

**12** Write each of the following in abbreviated form:

(a) 6·7777 . . . (b) 0·01001001001 . . . — 70 to 71

**13** Convert each of the following to fractional form in lowest terms:

(a) 0·4 (b) 3·68 (c) $1{\cdot}\dot{4}$ (d) – 6·1 — 68 to 73

**14** Write each of the following as a number raised to a power:

(a) $2^9 \times 2^2$ (b) $6^2 \div 5^2$ (c) $((-4)^4)^{-4}$ (d) $(3^{-5})^0$ — 78 to 89

**15** Find the value of each of the following to 3 dp:

(a) $11^{\frac{1}{4}}$ (b) $\sqrt[7]{3}$ (c) $(-81)^{\frac{1}{5}}$ (d) $(-81)^{\frac{1}{4}}$ — 90 to 94

**16** Express in standard form:

(a) 537·6 (b) 0·364 (c) 4902 (d) 0·000125 — 95 to 101

**17** Convert to preferred standard form:

(a) $6{\cdot}147 \times 10^7$ (b) $2{\cdot}439 \times 10^{-4}$ (c) $5{\cdot}286 \times 10^5$

(d) $4{\cdot}371 \times 10^{-7}$ — 102 to 104

**18** Determine the following product, giving the result in both standard form and preferred standard form:

$(6{\cdot}43 \times 10^3)(7{\cdot}35 \times 10^4)$ — 95 to 104

**19** Each of the following contains numbers obtained by measurement. Evaluate each to the appropriate level of accuracy:

(a) $18{\cdot}4^{1{\cdot}6} \times 0{\cdot}01$ (b) $\dfrac{7{\cdot}632 \times 2{\cdot}14 - 8{\cdot}32 \div 1{\cdot}1}{16{\cdot}04}$ — 105 to 108

**20** Express the following numbers in denary form:

(a) $1111{\cdot}11_2$ (b) $777{\cdot}701_8$ (c) $3\Lambda3{\cdot}9\Lambda1_{12}$

(d) $E02{\cdot}FAB_{16}$ — 112 to 129

**21** Convert $19{\cdot}872_{10}$ to the equivalent octal, binary, duodecimal and hexadecimal forms. — 130 to 152

## Further problems F.1

156

**1** Place the appropriate symbol < or > between each of the following pairs of numbers:

(a) – 4 – 11 (b) 7 – 13 (c) – 15 13

**2** Find the value of each of the following:

(a) $6 + 14 \div 7 - 2 \times 3$ (b) $(6 + 14) \div (7 - 2) \times 3$

**3** Round each number to the nearest 10, 100 and 1000:

(a) 3505 (b) 500 (c) – 2465 (d) – 9005

4 Calculate each of the following to:
(i) 5 sig fig;
(ii) 4 dp;
(iii) The required level of accuracy given that each number, other than those indicated in brackets, has been obtained by measurement.

(a) $\dfrac{3{\cdot}21^{2{\cdot}33}+(5{\cdot}77-3{\cdot}11)}{8{\cdot}32-2{\cdot}64\times\sqrt{2{\cdot}56}}$

(b) $\dfrac{3{\cdot}142\times 1{\cdot}95}{6}(3\times 5{\cdot}44^2+1{\cdot}95^2)$ (power 2, divisor 6, multiplier 3)

(c) $\dfrac{3{\cdot}142\times 1{\cdot}234}{12}(0{\cdot}424^2+0{\cdot}424\times 0{\cdot}951+0{\cdot}951^2)$ (power 2, divisor 12)

(d) $\sqrt{\dfrac{2\times 0{\cdot}577}{3{\cdot}142\times 2{\cdot}64}+\dfrac{2{\cdot}64^2}{3}}$ (power 2, divisor 3, multiplier 2)

(e) $\dfrac{3{\cdot}26+\sqrt{12{\cdot}13}}{14{\cdot}192-2{\cdot}4\times 1{\cdot}63^2}$ (power 2)

(f) $\dfrac{4{\cdot}62^2-(7{\cdot}16-2{\cdot}35)}{2{\cdot}63+1{\cdot}89\times\sqrt{73{\cdot}24}}$ (power 2)

5 Find the prime factorization for each of the following:
(a) 924 (b) 825 (c) 2310 (d) 35 530

6 Find the HCF and LCM for each of the following pairs of numbers:
(a) 9, 21 (b) 15, 85 (c) 66, 42 (d) 64, 360

7 Reduce each of the following fractions to their lowest terms:
(a) $\dfrac{6}{24}$ (b) $\dfrac{104}{48}$ (c) $-\dfrac{120}{15}$ (d) $-\dfrac{51}{7}$

8 Evaluate:

(a) $\dfrac{9}{2}-\dfrac{4}{5}\div\left(\dfrac{2}{3}\right)^2\times\dfrac{3}{11}$

(b) $\dfrac{\dfrac{3}{4}+\dfrac{7}{5}\div\dfrac{2}{9}\times\dfrac{1}{3}}{\dfrac{7}{3}-\dfrac{11}{2}\times\dfrac{2}{5}+\dfrac{4}{9}}$

(c) $\left(\dfrac{3}{4}+\dfrac{7}{5}\right)^2\div\left(\dfrac{7}{3}-\dfrac{11}{5}\right)^2$

(d) $\dfrac{\left(\dfrac{5}{2}\right)^3-\dfrac{2}{9}\div\left(\dfrac{2}{3}\right)^2\times\dfrac{3}{2}}{\dfrac{3}{11}+\left(\dfrac{11}{2}\times\dfrac{2}{5}\right)^2-\dfrac{7}{5}}$

9 Express each of the following as a fraction in its lowest terms:
(a) 36% (b) 17·5% (c) 8·7% (d) 72%

10 Express each of the following as a percentage accurate to 1 dp:
(a) $\dfrac{4}{5}$ (b) $\dfrac{3}{11}$ (c) $\dfrac{2}{9}$ (d) $\dfrac{1}{7}$
(e) $\dfrac{9}{19}$ (f) $\dfrac{13}{27}$ (g) $\dfrac{7}{101}$ (h) $\dfrac{199}{200}$

11 Find:
(a) 16% *of* 125 (b) 9·6% *of* 5·63
(c) 13·5% *of* (−13·5) (d) 0·13% *of* 92·66

▶

12 In each of the following the properties of a compound are given. In each case find A : B : C.

(a) $\frac{1}{5}$ of A, $\frac{2}{3}$ of B and the remainder of C;

(b) $\frac{3}{8}$ of A with B and C in the ratio 1 : 2;

(c) A, B and C are mixed according to the ratios A : B = 2 : 5 and B : C = 10 : 11;

(d) A, B and C are mixed according to the ratios A : B = 1 : 7 and B : C = 13 : 9.

13 Write each of the following in abbreviated form:

(a) 8·767676... (b) 212·211211211...

14 Convert each of the following to fractional form in lowest terms:

(a) 0·12 (b) 5·25 (c) $5{\cdot}\dot{3}0\dot{6}$ (d) $-9{\cdot}3$

15 Write each of the following as a number raised to a power:

(a) $8^4 \times 8^3$ (b) $2^9 \div 8^2$ (c) $(5^3)^5$ (d) $3^4 \div 9^2$

16 Find the value of each of the following to 3 dp:

(a) $17^{\frac{2}{5}}$ (b) $\sqrt[4]{13}$ (c) $(-5)^{\frac{2}{3}}$ (d) $\sqrt{(-5)^4}$

17 Convert each of the following to decimal form to 3 decimal places:

(a) $\frac{5}{21}$ (b) $-\frac{2}{17}$ (c) $\frac{8}{3}$ (d) $-\frac{32}{19}$

18 Express in standard form:

(a) 52·876 (b) 15 243 (c) 0·08765
(d) 0·0000492 (e) 436·2 (f) 0·5728

19 Rewrite in preferred standard form:

(a) 4285 (b) 0·0169 (c) $8{\cdot}526 \times 10^{-4}$
(d) $3{\cdot}629 \times 10^5$ (e) $1{\cdot}0073 \times 10^7$ (f) $5{\cdot}694 \times 10^8$

20 Evaluate the following, giving the result in standard form and in preferred standard form:

$$\frac{(4{\cdot}26 \times 10^4)(9{\cdot}38 \times 10^5)}{3{\cdot}179 \times 10^2}$$

21 Convert each of the following decimal numbers to (i) binary, (ii) octal, (iii) duodecimal and (iv) hexadecimal format:

(a) 1·83 (b) $3{\cdot}425 \times 10^2$

22 Convert each of the following octal numbers to (i) binary, (ii) decimal, (iii) duodecimal and (iv) hexadecimal format:

(a) 0·577 (b) 563

23 Convert each of the following duodecimal numbers to (i) binary, (ii) octal, (iii) decimal and (iv) hexadecimal format:

(a) 0·ΛX (b) 9Λ1

24 Convert each of the following binary numbers to (i) decimal, (ii) octal, (iii) duodecimal and (iv) hexadecimal format:

(a) 0·10011 (b) 111001100

25 Convert each of the following hexadecimal numbers to (i) binary, (ii) octal, (iii) duodecimal and (iv) decimal format:

(a) 0·F4B (b) 3A5

157 Now visit the companion website for this book at www.palgrave.com/stroud for more questions applying this mathematics to science and engineering.

# Introduction to algebra

## Learning outcomes

*When you have completed this Programme you will be able to:*

- ☐ Use alphabetic symbols to supplement the numerals and to combine these symbols using all the operations of arithmetic
- ☐ Simplify algebraic expressions by collecting like terms and by abstracting common factors from similar terms
- ☐ Remove brackets and so obtain alternative algebraic expressions
- ☐ Manipulate expressions involving powers and logarithms
- ☐ Multiply and divide algebraic expressions
- ☐ Manipulate algebraic fractions
- ☐ Factorize algebraic expressions using standard factorizations
- ☐ Factorize quadratic algebraic expressions

If you already feel confident about these why not try the quiz over the page? You can check your answers at the end of the book.

# Quiz F.2

**Frames**

**1** Simplify each of the following:

(a) $3pq + 5pr - 2qr + qp - 6rp$

(b) $5l^2mn + 2nl^2m - 3mln^2 + l^2nm + 4n^2ml - nm^2$

(c) $w^4 \div w^{-a} \times w^{-b}$

(d) $\dfrac{(s^{\frac{1}{3}})^{\frac{3}{4}} \times (t^{\frac{1}{4}})^{-1} \div (s^{\frac{2}{7}})^{-\frac{7}{4}}}{(s^{-\frac{1}{4}})^{-1} \div (t^{\frac{1}{2}})^4}$

Frames 1 to 22

**2** Remove the brackets in each of the following:

(a) $-4x(2x - y)(3x + 2y)$

(b) $(a - 2b)(2a - 3b)(3a - 4b)$

(c) $-\{-2[x - 3(y - 4)] - 5(z + 6)\}$

Frames 12 to 17

**3** Evaluate by calculator or by change of base where necessary (to 3 dp):

(a) $\log 0{\cdot}0101$

(b) $\ln 3{\cdot}47$

(c) $\log_2 3{\cdot}16$

Frames 23 to 32

**4** Express $\log F = \log G + \log m - \log\left(\dfrac{1}{M}\right) - 2\log r$ without logs.

**5** Rewrite $T = 2\pi\sqrt{\dfrac{l}{g}}$ in log form.

Frames 33 to 35

**6** Perform the following multiplications and simplify your results:

(a) $(n^2 + 2n - 3)(4n + 5)$

(b) $(v^3 - v^2 - 2)(1 - 3v + 2v^2)$

Frames 39 to 43

**7** Perform the following divisions:

(a) $(2y^2 - y - 10) \div (y + 2)$

(b) $\dfrac{q^3 - 8}{q - 2}$

(c) $\dfrac{2r^3 + 5r^2 - 4r - 3}{r^2 + 2r - 3}$

Frames 44 to 48

**8** Simplify each of the following:

(a) $\dfrac{p}{q^3} \div \dfrac{q}{p^3}$

(b) $\dfrac{a^2b}{2c} \times \dfrac{ac^2}{2b} \div \dfrac{b^2c}{2a}$

Frames 52 to 57

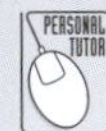

**9** Factorize the following:

(a) $18x^2y - 12xy^2$

(b) $x^3 + 4x^2y - 3xy^2 - 12y^3$

(c) $4(x - y)^2 - (x - 3y)^2$

(d) $12x^2 - 25x + 12$

Frames 61 to 78

# Algebraic expressions

**1**

*Think of a number*
*Add 15 to it*
*Double the result*
*Add this to the number you first thought of*
*Divide the result by 3*
*Take away the number you first thought of*

*The answer is 10*

Why?

*Check your answer in the next frame*

**2**

## Symbols other than numerals

A letter of the alphabet can be used to represent a number when the specific number is unknown and because the number is unknown (except, of course, to the person who thought of it) we shall represent the number by the letter $a$:

| | |
|---|---|
| Think of a number | $a$ |
| Add 15 to it | $a + 15$ |
| Double the result | $2 \times (a + 15) = (2 \times a) + (2 \times 15)$<br>$= (2 \times a) + 30$ |
| Add the result to the number you first thought of | $a + (2 \times a) + 30 = (3 \times a) + 30$ |
| Divide the result by 3 | $[(3 \times a) + 30] \div 3 = a + 10$ |
| Take away the number you first thought of | $a + 10 - a = 10$ |

The result is 10

*Next frame*

**3**

This little puzzle has demonstrated how:

*an unknown number can be represented by a letter of the alphabet which can then be manipulated just like an ordinary numeral within an arithmetic expression.*

So that, for example:

$a + a + a + a = 4 \times a$

$3 \times a - a = 2 \times a$

$8 \times a \div a = 8$

and

$a \times a \times a \times a \times a = a^5$

▶

If $a$ and $b$ represent two unknown numbers then we speak of the:

| | |
|---|---|
| sum of $a$ and $b$ | $a+b$ |
| difference of $a$ and $b$ | $a-b$ |
| product of $a$ and $b$ | $a \times b$, $a.b$ or simply $ab$ <br> (*we can and do suppress the multiplication sign*) |
| quotient of $a$ and $b$ | $a \div b$, $a/b$ or $\frac{a}{b}$ provided $b \neq 0$ |

and

| | |
|---|---|
| raising $a$ to the power $b$ | $a^b$ |

Using letters and numerals in this way is referred to as *algebra*.

*Now move to the next frame*

4

## Constants

In the puzzle of Frame 1 we saw how to use the letter $a$ to represent an unknown number – we call such a symbol a *constant*.

In many other problems we require a symbolism that can be used to represent not just one number but any one of a collection of numbers. Central to this symbolism is the idea of a variable.

*Next frame*

5

## Variables

We have seen that the operation of addition is commutative. That is, for example:

$$2+3=3+2$$

To describe this rule as applying to any pair of numbers and not just 2 and 3 we resort to the use of alphabetic characters $x$ and $y$ and write:

$$x+y=y+x$$

where $x$ and $y$ represent any two numbers. Used in this way, the letters $x$ and $y$ are referred to as *variables* because they each represent, not just one number, but any one of a collection of numbers.

So how would you write down the fact that multiplication is an associative operation? (Refer to Frame 12 of Programme F.1.)

*You can check your answer in the next frame*

6

$$\boxed{x(yz)=(xy)z=xyz}$$

where $x$, $y$ and $z$ represent numbers. Notice the suppression of the multiplication sign.

While it is not a hard and fast rule, it is generally accepted that letters from the beginning of the alphabet, i.e. $a, b, c, d, \ldots$ are used to represent constants and letters from the end of the alphabet, i.e. $\ldots v, w, x, y, z$ are used to represent variables. In any event, when a letter of the alphabet is used it should be made clear whether the letter stands for a constant or a variable.

*Now move on to the next frame*

7

## Rules of algebra

The rules of arithmetic that we met in the previous Programme for integers also apply to any type of number and we express this fact in the *rules of algebra* where we use variables rather than numerals as specific instances. The rules are:

### Commutativity

Two numbers $x$ and $y$ can be added or multiplied in any order without affecting the result. That is:

$x + y = y + x$ and

$xy = yx$

***Addition and multiplication are commutative operations***

The order in which two numbers are subtracted or divided *does* affect the result. That is:

$x - y \neq y - x$ unless $x = y$ and

$x \div y \neq y \div x, \left(\frac{x}{y} \neq \frac{y}{x}\right)$ unless $x = y$ and neither equals 0

***Subtraction and division are not commutative operations except in very special cases***

### Associativity

The way in which the numbers $x$, $y$ and $z$ are associated under addition or multiplication *does not* affect the result. That is:

$x + (y + z) = (x + y) + z = x + y + z$ and

$x(yz) = (xy)z = xyz$

***Addition and multiplication are associative operations***

The way in which the numbers are associated under subtraction or division *does* affect the result. That is:

$x - (y - z) \neq (x - y) - z$ unless $z = 0$ and

$x \div (y \div z) \neq (x \div y) \div z$ unless $z = 1$ and $y \neq 0$

***Subtraction and division are not associative operations except in very special cases***

### Distributivity

Multiplication is distributed over addition and subtraction from both the left and the right. For example:

$x(y + z) = xy + xz$ and $(x + y)z = xz + yz$

$x(y - z) = xy - xz$ and $(x - y)z = xz - yz$

Division is distributed over addition and subtraction from the right but not from the left. For example:

$(x + y) \div z = (x \div z) + (y \div z)$ but

$x \div (y + z) \neq (x \div y) + (x \div z)$

that is:

$\frac{x + y}{z} = \frac{x}{z} + \frac{y}{z}$ but $\frac{x}{y + z} \neq \frac{x}{y} + \frac{x}{z}$

***Take care here because it is a common mistake to get this wrong***

▶

## Rules of precedence

The familiar rules of precedence continue to apply when algebraic expressions involving mixed operations are to be manipulated.

*Next frame*

**8**

## Terms and coefficients

An algebraic expression consists of alphabetic characters and numerals linked together with the arithmetic operators. For example:

$8x - 3xy$

is an algebraic expression in the two variables $x$ and $y$. Each component of this expression is called a *term* of the expression. Here there are two terms, namely:

the $x$ term and the $xy$ term.

The numerals in each term are called the *coefficients* of the respective terms. So that:

8 is the coefficient of the $x$ term and $-3$ is the coefficient of the $xy$ term.

## Collecting like terms

Terms which have the same variables are called *like* terms and like terms can be collected together by addition or subtraction. For example:

$4x + 3y - 2z + 5y - 3x + 4z$ can be rearranged as $4x - 3x + 3y + 5y - 2z + 4z$ and simplified to:

$x + 8y + 2z$

Similarly, $4uv - 7uz - 6wz + 2uv + 3wz$ can be simplified to ...........

*Check your answer with the next frame*

**9**

$6uv - 7uz - 3wz$

*Next frame*

**10**

## Similar terms

In the algebraic expression:

$ab + ac$

both terms contain the letter $a$ and for this reason these terms, though not like terms, are called *similar* terms. Common symbols such as this letter $a$ are referred to as *common factors* and by using brackets these common factors can be *factored out*. For example, the common factor $a$ in this expression can be factored out to give:

$ab + ac = a(b + c)$ This process is known as *factorization.*

Numerical factors are factored out in the same way. For example, in the algebraic expression:

$3pq - 3qr$

the terms are similar, both containing the letter $q$. They also have a common coefficient 3 and this, as well as the common letter $q$, can be factored out to give:

$$\begin{aligned} 3pq - 3qr &= 3qp - 3qr \\ &= 3q(p - r) \end{aligned}$$

So the algebraic expression $9st - 3sv - 6sw$ simplifies to ...........

*The answer is in the next frame*

11

$$\boxed{3s(3t - v - 2w)}$$

Because

$$\begin{aligned} 9st - 3sv - 6sw &= 3s \times 3t - 3s \times v - 3s \times 2w \\ &= 3s(3t - v - 2w) \end{aligned}$$

*Next frame*

12

## Expanding brackets

Sometimes it will be desired to reverse the process of factorizing an expression by *removing* the brackets. This is done by:

(a) multiplying or dividing each term inside the bracket by the term outside the bracket, but

(b) if the term outside the bracket is negative then each term inside the bracket changes sign.

For example, the brackets in the expression:

$3x(y - 2z)$ are removed to give $3xy - 6xz$

and the brackets in the expression:

$-2y(2x - 4z)$ are removed to give $-4yx + 8yz$.

As a further example, the expression:

$$\frac{y+x}{8x} - \frac{y-x}{4x}$$

is an alternative form of $(y + x) \div 8x - (y - x) \div 4x$ and the brackets can be removed as follows:

$$\begin{aligned} \frac{y+x}{8x} - \frac{y-x}{4x} &= \frac{y}{8x} + \frac{x}{8x} - \frac{y}{4x} + \frac{x}{4x} \\ &= \frac{y}{8x} + \frac{1}{8} - \frac{y}{4x} + \frac{1}{4} \\ &= \frac{3}{8} - \frac{y}{8x} \end{aligned}$$

which can be written as $\frac{1}{8}\left(3 - \frac{y}{x}\right)$ or as $\frac{1}{8x}(3x - y)$

13

## Nested brackets

Whenever an algebraic expression contains brackets nested within other brackets the innermost brackets are removed first. For example:

$$\begin{aligned} 7(a - [4 - 5(b - 3a)]) &= 7(a - [4 - 5b + 15a]) \\ &= 7(a - 4 + 5b - 15a) \\ &= 7a - 28 + 35b - 105a \\ &= 35b - 98a - 28 \end{aligned}$$

So that the algebraic expression $4(2x + 3[5 - 2(x - y)])$ becomes, after the removal of the brackets ...........

*Next frame*

**14**

$$\boxed{24y - 16x + 60}$$

Because

$$\begin{aligned} 4(2x + 3[5 - 2(x - y)]) &= 4(2x + 3[5 - 2x + 2y]) \\ &= 4(2x + 15 - 6x + 6y) \\ &= 8x + 60 - 24x + 24y \\ &= 24y - 16x + 60 \end{aligned}$$

***At this point let us pause and summarize the main facts so far on algebraic expressions***

## Review summary

**15**

1 Alphabetic characters can be used to represent numbers and then be subjected to the arithmetic operations in much the same way as numerals.

2 An alphabetic character that represents a single number is called a *constant*.

3 An alphabetic character that represents any one of a collection of numbers is called a *variable*.

4 Some algebraic expressions contain terms multiplied by numerical coefficients.

5 Like terms contain identical alphabetic characters.

6 Similar terms have some but not all alphabetic characters in common.

7 Similar terms can be factorized by identifying their common factors and using brackets.

## Review exercise

**16**

1 Simplify each of the following by collecting like terms:

(a) $4xy + 3xz - 6zy - 5zx + yx$

(b) $-2a + 4ab + a - 4ba$

(c) $3rst - 10str + 8ts - 5rt + 2st$

(d) $2pq - 4pr + qr - 2rq + 3qp$

(e) $5lmn - 6ml + 7lm + 8mnl - 4ln$

2 Simplify each of the following by collecting like terms and factorizing:

(a) $4xy + 3xz - 6zy - 5zx + yx$

(b) $3rst - 10str + 8ts - 5rt + 2st$

(c) $2pq - 4pr + qr - 2rq + 3qp$

(d) $5lmn - 6ml + 7lm + 8mnl - 4ln$

3 Expand the following and then refactorize where possible:

(a) $8x(y - z) + 2y(7x + z)$ (b) $(3a - b)(b - 3a) + b^2$

(c) $-3(w - 7[x - 8(3 - z)])$ (d) $\dfrac{2a - 3}{4b} + \dfrac{3a + 2}{6b}$

17

1 (a) $4xy + 3xz - 6zy - 5zx + yx = 4xy + xy + 3xz - 5xz - 6yz$
$= 5xy - 2xz - 6yz$

Notice that the characters are written in alphabetic order.

(b) $-2a + 4ab + a - 4ba = -2a + a + 4ab - 4ab$
$= -a$

(c) $3rst - 10str + 8ts - 5rt + 2st = 3rst - 10rst + 8st + 2st - 5rt$
$= -7rst + 10st - 5rt$

(d) $2pq - 4pr + qr - 2rq + 3qp = 2pq + 3pq - 4pr + qr - 2qr$
$= 5pq - 4pr - qr$

(e) $5lmn - 6ml + 7lm + 8mnl - 4ln = 5lmn + 8lmn - 6lm + 7lm - 4ln$
$= 13lmn + lm - 4ln$

2 (a) $4xy + 3xz - 6zy - 5zx + yx = 5xy - 2xz - 6yz$
$= x(5y - 2z) - 6yz$ or
$= 5xy - 2z(x + 3y)$ or
$= y(5x - 6z) - 2xz$

(b) $3rst - 10str + 8ts - 5rt + 2st = -7rst + 10st - 5rt$
$= st(10 - 7r) - 5rt$ or
$= 10st - rt(7s + 5)$ or
$= t(10s - 7rs - 5r) = t(s[10 - 7r] - 5r)$

(c) $2pq - 4pr + qr - 2rq + 3qp = 5pq - 4pr - qr$
$= p(5q - 4r) - qr$ or
$= q(5p - r) - 4pr$ or
$= 5pq - r(4p + q)$

(d) $5lmn - 6ml + 7lm + 8mnl - 4ln = 13lmn + lm - 4ln$
$= l(13mn + m - 4n)$
$= l(m[13n + 1] - 4n)$ or
$= l(n[13m - 4] + m)$

3 (a) $8x(y - z) + 2y(7x + z) = 8xy - 8xz + 14xy + 2yz$
$= 22xy - 8xz + 2yz$
$= 2(x[11y - 4z] + yz)$

(b) $(3a - b)(b - 3a) + b^2 = 3a(b - 3a) - b(b - 3a) + b^2$
$= 3ab - 9a^2 - b^2 + 3ab + b^2$
$= 6ab - 9a^2$
$= 3a(2b - 3a)$

(c) $-3(w - 7[x - 8(3 - z)]) = -3(w - 7[x - 24 + 8z])$
$= -3(w - 7x + 168 - 56z)$
$= -3w + 21x - 504 + 168z$

(d) $\frac{2a - 3}{4b} + \frac{3a + 2}{6b} = \frac{2a}{4b} - \frac{3}{4b} + \frac{3a}{6b} + \frac{2}{6b}$
$= \frac{a}{2b} - \frac{3}{4b} + \frac{a}{2b} + \frac{1}{3b}$
$= \frac{a}{b} - \frac{5}{12b}$
$= \frac{1}{12b}(12a - 5)$

*So now on to the next topic*

# Powers and logarithms

18

## Powers

The use of *powers* (also called *indices* or *exponents*) provides a convenient form of algebraic shorthand. Repeated factors of the same base, for example $a \times a \times a \times a$ can be written as $a^4$, where the number 4 indicates the number of factors multiplied together. In general, the product of $n$ such factors $a$, where $a$ and $n$ are positive integers, is written $a^n$, where $a$ is called the *base* and $n$ is called the *index* or *exponent* or *power*. Any number multiplying $a^n$ is called the *coefficient* (as described in Frame 8)

$5a^3$ ← index or exponent or power; coefficient ↗ 5; base ↑ $a$

From the definitions above a number of rules of indices can immediately be established.

### Rules of indices

1 $a^m \times a^n = a^{m+n}$ e.g. $a^5 \times a^2 = a^{5+2} = a^7$

2 $a^m \div a^n = a^{m-n}$ e.g. $a^5 \div a^2 = a^{5-2} = a^3$

3 $(a^m)^n = a^{mn}$ e.g. $(a^5)^2 = a^5 \times a^5 = a^{10}$

These three basic rules lead to a number of important results.

4 $a^0 = 1$ because $a^m \div a^n = a^{m-n}$ and also $a^m \div a^n = \dfrac{a^m}{a^n}$

Then if $n = m$, $a^{m-m} = a^0$ and $\dfrac{a^m}{a^m} = 1$. So $a^0 = 1$

5 $a^{-m} = \dfrac{1}{a^m}$ because $a^{-m} = \dfrac{a^{-m} \times a^m}{a^m} = \dfrac{a^0}{a^m} = \dfrac{1}{a^m}$. So $a^{-m} = \dfrac{1}{a^m}$

6 $a^{\frac{1}{m}} = \sqrt[m]{a}$ because $\left(a^{\frac{1}{m}}\right)^m = a^{\frac{m}{m}} = a^1 = a$. So $a^{\frac{1}{m}} = \sqrt[m]{a}$

From this it follows that $a^{\frac{n}{m}} = \sqrt[m]{a^n}$ or $\left(\sqrt[m]{a}\right)^n$.

Make a note of any of these results that you may be unsure about.

*Then move on to the next frame*

19

So we have:

(a) $a^m \times a^n = a^{m+n}$

(b) $a^m \div a^n = a^{m-n}$

(c) $(a^m)^n = a^{mn}$

(d) $a^0 = 1$

(e) $a^{-m} = \dfrac{1}{a^m}$

(f) $a^{\frac{1}{m}} = \sqrt[m]{a}$

(g) $a^{\frac{n}{m}} = \left(\sqrt[m]{a}\right)^n$ or $\sqrt[m]{a^n}$

Now try to apply the rules:

$$\frac{6x^{-4} \times 2x^3}{8x^{-3}} = \ldots\ldots\ldots\ldots$$

**20**

$\boxed{\frac{3}{2}x^2}$

Because $\dfrac{6x^{-4} \times 2x^3}{8x^{-3}} = \dfrac{12}{8} \cdot \dfrac{x^{-4+3}}{x^{-3}} = \dfrac{12}{8} \cdot \dfrac{x^{-1}}{x^{-3}} = \dfrac{3}{2}x^{-1+3} = \dfrac{3}{2}x^2$

That was easy enough. In the same way, simplify:

$$E = (5x^2y^{-\frac{3}{2}}z^{\frac{1}{4}})^2 \times (4x^4y^2z)^{-\frac{1}{2}}$$
$$= \ldots\ldots\ldots\ldots$$

**21**

$$\boxed{\frac{25x^2}{2y^4}}$$

$$E = 25x^4y^{-3}z^{\frac{1}{2}} \times 4^{-\frac{1}{2}}x^{-2}y^{-1}z^{-\frac{1}{2}}$$
$$= 25x^4y^{-3}z^{\frac{1}{2}} \times \frac{1}{2}x^{-2}y^{-1}z^{-\frac{1}{2}}$$
$$= \frac{25}{2}x^2y^{-4}z^0 = \frac{25}{2}x^2y^{-4}.1 = \frac{25x^2}{2y^4}$$

And one more:

Simplify $F = \sqrt[3]{a^6b^3} \div \sqrt{\frac{1}{9}a^4b^6} \times \left(4\sqrt{a^6b^2}\right)^{-\frac{1}{2}}$ giving the result without fractional indices.

$$F = \ldots\ldots\ldots\ldots$$

**22**

$$\boxed{\frac{3}{2ab^2\sqrt{ab}}}$$

$$F = a^2b \div \frac{1}{3}a^2b^3 \times \frac{1}{(4a^3b)^{\frac{1}{2}}} = a^2b \times \frac{3}{a^2b^3} \times \frac{1}{2a^{\frac{3}{2}}b^{\frac{1}{2}}}$$
$$= \frac{3}{b^2} \times \frac{1}{2a^{\frac{3}{2}}b^{\frac{1}{2}}} = \frac{3}{b^2.2a(ab)^{\frac{1}{2}}} = \frac{3}{2ab^2\sqrt{ab}}$$

## Logarithms

**23**

Any real number can be written as another number raised to a power. For example:

$9 = 3^2$ and $27 = 3^3$

By writing numbers in the form of a number raised to a power some of the arithmetic operations can be performed in an alternative way. For example:

$$9 \times 27 = 3^2 \times 3^3$$
$$= 3^{2+3}$$
$$= 3^5$$
$$= 243$$

Here the process of multiplication is replaced by the process of relating numbers to powers and then adding the powers.

▶

If there were a simple way of relating numbers such as 9 and 27 to powers of 3 and then relating powers of 3 to numbers such as 243, the process of multiplying two numbers could be converted to the simpler process of adding two powers. In the past a system based on this reasoning was created. It was done using tables that were constructed of numbers and their respective powers.

In this instance:

| *Number* | *Power of 3* |
|---|---|
| 1 | 0 |
| 3 | 1 |
| 9 | 2 |
| 27 | 3 |
| 81 | 4 |
| 243 | 5 |
| ... | ... |

They were not called tables of powers but tables of *logarithms*. Nowadays, calculators have superseded the use of these tables but the logarithm remains an essential concept.

*Let's just formalize this*

**24**

If $a$, $b$ and $c$ are three real numbers where:

$a = b^c$ and $b > 1$

the power $c$ is called the *logarithm* of the number $a$ to the base $b$ and is written:

$c = \log_b a$ spoken as *$c$ is the log to the base $b$ of $a$*

For example, because

$25 = 5^2$

the power 2 is the logarithm of 25 to the base 5. That is:

$2 = \log_5 25$

So in each of the following what is the value of $x$, remembering that if $a = b^c$ then $c = \log_b a$?

(a) $x = \log_2 16$
(b) $4 = \log_x 81$
(c) $2 = \log_7 x$

*The answers are in the next frame*

**25**

(a) $x = 4$
(b) $x = 3$
(c) $x = 49$

Because

(a) If $x = \log_2 16$ then $2^x = 16 = 2^4$ and so $x = 4$
(b) If $4 = \log_x 81$ then $x^4 = 81 = 3^4$ and so $x = 3$
(c) If $2 = \log_7 x$ then $7^2 = x = 49$

*Move on to the next frame*

## Rules of logarithms

26

Since logarithms are powers, the rules that govern the manipulation of logarithms closely follow the rules of powers.

(a) If $x = a^b$ so that $b = \log_a x$ and
$y = a^c$ so that $c = \log_a y$ then:

$xy = a^b a^c = a^{b+c}$ hence $\log_a xy = b + c = \log_a x + \log_a y$. That is:

$\log_a xy = \log_a x + \log_a y$ *The log of a product equals the sum of the logs*

(b) Similarly $x \div y = a^b \div a^c = a^{b-c}$ so that $\log_a(x \div y) = b - c = \log_a x - \log_a y$ That is:

$\log_a(x \div y) = \log_a x - \log_a y$ *The log of a quotient equals the difference of the logs*

(c) Because $x^n = (a^b)^n = a^{bn}$, $\log_a(x^n) = bn = n\log_a x$. That is:

$\log_a(x^n) = n\log_a x$ *The log of a number raised to a power is the product of the power and the log of the number*

The following important results are also obtained from these rules:

(d) $\log_a 1 = 0$ because, from the laws of powers $a^0 = 1$. Therefore, from the definition of a logarithm $\log_a 1 = 0$

(e) $\log_a a = 1$ because $a^1 = a$ so that $\log_a a = 1$

(f) $\log_a a^x = x$ because $\log_a a^x = x\log_a a = x.1$ so that $\log_a a^x = x$

(g) $a^{\log_a x} = x$ because if we take the log of the left-hand side of this equation:

$\log_a a^{\log_a x} = \log_a x \log_a a = \log_a x$ so that $a^{\log_a x} = x$

(h) $\log_a b = \dfrac{1}{\log_b a}$ because, if $\log_b a = c$ then $b^c = a$ and so $b = \sqrt[c]{a} = a^{\frac{1}{c}}$

Hence, $\log_a b = \dfrac{1}{c} = \dfrac{1}{\log_b a}$. That is $\log_a b = \dfrac{1}{\log_b a}$

So, cover up the results above and complete the following

(a) $\log_a(x \times y) = \ldots\ldots\ldots\ldots$

(b) $\log_a(x \div y) = \ldots\ldots\ldots\ldots$

(c) $\log_a(x^n) = \ldots\ldots\ldots\ldots$

(d) $\log_a 1 = \ldots\ldots\ldots\ldots$

(e) $\log_a a = \ldots\ldots\ldots\ldots$

(f) $\log_a a^x = \ldots\ldots\ldots\ldots$

(g) $a^{\log_a x} = \ldots\ldots\ldots\ldots$

(h) $\dfrac{1}{\log_b a} = \ldots\ldots\ldots\ldots$

**27**

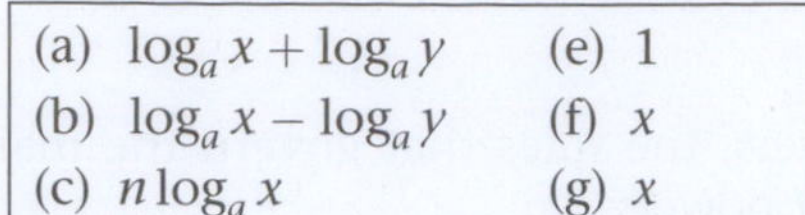

| | |
|---|---|
| (a) $\log_a x + \log_a y$ | (e) 1 |
| (b) $\log_a x - \log_a y$ | (f) $x$ |
| (c) $n \log_a x$ | (g) $x$ |
| (d) 0 | (h) $\log_a b$ |

Now try it with numbers. Complete the following:

(a) $\log_a(6{\cdot}788 \times 1{\cdot}043) = \ldots\ldots\ldots\ldots$

(b) $\log_a(19{\cdot}112 \div 0{\cdot}054) = \ldots\ldots\ldots\ldots$

(c) $\log_a(5{\cdot}889^{1{\cdot}2}) = \ldots\ldots\ldots\ldots$

(d) $\log_8 1 = \ldots\ldots\ldots\ldots$

(e) $\log_7 7 = \ldots\ldots\ldots\ldots$

(f) $\log_3 27 = \ldots\ldots\ldots\ldots$

(g) $12^{\log_{12} 4} = \ldots\ldots\ldots\ldots$

(h) $\dfrac{1}{\log_3 4} = \ldots\ldots\ldots\ldots$

**28**

| | |
|---|---|
| (a) $\log_a 6{\cdot}788 + \log_a 1{\cdot}043$ | (e) 1 |
| (b) $\log_a 19{\cdot}112 - \log_a 0{\cdot}054$ | (f) 3 |
| (c) $1{\cdot}2 \log_a 5{\cdot}889$ | (g) 4 |
| (d) 0 | (h) $\log_4 3$ |

*Next frame*

**29**

## Base 10 and base *e*

On a typical calculator there are buttons that provide access to logarithms to two different bases, namely 10 and the exponential number $e = 2{\cdot}71828\ldots$.

Logarithms to base 10 were commonly used in conjunction with tables for arithmetic calculations – they are called *common logarithms* and are written without indicating the base. For example:

$\log_{10} 1{\cdot}2345$ is normally written simply as $\log 1{\cdot}2345$

You will see it on your calculator as [log].

The logarithms to base *e* are called *natural logarithms* and are important for their mathematical properties. These also are written in an alternative form:

$\log_e 1{\cdot}2345$ is written as $\ln 1{\cdot}2345$

You will see it on your calculator as [ln].

So, use your calculator and complete the following (to 3 dp):

(a) $\log 5{\cdot}321 = \ldots\ldots\ldots\ldots$

(b) $\log 0{\cdot}278 = \ldots\ldots\ldots\ldots$

(c) $\log 1 = \ldots\ldots\ldots\ldots$

(d) $\log(-1{\cdot}005) = \ldots\ldots\ldots\ldots$

(e) $\ln 13{\cdot}45 = \ldots\ldots\ldots\ldots$

(f) $\ln 0{\cdot}278 = \ldots\ldots\ldots\ldots$

(g) $\ln 0{\cdot}00001 = \ldots\ldots\ldots\ldots$

(h) $\ln(-0{\cdot}001) = \ldots\ldots\ldots\ldots$

*The answers are in the next frame*

30

| (a) 0·726 | (e) 2·599 |
|---|---|
| (b) – 0·556 | (f) – 1·280 |
| (c) 0 | (g) – 11·513 |
| (d) ERROR | (h) ERROR |

Notice that for any base the:

logarithm of 1 is zero
logarithm of 0 is not defined
logarithm of a number greater than 1 is positive
logarithm of a number between 0 and 1 is negative
logarithm of a negative number cannot be evaluated as a real number.

*Move to the next frame*

31

## Change of base

In the previous two frames you saw that $\log 0{\cdot}278 \neq \ln 0{\cdot}278$, i.e. logarithms with different bases have different values. The different values are, however, related to each other as can be seen from the following:

Let $a = b^c$ so that $c = \log_b a$ and let $x = a^d$ so that $d = \log_a x$. Now:

$x = a^d = (b^c)^d = b^{cd}$ so that $cd = \log_b x$. That is:

$\log_b a \log_a x = \log_b x$

This is the change of base formula which relates the logarithms of a number relative to two different bases. For example:

$\log_e 0{\cdot}278 = -1{\cdot}280$ to 3 dp and
$\log_e 10 \times \log_{10} 0{\cdot}278 = 2{\cdot}303 \times (-0{\cdot}556) = -1{\cdot}280$ which confirms that:

$\log_e 10 \log_{10} 0{\cdot}278 = \log_e 0{\cdot}278$

Now, use your calculator to complete each of the following (to 3 dp):

(a) $\log_2 3{\cdot}66 = \ldots\ldots\ldots\ldots$ (c) $\log_{9{\cdot}9} 6{\cdot}35 = \ldots\ldots\ldots\ldots$

(b) $\log_{3{\cdot}4} 0{\cdot}293 = \ldots\ldots\ldots\ldots$ (d) $\log_{7{\cdot}34} 7{\cdot}34 = \ldots\ldots\ldots\ldots$

32

| (a) 1·872 | (b) −1·003 | (c) 0·806 | (d) 1 |
|---|---|---|---|

Because

(a) $(\log_{10} 2) \times (\log_2 3{\cdot}66) = \log_{10} 3{\cdot}66$ so that

$$\log_2 3{\cdot}66 = \frac{\log_{10} 3{\cdot}66}{\log_{10} 2} = \frac{0{\cdot}563\ldots}{0{\cdot}301\ldots} = 1{\cdot}872$$

(b) $(\log_{10} 3{\cdot}4) \times (\log_{3{\cdot}4} 0{\cdot}293) = \log_{10} 0{\cdot}293$ so that

$$\log_{3{\cdot}4} 0{\cdot}293 = \frac{\log_{10} 0{\cdot}293}{\log_{10} 3{\cdot}4} = \frac{-0{\cdot}533\ldots}{0{\cdot}531\ldots} = -1{\cdot}003$$

(c) $(\log_{10} 9{\cdot}9) \times (\log_{9{\cdot}9} 6{\cdot}35) = \log_{10} 6{\cdot}35$ so that

$$\log_{9{\cdot}9} 6{\cdot}35 = \frac{\log_{10} 6{\cdot}35}{\log_{10} 9{\cdot}9} = \frac{0{\cdot}802\ldots}{0{\cdot}995\ldots} = 0{\cdot}806$$

(d) $\log_{7{\cdot}34} 7{\cdot}34 = 1$ because for any base $a$, $\log_a a = 1$.

## 33 Logarithmic equations

The following four examples serve to show you how logarithmic expressions and equations can be manipulated.

**Example 1**

Simplify the following:

$\log_a x^2 + 3\log_a x - 2\log_a 4x$

*Solution*

$$\log_a x^2 + 3\log_a x - 2\log_a 4x = \log_a x^2 + \log_a x^3 - \log_a (4x)^2$$
$$= \log_a\left(\frac{x^2x^3}{16x^2}\right)$$
$$= \log_a\left(\frac{x^3}{16}\right)$$

**Example 2**

Solve the following for $x$:

$2\log_a x - \log_a(x-1) = \log_a(x-2)$

*Solution*

$$\text{LHS} = 2\log_a x - \log_a(x-1)$$
$$= \log_a x^2 - \log_a(x-1)$$
$$= \log_a\left(\frac{x^2}{x-1}\right)$$
$$= \log_a(x-2) \text{ so that } \frac{x^2}{x-1} = x-2. \text{ That is:}$$
$$x^2 = (x-2)(x-1) = x^2 - 3x + 2 \text{ so that } -3x + 2 = 0 \text{ giving } x = \frac{2}{3}$$

**Example 3**

Find $y$ in terms of $x$:

$5\log_a y - 2\log_a(x+4) = 2\log_a y + \log_a x$

*Solution*

$5\log_a y - 2\log_a(x+4) = 2\log_a y + \log_a x$ so that
$5\log_a y - 2\log_a y = \log_a x + 2\log_a(x+4)$ that is
$\log_a y^5 - \log_a y^2 = \log_a x + \log_a(x+4)^2$ that is
$\log_a\left(\frac{y^5}{y^2}\right) = \log_a y^3 = \log_a x(x+4)^2$ so that $y^3 = x(x+4)^2$ hence
$y = \sqrt[3]{x(x+4)^2}$

**Example 4**

For what values of $x$ is $\log_a(x-3)$ valid?

*Solution*

$\log_a(x-3)$ is valid for $x - 3 > 0$, that is $x > 3$

*Now you try some*

**34**

1 Simplify $2\log_a x - 3\log_a 2x + \log_a x^2$

2 Solve the following for $x$:

$$4\log_a \sqrt{x} - \log_a 3x = \log_a x^{-2}$$

3 Find $y$ in terms of $x$ where:

$$2\log_a y - 3\log_a(x^2) = \log_a \sqrt{y} + \log_a x$$

*Next frame for the answers*

**35**

1 $2\log_a x - 3\log_a 2x + \log_a x^2 = \log_a x^2 - \log_a(2x)^3 + \log_a x^2$

$$= \log_a\left(\frac{x^2x^2}{8x^3}\right)$$

$$= \log_a\left(\frac{x}{8}\right)$$

2 LHS $= 4\log_a \sqrt{x} - \log_a 3x$

$$= \log_a(\sqrt{x})^4 - \log_a 3x$$

$$= \log_a x^2 - \log_a 3x$$

$$= \log_a\left(\frac{x^2}{3x}\right)$$

$$= \log_a\left(\frac{x}{3}\right)$$

$= \log_a x^{-2}$ the right-hand side of the equation.

So that:

$x^{-2} = \dfrac{x}{3}$, that is $x^3 = 3$ giving $x = \sqrt[3]{3}$

3 $2\log_a y - 3\log_a(x^2) = \log_a \sqrt{y} + \log_a x$, that is

$\log_a y^2 - \log_a(x^2)^3 = \log_a y^{\frac{1}{2}} + \log_a x$ so that

$\log_a y^2 - \log_a y^{\frac{1}{2}} = \log_a(x^2)^3 + \log_a x$ giving

$\log_a \dfrac{y^2}{y^{\frac{1}{2}}} = \log_a x^6.x$. Consequently $y^{\frac{3}{2}} = x^7$ and so $y = \sqrt[3]{x^{14}}$

***At this point let us pause and summarize the main facts so far on powers and logarithms***

## Review summary

**36**

1 Rules of powers:

(a) $a^m \times a^n = a^{m+n}$ (b) $a^{-m} = \dfrac{1}{a^m}$

(c) $a^m \div a^n = a^{m-n}$ (d) $a^{\frac{1}{m}} = \sqrt[m]{a}$

(e) $(a^m)^n = a^{mn}$ (f) $a^{\frac{n}{m}} = \sqrt[m]{a^n}$ or $a^{\frac{n}{m}} = (\sqrt[m]{a})^n$

(g) $a^0 = 1$

▶

2 Rules of logarithms:

(a) $\log_a xy = \log_a x + \log_a y$ — *The log of a product equals the sum of the logs*

(b) $\log_a(x \div y) = \log_a x - \log_a y$ — *The log of a quotient equals the difference of the logs*

(c) $\log_a(x^n) = n\log_a x$ — *The log of a number raised to a power is the product of the power and the log of the number*

(d) $\log_a 1 = 0$

(e) $\log_a a = 1$ and $\log_a a^x = x$

(f) $a^{\log_a x} = x$

(g) $\log_a b = \dfrac{1}{\log_b a}$

3 Logarithms to base 10 are called *common logarithms* and are written as $\log x$.

4 Logarithms to base $e = 2{\cdot}71828\ldots$ are called *natural logarithms* and are written as $\ln x$.

---

## Review exercise

**37**

1 Simplify each of the following:

(a) $a^6 \times a^5$

(b) $x^7 \div x^3$

(c) $(w^2)^m \div (w^m)^3$

(d) $s^3 \div t^{-4} \times (s^{-3}t^{-2})^3$

(e) $\dfrac{8x^{-3} \times 3x^2}{6x^{-4}}$

(f) $(4a^3b^{-1}c)^2 \times (a^{-2}b^4c^{-2})^{\frac{1}{2}} \div \left[64(a^6b^4c^2)^{-\frac{1}{2}}\right]$

(g) $\sqrt[3]{8a^3b^6} \div \sqrt{\dfrac{1}{25}a^4b^7} \times \left(16\sqrt{a^4b^6}\right)^{-\frac{1}{2}}$

2 Express the following without logs:

(a) $\log K = \log P - \log T + 1{\cdot}3\log V$

(b) $\ln A = \ln P + rn$

3 Rewrite $R = r\sqrt{\dfrac{f+P}{f-P}}$ in log form.

4 Evaluate by calculator or by change of base where necessary (to 3 dp):

(a) $\log 5{\cdot}324$

(b) $\ln 0{\cdot}0023$

(c) $\log_4 1{\cdot}2$

**38**

1 (a) $a^6 \times a^5 = a^{6+5} = a^{11}$

(b) $x^7 \div x^3 = x^{7-3} = x^4$

(c) $(w^2)^m \div (w^m)^3 = w^{2m} \div w^{3m} = w^{2m} \times w^{-3m} = w^{-m}$

(d) $s^3 \div t^{-4} \times (s^{-3}t^{-2})^3 = s^3 \times t^4 \times s^{-9}t^{-6} = s^{-6}t^{-2}$

(e) $\dfrac{8x^{-3} \times 3x^2}{6x^{-4}} = \dfrac{24x^{-1}}{6x^{-4}} = 4x^3$

▶

(f) $(4a^3b^{-1}c)^2 \times (a^{-2}b^4c^{-2})^{\frac{1}{2}} \div 64(a^6b^4c^2)^{-\frac{1}{2}}$

$= (16a^6b^{-2}c^2) \times (a^{-1}b^2c^{-1}) \div 64(a^{-3}b^{-2}c^{-1})$

$= (16a^6b^{-2}c^2) \times (a^{-1}b^2c^{-1}) \times 64^{-1}(a^3b^2c^1)$

$= \dfrac{a^8b^2c^2}{4}$

(g) $\sqrt[3]{8a^3b^6} \div \sqrt{\dfrac{1}{25}a^4b^7} \times \left(16\sqrt{a^4b^6}\right)^{-\frac{1}{2}} = (2ab^2) \div \dfrac{a^2b^{\frac{7}{2}}}{5} \times \left(4ab^{\frac{3}{2}}\right)^{-1}$

$= (2ab^2) \times \dfrac{5}{a^2b^{\frac{7}{2}}} \times \dfrac{1}{4ab^{\frac{3}{2}}}$

$= \dfrac{5ab^2}{2a^2b^{\frac{7}{2}}ab^{\frac{3}{2}}}$

$= \dfrac{5}{2a^2b^3}$

2 (a) $K = \dfrac{PV^{1\cdot 3}}{T}$

(b) $A = Pe^{rn}$

3 $\log R = \log r + \dfrac{1}{2}(\log(f+P) - \log(f-P))$

4 (a) 0·726

(b) −6·075

(c) 0·132

*Now move to the next topic*

---

# Algebraic multiplication and division

## Multiplication

**39**

**Example 1**

$(x+2)(x+3) = x(x+3) + 2(x+3)$

$= x^2 + 3x + 2x + 6$

$= x^2 + 5x + 6$

*Now a slightly harder one*

---

**40**

**Example 2**

$(2x+5)(x^2+3x+4)$

Each term in the second expression is to be multiplied by $2x$ and then by 5 and the results added together, so we set it out thus:

| | |
|---|---|
| | $x^2 + 3x + 4$ |
| | $2x + 5$ |
| Multiply throughout by $2x$ | $2x^3 + 6x^2 + 8x$ |
| Multiply by 5 | $5x^2 + 15x + 20$ |
| Add the two lines | $2x^3 + 11x^2 + 23x + 20$ |

So $(2x+5)(x^2+3x+4) = 2x^3 + 11x^2 + 23x + 20$

Be sure to keep the same powers of the variable in the same column.

*Next frame*

**41** Now look at this one.

**Example 3**

Determine $(2x + 6)(4x^3 - 5x - 7)$

You will notice that the second expression is a cubic (highest power $x^3$), but that there is no term in $x^2$. In this case, we insert $0x^2$ in the working to keep the columns complete, that is:

$$\begin{array}{l} 4x^3 + 0x^2 - 5x - 7 \\ \underline{2x \ + 6 \qquad\qquad} \end{array}$$

which gives ...........

*Finish it*

**42**

$$\boxed{8x^4 + 24x^3 - 10x^2 - 44x - 42}$$

Here it is set out:

$$\begin{array}{l} 4x^3 + 0x^2 - 5x - 7 \\ 2x + 6 \\ \hline 8x^4 + \ 0x^3 - 10x^2 - 14x \\ \qquad\ 24x^3 + \ 0x^2 - 30x - 42 \\ \hline 8x^4 + 24x^3 - 10x^2 - 44x - 42 \\ \hline \end{array}$$

They are all done in the same way, so here is one more for practice.

**Example 4**

Determine the product $(3x - 5)(2x^3 - 4x^2 + 8)$

You can do that without any trouble.

The product is ...........

**43**

$$\boxed{6x^4 - 22x^3 + 20x^2 + 24x - 40}$$

All very straightforward:

$$\begin{array}{l} 2x^3 - 4x^2 + 0x + 8 \\ 3x - 5 \\ \hline 6x^4 - 12x^3 + \ 0x^2 + 24x \\ \quad\ - 10x^3 + 20x^2 + \ 0x - 40 \\ \hline 6x^4 - 22x^3 + 20x^2 + 24x - 40 \\ \hline \end{array}$$

**44** **Division**

Let us consider $(12x^3 - 2x^2 - 3x + 28) \div (3x + 4)$. The result of this division is called the *quotient* of the two expressions and we find the quotient by setting out the division in the same way as we do for the long division of numbers:

$$3x + 4 \overline{\big) 12x^3 - 2x^2 - 3x + 28}$$

▶

To make $12x^3$, $3x$ must be multiplied by $4x^2$, so we insert this as the first term in the quotient, multiply the divisor $(3x + 4)$ by $4x^2$, and subtract this from the first two terms:

$$\begin{array}{r|l} & 4x^2 \\ \hline 3x+4 & 12x^3 - 2x^2 - 3x + 28 \\ & 12x^3 + 16x^2 \\ \cline{2-2} & \quad - 18x^2 - 3x \end{array}$$

Bring down the next term $(-3x)$ and repeat the process

To make $-18x^2$, $3x$ must be multiplied by $-6x$, so do this and subtract as before, not forgetting to enter the $-6x$ in the quotient.

Do this and we get ...........

**45**

$$\begin{array}{r|l} & 4x^2 \quad - 6x \\ \hline 3x+4 & 12x^3 - 2x^2 - 3x + 28 \\ & 12x^3 + 16x^2 \\ \cline{2-2} & \quad - 18x^2 - 3x \\ & \quad - 18x^2 - 24x \\ \cline{2-2} & \qquad\qquad 21x \end{array}$$

Now bring down the next term and continue in the same way and finish it off.

So $(12x^3 - 2x^2 - 3x + 28) \div (3x + 4) =$ ...........

**46**

$4x^2 - 6x + 7$

As before, if an expression has a power missing, insert the power with zero coefficient. Now you can determine $(4x^3 + 13x + 33) \div (2x + 3)$

*Set it out as before and check the result with the next frame*

**47**

$2x^2 - 3x + 11$

Here it is:

$$\begin{array}{r|l} & 2x^2 - 3x + 11 \\ \hline 2x+3 & 4x^3 - 0x^2 + 13x + 33 \\ & 4x^3 + 6x^2 \\ \cline{2-2} & \quad - 6x^2 + 13x \\ & \quad - 6x^2 - 9x \\ \cline{2-2} & \qquad\qquad 22x + 33 \\ & \qquad\qquad 22x + 33 \\ \cline{2-2} & \qquad\qquad \bullet \quad \bullet \end{array}$$

So $(4x^3 + 13x + 33) \div (2x + 3) = 2x^2 - 3x + 11$

And one more.

Determine $(6x^3 - 7x^2 + 1) \div (3x + 1)$

Setting out as before, the quotient is ...........

48

$$\boxed{2x^2 - 3x + 1}$$

After inserting the $x$ term with zero coefficient, the rest is straightforward.

***At this point let us pause and summarize the main facts so far for multiplication and division of algebraic expressions***

## Review summary

49

1 *Multiplication of algebraic expressions*
Two algebraic expressions are multiplied together by successively multiplying the second expression by each term of the first expression.

2 *Division of algebraic expressions*
Two algebraic expressions are divided by setting out the division in the same way as we do for the long division of numbers.

## Review exercise

50

1 Perform the following multiplications and simplify your results:
(a) $(8x - 4)(4x^2 - 3x + 2)$
(b) $(2x + 3)(5x^3 + 3x - 4)$

2 Perform the following divisions:
(a) $(x^2 + 5x - 6) \div (x - 1)$
(b) $(x^2 - x - 2) \div (x + 1)$
(c) $(12x^3 - 11x^2 - 25) \div (3x - 5)$

51

1 (a)
$$\begin{aligned}(8x - 4)(4x^2 - 3x + 2) &= 8x(4x^2 - 3x + 2) - 4(4x^2 - 3x + 2)\\ &= 32x^3 - 24x^2 + 16x - 16x^2 + 12x - 8\\ &= 32x^3 - 40x^2 + 28x - 8\end{aligned}$$

(b)
$$\begin{aligned}(2x + 3)(5x^3 + 3x - 4) &= 2x(5x^3 + 3x - 4) + 3(5x^3 + 3x - 4)\\ &= 10x^4 + 6x^2 - 8x + 15x^3 + 9x - 12\\ &= 10x^4 + 15x^3 + 6x^2 + x - 12\end{aligned}$$

2 (a)
$$(x^2 + 5x - 6) \div (x - 1) = x - 1\,\overline{\big)\,x^2 + 5x - 6}$$

$$\begin{array}{r} x + 6 \\ x - 1 \,\overline{)\, x^2 + 5x - 6} \\ \underline{x^2 - \phantom{5}x \phantom{{}-6}} \\ 6x - 6 \\ \underline{6x - 6} \\ \underline{\bullet \quad \bullet} \end{array}$$

(b)

$$(x^2 - x - 2) \div (x + 1) = \begin{array}{r|l} & x - 2 \\ \hline x + 1 & x^2 - x - 2 \\ & x^2 + x \\ \hline & -2x - 2 \\ & -2x - 2 \\ \hline & \bullet \quad \bullet \\ \hline \end{array}$$

(c)

$$(12x^3 - 11x^2 - 25) \div (3x - 5) = \begin{array}{r|l} & 4x^2 + 3x + 5 \\ \hline 3x - 5 & 12x^3 - 11x^2 + 0x - 25 \\ & 12x^3 - 20x^2 \\ \hline & 9x^2 + 0x \\ & 9x^2 - 15x \\ \hline & 15x - 25 \\ & 15x - 25 \\ \hline & \bullet \quad \bullet \\ \hline \end{array}$$

# Algebraic fractions

52

A numerical fraction is represented by one integer divided by another. Division of symbols follows the same rules to create *algebraic fractions*. For example:

$5 \div 3$ can be written as the fraction $\dfrac{5}{3}$ so

$a \div b$ can be written as $\dfrac{a}{b}$

## Addition and subtraction

The addition and subtraction of algebraic fractions follow the same rules as the addition and subtraction of numerical fractions – the operations can only be performed when the denominators are the same. For example, just as:

$$\frac{4}{5} + \frac{3}{7} = \frac{4 \times 7}{5 \times 7} + \frac{3 \times 5}{7 \times 5}$$

$$= \frac{4 \times 7 + 3 \times 5}{5 \times 7}$$

$$= \frac{43}{35} \qquad \text{(where 35 is the LCM of 5 and 7)}$$

so:

$$\frac{a}{b} + \frac{c}{d} = \frac{a \times d}{b \times d} + \frac{c \times b}{d \times b}$$

$$= \frac{ad + cb}{bd} \text{ provided } b \neq 0 \text{ and } d \neq 0 \qquad \text{(where } bd \text{ is the LCM of } b \text{ and } d\text{)}$$

So that:

$$\frac{a}{b} - \frac{c}{d^2} + \frac{d}{a} = \ldots\ldots\ldots\ldots$$

*Answer in the next frame*

**53**

$$\boxed{\frac{a^2d^2 - abc + bd^3}{abd^2}}$$

Because

$$\frac{a}{b} - \frac{c}{d^2} + \frac{d}{a} = \frac{aad^2}{bad^2} - \frac{cab}{d^2ab} + \frac{dd^2b}{ad^2b}$$
$$= \frac{a^2d^2 - abc + bd^3}{abd^2}$$

where $abd^2$ is the LCM of $a$, $b$ and $d^2$.

*On now to the next frame*

**54** In the same way

$$\frac{2}{x+1} + \frac{4}{x+2} = \frac{2(x+2) + 4(x+1)}{(x+1)(x+2)}$$
$$= \frac{2x + 4 + 4x + 4}{(x+1)(x+2)}$$
$$= \frac{6x + 8}{x^2 + 3x + 2}$$

**55**

## Multiplication and division

Fractions are multiplied by multiplying their numerators and denominators separately. For example, just as:

$$\frac{5}{4} \times \frac{3}{7} = \frac{5 \times 3}{4 \times 7} = \frac{15}{28}$$

so:

$$\frac{a}{b} \times \frac{c}{d} = \frac{ac}{bd}$$

The *reciprocal* of a number is unity divided by the number. For example, the reciprocal of $a$ is $1/a$ and the reciprocal of $\frac{a}{b}$ is:

$$\frac{1}{a/b} = 1 \div \frac{a}{b} = 1 \times \frac{b}{a} = \frac{b}{a}$$ *the numerator and denominator in the divisor are interchanged*

*To divide by an algebraic fraction we multiply by its reciprocal.* For example:

$$\frac{a}{b} \div \frac{c}{d} = \frac{a}{b} \times \frac{d}{c} = \frac{ad}{bc}$$

So that $\frac{2a}{3b} \div \frac{a^2b}{6} =$ ...........

*Check with the next frame*

**56**

$$\boxed{\frac{4}{ab^2}}$$

Because

$$\frac{2a}{3b} \div \frac{a^2b}{6} = \frac{2a}{3b} \times \frac{6}{a^2b} = \frac{4}{ab^2}$$

▶

Try another one:

$$\frac{2a}{3b} \div \frac{a^2b}{6} \times \frac{ab}{2} = \ldots\ldots\ldots\ldots$$

*The answer is in the next frame*

---

57

$$\boxed{\frac{2}{b}}$$

Because

$$\frac{2a}{3b} \div \frac{a^2b}{6} \times \frac{ab}{2} = \frac{2a}{3b} \times \frac{6}{a^2b} \times \frac{ab}{2} = \frac{4}{ab^2} \times \frac{ab}{2} = \frac{2}{b}$$

Remember that by the rules of precedence we work through the expression from the left to the right so we perform the division before we multiply. If we were to multiply before dividing in the above expression we should obtain:

$$\begin{aligned}\frac{2a}{3b} \div \frac{a^2b}{6} \times \frac{ab}{2} &= \frac{2a}{3b} \div \frac{a^3b^2}{12} \\ &= \frac{2a}{3b} \times \frac{12}{a^3b^2} \\ &= \frac{24a}{3a^3b^3} = \frac{8}{a^2b^3}\end{aligned}$$

and this would be wrong.

---

## Review summary

58

1 The manipulation of algebraic fractions follows identical principles as those for arithmetic fractions.

2 Only fractions with identical denominators can be immediately added or subtracted.

3 Two fractions are multiplied by multiplying their respective numerators and denominators separately.

4 Two fractions are divided by multiplying the numerator fraction by the reciprocal of the divisor fraction.

---

## Review exercise

59

1 Perform the following multiplications and simplify your results:

(a) $(2a + 4b)(a - 3b)$ (b) $(9s^2 + 3)(s^2 - 4)$

2 Simplify each of the following into a single algebraic fraction:

(a) $\frac{ab}{c} + \frac{cb}{a}$ (b) $\frac{ab}{c} - 1$ (c) $\left(\frac{ab}{c} + \frac{ac}{b}\right) + \frac{bc}{a}$

3 Perform the following division:

$$\frac{a^3 + 8b^3}{a + 2b}$$

**60**

1 (a) $(2a+4b)(a-3b) = 2a(a-3b)+4b(a-3b) = 2a^2-2ab-12b^2$

(b) $(9s^2+3)(s^2-4) = 9s^2(s^2-4)+3(s^2-4)$

$$= 9s^4-36s^2+3s^2-12$$

$$= 9s^4-33s^2-12$$

2 (a) $\dfrac{ab}{c}+\dfrac{cb}{a}=\dfrac{aab}{ac}+\dfrac{cbc}{ac}=\dfrac{b(a^2+c^2)}{ac}$

(b) $\dfrac{ab}{c}-1=\dfrac{ab}{c}-\dfrac{c}{c}=\dfrac{ab-c}{c}$

(c) $\left(\dfrac{ab}{c}+\dfrac{ac}{b}\right)+\dfrac{bc}{a}=\dfrac{ab}{c}+\dfrac{ac}{b}+\dfrac{bc}{a}=\dfrac{a^2b^2+a^2c^2+b^2c^2}{abc}$

3

$$\frac{a^3+8b^3}{a+2b}=a+2b\;\overline{\big)\,a^3 \qquad\qquad +8b^3}$$

$$\begin{array}{rl}
 & a^2-2ab+4b^2 \\
a+2b\;\big) & a^3 \qquad\qquad +8b^3 \\
 & \underline{a^3+2a^2b} \\
 & -2a^2b \\
 & \underline{-2a^2b-4ab^2} \\
 & 4ab^2+8b^3 \\
 & \underline{4ab^2+8b^3} \\
 & \underline{\quad\bullet\qquad\bullet\quad}
\end{array}$$

# Factorization of algebraic expressions

**61**

An algebraic fraction can often be simplified by writing the numerator and denominator in terms of their factors and cancelling where possible. For example:

$$\frac{25ab^2-15a^2b}{40ab^2-24a^2b}=\frac{5ab(5b-3a)}{8ab(5b-3a)}$$

$$=\frac{5}{8}$$

This is an obvious example, but there are many uses for factorization of algebraic expressions in advanced processes.

## Common factors

The simplest form of factorization is the extraction of highest common factors (HCF) from an expression. For example, $(10x+8)$ can clearly be written $2(5x+4)$.

Similarly with $(35x^2y^2-10xy^3)$:

the HCF of the coefficients 35 and 10 is 5

the HCF of the powers of $x$ is $x$

the HCF of the powers of $y$ is $y^2$

So $(35x^2y^2-10xy^3)=5xy^2(7x-2y)$

In the same way: (a) $8x^4y^3+6x^3y^2=$ ...........

and (b) $15a^3b-9a^2b^2=$ ...........

62

(a) $2x^3y^2(4xy+3)$
(b) $3a^2b(5a-3b)$

## Common factors by grouping

Four-termed expressions can sometimes be factorized by grouping into two binomial expressions and extracting common factors from each.

For example: $2ac+6bc+ad+3bd$

$$= (2ac+6bc)+(ad+3bd) = 2c(a+3b)+d(a+3b)$$
$$= (a+3b)(2c+d)$$

Similarly:

$x^3-4x^2y+xy^2-4y^3 = \ldots\ldots\ldots\ldots$

63

$(x-4y)(x^2+y^2)$

Because

$$x^3-4x^2y+xy^2-4y^3 = (x^3-4x^2y)+(xy^2-4y^3)$$
$$= x^2(x-4y)+y^2(x-4y) = (x-4y)(x^2+y^2)$$

In some cases it might be necessary to rearrange the order of the original four terms. For example:

$$12x^2-y^2+3x-4xy^2 = 12x^2+3x-y^2-4xy^2$$
$$= (12x^2+3x)-(y^2+4xy^2) = 3x(4x+1)-y^2(1+4x)$$
$$= (4x+1)(3x-y^2)$$

Likewise, $20x^2-3y^2+4xy^2-15x = \ldots\ldots\ldots\ldots$

64

$(4x-3)(5x+y^2)$

Rearranging terms:

$$(20x^2-15x)+(4xy^2-3y^2) = 5x(4x-3)+y^2(4x-3)$$
$$= (4x-3)(5x+y^2)$$

65

## Useful products of two simple factors

A number of standard results are well-worth remembering for the products of simple factors of the form $(a+b)$ and $(a-b)$. These are:

(a) $(a+b)^2 = (a+b)(a+b) = a^2+ab+ba+b^2$

i.e. $(a+b)^2 = a^2+2ab+b^2$

(b) $(a-b)^2 = (a-b)(a-b) = a^2-ab-ba+b^2$

i.e. $(a-b)^2 = a^2-2ab+b^2$

(c) $(a-b)(a+b) = a^2+ab-ba-b^2$

i.e. $(a-b)(a+b) = a^2-b^2$ *the difference of two squares*

▶

For our immediate purpose, these results can be used in reverse:

$$a^2 + 2ab + b^2 = (a + b)^2$$
$$a^2 - 2ab + b^2 = (a - b)^2$$
$$a^2 - b^2 = (a - b)(a + b)$$

If an expression can be seen to be one of these forms, its factors can be obtained at once.

These expressions that involve the variables raised to the power 2 are examples of what are called *quadratic* expressions. If a quadratic expression can be seen to be one of these forms, its factors can be obtained at once.

*On to the next frame*

**66**

Remember

$$a^2 + 2ab + b^2 = (a + b)^2$$
$$a^2 - 2ab + b^2 = (a - b)^2$$
$$a^2 - b^2 = (a - b)(a + b)$$

**Example 1**

$$x^2 + 10x + 25 = (x)^2 + 2(x)(5) + (5)^2, \text{ like } a^2 + 2ab + b^2$$
$$= (x + 5)^2$$

So $x^2 + 10x + 25 = (x + 5)^2$

**Example 2**

$$4a^2 - 12a + 9 = (2a)^2 - 2(2a)(3) + (3)^2, \text{ like } a^2 - 2ab + b^2$$
$$= (2a - 3)^2$$

So $4a^2 - 12a + 9 = (2a - 3)^2$

**Example 3**

$$25x^2 - 16y^2 = (5x)^2 - (4y)^2$$
$$= (5x - 4y)(5x + 4y)$$

So $25x^2 - 16y^2 = (5x - 4y)(5x + 4y)$

Now can you factorize the following:

(a) $16x^2 + 40xy + 25y^2 = \ldots\ldots\ldots\ldots$
(b) $9x^2 - 12xy + 4y^2 = \ldots\ldots\ldots\ldots$
(c) $(2x + 3y)^2 - (x - 4y)^2 = \ldots\ldots\ldots\ldots$

**67**

(a) $(4x + 5y)^2$
(b) $(3x - 2y)^2$
(c) $(x + 7y)(3x - y)$

▶

## Quadratic expressions as the product of two simple factors

**1** $(x+g)(x+k) = x^2 + (g+k)x + gk$

The coefficient of the middle term is the sum of the two constants $g$ and $k$ and the last term is the product of $g$ and $k$.

**2** $(x-g)(x-k) = x^2 - (g+k)x + gk$

The coefficient of the middle term is minus the sum of the two constants $g$ and $k$ and the last term is the product of $g$ and $k$.

**3** $(x+g)(x-k) = x^2 + (g-k)x - gk$

The coefficient of the middle term is the difference of the two constants $g$ and $k$ and the last term is minus the product of $g$ and $k$.

*Now let's try some specific types of quadratic*

---

68

## Factorization of a quadratic expression, $ax^2 + bx + c$ when $a = 1$

If $a = 1$, the quadratic expression is similar to those you have just considered, that is $x^2 + bx + c$. From rules **1–3** in the previous frame you can see that the values of $f_1$ and $f_2$ in $(x+f_1)$ and $(x+f_2)$, the factors of the quadratic expression, will depend upon the signs of $b$ and $c$. Notice that $b$, $c$, $f_1$ and $f_2$ can be positive or negative and that:

*If c is positive*

(a) $f_1$ and $f_2$ are factors of $c$ and both have the sign of $b$

(b) the sum of $f_1$ and $f_2$ is $b$

*If c is negative*

(a) $f_1$ and $f_2$ are factors of $c$ and have opposite signs, the numerically larger having the sign of $b$

(b) the difference between $f_1$ and $f_2$ is $b$

*There are examples of this in the next frame*

---

69

**Example 1**

$x^2 + 5x + 6$

(a) Possible factors of 6 are (1, 6) and (2, 3), so $(\pm 1, \pm 6)$ and $(\pm 2, \pm 3)$ are possible choices for $f_1$ and $f_2$.

(b) $c$ is positive so the required factors add up to $b$, that is 5.

(c) $c$ is positive so the required factors have the sign of $b$, that is positive, therefore $(2, 3)$.

$$\text{So } x^2 + 5x + 6 = (x+2)(x+3)$$

▶

**Example 2**

$x^2 - 9x + 20$

(a) Possible factors of 20 are $(1, 20)$, $(2, 10)$ and $(4, 5)$, so $(\pm 1, \pm 20)$, $(\pm 2, \pm 10)$ and $(\pm 4, \pm 5)$ are possible choices for $f_1$ and $f_2$.

(b) $c$ is positive so the required factors add up to $b$, that is $-9$.

(c) $c$ is positive so the required factors have the sign of $b$, that is negative, therefore $(-4, -5)$.

$$\text{So } x^2 - 9x + 20 = (x - 4)(x - 5)$$

**Example 3**

$x^2 + 3x - 10$

(a) Possible factors of 10 are $(1, 10)$ and $(2, 5)$, so $(\pm 1, \pm 10)$ and $(\pm 2, \pm 5)$ are possible choices for $f_1$ and $f_2$.

(b) $c$ is negative so the required factors differ by $b$, that is 3.

(c) $c$ is negative so the required factors differ in sign, the numerically larger having the sign of $b$, that is positive, therefore $(-2, 5)$.

$$\text{So } x^2 + 3x - 10 = (x - 2)(x + 5)$$

**Example 4**

$x^2 - 2x - 24$

(a) Possible factors of 24 are (1, 24), (2, 12), (3, 8) and (4, 6), so $(\pm 1, \pm 24)$, $(\pm 2, \pm 12)$, $(\pm 3, \pm 8)$ and $(\pm 4, \pm 6)$ are possible choices for $f_1$ and $f_2$.

(b) $c$ is negative so the required factors differ by $b$, that is $-2$.

(c) $c$ is negative so the required factors differ in sign, the numerically larger having the sign of $b$, that is negative, therefore $(4, -6)$.

$$\text{So } x^2 - 2x - 24 = (x + 4)(x - 6)$$

Now, here is a short exercise for practice. Factorize each of the following into two linear factors:

(a) $x^2 + 7x + 12$

(b) $x^2 - 11x + 28$

(c) $x^2 - 3x - 18$

(d) $x^2 + 2x - 24$

(e) $x^2 - 2x - 35$

(f) $x^2 - 10x + 16$

*Finish all six and then check with the next frame*

---

**70**

(a) $(x + 3)(x + 4)$

(b) $(x - 4)(x - 7)$

(c) $(x - 6)(x + 3)$

(d) $(x + 6)(x - 4)$

(e) $(x - 7)(x + 5)$

(f) $(x - 2)(x - 8)$

## Factorization of a quadratic expression $ax^2 + bx + c$ when $a \neq 1$ **71**

If $a \neq 1$, the factorization is slightly more complicated, but still based on the same considerations as for the simpler examples already discussed.

To factorize such an expression into its linear factors, if they exist, we carry out the following steps.

(a) We obtain $|ac|$, i.e. the numerical value of the product $ac$ ignoring the sign of the product.
(b) We write down all the possible pairs of factors of $|ac|$.
(c) (i) *If c is positive,* we select the two factors of $|ac|$ whose sum is equal to $|b|$: both of these factors have the same sign as $b$.
(ii) *If c is negative,* we select the two factors of $|ac|$ which differ by the value of $|b|$: the numerically larger of these two factors has the same sign as that of $b$ and the other factor has the opposite sign.
(iii) In each case, denote the two factors so obtained by $f_1$ and $f_2$.
(d) Then $ax^2 + bx + c$ is now written $ax^2 + f_1x + f_2x + c$ and this is factorized by finding common factors by grouping – as in the previous work.

**Example 1**

To factorize $6x^2 + 11x + 3 \qquad (ax^2 + bx + c)$

In this case, $a = 6$; $b = 11$; $c = 3$. Therefore $|ac| = 18$

Possible factors of $18 = (1, 18)$, $(2, 9)$ and $(3, 6)$

$c$ is positive. So required factors, $f_1$ and $f_2$, add up to ...........

---

**72**

$|b|$, i.e. 11

So required factors are (2, 9).

$c$ is positive. Both factors have the same sign as $b$, i.e. positive.

So $f_1 = 2$; $f_2 = 9$; and $6x^2 + 11x + 3 = 6x^2 + 2x + 9x + 3$

which factorizes by grouping into ...........

---

**73**

$(2x + 3)(3x + 1)$

Because

$$\begin{aligned} 6x^2 + 11x + 3 &= 6x^2 + 2x + 9x + 3 \\ &= (6x^2 + 9x) + (2x + 3) \\ &= 3x(2x + 3) + 1(2x + 3) = (2x + 3)(3x + 1) \end{aligned}$$

Now this one.

**Example 2**

To factorize $3x^2 - 14x + 8 \qquad (ax^2 + bx + c)$

$a = 3$; $b = -14$; $c = 8$; $|ac| = 24$

Possible factors of 24 =(1, 24), (2, 12), (3, 8) and (4, 6)

$c$ is positive. So required factors total $|b|$, i.e. 14. Therefore (2, 12)

$c$ is positive. So factors have same sign as $b$, i.e. negative, $f_1 = -2$; $f_2 = -12$

So $3x^2 - 14x + 8 = 3x^2 - 2x - 12x + 8$
$= \ldots\ldots\ldots\ldots$

*Finish it off*

**74**

$(x-4)(3x-2)$

Because

$$\begin{aligned}3x^2-2x-12x+8 &= (3x^2-12x)-(2x-8)\\ &= 3x(x-4)-2(x-4)\\ 3x^2-14x+8 &= (x-4)(3x-2)\end{aligned}$$

And finally, this one.

**Example 3**

To factorize $8x^2+18x-5$ $\quad (ax^2+bx+c)$

Follow the routine as before and all will be well.

So $8x^2+18x-5=$ ...........

**75**

$(2x+5)(4x-1)$

In this case, $a=8$; $b=18$; $c=-5$; $|ac|=40$

Possible factors of $40 = (1, 40), (2, 20), (4, 10)$ and $(5, 8)$

$c$ is negative. So required factors differ by $|b|$, i.e. 18. Therefore (2, 20)

$c$ is negative. So numerically larger factor has sign of $b$, i.e. positive.

$c$ is negative. So signs of $f_1$ and $f_2$ are different, $f_1=20$; $f_2=-2$

$$\begin{aligned}\text{So } 8x^2+18x-5 &= 8x^2+20x-2x-5\\ &= 4x(2x+5)-1(2x+5)\\ &= (2x+5)(4x-1)\end{aligned}$$

*Next frame*

**76**

## Test for simple factors

Some quadratic equations are not capable of being written as the product of *simple factors* – that is, factors where all the coefficients are integers. To save time and effort, a quick test can be applied before the previous routine is put into action.

To determine whether $ax^2+bx+c$ can be factorized into two simple factors, first evaluate the expression $(b^2-4ac)$.

If $(b^2-4ac)$ *is a perfect square,* that is it can be written as $k^2$ for some integer $k$, $ax^2+bx+c$ can be factorized into two simple factors.

If $(b^2-4ac)$ *is not a perfect square,* no simple factors of $ax^2+bx+c$ exist.

**Example 1**

$3x^2-4x+5 \qquad a=3;\ b=-4;\ c=5$

$$\begin{aligned}b^2-4ac &= 16-4\times3\times5\\ &= 16-60=-44 \quad \text{(not a perfect square)}\end{aligned}$$

There are no simple factors of $3x^2-4x+5$

▶

Now test in the same way:

**Example 2**

$2x^2 + 5x - 3 \qquad a = 2;\ b = 5;\ c = -3$

$b^2 - 4ac = 25 - 4 \times 2 \times (-3) = 25 + 24 = 49 = 7^2$ (perfect square)

$2x^2 + 5x - 3$ can be factorized into simple factors.

Now as an exercise, determine whether or not each of the following could be expressed as the product of two simple factors:

(a) $4x^2 + 3x - 4$
(b) $6x^2 + 7x + 2$
(c) $3x^2 + x - 4$
(d) $7x^2 - 3x - 5$

---

**77**

(a) No simple factors
(b) Simple factors
(c) Simple factors
(d) No simple factors

Now we can link this test with the previous work. Work through the following short exercise: it makes useful revision.

Test whether each of the following could be expressed as the product of two simple factors and, where possible, determine those factors:

(a) $2x^2 + 7x + 3$
(b) $5x^2 - 4x + 6$
(c) $7x^2 - 5x - 4$
(d) $8x^2 + 2x - 3$

*Check the results with the next frame*

---

**78**

(a) $(2x + 1)(x + 3)$
(b) No simple factors
(c) No simple factors
(d) $(2x - 1)(4x + 3)$

Here is the working:

(a) $2x^2 + 7x + 3 \qquad a = 2;\ b = 7;\ c = 3$

$b^2 - 4ac = 49 - 4 \times 2 \times 3 = 49 - 24 = 25 = 5^2$. Factors exist.

$|ac| = 6$; possible factors of 6 are $(1, 6)$ and $(2, 3)$

$c$ is positive. Factors add up to 7, i.e.$(1, 6)$

Both factors have the same sign as $b$, i.e. positive.

So $f_1 = 1$ and $f_2 = 6$

$$\begin{aligned}2x^2 + 7x + 3 &= 2x^2 + x + 6x + 3\\ &= (2x^2 + x) + (6x + 3) = x(2x + 1) + 3(2x + 1)\\ &= (2x + 1)(x + 3)\end{aligned}$$

▶

(b) $5x^2 - 4x + 6 \qquad a = 5;\ b = -4;\ c = 6$

$b^2 - 4ac = 16 - 4 \times 5 \times 6 = 16 - 120 = -104$. Not a complete square.

Therefore, no simple factors exist.

(c) $7x^2 - 5x - 4 \qquad a = 7;\ b = -5;\ c = -4$

$b^2 - 4ac = 25 - 4 \times 7 \times (-4) = 25 + 112 = 137$. Not a complete square.

Therefore, no simple factors exist.

(d) $8x^2 + 2x - 3 \qquad a = 8;\ b = 2;\ c = -3$

$b^2 - 4ac = 4 - 4 \times 8 \times (-3) = 4 + 96 = 100 = 10^2$. Factors exist.

$|ac| = 24$; possible factors of 24 are $(1, 24)$, $(2, 12)$, $(3, 8)$ and $(4, 6)$

$c$ is negative. Factors differ by $|b|$, i.e. 2. So $(4, 6)$

$f_1$ and $f_2$ of opposite signs. Larger factor has the same sign as $b$, i.e. positive. $f_1 = 6;\ f_2 = -4$.

$$8x^2 + 2x - 3 = 8x^2 + 6x - 4x - 3$$

$$= (8x^2 - 4x) + (6x - 3) = 4x(2x - 1) + 3(2x - 1)$$

$$\text{So } 8x^2 + 2x - 3 = (2x - 1)(4x + 3)$$

***At this point let us pause and summarize the main facts so far on factorization of algebraic expressions***

---

## Review summary

**79**

*Factorization of algebraic expressions*

(a) Common factors of binomial expressions.

(b) Common factors of expressions by grouping.

Useful standard factors:

$$a^2 + 2ab + b^2 = (a + b)^2$$

$$a^2 - 2ab + b^2 = (a - b)^2$$

$$a^2 - b^2 = (a - b)(a + b) \quad \text{(Difference of two squares)}$$

*Factorization of quadratic expressions of the form* $ax^2 + bx + c$.
*Test for possibility of simple factors*: $(b^2 - 4ac)$ is a complete square.
*Determination of factors* of $ax^2 + bx + c$:

(a) Evaluate $|ac|$.

(b) Write down all possible factors of $|ac|$.

(c) *If c is positive*, select two factors of $|ac|$ with sum equal to $|b|$.
*If c is positive*, both factors have the same sign as $b$.

(d) *If c is negative*, select two factors of $|ac|$ that differ by $|b|$.
*If c is negative*, the factors have opposite signs, the numerically larger having the same sign as $b$.

(e) Let the required two factors be $f_1$ and $f_2$.
Then $ax^2 + bx + c = ax^2 + f_1x + f_2x + c$ and factorize this by the method of common factors by grouping.

---

##  Review exercise

**80**

1 Factorize the following:

(a) $18xy^3 - 8x^3y$

(b) $x^3 - 6x^2y - 2xy + 12y^2$

(c) $16x^2 - 24xy - 18x + 27y$

(d) $(x - 2y)^2 - (2x - y)^2$

(e) $x^2 + 7x - 30$

(f) $4x^2 - 36$

(g) $x^2 + 10x + 25$

(h) $3x^2 - 11x - 4$

**81**

1 (a) $18xy^3 - 8x^3y = 2xy(9y^2 - 4x^2)$
$= 2xy(3y - 2x)(3y + 2x)$

(b) $x^3 - 6x^2y - 2xy + 12y^2 = x^2(x - 6y) - 2y(x - 6y)$
$= (x^2 - 2y)(x - 6y)$

(c) $16x^2 - 24xy - 18x + 27y = (16x^2 - 24xy) - (18x - 27y)$
$= 8x(2x - 3y) - 9(2x - 3y)$
$= (8x - 9)(2x - 3y)$

(d) $(x - 2y)^2 - (2x - y)^2 = x^2 - 4xy + 4y^2 - 4x^2 + 4xy - y^2$
$= 3y^2 - 3x^2$
$= 3(y^2 - x^2)$
$= 3(y - x)(y + x)$

(e) $x^2 + 7x - 30 = x^2 + (10 - 3)x + (10) \times (-3)$
$= (x + 10)(x - 3)$

(f) $4x^2 - 36 = (2x)^2 - (6)^2$
$= (2x - 6)(2x + 6)$
$= 4(x - 3)(x + 3)$

(g) $x^2 + 10x + 25 = x^2 + (2 \times 5)x + 5^2$
$= (x + 5)^2$

(h) $3x^2 - 11x - 4 = 3x(x - 4) + (x - 4)$
$= (3x + 1)(x - 4)$

**82**

You have now come to the end of this Programme. A list of **Can You?** questions follows for you to gauge your understanding of the material in the Programme. You will notice that these questions match the **Learning outcomes** listed at the beginning of the Programme so go back and try the **Quiz** that follows them. After that try the **Test exercise**. *Work through these at your own pace, there is no need to hurry*. A set of **Further problems** provides additional valuable practice.

## Can You?

**83**

### Checklist F.2

Check this list before and after you try the end of Programme test.

On a scale of 1 to 5 how confident are you that you can: Frames

- Use alphabetic symbols to supplement the numerals and to combine these symbols using all the operations of arithmetic? 1 to 7
  *Yes* ☐ ☐ ☐ ☐ ☐ *No*
- Simplify algebraic expressions by collecting like terms and by abstracting common factors from similar terms? 8 to 11
  *Yes* ☐ ☐ ☐ ☐ ☐ *No*
- Remove brackets and so obtain alternative algebraic expressions? 12 to 14
  *Yes* ☐ ☐ ☐ ☐ ☐ *No*
- Manipulate expressions involving powers and logarithms? 18 to 38
  *Yes* ☐ ☐ ☐ ☐ ☐ *No*
- Multiply and divide two algebraic expressions? 39 to 51
  *Yes* ☐ ☐ ☐ ☐ ☐ *No*
- Manipulate algebraic fractions? 52 to 60
  *Yes* ☐ ☐ ☐ ☐ ☐ *No*
- Factorize algebraic expressions using standard factorizations? 61 to 70
  *Yes* ☐ ☐ ☐ ☐ ☐ *No*
- Factorize quadratic algebraic expressions? 71 to 78
  *Yes* ☐ ☐ ☐ ☐ ☐ *No*

## Test exercise F.2

**84**

Frames

1 Simplify each of the following:

(a) $2ab - 4ac + ba - 2cb + 3ba$

(b) $3x^2yz - zx^2y + 4yxz^2 - 2x^2zy + 3z^2yx - 3zy^2$

(c) $c^p \times c^{-q} \div c^{-2}$

(d) $\dfrac{\left(x^{\frac{1}{2}}\right)^{-\frac{2}{3}} \div \left(y^{\frac{3}{4}}\right)^{2} \times \left(x^{\frac{3}{5}}\right)^{-\frac{5}{3}}}{\left(x^{\frac{1}{4}}\right)^{-1} \times \left(y^{\frac{1}{3}}\right)^{6}}$ 1 to 22

2 Remove the brackets in each of the following:

(a) $2f(3g - 4h)(g + 2h)$

(b) $(5x - 6y)(2x + 6y)(5x - y)$

(c) $4\{3[p - 2(q - 3)] - 3(r - 2)\}$ 12 to 17

3 Evaluate by calculator or by change of base where necessary:

(a) $\log 0{\cdot}0270$ (b) $\ln 47{\cdot}89$ (c) $\log_7 126{\cdot}4$ 23 to 32

4 Rewrite the following in log form:

(a) $V = \dfrac{\pi h}{4}(D - h)(D + h)$ (b) $P = \dfrac{1}{16}(2d - 1)^2 N\sqrt{S}$

Frames

5 Express the following without logarithms:

(a) $\log x = \log P + 2\log Q - \log K - 3$

(b) $\log R = 1 + \frac{1}{3}\log M + 3\log S$

(c) $\ln P = \frac{1}{2}\ln(Q+1) - 3\ln R + 2$ 33 to 35

6 Perform the following multiplications and simplify your results:

(a) $(3x-1)(x^2-x-1)$

(b) $(a^2+2a+2)(3a^2+4a+4)$ 39 to 43

7 Simplify each of the following:

(a) $\frac{x^3}{y^2} \div \frac{x}{y^3}$

(b) $\frac{ab}{c} \div \frac{ac}{b} \times \frac{bc}{a}$ 44 to 48

8 Perform the following divisions:

(a) $(x^2+2x-3) \div (x-1)$

(b) $\frac{n^3+27}{n+3}$

(c) $\frac{3a^3+2a^2+1}{a+1}$ 49 to 60

9 Factorize the following:

(a) $36x^3y^2 - 8x^2y$

(b) $x^3 + 3x^2y + 2xy^2 + 6y^3$

(c) $4x^2 + 12x + 9$

(d) $(3x+4y)^2 - (2x-y)^2$

(e) $x^2 + 10x + 24$

(f) $x^2 - 10x + 16$

(g) $x^2 - 5x - 36$

(h) $6x^2 + 5x - 6$ 61 to 78

## Further problems F.2

**85**

1 Determine the following:

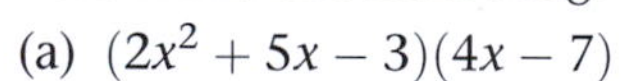

(a) $(2x^2+5x-3)(4x-7)$

(b) $(4x^2-7x+3)(5x+6)$

(c) $(5x^2-3x-4)(3x-5)$

(d) $(6x^3-5x^2-14x+12) \div (2x-3)$

(e) $(15x^3+46x^2-49) \div (5x+7)$

(f) $(18x^3+13x+14) \div (3x+2)$

2 Simplify the following, giving the result without fractional indices

$(x^2-1)^2 \times \sqrt{x+1} \div (x-1)^{\frac{3}{2}}$

3 Simplify:

(a) $\sqrt{a^{\frac{7}{3}}b^5c^{\frac{2}{3}}} \div \sqrt[3]{a^{\frac{1}{2}}b^3c^{-1}}$

(b) $\sqrt[3]{x^9y^{\frac{1}{3}}z^{\frac{1}{2}}} \times y^{\frac{8}{9}} \times (2^{-8}x^6y^2z^{\frac{1}{3}})^{-\frac{1}{2}}$

(c) $(6x^3y^{\frac{5}{2}}z^{\frac{1}{4}})^2 \div (9x^6y^4z^3)^{\frac{1}{2}}$

(d) $(x^2-y^2)^{\frac{1}{2}} \times (x-y)^{\frac{3}{2}} \times (x+y)^{-\frac{1}{2}}$

4 Evaluate:

(a) $\log 0{\cdot}008472$ (b) $\ln 25{\cdot}47$ (c) $\log_8 387{\cdot}5$

▶

**5** Express in log form:

(a) $f = \dfrac{1}{\pi d\sqrt{LC}}$

(b) $K = \dfrac{a^3 \times \sqrt{b}}{c^{\frac{1}{6}} d^{\frac{2}{5}}}$

**6** Rewrite the following without logarithms:

(a) $\log W = 2(\log A + \log w) - (\log 32 + 2\log\pi + 2\log r + \log c)$

(b) $\log S = \log K - \log 2 + 2\log\pi + 2\log n + \log y + \log r + 2\log L - 2\log h - \log g$

(c) $\ln I = \ln(2V) - \{\ln(KR + r) - \ln K + KL\}$

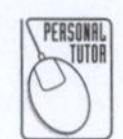

**7** Factorize the following:

(a) $15x^2y^2 + 20xy^3$

(b) $14a^3b - 12a^2b^2$

(c) $2x^2 + 3xy - 10x - 15y$

(d) $4xy - 7y^2 - 12x + 21y$

(e) $15x^2 + 8y + 20xy + 6x$

(f) $6xy - 20 + 15x - 8y$

(g) $9x^2 + 24xy + 16y^2$

(h) $16x^2 - 40xy + 25y^2$

(i) $25x^3y^4 - 16xy^2$

(j) $(2x + 5y)^2 - (x - 3y)^2$

**8** Find simple factors of the following where possible:

(a) $5x^2 + 13x + 6$

(b) $2x^2 - 11x + 12$

(c) $6x^2 - 5x - 6$

(d) $3x^2 + 7x - 4$

(e) $5x^2 - 19x + 12$

(f) $4x^2 - 6x + 9$

(g) $6x^2 - 5x - 7$

(h) $9x^2 - 18x + 8$

(i) $10x^2 + 11x - 6$

(j) $15x^2 - 19x + 6$

(k) $8x^2 + 2x - 15$

**86**

Now visit the companion website for this book at www.palgrave.com/stroud for more questions applying this mathematics to science and engineering.

# Expressions and equations

## Learning outcomes

When you have completed this Programme you will be able to:

- ☐ Numerically evaluate an algebraic expression by substituting numbers for variables
- ☐ Recognize the different types of equation
- ☐ Evaluate an independent variable
- ☐ Change the subject of an equation by transposition
- ☐ Evaluate polynomial expressions by 'nesting'
- ☐ Use the remainder and factor theorems to factorize polynomials
- ☐ Factorize fourth-order polynomials

If you already feel confident about these why not try the short quiz over the page? You can check your answers at the end of the book.

# Quiz F.3

**Frames**

1 Given $P = A\left(1+\frac{r}{100}\right)^n$ find $P$ to 2 dp given that $A = 12\,345{\cdot}66$, $r = 4{\cdot}65$ and $n = 6\frac{255}{365}$. (Frames 1 to 4)

2 Given $T = 2\pi\sqrt{\frac{l^2+4t^2}{3g(r-t)}}$ find:

(a) $l$ in terms of $T$, $t$, $r$ and $g$

(b) $r$ in terms of $T$, $t$, $l$ and $g$. (Frames 7 to 19)

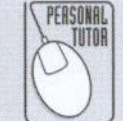

3 Write $f(x) = 7x^3 - 6x^2 + 4x + 1$ in nested form and find the value of $f(-2)$. (Frames 25 to 30)

4 Without dividing in full, determine the remainder of:

$(4x^4 + 3x^3 - 2x^2 - x + 7) \div (x+3)$ (Frames 31 to 33)

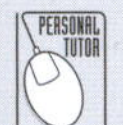

5 Factorize $6x^4 + 5x^3 - 39x^2 + 4x + 12$. (Frames 34 to 57)

# Expressions and equations

## Evaluating expressions

1

When numerical values are assigned to the variables and constants in an expression, the expression itself assumes a numerical value that is obtained by following the usual precedence rules. This process is known as *evaluating* the expression. For example, if $l = 2$ and $g = 9{\cdot}81$ then the expression:

$$2\pi\sqrt{\frac{l}{g}}$$

is evaluated as:

$$2\pi\sqrt{\frac{2}{9{\cdot}81}} = 2{\cdot}84 \text{ to 2 dp where } \pi = 3{\cdot}14159\ldots$$

So let's look at three examples:

**Example 1**

If $V = \frac{\pi h}{6}(3R^2 + h^2)$, determine the value of $V$ when $h = 2{\cdot}85$, $R = 6{\cdot}24$ and $\pi = 3{\cdot}142$.

Substituting the given values:

$$\begin{aligned} V &= \frac{3{\cdot}142 \times 2{\cdot}85}{6}(3 \times 6{\cdot}24^2 + 2{\cdot}85^2) \\ &= \frac{3{\cdot}142 \times 2{\cdot}85}{6}(3 \times 38{\cdot}938 + 8{\cdot}123) \\ &= \ldots\ldots\ldots\ldots \end{aligned}$$

*Finish it off*

2

$V = 186{\cdot}46$

**Example 2**

If $R = \frac{R_1 R_2}{R_1 + R_2}$, evaluate $R$ when $R_1 = 276$ and $R_2 = 145$.

That is easy enough, $R = \ldots\ldots\ldots\ldots$

3

$R = 95{\cdot}06$

Now let us deal with a more interesting one.

**Example 3**

If $V = \frac{\pi b}{12}(D^2 + Dd + d^2)$ evaluate $V$ to 3 sig fig when $b = 1{\cdot}46$, $D = 0{\cdot}864$, $d = 0{\cdot}517$ and $\pi = 3{\cdot}142$.

Substitute the values in the expressions and then apply the rules carefully. Take your time with the working: there are no prizes for speed!

$V = \ldots\ldots\ldots\ldots$

4

$$\boxed{V = 0{\cdot}558 \text{ to 3 sig fig}}$$

Here it is:

$$\begin{aligned} V &= \frac{3{\cdot}142 \times 1{\cdot}46}{12}(0{\cdot}864^2 + 0{\cdot}864 \times 0{\cdot}517 + 0{\cdot}517^2) \\ &= \frac{3{\cdot}142 \times 1{\cdot}46}{12}(0{\cdot}7465 + 0{\cdot}864 \times 0{\cdot}517 + 0{\cdot}2673) \\ &= \frac{3{\cdot}142 \times 1{\cdot}46}{12}(0{\cdot}7465 + 0{\cdot}4467 + 0{\cdot}2673) \\ &= \frac{3{\cdot}142 \times 1{\cdot}46}{12}(1{\cdot}4605) \\ &= 0{\cdot}5583\ldots \end{aligned}$$

$$\therefore V = 0{\cdot}558 \text{ to 3 sig fig}$$

5

## Equations

Because different values of the variables and constants produce different values for the expression, we assign these expression values to another variable and so form an *equation*. For example, the equation:

$$r = 2s^3 + 3t$$

states that the variable $r$ can be assigned values by successively assigning values to $s$ and to $t$, each time evaluating $2s^3 + 3t$. The variable $r$ is called the *dependent* variable and *subject* of the equation whose value *depends* on the values of the *independent* variables $s$ and $t$.

An *equation* is a statement of the equality of two expressions but there are different types of equation:

### Conditional equation

A *conditional equation*, usually just called an *equation*, is true only for certain values of the symbols involved. For example, the equation:

$$x^2 = 4$$

is an equation that is only true for each of the two values $x = +2$ and $x = -2$.

### Identity

An *identity* is a statement of equality of two expressions that is true for all values of the symbols for which both expressions are defined. For example, the equation:

$$2(5 - x) \equiv 10 - 2x$$

is true no matter what value is chosen for $x$ – it is an *identity*. The expression on the left is not just equal to the expression on the right, it is *equivalent* to it – one expression is an alternative form of the other. Hence the symbol $\equiv$ which stands for 'is equivalent to'.

### Defining equation

A *defining equation* is a statement of equality that defines an expression. For example:

$$a^2 \triangleq a \times a$$

Here the symbolism $a^2$ is defined to mean $a \times a$ where $\triangleq$ means 'is defined to be'.

▶

### Assigning equation

An *assigning equation* is a statement of equality that assigns a specific value to a variable. For example:

$p := 4$

Here, the value 4 is assigned to the variable $p$.

### Formula

A *formula* is a statement of equality that expresses a mathematical fact where all the variables, dependent and independent, are well-defined. For example, the equation:

$A = \pi r^2$

expresses the fact that the area $A$ of a circle of radius $r$ is given as $\pi r^2$.

The uses of $\equiv$, $\triangleq$ and $:=$ as connectives are often substituted by the $=$ sign. While it is not strictly correct to do so, it is acceptable.

So what type of equation is each of the following?

(a) $l = \dfrac{T^2 g}{4\pi^2}$ where $T$ is the periodic time and $l$ the length of a simple pendulum and where $g$ is the acceleration due to gravity

(b) $v = 23{\cdot}4$

(c) $4n = 4 \times n$

(d) $x^2 - 2x = 0$

(e) $\dfrac{r^3 - s^3}{r - s} = r^2 + rs + s^2$ where $r \neq s$

*The answers are in the next frame*

---

**6**

(a) Formula
(b) Assigning equation
(c) Defining equation
(d) Conditional equation
(e) Identity

Because

(a) It is a statement of a mathematical fact that relates the values of the variables $T$ and $l$ and the constant $g$ where $T$, $l$ and $g$ represent well-defined entities.

(b) It assigns the value 23·4 to the variable $v$.

(c) It defines the notation $4n$ whereby the multiplication sign is omitted.

(d) It is only true for certain values of the variables, namely $x = 0$ and $x = 2$.

(e) The left-hand side is an alternative form of the right-hand side, as can be seen by performing the division. Notice that the expression on the left is not defined when $r = s$ whereas the expression on the right is. We say that the equality only holds true for numerical values when *both* expressions are defined.

*Now to the next frame*

7

## Evaluating independent variables

Sometimes, the numerical values assigned to the variables and constants in a formula include a value for the dependent variable and exclude a value of one of the independent variables. In this case the exercise is to find the corresponding value of the independent variable. For example, given that:

$$T = 2\pi\sqrt{\frac{l}{g}} \text{ where } \pi = 3{\cdot}14 \text{ and } g = 9{\cdot}81$$

what is the length $l$ that corresponds to $T = 1{\cdot}03$? That is, given:

$$1{\cdot}03 = 6{\cdot}28\sqrt{\frac{l}{9{\cdot}81}}$$

find $l$. We do this by isolating $l$ on one side of the equation.

So we first divide both sides by 6·28 to give: $\dfrac{1{\cdot}03}{6{\cdot}28} = \sqrt{\dfrac{l}{9{\cdot}81}}$

*An equation is like a balance, so if any arithmetic operation is performed on one side of the equation the identical operation must be performed on the other side to maintain the balance.*

Square both sides to give: $\left(\dfrac{1{\cdot}03}{6{\cdot}28}\right)^2 = \dfrac{l}{9{\cdot}81}$ and now multiply both sides by 9·81 to give:

$$9{\cdot}81\left(\frac{1{\cdot}03}{6{\cdot}28}\right)^2 = l = 0{\cdot}264 \text{ to 3 sig fig}$$

So that if:

$$I = \frac{nE}{R + nr} \text{ and } n = 6,\ E = 2{\cdot}01,\ R = 12 \text{ and } I = 0{\cdot}98,$$

the corresponding value of $r$ is ...........

*Next frame*

8

$$\boxed{r = 0{\cdot}051}$$

Because

Given that $0{\cdot}98 = \dfrac{6 \times 2{\cdot}01}{12 + 6r} = \dfrac{12{\cdot}06}{12 + 6r}$ we see, by taking the reciprocal of each side, that:

$$\frac{1}{0{\cdot}98} = \frac{12 + 6r}{12{\cdot}06} \text{ and hence } \frac{12{\cdot}06}{0{\cdot}98} = 12 + 6r$$

after multiplying both sides by 12·06. Subtracting 12 from both sides yields:

$$\frac{12{\cdot}06}{0{\cdot}98} - 12 = 6r$$

and, dividing both sides by 6 gives:

$$\frac{1}{6}\left(\frac{12{\cdot}06}{0{\cdot}98} - 12\right) = r, \text{ giving } r = 0{\cdot}051 \text{ to 2 sig fig}$$

By this process of arithmetic manipulation the independent variable $r$ in the original equation has been *transposed* to become the dependent variable of a new equation, so enabling its value to be found.

▶

You will often encounter the need to transpose a variable in an equation so it is essential that you acquire the ability to do so. Furthermore, you will also need to transpose variables to obtain a new equation rather than just to find the numerical value of the transposed variable as you have done so far. In what follows we shall consider the transposition of variables algebraically rather then arithmetically.

## Transposition of formulas

9

The formula for the period of oscillation, $T$ seconds, of a pendulum is given by:

$$T = 2\pi\sqrt{\frac{l}{g}}$$

where $l$ is the length of the pendulum measured in metres, $g$ is the gravitational constant ($9{\cdot}81\ \mathrm{m\,s^{-2}}$) and $\pi = 3{\cdot}142$ to 4 sig fig. The single symbol on the left-hand side (LHS) of the formula – the dependent variable – is often referred to as the *subject of the formula*. We say that $T$ *is given in terms of* $l$. What we now require is a new formula where $l$ is the subject. That is, where $l$ is given in terms of $T$. To effect this transposition, keep in mind the following:

The formula is an equation, or balance. Whatever is done to one side must be done to the other.

To remove a symbol from the right-hand side (RHS) we carry out the opposite operation to that which the symbol is at present involved in. The 'opposites' are – *addition* and *subtraction*, *multiplication* and *division*, *powers* and *roots*.

In this case we start with:

$$T = 2\pi\sqrt{\frac{l}{g}}$$

To isolate $l$ we start by removing the $2\pi$ by dividing both sides by $2\pi$. This gives:

$$\frac{T}{2\pi} = \sqrt{\frac{l}{g}}$$

We next remove the square root sign by squaring both sides to give:

$$\frac{T^2}{4\pi^2} = \frac{l}{g}$$

Next we remove the $g$ on the RHS by multiplying both sides by $g$ to give:

$$\frac{gT^2}{4\pi^2} = l$$

Finally, we interchange sides to give:

$$l = \frac{gT^2}{4\pi^2}$$

because it is more usual to have the subject of the formula on the LHS.

*Now try a few examples*

**10**

**Example 1**

Transpose the formula $v = u + at$ to make $a$ the subject.

(a) Isolate the term involving $a$ by subtracting $u$ from both sides:

$$v - u = u + at - u = at$$

(b) Isolate the $a$ by dividing both sides by $t$:

$$\frac{v - u}{t} = \frac{at}{t} = a$$

(c) Finally, write the transposed equation with the new subject on the LHS,

$$a = \frac{v - u}{t}$$

*Let's just try another*

**11**

**Example 2**

Transpose the formula $a = \dfrac{2(ut - s)}{t^2}$ to make $u$ the subject.

(a) $u$ is part of the numerator on the RHS. Therefore first multiply both sides by $t^2$:

$$at^2 = 2(ut - s)$$

(b) We can now multiply out the bracket:

$$at^2 = 2ut - 2s$$

(c) Now we isolate the term containing $u$ by adding $2s$ to each side:

$$at^2 + 2s = 2ut$$

(d) $u$ is here multiplied by $2t$, therefore we divide each side by $2t$:

$$\frac{at^2 + 2s}{2t} = u$$

(e) Finally, write the transposed formula with the new subject on the LHS, i.e. $u = \dfrac{at^2 + 2s}{2t}$

Apply the procedure carefully and take one step at a time.

**Example 3**

Transpose the formula $d = 2\sqrt{h(2r - h)}$ to make $r$ the subject.

(a) First we divide both sides by 2:

$$\frac{d}{2} = \sqrt{h(2r - h)}$$

(b) To open up the expression under the square root sign, we ...........

**12**

square both sides

So $\dfrac{d^2}{4} = h(2r - h)$

(c) At present, the bracket expression is multiplied by $h$. Therefore, we ...........

**13**

divide both sides by $h$

So $\dfrac{d^2}{4h} = 2r - h$

(d) Next, we . . . . . . . . . . .

**14**

add $h$ to both sides

So $\dfrac{d^2}{4h} + h = 2r$

*Finish it off*

**15**

$$r = \frac{1}{2}\left\{\frac{d^2}{4h} + h\right\}$$

Of course, this could be written in a different form:

$$\begin{aligned} r &= \frac{1}{2}\left\{\frac{d^2}{4h} + h\right\} \\ &= \frac{1}{2}\left\{\frac{d^2 + 4h^2}{4h}\right\} \\ &= \frac{d^2 + 4h^2}{8h} \end{aligned}$$

All these forms are equivalent to each other.

Now, this one.

**Example 4**

Transpose $V = \dfrac{\pi h(3R^2 + h^2)}{6}$ to make $R$ the subject.

First locate the symbol $R$ in its present position and then take the necessary steps to isolate it. Do one step at a time.

$R =$ . . . . . . . . . . .

**16**

$$R = \sqrt{\frac{2V}{\pi h} - \frac{h^2}{3}}$$

Because

$$\begin{aligned} V &= \frac{\pi h(3R^2 + h^2)}{6} \\ 6V &= \pi h(3R^2 + h^2) \\ \frac{6V}{\pi h} &= 3R^2 + h^2 \end{aligned}$$

$$\frac{6V}{\pi h} - h^2 = 3R^2 \qquad \text{So } \frac{2V}{\pi h} - \frac{h^2}{3} = R^2$$

$$\text{Therefore } \sqrt{\frac{2V}{\pi h} - \frac{h^2}{3}} = R, \qquad R = \sqrt{\frac{2V}{\pi h} - \frac{h^2}{3}}$$

**17**

**Example 5**

This one is slightly different.

Transpose the formula $n = \dfrac{IR}{E - Ir}$ to make $I$ the subject.

In this case, you will see that the symbol $I$ occurs twice on the RHS. Our first step, therefore, is to move the denominator completely by multiplying both sides by $(E - Ir)$:

$$n(E - Ir) = IR$$

Then we can free the $I$ on the LHS by multiplying out the bracket:

$$nE - nIr = IR$$

Now we collect up the two terms containing $I$ on to the RHS:

$$\begin{aligned} nE &= IR + nIr \\ &= I(R + nr) \end{aligned}$$

$$\text{So } I = \frac{nE}{R + nr}$$

*Move on to the next frame*

**18**

**Example 6**

Here is one more, worked in very much the same way as the previous example, so you will have no trouble.

Transpose the formula $\dfrac{R}{r} = \sqrt{\dfrac{f + P}{f - P}}$ to make $f$ the subject.

Work right through it, using the rules and methods of the previous examples.

$f = \ldots\ldots\ldots\ldots$

**19**

$$\boxed{f = \frac{(R^2 + r^2)P}{R^2 - r^2}}$$

Here it is:

$$\frac{R^2}{r^2} = \frac{f + P}{f - P}$$

$$\frac{R^2}{r^2}(f - P) = f + P$$

$$R^2(f - P) = r^2(f + P)$$

$$R^2 f - R^2 P = r^2 f + r^2 P$$

$$R^2 f - r^2 f = R^2 P + r^2 P$$

$$f(R^2 - r^2) = P(R^2 + r^2)$$

$$\text{So } f = \frac{(R^2 + r^2)P}{R^2 - r^2}$$

## The evaluation process

20

A *system* is a process that is capable of accepting an *input*, *processing* the input and producing an *output*:

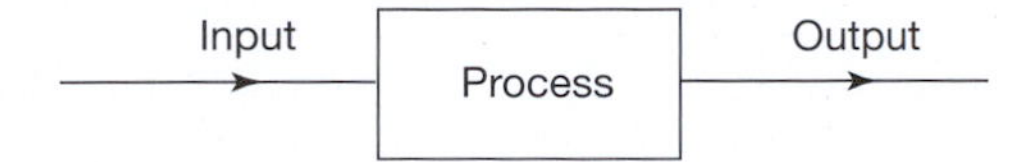


We can use this idea of a system to describe the way we evaluate an algebraic expression. For example, given the expression:

$3x - 4$

we evaluate it for $x = 5$, say, by multiplying 5 by 3 and then subtracting 4 to obtain the value 11; we *process* the *input* 5 to produce the *output* 11:

If we use the letter $x$ to denote the input and the letter $f$ to denote the process we denote the output as:

$f(x)$, that is '$f$ acting on $x$'

where the process $f$, represented by the box in the diagram, is:

*multiply x by 3 and then subtract 4*

How the evaluation is actually done, whether mentally, by pen and paper or by using a calculator is not important. What is important is that the prescription for evaluating it is given by the expression $3x - 4$ and that we can represent the *process* of executing this prescription by the label $f$.

The advantage of this notion is that it permits us to tabulate the results of evaluation in a meaningful way. For example, if:

$f(x) = 3x - 4$

then:

$f(5) = 15 - 4 = 11$

and in this way the corresponding values of the two variables are recorded.

So that, if $f(x) = 4x^3 - \dfrac{6}{2x}$ then:

(a) $f(3) =$ ...........

(b) $f(-4) =$ ...........

(c) $f(2/5) =$ ...........

(d) $f(-3{\cdot}24) =$ ........... (to 5 sig fig)

*Answers are in the next frame*

21

(a) $f(3) = 107$
(b) $f(-4) = -255{\cdot}25$
(c) $f(2/5) = -7{\cdot}244$
(d) $f(-3{\cdot}24) = -135{\cdot}12$

Because

(a) $f(3) = 4 \times 3^3 - \dfrac{6}{2 \times 3} = 108 - 1 = 107$

(b) $f(-4) = 4 \times (-4)^3 - \dfrac{6}{2 \times (-4)} = -256 + 0{\cdot}75 = -255{\cdot}25$

(c) $f(2/5) = 4 \times (2/5)^3 - \dfrac{6}{2 \times (2/5)} = 0{\cdot}256 - 7{\cdot}5 = -7{\cdot}244$

(d) $f(-3{\cdot}24) = 4 \times (-3{\cdot}24)^3 - \dfrac{6}{2 \times (-3{\cdot}24)} = -136{\cdot}05 + 0{\cdot}92593 = -135{\cdot}12$

***At this point let us pause and summarize the main facts so far***

## Review summary

22

1 An algebraic expression is evaluated by substituting numbers for the variables and constants in the expression and then using the arithmetic precedence rules.

2 Values so obtained can be assigned to a variable to form an equation. This variable is called the *subject* of the equation.

3 The subject of an equation is called the *dependent variable* and the variables within the expression are called the *independent variables*.

4 *There is more than one type of equation:*

Conditional equation
Identity
Defining equation
Assigning equation
Formula

5 Any one of the independent variables in a formula can be made the subject of a new formula obtained by transposing it with the dependent variable of the original formula.

6 Transposition is effected by performing identical arithmetic operations on both sides of the equation.

7 The evaluation process can be represented as a system capable of accepting an input, processing the input and producing an output. This has the advantage of enabling the results of continual evaluation in a meaningful way.

## Review exercise

23

1 Evaluate each of the following to 3 sig fig:

(a) $I = \dfrac{nE}{R + nr}$ where $n = 4$, $E = 1{\cdot}08$, $R = 5$ and $r = 0{\cdot}04$

(b) $A = P\left(1 + \dfrac{r}{100}\right)^n$ where $P = 285{\cdot}79$, $r = 5{\cdot}25$ and $n = 12$

(c) $P = A\dfrac{(nv/u)^{\frac{3}{2}}}{1 + (nv/u)^3}$ where $A = 40$, $u = 30$, $n = 2{\cdot}5$ and $v = 42{\cdot}75$

2 Transpose the formula $f = \dfrac{S(M - m)}{M + m}$ to make $m$ the subject.

3 Given $f(x) = 3x^2 - 4x$ find $f(2)$ and $f(-3)$.

24

1 (a) $I = \dfrac{4 \times 1{\cdot}08}{5 + 4 \times 0{\cdot}04} = \dfrac{4{\cdot}32}{5 + 0{\cdot}16} = 0{\cdot}837$

(b) $A = 285{\cdot}79\left(1 + \dfrac{5{\cdot}25}{100}\right)^{12} = 285{\cdot}79(1{\cdot}0525)^{12}$

$= 285{\cdot}79 \times 1{\cdot}84784\ldots = 528$

(c) $P = 40\dfrac{(2{\cdot}5 \times 42{\cdot}75 \div 30)^{\frac{3}{2}}}{1 + (2{\cdot}5 \times 42{\cdot}75 \div 30)^3} = 40\dfrac{3{\cdot}5625^{\frac{3}{2}}}{1 + 3{\cdot}5625^3} = 40\dfrac{3{\cdot}5625^{\frac{3}{2}}}{1 + 3{\cdot}5625^3}$

$= 40\dfrac{6{\cdot}7240\ldots}{46{\cdot}2131\ldots} = 5{\cdot}82$

2 $f = \dfrac{S(M - m)}{M + m}$ so $f(M + m) = S(M - m)$ thus $fM + fm = SM - Sm$, that is

$fm + Sm = SM - fM$. Factorizing yields $m(f + S) = M(S - f)$ giving $m = \dfrac{M(S - f)}{f + S}$.

3 $f(2) = 3(2)^2 - 4(2) = 12 - 8 = 4$ $\quad : f(2) = 4$

$f(-3) = 3(-3)^2 - 4(-3) = 27 + 12 = 39$ $\quad : f(-3) = 39$

*And now to the next topic*

# Polynomial equations

25

## Polynomial expressions

A *polynomial* in $x$ is an expression involving powers of $x$, normally arranged in descending (or sometimes ascending) powers. The degree of the polynomial is given by the highest power of $x$ occurring in the expression, for example:

$5x^4 + 7x^3 + 3x - 4$ is a polynomial of the 4th degree

and $2x^3 + 4x^2 - 2x + 7$ is a polynomial of the 3rd degree.

Polynomials of low degree often have alternative names:

$2x - 3$ is a polynomial of the 1st degree – or a *linear* expression.

$3x^2 + 4x + 2$ is a polynomial of the 2nd degree – or a *quadratic* expression.

A polynomial of the 3rd degree is often referred to as a *cubic* expression.

A polynomial of the 4th degree is often referred to as a *quartic* expression.

## Evaluation of polynomials

If $f(x) = 3x^4 - 5x^3 + 7x^2 - 4x + 2$, then evaluating $f(3)$ would involve finding the values of each term before finally totalling up the five individual values. This would mean recording the partial values – with the danger of including errors in the process.

This can be avoided by using the method known as *nesting* – so move to the next frame to see what it entails.

**26**

## Evaluation of a polynomial by nesting

Consider the polynomial $f(x) = 5x^3 + 2x^2 - 3x + 6$. To express this in *nested* form, write down the coefficient and one factor $x$ from the first term and add on the coefficient of the next term:

i.e. $5x + 2$

Enclose these in brackets, multiply by $x$ and add on the next coefficient:

i.e. $(5x + 2)x - 3$

Repeat the process: enclose the whole of this in square brackets, multiply by $x$ and add on the next coefficient:

i.e. $[(5x + 2)x - 3]x + 6$

So $f(x) = 5x^3 + 2x^2 - 3x + 6$

$= [(5x + 2)x - 3]x + 6$ in nested form.

Starting with the innermost brackets, we can now substitute the given value of $x$ and carry on in a linear fashion. No recording is required:

$f(4) = [(22)4 - 3]4 + 6$

$= [85]4 + 6 = 346$

So $f(4) = 346$

*Note*: The working has been set out here purely by way of explanation. Normally it would be carried out mentally.

So, in the same way,

$f(2) =$ ........... and $f(-1) =$ ...........

**27**

48; 6

*Notes*: (a) The terms of the polynomial must be arranged in descending order of powers.

(b) If any power is missing from the polynomial, it must be included with a zero coefficient before nesting is carried out.

Therefore, if $f(x) = 3x^4 + 2x^2 - 4x + 5$

(a) $f(x)$ in nested form $=$ ...........

(b) $f(2) =$ ...........

**28**

(a) $f(x) = \{[(3x + 0)x + 2]x - 4\}x + 5$

(b) $f(2) = 53$

*On to the next frame*

**29**

Now a short exercise. In each of the following cases, express the polynomial in nested form and evaluate the function for the given value of $x$:

(a) $f(x) = 4x^3 + 3x^2 + 2x - 4$ $[x = 2]$

(b) $f(x) = 2x^4 + x^3 - 3x^2 + 5x - 6$ $[x = 3]$

(c) $f(x) = x^4 - 3x^3 + 2x - 3$ $[x = 5]$

(d) $f(x) = 2x^4 - 5x^3 - 3x^2 + 4$ $[x = 4]$

*Results in the next frame*

**30**

| | |
|---|---|
| (a) $[(4x + 3)x + 2]x - 4$ | $f(2) = 44$ |
| (b) $\{[(2x + 1)x - 3]x + 5\}x - 6$ | $f(3) = 171$ |
| (c) $\{[(x - 3)x + 0]x + 2\}x - 3$ | $f(5) = 257$ |
| (d) $\{[(2x - 5)x - 3]x + 0\}x + 4$ | $f(4) = 148$ |

This method for evaluating polynomials will be most useful in the following work, so let us now move on to the next topic.

**31**

## Remainder theorem

The *remainder theorem* states that if a polynomial $f(x)$ is divided by $(x - a)$, the quotient will be a polynomial $g(x)$ of one degree less than the degree of $f(x)$, together with a remainder $R$ still to be divided by $(x - a)$. That is:

$$\frac{f(x)}{x - a} = g(x) + \frac{R}{x - a}$$

So $\quad f(x) = (x - a).g(x) + R$

When $x = a$

$$f(a) = 0.g(a) + R$$

i.e. $\quad f(a) = R$

That is:

*If $f(x)$ were to be divided by $(x - a)$, the remainder would be $f(a)$.*

So, $(x^3 + 3x^2 - 13x - 10) \div (x - 3)$ would give a remainder

$$R = f(3) = \dots\dots\dots\dots$$

**32**

5

Because

$$f(x) = x^3 + 3x^2 - 13x - 10$$
$$= [(x + 3)x - 13]x - 10$$

so $f(3) = 5$

We can verify this by actually performing the long division:

$$(x^3 + 3x^2 - 13x - 10) \div (x - 3) = \dots\dots\dots\dots$$

**33**

$$\boxed{x^2 + 6x + 5 \text{ with remainder } 5}$$

Here it is:

$$\begin{array}{rl}
 & \phantom{x^3 + {}} x^2 + \ \ 6x + \ \ 5 \\
x - 3 \,\big) & x^3 + 3x^2 - 13x - 10 \\
 & x^3 - 3x^2 \\
\hline
 & \qquad 6x^2 - 13x \\
 & \qquad 6x^2 - 18x \\
\hline
 & \qquad\qquad 5x - 10 \\
 & \qquad\qquad 5x - 15 \\
\hline
 & \qquad\qquad\qquad 5 \leftarrow \text{Remainder}
\end{array}$$

Now as an exercise, apply the remainder theorem to determine the remainder in each of the following cases.

(a) $(5x^3 - 4x^2 - 3x + 6) \div (x - 2)$
(b) $(4x^3 - 3x^2 + 5x - 3) \div (x - 4)$
(c) $(x^3 - 2x^2 - 3x + 5) \div (x - 5)$
(d) $(2x^3 + 3x^2 - x + 4) \div (x + 2)$
(e) $(3x^3 - 11x^2 + 10x - 12) \div (x - 3)$

*Finish all five and then check with the next frame*

**34**

(a) 24
(b) 225
(c) 65
(d) 2
(e) 0

## Factor theorem

We have seen in Frame 31 that the *remainder theorem* states that if a polynomial $f(x)$ is divided by $(x - a)$, the remainder is $f(a)$. If there is no remainder, that is, if $f(a) = 0$, then $(x - a)$ is a factor of $f(x)$.

For example, if $f(x) = x^3 + 2x^2 - 14x + 12 = [(x + 2)x - 14]x + 12$ and we substitute $x = 2$, $f(2) = 0$, so that division of $f(x)$ by $(x - 2)$ gives a zero remainder, i.e. $(x - 2)$ is a factor of $f(x)$. The remaining factor can be found by long division of $f(x)$ by $(x - 2)$.

$$f(x) = (x - 2)(\dots\dots\dots\dots)$$

**35**

$$\boxed{x^2 + 4x - 6}$$

$$\text{So } f(x) = (x - 2)(x^2 + 4x - 6)$$

The quadratic factor so obtained can sometimes be factorized further into two simple factors, so we apply the $(b^2 - 4ac)$ test – which we have used before. In this particular case $(b^2 - 4ac) = \dots\dots\dots\dots$

**36**

$$\boxed{40}$$

because $(b^2 - 4ac) = 16 - [4 \times 1 \times (-6)] = 16 + 24 = 40$. This is not a perfect square, so no simple factors exist. Therefore, we cannot factorize further.

So $f(x) = (x - 2)(x^2 + 4x - 6)$

As an example, test whether $(x - 3)$ is a factor of $f(x) = x^3 - 5x^2 - 2x + 24$ and, if so, determine the remaining factor (or factors).

$f(x) = x^3 - 5x^2 - 2x + 24 = [(x - 5)x - 2]x + 24$

$\therefore\ f(3) = \ldots\ldots\ldots\ldots$

**37**

$$\boxed{0}$$

There is no remainder. So $(x - 3)$ is a factor of $f(x)$. Long division now gives the remaining quadratic factor, so that

$f(x) = (x - 3)(\ldots\ldots\ldots\ldots)$

**38**

$$\boxed{x^2 - 2x - 8}$$

$$\begin{array}{r|l} & \quad x^2 - 2x - 8 \\ \hline x - 3 & x^3 - 5x^2 - 2x + 24 \\ & x^3 - 3x^2 \\ \hline & \quad - 2x^2 - 2x \\ & \quad - 2x^2 + 6x \\ \hline & \qquad\quad - 8x + 24 \\ & \qquad\quad - 8x + 24 \\ \hline & \qquad\qquad \cdot \quad \cdot \end{array}$$

$f(x) = (x - 3)(x^2 - 2x - 8)$

Now test whether $x^2 - 2x - 8$ can be factorized further.

$b^2 - 4ac = \ldots\ldots\ldots\ldots$

**39**

$$\boxed{36,\text{ i.e. } 6^2}$$

$(b^2 - 4ac) = 6^2$. Therefore there are simple factors of $x^2 - 2x - 8$. We have previously seen how to factorize a quadratic expression when such factors exist and, in this case:

$x^2 - 2x - 8 = (\ldots\ldots\ldots\ldots)(\ldots\ldots\ldots\ldots)$

**40**

$$\boxed{(x - 4)(x + 2)}$$

Collecting our results together:

$$\begin{aligned} f(x) &= x^3 - 5x^2 - 2x + 24 \\ &= (x - 3)(x^2 - 2x - 8) \\ &= (x - 3)(x - 4)(x + 2) \end{aligned}$$

▶

And now another example. Show that $(x-4)$ is a factor of $f(x)=x^3-6x^2-7x+60$ and, as far as possible, factorize $f(x)$ into simple factors.

*Work through it just as before and then check with the next frame*

**41**

$$f(x)=(x-4)(x+3)(x-5)$$

Here it is: $f(x)=x^3-6x^2-7x+60=[(x-6)x-7]x+60$
$f(4)=0$, so $(x-4)$ is a factor of $f(x)$.

$$\begin{array}{r|rrrr} & x^2 & -\ 2x & -\ 15 & \\ \hline x-4 & x^3 & -\ 6x^2 & -\ 7x & +60 \\ & x^3 & -\ 4x^2 & & \\ \hline & & -\ 2x^2 & -\ 7x & \\ & & -\ 2x^2 & +\ 8x & \\ \hline & & & -\ 15x & +60 \\ & & & -\ 15x & +60 \\ \hline & & & \cdot & \cdot \end{array}$$

$$f(x)=(x-4)(x^2-2x-15)$$

Now we attend to the quadratic factor. $(b^2-4ac)=64$, i.e. $8^2$. This is a complete square. Simple factors exist.

$$x^2-2x-15=(x+3)(x-5)$$
$$\text{so } f(x)=(x-4)(x+3)(x-5)$$

But how do we proceed if we are not given the first factor? We will attend to that in the next frame.

**42**

If we are not given the first factor, we proceed as follows:

(a) We write the cubic function in nested form.

(b) By trial and error, we substitute values of $x$, e.g. $x=1$, $x=-1$, $x=2$, $x=-2$ etc. until we find a substitution $x=k$ that gives a zero remainder. Then $(x-k)$ is a factor of $f(x)$.

After that, of course, we can continue as in the previous examples.

So, to factorize $f(x)=x^3+5x^2-2x-24$ as far as possible, we first write $f(x)$ in nested form,

i.e. . . . . . . . . . . .

**43**

$$f(x)=[(x+5)x-2]x-24$$

Now substitute values $x=k$ for $x$ until $f(k)=0$:

$f(1)=-20$ so $(x-1)$ is not a factor

$f(-1)=-18$ so $(x+1)$ is not a factor

$f(2)=0$ so $(x-2)$ *is* a factor of $f(x)$

Now you can work through the rest of it, giving finally that

$f(x)=$ . . . . . . . . . . .

**44**

$$\boxed{f(x) = (x-2)(x+3)(x+4)}$$

Because the long division gives $f(x) = (x-2)(x^2+7x+12)$ and factorizing the quadratic expression finally gives

$f(x) = (x-2)(x+3)(x+4)$

And now one more, entirely on your own:

Factorize $f(x) = 2x^3 - 9x^2 + 7x + 6$

There are no snags. Just take your time. Work through the same steps as before and you get:

$f(x) = \dots\dots\dots\dots$

**45**

$$\boxed{f(x) = (x-2)(x-3)(2x+1)}$$

$$f(x) = 2x^3 - 9x^2 + 7x + 6 = [(2x-9)x+7]x+6$$

$x = 2$ is the first substitution to give $f(x) = 0$. So $(x-2)$ is a factor.

Long division then leads to $f(x) = (x-2)(2x^2-5x-3)$.

$(b^2 - 4ac) = 49$, i.e. $7^2$, showing that simple factors exist for the quadratic.

In fact $2x^2 - 5x - 3 = (x-3)(2x+1)$

so $f(x) = (x-2)(x-3)(2x+1)$

**46**

## Factorization of fourth-order polynomials

The same method can be applied to polynomials of the fourth degree, provided that the given function has at least one simple factor.

For example, to factorize $f(x) = 2x^4 - x^3 - 8x^2 + x + 6$:

In nested form, $f(x) = \{[(2x-1)x-8]x+1\}x+6$

$f(1) = 0$, so $(x-1)$ *is* a factor.

$$\begin{array}{r|l}
 & 2x^3 + x^2 - 7x - 6 \\
\hline
x-1 & 2x^4 - x^3 - 8x^2 + x + 6 \\
 & 2x^4 - 2x^3 \\
\hline
 & \quad x^3 - 8x^2 \\
 & \quad x^3 - x^2 \\
\hline
 & \qquad -7x^2 + x \\
 & \qquad -7x^2 + 7x \\
\hline
 & \qquad\qquad -6x + 6 \\
 & \qquad\qquad -6x + 6 \\
\hline
 & \qquad\qquad \cdot \quad \cdot \\
\hline
\end{array}$$

So $f(x) = 2x^4 - x^3 - 8x^2 + x + 6$

$= (x-1)(2x^3 + x^2 - 7x - 6)$

$= (x-1).g(x)$

Now we can proceed to factorize $g(x) = (2x^3 + x^2 - 7x - 6)$ as we did with previous cubics:

$$g(x) = [(2x+1)x - 7]x - 6$$

$g(1) = -10$ so $(x-1)$ is not a factor of $g(x)$

$g(-1) = 0$ so $(x+1)$ *is* a factor of $g(x)$

Long division shows that $g(x) = (x+1)(2x^2 - x - 6)$

so $f(x) = (x-1)(x+1)(2x^2 - x - 6)$

Attending to the quadratic $(2x^2 - x - 6)$:

$(b^2 - 4ac) = 1 + 48 = 49 = 7^2$ There are simple factors.

In fact $2x^2 - x - 6 = (2x+3)(x-2)$

Finally, then, $f(x) = (x-1)(x+1)(x-2)(2x+3)$

*On to the next frame for another example*

**47**

Factorize $f(x) = x^4 + x^3 - 9x^2 + x + 10$.

First, in nested form, $f(x) = \ldots\ldots\ldots\ldots$

**48**

$$\boxed{f(x) = \{[(x+1)x - 9]x + 1\}x + 10}$$

Now we substitute $x = 1, -1, 2, \ldots$ from which we get

$f(1) = 4$ so $(x-1)$ is not a factor

$f(-1) = 0$ so $(x+1)$ *is* a factor

$$\begin{array}{r|llll}
 & x^3 & +0x^2 & -\;9x & +10 \\ \hline
x+1 & x^4 \;+x^3 & -9x^2 & +\;\;x & +10 \\
 & x^4 \;+x^3 & & & \\ \cline{2-2}
 & \bullet \quad \bullet & -9x^2 & +\;\;x & \\
 & & -9x^2 & -\;9x & \\ \cline{3-4}
 & & & 10x & +10 \\
 & & & 10x & +10 \\ \cline{4-5}
 & & & \bullet & \bullet
\end{array}$$

So $f(x) = (x+1)(x^3 + 0x^2 - 9x + 10) = (x+1).g(x)$

Then in nested form, $g(x) = \ldots\ldots\ldots\ldots$

**49**

$$\boxed{g(x) = [(x+0)x - 9]x + 10}$$

Now we hunt for factors by substituting $x = 1, -1, 2\ldots$ in $g(x)$

$g(1) = \ldots\ldots\ldots\ldots; g(-1) = \ldots\ldots\ldots\ldots; g(2) = \ldots\ldots\ldots\ldots$

**50**

$$\boxed{g(1) = 2;\ g(-1) = 18;\ g(2) = 0}$$

$g(2) = 0$ so $5(x-2)$ *is* a factor of $g(x)$

Long division $(x^3 + 0x^2 - 9x + 10) \div (x-2)$ gives the quotient $\ldots\ldots\ldots\ldots$

51

$$\boxed{x^2 + 2x - 5}$$

$f(x) = (x+1)(x-2)(x^2+2x-5)$, so we finally test the quadratic factor for simple factors and finish it off.

There are no linear factors of the quadratic

so $f(x) = (x+1)(x-2)(x^2+2x-5)$

52

One stage of long division can be avoided if we can find two factors of the original polynomial.

For example, factorize $f(x) = 2x^4 - 5x^3 - 15x^2 + 10x + 8$.

In nested form, $f(x)$ . . . . . . . . . . .

53

$$\boxed{f(x) = \{[(2x-5)x - 15]x + 10\}x + 8}$$

$f(1) = 0$ so $(x-1)$ *is* a factor of $f(x)$

$f(-1) = -10$ so $(x+1)$ is not a factor of $f(x)$

$f(2) = -40$ so $(x-2)$ is not a factor of $f(x)$

$f(-2) =$ . . . . . . . . . . .

54

$$\boxed{f(-2) = 0 \text{ so } (x+2) \textit{ is} \text{ a factor of } f(x)}$$

$$\begin{aligned} f(x) &= (x-1)(x+2)(ax^2+bx+c) \\ &= (x^2+x-2)(ax^2+bx+c) \end{aligned}$$

We can now find the quadratic factor by dividing $f(x)$ by $(x^2+x-2)$:

$$\begin{array}{r|l} & 2x^2 \quad - 7x \quad - 4 \\ \hline x^2+x-2 & 2x^4 - 5x^3 - 15x^2 + 10x + 8 \\ & 2x^4 + 2x^3 - 4x^2 \\ \hline & \quad - 7x^3 - 11x^2 + 10x \\ & \quad - 7x^3 - 7x^2 + 14x \\ \hline & \qquad - 4x^2 - 4x + 8 \\ & \qquad - 4x^2 - 4x + 8 \\ \hline & \qquad \cdot \quad \cdot \quad \cdot \end{array}$$

So $f(x) = (x-1)(x+2)(2x^2-7x-4)$

Finally test the quadratic factor for simple factors and finish it off.

$f(x) =$ . . . . . . . . . . .

**55**

$$\boxed{f(x) = (x-1)(x+2)(x-4)(2x+1)}$$

For the quadratic, $(2x^2 - 7x - 4)$, $b^2 - 4ac = 81 = 9^2$ so factors exist.

In fact, $2x^2 - 7x - 4 = (x-4)(2x+1)$

so $f(x) = (x-1)(x+2)(x-4)(2x+1)$

*Next frame*

**56**

Now one further example that you can do on your own. It is similar to the previous one.

Factorize $f(x) = 2x^4 - 3x^3 - 14x^2 + 33x - 18$.

Follow the usual steps and you will have no trouble.

$f(x) = \dots\dots\dots\dots$

**57**

$$\boxed{f(x) = (x-1)(x-2)(x+3)(2x-3)}$$

Here is the working:

$$\begin{aligned} f(x) &= 2x^4 - 3x^3 - 14x^2 + 33x - 18 \\ &= \{[(2x-3)x - 14]x + 33\}x - 18 \end{aligned}$$

$$\begin{array}{ll} f(1) = 0 & \text{so } (x-1) \textit{ is } \text{a factor of } f(x) \\ f(-1) = -60 & \text{so } (x+1) \text{ is not a factor of } f(x) \\ f(2) = 0 & \text{so } (x-2) \textit{ is } \text{a factor of } f(x) \end{array}$$

$$\begin{aligned} \text{So } f(x) &= (x-1)(x-2)(ax^2 + bx + c) \\ &= (x^2 - 3x + 2)(ax^2 + bx + c) \end{aligned}$$

$$\begin{array}{r|l} & 2x^2 + 3x - 9 \\ \hline x^2 - 3x + 2 & 2x^4 - 3x^3 - 14x^2 + 33x - 18 \\ & 2x^4 - 6x^3 + 4x^2 \\ \hline & 3x^3 - 18x^2 + 33x \\ & 3x^3 - 9x^2 + 6x \\ \hline & -9x^2 + 27x - 18 \\ & -9x^2 + 27x - 18 \\ \hline & \cdot \quad \cdot \quad \cdot \\ \hline \end{array}$$

So $f(x) = (x-1)(x-2)(2x^2 + 3x - 9)$

For $2x^2 + 3x - 9$, $(b^2 - 4ac) = 81 = 9^2$ Simple factors exist.

So $f(x) = (x-1)(x-2)(x+3)(2x-3)$

***At this point let us pause and summarize the main facts so far on evaluating polynomial equations, the remainder theorem and the factor theorem.***

## Review summary

**58**

1. A polynomial in $x$ is an expression involving powers of $x$, normally arranged in descending or ascending order. The degree of the polynomial is given by the highest power of $x$ occurring in the expression. Polynomials can be efficiently evaluated by method known as nesting.
2. *Remainder theorem*
   If a polynomial $f(x)$ is divided by $(x - a)$, the quotient will be a polynomial $g(x)$ of one degree less than that of $f(x)$, together with a remainder $R$ such that $R = f(a)$.
3. *Factor theorem*
   If $f(x)$ is a polynomial and substituting $x = a$ gives a remainder of zero, i.e. $f(a) = 0$, then $(x - a)$ is a factor of $f(x)$.

## Review exercise

**59**

1. Rewrite $f(x) = 6x^3 - 5x^2 + 4x - 3$ in nested form and determine the value of $f(2)$.
2. Without dividing in full, determine the remainder of
   $(x^4 - 2x^3 + 3x^2 - 4) \div (x - 2)$.
3. Factorize $6x^4 + x^3 - 25x^2 - 4x + 4$.

**60**

1. $f(x) = 6x^3 - 5x^2 + 4x - 3 = ((6x - 5)x + 4)x - 3$ so that
   $f(2) = ((12 - 5)2 + 4)2 - 3 = 33$
2. $f(x) = x^4 - 2x^3 + 3x^2 - 4 = (((x - 2)x + 3)x + 0)x - 4$ so that
   $f(2) = (((2 - 2)2 + 3)2 + 0)2 - 4 = 8 =$ remainder
3. $f(x) = 6x^4 + x^3 - 25x^2 - 4x + 4 = (((6x + 1)x - 25)x - 4)x + 4$
   $f(2) = 0$ so $x - 2$ *is* a factor, $f(-2) = 0$ so $x + 2$ *is* a factor.
   $(x - 2)(x + 2) = x^2 - 4$ and:

$$
\begin{array}{r|lllll}
 & & & 6x^2 & +\ x & -\ 1 \\
\cline{2-6}
x^2 - 4 & 6x^4 & +\ x^3 & -\ 25x^2 & -\ 4x & +\ 4 \\
 & 6x^4 & & -\ 24x^2 & & \\
\cline{2-4}
 & & x^3 & -\ x^2 & & \\
 & & x^3 & & -\ 4x & \\
\cline{3-5}
 & & & -\ x^2 & & +\ 4 \\
 & & & -\ x^2 & & +\ 4 \\
\cline{4-6}
 & & & \bullet & & \bullet \\
\cline{4-6}
\end{array}
$$

   Furthermore, $6x^2 + x - 1 = (2x + 1)(3x - 1)$ so that:
   $6x^4 + x^3 - 25x^2 - 4x + 4 = (x + 2)(x - 2)(2x + 1)(3x - 1)$

61

You have now come to the end of this Programme. A list of **Can You?** questions follows for you to gauge your understanding of the material in the Programme. You will notice that these questions match the **Learning outcomes** listed at the beginning of the Programme so go back and try the **Quiz** that follows them. After that try the **Test exercise**. *Work through them at your own pace, there is no need to hurry.* A set of **Further problems** provides additional valuable practice.

## Can You?

62

### Checklist F.3

Check this list before and after you try the end of Programme test.

On a scale of 1 to 5 how confident are you that you can: Frames

- Numerically evaluate an algebraic expression by substituting numbers for variables? 1 to 4
  *Yes* ☐ ☐ ☐ ☐ ☐ *No*
- Recognize the different types of equation? 5 to 6
  *Yes* ☐ ☐ ☐ ☐ ☐ *No*
- Evaluate an independent variable? 7 to 8
  *Yes* ☐ ☐ ☐ ☐ ☐ *No*
- Change the subject of an equation by transposition? 9 to 19
  *Yes* ☐ ☐ ☐ ☐ ☐ *No*
- Evaluate polynomial expressions by 'nesting'? 26 to 30
  *Yes* ☐ ☐ ☐ ☐ ☐ *No*
- Use the remainder and factor theorems to factorize polynomials? 31 to 45
  *Yes* ☐ ☐ ☐ ☐ ☐ *No*
- Factorize fourth-order polynomials? 46 to 57
  *Yes* ☐ ☐ ☐ ☐ ☐ *No*

## Test exercise F.3

63

PERSONAL TUTOR

Frames

1 Evaluate to 3 sig fig:

$\sqrt{\dfrac{2V}{\pi h}-\dfrac{h^2}{3}}$ where

(a) $V = 23{\cdot}05$ and $h = 2{\cdot}69$

(b) $V = 85{\cdot}67$ and $h = 5{\cdot}44$. 1 to 4

2 Given $q = \sqrt{(5p^2 - 1)^2 - 2}$ find $p$ in terms of $q$. 7 to 19

3 Express each of the following functions in nested form and determine the value of the function for the stated value of $x$:

(a) $f(x) = 2x^3 - 3x^2 + 5x - 4 \qquad f(3) = \ldots\ldots\ldots\ldots$

(b) $f(x) = 4x^3 + 2x^2 - 7x - 2 \qquad f(2) = \ldots\ldots\ldots\ldots$

(c) $f(x) = 3x^3 - 4x^2 + 8 \qquad f(4) = \ldots\ldots\ldots\ldots$

(d) $f(x) = 2x^3 + x - 5 \qquad f(5) = \ldots\ldots\ldots\ldots$

Frames 26 to 30

4 Determine the remainder that would occur if $(4x^3 - 5x^2 + 7x - 3)$ were divided by $(x - 3)$

31 to 33

5 Test whether $(x - 2)$ is a factor of

$f(x) = 2x^3 + 2x^2 - 17x + 10$.

If so, factorize $f(x)$ as far as possible.

34 to 45

6 Rewrite $f(x) = 2x^3 + 7x^2 - 14x - 40$ as the product of three linear factors.

34 to 45

7 Express the quartic $f(x) = 2x^4 - 7x^3 - 2x^2 + 13x + 6$ as the product of four linear factors.

46 to 57

---

## Further problems F.3

**64**

1 Evaluate the following, giving the results to 4 sig fig:

(a) $K = 14{\cdot}26 - 6{\cdot}38 + \sqrt{136{\cdot}5} \div (8{\cdot}72 + 4{\cdot}63)$

(b) $P = (21{\cdot}26 + 3{\cdot}74) \div 1{\cdot}24 + 4{\cdot}18^2 \times 6{\cdot}32$

(c) $Q = \dfrac{3{\cdot}26 + \sqrt{12{\cdot}13}}{14{\cdot}192 - 2{\cdot}4 \times 1{\cdot}63^2}$

2 Transpose each of the following formulas to make the symbol in the square brackets the subject:

(a) $V = IR$ $\qquad [I]$

(b) $v = u + at$ $\qquad [u]$

(c) $s = ut + \frac{1}{2}at^2$ $\qquad [u]$

(d) $f = \dfrac{1}{2\pi\sqrt{LC}}$ $\qquad [L]$

(e) $P = \dfrac{S(C - F)}{C}$ $\qquad [C]$

(f) $S = \sqrt{\dfrac{3D(L - D)}{8}}$ $\qquad [L]$

(g) $T = \dfrac{M - m}{1 + Mm}$ $\qquad [M]$

(h) $A = \pi r\sqrt{r^2 + h^2}$ $\qquad [h]$

(i) $V = \dfrac{\pi h}{6}(3R^2 + h^2)$ $\qquad [R]$

▶

3 Rewrite the following in nested form and, in each case, determine the value of the function for the value of $x$ stated:

(a) $f(x) = 5x^3 - 4x^2 + 3x - 12$ $[x = 2]$

(b) $f(x) = 3x^3 - 2x^2 - 5x + 3$ $[x = 4]$

(c) $f(x) = 4x^3 + x^2 - 6x + 2$ $[x = -3]$

(d) $f(x) = 2x^4 - 4x^3 + 2x^2 - 3x + 6$ $[x = 3]$

(e) $f(x) = x^4 - 5x^3 + 3x - 8$ $[x = 6]$

4 Without dividing in full, determine the remainder in each of the following cases:

(a) $(5x^3 + 4x^2 - 6x + 3) \div (x - 4)$

(b) $(3x^3 - 5x^2 + 3x - 4) \div (x - 5)$

(c) $(4x^3 + x^2 - 7x + 2) \div (x + 3)$

(d) $(2x^3 + 3x^2 - 4x + 5) \div (x + 4)$

(e) $(3x^4 - 2x^3 - 10x - 5) \div (x - 4)$

5 Factorize the following cubics as completely as possible:

(a) $x^3 + 6x^2 + 5x - 12$

(b) $2x^3 + 9x^2 - 11x - 30$

(c) $3x^3 - 4x^2 - 28x - 16$

(d) $3x^3 - x^2 + x + 5$

(e) $6x^3 - 5x^2 - 34x + 40$

6 Factorize the following quartics, expressing the results as the products of linear factors where possible:

(a) $f(x) = 2x^4 - 5x^3 - 15x^2 + 10x + 8$

(b) $f(x) = 3x^4 - 7x^3 - 25x^2 + 63x - 18$

(c) $f(x) = 4x^4 - 4x^3 - 35x^2 + 45x + 18$

(d) $f(x) = x^4 + 2x^3 - 6x^2 - 2x + 5$

(e) $f(x) = 6x^4 - 11x^3 - 35x^2 + 34x + 24$

(f) $f(x) = 2x^4 + 5x^3 - 20x^2 - 20x + 48$

Now visit the companion website for this book at www.palgrave.com/stroud for more questions applying this mathematics to science and engineering.

# Graphs

## Learning outcomes

When you have completed this Programme you will be able to:

- ☐ Construct a collection of ordered pairs of numbers from an equation
- ☐ Plot points associated with ordered pairs of numbers against Cartesian axes and generate graphs
- ☐ Appreciate the existence of asymptotes to curves and discontinuities
- ☐ Use an electronic spreadsheet to draw Cartesian graphs of equations
- ☐ Describe regions of the $x$–$y$ plane that are represented by inequalities
- ☐ Draw graphs of and algebraically manipulate the absolute value or modulus function

If you already feel confident about these why not try the quiz over the page? You can check your answers at the end of the book.

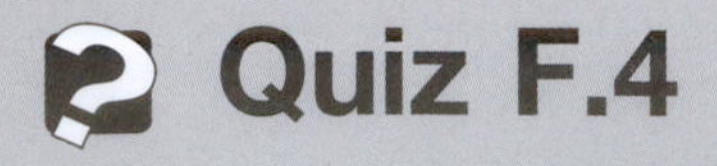

# Quiz F.4

**Frames**

1 Given the equation:

$$y^2 - x^2 = 1$$

(a) Transpose the equation to make $y$ the subject of the transposed equation.
(b) Construct ordered pairs of numbers corresponding to the even integer values of $x$ where $-10 \leq x \leq 10$.
(c) Plot the ordered pairs of numbers on a Cartesian graph and join the points plotted with a continuous curve. [1] to [8]

2 Plot the graph of:

(a) $y = \dfrac{1}{x^2}$ for $-3 \leq x \leq 3$ with intervals of 0·5

(b) $y = \begin{cases} -x : x \leq 0 \\ x \quad : x > 0 \end{cases}$ for $-3 \leq x \leq 4$ with intervals of 0·5 [9] to [16]

3 Use a spreadsheet to draw the Cartesian graphs of:

(a) $y = 2x^2 - 7x - 4$ for $-2 \leq x \leq 4{\cdot}9$ with step value 0·3

(b) $y = x^3 - x^2 + x - 1$ for $-2 \leq x \leq 4{\cdot}9$ with step value 0·3 [17] to [29]

4 Describe the region of the $x$–$y$ plane that corresponds to each of the following:

(a) $y > -x$

(b) $y \leq x - 3x^3$

(c) $x^2 + y^2 \leq 1$ [30] to [35]

5 What values of $x$ satisfy each of the following:

(a) $|x + 2| < 5$

(b) $|x + 2| > 5$

(c) $|2x - 3| < 4$

(d) $|8 - 5x| > 12$ [36] to [53]

# Graphs of equations

## Equations

1

A conditional equation is a statement of the equality of two expressions that is only true for restricted values of the symbols involved. For example, consider the equation:

$$x - y = 1$$

Transposing this equation to make $y$ the subject of a new equation gives:

$$y = x - 1$$

We evaluate this equation by freely selecting a value of the independent variable $x$ and then calculating the corresponding value of the dependent variable $y$. The value of $y$ is *restricted* to being the value obtained from the right-hand side of the equation.

We could equally have transposed the original equation to make $y$ the independent variable:

$$x = y + 1$$

but for historic reasons the variable $y$ is normally selected to be the subject of the transposed equation rather than $x$, so we shall concentrate on equations of the form $y =$ *some expression in* $x$.

So how are the values of $y$ related to the values of $x$ in the equation $x^2 + y^2 = 1$?

*See the next frame*

2

$$\boxed{y = \pm\sqrt{1 - x^2}}$$

Because

$y^2 = 1 - x^2$ subtracting $x^2$ from both sides

so that:

$y = \pm\sqrt{1 - x^2}$ (the symbol $\sqrt{x}$ refers to the *positive* square root of $x$ whereas $x^{\frac{1}{2}}$ refers to both the positive and the negative square roots of $x$)

Here we see that not only are the $y$-values restricted by the equation but our choice of value of the independent variable $x$ is also restricted. The value of $1 - x^2$ must not be negative otherwise we would be unable to find the square root.

The permitted values of $x$ are ............

*The answer is in the next frame*

3

$$\boxed{-1 \leq x \leq 1}$$

Because we demand that $1 - x^2$ must not be negative, it must be greater than or equal to zero. That is, $1 - x^2 \geq 0$ so that $1 \geq x^2$ (which can also be written as $x^2 \leq 1$). This inequality is only satisfied if $-1 \leq x \leq 1$. Notice also, that for each permitted value of $x$ there are two values of $y$, namely

$y = +\sqrt{1 - x^2}$ and $y = -\sqrt{1 - x^2}$.

*Move to the next frame*

4

## Ordered pairs of numbers

Consider the equation:

$y = x^2$

where, you will recall, $x$ is referred to as the independent variable and $y$ is the dependent variable. Evaluating such an equation enables a collection of ordered pairs of numbers to be constructed. For example, if we select $x = 2$ the corresponding value of the dependent variable $y$ is found to be $2^2 = 4$. From these two values the ordered pair of numbers (2, 4) can be constructed. It is called an *ordered* pair because the first number of the pair is always the value of the independent variable (here $x$) and the second number is the corresponding value of the dependent variable (here $y$).

So the collection of ordered pairs constructed from $y = x^2$ using successive integer values of $x$ from $-5$ to 5 is

. . . . . . . . . . . .

*See the next frame*

5

(−5, 25), (−4, 16), (−3, 9), (−2, 4), (−1, 1),
(0, 0), (1, 1), (2, 4), (3, 9), (4, 16), (5, 25)

## Cartesian axes

We can *plot* the ordered pair of numbers (2, 4) on a graph. On a sheet of graph paper, we can drasw two straight lines perpendicular to each other. On each line we mark off the integers so that the two lines intersect at their common zero points. We can then plot the ordered pair of numbers (2, 4) as a point in the plane referenced against the integers on the two lines thus:

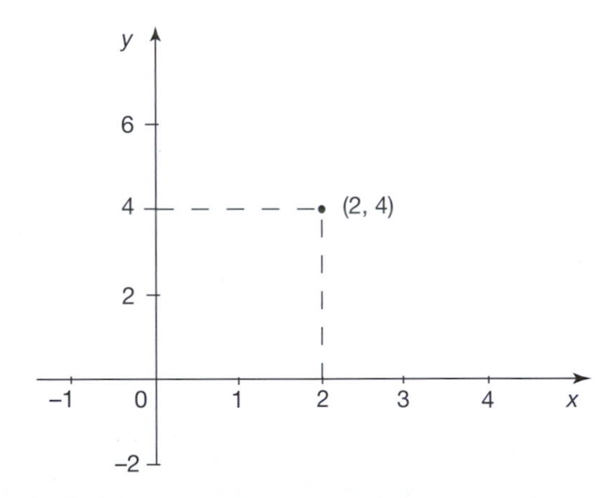


The arrangement of numbered lines is called the *Cartesian coordinate frame* and each line is called an *axis* (plural *axes*). The horizontal line is always taken to be the independent variable axis (here the $x$-axis) and the vertical line the dependent variable axis (here the $y$-axis). Notice that the scales of each axis need not be the same - the scales are chosen to make optimum use of the sheet of graph paper.

Now you try a couple of points. Plot the points (1, 1) and (−3, 9) against the same axes.

*The result is in the next frame*

6

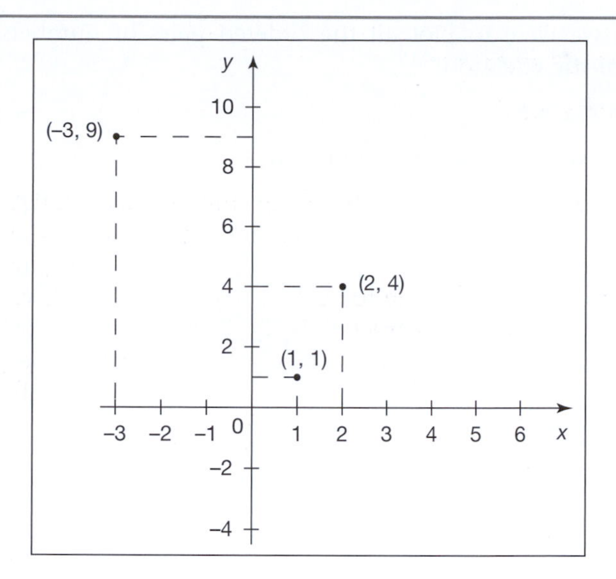


## Drawing a graph

For an equation in a single independent variable, we can construct a collection of ordered pairs. If we plot each pair as a point in the same Cartesian coordinate frame we obtain a collection of isolated points.

On a sheet of graph paper plot the points $(-5, 25)$, $(-4, 16)$, $(-3, 9)$, $(-2, 4)$, $(-1, 1)$, $(0, 0)$, $(1, 1)$, $(2, 4)$, $(3, 9)$, $(4, 16)$ and $(5, 25)$ obtained from the equation $y = x^2$.

*Check your graph with the next frame*

7

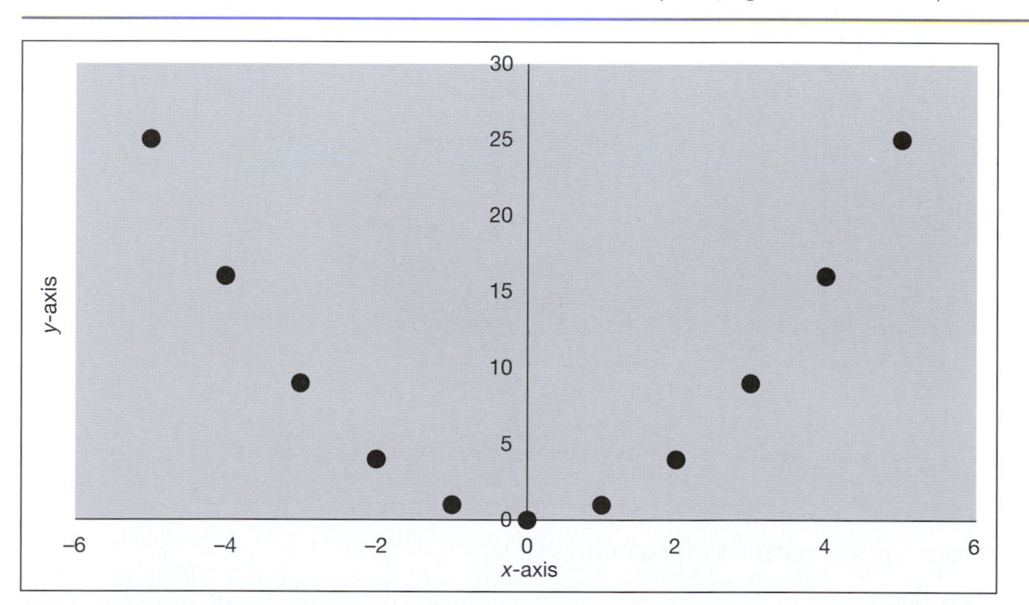


We have taken as our $x$-values only integers. We could have taken any value of $x$. In fact, there is an infinite number of possible choices of $x$ values and, hence, points. If we were able to plot this infinity of all possible points they would merge together to form a continuous line known as the *graph of the equation*. In practice, what we do is to plot a collection of isolated points and then join them up with a continuous line.

For example, if we were to plot all the ordered pairs of numbers that could be constructed from the equation:

$y = x^2$ for $-5 \leq x \leq 5$

we would end up with the shape given below:

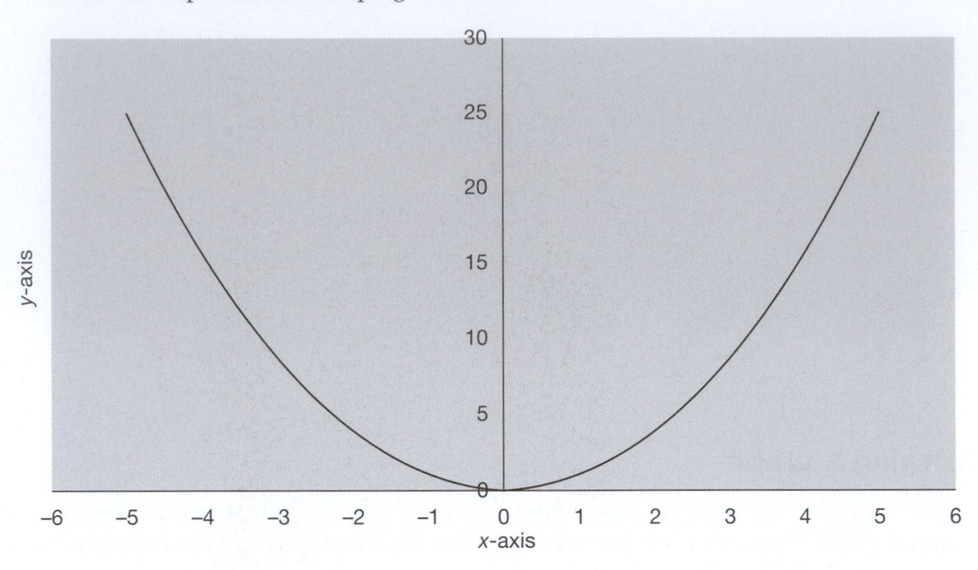


This is the graph of the equation $y = x^2$ for $-5 \leq x \leq 5$. We call the shape a *parabola*.

So what is the shape of the graph of $y = x + 1$ where $-4 \leq x \leq 4$? [Plot the graph using integer values of $x$.]

*The answer is in the next frame*

---

**8**

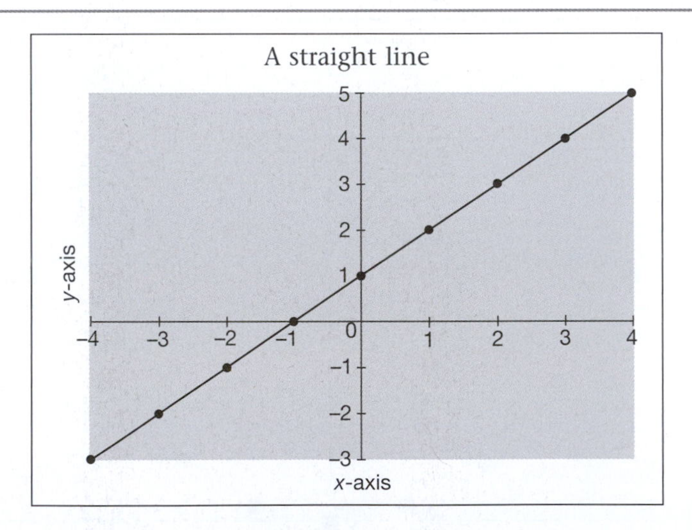


Because, if we construct the following table:

| $x$-value | $-4$ | $-3$ | $-2$ | $-1$ | 0 | 1 | 2 | 3 | 4 |
|---|---|---|---|---|---|---|---|---|---|
| $y$-value | $-3$ | $-2$ | $-1$ | 0 | 1 | 2 | 3 | 4 | 5 |

this gives rise to the following ordered pairs: $(-4, -3)$, $(-3, -2)$, $(-2, -1)$, $(-1, 0)$, $(0, 1)$, $(1, 2)$, $(2, 3)$, $(3, 4)$ and $(4, 5)$ which can be plotted as shown.

The plot we have obtained is the plot of just those ordered pairs of numbers that we have constructed, joined together by a continuous line.

*Next frame*

9

Try another one. Construct a table of $y$-values and then the graph of:

$y = x^3 - 2x^2 - x + 2$ for $-2 \leq x \leq 3$ with intervals of 0·5

*Check the next frame for the answer*

10

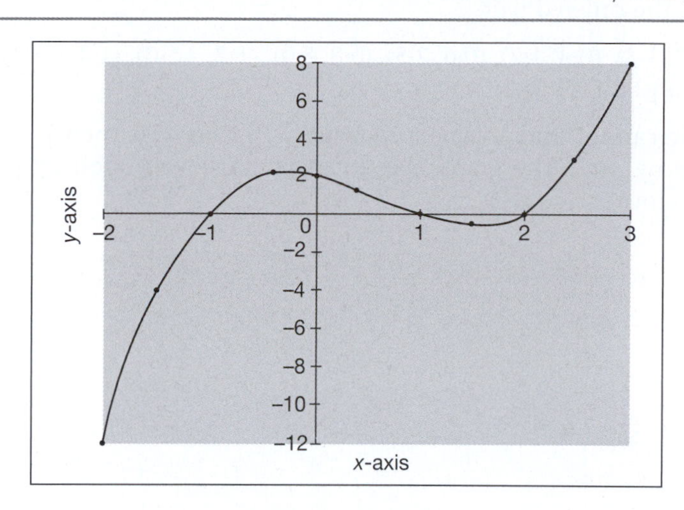


Because not all equations are polynomial equations we shall construct the ordered pairs by making use of a table rather than by *nesting* as we did in Programme F.3. Here is the table:

| $x$ | **−2** | **−1·5** | **−1** | **−0·5** | **0** | **0·5** | **1** | **1·5** | **2** | **2·5** | **3** |
|---|---|---|---|---|---|---|---|---|---|---|---|
| $x^3$ | −8 | −3·4 | −1 | −0·1 | 0 | 0·1 | 1 | 3·4 | 8 | 15·6 | 27 |
| $-2x^2$ | −8 | −4·5 | −2 | −0·5 | 0 | −0·5 | −2 | −4·5 | −8 | −12·5 | −18 |
| $-x$ | 2 | 1·5 | 1 | 0·5 | 0 | −0·5 | −1 | −1·5 | −2 | −2·5 | −3 |
| 2 | 2 | 2 | 2 | 2 | 2 | 2 | 2 | 2 | 2 | 2 | 2 |
| $y$ | **−12** | **−4·4** | **0** | **1·9** | **2** | **1·1** | **0** | **−0·6** | **0** | **2·6** | **8** |

Pairs: (−2, −12), (−1·5, −4·4), (−1, 0), (−0·5, 1·9), (0, 2), (0·5, 1·1), (1, 0), (1·5, −0·6), (2, 0), (2·5, 2·6), (3, 8)

How about the graph of:

$$y = \frac{1}{1-x} \text{ for } 0 \leq x \leq 2?$$

Select values of $x$ ranging from 0 to 2 with intervals of 0·2 and take care here with the values that are near to 1.

*The answer is in the next frame*

11

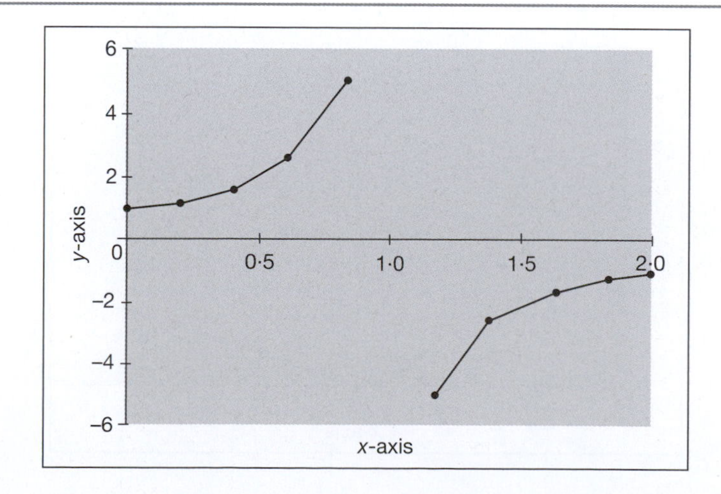


▶

Here is the table:

| $x$ | **0** | **0·2** | **0·4** | **0·6** | **0·8** | **1·2** | **1·4** | **1·6** | **1·8** | **2·0** |
|---|---|---|---|---|---|---|---|---|---|---|
| $1-x$ | 1·0 | 0·8 | 0·6 | 0·4 | 0·2 | −0·2 | −0·4 | −0·6 | −0·8 | −1·0 |
| $y$ | **1·0** | **1·3** | **1·7** | **2·5** | **5·0** | **−5·0** | **−2·5** | **−1·7** | **−1·3** | **−1·0** |

Giving rise to the ordered pairs:

(0, 1·0), (0·2, 1·3), (0·4, 1·7), (0·6, 2·5), (0·8, 5·0), (1·2, −5·0), (1·4, −2·5), (1·6, −1·7), (1·8, −1·3) and (2, −1·0)

Notice that we cannot find a value for $y$ when $x = 1$ because then $1 - x = 0$ and we cannot divide by zero. The graph above can be improved upon by plotting more points – see below:

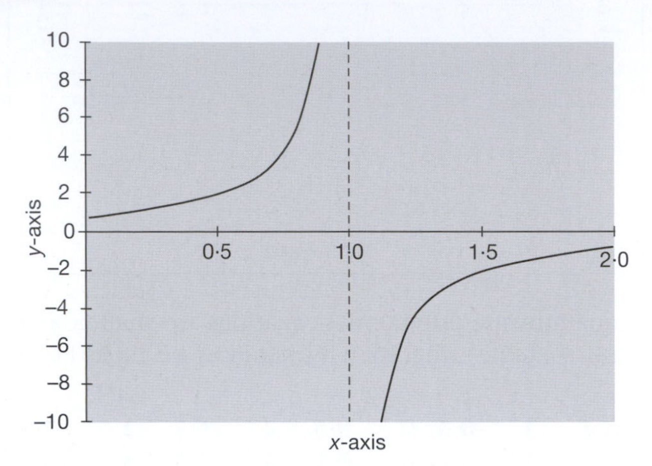


As $x$ approaches the value 1 from the left, the graph rises and approaches (but never crosses) the vertical line through $x = 1$. Also as $x$ approaches the value 1 from the right, the graph falls and approaches (but never crosses) the same vertical line. The vertical line through $x = 1$ is called a vertical *asymptote* to the graph. Not all asymptotes are vertical or even straight lines. Indeed, whenever a curve approaches a second curve without actually meeting or crossing it, the second curve is called an asymptote to the first curve.

Now, try finding the graph of the following:

$$y = \begin{cases} x^2 & \text{for } -5 \le x < 0 \\ x & \text{for } \quad\quad x \ge 0 \end{cases} \quad -5 \le x \le 5 \text{ with intervals of } 0{\cdot}5$$

*Check the next frame*

## 12

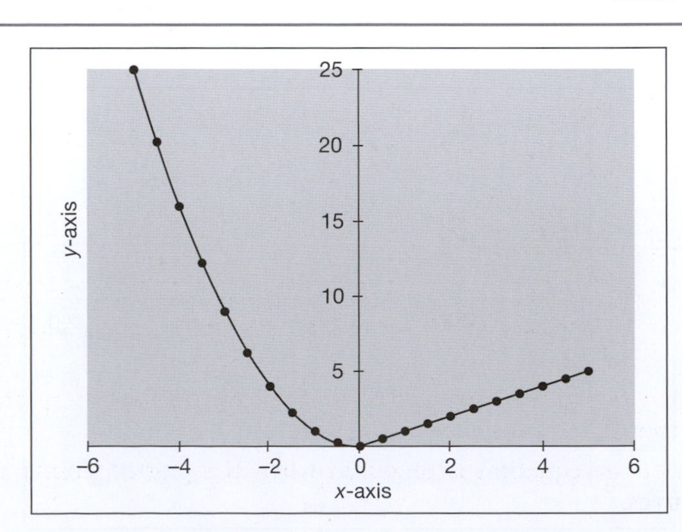

Because the equation:

$$y = \begin{cases} x^2 \text{ for } & -5 \le x < 0 \\ x \text{ for } & x \ge 0 \end{cases}$$

means that if $x$ is chosen so that its value lies between $-5$ and 0 then $y = x^2$ and that part of the graph is the parabola shape. If $x$ is greater than or equal to zero then $y = x$ and that part of the graph is the straight line. Notice that not all equations are of the simple form $y$ = *some expression in x*.

And finally, how about the graph of:

$$y = \begin{cases} 1 & \text{for } x \le 2 \\ -1 & \text{for } x > 2 \end{cases} \quad \text{for } -1 \le x \le 4$$

*Next frame*

13

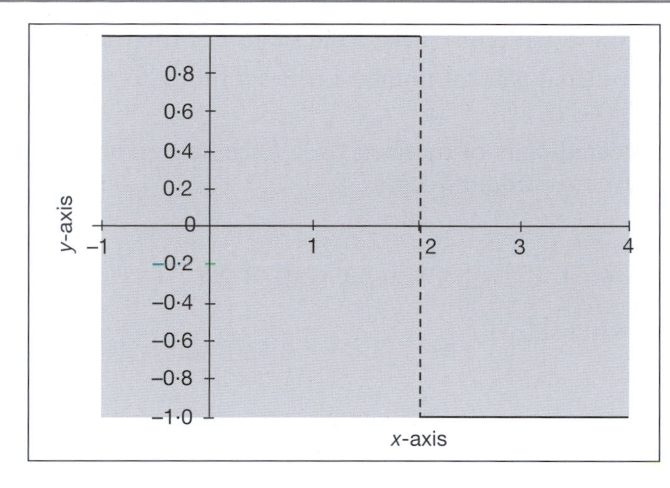


Because the equation:

$$y = \begin{cases} 1 & \text{for } x \le 2 \\ -1 & \text{for } x > 2 \end{cases}$$

means that no matter what the value is that is assigned to $x$, if it is less than or equal to 2 the value of $y$ is 1. If the value of $x$ is greater than 2 the corresponding value of $y$ is $-1$. Notice the gap between the two continuous straight lines. This is called a *discontinuity* – not all equations produce smooth continuous shapes so *care must be taken when joining points together with a continuous line*. The dashed line joining the two end points of the straight lines is only there as a visual guide, *it is not part of the graph*.

***At this point let us pause and summarize the main facts so far***

##  Review summary

14

1 Ordered pairs of numbers can be generated from an equation involving a single independent variable.

2 Ordered pairs of numbers generated from an equation can be plotted against a Cartesian coordinate frame.

3 The graph of the equation is produced when the plotted points are joined by smooth curves.

▶

4 Some equations have graphs that are given specific names, such as straight lines and parabolas.

5 Not all equations are of the simple form $y =$ *some expression in* $x$.

6 Not all graphs are smooth, unbroken lines. Some graphs consist of breaks called discontinuities.

---

## Review exercise

**15**

1 Given the equation:

$$x^2 + y^3 = 1$$

(a) Transpose the equation to make $y$ the subject of the transposed equation.

(b) Construct ordered pairs of numbers corresponding to the integer values of $x$ where $-5 \leq x \leq 5$.

(c) Plot the ordered pairs of numbers on a Cartesian graph and join the points plotted with a continuous curve.

2 Plot the graph of:

(a) $y = x^2 + \dfrac{1}{x}$ for $-3 \leq x \leq 3$ with intervals of 0·5.

(b) $y = \begin{cases} x^2 + x + 1 : x \leq 1 \\ 3 - x \quad : x > 1 \end{cases}$ for $-3 \leq x \leq 3$ with intervals of 0·5.

**16**

1 (a) $y = (1 - x^2)^{\frac{1}{3}}$

(b) $(-5, -2{\cdot}9)$, $(-4, -2{\cdot}5)$, $(-3, -2)$, $(-2, -1{\cdot}4)$, $(-1, 0)$, $(0, 1)$, $(1, 0)$, $(2, -1{\cdot}4)$, $(3, -2)$, $(4, -2{\cdot}5)$, $(5, -2{\cdot}9)$

(c)

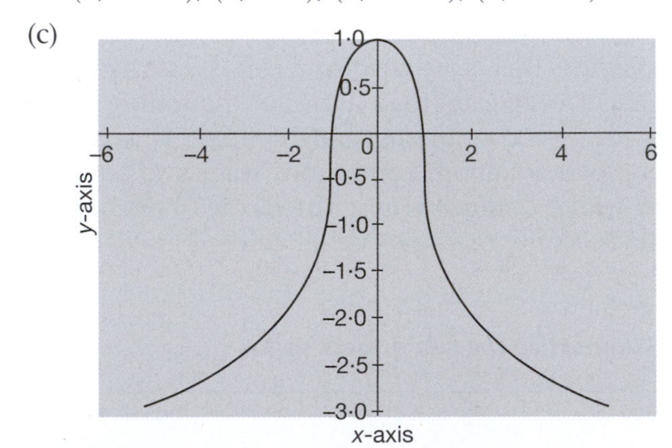


▶

2 (a) (b)

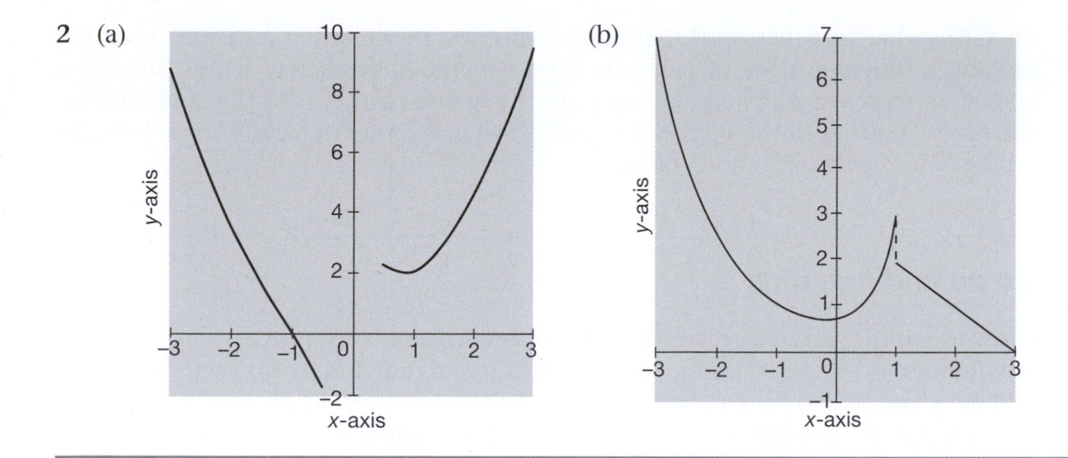


# Using a spreadsheet

## Spreadsheets

17

Electronic spreadsheets provide extensive graphing capabilities and their use is widespread. Because *Microsoft* products are the most widely used software products on a PC, all the examples will be based on the *Microsoft* spreadsheet *Excel*. It is expected that all later versions of *Excel* will support the handling characteristics of earlier versions with only a few minor changes. *If you have access to a computer algebra package you might ask your tutor how to use it to draw graphs.*

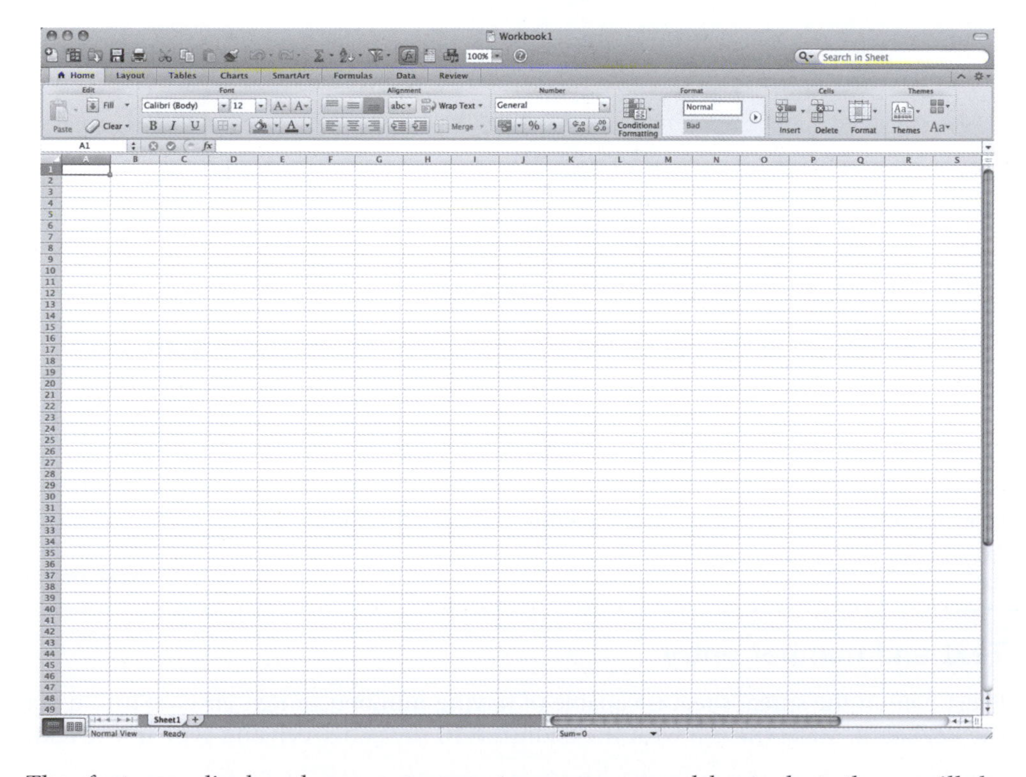


The features displayed are common to many spreadsheets but there will be differences in style between different products. Here the screen displays a worksheet headed by a selection of **icons** each for performing a given task when activated. If you place the mouse pointer over an icon a small box will appear to tell you what operation that icon performs. Below the selection of icons is a **tab bar**, each tab causing a different array of icons to appear in the space below when the tab is activated. To activate a particular tab place the mouse pointer over the tab and press the mouse button. In the body of the worksheet is an array of rectangles called **cells** arranged in regular rows and columns.

*Next frame*

18

## Rows and columns

Every electronic spreadsheet consists of a collection of cells arranged in a regular array of columns and rows. To enable the identification of individual cells each cell has an address given by the *column label* followed by the *row label*. In an *Excel* spreadsheet the columns are labelled alphabetically from **A** onwards and the rows are numbered from **1** onwards.

So that the cell with address **P123** is on the ......... column of the ......... row.

*Check with the next frame*

19

**P** (16th) column of the **123**rd row

Because the address consists of the column label **P** (16th letter of the alphabet) followed by the row number **123**.

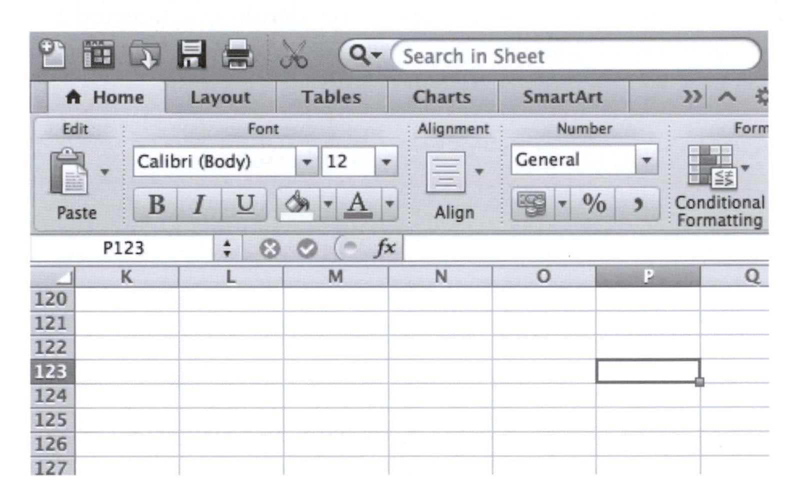


At any time one particular cell boundary is highlighted with a *cursor* and this cell is called the *active cell*. An alternative cell can become the active cell by pressing the cursor movement keys on the keyboard (←, ↑, → and ↓) or, alternatively, by pointing at a particular cell with the *mouse pointer* (✜) and then clicking the mouse button. Try it.

*Next frame*

20

## Text and number entry

Every cell on the spreadsheet is capable of having numbers or text entered into it via the keyboard. Make the cell with the address **B10** the active cell and type in the text:

Text

and then press Enter (↵). Now make cell **C15** the active cell and type in the number 12 followed by Enter.

*Next frame*

## Formulas

21

As well as text and numbers, each cell is capable of containing a formula. In an *Excel* spreadsheet every formula begins with the = (equals) sign when it is being entered at the keyboard. Move the cursor to cell **C16** and enter at the keyboard:

= 3*C15

followed by Enter. The * represents multiplication ($\times$) and the formula states that the contents of cell **C16** will be three times the contents of cell **C15**. You will notice that the number displayed in cell **C16** is three times the number displayed in **C15**. Now change the number in **C15** by overwriting to see the effect. The number in **C16** also changes to reflect the change in the number in **C15**.

*Next frame*

## Clearing entries

22

To clear an entry, point and click at the cell to be cleared. This makes it the active cell. Click the **Edit** command on the Command Bar to reveal a drop-down menu.

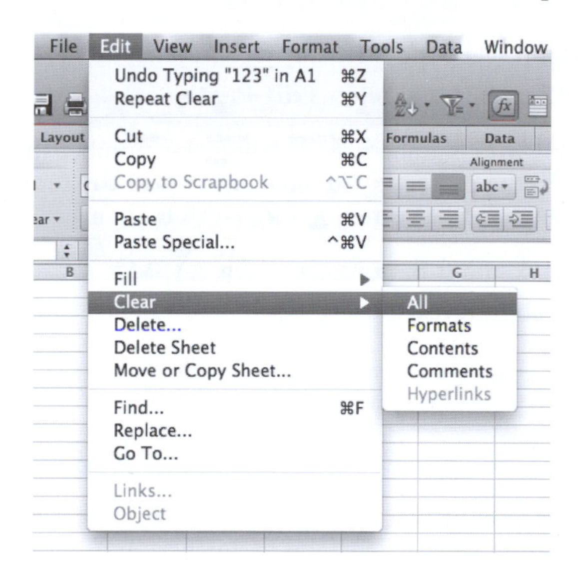


Select from this menu the option **Clear** to reveal a further drop-down menu. In this second menu select **All** and the cell contents are then cleared. Now make sure that all entries on the spreadsheet have been cleared because we want to use the spreadsheet to construct a graph.

*Let's now put all this together in the next frame*

## Construction of a Cartesian graph

23

Follow these instructions to plot the graph of $y = (x - 2)^3$:

Enter the number $-1$ in **A1**

Highlight the cells **A1** to **A21** by pointing at **A1**, holding down the mouse button, dragging the pointer to **A21** and then releasing the mouse button (all the cells from **A2** to **A21** change colour to indicate that they have been selected)

▶

Select the **Edit-Fill-Series** commands from the Command Bar and then in the *Series* window change the **Step value** from 1 to 0·3 and click the **OK** button:

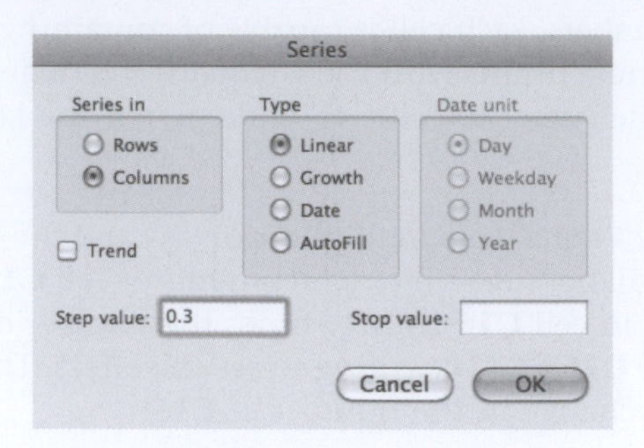


Cells **A1** to **A21** fill with single place decimals ranging from $-1$ to $+5$ with step value intervals of 0·3. These are the $x$-values, where $-1 \leq x \leq 5$.

In cell **B1**, type in the formula =(A1−2)^3 and then press Enter (the symbol ^ represents raising to a power). The number $-27$ appears in cell **B1** - that is $(-1-2)^3 = -27$ where $-1$ is the content of **A1**.

Activate cell **B1** and select the **Edit-Copy** commands

Highlight cells **B2** to **B21** and select **Edit-Paste**

Cells **B2** to **B21** fill with numbers, each being the number in the adjoining cell minus 2, all raised to the power 3; you have just copied the formula in **B1** into the cells **B2** to **B21**. These are the corresponding $y$-values.

Highlight the two columns of numbers - cells **A1:B21**

Click the Charts tab to reveal a collection of icons beneath the tab bar

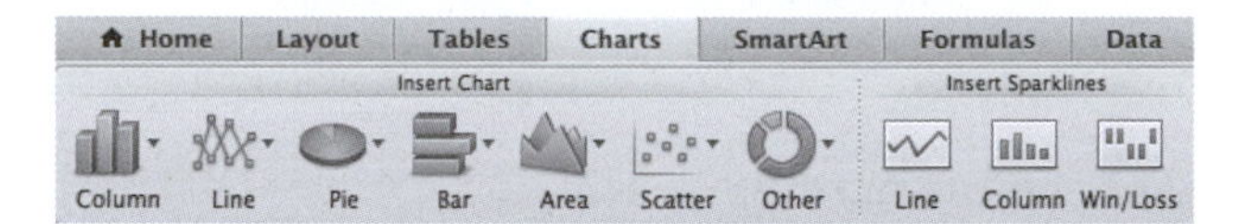


It is here that you choose the type of graph that you require. Select the scatter graph by clicking the Scatter icon whereupon a further drop-down window appears requesting a selection from a number of different types of scatter graphs.

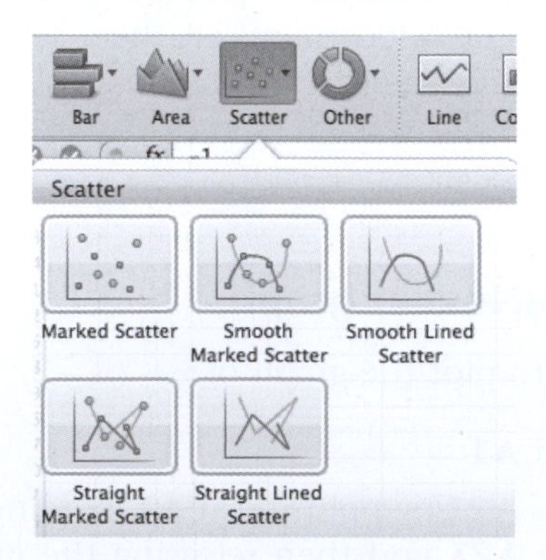


▶

Select Smooth Lined Scatter by clicking that icon (a scatter chart with data points connected by smooth lines without markers). A graph appears in its own window.

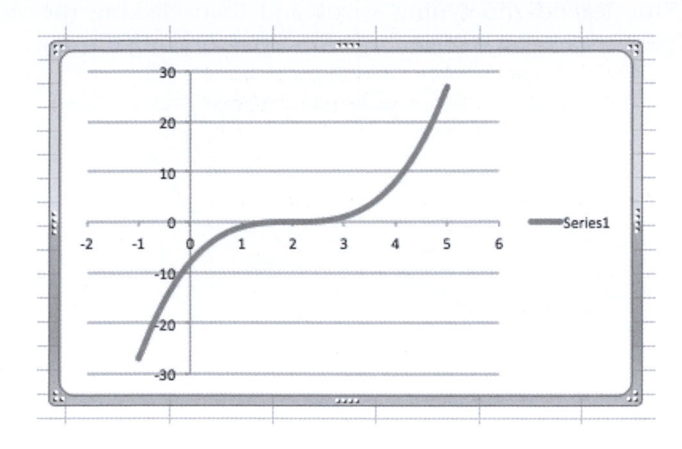


Now we need to consider the display.

*Next frame*

## Displays

**24**

There are numerous displays that can be selected for the graph depending upon the purpose of the display. For example, with the graph highlighted there is a further collection of icons under the heading Chart Quick Layouts just to the right of the icons used to select the type of graph.

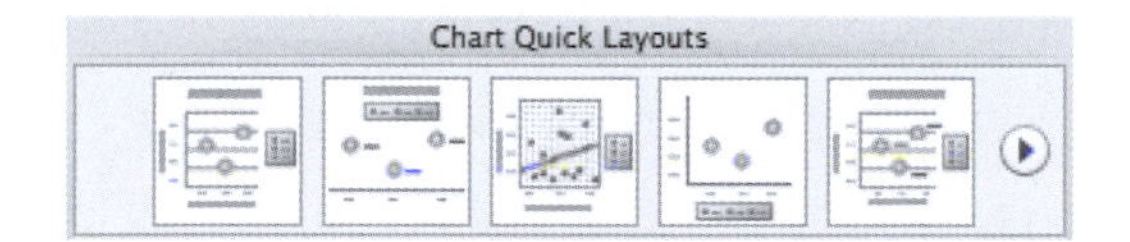


By moving the mouse pointer over each icon in turn you will see that they are numbered. By selecting Layout 10 the graph will change to:

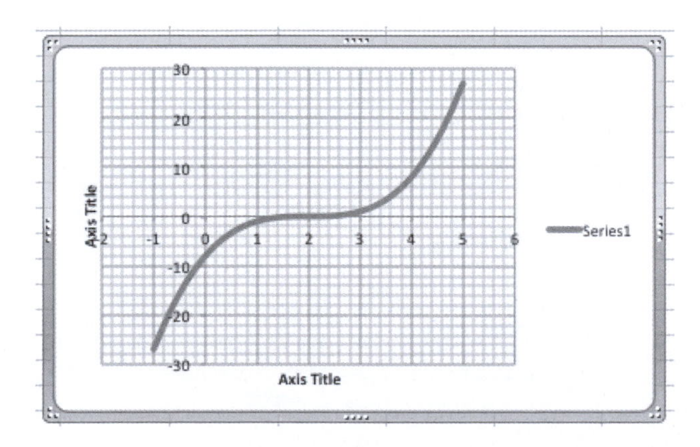

By clicking the legend Series 1 the text box containing the legend will highlight and by pressing backspace on your keyboard you can delete it. Also by clicking the bottom Axis Title legend and typing x-axis and then clicking the side Axis Title legend and typing y-axis you will end up with the following display:

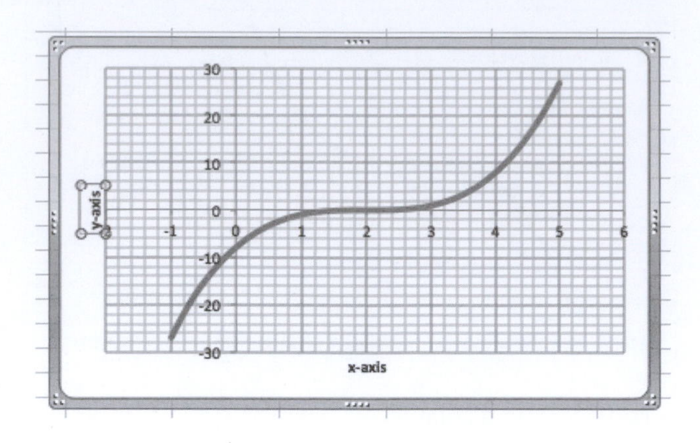


Finally, by placing the mouse pointer on one of the handles located at the centre of each side and at each corner and holding the mouse button down you can drag the handle and so change the size of the graph. Try it.

Now produce the graphs of the following equations. All you need to do is to change the formula in cell **B1** by activating it and then overtyping. Copy the new formula in **B1** down the **B** column and the graph will automatically update itself (*you do not have to clear the old graph, just change the formula*):

(a) $y = x^2 - 5x + 6$ Use * for multiplication so that $5x$ is entered as 5*$x$
(b) $y = x^2 - 6x + 9$
(c) $y = x^2 - x + 1$
(d) $y = x^3 - 6x^2 + 11x - 6$

25

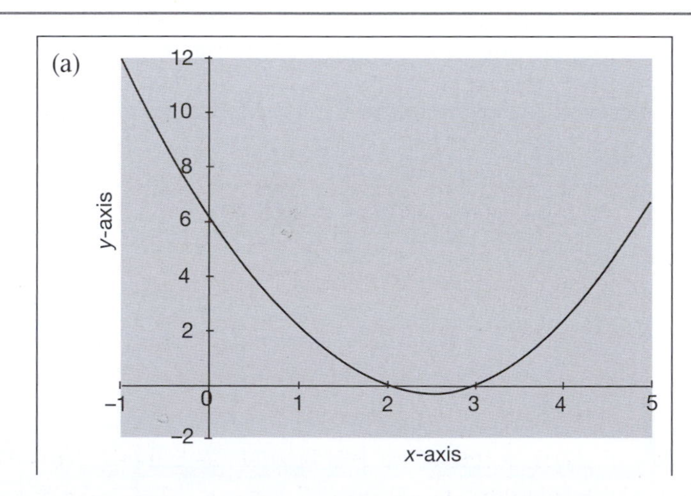


▶

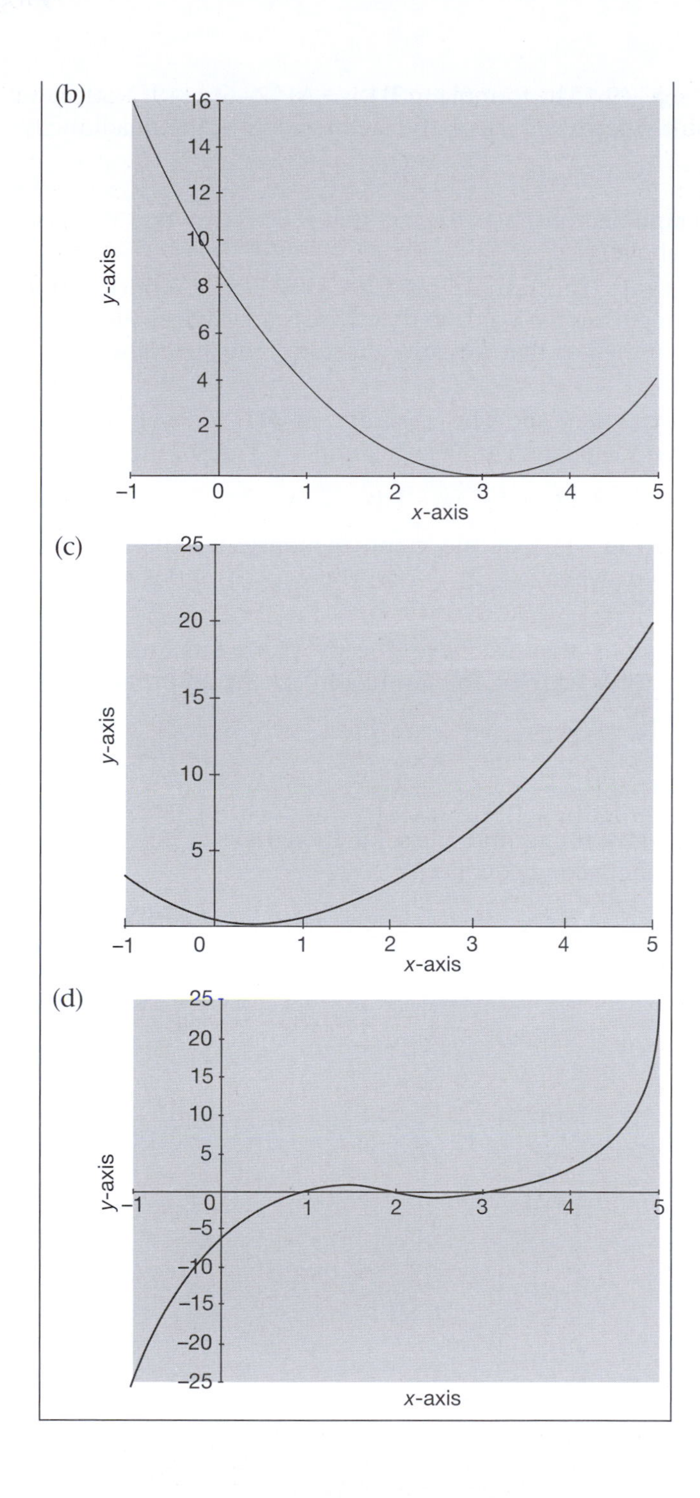


Because:

(a) $y = x^2 - 5x + 6$ The formula in **B1** is =A1^2–5*A1+6. This shape is the parabola – every quadratic equation has a graph in the shape of a parabola. Notice that $y = 0$ at those points where the curve coincides with the $x$-axis, namely when $x = 2$ and $x = 3$. Also, because we can factorize the quadratic [see Programme F.2, Frame 58 onwards]:

$$y = x^2 - 5x + 6 = (x - 2)(x - 3)$$

we can see that the graph demonstrates the fact that $y = 0$ when $x - 2 = 0$ or when $x - 3 = 0$ (see graph (a) in the box above).

▶

(b) $y = x^2 - 6x + 9$ The formula in **B1** is =A1^2–6*A1+9. Notice that $y = 0$ at just one point when $x = 3$. Here, the factorization of the quadratic is:

$$y = x^2 - 6x + 9 = (x - 3)(x - 3)$$

so the graph demonstrates the fact that $y = 0$ only when $x = 3$ (see graph (b) in the box above).

(c) $y = x^2 - x + 1$ The formula in **B1** is =A1^2–A1+1. Notice that the curve never touches or crosses the $x$-axis so that there is no value of $x$ for which $y = 0$. This is reflected in the fact that we cannot factorize the quadratic. (see graph (c) in the box above).

(d) $y = x^3 - 6x^2 + 11x - 6$ The formula in **B1** is =A1^3–6*A1^2+11*A1–6. Notice that $y = 0$ when $x = 1$, $x = 2$ and $x = 3$. Also, the cubic factorizes as:

$$y = x^3 - 6x^2 + 11x - 6 = (x - 1)(x - 2)(x - 3)$$

Again, we can see that the graph demonstrates the fact that $y = 0$ when $(x - 1) = 0$ or when $(x - 2) = 0$ or when $(x - 3) = 0$ (see graph (d) in the box above).

Let's try two graphs that we have already plotted manually. Repeat the same procedure by simply changing the formula in cell **B1** and then copying it into cells **B2** to **B21** to plot:

(a) $y = \dfrac{1}{1 - x}$ (b) $y = \begin{cases} 1 & \text{for } x \leq 2 \\ -1 & \text{for } x > 2 \end{cases}$

You will have to give the second one some thought as to how you are going to enter the formula for the second equation.

*Check your results in the next frame*

**26**

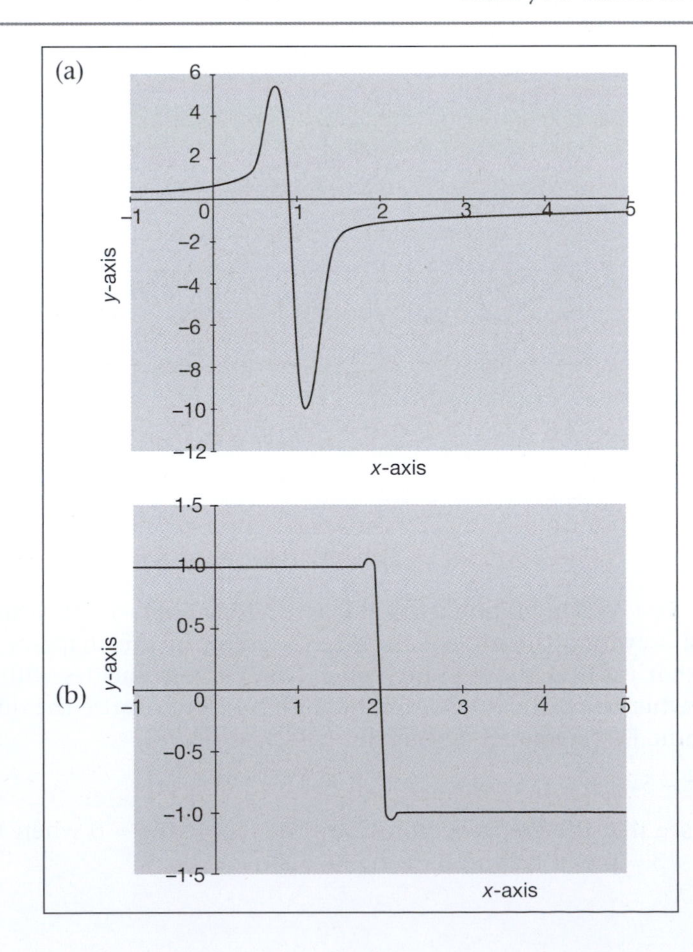

Because:

(a) $y = \dfrac{1}{1-x}$ The formula in **B1** is =1/(1−A1). Notice that the asymptotic behaviour is hidden by the joining of the two inside ends of the graph. See graph (a) in the box above. We can overcome this problem:

Make cell **A8** the active cell

On the Command Bar select **Insert-Rows**

An empty row appears above cell **A8** and the stray line on the graph disappears. (Clear the empty row using **Edit-Delete-Entire row**.)

(b) Here there are two formulas. The first formula in **B1** is =1 copied down to cell **B11** and the second formula in **B12** is =−1 copied down to **B21**.

Notice that the two sides of the graph are joined together when they should not be – see graph (b) in the box above. Make **A12** the active cell and insert a row to remove the stray line just as you did in part (a).

***At this point let us pause and summarize the main facts so far on using a spreadsheet***

## Review summary

27

1 A spreadsheet consists of an array of cells arranged in regular columns and rows.

2 Each cell has an address consisting of the column letter followed by the row number.

3 Each cell is capable of containing text, a number or a formula.

4 Cell entries are cleared by using the **Edit-Clear-All** sequence of commands.

5 To construct a graph:

Enter the range of $x$-values in the first column

Enter the corresponding collection of $y$-values in the second column

Use the Charts tab to construct the graph using the Scatter icon.

## Review exercise

28

1 For $x$ in the range $-2 \leq x \leq 6$ with a step value of 0·4 use a spreadsheet to draw the graphs of:

(a) $y = x^2 - 3x + 2$

(b) $y = x^2 - 1$

(c) $y = 4x^2 - 3x + 25$

(d) $y = -x^3 + 6x^2 - 8x$

29

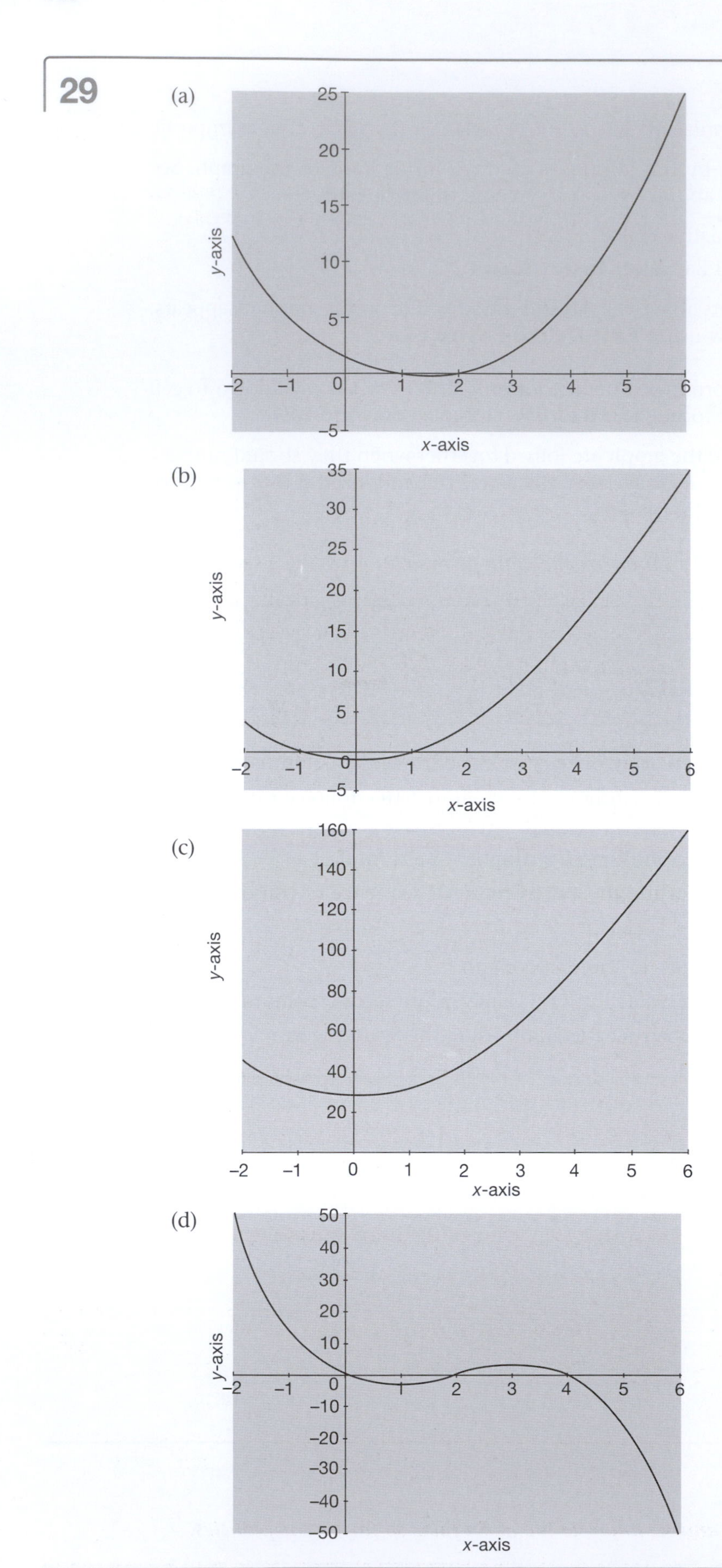


*Let's look now at an extension of these ideas*

# Inequalities

## Less than or greater than

30

You are familiar with the use of the two inequality symbols > and <. For example, $3 < 5$ and $-2 > -4$. We can also use them in algebraic expressions. For example:

$y > x$

The inequality tells us that whatever value is chosen for $x$ the corresponding value for $y$ is greater. Obviously there is an infinity of $y$-values greater than the chosen value for $x$, so the plot of an inequality produces an area rather than a line. For example, if we were to plot the graph of $y = x^2$ for $1 \le x \le 25$ we would obtain the graph shown below:

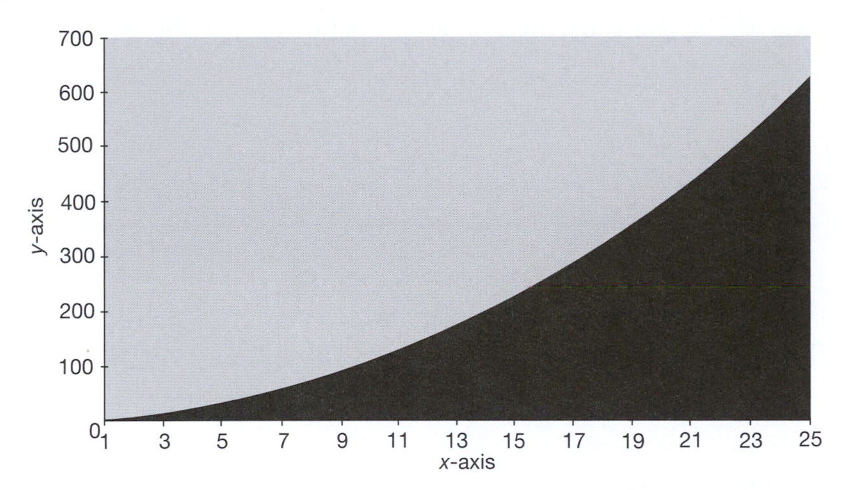


The graph is in the form of the curve $y = x^2$ acting as a separator for two different regions. The region above the line $y = x^2$ represents the plot of $y > x^2$ for $1 \le x \le 25$ because for every point in this region the $y$-value is greater than the square of the corresponding $x$-value. Similarly, the region below the line represents the plot of $y < x^2$ because for every point in this region the $y$-value is less than the square of the corresponding $x$-value.

So, without actually plotting it, what would be the region of the $x$–$y$ plane that corresponds to the inequality $y < x^3 - 2x^2$

*Check the next frame for the answer*

31

| The region of the graph below the curve $y = x^3 - 2x^2$ |
|---|

Because

For every point in this region the $y$-value is less than the corresponding $x$-value.

How about $y \ge 2x - 1$? The region of the $x$–$y$ plane that this describes is . . . . . . . . . . .

32

| The region of the graph on or above the line $y = 2x - 1$ |
|---|

Because

The inequality $y \ge 2x - 1$ means one of two conditions. Namely, $y > 2x - 1$ *or* $y = 2x - 1$ so that every point on or above the line satisfies one or the other of these two conditions.

***At this point let us pause and summarize the main facts so far on inequalities.***

## Review summary

33

1 The graph of an inequality is a region of the $x$–$y$ plane rather than a line or a curve.
2 Points above the graph of $y = f(x)$ are in the region described by $y > f(x)$.
3 Points below the graph of $y = f(x)$ are in the region described by $y < f(x)$.

## Review exercise

34

1 Describe the regions of the $x$–$y$ plane defined by each of the following inequalities:
(a) $y < 3x - 4$
(b) $y > -2x^2 + 1$
(c) $y \leq x^2 - 3x + 2$

35

1 (a) The region below the line $y = 3x - 4$.
(b) The region above the line $y = -2x^2 + 1$.
(c) The region below and on the line $y = x^2 - 3x + 2$.

## Absolute values

36

### Modulus

When numbers are plotted on a straight line the distance of a given number from zero is called the **absolute value** or **modulus** of that number. So the absolute value of $-5$ is 5 because it is 5 units distant from 0 and the absolute value of 3 is 3 because it is 3 units distant from 0.
So the absolute value of:

(a) $-2.41$ is ...........
(b) 13.6 is ...........

*Answers in the next frame*

37

(a) 2.41
(b) 13.6

Because:

$-2.41$ is 2.41 units distant from 0 and 13.6 is 13.6 units distant from 0.

And the modulus of:

(a) $-2.41$ is ...........
(b) 13.6 is ...........

*Answers in the next frame*

38

(a) 2.41
(b) 13.6

Because:

The modulus is just another name for the absolute value.

The absolute value of a number is denoted by placing the number between two vertical lines thus $|\dots|$. For example, $|-7.35|$ represents the absolute value of $-7.35$. The absolute value 7.35 of the negative number $-7.35$ is obtained arithmetically by multiplying the negative number by $-1$.

We can use the same notation algebraically. For example, the equation $y = |x|$, (read as '$y = \text{mod } x$') means that $y$ is equal to the absolute value of $x$. Where:

$|x| = x$ if $x \geq 0$ and
$|x| = -x$ if $x < 0$

so that:

$|6| = 6$ because $6 > 0$ and
$|-4| = -(-4) = 4$ because $-4 < 0$.

The absolute values of other expressions can similarly be found, for example

$|x+5| = x+5$ if $x+5 \geq 0$, that is $x \geq -5$ and
$|x+5| = -(x+5)$ if $x+5 < 0$, that is $x < -5$

So that $|x-3| = \dots\dots\dots$ if $x - 3 \geq 0$, that is $x \dots\dots\dots$ and
$|x-3| = \dots\dots\dots$ if $x - 3 < 0$, that is $x \dots\dots\dots$

*The answer is in the following frame*

39

$|x-3| = x-3$ if $x-3 \geq 0$, that is $x \geq 3$ and
$|x-3| = -(x-3)$ if $x-3 < 0$, that is $x < 3$

Now we shall look at the graphical properties of the absolute value function.

*Next frame*

## Graphs

40

Using the spreadsheet to plot the graph of $y = |x|$ we can take advantage of the built-in function **ABS** which finds the absolute value.

Fill cells **A1** to **A21** with numbers in the range $-5$ to 5 with Step Value 0.5. In cell **B1** type in **=ABS(A1)**.

Now click the cell **B1** to place the highlight there. Move the cursor so that it is over the small black square, bottom right of the highlight, hold down the left hand mouse button and drag the small square down until you reach cell **B21**. Release the mouse button and all the cells **B1 ... B21** fill with the appropriate copies of the formula. This is a much easier and quicker way to copy into a column of cells.

▶

Highlight cells **A1 . . . B21** and construct the graph of $y = |x|$ using the Chart Wizard and selecting the XY (Scatter) option to produce the following graph.

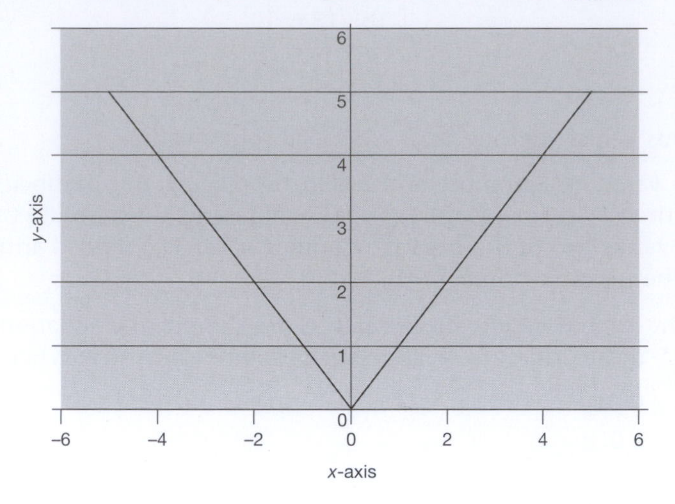


Notice that *Excel* does not do the best job possible with this graph. The point at the origin should be a sharp point whereas the graph produced by *Excel* appears to be rounded there. This demonstrates another danger of using *Excel* as anything other than a guide.

*Next frame*

## 41 Inequalities

Notice that a line drawn parallel to the $x$-axis, through the point $y = 2$ intersects the graph at $x = \pm 2$. This graphically demonstrates that if:

$y < 2$, that is $|x| < 2$ then $-2 < x < 2$ and if

$y > 2$, that is $|x| > 2$ then $x < -2$ or $x > 2$

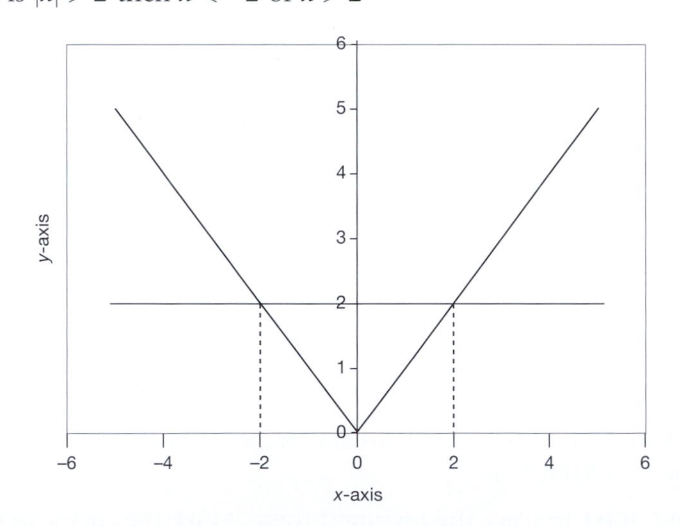


Indeed, if:

$|x| < a$ where $a > 0$ then $-a < x < a$ and if

$|x| > a$ where $a > 0$ then $x < -a$ or $x > a$

You try one. Draw the graph of $y = |x - 1|$ and ascertain from the graph those values of $x$ for which $|x - 1| < 3$ and those values of $x$ for which $|x - 1| > 3$.

*Next frame*

42

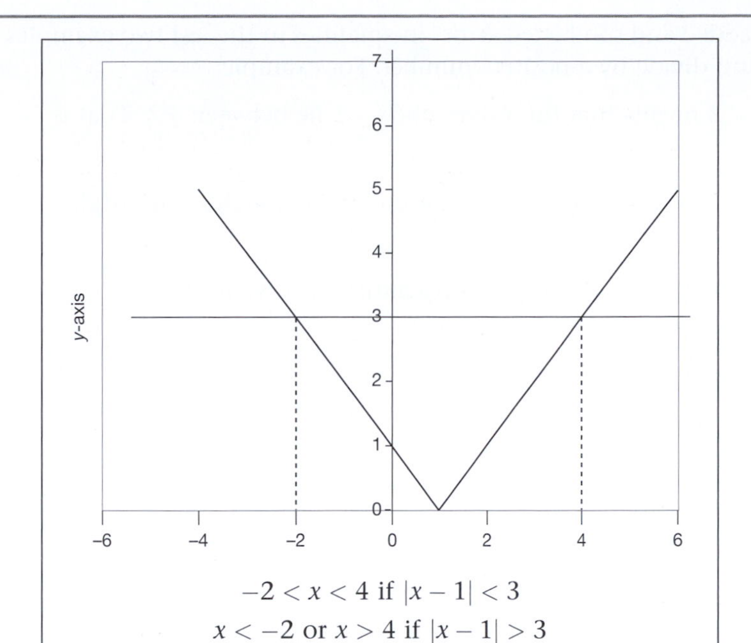


$-2 < x < 4$ if $|x - 1| < 3$

$x < -2$ or $x > 4$ if $|x - 1| > 3$

The graph has the same shape but is shifted to the right 1 unit along the $x$-axis. The values of $x$ for which $|x - 1| < 3$ are clearly $-2 < x < 4$ and for $|x - 1| > 3$ they are $x < -2$ or $x > 4$.

These graphical considerations are all very well but we do really need to be able to derive these results algebraically and this we now do before returning to an interactive graphical construction.

**Less-than inequalities**

If $|x| < 3$ then this means that the value of $x$ lie between $\pm 3$, that is $-3 < x < 3$. Now, extending this, the values of $x$ that satisfy the inequality $|x - 1| < 3$ are those where the values of $x - 1$ lie between $\pm 3$. That is:

$$-3 < x - 1 < 3$$

By adding 1 to both sides of each of the inequalities we find:

$$-3 + 1 < x < 3 + 1 \quad \text{that is} \quad -2 < x < 4$$

In general if $-a < x \pm b < a$ then $-a \mp b < x < a \mp b$

Now you try one. The values of $x$ that satisfy $|x + 4| < 1$ are ...........

*The result is in the next frame*

43

$$\boxed{-5 < x < -3}$$

Because:

$|x + 4| < 1$ means that the values of $x + 4$ lie between $\pm 1$. That is:

$$-1 < x + 4 < 1$$

By subtracting 4 from both sides of each of the two inequalities we find:

$$-1 - 4 < x < 1 - 4 \quad \text{that is} \quad -5 < x < -3$$

▶

Just as we added and subtracted across inequalities in the last two examples so we can multiply and divide by a positive number. For example:

$|5x - 1| < 2$ means that the values of $5x - 1$ lie between $\pm 2$. That is:

$-2 < 5x - 1 < 2$

By adding 1 to both sides of each of the two inequalities we find:

$-2 + 1 < 5x < 2 + 1$ that is $-1 < 5x < 3$

Now, dividing each of the two inequalities by 5 we find:

$$-\frac{1}{5} < x < \frac{3}{5}$$

Now you try one.

The values of $x$ that satisfy the inequality $\left|\frac{x}{2} + 6\right| < 9$ are ...........

*The result is in the next frame*

**44**

$$\boxed{-30 < x < 6}$$

Because:

$\left|\frac{x}{2} + 6\right| < 9$ means that the values of $\frac{x}{2} + 6$ lie between $\pm 9$. That is:

$$-9 < \frac{x}{2} + 6 < 9$$

By subtracting 6 from both sides of each of the two inequalities we find:

$-9 - 6 < \frac{x}{2} < 9 - 6$ that is $-15 < \frac{x}{2} < 3$

Finally, by multiplying by 2 across both sides of each of the two inequalities we find:

$-30 < x < 6$

Multiplying and dividing an inequality by a negative number raises an issue. For example, if

$x < 2$ then the value $x = 1$ will satisfy this inequality because $1 < 2$

If, however, we multiply both sides of the inequality by $-1$ we cannot assert that:

$-x < -2$ and that the value $x = 1$ will still satisfy the inequality because $-1 > -2$.

If, however, *when we multiply or divide by a negative number we switch the inequality* then this issue is resolved. Indeed, if:

$x < 2$ then multiplying by $-1$ gives $-x > -2$ and $x = 1$ satisfies both inequalities.

So, the values of $x$ that satisfy the inequality $|7 - 2x| < 9$ are ...........

*Next frame*

**45**

$-1 < x < 8$

Because:

The values of $x$ are such that $-9 < 7 - 2x < 9$. That is $-9 - 7 < -2x < 9 - 7$ so that $-16 < -2x < 2$ and so, by dividing both sides of the inequalities by $-2$, we see that $8 > x > -1$. *Notice the switch in the inequalities because we have divided both sides of the inequalities by a negative number.* This is then written as $-1 < x < 8$.

One more just to make sure.

The values of $x$ that satisfy the inequality $|5 - x/3| < 11$ are ...........

*Next frame*

**46**

$-18 < x < 48$

Because:

The values of $x$ are such that $-11 < 5 - x/3 < 11$. That is $-11 - 5 < -x/3 < 11 - 5$ so that $-16 < -x/3 < 6$ and so $48 > x > -18$. *Again, notice the switch in the inequalities because we have multiplied both sides of the inequality by a negative number, $-3$.* This is then written as $-18 < x < 48$.

### Greater-than inequalities

If $|x| > 4$ then this means that the $x > 4$ or $x < -4$. Now, extending this, the values of $x$ that satisfy the inequality $|x + 3| > 7$ are those where:

$x + 3 > 7$ or $x + 3 < -7$

That is:

$x > 7 - 3$ or $x < -7 - 3$

and so:

$x > 4$ or $x < -10$

In general, if $|x \pm b| > a$ then $x > a \mp b$ or $x < -a \mp b$

Now you try one.

The values of $x$ that satisfy the inequality $|x - 1| > 3$ are ...........

*Next frame*

**47**

$x > 4$ or $x < -2$

Because:

The values of $x$ are such that $x - 1 < -3$ or $x - 1 > 3$. That is $x < -3 + 1$ or $x > 3 + 1$ so that $x < -2$ or $x > 4$.

As we have already seen in Frame 44, multiplying and dividing an inequality by a negative number raises an issue which is resolved by *switching the inequality*. Therefore the values of $x$ that satisfy the inequality $|5 - x| > 2$ can be found as follows:

The values of $x$ that satisfy the inequality $|5 - x| > 2$ are those where:

$5 - x > 2$ or $5 - x < -2$

That is:

$-x > 2 - 5$ or $-x < -2 - 5$

and so:

$-x > -3$ or $-x < -7$

Multiplying by $-1$ then yields:

$x < 3$ or $x > 7$

Try one for yourself.

The values of $x$ that satisfy the inequality $|7 - 2x| > 9$ are ...........

*Next frame*

## 48

$x < -1$ or $x > 8$

Because:

The values of $x$ are such that $7 - 2x < -9$ or $7 - 2x > 9$. That is $-2x < -16$ or $-2x > 2$ so that $x > 8$ or $x < -1$. *Notice again the switch in the inequalities because we have divided by a negative number.*

Now we look at the construction of an interactive graph.

*Next frame*

## 49 Interaction

The spreadsheet can be used to demonstrate how changing various features of an equation affect the graph but before we can do this we need to know the difference between the cell addresses **A1** and **$A$1** when used within a formula. The first address is called a relative address and the spreadsheet interprets it not as an actual address but as an address relative to the cell in which it is written. For example, if **A1** forms part of a formula in cell **B1** then the spreadsheet understands **A1** to refer to the cell on the same row, namely row **1** but one column to the left. So if the formula were then copied into cell **B2** the reference **A1** would automatically be copied as **A2** – the cell on the same row as **B2** but one column to the left. The contents of cell **B2** would then use the contents of cell **A2** and not those of **A1**. The address **$A$1**, on the other hand, refers to the actual cell at that address. We shall see the effect of this now.

Fill cells **A1** to **A21** with numbers in the range $-5$ to 5 with Step Value 0·5.

In cell **B1** type in **= ABS($C$1*(A1+$D$1))+$E$1**.

That is $y = |a(x + b)| + c$. *We are going to use cells* **C1**, **D1** *and* **E1** *to vary the parameters a, b and c.*

In cell **C1** enter the number 1.

In cell **D1** enter the number 0.

In cell **E1** enter the number 0.

Copy the formula in **B1** into cells **B2 ... B21**.

Draw the graph.

▶

The result is the graph of $y = |x|$. That is $y = |a(x + b)| + c$ where $a = 1$, $b = 0$ and $c = 0$.

The graph has been created by using the fixed values of 1, 0 and 0 for $a$, $b$ and $c$ respectively located in cells **D1**, **C1** and **E1** and referring to them in the formula by using the absolute addresses **$C$1**, **$D$1** and **$E$1**. The relative address **A1** has been used in the formula to refer to the value of $x$; the variable values of which are located in cells **A1...A21**. If, on the Command Bar, you select **View** and **Formula Bar** you display the *Formula Bar*. Now move the cursor down column **B** from row **1** to row **21** and you will see in the Formula Bar that the relative address changes but that the absolute addresses do not.

Now:

(a) In cell **D1** enter the number 1 ($b = 2$). What happens?

(b) In cell **E1** enter the number 3 ($c = 3$). What happens?

(c) In cell **C1** enter the number 4 ($a = 4$). What happens?

*Next frame*

---

**50**

| |
|---|
| (a) The graph moves 2 units to the left |
| (b) The graph moves 3 units up |
| (c) The gradients of the arms of the graph increase |

This latter is noticeable from the change in the scale on the vertical axis. By changing the numbers in cells **C1**, **D1** and **E1** in this way you will be able to see the immediate effect on the graph.

---

## Review summary

**51**

1 *Absolute values*: When numbers are plotted on a straight line the distance of a given number from zero is called the absolute value or modulus of that number. The absolute value is denoted by placing the number between two vertical lines. For example $|-3| = 3$, which reads that the absolute value of $-3$ is 3.

2 *Graphs*: Using an Excel spreadsheet to plot a graph of absolute values necessitates the use of the **ABS** function. The graph of $y = |x|$ consists of two straight lines forming a 'V' with the point at the origin. The graph of $y = |x - a|$ consists of the same two straight lines forming a 'V' with the point $x = a$.

3 *Inequalities*: If, for $a > 0$,

(a) $|x| < a$ then $-a < x < a$

(b) $|x \pm b| < a$ then $-a \mp b < x < a \mp b$

(c) $|x| > a$ then $x > a$ or $x < -a$

(d) $|x \pm b| > a$ then $x > a \mp b$ or $x < -a \mp b$

(e) $-ax > b$ then $x < -\dfrac{b}{a}$

(f) $-ax < b$ then $x > -\dfrac{b}{a}$

4 *Interaction*: A spreadsheet has two types of address. The first is the relative address which is not an actual address, but an address relative to the cell in which it is written. For example if **A1** forms part of a formula in cell **B1** then the spreadsheet understands **A1** to refer to the cell in the same row (here row **1**) but one column to the left. If the formula were then copied to cell **B2** the reference **A1** would automatically be copied as **A2** (that is the same row but one column to the left).

The address **$A$1**, on the other hand, refers to the actual cell at that address. It is called an absolute address because it refers to an actual cell.

The spreadsheet can be made interactive by using relative and absolute addresses.

## Review exercise

**52**

1 Describe the region of the plane that corresponds to each of the following:

(a) $y > 1 - 4x$ (b) $y \leq \frac{3}{x} - 5x$ (c) $2x + 3y > 6$

2 What values of $x$ satisfy each of the following?

(a) $|2 + x| > 5$ (b) $|3 - x| < 4$ (c) $|3x - 2| \geq 6$

*Complete the questions. Take your time.*
*Look back at the Programme if necessary, but don't rush.*
*The answers and working are in the next frame.*

**53**

1 (a) $y > 1 - 4x$

The equation $y = 1 - 4x$ is the equation of a straight line with gradient $-4$ that crosses the $y$-axis at $y = 1$. The inequality $y > 1 - 4x$ defines points for a given $x$-value whose $y$-values are greater than the $y$-value of the point that lies on the line. Therefore the inequality defines the region above the line.

(b) $y \leq \frac{3}{x} - 5x$

The equation $y = \frac{3}{x} - 5x$ is the equation of a curve. The inequality $y \leq \frac{3}{x} - 5x$ defines points for a given $x$-value whose $y$-values are less than or equal to the $y$-value of the point that lies on the line. Therefore the inequality defines the region on and below the curve.

(c) $2x + 3y > 6$

The inequality $2x + 3y > 6$ can be rewritten as follows:

subtracting $2x$ from both sides gives $3y > 6 - 2x$

dividing both sides by 3 gives $y > 2 - (2/3)x$

The equation $2x + 3y = 6$ is, therefore, the equation of a straight line with gradient $-2/3$ that crosses the $y$-axis at $y = 2$.

This inequality defines points for a given $x$-value whose $y$-values are greater than the $y$-value of the point that lies on the line. Therefore the inequality defines the region above the line.

▶

2 (a) $|2+x| > 5$ can be written as:

$(2+x) > 5$ or $(2+x) < -5$ that is $x > 5-2$

or $x < -5-2$ so $x > 3$ or $x < -7$

(b) $|3-x| < 4$ can be written as:

$-4 < (3-x) < 4$ that is $-4-3 < -x < 4-3$

so $-7 < -x < 1$, therefore $-7 < -x$ and $-x < 1$

so that $7 > x$ and $x > -1$ therefore $-1 < x < 7$

(c) $|3x-2| \geq 6$ can be written as:

$(3x-2) \geq 6$ or $(3x-2) \leq -6$ that is $x \geq \dfrac{6+2}{3}$ or $x \leq \dfrac{-6+2}{3}$

so $x \geq \dfrac{8}{3}$ or $x \leq -\dfrac{4}{3}$

---

54

You have now come to the end of this Programme. A list of **Can You?** questions follows for you to gauge your understanding of the material in the Programme. You will notice that these questions match the **Learning outcomes** listed at the beginning of the Programme so go back and try the **Quiz** that follows them. After that try the **Test exercise**. *Work through them at your own pace, there is no need to hurry.* A set of **Further problems** provides additional valuable practice.

---

## Can You?

### Checklist F.4

55

Check this list before and after you try the end of Programme test.

On a scale of 1 to 5 how confident are you that you can: | Frames

- Construct a collection of ordered pairs of numbers from an equation? — Frames 1 to 4
  *Yes* ☐ ☐ ☐ ☐ ☐ *No*
- Plot points associated with ordered pairs of numbers against Cartesian axes and generate graphs? — Frames 5 to 10
  *Yes* ☐ ☐ ☐ ☐ ☐ *No*
- Appreciate the existence of asymptotes to curves and discontinuities? — Frames 11 to 16
  *Yes* ☐ ☐ ☐ ☐ ☐ *No*
- Use an electronic spreadsheet to draw Cartesian graphs of equations? — Frames 17 to 29
  *Yes* ☐ ☐ ☐ ☐ ☐ *No*
- Describe regions of the $x$–$y$ plane that are represented by inequalities? — Frames 30 to 35
  *Yes* ☐ ☐ ☐ ☐ ☐ *No*
- Draw graphs of and algebraically manipulate the absolute value or modulus function? — Frames 36 to 53
  *Yes* ☐ ☐ ☐ ☐ ☐ *No*

# Test exercise F.4

56

Frames

1 Given the equation:

$$x^2 + y^2 = 1$$

(a) Transpose the equation to make $y$ the subject of the transposed equation.

(b) Construct ordered pairs of numbers corresponding to the integer values of $x$ where $-1 \leq x \leq 1$ in intervals of 0·2.

(c) Plot the ordered pairs of numbers on a Cartesian graph and join the points plotted with a continuous curve.

1 to 8

2 Plot the graph of:

(a) $y = x - \dfrac{1}{x}$ for $-3 \leq x \leq 3$.

What is happening near to $x = 0$?

(b) $y = \begin{cases} 2 - x & : x \leq 0 \\ x^2 + 2 & : x > 0 \end{cases}$ for $-4 \leq x \leq 4$

9 to 16

3 Use a spreadsheet to draw the Cartesian graphs of:

(a) $y = 3x^2 + 7x - 6$ for $-4 \leq x \leq 5{\cdot}2$ with step value 0·4

(b) $y = x^3 + x^2 - x + 1$ for $-2 \leq x \leq 4{\cdot}9$ with step value 0·3

17 to 29

4 Describe the region of the $x$–$y$ plane that corresponds to each of the following:

(a) $y < 2 - 3x$ (b) $y > x - \dfrac{2}{x}$ (c) $x + y \leq 1$

30 to 35

5 What values of $x$ satisfy each of the following?

(a) $|x + 8| < 13$

(b) $|x + 8| > 13$

(c) $|4x - 5| < 8$

(d) $|6 - 3x| > 14$

36 to 53

# Further problems F.4

57

1 Given $x^2 - y^2 = 0$ find the equation giving $y$ in terms of $x$ and plot the graph of this equation for $-3 \leq x \leq 3$.

2 Using a spreadsheet plot the graph of:

$$y = x^3 + 10x^2 + 10x - 1$$

for $-10 \leq x \leq 2$ with a step value of 0·5.

3 Using a spreadsheet plot the graph of:

$$y = \frac{x^3}{1 - x^2}$$

for $-2 \leq x \leq 2{\cdot}6$ with a step value of 0·2. Draw a sketch of this graph on a sheet of graph paper indicating discontinuities and asymptotic behaviour more accurately.

4 Given the equation $x^2 + y^2 = 4$ transpose it to find $y$ in terms of $x$. With the aid of a spreadsheet describe the shape that this equation describes.

5 Given the equation:

$$\left(\frac{x}{2}\right)^2 + \left(\frac{y}{4}\right)^2 = 1$$

transpose it to find $y$ in terms of $x$. With the aid of a spreadsheet describe the shape that this equation describes.

6 Given the equation $x^2 + y^2 + 2x + 2y + 1 = 0$ transpose it to find $y$ in terms of $x$. With the aid of a spreadsheet describe the shape that this equation describes.

7 Describe the region of the $x$–$y$ plane that corresponds to each of the following:

(a) $y \geq 2x + 4$

(b) $y < 3 - x$

(c) $3x - 4y \geq 1$

(d) $x^2 + y^2 > 2$

8 Draw the graph of $f(x) = |3x - 4| + 2$ and describe the differences between that graph and the graph of $g(x) = |4 - 3x| + 2$.

9 What values of $x$ between 0 and $2\pi$ radians satisfy each of the following?

(a) $|\sin x| < 0.5$ (b) $|\cos x| > 0.5$

10 Show that $|a - b| \leq c$ is equivalent to $b - c \leq a \leq b + c$.

11 Verify the rule that for two real numbers $x$ and $y$ then $|x + y| \leq |x| + |y|$.

12 Given that for two real numbers $x$ and $y$, $|x + y| \leq |x| + |y|$, deduce that $|x - y| \geq |x| - |y|$.

Now visit the companion website for this book at www.palgrave.com/stroud for more questions applying this mathematics to science and engineering.

58

# Linear equations

## Learning outcomes

When you have completed this Programme you will be able to:

- ☐ Solve any linear equation
- ☐ Solve simultaneous linear equations in two unknowns
- ☐ Solve simultaneous linear equations in three unknowns

If you already feel confident about these why not try the quiz over the page? You can check your answers at the end of the book.

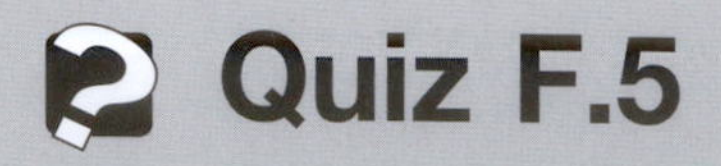

# Quiz F.5

**Frames**

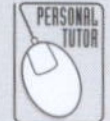

**1** Solve for $x$:

(a) $3(x-1)+2(3-2x)=6(x+4)-7$

(b) $\dfrac{x-2}{3}-\dfrac{3-x}{2}=1+\dfrac{4-5x}{4}$

(c) $\dfrac{4}{x-1}+\dfrac{2}{x+1}=\dfrac{6}{x}$ 1 to 6

**2** Solve the following pair of simultaneous equations for $x$ and $y$:

$$3x-4y=7$$
$$5x-2y=7$$

7 to 11

**3** Solve the following set of three equations in three unknowns:

$$x-2y+3z=10$$
$$3x-2y+z=2$$
$$4x+5y+2z=29$$

12 to 20

# Linear equations

**1**

A linear equation in a single variable (unknown) involves powers of the variable no higher than the first. A linear equation is also referred to as a simple equation.

## Solution of simple equations

The solution of simple equations consists essentially of simplifying the expressions on each side of the equation to obtain an equation of the form $ax + b = cx + d$ giving $ax - cx = d - b$ and

hence $x = \dfrac{d-b}{a-c}$ provided $a \neq c$.

**Example**

If $\quad 6x + 7 + 5x - 2 + 4x - 1 = 36 + 7x$

then $\quad 15x + 4 = 7x + 36$

so $8x = 32$ therefore $x = 4$

Similarly:

if $\quad 5(x-1) + 3(2x+9) - 2 = 4(3x-1) + 2(4x+3)$

then $\quad x = \dots\dots\dots\dots$

**2**

$x = 2$

Because

$$5(x-1) + 3(2x+9) - 2 = 4(3x-1) + 2(4x+3)$$

$$\text{so } 5x - 5 + 6x + 27 - 2 = 12x - 4 + 8x + 6$$

$$11x + 20 = 20x + 2$$

$$\text{so } 18 = 9x \qquad \text{therefore } x = 2$$

Equations which appear not to be simple equations sometimes develop into simple equations during simplification. For example:

$$(4x+1)(x+3) - (x+5)(x-3) = (3x+1)(x-4)$$

$$(4x^2 + 13x + 3) - (x^2 + 2x - 15) = 3x^2 - 11x - 4$$

$$\text{so } 3x^2 + 11x + 18 = 3x^2 - 11x - 4$$

$3x^2$ can now be subtracted from each side, giving:

$$11x + 18 = -11x - 4$$

$$\text{so } 22x = -22 \qquad \text{therefore } x = -1$$

It is always wise to check the result by substituting this value for $x$ in the original equation:

$$\text{LHS} = (4x+1)(x+3) - (x+5)(x-3)$$

$$= (-3)(2) - (4)(-4) = -6 + 16 = 10$$

$$\text{RHS} = (3x+1)(x-4) = (-2)(-5) = 10 \quad \therefore \text{LHS} = \text{RHS}$$

So, in the same way, solving the equation:

$$(4x+3)(3x-1) - (5x-3)(x+2) = (7x+9)(x-3)$$

we have $x = \dots\dots\dots\dots$

**3**

$$\boxed{x = -3}$$

Simplification gives $7x^2 - 2x + 3 = 7x^2 - 12x - 27$

Hence $10x = -30 \quad \therefore x = -3$

Where simple equations involve algebraic fractions, the first step is to eliminate the denominators by multiplying throughout by the LCM (Lowest Common Multiple) of the denominators. For example, to solve:

$$\frac{x+2}{2} - \frac{x+5}{3} = \frac{2x-5}{4} + \frac{x+3}{6}$$

The LCM of 2, 3, 4 and 6 is 12:

$$\frac{12(x+2)}{2} - \frac{12(x+5)}{3} = \frac{12(2x-5)}{4} + \frac{12(x+3)}{6}$$

$$6(x+2) - 4(x+5) = 3(2x-5) + 2(x+3)$$

$$\therefore x = \ldots\ldots\ldots\ldots$$

**4**

$$\boxed{x = \frac{1}{6}}$$

That was easy enough. Now let us look at this one.

To solve $\dfrac{4}{x-3} + \dfrac{2}{x} = \dfrac{6}{x-5}$

Here, the LCM of the denominators is $x(x-3)(x-5)$. So, multiplying throughout by the LCM, we have

$$\frac{4x(x-3)(x-5)}{x-3} + \frac{2x(x-3)(x-5)}{x} = \frac{6x(x-3)(x-5)}{x-5}$$

After cancelling where possible:

$$4x(x-5) + 2(x-3)(x-5) = 6x(x-3)$$

so, finishing it off,

$$x = \ldots\ldots\ldots\ldots$$

**5**

$$\boxed{x = \frac{5}{3}}$$

Because $4x^2 - 20x + 2(x^2 - 8x + 15) = 6x^2 - 18x$

$$4x^2 - 20x + 2x^2 - 16x + 30 = 6x^2 - 18x$$

$$\therefore 6x^2 - 36x + 30 = 6x^2 - 18x$$

Subtracting the $6x^2$ term from both sides:

$$-36x + 30 = -18x \quad \therefore 30 = 18x \quad \therefore x = \frac{5}{3}$$

Working in the same way:

$$\frac{3}{x-2} + \frac{5}{x-3} - \frac{8}{x+3} = 0$$

gives the solution

$$x = \ldots\ldots\ldots\ldots$$

6

$$\boxed{x = \frac{7}{3}}$$

The LCM of the denominators is $(x - 2)(x - 3)(x + 3)$, so

$$\frac{3(x-2)(x-3)(x+3)}{x-2} + \frac{5(x-2)(x-3)(x+3)}{x-3} - \frac{8(x-2)(x-3)(x+3)}{x+3} = 0$$

$$\therefore\ 3(x-3)(x+3) + 5(x-2)(x+3) - 8(x-2)(x-3) = 0$$

$$3(x^2-9) + 5(x^2+x-6) - 8(x^2-5x+6) = 0$$

$$3x^2 - 27 + 5x^2 + 5x - 30 - 8x^2 + 40x - 48 = 0$$

$$\therefore\ 45x - 105 = 0$$

$$\therefore\ x = \frac{105}{45} = \frac{7}{3} \quad \therefore\ x = \frac{7}{3}$$

*Now on to Frame 7*

## Simultaneous linear equations with two unknowns

7

A linear equation in two variables has an infinite number of solutions. For example, the two-variable linear equation $y - x = 3$ can be transposed to read:

$$y = x + 3$$

Any one of an infinite number of $x$-values can be substituted into this equation and each one has a corresponding $y$-value. However, for two such equations there may be just one pair of $x$- and $y$-values that satisfy both equations *simultaneously*.

### 1 Solution by substitution

To solve the pair of equations

$$5x + 2y = 14 \qquad (1)$$

$$3x - 4y = 24 \qquad (2)$$

From (1): $5x + 2y = 14 \ \therefore\ 2y = 14 - 5x \ \therefore\ y = 7 - \dfrac{5x}{2}$

If we substitute this for $y$ in (2), we get:

$$3x - 4\left(7 - \frac{5x}{2}\right) = 24$$

$$\therefore\ 3x - 28 + 10x = 24$$

$$13x = 52 \quad \therefore\ x = 4$$

If we now substitute this value for $x$ in the other original equation, i.e. (1), we get:

$$5(4) + 2y = 14$$

$$20 + 2y = 14 \quad \therefore\ 2y = -6 \quad \therefore\ y = -3$$

$\therefore$ We have $x = 4$, $y = -3$

As a check, we can substitute both these values in (1) and (2):

(1) $5x + 2y = 5(4) + 2(-3) = 20 - 6 = 14$ ✓

(2) $3x - 4y = 3(4) - 4(-3) = 12 + 12 = 24$ ✓

$\therefore\ x = 4$, $y = -3$ is the required solution.

▶

Another example:

To solve

$$3x + 4y = 9 \quad (1)$$
$$2x + 3y = 8 \quad (2)$$

Proceeding as before, determine the values of $x$ and $y$ and check the results by substitution in the given equations.

$x = \ldots\ldots\ldots\ldots, \quad y = \ldots\ldots\ldots\ldots$

**8**

$$\boxed{x = -5, y = 6}$$

Because

(1) $\quad 3x + 4y = 9 \quad \therefore \quad 4y = 9 - 3x \;\text{ so }\; y = \frac{9}{4} - \frac{3x}{4}$

Substituting this for $y$ in (2) yields:

$$2x + 3\left(\frac{9}{4} - \frac{3x}{4}\right) = 8 \;\text{ that is }\; 2x + \frac{27}{4} - \frac{9x}{4} = 8 \;\therefore\; -\frac{x}{4} = \frac{5}{4} \;\therefore\; x = -5$$

Substituting this value for $x$ in (1) yields:

$$-15 + 4y = 9 \;\therefore\; 4y = 24 \;\therefore\; y = 6$$

### 2 Solution by equating coefficients

To solve $\quad 3x + 2y = 16 \quad (1)$
$\qquad\qquad 4x - 3y = 10 \quad (2)$

If we multiply both sides of (1) by 3 (the coefficient of $y$ in (2)) and we multiply both sides of (2) by 2 (the coefficient of $y$ in (1)) then we have

$$9x + 6y = 48$$
$$8x - 6y = 20$$

If we now add these two lines together, the $y$-term disappears:

$$\therefore\; 17x = 68 \quad \therefore\; x = 4$$

Substituting this result, $x = 4$, in either of the original equations, provides the value of $y$.

In (1) $\quad 3(4) + 2y = 16 \quad \therefore\; 12 + 2y = 16 \quad \therefore\; y = 2$

$$\therefore\; x = 4, y = 2$$

Check in (2): $\quad 4(4) - 3(2) = 16 - 6 = 10 \quad \checkmark$

Had the $y$-terms been of the same sign, we should, of course, have subtracted one line from the other to eliminate one of the variables.

Another example:

To solve $\quad 3x + y = 18 \quad (1)$
$\qquad\qquad 4x + 2y = 21 \quad (2)$

Working as before:

$x = \ldots\ldots\ldots\ldots, \quad y = \ldots\ldots\ldots\ldots$

**9**

$$\boxed{x = 7{\cdot}5,\ y = -4{\cdot}5}$$

$$\begin{array}{lll} (1) & \times 2 & 6x + 2y = 36 \\ (2) & & 4x + 2y = 21 \\ \hline \text{Subtract:} & & 2x \quad\quad = 15 \quad \therefore\ x = 7{\cdot}5 \end{array}$$

Substitute $x = 7{\cdot}5$ in (1):

$$3(7{\cdot}5) + y = 18 \qquad 22{\cdot}5 + y = 18$$

$$\therefore\ y = 18 - 22{\cdot}5 = -4{\cdot}5 \qquad \therefore\ y = -4{\cdot}5$$

Check in (2): $4x + 2y = 4(7{\cdot}5) + 2(-4{\cdot}5) = 30 - 9 = 21$ ✓

$\therefore\ x = 7{\cdot}5,\ y = -4{\cdot}5$

**10**

And one more – on your own.

Solve $7x - 4y = 23$ (1)

$4x - 3y = 11$ (2)

Working as before:

$x = \dots\dots\dots\dots,\quad y = \dots\dots\dots\dots$

**11**

$$\boxed{x = 5,\ y = 3}$$

$$\begin{array}{lll} (1) & \times 3 & 21x - 12y = 69 \\ (2) & \times 4 & 16x - 12y = 44 \\ \hline \text{Subtract:} & & 5x \quad\quad = 25 \quad \therefore\ x = 5 \end{array}$$

Substitute in (2): $20 - 3y = 11 \quad \therefore\ 3y = 9 \quad \therefore\ y = 3$

$\therefore\ x = 5,\ y = 3$

Check in (1): $7x - 4y = 35 - 12 = 23$ ✓

**12**

## Simultaneous linear equations with three unknowns

With three unknowns and three equations the method of solution is just an extension of the work with two unknowns.

### Example 1

To solve

$$3x + 2y - z = 19 \quad (1)$$
$$4x - y + 2z = 4 \quad (2)$$
$$2x + 4y - 5z = 32 \quad (3)$$

We take a pair of equations and eliminate one of the variables using the method in Frame 7:

$$3x + 2y - z = 19 \quad (1)$$
$$4x - y + 2z = 4 \quad (2)$$

$$\begin{array}{ll} (1) \times 2 & 6x + 4y - 2z = 38 \\ (2) & 4x - y + 2z = 4 \\ \hline \text{Add:} & 10x + 3y \quad\quad = 42 \quad (4) \end{array}$$

▶

Now take another pair, e.g. (1) and (3):

$$
\begin{array}{lrl}
(1)\times 5 & 15x + 10y - 5z = 95 & \\
(3) & 2x + 4y - 5z = 32 & \\
\hline
\text{Subtract:} & 13x + 6y = 63 & (5)
\end{array}
$$

We can now solve equations (4) and (5) for values of $x$ and $y$ in the usual way.

$x = \dots\dots\dots\dots$, $y = \dots\dots\dots\dots$

## 13

$x = 3, y = 4$

$$
\begin{array}{lrr}
\text{Because} & 10x + 3y = 42 & (4) \\
 & 13x + 6y = 63 & (5) \\
(4)\times 2 & 20x + 6y = 84 & \\
(5) & 13x + 6y = 63 & \\
\hline
\text{Subtract:} & 7x = 21 & \therefore\ x = 3
\end{array}
$$

Substitute in (4): $30 + 3y = 42 \quad \therefore\ 3y = 12 \quad \therefore\ y = 4$

We substitute these values in one of the original equations to obtain the value of $z$:

e.g. (2) $4x - y + 2z = 12 - 4 + 2z = 4$

$\therefore\ 2z = -4 \quad \therefore\ z = -2$

$\therefore\ x = 3, y = 4, z = -2$

Finally, substitute all three values in the other two original equations as a check procedure:

(1) $3x + 2y - z = 9 + 8 + 2 = 19$

(3) $2x + 4y - 5z = 6 + 16 + 10 = 32$ – so all is well.

The working is clearly longer than with only two unknowns, but the method is no more difficult. Here is another.

**Example 2**

Solve

$$
\begin{array}{rr}
5x - 3y - 2z = 31 & (1) \\
2x + 6y + 3z = 4 & (2) \\
4x + 2y - z = 30 & (3)
\end{array}
$$

Work through it in just the same way and, as usual, check the results.

$x = \dots\dots\dots\dots$, $y = \dots\dots\dots\dots$, $z = \dots\dots\dots\dots$

## 14

$x = 5, y = 2, z = -6$

$$
\begin{array}{lrr}
(1)\times 3 & 15x - 9y - 6z = 93 & \\
(2)\times 2 & 4x + 12y + 6z = 8 & \\
\hline
 & 19x + 3y = 101 & (4) \\
(1) & 5x - 3y - 2z = 31 & \\
(3)\times 2 & 8x + 4y - 2z = 60 & \\
\hline
 & 3x + 7y = 29 & (5)
\end{array}
$$

Solving (4) and (5) and substituting back gives the results above:

$x = 5, \quad y = 2, \quad z = -6$

**15**

Sometimes, the given equations need to be simplified before the method of solution can be carried out.

## Pre-simplification

**Example 1**

Solve the pair of equations:

$$2(x+2y)+3(3x-y)=38 \quad (1)$$
$$4(3x+2y)-3(x+5y)=-8 \quad (2)$$

$$2x+4y+9x-3y=38 \quad \therefore\ 11x+y=38 \quad (3)$$
$$12x+8y-3x-15y=-8 \quad \therefore\ 9x-7y=-8 \quad (4)$$

The pair of equations can now be solved in the usual manner.

$x=$ ......., $y=$ ............

**16**

$$\boxed{x=3,\ y=5}$$

Because (3) × 7 $\quad 77x+7y=266$

(4) $\quad 9x-7y=-8$

$86x = 258 \quad \therefore\ x=3$

Substitute in (3): $33+y=38 \quad \therefore\ y=5$

Check in (4): $9(3)-7(5)=27-35=-8$ ✓

$\therefore\ x=3,\ y=5$

**Example 2**

$$\frac{2x-1}{5}+\frac{x-2y}{10}=\frac{x+1}{4} \quad (1)$$
$$\frac{3y+2}{3}+\frac{4x-3y}{2}=\frac{5x+4}{4} \quad (2)$$

For (1) LCM = 20 $\quad \therefore\ \dfrac{20(2x-1)}{5}+\dfrac{20(x-2y)}{10}=\dfrac{20(x+1)}{4}$

$$4(2x-1)+2(x-2y)=5(x+1)$$
$$8x-4+2x-4y=5x+5$$
$$\therefore\ 5x-4y=9 \quad (3)$$

Similarly for (2), the simplified equation is ...........

**17**

$$\boxed{9x - 6y = 4}$$

Because

$$\frac{3y+2}{3} + \frac{4x-3y}{2} = \frac{5x+4}{4} \qquad \text{LCM} = 12$$

$$\therefore \quad \frac{12(3y+2)}{3} + \frac{12(4x-3y)}{2} = \frac{12(5x+4)}{4}$$

$$4(3y+2) + 6(4x-3y) = 3(5x+4)$$

$$12y + 8 + 24x - 18y = 15x + 12$$

$$\therefore \; 24x - 6y + 8 = 15x + 12$$

$$\therefore \; 9x - 6y = 4 \qquad (4)$$

So we have

$$5x - 4y = 9 \qquad (3)$$

$$9x - 6y = 4 \qquad (4)$$

Finishing off, we get

$x = \ldots\ldots\ldots\ldots$, $y = \ldots\ldots\ldots\ldots$

**18**

$$\boxed{x = -19/3,\; y = -61/6}$$

Because

$9 \times (5x - 4y = 9)$ gives $45x - 36y = 81$

$5 \times (9x - 6y = 4)$ gives $45x - 30y = 20$

Subtracting gives $-6y = 61$ so that $y = -\dfrac{61}{6}$. Substitution then gives:

$$5x = 9 + 4y = 9 - \frac{244}{6} = -\frac{190}{6},$$

so $\quad x = -\dfrac{190}{30} = -\dfrac{19}{3}$

That was easy enough.

**Example 3**

$$5(x+2y) - 4(3x+4z) - 2(x+3y-5z) = 16$$

$$2(3x-y) + 3(x-2z) + 4(2x-3y+z) = -16$$

$$4(y-2z) + 2(2x-4y-3) - 3(x+4y-2z) = -62$$

Simplifying these three equations, we get

$\ldots\ldots\ldots\ldots$

**19**

$$\boxed{\begin{aligned} -9x + 4y - 6z &= 16 \\ 17x - 14y - 2z &= -16 \\ x - 16y - 2z &= -56 \end{aligned}}$$

Solving this set of three by equating coefficients, we obtain

$x = \ldots\ldots\ldots\ldots$, $y = \ldots\ldots\ldots\ldots$, $z = \ldots\ldots\ldots\ldots$

20

$$x = 2,\ y = 4,\ z = -3$$

Here is the working as a check:

$$-9x + 4y - 6z = 16 \quad (1)$$
$$17x - 14y - 2z = -16 \quad (2)$$
$$x - 16y - 2z = -56 \quad (3)$$

(1) $\quad -9x + 4y - 6z = 16$

(2) × 3 $\quad 51x - 42y - 6z = -48$

Subtracting: $\quad 60x - 46y = -64 \quad \therefore 30x - 23y = -32 \quad (4)$

(2) $\quad 17x - 14y - 2z = -16$

(3) $\quad x - 16y - 2z = -56$

Subtracting: $\quad 16x + 2y = 40 \quad \therefore 8x + y = 20 \quad (5)$

$\therefore$ (4) $\quad 30x - 23y = -32$

(5) × 23 $\quad 184x + 23y = 460$

Add: $\quad 214x = 428 \quad \therefore x = 2$

Substitute in (5): $\quad 16 + y = 20 \quad \therefore y = 4$

Substitute for $x$ and $y$ in (3):

$$2 - 64 - 2z = -56$$
$$-62 - 2z = -56 \quad \therefore z + 31 = 28$$
$$\therefore z = -3$$

Check by substituting all values in (1):

$$-9(2) + 4(4) - 6(-3) = -18 + 16 + 18 = 16 \quad ✓$$
$$\therefore x = 2,\ y = 4,\ z = -3$$

## Review summary

21

1 A *linear equation* involves powers of the variables no higher than the first. A linear equation in a single variable is also referred to as a *simple equation*.

2 Simple equations are solved by isolating the variable on one side of the equation.

3 A linear equation in two variables has an infinite number of solutions. For two such equations there may be only one solution that satisfies both of them simultaneously.

4 Simultaneous equations can be solved:

(a) by substitution

(b) by elimination.

##  Review exercise

**22**

1 Solve the following linear equations:

(a) $2(x-1)-4(x+2)=3(x+5)+(x-1)$

(b) $\dfrac{x-1}{2}-\dfrac{x+1}{2}=5-\dfrac{x+2}{4}$

(c) $\dfrac{3}{x+2}-\dfrac{5}{x}=-\dfrac{2}{x-1}$

2 Solve the following pairs of simultaneous equations:

(a) $x-y=2$ by elimination
$2x+3y=9$

(b) $4x+2y=10$ by substitution
$3x-5y=1$

3 Solve the following set of three equations in three unknowns:

$x+y+z=6$
$2x-y+3z=9$
$x+2y-3z=-4$

4 Simplify and solve the following set of simultaneous equations:

$2(x+2y)-3(x+2z)-4(y-2z)=-1$
$3(2x-y)+2(3y-4z)-5(3x-2z)=-7$
$-(x-y)+(x-z)+(y+z)=-2$

**23**

1 (a) $2(x-1)-4(x+2)=3(x+5)+(x-1)$.

Eliminate the brackets to give $2x-2-4x-8=3x+15+x-1$.
Simplify each side to give $-2x-10=4x+14$.
Add $2x-14$ to both sides of the equation to give $-24=6x$ so that $x=-4$.

(b) $\dfrac{x-1}{2}-\dfrac{x+1}{2}=5-\dfrac{x+2}{4}$.

Multiply throughout by 4 to give $2(x-1)-2(x+1)=20-x-2$. Simplify to give $-4=18-x$, therefore $x=22$.

(c) $\dfrac{3}{x+2}-\dfrac{5}{x}=-\dfrac{2}{x-1}$.

Multiply throughout by $x(x+2)(x-1)$ to give
$3x(x-1)-5(x+2)(x-1)=-2x(x+2)$.
Eliminate the brackets to give $3x^2-3x-5x^2-5x+10=-2x^2-4x$.
Simplify to give $-2x^2-8x+10=-2x^2-4x$.
Add $2x^2$ to both sides to give $-8x+10=-4x$. Add $8x$ to both sides of the equation to give $10=4x$ so that $x=\dfrac{5}{2}$.

2 (a)

| | |
|---|---|
| $x-y=2$ | [1] |
| $2x+3y=9$ | [2] Multiply [1] by $-2$ |
| $-2x+2y=-4$ | Add [2] to give $5y=5$ so that $y=1$. Substitution in [1] gives $x=3$ |

▶

(b) $4x + 2y = 10$ [1]

$3x - 5y = 1$ [2] Subtract $4x$ from both sides of [1] and then divide by 2 to give: $y = 5 - 2x$. Substitute this expression for $y$ in [2] to give $3x - 5(5 - 2x) = 1$. Simplify to give $13x - 25 = 1$ so that $x = 2$ and $y = 5 - 4 = 1$.

**3** $x + y + z = 6$ [1]

$2x - y + 3z = 9$ [2]

$x + 2y - 3z = -4$ [3] Add [1] and [3] and subtract [2] to give

$(2x + 3y - 2z) - (2x - y + 3z) = (6 - 4) - 9$. That is

$4y - 5z = -7$ [4] Subtract [1] from [3] to give

$y - 4z = -10$ [5] Multiply this by 4 to give

$4y - 16z = -40$ [6] Subtract [6] from [4] to give $11z = 33$ so that $z = 3$. Substitute this value in [5] to give $y = -10 + 12 = 2$ so from [1], $x = 6 - 3 - 2 = 1$.

**4** $2(x + 2y) - 3(x + 2z) - 4(y - 2z) = -1$ [1]

$3(2x - y) + 2(3y - 4z) - 5(3x - 2z) = -7$ [2]

$-(x - y) + (x - z) + (y + z) = -2$ [3] Eliminate the brackets to give

$-x + 2z = -1$ [4]

$-9x + 3y + 2z = -7$ [5]

$2y = -2$ [6] Thus $y = -1$. Substituting this value in [5] gives

$-9x + 2z = -4$ Subtracting [4] gives $-8x = -3$ so that $x = \frac{3}{8}$. Substitution in [4] gives $z = -\frac{5}{16}$.

---

**24**

You have now come to the end of this Programme. A list of **Can You?** questions follows for you to gauge your understanding of the material in the Programme. You will notice that these questions match the **Learning outcomes** listed at the beginning so go back and try the **Quiz** that follows them. After that try the **Test exercise**. *Work through them at your own pace, there is no need to hurry*. A set of **Further problems** provides additional valuable practice.

---

## Can You?

### Checklist F.5

**25**

Check this list before and after you try the end of Programme test.

On a scale of 1 to 5 how confident are you that you can:

| | | Frames |
|---|---|---|
| • Solve any linear equation? | *Yes* ☐ ☐ ☐ ☐ ☐ *No* | 1 to 6 |
| • Solve simultaneous linear equations in two unknowns? | *Yes* ☐ ☐ ☐ ☐ ☐ *No* | 7 to 11 |
| • Solve simultaneous linear equations in three unknowns? | *Yes* ☐ ☐ ☐ ☐ ☐ *No* | 12 to 20 |

## Test exercise F.5

**26**

Frames

1 Solve the following linear equations:

(a) $4(x+5)-6(2x+3)=3(x+14)-2(5-x)+9$

(b) $\dfrac{2x+1}{3}-\dfrac{2x+5}{5}=2+\dfrac{x-1}{6}$

(c) $\dfrac{2}{x-2}+\dfrac{3}{x}=\dfrac{5}{x-4}$

(d) $(4x-3)(3x-1)-(7x+2)(x+1)=(5x+1)(x-2)-10$ 1 to 6

2 Solve the following pairs of simultaneous equations:

(a) $2x+3y=7$
$5x-2y=8$ by substitution

(b) $4x+2y=5$
$3x+y=9$ by elimination 7 to 11

3 Solve the following set of three equations in three unknowns:

$$2x+3y-z=-5$$
$$x-4y+2z=21$$
$$5x+2y-3z=-4$$ 12 to 14

4 Simplify and solve the following set of simultaneous equations:

$$4(x+3y)-2(4x+3z)-3(x-2y-4z)=17$$
$$2(4x-3y)+5(x-4z)+4(x-3y+2z)=23$$
$$3(y+4z)+4(2x-y-z)+2(x+3y-2z)=5$$ 15 to 20

## Further problems F.5

**27**

Solve the following equations, numbered **1** to **15**:

1 $(4x+5)(4x-3)+(5-4x)(5+4x)=(3x-2)(3x+2)-(9x-5)(x+1)$

2 $(3x-1)(x-3)-(4x-5)(3x-4)=6(x-7)-(3x-5)^2$

3 $\dfrac{4x+7}{3}-\dfrac{x-7}{5}=\dfrac{x+6}{9}$

4 $\dfrac{2x-5}{2}+\dfrac{5x+2}{6}=\dfrac{2x+3}{4}-\dfrac{x+5}{3}$

5 $(3x+1)(x-2)-(5x-2)(2x-1)=16-(7x+3)(x+2)$

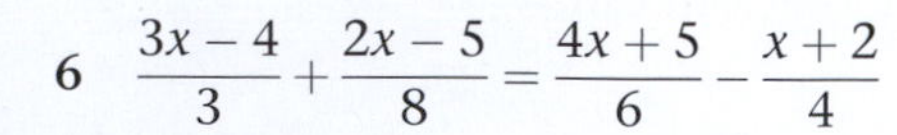

6 $\dfrac{3x-4}{3}+\dfrac{2x-5}{8}=\dfrac{4x+5}{6}-\dfrac{x+2}{4}$

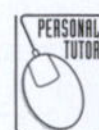

7 $8x-5y=10$
$6x-4y=11$

8 $\dfrac{2}{x+3}+\dfrac{5}{x}=\dfrac{7}{x-2}$

▶

**9** $(3-4x)(3+4x) = 3(3x-2)(x+1) - (5x-3)(5x+3)$

**10** $3x - 2y + z = 2$
$x - 3y + 2z = 1$
$2x + y - 3z = -5$

**11** $2x - y + 3z = 20$
$x - 6y - z = -41$
$3x + 6y + 2z = 70$

**12** $3x + 2y + 5z = 2$
$5x + 3y - 2z = 4$
$2x - 5y - 3z = 14$

**13** $5x - 6y + 3z = -9$
$2x - 3y + 2z = -5$
$3x - 7y + 5z = -16$

**14** $\dfrac{3x+2}{4} - \dfrac{x+2y}{2} = \dfrac{x-3}{12}$
$\dfrac{2y+1}{5} + \dfrac{x-3y}{4} = \dfrac{3x+1}{10}$

**15** $\dfrac{x-5}{x+5} + \dfrac{x-7}{x+7} = 2$

**16** Simplify each side separately and hence determine the solution:
$$\frac{x-2}{x-4} - \frac{x-4}{x-6} = \frac{x-1}{x-3} - \frac{x-3}{x-5}$$

**17** Solve the simultaneous equations:
$$\frac{3x-2y}{2} = \frac{2x+y}{7} + \frac{3}{2}$$
$$7 - \frac{2x-y}{6} = x + \frac{y}{4}$$

**18** Writing $u$ for $\dfrac{1}{x+8y}$ and $v$ for $\dfrac{1}{8x-y}$, solve the following equations for $u$ and $v$, and hence determine the values of $x$ and $y$:
$$\frac{2}{x+8y} - \frac{1}{8x-y} = 4; \qquad \frac{1}{x+8y} + \frac{2}{8x-y} = 7$$

**19** Solve the pair of equations:
$$\frac{6}{x-2y} - \frac{15}{x+y} = 0{\cdot}5$$
$$\frac{12}{x-2y} - \frac{9}{x+y} = -0{\cdot}4$$

**20** Solve:
$$\frac{4}{x+1} + \frac{7}{4(2-y)} = 9$$
$$\frac{3}{2(x+1)} + \frac{7}{2-y} = 7$$

Now visit the companion website for this book at www.palgrave.com/stroud for more questions applying this mathematics to science and engineering.

28

# Polynomial equations

## Learning outcomes

When you have completed this Programme you will be able to:

- ☐ Solve quadratic equations by factors, by completing the square and by formula
- ☐ Solve cubic equations with at least one linear factor
- ☐ Solve fourth-order equations with at least two linear factors

If you already feel confident about these why not try the quiz over the page? You can check your answers at the end of the book.

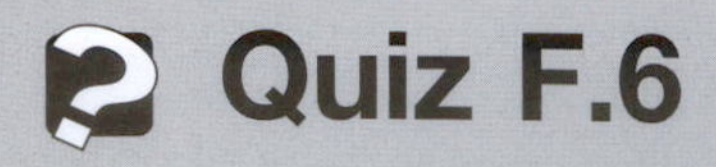

# Quiz F.6

**Frames**

1 Solve $x$ by factorizing:

$6x^2 - 13x + 6 = 0$ — 2 to 7

2 Solve for $x$ (accurate to 3 dp) by completing the square:

$3x^2 + 4x - 1 = 0$ — 8 to 13

3 Solve $x$ (accurate to 3 dp) by using the formula:

$x^2 - 3x + 1 = 0$ — 14 to 17

4 Solve for $x$:

$x^3 - 8x^2 + 18x - 9 = 0$ — 18 to 26

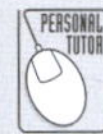

5 Solve for $x$:

$2x^4 - 9x^3 + 12x^2 - 3x - 2 = 0$ — 27 to 33

# Polynomial equations

**1**

In Programme F.3 we looked at polynomial equations. In particular we selected a value for the variable $x$ in a polynomial expression and found the resulting value of the polynomial expression. In other words, we *evaluated* the polynomial expression. Here we reverse the process by giving the polynomial expression a value of zero and finding those values of $x$ which satisfy the resulting equation. We start with quadratic equations.

*On now to the second frame*

**2**

## Quadratic equations

### 1 Solution by factors

We dealt at some length in Programmes F.2 and F.3 with the representation of a quadratic expression as a product of two simple linear factors, where such factors exist.

For example, $x^2 + 5x - 14$ can be factorized into $(x + 7)(x - 2)$. The equation $x^2 + 5x - 14 = 0$ can therefore be written in the form:

$$(x + 7)(x - 2) = 0$$

and this equation is satisfied if either factor has a zero value.

$\therefore\ x + 7 = 0$ or $x - 2 = 0$

i.e. $x = -7$ or $x = 2$

are the solutions of the given equation $x^2 + 5x - 14 = 0$.

By way of revision, then, you can solve the following equations with no trouble:

(a) $x^2 - 9x + 18 = 0$

(b) $x^2 + 11x + 28 = 0$

(c) $x^2 + 5x - 24 = 0$

(d) $x^2 - 4x - 21 = 0$

**3**

(a) $x = 3$ or $x = 6$
(b) $x = -4$ or $x = -7$
(c) $x = 3$ or $x = -8$
(d) $x = 7$ or $x = -3$

Not all quadratic expressions can be factorized as two simple linear factors. Remember that the test for the availability of factors with $ax^2 + bx + c$ is to calculate whether $(b^2 - 4ac)$ is ...........

**4**

a perfect square

The test should always be applied at the begining to see whether, in fact, simple linear factors exist.

For example, with $x^2 + 8x + 15$, $(b^2 - 4ac) = $ ...........

**5**

$4$, i.e. $2^2$

So $x^2 + 8x + 15$ can be written as the product of two simple linear factors.
But with $x^2 + 8x + 20$, $a = 1, b = 8, c = 20$
and $(b^2 - 4ac) = 8^2 - 4 \times 1 \times 20 = 64 - 80 = -16$, which is not a perfect square.
So $x^2 + 8x + 20$ cannot be written as the product of two simple linear factors.

*On to the next frame*

**6**

If the coefficient of the $x^2$ term is other than unity, the factorization process is a trifle more involved, but follows the routine already established in Programme F.2. You will remember the method.

**Example 1**

To solve $2x^2 - 3x - 5 = 0$

In this case, $ax^2 + bx + c = 0$, $a = 2$, $b = -3$, $c = -5$.

(a) Test for simple factors:

$$(b^2 - 4ac) = (-3)^2 - 4 \times 2 \times (-5)$$
$$= 9 + 40 = 49 = 7^2 \qquad \therefore \text{ simple factors exist.}$$

(b) $|ac| = 10$. Possible factors of 10 are (1, 10) and (2, 5).
$c$ is negative: $\therefore$ factors differ by $|b|$, i.e. 3
$\therefore$ Required factors are $(2, 5)$

$c$ is negative: $\therefore$ factors are of different sign, the numerically larger having the sign of $b$, i.e. negative.

$$\therefore\ 2x^2 - 3x - 5 = 2x^2 + 2x - 5x - 5$$
$$= 2x(x + 1) - 5(x + 1)$$
$$= (x + 1)(2x - 5)$$

$\therefore$ the equation $2x^2 - 3x - 5 = 0$ becomes $(x + 1)(2x - 5) = 0$

$\therefore\ x + 1 = 0$ or $2x - 5 = 0$
$\therefore\ x = -1$ or $x = 2{\cdot}5$

**Example 2**

In the same way, the equation:

$3x^2 + 14x + 8 = 0$

has solutions

$x = \ldots\ldots\ldots\ldots$ or $x = \ldots\ldots\ldots\ldots$

**7**

$$x = -4 \text{ or } x = -\frac{2}{3}$$

Because $3x^2 + 14x + 8 = 0 \quad a = 3, b = 14, c = 8$

Test for simple factors: $(b^2 - 4ac) = 14^2 - 4 \times 3 \times 8 = 196 - 96 = 100 = 10^2$
$= \text{perfect square}, \therefore$ simple factors exist.

$|ac| = 24$. Possible factors of 24 are $(1, 24)$, $(2, 12)$, $(3, 8)$ and $(4, 6)$.
$c$ is positive: $\therefore$ factors add up to $|b|$, i.e. 14 $\therefore$ $(2, 12)$
$c$ is positive: $\therefore$ both factors have same sign as $b$, i.e. positive.

$$\begin{aligned}\therefore\ 3x^2 + 14x + 8 &= 3x^2 + 2x + 12x + 8 = 0\\ &= x(3x + 2) + 4(3x + 2) = 0\\ &= (3x + 2)(x + 4) = 0\end{aligned}$$

$\therefore\ x + 4 = 0$ or $3x + 2 = 0 \ \therefore\ x = -4$ or $x = -\frac{2}{3}$

They are all done in the same way – provided that simple linear factors exist, so always test to begin with by evaluating $(b^2 - 4ac)$.

Here is another: work right through it on your own.

Solve the equation $4x^2 - 16x + 15 = 0$

$x = \ldots\ldots\ldots\ldots$ or $x = \ldots\ldots\ldots\ldots$

**8**

$$x = \frac{3}{2} \text{ or } x = \frac{5}{2}$$

## 2 Solution by completing the square

We have already seen that some quadratic equations are incapable of being factorized into two simple factors. In such cases, another method of solution must be employed. The following example will show the procedure.

Solve $x^2 - 6x - 4 = 0 \qquad a = 1, b = -6, c = -4.$

$(b^2 - 4ac) = 36 - 4 \times 1 \times (-4) = 36 + 16 = 52$. Not a perfect square.
$\therefore$ No simple factors.

So: $\qquad x^2 - 6x - 4 = 0$

Add 4 to both sides: $\qquad x^2 - 6x = 4$

Add to each side the square of half the coefficient of $x$:

$x^2 - 6x + (-3)^2 = 4 + (-3)^2$

$x^2 - 6x + 9 = 4 + 9 = 13$

$\therefore\ (x - 3)^2 = 13$

$\therefore\ x - 3 = \pm\sqrt{13}$. Remember to include the two signs.

$\therefore\ x = 3 \pm \sqrt{13} = 3 \pm 3{\cdot}6056 \ \therefore\ x = 6{\cdot}606$ or $x = -0{\cdot}606$

**9**

Now this one. Solve $x^2 + 8x + 5 = 0$ by the method of completing the square. First we take the constant term to the right-hand side:

$$x^2 + 8x + 5 = 0$$
$$x^2 + 8x \quad = -5$$

Then we add to both sides ...........

**10**

the square of half the coefficient of $x$

$$x^2 + 8x + 4^2 = -5 + 4^2$$
$$\therefore\ x^2 + 8x + 16 = -5 + 16 = 11$$
$$\therefore\ (x + 4)^2 = 11$$

And now we can finish it off, finally getting:

$x = \ldots\ldots\ldots\ldots$ or $x = \ldots\ldots\ldots\ldots$

**11**

$x = -0{\cdot}683$ or $x = -7{\cdot}317$

If the coefficient of the squared term is not unity, the first step is to divide both sides of the equation by the existing coefficient.

For example: $2x^2 + 10x - 7 = 0$

Dividing throughout by 2: $\quad x^2 + 5x - 3{\cdot}5 = 0$

We then proceed as before, which finally gives:

$x = \ldots\ldots\ldots\ldots$ or $x = \ldots\ldots\ldots\ldots$

**12**

$x = 0{\cdot}622$ or $x = -5{\cdot}622$

Because

$$x^2 + 5x - 3{\cdot}5 = 0$$
$$\therefore\ x^2 + 5x = 3{\cdot}5$$
$$x^2 + 5x + 2{\cdot}5^2 = 3{\cdot}5 + 6{\cdot}25 = 9{\cdot}75$$
$$\therefore\ (x + 2{\cdot}5)^2 = 9{\cdot}75$$
$$\therefore\ x + 2{\cdot}5 = \pm\sqrt{9{\cdot}75} = \pm 3{\cdot}122$$
$$\therefore\ x = -2{\cdot}5 \pm 3{\cdot}122 \ \therefore\ x = 0{\cdot}622 \text{ or } x = -5{\cdot}622$$

One more. Solve the equation $4x^2 - 16x + 3 = 0$ by completing the square.

$x = \ldots\ldots\ldots\ldots$ or $x = \ldots\ldots\ldots\ldots$

13

$$\boxed{x = 0{\cdot}197 \text{ or } x = 3{\cdot}803}$$

Because

$$4x^2 - 16x + 3 = 0 \qquad \therefore\ x^2 - 4x + 0{\cdot}75 = 0$$
$$\therefore\ x^2 - 4x = -0{\cdot}75$$
$$x^2 - 4x + (-2)^2 = -0{\cdot}75 + (-2)^2$$
$$x^2 - 4x + 4 = -0{\cdot}75 + 4 = 3{\cdot}25$$
$$\therefore\ (x-2)^2 = 3{\cdot}25$$
$$\therefore\ x - 2 = \pm\sqrt{3{\cdot}25} = \pm 1{\cdot}8028$$
$$\therefore\ x = 2 \pm 1{\cdot}803 \ \therefore\ x = 0{\cdot}197 \text{ or } x = 3{\cdot}803$$

*On to the next topic*

14

### 3 Solution by formula

We can establish a formula for the solution of the general quadratic equation $ax^2 + bx + c = 0$ which is based on the method of completing the square:

$$ax^2 + bx + c = 0$$

Dividing throughout by the coefficient of $x$, i.e. $a$:

$$x^2 + \frac{b}{a}x + \frac{c}{a} = 0$$

Subtracting $\frac{c}{a}$ from each side gives $x^2 + \frac{b}{a}x = -\frac{c}{a}$

We then add to each side the square of half the coefficient of $x$:

$$x^2 + \frac{b}{a}x + \left(\frac{b}{2a}\right)^2 = -\frac{c}{a} + \left(\frac{b}{2a}\right)^2$$
$$x^2 + \frac{b}{a}x + \frac{b^2}{4a^2} = \frac{b^2}{4a^2} - \frac{c}{a}$$
$$\left(x + \frac{b}{2a}\right)^2 = \frac{b^2 - 4ac}{4a^2}$$
$$\therefore\ x + \frac{b}{2a} = \pm\sqrt{\frac{b^2 - 4ac}{4a^2}} = \pm\frac{\sqrt{b^2 - 4ac}}{2a} \quad \therefore\ x = -\frac{b}{2a} \pm \frac{\sqrt{b^2 - 4ac}}{2a}$$
$$\therefore\ \text{If } ax^2 + bx + c = 0,\ x = \frac{-b \pm \sqrt{b^2 - 4ac}}{2a}$$

Substituting the values of $a$, $b$ and $c$ for any particular quadratic equation gives the solutions of the equation.

*Make a note of the formula: it is important*

15

As an example, we shall solve the equation $2x^2 - 3x - 4 = 0$.

Here $a = 2$, $b = -3$, $c = -4$ and $x = \dfrac{-b \pm \sqrt{b^2 - 4ac}}{2a}$

$$x = \frac{3 \pm \sqrt{9 - 4 \times 2 \times (-4)}}{4} = \frac{3 \pm \sqrt{9 + 32}}{4} = \frac{3 \pm \sqrt{41}}{4}$$
$$= \frac{3 \pm 6{\cdot}403}{4} = \frac{-3{\cdot}403}{4} \text{ or } \frac{9{\cdot}403}{4}$$
$$\therefore\ x = -0{\cdot}851 \text{ or } x = 2{\cdot}351$$

▶

It is just a case of careful substitution. You need, of course, to remember the formula. For

$ax^2 + bx + c + 0 \quad x = \ldots\ldots\ldots\ldots$

**16**

$$x = \frac{-b \pm \sqrt{b^2 - 4ac}}{2a}$$

As an exercise, use the formula method to solve the following:

(a) $5x^2 + 12x + 3 = 0$ (c) $x^2 + 15x - 7 = 0$
(b) $3x^2 - 10x + 4 = 0$ (d) $6x^2 - 8x - 9 = 0$

**17**

(a) $x = -2{\cdot}117$ or $x = -0{\cdot}283$ (c) $x = 0{\cdot}453$ or $x = -15{\cdot}453$
(b) $x = 0{\cdot}465$ or $x = 2{\cdot}869$ (d) $x = -0{\cdot}728$ or $x = 2{\cdot}061$

At this stage of our work, we can with advantage bring together a number of the items that we have studied earlier.

**18**

## Cubic equations having at least one simple linear factor

In Programme F.3, we dealt with the factorization of cubic polynomials, with application of the remainder theorem and factor theorem, and the evaluation of the polynomial functions by nesting. These we can now re-apply to the solution of cubic equations.

*So move on to the next frame for some examples*

**19**

**Example 1**

To solve the cubic equation:

$2x^3 - 11x^2 + 18x - 8 = 0$

The first step is to find a linear factor of the cubic expression:

$f(x) = 2x^3 - 11x^2 + 18x - 8$

by application of the remainder theorem. To facilitate the calculation, we first write $f(x)$ in nested form:

$f(x) = [(2x - 11)x + 18]x - 8$

Now we seek a value for $x$ $(x = k)$ which gives a zero remainder on division by $(x - k)$. We therefore evaluate $f(1)$, $f(-1)$, $f(2)\ldots$ etc.

$f(1) = 1 \;\therefore\; (x - 1)$ is not a factor of $f(x)$
$f(-1) = -39 \;\therefore\; (x + 1)$ is not a factor of $f(x)$
$f(2) = 0 \;\therefore\; (x - 2)$ *is* a factor of $f(x)$

We therefore divide $f(x)$ by $(x - 2)$ to determine the remaining factor, which is $\ldots\ldots\ldots\ldots$

**20**

$$\boxed{2x^2 - 7x + 4}$$

Because

$$\begin{array}{r|l} & 2x^2 \quad - \quad 7x \; + 4 \\ \hline x-2 & 2x^3 \; - 11x^2 \; + 18x \; - 8 \\ & 2x^3 \; - \;\; 4x^2 \\ \hline & \quad\;\; - \;\; 7x^2 \; + 18x \\ & \quad\;\; - \;\; 7x^2 \; + 14x \\ \hline & \qquad\qquad\quad 4x \; - 8 \\ & \qquad\qquad\quad 4x \; - 8 \\ \hline & \qquad\qquad\quad \bullet \quad \bullet \end{array}$$

$\therefore\ f(x) = (x-2)(2x^2 - 7x + 4)$ and the cubic equation is now written:

$(x-2)(2x^2 - 7x + 4) = 0$

which gives $x - 2 = 0$ or $2x^2 - 7x + 4 = 0$.

$\therefore\ x = 2$ and the quadratic equation can be solved in the usual way giving

$x = \ldots\ldots\ldots\ldots$ or $x = \ldots\ldots\ldots\ldots$

**21**

$$\boxed{x = 0{\cdot}719 \text{ or } x = 2{\cdot}781}$$

$$2x^2 - 7x + 4 = 0 \ \therefore\ x = \frac{7 \pm \sqrt{49 - 32}}{4} = \frac{7 \pm \sqrt{17}}{4} = \frac{7 \pm 4{\cdot}1231}{4}$$

$$= \frac{2{\cdot}8769}{4} \text{ or } \frac{11{\cdot}1231}{4}$$

$$\therefore\ x = 0{\cdot}719 \text{ or } x = 2{\cdot}781$$

$\therefore\ 2x^3 - 11x^2 + 18x - 8 = 0$ has the solutions

$$x = 2,\ x = 0{\cdot}719,\ x = 2{\cdot}781$$

The whole method depends on the given expression in the equation having at least one linear factor.

Here is another.

**Example 2**

Solve the equation $3x^3 + 12x^2 + 13x + 4 = 0$

First, in nested form $f(x) = \ldots\ldots\ldots\ldots$

**22**

$$\boxed{f(x) = [(3x + 12)x + 13]x + 4}$$

Now evaluate $f(1), f(-1), f(2), \ldots$

$f(1) = 32 \quad \therefore\ (x-1)$ is not a factor of $f(x)$

$f(-1) = \ldots\ldots\ldots\ldots$

**23**

$$\boxed{f(-1) = 0}$$

$\therefore\ (x+1)$ *is* a factor of $f(x)$. Then, by long division, the remaining factor of $f(x)$ is $\ldots\ldots\ldots\ldots$

**24**

$$\boxed{3x^2 + 9x + 4}$$

$\therefore$ The equation $3x^3 + 12x^2 + 13x + 4 = 0$ can be written

$$(x + 1)(3x^2 + 9x + 4) = 0$$

so that $x + 1 = 0$ or $3x^2 + 9x + 4 = 0$

which gives $x = -1$ or $x =$ ........... or $x =$ ...........

**25**

$$\boxed{x = -2{\cdot}457 \text{ or } x = -0{\cdot}543}$$

$$3x^2 + 9x + 4 = 0 \quad \therefore \ x = \frac{-9 \pm \sqrt{81 - 48}}{6} = \frac{-9 \pm \sqrt{33}}{6}$$

$$= \frac{-9 \pm 5{\cdot}7446}{6}$$

$$= \frac{-14{\cdot}7446}{6} \text{ or } \frac{-3{\cdot}2554}{6}$$

$$= -2{\cdot}4574 \text{ or } -0{\cdot}5426$$

The complete solutions of $3x^3 + 12x^2 + 13x + 4 = 0$ are

$x = -1$, $x = -2{\cdot}457$, $x = -0{\cdot}543$

And now one more.

**Example 3**

Solve the equation $5x^3 + 2x^2 - 26x - 20 = 0$

Working through the method step by step, as before:

$x =$ ........... or $x =$ ........... or $x =$ ...........

**26**

$$\boxed{x = -2 \text{ or } x = -0{\cdot}825 \text{ or } x = 2{\cdot}425}$$

Here is the working, just as a check:

$$f(x) = 5x^3 + 2x^2 - 26x - 20 = 0$$

$$\therefore \ f(x) = [(5x + 2)x - 26]x - 20$$

$f(1) = -39 \quad \therefore \ (x - 1)$ is not a factor of $f(x)$

$f(-1) = 3 \quad \therefore \ (x + 1)$ is not a factor of $f(x)$

$f(2) = -24 \quad \therefore \ (x - 2)$ is not a factor of $f(x)$

$f(-2) = 0 \quad \therefore \ (x + 2)$ *is* a factor of $f(x)$:

$$\begin{array}{r|l}
 & 5x^2 - 8x - 10 \\ \hline
x + 2 & 5x^3 + 2x^2 - 26x - 20 \\
 & 5x^3 + 10x^2 \\ \hline
 & \quad - 8x^2 - 26x \\
 & \quad - 8x^2 - 16x \\ \hline
 & \qquad - 10x - 20 \\
 & \qquad - 10x - 20 \\ \hline
 & \qquad \bullet \quad \bullet \\ \hline
\end{array}$$

$f(x) = (x+2)(5x^2 - 8x - 10) = 0 \quad \therefore\ x + 2 = 0$ or $5x^2 - 8x - 10 = 0$

$$\therefore\ x = -2 \text{ or } x = \frac{8 \pm \sqrt{64+200}}{10} = \frac{8 \pm \sqrt{264}}{10}$$

$$= \frac{8 \pm 16{\cdot}248}{10} = -0{\cdot}825 \text{ or } 2{\cdot}425$$

$\therefore\ x = -2$ or $x = -0{\cdot}825$ or $x = 2{\cdot}425$

*Next frame*

## Fourth-order equations having at least two linear factors

**27**

The method here is practically the same as that for solving cubic equations which we have just considered. The only difference is that we have to find two simple linear factors of $f(x)$. An example will reveal all.

### Example 1

To solve the equation $4x^4 - 19x^3 + 24x^2 + x - 10 = 0$

(a) As before, express the polynomial in nested form:

$$f(x) = \{[(4x-19)x + 24]x + 1\}x - 10$$

(b) Determine $f(1)$, $f(-1)$, $f(2)$ etc.

$f(1) = 0 \quad \therefore\ (x-1)$ *is* a factor of $f(x)$:

$$\begin{array}{r|llllll}
 & 4x^3 & -15x^2 & +\ 9x & +10 & \\
\hline
x-1 & 4x^4 & -19x^3 & +24x^2 & +\ x & -10 \\
 & 4x^4 & -\ 4x^3 & & & \\
\hline
 & & -15x^3 & +24x^2 & & \\
 & & -15x^3 & +15x^2 & & \\
\hline
 & & & 9x^2 & +\ x & \\
 & & & 9x^2 & -\ 9x & \\
\hline
 & & & & 10x & -10 \\
 & & & & 10x & -10 \\
\hline
 & & & & \bullet & \bullet
\end{array}$$

$\therefore\ f(x) = (x-1)(4x^3 - 15x^2 + 9x + 10) = 0$

$\therefore\ x = 1$ or $4x^3 - 15x^2 + 9x + 10 = 0$, i.e. $F(x) = 0$

$F(x) = [(4x - 15)x + 9]x + 10$

$F(1) = 8 \quad \therefore\ (x-1)$ is not a factor of $F(x)$

$F(-1) = -18 \quad \therefore\ (x+1)$ is not a factor of $F(x)$

$F(2) = 0 \quad \therefore\ (x-2)$ *is* a factor of $F(x)$

Division of $F(x)$ by $(x-2)$ by long division gives $(4x^2 - 7x - 5)$:

$F(x) = (x-2)(4x^2 - 7x - 5)$

$\therefore\ f(x) = (x-1)(x-2)(4x^2 - 7x - 5) = 0$

$\therefore\ x = 1,\ x = 2$, or $4x^2 - 7x - 5 = 0$

Solving the quadratic by formula gives solutions $x = -0{\cdot}545$ or $x = 2{\cdot}295$.

$\therefore\ x = 1,\ \ x = 2,\ \ x = -0{\cdot}545,\ \ x = 2{\cdot}295$

*Now let us deal with another example*

**28**

**Example 2**

Solve $2x^4 - 4x^3 - 23x^2 - 11x + 6 = 0$

$f(x) = \{[(2x-4)x-23]x-11\}x+6$

$f(1) = -30 \quad \therefore \ (x-1)$ is not a factor of $f(x)$

$f(-1) = 0 \quad \therefore \ (x+1)$ *is* a factor of $f(x)$:

$$
\begin{array}{r|rrrrr}
 & 2x^3 & -\ 6x^2 & -\ 17x & +\ 6 & \\
\hline
x+1 & 2x^4 & -\ 4x^3 & -\ 23x^2 & -\ 11x & +\ 6 \\
 & 2x^4 & +\ 2x^3 & & & \\
\hline
 & & -\ 6x^3 & -\ 23x^2 & & \\
 & & -\ 6x^3 & -\ 6x^2 & & \\
\hline
 & & & -\ 17x^2 & -\ 11x & \\
 & & & -\ 17x^2 & -\ 17x & \\
\hline
 & & & & 6x & +\ 6 \\
 & & & & 6x & +\ 6 \\
\hline
 & & & & \cdot & \cdot \\
\hline
\end{array}
$$

$\therefore \ f(x) = (x+1)(2x^3 - 6x^2 - 17x + 6) = 0$

$\therefore \ x = -1$ or $2x^3 - 6x^2 - 17x + 6 = 0$, i.e. $F(x) = 0$

$F(x) = [(2x-6)x-17]x+6$

$F(1) = -15 \quad \therefore \ (x-1)$ is not a factor of $F(x)$

Now you can continue and finish it off:

$x = -1, \ x = \ldots\ldots\ldots\ldots, \ x = \ldots\ldots\ldots\ldots, \ x = \ldots\ldots\ldots\ldots$

**29**

$$\boxed{x = -2, \ x = 0{\cdot}321, \ x = 4{\cdot}679}$$

Because

$F(-1) = 15 \quad \therefore \ (x+1)$ is not a factor of $F(x)$

$F(2) \ = -36 \quad \therefore \ (x-2)$ is not a factor of $F(x)$

$F(-2) = 0 \quad \therefore \ (x+2)$ *is* a factor of $F(x)$

Division of $F(x)$ by $(x+2)$ by long division gives:

$F(x) = (x+2)(2x^2 - 10x + 3)$

$\therefore \ f(x) = (x+1)(x+2)(2x^2 - 10x + 3) = 0$

$$\therefore \ x = -1, \ x = -2, \ x = \frac{10 \pm \sqrt{100-24}}{4} = \frac{10 \pm \sqrt{76}}{4}$$

$$= \frac{10 \pm 8{\cdot}7178}{4} = \frac{1{\cdot}2822}{4} \text{ or } \frac{18{\cdot}7178}{4}$$

$$= 0{\cdot}3206 \text{ or } 4{\cdot}6794$$

$$\therefore \ x = -1, \ x = -2, \ x = 0{\cdot}321, \ x = 4{\cdot}679$$

**30**

**Example 3**

Solve $f(x) = 0$ when $f(x) = 3x^4 + 2x^3 - 15x^2 + 12x - 2$

$$f(x) = 3x^4 + 2x^3 - 15x^2 + 12x - 2$$
$$= \{[(3x + 2)x - 15]x + 12\}x - 2$$

Now you can work through the solution, taking the same steps as before. There are no snags, so you will have no trouble:

$$x = \ldots\ldots\ldots\ldots$$

---

**31**

$$\boxed{x = 1,\ x = 1,\ x = -2{\cdot}897,\ x = 0{\cdot}230}$$

Because $f(1) = 0$, $\therefore$ $(x - 1)$ *is* a factor of $f(x)$:

$$\begin{array}{r|rrrrr}
 & 3x^3 & +\ 5x^2 & -\ 10x & +2 & \\
\hline
x-1 & 3x^4 & +2x^3 & -15x^2 & +12x & -2 \\
 & 3x^4 & -3x^3 & & & \\
\hline
 & & 5x^3 & -15x^2 & & \\
 & & 5x^3 & -\ \ 5x^2 & & \\
\hline
 & & & -10x^2 & +12x & \\
 & & & -10x^2 & +10x & \\
\hline
 & & & & 2x & -2 \\
 & & & & 2x & -2 \\
\hline
 & & & & \bullet & \bullet
\end{array}$$

$\therefore\ f(x) = (x - 1)(3x^3 + 5x^2 - 10x + 2) = (x - 1) \times F(x) = 0$

$F(x) = [(3x + 5)x - 10]x + 2$

$F(1) = 0 \quad \therefore\ (x - 1)$ *is* a factor of $F(x)$:

$$\begin{array}{r|rrrr}
 & 3x^2 & +\ 8x & -\ 2 & \\
\hline
x-1 & 3x^3 & +5x^2 & -10x & +2 \\
 & 3x^3 & -3x^2 & & \\
\hline
 & & 8x^2 & -10x & \\
 & & 8x^2 & -\ \ 8x & \\
\hline
 & & & -\ \ 2x & +2 \\
 & & & -\ \ 2x & +2 \\
\hline
 & & & \bullet & \bullet
\end{array}$$

$\therefore\ f(x) = (x - 1)(x - 1)(3x^2 + 8x - 2) = 0$

Solving the quadratic by formula gives $x = -2{\cdot}8968$ or $x = 0{\cdot}2301$

$\therefore\ x = 1,\ x = 1,\ x = -2{\cdot}897,\ x = 0{\cdot}230$

*All correct?*

**32**

Finally, one more for good measure.

**Example 4**

Solve the equation $2x^4 + 3x^3 - 13x^2 - 6x + 8 = 0$

Work through it using the same method as before.

$x = \dots\dots\dots\dots$

**33**

$$x = -1,\ x = 2,\ x = 0{\cdot}637,\ x = -3{\cdot}137$$

Here is an outline of the solution:

$$f(x) = 2x^4 + 3x^3 - 13x^2 - 6x + 8$$
$$= \{[(2x+3)x - 13]x - 6\}x + 8$$
$$f(-1) = 0 \quad \therefore \ (x+1) \text{ is a factor of } f(x).$$

Division of $f(x)$ by $(x+1)$ gives:

$$f(x) = (x+1)(2x^3 + x^2 - 14x + 8) = (x+1) \times F(x)$$
$$F(x) = 2x^3 + x^2 - 14x + 8 = [(2x+1)x - 14]x + 8$$
$$F(2) = 0 \quad \therefore \ (x-2) \text{ is a factor of } F(x).$$

Division of $F(x)$ by $(x-2)$ gives $F(x) = (x-2)(2x^2 + 5x - 4)$.

$$\therefore \ f(x) = (x+1)(x-2)(2x^2 + 5x - 4) = 0$$

Solution of the quadratic by formula shows $x = 0{\cdot}637$ or $x = -3{\cdot}137$.

$$\therefore \ x = -1,\ x = 2,\ x = 0{\cdot}637,\ x = -3{\cdot}137$$

**34**

The methods we have used for solving cubic and fourth-order equations have depended on the initial finding of one or more simple factors by application of the remainder theorem.

There are, however, many cubic and fourth-order equations which have no simple factors and other methods of solution are necessary. These are more advanced and will be dealt with later in the course.

## Review summary

**35**

1 *Quadratic equations* $ax^2 + bx + c = 0$

(a) *Solution by factors*

Test for availability of factors: evaluate $(b^2 - 4ac)$.
If the value of $(b^2 - 4ac)$ is a perfect square, simple linear factors exist. If $(b^2 - 4ac)$ is not a perfect square, there are no simple linear factors.

(b) *Solution by completing the square*

(i) Remove the constant term to the RHS.

(ii) Add to both sides the square of half the coefficient of $x$. This completes the square on the LHS.

(iii) Take the square root of each side – including both signs.

(iv) Simplify the results to find the values of $x$.

▶

(c) *Solution by formula*

Evaluate $x = \dfrac{-b \pm \sqrt{b^2 - 4ac}}{2a}$, to obtain two values for $x$.

2 *Cubic equations* – with at least one linear factor

(a) Rewrite the polynomial function, $f(x)$, in nested form.

(b) Apply the remainder theorem by substituting values for $x$ until a value, $x = k$, gives zero remainder. Then $(x - k)$ is a factor of $f(x)$.

(c) By long division, determine the remaining quadratic factor of $f(x)$, i.e. $(x - k)(ax^2 + bx + c) = 0$. Then $x = k$, and $ax^2 + bx + c = 0$ can be solved by the usual methods. $\therefore\ x = k,\ x = x_1,\ x = x_2$.

3 *Fourth-order equations* – with at least two linear factors.

(a) Find the first linear factor as in section 2 above.

(b) Divide by $(x - k)$ to obtain the remaining cubic expression. Then $f(x) = (x - k) \times F(x)$ where $F(x)$ is a cubic.

(c) $f(x) = 0 \quad \therefore\ (x - k) \times F(x) = 0 \quad \therefore\ x = k$ or $F(x) = 0$.

(d) The cubic $F(x)$ is now solved as in section 2 above, giving:

$$(x - m)(ax^2 + bx + c) = 0$$

$$\therefore\ x = k,\ x = m \text{ and } ax^2 + bx + c = 0$$

giving the four solutions $x = k$, $x = m$, $x = x_1$ and $x = x_2$.

---

## Review exercise

**36**

1 Solve for $x$ by factorizing: $12x^2 - 5x - 2 = 0$

2 Solve for $x$ (accurate to 3 dp) by completing the square:

$3x^2 - 4x - 2 = 0$

3 Solve for $x$ (accurate to 3 dp) by using the formula:

$x^2 - 3x + 1 = 0$

4 Solve for $x$: $x^3 - 5x^2 + 7x - 2 = 0$

5 Solve for $x$: $x^4 + x^3 - 2x^2 - x + 1 = 0$

---

**37**

1 $12x^2 - 5x - 2 = 0$

Here, $ax^2 + bx + c = 0$ where $a = 12$, $b = -5$, $c = -2$.

(a) Test for factors: $(b^2 - 4ac) = (-5)^2 - [4 \times 12 \times (-2)] = 25 + 96 = 121$ so that factors exist.

(b) $|ac| = 24$. Possible factors of 24 are (1, 24), (2, 12), (3, 8) and (4, 6).
$c$ is negative $\therefore$ factors differ by $|b| = 5$ $\therefore$ Required factors are (3, 8)
$c$ is negative $\therefore$ factors of different sign, the numerically larger having the same sign of $b$, that is negative, therefore:

$$\begin{aligned} 12x^2 - 5x - 2 &= 12x^2 - 8x + 3x - 2 \\ &= 4x(3x - 2) + (3x - 2) \\ &= (4x + 1)(3x - 2) = 0 \end{aligned}$$

So that $(4x + 1) = 0$ or $(3x - 2) = 0$. That is $x = -1/4$ or $x = 2/3$.

▶

2 $3x^2 - 4x - 2 = 0$, that is $x^2 - (4/3)x - (2/3) = 0$.

Here $a = 1$, $b = -4/3$, $c = -2/3$. So that $(b^2 - 4ac) = \frac{16}{9} + 4 \times 1 \times \frac{2}{3} = \frac{40}{9}$ – not a perfect square so we proceed by completing the square:

$$\begin{aligned} x^2 - (4/3)x - 2/3 &= 0 \\ x^2 - (4/3)x &= 2/3 \\ x^2 - (4/3)x + (2/3)^2 &= 2/3 + (2/3)^2 \\ x^2 - (4/3)x + 4/9 &= 10/9 \\ (x - 2/3)^2 &= 10/9 \end{aligned}$$

So that $x - 2/3 = \pm\sqrt{10/9}$ giving $x = 2/3 \pm \sqrt{10/9} = 0{\cdot}6667 \pm 1{\cdot}0541$ to 4 dp. Therefore $x = 1{\cdot}721$ or $x = -0{\cdot}387$.

3 $x^2 - 3x + 1 = 0$. Here $a = 1$, $b = -3$, $c = 1$ where:

$$x = \frac{-b \pm \sqrt{b^2 - 4ac}}{2a}$$

$$\text{so that } x = \frac{3 \pm \sqrt{9 - 4}}{2} = \frac{3 \pm \sqrt{5}}{2} = 2{\cdot}618 \text{ or } 0{\cdot}382$$

4 $x^3 - 5x^2 + 7x - 2 = 0$.

$$\begin{aligned} \text{Let } f(x) &= x^3 - 5x^2 + 7x - 2 \\ &= ((x - 5)x + 7)x - 2 \text{ so} \\ f(2) &= ((2 - 5)2 + 7)2 - 2 = 0 \end{aligned}$$

Thus $x - 2$ is a factor of $f(x)$ and so $x = 2$ is a solution of the cubic. To find the other two we first need to divide the cubic by this factor:

$$\frac{x^3 - 5x^2 + 7x - 2}{x - 2} = x^2 - 3x + 1$$

That is $x^3 - 5x^2 + 7x - 2 = (x - 2)(x^2 - 3x + 1) = 0$. So, from the previous question, the complete solution is $x = 2$, $x = 2{\cdot}6180$ or $x = 0{\cdot}3820$ to 4 dp.

5 $x^4 + x^3 - 2x^2 - x + 1 = 0$.

$$\begin{aligned} \text{Let } f(x) &= x^4 + x^3 - 2x^2 - x + 1 \\ &= (((x + 1)x - 2)x - 1)x + 1 \text{ so} \\ f(1) &= (((1 + 1)1 - 2)1 - 1)1 + 1 = 0 \text{ and} \\ f(-1) &= ((([-1] + 1)[-1] - 2)[-1] - 1)[-1] + 1 = 0 \end{aligned}$$

Thus $x - 1$ and $x + 1$ are factors of $f(x)$. Now $(x - 1)(x + 1) = x^2 - 1$ and:

$$\frac{x^4 + x^3 - 2x^2 - x + 1}{x^2 - 1} = x^2 + x - 1$$

so that $x^4 + x^3 - 2x^2 - x + 1 = (x^2 - 1)(x^2 + x - 1)$ giving the solution of $x^4 + x^3 - 2x^2 - x + 1 = 0$ as $x = \pm 1$, $x = 0{\cdot}618$ or $x = -1{\cdot}618$.

**38**

You have now come to the end of this Programme. A list of **Can You?** questions follows for you to gauge your understanding of the material in the Programme. You will notice that these questions match the **Learning outcomes** listed at the begining of the Programme so go back and try the **Quiz** that follows them. After that try the **Test exercise**. *Work through them at your own pace, there is no need to hurry.* A set of **Further problems** provides additional valuable practice.

# Can You?

## Checklist F.6 39

Check this list before you try the end of Programme test.

On a scale of 1 to 5 how confident are you that you can: Frames

- Solve quadratic equations by factors, by completing the square and by formula? 2 to 17
  *Yes* ☐ ☐ ☐ ☐ ☐ *No*
- Solve cubic equations with at least one linear factor? 18 to 26
  *Yes* ☐ ☐ ☐ ☐ ☐ *No*
- Solve fourth-order equations with at least two linear factors? 27 to 33
  *Yes* ☐ ☐ ☐ ☐ ☐ *No*

# Test exercise F.6

**40**

1 Solve the following equations by the method of factors: Frames

(a) $x^2 + 11x + 18 = 0$ (d) $2x^2 + 13x + 20 = 0$

(b) $x^2 - 13x + 42 = 0$ (e) $3x^2 + 5x - 12 = 0$

(c) $x^2 + 4x - 21 = 0$ (f) $5x^2 - 26x + 24 = 0$ 2 to 7

2 Solve the following by completing the square:

(a) $2x^2 - 4x - 3 = 0$ (c) $3x^2 + 12x - 18 = 0$

(b) $5x^2 + 10x + 2 = 0$ 8 to 13

3 Solve by means of the formula:

(a) $4x^2 + 11x + 2 = 0$ (c) $5x^2 - 9x - 4 = 0$

(b) $6x^2 - 3x - 5 = 0$ 14 to 17

4 Solve the cubic equation:

$2x^3 + 11x^2 + 6x - 3 = 0$ 18 to 26

5 Solve the fourth-order equation:

$4x^4 - 18x^3 + 23x^2 - 3x - 6 = 0$ 27 to 33

# Further problems F.6

**41**

1 Solve the following by the method of factors:

(a) $x^2 + 3x - 40 = 0$ (c) $x^2 + 10x + 24 = 0$

(b) $x^2 - 11x + 28 = 0$ (d) $x^2 - 4x - 45 = 0$

2 Solve the following:

(a) $4x^2 - 5x - 6 = 0$ (d) $7x^2 + 4x - 5 = 0$

(b) $3x^2 + 7x + 3 = 0$ (e) $6x^2 - 15x + 7 = 0$

(c) $5x^2 + 8x + 2 = 0$ (f) $8x^2 + 11x - 3 = 0$

▶

3 Solve the following cubic equations:

(a) $5x^3 + 14x^2 + 7x - 2 = 0$
(b) $4x^3 + 7x^2 - 6x - 5 = 0$
(c) $3x^3 - 2x^2 - 21x - 10 = 0$
(d) $2x^3 + 4x^2 - 3x - 3 = 0$
(e) $4x^3 + 2x^2 - 17x - 6 = 0$
(f) $4x^3 - 7x^2 - 17x + 6 = 0$

4 Solve the following fourth-order equations:

(a) $2x^4 - 4x^3 - 23x^2 - 11x + 6 = 0$
(b) $5x^4 + 8x^3 - 8x^2 - 8x + 3 = 0$
(c) $4x^4 - 3x^3 - 30x^2 + 41x - 12 = 0$
(d) $2x^4 + 14x^3 + 33x^2 + 31x + 10 = 0$
(e) $3x^4 - 6x^3 - 25x^2 + 44x + 12 = 0$
(f) $5x^4 - 12x^3 - 6x^2 + 17x + 6 = 0$
(g) $2x^4 - 16x^3 + 37x^2 - 29x + 6 = 0$

**42**

Now visit the companion website for this book at www.palgrave.com/stroud for more questions applying this mathematics to science and engineering.

# Binomials

## Learning outcomes

When you have completed this Programme you will be able to:

- ☐ Define $n!$ and recognize that there are $n!$ different arrangements of $n$ different items
- ☐ Evaluate $n!$ for moderately sized $n$ using a calculator
- ☐ Manipulate expressions containing factorials
- ☐ Recognize that there are $\dfrac{n!}{(n-r)!r!}$ combinations of $r$ identical items in $n$ locations
- ☐ Recognize simple properties of combinatorial coefficients
- ☐ Construct Pascal's triangle
- ☐ Write down the binomial expansion for natural number powers
- ☐ Obtain specific terms in the binomial expansion using the general term
- ☐ Use the sigma notation
- ☐ Recognize and reproduce the expansion for $e^x$ where $e$ is the exponential number

If you already feel confident about these why not try the short quiz over the page? You can check your answers at the end of the book.

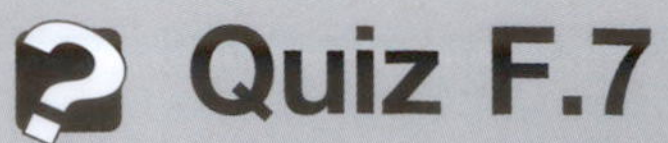

# Quiz F.7

**Frames**

1 How many combinations are there of 7 identical bottles of wine arranged in a wine rack with spaces for 12 bottles? — 1 to 3

2 Find the value of:

(a) $6!$ (b) $11!$ (c) $\dfrac{15!}{12!}$ (d) $(13-7)!$ (e) $\dfrac{0!}{3!}$ — 4 to 9

3 Evaluate each of the following:

(a) ${}^{4}C_{2}$ (b) ${}^{10}C_{6}$ (c) ${}^{89}C_{0}$ (d) ${}^{36}C_{35}$ — 10 to 18

4 Expand $(3a+4b)^4$ as a binomial series. — 22 to 31

5 In the binomial expansion of $(1-x/2)^9$ written in terms of ascending powers of $x$, find:

(a) the 4th term

(b) the coefficient of $x^5$. — 32 to 33

6 Evaluate:

(a) $\displaystyle\sum_{r=1}^{20} r$ (b) $\displaystyle\sum_{r=1}^{n}(2r+3)$ — 37 to 47

7 Determine the 6th term and the sum of the first 10 terms of the series: $2+4+6+8+\ldots\ldots\ldots\ldots$ — 37 to 47

8 Using the series expansion of $e^x$ find $e^{-0{\cdot}45}$ accurate to 3 decimal places. — 48 to 50

# Factorials and combinations

## Factorials

**1**

Answer this question. How many different three-digit numbers can you construct using the three numerals 5, 7 and 8 once each?

*The answer is in the next frame*

**2**

6

They are: 578 587
758 785
857 875

Instead of listing them like this you can work it out. There are 3 choices for the first numeral and for each choice there are a further 2 choices for the second numeral. That is, there are:

$3 \times 2 = 6$ choices of first and second numeral combined

The third numeral is then the one that is left.

So, how many four-digit numbers can be constructed using the numerals 3, 5, 7 and 9 once each?

*Answer in the next frame*

**3**

$4 \times 3 \times 2 \times 1 = 24$

Because

The first numeral can be selected in one of 4 ways, each selection leaving 3 ways to select the second numeral. So there are:

$4 \times 3 = 12$ ways of selecting the first two numerals

Each combination of the first two numerals leaves 2 ways to select the third numeral. So there are:

$4 \times 3 \times 2 = 24$ ways of selecting the first three numerals

The last numeral is the one that is left so there is only one choice for that. Therefore, there are:

$4 \times 3 \times 2 \times 1 = 24$ choices of of first, second, third and fourth numeral combined.

Can you see the pattern here? If you have $n$ different items then you can form

$n \times (n - 1) \times (n - 2) \times \ldots \times 2$

different arrangements using each item just once.

This type of product of decreasing natural numbers occurs quite often in mathematics so a general notation has been devised. For example, the product:

$3 \times 2 \times 1$

is called *3-factorial* and is written as 3!

So the value of 5! is ...........

*Next frame*

**4**

$\boxed{120}$

Because

$5! = 5 \times 4 \times 3 \times 2 \times 1 = 120$

The factorial expression can be generalized to the factorial of an arbitrary natural number $n$ as:

$n! = n \times (n-1) \times (n-2) \times \ldots \times 2 \times 1$

To save time a calculator can be used to evaluate $n!$ for moderately sized $n$. For example, on your calculator:

Enter the number 9
Press the $x!$ key followed by =
The display changes to 362880. that is:

$9! = 9 \times 8 \times 7 \times 6 \times 5 \times 4 \times 3 \times 2 \times 1 = 362\,880$

So the value of 11! is . . . . . . . . . . .

**5**

$\boxed{39\,916\,800}$

For reasons that will soon become clear the value of 0! is defined to be 1. Try it on your calculator.

Try a few examples. Evaluate each of the following:

(a) $6!$ (b) $\dfrac{8!}{3!}$ (c) $(7-2)!$ (d) $\dfrac{3!}{0!}$ (e) $\dfrac{5!}{(5-2)!2!}$

**6**

| (a) 720 | (b) 6720 | (c) 120 | (d) 6 | (e) 10 |
|---|---|---|---|---|

Because

(a) $6! = 6 \times 5 \times 4 \times 3 \times 2 \times 1 = 720$

(b) $\dfrac{8!}{3!} = \dfrac{8 \times 7 \times 6 \times 5 \times 4 \times 3 \times 2 \times 1}{3 \times 2 \times 1} = 8 \times 7 \times 6 \times 5 \times 4 = 6720$

(c) $(7-2)! = 5! = 120$

(d) $\dfrac{3!}{0!} = \dfrac{6}{1} = 6$ Notice that we have used the fact that $0! = 1$

(e) $\dfrac{5!}{(5-2)!2!} = \dfrac{5!}{3!2!} = \dfrac{5 \times 4 \times 3 \times 2 \times 1}{(3 \times 2 \times 1) \times (2 \times 1)} = \dfrac{5 \times 4}{(2 \times 1)} = 10$

Now try some for general $n$. Because

$$\begin{aligned} n! &= n \times (n-1) \times (n-2) \times \ldots \times 2 \times 1 \\ &= n \times (n-1)! \end{aligned}$$

then:

$$\frac{n!}{(n-1)!} = \frac{n \times (n-1)!}{(n-1)!} = n$$

Also note that while

$$2n! = 2 \times n!, (2n)! = (2n) \times (2n-1) \times (2n-2) \times \ldots \times 2 \times 1$$

so that, for example:

$$\begin{aligned}\frac{(2n+1)!}{(2n-1)!} &= \frac{(2n+1) \times (2n+1-1) \times (2n+1-2) \times \ldots \times 2 \times 1}{(2n-1) \times (2n-1-1) \times (2n-1-2) \times \ldots \times 2 \times 1}\\ &= \frac{(2n+1) \times (2n) \times (2n-1) \times \ldots \times 2 \times 1}{(2n-1) \times (2n-2) \times (2n-3) \times \ldots \times 2 \times 1}\\ &= (2n+1) \times (2n)\end{aligned}$$

So simplify each of these:

(a) $\dfrac{n!}{(n+1)!}$ (b) $\dfrac{(n+1)!}{(n-1)!}$ (c) $\dfrac{(2n)!}{(2n+2)!}$

*Answers in the next frame*

---

**7**

(a) $\dfrac{1}{n+1}$

(b) $n(n+1)$

(c) $\dfrac{1}{(2n+2) \times (2n+1)}$

Because

(a) $\dfrac{n!}{(n+1)!} = \dfrac{n \times (n-1) \times (n-2) \times \ldots \times 2 \times 1}{(n+1) \times n \times (n-1) \times (n-2) \times \ldots \times 2 \times 1} = \dfrac{1}{(n+1)}$

(b) $\dfrac{(n+1)!}{(n-1)!} = \dfrac{(n+1) \times (n) \times (n-1)!}{(n-1)!} = (n+1) \times (n) = n(n+1)$

(c) $\dfrac{(2n)!}{(2n+2)!} = \dfrac{(2n)!}{(2n+2) \times (2n+1) \times (2n)!} = \dfrac{1}{(2n+2) \times (2n+1)}$

---

**8**

Try some more. Write each of the following in factorial form:

(a) $4 \times 3 \times 2 \times 1$

(b) $6 \times 5 \times 4$

(c) $\dfrac{(7 \times 6) \times (3 \times 2 \times 1)}{2}$

*Next frame for the answers*

---

**9**

(a) $4!$ (b) $\dfrac{6!}{3!}$ (c) $\dfrac{7! \times 3!}{5! \times 2!}$

Because

(a) $4 \times 3 \times 2 \times 1 = 4!$ by the definition of the factorial

(b) $6 \times 5 \times 4 = \dfrac{6 \times 5 \times 4 \times 3 \times 2 \times 1}{3 \times 2 \times 1} = \dfrac{6!}{3!}$

(c) $$\begin{aligned}\frac{(7 \times 6) \times (3 \times 2 \times 1)}{2} &= \frac{(7 \times 6 \times 5 \times 4 \times 3 \times 2 \times 1) \times (3 \times 2 \times 1)}{(5 \times 4 \times 3 \times 2 \times 1) \times (2 \times 1)}\\ &= \frac{7! \times 3!}{5! \times 2!}\end{aligned}$$

*Now let's use these ideas. Next frame*

**10**

## Combinations

In mathematics, arranging (also called permuting) and combining are two related but distinct activities. Items are combined in no particular order but arrangements (also called permutations) are all about order. For example, there is only one combination of the first three letters of the alphabet if we are not concerned about their order whereas there are six arrangements if we are, namely:

*abc*, *acb*, *bac*, *bca*, *cab* and *cba*

In an arrangement the order in which items are listed is relevant but in a combination the order is irrelevant. So that, whilst *abc* and *cba* form distinct arrangements, they both form the same combination if we are not concerned about the order of the letters.

Let's assume you have two identical pegs and a board with five holes each capable of taking a peg. What we want to know is how many combinations there are of two pegs amongst the five holes in the board. One possible arrangement could have the pegs in the second and fifth holes:

Another arrangement could have the pegs in the first and fourth holes:

How many arrangements are there of two pegs in five holes?

*The answer is in the next frame*

**11**

$$5 \times 4 = 20$$

Because

There are 5 holes into which the first peg can be placed and for each first selection there are 4 holes left in which to insert the second peg. This gives a total of $5 \times 4 = 20$ possible two-peg arrangements.

However, not all two-peg arrangements are different. For example, if you made

*your first choice as hole 2 and your second choice as hole 5*

this would be the same as making

*your first choice as hole 5 and your second choice as hole 2*

because the two pegs are identical. So every two-peg arrangement is duplicated because the order in which the two holes are selected in this problem is not relevant.

How many different two-peg arrangements are there?

**12**

$$\boxed{\frac{5 \times 4}{2} = 10}$$

Because each two-peg arrangement is duplicated there are not 20 but 10 different arrangements. List them:

| | | | |
|---|---|---|---|
| hole 1, hole 2 | hole 1, hole 3 | hole 1, hole 4 | hole 1, hole 5 |
| | hole 2, hole 3 | hole 2, hole 4 | hole 2, hole 5 |
| | | hole 3, hole 4 | hole 3, hole 5 |
| | | | hole 4, hole 5 |

*There are 10 different ways of combining 2 identical pegs amongst 5 holes.*

The expression $\frac{5 \times 4}{2}$ can be written in factorial form as follows:

$$\frac{5 \times 4}{2} = \frac{5 \times 4}{2 \times 1} = \frac{5 \times 4 \times 3 \times 2 \times 1}{(3 \times 2 \times 1)(2 \times 1)} = \frac{5!}{3!2!}$$

or better still:

$$\frac{5 \times 4}{2} = \frac{5!}{3!2!} = \frac{5!}{(5-2)!2!}$$

because this just contains the numbers 5 and 2 and will link to a general notation in a later Frame.

*There are* $\frac{5!}{(5-2)!2!}$ *different ways of combining 2 identical pegs amongst 5 holes.*

So now, if you were to combine 3 identical pegs amongst the 5 holes how many three-peg arrangements could you make? (Some arrangements will contain the same pegs but in a different order.)

*Next frame*

**13**

$$\boxed{5 \times 4 \times 3 = 60}$$

Because

There are 5 holes from which to make a first selection and for each first selection there are 4 holes left from which to make the second selection and then 3 holes left from which to make the third selection. This gives a total of $5 \times 4 \times 3 = 60$ possible three-peg arrangements.

However, within the 60 arrangements every three-peg arrangement is repeated but with the pegs in a different order. In this problem the order in which the pegs are arranged is irrelevant because the pegs are identical. So we need to find out how many times each three-peg arrangement is repeated. To find the number of repeats consider any single three-peg selection and work out how many different arrangements of these three pegs can be made. Think of the three-peg arrangement as having the pegs in the first, second and third holes. The first hole can be taken by any one of the 3 pegs leaving the second hole to be taken by one of the remaining 2 pegs. Finally, the third hole is taken by the 1 peg that is left so there are:

$$3 \times 2 \times 1 = 6$$

different arrangements of any single three-peg combination. Consequently the total number of combinations of three pegs is:

. . . . . . . . . . . .

*The answer is in the next frame*

**14**

$$\boxed{\frac{5 \times 4 \times 3}{3 \times 2 \times 1} = 10}$$

Because

Any one three-peg arrangement can be rearranged within itself $3 \times 2 \times 1 = 6$ times. This means that every combination of three specific pegs appears 6 times in the list of 60 possible three-peg arrangements. Therefore there are:

$$\frac{5 \times 4 \times 3}{3 \times 2 \times 1} = \frac{60}{6} = 10$$

combinations of 3 pegs amongst the 5 holes.

Written in factorial form

$$\frac{5 \times 4 \times 3}{3 \times 2 \times 1} = \ldots\ldots\ldots\ldots$$

*Next frame*

**15**

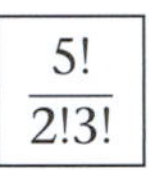

Because

$$\frac{5 \times 4 \times 3}{3 \times 2 \times 1} = \frac{5 \times 4 \times 3 \times 2 \times 1}{(2 \times 1)(3 \times 2 \times 1)}$$
$$= \frac{5!}{2!3!}$$

Or, better still $\dfrac{5!}{2!3!} = \dfrac{5!}{(5-3)!3!}$ because it only contains the numbers mentioned in the problem, namely 5 and 3.

Can you see a pattern emerging here?

There are $\dfrac{5!}{(5-2)!2!}$ different combinations of *two* identical items in *five* different places.

There are $\dfrac{5!}{(5-3)!3!}$ different combinations of *three* identical items in *five* different places.

So if you have $r$ identical items to be located in $n$ different places where $n \geq r$, the number of combinations is

$\ldots\ldots\ldots\ldots$

*The answer is in the next frame*

**16**

$$\frac{n!}{(n-r)!r!}$$

Because

The first item can be located in any one of $n$ places.
The second item can be located in any one of the remaining $n-1$ places.
The third item can be located in any one of the remaining $n-2$ places.

⋮

The $r$th item can be located in any one of the remaining $n-(r-1)$ places.

This means that there are $n(n-1)(n-2)\dots(n-(r-1)) = \frac{n!}{(n-r)!}$ arrangements.
However, every arrangement is repeated $r!$ times. (Any one arrangement can be rearranged within itself $r!$ times.) So the total number of combinations is given as:

$$\frac{n!}{(n-r)!r!}$$

This particular ratio of factorials is called a *combinatorial coefficient* and is denoted by $^nC_r$:

$$^nC_r = \frac{n!}{(n-r)!r!} \text{ where } 0 \leq r \leq n$$

So, evaluate each of the following:

(a) $^6C_3$ (b) $^7C_2$ (c) $^4C_4$ (d) $^3C_0$ (e) $^5C_1$

**17**

(a) 20 (b) 21 (c) 1 (d) 1 (e) 5

Because

(a) $^6C_3 = \frac{6!}{(6-3)!3!} = \frac{6!}{3!3!} = \frac{720}{36} = 20$

(b) $^7C_2 = \frac{7!}{(7-2)!2!} = \frac{7!}{5!2!} = \frac{5040}{120 \times 2} = 21$

(c) $^4C_4 = \frac{4!}{(4-4)!4!} = \frac{4!}{0!4!} = \frac{4!}{4!} = 1$ Remember $0! = 1$

(d) $^3C_0 = \frac{3!}{(3-0)!0!} = \frac{3!}{3!0!} = 1$

(e) $^5C_1 = \frac{5!}{(5-1)!1!} = \frac{5!}{4!1!} = \frac{5!}{4!} = 5$

## Three properties of combinatorial coefficients

**18**

**1** $^nC_n = {^nC_0} = 1$

This is quite straightforward to prove:

$$^nC_n = \frac{n!}{(n-n)!n!} = \frac{n!}{0!n!} = \frac{n!}{n!} = 1 \text{ and } {^nC_0} = \frac{n!}{(n-0)!0!} = \frac{n!}{n!0!} = \frac{n!}{n!} = 1$$

**2** $^nC_{n-r} = {^nC_r}$. This is a little more involved:

$$^nC_{n-r} = \frac{n!}{(n-[n-r])!(n-r)!} = \frac{n!}{(n-n+r)!(n-r)!} = \frac{n!}{r!(n-r)!} = {^nC_r}$$

▶

3 $^nC_r + {^nC_{r+1}} = {^{n+1}C_{r+1}}$

This requires some care to prove:

$$^nC_r + {^nC_{r+1}} = \frac{n!}{(n-r)!r!} + \frac{n!}{(n-[r+1])!(r+1)!}$$

$$= \frac{n!}{(n-r)!r!} + \frac{n!}{(n-r-1)!(r+1)!}$$

$$= \frac{n!}{(n-r-1)!r!}\left(\frac{1}{n-r} + \frac{1}{r+1}\right)$$ Taking out common factors where $(n-r)! = (n-r)(n-r-1)!$ and $(r+1)! = (r+1)r!$

$$= \frac{n!}{(n-r-1)!r!}\left(\frac{r+1+n-r}{(n-r)(r+1)}\right)$$ Adding the two fractions together

$$= \frac{n!}{(n-r)!(r+1)!}(n+1)$$ Because $(n-r-1)!(n-r) = (n-r)!$ and $r!(r+1) = (r+1)!$

$$= \frac{(n+1)!}{(n-r)!(r+1)!}$$ Because $n!(n+1) = (n+1)!$

$$= \frac{(n+1)!}{([n+1]-[r+1])!(r+1)!}$$ Because $[n+1]-[r+1] = n-r$

$$= {^{n+1}C_{r+1}}$$

***At this point let us pause and summarize the main facts on factorials and combinations***

---

## Review summary

**19**

1 The product $n \times (n-1) \times (n-2) \times \ldots \times 2 \times 1$ is called $n$-factorial and is denoted by $n!$

2 The number of different arrangements of $n$ different items is $n!$

3 The number of combinations of $r$ identical items among $n$ locations is given by the combinatorial coefficient $^nC_r = \dfrac{n!}{(n-r)!r!}$ where $0 \leq r \leq n$

4 $^nC_n = {^nC_0} = 1$

5 $^nC_{n-r} = {^nC_r}$

6 $^nC_r + {^nC_{r+1}} = {^{n+1}C_{r+1}}$

---

## Review exercise

**20**

1 Find the value of:

(a) $5!$ (b) $10!$ (c) $\dfrac{13!}{8!}$ (d) $(9-3)!$ (e) $0!$

2 How many combinations are there of 9 identical umbrellas on a rack of 15 coat hooks?

3 Evaluate each of the following:

(a) $^3C_1$ (b) $^9C_4$ (c) $^{100}C_0$ (d) $^{50}C_{50}$

4 Without evaluating, explain why each of the following equations is true:

(a) $^{15}C_5 = {^{15}C_{10}}$ (b) $^5C_3 + {^5C_4} = {^6C_4}$

21

1 (a) $5! = 5 \times 4 \times 3 \times 2 \times 1 = 120$
(b) $10! = 3\,628\,800$ (use a calculator)
(c) $\dfrac{13!}{8!} = 13 \times 12 \times 11 \times 10 \times 9 = 154\,440$
(d) $(9-3)! = 6! = 720$
(e) $0! = 1$

2 There are ${}^{15}C_9$ combinations of 9 identical umbrellas on a rack of 15 coat hooks, where:

$${}^{15}C_9 = \frac{15!}{(15-9)!9!} = \frac{15!}{6!9!} = 5005$$

3 (a) ${}^{3}C_1 = \dfrac{3!}{(3-1)!1!} = \dfrac{3!}{2!1!} = 3$
(b) ${}^{9}C_4 = \dfrac{9!}{(9-4)!4!} = \dfrac{9!}{5!4!} = 126$
(c) ${}^{100}C_0 = 1$
(d) ${}^{50}C_{50} = 1$

4 (a) ${}^{15}C_5 = {}^{15}C_{10}$ because ${}^{n}C_{n-r} = {}^{n}C_r$ where $n = 15$ and $r = 10$
(b) ${}^{5}C_3 + {}^{5}C_4 = {}^{6}C_4$ because ${}^{n}C_r + {}^{n}C_{r+1} = {}^{n+1}C_{r+1}$ where $n = 5$ and $r = 3$

*The next topic is related to this last one*

# Binomial expansions

22

## Pascal's triangle

The following triangular array of combinatorial coefficients can be constructed where the *superscript* to the left of each coefficient indicates the row number and the *subscript* to the right indicates the column number:

| *Row* | *Column* | | | | |
|---|---|---|---|---|---|
| | *0* | *1* | *2* | *3* | *4* |
| *0* | ${}^{0}C_0$ | | | | |
| *1* | ${}^{1}C_0$ | ${}^{1}C_1$ | | | |
| *2* | ${}^{2}C_0$ | ${}^{2}C_1$ | ${}^{2}C_2$ | | |
| *3* | ${}^{3}C_0$ | ${}^{3}C_1$ | ${}^{3}C_2$ | ${}^{3}C_3$ | |
| *4* | ...... | ...... | ...... | ...... | ...... |

Follow the pattern and fill in the next row.

*The answer is in the following frame*

23

${}^{4}C_0$, ${}^{4}C_1$, ${}^{4}C_2$, ${}^{4}C_3$ and ${}^{4}C_4$

Because

The superscript indicates row 4 and the subscripts indicate columns 0 to 4.

The pattern devised in this array can be used to demonstrate the third property of combinatorial coefficients that you considered in Frame 18, namely that:

$${}^{n}C_r + {}^{n}C_{r+1} = {}^{n+1}C_{r+1}$$

▶

In the following array, arrows have been inserted to indicate that any coefficient is equal to the coefficient immediately above it plus the one above it and to the left.

| *Row* | *Column* | | | | | |
|---|---|---|---|---|---|---|
| | *0* | *1* | *2* | *3* | *4* | |
| *0* | ${}^0C_0$ | | | | | |
| *1* | ${}^1C_0$ | ${}^1C_1$ | | | | |
| *2* | ${}^2C_0$ | ${}^2C_1$ | ${}^2C_2$ | | | ${}^1C_0 + {}^1C_1 = {}^{1+1}C_{0+1} = {}^2C_1$ |
| *3* | ${}^3C_0$ | ${}^3C_1$ | ${}^3C_2$ | ${}^3C_3$ | | ${}^2C_0 + {}^2C_1 = {}^{2+1}C_{0+1} = {}^3C_1$ etc. |
| *4* | ${}^4C_0$ | ${}^4C_1$ | ${}^4C_2$ | ${}^4C_3$ | ${}^4C_4$ | ${}^3C_0 + {}^3C_1 = {}^{3+1}C_{0+1} = {}^4C_1$ etc. |

Now you have already seen from the first property of the combinatorial coefficients in Frame 18 that ${}^nC_n = {}^nC_0 = 1$ so the values of some of these coefficients can be filled in immediately:

| *Row* | *Column* | | | | |
|---|---|---|---|---|---|
| | *0* | *1* | *2* | *3* | *4* |
| *0* | 1 | | | | |
| *1* | 1 | 1 | | | |
| *2* | 1 | ${}^2C_1$ | 1 | | |
| *3* | 1 | ${}^3C_1$ | ${}^3C_2$ | 1 | |
| *4* | 1 | ${}^4C_1$ | ${}^4C_2$ | ${}^4C_3$ | 1 |

Fill in the numerical values of the remaining combinatorial coefficients using the fact that ${}^nC_r + {}^nC_{r+1} = {}^{n+1}C_{r+1}$: that is, any coefficient is equal to the sum of the number immediately above it and the one above it and to the left.

*The answer is in Frame 24*

## 24

| *Row* | *Column* | | | | | |
|---|---|---|---|---|---|---|
| | *0* | *1* | *2* | *3* | *4* | |
| *0* | 1 | | | | | |
| *1* | 1 | 1 | | | | |
| *2* | 1 | 2 | 1 | | | $1 + 1 = 2$ |
| *3* | 1 | 3 | 3 | 1 | | $1 + 2 = 3,\ 2 + 1 = 3$ |
| *4* | 1 | 4 | 6 | 4 | 1 | $1 + 3 = 4,\ 3 + 3 = 6,\ 3 + 1 = 4$ |

This is called *Pascal's triangle.*

The numbers on row 5, reading from left to right are ...........

*Answers are in the next frame*

## 25

**1, 5, 10, 10, 5, 1**

Because

The first number is ${}^5C_0 = 1$ and then $1 + 4 = 5$, $4 + 6 = 10$, $6 + 4 = 10$, $4 + 1 = 5$. Finally, ${}^5C_5 = 1$.

*Now let's move on to a related topic in the next frame*

## Binomial expansions

26

A *binomial* is the sum of a pair of numbers raised to a power. In this Programme we shall only consider natural number powers, namely, binomials of the form:

$(a+b)^n$

where $n$ is a natural number. In particular, look at the following expansions:

$(a+b)^1 = a+b$ Note the coefficients 1, 1

$(a+b)^2 = a^2+2ab+b^2$ Note the coefficients 1, 2, 1

$(a+b)^3 = a^3+3a^2b+3ab^2+b^3$ Note the coefficients 1, 3, 3, 1

So what is the expansion of $(a+b)^4$ and what are the coefficients?

*Next frame for the answer*

27

$(a+b)^4 = a^4+4a^3b+6a^2b^2+4ab^3+b^4$
Coefficients 1, 4, 6, 4, 1

Because

$$\begin{aligned}(a+b)^4 &= (a+b)^3(a+b)\\ &= (a^3+3a^2b+3ab^2+b^3)(a+b)\\ &= a^4+3a^3b+3a^2b^2+ab^3+a^3b+3a^2b^2+3ab^3+b^4\\ &= a^4+4a^3b+6a^2b^2+4ab^3+b^4\end{aligned}$$

Can you see the connection with Pascal's triangle?

| *Row* | *Column* 0 | 1 | 2 | 3 | 4 | |
|---|---|---|---|---|---|---|
| 0 | 1 | | | | | |
| 1 | 1 | 1 | | | | Coefficients of $(a+b)^1$ |
| 2 | 1 | 2 | 1 | | | Coefficients of $(a+b)^2$ |
| 3 | 1 | 3 | 3 | 1 | | Coefficients of $(a+b)^3$ |
| 4 | 1 | 4 | 6 | 4 | 1 | Coefficients of $(a+b)^4$ |

So what are the coefficients of $(a+b)^5$?

*Next frame for the answer*

28

1, 5, 10, 10, 5, 1

Because

The values in the next row of Pascal's triangle are 1, 5, 10, 10, 5, 1 and the numbers in row 5 are also the values of the coefficients of the binomial expansion:

$$\begin{aligned}(a+b)^5 &= a^5+5a^4b+10a^3b^2+10a^2b^3+5ab^4+b^5\\ &= 1a^5b^0+5a^4b^1+10a^3b^2+10a^2b^3+5a^1b^4+1a^0b^5\end{aligned}$$

Notice how in this expansion, as the power of $a$ decreases, the power of $b$ increases so that in each term the two powers add up to 5.

▶

Because the numbers in row 5 are also the values of the appropriate combinatorial coefficients this expansion can be written as:

$$(a+b)^5 = {}^5C_0a^5b^0 + {}^5C_1a^4b^1 + {}^5C_2a^3b^2 + {}^5C_3a^2b^3 + {}^5C_4a^1b^4 + {}^5C_5a^0b^5$$

The general binomial expansion for natural number $n$ is then given as:

$$\begin{aligned}(a+b)^n &= {}^nC_0a^nb^0 + {}^nC_1a^{n-1}b^1 + {}^nC_2a^{n-2}b^2 + \ldots + {}^nC_na^{n-n}b^n \\ &= a^n + {}^nC_1a^{n-1}b + {}^nC_2a^{n-2}b^2 + \ldots + b^n\end{aligned}$$

Now ${}^nC_1 = \dfrac{n!}{(n-1)!1!} = n$ and ${}^nC_2 = \dfrac{n!}{(n-2)!2!} = \dfrac{n(n-1)}{2!}$ so ${}^nC_3 = \ldots\ldots\ldots\ldots$

*Next frame for the answer*

**29**

$$\boxed{\frac{n(n-1)(n-2)}{3!}}$$

Because

$${}^nC_3 = \frac{n!}{(n-3)!3!} = \frac{n(n-1)(n-2)}{3!}$$

From this, the binomial expansion can be written as:

$$(a+b)^n = a^n + na^{n-1}b + \frac{n(n-1)}{2!}a^{n-2}b^2 + \frac{n(n-1)(n-2)}{3!}a^{n-3}b^3 + \ldots + b^n$$

So use this form of the binomial expansion to expand $(a+b)^6$.

*Next frame for the answer*

**30**

$$\boxed{(a+b)^6 = a^6 + 6a^5b + 15a^4b^2 + 20a^3b^3 + 15a^2b^4 + 6ab^5 + b^6}$$

Because

$$\begin{aligned}(a+b)^6 &= a^6 + 6a^5b + \frac{6\times5}{2!}a^4b^2 + \frac{6\times5\times4}{3!}a^3b^3 \\ &\quad + \frac{6\times5\times4\times3}{4!}a^2b^4 + \frac{6\times5\times4\times3\times2}{5!}ab^5 \\ &\quad + \frac{6\times5\times4\times3\times2\times1}{6!}b^6 \\ &= a^6 + 6a^5b + \frac{6\times5}{2}a^4b^2 + \frac{6\times5\times4}{6}a^3b^3 \\ &\quad + \frac{6\times5\times4\times3}{4\times3\times2\times1}a^2b^4 + \frac{6\times5\times4\times3\times2}{5\times4\times3\times2\times1}ab^5 + \frac{6!}{6!}b^6 \\ &= a^6 + 6a^5b + 15a^4b^2 + 20a^3b^3 + 15a^2b^4 + 6ab^5 + b^6\end{aligned}$$

Notice that 1, 6, 15, 20, 15, 6 and 1 are the numbers in row 6 of Pascal's triangle.

There are now two ways of obtaining the binomial expansion of $(a+b)^n$:

1 Use Pascal's triangle. This is appropriate when $n$ is small

2 Use the combinatorial coefficients. This is appropriate when $n$ is large

So, expand each of the following binomials:

(a) $(1+x)^7$ using Pascal's triangle

(b) $(3-2x)^4$ using the combinatorial coefficients

*Next frame for the answer*

31

(a) $(1+x)^7 = 1 + 7x + 21x^2 + 35x^3 + 35x^4 + 21x^5 + 7x^6 + x^7$
(b) $(3-2x)^4 = 81 - 216x + 216x^2 - 96x^3 + 16x^4$

Because

(a) Using Pascal's triangle:

$$\begin{aligned}(1+x)^7 &= 1 + (1+6)x + (6+15)x^2 + (15+20)x^3 \\ &\quad + (20+15)x^4 + (15+6)x^5 + (6+1)x^6 + x^7 \\ &= 1 + 7x + 21x^2 + 35x^3 + 35x^4 + 21x^5 + 7x^6 + x^7\end{aligned}$$

(b) Using the general form of the binomial expansion:

$$\begin{aligned}(3-2x)^4 &= (3+[-2x])^4 \\ &= 3^4 + 4 \times 3^3 \times (-2x) + \frac{4 \times 3}{2!} \times 3^2 \times (-2x)^2 \\ &\quad + \frac{4 \times 3 \times 2}{3!} \times 3 \times (-2x)^3 + \frac{4 \times 3 \times 2 \times 1}{4!} \times (-2x)^4 \\ &= 3^4 + 3^3 \times (-8x) + 6 \times 3^2 \times (4x^2) + 4 \times 3 \times (-8x^3) + (16x^4) \\ &= 81 - 216x + 216x^2 - 96x^3 + 16x^4\end{aligned}$$

## The general term of the binomial expansion

32

In Frame 28 we found that the binomial expansion of $(a+b)^n$ is given as:

$$(a+b)^n = {}^nC_0 a^n b^0 + {}^nC_1 a^{n-1} b^1 + {}^nC_2 a^{n-2} b^2 + \ldots + {}^nC_n a^{n-n} b^n$$

Each term of this expansion looks like ${}^nC_r a^{n-r} b^r$ where the value of $r$ ranges progressively from $r = 0$ to $r = n$ (there are $n+1$ terms in the expansion). Because the expression ${}^nC_r a^{n-r} b^r$ is typical of each and every term in the expansion we call it the *general term* of the expansion.

Any *specific* term can be derived from the general term. For example, consider the case when $r = 2$. The general term then becomes:

$${}^nC_2 a^{n-2} b^2 = \frac{n!}{(n-2)!2!} a^{n-2} b^2 = \frac{n(n-1)}{2!} a^{n-2} b^2$$

and this is the *third* term of the expansion:

The 3rd term is obtained from the general term by letting $r = 2$.

Consequently, we can say that:

${}^nC_r a^{n-r} b^r$ represents the $(r+1)$th term in the expansion for $0 \le r \le n$

Let's look at an example. To find the 10th term in the binomial expansion of $(1+x)^{15}$ written in ascending powers of $x$, we note that $a = 1$, $b = x$, $n = 15$ and $r + 1 = 10$ so $r = 9$. This gives the 10th term as:

$$\begin{aligned}{}^{15}C_9 1^{15-9} x^9 &= \frac{15!}{(15-9)!9!} 1^{15-9} x^9 \\ &= \frac{15!}{6!9!} x^9 \\ &= 5005x^9 \quad \text{obtained using a calculator}\end{aligned}$$

Try this one yourself. The 8th term in the binomial expansion of

$\left(2 - \frac{x}{3}\right)^{12}$ is . . . . . . . . . . .

*The answer is in the next frame*

**33**

$$\boxed{-\frac{2816}{243}x^7}$$

Because

Here $a = 2, b = -x/3, n = 12$ and $r + 1 = 8$ so $r = 7$. The 8th term is:

$$\begin{aligned}
{}^{12}C_7 2^{12-7}(-x/3)^7 &= \frac{12!}{(12-7)!7!}2^5(-x/3)^7 \\
&= \frac{12!}{5!7!}32 \times \frac{x^7}{(-3)^7} \\
&= \frac{792 \times 32}{-2187}x^7 \\
&= -\frac{25\,344}{2187}x^7 \\
&= -\frac{2816}{243}x^7
\end{aligned}$$

***At this point let us pause and summarize the main facts on binomial expansions***

##  Review summary

**34**

1 Row $n$ in Pascal's triangle contains the coefficients of the binomial expansion of $(a+b)^n$.

2 An alternative form of the binomial expansion of $(a+b)^n$ is given in terms of combinatorial coefficients as:

$$\begin{aligned}
(a+b)^n &= {}^nC_0a^nb^0 + {}^nC_1a^{n-1}b^1 + {}^nC_2a^{n-2}b^2 + \ldots + {}^nC_na^{n-n}b^n \\
&= a^n + na^{n-1}b + \frac{n(n-1)}{2!}a^{n-2}b^2 + \frac{n(n-1)(n-2)}{3!}a^{n-3}b^3 \ldots + b^n
\end{aligned}$$

##  Review exercise

**35**

1 Using Pascal's triangle write down the binomial expansion of:

(a) $(a+b)^6$ (b) $(a+b)^7$

2 Write down the binomial expansion of each of the following:

(a) $(1+x)^5$ (b) $(2+3x)^4$ (c) $(2-x/2)^3$

3 In the binomial expansion of $\left(2-\frac{3}{x}\right)^8$ written in terms of descending powers of $x$, find:

(a) the 4th term (b) the coefficient of $x^{-4}$

**36**

1 From Pascal's triangle:

(a) $(a+b)^6 = a^6b^0 + 6a^5b^1 + 15a^4b^2 + 20a^3b^3 + 15a^2b^4 + 6a^1b^5 + a^0b^6$
$= a^6 + 6a^5b + 15a^4b^2 + 20a^3b^3 + 15a^2b^4 + 6ab^5 + b^6$

(b) $(a+b)^7 = a^7 + 7a^6b + 21a^5b^2 + 35a^4b^3 + 35a^3b^4 + 21a^2b^5 + 7ab^6 + b^7$

2 (a) $(1+x)^5 = 1^5 + 5\times 1^4x^1 + 10\times 1^3x^2 + 10\times 1^2x^3 + 5\times 1^1x^4 + x^5$
$= 1 + 5x + 10x^2 + 10x^3 + 5x^4 + x^5$

(b) $(2+3x)^4 = 2^4 + 4\times 2^3(3x)^1 + 6\times 2^2(3x)^2 + 4\times 2^1(3x)^3 + (3x)^4$
$= 16 + 96x + 216x^2 + 216x^3 + 81x^4$

(c) $(2-x/2)^3 = 2^3 + 3\times 2^2(-x/2)^1 + 3\times 2^1(-x/2)^2 + (-x/2)^3$
$= 8 - 6x + 3x^2/2 - x^3/8$

3 In the binomial expansion of $\left(2-\dfrac{3}{x}\right)^8$:

(a) The 4th term is derived from the general term ${}^nC_ra^{n-r}b^r$ where $a=2$, $b=-3/x$, $n=8$ and $r+1=4$ so $r=3$. That is, the 4th term is:

$$\begin{aligned}{}^8C_32^{8-3}(-3/x)^3 &= \frac{8!}{(8-3)!3!}2^5(-3/x)^3\\ &= \frac{8!}{5!3!}\times 32\times(-27/x^3)\\ &= -\frac{56}{x^3}\times 864 = -\frac{48\,384}{x^3}\end{aligned}$$

(b) The coefficient of $x^{-4}$ is derived from the general term when $r=4$. That is: ${}^8C_42^{8-4}(-3/x)^4$, giving the coefficient as:

$$\begin{aligned}{}^8C_42^{8-4}(-3)^4 &= \frac{8!}{(8-4)!4!}2^4(-3)^4\\ &= 70\times 16\times 81 = 90\,720\end{aligned}$$

## The $\sum$ (sigma) notation

**37**

The binomial expansion of $(a+b)^n$ is given as a sum of terms:

$$(a+b)^n = {}^nC_0a^nb^0 + {}^nC_1a^{n-1}b^1 + {}^nC_2a^{n-2}b^2 + \ldots + {}^nC_na^{n-n}b^n$$

Instead of writing down each term in the sum in this way a shorthand notation has been devised. We write down the general term and then use the Greek letter $\sum$ (sigma) to denote the sum. That is:

$$\begin{aligned}(a+b)^n &= {}^nC_0a^nb^0 + {}^nC_1a^{n-1}b^1 + {}^nC_2a^{n-2}b^2 + \ldots + {}^nC_na^{n-n}b^n\\ &= \sum_{r=0}^{n}{}^nC_ra^{n-r}b^r\end{aligned}$$

Here the Greek letter $\sum$ denotes a *sum of terms* where the typical term is ${}^nC_ra^{n-r}b^r$ and where the value of $r$ ranges in integer steps from $r=0$, as indicated at the bottom of the sigma, to $r=n$, as indicated on the top of the sigma.

▶

One immediate benefit of this notation is that it permits further properties of the combinatorial coefficients to be proved. For example:

$\sum_{r=0}^{n} {}^nC_r = 2^n$ The sum of the numbers in any row of Pascal's triangle is equal to 2 raised to the power of the row number

This is easily proved using the fact that:

$(a+b)^n = \sum_{r=0}^{n} {}^nC_r a^{n-r} b^r$. When $a = 1$ and $b = 1$, then:

$$(1+1)^n = 2^n = \sum_{r=0}^{n} {}^nC_r 1^{n-r} 1^r = \sum_{r=0}^{n} {}^nC_r$$

## 38 General terms

It is necessary for you to acquire the ability to form the general term from a sum of specific terms and so write the sum of specific terms using the sigma notation. To begin, consider the sum of the first $n$ even numbers:

$2 + 4 + 6 + 8 + \ldots\ldots\ldots\ldots$

Every even integer is divisible by 2 so every even integer can be written in the form $2r$ where $r$ is some integer. For example:

$8 = 2 \times 4$ so here $8 = 2r$ where $r = 4$

Every odd integer can be written in the form $2r - 1$ or as $2r + 1$, as an even integer minus or plus 1. For example:

$13 = 14 - 1 = 2 \times 7 - 1$ so that $13 = 2r - 1$ where $r = 7$

Alternatively:

$13 = 12 + 1 = 2 \times 6 + 1$ so that $13 = 2r + 1$ where $r = 6$

Writing (a) 16, 248, $-32$ each in the form $2r$, give the value of $r$ in each case.
(b) 21, 197, $-23$ each in the form $2r - 1$, give the value of $r$ in each case.

*The answer is in the next frame*

## 39

| (a) $r = 8$ | (b) $r = 11$ |
|---|---|
| $r = 124$ | $r = 99$ |
| $r = -16$ | $r = -11$ |

Because

(a) $16 = 2 \times 8 = 2r$ where $r = 8$
$248 = 2 \times 124 = 2r$ where $r = 124$
$-32 = 2 \times (-16)$ where $r = -16$

(b) $21 = 22 - 1 = 2 \times 11 - 1 = 2r - 1$ where $r = 11$
$197 = 198 - 1 = 2 \times 99 - 1 = 2r - 1$ where $r = 99$
$-23 = -22 - 1 = 2 \times (-11) - 1 = 2r - 1$ where $r = -11$

*Next frame*

40

We saw in Frame 37 that we can use the sigma notation to denote sums of general terms. We shall now use the notation to denote sums of terms involving integers. For example, in the sum of the odd natural numbers:

$1 + 3 + 5 + 7 + 9 + \dots\dots\dots\dots$

the general term can now be denoted by $2r - 1$ where $r \geq 1$. The symbol $\sum$ can then be used to denote a sum of terms of which the general term is typical:

$1 + 3 + 5 + 7 + 9 + \dots\dots\dots\dots = \sum(2r - 1)$

We can now also denote the range of terms over which we wish to extend the sum by inserting the appropriate values of the *counting number* $r$ below and above the sigma sign. For example:

$\sum_{r=1}^{7}(2r - 1)$ indicates the sum of 7 terms where $r$ ranges from $r = 1$ to $r = 7$.

That is:

$$\sum_{r=1}^{7}(2r - 1) = 1 + 3 + 5 + 7 + 9 + 11 + 13 = 49$$

Now you try some. In each of the following write down the general term and then write down the sum of the first 10 terms using the sigma notation:

(a) $2 + 4 + 6 + 8 + \dots$　　(b) $1 + \frac{1}{2} + \frac{1}{3} + \frac{1}{4} + \dots$

(c) $1 + 8 + 27 + 64 + \dots$　　(d) $-1 + 2 - 3 + 4 - 5 + \dots$

(e) $\frac{1}{0!} + \frac{1}{1!} + \frac{1}{2!} + \frac{1}{3!} + \dots$

*Answers in the next frame*

41

(a) $2r, \sum_{r=1}^{10} 2r$　　(b) $\frac{1}{r}, \sum_{r=1}^{10} \frac{1}{r}$

(c) $r^3, \sum_{r=1}^{10} r^3$　　(d) $(-1)^r r, \sum_{r=1}^{10} (-1)^r r$

(e) $\frac{1}{r!}, \sum_{r=0}^{9} \frac{1}{r!}$

Because

(a) This is the sum of the first 10 even numbers and every number is divisible by 2 so the general even number can be denoted by $2r$, giving the sum as $\sum_{r=1}^{10} 2r$.

(b) This is the sum of the first 10 reciprocals and the general reciprocal can be denoted by $\frac{1}{r}$ where $r \neq 0$. This gives the sum as $\sum_{r=1}^{10} \frac{1}{r}$.

(c) This is the sum of the first 10 numbers cubed and every number can be denoted by $r^3$, giving the sum as $\sum_{r=1}^{10} r^3$.

▶

(d) Here, every odd number is preceded by a minus sign. This can be denoted by $(-1)^r$ because when $r$ is even $(-1)^r = 1$ and when $r$ is odd $(-1)^r = -1$. This permits the general term to be written as $(-1)^r r$ and the sum is $\sum_{r=1}^{10}(-1)^r r$.

(e) This is the sum of the first 10 reciprocal factorials and the general reciprocal factorial can be denoted by $\frac{1}{r!}$, giving the sum as $\sum_{r=0}^{9}\frac{1}{r!}$. Notice that the sum of the first 10 terms starts with $r = 0$ and ends with $r = 9$.

## 42

If the sum of terms is required up to some final but unspecified value of the counting variable $r$, say $r = n$, then the symbol $n$ is placed on the top of the sigma. For example, the sum of the first $n$ terms of the series with general term $r^2$ is given by:

$$\sum_{r=1}^{n} r^2 = 1^2 + 2^2 + 3^2 + \ldots + n^2$$

Try some examples. In each of the following write down the general term and then write down the sum of the first $n$ terms using the sigma notation:

(a) $\frac{5}{2}+\frac{5}{4}+\frac{5}{6}+\frac{5}{8}+\ldots$ (b) $1+\frac{1}{3^2}+\frac{1}{5^2}+\frac{1}{7^2}+\ldots$

(c) $2-4+6-8+\ldots$ (d) $1-3+9-27+\ldots$

(e) $1-1+1-1+\ldots$

*Answers in the next frame*

## 43

(a) $\frac{5}{2r}, \sum_{r=1}^{n}\frac{5}{2r}$ (b) $\frac{1}{(2r-1)^2}, \sum_{r=1}^{n}\frac{1}{(2r-1)^2}$

(c) $(-1)^{r+1}2r, \sum_{r=1}^{n}(-1)^{r+1}2r$ (d) $(-1)^r 3^r, \sum_{r=0}^{n-1}(-1)^r 3^r$

(e) $(-1)^r, \sum_{r=0}^{n-1}(-1)^r$

Because

(a) Each term is of the form of 5 divided by an even number.

(b) Each term is of the form of the reciprocal of an odd number squared. Notice that the first term could be written as $\frac{1}{1^2}$ to maintain the pattern.

(c) Here the alternating sign is positive for every odd term ($r$ odd) and negative for every even term ($r$ even). Consequently, to force a positive sign for $r$ odd we must raise $-1$ to an even power – hence $r + 1$ which is even when $r$ is odd and odd when $r$ is even.

(d) Here the counting starts at $r = 0$ for the first term so while odd terms are preceded by a minus sign the value of $r$ is even. Also the $n$th term corresponds to $r = n - 1$.

(e) Again, the $n$th term corresponds to $r = n - 1$ as the first term corresponds to $r = 0$.

## The sum of the first *n* natural numbers

**44**

Consider the sum of the first $n$ non-zero natural numbers:

$\sum_{r=1}^{n} r = 1 + 2 + 3 + \ldots + n$. This can equally well be written as:

$\sum_{r=1}^{n} r = n + (n-1) + (n-2) + \ldots + 1$ starting with $n$ and working backwards.

If these two are added together term by term then:

$$2\sum_{r=1}^{n} r = (1+n) + (2+n-1) + (3+n-2) + \ldots + (n+1)$$

That is:

$$2\sum_{r=1}^{n} r = (n+1) + (n+1) + (n+1) + \ldots + (n+1) \qquad (n+1) \text{ added } n \text{ times}$$

That is:

$2\sum_{r=1}^{n} r = n(n+1)$ so that $\sum_{r=1}^{n} r = \dfrac{n(n+1)}{2}$ the sum of the first $n$ non-zero natural numbers.

*This is a useful formula to remember so make a note of it in your workbook.*

*Now for two rules to be used when manipulating sums. Next frame*

## Rules for manipulating sums

**45**

**Rule 1**

If $f(r)$ is some general term and $k$ is a constant then:

$$\begin{aligned}\sum_{r=1}^{n} kf(r) &= kf(1) + kf(2) + kf(3) + \ldots + kf(n)\\ &= k(f(1) + f(2) + f(3) + \ldots + f(n))\\ &= k\sum_{r=1}^{n} f(r)\end{aligned}$$

*Common constants can be factored out of the sigma.*

In particular, when $f(r) = 1$ for all values of $r$:

$$\begin{aligned}\sum_{r=1}^{n} k &= k\sum_{r=1}^{n} 1\\ &= k(1 + 1 + \ldots + 1) \qquad k \text{ multiplied by 1 added to itself } n \text{ times}\\ &= kn\end{aligned}$$

**Rule 2**

If $f(r)$ and $g(r)$ are two general terms then:

$$\begin{aligned}\sum_{r=1}^{n}(f(r) + g(r)) &= \{f(1) + g(1) + f(2) + g(2) + \ldots\}\\ &= \{f(1) + f(2) + \ldots\} + \{g(1) + g(2) + \ldots\}\\ &= \sum_{r=1}^{n} f(r) + \sum_{r=1}^{n} g(r)\end{aligned}$$

*Now for two worked examples*

**46**

**Example 1**

Find the value of the first 100 natural numbers (excluding zero).

*Solution*

$$\begin{aligned}1+2+3+\ldots+100 &= \sum_{r=1}^{100} r\\ &= \frac{100(100+1)}{2} \quad \text{using the formula in Frame 44}\\ &= 5050\end{aligned}$$

**Example 2**

Find the value of $\sum_{r=1}^{n}(6r+5)$

*Solution*

$$\begin{aligned}\sum_{r=1}^{n}(6r+5) &= \sum_{r=1}^{n}6r+\sum_{r=1}^{n}5 \quad \text{by Rule 2}\\ &= 6\sum_{r=1}^{n}r+\sum_{r=1}^{n}5 \quad \text{by Rule 1}\\ &= 6\frac{n(n+1)}{2}+5n \quad \text{using the formulas in Frames 44 and 45}\\ &= 3(n^2+n)+5n\\ &= 3n^2+8n\\ &= n(3n+8)\end{aligned}$$

So the values of:

(a) $\sum_{r=1}^{50} r$

(b) $\sum_{r=1}^{n}(8r-7)$ are ...........

*Answers in next frame*

**47**

(a) 1275
(b) $n(4n-3)$

Because

(a) $\sum_{r=1}^{50} r = \frac{50\times 51}{2} = 1275$

(b) $\sum_{r=1}^{n}(8r-7) = 8\sum_{r=1}^{n}r - \sum_{r=1}^{n}7 = 4n^2-3n = n(4n-3)$

*Now for the last topic*

## The exponential number e

48

The binomial expansion of $\left(1+\frac{1}{n}\right)^n$ is given as:

$$\left(1+\frac{1}{n}\right)^n = 1+n\left(\frac{1}{n}\right)+\frac{n(n-1)}{2!}\left(\frac{1}{n}\right)^2+\frac{n(n-1)(n-2)}{3!}\left(\frac{1}{n}\right)^3+\ldots+\left(\frac{1}{n}\right)^n$$
$$= 1+\frac{n}{n}+\frac{n(n-1)}{2!n^2}+\frac{n(n-1)(n-2)}{3!n^3}+\ldots+\frac{1}{n^n}$$
$$= 1+1+\frac{(1-1/n)}{2!}+\frac{(1-1/n)(1-2/n)}{3!}+\ldots+\frac{1}{n^n}$$

This expansion is true for any natural number value of $n$, large or small, but when $n$ is a large natural number then $\frac{1}{n}$ is a small number. If we now let the value of $n$ increase then, as it does so, the value of $\frac{1}{n}$ decreases. Indeed, the larger the value of $n$ becomes, the closer $\frac{1}{n}$ becomes to zero and the closer the expansion above comes to the sum of terms:

$$1+1+\frac{(1-0)}{2!}+\frac{(1-0)(1-0)}{3!}+\ldots = 1+1+\frac{1}{2!}+\frac{1}{3!}+\ldots$$
$$= \frac{1}{0!}+\frac{1}{1!}+\frac{1}{2!}+\frac{1}{3!}+\ldots$$

Here the *ellipsis* (...) at the end of the expansion means that the expansion never ends – we say that it has an *infinite number of terms*. Indeed, we can use the sigma notation here and write:

$$\frac{1}{0!}+\frac{1}{1!}+\frac{1}{2!}+\frac{1}{3!}+\ldots = \sum_{r=0}^{\infty}\frac{1}{r!}$$

Notice the symbol for infinity ($\infty$) at the top of the sigma; this denotes the fact that the sum is a sum of an infinite number of terms.

It can be shown that this sum is a finite number which is denoted by $e$, the exponential number, whose value is 2·7182818...
That is:

$$\sum_{r=0}^{\infty}\frac{1}{r!} = e$$

You will find this number on your calculator:

Press the $e^x$ key

Enter the number 1 followed by = and the value of $e$ is displayed.

In Part II we shall show that:

$$e^x = \sum_{r=0}^{\infty}\frac{x^r}{r!} = 1+x+\frac{x^2}{2!}+\frac{x^3}{3!}+\frac{x^4}{4!}+\ldots$$

*Next frame*

**49**

Use the series expansion to find the value of $e^{0\cdot1}$ accurate to 3 sig fig.

Since:

$$e^x = 1 + x + \frac{x^2}{2!} + \frac{x^3}{3!} + \frac{x^4}{4!} + \ldots \text{ then:}$$

$$e^{0\cdot1} = 1 + 0\cdot1 + \frac{(0\cdot1)^2}{2!} + \frac{(0\cdot1)^3}{3!} + \frac{(0\cdot1)^4}{4!} + \ldots$$

$$= 1 + 0\cdot1 + 0\cdot005 + 0\cdot0001\dot{6} + \ldots$$ subsequent terms will not affect the value to 3 sig fig

$$= 1\cdot11 \text{ to 3 sig fig}$$

So, using the series expansion of $e^x$, it is found that the value of $e^{-0\cdot25}$ to 3 dp is . . . . . . . . . . .

*Next frame for the answer*

**50**

0·779

Because

$$e^{-0\cdot25} = 1 + (-0\cdot25) + \frac{(-0\cdot25)^2}{2!} + \frac{(-0\cdot25)^3}{3!} + \ldots$$

$$= 1 - 0\cdot25 + 0\cdot03125 - 0\cdot0026 + \ldots$$ subsequent terms will not affect the value to 3 dp

$$= 1\cdot03125 - 0\cdot2526$$

$$= 0\cdot779 \text{ to 3 dp}$$

Check this answer using your calculator.

***At this point let us pause and summarize the main facts on the sigma notation and the series expansion of $e^x$.***

##  Review summary

**51**

1 The sigma notation is used as a shorthand notation for the sum of a number of terms, each term typified by a general term:

$$\sum_{r=1}^{n} f(r) = f(1) + f(2) + f(3) + \ldots + f(n)$$

2 The binomial expansion can be written using the sigma notation as:

$$(a+b)^n = \sum_{r=0}^{n} {}^nC_r a^{n-r} b^r$$

3 The exponential expression $e^x$ is given as a sum of an infinite number of terms:

$$e^x = \sum_{r=0}^{\infty} \frac{x^r}{r!} + 1 + x + \frac{x^2}{2!} + \frac{x^3}{3!} + \frac{x^4}{4!} + \ldots$$

## Review exercise

**52**

1 Evaluate:

(a) $\sum_{r=1}^{36} r$

(b) $\sum_{r=1}^{n}(4r+2)$

2 Determine the 5th term and the sum of the first 20 terms of the series: $3+6+9+12+\ldots$

3 Using the series expansion of $e^x$ find $e^{0\cdot 23}$ accurate to 3 decimal places.

**53**

1 (a) $\sum_{r=1}^{n} r = \dfrac{n(n+1)}{2}$ so that $\sum_{r=1}^{36} r = \dfrac{36(37)}{2} = 666$

(b) $\sum_{r=1}^{n}(4r+2) = \sum_{r=1}^{n} 4r + \sum_{r=1}^{n} 2$

$$= 4\sum_{r=1}^{n} r + (2+2+\ldots+2) \quad \text{2 added to itself } n \text{ times}$$

$$= 4\frac{n(n+1)}{2} + 2n$$

$$= 2n^2 + 4n$$

$$= 2n(n+2)$$

2 The 5th term of $3+6+9+12+\ldots$ is 15 because the general term is $3r$. The sum of the first 20 terms of the series is then:

$$\sum_{r=1}^{20} 3r = 3\sum_{r=1}^{20} r$$

$$= 3\frac{20(21)}{2}$$

$$= 630$$

3 Given that:

$$e^x = \sum_{r=0}^{\infty} \frac{x^r}{r!} = 1 + x + \frac{x^2}{2!} + \frac{x^3}{3!} + \ldots \text{ then:}$$

$$e^{0\cdot 23} = \sum_{r=1}^{\infty} \frac{(0\cdot 23)^r}{r!} = 1 + 0\cdot 23 + \frac{(0\cdot 23)^2}{2!} + \frac{(0\cdot 23)^3}{3!} + \frac{(0\cdot 23)^4}{4!} + \ldots$$

$$= 1 + 0\cdot 23 + 0\cdot 02645 + 0\cdot 002028 + 0\cdot 000117 + \ldots$$

$$= 1\cdot 259 \text{ to 3 dp}$$

**54**

You have now come to the end of this Programme. A list of **Can You?** questions follows for you to gauge your understanding of the material in the Programme. You will notice that these questions match the **Learning outcomes** listed at the beginning of the Programme so go back and try the **Quiz** that follows them. After that try the **Test exercise**. *Work through these at your own pace, there is no need to hurry.* A set of **Further problems** provides additional valuable practice.

## Can You?

**55**

### Checklist F.7

Check this list before and after you try the end of Programme test.

On a scale of 1 to 5 how confident are you that you can: Frames

- Define $n!$ and recognize that there are $n!$ different arrangements of $n$ different items? [1] to [3]
  *Yes* ☐ ☐ ☐ ☐ ☐ *No*
- Evaluate $n!$ for moderately sized $n$ using a calculator? [4] to [5]
  *Yes* ☐ ☐ ☐ ☐ ☐ *No*
- Manipulate expressions containing factorials? [6] to [9]
  *Yes* ☐ ☐ ☐ ☐ ☐ *No*
- Recognize that there are $\dfrac{n!}{(n-r)!r!}$ combinations of $r$ identical items in $n$ locations? [10] to [17]
  *Yes* ☐ ☐ ☐ ☐ ☐ *No*
- Recognize simple properties of combinatorial coefficients? [18]
  *Yes* ☐ ☐ ☐ ☐ ☐ *No*
- Construct Pascal's triangle? [22] to [25]
  *Yes* ☐ ☐ ☐ ☐ ☐ *No*
- Write down the binomial expansion for natural number powers? [26] to [31]
  *Yes* ☐ ☐ ☐ ☐ ☐ *No*
- Obtain specific terms in the binomial expansion using the general term? [32] to [33]
  *Yes* ☐ ☐ ☐ ☐ ☐ *No*
- Use the sigma notation? [37] to [47]
  *Yes* ☐ ☐ ☐ ☐ ☐ *No*
- Recognize and reproduce the expansion for $e^x$ where $e$ is the exponential number? [48] to [50]
  *Yes* ☐ ☐ ☐ ☐ ☐ *No*

## Test exercise F.7

**56**

Frames

1 How many combinations are there of 6 different numbers selected from the numbers 1 to 49 if the order in which the selection is made does not matter? [1] to [3]

2 Find the value of: [4] to [9]

(a) $8!$ (b) $10!$ (c) $\dfrac{17!}{14!}$ (d) $(15-11)!$ (e) $\dfrac{4!}{0!}$

3 Evaluate each of the following: [10] to [18]

(a) ${}^{8}C_{3}$ (b) ${}^{15}C_{12}$ (c) ${}^{159}C_{158}$ (d) ${}^{204}C_{0}$

4 Expand $(2a-5b)^7$ as a binomial series. [22] to [31]

▶

5 In the binomial expansion of $(1 + 10/x)^{10}$ written in terms of descending powers of $x$, find: Frames

(a) the 8th term (b) the coefficient of $x^{-8}$ 32 to 33

6 Evaluate:

(a) $\sum_{r=1}^{45} r$ (b) $\sum_{r=1}^{n} (9 - 3r)$ 37 to 47

7 Determine the 5th term and the sum of the first 20 terms of the series: $1 + 3 + 5 + 7 + \ldots$ 37 to 47

8 Using the series expansion of $e^x$ find $e^{-2}$ accurate to 3 decimal places. 48 to 50

## Further problems F.7

57

1 Given a row of 12 hat pegs, in how many combinations can:

(a) 5 identical red hardhats be hung?

(b) 5 identical red and 4 identical blue hardhats be hung?

(c) 5 identical red, 4 identical blue and 2 identical white hardhats be hung?

2 Show that:

(a) ${}^{n+1}C_1 - {}^{n}C_1 = 1$

(b) $\sum_{r=0}^{n} (-1)^r \, {}^{n}C_r = 0$, where $n > 0$

(c) $\sum_{r=0}^{n} {}^{n}C_r \, 2^r = 3^n$

3 Write out the binomial expansions of:

(a) $(1 - 3x)^4$ (b) $(2 + x/2)^5$ (c) $\left(1 - \frac{1}{x}\right)^5$ (d) $(x + 1/x)^6$

4 Evaluate:

(a) $\sum_{r=1}^{16} (5r - 7)$ (b) $\sum_{r=1}^{n} (4 - 5r)$

5 Using the first 6 terms of the series for $e^x$, determine approximate values of $e^2$ and $\sqrt{e}$ to 4 sig fig.

6 Expand $e^{3x}$ and $e^{-2x}$ as a series of ascending powers of $x$ using the first 6 terms in each case. Hence determine a series for $e^{3x} - e^{-2x}$.

7 Using the fact that $a^x = e^{x \ln a}$ and the series expansion of $e^{kx}$, evaluate $2^{-3 \cdot 4}$ accurate to 3 dp. (This will take 12 terms.)

Now visit the companion website for this book at www.palgrave.com/stroud for more questions applying this mathematics to science and engineering.

58

# Partial fractions

## Learning outcomes

When you have completed this Programme you will be able to:

- ☐ Factorize the denominator of an algebraic fraction into its factors
- ☐ Separate an algebraic fraction into its partial fractions
- ☐ Recognize the rules of partial fractions

If you already feel confident about these why not try the quiz over the page? You can check your answers at the end of the book.

# Quiz F.8

**Frames**

Express each of the following in partial fraction form:

1 $\dfrac{3x+9}{x^2+8x+12}$ — Frames 1 to 7

2 $\dfrac{x^2+x+1}{x^2+3x+2}$ — Frames 8 to 21

3 $\dfrac{7x^2+6x+5}{(x+1)(x^2+x+1)}$ — Frames 25 to 31

4 $\dfrac{5x+6}{(x-1)^2}$ — Frames 32 to 34

5 $\dfrac{2x^3-5x+13}{(x+4)^2}$ — Frames 8 to 34

# Partial fractions

**1**

To simplify an arithmetical expression consisting of a number of fractions, we first convert the individual fractions to a new form having a common denominator which is the LCM of the individual denominators.

With $\frac{2}{5}-\frac{3}{4}+\frac{1}{2}$ the LCM of the denominators, 5, 4 and 2, is 20.

$$\therefore \frac{2}{5}-\frac{3}{4}+\frac{1}{2}=\frac{8-15+10}{20}$$
$$=\frac{3}{20}$$

In just the same way, algebraic fractions can be combined by converting them to a new denominator which is the LCM of the individual denominators.

For example, with $\frac{2}{x-3}-\frac{4}{x-1}$, the LCM of the denominators is $(x-3)(x-1)$.

Therefore:

$$\frac{2}{x-3}-\frac{4}{x-1}=\frac{2(x-1)-4(x-3)}{(x-3)(x-1)}$$
$$=\frac{2x-2-4x+12}{(x-3)(x-1)}$$
$$=\frac{10-2x}{(x-3)(x-1)}$$

In practice, the reverse process is often required. That is, presented with a somewhat cumbersome algebraic fraction there is a need to express this as a number of simpler component fractions.

From the previous example:

$$\frac{2}{x-3}-\frac{4}{x-1}=\frac{10-2x}{(x-3)(x-1)}$$

The two simple fractions on the left-hand side are called the *partial fractions* of the expression on the right-hand side. What follows describes how these partial fractions can be obtained from the original fraction.

*So, on to the next frame*

**2**

Let us consider a simple case and proceed step by step.
To separate:

$$\frac{8x-28}{x^2-6x+8}$$

into its partial fractions we must first factorize the denominator into its prime factors. You can do this:

$$x^2-6x+8=(\ldots\ldots\ldots\ldots)(\ldots\ldots\ldots\ldots)$$

**3**

$$\boxed{(x-2)(x-4)}$$

Therefore:

$$\frac{8x-28}{x^2-6x+8}=\frac{8x-28}{(x-2)(x-4)}$$

We now assume that each simple factor in the denominator gives rise to a single partial fraction. That is, we assume that we can write:

$$\frac{8x-28}{(x-2)(x-4)}=\frac{A}{x-2}+\frac{B}{x-4}$$

where $A$ and $B$ are constants. We shall now show that this assumption is valid by finding the values of $A$ and $B$. First we add the two partial fractions on the RHS to give:

$$\frac{8x-28}{(x-2)(x-4)}=\ldots\ldots\ldots\ldots$$

*The answer is in the next frame*

**4**

$$\boxed{\frac{8x-28}{x^2-6x+8}=\frac{A(x-4)+B(x-2)}{(x-2)(x-4)}}$$

Because

$$\frac{A}{x-2}+\frac{B}{x-4}=\frac{A(x-4)}{(x-2)(x-4)}+\frac{B(x-2)}{(x-2)(x-4)}$$
$$=\frac{A(x-4)+B(x-2)}{(x-2)(x-4)}$$

The equation $\dfrac{8x-28}{x^2-6x+8}=\dfrac{A(x-4)+B(x-2)}{(x-2)(x-4)}$ is an identity because the RHS is just an alternative way of writing the LHS. Also, because the denominator on the RHS is an alternative way of writing the denominator on the LHS, the same must be true of the numerators. Consequently, equating the numerators we find that …………

*Next frame*

**5**

$$\boxed{8x-28\equiv A(x-4)+B(x-2)}$$

Because this is an identity it must be true for all values of $x$. It is convenient to choose a value of $x$ that makes one of the brackets zero. For example:

Letting $x=4$ gives $B=\ldots\ldots\ldots\ldots$

**6**

$$\boxed{B=2}$$

Because

$32-28=A(0)+B(2)$ so that $4=2B$ giving $B=2$.

Similarly if we let $x=2$ then $A=\ldots\ldots\ldots\ldots$

**7**

$$\boxed{A = 6}$$

Because

$16 - 28 = A(-2) + B(0)$ so that $-12 = -2A$ giving $A = 6$.

Therefore:

$$\frac{8x - 28}{(x-2)(x-4)} = \frac{6}{x-2} + \frac{2}{x-4}$$

the required partial fraction breakdown.

This example has demonstrated the basic process whereby the partial fractions of a given rational expression can be obtained. There is, however, one important proviso that has not been mentioned:

*To effect the partial fraction breakdown of a rational algebraic expression it is necessary for the degree of the numerator to be less than the degree of the denominator.*

If, in the original algebraic rational expression, the degree of the numerator is not less than the degree of the denominator then we divide out by long division. This gives a polynomial with a rational remainder where the remainder has a numerator with degree less than the denominator. The remainder can then be broken down into its partial fractions. In the following frames we consider some examples of this type.

**8**

**Example 1**

Express $\dfrac{x^2 + 3x - 10}{x^2 - 2x - 3}$ in partial fractions.

The first consideration is ...........

**9**

Is the numerator of lower degree than the denominator?

No, it is not, so we have to divide out by long division:

$$\begin{array}{r|l} & \;\;\;\;\;\;\;\;\;\;\;\;\;1 \\ x^2 - 2x - 3 & x^2 + 3x - 10 \\ & x^2 - 2x - 3 \\ \hline & \;\;\;\;\;\;5x - 7 \end{array}$$

$$\therefore \frac{x^2 + 3x - 10}{x^2 - 2x - 3} = 1 + \frac{5x - 7}{x^2 - 2x - 3}$$

Now we factorize the denominator into its prime factors, which gives ...........

**10**

$$\boxed{(x+1)(x-3)}$$

$$\therefore \frac{x^2 + 3x - 10}{x^2 - 2x - 3} = 1 + \frac{5x - 7}{(x+1)(x-3)}$$

The remaining fraction will give partial fractions of the form:

$$\frac{5x - 7}{(x+1)(x-3)} = \frac{A}{x+1} + \frac{B}{x-3}$$

Multiplying both sides by the denominator $(x+1)(x-3)$:

$5x - 7 =$ ...........

**11**

$A(x-3)+B(x+1)$

$5x-7 \equiv A(x-3)+B(x+1)$ is an identity, since the RHS is the LHS merely written in a different form. Therefore the statement is true for any value of $x$ we choose to substitute.

As was said previously, it is convenient to select a value for $x$ that makes one of the brackets zero. So if we put $x=3$ in both sides of the identity, we get ...........

**12**

$15-7=A(0)+B(4)$

i.e. $8=4B \quad \therefore B=2$

Similarly, if we substitute $x=-1$, we get ...........

**13**

$-5-7=A(-4)+B(0)$

$\therefore\ -12=-4A \qquad \therefore A=3$

$$\therefore \frac{5x-7}{(x+1)(x-3)}=\frac{3}{x+1}+\frac{2}{x-3}$$

So, collecting our results together:

$$\frac{x^2+3x-10}{x^2-2x-3}=1+\frac{5x-7}{x^2-2x-3}=1+\frac{5x-7}{(x+1)(x-3)}$$
$$=1+\frac{3}{x+1}+\frac{2}{x-3}$$

**Example 2**

Express $\dfrac{2x^2+18x+31}{x^2+5x+6}$ in partial fractions.

The first step is ...........

**14**

to divide the numerator by the denominator

since the numerator is not of lower degree than that of the denominator.

$$\therefore \frac{2x^2+18x+31}{x^2+5x+6}=2+\frac{8x+19}{x^2+5x+6}.$$

Now we attend to $\dfrac{8x+19}{x^2+5x+6}$.

Factorizing the denominator, we have ...........

**15**

$$\frac{8x+19}{(x+2)(x+3)}$$

so the form of the partial fractions will be ...........

16

$$\boxed{\frac{A}{x+2}+\frac{B}{x+3}}$$

i.e. $\dfrac{8x+19}{(x+2)(x+3)}=\dfrac{A}{x+2}+\dfrac{B}{x+3}$

You can now multiply both sides by the denominator $(x+2)(x+3)$ and finish it off:

$$\frac{2x^2+18x+31}{x^2+5x+6}=\dots\dots\dots$$

17

$$\boxed{2+\frac{3}{x+2}+\frac{5}{x+3}}$$

*So move on to the next frame*

18

**Example 3**

Express $\dfrac{3x^3-x^2-13x-13}{x^2-x-6}$ in partial fractions.

Applying the rules, we first divide out:

$$\frac{3x^3-x^2-13x-13}{x^2-x-6}=\dots\dots\dots$$

19

$$\boxed{3x+2+\frac{7x-1}{x^2-x-6}}$$

Now we attend to $\dfrac{7x-1}{x^2-x-6}$ in the normal way.

*Finish it off*

20

$$\boxed{3x+2+\frac{3}{x+2}+\frac{4}{x-3}}$$

Because

$$\frac{7x-1}{(x+2)(x-3)}=\frac{A}{x+2}+\frac{B}{x-3}$$

$$\therefore\ 7x-1=A(x-3)+B(x+2)$$

$x=3 \qquad 20=A(0)+B(5) \qquad \therefore B=4$

$x=-2 \qquad -15=A(-5)+B(0) \qquad \therefore A=3$

Remembering to include the polynomial part:

$$\therefore\ \frac{3x^3-x^2-13x-13}{x^2-x-6}=3x+2+\frac{3}{x+2}+\frac{4}{x-3}$$

Now one more entirely on your own just like the last one.

**Example 4**

Express $\dfrac{2x^3+3x^2-54x+50}{x^2+2x-24}$ in partial fractions.

Work right through it: then check with the next frame.

$$\frac{2x^3+3x^2-54x+50}{x^2+2x-24}=\dots\dots\dots$$

**21**

$$2x - 1 + \frac{1}{x-4} - \frac{5}{x+6}$$

Here it is:

$$\begin{array}{r|l} & 2x \quad - \quad 1 \\ \hline x^2 + 2x - 24 & 2x^3 + 3x^2 - 54x + 50 \\ & 2x^3 + 4x^2 - 48x \\ \hline & \quad - x^2 - 6x + 50 \\ & \quad - x^2 - 2x + 24 \\ \hline & \qquad\quad - 4x + 26 \end{array}$$

$$\therefore \frac{2x^3 + 3x^2 - 54x + 50}{x^2 + 2x - 24} = 2x - 1 - \frac{4x - 26}{x^2 + 2x - 24}$$

$$\frac{4x - 26}{(x-4)(x+6)} = \frac{A}{x-4} + \frac{B}{x+6} \qquad \therefore 4x - 26 = A(x+6) + B(x-4)$$

$$x = 4 \qquad -10 = A(10) + B(0) \qquad \therefore A = -1$$

$$x = -6 \qquad -50 = A(0) + B(-10) \qquad \therefore B = 5$$

$$\therefore \frac{2x^3 + 3x^2 - 54x + 50}{x^2 + 2x - 24} = 2x - 1 - \left\{-\frac{1}{x-4} + \frac{5}{x+6}\right\}$$

$$= 2x - 1 + \frac{1}{x-4} - \frac{5}{x+6}$$

***At this point let us pause and summarize the main facts so far on the breaking into partial fractions of rational algebraic expressions with denominators in the form of a product of two simple factors***

## Review summary

**22**

1 To effect the partial fraction breakdown of a rational algebraic expression it is necessary for the degree of the numerator to be less than the degree of the denominator. In such an expression whose denominator can be expressed as a product of simple prime factors, each of the form $ax + b$:

(a) Write the rational expression with the denominator given as a product of its prime factors.

(b) Each factor then gives rise to a partial fraction of the form $\frac{A}{ax+b}$ where $A$ is a constant whose value is to be determined.

(c) Add the partial fractions together to form a single algebraic fraction whose numerator contains the unknown constants and whose denominator is identical to that of the original expression.

(d) Equate the numerator so obtained with the numerator of the original algebraic fraction.

(e) By substituting appropriate values of $x$ in this equation determine the values of the unknown constants.

2 If, in the original algebraic rational expression, the degree of the numerator is not less than the degree of the denominator then we divide out by long division. This gives a polynomial with a rational remainder where the remainder has a numerator with degree less than the denominator. The remainder can then be broken down into its partial fractions.

## Review exercise

23

Express the following in partial fractions:

1 $\dfrac{x+7}{x^2-7x+10}$

3 $\dfrac{3x^2-8x-63}{x^2-3x-10}$

2 $\dfrac{10x+37}{x^2+3x-28}$

4 $\dfrac{2x^2+6x-35}{x^2-x-12}$

24

1 $\dfrac{x+7}{x^2-7x+10}=\dfrac{x+7}{(x-2)(x-5)}=\dfrac{A}{x-2}+\dfrac{B}{x-5}$

$\therefore x+7=A(x-5)+B(x-2)$

$x=5 \quad 12=A(0)+B(3) \quad \therefore B=4$

$x=2 \quad 9=A(-3)+B(0) \quad \therefore A=-3$

$\dfrac{x+7}{x^2-7x+10}=\dfrac{4}{x-5}-\dfrac{3}{x-2}$

2 $\dfrac{10x+37}{x^2+3x-28}=\dfrac{10x+37}{(x-4)(x+7)}=\dfrac{A}{x-4}+\dfrac{B}{x+7}$

$\therefore 10x+37=A(x+7)+B(x-4)$

$x=-7 \quad -33=A(0)+B(-11) \quad \therefore B=3$

$x=4 \quad 77=A(11)+B(0) \quad \therefore A=7$

$\therefore \dfrac{10x+37}{x^2+3x-28}=\dfrac{7}{x-4}+\dfrac{3}{x+7}$

3 $\dfrac{3x^2-8x-63}{x^2-3x-10}=3+\dfrac{x-33}{(x+2)(x-5)}$

$\dfrac{x-33}{(x+2)(x-5)}=\dfrac{A}{x+2}+\dfrac{B}{x-5}$

$\therefore x-33=A(x-5)+B(x+2)$

$x=5 \quad -28=A(0)+B(7) \quad \therefore B=-4$

$x=-2 \quad -35=A(-7)+B(0) \quad \therefore A=5$

$\therefore \dfrac{3x^2-8x-63}{x^2-3x-10}=3+\dfrac{5}{x+2}-\dfrac{4}{x-5}$

4 $\dfrac{2x^2+6x-35}{x^2-x-12}=2+\dfrac{8x-11}{(x+3)(x-4)}$

$\dfrac{8x-11}{(x+3)(x-4)}=\dfrac{A}{x+3}+\dfrac{B}{x-4}$

$8x-11=A(x-4)+B(x+3)$

$x=4 \quad 21=A(0)+B(7) \quad \therefore B=3$

$x=-3 \quad -35=A(-7)+B(0) \quad \therefore A=5$

$\therefore \dfrac{2x^2+6x-35}{x^2-x-12}=2+\dfrac{5}{x+3}+\dfrac{3}{x-4}$

# Denominators with repeated and quadratic factors

**25**

Now let's look at a rational algebraic fraction where the denominator contains a quadratic factor that will not factorize into two simple factors.

Express $\dfrac{15x^2 - x + 2}{(x-5)(3x^2+4x-2)}$ in partial fractions.

Here the degree of the numerator is less than the degree of the denominator so no initial division is required. However, the denominator contains a quadratic factor that cannot be factorized further into simple factors. The usual test confirms this because $(b^2 - 4ac) = 16 - 4 \times 3 \times (-2) = 40$ which is not a perfect square. In this situation there is a rule that applies:

*An irreducible quadratic factor in the denominator of the original rational expression of the form $(ax^2 + bx + c)$ gives rise to a partial fraction of the form $\dfrac{Ax+B}{ax^2+bx+c}$*

$$\therefore \frac{15x^2 - x + 2}{(x-5)(3x^2+4x-2)} = \frac{A}{x-5} + \frac{Bx+C}{3x^2+4x-2}$$

Multiplying throughout by the denominator:

$$\begin{aligned} 15x^2 - x + 2 &= A(3x^2+4x-2) + (Bx+C)(x-5) \\ &= 3Ax^2 + 4Ax - 2A + Bx^2 + Cx - 5Bx - 5C \end{aligned}$$

Collecting up like terms on the RHS gives:

$15x^2 - x + 2 = \dots\dots\dots\dots$

**26**

$$\boxed{15x^2 - x + 2 = (3A+B)x^2 + (4A - 5B + C)x - 2A - 5C}$$

This is an identity, so we can equate coefficients of like terms on each side:

| | | | |
|---|---|---|---|
| | $[x^2]$ | $15 = 3A + B$ | (1) |
| constant term | [CT] | $2 = -2A - 5C$ | (2) |
| | $[x]$ | $-1 = 4A - 5B + C$ | (3) |

From (1): $B = 15 - 3A$

From (2): $2 = -2A - 5C \quad \therefore 5C = -2A - 2 \quad \therefore C = \dfrac{-(2A+2)}{5}$

Substituting for $B$ and $C$ in (3), we have:

$$\begin{aligned} -1 &= 4A - 5(15 - 3A) - \frac{2A+2}{5} \\ \therefore -5 &= 20A - 25(15 - 3A) - (2A + 2) \\ -5 &= 20A - 375 + 75A - 2A - 2 = 93A - 377 \\ \therefore 93A &= 377 - 5 = 372 \quad \therefore A = 4 \end{aligned}$$

Sub. in (1): $15 = 12 + B \quad \therefore B = 3$

Sub. in (3): $-1 = 16 - 15 + C \quad \therefore C = -2$

$$\therefore \frac{15x^2 - x + 2}{(x-5)(3x^2+4x-2)} = \frac{4}{x-5} + \frac{3x-2}{3x^2+4x-2}$$

*Now on to another example*

**27**

Express $\dfrac{7x^2 - 18x - 7}{(x-4)(2x^2 - 6x + 3)}$ in partial fractions.

Here again, for the factor $(2x^2 - 6x + 3)$, $(b^2 - 4ac) = 36 - 24 = 12$ which is not a perfect square. $\therefore$ $(2x^2 - 6x + 3)$ is irreducible.

$\therefore$ The partial fractions of $\dfrac{7x^2 - 18x - 7}{(x-4)(2x^2 - 6x + 3)}$ will be of the form ...........

**28**

$$\frac{A}{x-4} + \frac{Bx + C}{2x^2 - 6x + 3}$$

Multiplying throughout by the complete denominator:

$7x^2 - 18x - 7 =$ ...........

**29**

$$A(2x^2 - 6x + 3) + (Bx + C)(x - 4)$$

Then multiply out and collect up like terms, and that gives:

$7x^2 - 18x - 7 =$ ...........

**30**

$$7x^2 - 18x - 7 = (2A + B)x^2 - (6A + 4B - C)x + 3A - 4C$$

Now you can equate coefficients of like terms on each side and finish it. The required partial fractions for

$\dfrac{7x^2 - 18x - 7}{(x-4)(2x^2 - 6x + 3)} =$ ...........

**31**

$$\frac{3}{x-4} + \frac{x + 4}{2x^2 - 6x + 3}$$

Because

$[x^2] \quad 7 = 2A + B \quad \therefore B = 7 - 2A \qquad (1)$

$[CT] \quad -7 = 3A - 4C \quad \therefore C = \dfrac{3A + 7}{4} \qquad (2)$

$[x] \quad -18 = -\left(6A + 28 - 8A - \dfrac{3A + 7}{4}\right)$

$$\therefore 72 = 24A + 112 - 32A - 3A - 7$$
$$= -11A + 105$$
$$\therefore 11A = 33$$
$$\therefore A = 3$$

Substitution in (1) and (2) gives $B = 1$ and $C = 4$.

$$\therefore \frac{7x^2 - 18x - 7}{(x-4)(2x^2 - 6x + 3)} = \frac{3}{x-4} + \frac{x + 4}{2x^2 - 6x + 3}$$

*Next frame*

**32**

Now let's look at a rational algebraic fraction where the denominator contains a repeated simple factor.

Express $\dfrac{35x-14}{(7x-2)^2}$ in partial fractions.

Again, there is a rule that applies:

*Repeated factors in the denominator of the algebraic expression of the form* $(ax+b)^2$ *give partial fractions of the form* $\dfrac{A}{ax+b}+\dfrac{B}{(ax+b)^2}$. *Similarly* $(ax+b)^3$ *gives rise to partial fractions of the form:*

$$\frac{A}{ax+b}+\frac{B}{(ax+b)^2}+\frac{C}{(ax+b)^3}$$

Consequently, we write:

$$\frac{35x-14}{(7x-2)^2}=\frac{A}{7x-2}+\frac{B}{(7x-2)^2}$$

Then we multiply throughout as usual by the original denominator:

$$\begin{aligned}35x-14&=A(7x-2)+B\\&=7Ax-2A+B\end{aligned}$$

Now we simply equate coefficients and $A$ and $B$ are found:

$$\frac{35x-14}{(7x-2)^2}=\ldots\ldots\ldots\ldots$$

**33**

$$\boxed{\frac{5}{7x-2}-\frac{4}{(7x-2)^2}}$$

Similarly:

$\dfrac{42x+44}{(6x+5)^2}$ in partial fractions $=\ldots\ldots\ldots\ldots$

*Complete it*

**34**

$$\boxed{\frac{7}{6x+5}+\frac{9}{(6x+5)^2}}$$

$$\frac{42x+44}{(6x+5)^2}=\frac{A}{6x+5}+\frac{B}{(6x+5)^2}$$

$$\therefore\ 42x+44=A(6x+5)+B=6Ax+5A+B$$

$$[x]\quad 42=6A\quad \therefore A=7$$

$$[\text{CT}]\quad 44=5A+B=35+B\quad \therefore B=9$$

$$\therefore\ \frac{42x+44}{(6x+5)^2}=\frac{7}{6x+5}+\frac{9}{(6x+5)^2}$$

And now this one:

Express $\dfrac{18x^2+3x+6}{(3x+1)^3}$ in partial fractions.

*Complete the work with that one and then check with the next frame*

35

$$\frac{2}{3x+1} - \frac{3}{(3x+1)^2} + \frac{7}{(3x+1)^3}$$

Here $\dfrac{18x^2+3x+6}{(3x+1)^3} = \dfrac{A}{3x+1} + \dfrac{B}{(3x+1)^2} + \dfrac{C}{(3x+1)^3}$

$$\begin{aligned}\therefore\ 18x^2+3x+6 &= A(3x+1)^2 + B(3x+1) + C \\ &= A(9x^2+6x+1) + B(3x+1) + C \\ &= 9Ax^2 + 6Ax + A + 3Bx + B + C \\ &= 9Ax^2 + (6A+3B)x + (A+B+C)\end{aligned}$$

Equating coefficients:

$[x^2]$ $\quad 18 = 9A \quad \therefore A = 2$

$[x]$ $\quad 3 = 6A + 3B \quad 3 = 12 + 3B \quad 3B = -9 \quad \therefore B = -3$

[CT] $\quad 6 = A + B + C \quad 6 = 2 - 3 + C \quad \therefore C = 7$

$$\therefore \frac{18x^2+3x+6}{(3x+1)^3} = \frac{2}{3x+1} - \frac{3}{(3x+1)^2} + \frac{7}{(3x+1)^3}$$

Now determine the partial fractions of $\dfrac{20x^2-54x+35}{(2x-3)^3}$

The working is just the same as with the previous example:

$$\frac{20x^2-54x+35}{(2x-3)^3} = \ldots\ldots\ldots\ldots$$

36

$$\frac{5}{2x-3} + \frac{3}{(2x-3)^2} - \frac{1}{(2x-3)^3}$$

Because

$$\frac{20x^2-54x+35}{(2x-3)^3} = \frac{A}{2x-3} + \frac{B}{(2x-3)^2} + \frac{C}{(2x-3)^3}$$

$$\therefore\ 20x^2 - 54x + 35 = A(2x-3)^2 + B(2x-3) + C$$

Multiplying out and collecting up like terms:

$$20x^2 - 54x + 35 = 4Ax^2 - (12A - 2B)x + (9A - 3B + C)$$

Then, equating coefficients in the usual way:

$A = 5;\ B = 3;\ C = -1$

$$\therefore \frac{20x^2-54x+35}{(2x-3)^3} = \frac{5}{2x-3} + \frac{3}{(2x-3)^2} - \frac{1}{(2x-3)^3}$$

*Next frame*

37

Let us now consider the case where the denominator is a cubic expression with different linear factors.

Express $\dfrac{10x^2+7x-42}{(x-2)(x+4)(x-1)}$ in partial fractions.

Here the factors are all different, so the partial fractions will be ............

**38**

$$\boxed{\frac{A}{x-2}+\frac{B}{x+4}+\frac{C}{x-1}}$$

$$\frac{10x^2+7x-42}{(x-2)(x+4)(x-1)}=\frac{A}{x-2}+\frac{B}{x+4}+\frac{C}{x-1}$$

$$\therefore\ 10x^2+7x-42=A(x+4)(x-1)+B(x-2)(x-1)+C(x+4)(x-2)$$

$$=A(x^2+3x-4)+B(x^2-3x+2)+C(x^2+2x-8)$$

Multiplying out and collecting up like terms:

$10x^2+7x-42=\ldots\ldots\ldots\ldots$

**39**

$$\boxed{(A+B+C)x^2+(3A-3B+2C)x-(4A-2B+8C)}$$

Equating coefficients:

$[x^2]\quad 10=A+B+C \quad (1)$

$[x]\quad 7=3A-3B+2C \quad (2)$

$[\text{CT}]\quad -42=-4A+2B-8C \quad (3)$

Solving these three simultaneous equations provides the values of $A$, $B$ and $C$:

$(1)\times 2\quad 2A+2B+2C=20$

$(2)\quad 3A-3B+2C=\ \ 7$

$-A+5B\quad =13 \quad (4)$

$(2)\times 4\quad 12A-12B+8C=\ \ 28$

$(3)\quad -4A+\ \ 2B-8C=-42$

$8A-10B\quad =-14 \quad (5)$

Now we can solve (4) and (5) to find $A$ and $B$, and then substitute in (1) to find $C$. So finally:

$$\frac{10x^2+7x-42}{(x-2)(x+4)(x-1)}=\ldots\ldots\ldots\ldots$$

40

$$\boxed{\frac{2}{x-2}+\frac{3}{x+4}+\frac{5}{x-1}}$$

In this latest example, the denominator has conveniently been given as the product of three linear factors. It may well be that this could be given as a cubic expression, in which case factorization would have to be carried out using the remainder theorem before further progress could be made. Here then is an example which brings us to the peak of this programme.

Determine the partial fractions of $\dfrac{8x^2-14x-10}{x^3-4x^2+x+6}$.

First we see that no initial division is necessary. Then we have to factorize the denominator into its prime factors, as we did in Programme F.3.

So, putting $f(x)=x^3-4x^2+x+6$, we determine the three simple factors of $f(x)$, if they exist.

These are ...........

41

$$\boxed{(x+1)(x-2)(x-3)}$$

Because

$f(x)=x^3-4x^2+x+6=[(x-4)x+1]x+6$ in nested form.

$f(1)=4 \quad \therefore (x-1)$ is not a factor

$f(-1)=0 \quad \therefore (x+1)$ is a factor of $f(x)$:

$$\begin{array}{r|l}
 & x^2 - 5x + 6 \\
\hline
x+1 & x^3 - 4x^2 + x + 6 \\
 & x^3 + x^2 \\
 & \quad -5x^2 + x \\
 & \quad -5x^2 - 5x \\
 & \qquad 6x + 6 \\
 & \qquad 6x + 6 \\
 & \qquad \cdot \quad \cdot
\end{array}$$

$$\therefore f(x)=(x+1)(x^2-5x+6)$$
$$=(x+1)(x-2)(x-3)$$
$$\therefore \frac{8x^2-14x-10}{x^3-4x^2+x+6}=\frac{8x^2-14x-10}{(x+1)(x-2)(x-3)}$$

and now we can proceed as in the previous example. Work right through it and then check the results with the following frame.

$$\frac{8x^2-14x-10}{(x+1)(x-2)(x-3)}= \ldots\ldots\ldots\ldots$$

42

$$\boxed{\frac{1}{x+1}+\frac{2}{x-2}+\frac{5}{x-3}}$$

Because

$$\frac{8x^2-14x-10}{(x+1)(x-2)(x-3)}=\frac{A}{x+1}+\frac{B}{x-2}+\frac{C}{x-3}$$

$$\begin{aligned}\therefore\ 8x^2-14x-10 &= A(x-2)(x-3)+B(x+1)(x-3)+C(x+1)(x-2)\\ &= A(x^2-5x+6)+B(x^2-2x-3)+C(x^2-x-2)\\ &= (A+B+C)x^2-(5A+2B+C)x+(6A-3B-2C)\end{aligned}$$

$[x^2]$ $\quad A+B+C=8$ (1)
$[x]$ $\quad 5A+2B+C=14$ (2)
[CT] $\quad 6A-3B-2C=-10$ (3)

(1) × 2 $\quad 2A+2B+2C=16$
(3) $\quad 6A-3B-2C=-10$

$8A-B=6$ (4)

(2) $\quad 5A+2B+C=14$
(1) $\quad A+B+C=8$

$4A+B=6$ (5)

(5) $\quad 4A+B=6$
(6) $\quad 8A-B=6$

$12A=12 \quad \therefore A=1$

(5) $\quad 4A+B=6 \quad 4+B=6 \quad \therefore B=2$
(1) $\quad A+B+C=8 \quad 1+2+C=8 \quad \therefore C=5$

$$\begin{aligned}\therefore\ \frac{8x^2-14x-10}{x^3-4x^2+x+6} &= \frac{8x^2-14x-10}{(x+1)(x-2)(x-3)}\\ &= \frac{1}{x+1}+\frac{2}{x-2}+\frac{5}{x-3}\end{aligned}$$

***At this point let us pause and summarize the main facts on the breaking into partial fractions of rational algebraic expressions with denominators containing irreducible quadratic factors or repeated simple factors***

## Review summary

43

To effect the partial fraction breakdown of a rational algebraic expression it is necessary for the degree of the numerator to be less than the degree of the denominator. In such an expression whose denominator contains:

1 A quadratic factor of the form $ax^2+bx+c$ which cannot be expressed as a product of simple factors, the partial fraction breakdown gives rise to a partial fraction of the form $\dfrac{Ax+B}{ax^2+bx+c}$

2 Repeated factors of the form $(ax+b)^2$, the partial fraction breakdown gives rise to a partial fraction of the form $\dfrac{A}{ax+b}+\dfrac{B}{(ax+b)^2}$. Similarly $(ax+b)^3$ gives rise to partial fractions of the form

$$\frac{A}{ax+b}+\frac{B}{(ax+b)^2}+\frac{C}{(ax+b)^3}$$

## Review exercise

**44**

Express in partial fractions:

1 $\dfrac{32x^2-28x-5}{(4x-3)^3}$

2 $\dfrac{9x^2+48x+18}{(2x+1)(x^2+8x+3)}$

3 $\dfrac{12x^2+36x+6}{x^3+6x^2+3x-10}$

**45**

1 $\dfrac{2}{4x-3}+\dfrac{5}{(4x-3)^2}-\dfrac{8}{(4x-3)^3}$

2 $\dfrac{5}{2x+1}+\dfrac{2x+3}{x^2+8x+3}$

3 $\dfrac{3}{x-1}+\dfrac{2}{x+2}+\dfrac{7}{x+5}$

**46**

You have now come to the end of this Programme. A list of **Can You?** questions follows for you to gauge your understanding of the material in the Programme. You will notice that these questions match the **Learning outcomes** listed at the beginning of the Programme so go back and try the **Quiz** that follows them. After that try the **Test exercise**. *Work through them at your own pace, there is no need to hurry.* A set of **Further problems** provides additional valuable practice.

## Can You?

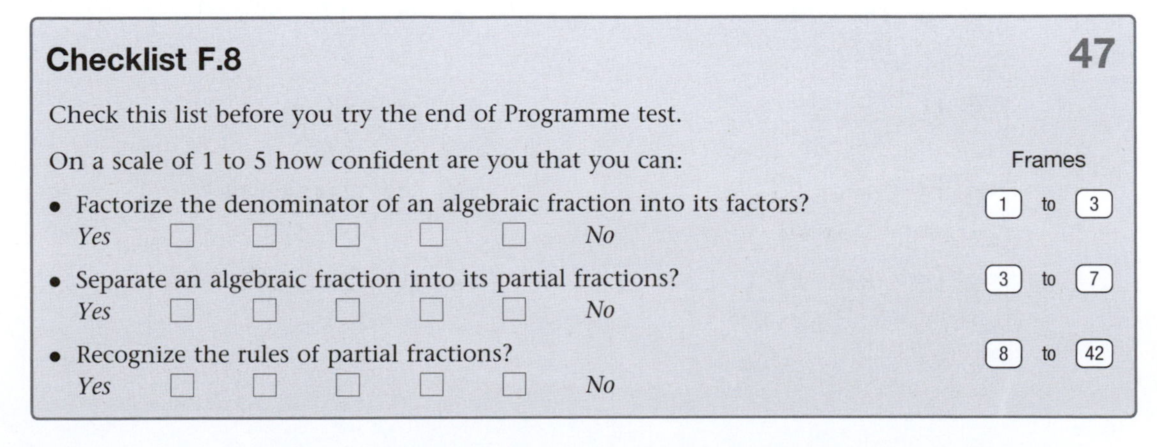

### Checklist F.8

**47**

Check this list before you try the end of Programme test.

On a scale of 1 to 5 how confident are you that you can: Frames

- Factorize the denominator of an algebraic fraction into its factors? 1 to 3
  *Yes* ☐ ☐ ☐ ☐ ☐ *No*
- Separate an algebraic fraction into its partial fractions? 3 to 7
  *Yes* ☐ ☐ ☐ ☐ ☐ *No*
- Recognize the rules of partial fractions? 8 to 42
  *Yes* ☐ ☐ ☐ ☐ ☐ *No*

## Test exercise F.8

**48**

Express each of the following in partial fractions: Frames

1 $\dfrac{x-14}{x^2-10x+24}$ — Frames 1 to 7

2 $\dfrac{13x-7}{10x^2-11x+3}$ — Frames 1 to 7

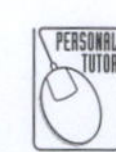

3 $\dfrac{4x^2+9x-73}{x^2+x-20}$ — Frames 8 to 21

4 $\dfrac{6x^2+19x-11}{(x+1)(x^2+5x-2)}$ — Frames 25 to 31

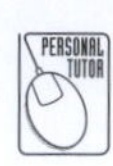

5 $\dfrac{10x-13}{(2x-3)^2}$ — Frames 32 to 34

6 $\dfrac{3x^2-34x+97}{(x-5)^3}$ — Frames 35 to 36

7 $\dfrac{9x^2-16x+34}{(x+4)(2x-3)(3x+1)}$ — Frames 37 to 40

8 $\dfrac{8x^2+27x+13}{x^3+4x^2+x-6}$ — Frames 41 to 42

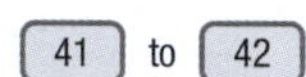

## Further problems F.8

**49**

Express each of the following in partial fractions:

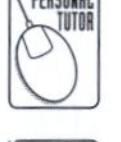

1 $\dfrac{7x+36}{x^2+12x+32}$

2 $\dfrac{5x-2}{x^2-3x-28}$

3 $\dfrac{x+7}{x^2-7x+10}$

4 $\dfrac{3x-9}{x^2-3x-18}$

5 $\dfrac{7x-9}{2x^2-7x-15}$

6 $\dfrac{14x}{6x^2-x-2}$

7 $\dfrac{13x-7}{10x^2-11x+3}$

8 $\dfrac{7x-7}{6x^2+11x+3}$

9 $\dfrac{18x+20}{(3x+4)^2}$

10 $\dfrac{35x+17}{(5x+2)^2}$

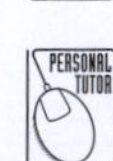

11 $\dfrac{12x-16}{(4x-5)^2}$

12 $\dfrac{5x^2-13x+5}{(x-2)^3}$

13 $\dfrac{75x^2+35x-4}{(5x+2)^3}$

14 $\dfrac{64x^2-148x+78}{(4x-5)^3}$

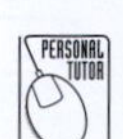

15 $\dfrac{8x^2+x-3}{(x+2)(x-1)^2}$

16 $\dfrac{4x^2-24x+11}{(x+2)(x-3)^2}$

 **17** $\dfrac{14x^2 + 31x + 5}{(x-1)(2x+3)^2}$

**18** $\dfrac{4x^2 - 47x + 141}{x^2 - 13x + 40}$

 **19** $\dfrac{5x^2 - 77}{x^2 - 2x - 15}$

**20** $\dfrac{8x^2 - 19x - 24}{(x-2)(x^2 - 2x - 5)}$

 **21** $\dfrac{5x^2 + 9x - 1}{(2x+3)(x^2 + 5x + 2)}$

**22** $\dfrac{7x^2 - 18x - 7}{(3x-1)(2x^2 - 4x - 5)}$

 **23** $\dfrac{11x + 23}{(x+1)(x+2)(x+3)}$

**24** $\dfrac{5x^2 + 28x + 47}{(x-1)(x+3)(x+4)}$

 **25** $\dfrac{8x^2 - 60x - 43}{(4x+1)(2x+3)(3x-2)}$

**26** $\dfrac{74x^2 - 39x - 6}{(2x-1)(3x+2)(4x-3)}$

 **27** $\dfrac{16x^2 - x + 3}{6x^3 - 5x^2 - 2x + 1}$

**28** $\dfrac{11x^2 + 11x - 32}{6x^3 + 5x^2 - 16x - 15}$

 **29** $\dfrac{4x^2 - 3x - 4}{6x^3 - 29x^2 + 46x - 24}$

**30** $\dfrac{2x^2 + 85x + 36}{20x^3 + 47x^2 + 11x - 6}$

---

 Now visit the companion website for this book at www.palgrave.com/stroud for more questions applying this mathematics to science and engineering.

**50**

---

# Trigonometry

## Learning outcomes

*When you have completed this Programme you will be able to:*

- ☐ Convert angles measured in degrees, minutes and seconds into decimal degrees
- ☐ Convert degrees into radians and vice versa
- ☐ Use a calculator to determine the values of trigonometric ratios for any acute angle
- ☐ Verify trigonometric identities

If you already feel confident about these why not try the quiz over the page? You can check your answers at the end of the book.

# Quiz F.9

**Frames**

1 Convert the angle $253°18'42''$ to decimal degree format. — 1 to 2

2 Convert the angle $73{\cdot}415°$ to degrees, minutes and seconds. — 2 to 3

3 Convert the following to radians to 2 dp:
(a) $47°$ (b) $12{\cdot}61°$ (c) $135°$ (as a multiple of $\pi$) — 4 to 6

4 Convert the following to degrees to 2 dp:
(a) 4·621 rad (b) $9\pi/4$ rad (c) $13\pi/5$ rad — 6 to 8

5 Find the value of each of the following to 4 dp:
(a) $\cos 24°$ (b) $\sin 5\pi/12$ (c) $\cot \pi/3$
(d) $\operatorname{cosec} 17{\cdot}9°$ (e) $\sec 5{\cdot}42°$ (f) $\tan 3{\cdot}24$ rad — 9 to 19

6 Given one side and the hypotenuse of a right-angled triangle as 5·6 cm and 12·3 cm, find the length of the other side. — 20 to 23

7 Show that the triangle with sides 7 cm, 24 cm and 25 cm is a right-angled triangle. — 20 to 23

8 A ship sails 12 km due north of a port and then sails 14 km due east. How far is the ship from the port? How much further east will it have sailed when it is 30 km from the port, assuming it keeps on the same course? — 24 to 27

9 Verify each of the following trigonometric identities:
(a) $(\sin\theta - \cos\theta)^2 + (\sin\theta + \cos\theta)^2 \equiv 2$
(b) $(1 - \cos\theta)^{\frac{1}{2}}(1 + \cos\theta)^{\frac{1}{2}} \equiv \sin\theta$
(c) $\tan\theta + \sec\theta \equiv \dfrac{1}{\sec\theta - \tan\theta}$
(d) $\cos\theta - \cos\phi \equiv -2\sin\dfrac{\theta+\phi}{2}\sin\dfrac{\theta-\phi}{2}$ — 31 to 40

10 Show that:
(a) $\tan 75° = \dfrac{\sqrt{3}+1}{\sqrt{3}-1}$
(b) $\sin 15° = \dfrac{\sqrt{3}-1}{2\sqrt{2}}$ — 38 to 39

# Angles

## Rotation

**1**

When a straight line is rotated about a point it sweeps out an angle that can be measured either in *degrees* or in *radians*. By convention a straight line rotating through a *full angle* and returning to its starting position is said to have rotated through 360 degrees – $360^{\circ}$ – where each degree is subdivided into 60 minutes – $60'$ – and each minute further subdivided into 60 seconds – $60''$. A *straight angle* is half of this, namely $180^{\circ}$ and a *right angle* is half of this again, namely $90^{\circ}$. Any angle less than $90^{\circ}$ is called an *acute* angle and any angle greater than $90^{\circ}$ is called an *obtuse* angle.

An angle that is measured in degrees, minutes and seconds can be converted to a decimal degree as follows:

$$\begin{aligned}45^{\circ}36'18'' &= 45^{\circ} + \left(\frac{36}{60}\right)^{\circ} + \left(\frac{18}{60 \times 60}\right)^{\circ}\\ &= (45 + 0{\cdot}6 + 0{\cdot}005)^{\circ}\\ &= 45{\cdot}605^{\circ}\end{aligned}$$

That was easy, so the decimal form of $53^{\circ}29'7''$ to 3 dp is ...........

*The answer is in the next frame*

---

**2**

$$\boxed{53{\cdot}485^{\circ}}$$

Because

$$\begin{aligned}53^{\circ}29'7'' &= 53^{\circ} + \left(\frac{29}{60}\right)^{\circ} + \left(\frac{7}{60 \times 60}\right)^{\circ}\\ &= (53 + 0{\cdot}48\dot{3} + 0{\cdot}0019\dot{4})^{\circ}\\ &= 53{\cdot}485^{\circ} \text{ to 3 dp}\end{aligned}$$

How about the other way? For example, to convert $18{\cdot}478^{\circ}$ to degrees, minutes and seconds we proceed as follows:

$18{\cdot}478^{\circ} = 18^{\circ} + (0{\cdot}478 \times 60)'$ Multiply the fractional part of the degree by 60

$= 18^{\circ} + 28{\cdot}68'$

$= 18^{\circ} + 28' + (0{\cdot}68 \times 60)''$ Multiply the fractional part of the minute by 60

$= 18^{\circ} + 28' + 40{\cdot}8''$

$= 18^{\circ}28'41''$ to the nearest second

So that $236{\cdot}986^{\circ} =$ ........... (in degrees, minutes and seconds)

*Next frame*

**3**

$$\boxed{236°59'10''}$$

Because

$$\begin{aligned}236{\cdot}986° &= 236° + (0{\cdot}986 \times 60)' \\ &= 236° + 59{\cdot}16' \\ &= 236° + 59' + (0{\cdot}16 \times 60)'' \\ &= 236° + 59' + 9{\cdot}6'' \\ &= 236°59'10'' \text{ to the nearest second}\end{aligned}$$

*Move now to the next frame*

**4**

## Radians

An alternative unit of measure of an angle is the radian. If a straight line of length $r$ rotates about one end so that the other end describes an arc of length $r$, the line is said to have rotated through 1 radian – 1 rad.

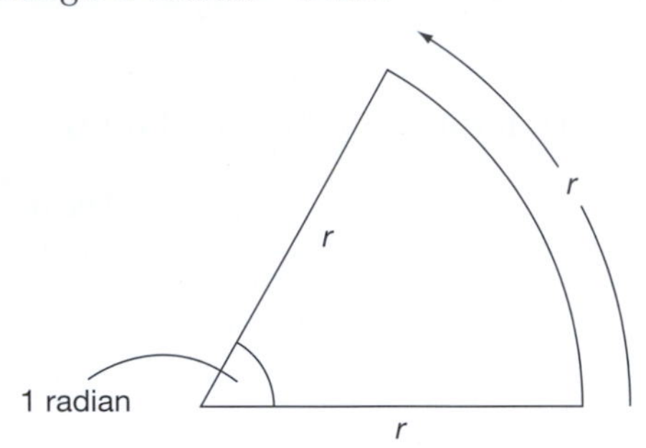


Because the arc described when the line rotates through a full angle is the circumference of a circle which measures $2\pi r$, the number of radians in a full angle is $2\pi$ rad. Consequently, relating degrees to radians we see that:

$$\begin{aligned}360° &= 2\pi \text{ rad} \\ &= 6{\cdot}2831\ldots \text{ rad}\end{aligned}$$

So that $1° = \ldots\ldots\ldots\ldots$ rad (to 3 sig fig)

*The answer is in the next frame*

**5**

$$\boxed{0{\cdot}0175 \text{ rad}}$$

Because

$$360° = 2\pi \text{ rad, so } 1° = \frac{2\pi}{360} = \frac{\pi}{180} = 0{\cdot}0175 \text{ rad to 3 sig fig}$$

Often, when degrees are transformed to radians they are given as multiples of $\pi$. For example:

$$360° = 2\pi \text{ rad, so that } 180° = \pi \text{ rad, } 90° = \pi/2 \text{ rad, } 45° = \pi/4 \text{ rad and so on}$$

So, 30°, 120° and 270° are given in multiples of $\pi$ as ……, ……, …….

*Answers in the next frame*

6

$\pi/6$ rad, $2\pi/3$ rad, $3\pi/2$ rad

Because

$$30^\circ = 180^\circ/6 = \pi/6 \text{ rad}$$
$$120^\circ = 2 \times 60^\circ = 2 \times (180^\circ/3) = 2\pi/3 \text{ rad}$$
$$270^\circ = 3 \times 90^\circ = 3 \times (180^\circ/2) = 3\pi/2 \text{ rad}$$

Also, 1 rad = ........... degrees (to 3 dp)

*Check your answer in the next frame*

7

$57{\cdot}296^\circ$

Because

$$2\pi \text{ rad} = 360^\circ, \text{ so } 1 \text{ rad} = \frac{360}{2\pi} = \frac{180}{\pi} = 57{\cdot}296^\circ$$

So, the degree equivalents of 2·34 rad, $\pi/3$ rad, $5\pi/6$ rad and $7\pi/4$ rad are ...........

*Check with the next frame*

8

$134{\cdot}1^\circ$ to 1 dp, $60^\circ$, $150^\circ$ and $315^\circ$

Because

$$2{\cdot}34 \text{ rad} = \left(2{\cdot}34 \times \frac{180}{\pi}\right)^\circ = 134{\cdot}1^\circ \text{ to 1 dp}$$
$$\pi/3 \text{ rad} = \left(\frac{\pi}{3} \times \frac{180}{\pi}\right)^\circ = 60^\circ$$
$$5\pi/6 \text{ rad} = \left(\frac{5\pi}{6} \times \frac{180}{\pi}\right)^\circ = 150^\circ$$
$$7\pi/4 \text{ rad} = \left(\frac{7\pi}{4} \times \frac{180}{\pi}\right)^\circ = 315^\circ$$

*Move to the next frame*

9

## Triangles

All triangles possess shape and size. The shape of a triangle is governed by the three angles (which always add up to 180°) and the size by the lengths of the three sides. Two triangles can possess the same shape – possess the same angles – but be of different sizes. We say that two such triangles are *similar*. It is the similarity of figures of different sizes that permits an artist to draw a picture of a scene that looks like the real thing – the lengths of the corresponding lines in the picture and the scene are obviously different but the corresponding angles in the picture and the scene are the same.

▶

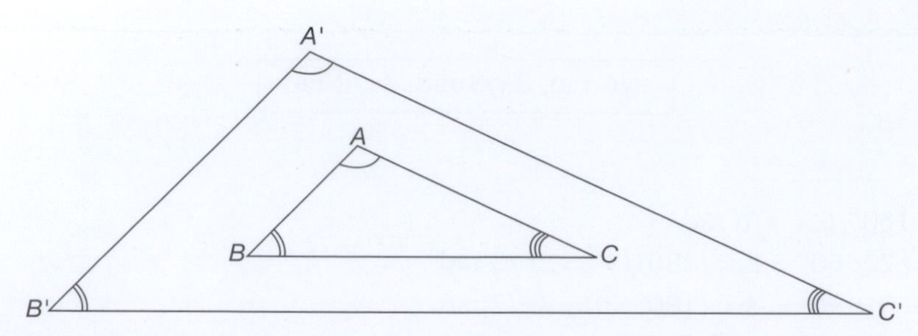


A significant feature of similar figures is that lengths of corresponding sides are all in the same ratio so that, for example, in the similar triangles $ABC$ and $A'B'C'$ in the figure:

$$\frac{AB}{A'B'} = \frac{AC}{A'C'} = \frac{BC}{B'C'}$$

So from a knowledge of the ratios of the sides of a given triangle we can make deductions about any triangle that is similar to it. For example, if in triangle $ABC$ of the above figure:

$AB = 2$ cm, $AC = 5$ cm and $BC = 4$ cm

and in triangle $A'B'C'$, $A'B' = 3$ cm, the length of $A'C'$ can be found as follows:

Since $\frac{AB}{A'B'} = \frac{AC}{A'C'}$ and $\frac{AB}{A'B'} = \frac{2}{3}$ then $\frac{AC}{A'C'} = \frac{5}{A'C'} = \frac{2}{3}$ giving

$$A'C' = \frac{5 \times 3}{2} = 7{\cdot}5 \text{ cm}$$

This means that the length of $B'C' = \ldots\ldots\ldots\ldots$

*Check your answer in the next frame*

---

**10**

6 cm

Because

$$\frac{AB}{A'B'} = \frac{BC}{B'C'} = \frac{2}{3} \text{ then } \frac{4}{B'C'} = \frac{2}{3} \text{ so that } B'C' = \frac{4 \times 3}{2} = 6 \text{ cm}$$

Ratios between side lengths of one given triangle are also equal to the corresponding ratios between side lengths of a similar triangle. We can prove this using the figure above where:

$$\frac{AB}{A'B'} = \frac{AC}{A'C'}$$

By multiplying both sides of this equation by $\frac{A'B'}{AC}$ we find that:

$$\frac{AB}{A'B'} \times \frac{A'B'}{AC} = \frac{AC}{A'C'} \times \frac{A'B'}{AC}, \text{ that is } \frac{AB}{AC} = \frac{A'B'}{A'C'}$$

so that the ratio between sides $AB$ and $AC$ in the smaller triangle is equal to the ratio between the two corresponding sides $A'B'$ and $A'C'$ in the larger, similar triangle.

So that $\frac{AB}{BC} = \ldots\ldots\ldots\ldots$

*Answer in the next frame*

**11**

$$\boxed{\frac{A'B'}{B'C'}}$$

Because

$\dfrac{AB}{A'B'} = \dfrac{BC}{B'C'}$ then multiplying both sides of this equation by $\dfrac{A'B'}{BC}$ we find that:

$\dfrac{AB}{A'B'} \times \dfrac{A'B'}{BC} = \dfrac{BC}{B'C'} \times \dfrac{A'B'}{BC}$, that is $\dfrac{AB}{BC} = \dfrac{A'B'}{B'C'}$

Similarly, $\dfrac{AC}{BC} =$ ...........

*Next frame*

**12**

$$\boxed{\frac{A'C'}{B'C'}}$$

Because

$\dfrac{AC}{A'C'} = \dfrac{BC}{B'C'}$ then multiplying both sides of this equation by $\dfrac{A'C'}{BC}$ gives:

$\dfrac{AC}{A'C'} \times \dfrac{A'C'}{BC} = \dfrac{BC}{B'C'} \times \dfrac{A'C'}{BC}$, that is $\dfrac{AC}{BC} = \dfrac{A'C'}{B'C'}$

All triangles whose corresponding ratios of side lengths are equal have the same shape – they are similar triangles because corresponding angles are equal. Consequently, whilst the lengths of the sides of a triangle dictate the size of the triangle, the *ratios* of the side lengths dictate the shape – the angles of the triangle.

Because we need to know the properties of similar triangles we shall now link these ratios of side lengths to specific angles by using a right-angled triangle; the ratios are then called the *trigonometric ratios*.

*On now to the next frame*

## Trigonometric ratios

**13**

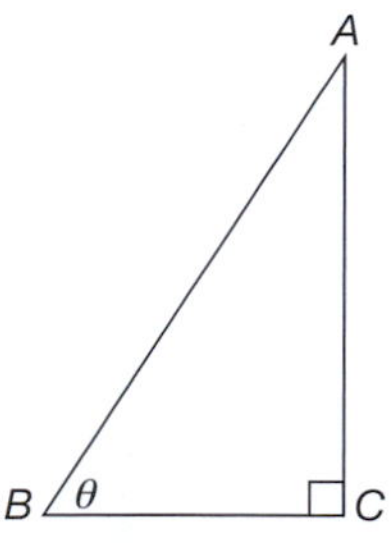


Triangle $ABC$ has a right angle at $C$, as denoted by the small square. Because of this the triangle $ABC$ is called a right-angled triangle. In this triangle the angle at B is denoted by $\theta$ [theta] where side $AC$ is *opposite* $\theta$, side $BC$ is *adjacent* to $\theta$ and side $AB$ is called the *hypotenuse*, we define the trigonometric ratios as:

*sine* of angle $\theta$ as $\dfrac{\text{opposite}}{\text{hypotenuse}} = \dfrac{AC}{AB}$ – this ratio is denoted by $\sin\theta$

*cosine* of angle $\theta$ as $\dfrac{\text{adjacent}}{\text{hypotenuse}} = \dfrac{BC}{AB}$ – this ratio is denoted by $\cos\theta$

*tangent* of angle $\theta$ as $\dfrac{\text{opposite}}{\text{adjacent}} = \dfrac{AC}{BC}$ – this ratio is denoted by $\tan\theta$

▶

Every angle possesses its respective set of values for the trigonometric ratios and these are most easily found by using a calculator. For example, with the calculator in degree mode, press the *sin* key and enter 58 = to display 0·84804... which is the value of sin 58° (that is the ratio of the opposite side over the hypotenuse of all right-angled triangles with an angle of 58°).

Now, with your calculator in radian mode, press the *sin* key and enter 2 = to display 0·90929... which is the value of sin 2 rad – ordinarily we shall omit the rad and just write sin 2. Similar results are obtained using the *cos* key to find the cosine of an angle and the *tan* key to find the tangent of an angle.

Use a calculator in degree mode to find to 4 dp the values of:

(a) sin 27°
(b) cos 84°
(c) tan 43°

*The answers are in the next frame*

## 14

(a) 0·4540
(b) 0·1045
(c) 0·9325

That was easy enough. Now use a calculator in radian mode to find to 4 dp the values of the following where the angles are measured in radians:

(a) cos 1·321
(b) tan 0·013
(c) sin $\pi/6$

*Check with the next frame*

## 15

(a) 0·2472
(b) 0·0130
(c) 0·5000

We can now use these ratios to find unknowns. For example (see figure), a ladder of length 3 m leans against a vertical wall at an angle of 56° to the horizontal.

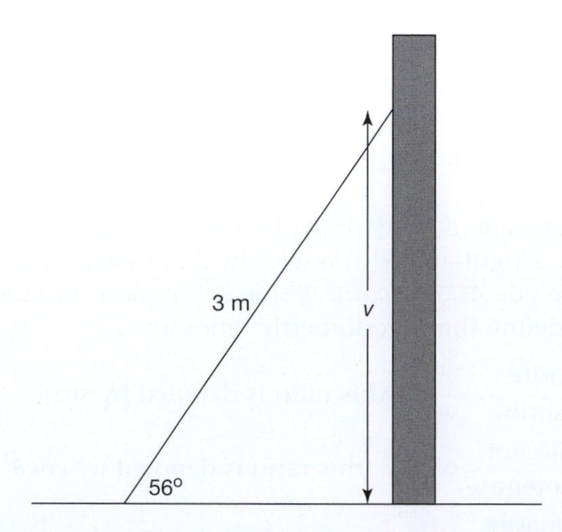


▶

The vertical height of the ladder can now be found as follows.

Dividing the vertical height $v$ (the opposite) by the length of the ladder (the hypotenuse) gives the sine of the angle of inclination 56°. That is:

$$\frac{\text{vertical height}}{\text{length of ladder}} = \sin 56°.$$

That is $\frac{v}{3} = 0{\cdot}82903\ldots$ giving the vertical height $v$ as

$3 \times 0{\cdot}82903\ldots = 2{\cdot}49$ m (to 3 sig fig)

So if a ladder of length $L$ leans against a wall at an angle of 60° to the horizontal with the top of the ladder 4·5 m above the ground, the length of the ladder is:

$L = \ldots\ldots$

*The answer is in the next frame*

---

**16**

5·20 m

Because

$$\frac{\text{vertical height}}{L} = \frac{4{\cdot}5}{L} = \sin 60° = 0{\cdot}8660\ldots$$

so that $L = \dfrac{4{\cdot}5}{0{\cdot}8660} = 5{\cdot}20$ m (to 2 dp)

*Next frame*

---

**17**

## Reciprocal ratios

In addition to the three trigonometrical ratios there are three *reciprocal ratios*, namely the cosecant (cosec), the secant (sec) and the cotangent (cot) where:

$$\operatorname{cosec}\theta = \frac{1}{\sin\theta},\ \sec\theta = \frac{1}{\cos\theta} \text{ and } \cot\theta = \frac{1}{\tan\theta} \equiv \frac{\cos\theta}{\sin\theta}$$

The values of these for a given angle can also be found using a calculator by finding the appropriate trigonometric ratio and then pressing the *reciprocal* key – the $x^{-1}$ key.

So that, to 4 dp:

(a) $\cot 12° = \ldots\ldots\ldots\ldots$
(b) $\sec 37° = \ldots\ldots\ldots\ldots$
(c) $\operatorname{cosec} 71° = \ldots\ldots\ldots\ldots$

*Next frame*

---

**18**

(a) 4·7046
(b) 1·2521
(c) 1·0576

Because

(a) $\tan 12° = 0{\cdot}21255\ldots$ and the reciprocal of that is 4·7046 to 4 dp

(b) $\cos 37° = 0{\cdot}79863\ldots$ and the reciprocal of that is 1·2521 to 4 dp

(c) $\sin 71° = 0{\cdot}94551\ldots$ and the reciprocal of that is 1·0576 to 4 dp

▶

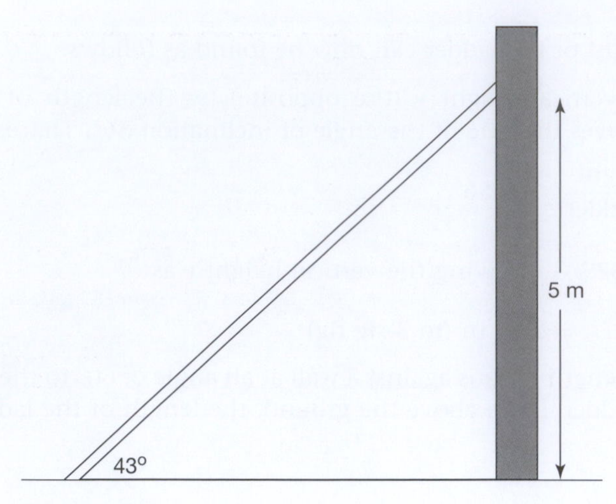


To strengthen a vertical wall a strut has to be placed 5 m up the wall and inclined at an angle of 43° to the ground. To do this the length of the strut must be ...........

*Check the next frame*

**19**

7·33 m

Because

$$\frac{\text{length of strut}}{5} = \frac{1}{\sin 43^\circ} = \operatorname{cosec} 43^\circ$$

that is $\frac{L}{5} = 1{\cdot}4662\ldots$

giving $L = 7{\cdot}33$ to 2 dp

*Now go to the next frame*

**20**

## Pythagoras' theorem

All right-angled triangles have a property in common that is expressed in Pythagoras' theorem:

*The square on the hypotenuse of a right-angled triangle is equal to the sum of the squares on the other two sides*

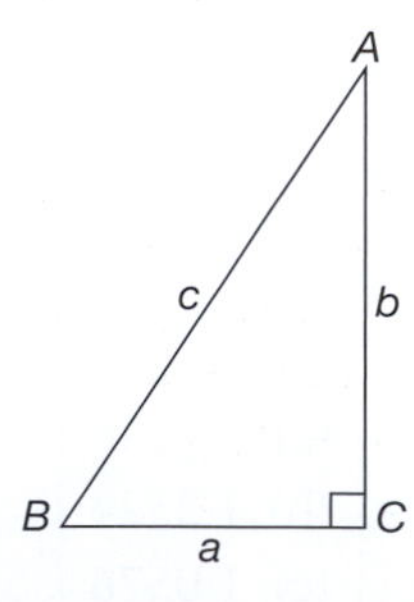


So in the figure:

$a^2 + b^2 = c^2$

Notice how the letter for each side length corresponds to the opposite angle ($a$ is opposite angle $A$ etc.); this is the common convention.
So, if a right-angled triangle has a hypotenuse of length 8 and one other side of length 3, the length of the third side to 3 dp is ...........

*Check your answer in the next frame*

21

7·416

Because

If $a$ represents the length of the third side then:

$$a^2 + 3^2 = 8^2 \text{ so } a^2 = 64 - 9 = 55 \text{ giving } a = 7{\cdot}416 \text{ to 3 dp}$$

Here's another. Is the triangle with sides 7, 24 and 25 a right-angled triangle?

*Answer in the next frame*

22

Yes

Because

Squaring the lengths of the sides gives:

$7^2 = 49$, $24^2 = 576$ and $25^2 = 625$.

Now, $49 + 576 = 625$ so that $7^2 + 24^2 = 25^2$

The sum of the squares of the lengths of the two smaller sides is equal to the square on the longest side. Because the lengths satisfy Pythagoras' theorem, the triangle is right-angled.

How about the triangle with sides 5, 11 and 12? Is this a right-angled triangle?

*Check in the next frame*

23

No

Because

$5^2 = 25$ and $11^2 = 121$ so $5^2 + 11^2 = 146 \neq 12^2$. The squares of the smaller sides do not add up to the square of the longest side so the triangle does not satisfy Pythagoras' theorem and so is not a right-angled triangle.

*Next frame*

24

## Special triangles

Two right-angled triangles are of special interest because the trigonometric ratios of their angles can be given in surd or fractional form. The first is the right-angled *isosceles* triangle (an isosceles triangle is any triangle with two sides of equal length). Since the angles of any triangle add up to 180°, the angles in a right-angled isosceles triangle are 90°, 45° and 45° (or, in radians, $\pi/2$, $\pi/4$ and $\pi/4$) with side lengths, therefore, in the ratio $1 : 1 : \sqrt{2}$ (by Pythagoras' theorem).

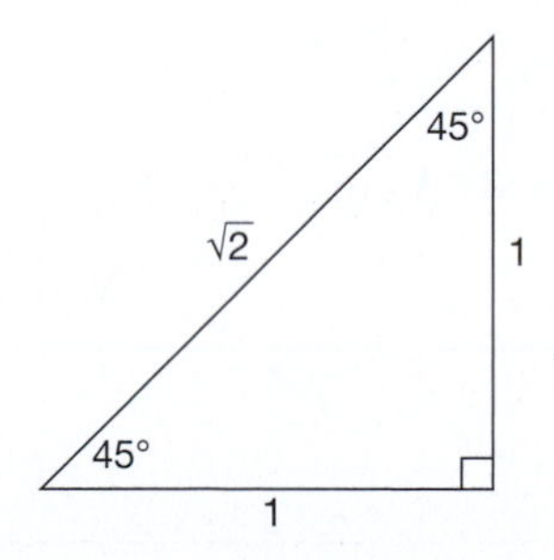


▶

Here we see that:

$$\sin 45^\circ = \cos 45^\circ = \frac{1}{\sqrt{2}} \text{ and } \tan 45^\circ = 1$$

Or, measuring the angles in radians:

$$\sin \pi/4 = \cos \pi/4 = \frac{1}{\sqrt{2}} \text{ and } \tan \pi/4 = 1$$

Now, a problem using these ratios:

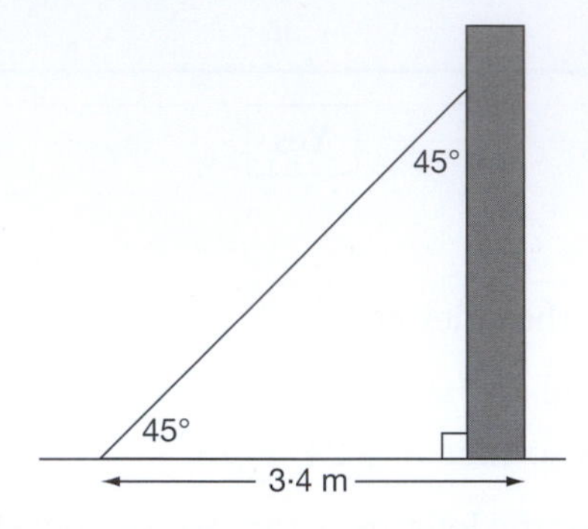


A prop in the form of an isosceles triangle constructed out of timber is placed against a vertical wall. If the length of the side along the horizontal ground is 3·4 m the length of the hypotenuse to 2 dp is obtained as follows:

$$\frac{\text{ground length}}{\text{hypotenuse}} = \frac{3{\cdot}4}{\text{hypotenuse}} = \cos 45^\circ = \frac{1}{\sqrt{2}}$$

so that:

$$\text{hypotenuse} = \sqrt{2} \times 3{\cdot}4 = 4{\cdot}81 \text{ m}$$

Now one for you to try.

A bicycle frame is in the form of an isosceles triangle with the horizontal crossbar forming the hypotenuse. If the crossbar is 53 cm long, the length of each of the other two sides to the nearest mm is ...........

*The answer is in the next frame*

**25**

37·5 cm

Because

$$\frac{\text{side length}}{\text{hypotenuse}} = \frac{\text{side length}}{53}$$
$$= \cos 45^\circ$$
$$= \frac{1}{\sqrt{2}} = 0{\cdot}7071\ldots$$

so that:

$$\text{side length} = 53 \times 0{\cdot}7071 = 37{\cdot}5 \text{ cm}$$

*Next frame for some more surd forms*

## Half equilateral



An equilateral triangle is a triangle whose sides are all the same length and whose angles are all equal to 60° (or, in radians, $\pi/3$).

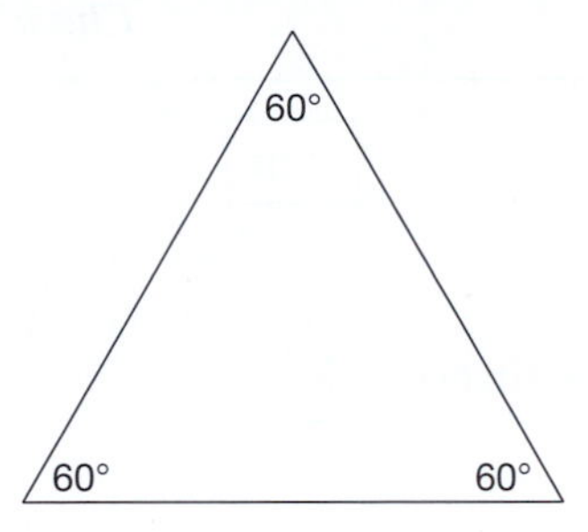


The second right-angled triangle of interest is the *half equilateral* triangle with side lengths (again, by Pythagoras) in the ratio $1 : \sqrt{3} : 2$.

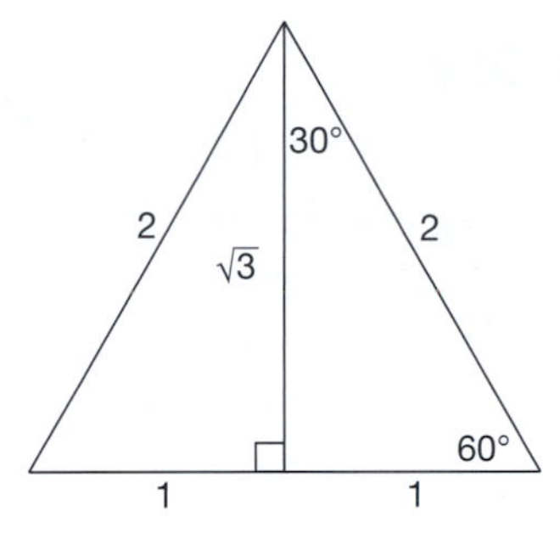


Here we see that:

$$\sin 30^\circ = \cos 60^\circ = \frac{1}{2}, \ \sin 60^\circ = \cos 30^\circ = \frac{\sqrt{3}}{2} \text{ and } \tan 60^\circ = \frac{1}{\tan 30^\circ} = \sqrt{3}$$

Again, if we measure the angles in radians:

$$\sin \pi/6 = \cos \pi/3 = \frac{1}{2}, \ \sin \pi/3 = \cos \pi/6 = \frac{\sqrt{3}}{2} \text{ and } \tan \pi/3 = \frac{1}{\tan \pi/6} = \sqrt{3}$$

Here's an example using these new ratios.

A tree casts a horizontal shadow $8\sqrt{3}$ m long. If a line were to be drawn from the end of the shadow to the top of the tree it would be inclined to the horizontal at 60°. The height of the tree is obtained as follows:

$$\frac{\text{height of tree}}{\text{length of shadow}} = \tan 60^\circ = \sqrt{3}$$

so that

$$\text{height of tree} = \sqrt{3} \times \text{ length of shadow } = \sqrt{3} \times 8\sqrt{3} = 8 \times 3 = 24 \text{ m}$$

▶

Now try this one.

When a small tent is erected the front forms an equilateral triangle. If the tent pole is $\sqrt{3}$ m long, the lengths of the sides of the tent are both …………

*Check your answer in the next frame*

**27**

2 m

Because

$$\frac{\text{length of tent pole}}{\text{length of tent side}} = \frac{\sqrt{3}}{L} = \sin 60^\circ = \frac{\sqrt{3}}{2}$$

so that $L = 2$ m.

##  Review summary

**28**

1 Angles can be measured in degrees, minutes and seconds or radians where a full angle has a magnitude of $360^\circ$ or $2\pi$ radians.

2 Similar triangles have the same shape – the same angles – but different sizes.

3 Ratios of the sides of one triangle have the same values as ratios of the corresponding sides in a similar triangle.

4 The trigonometric ratios are defined within a right-angled triangle. They are:

$$\sin\theta = \frac{\text{opposite}}{\text{hypotenuse}}$$

$$\cos\theta = \frac{\text{adjacent}}{\text{hypotenuse}}$$

$$\tan\theta = \frac{\sin\theta}{\cos\theta} = \frac{\text{opposite}}{\text{adjacent}}$$

and their reciprocal ratios are:

$$\operatorname{cosec}\theta = 1/\sin\theta$$

$$\sec\theta = 1/\cos\theta$$

$$\cot\theta = 1/\tan\theta$$

5 Pythagoras' theorem states that:

*The square on the hypotenuse of a right-angled triangle is equal to the sum of the squares on the other two sides*

$$a^2 + b^2 = c^2$$

where $a$ and $b$ are the lengths of the two smaller sides and $c$ is the length of the hypotenuse.

6 The right-angled isosceles triangle has angles $\pi/2$, $\pi/4$ and $\pi/4$, and sides in the ratio $1 : 1 : \sqrt{2}$.

7 The right-angled half equilateral triangle has angles $\pi/2$, $\pi/3$ and $\pi/6$, and sides in the ratio $1 : \sqrt{3} : 2$.

##  Review exercise

29

1 Convert the angle $164°49'13''$ to decimal degree format.

2 Convert the angle $87{\cdot}375°$ to degrees, minutes and seconds.

3 Convert the following to radians to 2 dp:

(a) $73°$ (b) $18{\cdot}34°$ (c) $240°$

4 Convert the following to degrees to 2 dp:

(a) $3{\cdot}721$ rad (b) $7\pi/6$ rad (c) $11\pi/12$ rad

5 Find the value of each of the following to 4 dp:

(a) $\sin 32°$ (b) $\cos \pi/12$ (c) $\tan 2\pi/5$

(d) $\sec 57{\cdot}8°$ (e) $\operatorname{cosec} 13{\cdot}33°$ (f) $\cot 0{\cdot}99$ rad

6 Given one side and the hypotenuse of a right-angled triangle as 5·6 cm and 12·3 cm respectively, find the length of the other side.

7 Show that the triangle with sides 9 m, 40 m and 41 m is a right-angled triangle.

8 A rod of length $7\sqrt{2}$ cm is inclined to the horizontal at an angle of $\pi/4$ radians. A shadow is cast immediately below it from a lamp directly overhead. What is the length of the shadow? What is the new length of the shadow if the rod's inclination is changed to $\pi/3$ to the vertical?

30

1 $164{\cdot}8203°$ to 4 dp.

2 $87°22'30''$

3 (a) $1{\cdot}27$ rad (b) $0{\cdot}32$ (c) $4\pi/3$ rad $= 4{\cdot}19$ rad

4 (a) $213{\cdot}20°$ (b) $210°$ (c) $165°$

5 (a) $0{\cdot}5299$ (b) $0{\cdot}9659$ (c) $3{\cdot}0777$

(d) $1{\cdot}8766$ (e) $4{\cdot}3373$ (f) $0{\cdot}6563$

6 If the sides are $a$, $b$ and $c$ where $c$ is the hypotenuse then $a^2 + b^2 = c^2$. That is, $(5{\cdot}6)^2 + b^2 = (12{\cdot}3)^2$ so that $b = \sqrt{(12{\cdot}3)^2 - (5{\cdot}6)^2} = 11{\cdot}0$ to 1 dp.

7 $40^2 + 9^2 = 1681 = 41^2$ thereby satisfying Pythagoras' theorem.

8 If $l$ is the length of the shadow then $\dfrac{l}{7\sqrt{2}} = \cos \pi/4 = \dfrac{1}{\sqrt{2}}$ so that $l = 7$ cm.

If the angle is $\pi/3$ to the vertical then $\dfrac{l}{7\sqrt{2}} = \sin \pi/3 = \dfrac{\sqrt{3}}{2}$ so that

$$l = \frac{7\sqrt{3}\sqrt{2}}{2} = 7\sqrt{\frac{3}{2}} = 8{\cdot}6 \text{ cm.}$$

*On now to the next topic*

# Trigonometric identities

## 31 The fundamental identity

Given the right-angled triangle of the figure with vertices $A$, $B$ and $C$, sides opposite the vertices of $a$, $b$ and hypotenuse $c$ and angle $\theta$ at $B$ then:

$$a^2 + b^2 = c^2$$

Dividing both sides by $c^2$ gives:

$$\left(\frac{a}{c}\right)^2 + \left(\frac{b}{c}\right)^2 = 1$$

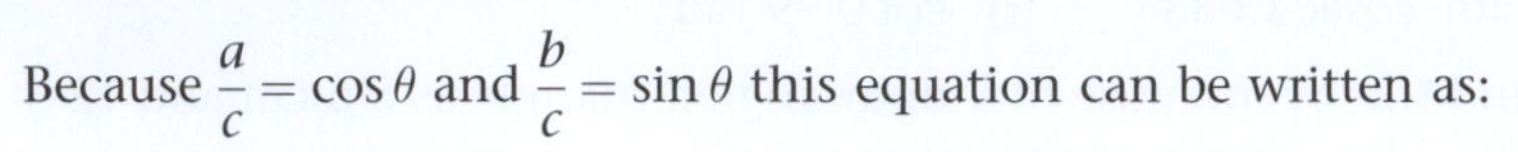

Because $\frac{a}{c} = \cos\theta$ and $\frac{b}{c} = \sin\theta$ this equation can be written as:

$$\cos^2\theta + \sin^2\theta = 1$$

where the notation $\cos^2\theta = (\cos\theta)^2$ and $\sin^2\theta = (\sin\theta)^2$. Since this equation is true for any angle $\theta$ the equation is in fact an identity:

$$\cos^2\theta + \sin^2\theta \equiv 1$$

and is called the *fundamental trigonometrical identity*.

For example, to show that the triangle with sides 3 cm, 4 cm and 5 cm is a right-angled triangle it is sufficient to show that it satisfies the fundamental trigonometrical identity. That is, taking the side of length 3 cm to be adjacent to $\theta$ (the side with length 5 cm is obviously the hypotenuse as it is the longest side) then:

$$\cos\theta = \frac{3}{5} \text{ and } \sin\theta = \frac{4}{5} \text{ and so}$$

$$\cos^2\theta + \sin^2\theta = \left(\frac{3}{5}\right)^2 + \left(\frac{4}{5}\right)^2 = \frac{9}{25} + \frac{16}{25}$$

$$= \frac{25}{25} = 1$$

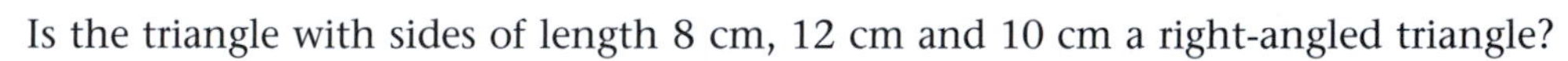

Is the triangle with sides of length 8 cm, 12 cm and 10 cm a right-angled triangle?

*The answer is in the next frame*

## 32

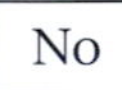

No

Because

$$\text{Letting } \cos\theta = \frac{8}{12} \text{ and } \sin\theta = \frac{10}{12}, \text{ then}$$

$$\cos^2\theta + \sin^2\theta = \left(\frac{8}{12}\right)^2 + \left(\frac{10}{12}\right)^2$$

$$= \frac{64}{144} + \frac{100}{144} = \frac{164}{144} \neq 1$$

Since the fundamental trigonometric identity is not satisfied this is not a right-angled triangle.

*Move to the next frame*

33

## Two more identities

Two more identities can be derived directly from the fundamental identity; dividing both sides of the fundamental identity by $\cos^2\theta$ gives the identity ...........

*Check your answer in the next frame*

34

$$\boxed{1+\tan^2\theta \equiv \sec^2\theta}$$

Because

$$\frac{\cos^2\theta}{\cos^2\theta}+\frac{\sin^2\theta}{\cos^2\theta}\equiv\frac{1}{\cos^2\theta}$$

that is $1+\tan^2\theta \equiv \sec^2\theta$

Dividing the fundamental identity by $\sin^2\theta$ gives a third identity ...........

*Next frame*

35

$$\boxed{\cot^2\theta+1\equiv \operatorname{cosec}^2\theta}$$

Because

$$\frac{\cos^2\theta}{\sin^2\theta}+\frac{\sin^2\theta}{\sin^2\theta}\equiv\frac{1}{\sin^2\theta}$$

that is $\cot^2\theta+1\equiv \operatorname{cosec}^2\theta$

Using these three identities and the definitions of the trigonometric ratios it is possible to demonstrate the validity of other identities. For example, to demonstrate the validity of the identity:

$$\frac{1}{1-\cos\theta}+\frac{1}{1+\cos\theta}\equiv 2\operatorname{cosec}^2\theta$$

we start with the left-hand side of this identity and demonstrate that it is equivalent to the right-hand side:

$$\begin{aligned}
\text{LHS} &= \frac{1}{1-\cos\theta}+\frac{1}{1+\cos\theta}\\
&\equiv \frac{1+\cos\theta+1-\cos\theta}{(1-\cos\theta)(1+\cos\theta)} && \text{Adding the two fractions together}\\
&\equiv \frac{2}{1-\cos^2\theta}\\
&\equiv \frac{2}{\sin^2\theta} && \text{From the fundamental identity}\\
&\equiv 2\operatorname{cosec}^2\theta\\
&= \text{RHS}
\end{aligned}$$

Try this one. Show that:

$$\tan\theta+\cot\theta\equiv\sec\theta\operatorname{cosec}\theta$$

*Next frame*

**36**

We proceed as follows:

$$\begin{aligned} \text{LHS} &= \tan\theta + \cot\theta \\ &\equiv \frac{\sin\theta}{\cos\theta} + \frac{\cos\theta}{\sin\theta} && \text{Writing explicitly in terms of sines and cosines} \\ &\equiv \frac{\sin^2\theta + \cos^2\theta}{\cos\theta\sin\theta} && \text{Adding the two fractions together} \\ &\equiv \frac{1}{\cos\theta\sin\theta} && \text{Since } \sin^2\theta + \cos^2\theta = 1 \text{ (the fundamental identity)} \\ &\equiv \sec\theta\,\text{cosec}\,\theta \\ &= \text{RHS} \end{aligned}$$

So demonstrate the validity of each of the following identities:

(a) $\tan^2\theta - \sin^2\theta \equiv \sin^4\theta\sec^2\theta$

(b) $\dfrac{1+\sin\theta}{\cos\theta} \equiv \dfrac{\cos\theta}{1-\sin\theta}$

Take care with the second one – it is done by performing an operation on both sides first.

*The answers are in the next frame.*

**37**

(a)
$$\begin{aligned} \text{LHS} &= \tan^2\theta - \sin^2\theta \\ &\equiv \frac{\sin^2\theta}{\cos^2\theta} - \sin^2\theta && \text{Writing explicitly in terms of sines and cosines} \\ &\equiv \sin^2\theta\sec^2\theta - \sin^2\theta \\ &\equiv \sin^2\theta(\sec^2\theta - 1) && \text{Factorizing out the } \sin^2\theta \\ &\equiv \sin^2\theta\tan^2\theta && \text{Using the identity } 1 + \tan^2\theta \equiv \sec^2\theta \\ &\equiv \sin^2\theta\frac{\sin^2\theta}{\cos^2\theta} \\ &\equiv \sin^4\theta\sec^2\theta \\ &= \text{RHS} \end{aligned}$$

(b) $\dfrac{1+\sin\theta}{\cos\theta} \equiv \dfrac{\cos\theta}{1-\sin\theta}$

Multiplying both sides by $\cos\theta(1-\sin\theta)$ transforms the identity into:

$(1-\sin\theta)(1+\sin\theta) \equiv \cos^2\theta$. From this we find that:

$$\begin{aligned} \text{LHS} &= (1-\sin\theta)(1+\sin\theta) \\ &\equiv 1 - \sin^2\theta \\ &\equiv \cos^2\theta && \text{since } \cos^2\theta + \sin^2\theta \equiv 1 \\ &= \text{RHS} \end{aligned}$$

*Move on now to the next frame*

**38**

## Identities for compound angles

The trigonometric ratios of the sum or difference of two angles can be given in terms of the ratios of the individual angles. For example, the cosine of a sum of angles is given by:

$$\cos(\theta+\phi) \equiv \cos\theta\cos\phi - \sin\theta\sin\phi$$

▶

To demonstrate the validity of this, consider the following figure:

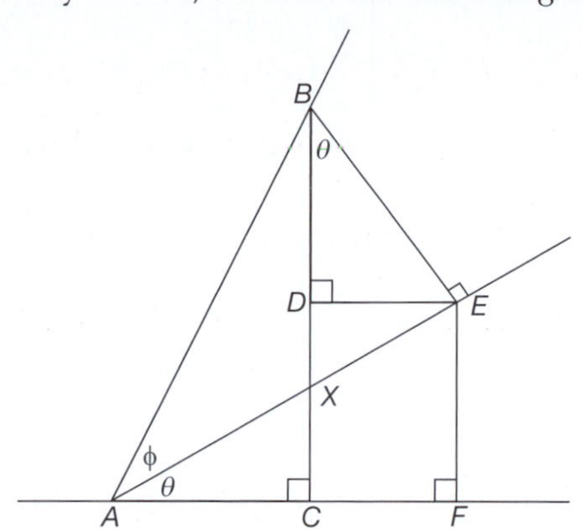


(Notice that in triangles $AXC$ and $BXE$, $\angle C = \angle E$ as both are right angles, and $\angle AXC = \angle BXE$ as they are equal and opposite. Consequently, the third angles must also be equal so that $\angle EBX = \angle CAX = \theta$.)

Hence we see that:

$$\cos(\theta + \phi) = \frac{AC}{AB} \quad \text{Adjacent over hypotenuse}$$
$$= \frac{AF - CF}{AB}$$
$$= \frac{AF - DE}{AB} \quad \text{Because } DE = CF$$
$$= \frac{AF}{AB} - \frac{DE}{AB} \quad \text{Separating out the fraction}$$

Now, $\cos\theta = \frac{AF}{AE}$ so that $AF = AE\cos\theta$. Similarly, $\sin\theta = \frac{DE}{BE}$ so that $DE = BE\sin\theta$. This means that:

$$\cos(\theta + \phi) = \frac{AF}{AB} - \frac{DE}{AB}$$

$= \frac{AE\cos\theta}{AB} - \frac{BE\sin\theta}{AB}$. Now, $\frac{AE}{AB} = \cos\phi$ and $\frac{BE}{AB} = \sin\phi$, therefore

$$\cos(\theta + \phi) \equiv \cos\theta\cos\phi - \sin\theta\sin\phi$$

A similar identity can be demonstrated for the difference of two angles, namely:

$$\cos(\theta - \phi) \equiv \cos\theta\cos\phi + \sin\theta\sin\phi$$

Using these identities it is possible to obtain the cosine of angles other than 30°, 60° and 45° in surd form. For example:

$$\cos 75^\circ = \cos(45^\circ + 30^\circ) \quad \text{Expressing } 75^\circ \text{ in angles where we know the surd form for the trigonometric ratios}$$
$$= \cos 45^\circ \cos 30^\circ - \sin 45^\circ \sin 30^\circ \quad \text{Using the new formula}$$
$$= \frac{1}{\sqrt{2}} \times \frac{\sqrt{3}}{2} - \frac{1}{\sqrt{2}} \times \frac{1}{2}$$
$$= \frac{\sqrt{3} - 1}{2\sqrt{2}}$$

So the value of $\cos 15^\circ$ in surd form is ...........

*The answer is in the next frame*

39

$$\boxed{\frac{1+\sqrt{3}}{2\sqrt{2}}}$$

Because

$$\begin{aligned}\cos 15^\circ &= \cos(60^\circ - 45^\circ)\\ &= \cos 60^\circ \cos 45^\circ + \sin 60^\circ \sin 45^\circ\\ &= \frac{1}{2}\times\frac{1}{\sqrt{2}}+\frac{\sqrt{3}}{2}\times\frac{1}{\sqrt{2}}\\ &= \frac{1+\sqrt{3}}{2\sqrt{2}}\end{aligned}$$

Just as it is possible to derive the cosine of a sum of angles, it is also possible to derive other trigonometric ratios of sums and differences of angles. In the next frame a list of such identities is given for future reference.

40

## Trigonometric formulas

### Sums and differences of angles

$$\cos(\theta+\phi) \equiv \cos\theta\cos\phi - \sin\theta\sin\phi \qquad \sin(\theta+\phi) \equiv \sin\theta\cos\phi + \cos\theta\sin\phi$$

$$\cos(\theta-\phi) \equiv \cos\theta\cos\phi + \sin\theta\sin\phi \qquad \sin(\theta-\phi) \equiv \sin\theta\cos\phi - \cos\theta\sin\phi$$

$$\tan(\theta+\phi) \equiv \frac{\sin(\theta+\phi)}{\cos(\theta+\phi)} \equiv \frac{\sin\theta\cos\phi+\cos\theta\sin\phi}{\cos\theta\cos\phi-\sin\theta\sin\phi}$$

Now divide numerator and denominator by $\cos\theta\cos\phi$

$$\equiv \frac{\tan\theta+\tan\phi}{1-\tan\theta\tan\phi}$$

$$\tan(\theta-\phi) \equiv \frac{\tan\theta-\tan\phi}{1+\tan\theta\tan\phi}$$

### Double angles

Double angle formulas come from the above formulas for sums when $\theta=\phi$:

$$\sin 2\theta \equiv 2\sin\theta\cos\theta$$

$$\cos 2\theta \equiv \cos^2\theta - \sin^2\theta \equiv 2\cos^2\theta - 1 \equiv 1 - 2\sin^2\theta$$

$$\tan 2\theta \equiv \frac{2\tan\theta}{1-\tan^2\theta}$$

For future reference we now list identities for sums, differences and products of the trigonometric ratios. Each of these can be proved by using the earlier identities and showing that RHS $\equiv$ LHS (rather than showing LHS $\equiv$ RHS as we have done hitherto).

### Sums and differences of ratios

$$\sin\theta + \sin\phi \equiv 2\sin\frac{\theta+\phi}{2}\cos\frac{\theta-\phi}{2}$$

$$\sin\theta - \sin\phi \equiv 2\cos\frac{\theta+\phi}{2}\sin\frac{\theta-\phi}{2}$$

$$\cos\theta + \cos\phi \equiv 2\cos\frac{\theta+\phi}{2}\cos\frac{\theta-\phi}{2}$$

$$\cos\theta - \cos\phi \equiv -2\sin\frac{\theta+\phi}{2}\sin\frac{\theta-\phi}{2}$$

### Products of ratios

$$2\sin\theta\cos\phi \equiv \sin(\theta+\phi) + \sin(\theta-\phi)$$

$$2\cos\theta\cos\phi \equiv \cos(\theta+\phi) + \cos(\theta-\phi)$$

$$2\sin\theta\sin\phi \equiv \cos(\theta-\phi) - \cos(\theta+\phi)$$

##  Review summary

41

1 The fundamental trigonometric identity is $\cos^2\theta + \sin^2\theta \equiv 1$ and is derived from *Pythagoras*' theorem.

2 Trigonometric identities can be verified using both the fundamental identity and the definitions of the trigonometric ratios.

## Review exercise

42

1 Use the fundamental trigonometric identity to show that:

(a) the triangle with sides 5 cm, 12 cm and 13 cm is a right-angled triangle.

(b) the triangle with sides 7 cm, 15 cm and 16 cm is not a right-angled triangle.

2 Verify each of the following identities:

(a) $1 - \dfrac{\sin\theta\tan\theta}{1+\sec\theta} \equiv \cos\theta$ (b) $\sin\theta + \sin\phi \equiv 2\sin\dfrac{\theta+\phi}{2}\cos\dfrac{\theta-\phi}{2}$

43

1 (a) $5^2 + 12^2 = 25 + 144 = 169 = 13^3$

(b) $7^2 + 15^2 = 49 + 225 = 274 \neq 16^2$

2 (a)

$$\begin{aligned}
\text{LHS} &= 1 - \frac{\sin\theta\tan\theta}{1+\sec\theta}\\
&\equiv \frac{1+\sec\theta - \sin\theta\tan\theta}{1+\sec\theta}\\
&\equiv \frac{\cos\theta + 1 - \sin^2\theta}{\cos\theta + 1} \quad \text{multiplying top and bottom by } \cos\theta\\
&\equiv \frac{\cos\theta + \cos^2\theta}{\cos\theta + 1}\\
&\equiv \frac{\cos\theta(1+\cos\theta)}{\cos\theta + 1}\\
&\equiv \cos\theta\\
&= \text{RHS}
\end{aligned}$$

(b)

$$\begin{aligned}
\text{RHS} &= 2\sin\frac{\theta+\phi}{2}\cos\frac{\theta-\phi}{2}\\
&\equiv 2\left(\sin\frac{\theta}{2}\cos\frac{\phi}{2} + \sin\frac{\phi}{2}\cos\frac{\theta}{2}\right)\left(\cos\frac{\theta}{2}\cos\frac{\phi}{2} + \sin\frac{\phi}{2}\sin\frac{\theta}{2}\right)\\
&\equiv 2\left(\sin\frac{\theta}{2}\cos\frac{\theta}{2}\cos^2\frac{\phi}{2} + \sin\frac{\phi}{2}\cos\frac{\phi}{2}\cos^2\frac{\theta}{2}\right.\\
&\qquad \left. + \sin\frac{\phi}{2}\cos\frac{\phi}{2}\sin^2\frac{\theta}{2} + \sin^2\frac{\phi}{2}\sin\frac{\theta}{2}\cos\frac{\theta}{2}\right)\\
&\equiv \sin\theta\cos^2\frac{\phi}{2} + \sin\phi\cos^2\frac{\theta}{2} + \sin\phi\sin^2\frac{\theta}{2} + \sin^2\frac{\phi}{2}\sin\theta\\
&\equiv \sin\theta\cos^2\frac{\phi}{2} + \sin^2\frac{\phi}{2}\sin\theta + \sin\phi\cos^2\frac{\theta}{2} + \sin\phi\sin^2\frac{\theta}{2}\\
&\equiv \sin\theta\left(\cos^2\frac{\phi}{2} + \sin^2\frac{\phi}{2}\right) + \sin\phi\left(\cos^2\frac{\theta}{2} + \sin^2\frac{\theta}{2}\right)\\
&\equiv \sin\theta + \sin\phi\\
&= \text{LHS}
\end{aligned}$$

**44**

You have now come to the end of this Programme. A list of **Can You?** questions follows for you to gauge your understanding of the material in the Programme. You will notice that these questions match the **Learning outcomes** listed at the beginning of the Programme so go back and try the **Quiz** that follows them. After that try the **Test exercise**. *Work through these at your own pace, there is no need to hurry.* A set of **Further problems** provides additional valuable practice.

## Can You?

**45**

### Checklist F.9

Check this list before and after you try the end of Programme test.

On a scale of 1 to 5 how confident are you that you can: Frames

- Convert angles measured in degrees, minutes and seconds into decimal degrees? 1 to 3
  *Yes* ☐ ☐ ☐ ☐ ☐ *No*
- Convert degrees into radians and vice versa? 4 to 8
  *Yes* ☐ ☐ ☐ ☐ ☐ *No*
- Use a calculator to determine the values of trigonometric ratios for any angle? 9 to 27
  *Yes* ☐ ☐ ☐ ☐ ☐ *No*
- Verify trigonometric identities? 31 to 43
  *Yes* ☐ ☐ ☐ ☐ ☐ *No*

## Test exercise F.9

**46**

Frames

1. Convert the angle $39°57'2''$ to decimal degree format. 1 to 2
2. Convert the angle $52{\cdot}505°$ to degrees, minutes and seconds. 2 to 3
3. Convert the following to radians to 2 dp: 4 to 6
   (a) $84°$ (b) $69{\cdot}12°$ (c) $240°$ (as a multiple of $\pi$)
4. Convert the following to degrees to 2 dp: 6 to 8
   (a) 2·139 rad (b) $5\pi/3$ rad (c) $9\pi/10$ rad
5. Find the value of each of the following to 4 dp: 9 to 19
   (a) $\cos 18°$ (b) $\sin \pi/11$ (c) $\cos 2\pi/7$
   (d) $\cot 48{\cdot}7°$ (e) cosec 1·04 rad (f) sec 0·85 rad
6. Given one side and the hypotenuse of a right-angled triangle as 4·3 cm and 11·2 cm, find the length of the other side. 20 to 23
7. Show that the triangle with sides 9 cm, 12 cm and 15 cm is a right-angled triangle. 20 to 23
8. A triangle has its three sides in the ratio 1 : 0·6 : 0·8. Is it a right-angled triangle? 24 to 27

**9** Verify each of the following trigonometric identities: Frames

(a) $\dfrac{(\cos\theta - \sin\theta)^2}{\cos\theta} \equiv \sec\theta - 2\sin\theta$

(b) $\dfrac{\operatorname{cosec}\theta\sec\theta}{\cot\theta} \equiv 1 + \tan^2\theta$

(c) $2\sin\theta\cos\phi \equiv \sin(\theta+\phi) + \sin(\theta-\phi)$ 31 to 40

**10** Show that:

(a) $\sin 75^\circ = \dfrac{1+\sqrt{3}}{2\sqrt{2}}$ (b) $\tan 15^\circ = \dfrac{\sqrt{3}-1}{\sqrt{3}+1}$ 38 to 39

---

## Further problems F.9

**47**

1 Convert the angle $81^\circ 18' 23''$ to decimal degree format.

2 Convert the angle $63{\cdot}216^\circ$ to degrees, minutes and seconds.

3 Convert the following to radians to 2 dp:

(a) $31^\circ$ (b) $48{\cdot}15^\circ$ (c) $225^\circ$ (as a multiple of $\pi$)

4 Convert the following to degrees to 2 dp:

(a) $1{\cdot}784$ rad (b) $3\pi/4$ rad (c) $4\pi/5$ rad

5 Find the value of each of the following to 4 dp:

(a) $\tan 27^\circ$ (b) $\sin \pi/5$ (c) $\tan 4\pi/9$

(d) $\sec 89{\cdot}2^\circ$ (e) $\operatorname{cosec} 0{\cdot}04^\circ$ (f) $\cot 1{\cdot}18$ rad

6 Given one side and the hypotenuse of a right-angled triangle as 6·4 cm and 9·1 cm, find the length of the other side.

7 Show that the triangle with sides 5 cm, 11 cm and 12 cm is not a right-angled triangle.

8 What is the length of the diagonal of a square of side length $\sqrt{2}$?

9 Verify each of the following trigonometric identities:

(a) $\dfrac{\cos\theta - 1}{\sec\theta + \tan\theta} + \dfrac{\cos\theta + 1}{\sec\theta - \tan\theta} \equiv 2(1+\tan\theta)$

(b) $\sin^3\theta - \cos^3\theta \equiv (\sin\theta - \cos\theta)(1 + \sin\theta\cos\theta)$

(c) $\operatorname{cosec}^2\theta - \operatorname{cosec}\theta \equiv \dfrac{\cot^2\theta}{1+\sin\theta}$

(d) $\cot\theta\cos\theta + \tan\theta\sin\theta \equiv (\operatorname{cosec}\theta + \sec\theta)(1 - \sin\theta\cos\theta)$

(e) $\dfrac{\cos\theta + \sin\theta}{\cos\theta - \sin\theta} \equiv 1 + \dfrac{2\tan\theta}{1-\tan\theta}$

(f) $(\sin\theta - \cos\theta)^2 + (\sin\theta + \cos\theta)^2 \equiv 2$

(g) $\sqrt{\dfrac{1+\tan^2\theta}{1+\cot^2\theta}} \equiv \tan\theta$

---

**48**

Now visit the companion website for this book at www.palgrave.com/stroud for more questions applying this mathematics to science and engineering.

# Functions

## Learning outcomes

When you have completed this Programme you will be able to:

- ☐ Identify a function as a rule and recognize rules that are not functions
- ☐ Determine the domain and range of a function
- ☐ Construct the inverse of a function and draw its graph
- ☐ Construct compositions of functions and de-construct them into their component functions

If you already feel confident about these why not try the quiz over the page? You can check your answers at the end of the book.

# Quiz F.10

**Frames**

1 Which of the following equations expresses a rule that is a function?

(a) $y = 1 - x^2$

(b) $y = -\sqrt{x^4}$

(c) $y = x^{\frac{1}{6}}$ — 1 to 4

2 Given the two functions $f$ and $g$ expressed by:

$$f(x) = \frac{1}{4 - x} \text{ for } 0 \leq x < 4 \quad \text{and} \quad g(x) = x - 3 \text{ for } 0 < x \leq 5$$

find the domain and range of functions $h$ and $k$ where:

(a) $h(x) = f(x) + 3g(x)$

(b) $k(x) = \dfrac{f(x)}{2g(x)}$ — 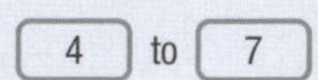 4 to 7

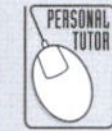

3 Use your spreadsheet to draw each of the following and their inverses. Is the inverse a function?

(a) $y = x^5$

(b) $y = -3x^2$

(c) $y = \sqrt{1 - x^2}$ — 8 to 16

4 Given that $a(x) = -2x$, $b(x) = x^3$, $c(x) = x - 1$ and $d(x) = \sqrt{x}$ find:

(a) $f(x) = a[b(c[d(x)])]$

(b) $f(x) = a(a[d(x)])$

(c) $f(x) = b[c(b[c(x)])]$ — 17 to 21

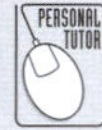

5 Given that $f(x) = (3x + 4)^2 + 4$, decompose $f$ into its component functions and find its inverse. Is the inverse a function? — 22 to 23

## Processing numbers

**1**

The equation that states that $y$ *is equal to some expression in* $x$, written as:

$y = f(x)$

has been described with the words '$y$ *is a function of* $x$'. Despite being widely used and commonly accepted, this description is not strictly correct as will be seen in Frame 3. Put simply, for all the functions that you have considered so far, both $x$ and $y$ are *numbers*.

Take out your calculator and enter the number:

5 this is $x$, the *input* number

Now press the $x^2$ key and the display changes to:

25 this is $y$, the *output* number where $y = x^2$

The *function* is a *rule* embodied in a *set of instructions* within the calculator that changed the 5 to 25, activated by you pressing the $x^2$ key. A diagram can be constructed to represent this:

The box labelled $f$ represents the function. The notation ^2 inside the box means *raising to the power 2* and describes the rule – what the set of instructions will do when activated. The diagram tells you that the input number $x$ is *processed* by the function $f$ to produce the output number $y = f(x)$. So that $y = f(x)$ is the *result* of function $f$ acting on $x$.

So, use diagrams and describe the functions appropriate to each of the following equations:

(a) $y = \dfrac{1}{x}$ (b) $y = x - 6$ (c) $y = 4x$ (d) $y = x^{1/3}$

*Just follow the reasoning above, the answers are in the next frame*

**2**

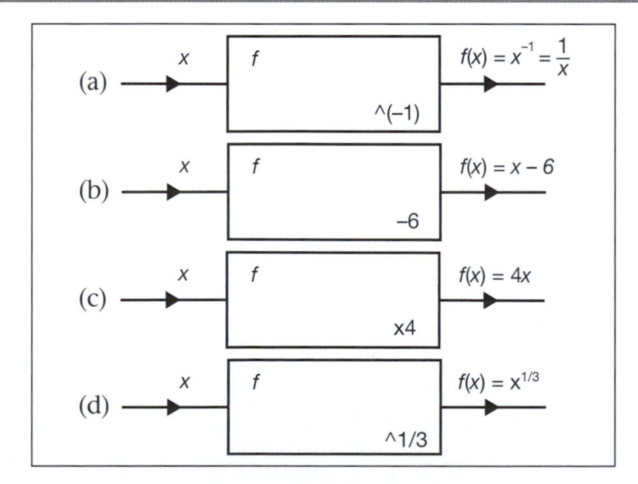


(a) Function $f$ produces the reciprocal of the input
(b) Function $f$ subtracts 6 from the input
(c) Function $f$ multiplies the input by 4
(d) Function $f$ produces the cube root of the input

*Let's now expand this idea*

## 3 Functions are rules but not all rules are functions

A function of a variable $x$ is a *rule* that describes how a value of the variable $x$ is manipulated to generate a value of the variable $y$. The rule is often expressed in the form of an equation $y = f(x)$ with the proviso that for any specific input $x$ there is a unique value for $y$. Different outputs are associated with different inputs – the function is said to be *single valued* because for a given input there is only one output. For example, the equation:

$$y = 2x + 3$$

expresses the rule '*multiply the value of x by two and add three*' and this rule is the function. On the other hand, the equation:

$$y = x^{\frac{1}{2}} \text{ which is the same as } y = \pm\sqrt{x}$$

expresses the rule '*take the positive and negative square roots of the value of x*'. This rule is not a function because to each value of the input $x > 0$ there are two different values of output $y$. The graph of $y = \pm\sqrt{x}$ illustrates this quite clearly:

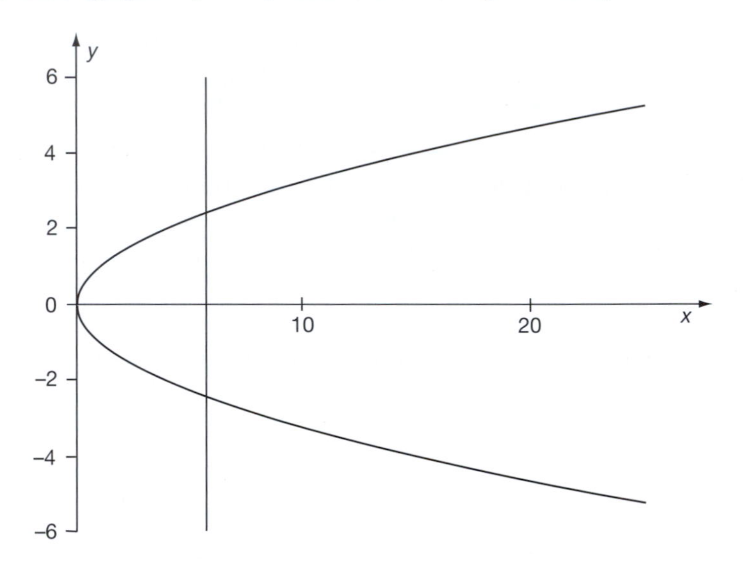


If a vertical line is drawn through the $x$-axis (for $x > 0$) it intersects the graph at more than one point. The fact that for $x = 0$ the vertical line intersects the graph at only *one* point does not matter – that there are other points where the vertical line intersects the graph in more than one point is sufficient to bar this from being the graph of a function. Notice that $y = x^{\frac{1}{2}}$ has no real values of $y$ for $x < 0$.

Also note that your calculator only gives a single answer to $x^{\frac{1}{2}}$ because it is, in fact, calculating $\sqrt{x}$.

So, which of the following equations express rules that are functions?

(a) $y = 5x^2 + 2x^{-\frac{1}{4}}$

(b) $y = 7x^{\frac{1}{3}} - 3x^{-1}$

*Next frame*

4

(a) $y = 5x^2 + 2x^{-\frac{1}{4}}$ does not
(b) $y = 7x^{\frac{1}{3}} - 3x^{-1}$ does

(a) $y = 5x^2 + 2x^{-\frac{1}{4}}$ does not express a function because to each value of $x$ ($x > 0$) there are two values of $x^{-\frac{1}{4}}$, positive and negative because $x^{-\frac{1}{4}} \equiv (x^{-\frac{1}{2}})^{\frac{1}{2}} \equiv \pm\sqrt{x^{-\frac{1}{2}}}$. Indeed, *any* even root produces two values.
(b) $y = 7x^{\frac{1}{3}} - 3x^{-1}$ does express a function because to each value of $x$ ($x \neq 0$) there is just one value of $y$.

All the input numbers $x$ that a function can process are collectively called the function's *domain*. The complete collection of numbers $y$ that correspond to the numbers in the domain is called the *range* (or *co-domain*) of the function. For example, if:

$y = \sqrt{1 - x^2}$ where both $x$ and $y$ are real numbers

the domain is $-1 \leq x \leq 1$ because these are the only values of $x$ for which $y$ has a real value. The range is $0 \leq y \leq 1$ because 0 and 1 are the minimum and maximum values of $y$ over the domain. Other functions may, for some purpose or other, be defined on a restricted domain. For example, if we specify:

$y = x^3, \quad -2 \leq x < 3$ (the function is defined only for the restricted set of $x$-values given)

the domain is given as $-2 \leq x < 3$ and the range as $-8 \leq y < 27$ because $-8$ and 27 are the minimum and maximum values of $y$ over the domain.
So the domains and ranges of each of the following are:

(a) $y = x^3 \quad -5 \leq x < 4$ (b) $y = x^4$ (c) $y = \dfrac{1}{(x-1)(x+2)} \quad 0 \leq x \leq 6$

*The answers are in the next frame*

5

(a) $y = x^3 \quad -5 \leq x < 4$
domain $-5 \leq x < 4$, range $-125 \leq y < 64$
(b) $y = x^4$
domain $-\infty < x < \infty$, range $0 \leq y < \infty$
(c) $y = \dfrac{1}{(x-1)(x+2)}, \quad 0 \leq x \leq 6$
domain $0 \leq x < 1$ and $1 < x \leq 6$,
range $-\infty < y \leq -0{\cdot}5$, $0{\cdot}025 \leq y < \infty$

Because

(a) The domain is given as $-5 \leq x < 4$ and the range as $-125 \leq y < 64$ because $-125$ and 64 are the minimum and maximum values of $y$ over the domain.
(b) The domain is not given and is assumed to consist of all finite values of $x$, that is, $-\infty < x < \infty$. The range values are all positive because of the even power.
(c) The domain is $0 \leq x < 1$ and $1 < x \leq 6$ since $y$ is not defined when $x = 1$ where there is a vertical asymptote. To the left of the asymptote ($0 \leq x < 1$) the $y$-values range from $y = -0{\cdot}5$ when $x = 0$ and increase negatively towards $-\infty$ as $x \to 1$. To the right of the asymptote $1 < x \leq 6$ the $y$-values range from infinitely large and positive to $0{\cdot}025$ when $x = 6$. If you plot the graph on your spreadsheet this will be evident.

*Next frame*

**6**

## Functions and the arithmetic operations

Functions can be combined under the action of the arithmetic operators provided care is taken over their common domains. For example:

If $f(x) = x^2 - 1, \quad -2 \le x < 4$ and $g(x) = \dfrac{2}{x+3}, \quad 0 < x \le 5$ then, for example

(a) $h(x) = f(x) + g(x) = x^2 - 1 + \dfrac{2}{x+3}, \quad 0 < x < 4$

because $g(x)$ is not defined for $-2 \le x \le 0$ and $f(x)$ is not defined for $4 \le x \le 5$ so $0 < x < 4$ is the common domain between them.

(b) $k(x) = \dfrac{g(x)}{f(x)} = \dfrac{2}{(x+3)(x^2-1)}, \quad 0 < x < 4$ and $x \ne 1$

because $g(x)$ is not defined for $-2 \le x \le 0$, $f(x)$ is not defined for $4 \le x \le 5$ and $k(x)$ is not defined when $x = 1$.

So if:

$$f(x) = \frac{2x}{x^3 - 1}, \text{ where } -3 < x < 3 \text{ and } x \ne 1 \text{ and}$$

$$g(x) = \frac{4x-8}{x+5}, 0 < x \le 6 \text{ then } h(x) = \frac{f(x)}{g(x)} \text{ is } \ldots\ldots\ldots\ldots$$

*The answer is in the next frame*

**7**

$$h(x) = \frac{f(x)}{g(x)} = \frac{2x(x+5)}{(x^3-1)(4x-8)} \text{ where } 0 < x < 3, x \ne 1 \text{ and } x \ne 2$$

Because when $x = 1$ or 2, $h(x)$ is not defined; when $-3 < x \le 0$, $g(x)$ is not defined; and when $3 \le x \le 6$, $f(x)$ is not defined.

**8**

## Inverses of functions

The process of generating the output of a function is assumed to be reversible so that what has been constructed can be de-constructed. The effect can be described by reversing the flow of information through the diagram so that, for example, if:

$y = f(x) = x + 5$

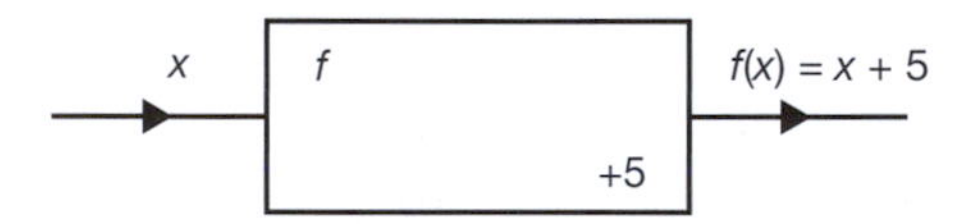


the flow is reversed by making the output the input and *retrieving the original input as the new output*:

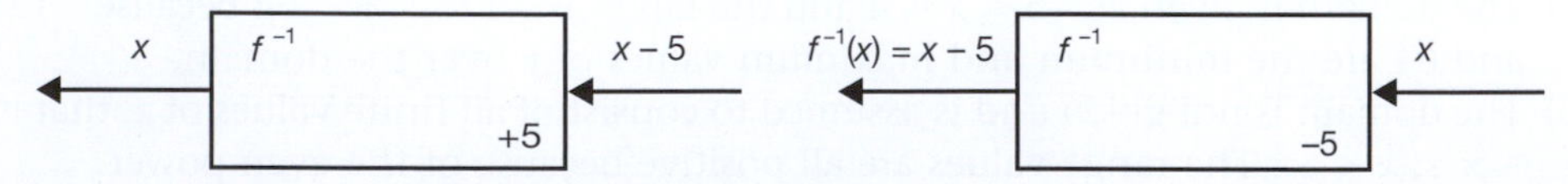


The reverse process is different because instead of adding 5 to the input, 5 is now subtracted from the input. The rule that describes the reversed process is called the *inverse of the function* which is labelled as either $f^{-1}$ or *arcf*. That is:

$f^{-1}(x) = x - 5$

The notation $f^{-1}$ is very commonly used but care must be taken to remember that the $-1$ does not mean that it is in any way related to the reciprocal of $f$.

Try some. Find $f^{-1}(x)$ in each of the following cases:

(a) $f(x) = 6x$ (b) $f(x) = x^3$ (c) $f(x) = \frac{x}{2}$

*Draw the diagram, reverse the flow and find the inverse of the function in each case*

---

**9**

(a) $f^{-1}(x) = \frac{x}{6}$
(b) $f^{-1}(x) = x^{\frac{1}{3}}$
(c) $f^{-1}(x) = 2x$

Because

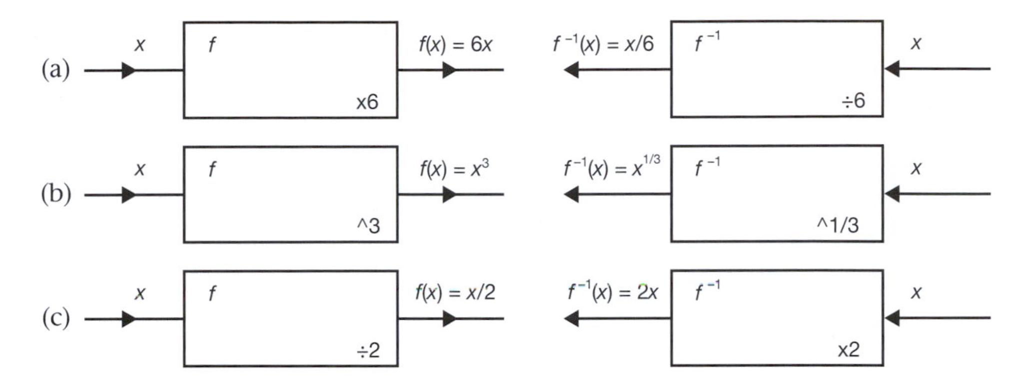


The inverses of the arithmetic operations are just as you would expect:

*addition and subtraction are inverses of each other*
*multiplication and division are inverses of each other*
*raising to the power k and raising to a power 1/k are inverses of each other*

Now, can you think of two functions that are each identical to their inverse?

*Think carefully*

---

**10**

$f(x) = x$ and $f(x) = \frac{1}{x}$

Because the function with output $f(x) = x$ does not alter the input at all so the inverse does not either, and the function with output $f(x) = \frac{1}{x}$ is its own inverse because the reciprocal of the reciprocal of a number is the number:

$$\frac{1}{1/x} = x$$

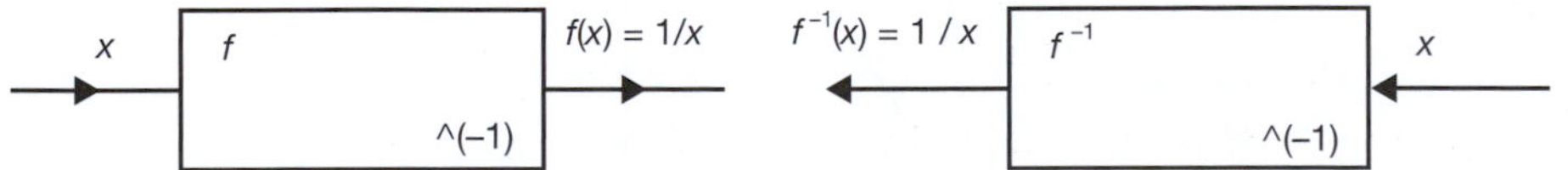


*Let's progress*

## 11 Graphs of inverses

The diagram of the inverse of a function can be drawn by reversing the flow of information and this is the same as interchanging the contents of each ordered pair generated by the function. As a result, when the ordered pairs generated by the inverse of a function are plotted, the graph takes up the shape of the original function but reflected in the line $y = x$. Let's try it. Use your spreadsheet to plot $y = x^3$ and the inverse $y = x^{\frac{1}{3}}$. If you are unfamiliar with the use of a spreadsheet, read Programme F.4 first where the spreadsheet is introduced.

*What you are about to do is a little involved, so follow the instructions to the letter and take it slowly and carefully*

## 12 The graph of $y = x^3$

Open up your spreadsheet

Enter $-1{\cdot}1$ in cell **A1**
Highlight **A1** to **A24**
Click **Edit-Fill-Series** and enter the **step value** as $0{\cdot}1$

The cells **A1** to **A24** then fill with the numbers $-1{\cdot}1$ to $1{\cdot}2$.

In cell **B1** enter the formula **=A1^3** and press **Enter**

Cell **B1** now contains the cube of the contents of cell **A1**

Make **B1** the active cell
Click **Edit-Copy** This copies the contents of B1 to the Clipboard
Highlight **B2** to **B24**
Click **Edit-Paste** This pastes the contents of the Clipboard to B2 to B24

Each of the cells **B1** to **B24** contains the cube of the contents of the adjacent cell in the **A** column.

Highlight the block of cells **A1** to **B24**
Click the *Charts* tab and construct a Smooth Lined Scatter graph. [See Frame 23 of Programme F.4, Graphs.]

The graph you obtain will look like this:

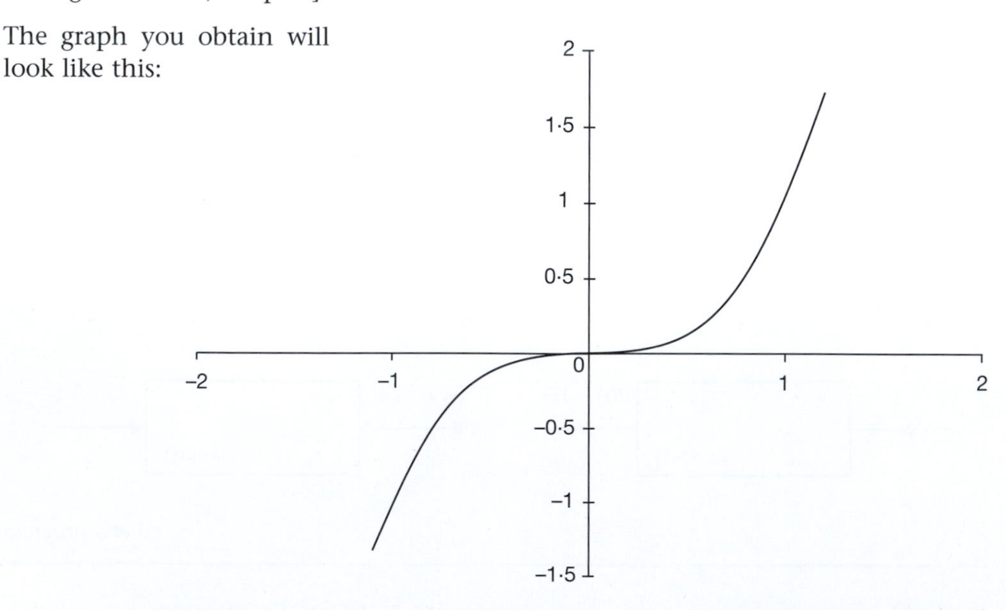


*Now for the graph of $y = x^{\frac{1}{3}}$*

## The graph of $y = x^{1/3}$

13

Keep the data you already have on the spreadsheet, you are going to use it:

Highlight cells **A1** to **A24**
Click **Edit-Copy** This copies the contents of A1 to A24 to the Clipboard
Place the cursor in cell **B26**
Click **Edit-Paste** This pastes the contents of the Clipboard to B26 to B49

The cells **B26** to **B49** then fill with the same values as those in cells **A1** to **A24**

Highlight cells **B1** to **B24**
Click **Edit-Copy** This copies the contents of B1 to B24 to the Clipboard
Place the cursor in cell **A26**
Click **Edit-Paste Special**
In the *Paste Special* window select **Values** and click **OK**

The cells **A26** to **A49** then fill with the same values as those in cells **B1** to **B24**. Because the cells **B1** to **B24** contain formulas, using **Paste Special** rather than simply **Paste** ensures that you copy the values rather than the formulas.

What you now have are the original ordered pairs for the first function reversed in readiness to draw the graph of the inverse of the function.

*Notice that row 25 is empty. This is essential because later on you are going to obtain a plot of two curves on the same graph.*

For now you must first clear away the old graph:

Click the boundary of the graph to display the handles
Click **Edit-Clear-All**

and the graph disappears. Now, to draw the new graph:

Highlight the block of cells **A26** to **B49**
Click the *Charts* tab and construct a Smooth Lined Scatter graph. [See Frame 23 of Programme F.4, Graphs.]

The graph you obtain will look like that depicted below. Same shape as the previous one but a different orientation.

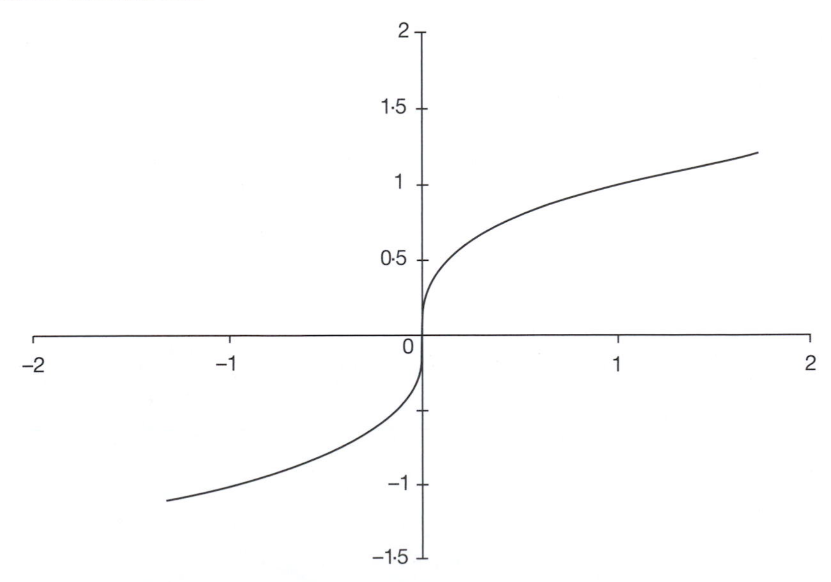


*Now for both the graphs of $y = x^3$ and $y = x^{\frac{1}{3}}$ together*

## 14 The graphs of $y = x^3$ and $y = x^{1/3}$ plotted together

Clear away the graph you have just drawn. Then:

Highlight the block of cells **A1** to **B49**
Click the *Charts* tab and construct a Smooth Lined Scatter graph.

The graph you obtain will look like that depicted to the right:

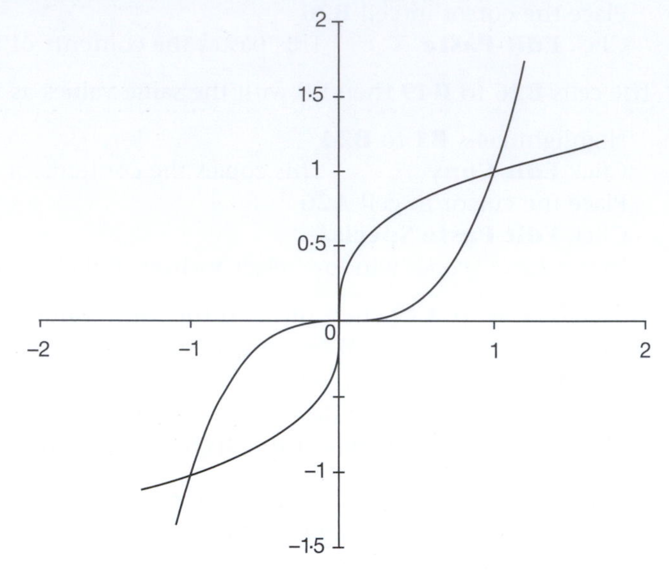


Now you can see that the two graphs are each a reflection of the other in the line $y = x$. To firmly convince yourself of this:

Place the cursor in cell **A51** and enter the number $-1{\cdot}1$
Enter the number $-1{\cdot}1$ in cell **B51**
Enter the number $1{\cdot}2$ in cell **A52**
Enter the number $1{\cdot}2$ in cell **B52**

You now have two points with which to plot the straight line $y = x$. *Notice again, row 50 this time is empty.*

Clear away the last graph. Then:

Highlight the block of cells **A1** to **B52**
Click the *Charts* tab and construct a Smooth Lined Scatter graph.

The graph you obtain will look like that depicted:

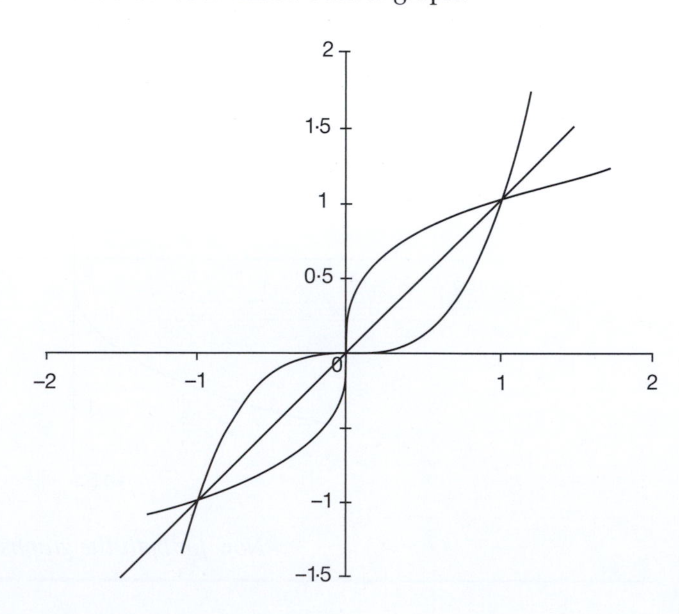


▶

As you can see, the graphs are symmetric about the sloping line $y = x$. We say they are **reflection symmetric about $y=x$** because each one could be considered as a reflection of the other in a double-sided mirror lying along this line.

Now you try one. Use the spreadsheet to plot the graphs of $y = x^2$ and its inverse $y = x^{\frac{1}{2}}$. You do not need to start from scratch, just use the sheet you have already used and change the contents of cell **B1** to the formula **=A1^2**, copy this down the **B** column to **B24** and then **Paste Special** these values into cells **A26** to **A49**.



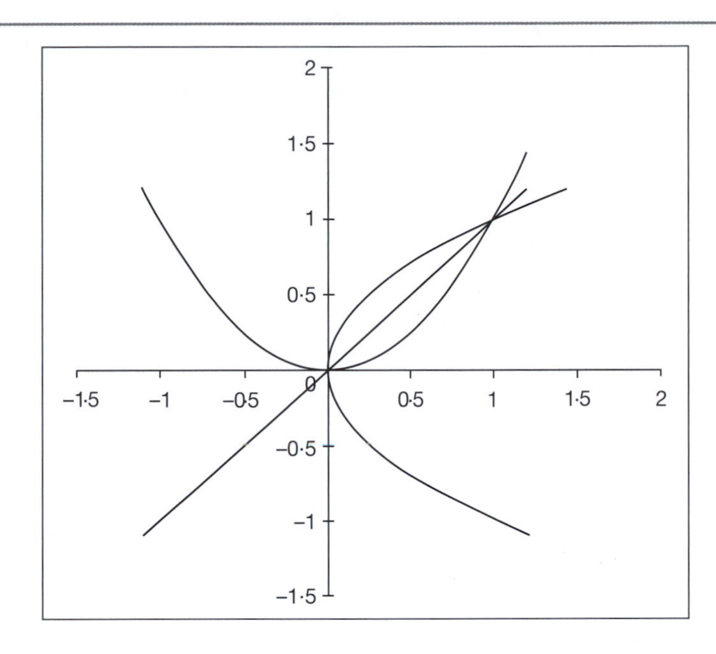


The graph of the inverse of the square function is a parabola on its side. However, as you have seen earlier, this is not a graph of a function. If, however, the bottom branch of this graph is removed, what is left is the graph of the function expressed by $y = \sqrt{x}$ which is called the *inverse function* because it is single valued.

Plot the graph of $y = x^4 - x^2 + 1$ by simply changing the formula in **B1** and copying it into cells **B2** to **B24**. So:

(a) What does the inverse of the function look like?

(b) Is the inverse of the function the inverse function?

16

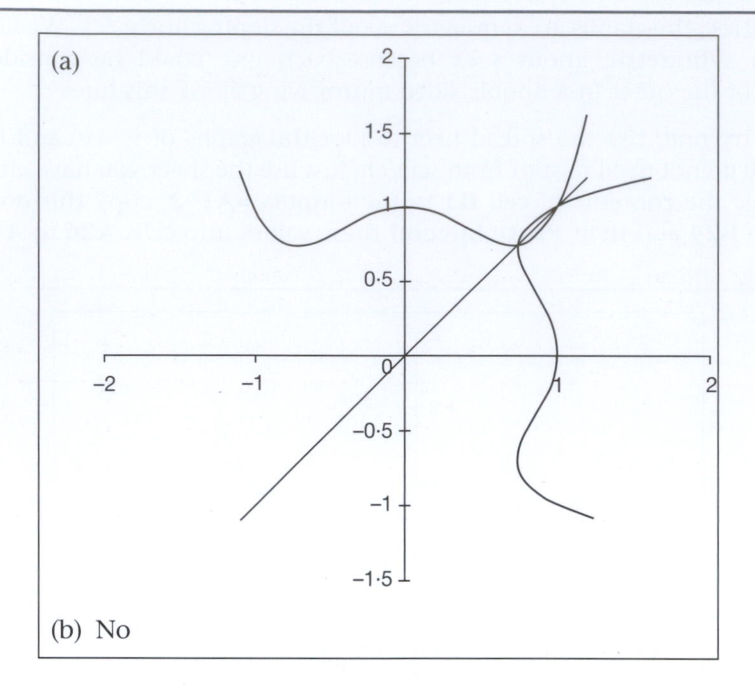


(b) No

The answer to (b) is ‘no’ because the inverse of the function is not single valued so it cannot be a function. The inverse function would have to be obtained by removing parts of the inverse of the function to obtain a function that was single valued.

***At this point let us pause and summarize the main facts so far on functions and their inverses***

## Review summary

17

1 A function is a rule expressed in the form $y = f(x)$ with the proviso that for each value of $x$ there is a unique value of $y$.

2 The collection of permitted input values to a function is called the *domain* of the function and the collection of corresponding output values is called the *range*.

3 The inverse of a function is a rule that associates range values to domain values of the original function.

## Review exercise

18

1 Which of the following equations expresses a rule that is a function:

(a) $y = 6x - 2$

(b) $y = \sqrt{x^3}$

(c) $y = \left(\dfrac{3x}{x^2 + 3}\right)^{\frac{5}{2}}$

▶

2 Given the two functions $f$ and $g$ expressed by:

$$f(x) = 2x - 1 \text{ for } -2 < x < 4 \text{ and } g(x) = \frac{4}{x-2} \text{ for } 3 < x < 5,$$

find the domain and range of:

(a) $h(x) = f(x) - g(x)$

(b) $k(x) = -\dfrac{2f(x)}{g(x)}$

3 Use your spreadsheet to draw each of the following and their inverses. Is the inverse a function?

(a) $y = x^6$ Use the data from the text and just change the formula

(b) $y = -3x$ Use the data from the text and just change the formula

(c) $y = \sqrt{x^3}$ Enter 0 in cell **A1** and **Edit-Fill-Series** with *step value* 0·1

---

**19**

1 (a) $y = 6x - 2$ expresses a rule that is a function because to each value of $x$ there is only one value of $y$.

(b) $y = \sqrt{x^3}$ expresses a rule that is a function because to each value of $x$ there is only one value of $y$. The surd sign $\sqrt{\ }$ stands for the positive square root.

(c) $y = \left(\dfrac{3x}{x^2+3}\right)^{\frac{5}{2}}$ expresses a rule that is not a function because to each positive value of the bracket there are two values of $y$. The power 5/2 represents raising to the power 5 and taking the square root, and there are always two square roots to each positive number.

2 (a) $h(x) = f(x) - g(x) = 2x - 1 - \dfrac{4}{x-2}$ for $3 < x < 4$ because $g(x)$ is not defined for $-2 < x \leq 3$ and $f(x)$ is not defined for $4 \leq x < 5$. Range $1 < h(x) < 5$.

(b) $k(x) = -\dfrac{2f(x)}{g(x)} = -\dfrac{(2x-1)(x-2)}{2}$ for $3 < x < 4$.

Range $-7 < k(x) < -5/2$.

3 (a) $y = x^6$ has an inverse $y = x^{\frac{1}{6}}$. This does not express a function because there are always two values to an even root (see Frame 4).

(b) $y = -3x$ has an inverse $y = -\dfrac{x}{3}$. This does express a function because there is only one value of $y$ to each value of $x$.

(c) $y = \sqrt{x^3}$ has an inverse $y = x^{\frac{2}{3}}$ because $\sqrt{x^3}$ represents the positive value of $y = x^{\frac{3}{2}}$. The inverse does express a function.

*Now let's move on*

---

# Composition

## 20 Function of a function

Chains of functions can be built up where the output from one function forms the input to the next function in the chain. Take out your calculator again and this time press the key:

$\boxed{x^{-1}}$ the reciprocal key.

Now enter 4 = and the display changes to:

0·25 the reciprocal of 4

Now press the $\boxed{x^2}$ key followed by = and the display changes to:

0·0625 the square of 0·25

Here, the number 4 was the input to the reciprocal function and the number 0·25 was the output. This same number 0·25 was then the input to the squaring function with output 0·0625. This can be represented by the following diagram:

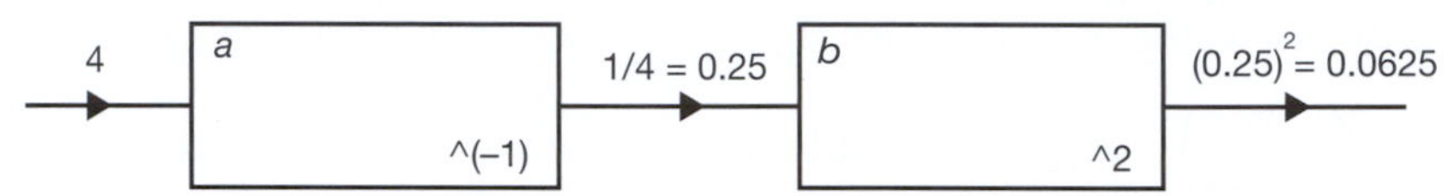


Notice that the two functions have been named *a* and *b*, but any letter can be used to label a function.

At the same time the *total* processing by $f$ could be said to be that the number 4 was input and the number 0·0625 was output:

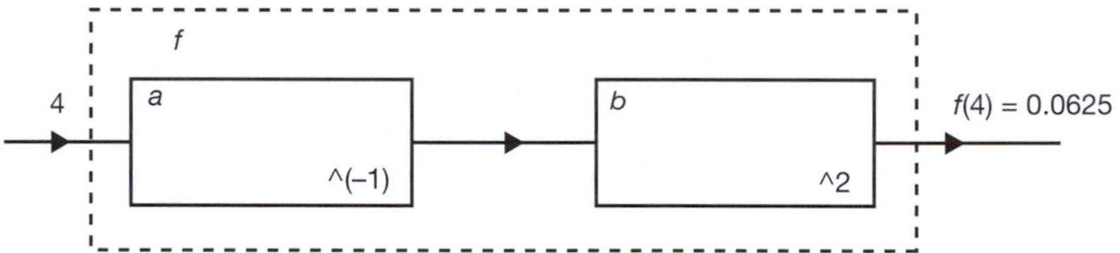


So the function $f$ is *composed* of the two functions $a$ and $b$ where $a(x) = \frac{1}{x}$, $b(x) = x^2$ and $f(x) = \left(\frac{1}{x}\right)^2$. It is said that $f$ is the *composition* of $a$ and $b$, written as:

$$f = b \circ a$$

and read as *b of a*. Notice that the functions $a$ and $b$ are written down algebraically in the reverse order from the order in which they are given in the diagram. This is because in the diagram the input to the composition enters on the left, whereas algebraically the input is placed to the right:

$$f(x) = b \circ a(x)$$

So that $f(x) = b \circ a(x)$, which is read as *f of x equals b of a of x*. An alternative notation, more commonly used, is:

$$f(x) = b[a(x)]$$

and $f$ is described as being a *function of a function*.

Now you try. Given that $a(x) = x + 3$, $b(x) = 4x$ find the functions $f$ and $g$ where:

(a) $f(x) = b[a(x)]$
(b) $g(x) = a[b(x)]$

*Stick with what you know, draw the boxes and see what you find*

21

(a) $f(x) = 4x + 12$
(b) $g(x) = 4x + 3$

Because

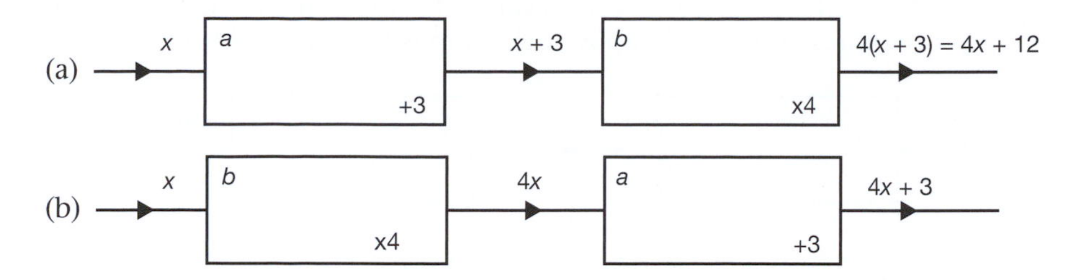


Notice how these two examples show that $b[a(x)]$ is different from $a[b(x)]$. That is, the order of composition matters.

Now, how about something a little more complicated? Given the three functions $a, b$ and $c$ where $a(x) = x^3$, $b(x) = 2x$ and $c(x) = x - 5$, find each of the following as expressions in $x$:

(a) $f(x) = a(b[c(x)])$

(b) $g(x) = c(a[b(x)])$

(c) $h(x) = a(a[b(x)])$

*Remember, draw the boxes and follow the logic*

22

(a) $f(x) = 8(x - 5)^3$
(b) $g(x) = 8x^3 - 5$
(c) $h(x) = 512x^9$

Because

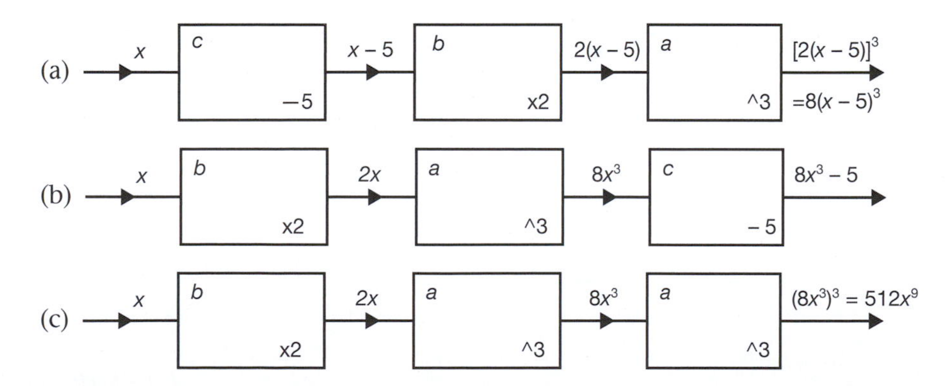


How about working the other way? Given the expression $f(x)$ for the output from a composition of functions, how do you decompose it into its component functions? This is particularly easy because you already know how to do it even though you may not yet realize it.

Let's look at a specific example first.

▶

Given the output from a composition of functions as $f(x) = 6x - 4$ ask yourself how, given a calculator, would you find the value of $f(2)$? You would:

| | | | |
|---|---|---|---|
| enter the number 2 | the input | $x$ | |
| *multiply by 6* to give 12 | the first function | $a(x) = 6x$ | input times 6 |
| *subtract 4* to give 8 | the second function | $b[a(x)] = 6x - 4$ | input minus 4 |

so that $f(x) = b[a(x)]$. The very act of using a calculator to enumerate the output from a composition requires you to decompose the composition automatically as you go.

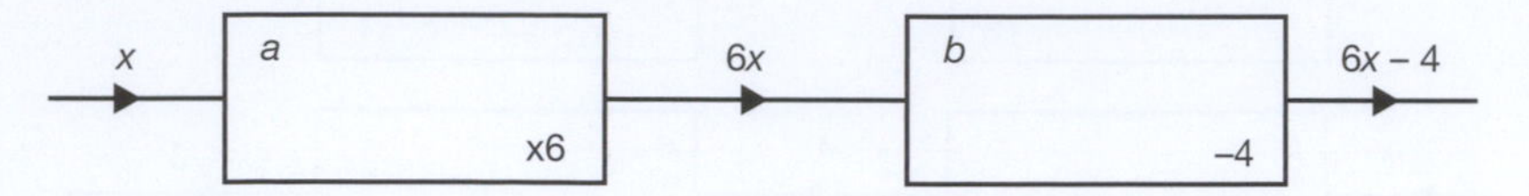


Try it yourself. Decompose the composition with output $f(x) = (x + 5)^4$.

*Get your calculator out and find the output for a specific input*

**23**

$$f(x) = b[a(x)] \text{ where } a(x) = x + 5 \text{ and } b(x) = x^4$$

Because

x
a
+5
x + 5
b
^4
(x + 5)⁴

Notice that this decomposition is not unique. You could have defined $b(x) = x^2$ in which case the composition would have been $f(x) = b(b[a(x)])$.

Just to make sure you are clear about this, decompose the composition with output $f(x) = 3(2x + 7)^4$.

*Use your calculator and take it steady, there are four functions here*

**24**

$$f(x) = d[c(b[a(x)])] \text{ where } a(x) = 2x$$
$$b(x) = x + 7,\ c(x) = x^4 \text{ and } d(x) = 3x$$

Because

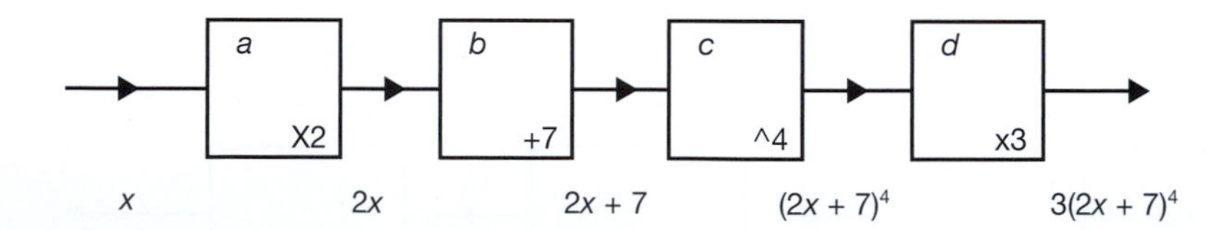


*Let's keep going*

**25**

## Inverses of compositions

As has been stated before, the diagram of the inverse of a function can be drawn as the function with the information flowing through it in the reverse direction, and the same applies to a composition.

For example, consider the function $f$ with output $f(x) = (3x - 5)^{\frac{1}{3}}$. By decomposing $f$ you find that:

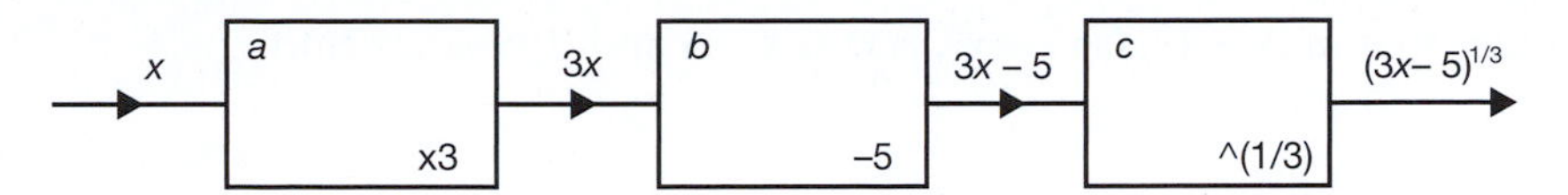


where $a(x) = 3x$, $b(x) = x - 5$ and $c(x) = x^{\frac{1}{3}}$, so $f(x) = c(b[a(x)])$. Each of the three functions in the composition has an inverse, namely:

$$a^{-1}(x) = \frac{x}{3}$$
$$b^{-1}(x) = x + 5$$
$$c^{-1}(x) = x^3$$

By reversing the flow of information through the diagram we find that:

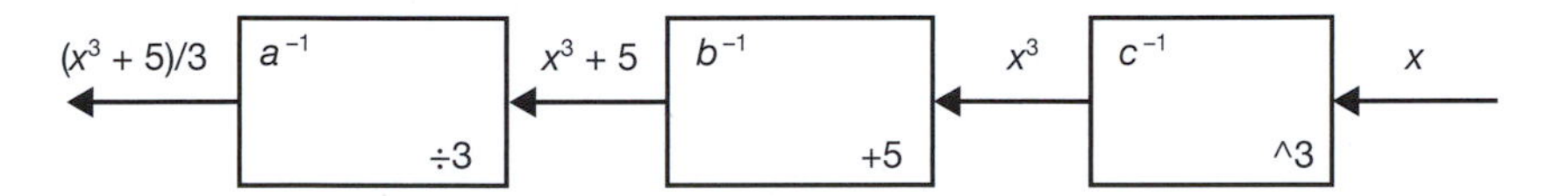


so that

$$f^{-1}(x) = a^{-1}(b^{-1}[c^{-1}(x)]) = (x^3 + 5)/3$$

Notice the *reversal* of the order of the components:

$$f(x) = c(b[a(x)]), \qquad f^{-1}(x) = a^{-1}(b^{-1}[c^{-1}(x)])$$

Now you try this one. Find the inverse of the function $f$ with output $f(x) = \left(\frac{x+2}{4}\right)^5$.

*Answer in the next frame*

---

**26**

$$\boxed{f^{-1}(x) = 4x^{\frac{1}{5}} - 2}$$

Because

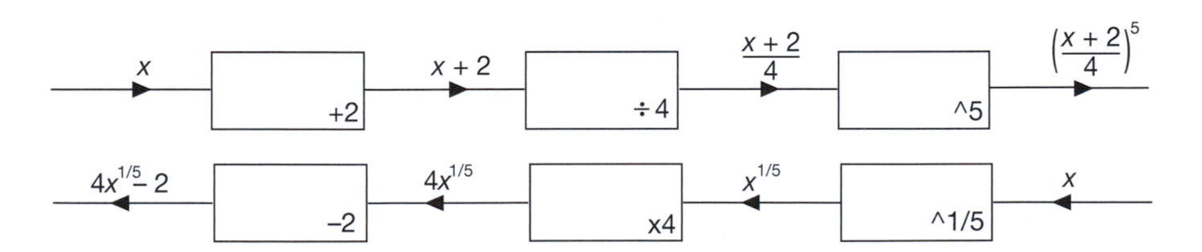


---

## Review summary

**27**

*Composition of functions*

Chains of functions can be built up where the output from one function forms the input to the next function in the chain.

## Review exercise

**28**

1 Given that $a(x) = 4x$, $b(x) = x^2$, $c(x) = x - 5$ and $d(x) = \sqrt{x}$ find:

(a) $f(x) = a[b(c[d(x)])]$

(b) $f(x) = a(a[d(x)])$

(c) $f(x) = b[c(b[c(x)])]$

2 Given that $f(x) = (2x-3)^3 - 3$, decompose $f$ into its component functions and find its inverse. Is the inverse a function?

*Take it steady – you will find the solutions in the next frame*

**29**

1 (a) $f(x) = a[b(c[d(x)])] = 4(\sqrt{x} - 5)^2$

(b) $f(x) = a(a[d(x)]) = 16\sqrt{x}$

(c) $f(x) = b[c(b[c(x)])] = ((x-5)^2 - 5)^2 = x^4 - 20x^3 + 140x^2 - 400x + 400$

2 $f = b \circ c \circ b \circ a$ so that $f(x) = b[c(b[a(x)])]$ where $a(x) = 2x, b(x) = x - 3$ and $c(x) = x^3$. The inverse is $f^{-1}(x) = a^{-1}[b^{-1}(c^{-1}[b^{-1}(x)])]$ so that:

$f^{-1}(x) = a^{-1}[b^{-1}(c^{-1}[b^{-1}(x)])] = \dfrac{(x+3)^{\frac{1}{3}} + 3}{2}$ where $a^{-1}(x) = x/2$,

$b^{-1}(x) = x + 3$ and $c^{-1}(x) = x^{\frac{1}{3}}$. The inverse is a function.

So far our work on functions has centred around *algebraic functions*. This is just one category of function. In the next Programme we shall move on and consider other types of function and their specific properties.

*Next frame*

**30**

You have now come to the end of this Programme. A list of **Can You?** questions follows for you to gauge your understanding of the material in the Programme. These questions match the **Learning outcomes** listed at the beginning of the Programme so go back and try the **Quiz** that follows them. After that try the **Test exercise**. *Work through them at your own pace, there is no need to hurry*. A set of **Further problems** provides additional valuable practice.

## Can You?

**31**

### Checklist F.10

Check this list before and after you try the end of Programme test.

On a scale of 1 to 5 how confident are you that you can: Frames

- Identify a function as a rule and recognize rules that are not functions? 1 to 4
  *Yes* ☐ ☐ ☐ ☐ ☐ *No*
- Determine the domain and range of a function? 4 to 7
  *Yes* ☐ ☐ ☐ ☐ ☐ *No*
- Construct the inverse of a function and draw its graph? 8 to 16
  *Yes* ☐ ☐ ☐ ☐ ☐ *No*
- Construct compositions of functions and de-construct them into their component functions? 17 to 23
  *Yes* ☐ ☐ ☐ ☐ ☐ *No*

## Test exercise F.10

32

Frames

1 Which of the following equations expresses a rule that is a function?

(a) $y = -x^{\frac{3}{2}}$ (b) $y = x^2 + x + 1$ (c) $y = (\sqrt{x})^3$ — 1 to 4

2 Given the two functions $f$ and $g$ expressed by:

$f(x) = \dfrac{1}{x-2}$ for $2 < x \leq 4$ and $g(x) = x - 1$ for $0 \leq x < 3$

find the domain and range of functions $h$ and $k$ where:

(a) $h(x) = 2f(x) - 3g(x)$ (b) $k(x) = -\dfrac{3f(x)}{5g(x)}$ — 4 to 7

3 Use your spreadsheet to draw each of the following and their inverses. Is the inverse a function?

(a) $y = 3x^4$ (b) $y = -x^3$ (c) $y = 3 - x^2$ — 8 to 16

4 Given that $a(x) = 5x$, $b(x) = x^4$, $c(x) = x + 3$ and $d(x) = \sqrt{x}$ find:

(a) $f(x) = a[b(c[d(x)])]$ (b) $f(x) = a(a[d(x)])$

(c) $f(x) = b[c(b[c(x)])]$

17 to 21

5 Given that $f(x) = (5x - 4)^3 - 4$ decompose $f$ into its component functions and find its inverse. Is the inverse a function?

22 to 23

## Further problems F.10

33

1 Pairs of real numbers $(x, y)$ are generated by the rule $y =$ some expression in $x$. In each of the following collections of pairs of numbers is the corresponding rule that generated them a function or not? Is the inverse?

(a) $\{(1, 2), (2, 3), (3, 4), (4, 5)\}$ (b) $\{(-5, 2), (-4, 3), (-3, 2), (-2, 2)\}$

(c) $\{(11, 12)\}$ (d) $\{(1, 2), (1, 3), (2, 4), (3, 5)\}$

(e) $\{(1, 2), (2, 1)\}$ (f) $\{(4, 2), (4, -2), (9, 3), (9, -3)\}$

2 Each of the following graphs displays a plot of a collection of ordered pairs of real numbers where the horizontal line contains the domain points and the vertical line contains the range points. Which graphs are graphs of functions and which are not?

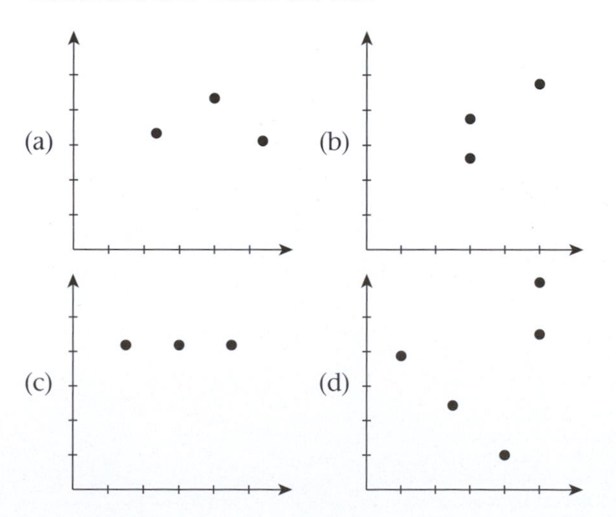


▶

3 Given the two functions $f$ and $g$ expressed by:

$f(x) = \sqrt{x^3}$ and $g(x) = \dfrac{1}{\sqrt{x}}$

find the domain and range of $h$ where:

(a) $h(x) = f(x) - g(x)$ (b) $h(x) = 2f(x) + 3g(x)$

(c) $h(x) = \dfrac{f(x)}{g(x)}$ (d) $h(x) = f(x)g(x)$

4 Use your spreadsheet to draw each of the following. Is the inverse a function?

(a) $f(x) = \sqrt{x^3}$ (b) $f(x) = \dfrac{1}{\sqrt{x}}$ (c) $f(x) = \dfrac{1}{x} + 1 + x$

(d) $f(x) = \dfrac{5x^5 - 3x^3 + x}{4x^4 - 2x^2 + 1}$ (e) $f(x) = \dfrac{x^2 - 1}{x + 1}$ (f) $f(x) = \dfrac{x^3 + 1}{x + 1}$

5 Given that $a(x) = x - 5$, $b(x) = 3x$, $c(x) = -\dfrac{x}{2}$ and $d(x) = \sqrt{x}$ find:

(a) $a(b[c([d(x)])])$ (b) $a(b[a(x)])$

(c) $a(a[a([a(x)])])$ (d) $d(d[d(x)])$

6 For each of the following decompose $f$ into its component functions and find its inverse. Is the inverse a function?

(a) $f(x) = 5(3x - 1)^3$ (b) $f(x) = [5(3x - 1)]^3$

(c) $f(x) = 5x^3 - 1$ (d) $f(x) = (x - 1)^3$

7 Find a function that is identical to its inverse?

8 Given the function $f$ whose input values $x$ and output values $y$ are related via the equation

$y = \dfrac{x}{x - 1}$

it is clear that $x = 1$ is not in the domain of $f$. Is $y = 1$ in the range of $f$?

34

Now visit the companion website for this book at www.palgrave.com/stroud for more questions applying this mathematics to science and engineering.

# Trigonometric and exponential functions

## Learning outcomes

When you have completed this Programme you will be able to:

- ☐ Develop the trigonometric functions from the trigonometric ratios
- ☐ Find the period, amplitude and phase of a periodic function
- ☐ Distinguish between the inverse of a trigonometric function and the inverse trigonometric function
- ☐ Solve trigonometric equations using the inverse trigonometric functions and trigonometric identities
- ☐ Recognize that the exponential function and the natural logarithmic function are mutual inverses and solve indicial and logarithmic equations
- ☐ Find the even and odd parts of a function when they exist
- ☐ Construct the hyperbolic functions from the odd and even parts of the exponential function
- ☐ Evaluate limits of simple functions

If you already feel confident about these why not try the quiz over the page? You can check your answers at the end of the book.

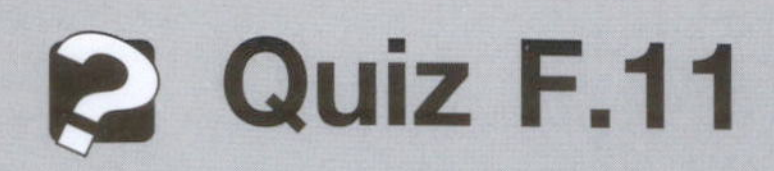

# Quiz F.11

**Frames**

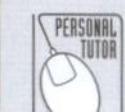

1 Give the value of:
(a) $\sin 0°$ (b) $\cos 90°$ (c) $\tan 90°$ — 2 to 7

2 Use a calculator to find the value of each of the following:
(a) $\cos(-5\pi/3)$ (b) $\sec(115°)$ (c) $\tan(-13°)$ — 2 to 9

3 Find the period, amplitude and phase of each of the following:
(a) $f(\theta) = 2\cos 6\theta$ (b) $f(\theta) = -2\tan(2\theta - 2)$
(c) $f(\theta) = \cos(\pi - \theta)$ — 10 to 22

4 A function is defined by the following prescription:

$$f(x) = 4 - x^2, \quad 0 \le x < 2, \quad f(x \pm 2) = f(x)$$

Plot a graph of this function for $-6 \le x \le 6$ and find:
(a) the period
(b) the amplitude
(c) the phase of $f(x) + 2$ with respect to $f(x)$ — 10 to 22

5 Solve the following trigonometric equations:
(a) $\cos(x + \pi/3) + \cos(x - \pi/4) = 0$
(b) $\sin^2 x - 4\sin x \cos x + 3\cos^2 x = 0$
(c) $9\cos x - 4\sin x = 2\sqrt{2}$ for $0 \le x \le \pi$ — 23 to 30

6 Find the value of $x$ corresponding to each of the following:
(a) $2^{-2x} = 1$ (b) $\exp(-4x) = e^3$ (c) $e^{\frac{1}{x}} = 2{\cdot}34$
(d) $\log_2 x = 0{\cdot}4$ (e) $\ln x = 2$ — 34 to 41

7 Solve for $x$:
(a) $2^{x-5}5^{x+1} = 62{\cdot}5$
(b) $e^{2x} - e^{x+3} - e^{x+1} + e^4 = 0$
(c) $\log(x^2) = \log\left(\frac{1}{x}\right) - \log 5$
(d) $\log_4(3 - x^2) = -6{\cdot}2$ — 34 to 41

8 Find the even and odd parts of $f(x) = x(x^2 + x + 1)$. — 42 to 47

9 Evaluate each of the following limits:
(a) $\underset{x\to 3}{Lim}\left(\frac{x^2 - 9}{x - 3}\right)$
(b) $\underset{x\to 1/3}{Lim}(3x^2 - 2x + 1)$
(c) $\underset{x\to -1}{Lim}\left(\frac{(x+1)e^{3x+2}}{x^2 - 1}\right)$ — 48 to 50

# Introduction

1

In Programme F.10 the idea of a function was introduced. During the course of the Programme we looked particularly at functions whose outputs could be represented by algebraic expressions – we call these *algebraic functions*. In this Programme we shall be extending our work on functions by considering trigonometric and exponential functions.

# Trigonometric functions

2

## Rotation

In Programme F.9 the trigonometric ratios were defined for the two acute angles in a right-angled triangle. These definitions can be extended to form *trigonometric functions* that are valid for *any* angle and yet retain all the properties of the original ratios. Start with the circle generated by the end point $A$ of a straight line $OA$ of unit length rotating anticlockwise about the end $O$ as shown in the diagram:

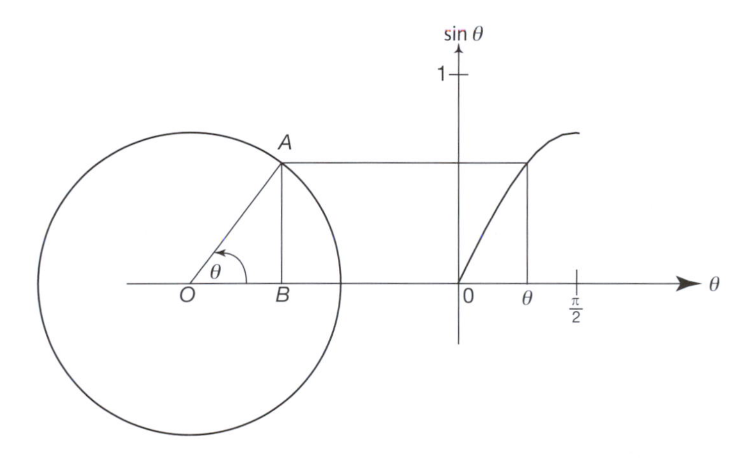


For angles $\theta$ where $0 < \theta < \pi/2$ radians you already know that:

$$\sin\theta = \frac{AB}{OA} = AB \text{ since } OA = 1$$

That is, the value of the trigonometric ratio $\sin\theta$ is equal to the height of $A$ above $B$. The *sine* function with output $\sin\theta$ is now defined as the height of $A$ above $B$ *for any angle* $\theta$ $(0 \leq \theta < \infty)$.

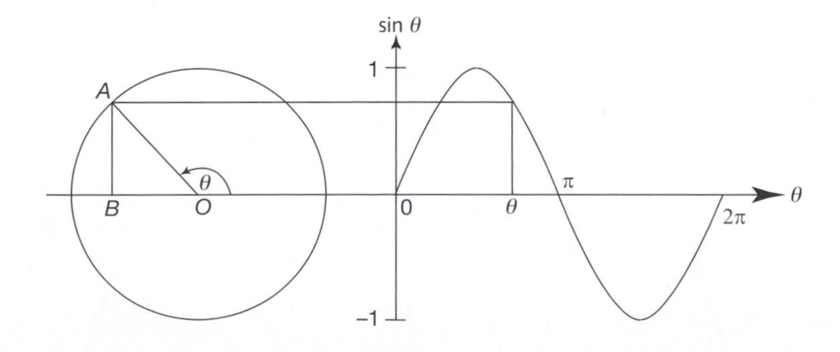


▶

Notice that when $A$ is *below* $B$ the height is *negative*. The definition of the sine function can be further extended by taking into account negative angles, which represent a clockwise rotation of the line $OA$ giving the complete graph of the sine function as in the diagram below:

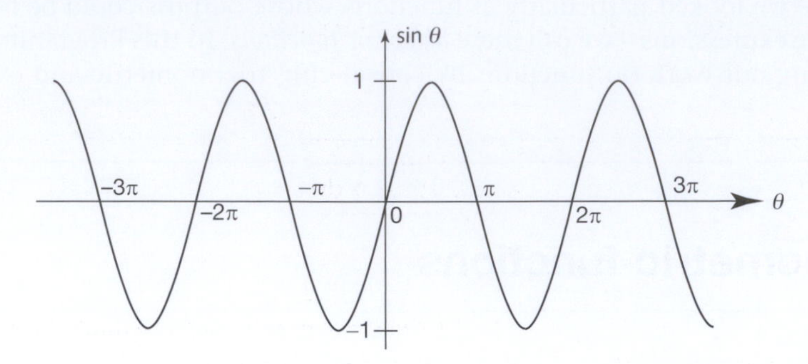


As you can see from this diagram, the value of $\sin\theta$ ranges from $+1$ to $-1$ depending upon the value of $\theta$. You can reproduce this graph using a spreadsheet – in cells **A1** to **A21** enter the numbers $-10$ to 10 in steps of 1 and in cell **B1** enter the formula **=sin(A1)** and copy this into cells **B2** to **B21**. Use the *Charts* tab to draw the graph (see Programme F.4).

Just as before, you can use a calculator to find the values of the sine of an angle. So the sine of $153°$ is ...........

*Remember to put your calculator in degree mode*

**3**

0·4540 to 4 dp

and the sine of $-\pi/4$ radians is ...........

*Remember to put your calculator in radian mode*

**4**

$-0{\cdot}7071$ to 4 dp

By the same reasoning, referring back to the first diagram in Frame 2, for angles $\theta$ where $0 < \theta < \pi/2$ radians you already know that:

$$\cos\theta = \frac{OB}{OA} = OB \text{ since } OA = 1$$

This time, the value of the trigonometric ratio $\cos\theta$ is equal to the distance from $O$ to $B$. The *cosine* function with output $\cos\theta$ is now defined as the distance from $O$ to $B$ *for any angle* $\theta$ $(-\infty < \theta < \infty)$.

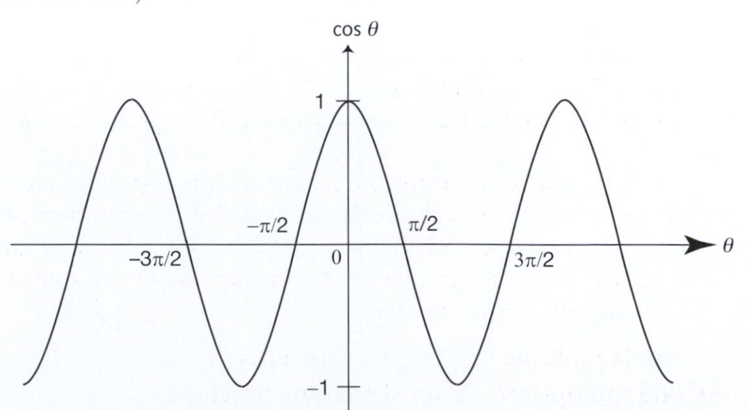


▶

Notice that when $B$ is to the left of $O$ the distance from $O$ to $B$ is negative.

Again, you can reproduce this graph using a spreadsheet – in cells **A1** to **A21** enter the numbers $-10$ to 10 in steps of 1 and in cell **B1** enter the formula **=cos (A1)** and copy this into cells **B2** to **B21**. Use the *Charts* tab to draw the graph.

A calculator is used to find the values of the cosine of an angle. So the cosine of $-272°$ is ...........

*Remember to put your calculator in degree mode*

**5**

0·0349 to 4 dp

and the cosine of $2\pi/3$ radians is ...........

*Remember to put your calculator in radian mode*

**6**

$-0{\cdot}5$

*Now to put these two functions together*

**7**

## The tangent

The third basic trigonometric function, the tangent, is defined as the ratio of the sine to the cosine:

$$\tan\theta = \frac{\sin\theta}{\cos\theta}$$

Because $\cos\theta = 0$ whenever $\theta$ is an odd multiple of $\pi/2$, the tangent is not defined at these points. Instead the graph has vertical asymptotes as seen below:

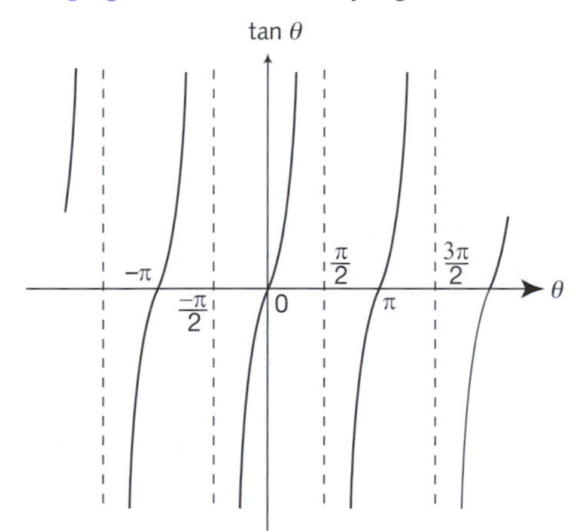


You can plot a single branch of the tangent function using your spreadsheet. Enter $-1{\cdot}5$ in cell **A1** and **Edit-Fill-Series** down to **A21** with *step value* 0·15, then use the function **= tan (A1)** in cell **B1** and copy down to **B21**. Use the *Charts* tab to create the graph.

Make a note of the diagram in this frame and the diagrams in Frames 2 and 4. It is essential that you are able to draw sketch graphs of these functions.

For the sine and cosine functions the repeated sinusoidal wave pattern is easily remembered, all you then have to remember is that each rises and falls between $+1$ and $-1$ and crosses the horizontal axis:

(a) every whole multiple of $\pi$ for the sine function
(b) every odd multiple of $\pi/2$ for the cosine function

▶

The repeated *branch* pattern of the tangent function is also easily remembered, all you then have to remember is that it rises from $-\infty$ to $+\infty$, crosses the horizontal axis every even multiple of $\pi/2$ and has a vertical asymptote every odd multiple of $\pi/2$.

Just as before, you can use a calculator to find the values of the tangent of an angle. So the tangent of 333° is ...........

*Remember to put your calculator in degree mode*

**8**

$-0{\cdot}5095$ to 4 dp

and the tangent of $-6\pi/5$ radians is ...........

*Remember to put your calculator in radian mode*

**9**

$-0{\cdot}7265$ to 4 dp

*Now to look at some common properties of these trigonometric functions*

**10**

## Period

As can be seen from the graph in Frame 7, the tangent function $f(\theta) = \tan\theta$ repeats its output values every 180° ($\pi$ radians). Use your calculator in degree mode to verify that:

$$\begin{aligned} f(45) &= \tan 45^\circ \;\; = 1 \\ f(45+180) &= \tan 225^\circ = 1 \\ f(45+360) &= \tan 405^\circ = 1 \\ f(45+540) &= \tan 585^\circ = 1 \end{aligned}$$

We can write this fact in the form of an equation:

$\tan(45 + 180n) = \tan 45^\circ$ where $n = 1, 2, 3, \ldots$

or, using radians:

$\tan(\pi/4 + n\pi) = \tan \pi/4$ where $n = 1, 2, 3, \ldots$

Indeed, as can be seen from the graph in Frame 7, the tangent function continually repeats the output value corresponding to each input value of $\theta$ in the interval $-\pi/2 < \theta < \pi/2$ which is of width $\pi$. That is:

$\tan(\theta + n\pi) = \tan\theta$ where $n = 1, 2, 3, \ldots$

The sine function $f(\theta) = \sin\theta$ is periodic with period ...........

*Answer in the next frame*

**11**

$2\pi$

Because

As is evident from the graph of $f(\theta) = \sin\theta$ in Frame 2 the sinusoidal wave pattern consists of a repeated wave of width $2\pi$. The sine function therefore continually repeats the output value corresponding to each input value of $\theta$ in the inveral $0 \le \theta < 2\pi$.

Similarly, the cosine function $f(\theta) = \cos\theta$ is periodic with period $2\pi$ as can be seen in the graph in Frame 4.

▶

Finding the periods of trigonometric functions with more involved outputs requires some manipulation. For example, consider $f(\theta) = \cos 3\theta$. Here we see that, for example:

$$f(\theta) = \cos 3\theta = 1 \text{ when } 3\theta = 0,\ 2\pi,\ 4\pi,\ 6\pi, \ldots,\ 2n\pi$$

that is when:

$$\theta = 0,\ 2\pi/3,\ 4\pi/3,\ 6\pi/3\ (2\pi),\ \ldots,\ 2n\pi/3$$

So that $f(\theta + 2\pi/3) = \cos 3(\theta + 2\pi/3) = \cos(3\theta + 2\pi) = \cos 3\theta = f(\theta)$. That is, the period of $f(\theta) = \cos 3\theta$ is $2\pi/3$. The output of $f(\theta) = \cos 3\theta$ certainly repeats itself over $2\pi$ radians but within $2\pi$ the basic sinusoidal shape is repeated three times.

So the period of $\cos 4\theta$ is ...........

---

**12**

$$\boxed{\frac{\pi}{2}}$$

Because

$$\cos 4\theta = \cos(4\theta + 2\pi) = \cos 4\left(\theta + \frac{2\pi}{4}\right) = \cos 4\left(\theta + \frac{\pi}{2}\right)$$

And the period of $\tan 5\theta =$ ...........

*Answer in the next frame*

---

**13**

$$\boxed{\frac{\pi}{5}}$$

Because

$$\tan 5\theta = \tan(5\theta + \pi) = \tan 5(\theta + \pi/5)$$

Now, try another one. The period of $\sin(\theta/3) =$ ...........

*Just follow the same procedure. The answer may surprise you*

---

**14**

$$\boxed{6\pi}$$

Because

$$\sin(\theta/3) = \sin(\theta/3 + 2\pi) = \sin\frac{1}{3}(\theta + 6\pi)$$

The answer is not $2\pi$ because the basic sinusoidal shape is only completed over the interval of $6\pi$ radians. If you are still not convinced of all this, use the spreadsheet to plot their graphs. Just one more before moving on.

The period of $\cos(\theta/2 + \pi/3) =$ ...........

*Just follow the procedure*

---

**15**

$$\boxed{4\pi}$$

Because

$$\cos(\theta/2 + \pi/3) = \cos(\theta/2 + \pi/3 + 2\pi) = \cos\left(\frac{1}{2}[\theta + 4\pi] + \pi/3\right)$$

The $\pi/3$ has no effect on the period, it just shifts the basic sinusoidal shape $\pi/3$ radians to the left.

*Move on*

---

16

## Amplitude

Every periodic function possesses an *amplitude* that is given as *the difference between the maximum value and the average value of the output taken over a single period.* For example, the average value of the output from the cosine function is zero (it ranges between +1 and −1) and the maximum value of the output is 1, so the amplitude is $1 - 0 = 1$.

So the amplitude of $4\cos(2\theta - 3)$ is ...........

*Next frame*

17

4

Because

The maximum and minimum values of the cosine function are +1 and −1 respectively, so the output here ranges from +4 to −4 with an average of zero. The maximum value is 4 so that the amplitude is $4 - 0 = 4$.

Periodic functions are not always trigonometric functions. For example, the function with the graph shown in the diagram below is also periodic:

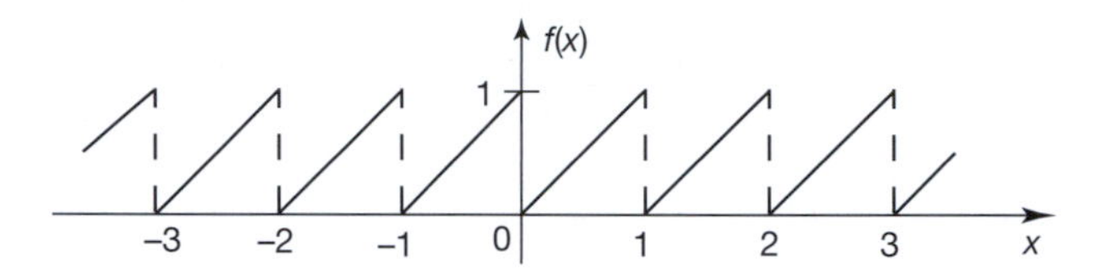


The straight line branch between $x = 0$ and $x = 1$ repeats itself indefinitely. For $0 \leq x < 1$ the output from $f$ is given as $f(x) = x$. The output from $f$ for $1 \leq x < 2$ matches the output for $0 \leq x < 1$. That is:

$f(x + 1) = f(x)$ for $0 \leq x < 1$

So for example, $f(1{\cdot}5) = f(0{\cdot}5 + 1) = f(0{\cdot}5) = 0{\cdot}5$

The output from $f$ for $2 \leq x < 3$ also matches the output for $0 \leq x < 1$. That is:

$f(x + 2) = f(x)$ for $0 \leq x < 1$

So that, for example, $f(2{\cdot}5) = f(0{\cdot}5 + 2) = f(0{\cdot}5) = 0{\cdot}5$

This means that we can give the prescription for the function as:

$f(x) = x$ for $0 \leq x < 1$

$f(x + n) = f(x)$ for any integer $n$

For a periodic function of this type with period $P$ where the first branch of the function is given for $a \leq x < a + P$ we can say that:

$f(x) =$ some expression in $x$ for $a \leq x < a + P$

$f(x + nP) = f(x)$

Because of its shape, the specific function we have considered is called a *sawtooth wave.*

The amplitude of this sawtooth wave is ...........

*Remember the definition of amplitude*

18

$$\boxed{\frac{1}{2}}$$

Because

The amplitude is given as the *difference between the maximum value and the average value of the output taken over a single period*. Here the maximum value of the output is 1 and the average output is $\frac{1}{2}$, so the amplitude is $1 - 1/2 = 1/2$.

*Next frame*

19

## Phase difference

The phase difference of a periodic function is the interval of the input by which the output leads or lags behind the *reference function*. For example, the plots of $y = \sin x$ and $y = \sin(x + \pi/4)$ on the same graph are shown below:

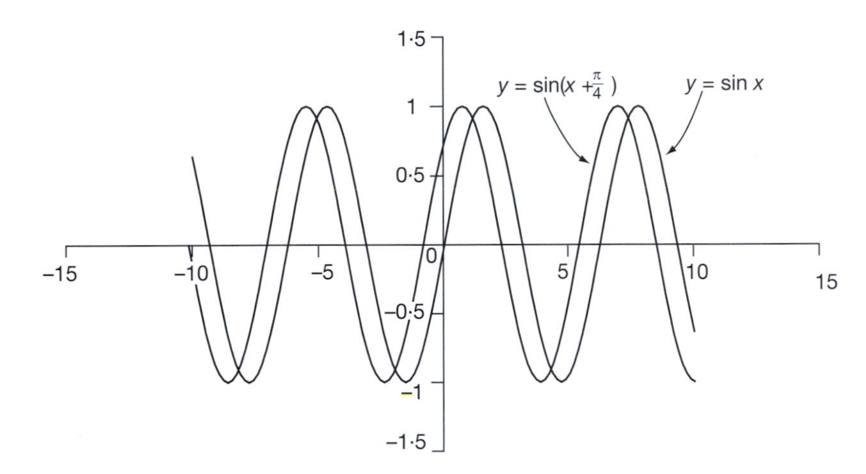


The diagram shows that $y = \sin(x + \pi/4)$ has the identical shape to $y = \sin x$ but is *leading* it by $\pi/4$ radians. It might appear to lag behind when you look at the diagram but it is, in fact, leading because when $x = 0$ then $\sin(x + \pi/4)$ already has the value $\sin \pi/4$, whereas $\sin x$ only has the value $\sin 0$. It is said that $y = \sin(x + \pi/4)$ leads with a *phase difference* of $\pi/4$ radians relative to the reference function $y = \sin x$. A function with a negative phase difference is said to *lag behind* the reference function. So that $y = \sin(x - \pi/4)$ lags behind $y = \sin x$ with a phase difference of $-\pi/4$.

So the phase difference of $y = \sin(x - \pi/6)$ relative to $y = \sin x$ is ...........

*Next frame*

20

$$\boxed{-\pi/6 \text{ radians}}$$

Because

The graph of $y = \sin(x - \pi/6)$ *lags behind* $y = \sin x$ by $\pi/6$ radians.

The phase difference of $y = \cos x$ relative to the reference function $y = \sin x$ is ...........

*Think how $\cos x$ relates to $\sin x$*

21

$\pi/2$ radians

Because

$\cos x = \sin(x + \pi/2)$ and $\sin(x + \pi/2)$ *leads* $\sin x$ by $\pi/2$ radians.

Finally, try this. The phase difference of $y = \sin(3x + \pi/8)$ relative to $y = \sin 3x$ is ...........

*Take care to compare like with like – plot the graph if necessary*

22

$\pi/24$ radians

Because

$\sin(3x + \pi/8) = \sin(3[x + \pi/24])$ and $\sin(3[x + \pi/24])$ *leads* $\sin 3x$ by $\pi/24$ radians.

*Now for inverse trigonometric functions*

23

## Inverse trigonometric functions

If the graph of $y = \sin x$ is reflected in the line $y = x$, the graph of the inverse of the sine function is what results:

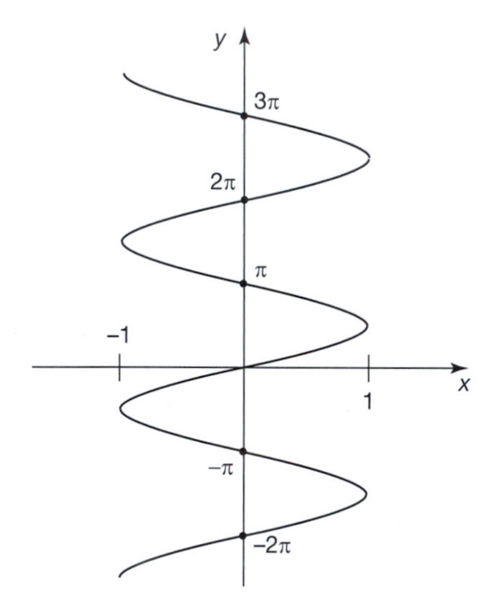


However, as you can see, this is not a function because there is more than one value of $y$ corresponding to a given value of $x$. If you cut off the upper and lower parts of the graph you obtain a single-valued function and it is this that is the *inverse sine function*:

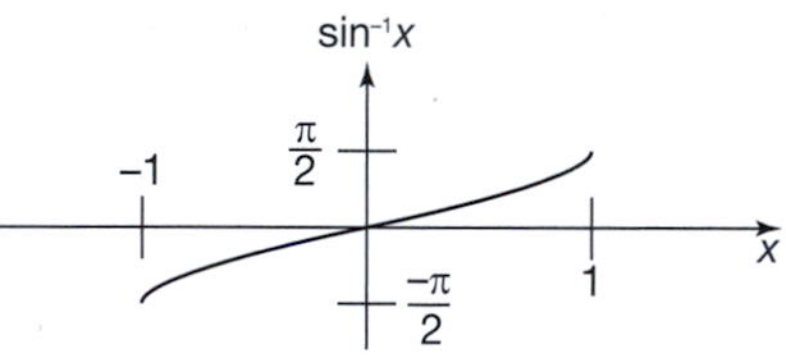


In a similar manner you can obtain the *inverse cosine function* and the *inverse tangent function*:

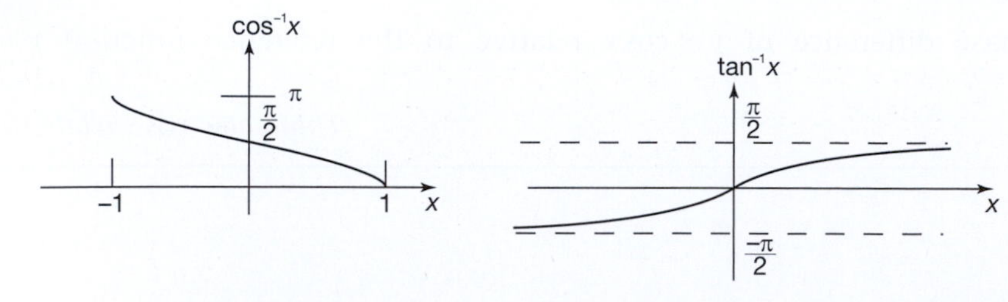

As in the case of the trigonometric functions, the values of these inverse functions are found using a calculator.

So that:

(a) $\sin^{-1}(0{\cdot}5) = \ldots\ldots\ldots$

(b) $\tan^{-1}(-3{\cdot}5) = \ldots\ldots\ldots$

(c) $\sec^{-1}(10) = \ldots\ldots\ldots$ (refer to Programme F.9 for the definition of $\sec\theta$)

*Next frame for the answer*

**24**

(a) $30^\circ$ (b) $-74{\cdot}05^\circ$ (c) $84{\cdot}26^\circ$

Because

(c) If $\sec^{-1}(10) = \theta$ then $\sec\theta = 10 = \dfrac{1}{\cos\theta}$ so that $\cos\theta = 0{\cdot}1$ and

$\theta = \cos^{-1}(0{\cdot}1) = 84{\cdot}26^\circ$

So remember $\sec^{-1}\theta = \cos^{-1}\dfrac{1}{\theta}$.

Similar results are obtained for $\text{cosec}^{-1}\theta$ and $\cot^{-1}\theta$:

$$\text{cosec}^{-1}\theta = \sin^{-1}\frac{1}{\theta} \text{ and } \cot^{-1}\theta = \tan^{-1}\frac{1}{\theta}$$

*Now to use these functions and their inverses to solve equations*

**25**

## Trigonometric equations

A simple trigonometric equation is one that involves just a single trigonometric expression. For example, the equation:

$\sin 3x = 0$ is a simple trigonometric equation

The solution of this equation can be found from inspecting the graph of the sine function $\sin\theta$ which crosses the $\theta$-axis whenever $\theta$ is an integer multiple of $\pi$. That is, $\sin n\pi = 0$ where $n$ is an integer. This means that the solutions to $\sin 3x = 0$ are found when:

$$3x = n\pi \text{ so that } x = \frac{n\pi}{3},\ n = 0,\ \pm 1,\ \pm 2,\ \ldots$$

So the values of $x$ that satisfy the simple trigonometric equation:

$\cos 2x = 1$ are $\ldots\ldots\ldots\ldots$

*Next frame*

**26**

$x = n\pi,\ n = 0,\ \pm 1,\ \pm 2,\ \ldots$

Because

From the graph of the cosine function you can see that it rises to its maximum $\cos\theta = 1$ whenever $\theta$ is an even multiple of $\pi$, that is $\theta = 0,\ \pm 2\pi,\ \pm 4\pi$. Consequently, $\cos 2x = 1$ when $2x = 2n\pi$ so that $x = n\pi,\ n = 0,\ \pm 1,\ \pm 2,\ \ldots$

▶

Just look at another. Consider the equation:

$$2\sin 3x = \sqrt{2}$$

This can be rewritten as

$$\sin 3x = \frac{\sqrt{2}}{2} = \frac{1}{\sqrt{2}}.$$

From what you know about a right-angled isosceles triangle, you can say that when $\theta = \frac{\pi}{4}$ then $\sin\theta = \frac{1}{\sqrt{2}}$. However if you look at at the graph of the sine function you can see that between $\theta = 0$ and $\theta = 2\pi$ there are two values of $\theta$ where $\sin\theta = \frac{1}{\sqrt{2}}$, namely $\theta = \frac{\pi}{4}$ and $\frac{3\pi}{4}$.

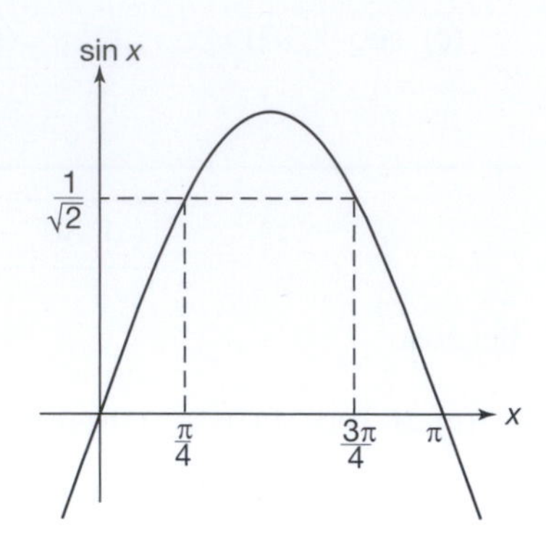


Consequently:

$$\sin 3x = \frac{1}{\sqrt{2}} \text{ when } 3x = \frac{\pi}{4} \pm 2n\pi \text{ and } 3x = \frac{3\pi}{4} \pm 2n\pi,\ n = 0,\ \pm 1,\ \pm 2,\ \ldots$$

So the values of $x$ that satisfy $2\sin 3x = \sqrt{2}$ are:

$$x = \frac{\pi}{12} \pm \frac{2n\pi}{3} \text{ and } x = \frac{\pi}{4} \pm \frac{2n\pi}{3} \text{ where } n = 0,\ \pm 1,\ \pm 2,\ \ldots$$

So, the values of $x$ that satisfy $\cos 4x = \frac{1}{2}$ are ...........

*Next frame*

**27**

$$x = \frac{\pi}{12} \pm \frac{n\pi}{2},\ \text{and}\ \frac{5\pi}{12} \pm \frac{n\pi}{2},\ n = 0,\ \pm 1,\ \pm 2,\ \ldots$$

Because

From the graph of the cosine function you see that when $\cos\theta = \frac{1}{2}$

$$\theta = \frac{\pi}{3} \pm 2n\pi \text{ or } \theta = \frac{5\pi}{3} \pm 2n\pi$$

where $n = 0,\ \pm 1,\ \pm 2,\ \ldots$

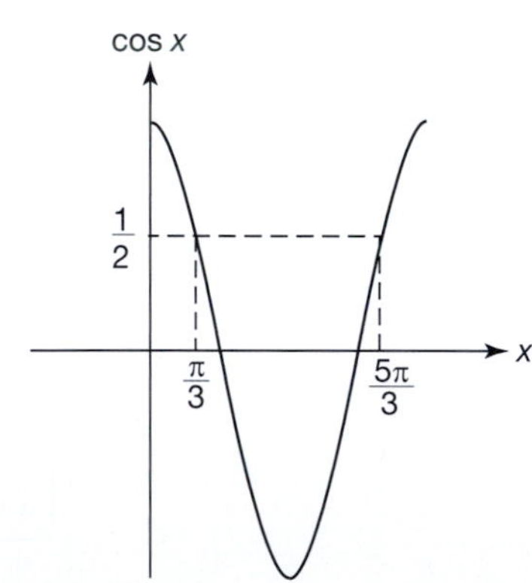


So that when $\cos 4x = \frac{1}{2}$, $4x = \frac{\pi}{3} \pm 2n\pi$ or $4x = \frac{5\pi}{3} \pm 2n\pi$. That is:

$$x = \frac{\pi}{12} \pm \frac{n\pi}{2} \text{ and } x = \frac{5\pi}{12} \pm \frac{n\pi}{2} \text{ where } n = 0,\ \pm 1,\ \pm 2,\ \ldots$$

## Equations of the form $a\cos x + b\sin x = c$

A plot of $f(x) = a\cos x + b\sin x$ against $x$ will produce a sinusoidal graph. Try it. Use your spreadsheet to plot:

$f(x) = 3\cos x + 4\sin x$ against $x$ for $-10 \le x \le 10$ with *step value* 1. The result is shown below:

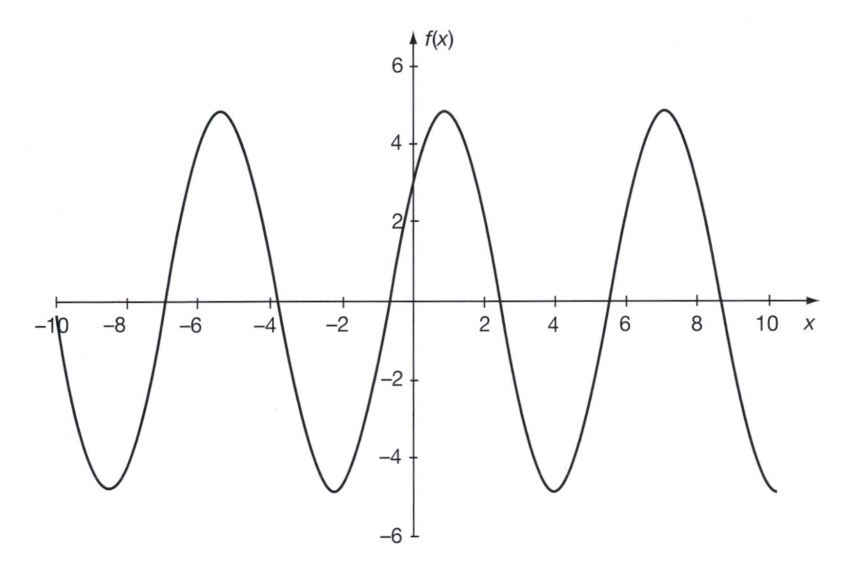


This sinusoidal shape possesses an amplitude and a phase, so its equation must be of the form:

$$f(x) = R\sin(x + \theta)$$

or $f(x) = R\cos(x + \phi)$

Either form will suffice – we shall select the first one to find solutions to the equation:

$$3\cos x + 4\sin x = 5$$

That is:

$$R\sin(x + \theta) = 5$$

The left-hand side can be expanded to give:

$$R\sin\theta\cos x + R\cos\theta\sin x = 5$$

Comparing this equation with the equation $3\cos x + 4\sin x = 5$ enables us to say that:

$3 = R\sin\theta$ and $4 = R\cos\theta$

Now:

$$\begin{aligned} R^2\sin^2\theta + R^2\cos^2\theta &= R^2 \\ &= 3^2 + 4^2 = 25 \\ &= 5^2 \end{aligned}$$

so that $R = 5$. This means that $5\sin(\theta + x) = 5$ so that:

$\sin(\theta + x) = 1$ with solution $\theta + x = \dfrac{\pi}{2} \pm 2n\pi$

Thus:

$$x = \frac{\pi}{2} - \theta \pm 2n\pi$$

▶

Now, $\dfrac{R\sin\theta}{R\cos\theta} = \tan\theta = \dfrac{3}{4}$ so that $\theta = \tan^{-1}\left(\dfrac{3}{4}\right) = 0{\cdot}64$ rad. This gives the solution to the original equation as:

$$x = \frac{\pi}{2} - 0{\cdot}64 \pm 2n\pi = 0{\cdot}93 \pm 2n\pi \text{ radians to 2 dp}$$

Notice that if we had assumed a form $f(x) = R\cos(x + \phi)$ the end result would have been the same.

**29**

Try this one yourself.

The solutions to the equation $\sin x - \sqrt{2}\cos x = 1$ are ...........

*Next frame for the answer*

**30**

$$\boxed{x = 1{\cdot}571 \pm 2n\pi \text{ radians}}$$

Because

Letting $R\sin(x + \theta) = 1$ we find, by expanding the sine, that:

$$R\sin x\cos\theta + R\sin\theta\cos x = 1$$

Comparing this equation with $\sin x - \sqrt{2}\cos x = 1$ enables us to say that:

$$R\cos\theta = 1 \text{ and } R\sin\theta = -\sqrt{2}$$

where $R^2\cos^2\theta + R^2\sin^2\theta = R^2 = 1^2 + (-\sqrt{2})^2 = 3$ so that $R = \sqrt{3}$.

This means that $\sqrt{3}\sin(x + \theta) = 1$ so that $\sin(x + \theta) = \dfrac{1}{\sqrt{3}}$ giving:

$$x + \theta = \sin^{-1}\left(\frac{1}{\sqrt{3}}\right) = 0{\cdot}6155 \pm 2n\pi \text{ radians}$$

That is, $x = 0{\cdot}6155 \pm 2n\pi - \theta$ rad. Now:

$$\frac{R\sin\theta}{R\cos\theta} = \tan\theta = -\sqrt{2} \text{ so that } \theta = \tan^{-1}\left(-\sqrt{2}\right) = -0{\cdot}9553 \text{ rad,}$$

giving the final solution as $x = 1{\cdot}571 \pm 2n\pi$ radians.

***At this point let us pause and summarize the main facts so far on trigonometric functions and equations***

## Review summary

**31**

1. The definitions of the trigonometric ratios, valid for angles greater than 0° and less than 90°, can be extended to the trigonometric functions valid for any angle.
2. The trigonometric functions possess periods, amplitudes and phases.
3. The inverse trigonometric functions have restricted ranges.

##  Review exercise

32

1 Use a calculator to find the value of each of the following (take care to ensure that your calculator is in the correct mode):
(a) $\sin(3\pi/4)$ (b) $\operatorname{cosec}(-\pi/13)$ (c) $\tan(125°)$
(d) $\cot(-30°)$ (e) $\cos(-5\pi/7)$ (f) $\sec(18\pi/11)$

2 Find the period, amplitude and phase (in radians) of each of the following:
(a) $f(\theta) = 3\sin 9\theta$ (b) $f(\theta) = -7\cos(5\theta - 3)$
(c) $f(\theta) = \tan(2 - \theta)$ (d) $f(\theta) = -\cot(3\theta - 4)$

3 A function is defined by the following prescription:

$$f(x) = -x + 4, \qquad 0 \leq x < 3, \qquad f(x+3) = f(x)$$

Plot a graph of this function for $-9 \leq x < 9$ and find:
(a) the period
(b) the amplitude
(c) the phase of $f(x) + 2$ with respect to $f(x)$

4 Solve the following trigonometric equations:
(a) $\tan 4x = 1$
(b) $\sin(x + 2\pi) + \sin(x - 2\pi) = \dfrac{1}{2}$
(c) $2\cot\theta + 3\cot\phi = 1{\cdot}4$
$\cot\theta - \cot\phi = 0{\cdot}2$
(d) $18\cos^2 x + 3\cos x - 1 = 0$
(e) $3\sin^2 x - \cos^2 x = \sin 2x$
(f) $\cos x + \sqrt{3}\sin x = \sqrt{2}$ for $0 \leq x \leq 2\pi$

33

1 Using your calculator you will find:
(a) 0·7071 (b) – 4·1786 (c) – 1·4281
(d) – 1·7321 (e) – 0·6235 (f) 2·4072

2 (a) $3\sin 9\theta = 3\sin(9\theta + 2\pi) = 3\sin 9(\theta + 2\pi/9)$ so the period of $f(\theta)$ is $2\pi/9$ and the phase is 0. The maximum value of $f(\theta)$ is 3 and the average value is 0, so the amplitude of $f(\theta)$ is 3.

(b) $-7\cos(5\theta - 3) = -7\cos(5\theta - 3 + 2\pi) = -7\cos 5(\theta + 2\pi/5 - 3/5)$ so the period of $f(\theta)$ is $2\pi/5$, the phase is $-3/5$ and the amplitude is 7.

(c) $\tan(2 - \theta) = \tan(2 - \theta + \pi) = \tan(-\theta + 2 + \pi)$ so the period of $f(\theta)$ is $\pi$ and $f(\theta)$ leads $\tan(-\theta)$ by the phase 2 with an infinite amplitude.

(d) $-\cot(3\theta - 4) = \cot(-3\theta + 4 + \pi) = \cot 3(-\theta + 4/3 + \pi/3)$ so the period of $f(\theta)$ is $\pi/3$ and $f(\theta)$ leads $\cot(-3\theta)$ by the phase 4/3 with an infinite amplitude.

3 (a) 3 (b) 1·5 (c) 0

4 (a) If $\tan\theta = 1$ then $\theta = \dfrac{\pi}{4} \pm n\pi$ radians. Since $\tan 4x = 1$ then $4x = \dfrac{\pi}{4} \pm n\pi$ so that $x = \dfrac{\pi}{16} \pm n\dfrac{\pi}{4}$.

▶

(b) LHS $= (\sin x\cos 2\pi + \sin 2\pi\cos x) + (\sin x\cos 2\pi - \sin 2\pi\cos x)$

$= 2\sin x\cos 2\pi$

$= 2\sin x$ because $\cos 2\pi = 1$

$= 1/2$ right-hand side

Therefore $\sin x = 1/4$ so $x = \sin^{-1}\left(\frac{1}{4}\right) = 0{\cdot}2527$ or $2{\cdot}8889 \pm 2n\pi$ radians.

(c) Multiplying the second equation by 3 yields:

$2\cot\theta + 3\cot\phi = 1{\cdot}4$

$3\cot\theta - 3\cot\phi = 0{\cdot}6$ adding yields $5\cot\theta = 2{\cdot}0$ so that

$\theta = \cot^{-1}\left(\frac{2}{5}\right) = 1{\cdot}1903 \pm n\pi$ radians.

Also, substituting $\cot\theta = 0{\cdot}4$ into the first equation gives

$\cot\phi = \frac{1{\cdot}4 - 0{\cdot}8}{3} = 0{\cdot}2$ so that $\phi = \cot^{-1}(0{\cdot}2) = 1{\cdot}3734 \pm n\pi$ radians.

(d) This equation factorizes as $(6\cos x - 1)(3\cos x + 1) = 0$ so that:

$\cos x = 1/6$ or $-1/3$. Thus $x = \pm 1{\cdot}4033 \pm 2n\pi$ radians or $x = \pm 1{\cdot}9106 \pm 2n\pi$ radians.

(e) This equation can be written as $3\sin^2 x - \cos^2 x = 2\sin x\cos x$.

That is: $3\sin^2 x - 2\sin x\cos x - \cos^2 x = 0$.
That is $(3\sin x + \cos x)(\sin x - \cos x) = 0$
so that $3\sin x + \cos x = 0$ or $\sin x - \cos x = 0$.
If $3\sin x + \cos x = 0$ then $\tan x = -1/3$ and so $x = -0{\cdot}3218 \pm n\pi$
and if $\sin x - \cos x = 0$ then $\tan x = 1$ and so $x = \pi/4 \pm n\pi$.

(f) To solve $\cos x + \sqrt{3}\sin x = \sqrt{2}$, write $R\sin\theta = 1$ and $R\cos\theta = \sqrt{3}$. The equation then becomes:

$R\sin\theta\cos x + R\cos\theta\sin x = \sqrt{2}$

That is:

$R\sin(\theta + x) = \sqrt{2}$

Now:

$R^2\sin^2\theta + R^2\cos^2\theta = R^2 = 1^2 + (\sqrt{3})^2 = 4$

so that $R = 2$. This means that $2\sin(\theta + x) = \sqrt{2}$ so that:

$\sin(\theta + x) = \frac{1}{\sqrt{2}}$ with solution $\theta + x = \frac{\pi}{4} \pm 2n\pi$ or $\theta + x = \frac{3\pi}{4} \pm 2n\pi$

Thus:

$\theta = \frac{\pi}{4} - x \pm 2n\pi$ or $\theta = \frac{3\pi}{4} - x \pm 2n\pi$

Now $\frac{R\sin\theta}{R\cos\theta} = \tan\theta = \frac{1}{\sqrt{3}}$ so that $\theta = \tan^{-1}\left(\frac{1}{\sqrt{3}}\right) = \frac{\pi}{6}$ or $\frac{7\pi}{6}$ rad. This gives the solution to the original equation as:

$x = \frac{\pi}{4} - \frac{\pi}{6} = \frac{\pi}{12}$ or $x = \frac{3\pi}{4} - \frac{\pi}{6} = \frac{7\pi}{12}$ within the range $0 \le x \le 2\pi$.

# Exponential and logarithmic functions

## Exponential functions



The exponential function is expressed by the equation:

$y = e^x$ or $y = \exp(x)$

where $e$ is the exponential number 2·7182818... . The graph of this function lies entirely above the $x$-axis as does the graph of its reciprocal $y = e^{-x}$, as can be seen in the diagram:

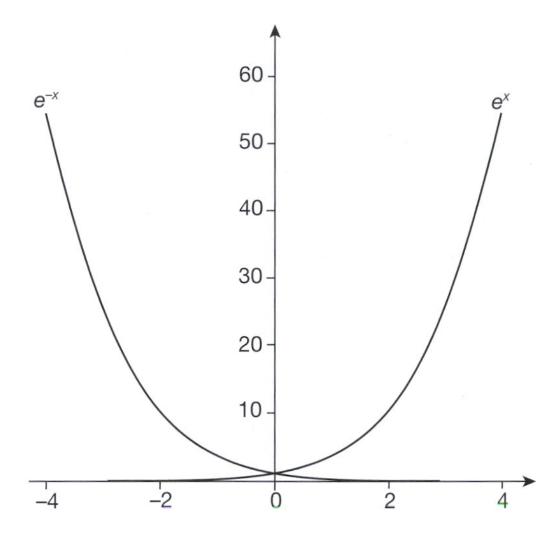


The value of $e^x$ can be found to any level of precision desired from the series expansion:

$$e^x = 1 + x + \frac{x^2}{2!} + \frac{x^3}{3!} + \frac{x^4}{4!} + \ldots$$

In practice a calculator is used. The general exponential function is given by:

$y = a^x$ where $a > 0$

and because $a = e^{\ln a}$ the general exponential function can be written in the form:

$y = e^{x \ln a}$

Because $\ln a < 1$ when $a < e$ you can see that the graph increases less quickly than the graph of $e^x$ and if $a > e$ it grows faster.

The inverse exponential function is the logarithmic function expressed by the equation:

$y = \log_a x$

with the graph shown in the diagram:

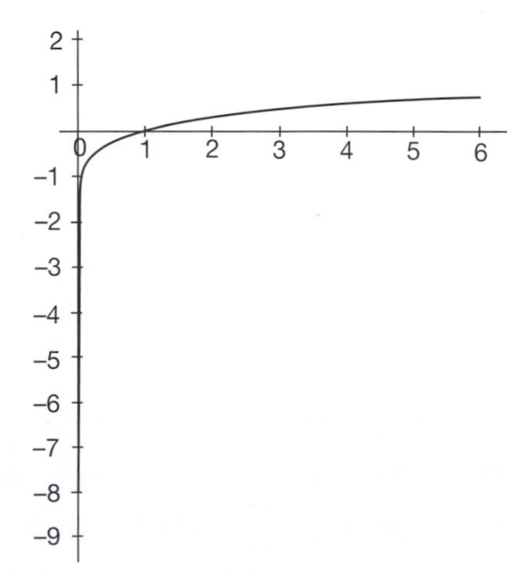


When the base $a$ of the logarithmic function takes on the value of the exponential number $e$ the notation $y = \ln x$ is used.

## Indicial equations

An *indicial equation* is an equation where the variable appears as an index and the solution of such an equation requires the application of logarithms.

**Example 1**

Here is a simple case. We have to find the value of $x$, given that $12^{2x} = 35{\cdot}4$.

Taking logs of both sides – and using $\log(A^n) = n\log A$ we have

$$
\begin{aligned}
&(2x)\log 12 = \log 35{\cdot}4\\
\text{i.e.}\quad &(2x)1{\cdot}0792 = 1{\cdot}5490\\
&2{\cdot}1584x = 1{\cdot}5490\\
&\therefore x = \frac{1{\cdot}5490}{2{\cdot}1584} = 0{\cdot}71766\\
&\therefore x = 0{\cdot}7177 \text{ to 4 sig fig}
\end{aligned}
$$

**Example 2**

Solve the equation $4^{3x-2} = 26^{x+1}$

The first line in this solution is ...........

**35**

$$\boxed{(3x-2)\log 4 = (x+1)\log 26}$$

$\therefore (3x-2)0{\cdot}6021 = (x+1)1{\cdot}4150$

Multiplying out and collecting up, we eventually get

$x =$ ........... to 4 sig fig

**36**

$$\boxed{6{\cdot}694}$$

Because we have

$$
\begin{aligned}
(3x-2)0{\cdot}6021 &= (x+1)1{\cdot}4150\\
1{\cdot}8063x - 1{\cdot}2042 &= 1{\cdot}4150x + 1{\cdot}4150\\
(1{\cdot}8063 - 1{\cdot}4150)x &= (1{\cdot}4150 + 1{\cdot}2042)\\
0{\cdot}3913x &= 2{\cdot}6192\\
\therefore x &= \frac{2{\cdot}6192}{0{\cdot}3913} = 6{\cdot}6936\\
\therefore x &= 6{\cdot}694 \text{ to 4 sig fig}
\end{aligned}
$$

Care must be taken to apply the rules of logarithms rigidly.

*Now we will deal with another example*

**37**

**Example 3**

Solve the equation $5{\cdot}4^{x+3} \times 8{\cdot}2^{2x-1} = 4{\cdot}8^{3x}$

We recall that $\log(A \times B) = \log A + \log B$

Therefore, we have $\log\{5{\cdot}4^{x+3}\} + \log\{8{\cdot}2^{2x-1}\} = \log\{4{\cdot}8^{3x}\}$

i.e. $\quad (x+3)\log 5{\cdot}4 + (2x-1)\log 8{\cdot}2 = 3x\log 4{\cdot}8$

You can finish it off, finally getting

$x =$ ........... to 4 sig fig

**38**

$\boxed{2{\cdot}485}$

Here is the working as a check:

$$(x+3)0{\cdot}7324 + (2x-1)0{\cdot}9138 = (3x)0{\cdot}6812$$
$$0{\cdot}7324x + 2{\cdot}1972 + 1{\cdot}8276x - 0{\cdot}9138 = 2{\cdot}0436x$$
$$(0{\cdot}7324 + 1{\cdot}8276)x + (2{\cdot}1972 - 0{\cdot}9138) = 2{\cdot}0436x$$
$$\therefore\ 2{\cdot}5600x + 1{\cdot}2834 = 2{\cdot}0436x$$
$$(2{\cdot}5600 - 2{\cdot}0436)x = -1{\cdot}2834$$
$$0{\cdot}5164x = -1{\cdot}2834$$
$$x = -\frac{1{\cdot}2834}{0{\cdot}5164} = -2{\cdot}4853 \quad \therefore x = -2{\cdot}485 \text{ to 4 sig fig}$$

Finally, here is one to do all on your own.

**Example 4**

Solve the equation $7(14{\cdot}3^{x+5}) \times 6{\cdot}4^{2x} = 294$

Work right through it, giving the result to 4 sig fig.

$x = \ldots\ldots\ldots\ldots$

**39**

$\boxed{-1{\cdot}501}$

Check the working:

$$7(14{\cdot}3^{x+5}) \times 6{\cdot}4^{2x} = 294$$
$$\therefore\ \log 7 + (x+5)\log 14{\cdot}3 + (2x)\log 6{\cdot}4 = \log 294$$
$$0{\cdot}8451 + (x+5)1{\cdot}1553 + (2x)0{\cdot}8062 = 2{\cdot}4683$$
$$0{\cdot}8451 + 1{\cdot}1553x + 5{\cdot}7765 + 1{\cdot}6124x = 2{\cdot}4683$$
$$(1{\cdot}1553 + 1{\cdot}6124)x + (0{\cdot}8451 + 5{\cdot}7765) = 2{\cdot}4683$$
$$2{\cdot}7677x + 6{\cdot}6216 = 2{\cdot}4683$$
$$2{\cdot}7677x = 2{\cdot}4683 - 6{\cdot}6216 = -4{\cdot}1533$$
$$\therefore\ x = -\frac{4{\cdot}1533}{2{\cdot}7677} = -1{\cdot}5006$$
$$x = -1{\cdot}501 \text{ to 4 sig fig}$$

**40**

Some indicial equations may need a little manipulation before the rules of logarithms can be applied.

**Example 5**

Solve the equation $2^{2x} - 6 \times 2^x + 8 = 0$

Because $2^{2x} = (2^x)^2$ this is an equation that is quadratic in $2^x$. We can, therefore, write $y = 2^x$ and substitute $y$ for $2^x$ in the equation to give:

$$y^2 - 6y + 8 = 0$$

which factorizes to give:

$$(y-2)(y-4) = 0$$

so that $y = 2$ or $y = 4$. That is $2^x = 2$ or $2^x = 4$ so that $x = 1$ or $x = 2$.

Try this one. Solve $2 \times 3^{2x} - 6 \times 3^x + 4 = 0$.

*The answer is in Frame 41*

**41**

$x = 0{\cdot}631$ to 3 dp or $x = 0$

Because $3^{2x} = (3^x)^2$ this is an equation that is quadratic in $3^x$. We can, therefore, write $y = 3^x$ and substitute $y$ for $3^x$ in the equation to give:

$$2y^2 - 6y + 4 = 0$$

which factorizes to give:

$$(2y - 4)(y - 1) = 0$$

so that $y = 2$ or $y = 1$. That is $3^x = 2$ or $3^x = 1$ so that $x \log 3 = \log 2$ or $x = 0$. That is $x = \dfrac{\log 2}{\log 3} = 0{\cdot}631$ to 3 dp or $x = 0$.

***At this point let us pause and summarize the main facts so far on exponential and logarithmic functions***

## Review summary

**42**

1 Any function of the form $f(x) = a^x$ is called an *exponential* function.

2 The exponential function $f(x) = a^x$ and the logarithmic function $g(x) = \log_a x$ are *mutual inverses*: $f^{-1}(x) = g(x)$ and $g^{-1}(x) = f(x)$.

## Review exercise

**43**

1 Solve the following indicial equations giving the results to 4 dp:

(a) $13^{3x} = 8{\cdot}4$ (b) $2{\cdot}8^{2x+1} \times 9{\cdot}4^{2x-1} = 6{\cdot}3^{4x}$

(c) $7^{2x} - 9 \times 7^x + 14 = 0$

**44**

1 (a) $13^{3x} = 8{\cdot}4$

Taking logs of both sides gives: $3x \log 13 = \log 8{\cdot}4$ so that:

$$x = \frac{\log 8{\cdot}4}{3 \log 13} = \frac{0{\cdot}9242\ldots}{3{\cdot}3418\ldots} = 0{\cdot}277 \text{ to 3 dp}$$

(b) $2{\cdot}8^{2x+1} \times 9{\cdot}4^{2x-1} = 6{\cdot}3^{4x}$

Taking logs:

$$(2x + 1) \log 2{\cdot}8 + (2x - 1) \log 9{\cdot}4 = 4x \log 6{\cdot}3$$

Factorizing $x$ gives:

$$x(2 \log 2{\cdot}8 + 2 \log 9{\cdot}4 - 4 \log 6{\cdot}3) = \log 9{\cdot}4 - \log 2{\cdot}8$$

So that:

$$x = \frac{\log 9{\cdot}4 - \log 2{\cdot}8}{2 \log 2{\cdot}8 + 2 \log 9{\cdot}4 - 4 \log 6{\cdot}3}$$

$$= \frac{0{\cdot}9731\ldots - 0{\cdot}4471\ldots}{2(0{\cdot}4471\ldots) + 2(0{\cdot}9731\ldots) - 4(0{\cdot}7993\ldots)} = -1{\cdot}474 \text{ to 3 dp}$$

▶

(c) $7^{2x} + 9 \times 7^x - 14 = 0$

This equation is quadratic in $7^x$ so let $y = 7^x$ and rewrite the equation as:

$y^2 - 9y + 14 = 0$ which factorizes to $(y - 2)(y - 7) = 0$
with solution: $y = 2$ or $y = 7$ that is $7^x = 2$ or $7^x = 7$

so that $x = \dfrac{\log 2}{\log 7} = 0{\cdot}356$ to 3 dp or $x = 1$

---

# Odd and even functions

**45**

If, by replacing $x$ by $-x$ in $f(x)$ the expression does not change its value, $f$ is called an *even* function. For example, if:

$f(x) = x^2$ then $f(-x) = (-x)^2 = x^2 = f(x)$ so that $f$ is an even function.

On the other hand, if $f(-x) = -f(x)$ then $f$ is called an *odd* function. For example, if:

$f(x) = x^3$ then $f(-x) = (-x)^3 = -x^3 = -f(x)$ so that $f$ is an odd function .

Because $\sin(-\theta) = -\sin\theta$ the sine function is an odd function and because $\cos(-\theta) = \cos\theta$ the cosine function is an even function. Notice how the graph of the cosine function is reflection symmetric about the vertical axis through $\theta = 0$ in the diagram in Frame 4. All even functions possess this type of symmetry. The graph of the sine function is rotation symmetric about the origin as it goes into itself under a rotation of 180° about this point as can be seen from the third diagram in Frame 2. All odd functions possess this type of antisymmetry. Notice also that $\tan(-\theta) = -\tan\theta$ so that the tangent function, like the sine function, is odd and has an antisymmetric graph (see the diagram in Frame 7).

---

## Odd and even parts

**46**

Not every function is either even or odd but many can be written as the sum of an even part and an odd part. If, given $f(x)$ where $f(-x)$ is also defined then:

$f_e(x) = \dfrac{f(x) + f(-x)}{2}$ is even and $f_o(x) = \dfrac{f(x) - f(-x)}{2}$ is odd.

Furthermore $f_e(x)$ is called the *even part* of $f(x)$ and

$f_o(x)$ is called the *odd part* of $f(x)$.

For example, if $f(x) = 3x^2 - 2x + 1$ then $f(-x) = 3(-x)^2 - 2(-x) + 1 = 3x^2 + 2x + 1$ so that the even and odd parts of $f(x)$ are:

$$f_e(x) = \frac{(3x^2 - 2x + 1) + (3x^2 + 2x + 1)}{2} = 3x^2 + 1 \text{ and}$$

$$f_o(x) = \frac{(3x^2 - 2x + 1) - (3x^2 + 2x + 1)}{2} = -2x$$

So, the even and odd parts of $f(x) = x^3 - 2x^2 - 3x + 4$ are ...........

*Apply the two formulas; the answer is in the next frame*

---

**47**

$$f_e(x) = -2x^2 + 4$$
$$f_o(x) = x^3 - 3x$$

Because

$$f_e(x) = \frac{f(x) + f(-x)}{2}$$
$$= \frac{(x^3 - 2x^2 - 3x + 4) + ((-x)^3 - 2(-x)^2 - 3(-x) + 4)}{2}$$
$$= \frac{x^3 - 2x^2 - 3x + 4 - x^3 - 2x^2 + 3x + 4}{2} = \frac{-2x^2 + 4 - 2x^2 + 4}{2}$$
$$= -2x^2 + 4$$

and

$$f_o(x) = \frac{f(x) - f(-x)}{2}$$
$$= \frac{(x^3 - 2x^2 - 3x + 4) - ((-x)^3 - 2(-x)^2 - 3(-x) + 4)}{2}$$
$$= \frac{x^3 - 2x^2 - 3x + 4 + x^3 + 2x^2 - 3x - 4}{2} = \frac{x^3 - 3x + x^3 - 3x}{2}$$
$$= x^3 - 3x$$

Make a note here that *even polynomial functions consist of only even powers* and *odd polynomial functions consist of only odd powers.*

So the odd and even parts of $f(x) = x^3(x^2 - 3x + 5)$ are ...........

*Answer in next frame*

**48**

$$f_e(x) = -3x^4 \qquad f_o(x) = x^5 + 5x^3$$

Because

Even polynomial functions consist of only even powers and odd polynomial functions consist of only odd powers. Consequently, the even part of $f(x)$ consists of even powers only and the odd part of $f(x)$ consists of odd powers only.

Now try this. The even and odd parts of $f(x) = \dfrac{1}{x-1}$ are ...........

*Next frame*

**49**

$$f_e(x) = \frac{1}{(x-1)(x+1)}$$
$$f_o(x) = \frac{x}{(x-1)(x+1)}$$

Because

$$f_e(x) = \frac{f(x) + f(-x)}{2}$$
$$= \frac{1}{2}\left(\frac{1}{(x-1)} + \frac{1}{(-x-1)}\right) = \frac{1}{2}\left(\frac{1}{(x-1)} - \frac{1}{(x+1)}\right)$$
$$= \frac{1}{2}\frac{x + 1 - (x-1)}{(x-1)(x+1)} = \frac{1}{2}\frac{2}{(x-1)(x+1)}$$
$$= \frac{1}{(x-1)(x+1)}$$

▶

and

$$f_o(x) = \frac{f(x) - f(-x)}{2}$$
$$= \frac{1}{2}\left(\frac{1}{(x-1)} - \frac{1}{(-x-1)}\right) = \frac{1}{2}\left(\frac{1}{(x-1)} + \frac{1}{(x+1)}\right)$$
$$= \frac{1}{2}\frac{x+1+x-1}{(x-1)(x+1)} = \frac{1}{2}\frac{2x}{(x-1)(x+1)}$$
$$= \frac{x}{(x-1)(x+1)}$$

*Next frame*

50

## Odd and even parts of the exponential function

The exponential function is neither odd nor even but it can be written as a sum of an odd part and an even part.

That is, $\exp_e(x) = \frac{\exp(x) + \exp(-x)}{2}$ and $\exp_o(x) = \frac{\exp(x) - \exp(-x)}{2}$. These two functions are known as the *hyperbolic cosine* and the *hyperbolic sine* respectively:

$$\cosh x = \frac{e^x + e^{-x}}{2} \quad \text{and} \quad \sinh x = \frac{e^x - e^{-x}}{2}$$

Using these two functions the hyperbolic tangent can also be defined:

$$\tanh x = \frac{e^x - e^{-x}}{e^x + e^{-x}}$$

No more will be said about these hyperbolic trigonometric functions here. Instead they will be looked at in more detail in Programme 3 of Part II.

The logarithmic function $y = \log_a x$ is neither odd nor even and indeed does not possess even and odd parts because $\log_a(-x)$ is not defined.

51

## Limits of functions

There are times when a function is not defined for a particular value of $x$, say $x = x_0$ but it is defined for values of $x$ that are arbitrarily close to $x_0$. For example, the expression:

$$f(x) = \frac{x^2 - 1}{x - 1}$$

is not defined when $x = 1$ because at that point the denominator is zero and division by zero is not defined. However, we note that:

$$f(x) = \frac{x^2 - 1}{x - 1} = \frac{(x-1)(x+1)}{x-1} = x + 1 \text{ provided that } x \neq 1$$

We still cannot permit the value $x = 1$ because to do so would mean that the cancellation of the $x - 1$ factor would be division by zero. But we can say that as the value of $x$ approaches 1, the value of $f(x)$ approaches 2. Clearly the value of $f(x)$ never actually attains the value of 2 but it does get as close to it as you wish it to be by selecting a value of $x$ sufficiently close to 1. We say that the limit of $\frac{x^2 - 1}{x - 1}$ as $x$ approaches 1 is 2 and we write:

$$\operatorname*{Lim}_{x \to 1}\left(\frac{x^2 - 1}{x - 1}\right) = 2$$

You try one, $\operatorname*{Lim}_{x \to -3}\left(\frac{x^2 - 9}{x + 3}\right) = \ldots\ldots\ldots\ldots$

*The answer is in the next frame*

**52**

$\boxed{-6}$

Because

$$\begin{aligned} f(x) &= \frac{x^2 - 9}{x + 3} \\ &= \frac{(x + 3)(x - 3)}{x + 3} \\ &= x - 3 \text{ provided } x \neq -3 \end{aligned}$$

So that:

$$\underset{x \to -3}{Lim} \left( \frac{x^2 - 9}{x + 3} \right) = -6$$

## The rules of limits

Here we list the rules for evaluating the limits of expressions involving a real variable $x$ whose value approaches a fixed value $x_0$. They are what you would expect with no surprises.

If $\underset{x \to x_0}{Lim} f(x) = A$ and $\underset{x \to x_0}{Lim} g(x) = B$ then:

$$\underset{x \to x_0}{Lim} (f(x) \pm g(x)) = \underset{x \to x_0}{Lim} f(x) \pm \underset{x \to x_0}{Lim} g(x) = A \pm B$$

$$\underset{x \to x_0}{Lim} (f(x)g(x)) = \underset{x \to x_0}{Lim} f(x) \, \underset{x \to x_0}{Lim} g(x) = AB$$

$$\underset{x \to x_0}{Lim} \frac{f(x)}{g(x)} = \frac{\underset{x \to x_0}{Lim} f(x)}{\underset{x \to x_0}{Lim} g(x)} = \frac{A}{B} \text{ provided } \underset{x \to x_0}{Lim} g(x) = B \neq 0$$

$$\underset{x \to x_0}{Lim} f(g(x)) = f\left( \underset{x \to x_0}{Lim} g(x) \right) = f(B) \text{ provided } f(g(x)) \text{ is continuous at } x_0$$

So you try these for yourself:

(a) $\underset{x \to \pi}{Lim} (x^2 - \sin x)$

(b) $\underset{x \to \pi}{Lim} (x^2 \sin x)$

(c) $\underset{x \to \pi/4}{Lim} \dfrac{(\tan x)}{(\sin x)}$

(d) $\underset{x \to 1}{Lim} \cos(x^2 - 1)$

*The answers are in the next frame*

**53**

| (a) $\pi^2$ | (b) 0 | (c) $\sqrt{2}$ | (d) 1 |
|---|---|---|---|

Because:

(a) $\underset{x \to \pi}{Lim} (x^2 - \sin x) = \underset{x \to \pi}{Lim} x^2 - \underset{x \to \pi}{Lim} \sin x = \pi^2 - 0 = \pi^2$

(b) $\underset{x \to \pi}{Lim} (x^2 \sin x) = \underset{x \to \pi}{Lim} (x^2) \underset{x \to \pi}{Lim} (\sin x) = \pi^2 \times 0 = 0$

(c) $\underset{x \to \pi/4}{Lim} \dfrac{(\tan x)}{(\sin x)} = \dfrac{\underset{x \to \pi/4}{Lim} \tan x}{\underset{x \to \pi/4}{Lim} \sin x} = \dfrac{1}{1/\sqrt{2}} = \sqrt{2}$

(d) $\underset{x \to 1}{Lim} \cos(x^2 - 1) = \cos\left( \underset{x \to 1}{Lim} (x^2 - 1) \right) = \cos 0 = 1$

▶

All fairly straightforward for these simple problems. Difficulties do occur for the limits of quotients when both the numerator and the denominator are simultaneously zero at the limit point but those problems we shall leave for later.

***At this point let us pause and summarize the main facts so far on odd and even functions***

## Review summary

54

1 If $f(-x) = f(x)$ then $f$ is called an *even* function and if $f(-x) = -f(x)$ then $f$ is called an *odd* function.

2 If $f_e(x) = \dfrac{f(x) + f(-x)}{2}$ can be defined it is called the *even* part of $f(x)$ and if $f_o(x) = \dfrac{f(x) - f(-x)}{2}$ can be defined it is called the *odd* part of $f(x)$.

3 The limit of a sum (difference) is equal to the sum (difference) of the limits.
The limit of a product is equal to the product of the limits.
The limit of a quotient is equal to the quotient of the limits provided the denominator limit is not zero.

## Review exercise

55

1 Which of the following are odd functions and which are even? For those that are neither odd nor even find their odd and even parts.

(a) $f(x) = x \sin x$ (b) $f(x) = x^2 e^x$ (c) $f(x) = \dfrac{x}{x^2 - 1}$ (d) $f(x) = x^4 \ln x$

2 Evaluate

(a) $\underset{x \to -1}{Lim}\left(\dfrac{x^2 + 2x + 1}{x^2 + 3x + 2}\right)$ (b) $\underset{x \to \pi/4}{Lim}\,(3x - \tan x)$ (c) $\underset{x \to \pi/2}{Lim}\,(2x^2 \cos[3x - \pi/2])$

56

1 (a) $f(x) = x \sin x$ and $f(-x) = (-x)\sin(-x)$

$$= x \sin x$$

$$= f(x) \quad \text{so } f(x) \text{ is even}$$

(b) $f(x) = x^2 e^x$ and $f(-x) = (-x)^2 e^{-x}$

$$= x^2 e^{-x}$$

$$\neq f(x) \text{ or } -f(x) \quad \text{so } f(x) \text{ is neither odd nor even}$$

$$f_e(x) = \frac{f(x) + f(-x)}{2} = \frac{x^2 e^x + x^2 e^{-x}}{2} = x^2 \frac{e^x + e^{-x}}{2} = x^2 \cosh x$$

and

$$f_o(x) = \frac{f(x) - f(-x)}{2} = \frac{x^2 e^x - x^2 e^{-x}}{2} = x^2 \frac{e^x - e^{-x}}{2} = x^2 \sinh x$$

(c) $f(x) = \dfrac{x}{x^2 - 1}$ and $f(-x) = \dfrac{-x}{(-x)^2 - 1}$

$$= -\frac{x}{x^2 - 1}$$

$$= -f(x) \quad \text{so } f(x) \text{ is odd}$$

▶

(d) $f(x) = x^4 \ln x$ and $f(-x) = (-x)\ln(-x)$ but $\ln(-x)$ is not defined so not only is $f(x)$ neither odd nor even but it does not have odd or even parts.

2 (a) $$\underset{x\to -1}{Lim}\left(\frac{x^2+2x+1}{x^2+3x+2}\right) = \underset{x\to -1}{Lim}\left(\frac{(x+1)^2}{(x+1)(x+2)}\right)$$

$$= \underset{x\to -1}{Lim}\left(\frac{x+1}{x+2}\right)$$ The cancellation is permitted since $x \neq -1$

$$= \frac{0}{1} = 0$$

(b) $$\underset{x\to \pi/4}{Lim}\,(3x - \tan x) = \underset{x\to \pi/4}{Lim}\,(3x) - \underset{x\to \pi/4}{Lim}\,(\tan x)$$

$$= \frac{3\pi}{4} - \tan\frac{\pi}{4} = \frac{3\pi}{4} - 1$$

(c) $$\underset{x\to \pi/2}{Lim}\,(2x^2\cos[3x - \pi/2]) = \underset{x\to \pi/2}{Lim}\,(2x^2)\,\underset{x\to \pi/2}{Lim}\,(\cos[3x - \pi/2])$$

$$= \frac{2\pi^2}{4} \times \cos \underset{x\to \pi/2}{Lim}\,[3x - \pi/2]$$

$$= \frac{2\pi^2}{4} \times \cos[3\pi/2 - \pi/2]$$

$$= \frac{2\pi^2}{4} \times \cos\pi = -\frac{\pi^2}{2}$$

**57**

You have now come to the end of this Programme. A list of **Can You?** questions follows for you to gauge your understanding of the material in the Programme. These questions match the **Learning outcomes** listed at the beginning of the Programme so go back and try the **Quiz** that follows them. After that try the **Test exercise**. *Work through them at your own pace, there is no need to hurry*. A set of **Further problems** provides additional valuable practice.

## Can You?

**58**

### Checklist F.11

Check this list before and after you try the end of Programme test.

On a scale of 1 to 5 how confident are you that you can: Frames

- Develop the trigonometric functions from the trigonometric ratios? 2 to 9
  *Yes* ☐ ☐ ☐ ☐ ☐ *No*
- Find the period, amplitude and phase of a periodic function? 10 to 22
  *Yes* ☐ ☐ ☐ ☐ ☐ *No*
- Distinguish between the inverse of a trigonometric function and the inverse trigonometric function? 23 to 24
  *Yes* ☐ ☐ ☐ ☐ ☐ *No*
- Solve trigonometric equations using the inverse trigonometric functions? 25 to 30
  *Yes* ☐ ☐ ☐ ☐ ☐ *No*
- Recognize that the exponential function and the natural logarithmic function are mutual inverses and solve indicial and logarithmic equations? 34 to 41
  *Yes* ☐ ☐ ☐ ☐ ☐ *No*

- Find the even and odd parts of a function when they exist? 42 to 46
  *Yes* ☐ ☐ ☐ ☐ ☐ *No*
- Construct the hyperbolic functions from the odd and even parts of the exponential function? 47
  *Yes* ☐ ☐ ☐ ☐ ☐ *No*
- Evaluate limits of simple functions? 48 to 50
  *Yes* ☐ ☐ ☐ ☐ ☐ *No*

## Test exercise F.11

59

Frames

1 Give the value of: (Frames 2 to 7)
(a) $\sin 90^\circ$ (b) $\cos 0^\circ$ (c) $\cot 90^\circ$

2 Use a calculator to find the value of each of the following: (Frames 2 to 9)
(a) $\sin(-320^\circ)$ (b) $\operatorname{cosec}(\pi/11)$ (c) $\cot(-\pi/2)$

3 Find the period, amplitude and phase of each of the following: (Frames 10 to 22)
(a) $f(\theta) = 4\cos 7\theta$ (b) $f(\theta) = -2\sin(2\theta - \pi/2)$
(c) $f(\theta) = \sec(3\theta + 4)$

4 A function is defined by the following prescription: (Frames 10 to 22)
$f(x) = 9 - x^2, \quad 0 \le x < 3$
$f(x+3) = f(x)$
Plot a graph of this function for $-6 \le x \le 6$ and find:
(a) the period
(b) the amplitude
(c) the phase of $f(x) + 2$ with respect to $f(x)$

5 Solve the following trigonometric equations: (Frames 23 to 30)
(a) $\tan(x + \pi) + \tan(x - \pi) = 0$
(b) $6\sin^2 x - 3\sin^2 2x + \cos^2 x = 0$
(c) $3\sin x - 5\cos x = 2$

6 Find the value of $x$ corresponding to each of the following: (Frames 34 to 41)
(a) $4^{-x} = 1$ (b) $\exp(3x) = e$
(c) $e^x = 54{\cdot}32$ (d) $\log_5 x = 4$
(e) $\log_{10}(x-3) = 0{\cdot}101$ (f) $\ln 100 = x$

7 Solve for $x$: (Frames 34 to 41)
(a) $4^{x+2}5^{x+1} = 32\,000$ (b) $e^{2x} - 5e^x + 6 = 0$
(c) $\log_x 36 = 2$ (d) $\frac{1}{2}\log(x^2) = 5\log 2 - 4\log 3$
(e) $\ln(x^{1/2}) = \ln x + \ln 3$

8 Find the even and odd parts of $f(x) = a^x$. (Frames 42 to 47)

9 Evaluate: (Frames 48 to 50)

(a) $\underset{x\to 4}{Lim}\left(\dfrac{x^2 - 8x + 16}{x^2 - 5x + 4}\right)$ (b) $\underset{x\to 1}{Lim}\,(6x^{-1} + \ln x)$

(c) $\underset{x\to 1/2}{Lim}\,[(3x^3 - 1)/(4 - 2x^2)]$ (d) $\underset{x\to -3}{Lim}\,\dfrac{(1 + 9x^{-2})}{(3x - x^2)}$

(e) $\underset{x\to 1}{Lim}\left(\tan^{-1}\left[\dfrac{x^2 - 1}{2(x-1)}\right]\right)$

# Further problems F.11

**60**

1 Do the graphs of $f(x) = 3\log x$ and $g(x) = \log(3x)$ intersect?

2 Let $f(x) = \ln\left(\dfrac{1+x}{1-x}\right)$.

(a) Find the domain and range of $f$.

(b) Show that the new function $g$ formed by replacing $x$ in $f(x)$ by $\dfrac{2x}{1+x^2}$ is given by $g(x) = 2f(x)$.

3 Describe the graph of $x^2 - 9y^2 = 0$.

4 Two functions $C$ and $S$ are defined as the even and odd parts of $f$ where $f(x) = a^x$. Show that:

(a) $[C(x)]^2 - [S(x)]^2 = 1$ (b) $S(2x) = 2S(x)C(x)$

5 Show by using a diagram that, for functions $f$ and $g$, $(f \circ g)^{-1} = g^{-1} \circ f^{-1}$.

6 Is it possible to find values of $a$ and $x$ such that $\log_a(x) = a^x$ for $a > 1$?

7 Given the three functions $a$, $b$ and $c$ where $a(x) = 6x$, $b(x) = x - 2$ and $c(x) = x^3$ find the inverse of:

(a) $f(x) = a(b[c(x)])$ (b) $f = c \circ b \circ c$ (c) $f = b \circ c \circ a \circ b \circ c$

8 Use your spreadsheet to plot $\sin\theta$ and $\sin 2\theta$ on the same graph. Plot from $\theta = -5$ to $\theta = +5$ with a step value of 0·5.

9 The square sine wave with period 2 is given by the prescription:

$$f(x) = \begin{cases} 1 & 0 \le x < 1 \\ -1 & 1 \le x < 2 \end{cases}$$

Plot this wave for $-4 \le x \le 4$ on a sheet of graph paper.

10 The absolute value of $x$ is given as:

$$|x| = \begin{cases} x & \text{if } x \ge 0 \\ -x & \text{if } x < 0 \end{cases}$$

(a) Plot the graph of $y = |x|$ for $-2 \le x \le 2$.

(b) Find the derivative of $y$.

(c) Does the derivative exist at $x = 0$?

11 Use the spreadsheet to plot the rectified sine wave $f(x) = |\sin x|$ for $-10 \le x \le 10$ with step value 1.

12 Use your spreadsheet to plot $f(x) = \dfrac{\sin x}{x}$ for $-40 \le x \le 40$ with step value 4. (You will have to enter the value $f(0) = 1$ specifically in cell **B11**.)

13 Solve the following giving the results to 4 sig fig:

(a) $6\{8^{3x+2}\} = 5^{2x-7}$ (b) $4{\cdot}5^{1-2x} \times 6{\cdot}2^{3x+4} = 12{\cdot}7^{5x}$

(c) $5\{17{\cdot}2^{x+4}\} \times 3\{8{\cdot}6^{2x}\} = 4{\cdot}7^{x-1}$

14 Evaluate each of the following limits:

(a) $\underset{x\to\pi/2}{Lim}\left(\dfrac{(x^2 - \pi/4)\sin(\cos x)}{x - \pi/2}\right)$ (b) $\underset{x\to-1}{Lim}\ln\left(\exp\left[\dfrac{3x^2+2x-1}{x+1}\right]\right)$

(c) $\underset{x\to 2+\sqrt{3}}{Lim}\cos\left(\sin^{-1}\left(\dfrac{x-2}{x-\sqrt{3}}\right)\right)$

**61** Now visit the companion website for this book at www.palgrave.com/stroud for more questions applying this mathematics to science and engineering.

# Differentiation

## Learning outcomes

When you have completed this Programme you will be able to:

- ☐ Determine the gradient of a straight-line graph
- ☐ Evaluate from first principles the gradient at a point on a quadratic curve
- ☐ Differentiate powers of $x$ and polynomials
- ☐ Evaluate second derivatives and use tables of standard derivatives
- ☐ Differentiate products and quotients of expressions
- ☐ Differentiate using the chain rule for a 'function of a function'
- ☐ Use the Newton–Raphson method to obtain a numerical solution to an equation

If you already feel confident about these why not try the quiz over the page? You can check your answers at the end of the book.

# Quiz F.12

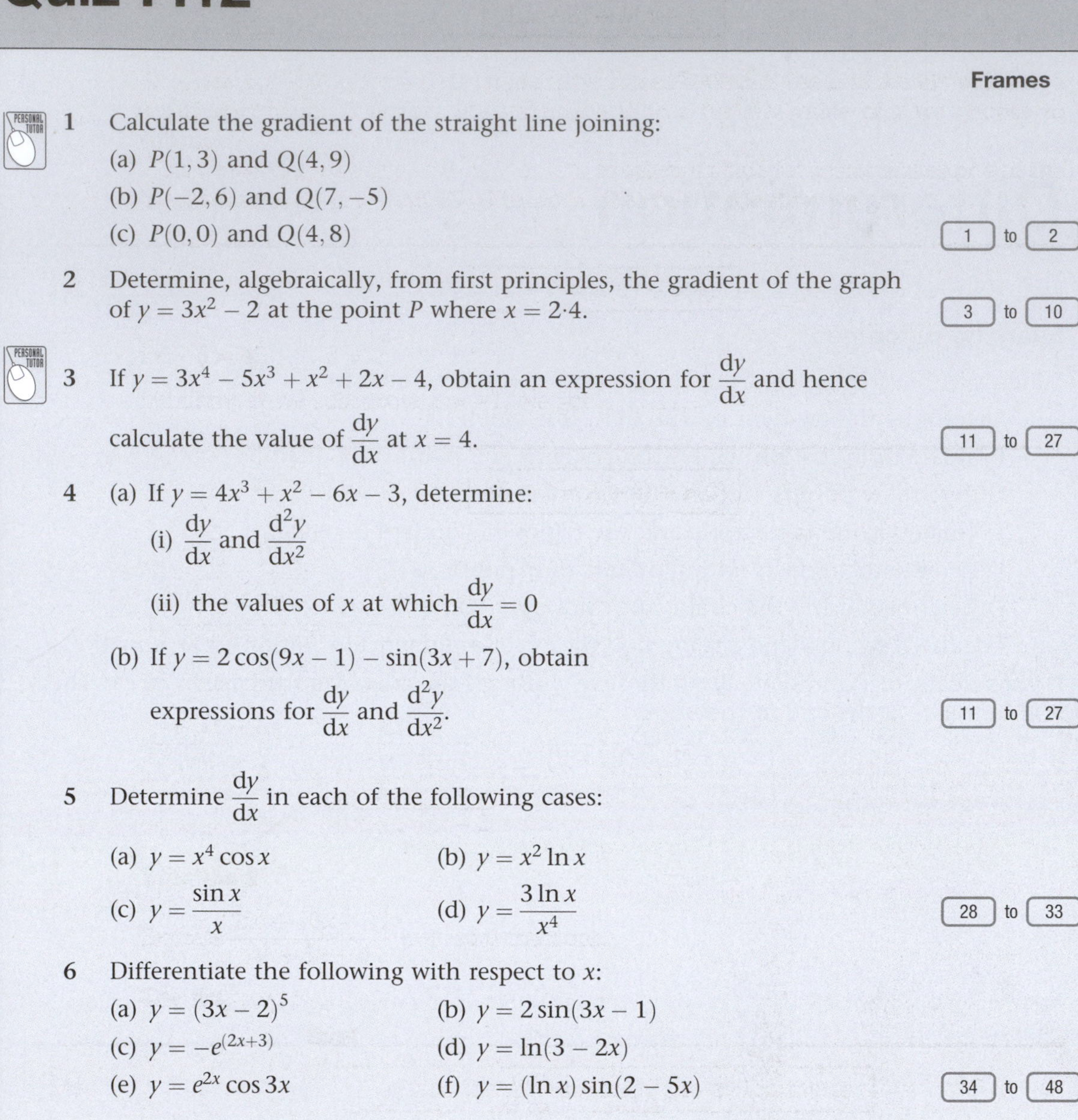

**Frames**

1. Calculate the gradient of the straight line joining:
   (a) $P(1, 3)$ and $Q(4, 9)$
   (b) $P(-2, 6)$ and $Q(7, -5)$
   (c) $P(0, 0)$ and $Q(4, 8)$ — 1 to 2

2. Determine, algebraically, from first principles, the gradient of the graph of $y = 3x^2 - 2$ at the point $P$ where $x = 2{\cdot}4$. — 3 to 10

3. If $y = 3x^4 - 5x^3 + x^2 + 2x - 4$, obtain an expression for $\dfrac{dy}{dx}$ and hence calculate the value of $\dfrac{dy}{dx}$ at $x = 4$. — 11 to 27

4. (a) If $y = 4x^3 + x^2 - 6x - 3$, determine:
   (i) $\dfrac{dy}{dx}$ and $\dfrac{d^2y}{dx^2}$
   (ii) the values of $x$ at which $\dfrac{dy}{dx} = 0$
   (b) If $y = 2\cos(9x - 1) - \sin(3x + 7)$, obtain expressions for $\dfrac{dy}{dx}$ and $\dfrac{d^2y}{dx^2}$. — 11 to 27

5. Determine $\dfrac{dy}{dx}$ in each of the following cases:
   (a) $y = x^4 \cos x$ (b) $y = x^2 \ln x$
   (c) $y = \dfrac{\sin x}{x}$ (d) $y = \dfrac{3 \ln x}{x^4}$ — 28 to 33

6. Differentiate the following with respect to $x$:
   (a) $y = (3x - 2)^5$ (b) $y = 2\sin(3x - 1)$
   (c) $y = -e^{(2x+3)}$ (d) $y = \ln(3 - 2x)$
   (e) $y = e^{2x} \cos 3x$ (f) $y = (\ln x)\sin(2 - 5x)$ — 34 to 48

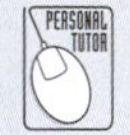

7. Use the Newton–Raphson method to solve the equation $x^3 - 3 = 0$, accurate to 6 decimal places given that $x = 1{\cdot}4$ is the solution accurate to 1 decimal place. — 49 to 71

# Gradients

## The gradient of a straight-line **1**

The *gradient* of the sloping straight line shown in the figure is defined as:

$$\frac{\text{the vertical distance the line rises or falls between two points } P \text{ and } Q}{\text{the horizontal distance between } P \text{ and } Q}$$

where $P$ is a point to the left of point $Q$ on the straight line $AB$ which slopes upwards from left to right. The changes in the $x$- and $y$-values of the points $P$ and $Q$ are denoted by $\mathrm{d}x$ and $\mathrm{d}y$ respectively. So the gradient of this line is given as:

$$\frac{\mathrm{d}y}{\mathrm{d}x}$$

We could have chosen any pair of points on the straight line for $P$ and $Q$ and by similar triangles this ratio would have worked out to the same value:

*the gradient of a straight line is constant throughout its length*

Its value is denoted by the symbol $m$.

Therefore $m = \dfrac{\mathrm{d}y}{\mathrm{d}x}$.

For example, if, for some line (see the figure below), $P$ is the point (2, 3) and $Q$ is the point (6, 4), then $P$ is to the left and *below* the point $Q$. In this case:

$\mathrm{d}y$ = the change in the $y$-values $= 4 - 3 = 1$ and
$\mathrm{d}x$ = the change in the $x$-values $= 6 - 2 = 4$ so that:

$m = \dfrac{\mathrm{d}y}{\mathrm{d}x} = \dfrac{1}{4} = 0{\cdot}25$. The sloping line *rises vertically* from left to right by 0·25 unit for every 1 unit horizontally.

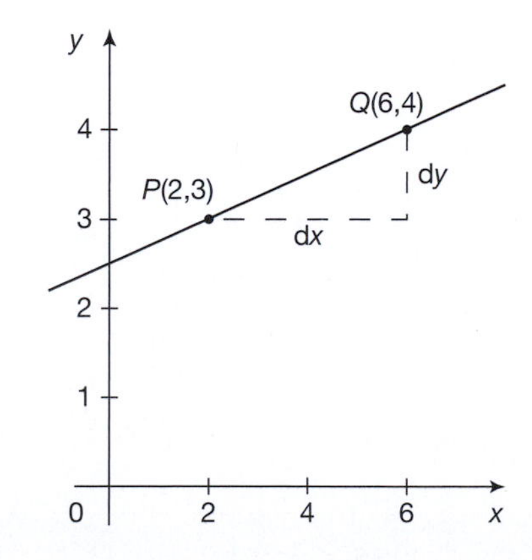


▶

If, for some other line (see the figure below), $P$ is the point (3, 5) and Q is the point (7, 1), then $P$ is to the left and *above* the point Q. In this case:

dy = the change in the $y$-values = 1 – 5 = –4 and
dx = the change in the $x$-values = 7 – 3 = 4 so that:

$m = \frac{dy}{dx} = \frac{-4}{4} = -1$. The sloping line *falls vertically* from left to right by 1 unit for every 1 unit horizontally

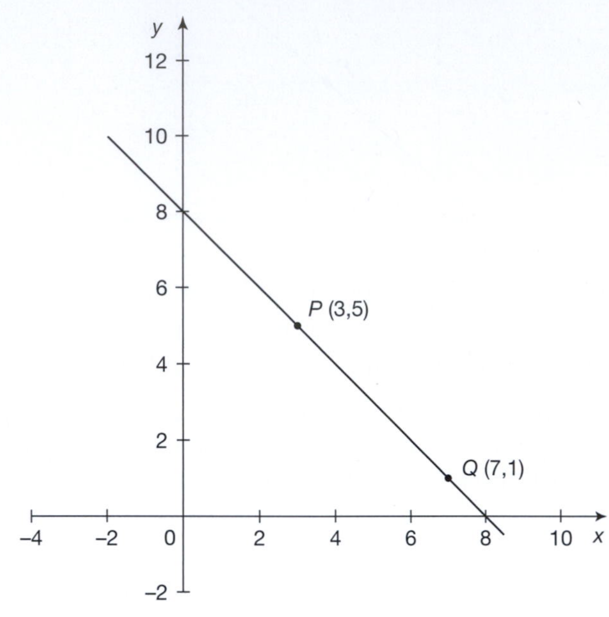


So, lines going *up* to the right have a *positive* gradient, lines going *down* to the right have a *negative* gradient.

Try the following exercises. Determine the gradients of the straight lines joining:

1 $P$ (3, 7) and Q (5, 8)
2 $P$ (2, 4) and Q (6, 9)
3 $P$ (1, 6) and Q (4, 4)
4 $P$ (–3, 6) and Q (5, 2)
5 $P$ (–2, 4) and Q (3, –2)

*Draw a diagram in each case*

**2**

Here they are:

**1**

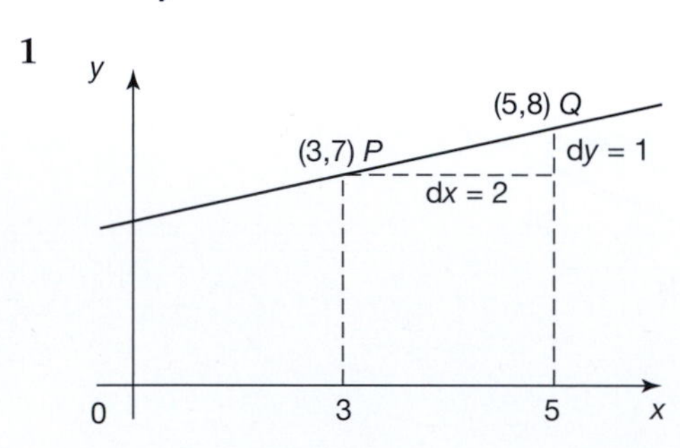


$m = \frac{1}{2} = 0{\cdot}5$

▶

**2**

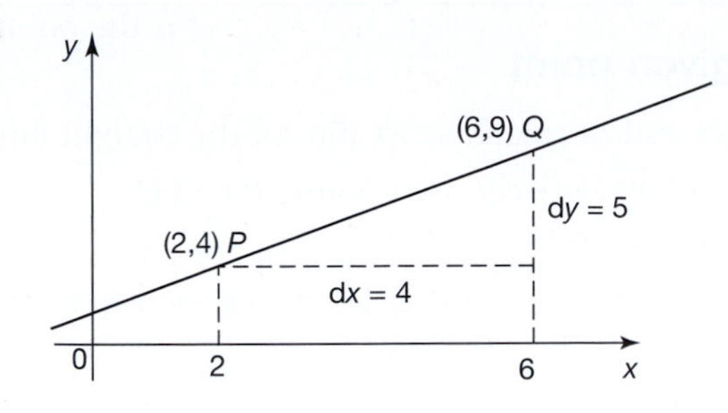


$$m = \frac{5}{4} = 1{\cdot}25$$

**3**

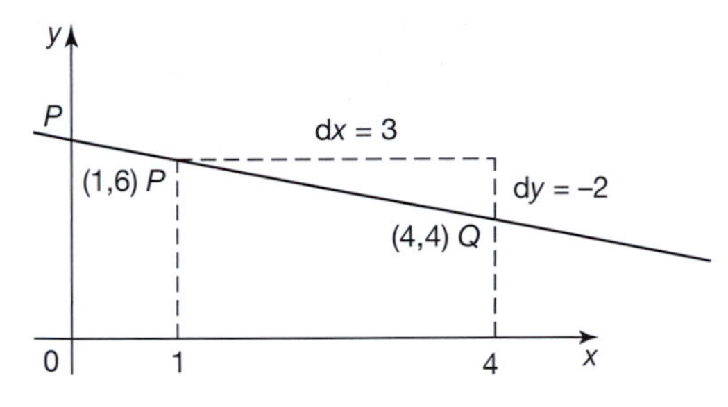


$$m = -\frac{2}{3} = -0{\cdot}667$$

**4**

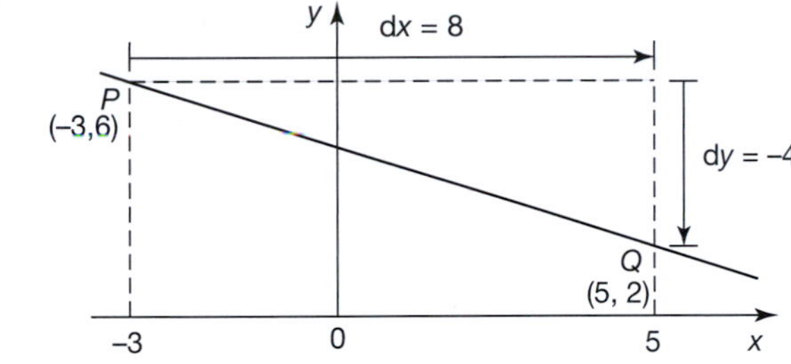


$$m = -\frac{1}{2} = -0{\cdot}5$$

**5**

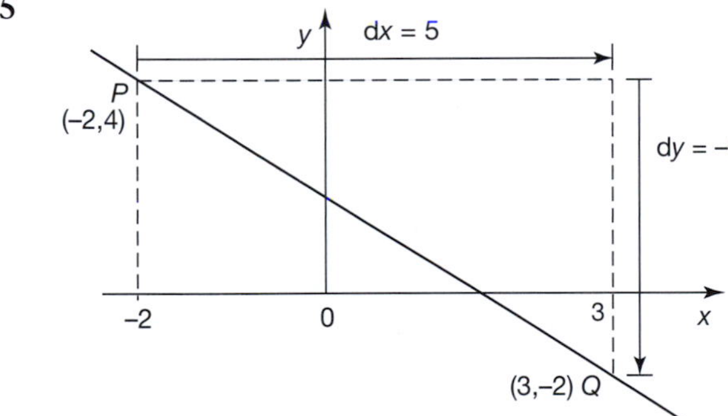


$$m = -\frac{6}{5} = -1{\cdot}2$$

Now let us extend these ideas to graphs that are not straight lines.

*On then to the next frame*

3

## The gradient of a curve at a given point

If we take two points $P$ and $Q$ on a curve and calculate, as we did for the straight line:

$$\frac{\text{the vertical distance the curve rises or falls between two points } P \text{ and } Q}{\text{the horizontal distance between } P \text{ and } Q}$$

the result will depend upon the choice of points $P$ and $Q$ as can be seen from the figure below:

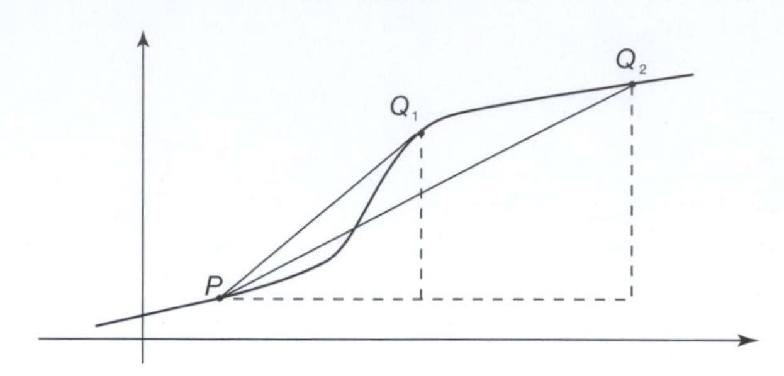


This is because the gradient of the curve varies along its length, as anyone who has climbed a hill will appreciate. Because of this the gradient of a curve is not defined *between two points* as in the case of a straight line but *at a single point*. The gradient of a curve at a given point is defined to be the gradient of the straight line that just touches the curve at that point – the *gradient of the tangent.*

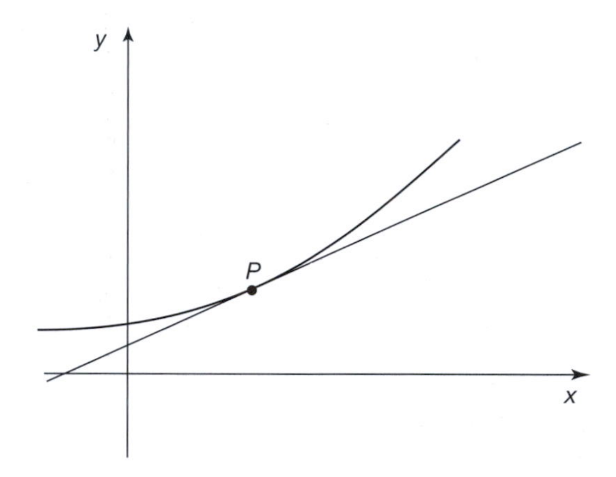


*The gradient of the curve at P is equal to the gradient of the tangent to the curve at P.*

**This is a straightforward but very important definition because all of the differential calculus depends upon it. Make a note of it in your workbook.**

For example, find the gradient of the curve of $y = 2x^2 + 5$ at the point $P$ at which $x = 1{\cdot}5$.

First we must compile a table giving the $y$-values of $y = 2x^2 + 5$ at $0{\cdot}5$ intervals of $x$ between $x = 0$ and $x = 3$.

*Complete the table and then move on to the next frame*

4

| $x$ | 0 | 0·5 | 1·0 | 1·5 | 2·0 | 2·5 | 3·0 |
|---|---|---|---|---|---|---|---|
| $x^2$ | 0 | 0·25 | 1·0 | 2·25 | 4·0 | 6·25 | 9·0 |
| $2x^2$ | 0 | 0·5 | 2·0 | 4·5 | 8·0 | 12·5 | 18·0 |
| $y = 2x^2 + 5$ | 5 | 5·5 | 7·0 | 9·5 | 13·0 | 17·5 | 23·0 |

Then we can plot the graph accurately, using these results, and mark the point $P$ on the graph at which $x = 1{\cdot}5$.

*So do that*

5

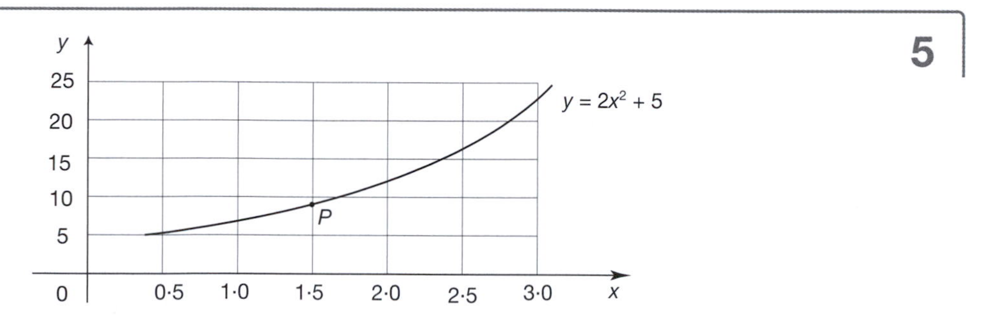


Now we bring a ruler up to the point $P$ and adjust the angle of the ruler by eye until it takes on the position of the tangent to the curve. Then we carefully draw the tangent.

6

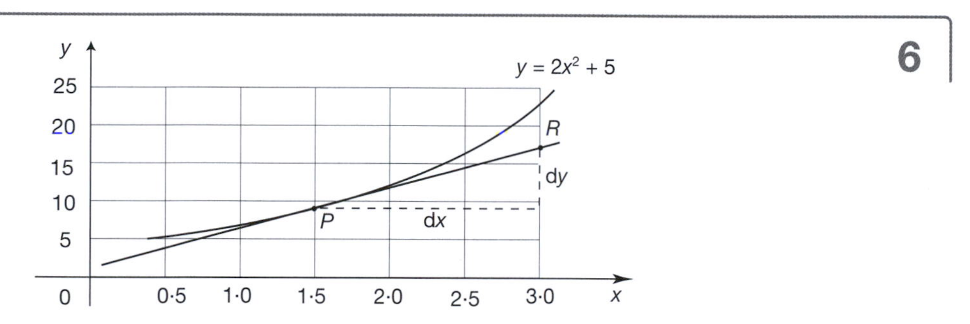


Now select a second point on the tangent at $R$, for example $R$ at $x = 3{\cdot}0$ where at $P$, $x = 1{\cdot}5$. Determine d$x$ and d$y$ between them and hence calculate the gradient of the tangent, i.e. the gradient of the curve at $P$.

This gives $m =$ ...........

7

$m = 5{\cdot}9$

Because, at $x = 1{\cdot}5$, $y = 9{\cdot}3$ (gauged from the graph)
$x = 3{\cdot}0$, $y = 18{\cdot}1$ (again, gauged from the graph)

Therefore d$x = 1{\cdot}5$ and d$y = 8{\cdot}8$.

Therefore for the tangent $m = \dfrac{dy}{dx} = \dfrac{8{\cdot}8}{1{\cdot}5} = 5{\cdot}9$ to 2 sig fig.

*Therefore the gradient of the curve at P is approximately 5·9.*

***Note*: Your results may differ slightly from those given here because the value obtained is the result of practical plotting and construction and may, therefore, contain minor differences. However, the description given is designed to clarify the method.**

8

## Algebraic determination of the gradient of a curve

The limitations of the practical construction in the previous method call for a more accurate method of finding the gradient, so let us start afresh from a different viewpoint and prove a general rule.

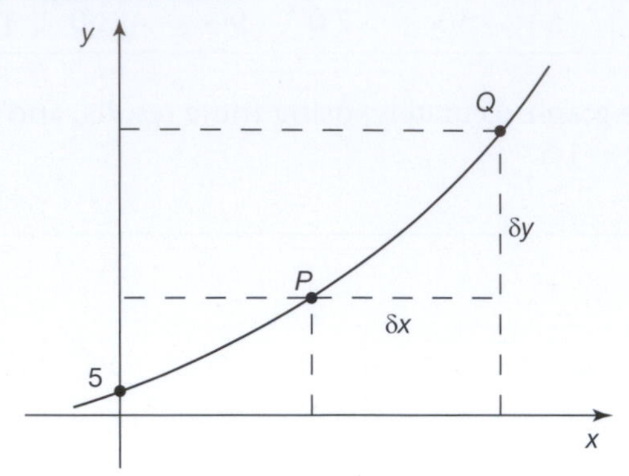


Let $P$ be a fixed point $(x, y)$ on the curve $y = 2x^2 + 5$, and $Q$ be a neighbouring point, with coordinates $(x + \delta x, y + \delta y)$.

Now draw two straight lines, one through the points $P$ and $Q$ and the other tangential to the curve at point $P$.

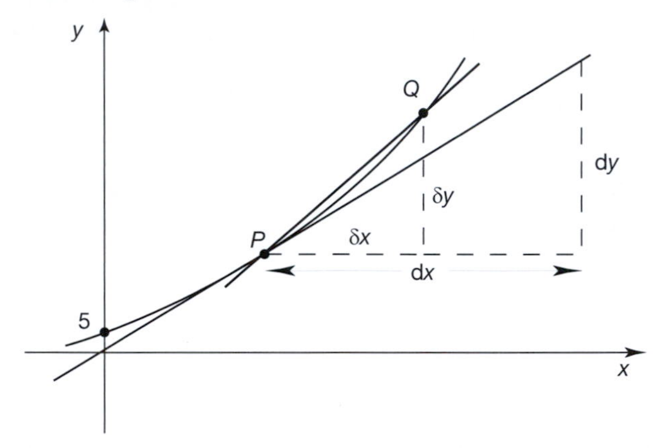


*Notice* that we use $\delta x$ and $\delta y$ to denote the respective differences in the $x$- and $y$-values of points $P$ and $Q$ on the curve. We reserve the notation $\mathrm{d}x$ and $\mathrm{d}y$ for the differences in the $x$- and $y$-values of two points on a *straight line* and particularly for a tangent. The quantities $\mathrm{d}x$ and $\mathrm{d}y$ are called *differentials*.

*Next frame*

9

At Q: $y + \delta y = 2(x + \delta x)^2 + 5$

$= 2(x^2 + 2x.\delta x + [\delta x]^2) + 5$ Expanding the bracket

$= 2x^2 + 4x.\delta x + 2[\delta x]^2 + 5$ Multiplying through by 2

Subtracting $y$ from both sides:

$$y + \delta y - y = 2x^2 + 4x.\delta x + 2[\delta x]^2 + 5 - (2x^2 + 5)$$
$$= 4x.\delta x + 2[\delta x]^2$$

Therefore: $\delta y = 4x.\delta x + 2[\delta x]^2$

$\delta y$ is the *vertical* distance between point $P$ and point $Q$ and this has now been given in terms of $x$ and $\delta x$ where $\delta x$ is the *horizontal* distance between point $P$ and point $Q$.

▶

Now, if we divide both sides by $\delta x$:

$$\frac{\delta y}{\delta x} = 4x + 2.\delta x$$

This expression, giving the change in vertical distance per unit change in horizontal distance, is the *gradient* of the straight line through $P$ and $Q$.

If the line through $P$ and $Q$ now rotates clockwise about $P$, the point $Q$ moves down the curve and approaches $P$. Also, both $\delta x$ and $\delta y$ approach 0 ($\delta x \to 0$ and $\delta y \to 0$). However, *their ratio*, which is the gradient of $PQ$, approaches the gradient of the tangent at $P$:

$$\frac{\delta y}{\delta x} \to \text{the gradient of the tangent at } P = \frac{dy}{dx}$$

Therefore $\dfrac{dy}{dx} = \ldots\ldots\ldots\ldots$

**10**

$$\boxed{\frac{dy}{dx} = 4x}$$

Because

$$\frac{\delta y}{\delta x} = 4x + 2.\delta x \text{ and as } \delta x \to 0 \text{ so } \frac{\delta y}{\delta x} \to 4x + 2 \times 0 = 4x$$

This is a general result giving the slope of the curve at any point on the curve $y = 2x^2 + 5$.

$\therefore$ At $x = 1{\cdot}5$ $\dfrac{dy}{dx} = 4(1{\cdot}5) = 6$

$\therefore$ The real slope of the curve $y = 2x^2 + 5$ at $x = 1{\cdot}5$ is 6.

The graphical solution previously obtained, i.e. 5·9, is an approximation.

The expression $\dfrac{dy}{dx}$ is called *the derivative of y with respect to x* because it is *derived* from the expression for $y$. The process of finding the derivative is called *differentiation* because it involves manipulating differences in coordinate values.

## Derivative of powers of *x*

**11**

### Two straight lines

(a) $y = c$ (constant)

The graph of $y = c$ is a straight line parallel to the $x$-axis. Therefore its gradient is zero.

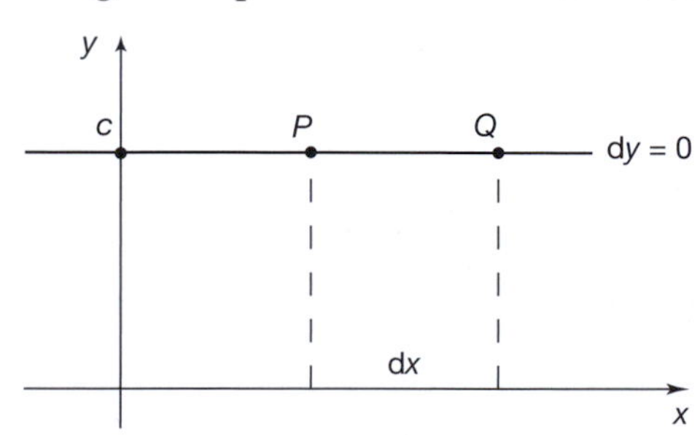


$$y = c,\ dy = 0 \text{ therefore } \frac{dy}{dx} = \frac{0}{dx} = 0$$

$$\text{If } y = c,\ \frac{dy}{dx} = 0$$

*Move to the next frame*

**12**

(b) $y = ax$

So for a point further up the line

$$y + \mathrm{d}y = a(x + \mathrm{d}x)$$
$$= ax + a.\mathrm{d}x$$

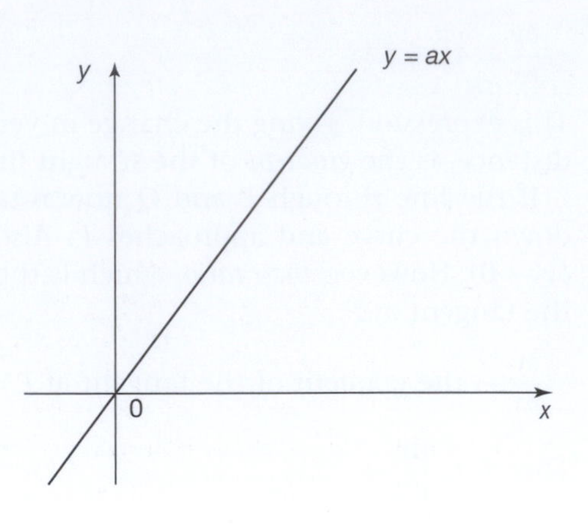


Subtract $y$:

$$\mathrm{d}y = ax + a.\mathrm{d}x - ax \text{ where } y = ax$$
$$= a.\mathrm{d}x$$

Therefore:

$$\frac{\mathrm{d}y}{\mathrm{d}x} = a$$

Meaning that the gradient of a straight line is constant along its length.

$$\text{If } y = ax, \ \frac{\mathrm{d}y}{\mathrm{d}x} = a$$

In particular, if $a = 1$ then $y = x$ and $\frac{\mathrm{d}y}{\mathrm{d}x} = 1$

*And now for curves*

**13**

**Two curves**

(a) $y = x^2$

At Q, $y + \delta y = (x + \delta x)^2$

$$y + \delta y = x^2 + 2x.\delta x + (\delta x)^2$$
$$y = x^2$$
$$\therefore \delta y = 2x.\delta x + (\delta x)^2$$
$$\therefore \frac{\delta y}{\delta x} = 2x + \delta x$$

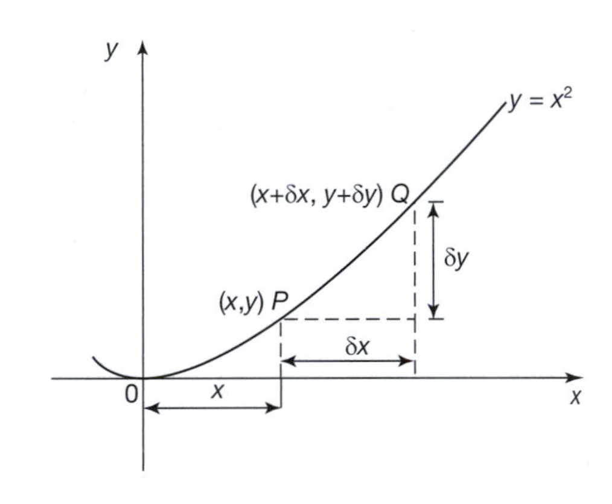


If $\delta x \to 0$, $\frac{\delta y}{\delta x} \to \frac{\mathrm{d}y}{\mathrm{d}x}$ and $\therefore \frac{\mathrm{d}y}{\mathrm{d}x} = 2x$

$\therefore$ If $y = x^2$, $\frac{\mathrm{d}y}{\mathrm{d}x} = 2x$

*Remember that* $\frac{\mathrm{d}y}{\mathrm{d}x}$ *is the gradient of the tangent to the curve.*

**14**

(b) $y = x^3$

At Q, $y + \delta y = (x + \delta x)^3$

$$= x^3 + 3x^2.\delta x + 3x(\delta x)^2 + (\delta x)^3$$

Now subtract $y = x^3$ from each side and finish it off as before.

$\therefore$ If $y = x^3$, $\frac{\mathrm{d}y}{\mathrm{d}x} = \ldots\ldots\ldots\ldots$

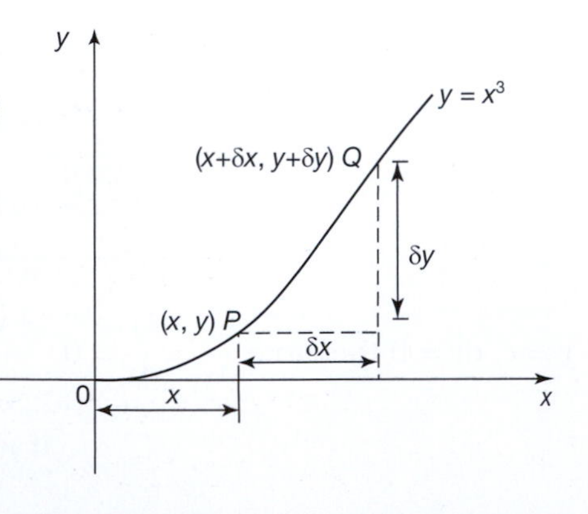

**15**

$3x^2$

Because

$$\delta y = 3x^2.\delta x + 3x(\delta x)^2 + (\delta x)^3$$

$$\therefore \frac{\delta y}{\delta x} = 3x^2 + 3x(\delta x) + (\delta x)^2$$

Then, if $\delta x \to 0$, $\frac{\delta y}{\delta x} \to \frac{dy}{dx}$ $\therefore \frac{dy}{dx} = 3x^2$

$\therefore$ If $y = x^3$, $\frac{dy}{dx} = 3x^2$

Continuing in the same manner, we should find that:

when $y = x^4$, $\frac{dy}{dx} = 4x^3$

and when $y = x^5$, $\frac{dy}{dx} = 5x^4$

If we now collect these results together and look for a pattern, we have the evidence given in the next frame.

*So move on*

**16**

| $y$ | $\frac{dy}{dx}$ |
|---|---|
| $c$ | $0$ |
| $x$ | $1$ |
| $x^2$ | $2x$ |
| $x^3$ | $3x^2$ |
| $x^4$ | $4x^3$ |
| $x^5$ | $5x^4$ |

We soon see a clear pattern emerging.

For a power of $x$:

In the derivative, the old index becomes a coefficient and the new index is one less than the index in the original power.

i.e. If $y = x^n$, $\frac{dy}{dx} = nx^{n-1}$

While this has only been demonstrated to be valid for $n = 0, 1, 2, 3$ it is in fact true for any value of $n$.

We have established that if $y = a$ ($a$ constant) then $\frac{dy}{dx} = 0$. It can also be established that if $y = ax^n$ then $\frac{dy}{dx} = anx^{n-1}$. For example, if $y = ax^4$ then $\frac{dy}{dx} = a \times 4x^3 = 4ax^3$.

We can prove these results by using the Binomial theorem:

$$(a+b)^n = a^n + na^{n-1}b + \frac{n(n-1)}{2!}a^{n-2}b^2 + \frac{n(n-1)(n-2)}{3!}a^{n-3}b^3 + \dots$$

If $y = x^n$, $y + \delta y = (x + \delta x)^n$

$$\therefore y + \delta y = x^n + nx^{n-1}(\delta x) + \frac{n(n-1)}{2!}x^{n-2}(\delta x)^2 + \frac{n(n-1)(n-2)}{3!}x^{n-3}(\delta x)^3 + \dots$$

$$y = x^n$$

$$\therefore \delta y = nx^{n-1}(\delta x) + \frac{n(n-1)}{2!}x^{n-2}(\delta x)^2 + \frac{n(n-1)(n-2)}{3!}x^{n-3}(\delta x)^3 + \dots$$

$$\frac{\delta y}{\delta x} = nx^{n-1} + \frac{n(n-1)}{2!}x^{n-2}(\delta x) + \frac{n(n-1)(n-2)}{3!}x^{n-3}(\delta x)^2 + \dots$$

▶

If $\delta x \to 0$, $\dfrac{\delta y}{\delta x} \to \dfrac{dy}{dx}$ and all terms on the RHS, except the first $\to 0$.

$\therefore$ If $\delta x \to 0$, $\dfrac{dy}{dx} = nx^{n-1} + 0 + 0 + 0 + \dots$

$\therefore$ If $y = x^n$, $\dfrac{dy}{dx} = nx^{n-1}$

which is, of course, the general form of the results we obtained in the examples above.

*Make a note of this important result*

**17**

## Differentiation of polynomials

To *differentiate a polynomial*, we differentiate each term in turn.

e.g. If $y = x^4 + 5x^3 - 4x^2 + 7x - 2$

$$\frac{dy}{dx} = 4x^3 + 5 \times 3x^2 - 4 \times 2x + 7 \times 1 - 0$$

$$\therefore \frac{dy}{dx} = 4x^3 + 15x^2 - 8x + 7$$

### Example

If $y = 2x^5 + 4x^4 - x^3 + 3x^2 - 5x + 7$, find an expression for $\dfrac{dy}{dx}$ and the value of $\dfrac{dy}{dx}$ at $x = 2$.

So, first of all, $\dfrac{dy}{dx} = \dots\dots\dots\dots$

**18**

$$\boxed{\frac{dy}{dx} = 10x^4 + 16x^3 - 3x^2 + 6x - 5}$$

Then, expressing the RHS in nested form and substituting $x = 2$, we have:

At $x = 2$, $\dfrac{dy}{dx} = \dots\dots\dots\dots$

**19**

$$\boxed{283}$$

Because

$$\frac{dy}{dx} = f(x) = \{[(10x + 16)x - 3]x + 6\}x - 5$$

$\therefore f(2) = 283$

Now, as an exercise, determine an expression for $\dfrac{dy}{dx}$ in each of the following cases and find the value of $\dfrac{dy}{dx}$ at the stated value of $x$:

1. $y = 3x^4 - 7x^3 + 4x^2 + 3x - 4$ $[x = 2]$
2. $y = x^5 + 2x^4 - 4x^3 - 5x^2 + 8x - 3$ $[x = -1]$
3. $y = 6x^3 - 7x^2 + 4x + 5$ $[x = 3]$

**20**

| | | |
|---|---|---|
| 1 | $\frac{dy}{dx} = 12x^3 - 21x^2 + 8x + 3.$ | At $x = 2$, $\frac{dy}{dx} = 31$ |
| 2 | $\frac{dy}{dx} = 5x^4 + 8x^3 - 12x^2 - 10x + 8.$ | At $x = -1$, $\frac{dy}{dx} = 3$ |
| 3 | $\frac{dy}{dx} = 18x^2 - 14x + 4.$ | At $x = 3$, $\frac{dy}{dx} = 124$ |

*Now on to the next topic*

**21**

## Second and higher derivatives – an alternative notation

If $y = 2x^2 - 5x + 3$, then $\frac{dy}{dx} = 4x - 5$. This double statement can be written as a single statement by putting $2x^2 - 5x + 3$ in place of $y$ in $\frac{dy}{dx}$.

i.e. $\frac{d}{dx}(2x^2 - 5x + 3) = 4x - 5$

In the same way, $\frac{d}{dx}(4x^3 - 7x^2 + 2x - 5) = \ldots\ldots\ldots\ldots$

**22**

$12x^2 - 14x + 2$

Either of the two methods is acceptable: it is just a case of which is the more convenient in any situation.

If $y = 2x^4 - 5x^3 + 3x^2 - 2x + 4$, then, by the previous method:

$$\frac{dy}{dx} = \frac{d}{dx}(2x^4 - 5x^3 + 3x^2 - 2x + 4) = 8x^3 - 15x^2 + 6x - 2$$

This expression for $\frac{dy}{dx}$ is itself a polynomial in powers of $x$ and can be differentiated in the same way as before, i.e. we can find the derivative of $\frac{dy}{dx}$.

$\frac{d}{dx}\left(\frac{dy}{dx}\right)$ is written $\frac{d^2y}{dx^2}$ and is the *second derivative of y with respect to x* (spoken as 'dee two $y$ by dee $x$ squared').

So, in this example, we have:

$$y = 2x^4 - 5x^3 + 3x^2 - 2x + 4$$

$$\frac{dy}{dx} = 8x^3 - 15x^2 + 6x - 2$$

$$\frac{d^2y}{dx^2} = 24x^2 - 30x + 6$$

We could, if necessary, find the third derivative of $y$ in the same way:

$\frac{d^3y}{dx^3} = \ldots\ldots\ldots\ldots$

**23**

$$\boxed{48x - 30}$$

Similarly, if $y = 3x^4 + 2x^3 - 4x^2 + 5x + 1$

$$\frac{dy}{dx} = \ldots\ldots\ldots\ldots$$

$$\frac{d^2y}{dx^2} = \ldots\ldots\ldots\ldots$$

**24**

$$\frac{dy}{dx} = 12x^3 + 6x^2 - 8x + 5$$

$$\frac{d^2y}{dx^2} = 36x^2 + 12x - 8$$

***At this point let us pause and summarize derivatives of powers of x and polynomials***

## Review summary

**25**

1 *Gradient of a straight line graph* $m = \dfrac{dy}{dx}$

2 *Gradient of a curve at a given point P at* $(x, y)$

$\dfrac{dy}{dx}$ = gradient of the tangent to the curve at $P$.

3 *Derivatives of powers of x*

(a) $y = c$ (constant), $\dfrac{dy}{dx} = 0$

(b) $y = x^n$, $\dfrac{dy}{dx} = nx^{n-1}$

(c) $y = ax^n$, $\dfrac{dy}{dx} = anx^{n-1}$

4 *Differentiation of polynomials* – differentiate each term in turn.

5 *Second and higher derivatives* – Just as $y$ = polynomial in $x$ can be differentiated to form the derivative $\dfrac{dy}{dx}$ so $\dfrac{dy}{dx}$ can also be differentiated to form $\dfrac{d}{dx}\left(\dfrac{dy}{dx}\right) = \dfrac{d^2y}{dx^2}$, the second derivative with respect to $x$ of $y$. Further differentiation yields the third derivative

$$\frac{d}{dx}\left(\frac{d^2y}{dx^2}\right) = \frac{d^3y}{dx^3}$$

and so on.

##  Review exercise

26

1 Calculate the gradient of the straight line joining:

(a) $P(4, 0)$ and $Q(7, 3)$

(b) $P(5, 6)$ and $Q(9, 2)$

(c) $P(-4, -7)$ and $Q(1, 3)$

(d) $P(0, 5)$ and $Q(5, -6)$

2 Determine, algebraically, from first principles, the gradient of the graph of $y = 5x^2 + 2$ at the point $P$ where $x = -1{\cdot}6$.

3 If $y = -2x^4 - 3x^3 + 4x^2 - x + 5$, obtain an expression for $\frac{dy}{dx}$ and hence calculate the value of $\frac{dy}{dx}$ at $x = -3$.

4 Given $y = 7x^3 - 4x^2 + 5x - 9$ find $\frac{d^2y}{dx^2}$ and $\frac{d^3y}{dx^3}$.

27

1 (a) $P(4, 0)$ and $Q(7, 3)$:

$$dy = 3 - 0 = 3,\ dx = 7 - 4 = 3 \text{ so } \frac{dy}{dx} = \frac{3}{3} = 1$$

(b) $P(5, 6)$ and $Q(9, 2)$:

$$dy = 2 - 6 = -4,\ dx = 9 - 5 = 4 \text{ so } \frac{dy}{dx} = \frac{-4}{4} = -1$$

(c) $P(-4, -7)$ and $Q(1, 3)$:

$$dy = 3 - (-7) = 10,\ dx = 1 - (-4) = 5 \text{ so } \frac{dy}{dx} = \frac{10}{5} = 2$$

(d) $P(0, 5)$ and $Q(5, -6)$:

$$dy = -6 - 5 = -11,\ dx = 5 - 0 = 5 \text{ so } \frac{dy}{dx} = \frac{-11}{5} = -2{\cdot}2$$

2 $y = 5x^2 + 2$ so that:

$$\begin{aligned} y + \delta y &= 5(x + \delta x)^2 + 2 \\ &= 5(x^2 + 2x\delta x + [\delta x]^2) + 2 \\ &= 5x^2 + 10x\delta x + 5[\delta x]^2 + 2 \text{ so that:} \\ y + \delta y - y &= 5x^2 + 10x\delta x + 5[\delta x]^2 + 2 - (5x^2 + 2) \\ &= 10x\delta x + 5[\delta x]^2 \\ &= \delta y \end{aligned}$$

Hence:

$$\frac{\delta y}{\delta x} = \frac{10x\delta x + 5[\delta x]^2}{\delta x} = 10x + 5\delta x \text{ therefore } \frac{dy}{dx} = 10x. \text{ When } x = -1{\cdot}6:$$

$$\frac{dy}{dx} = -16$$

▶

3 If $y = -2x^4 - 3x^3 + 4x^2 - x + 5$ then, differentiating term by term:

$$\frac{dy}{dx} = -2 \times 4x^3 - 3 \times 3x^2 + 4 \times 2x - 1$$
$$= -8x^3 - 9x^2 + 8x - 1 = ((-8x - 9)x + 8)x - 1. \text{ When } x = -3:$$
$$\frac{dy}{dx} = ((-8(-3) - 9)(-3) + 8)(-3) - 1$$
$$= 110$$

4 Given that $y = 7x^3 - 4x^2 + 5x - 9$ then

$$\frac{dy}{dx} = \frac{d}{dx}(7x^3 - 4x^2 + 5x - 9)$$
$$= 21x^2 - 8x + 5$$
$$\frac{d^2y}{dx^2} = \frac{d}{dx}(21x^2 - 8x + 5)$$
$$= 42x - 8$$
$$\frac{d^3y}{dx^3} = \frac{d}{dx}(42x - 8)$$
$$= 42$$

## Standard derivatives and rules

**28**

Let us first establish a limiting value that we shall need in the future when we come to differentiate the trigonometric functions.

### Limiting value of $\frac{\sin\theta}{\theta}$ as $\theta \to 0$

$P$ is a point on the circumference of a circle, centre $O$ and radius $r$. $PT$ is a tangent to the circle at $P$.

Use your trigonometry to show that:

$h = r\sin\theta$ and $H = r\tan\theta$

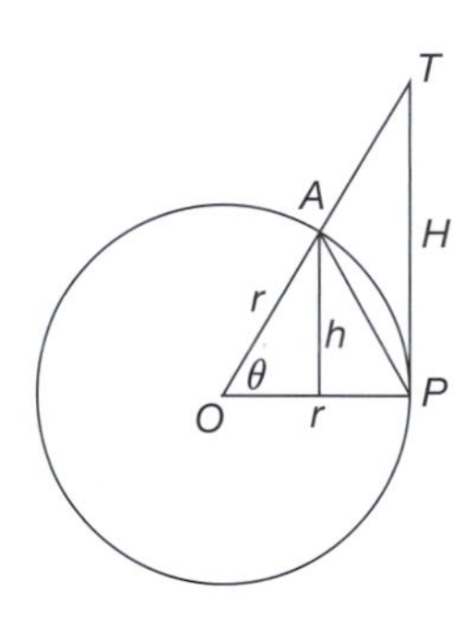


*Recollect*:

Area of a triangle $= \frac{1}{2} \times (\text{base}) \times (\text{height})$

Area of a circle $= \pi r^2 = \frac{1}{2} r^2 .2\pi$, so

Area of a sector $= \frac{1}{2} r^2 .\theta$ ($\theta$ in radians)

$\triangle POA$: area $= \frac{1}{2} r.h = \frac{1}{2} r.r\sin\theta = \frac{1}{2} r^2 \sin\theta$

Sector $POA$: area $= \frac{1}{2} r^2 \theta$

$\triangle POT$: area $= \frac{1}{2} r.H = \frac{1}{2} r.r\tan\theta = \frac{1}{2} r^2 \tan\theta$

▶

In terms of area: $\triangle POA <$ sector $POA < \triangle POT$

$$\therefore \frac{1}{2}r^2 \sin\theta < \frac{1}{2}r^2\theta < \frac{1}{2}r^2 \tan\theta$$

so dividing through by $\frac{1}{2}r^2$ gives

$$\sin\theta < \theta < \tan\theta \text{ that is } \sin\theta < \theta < \frac{\sin\theta}{\cos\theta}$$

and taking the reciprocals reverses the inequalities so

$$\frac{1}{\sin\theta} > \frac{1}{\theta} > \frac{\cos\theta}{\sin\theta}$$

Multiplying throughout by $\sin\theta$: $1 > \frac{\sin\theta}{\theta} > \cos\theta$

When $\theta \to 0$, $\cos\theta \to 1$. $\therefore$ The limiting value of $\frac{\sin\theta}{\theta}$ is bounded on both sides by the value 1.

$$\therefore \text{Limiting value of } \frac{\sin\theta}{\theta} \text{ as } \theta \to 0 = 1$$

*Make a note of this result. We shall certainly meet it again in due course*

## Standard derivatives

29

So far, we have found derivatives of polynomials using the standard derivative $\frac{d}{dx}(x^n) = nx^{n-1}$. Derivatives of trigonometric expressions can be established by using a number of trigonometrical formulas. We shall deal with some of these in the next few frames.

### 1 Derivative of $y = \sin x$

If $y = \sin x$, $y + \delta y = \sin(x + \delta x)$ $\therefore \delta y = \sin(x + \delta x) - \sin x$

We now apply the trigonometrical formula:

$$\sin A - \sin B = 2\cos\frac{A+B}{2}\sin\frac{A-B}{2} \quad \text{where } A = x + \delta x \text{ and } B = x$$

$$\therefore \delta y = 2\cos\left(\frac{2x+\delta x}{2}\right).\sin\left(\frac{\delta x}{2}\right) = 2\cos\left(x + \frac{\delta x}{2}\right).\sin\left(\frac{\delta x}{2}\right)$$

$$\therefore \frac{\delta y}{\delta x} = \frac{2\cos\left(x + \frac{\delta x}{2}\right).\sin\left(\frac{\delta x}{2}\right)}{\delta x}$$

$$= \frac{\cos\left(x + \frac{\delta x}{2}\right).\sin\left(\frac{\delta x}{2}\right)}{\frac{\delta x}{2}}$$

$$= \cos\left(x + \frac{\delta x}{2}\right).\frac{\sin\left(\frac{\delta x}{2}\right)}{\frac{\delta x}{2}}$$

When $\delta x \to 0$, $\frac{\delta y}{\delta x} \to \frac{dy}{dx}$ and $\frac{dy}{dx} \to \cos x.1$ using the result of Frame 28

$$\therefore \text{If } y = \sin x, \ \frac{dy}{dx} = \cos x$$

▶

**2 Derivative of $y = \cos x$**

This is obtained in much the same way as for the previous case.

If $y = \cos x$, $y + \delta y = \cos(x + \delta x)$ $\therefore$ $\delta y = \cos(x + \delta x) - \cos x$

Now we use the formula $\cos A - \cos B = -2\sin\frac{A+B}{2}\sin\frac{A-B}{2}$

$$\therefore\ \delta y = -2\sin\left(\frac{2x+\delta x}{2}\right)\cdot\sin\left(\frac{\delta x}{2}\right)$$
$$= -2\sin\left(x+\frac{\delta x}{2}\right).\sin\left(\frac{\delta x}{2}\right)$$
$$\therefore\ \frac{\delta y}{\delta x} = \frac{-2\sin\left(x+\frac{\delta x}{2}\right).\sin\left(\frac{\delta x}{2}\right)}{\delta x} = -\sin\left(x+\frac{\delta x}{2}\right).\frac{\sin\left(\frac{\delta x}{2}\right)}{\frac{\delta x}{2}}$$

As $\delta x \to 0$, $\frac{\delta y}{\delta x} \to (-\sin x).(1)$ $\therefore$ If $y = \cos x$, $\frac{dy}{dx} = -\sin x$ using the result of Frame 28

**30**

At this stage, there is one more derivative that we should determine.

**3 Derivative of $y = e^x$**

We have already discussed the series representation of $e^x$ in Frame 48 of Programme F.7, so:

$$y = e^x = 1 + x + \frac{x^2}{2!} + \frac{x^3}{3!} + \frac{x^4}{4!} + \ldots$$

If we differentiate each power of $x$ on the RHS, this gives

$$\frac{dy}{dx} = 0 + 1 + \frac{2x}{2!} + \frac{3x^2}{3!} + \frac{4x^3}{4!} + \ldots$$
$$= 1 + x + \frac{x^2}{2!} + \frac{x^3}{3!} + \ldots$$
$$= e^x$$

$\therefore$ If $y = e^x$, $\frac{dy}{dx} = e^x$

Note that $e^x$ is a special function in which the derivative is equal to the function itself.

So, we have obtained some important standard results:

(a) If $y = x^n$, $\frac{dy}{dx} = n.x^{n-1}$

(b) If $y = c$ (a constant), $\frac{dy}{dx} = 0$

(c) A constant factor is unchanged, for example:

if $y = a.x^n$, $\frac{dy}{dx} = a.n.x^{n-1}$

(d) If $y = \sin x$, $\frac{dy}{dx} = \cos x$

(e) If $y = \cos x$, $\frac{dy}{dx} = -\sin x$

(f) if $y = e^x$, $\frac{dy}{dx} = e^x$

▶

Now cover up the list of results above and complete the following table:

| $y$ | $\frac{dy}{dx}$ |
|---|---|
| 14 | |
| $\cos x$ | |
| $x^n$ | |

| $y$ | $\frac{dy}{dx}$ |
|---|---|
| $e^x$ | |
| $5x^3$ | |
| $\sin x$ | |

*You can check these in the next frame*

31

| $y$ | $\frac{dy}{dx}$ |
|---|---|
| 14 | 0 |
| $\cos x$ | $-\sin x$ |
| $x^n$ | $nx^{n-1}$ |

| $y$ | $\frac{dy}{dx}$ |
|---|---|
| $e^x$ | $e^x$ |
| $5x^3$ | $15x^2$ |
| $\sin x$ | $\cos x$ |

*On to the next frame*

32

## Derivative of a product of functions

Let $y = uv$, where $u$ and $v$ are functions of $x$.

If $x$ is changed to $x + \delta x$ then $u$ will be changed to $u + \delta u$ and $v$ will be changed to $v + \delta v$. As a result, $y$ will be changed to $y + \delta y$ where:

$$\begin{aligned} y + \delta y &= (u + \delta u)(v + \delta v) \\ &= uv + u\delta v + v\delta u + \delta u\delta v \\ &= y + u\delta v + v\delta u + \delta u\delta v \end{aligned}$$

Therefore $\delta y = u\delta v + v\delta u + \delta u\delta v$. Dividing through by $\delta x$ we find:

$$\frac{\delta y}{\delta x} = u\frac{\delta v}{\delta x} + v\frac{\delta u}{\delta x} + \delta u\frac{\delta v}{\delta x}$$

As $\delta x \to 0$, so $\frac{\delta y}{\delta x} \to \frac{dy}{dx}$, $\frac{\delta u}{\delta x} \to \frac{du}{dx}$, $\frac{\delta v}{\delta x} \to \frac{dv}{dx}$ and $\delta u \to 0$

therefore $\frac{dy}{dx} = u\frac{dv}{dx} + v\frac{du}{dx} + 0\frac{dv}{dx}$.

$$\text{So, if } y = uv, \quad \frac{dy}{dx} = u\frac{dv}{dx} + v\frac{du}{dx}$$

That is:

*To differentiate a product of two functions:*

*Put down the first (differentiate the second) + put down the second (differentiate the first)*

**Example 1**

$$\begin{aligned} y &= x^3 . \sin x \\ \frac{dy}{dx} &= x^3(\cos x) + \sin x(3x^2) \\ &= x^3 . \cos x + 3x^2 . \sin x = x^2(x\cos x + 3\sin x) \end{aligned}$$

▶

**Example 2**

$$y = x^4 . \cos x$$

$$\frac{dy}{dx} = x^4(-\sin x) + \cos x.(4x^3)$$

$$= -x^4 \sin x + 4x^3 \cos x = x^3(4\cos x - x \sin x)$$

**Example 3**

$$y = x^5 . e^x$$

$$\frac{dy}{dx} = x^5 e^x + e^x 5x^4 = x^4 e^x (x + 5)$$

In the same way, as an exercise, now you can differentiate the following:

1 $y = e^x . \sin x$

2 $y = 4x^3 . \sin x$

3 $y = e^x . \cos x$

4 $y = \cos x . \sin x$

5 $y = 3x^3 . e^x$

6 $y = 2x^5 . \cos x$

*Finish all six and then check with the next frame*

**33**

1 $\frac{dy}{dx} = e^x . \cos x + e^x . \sin x = e^x(\cos x + \sin x)$

2 $\frac{dy}{dx} = 4x^3 . \cos x + 12x^2 . \sin x = 4x^2(x \cos x + 3 \sin x)$

3 $\frac{dy}{dx} = e^x(-\sin x) + e^x \cos x = e^x(\cos x - \sin x)$

4 $\frac{dy}{dx} = \cos x . \cos x + \sin x(-\sin x) = \cos^2 x - \sin^2 x$

5 $\frac{dy}{dx} = 3x^3 . e^x + 9x^2 . e^x = 3x^2 e^x (x + 3)$

6 $\frac{dy}{dx} = 2x^5(-\sin x) + 10x^4 . \cos x = 2x^4(5 \cos x - x \sin x)$

*Now we will see how to deal with the quotient of two functions*

**34**

## Derivative of a quotient of functions

Let $y = \frac{u}{v}$, where $u$ and $v$ are functions of $x$.

$$\text{Then } y + \delta y = \frac{u + \delta u}{v + \delta v}$$

$$\therefore \delta y = \frac{u + \delta u}{v + \delta v} - \frac{u}{v}$$

$$= \frac{(u + \delta u)v - u(v + \delta v)}{(v + \delta v)v}$$

$$= \frac{uv + v.\delta u - uv - u.\delta v}{v(v + \delta v)}$$

$$= \frac{v.\delta u - u.\delta v}{v^2 + v.\delta v}$$

$$\therefore \frac{\delta y}{\delta x} = \frac{v\frac{\delta u}{\delta x} - u\frac{\delta v}{\delta x}}{v^2 + v.\delta v}$$

▶

If $\delta x \to 0$, $\delta u \to 0$ and $\delta v \to 0$

$$\therefore \frac{dy}{dx} = \frac{v\frac{du}{dx} - u\frac{dv}{dx}}{v^2}$$

$$\therefore \text{If } y = \frac{u}{v}, \quad \frac{dy}{dx} = \frac{v\frac{du}{dx} - u\frac{dv}{dx}}{v^2}$$

$\therefore$ *To differentiate a quotient of two functions:*

*[Put down the bottom (differentiate the top) – put down the top (differentiate the bottom)] all over the bottom squared.*

$\therefore$ If $y = \frac{\sin x}{x^2}$, $\frac{dy}{dx} = \ldots\ldots\ldots\ldots$

---

**35**

$$\boxed{\frac{x\cos x - 2\sin x}{x^3}}$$

Because

$$y = \frac{\sin x}{x^2} \quad \therefore \frac{dy}{dx} = \frac{x^2\cos x - \sin x.(2x)}{(x^2)^2}$$

$$= \frac{x^2\cos x - 2x\sin x}{x^4} = \frac{x\cos x - 2\sin x}{x^3}$$

Another example: $y = \frac{5e^x}{\cos x}$

$$y = \frac{u}{v}, \quad \frac{dy}{dx} = \frac{v\frac{du}{dx} - u\frac{dv}{dx}}{v^2}$$

$$= \frac{\cos x.(5e^x) - 5e^x(-\sin x)}{\cos^2 x}$$

$$\therefore \frac{dy}{dx} = \frac{5e^x(\cos x + \sin x)}{\cos^2 x}$$

Now let us deal with this one:

$$y = \frac{\sin x}{\cos x}$$

$$\frac{dy}{dx} = \frac{\cos x.\cos x - \sin x(-\sin x)}{\cos^2 x}$$

$$= \frac{\cos^2 x + \sin^2 x}{\cos^2 x} \qquad \text{But } \sin^2 x + \cos^2 x = 1$$

$$= \frac{1}{\cos^2 x} \qquad \text{Also } \frac{1}{\cos x} = \sec x$$

$$= \sec^2 x \qquad \text{and } \frac{\sin x}{\cos x} = \tan x$$

$\therefore$ If $y = \tan x$, $\frac{dy}{dx} = \sec^2 x$

This is another one for our list of standard derivatives, so make a note of it.

We now have the following standard derivatives:

| $y$ | $\frac{dy}{dx}$ |
|---|---|
| $x^n$ | |
| $c$ | |
| $a.x^n$ | |
| $\sin x$ | |
| $\cos x$ | |
| $\tan x$ | |
| $e^x$ | |

*Fill in the results in the right-hand column and then check with the next frame*

**36**

| $y$ | $\frac{dy}{dx}$ |
|---|---|
| $x^n$ | $n.x^{n-1}$ |
| $c$ | $0$ |
| $a.x^n$ | $an.x^{n-1}$ |
| $\sin x$ | $\cos x$ |
| $\cos x$ | $-\sin x$ |
| $\tan x$ | $\sec^2 x$ |
| $e^x$ | $e^x$ |

Also **1** If $y = uv$, $\frac{dy}{dx} = \ldots\ldots\ldots\ldots$

and **2** If $y = \frac{u}{v}$, $\frac{dy}{dx} = \ldots\ldots\ldots\ldots$

**37**

$$1 \quad y = uv, \quad \frac{dy}{dx} = u\frac{dv}{dx} + v\frac{du}{dx}$$

$$2 \quad y = \frac{u}{v}, \quad \frac{dy}{dx} = \frac{v\frac{du}{dx} - u\frac{dv}{dx}}{v^2}$$

Now here is an exercise covering the work we have been doing. In each of the following functions, determine an expression for $\frac{dy}{dx}$:

**1** $y = x^2 \cos x$ **2** $y = e^x \sin x$ **3** $y = \frac{4e^x}{\sin x}$ **4** $y = \frac{\cos x}{x^4}$

**5** $y = 5x^3.e^x$ **6** $y = 4x^2.\tan x$ **7** $y = \frac{\cos x}{\sin x}$ **8** $y = \frac{\tan x}{e^x}$

**9** $y = x^5.\sin x$ **10** $y = \frac{3x^2}{\cos x}$

*All the solutions are shown in the next frame*

38

1
$$y = x^2 . \cos x$$
$$\frac{dy}{dx} = x^2(-\sin x) + \cos x.(2x)$$
$$= x(2\cos x - x\sin x)$$

2
$$y = e^x . \sin x$$
$$\frac{dy}{dx} = e^x . \cos x + \sin x(e^x)$$
$$= e^x(\sin x + \cos x)$$

3
$$y = \frac{4e^x}{\sin x}$$
$$\frac{dy}{dx} = \frac{\sin x.(4e^x) - 4e^x . \cos x}{\sin^2 x}$$
$$= \frac{4e^x(\sin x - \cos x)}{\sin^2 x}$$

4
$$y = \frac{\cos x}{x^4}$$
$$\frac{dy}{dx} = \frac{x^4(-\sin x) - \cos x(4x^3)}{x^8}$$
$$= \frac{-x\sin x - 4\cos x}{x^5}$$

5
$$y = 5x^3 e^x$$
$$\frac{dy}{dx} = 5x^3 .e^x + e^x .15x^2$$
$$= 5x^2 e^x(x+3)$$

6
$$y = 4x^2 \tan x$$
$$\frac{dy}{dx} = 4x^2 . \sec^2 x + \tan x.(8x)$$
$$= 4x(x. \sec^2 x + 2\tan x)$$

7
$$y = \frac{\cos x}{\sin x}$$
$$\frac{dy}{dx} = \frac{\sin x(-\sin x) - \cos x. \cos x}{\sin^2 x}$$
$$= \frac{-(\sin^2 x + \cos^2 x)}{\sin^2 x}$$
$$= -\frac{1}{\sin^2 x} = -\text{cosec}^2 x$$

8
$$y = \frac{\tan x}{e^x}$$
$$\frac{dy}{dx} = \frac{e^x . \sec^2 x - \tan x.e^x}{e^{2x}}$$
$$= \frac{\sec^2 x - \tan x}{e^x}$$

9
$$y = x^5 . \sin x$$
$$\frac{dy}{dx} = x^5 . \cos x + \sin x.(5x^4)$$
$$= x^4(x\cos x + 5\sin x)$$

10
$$y = \frac{3x^2}{\cos x}$$
$$\frac{dy}{dx} = \frac{\cos x.6x - 3x^2(-\sin x)}{\cos^2 x}$$
$$= \frac{3x(2\cos x + x\sin x)}{\cos^2 x}$$

*Check the results and then move on to the next topic*

39

## Derivative of a function of a function

If $y = \sin x$, $y$ is a function of the angle $x$, since the value of $y$ depends on the value given to $x$.

If $y = \sin(2x - 3)$, $y$ is a function of the angle $(2x - 3)$ which is itself a function of $x$.

Therefore, $y$ is a function of (a function of $x$) and is said to be a *function of a function* of $x$.

To differentiate a function of a function, we must first introduce the *chain rule*.

▶

With the example, $y = \sin(2x - 3)$, we put $u = (2x - 3)$

i.e. $y = \sin u$ where $u = 2x - 3$.

If $x$ has an increase $\delta x$, $u$ will have an increase $\delta u$ and then $y$ will have an increase $\delta y$, i.e. $x \to x + \delta x$, $u \to u + \delta u$ and $y \to y + \delta y$.

At this stage, the increases $\delta x$, $\delta u$ and $\delta y$ are all finite values and therefore we can say that

$$\frac{\delta y}{\delta x} = \frac{\delta y}{\delta u} \times \frac{\delta u}{\delta x}$$

because the $\delta u$ in $\frac{\delta y}{\delta u}$ cancels the $\delta u$ in $\frac{\delta u}{\delta x}$.

If now $\delta x \to 0$, then $\delta u \to 0$ and $\delta y \to 0$

Also $\frac{\delta y}{\delta x} \to \frac{dy}{dx}$, $\frac{\delta y}{\delta u} \to \frac{dy}{du}$ and $\frac{\delta u}{\delta x} \to \frac{du}{dx}$, and the previous statement now becomes

$$\frac{dy}{dx} = \frac{dy}{du} \cdot \frac{du}{dx}.$$

This is the *chain rule* and is particularly useful when determining the derivatives of functions of a function.

**Example 1**

To differentiate $y = \sin(2x - 3)$

Put $u = (2x - 3)$ $\quad \therefore y = \sin u$

$$\therefore \frac{du}{dx} = 2 \text{ and } \frac{dy}{du} = \cos u$$

$$\frac{dy}{dx} = \frac{dy}{du} \cdot \frac{du}{dx} = \cos u.(2) = 2\cos u = 2\cos(2x - 3)$$

$$\therefore \text{If } y = \sin(2x - 3),\ \frac{dy}{dx} = 2\cos(2x - 3)$$

*Further examples follow*

**40**

**Example 2**

If $y = (3x + 5)^4$, determine $\frac{dy}{dx}$

$y = (3x + 5)^4$. Put $u = (3x + 5)$. $\therefore y = u^4$

$$\therefore \frac{dy}{du} = 4u^3 \text{ and } \frac{du}{dx} = 3. \quad \frac{dy}{dx} = \frac{dy}{du} \cdot \frac{du}{dx}$$

$$\therefore \frac{dy}{dx} = 4u^3.(3) = 12u^3 = 12(3x + 5)^3$$

$$\therefore \text{If } y = (3x + 5)^4,\ \frac{dy}{dx} = 12(3x + 5)^3$$

And in just the same way:

**Example 3**

If $y = \tan(4x + 1)$, $\frac{dy}{dx} = \ldots\ldots\ldots\ldots$

41

$$\boxed{4\sec^2(4x+1)}$$

Because

$y = \tan(4x+1) \quad \therefore \text{ Put } u = 4x+1 \quad \therefore \dfrac{du}{dx} = 4$

$y = \tan u \quad \therefore \dfrac{dy}{du} = \sec^2 u$

$\dfrac{dy}{dx} = \dfrac{dy}{du} \cdot \dfrac{du}{dx} = \sec^2 u.(4) = 4\sec^2(4x+1)$

$\therefore$ If $y = \tan(4x+1)$, $\dfrac{dy}{dx} = 4\sec^2(4x+1)$

And now this one:

**Example 4**

If $y = e^{5x}$, $\dfrac{dy}{dx} = \dots\dots\dots\dots$

42

$$\boxed{5e^{5x}}$$

Because

$y = e^{5x} \quad \therefore \text{ Put } u = 5x \quad \therefore \dfrac{du}{dx} = 5$

$y = e^{u} \quad \therefore \dfrac{dy}{du} = e^{u}$

$\dfrac{dy}{dx} = \dfrac{dy}{du} \cdot \dfrac{du}{dx} = e^{u}.(5) = 5e^{u} = 5e^{5x}$

$\therefore$ If $y = e^{5x}$, $\dfrac{dy}{dx} = 5e^{5x}$

Many of these functions can be differentiated at sight by slight modification to our list of standard derivatives:

| $F$ is a function of $x$ | | | |
|---|---|---|---|
| $y$ | $\dfrac{dy}{dx}$ | $y$ | $\dfrac{dy}{dx}$ |
| $F^n$ | $nF^{n-1}.\dfrac{dF}{dx}$ | $\cos F$ | $-\sin F.\dfrac{dF}{dx}$ |
| $a.F^n$ | $a.nF^{n-1}.\dfrac{dF}{dx}$ | $\tan F$ | $\sec^2 F.\dfrac{dF}{dx}$ |
| $\sin F$ | $\cos F.\dfrac{dF}{dx}$ | $e^F$ | $e^F.\dfrac{dF}{dx}$ |

*Let us now apply these results*

**43**

Here are four examples:

1 $y = \cos(2x-1)$, $\dfrac{dy}{dx} = -\sin(2x-1)\times 2 = -2\sin(2x-1)$

2 $y = e^{(3x+4)}$, $\dfrac{dy}{dx} = e^{(3x+4)} \times 3 = 3.e^{(3x+4)}$

3 $y = (5x-2)^3$, $\dfrac{dy}{dx} = 3(5x-2)^2 \times 5 = 15(5x-2)^2$

4 $y = 4.e^{\sin x}$, $\dfrac{dy}{dx} = 4.e^{\sin x} \times \cos x = 4\cos x.e^{\sin x}$

In just the same way, as an exercise, differentiate the following:

| | |
|---|---|
| 1 $y = \sin(4x+3)$ | 4 $y = \tan 5x$ |
| 2 $y = (2x-5)^4$ | 5 $y = e^{2x-3}$ |
| 3 $y = \sin^3 x$ | 6 $y = 4\cos(7x+2)$ |

**44**

| | |
|---|---|
| 1 $\dfrac{dy}{dx} = 4.\cos(4x+3)$ | 4 $\dfrac{dy}{dx} = 5.\sec^2 5x$ |
| 2 $\dfrac{dy}{dx} = 8.(2x-5)^3$ | 5 $\dfrac{dy}{dx} = 2.e^{2x-3}$ |
| 3 $\dfrac{dy}{dx} = 3.\sin^2 x\cos x$ | 6 $\dfrac{dy}{dx} = -28.\sin(7x+2)$ |

Now let us consider the derivative of $y = \ln x$:

If $y = \ln x$ then $x = e^y$

Differentiating with respect to $x$: $\dfrac{d}{dx}(x) = \dfrac{d}{dx}(e^y)$ we find that the derivative on the LHS is equal to 1 and, by the chain rule, the derivative on the RHS is $\dfrac{d}{dy}(e^y)\dfrac{dy}{dx} = e^y\dfrac{dy}{dx}$ so that:

$1 = e^y\dfrac{dy}{dx}$ and, since $x = e^y$ this can be written as:

$1 = x\dfrac{dy}{dx}$ therefore $\dfrac{dy}{dx} = \dfrac{1}{x}$. Therefore if $y = \ln x$, $\dfrac{dy}{dx} = \dfrac{1}{x}$

We can add this to our list of standard derivatives. Also, if $F$ is a function of $x$ then:

if $y = \ln F$ we have $\dfrac{dy}{dx} = \dfrac{1}{F}\cdot\dfrac{dF}{dx}$.

Here is an example:

If $y = \ln(3x-5)$, $\dfrac{dy}{dx} = \dfrac{1}{3x-5}.3 = \dfrac{3}{3x-5}$

and if $y = \ln(\sin x)$, $\dfrac{dy}{dx} = \dfrac{1}{\sin x}.\cos x = \cot x$

***There is one further standard derivative to be established at this stage, so move on to the next frame***

45

## Derivative of $y = a^x$

We already know that if $y = e^x$, $\dfrac{dy}{dx} = e^x$

and that if $y = e^F$, $\dfrac{dy}{dx} = e^F.\dfrac{dF}{dx}$

Then, if $y = a^x$, we can write $a = e^k$ and then $y = a^x = (e^k)^x = e^{kx}$

$$\therefore \frac{dy}{dx} = e^{kx}.\frac{d}{dx}(kx) = e^{kx}(k) = k.e^{kx}$$

But, $e^{kx} = a^x$ and $k = \ln a$ $\quad\therefore$ If $y = a^x$, $\dfrac{dy}{dx} = a^x \ln a$

We can add this result to our list for future reference.

***For the moment we shall just pause and review the work we have done so far on standard derivatives and the rules of differentiation***

# Review summary

46

1 *Standard derivatives*

| $y$ | $\dfrac{dy}{dx}$ | $y$ | $\dfrac{dy}{dx}$ |
|---|---|---|---|
| $x^n$ | $n.x^{n-1}$ | $\tan x$ | $\sec^2 x$ |
| $c$ | $0$ | $e^x$ | $e^x$ |
| $\sin x$ | $\cos x$ | $\ln x$ | $\dfrac{1}{x}$ |
| $\cos x$ | $-\sin x$ | $a^x$ | $a^x \ln a$ |

2 *Rules of derivatives*

(a) of product $y = uv$, $\dfrac{dy}{dx} = u\dfrac{dv}{dx} + v\dfrac{du}{dx}$

(b) of quotient $y = \dfrac{u}{v}$, $\dfrac{dy}{dx} = \dfrac{v\dfrac{du}{dx} - u\dfrac{dv}{dx}}{v^2}$

3 *Chain rule*

| *F* is a function of *x* | | | |
|---|---|---|---|
| $y$ | $\dfrac{dy}{dx}$ | $y$ | $\dfrac{dy}{dx}$ |
| $F^n$ | $nF^{n-1}.\dfrac{dF}{dx}$ | $\tan F$ | $\sec^2 F.\dfrac{dF}{dx}$ |
| $\sin F$ | $\cos F.\dfrac{dF}{dx}$ | $e^F$ | $e^F.\dfrac{dF}{dx}$ |
| $\cos F$ | $-\sin F.\dfrac{dF}{dx}$ | $\ln F$ | $\dfrac{1}{F}.\dfrac{dF}{dx}$ |

## Review exercise

**47**

1 Determine $\frac{dy}{dx}$ in each of the following cases:

(a) $y = x^3 \tan x$ (b) $y = x^2 e^x$

(c) $y = \frac{2e^x}{x^2}$ (d) $y = \frac{x^3}{\sin x}$

2 Differentiate the following with respect to $x$:

(a) $y = (4x + 3)^6$ (b) $y = \tan(2x + 3)$

(c) $y = \ln(3x - 4)$ (d) $y = -2e^{(1-3x)}$

(e) $y = e^{-3x}\sin(2x)$ (f) $y = \tan^2 x$

3 (a) If $y = x^3 + 2x^2 - 3x - 4$, determine:

(i) $\frac{dy}{dx}$ and $\frac{d^2y}{dx^2}$

(ii) the values of $x$ at which $\frac{dy}{dx} = 0$.

(b) If $y = 2\cos(x + 1) - 3\sin(x - 1)$, obtain expressions for $\frac{dy}{dx}$ and $\frac{d^2y}{dx^2}$.

**48**

1 (a) $y = x^3 \tan x$. Applying the product rule we find that:

$$\frac{dy}{dx} = 3x^2 \tan x + x^3 \sec^2 x$$
$$= x^2(3\tan x + x\sec^2 x)$$

(b) $y = x^2 e^x$. Applying the product rule we find that:

$$\frac{dy}{dx} = 2xe^x + x^2 e^x$$
$$= xe^x(2 + x)$$

(c) $y = \frac{2e^x}{x^2}$. Applying the quotient rule we find that:

$$\frac{dy}{dx} = \frac{2e^x x^2 - 2e^x 2x}{[x^2]^2}$$
$$= 2e^x\left(\frac{x^2 - 2x}{x^4}\right)$$
$$= 2e^x\left(\frac{x - 2}{x^3}\right)$$

(d) $y = \frac{x^3}{\sin x}$. Applying the quotient rule we find that:

$$\frac{dy}{dx} = \frac{3x^2 \sin x - x^3 \cos x}{[\sin x]^2}$$
$$= x^2\left(\frac{3\sin x - x\cos x}{\sin^2 x}\right)$$

▶

2 (a) $y = (4x+3)^6$. Applying the chain rule we find that:

$$\frac{dy}{dx} = 6(4x+3)^5 \times 4$$
$$= 24(4x+3)^5$$

(b) $y = \tan(2x+3)$. Applying the chain rule we find that:

$$\frac{dy}{dx} = \sec^2(2x+3) \times 2$$
$$= 2\sec^2(2x+3)$$

(c) $y = \ln(3x-4)$. Applying the chain rule we find that:

$$\frac{dy}{dx} = \frac{1}{3x-4} \times 3$$
$$= \frac{3}{3x-4}$$

(d) $y = -2e^{(1-3x)}$. Applying the chain rule we find that:

$$\frac{dy}{dx} = -2e^{(1-3x)} \times (-3)$$
$$= 6e^{(1-3x)}$$

(e) $y = e^{-3x}\sin(2x)$. Applying the chain rule and product rule combined we find that:

$$\frac{dy}{dx} = -3e^{-3x} \times \sin(2x) + e^{-3x} \times 2\cos(2x)$$
$$= e^{-3x}(2\cos(2x) - 3\sin(2x))$$

(f) $y = \tan^2 x$. Applying the chain rule we find that:

$$\frac{dy}{dx} = (2\tan x) \times \sec^2 x$$
$$= 2\tan x \sec^2 x$$

3 (a) If $y = x^3 + 2x^2 - 3x - 4$, then:

(i) $\frac{dy}{dx} = 3x^2 + 4x - 3$ and $\frac{d^2y}{dx^2} = \frac{d}{dx}\left(\frac{dy}{dx}\right) = 6x + 4$.

(ii) $\frac{dy}{dx} = 0$ when $3x^2 + 4x - 3 = 0$. That is when:

$$x = \frac{-4 \pm \sqrt{16 - 4 \times 3 \times (-3)}}{2 \times 3}$$
$$= \frac{-4 \pm \sqrt{52}}{6} = -1{\cdot}869 \text{ or } 0{\cdot}535 \text{ to 3 dp.}$$

(b) If $y = 2\cos(x+1) - 3\sin(x-1)$, then:

$$\frac{dy}{dx} = -2\sin(x+1) - 3\cos(x-1) \text{ and}$$
$$\frac{d^2y}{dx^2} = -2\cos(x+1) + 3\sin(x-1) = -y$$

# Newton–Raphson iterative method

**49**

To complete the topic of differentiation at this stage we shall look at a method that uses the differential calculus to find numerical solutions to equations, but first we must attend to an alternative notation.

## Notation

So far we have considered $y$ to be an expression involving the variable $x$ with a derivative denoted by:

$$\frac{dy}{dx}$$

This notation has a substantial history dating back to the creation of the calculus in the 17th century. However, it is rather cumbersome when the mathematics becomes more involved so an alternative notation has been devised. Consider $y$ to be an expression involving the variable $x$. That is:

$$y = f(x)$$

the derivative can then be written as

$$\frac{dy}{dx} = f'(x)$$

which we refer to as '*f primed of x*'. Referring back to Frame 22 this notation can easily be extended to cater for higher order derivatives. For example

$\frac{d^2y}{dx^2} = f''(x)$ the second derivative is '$f$ double primed'

$\frac{d^3y}{dx^3} = f'''(x)$ the third derivative is '$f$ triple primed' and so on.

This is the notation we shall use for the next topic.

*Move on to the final topic of this Programme in the next frame*

**50**

Consider the graph of $y = f(x)$ as shown. The $x$-value at the point A, where the graph crosses the $x$-axis, gives a solution of the equation $f(x) = 0$.

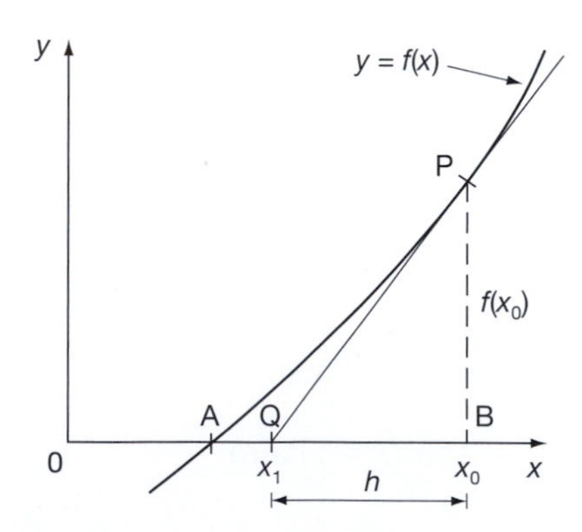


If P is a point on the curve near to A, then $x = x_0$ is an approximate value of the root of $f(x) = 0$, the error of the approximation being given by AB.

Let PQ be the tangent to the curve at P, crossing the $x$-axis at Q $(x_1, 0)$. Then $x = x_1$ is a better approximation to the required root.

From the diagram, $\frac{PB}{QB} = \left[\frac{dy}{dx}\right]_P$ i.e. the value of the derivative of $y$ at the point P, $x = x_0$.

$$\therefore \frac{PB}{QB} = f'(x_0) \quad \text{and} \quad PB = f(x_0) \quad [f'(x_0) \text{ is the derivative of } f(x) \text{ at } x = x_0]$$

$$\therefore QB = \frac{PB}{f'(x_0)} = \frac{f(x_0)}{f'(x_0)} = h \text{ (say)}$$

$$x_1 = x_0 - h \qquad \therefore x_1 = x_0 - \frac{f(x_0)}{f'(x_0)}$$

If we begin, therefore, with an approximate value ($x_0$) of the root, we can determine a better approximation ($x_1$). Naturally, the process can be repeated to improve the result still further. Let us see this in operation.

*On to the next frame*

---

**51**

**Example 1**

The equation $x^3 - 3x - 4 = 0$ is of the form $f(x) = 0$ where $f(1) < 0$ and $f(3) > 0$ so there is a solution to the equation between 1 and 3. We shall take this to be 2, by bisection. Find a better approximation to the root.

We have $f(x) = x^3 - 3x - 4 \quad \therefore f'(x) = 3x^2 - 3$

If the first approximation is $x_0 = 2$, then

$$f(x_0) = f(2) = -2 \quad \text{and} \quad f'(x_0) = f'(2) = 9$$

A better approximation $x_1$ is given by

$$x_1 = x_0 - \frac{f(x_0)}{f'(x_0)} = x_0 - \frac{{x_0}^3 - 3x_0 - 4}{3{x_0}^2 - 3}$$

$$x_1 = 2 - \frac{(-2)}{9} = 2{\cdot}22$$

$$\therefore x_0 = 2; \quad x_1 = 2{\cdot}22$$

If we now start from $x_1$ we can get a better approximation still by repeating the process.

$$x_2 = x_1 - \frac{f(x_1)}{f'(x_1)} = x_1 - \frac{{x_1}^3 - 3x_1 - 4}{3{x_1}^2 - 3}$$

Here $x_1 = 2{\cdot}22 \qquad f(x_1) = \dots\dots\dots; \quad f'(x_1) = \dots\dots\dots$

---

**52**

$$\boxed{f(x_1) = 0{\cdot}281; \quad f'(x_1) = 11{\cdot}785}$$

Then $x_2 = \dots\dots\dots$

---

**53**

$$\boxed{x_2 = 2{\cdot}196}$$

Because

$$x_2 = 2{\cdot}22 - \frac{0{\cdot}281}{11{\cdot}79} = 2{\cdot}196$$

Using $x_2 = 2{\cdot}196$ as a starter value, we can continue the process until successive results agree to the desired degree of accuracy.

$x_3 = \dots\dots\dots$

---

**54**

$$\boxed{x_3 = 2{\cdot}196}$$

Because

$$f(x_2) = f(2{\cdot}196) = 0{\cdot}002026; \qquad f'(x_2) = f'(2{\cdot}196) = 11{\cdot}467$$

$$\therefore\ x_3 = x_2 - \frac{f(x_2)}{f'(x_2)} = 2{\cdot}196 - \frac{0{\cdot}00203}{11{\cdot}467} = 2{\cdot}196 \text{ (to 4 sig. fig.)}$$

The process is simple but effective and can be repeated again and again. Each repetition, or *iteration*, usually gives a result nearer to the required root $x = x_A$.

In general $x_{n+1} =$ ...........

**55**

$$\boxed{x_{n+1} = x_n - \frac{f(x_n)}{f'(x_n)}}$$

## Tabular display of results

Open your spreadsheet and in cells **A1** to **D1** enter the headings $n$, $x$, $f(x)$ and $f'(x)$

Fill cells **A2** to **A6** with the numbers 0 to 4

In cell **B2** enter the value for $x_0$, namely 2

In cell **C2** enter the formula for $f(x_0)$, namely **B2^3 – 3*B2 – 4** and copy into cells **C3** to **C6**

In cell **D2** enter the formula for $f'(x_0)$, namely **3*B2^2 – 3** and copy into cells **D3** to **D6**

In cell **B3** enter the formula for $x_1$, namely **B2 – C2/D2** and copy into cells **B4** to **B6**.

The final display is ...........

**56**

| $n$ | $x$ | $f(x)$ | $f'(x)$ |
|---|---|---|---|
| 0 | 2 | −2 | 9 |
| 1 | 2·222222 | 0·30727 | 11·81481 |
| 2 | 2·196215 | 0·004492 | 11·47008 |
| 3 | 2·195823 | 1·01E-06 | 11·46492 |
| 4 | 2·195823 | 5·15E-14 | 11·46492 |

As soon as the number in the second column is repeated then we know that we have arrived at that particular level of accuracy. The required root is therefore $x = 2{\cdot}195823$ to 6 dp. Save the spreadsheet so that it can be used as a template for other such problems.

Now let us have another example.

*Next frame*

**57**

### Example 2

The equation $x^3 + 2x^2 - 5x - 1 = 0$ is of the form $f(x) = 0$ where $f(1) < 0$ and $f(2) > 0$ so there is a solution to the equation between 1 and 2. We shall take this to be $x = 1{\cdot}5$. Use the Newton–Raphson method to find the root to six decimal places.

Use the previous spreadsheet as a template and make the following amendments

In cell **B2** enter the number ...........

**58**

1·5

Because

That is the value of $x_0$ that is used to start the iteration

In cell **C2** enter the formula . . . . . . . . . . .

**59**

**B2^3 + 2*B2^2 – 5*B2 – 1**

Because

That is the value of $f(x_0) = {x_0}^3 + 2{x_0}^2 - 5x_0 - 1$. Copy the contents of cell **C2** into cells **C3** to **C5**.

In cell **D2** enter the formula . . . . . . . . . . .

**60**

**3*B2^2 + 4*B2 – 5**

Because

That is the value of $f'(x_0) = 3{x_0}^2 + 4x_0 - 5$. Copy the contents of cell **D2** into cells **D3** to **D5**.

In cell **B2** the formula remains the same as . . . . . . . . . . .

**61**

**B2 – C2/D2**

The final display is then . . . . . . . . . . .

**62**

| $n$ | $x$ | $f(x)$ | $f'(x)$ |
|---|---|---|---|
| 0 | 1·5 | −0·625 | 7·75 |
| 1 | 1·580645 | 0·042798 | 8·817898 |
| 2 | 1·575792 | 0·000159 | 8·752524 |
| 3 | 1·575773 | 2·21E-09 | 8·75228 |

We cannot be sure that the value 1·575773 is accurate to the sixth decimal place so we must extend the table.

Highlight cells **A5** to **D5**, click **Edit** on the Command bar and select **Copy** from the drop-down menu.

Place the cell highlight in cell **A6**, click **Edit** and then **Paste**.

The sixth row of the spreadsheet then fills to produce the display

| $n$ | $x$ | $f(x)$ | $f'(x)$ |
|---|---|---|---|
| 0 | 1·5 | −0·625 | 7·75 |
| 1 | 1·580645 | 0·042798 | 8·817898 |
| 2 | 1·575792 | 0·000159 | 8·752524 |
| 3 | 1·575773 | 2·21E-09 | 8·75228 |
| 4 | 1·575773 | −8·9E-16 | 8·75228 |

And the repetition of the $x$-value ensures that the solution $x = 1{\cdot}575773$ is indeed accurate to 6 dp.

Now do one completely on your own.

*Next frame*

63

**Example 3**

The equation $2x^3 - 7x^2 - x + 12 = 0$ has a root near to $x = 1{\cdot}5$. Use the Newton-Raphson method to find the root to six decimal places.

The spreadsheet solution produces …………

64

$x = 1{\cdot}686141$ to 6 dp

Because

Fill cells **A2** to **A6** with the numbers 0 to 4

In cell **B2** enter the value for $x_0$, namely 1·5

In cell **C2** enter the formula for $f(x_0)$, namely **2*B2^3 – 7*B2^2 – B2 + 12** and copy into cells **C3** to **C6**

In cell **D2** enter the formula for $f'(x_0)$, namely **6*B2^2 – 14*B2 – 1** and copy into cells **D3** to **D6**

In cell **B3** enter the formula for $x_1$, namely **B2 – C2/D2** and copy into cells **B4** to **B6**.

The final display is …………

65

| $n$ | $x$ | $f(x)$ | $f'(x)$ |
|---|---|---|---|
| 0 | 1·5 | 1·5 | −8·5 |
| 1 | 1·676471 | 0·073275 | −7·60727 |
| 2 | 1·686103 | 0·000286 | −7·54778 |
| 3 | 1·686141 | 4·46E-09 | −7·54755 |
| 4 | 1·686141 | 0 | −7·54755 |

As soon as the number in the second column is repeated then we know that we have arrived at that particular level of accuracy. The required root is therefore $x = 1{\cdot}686141$ to 6 dp.

**First approximations**

The whole process hinges on knowing a 'starter' value as first approximation. If we are not given a hint, this information can be found by either

(a) applying the remainder theorem if the function is a polynomial
(b) drawing a sketch graph of the function.

**Example 4**

Find the real root of the equation $x^3 + 5x^2 - 3x - 4 = 0$ correct to six significant figures.

Application of the remainder theorem involves substituting $x = 0$, $x = \pm 1$, $x = \pm 2$, etc. until two adjacent values give a change in sign.

$$f(x) = x^3 + 5x^2 - 3x - 4$$
$$f(0) = -4; \quad f(1) = -1; \quad f(-1) = 3$$

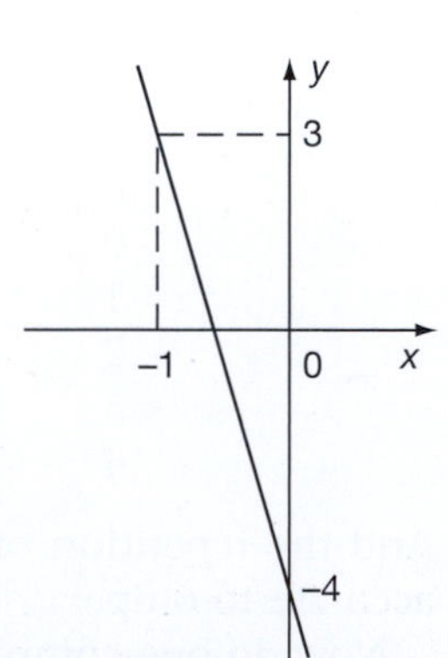


The sign changes from $f(0)$ to $f(-1)$. There is thus a root between $x = 0$ and $x = -1$.

Therefore choose $x = -0{\cdot}5$ as the first approximation and then proceed as before.

Complete the table and obtain the root $x =$ …………

**64**

$x = -0{\cdot}675527$

The final spreadsheet display is

| $n$ | $x$ | $f(x)$ | $f'(x)$ |
|---|---|---|---|
| 0 | −0·5 | −1·375 | −7·25 |
| 1 | −0·689655 | 0·11907 | −8·469679 |
| 2 | −0·675597 | 0·000582 | −8·386675 |
| 3 | −0·675527 | 1·43E-08 | −8·386262 |
| 4 | −0·675527 | 0 | −8·386262 |

**67**

**Example 5**

Solve the equation $e^x + x - 2 = 0$ giving the root to 6 significant figures.

It is sometimes more convenient to obtain a first approximation to the required root from a sketch graph of the function, or by some other graphical means.

In this case, the equation can be rewritten as $e^x = 2 - x$ and we therefore sketch graphs of $y = e^x$ and $y = 2 - x$.

| $x$ | 0·2 | 0·4 | 0·6 | 0·8 | 1 |
|---|---|---|---|---|---|
| $e^x$ | 1·22 | 1·49 | 1·82 | 2·23 | 2·72 |
| $2 - x$ | 1·8 | 1·6 | 1·4 | 1·2 | 1 |

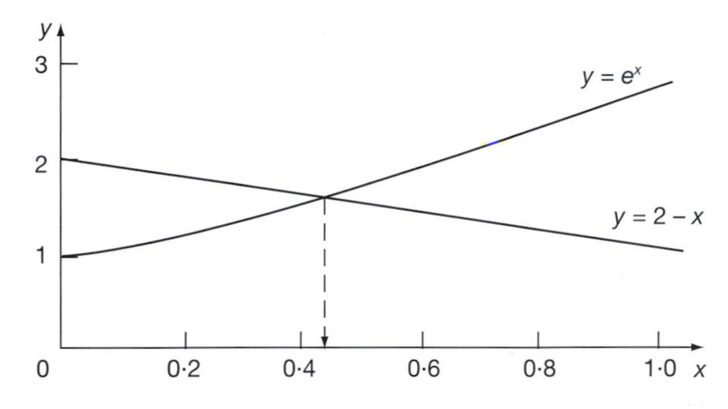


It can be seen that the two curves cross over between $x = 0{\cdot}4$ and $x = 0{\cdot}6$.

Approximate root $x = 0{\cdot}4$

$$f(x) = e^x + x - 2 \qquad f'(x) = e^x + 1$$

$x = \ldots\ldots\ldots\ldots$

*Finish it off using EXP(x) for $e^x$*

**68**

$x = 0{\cdot}442854$

The final spreadsheet display is

| $n$ | $x$ | $f(x)$ | $f'(x)$ |
|---|---|---|---|
| 0 | 0·4 | −0·10818 | 2·491825 |
| 1 | 0·443412 | 0·001426 | 2·558014 |
| 2 | 0·442854 | 2·42E-07 | 2·557146 |
| 3 | 0·442854 | 7·11E-15 | 2·557146 |

*Note*: There are times when the normal application of the Newton–Raphson method fails to converge to the required root. This is particularly so when $f'(x_0)$ is very small.

## Review summary

69

1 *Newton–Raphson method*

If $x = x_0$ is an approximate solution to the equation $f(x) = 0$ then $x_1 = x_0 - \dfrac{f(x_0)}{f'(x_0)}$ gives a more accurate value, where $f'(x_0)$ is the derivative of $f(x)$ at $x = x_0$. This defines an iterative procedure $x_{n+1} = x_n - \dfrac{f(x_n)}{f'(x_n)}$ for $n \geq 0$ to find progressively more accurate values.

## Review exercise

70

1 Use the Newton–Raphson method to solve the equation $x^2 - 5 = 0$, accurate to 6 decimal places given that $x = 2{\cdot}2$ is the solution accurate to 1 decimal place.

71

1 Let $f(x) = x^2 - 5$ so that $f'(x) = 2x$. The iterative solution is then given as:

$$x_{n+1} = x_n - \frac{f(x_n)}{f'(x_n)} \text{ for } n \geq 0$$

We are given that $x = 2{\cdot}2$ is the solution accurate to 1 decimal place so we can set up our tabular display as follows where cell **C2** contains the formula **=B2^2-5** and cell **D2** contains the formula **=2*B2**:

| $n$ | $x$ | $f(x)$ | $f'(x)$ |
|---|---|---|---|
| 0 | 2·2000000 | −0·1600000 | 4·4000000 |
| 1 | 2·2363636 | 0·0013223 | 4·4727273 |
| 2 | 2·2360680 | 0·0000001 | 4·4721360 |
| 3 | 2·2360680 | 0.0000000 | 4·4721360 |

Giving the solution as 2·236068 accurate to 6 decimal places.

72

You have now come to the end of this Programme. A list of **Can You?** questions follows for you to gauge your understanding of the material in the Programme. You will notice that these questions match the **Learning outcomes** listed at the beginning of the Programme so go back and try the **Quiz** that follows them. After that try the **Test exercise**. *Work through these at your own pace, there is no need to hurry.* A set of **Further problems** provides additional valuable practice.

## Can You?

73

### Checklist F.12

Check this list before and after you try the end of Programme test.

**On a scale of 1 to 5 how confident are you that you can:** Frames

- Determine the gradient of a straight-line graph? 1 to 2
  *Yes* ☐ ☐ ☐ ☐ ☐ *No*
- Evaluate from first principles the gradient at a point on a quadratic curve? 3 to 10
  *Yes* ☐ ☐ ☐ ☐ ☐ *No*

- Differentiate powers of $x$ and polynomials? — Frames 11 to 27
  *Yes* ☐ ☐ ☐ ☐ ☐ *No*
- Evaluate second derivatives and use tables of standard derivatives? — Frames 11 to 27
  *Yes* ☐ ☐ ☐ ☐ ☐ *No*
- Differentiate products and quotients of expressions? — Frames 28 to 33
  *Yes* ☐ ☐ ☐ ☐ ☐ *No*
- Differentiate using the chain rule for a function of a function? — Frames 34 to 48
  *Yes* ☐ ☐ ☐ ☐ ☐ *No*
- Use the Newton–Raphson method to obtain a numerical solution to an equation? — Frames 49 to 71
  *Yes* ☐ ☐ ☐ ☐ ☐ *No*

## Test exercise F.12

**74**

Frames

1 Calculate the slope of the straight line joining:
  (a) $P\ (3, 5)$ and $Q\ (6, 9)$  (c) $P\ (-3, 8)$ and $Q\ (4, 2)$
  (b) $P(2, 6)$ and $Q\ (7, 4)$  (d) $P\ (-2, 5)$ and $Q\ (3, -8)$ — Frames 1 to 2

2 Determine algebraically, from first principles, the slope of the graph of $y = 3x^2 + 4$ at the point $P$ where $x = 1{\cdot}2$. — Frames 3 to 10

3 If $y = x^4 + 5x^3 - 6x^2 + 7x - 3$, obtain an expression for $\dfrac{dy}{dx}$ and hence calculate the value of $\dfrac{dy}{dx}$ at $x = -2$. — Frames 11 to 27

4 (a) If $y = 2x^3 - 11x^2 + 12x - 5$, determine:
  (i) $\dfrac{dy}{dx}$ and $\dfrac{d^2y}{dx^2}$  (ii) the values of $x$ at which $\dfrac{dy}{dx} = 0$.
  (b) If $y = 3\sin(2x + 1) + 4\cos(3x - 1)$, obtain expressions for $\dfrac{dy}{dx}$ and $\dfrac{d^2y}{dx^2}$. — Frames 11 to 27

5 Determine $\dfrac{dy}{dx}$ in each of the following cases:
  (a) $y = x^2 . \sin x$  (b) $y = x^3 . e^x$
  (c) $y = \dfrac{\cos x}{x^2}$  (d) $y = \dfrac{2e^x}{\tan x}$ — Frames 28 to 33

6 Differentiate the following with respect to $x$:
  (a) $y = (5x + 2)^4$  (d) $y = 5\cos(2x + 3)$
  (b) $y = \sin(3x + 2)$  (e) $y = \cos^3 x$
  (c) $y = e^{(4x-1)}$  (f) $y = \ln(4x - 5)$ — Frames 34 to 48

7 Use the Newton–Raphson method to solve the equation $4 - 2x - x^3 = 0$, accurate to 6 decimal places. — Frames 49 to 71

# Further problems F.12

75

1 Determine algebraically from first principles, the slope of the following graphs at the value of $x$ indicated:

(a) $y = 4x^2 - 7$ at $x = -0{\cdot}5$
(b) $y = 2x^3 + x - 4$ at $x = 2$
(c) $y = 3x^3 - 2x^2 + x - 4$ at $x = -1$

2 Differentiate the following and calculate the value of $dy/dx$ at the value of $x$ stated:

(a) $y = 2x^3 + 4x^2 - 2x + 7$ $[x = -2]$
(b) $y = 3x^4 - 5x^3 + 4x^2 - x + 4$ $[x = 3]$
(c) $y = 4x^5 + 2x^4 - 3x^3 + 7x^2 - 2x + 3$ $[x = 1]$

Differentiate the functions given in questions **3**, **4** and **5**:

3 (a) $y = x^5 \sin x$ (b) $y = e^x \cos x$ (c) $y = x^3 \tan x$ (d) $y = x^4 \cos x$ (e) $y = 5x^2 \sin x$ (f) $y = 2e^x \ln x$

4 (a) $y = \dfrac{\cos x}{x^3}$ (b) $y = \dfrac{\sin x}{2e^x}$ (c) $y = \dfrac{\cos x}{\tan x}$ (d) $y = \dfrac{4x^3}{\cos x}$ (e) $y = \dfrac{\tan x}{e^x}$ (f) $y = \dfrac{\ln x}{x^3}$

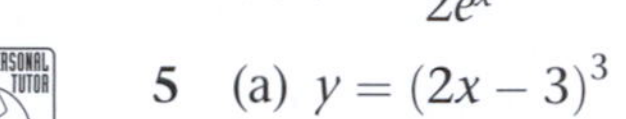

5 (a) $y = (2x - 3)^3$ (b) $y = e^{3x+2}$ (c) $y = 4\cos(3x + 1)$ (d) $y = \ln(x^2 + 4)$ (e) $y = \sin(2x - 3)$ (f) $y = \tan(x^2 - 3)$ (g) $y = 5(4x + 5)^2$ (h) $y = 6e^{x^2+2}$ (i) $y = \ln(3x^2)$ (j) $y = 3\sin(4 - 5x)$

6 Use the Newton–Raphson iterative method to solve the following.

(a) Show that a root of the equation $x^3 + 3x^2 + 5x + 9 = 0$ occurs between $x = -2$ and $x = -3$. Evaluate the root to four significant figures.

(b) Show graphically that the equation $e^{2x} = 25x - 10$ has two real roots and find the larger root correct to four significant figures.

(c) Verify that the equation $x - \cos x = 0$ has a root near to $x = 0{\cdot}8$ and determine the root correct to three significant figures.

(d) Obtain graphically an approximate root of the equation $2\ln x = 3 - x$. Evaluate the root correct to four significant figures.

(e) Verify that the equation $x^4 + 5x - 20 = 0$ has a root at approximately $x = 1{\cdot}8$. Determine the root correct to five significant figures.

(f) Show that the equation $x + 3\sin x = 2$ has a root between $x = 0{\cdot}4$ and $x = 0{\cdot}6$. Evaluate the root correct to four significant figures.

(g) The equation $2\cos x = e^x - 1$ has a real root between $x = 0{\cdot}8$ and $x = 0{\cdot}9$. Evaluate the root correct to four significant figures.

(h) The equation $20x^3 - 22x^2 + 5x - 1 = 0$ has a root at approximately $x = 0{\cdot}8$. Determine the value of the root correct to four significant figures.

76

Now visit the companion website for this book at www.palgrave.com/stroud for more questions applying this mathematics to science and engineering.

# Integration

## Learning outcomes

When you have completed this Programme you will be able to:

- ☐ Appreciate that integration is the reverse process of differentiation
- ☐ Recognize the need for a constant of integration
- ☐ Evaluate indefinite integrals of standard forms
- ☐ Evaluate indefinite integrals of polynomials
- ☐ Evaluate indefinite integrals of 'functions of a linear function of $x$'
- ☐ Integrate by partial fractions
- ☐ Appreciate that a definite integral is a measure of the area under a curve
- ☐ Evaluate definite integrals of standard forms
- ☐ Use the definite integral to find areas between a curve and the horizontal axis
- ☐ Use the definite integral to find areas between a curve and a given straight line

If you already feel confident about these why not try the quiz over the page? You can check your answers at the end of the book.

# Quiz F.13

**Frames**

1 Determine the following integrals:

(a) $\int x^7\,dx$ (b) $\int 4\cos x\,dx$ (c) $\int 2e^x\,dx$

(d) $\int 12\,dx$ (e) $\int x^{-4}\,dx$ (f) $\int 6^x\,dx$

(g) $\int 9x^{\frac{1}{3}}\,dx$ (h) $\int 4\sec^2 x\,dx$ (i) $\int 9\sin x\,dx$

(j) $\int \frac{8}{x}\,dx$

1 to 5

2 Determine the following integrals:

(a) $I = \int (x^3 - x^2 + x - 1)\,dx$

(b) $I = \int (4x^3 - 9x^2 + 8x - 2)\,dx$ given that $I = \frac{11}{16}$ when $x = \frac{1}{2}$.

6 to 11

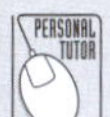

3 Determine the following integrals:

(a) $\int (5x-1)^4\,dx$ (b) $\int \frac{\sin(6x-1)}{3}\,dx$

(c) $\int \sqrt{4-2x}\,dx$ (d) $\int 2e^{3x+2}\,dx$

(e) $\int 5^{1-x}\,dx$ (f) $\int \frac{3}{2x-3}\,dx$

(g) $\int \frac{\sec^2(2-5x)}{5}\,dx$

12 to 15

4 Integrate by partial fractions each of the following integrals:

(a) $\int \frac{5x}{6x^2 - x - 1}\,dx$ (b) $\int \frac{14x+1}{2-7x-4x^2}\,dx$

(c) $\int \frac{1-9x}{1-9x^2}\,dx$

16 to 25

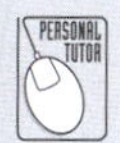

5 Find the area enclosed between the $x$-axis and the curve $y = e^x$ between $x = 1$ and $x = 2$, giving your answer in terms of $e$.

26 to 36

6 Find the area enclosed between the curve $y = e^x$ and the straight line $y = 1 - x$ between $x = 1$ and $x = 2$, giving your answer to 3 dp.

37 to 42

# Integration

1

*Integration* is the reverse process of differentiation. When we differentiate we start with an expression and proceed to find its derivative. When we integrate, we start with the derivative and then find the expression from which it has been derived.

For example, $\frac{d}{dx}(x^4) = 4x^3$. Therefore, the integral of $4x^3$ with respect to $x$ we know to be $x^4$. This is written:

$$\int 4x^3\, dx = x^4$$

The symbols $\int f(x)\, dx$ denote the *integral of $f(x)$ with respect to the variable $x$*; the symbol $\int$ was developed from a capital S which was used in the 17th century when the ideas of the calculus were first devised. The expression $f(x)$ to be integrated is called the *integrand* and the differential $dx$ is usefully there to assist in the evaluation of certain integrals, as we shall see in a later Programme.

## Constant of integration

$$\text{So} \quad \frac{d}{dx}(x^4) = 4x^3 \quad \therefore \quad \int 4x^3.dx = x^4$$

$$\text{Also} \quad \frac{d}{dx}(x^4 + 2) = 4x^3 \quad \therefore \quad \int 4x^3.dx = x^4 + 2$$

$$\text{and} \quad \frac{d}{dx}(x^4 - 5) = 4x^3 \quad \therefore \quad \int 4x^3.dx = x^4 - 5$$

In these three examples we happen to know the expressions from which the derivative $4x^3$ was derived. But any constant term in the original expression becomes zero in the derivative and all trace of it is lost. So if we do not know the history of the derivative $4x^3$ we have no evidence of the value of the constant term, be it 0, +2, −5 or any other value. We therefore acknowledge the presence of such a constant term of some value by adding a symbol $C$ to the result of the integration:

$$\text{i.e.} \quad \int 4x^3.dx = x^4 + C$$

$C$ is called the *constant of integration* and must always be included.

Such an integral is called an *indefinite integral* since normally we do not know the value of $C$. In certain circumstances, however, the value of $C$ might be found if further information about the integral is available.

For example, to determine $I = \int 4x^3.dx$, given that $I = 3$ when $x = 2$. As before:

$$I = \int 4x^3\, dx = x^4 + C$$

But $I = 3$ when $x = 2$ so that $3 = 2^4 + C = 16 + C \;\therefore\; C = -13$.
So, in this case $I = x^4 - 13$.

*Next frame*

## Standard integrals

2

Every derivative written in reverse gives an integral.

$$\text{e.g.} \quad \frac{d}{dx}(\sin x) = \cos x \;\therefore\; \int \cos x.dx = \sin x + C$$

It follows, therefore, that our list of standard derivatives provides a source of standard integrals.

▶

(a) $\dfrac{d}{dx}(x^n) = nx^{n-1}$. Replacing $n$ by $(n+1)$, $\dfrac{d}{dx}(x^{n+1}) = (n+1)x^n$

$\therefore \quad \dfrac{d}{dx}\left(\dfrac{x^{n+1}}{n+1}\right) = x^n \qquad \therefore \quad \int x^n.dx = \dfrac{x^{n+1}}{n+1} + C$

This is true except when $n = -1$, for then we should be dividing by 0.

(b) $\dfrac{d}{dx}(\sin x) = \cos x \qquad \therefore \quad \int \cos x.dx = \sin x + C$

(c) $\dfrac{d}{dx}(\cos x) = -\sin x \qquad \therefore \quad \dfrac{d}{dx}(-\cos x) = \sin x$

$\therefore \quad \int \sin x.dx = -\cos x + C$

(d) $\dfrac{d}{dx}(\tan x) = \sec^2 x \qquad \therefore \quad \int \sec^2 x.dx = \tan x + C$

(e) $\dfrac{d}{dx}(e^x) = e^x \qquad \therefore \quad \int e^x.dx = e^x + C$

(f) $\dfrac{d}{dx}(\ln x) = \dfrac{1}{x} \qquad \therefore \quad \int \dfrac{1}{x}dx = \ln x + C$

(g) $\dfrac{d}{dx}(a^x) = a^x.\ln a \qquad \therefore \quad \int a^x.dx = \dfrac{a^x}{\ln a} + C$

As with differentiation, a constant coefficient remains unchanged

e.g. $\int 5.\cos x.dx = 5\sin x + C$, etc.

Collecting the results together, we have:

| $f(x)$ | $\int f(x).dx$ | |
|---|---|---|
| $x^n$ | $\dfrac{x^{n+1}}{n+1} + C$ | $(n \neq -1)$ |
| $1$ | $x + C$ | |
| $a$ | $ax + C$ | |
| $\sin x$ | $-\cos x + C$ | |
| $\cos x$ | $\sin x + C$ | |
| $\sec^2 x$ | $\tan x + C$ | |
| $e^x$ | $e^x + C$ | |
| $a^x$ | $a^x/\ln a + C$ | |
| $\dfrac{1}{x}$ | $\ln x + C$ | |

***At this point let us pause and summarize the main facts so far***

##  Review summary

3

1 Integration is the reverse process of differentiation. Given a derivative we have to find the expression from which it was derived.

2 Because constants have a zero derivative we find that when we reverse the process of differentiation we must introduce an integration constant into our result.

3 Such integrals are called indefinite integrals.

##  Review exercise

4

1 Determine the following integrals:

(a) $\int x^6\,dx$ (b) $\int 3e^x\,dx$ (c) $\int \frac{6}{x}\,dx$ (d) $\int 5\sin x\,dx$ (e) $\int \sec^2 x\,dx$

(f) $\int 8\,dx$ (g) $\int x^{\frac{1}{2}}\,dx$ (h) $\int 2\cos x\,dx$ (i) $\int x^{-3}\,dx$ (j) $\int 4^x\,dx$

2 (a) Determine $I = \int 4x^2.dx$, given that $I = 25$ when $x = 3$.

(b) Determine $I = \int 5.dx$, given that $I = 16$ when $x = 2$.

(c) Determine $I = \int 2.\cos x.dx$, given that $I = 7$ when $x = \frac{\pi}{2}$ (radians).

(d) Determine $I = \int 2.e^x.dx$, given that $I = 50{\cdot}2$ when $x = 3$.

*Complete and then check with the next frame*

5

1 (a) $\int x^6\,dx = \frac{x^7}{7} + C$  Check by differentiating the result – the derivative of the result will give the integrand

$$\frac{d}{dx}\left(\frac{x^7}{7} + C\right) = \frac{7x^6}{7} + 0 = x^6 \text{ the integrand}$$

(b) $\int 3e^x\,dx = 3e^x + C$  The derivative of the result is $3e^x + 0 = 3e^x$ – the integrand

(c) $\int \frac{6}{x}\,dx = 6\ln x + C$  The derivative of the result is $\frac{6}{x} + 0 = \frac{6}{x}$

(d) $\int 5\sin x\,dx = -5\cos x + C$  The derivative of the result is $-5(-\sin x) + 0 = 5\sin x$

(e) $\int \sec^2 x\,dx = \tan x + C$  The derivative of the result is $\sec^2 x + 0 = \sec^2 x$

(f) $\int 8\,dx = 8x + C$  The derivative of the result is $8 + 0 = 8$

(g) $\int x^{\frac{1}{2}}\,dx = \frac{x^{\frac{1}{2}+1}}{\frac{1}{2}+1} + C = \frac{x^{\frac{3}{2}}}{\frac{3}{2}} + C = \frac{2x^{\frac{3}{2}}}{3} + C$  The derivative of the result is $\frac{3}{2} \times \frac{2x^{\frac{3}{2}-1}}{3} + 0 = x^{\frac{1}{2}}$

▶

(h) $\int 2\cos x\,dx = 2\sin x + C$ The derivative of the result is $2\cos x + 0 = 2\cos x$

(i) $\int x^{-3}\,dx = \frac{x^{-3+1}}{-3+1} + C$ The derivative of the result is

$= -\frac{x^{-2}}{2} + C$ $\quad -(-2)\frac{x^{-2-1}}{2} + 0 = x^{-3}$

(j) $\int 4^x\,dx = \frac{4^x}{\ln 4} + C$ Recall the standard derivative $\frac{d}{dx}(a^x) = a^x \ln a$

2 (a) $I = \int 4x^2.dx = 4\frac{x^3}{3} + C$

$\therefore\ 25 = 36 + C\ \therefore\ C = -11$

$\therefore\ \int 4x^2.dx = \frac{4x^3}{3} - 11$

(b) $I = \int 5.dx = 5x + C$

$16 = 10 + C\ \therefore\ C = 6$

$\therefore\ \int 5.dx = 5x + 6$

(c) $I = \int 2.\cos x.dx = 2\sin x + C$

$\therefore\ 7 = 2 + C\ \therefore\ C = 5$

$\therefore\ \int 2.\cos x.dx = 2\sin x + 5$

(d) $I = \int 2.e^x.dx = 2e^x + C$

$\therefore\ 50{\cdot}2 = 2e^3 + C = 40{\cdot}2 + C$

$\therefore\ C = 10$

$\therefore\ \int 2.e^x.dx = 2e^x + 10$

---

## Integration of polynomial expressions

**6**

In the previous Programme we differentiated a polynomial expression by dealing with the separate terms, one by one. It is not surprising, therefore, that we do much the same with the integration of polynomial expressions.

*Polynomial expressions* are integrated term by term with the individual constants of integration consolidated into one constant $C$ for the whole expression. For example:

$$\int (4x^3 + 5x^2 - 2x + 7)\,dx$$

$$= x^4 + \frac{5x^3}{3} - x^2 + 7x + C$$

So, what about this one? If $I = \int (8x^3 - 3x^2 + 4x - 5)\,dx$, determine the value of $I$ when $x = 3$, given that at $x = 2$, $I = 26$.

First we must determine the function for $I$, so carrying out the integration, we get

$I = \ldots\ldots\ldots\ldots$

**7**

$$I = 2x^4 - x^3 + 2x^2 - 5x + C$$

Now we can calculate the value of $C$ since we are told that when $x = 2$, $I = 26$. So, expressing the function for $I$ in nested form, we have

$I = \ldots\ldots\ldots\ldots$

**8**

$I = \{[(2x-1)x+2]x-5\}x + C$

Substituting $x = 2$:

$26 = \ldots\ldots\ldots\ldots$

**9**

$22 + C$

We have $26 = 22 + C \quad \therefore \; C = 4$

$\therefore \; I = \{[(2x-1)x+2]x-5\}x + 4$

Finally, all we now have to do is to put $x = 3$ in this expression which gives us that, when $x = 3$,

$I = \ldots\ldots\ldots$

**10**

142

Just take one step at a time. There are no snags.

Now here is another of the same type. Determine the value of

$I = \int (4x^3 - 6x^2 - 16x + 4)\,dx$ when $x = -2$, given that at $x = 3$, $I = -13$.

As before:

(a) Perform the integration.

(b) Express the resulting function in nested form.

(c) Evaluate the constant of integration, using the fact that when $x = 3$, $I = -13$.

(d) Determine the value of $I$ when $x = -2$.

The method is just the same as before, so work through it.

$\therefore$ When $x = -2, \quad I = \ldots\ldots\ldots\ldots$

**11**

12

Here is a check on the working:

(a) $I = x^4 - 2x^3 - 8x^2 + 4x + C$

(b) In nested form, $I = \{[(x-2)x-8]x+4\}x + C$

(c) At $x = 3$, $I = -13 = -33 + C \;\; \therefore \; C = 20$

$\therefore \; I = \{[(x-2)x-8]x+4\}x + 20$

(d) $\therefore$ When $x = -2, \quad I = 12$

It is all very straightforward.

*Now let us move on to something slightly different*

## 12 Functions of a linear function of x

It is often necessary to integrate any one of the expressions shown in our list of standard integrals when the variable $x$ is replaced by a linear expression in $x$. That is, of the form $ax + b$. For example, $y = \int (3x+2)^4\,dx$ is of the same structure as $y = \int x^4\,dx$ except that $x$ is replaced by the linear expression $3x + 2$. Let us put $u = 3x + 2$ then:

$$\int (3x+2)^4\,dx \text{ becomes } \int u^4\,dx$$

We now have to change the variable $x$ in d$x$ before we can progress. Now, $u = 3x + 2$ so that:

$$\frac{du}{dx} = 3$$

That is: $du = 3\,dx$ or, alternatively $dx = \frac{du}{3}$

We now find that our integral can be determined as:

$$y = \int u^4\,dx = \int u^4 \frac{du}{3} = \frac{1}{3}\left(\frac{u^5}{5}\right) + C = \frac{1}{3}\left(\frac{(3x+2)^5}{5}\right) + C$$

That is: $y = \frac{(3x+2)^5}{15} + C$

To integrate a 'function of a linear function of x', simply replace x in the corresponding standard result by the linear expression and divide by the coefficient of x in the linear expression.

Here are three examples:

(1) $\int (4x-3)^2.dx$ [Standard integral $\int x^2.dx = \frac{x^3}{3} + C$]

$$\therefore \int (4x-3)^2.dx = \frac{(4x-3)^3}{3} \times \frac{1}{4} + C = \frac{(4x-3)^3}{12} + C$$

(2) $\int \cos 3x.dx$ [Standard integral $\int \cos x.dx = \sin x + C$]

$$\therefore \int \cos 3x.dx = \sin 3x.\frac{1}{3} + C = \frac{\sin 3x}{3} + C$$

(3) $\int e^{5x+2}.dx$ [Standard integral $\int e^x.dx = e^x + C$]

$$\therefore \int e^{5x+2}.dx = e^{5x+2}\,\frac{1}{5} + C = \frac{e^{5x+2}}{5} + C$$

Just refer to the basic standard integral of the same form, replace $x$ in the result by the linear expression and finally divide by the coefficient of $x$ in the linear expression – and remember the constant of integration.

***At this point let us pause and summarize the main facts dealing with the integration of polynomial expressions and 'functions of a linear function of x'***

## Review summary

13

1 *Integration of polynomial expressions*
Integrate term by term and combine the individual constants of integration into one symbol.

2 *Integration of 'functions of a linear function of x'*
Replace $x$ in the corresponding standard integral by the linear expression and divide the result by the coefficient of $x$ in the linear expression.

## Review exercise

14

1 Determine the following integrals:

(a) $I = \int (2x^3 - 5x^2 + 6x - 9)\,dx$

(b) $I = \int (9x^3 + 11x^2 - x - 3)\,dx$, given that when $x = 1$, $I = 2$.

2 Determine the following integrals:

(a) $\int (1-4x)^2.dx$

(b) $\int 4.e^{5x-2}.dx$

(c) $\int 3.\sin(2x+1).dx$

(d) $\int (3-2x)^{-5}.dx$

(e) $\int \dfrac{7}{2x-5}.dx$

(f) $\int \cos(1-3x).dx$

(g) $\int 2^{3x-1}.dx$

(h) $\int 6\sec^2(2+3x).dx$

(i) $\int \sqrt{3-4x}.dx$

(j) $\int 5.e^{1-3x}.dx$

15

1 (a) $I = \int (2x^3 - 5x^2 + 6x - 9)\,dx = 2\dfrac{x^4}{4} - 5\dfrac{x^3}{3} + 6\dfrac{x^2}{2} - 9x + C$

$$= \frac{x^4}{2} - \frac{5}{3}x^3 + 3x^2 - 9x + C$$

(b) $I = \int (9x^3 + 11x^2 - x - 3)\,dx = 9\dfrac{x^4}{4} + 11\dfrac{x^3}{3} - \dfrac{x^2}{2} - 3x + C$

$$= \left(\left(\left(\frac{9x}{4} + \frac{11}{3}\right)x - \frac{1}{2}\right)x - 3\right)x + C$$

Given $I = 2$ when $x = 1$ we find that:

$$2 = \frac{29}{12} + C$$

So that $C = -\dfrac{5}{12}$ and $I = 9\dfrac{x^4}{4} + 11\dfrac{x^3}{3} - \dfrac{x^2}{2} - 3x - \dfrac{5}{12}$

▶

2 (a) $\int (1-4x)^2\,dx$ [Standard integral $\int x^2\,dx = \frac{x^3}{3} + C$]

Therefore, $\int (1-4x)^2\,dx = \frac{(1-4x)^3}{3} \times \frac{1}{(-4)} + C = -\frac{(1-4x)^3}{12} + C$

(b) $\int 4e^{5x-2}\,dx$ [Standard integral $\int e^x\,dx = e^x + C$]

Therefore, $\int 4e^{5x-2}\,dx = 4e^{5x-2} \times \frac{1}{5} + C = \frac{4}{5}e^{5x-2} + C$

(c) $\int 3\sin(2x+1)\,dx$ [Standard integral $\int \sin x\,dx = -\cos x + C$]

Therefore, $\int 3\sin(2x+1)\,dx = 3(-\cos(2x+1)) \times \frac{1}{2} + C$

$= -\frac{3}{2}\cos(2x+1) + C$

(d) $\int (3-2x)^{-5}\,dx$ [Standard integral $\int x^{-5}\,dx = \frac{-x^{-4}}{4} + C$]

Therefore, $\int (3-2x)^{-5}\,dx = -\frac{(3-2x)^{-4}}{4} \times \frac{1}{(-2)} + C = \frac{(3-2x)^{-4}}{8} + C$

(e) $\int \frac{7}{2x-5}\,dx$ [Standard integral $\int \frac{1}{x}\,dx = \ln x + C$]

Therefore, $\int \frac{7}{2x-5}\,dx = 7\ln(2x-5) \times \frac{1}{2} + C = \frac{7}{2}\ln(2x-5) + C$

(f) $\int \cos(1-3x)\,dx$ [Standard integral $\int \cos x\,dx = \sin x + C$]

Therefore, $\int \cos(1-3x)\,dx = \sin(1-3x) \times \frac{1}{(-3)} + C$

$= -\frac{\sin(1-3x)}{3} + C$

(g) $\int 2^{3x-1}\,dx$ [Standard integral $\int 2^x\,dx = \frac{2^x}{\ln 2} + C$]

Therefore, $\int 2^{3x-1}\,dx = \frac{2^{3x-1}}{\ln 2} \times \frac{1}{3} + C = \frac{2^{3x-1}}{3\ln 2} + C$

(h) $\int 6\sec^2(2+3x)\,dx$ [Standard integral $\int \sec^2 x\,dx = \tan x + C$]

Therefore, $\int 6\sec^2(2+3x)\,dx = 6\tan(2+3x) \times \frac{1}{3} + C$

$= 2\tan(2+3x) + C$

(i) $\int \sqrt{3-4x}\,dx$ [Standard integral $\int \sqrt{x}\,dx = \frac{x^{\frac{3}{2}}}{3/2} + C$]

Therefore, $\int \sqrt{3-4x}\,dx = \frac{2(3-4x)^{\frac{3}{2}}}{3} \times \frac{1}{(-4)} + C = -\frac{(3-4x)^{\frac{3}{2}}}{6} + C$

(j) $\int 5e^{1-3x}\,dx$ [Standard integral $\int 5e^x\,dx = 5e^x + C$]

Therefore, $\int 5e^{1-3x}\,dx = 5e^{1-3x} \times \frac{1}{(-3)} + C = -\frac{5}{3}e^{1-3x} + C$

## Integration by partial fractions

**16**

Expressions such as $\int \frac{7x+8}{2x^2+11x+5}\,dx$ do not appear in our list of standard integrals, but do, in fact, occur in many mathematical applications. We saw in Programme F.8 that such an expression as $\frac{7x+8}{2x^2+11x+5}$ can be expressed in partial fractions which are simpler in structure.

In fact, $\frac{7x+8}{2x^2+11x+5} = \frac{7x+8}{(x+5)(2x+1)} = \frac{3}{x+5} + \frac{1}{2x+1}$ so that

$$\int \frac{7x+8}{2x^2+11x+5}\,dx = \int \frac{3}{x+5}\,dx + \int \frac{1}{2x+1}\,dx$$

These partial fractions are 'functions of a linear function of $x$', based on the standard integral $\int \frac{1}{x}\,dx$, so the result is clear:

$$\int \frac{7x+8}{2x^2+11x+5}\,dx = \int \frac{7x+8}{(x+5)(2x-1)}\,dx = \int \frac{3}{x+5}\,dx + \int \frac{1}{2x+1}\,dx$$
$$= 3\ln(x+5) + \frac{1}{2}\ln(2x+1) + C$$

You will recall the Rules of Partial Fractions which we listed earlier and used in Programme F.8, so let us apply them in this example. We will only deal with simple linear denominators at this stage.

Determine $\int \frac{3x^2+18x+3}{3x^2+5x-2}\,dx$ by partial fractions.

The first step is ...........

---

**17**

to divide out

because the numerator is not of lower degree than that of the denominator

So $\frac{3x^2+18x+3}{3x^2+5x-2} = $ ...........

---

**18**

$$1 + \frac{13x+5}{3x^2+5x-2}$$

The denominator factorizes into $(3x-1)(x+2)$ so the partial fractions of

$\frac{13x+5}{(3x-1)(x+2)} = $ ...........

---

**19**

$$\boxed{\frac{4}{3x-1}+\frac{3}{x+2}}$$

Because

$$\frac{13x+5}{(3x-1)(x+2)}=\frac{A}{3x-1}+\frac{B}{x+2}$$

$$\begin{aligned}\therefore\ 13x+5&=A(x+2)+B(3x-1)\\&=Ax+2A+3Bx-B\\&=(A+3B)x+(2A-B)\end{aligned}$$

$$\begin{array}{llll}[x] & \therefore\ A+3B=13 & A+3B=13 & \\ [CT] & 2A-B=5 & \underline{6A-3B=15} & \\ & & 7A\quad\ =28 & \therefore\ A=4\\ & \therefore\ 4+3B=13 & \therefore\ 3B=9 & \therefore\ B=3\end{array}$$

$$\therefore\ \frac{13x+5}{(3x-1)(x+2)}=\frac{4}{3x-1}+\frac{3}{x+2}$$

$$\therefore\ \int\frac{3x^2+18x+3}{3x^2+5x-2}\,dx=\int\left(1+\frac{4}{3x-1}+\frac{3}{x+2}\right)dx$$

$$=\ldots\ldots\ldots$$

*Finish it*

**20**

$$\boxed{I=x+\frac{4.\ln(3x-1)}{3}+3.\ln(x+2)+C}$$

Now you can do this one in like manner:

$$\int\frac{4x^2+26x+5}{2x^2+9x+4}\,dx=\ldots\ldots\ldots\ldots$$

*Work right through it and then check the solution with the next frame*

**21**

$$\boxed{2x+5\ln(x+4)-\ln(2x+1)+C}$$

Here is the working:

$$\frac{4x^2+26x+5}{2x^2+9x+4}=2+\frac{8x-3}{2x^2+9x+4}$$

$$\frac{8x-3}{2x^2+9x+4}=\frac{8x-3}{(x+4)(2x+1)}=\frac{A}{x+4}+\frac{B}{2x+1}$$

$$\therefore\ 8x-3=A(2x+1)+B(x+4)=(2A+B)x+(A+4B)$$

$$\begin{array}{lll}\therefore\ 2A+B=8 & 8A+4B=32 & \\ A+4B=-3 & \underline{A+4B=-3} & \\ & \therefore\ 7A\quad\ =35 & \therefore\ A=5\\ 10+B=8 & & \therefore\ B=-2\end{array}$$

$$\therefore\ \int\frac{4x^2+26x+5}{2x^2+9x+4}\,dx=\int\left(2+\frac{5}{x+4}-\frac{2}{2x+1}\right)dx$$

$$=2x+5\ln(x+4)-\frac{2\ln(2x+1)}{2}+C$$

$$=2x+5\ln(x+4)-\ln(2x+1)+C$$

▶

And finally this one:

Determine $I = \int \frac{16x+7}{6x^2+x-12}\,dx$ by partial fractions.

$I = \dots\dots\dots$

22

$$\ln(2x+3) + \frac{5}{3}\ln(3x-4) + C$$

$$\frac{16x+7}{6x^2+x-12} = \frac{16x+7}{(2x+3)(3x-4)} = \frac{A}{2x+3} + \frac{B}{3x-4}$$

$$\therefore\ 16x+7 = A(3x-4) + B(2x+3)$$
$$= (3A+2B)x - (4A-3B)$$

Equating coefficients gives $A = 2$ and $B = 5$

$$\therefore\ \int \frac{16x+7}{6x^2+x-12}\,dx = \int\left(\frac{2}{2x+3} + \frac{5}{3x-4}\right)dx$$
$$= \ln(2x+3) + \frac{5}{3}\ln(3x-4) + C$$

***At this point let us pause and summarize the main facts dealing with integration by partial fractions***

## Review summary

23

*Integration by partial fractions*

Algebraic fractions can often be expressed in terms of partial fractions. This renders integration of such algebraic fractions possible, the integration of each partial fraction

$$\int \frac{A}{ax+b}\,dx = A\frac{\ln(ax+b)}{a} + C$$

## Review exercise

24

**1** Integrate by partial fractions each of the following integrals:

(a) $\int \frac{5x+2}{3x^2+x-4}\,dx$ (b) $\int \frac{x+1}{4x^2-1}\,dx$ (c) $\int \frac{3x}{1+x-2x^2}\,dx$

25

**1** (a) $\frac{5x+2}{3x^2+x-4} = \frac{5x+2}{(3x+4)(x-1)} = \frac{A}{3x+4} + \frac{B}{x-1}$ therefore

$$5x+2 = A(x-1) + B(3x+4)$$
$$= (A+3B)x + (-A+4B) \text{ so that}$$

$A + 3B = 5$

$-A + 4B = 2$ therefore adding we find that $7B = 7$ so $B = 1$ and $A = 2$.

▶

Therefore: $\displaystyle\int \frac{5x+2}{3x^2+x-4}\,dx = \int \frac{2}{3x+4}\,dx + \int \frac{1}{x-1}\,dx$

$\displaystyle = \frac{2}{3}\ln(3x+4) + \ln(x-1) + C$

(b) $\displaystyle \frac{x+1}{4x^2-1} = \frac{x+1}{(2x+1)(2x-1)} = \frac{A}{2x+1} + \frac{B}{2x-1}$ therefore

$x+1 = A(2x-1) + B(2x+1)$

$= (2A+2B)x + (-A+B)$ so that

$2A+2B = 1 \qquad 2A+2B = 1$

$-A+B = 1 \qquad -2A+2B = 2$ therefore adding we find that $4B = 3$ so $B = 3/4$ and $A = -1/4$.

Therefore:

$\displaystyle\int \frac{x+1}{4x^2-1}\,dx = -\int \frac{1/4}{2x+1}\,dx + \int \frac{3/4}{2x-1}\,dx$

$\displaystyle = -\frac{1}{8}\ln(2x+1) + \frac{3}{8}\ln(2x-1) + C$

(c) $\displaystyle \frac{3x}{1+x-2x^2} = \frac{3x}{(1-x)(1+2x)} = \frac{A}{1-x} + \frac{B}{1+2x}$ therefore

$3x = A(1+2x) + B(1-x)$

$= (2A-B)x + (A+B)$ so that

$2A - B = 3$

$A + B = 0$ therefore, adding we find that $3A = 3$ so $A = 1$ and $B = -1$.

Therefore:

$\displaystyle\int \frac{3x}{1+x-2x^2}\,dx = \int \frac{1}{1-x}\,dx - \int \frac{1}{1+2x}\,dx$

$\displaystyle = -\ln(1-x) - \frac{1}{2}\ln(1+2x) + C$

*Now on to something different*

## Areas under curves

**26**

Consider the area $A$ of the figure bounded by the curve $y = f(x)$, the $x$-axis and the two vertical lines through $x = a$ and $x = b$ (where $b > a$).

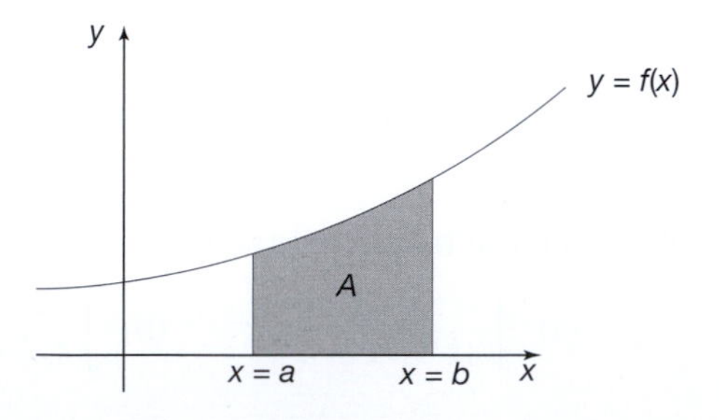


To evaluate the area $A$ you need to consider the total area between the same curve and the $x$-axis from the left up to some arbitrary point $P$ on the curve with coordinates $(x, y)$ which we shall denote by $A_x$.

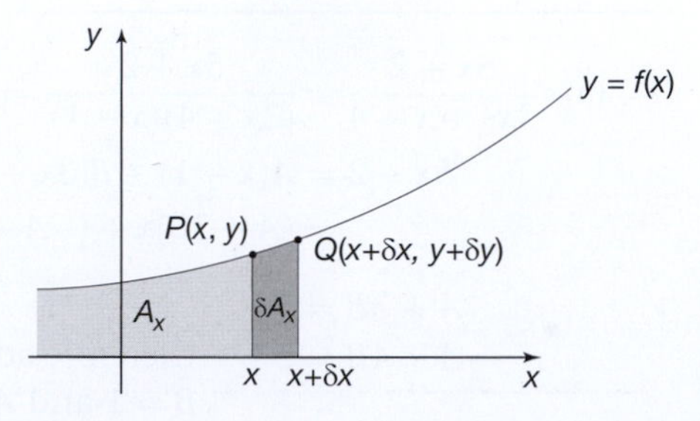

Area $\delta A_x$ is the area enclosed by the strip under the arc $PQ$ where $Q$ has the coordinates $(x+\delta x, y+\delta y)$. If the strip is approximated by a rectangle of height $y$ and width $\delta x$ then $\delta A_x \approx y.\delta x$. This means that:

$$\frac{\delta A_x}{\delta x} \approx y$$

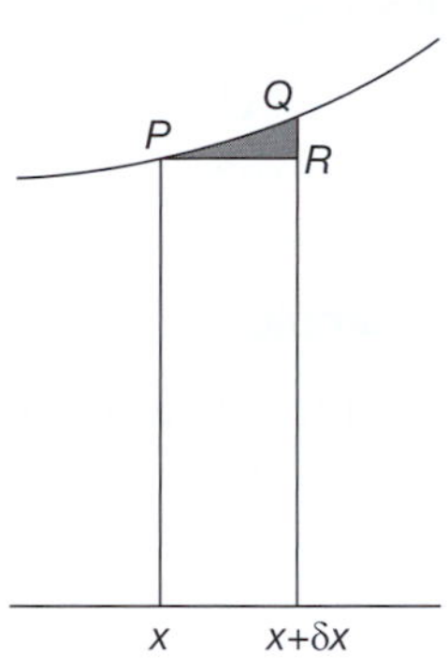


The error in this approximation is given by the area of $PQR$ in the figure to the right, where the strip has been magnified.

If the width of the strip is reduced then the error is accordingly reduced. Also, if $\delta x \to 0$ then $\delta A_x \to 0$ and:

$$\frac{\delta A_x}{\delta x} \to \frac{dA_x}{dx} \text{ so that, in the limit, } \frac{dA_x}{dx} = y$$

Consequently, because integration is the reverse process of differentiation it is seen that:

$$A_x = \int y\,dx$$

*The total area between the curve and the x-axis up to the point P is given by the indefinite integral.*

If $x = b$ then $A_b = \int\limits_{(x=b)} y\,dx$ (the value of the integral and hence the area up to $b$) and

if $x = a$ then $A_a = \int\limits_{(x=a)} y\,dx$ (the value of the integral and hence the area up to $a$).

Because $b > a$, the difference in these two areas $A_b - A_a$ gives the required area $A$. That is:

$$A = \int\limits_{(x=b)} y\,dx - \int\limits_{(x=a)} y\,dx \text{ which is written } A = \int_a^b y\,dx$$

The numbers $a$ and $b$ are called the *limits* of the integral where the right-hand limit is at the top of the integral sign and the left-hand limit is at the bottom. Such an integral with limits is called a *definite integral*. Notice that in the subtraction process when the integral is evaluated, the constant of integration disappears leaving the numerical value of the area.

### Example 1

To determine the area bounded by the curve $y = 3x^2 + 6x + 8$, the $x$-axis and the ordinates $x = 1$ and $x = 3$.

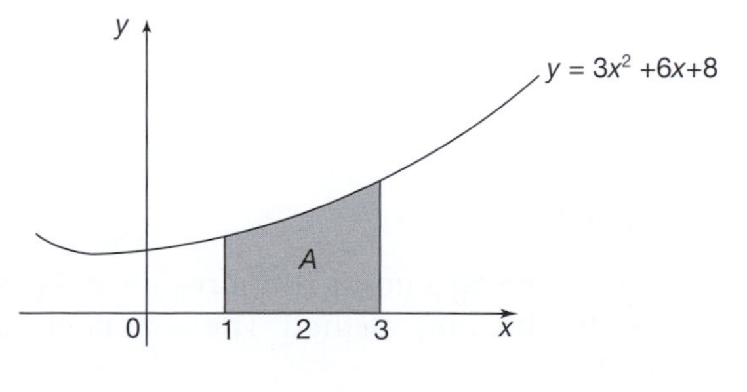


$$A = \int_1^3 y\,dx = \int_1^3 (3x^2 + 6x + 8)\,dx$$

$$= \left[x^3 + 3x^2 + 8x\right]_1^3$$

Note that we enclose the expression in square brackets with the limits attached.

Now calculate the values at the upper and lower limits and subtract the second from the first which gives $A = \ldots\ldots\ldots$

**27**

66 unit$^2$

Because

$$A = \left[x^3 + 3x^2 + 8x\right]_1^3$$
$$= \{27 + 27 + 24\} - \{1 + 3 + 8\} = 78 - 12 = 66 \text{ unit}^2$$

**Example 2**

Find the area bounded by the curve $y = 3x^2 + 14x + 15$, the $x$-axis and ordinates at $x = -1$ and $x = 2$.

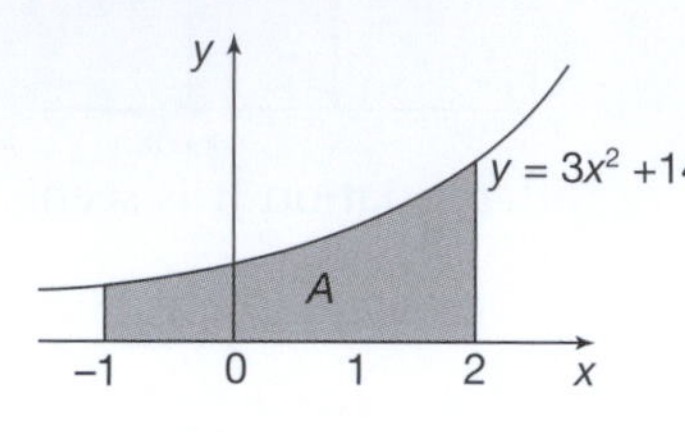


$$A = \int_{-1}^{2} y\,dx = \int_{-1}^{2} (3x^2 + 14x + 15)\,dx$$
$$= \left[x^3 + 7x^2 + 15x\right]_{-1}^{2}$$
$$\therefore\ A = \{8 + 28 + 30\} - \{-1 + 7 - 15\}$$
$$= 66 - (-9) = 75 \text{ unit}^2$$

**Example 3**

Calculate the area bounded by the curve $y = -6x^2 + 24x + 10$, the $x$-axis and the ordinates $x = 0$ and $x = 4$.

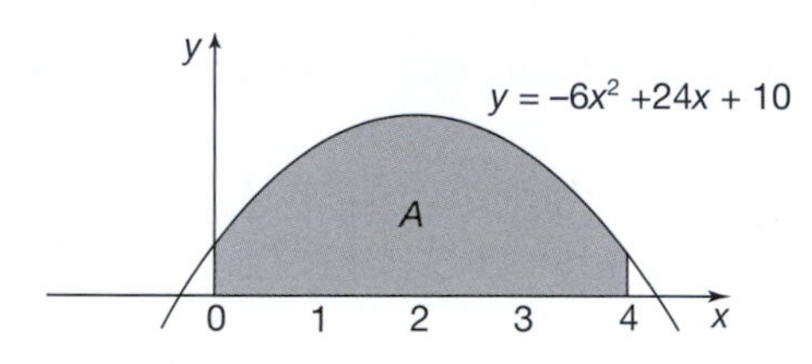


$$A = \int_0^4 y\,dx = \int_0^4 (-6x^2 + 24x + 10)\,dx$$
$$= \ldots\ldots\ldots\ldots$$

**28**

104 unit$^2$

Because

$$A = \left[-2x^3 + 12x^2 + 10x\right]_0^4 = 104 - 0 \quad \therefore\ A = 104 \text{ unit}^2$$

And now:

**Example 4**

Determine the area under the curve $y = e^x$ between $x = -2$ and $x = 3$.
Do this by the same method. The powers of $e$ are available from most calculators or from tables.

$A = \ldots\ldots\ldots\ldots$

29

19·95 unit$^2$

As a check:

$$A = \int_{-2}^{3} y\,dx = \int_{-2}^{3} e^x\,dx = [e^x]_{-2}^{3} = \{e^3\} - \{e^{-2}\}$$

$e^3 = 20{\cdot}09$ and $e^{-2} = 0{\cdot}14$

$\therefore\ A = 20{\cdot}09 - 0{\cdot}14 = 19{\cdot}95$ unit$^2$

***At this point let us pause and summarize the main facts dealing with areas beneath curves***

## Review summary

30

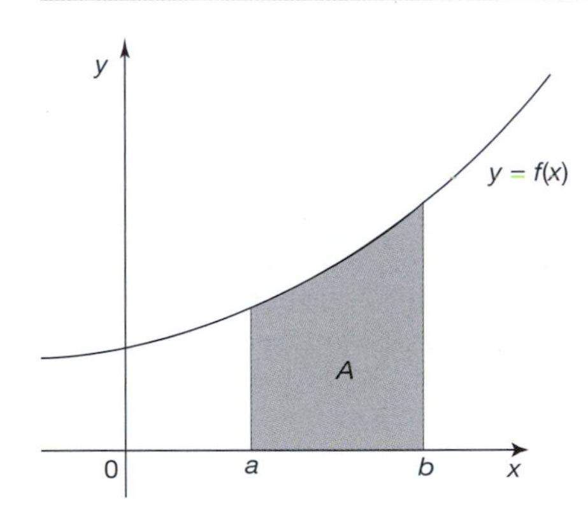


Area $A$, bounded by the curve $y = f(x)$, the $x$-axis and ordinates $x = a$ and $x = b$, is given by:

$$A = \int_{a}^{b} y\,dx$$

## Review exercise

31

1 Find the area bounded by $y = 5 + 4x - x^2$, the $x$-axis and the ordinates $x = 1$ and $x = 4$.

2 Calculate the area under the curve $y = 2x^2 + 4x + 3$, between $x = 2$ and $x = 5$.

3 Determine the area bounded by $y = x^2 - 2x + 3$, the $x$-axis and ordinates $x = -1$ and $x = 3$.

*Finish all three and then check with the next frame*

32

**1**. 24 unit$^2$, **2**. 129 unit$^2$, **3**. $13\frac{1}{3}$ unit$^2$

Here is the working:

**1** $$A = \int_{1}^{4} y\,dx = \int_{1}^{4} (5 + 4x - x^2)\,dx = \left[5x + 2x^2 - \frac{x^3}{3}\right]_1^4$$

$$= \left\{20 + 32 - \frac{64}{3}\right\} - \left\{5 + 2 - \frac{1}{3}\right\}$$

$$= 30\frac{2}{3} - 6\frac{2}{3} = 24 \text{ unit}^2$$

▶

2 $A = \int_2^5 y\,dx = \int_2^5 (2x^2 + 4x + 3)\,dx = \left[2\frac{x^3}{3} + 2x^2 + 3x\right]_2^5$

$= \left\{\frac{250}{3} + 50 + 15\right\} - \left\{\frac{16}{3} + 8 + 6\right\} = 148\frac{1}{3} - 19\frac{1}{3} = 129 \text{ unit}^2$

3 $A = \int_{-1}^3 y\,dx = \int_{-1}^3 (x^2 - 2x + 3)\,dx = \left[\frac{x^3}{3} - x^2 + 3x\right]_{-1}^3$

$= \left\{9 - 9 + 9\right\} - \left\{-\frac{1}{3} - 1 - 3\right\}$

$= 9 - \left\{-4\frac{1}{3}\right\} = 13\frac{1}{3} \text{ unit}^2.$

Notice that in all these definite integrals, we omit the constant of integration because we know it will disappear at the subtraction stage.

In an indefinite integral, however, it must always be included.

*Now move on to the next section*

## Integration as a summation

**33**

We have identified the value of a definite integral as the area beneath a curve. However, some definite integrals have a negative value so how can we link this to an area because all areas are *positive quantities*? Before we can make this link we must consider the determination of area in a slightly different manner.

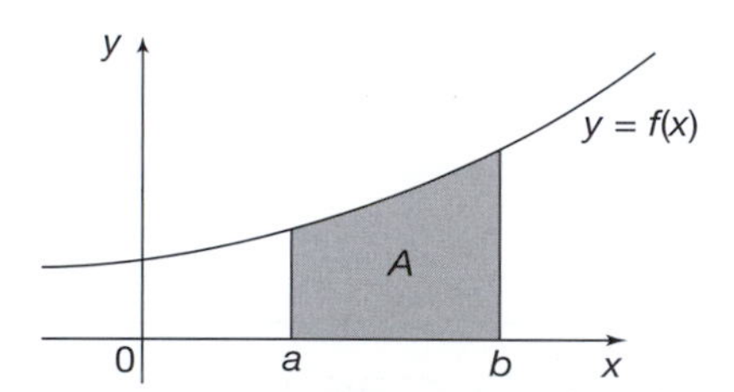


We have already seen that the area $A$ under a curve $y = f(x)$ between $x = a$ and $x = b$ is given by the definite integral:

$$A = \int_a^b y\,dx$$

Let's look at this area a little more closely:

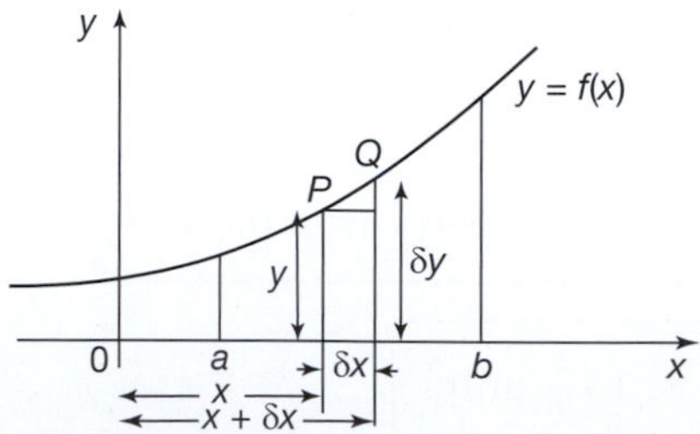


Let $P$ be the point $(x, y)$ on the curve and $Q$ a similar point $(x + \delta x,\ y + \delta y)$. The approximate area $\delta A$ of the strip under the arc $PQ$ is given by

$$\delta A \approx y.\delta x$$

As we have indicated earlier, the error in the approximation is the area above the rectangle.

If we divide the whole figure between $x = a$ and $x = b$ into a number of such strips, the total area is approximately the sum of the areas of all rectangles $y.\delta x$.

i.e. $A \approx$ the sum of all rectangles $y.\delta x$ between $x = a$ and $x = b$. This can be written

$A \approx \sum_{x=a}^{x=b} y.\delta x$ where the symbol $\Sigma$ represents '*the sum of all terms of the form …*'

▶

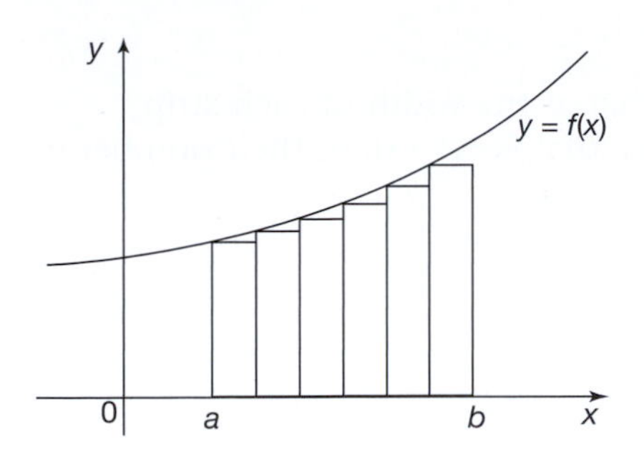


If we make the strips narrower, there will be more of them to cover the whole figure, but the total error in the approximation diminishes.

If we continue the process, we arrive at an infinite number of minutely narrow rectangles, each with an area too small to exist alone.

Then, in the limit as $\delta x \to 0$, $A = \lim_{\delta x \to 0} \sum_{x=a}^{x=b} y.\delta x$

But we already know that $A = \int_a^b y\,dx \quad \therefore \quad \lim_{\delta x \to 0} \sum_{x=a}^{x=b} y.\delta x = A = \int_a^b y\,dx$

*Let us consider an example*

## 34

To illustrate this, consider the area $A$ beneath the straight line $y = x$, above the $x$-axis and between the values $x = 2$ and $x = 4$ as shown in the figure below.

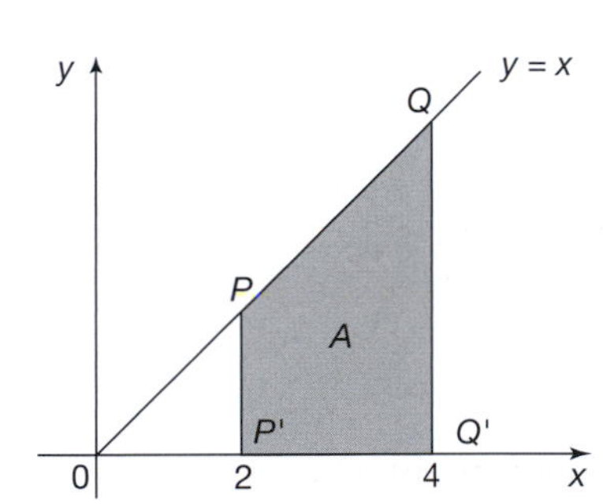


The area of a triangle is

$\frac{1}{2} \times$ base $\times$ height

and area $A$ is equal to the difference in the areas of the two triangles $OQQ'$ and $OPP'$ so that:

$$A = \frac{1}{2} \times 4 \times 4 - \frac{1}{2} \times 2 \times 2$$
$$= 8 - 2 = 6 \text{ units}^2$$

This value will now be confirmed by dividing the area into equal strips, summing their areas and then taking the limit as the width of the strips goes to zero.

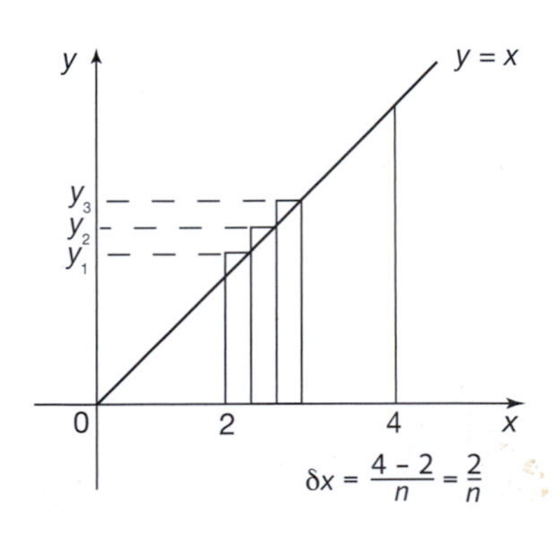


In the figure, the area has been subdivided into $n$ strips each of width $\delta x$ where:

$$\delta x = \frac{4-2}{n} = \frac{2}{n}$$

The strip heights are given as:

$$y_1 = y(2 + \delta x) = y\left(2 + \frac{2}{n}\right) = 2 + \frac{2}{n}$$

$$y_2 = y(2 + 2\delta x) = y\left(2 + 2 \times \frac{2}{n}\right) = 2 + 2 \times \frac{2}{n}$$

$$y_3 = y(2 + 3\delta x) = y\left(2 + 3 \times \frac{2}{n}\right) = 2 + 3 \times \frac{2}{n}$$

......

$$y_r = y(2 + r\delta x) = y\left(2 + r \times \frac{2}{n}\right) = 2 + r \times \frac{2}{n}$$

......

$$y_n = y(2 + n\delta x) = y\left(2 + n \times \frac{2}{n}\right) = 2 + n \times \frac{2}{n} = 4$$

▶

Consequently:

$$\underset{\delta x\to 0}{Lim}\sum_{x=2}^{x=4} y.\delta x = \underset{n\to\infty}{Lim}\sum_{r=1}^{n}\left(2+\frac{2r}{n}\right).\frac{2}{n}$$

Notice that as the width of each strip decreases, that is $\delta x \to 0$, so their number $n$ increases, $n \to \infty$

$$= \underset{n\to\infty}{Lim}\sum_{r=1}^{n}\left(\frac{4}{n}+\frac{4r}{n^2}\right)$$

$$= \underset{n\to\infty}{Lim}\sum_{r=1}^{n}\frac{4}{n} + \underset{n\to\infty}{Lim}\sum_{r=1}^{n}\frac{4r}{n^2}$$

$$= \underset{n\to\infty}{Lim}\frac{4}{n}\sum_{r=1}^{n}1 + \underset{n\to\infty}{Lim}\frac{4}{n^2}\sum_{r=1}^{n}r$$

$$= \underset{n\to\infty}{Lim}\left(\frac{4}{n}\times n\right) + \underset{n\to\infty}{Lim}\left(\frac{4}{n^2}\times\frac{n(n+1)}{2}\right)$$

$$= 4 + 2 = 6 \text{ units}^2$$

Notice the use of $\sum_{r=1}^{n} 1 = n$ and $\sum_{r=1}^{n} r = \frac{n(n+1)}{2}$ from Programme F.7.

Now, without following the above procedure, but by integrating normally, find the area bounded by the curve $y = x^2 - 9$ and the $x$-axis and between $x = -3$ and $x = 3$.

## 35

$A = -36 \text{ unit}^2$

Because

$$A = \int_{-3}^{3} (x^2 - 9)\,dx$$

$$= \left[\frac{x^3}{3} - 9x\right]_{-3}^{3}$$

$$= (9 - 27) - (-9 + 27)$$

$$= -36$$

Simple enough, but what is meant by a negative area?

*Have you any suggestions?*

## 36

The area that lies beneath the $x$-axis

If we sketch the figure we get:

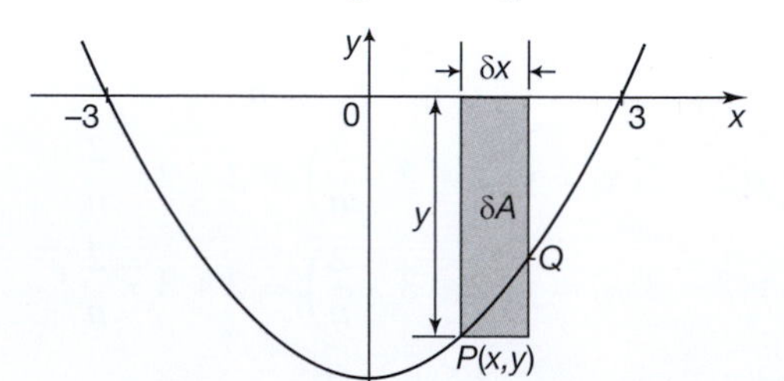


As before, the area of the strip

$$\delta A \approx y\,\delta x$$

and the total area

$$A \approx \sum_{x=-3}^{x=3} y\,\delta x$$

But, for all such strips across the figure, $y$ is negative and $\delta x$ a positive value.

Therefore, in this case, $y.\delta x$ is negative and the sum of all such quantities gives a negative total, even when $\delta x \to 0$.

So $\int_{-3}^{3} y\,dx$ has a negative value, namely *minus the value of the enclosed area.*

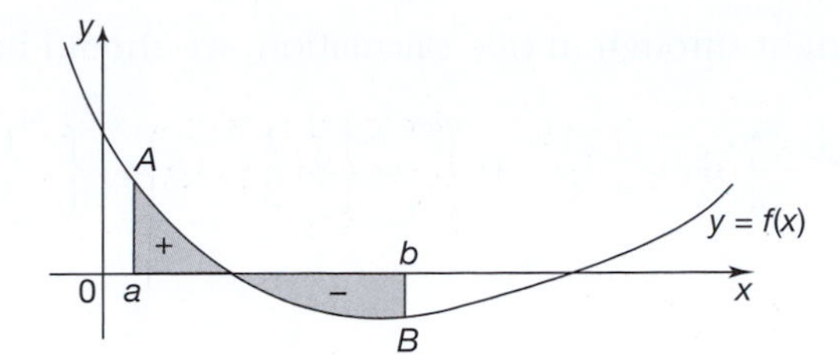


The trouble comes when part of the area to be calculated lies above the $x$-axis and part below. In that case, integration from $x = a$ to $x = b$ gives the algebraic sum of the two parts.

It is always wise, therefore, to sketch the figure of the problem before carrying out the integration and, if necessary, to calculate the positive and negative parts separately and to add the numerical values of each part to obtain the physical area between the limits stated.

As an example, we will determine the physical area of the figure bounded by the curve $y = x^2 - 4$, the $x$-axis and ordinates at $x = -1$ and $x = 4$.

The curve $y = x^2 - 4$ is, of course, the parabola $y = x^2$ lowered 4 units on the $y$-axis. It crosses the $x$-axis when $y = 0$, i.e. $x^2 - 4 = 0$, $x^2 = 4 \therefore x = \pm 2$.

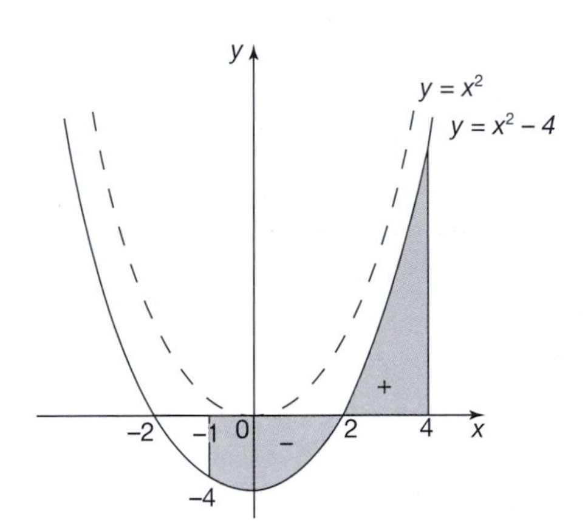


The figure of our problem extends from $x = -1$ to $x = 4$, which includes an area *beneath* the $x$-axis between $x = -1$ and $x = 2$ and an area *above* the $x$-axis between $x = 2$ and $x = 4$. Therefore, we calculate the physical area enclosed in two parts and add the results.

So let $I_1 = \int_{-1}^{2} y\,dx$ and $I_2 = \int_{2}^{4} y\,dx$. Then:

$$I_1 = \int_{-1}^{2} (x^2 - 4)\,dx = \left[\frac{x^3}{3} - 4x\right]_{-1}^{2} = \left\{\frac{8}{3} - 8\right\} - \left\{-\frac{1}{3} + 4\right\}$$

$$= -9 \text{ so } A_1 = 9 \text{ units}^2$$

and

$$I_2 = \int_{2}^{4} y\,dx = \left[\frac{x^3}{3} - 4x\right]_{2}^{4} = \left\{\frac{64}{3} - 16\right\} - \left\{\frac{8}{3} - 8\right\}$$

$$= 10\frac{2}{3} \text{ units}^2$$

Consequently, $A_2 = I_2 = 10\frac{2}{3}$ units$^2$ and so:

$$A = A_1 + A_2 = 19\frac{2}{3} \text{ units}^2$$

▶

Had we integrated right through in one calculation, we should have obtained:

$$I = \int_{-1}^{4} y\,dx = \int_{-1}^{4} (x^2 - 4)\,dx = \left[\frac{x^3}{3} - 4x\right]_{-1}^{4} = \left\{21\frac{1}{3} - 16\right\} - \left\{-\frac{1}{3} + 4\right\}$$

$$= 1\frac{2}{3} \text{ units}^2$$

which, though it does give the correct value of the definite integral, does not give the correct value of the total area enclosed.

*On to the next frame*

37

## The area between a curve and an intersecting line

To find the area enclosed by the curve $y = 25 - x^2$ and the straight line $y = x + 13$.

First we must develop the figure. We know that $y = x^2$ is a normal parabola:

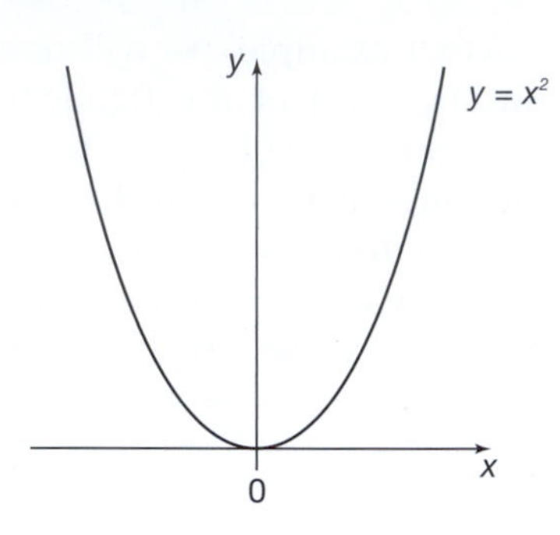


Then $y = -x^2$ is an inverted parabola:

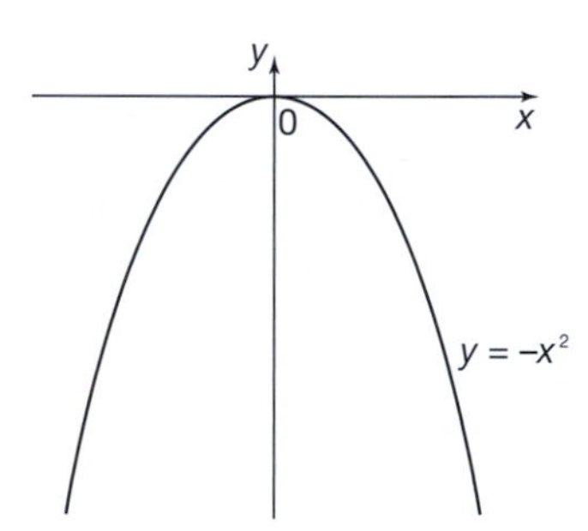


$y = 25 - x^2$ is the inverted parabola raised 25 units on the $y$-scale.

Also when $y = 0$, $x^2 = 25$, so $x = \pm 5$.

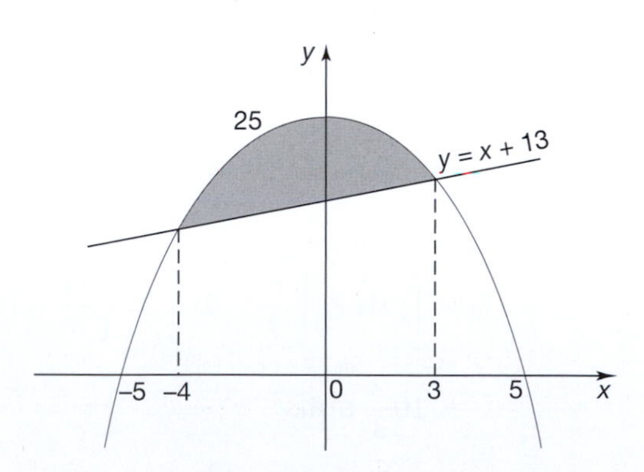


$y = x + 13$ is a straight line crossing $y = 25 - x^2$ when

$x + 13 = 25 - x^2$, i.e. $x^2 + x - 12 = 0$

$\therefore\ (x - 3)(x + 4) = 0 \quad \therefore\ x = -4$ or $3$

So the area $A$ we need is the part shaded.

*On to the next frame*

38

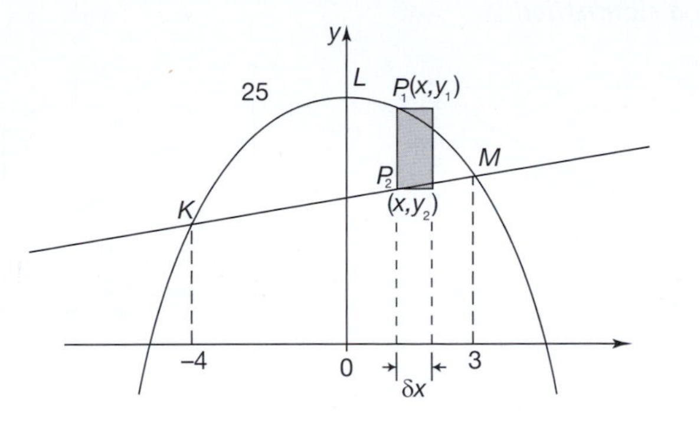


Let $P_1(x, y_1)$ be a point on $y_1 = 25 - x^2$ and $P_2(x, y_2)$ the corresponding point on $y_2 = x + 13$.

Then area of strip

$P_1P_2 \approx (y_1 - y_2).\delta x$

Then the area of the figure

$$KLM \approx \sum_{x=-4}^{x=3} (y_1 - y_2).\delta x$$

$$\text{If } \delta x \to 0,\ A = \int_{-4}^{3} (y_1 - y_2)\,dx$$

$$\therefore\ A = \int_{-4}^{3} (25 - x^2 - x - 13)\,dx = \int_{-4}^{3} (12 - x - x^2)\,dx$$

which you can now finish off, to give $A = \ldots\ldots\ldots$

39

$57{\cdot}2$ unit$^2$

$$A = \int_{-4}^{2} (12 - x - x^2)\,dx = \left[12x - \frac{x^2}{2} - \frac{x^3}{3}\right]_{-4}^{3}$$

$$= \left\{36 - \frac{9}{2} - 9\right\} - \left\{-48 - 8 + \frac{64}{3}\right\}$$

$$= 22{\cdot}5 + 34{\cdot}67 = 57{\cdot}17 \;\therefore\; A = 57{\cdot}2 \text{ unit}^2$$

***At this point let us pause and summarize the main facts dealing with areas and the definite integral***

## Review summary

40

1 *Definite integral*

An integral with limits (e.g. $x = a$ and $x = b$) is called a definite integral. The constant of integration $C$ in such cases will always disappear at the subtraction stage, since

$$\int_a^b y\,dx = \int_{x=b} y\,dx - \int_{x=a} y\,dx$$

▶

2 *Integration as a summation*

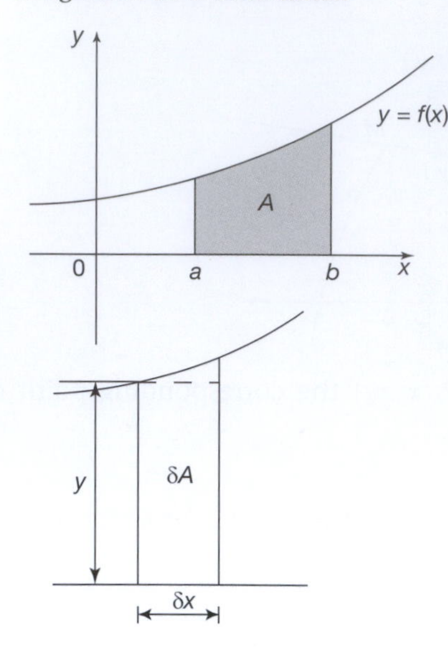


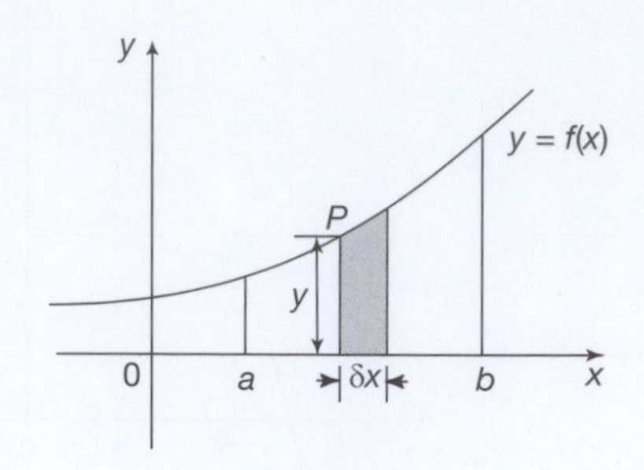


$\delta A \approx y.\delta x \qquad \therefore\ A \approx \sum_{x=a}^{b} y.\delta x$

If $\delta x \to 0 \qquad A = \int_a^b y.\mathrm{d}x$

---

## Review exercise

**41**

1 Evaluate each of the following definite integrals:

(a) $\int_2^4 3x^5\,\mathrm{d}x$ (b) $\int_0^{\pi/2} (\sin x - \cos x)\,\mathrm{d}x$ (c) $\int_0^1 e^{2x}\,\mathrm{d}x$ (d) $\int_{-1}^0 x^3\,\mathrm{d}x$

2 Find the area enclosed between the $x$-axis and the curves:

(a) $y = x^3 + 2x^2 + x + 1$ between $x = -1$ and $x = 2$

(b) $y = x^2 - 25$ for $-5 \leq x \leq 5$.

3 Find the area enclosed between the curve $y = x^3$ and the straight line $y = x$.

**42**

1 (a) $\int_2^4 3x^5\,\mathrm{d}x = \left[3\dfrac{x^6}{6}\right]_2^4$

$= \left\{3\dfrac{4^6}{6}\right\} - \left\{3\dfrac{2^6}{6}\right\}$

$= 2048 - 32 = 2016$

(b) $\int_0^{\pi/2} (\sin x - \cos x)\,\mathrm{d}x = [-\cos x - \sin x]_0^{\pi/2}$

$= \left\{-\cos\dfrac{\pi}{2} - \sin\dfrac{\pi}{2}\right\} - \{-\cos 0 - \sin 0\}$

$= \{-0 - 1\} - \{-1 - 0\}$

$= 0$

▶

(c) $$\int_0^1 e^{2x}\,dx = \left[\frac{e^{2x}}{2}\right]_0^1$$

$$= \left\{\frac{e^2}{2}\right\} - \left\{\frac{e^0}{2}\right\}$$

$$= \frac{e^2}{2} - \frac{1}{2}$$

$$= \frac{e^2 - 1}{2}$$

(d) $$\int_{-1}^{0} x^3\,dx = \left[\frac{x^4}{4}\right]_{-1}^{0}$$

$$= \{0^4/4\} - \{(-1)^4/4\} = -1/4$$

**2** (a) The graph of $y = x^3 + 2x^2 + x + 1$ between $x = -1$ and $x = 2$ lies entirely above the $x$-axis

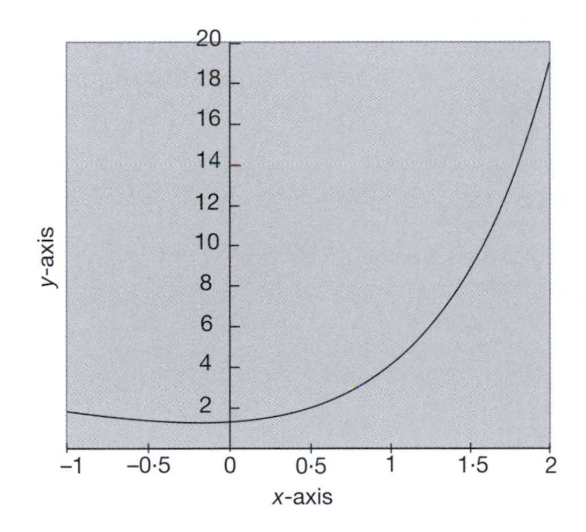


so that the area enclosed between the curve, the $x$-axis and between $x = -1$ and $x = 2$ is given by:

$$\int_{-1}^{2} (x^3 + 2x^2 + x + 1)\,dx$$

$$= \left[\frac{x^4}{4} + 2\frac{x^3}{3} + \frac{x^2}{2} + x\right]_{-1}^{2}$$

$$= \left\{\frac{2^4}{4} + 2.\frac{2^3}{3} + \frac{2^2}{2} + 2\right\} - \left\{\frac{(-1)^4}{4} - 2.\frac{(-1)^3}{3} + \frac{(-1)^2}{2} - 1\right\}$$

$$= \left\{4 + \frac{16}{3} + 2 + 2\right\} - \left\{\frac{1}{4} - \frac{2}{3} + \frac{1}{2} - 1\right\}$$

$$= 13\tfrac{1}{3} + \tfrac{11}{12}$$

$$= 14\tfrac{1}{4}$$

▶

(b) The graph of $y = x^2 - 25$ for $-5 \leq x \leq 5$ lies entirely below the $x$-axis

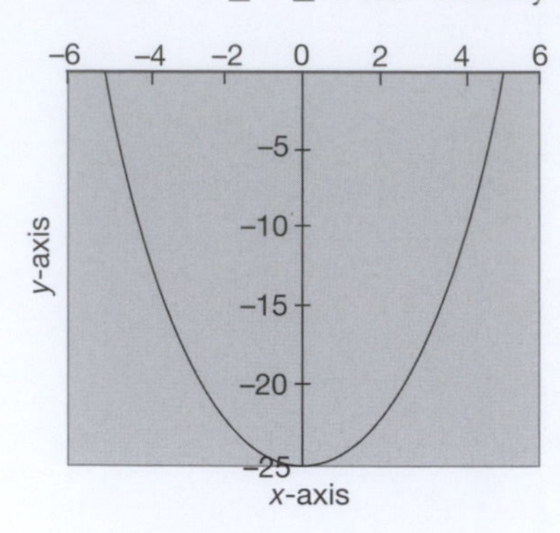


so that the area enclosed between the curve and the $x$-axis for $-5 \leq x \leq 5$ is given by $-I$ where:

$$
\begin{aligned}
I &= \int_{-5}^{5} (x^2 - 25)\,dx \\
&= \left[\frac{x^3}{3} - 25x\right]_{-5}^{5} \\
&= \left\{\frac{5^3}{3} - 25 \times 5\right\} - \left\{\frac{(-5)^3}{3} - 25 \times (-5)\right\} \\
&= \left\{\frac{125}{3} - 125\right\} - \left\{-\frac{125}{3} + 125\right\} \\
&= -\frac{500}{3}
\end{aligned}
$$

So the enclosed area is $-I = \dfrac{500}{3}$ units$^2$

3 The curve $y = x^3$ and the line $y = x$ intersect when $x^3 = x$, that is when $x^3 - x = 0$. Factorizing we see that this means $x(x^2 - 1) = 0$ which is satisfied when $x = 0$, $x = 1$ and $x = -1$.

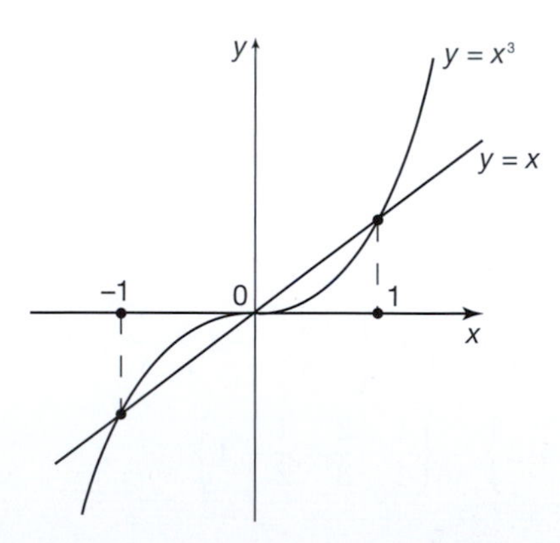


From the figure we can see that the area enclosed between the curve $y = x^3$ and the straight line $y = x$ is in two parts, one above the $x$-axis and the other below. From the figure it is easily seen that both parts are the same, so we only need find the area between $x = 0$ and $x = 1$ and then double it to find the total area enclosed.

▶

The area enclosed between $x = 0$ and $x = 1$ is equal to the area beneath the line $y = x$ minus the area beneath the curve $y = x^3$. That is:

$$\int_0^1 x\,dx - \int_0^1 x^3\,dx = \int_0^1 (x - x^3)\,dx$$

$$= \left[\frac{x^2}{2} - \frac{x^4}{4}\right]_0^1$$

$$= \left\{\frac{1^2}{2} - \frac{1^4}{4}\right\} - \{0\} = \frac{1}{4}$$

The total area enclosed is then twice this, namely $\frac{1}{2}$ unit$^2$.

**43**

You have now come to the end of this Programme. A list of **Can You?** questions follows for you to gauge your understanding of the material in the Programme. You will notice that these questions match the **Learning outcomes** listed at the beginning of the Programme so go back and try the **Quiz** that follows them. After that try the **Test exercise**. *Work through these at your own pace, there is no need to hurry.* A set of **Further problems** provides additional valuable practice.

## Can You?

### Checklist F.13

**44**

Check this list before and after you try the end of Programme test.

On a scale of 1 to 5 how confident are you that you can:

| | | | | | | | Frames |
|---|---|---|---|---|---|---|---|
| • Appreciate that integration is the reverse process of differentiation? | | | | | | | 1 |
| *Yes* | ☐ | ☐ | ☐ | ☐ | ☐ | *No* | |
| • Recognize the need for a constant of integration? | | | | | | | 1 to 2 |
| *Yes* | ☐ | ☐ | ☐ | ☐ | ☐ | *No* | |
| • Evaluate indefinite integrals of standard forms? | | | | | | | 2 |
| *Yes* | ☐ | ☐ | ☐ | ☐ | ☐ | *No* | |
| • Evaluate indefinite integrals of polynomials? | | | | | | | 6 to 11 |
| *Yes* | ☐ | ☐ | ☐ | ☐ | ☐ | *No* | |
| • Evaluate indefinite integrals of 'functions of a linear function of $x$'? | | | | | | | 12 |
| *Yes* | ☐ | ☐ | ☐ | ☐ | ☐ | *No* | |
| • Integrate by partial fractions? | | | | | | | 16 to 22 |
| *Yes* | ☐ | ☐ | ☐ | ☐ | ☐ | *No* | |
| • Appreciate that a definite integral is a measure of an area under a curve? | | | | | | | 26 |
| *Yes* | ☐ | ☐ | ☐ | ☐ | ☐ | *No* | |
| • Evaluate definite integrals of standard forms? | | | | | | | 27 to 29 |
| *Yes* | ☐ | ☐ | ☐ | ☐ | ☐ | *No* | |
| • Use the definite integral to find areas between a curve and the horizontal axis? | | | | | | | 33 to 36 |
| *Yes* | ☐ | ☐ | ☐ | ☐ | ☐ | *No* | |
| • Use the definite integral to find areas between a curve and a given straight line? | | | | | | | 37 to 39 |
| *Yes* | ☐ | ☐ | ☐ | ☐ | ☐ | *No* | |

# Test exercise F.13

45

1 Determine the following integrals: Frames

(a) $\int x^4\,dx$ (f) $\int 5^x\,dx$

(b) $\int 3\sin x\,dx$ (g) $\int x^{-4}\,dx$

(c) $\int 4.e^x\,dx$ (h) $\int \frac{4}{x}\,dx$

(d) $\int 6\,dx$ (i) $\int 3\cos x\,dx$

(e) $\int 3x^{\frac{1}{2}}\,dx$ (j) $\int 2\sec^2 x\,dx$ 1 to 5

2 (a) Determine $\int (8x^3 + 6x^2 - 5x + 4)\,dx$.

(b) If $I = \int (4x^3 - 3x^2 + 6x - 2)\,dx$, determine the value of $I$ when $x = 4$, given that when $x = 2$, $I = 20$. 6 to 11

3 Determine:

(a) $\int 2\sin(3x+1)\,dx$ (e) $\int 4^{2x-3}\,dx$

(b) $\int \sqrt{5-2x}\,dx$ (f) $\int 6\cos(1-2x)\,dx$

(c) $\int 6.e^{1-3x}\,dx$ (g) $\int \frac{5}{3x-2}\,dx$

(d) $\int (4x+1)^3\,dx$ (h) $\int 3\sec^2(1+4x)\,dx$ 12 to 15

4 Express $\frac{2x^2+14x+10}{2x^2+9x+4}$ in partial fractions and hence determine

$$\int \frac{2x^2+14x+10}{2x^2+9x+4}\,dx$$

16 to 25

5 (a) Evaluate the area under the curve $y = 5.e^x$ between $x = 0$ and $x = 3$.

(b) Show that the straight line $y = -2x + 28$ crosses the curve $y = 36 - x^2$ at $x = -2$ and $x = 4$. Hence, determine the area enclosed by the curve $y = 36 - x^2$ and the line $y = -2x + 28$ 26 to 42

## Further problems F.13

46

1 Determine the following:

(a) $\int (5-6x)^2\,dx$ (e) $\int \sqrt{5-3x}\,dx$ (h) $\int 4\cos(1-2x)\,dx$

(b) $\int 4\sin(3x+2)\,dx$ (f) $\int 6.e^{3x+1}\,dx$ (i) $\int 3\sec^2(4-3x)\,dx$

(c) $\int \frac{5}{2x+3}\,dx$ (g) $\int (4-3x)^{-2}\,dx$ (j) $\int 8.e^{3-4x}\,dx$

(d) $\int 3^{2x-1}\,dx$

2 Determine $\int \left(6.e^{3x-5}+\frac{4}{3x-2}-5^{2x+1}\right)dx$.

3 If $I=\int (8x^3+3x^2-6x+7)\,dx$, determine the value of $I$ when $x=-3$, given that when $x=2$, $I=50$.

4 Determine the following using partial fractions:

(a) $\int \frac{6x+1}{4x^2+4x-3}\,dx$ (e) $\int \frac{6x^2-2x-2}{6x^2-7x+2}\,dx$

(b) $\int \frac{x+11}{x^2-3x-4}\,dx$ (f) $\int \frac{4x^2-9x-19}{2x^2-7x-4}\,dx$

(c) $\int \frac{3x-17}{12x^2-19x+4}\,dx$ (g) $\int \frac{6x^2+2x+1}{2x^2+x-6}\,dx$

(d) $\int \frac{20x+2}{8x^2-14x-15}\,dx$ (h) $\int \frac{4x^2-2x-11}{2x^2-7x-4}\,dx$

5 Determine the area bounded by the curve $y=f(x)$, the $x$-axis and the stated ordinates in the following cases:

(a) $y=x^2-3x+4$, $x=0$ and $x=4$

(b) $y=3x^2+5$, $x=-2$ and $x=3$

(c) $y=5+6x-3x^2$, $x=0$ and $x=2$

(d) $y=-3x^2+12x+10$, $x=1$ and $x=4$

(e) $y=x^3+10$, $x=-1$ and $x=2$

6 In each of the following, determine the area enclosed by the given boundaries:

(a) $y=10-x^2$ and $y=x^2+2$

(b) $y=x^2-4x+20$, $y=3x$, $x=0$ and $x=4$

(c) $y=4e^{2x}$, $y=4e^{-x}$, $x=1$ and $x=3$

(d) $y=5e^x$, $y=x^3$, $x=1$ and $x=4$

(e) $y=20+2x-x^2$, $y=\frac{e^x}{2}$, $x=1$ and $x=3$

Now visit the companion website for this book at www.palgrave.com/stroud for more questions applying this mathematics to science and engineering.

47

# PART II

# Complex numbers 1

## Learning outcomes

When you have completed this Programme you will be able to:

- ☐ Recognize $j$ as standing for $\sqrt{-1}$ and be able to reduce powers of $j$ to $\pm j$ or $\pm 1$
- ☐ Recognize that all complex numbers are in the form (real part) $+ j$(imaginary part)
- ☐ Add, subtract and multiply complex numbers
- ☐ Find the complex conjugate of a complex number
- ☐ Divide complex numbers
- ☐ State the conditions for the equality of two complex numbers
- ☐ Draw complex numbers and recognize the parallelogram law of addition
- ☐ Convert a complex number from Cartesian to polar form and vice versa
- ☐ Write a complex number in its exponential form
- ☐ Obtain the logarithm of a complex number

# Introduction

1

## Ideas and symbols

The numerals and their associated arithmetic symbols were devised so that we could make written calculations and records of quantities and measurements. When a grouping of symbols such as $\sqrt{-1}$ arises to which there is no corresponding quantity we must ask ourselves why? Why does the grouping of symbols occur if there is no quantity associated with it? Often, the only way to answer such a question is to accept the grouping of symbols and carry on manipulating with it to see if any new ideas are forthcoming that will help answer the question. This we shall now do.

By the way, because $\sqrt{-1}$ can have no quantity associated with it we call it an **imaginary** number. This is to distinguish it from those numbers to which we can associate quantities; these we call **real** numbers. This is a most unfortunate choice of words because it gives the idea that those numbers we call 'real' are somehow more concrete and proper than their ethereal and elusive cousins that we call 'imaginary'. In fact all numbers are constructs of the human mind and imagination; no one type is more concrete or proper than any other and the numerals we use to describe numbers of any type are just symbols that permit us to manipulate numbers and so communicate ideas of number from one mind to another. When we put those numbers to use as in counting or measuring it is easy to confuse the idea of number and the real world to which that number is applied.

In an earlier Programme it was stated that there are some quadratics that cannot be factorized into two linear factors. In fact all quadratics can be factorized into two linear factors but with the present state of your knowledge you cannot always find them. This will now be remedied.

*Read on and see how*

# The symbol *j*

2

## Quadratic equations

The solutions of the quadratic equation:

$$x^2 - 1 = 0$$

are:

$x = \dots\dots\dots\dots$ and $x = \dots\dots\dots\dots$

*The answers are in the next frame*

3

$$\boxed{x = 1 \text{ and } x = -1}$$

Because:

$x^2 - 1 = 0$ can be written as $x^2 = 1$ by adding 1 to each side of the equation. The solution is then given as:

$$\begin{aligned} x &= 1^{\frac{1}{2}} = \pm\sqrt{1} \\ &= 1 \text{ or } -1 \end{aligned}$$

▶

Try this next one.

The solutions to the quadratic equation:

$x^2 + 1 = 0$

are:

$x =$ ........... and ...........

*Again, the answers are in the next frame*

**4**

$$x = \sqrt{-1} \text{ and } x = -\sqrt{-1}$$

Because

$x^2 + 1 = 0$ can be written as $x^2 = -1$ by subtracting 1 from each side of the equation. The solution is then given as:

$$x = (-1)^{\frac{1}{2}} = \pm\sqrt{-1} = \sqrt{-1} \text{ or } -\sqrt{-1}$$

This notation is rather clumsy so we denote $\sqrt{-1}$ by the symbol $j$. This means that we can write the solution of the equation $x^2 + 1 = 0$ as:

........... and ...........

*Next frame*

**5**

$$x = j \text{ and } x = -j$$

Because

$j = \sqrt{-1}$ and so $-j = -\sqrt{-1}$

*Let's take a closer look at quadratic equations in the next frame*

**6**

We have already seen that the general quadratic equation is of the form:

$ax^2 + bx + c = 0$ with solution $x = \dfrac{-b \pm \sqrt{b^2 - 4ac}}{2a}$

For example, if:

$2x^2 + 9x + 7 = 0$, then we have $a = 2$, $b = 9$, and $c = 7$ and so:

$$x = \frac{-9 \pm \sqrt{81 - 56}}{4} = \frac{-9 \pm \sqrt{25}}{4} = \frac{-9 \pm 5}{4}$$

$$\therefore\ x = -\frac{4}{4} \text{ or } -\frac{14}{4}$$

$$\therefore\ x = -1 \text{ or } -3{\cdot}5$$

That was straightforward enough, but if we solve the equation $5x^2 - 6x + 5 = 0$ in the same way, we get

$$x = \frac{6 \pm \sqrt{36 - 100}}{10} = \frac{6 \pm \sqrt{-64}}{10}$$

and the next stage is now to determine the square root of $-64$.

Is it (a) 8
(b) $-8$
(c) neither?

*Next frame*

**7**

neither

It is, of course, neither, since +8 and −8 are the square roots of 64 and not of −64. In fact:

$$\sqrt{-64} = \sqrt{-1 \times 64} = \sqrt{-1} \times \sqrt{64} = j8 \text{ an } \textit{imaginary} \text{ number}$$

Similarly:

$$\sqrt{-36} = \sqrt{-1}\sqrt{36} = j6$$
$$\sqrt{-7} = \sqrt{-1}\sqrt{7} = j2{\cdot}646$$

So $\sqrt{-25}$ can be written as ...........

**8**

$j5$

We now have a way of finishing off the quadratic equation we started in Frame 6.

$$5x^2 - 6x + 5 = 0 \text{ therefore } x = \frac{6 \pm \sqrt{36 - 100}}{10} = \frac{6 \pm \sqrt{-64}}{10}$$

$$\therefore\ x = \frac{6 \pm j8}{10} \quad \therefore\ x = 0{\cdot}6 \pm j0{\cdot}8$$

$$\therefore\ x = 0{\cdot}6 + j0{\cdot}8 \text{ or } x = 0{\cdot}6 - j0{\cdot}8$$

Just try one yourself to make sure that you have followed all this. The solution of the quadratic $x^2 + x + 1 = 0$ equation is

$x =$ ........... and $x =$ ...........

*Next frame*

**9**

$$x = \frac{-1 + j\sqrt{3}}{2} \text{ and } x = \frac{-1 - j\sqrt{3}}{2}$$

Because:

$x^2 + x + 1 = 0$ so that $a = 1$, $b = 1$ and $c = 1$

Therefore $x = \dfrac{-1 \pm \sqrt{1 - 4}}{2} = \dfrac{-1 \pm j\sqrt{3}}{2}$,

that is $x = \dfrac{-1 + j\sqrt{3}}{2}$ or $x = \dfrac{-1 - j\sqrt{3}}{2}$

Now let's look a little closer at that number $j$.

*Move to the next frame*

# Powers of $j$

## Positive integer powers

10

Because $j$ represents $\sqrt{-1}$, it is clear that:

$$j^2 = -1$$
$$j^3 = (j^2)j = -j$$
$$j^4 = (j^2)^2 = (-1)^2 = 1$$
$$j^5 = (j^4)j = j$$

Note especially the third result: $j^4 = 1$. Every time a factor $j^4$ occurs, it can be replaced by the factor 1, so that $j$ raised to a positive power is reduced to one of the four results $\pm j$ or $\pm 1$. For example:

$$j^9 = (j^4)^2 j = (1)^2 j = 1.j = j$$
$$j^{20} = (j^4)^5 = (1)^5 = 1$$
$$j^{30} = (j^4)^7 j^2 = (1)^7(-1) = 1(-1) = -1$$
$$j^{15} = (j^4)^3 j^3 = (1)^3(-j) = -j$$

So, in the same way:

(a) $j^{42} = \dots\dots\dots\dots$

(b) $j^{12} = \dots\dots\dots\dots$

(c) $j^{11} = \dots\dots\dots\dots$

*The answers are in the next frame*

11

(a) $j^{42} = -1$
(b) $j^{12} = 1$
(c) $j^{11} = -j$

Because:

$$j^{42} = (j^4)^{10} j^2 = (1)^{10}(-1) = -1$$
$$j^{12} = (j^4)^3 = (1)^3 = 1$$
$$j^{11} = (j^4)^2 j^3 = (1)^2(-j) = -j$$

*Now for negative powers in the next frame*

## Negative integer powers

12

Negative powers follow from the reciprocal of $j$. Because

$$j^2 = -1$$

we see that by dividing both sides by $j$ that $j = -\dfrac{1}{j} = -j^{-1}$ so that:

$$j^{-1} = -j$$
$$j^{-2} = (j^2)^{-1} = (-1)^{-1} = -1$$
$$j^{-3} = (j^{-2})\,j^{-1} = (-1)(-j) = j$$
$$j^{-4} = (j^{-2})^2 = (-1)^2 = 1$$

▶

Note again the last result. Every time a factor $j^{-4}$ occurs, it can be replaced by the number 1, so that $j$ raised to a negative power is reduced to one of the four results $\pm j$ or $\pm 1$. For example:

$$j^{-8} = (j^{-4})^2 = (1)^2 = 1$$

$$j^{-19} = (j^{-4})^4\, j^{-3} = (1)^4(j) = j$$

$$j^{-30} = (j^{-4})^7\, j^{-2} = (1)^7(-1) = -1$$

$$j^{-15} = (j^{-4})^3\, j^{-3} = (1)^3(-j) = -j$$

So, in the same way:

(a) $j^{-32} =$ ...........
(b) $j^{-13} =$ ...........
(c) $j^{-23} =$ ...........

*The answers are in the next frame*

**13**

(a) $j^{-32} = 1$
(b) $j^{-13} = -j$
(c) $j^{-23} = j$

Because:

$$j^{-32} = (j^{-4})^8 = (1)^8 = 1$$

$$j^{-13} = (j^{-4})^3\, j^{-1} = (1)^3(-j) = -j$$

$$j^{-23} = (j^{-4})^5\, j^{-3} = (1)^5(j) = j$$

# Complex numbers

**14**

In Frame 8 we saw that the quadratic equation $5x^2 - 6x + 5 = 0$ had two solutions, one of which was $x = 0.6 + j0.8$. This consists of two separate terms, 0.6 and $j0.8$ which cannot be combined any further – it is a **mixture** of the real number 0.6 and the imaginary number $j0.8$. This mixture is called a **complex** number.

The word 'complex' means either:

(a) complicated, or
(b) mixture

*Answer in the next frame*

**15**

mixture

A complex number is a mixture of a real number and an imaginary number. The symbol $z$ is used to denote a complex number.

In the complex number $z = 3 + j5$ the number 3 is called the *real part* of $z$ and is denoted as $\text{Re}(z)$. The number 5 is called the *imaginary part* of $z$ and is denoted $\text{Im}(z)$.

So $\text{Re}(3 + j5) = 3$ and $\text{Im}(3 + j5) = 5$

▶

**Notice**: the imaginary part of a complex number is a real number so that:

$\text{Im}(3 + j5) = 5$ and not $j5$

So, a complex number $z = \text{Re}(z) + j\,\text{Im}(z)$.

In the complex number $z = 2 + j7$ the real and imaginary parts are

$\text{Re}(z) = \ldots\ldots\ldots\ldots$ and $\text{Im}(z) = \ldots\ldots\ldots\ldots$

*Go to the next frame*

---

**16**

$\text{Re}(z) = 2$ and $\text{Im}(z) = 7$ (*not* $j7$)

Complex numbers have many applications in engineering and science. To use them, we must know how to carry out the usual arithmetical operations.

## Addition and subtraction

This is easy, as a few examples will show:

$(4 + j5) + (3 - j2)$

Although the real and imaginary parts cannot be combined, we can remove the brackets and total up terms of the same kind:

$$\begin{aligned}(4 + j5) + (3 - j2) &= 4 + j5 + 3 - j2 = (4 + 3) + j(5 - 2)\\ &= 7 + j3\end{aligned}$$

Another example:

$$\begin{aligned}(4 + j7) - (2 - j5) &= 4 + j7 - 2 + j5 = (4 - 2) + j(7 + 5)\\ &= 2 + j12\end{aligned}$$

So in general, $(a + jb) + (c + jd) = (a + c) + j(b + d)$
Now you do this one:

$(5 + j7) + (3 - j4) - (6 - j3) = \ldots\ldots\ldots\ldots$

---

**17**

$2 + j6$

since $(5 + j7) + (3 - j4) - (6 - j3)$

$$\begin{aligned}&= 5 + j7 + 3 - j4 - 6 + j3\\ &= (5 + 3 - 6) + j(7 - 4 + 3) = 2 + j6\end{aligned}$$

Now you do these in the same way:

(a) $(6 + j5) - (4 - j3) + (2 - j7) = \ldots\ldots\ldots\ldots$

and (b) $(3 + j5) - (5 - j4) - (-2 - j3) = \ldots\ldots\ldots\ldots$

---

**18**

(a) $4 + j$ (b) $j12$

Here is the working:

(a) $(6 + j5) - (4 - j3) + (2 - j7)$

$$\begin{aligned}&= 6 + j5 - 4 + j3 + 2 - j7\\ &= (6 - 4 + 2) + j(5 + 3 - 7) = 4 + j\end{aligned}$$

▶

(b) $(3+j5)-(5-j4)-(-2-j3)$
$=3+j5-5+j4+2+j3$ (Take care with signs!)
$=(3-5+2)+j(5+4+3)$
$=0+j12=j12$

This is very easy then, so long as you remember that the real and the imaginary parts must be treated quite separately – just like $x$'s and $y$'s in an algebraic expression.

*On to Frame 19*

**19**

## Multiplication

Take as an example: $(3+j4)(2+j5)$

These are multiplied together in just the same way as you would determine the product $(3x+4y)(2x+5y)$.

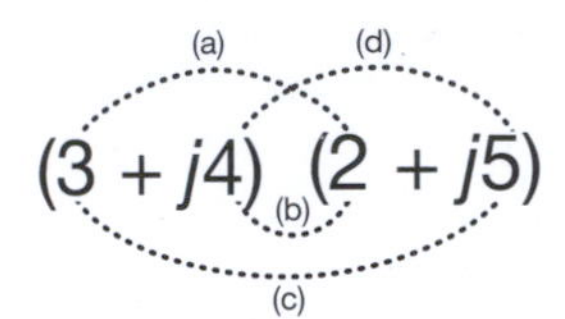


Form the product terms of

(a) the two left-hand terms
(b) the two inner terms
(c) the two outer terms
(d) the two right-hand terms

$$=6+j8+j15+j^2 20$$
$$=6+j23-20 \quad (\text{since } j^2=-1)$$
$$=-14+j23$$

Likewise, $(4-j5)(3+j2)$ . . . . . . . . . . .

**20**

$$\boxed{22-j7}$$

Because

$$(4-j5)(3+j2)=12-j15+j8-j^2 10$$
$$=12-j7+10 \quad (j^2=-1)$$
$$=22-j7$$

If the expression contains more than two factors, we multiply the factors together in stages:

$$(3+j4)(2-j5)(1-j2)=(6+j8-j15-j^2 20)(1-j2)$$
$$=(6-j7+20)(1-j2)$$
$$=(26-j7)(1-j2)$$
$$= \ldots\ldots\ldots$$

*Finish it off*

**21**

$$\boxed{12-j59}$$

Because

$$(26-j7)(1-j2)=26-j7-j52+j^2 14$$
$$=26-j59-14=12-j59$$

Note that when we are dealing with complex numbers, the result of our calculations is also, in general, a complex number.

▶

Now you do this one on your own.

$(5 + j8)(5 - j8) = \ldots\ldots\ldots\ldots$

**22**

89

Here it is:

$$\begin{aligned}(5 + j8)(5 - j8) &= 25 + j40 - j40 - j^2 64 \\ &= 25 + 64 \\ &= 89\end{aligned}$$

In spite of what we said above, here we have a result containing no $j$ term. The result is therefore entirely real.

This is rather an exceptional case. Look at the two complex numbers we have just multiplied together. Can you find anything special about them? If so, what is it?

*When you have decided, move on to the next frame*

**23**

They are identical except for the middle sign in the brackets, i.e. $(5 + j8)$ and $(5 - j8)$

A pair of complex numbers like these are called *conjugate* complex numbers and *the product of two conjugate complex numbers is always entirely real.*

Look at it this way:

$(a + b)(a - b) = a^2 - b^2$ Difference of two squares

Similarly

$$\begin{aligned}(5 + j8)(5 - j8) &= 5^2 - (j8)^2 = 5^2 - j^2 8^2 \\ &= 5^2 + 8^2 \quad \text{Sum of two squares} \\ &= 25 + 64 = 89\end{aligned}$$

Without actually working it out, will the product of $(7 - j6)$ and $(4 + j3)$ be

(a) a real number

(b) an imaginary number

(c) a complex number?

**24**

A complex number

since $(7 - j6)(4 + j3)$ is a product of two complex numbers which are *not* conjugate complex numbers or multiples of conjugates.

**Remember:** Conjugate complex numbers are identical except for the signs in the middle of the brackets.

| | | | |
|---|---|---|---|
| | $(4 + j5)$ and $(4 - j5)$ | *are* | conjugate complex numbers |
| | $(a + jb)$ and $(a - jb)$ | *are* | conjugate complex numbers |
| but | $(6 + j2)$ and $(2 + j6)$ | *are not* | conjugate complex numbers |
| | $(5 - j3)$ and $(-5 + j3)$ | *are not* | conjugate complex numbers |

So what must we multiply $(3 - j2)$ by, to produce a result that is entirely real?

## 25

$(3 + j2)$ or a multiple of it.

because the conjugate of $(3 - j2)$ is identical to it, except for the middle sign, i.e. $(3 + j2)$, and we know that the product of two *conjugate* complex numbers is always real.

Here are two examples:

$$(3 - j2)(3 + j2) = 3^2 - (j2)^2 = 9 - j^2 4$$
$$= 9 + 4 = 13$$
$$(2 + j7)(2 - j7) = 2^2 - (j7)^2 = 4 - j^2 49$$
$$= 4 + 49 = 53$$

. . . and so on.

Complex numbers of the form $(a + jb)$ and $(a - jb)$ are called . . . . . . . . . . . complex numbers.

## 26

conjugate

Now you should have no trouble with these:

(a) Write down the following products

(i) $(4 - j3)(4 + j3)$ (ii) $(4 + j7)(4 - j7)$

(iii) $(a + jb)(a - jb)$ (iv) $(x - jy)(x + jy)$

(b) Multiply $(3 - j5)$ by a suitable factor to give a product that is entirely real.

*When you have finished, move on to Frame 27*

## 27

Here are the results in detail:

(a) (i) $(4 - j3)(4 + j3) = 4^2 - j^2 3^2 = 16 + 9 = \boxed{25}$

(ii) $(4 + j7)(4 - j7) = 4^2 - j^2 7^2 = 16 + 49 = \boxed{65}$

(iii) $(a + jb)(a - jb) = a^2 - j^2 b^2 = \boxed{a^2 + b^2}$

(iv) $(x - jy)(x + jy) = x^2 - j^2 y^2 = \boxed{x^2 + y^2}$

(b) To obtain a real product, we can multiply $(3 - j5)$ by its conjugate, i.e. $(3 + j5)$, giving:

$$(3 - j5)(3 + j5) = 3^2 - j^2 5^2 = 9 + 25 = \boxed{34}$$

*Now move on to the next frame*

## 28

### Division

Now let us deal with division.

Division of a complex number by a real number is easy enough:

$$\frac{5 - j4}{3} = \frac{5}{3} - j\frac{4}{3} = 1{\cdot}67 - j1{\cdot}33 \text{ (to 2 dp)}$$

But how do we manage with $\dfrac{7 - j4}{4 + j3}$?

▶

If we could, somehow, convert the denominator into a real number, we could divide out as in the example above. So our problem is really, how can we convert $(4 + j3)$ into a completely real denominator – and this is where our last piece of work comes in.

We know that we can convert $(4 + j3)$ into a completely real number by multiplying it by its ...........

---

**29**

conjugate

i.e. the same complex number but with the opposite sign in the middle, in this case $(4 - j3)$.

But if we multiply the denominator by $(4 - j3)$, we must also multiply the numerator by the same factor:

$$\frac{7 - j4}{4 + j3} = \frac{(7 - j4)(4 - j3)}{(4 + j3)(4 - j3)} = \frac{28 - j37 - 12}{16 + 9} = \frac{16 - j37}{25}$$

$$\frac{16}{25} - j\frac{37}{25} = 0{\cdot}64 - j1{\cdot}48$$

and the job is done. To divide one complex number by another, therefore, we multiply numerator and denominator by the conjugate of the denominator. This will convert the denominator into a real number and the final step can then be completed.

Thus, to simplify $\frac{4 - j5}{1 + j2}$, we shall multiply top and bottom by ...........

---

**30**

the conjugate of the denominator, i.e. $(1 - j2)$

If we do that, we get:

$$\frac{4 - j5}{1 + j2} = \frac{(4 - j5)(1 - j2)}{(1 + j2)(1 - j2)} = \frac{4 - j13 - 10}{1 + 4}$$

$$= \frac{-6 - j13}{5} = \frac{-6}{5} - j\frac{13}{5} = -1{\cdot}2 - j2{\cdot}6$$

Now here is one for you to do: Simplify $\frac{3 + j2}{1 - j3}$

*When you have done it, move on to the next frame*

---

**31**

$-0{\cdot}3 + j1{\cdot}1$

Because

$$\frac{3 + j2}{1 - j3} = \frac{(3 + j2)(1 + j3)}{(1 - j3)(1 + j3)} = \frac{3 + j11 - 6}{1 + 9} = \frac{-3 + j11}{10} = -0{\cdot}3 + j1{\cdot}1$$

Now do these in the same way:

(a) $\frac{4 - j5}{2 - j}$ (b) $\frac{3 + j5}{5 - j3}$ (c) $\frac{(2 + j3)(1 - j2)}{3 + j4}$

*When you have worked these, move on to Frame 32 to check your results*

**32**

Here are the solutions in detail:

(a) $\dfrac{4-j5}{2-j}=\dfrac{(4-j5)(2+j)}{(2-j)(2+j)}=\dfrac{8-j6+5}{4+1}=\dfrac{13-j6}{5}=\boxed{2{\cdot}6-j1{\cdot}2}$

(b) $\dfrac{3+j5}{5-j3}=\dfrac{(3+j5)(5+j3)}{(5-j3)(5+j3)}=\dfrac{15+j34-15}{25+9}=\dfrac{j34}{34}=\boxed{j}$

(c) $\dfrac{(2+j3)(1-j2)}{(3+j4)}=\dfrac{2-j+6}{3+j4}=\dfrac{8-j}{3+j4}$

$$=\frac{(8-j)(3-j4)}{(3+j4)(3-j4)}$$

$$=\frac{24-j35-4}{9+16}=\frac{20-j35}{25}$$

$$=\boxed{0{\cdot}8-j1{\cdot}4}$$

*And now you know how to add, subtract, multiply and divide complex numbers*

**33**

## Equal complex numbers

Now let us see what we can find out about two complex numbers which we are told are equal.

| | |
|---|---|
| Let the numbers be | $a+jb$ and $c+jd$ |
| Then we have | $a+jb=c+jd$ |
| Rearranging terms, we get | $a-c=j(d-b)$ |

In this last statement, the quantity on the left-hand side is entirely real, while that on the right-hand side is entirely imaginary, i.e. a real quantity equals an imaginary quantity! This seems contradictory and in general it just cannot be true. But there is *one* special case for which the statement can be true. That is when ...........

**34**

each side is zero

$a-c=j(d-b)$

can be true only if $a-c=0$, i.e. $a=c$
and if $d-b=0$, i.e. $b=d$

So we get this important result:
If two complex numbers are equal

(a) the two real parts are equal

(b) the two imaginary parts are equal

For example, if $x+jy=5+j4$, then we know $x=5$ and $y=4$

and if $a+jb=6-j3$, then $a=$ ........... and $b=$ ...........

**35**

$a = 6$ and $b = -3$

Be careful to include the sign!

Now what about this one?

If $(a + b) + j(a - b) = 7 + j2$, find the values of $a$ and $b$.

Well now, following our rule about two equal complex numbers, what can we say about $(a + b)$ and $(a - b)$?

**36**

$a + b = 7$ and $a - b = 2$

since the two real parts are equal and the two imaginary parts are equal.

This gives you two simultaneous equations, from which you can determine the values of $a$ and $b$.

So what are they?

**37**

$a = 4{\cdot}5$; $b = 2{\cdot}5$

For $\left.\begin{array}{l} a + b = 7 \\ a - b = 2 \end{array}\right\} \begin{array}{ll} 2a = 9 & \therefore\ a = 4{\cdot}5 \\ 2b = 5 & \therefore\ b = 2{\cdot}5 \end{array}$

We see then that an equation involving complex numbers leads to a pair of simultaneous equations by putting

(a) the two real parts equal

(b) the two imaginary parts equal

This is quite an important point to remember.

## Review exercise

**38**

1 Simplify (a) $j^{12}$ (b) $j^{10}$ (c) $j^{-7}$

2 Simplify:

(a) $(5 - j9) - (2 - j6) + (3 - j4)$

(b) $(6 - j3)(2 + j5)(6 - j2)$

(c) $(4 - j3)^2$

(d) $(5 - j4)(5 + j4)$

3 Multiply $(4 - j3)$ by an appropriate factor to give a product that is entirely real. What is the result?

4 Simplify $\dfrac{3 - j2}{1 + j}$ into the form $a + jb$

5 If $\dfrac{2 - a - j5}{1 - jb} = 3 - j5$ find $a$ and $b$

*When you have completed this exercise, move on to Frame 39*

**39**

Here are the results. Check yours.

1 (a) $j^{12} = (j^4)^3 = 1^3 = \boxed{1}$

(b) $j^{10} = (j^4)^2 j^2 = 1^2(-1) = \boxed{-1}$

(c) $j^{-7} = (j^{-4})^2 j = (1)^2 j = \boxed{j}$

2 (a) $(5 - j9) - (2 - j6) + (3 - j4)$

$= 5 - j9 - 2 + j6 + 3 - j4$

$= (5 - 2 + 3) + j(6 - 9 - 4) = \boxed{6 - j7}$

(b) $(6 - j3)(2 + j5)(6 - j2)$

$= (12 - j6 + j30 - j^2 15)(6 - j2)$

$= (27 + j24)(6 - j2)$

$= 162 + j144 - j54 + 48 = \boxed{210 + j90}$

(c) $(4 - j3)^2 = 16 - j24 - 9$

$= \boxed{7 - j24}$

(d) $(5 - j4)(5 + j4)$

$= 25 - j^2 16 = 25 + 16 = \boxed{41}$

3 A suitable factor is the conjugate of the given complex number:

$(4 - j3)(4 + j3) = 16 + 9 = \boxed{25}$

4
$$\frac{3 - j2}{1 + j} = \frac{(3 - j2)(1 - j)}{(1 + j)(1 - j)}$$
$$= \frac{3 - j3 - j2 - 2}{1 + 1}$$
$$= \frac{1 - j5}{2}$$
$$= \frac{1}{2} - j\frac{5}{2}$$

5
$$\frac{2 - a - j5}{1 - jb} = 3 - j5 \quad \text{so} \quad 2 - a - j5 = (3 - j5)(1 - jb)$$
$$= 3 - j3b - j5 - 5b$$
$$= 3 - 5b - j(5 + 3b)$$

Therefore $2 - a - j5 = 3 - 5b - j(5 + 3b)$ so

$$3 - 5b = 2 - a$$
$$5 + 3b = 5$$

$$9 - 15b = 6 - 3a$$
$$25 + 15b = 25$$
$$34 \qquad = 31 - 3a$$

so $a = -1$ and $b = 0$

All correct?

*Now move on to the next frame to continue the Programme*

# Graphical representation of a complex number

40

## Argand diagram

So far we have looked at the algebra of complex numbers, now we want to look at the geometry. A real number can be graphically represented as a point on a line – the *real line*. By using a Cartesian coordinate system a pair of real numbers can be graphically represented by a point in the plane. In 1806, the French mathematician Jean-Robert Argand devised a means of representing a complex number using the same Cartesian coordinate system. He plotted the pair of real numbers $(a, b)$ of $z = a + jb$ as a point in the plane and then joined that point to the origin with a straight line.

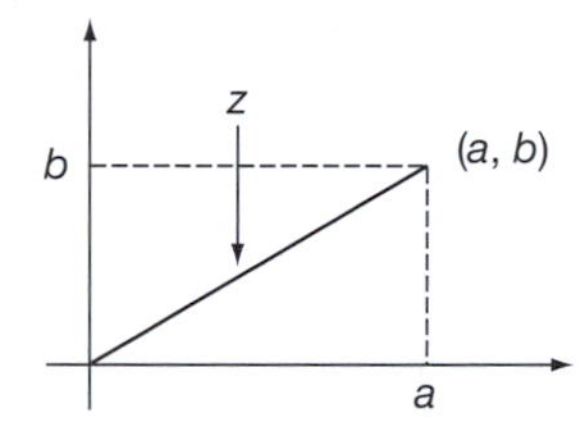


This straight line is then the graphical representation of the complex number $z = a + jb$ and the plane it is plotted against is referred to as the **complex plane**. The entire diagram is called an Argand diagram. The two axes are referred to as the real and imaginary axes respectively but note that it is *real numbers* that are plotted on both axes.

Draw an Argand diagram to represent the complex numbers:

(a) $z_1 = 2 + j3$ (b) $z_2 = -3 + j2$

(c) $z_3 = 4 - j3$ (d) $z_4 = -4 - j5$

*Label each one clearly*

41

Here they are. Check yours.

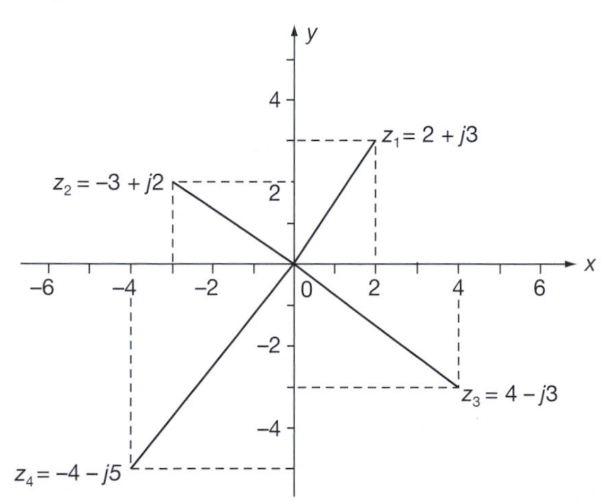


Note once again that the end of each complex number is plotted very much like plotting $x$ and $y$ coordinates.

The real part corresponds to the $x$-value.

The imaginary part corresponds to the $y$-value.

*Move on to Frame 42*

**42**

## Graphical addition of complex numbers

If we draw the two complex numbers $z_1 = 5 + j2$ and $z_2 = 2 + j3$ on an Argand diagram we can then construct the parallelogram formed from the two adjacent sides $z_1$ and $z_2$ as shown.

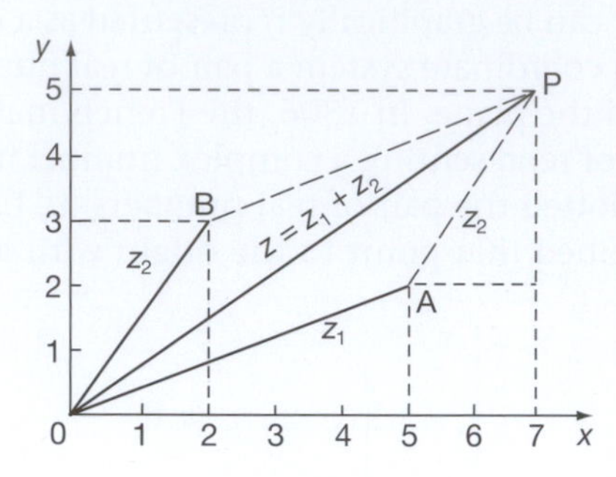


If the diagonal of the parallelogram OP represents the complex number $z = a + jb$, the values of $a$ and $b$ are, from the diagram:

$a = \dots\dots\dots\dots$

$b = \dots\dots\dots\dots$

*Next frame*

**43**

$$a = 5 + 2 = 7 \qquad b = 2 + 3 = 5$$

Therefore $OP = z = 7 + j5$.

$$\begin{aligned}\text{Since } z_1 + z_2 &= (5 + j2) + (2 + j3)\\ &= 7 + j5\\ &= z\end{aligned}$$

we can see that on an Argand diagram the sum of two complex numbers $z_1 + z_2$ is given by the ........... of the parallelogram formed from the two adjacent sides $z_1$ and $z_2$.

*Next frame*

**44**

diagonal

How do we do subtraction by similar means? We do this rather craftily without learning any new methods. The trick is simply this:

$$z_1 - z_2 = z_1 + (-z_2)$$

That is, we draw the lines representing $z_1$ and the *negative* of $z_2$ and add them as before. The negative of $z_2$ is simply a line with the same magnitude (or length) as $z_2$ but in the opposite direction.

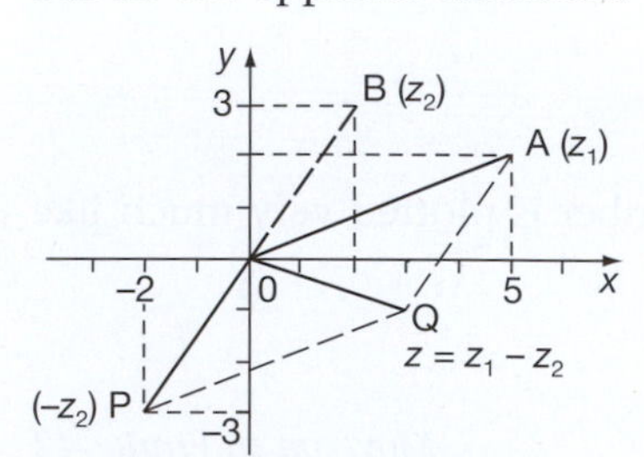


e.g. If $z_1 = 5 + j2$ and $z_2 = 2 + j3$

$$\begin{aligned}OA &= z_1 = 5 + j2\\ OP &= -z_2 = -(2 + j3)\\ \text{Then } OQ &= z_1 + (-z_2)\\ &= z_1 - z_2\end{aligned}$$

Determine on an Argand diagram

$$(4 + j2) + (-2 + j3) - (-1 + j6) = \dots\dots\dots\dots$$

45

$\boxed{3 - j}$

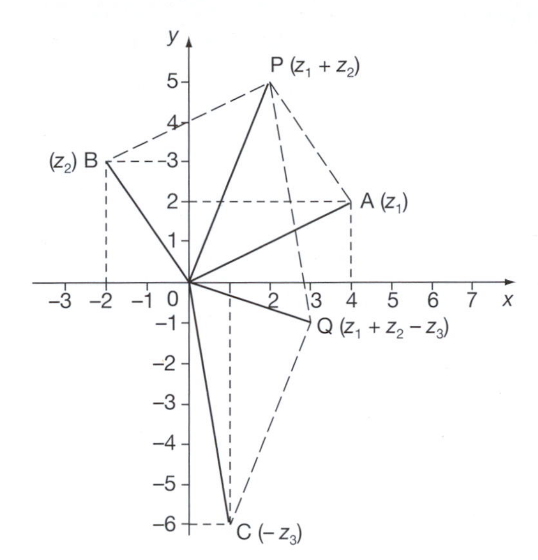


$$\text{OA} = z_1 = 4 + j2$$
$$\text{OB} = z_2 = -2 + j3$$
$$\text{OC} = -z_3 = 1 - j6$$

Then $\text{OP} = z_1 + z_2$

$$\text{OQ} = z_1 + z_2 - z_3 = 3 - j$$

## Polar form of a complex number

46

It is convenient sometimes to express a complex number $a + jb$ in a different form. On an Argand diagram, let OP be a complex number $a + jb$. Let $r =$ length of the complex number and $\theta$ the angle made with the $x$-axis.

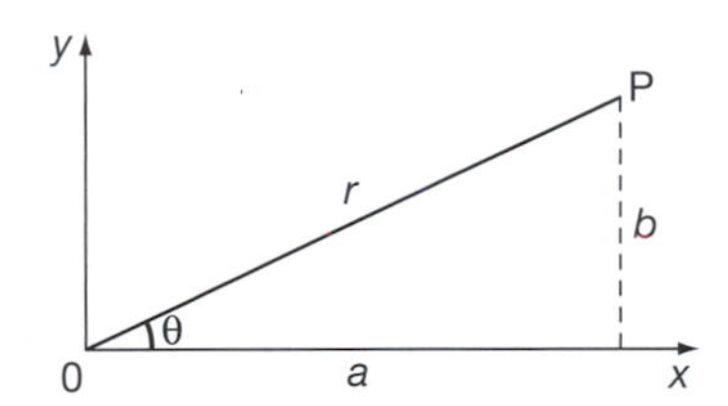


Then $\quad r^2 = a^2 + b^2 \qquad r = \sqrt{a^2 + b^2}$

and $\quad \tan\theta = \dfrac{b}{a} \qquad \theta = \tan^{-1}\dfrac{b}{a}$

Also $\quad a = r\cos\theta$ and $b = r\sin\theta$

Since $z = a + jb$, this can be written

$$z = r\cos\theta + jr\sin\theta$$

i.e. $z = r(\cos\theta + j\sin\theta)$

This is called the *polar form* of the complex number $a + jb$, where

$$r = \sqrt{a^2 + b^2}$$

and $\theta = \tan^{-1}\dfrac{b}{a}$

Let us take a numerical example.

*Next frame*

## 47

To express $z = 4 + j3$ in polar form.

First draw a sketch diagram (that always helps).

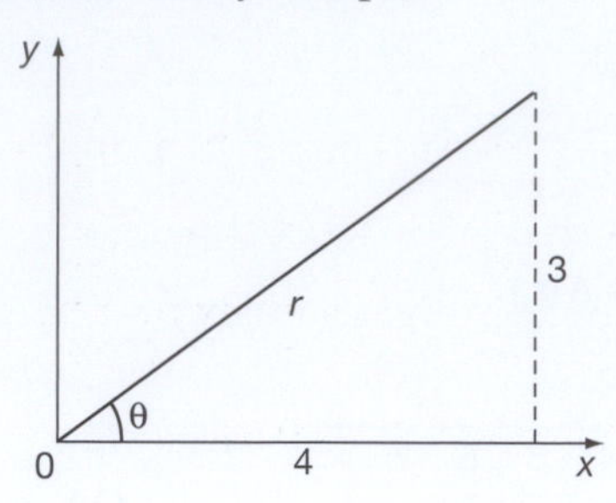


We can see that:

(a) $r^2 = 4^2 + 3^2 = 16 + 9 = 25$

$r = 5$

(b) $\tan\theta = \dfrac{3}{4} = 0{\cdot}75$

$\theta = 36°52'$

$z = a + jb = r(\cos\theta + j\sin\theta)$

So in this case $z = 5(\cos 36°52' + j\sin 36°52')$

Now here is one for you to do.

Find the polar form of the complex number $(2 + j3)$.

*When you have finished it, consult the next frame*

## 48

$$z = 3{\cdot}606(\cos 56°19' + j\sin 56°19')$$

Here is the working:

$z = 2 + j3 = r(\cos\theta + j\sin\theta)$

$r^2 = 4 + 9 = 13 \qquad r = 3{\cdot}606$ (to 3 dp)

$\tan\theta = \dfrac{3}{2} = 1{\cdot}5 \qquad \theta = 56°19'$

$z = 3{\cdot}606(\cos 56°19' + j\sin 56°19')$

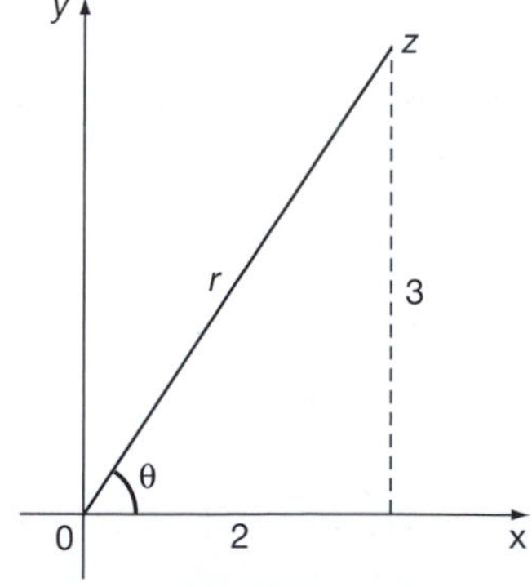


We have special names for the values of $r$ and $\theta$:

$z = a + jb = r(\cos\theta + j\sin\theta)$

(a) $r$ is called the *modulus* of the complex number $z$ and is often abbreviated to 'mod $z$' or indicated by $|z|$.

Thus if $z = 2 + j5$, then $|z| = \sqrt{2^2 + 5^2} = \sqrt{4 + 25} = \sqrt{29}$

(b) $\theta$ is called the *argument* of the complex number and can be abbreviated to 'arg $z$'.

So if $z = 2 + j5$, then $\arg z = \dots\dots\dots\dots$

**49**

$\arg z = 68°12'$

$z = 2 + j5$. Then $\arg z = \theta = \tan^{-1}\dfrac{5}{2} = 68°12'$

*Warning*: In finding $\theta$, there are of course two angles between $0°$ and $360°$, the tangent of which has the value $\dfrac{b}{a}$. We must be careful to use the angle in the correct quadrant. *Always* draw a sketch of the complex number to ensure you have the right one.

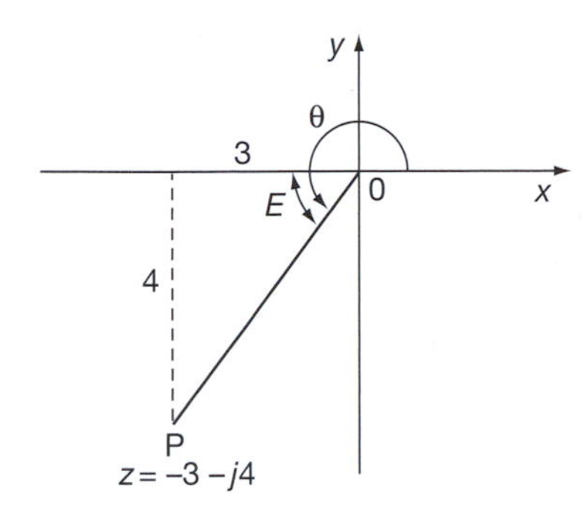


e.g. Find $\arg z$ when $z = -3 - j4$

$\theta$ is measured from O$x$ to OP. We first find $E$, the equivalent acute angle from the triangle shown:

$\tan E = \dfrac{4}{3} = 1{\cdot}333\ldots \qquad \therefore\ E = 53°8'$

Then in this case:

$\theta = 180° + E = 233°8' \qquad \arg z = 233°8'$

Now you find $\arg(-5 + j2)$

*Move on when you have finished*

**50**

$\arg z = 158°12'$

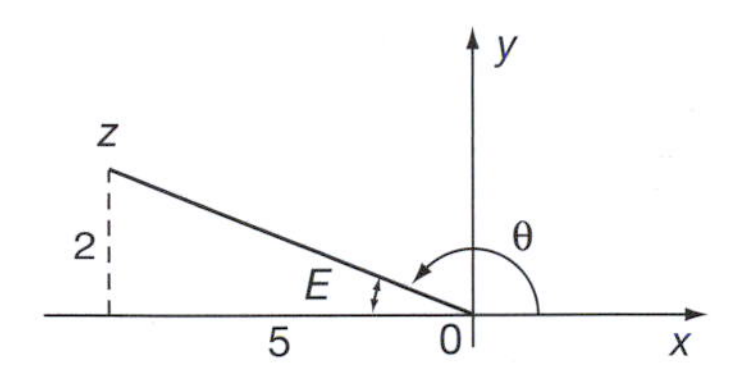


$z = -5 + j2$

$\tan E = \dfrac{2}{5} = 0{\cdot}4 \qquad \therefore\ E = 21°48'$

In this particular case, $\theta = 180° - E$

$\therefore\ \theta = 158°12'$

Complex numbers in polar form are always of the same shape and differ only in the actual values of $r$ and $\theta$. We often use the shorthand version $r\angle\theta$ to denote the polar form.

e.g. If $z = -5 + j2$, $r = \sqrt{25 + 4} = \sqrt{29} = 5{\cdot}385$ and from above $\theta = 158°12'$

$\therefore$ The full polar form is $z = 5{\cdot}385(\cos 158°12' + j\sin 158°12')$ and this can be shortened to $z = 5{\cdot}385\angle 158°12'$

Express in shortened form, the polar form of $(4 - j3)$

Do not forget to draw a sketch diagram first.

**51**

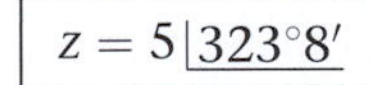


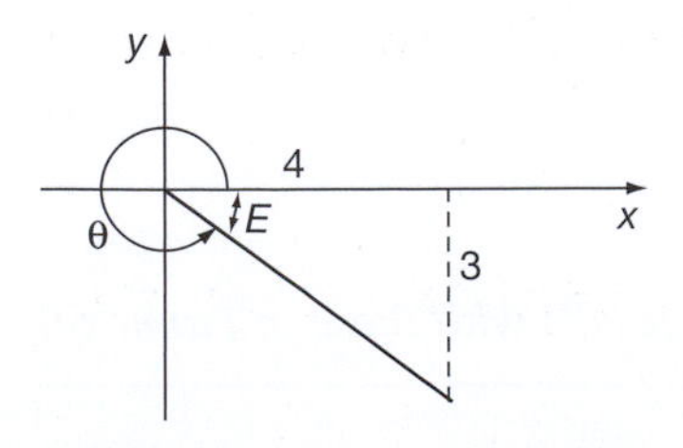


$r = \sqrt{4^2 + 3^2} \qquad r = 5$

$\tan E = 0{\cdot}75 \qquad \therefore\ E = 36°52'$

$\therefore\ \theta = 360° - E = 323°8'$

$\therefore\ z = 5(\cos 323°8' + j\sin 323°8')$

$= 5\angle 323°8'$

▶

Of course, given a complex number in polar form, you can convert it into basic form $a + jb$ simply by evaluating the cosine and the sine and multiplying by the value of $r$.

e.g. $z = 5(\cos 35° + j\sin 35°) = 5(0{\cdot}8192 + j0{\cdot}5736)$

$z = 4{\cdot}0958 + j2{\cdot}8679$ (to 4 dp)

Now you do this one.

Express in the form $a + jb$, $4(\cos 65° + j\sin 65°)$

## 52

$$z = 1{\cdot}6905 + j3{\cdot}6252$$

Because

$$\begin{aligned} z &= 4(\cos 65° + j\sin 65°) \\ &= 4(0{\cdot}4226 + j0{\cdot}9063) \\ &= 1{\cdot}6905 + j3{\cdot}6252 \text{ (to 4 dp)} \end{aligned}$$

If the argument is greater than 90°, care must be taken in evaluating the cosine and sine to include the appropriate signs.

e.g. If $z = 2(\cos 210° + j\sin 210°)$ the complex number lies in the third quadrant.

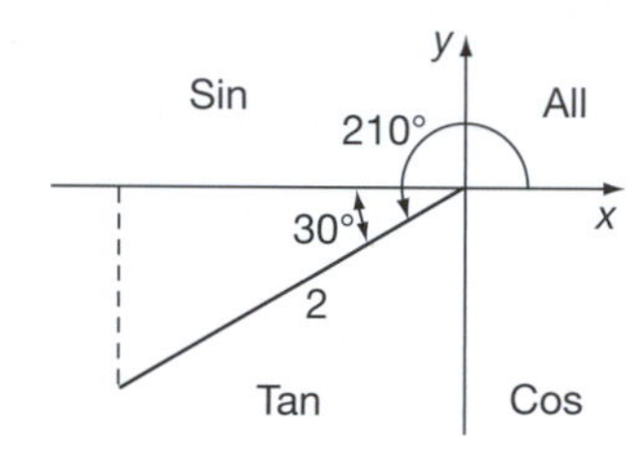


$\cos 210° = -\cos 30°$

$\sin 210° = -\sin 30°$

Then $z = 2(-\cos 30° - j\sin 30°)$

$= 2(-0{\cdot}8660 - j0{\cdot}5)$

$= -1{\cdot}732 - j$

Here you are. What about this one?

Express $z = 5(\cos 140° + j\sin 140°)$ in the form $a + jb$

What do you make it?

## 53

$$z = -3{\cdot}8302 + j3{\cdot}2139$$

Here are the details:

$\cos 140° = -\cos 40°$

$\sin 140° = \sin 40°$

$$\begin{aligned} z &= 5(\cos 140° + j\sin 140°) \\ &= 5(-\cos 40° + j\sin 40°) \\ &= 5(-0{\cdot}7660 + j0{\cdot}6428) \\ &= -3{\cdot}830 + j3{\cdot}214 \text{ (to 3 dp)} \end{aligned}$$

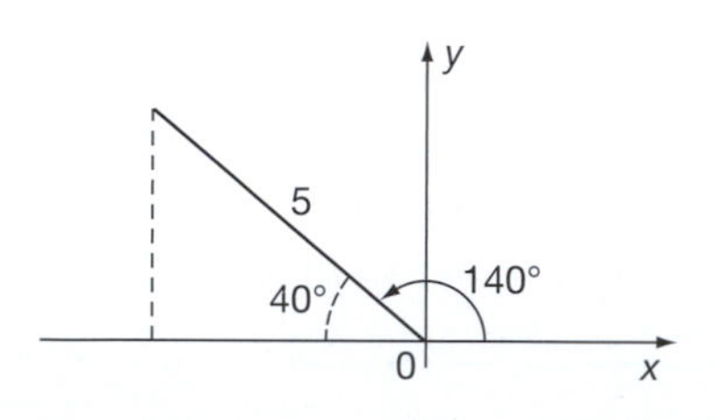


Fine. Now by way of revision, work out the following:

(a) Express $-5 + j4$ in polar form

(b) Express $3\underline{|300°}$ in the form $a + jb$

*When you have finished both of them, check your results with those in Frame 54*

**54**

Here is the working:

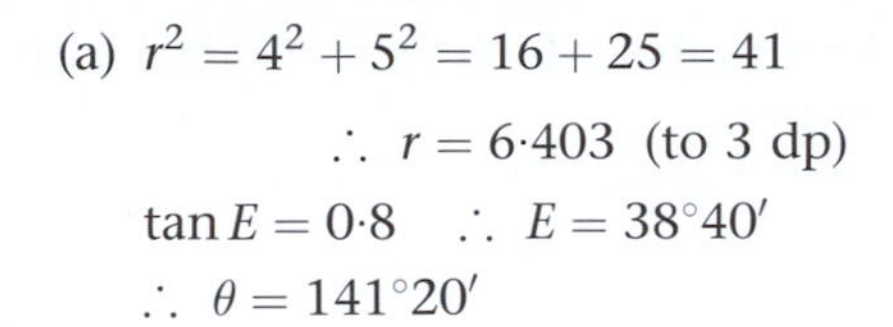

(a) $r^2 = 4^2 + 5^2 = 16 + 25 = 41$

$\therefore\ r = 6{\cdot}403$ (to 3 dp)

$\tan E = 0{\cdot}8 \quad \therefore\ E = 38^\circ 40'$

$\therefore\ \theta = 141^\circ 20'$

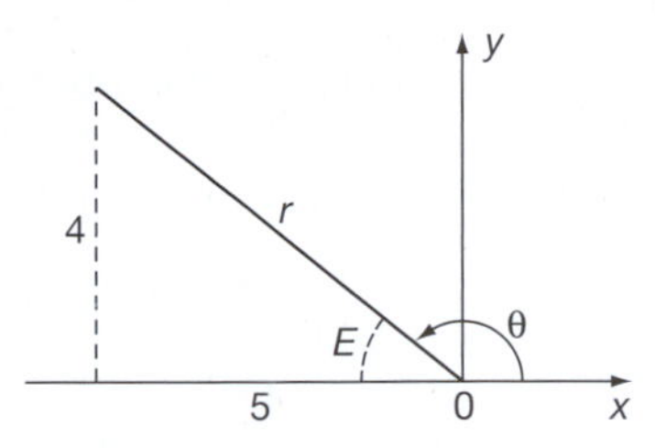


$$-5 + j4 = 6{\cdot}403(\cos 141^\circ 20' + j \sin 141^\circ 20') = \boxed{6{\cdot}403\underline{|141^\circ 20'}}$$

(b) $3\underline{|300^\circ} = 3(\cos 300^\circ + j \sin 300^\circ)$

$\cos 300^\circ = \cos 60^\circ$

$\sin 300^\circ = -\sin 60^\circ$

$3\underline{|300^\circ} = 3(\cos 60^\circ - j \sin 60^\circ)$

$= 3(0{\cdot}500 - j0{\cdot}866)$ (to 3 dp)

$= \boxed{1{\cdot}500 - j2{\cdot}598}$

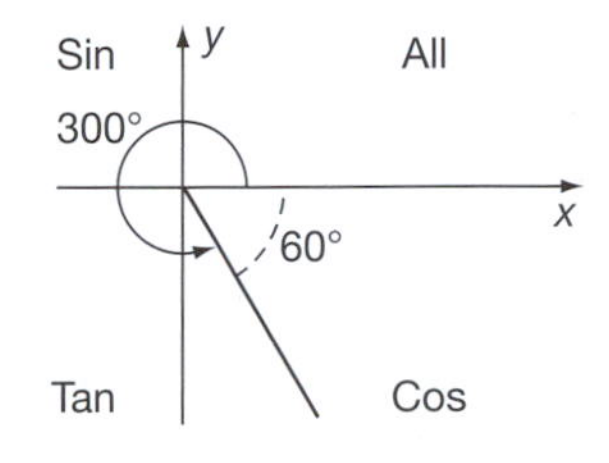


*Move to Frame 55*

**55**

We see then that there are two ways of expressing a complex number:

(a) in standard form $\quad z = a + jb$

(b) in polar form $\quad z = r(\cos\theta + j \sin\theta)$

where $\quad r = \sqrt{a^2 + b^2}$

and $\quad \theta = \tan^{-1}\dfrac{b}{a}$

If we remember the simple diagram, we can easily convert from one system to the other:

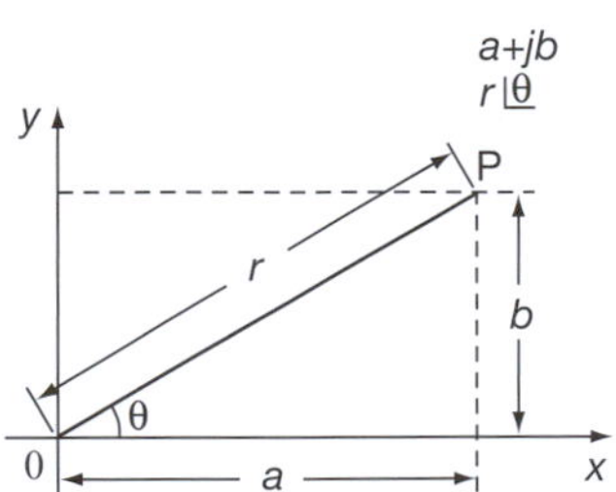


*So on now to Frame 56*

## Exponential form of a complex number

**56**

There is still another way of expressing a complex number which we must deal with, for it too has its uses. We shall arrive at it this way:

Many functions can be expressed as series. for example,

$$e^x = 1 + x + \frac{x^2}{2!} + \frac{x^3}{3!} + \frac{x^4}{4!} + \frac{x^5}{5!} + \dots\dots\dots$$

$$\sin x = x - \frac{x^3}{3!} + \frac{x^5}{5!} - \frac{x^7}{7!} + \frac{x^9}{9!} + \dots\dots\dots$$

$$\cos x = 1 - \frac{x^2}{2!} + \frac{x^4}{4!} - \frac{x^6}{6!} + \dots\dots\dots$$

You no doubt have hazy recollections of these series. You had better make a note of them since they have turned up here.

**57**

If we now take the series for $e^x$ and write $j\theta$ in place of $x$, we get:

$$\begin{aligned} e^{j\theta} &= 1 + j\theta + \frac{(j\theta)^2}{2!} + \frac{(j\theta)^3}{3!} + \frac{(j\theta)^4}{4!} + \dots\dots\dots \\ &= 1 + j\theta + \frac{j^2\theta^2}{2!} + \frac{j^3\theta^3}{3!} + \frac{j^4\theta^4}{4!} + \dots\dots\dots \\ &= 1 + j\theta - \frac{\theta^2}{2!} - \frac{j\theta^3}{3!} + \frac{\theta^4}{4!} + \dots\dots\dots \\ &= \left(1 - \frac{\theta^2}{2!} + \frac{\theta^4}{4!} - \dots\dots\dots\right) + j\left(\theta - \frac{\theta^3}{3!} + \frac{\theta^5}{5!} - \dots\dots\dots\right) \\ &= \cos\theta + j\sin\theta \end{aligned}$$

Therefore, $r(\cos\theta + j\sin\theta)$ can now be written as $re^{j\theta}$. This is called the *exponential form* of the complex number. It can be obtained from the polar quite easily since the $r$ value is the same and the angle $\theta$ is the same in both. It is important to note, however, that in the exponential form, the angle must be in *radians*.

*Move on to the next frame*

**58**

The three ways of expressing a complex number are therefore:

(a) $z = a + jb$ Cartesian form

(b) $z = r(\cos\theta + j\sin\theta)$ Polar form

(c) $z = r.e^{j\theta}$ Exponential form

Remember that the exponential form is obtained from the polar form:

(a) the $r$ value is the same in each case

(b) the angle is also the same in each case, but in the exponential form the angle must be in radians.

So, knowing that, change the polar form $5(\cos 60^\circ + j\sin 60^\circ)$ into the exponential form.

*Then turn to Frame 59*

59

$5e^{j\frac{\pi}{3}}$

Because we have $5(\cos 60° + j\sin 60°)$ $\quad r = 5$

$\theta = 60° = \frac{\pi}{3}$ radians

$\therefore$ Exponential form is $5e^{j\frac{\pi}{3}}$

And now a word about negative angles:

We know $e^{j\theta} = \cos\theta + j\sin\theta$

If we replace $\theta$ by $-\theta$ in this result, we get

$$e^{-j\theta} = \cos(-\theta) + j\sin(-\theta)$$
$$= \cos\theta - j\sin\theta$$

So we have

$$\left.\begin{aligned} e^{j\theta} &= \cos\theta + j\sin\theta \\ e^{-j\theta} &= \cos\theta - j\sin\theta \end{aligned}\right\}$$

*Make a note of these*

60

There is one operation that we have been unable to carry out with complex numbers before this. That is to find the logarithm of a complex number. The exponential form now makes this possible, since the exponential form consists only of products and powers.

For, if we have:

$$z = re^{j\theta}$$

then we can say:

$$\ln z = \ln r + j\theta$$

e.g. If $z = 6{\cdot}42e^{j1{\cdot}57}$ then

$$\ln z = \ln 6{\cdot}42 + j1{\cdot}57$$
$$= 1{\cdot}86 + j1{\cdot}57 \text{ (to 2 dp)}$$

and the result is once again a complex number.

And if $z = 3{\cdot}8e^{-j0{\cdot}236}$, then $\ln z =$ ............ to 3 dp

61

$$\ln z = \ln 3{\cdot}8 - j0{\cdot}236 = \boxed{1{\cdot}335 - j0{\cdot}236}$$

Finally, here is an example of a rather different kind. Once you have seen it done, you will be able to deal with others of this kind. Here it is.

Express $e^{1-j\pi/4}$ in the form $a + jb$

Well now, we can write:

$$e^{1-j\pi/4} \text{ as } e^1 e^{-j\pi/4}$$
$$= e(\cos\pi/4 - j\sin\pi/4)$$
$$= e\left\{\frac{1}{\sqrt{2}} - j\frac{1}{\sqrt{2}}\right\}$$
$$= \frac{e}{\sqrt{2}}(1 - j)$$

▶

This brings us to the end of this Programme, except for the **Can You?** checklist and the **Test exercise**. Before you do them, read down the **Review summary** that follows in the next frame and revise any points on which you are not completely sure.

*Move to Frame 62*

## Review summary

**62**

1 *Powers of j*

$$j = \sqrt{-1}, \quad j^2 = -1, \quad j^3 = -j, \quad j^4 = 1.$$
$$j^{-1} = -j, \quad j^{-2} = -1, \quad j^{-3} = j, \quad j^{-4} = 1$$

2 *Complex numbers*

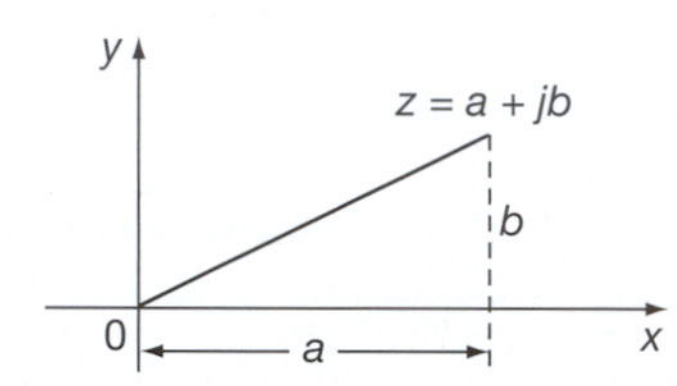


$a$ =real part
$b$ =imaginary part

3 *Conjugate complex numbers* $(a + jb)$ and $(a - jb)$
The product of two conjugate numbers is always real:

$$(a + jb)(a - jb) = a^2 + b^2$$

4 *Equal complex numbers*
If $a + jb = c + jd$, then $a = c$ and $b = d$.

5 *Polar form of a complex number*

$$z = a + jb$$
$$= r(\cos\theta + j\sin\theta)$$
$$= r\underline{|\theta}$$

$$r = \sqrt{a^2 + b^2}; \quad \theta = \tan^{-1}\left\{\frac{b}{a}\right\}$$

also $a = r\cos\theta$; $b = r\sin\theta$

$r =$ the modulus of $z$ written 'mod $z$' or $|z|$

$\theta =$ the argument of $z$, written 'arg $z$'

6 *Exponential form of a complex number*

$$\left.\begin{aligned} z &= r(\cos\theta + j\sin\theta) = re^{j\theta} \\ \text{and} \quad & r(\cos\theta - j\sin\theta) = re^{-j\theta} \end{aligned}\right\} \theta \text{ in radians}$$

7 *Logarithm of a complex number*

$$z = re^{j\theta} \quad \therefore \quad \ln z = \ln r + j\theta$$
$$\text{or if} \quad z = re^{-j\theta} \quad \therefore \quad \ln z = \ln r - j\theta$$

## Can You?

### Checklist 1

63

Check this list before and after you try the end of Programme test.

On a scale of 1 to 5 how confident are you that you can:

| | Frames |
|---|---|
| • Recognize $j$ as standing for $\sqrt{-1}$ and be able to reduce the powers of $j$ to $\pm j$ or $\pm 1$?<br>*Yes* ☐ ☐ ☐ ☐ ☐ *No* | 1 to 13 |
| • Recognize that all complex numbers are in the form (real part) + $j$(imaginary part)?<br>*Yes* ☐ ☐ ☐ ☐ ☐ *No* | 14 to 15 |
| • Add, subtract and multiply complex numbers?<br>*Yes* ☐ ☐ ☐ ☐ ☐ *No* | 16 to 22 |
| • Find the complex conjugate of a complex number?<br>*Yes* ☐ ☐ ☐ ☐ ☐ *No* | 23 to 27 |
| • Divide complex numbers?<br>*Yes* ☐ ☐ ☐ ☐ ☐ *No* | 28 to 32 |
| • State the conditions for the equality of two complex numbers?<br>*Yes* ☐ ☐ ☐ ☐ ☐ *No* | 33 to 39 |
| • Draw complex numbers and recognize the parallel law of addition?<br>*Yes* ☐ ☐ ☐ ☐ ☐ *No* | 40 to 45 |
| • Convert a complex number from Cartesian to polar form and vice versa?<br>*Yes* ☐ ☐ ☐ ☐ ☐ *No* | 46 to 55 |
| • Write a complex number in its exponential form?<br>*Yes* ☐ ☐ ☐ ☐ ☐ *No* | 56 to 59 |
| • Obtain the logarithm of a complex number?<br>*Yes* ☐ ☐ ☐ ☐ ☐ *No* | 60 |

## Test exercise 1

64

You will find the questions quite straightforward and easy.

1 Simplify: (a) $j^3$ (b) $j^5$ (c) $j^{12}$ (d) $j^{14}$.

2 Express in the form $a + jb$:

(a) $(4 - j7)(2 + j3)$ (b) $(-1 + j)^2$

(c) $(5 + j2)(4 - j5)(2 + j3)$ (d) $\dfrac{4 + j3}{2 - j}$

3 Find the values of $x$ and $y$ that satisfy the equation:
$(x + y) + j(x - y) = 14{\cdot}8 + j6{\cdot}2$

4 Express in polar form:
(a) $3 + j5$ (b) $-6 + j3$ (c) $-4 - j5$

5 Express in the form $a + jb$:
(a) $5(\cos 225^\circ + j \sin 225^\circ)$ (b) $4\underline{|330^\circ}$

▶

6 Express in exponential form:
(a) $z_1 = 10\underline{|37°15'}$ and (b) $z_2 = 10\underline{|322°45'}$
Hence find $\ln z_1$ and $\ln z_2$.

7 Express $z = e^{1+j\pi/2}$ in the form $a + jb$.

*Now you are ready to start Part 2 of the work on complex numbers*

---

## Further problems 1

**65**

1 Simplify:

(a) $(5 + j4)(3 + j7)(2 - j3)$ (b) $\dfrac{(2 - j3)(3 + j2)}{(4 - j3)}$ (c) $\dfrac{\cos 3x + j\sin 3x}{\cos x + j\sin x}$

2 Express $\dfrac{2 + j3}{j(4 - j5)} + \dfrac{2}{j}$ in the form $a + jb$.

3 If $z = \dfrac{1}{2 + j3} + \dfrac{1}{1 - j2}$, express $z$ in the form $a + jb$.

4 If $z = \dfrac{2 + j}{1 - j}$, find the real and imaginary parts of the complex number $z + \dfrac{1}{z}$.

5 Simplify $(2 + j5)^2 + \dfrac{5(7 + j2)}{3 - j4} - j(4 - j6)$, expressing the result in the form $a + jb$.

6 If $z_1 = 2 + j$, $z_2 = -2 + j4$ and $\dfrac{1}{z_3} = \dfrac{1}{z_1} + \dfrac{1}{z_2}$, evaluate $z_3$ in the form $a + jb$. If $z_1$, $z_2$, $z_3$ are represented on an Argand diagram by the points P, Q, R, respectively, prove that R is the foot of the perpendicular from the origin on to the line PQ.

7 Points A, B, C, D, on an Argand diagram, represent the complex numbers $9 + j$, $4 + j13$, $-8 + j8$, $-3 - j4$ respectively. Prove that ABCD is a square.

8 If $(2 + j3)(3 - j4) = x + jy$, evaluate $x$ and $y$.

9 If $(a + b) + j(a - b) = (2 + j5)^2 + j(2 - j3)$, find the values of $a$ and $b$.

10 If $x$ and $y$ are real, solve the equation:

$$\frac{jx}{1 + jy} = \frac{3x + j4}{x + 3y}$$

11 If $z = \dfrac{a + jb}{c + jd}$, where $a$, $b$, $c$ and $d$ are real quantities, show that (a) if $z$ is real then $\dfrac{a}{b} = \dfrac{c}{d}$ and (b) if $z$ is entirely imaginary then $\dfrac{a}{b} = -\dfrac{d}{c}$.

12 Given that $(a + b) + j(a - b) = (1 + j)^2 + j(2 + j)$, obtain the values of $a$ and $b$.

13 Express $(-1 + j)$ in the form $re^{j\theta}$ where $r$ is positive and $-\pi < \theta < \pi$.

14 Find the modulus of $z = (2 - j)(5 + j12)/(1 + j2)^3$.

15 If $x$ is real, show that $(2 + j)e^{(1+j3)x} + (2 - j)e^{(1-j3)x}$ is also real.

▶

**16** Given that $z_1 = R_1 + R + j\omega L$; $z_2 = R_2$; $z_3 = \dfrac{1}{j\omega C_3}$; and $z_4 = R_4 + \dfrac{1}{j\omega C_4}$; and also that $z_1 z_3 = z_2 z_4$, express $R$ and $L$ in terms of the real constants $R_1$, $R_2$, $R_4$, $C_3$ and $C_4$.

**17** If $z = x + jy$, where $x$ and $y$ are real, and if the real part of $(z + 1)/(z + j)$ is equal to 1, show that the point $z$ lies on a straight line in the Argand diagram.

**18** When $z_1 = 2 + j3$, $z_2 = 3 - j4$, $z_3 = -5 + j12$, then $z = z_1 + \dfrac{z_2 z_3}{z_2 + z_3}$. If $E = Iz$, find $E$ when $I = 5 + j6$.

**19** If $\dfrac{R_1 + j\omega L}{R_3} = \dfrac{R_2}{R_4 - j\frac{1}{\omega C}}$, where $R_1$, $R_2$, $R_3$, $R_4$, $\omega$, $L$ and $C$ are real, show that

$$L = \frac{CR_2 R_3}{\omega^2 C^2 R_4^2 + 1}$$

**20** If $z$ and $\bar{z}$ are conjugate complex numbers, find two complex numbers, $z = z_1$ and $z = z_2$, that satisfy the equation:

$$3z\bar{z} + 2(z - \bar{z}) = 39 + j12$$

On an Argand diagram, these two numbers are represented by the points P and Q. If R represents the number $j1$, show that the angle PRQ is a right angle.

Now visit the companion website for this book at www.palgrave.com/stroud for more questions applying this mathematics to science and engineering.

**66**

# Complex numbers 2

## Learning outcomes

When you have completed this Programme you will be able to:

- ☐ Use the shorthand form for a complex number in polar form
- ☐ Write complex numbers in polar form using negative angles
- ☐ Multiply and divide complex numbers in polar form
- ☐ Use DeMoivre's theorem
- ☐ Find the roots of a complex number
- ☐ Demonstrate trigonometric identities of multiple angles using complex numbers
- ☐ Solve loci problems using complex numbers

# Polar-form calculations

1

In Part 1 of this programme on complex numbers, we discovered how to manipulate them in adding, subtracting, multiplying and dividing. We also finished Part 1 by seeing that a complex number $a + jb$ can also be expressed in polar form, which is always of the form $r(\cos\theta + j\sin\theta)$.

You will remember that values of $r$ and $\theta$ can easily be found from the diagram of the given complex number:

$$r^2 = a^2 + b^2 \quad \therefore r = \sqrt{a^2 + b^2}$$

$$\text{and } \tan\theta = \frac{b}{a} \quad \therefore \theta = \tan^{-1}\frac{b}{a}$$

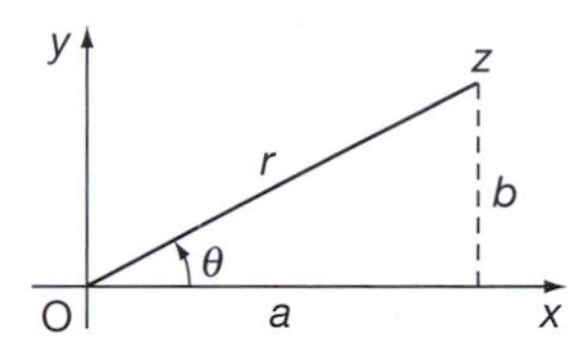


To be sure that you have taken the correct value of $\theta$, always *draw a sketch diagram* to see which quadrant the complex number is in.

Remember that $\theta$ is always measured from . . . . . . . . . . .

2

Ox i.e. the positive $x$-axis.

Right. As a warming-up exercise, do the following:

Express $z = 12 - j5$ in polar form

Do not forget the sketch diagram. It ensures that you get the correct value for $\theta$.

*When you have finished, and not before, move on to Frame 3 to check your result*

3

$13(\cos 337°23' + j\sin 337°23')$

Here it is, worked out in full:

$$r^2 = 12^2 + 5^2 = 144 + 25 = 169$$

$$\therefore r = 13$$

$$\tan E = \frac{5}{12} = 0{\cdot}4167 \quad \therefore E = 22°37'$$

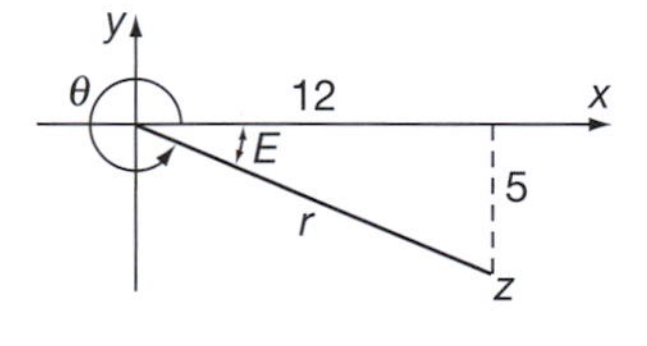


In this case, $\theta = 360° - E = 360° - 22°37'$

$$\therefore \theta = 337°23'$$

$$z = r(\cos\theta + j\sin\theta)$$

$$= 13(\cos 337°23' + j\sin 337°23')$$

Did you get that right? Here is one more, done in just the same way:

Express $-5 - j4$ in polar form.

Diagram first of all! Then you cannot go wrong.

*When you have the result, on to Frame 4*

**4**

$$\boxed{z = 6{\cdot}403(\cos 218°40' + j\sin 218°40')}$$

Here is the working: check yours.

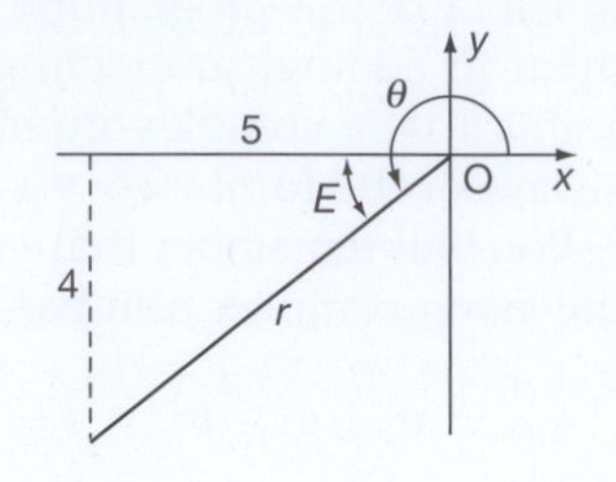


$$r^2 = 5^2 + 4^2 = 25 + 16 = 41$$
$$\therefore r = \sqrt{41} = 6{\cdot}403 \text{ (to 3 dp)}$$
$$\tan E = \frac{4}{5} = 0{\cdot}8$$
$$\therefore E = 38°40'$$

In this case, $\theta = 180° + E = 218°40'$

So $z = -5 - j4$
$= 6{\cdot}403(\cos 218°40' + j\sin 218°40')$

Since every complex number in polar form is of the same structure, i.e. $r(\cos\theta + j\sin\theta)$ and differs from another complex number simply by the values of $r$ and $\theta$, we have a shorthand method of quoting the result in polar form. Do you remember what it is? The shorthand way of writing the result above,

i.e. $6{\cdot}403(\cos 218°40' + j\sin 218°40')$ is ...........

**5**

$$\boxed{6{\cdot}403\underline{|218°40'}}$$

Correct. Likewise:

| | | |
|---|---|---|
| $5{\cdot}72(\cos 322°15' + j\sin 322°15')$ | is written | $5{\cdot}72\underline{|322°15'}$ |
| $5(\cos 105° + j\sin 105°)$ | is written | $5\underline{|105°}$ |
| $3{\cdot}4\left(\cos\frac{\pi}{6} + j\sin\frac{\pi}{6}\right)$ | is written | $3{\cdot}4\underline{\left|\frac{\pi}{6}\right.}$ |

They are all complex numbers in polar form. They are all the same structure and differ one from another simply by the values of ........... and ...........

**6**

$\boxed{r}$ and $\boxed{\theta}$

Now let us consider the following example.

Express $z = 4 - j3$ in polar form.

First the diagram:

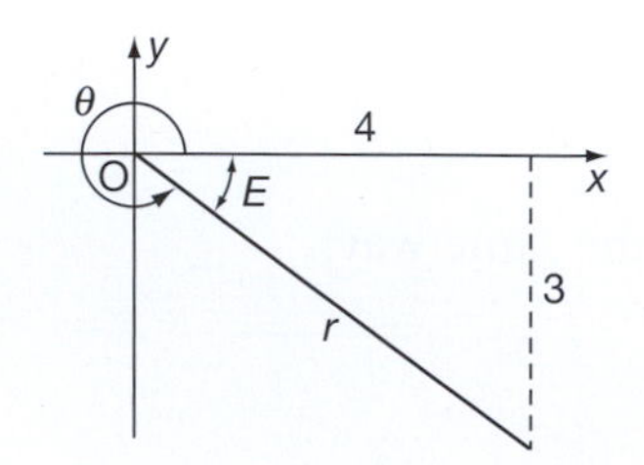


From this:

$$r = 5$$
$$\tan E = \frac{3}{4} = 0{\cdot}75$$
$$\therefore E = 36°52'$$
$$\theta = 360° - 36°52' = 323°8'$$
$$z = 4 - j3 = 5(\cos 323°8' + j\sin 323°8')$$

or in shortened form,

$z =$ ...........

**7**

$z = 5\underline{|323°8'}$

In the previous example, we have:

$z = 5(\cos 323°8' + j\sin 323°8')$

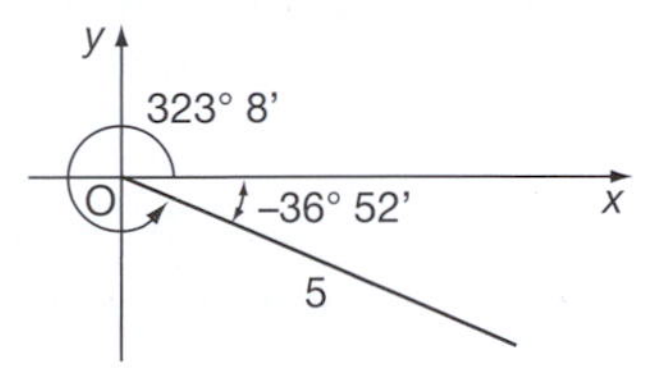


But the direction of the complex number, measured from O$x$, could be given as $-36°52'$, the minus sign showing that we are measuring the angle in the opposite sense from the usual positive direction.

We could write $z = 5(\cos[-36°52'] + j\sin[-36°52'])$. But you alreay know that $\cos[-\theta] = \cos\theta$ and $\sin[-\theta] = -\sin\theta$.

$z = 5(\cos 36°52' - j\sin 36°52')$

i.e. very much like the polar form but with a minus sign in the middle. This comes about whenever we use negative angles. In the same way:

$$\begin{aligned} z &= 4(\cos 250° + j\sin 250°) \\ &= 4(\cos[-110°] + j\sin[-110°]) \\ &= 4(\dots\dots\dots\dots) \end{aligned}$$

**8**

$z = 4(\cos 110° - j\sin 110°)$

since $\cos(-110°) = \cos 110°$
and $\sin(-110°) = -\sin 110°$

It is sometimes convenient to use this form when the value of $\theta$ is greater than 180°, i.e. in the 3rd and 4th quadrants.

Here are some examples:

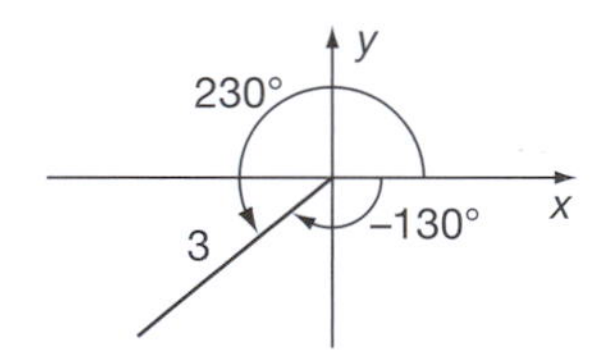


$$\begin{aligned} z &= 3(\cos 230° + j\sin 230°) \\ &= 3(\cos 130° - j\sin 130°) \end{aligned}$$

Similarly

$z = 3(\cos 300° + j\sin 300°) = 3(\cos 60° - j\sin 60°)$

$z = 4(\cos 290° + j\sin 290°) = 4(\cos 70° - j\sin 70°)$

$z = 2(\cos 215° + j\sin 215°) = 2(\cos 145° - j\sin 145°)$

and

$z = 6(\cos 310° + j\sin 310°) = \dots\dots\dots\dots$

**9**

$z = 6(\cos 50° - j\sin 50°)$

Because

$\cos 310° = \cos 50°$
and $\sin 310° = -\sin 50°$

▶

A moment ago we agreed that the minus sign comes about by the use of negative angles. To convert a complex number given in this way back into proper polar form, i.e. with a '+' in the middle, we simply work back the way we came. A complex number with a negative sign in the middle is equivalent to the same complex number with a positive sign, but with the angles made negative.

e.g. $z = 4(\cos 30^\circ - j\sin 30^\circ)$
$= 4(\cos[-30^\circ] + j\sin[-30^\circ])$
$= 4(\cos 330^\circ + j\sin 330^\circ)$

and we are back in proper polar form.

You do this one.

Convert $z = 5(\cos 40^\circ - j\sin 40^\circ)$ into proper polar form.

*Then on to Frame 10*

## 10

$$\boxed{z = 5(\cos 320^\circ + j\sin 320^\circ)}$$

Because

$$z = 5(\cos 40^\circ - j\sin 40^\circ) = 5(\cos[-40^\circ] + j\sin[-40^\circ])$$
$$= 5(\cos 320^\circ + j\sin 320^\circ)$$

Here is another for you to do.

Express $z = 4(\cos 100^\circ - j\sin 100^\circ)$ in proper polar form.

Do not forget, it all depends on the use of negative angles.

## 11

$$\boxed{z = 4(\cos 260^\circ + j\sin 260^\circ)}$$

Because

$$z = 4(\cos 100^\circ - j\sin 100^\circ) = 4(\cos[-100^\circ] + j\sin[-100^\circ])$$
$$= 4(\cos 260^\circ + j\sin 260^\circ)$$

We ought to see how this modified polar form affects our shorthand notation.

Remember, $5(\cos 60^\circ + j\sin 60^\circ)$ is written $5\underline{|60^\circ}$

How then shall we write $5(\cos 60^\circ - j\sin 60^\circ)$?

We know that this really stands for $5(\cos[-60^\circ] + j\sin[-60^\circ])$ so we could write $5\underline{|-60^\circ}$. But instead of using the negative angle we use a different symbol,

i.e. $5\underline{|-60^\circ}$ becomes $5\overline{|60^\circ}$.

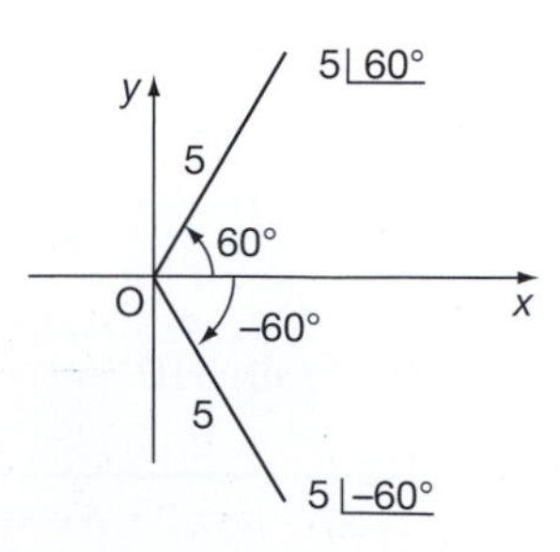


Similarly, $3(\cos 45^\circ - j\sin 45^\circ) = 3\underline{|-45^\circ} = \dots\dots\dots\dots$

**12**

$$3\overline{|45^\circ}$$

This is easy to remember,

for the sign $\underline{|\;\;\;}$ (the first-quadrant sign) resembles the first quadrant and indicates

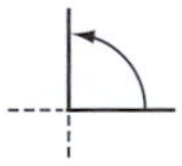

measuring angles ↺ i.e. in the positive direction,

while the sign $\overline{|\;\;\;}$ (the fourth-quadrant sign) resembles the fourth quadrant and indicates

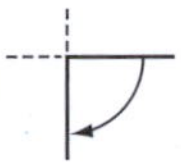

measuring angles ↻ i.e. in the negative direction.

For example:

$(\cos 15^\circ + j\sin 15^\circ)$ is written $\underline{|15^\circ}$

but $(\cos 15^\circ - j\sin 15^\circ)$, which is really $(\cos[-15^\circ] + j\sin[-15^\circ])$, is written $\overline{|15^\circ}$

So how do we write

(a) $(\cos 120^\circ + j\sin 120^\circ)$

and (b) $(\cos 135^\circ - j\sin 135^\circ)$

in the shorthand way?

---

**13**

(a) $\underline{|120^\circ}$ (b) $\overline{|135^\circ}$

The polar form at first sight seems to be a complicated way of representing a complex number. However it is very useful, as we shall see. Suppose we multiply together two complex numbers in this form:

Let $z_1 = r_1(\cos\theta_1 + j\sin\theta_1)$ and $z_2 = r_2(\cos\theta_2 + j\sin\theta_2)$

Then $z_1z_2 = r_1(\cos\theta_1 + j\sin\theta_1)r_2(\cos\theta_2 + j\sin\theta_2)$

$$= r_1r_2(\cos\theta_1\cos\theta_2 + j\sin\theta_1\cos\theta_2 + j\cos\theta_1\sin\theta_2 + j^2\sin\theta_1\sin\theta_2)$$

Rearranging the terms and remembering that $j^2 = -1$, we get

$$z_1z_2 = r_1r_2[(\cos\theta_1\cos\theta_2 - \sin\theta_1\sin\theta_2) + j(\sin\theta_1\cos\theta_2 + \cos\theta_1\sin\theta_2)]$$

Now the brackets $(\cos\theta_1\cos\theta_2 - \sin\theta_1\sin\theta_2)$

and $(\sin\theta_1\cos\theta_2 + \cos\theta_1\sin\theta_2)$

ought to ring a bell. What are they?

---

**14**

$$\cos\theta_1\cos\theta_2 - \sin\theta_1\sin\theta_2 = \cos(\theta_1 + \theta_2)$$
$$\sin\theta_1\cos\theta_2 + \cos\theta_1\sin\theta_2 = \sin(\theta_1 + \theta_2)$$

In that case, $z_1z_2 = r_1r_2[\cos(\theta_1 + \theta_2) + j\sin(\theta_1 + \theta_2)]$.
Note this important result. We have just shown that

$$r_1(\cos\theta_1 + j\sin\theta_1).r_2(\cos\theta_2 + j\sin\theta_2) = r_1r_2[\cos(\theta_1 + \theta_2) + j\sin(\theta_1 + \theta_2)]$$

i.e. To multiply together two complex numbers in polar form,

(a) multiply the $r$'s together,

(b) add the angles, $\theta$, together.

▶

It is just as easy as that!

e.g. $2(\cos 30^\circ + j\sin 30^\circ) \times 3(\cos 40^\circ + j\sin 40^\circ)$
$= 2 \times 3(\cos[30^\circ + 40^\circ] + j\sin[30^\circ + 40^\circ])$
$= 6(\cos 70^\circ + j\sin 70^\circ)$

So if we multiply together $5(\cos 50^\circ + j\sin 50^\circ)$ and $2(\cos 65^\circ + j\sin 65^\circ)$ we get ............

**15**

$10(\cos 115^\circ + j\sin 115^\circ)$

Remember, multiply the $r$'s; add the $\theta$'s.

Here you are then; all done the same way:

(a) $2(\cos 120^\circ + j\sin 120^\circ) \times 4(\cos 20^\circ + j\sin 20^\circ)$
$= 8(\cos 140^\circ + j\sin 140^\circ)$

(b) $a(\cos\theta + j\sin\theta) \times b(\cos\phi + j\sin\phi)$
$= ab(\cos[\theta + \phi] + j\sin[\theta + \phi])$

(c) $6(\cos 210^\circ + j\sin 210^\circ) \times 3(\cos 80^\circ + j\sin 80^\circ)$
$= 18(\cos 290^\circ + j\sin 290^\circ)$

(d) $5(\cos 50^\circ + j\sin 50^\circ) \times 3(\cos[-20^\circ] + j\sin[-20^\circ])$
$= 15(\cos 30^\circ + j\sin 30^\circ)$

Have you got it? No matter what the angles are, all we do is:

(a) multiply the moduli,
(b) add the arguments.

So therefore, $4(\cos 35^\circ + j\sin 35^\circ) \times 3(\cos 20^\circ + j\sin 20^\circ) =$ ...........

**16**

$12(\cos 55^\circ + j\sin 55^\circ)$

Now let us see if we can discover a similar set of rules for division.

We already know that to simplify $\dfrac{5 + j6}{3 + j4}$ we first obtain a denominator that is entirely real by multiplying top and bottom by ...........

**17**

the conjugate of the denominator, i.e. $3 - j4$

Right. Then let us do the same with $\dfrac{r_1(\cos\theta_1 + j\sin\theta_1)}{r_2(\cos\theta_2 + j\sin\theta_2)}$

$$\frac{r_1(\cos\theta_1 + j\sin\theta_1)}{r_2(\cos\theta_2 + j\sin\theta_2)} = \frac{r_1(\cos\theta_1 + j\sin\theta_1)(\cos\theta_2 - j\sin\theta_2)}{r_2(\cos\theta_2 + j\sin\theta_2)(\cos\theta_2 - j\sin\theta_2)}$$
$$= \frac{r_1}{r_2}\frac{(\cos\theta_1\cos\theta_2 + j\sin\theta_1\cos\theta_2 - j\cos\theta_1\sin\theta_2 + \sin\theta_1\sin\theta_2)}{(\cos^2\theta_2 + \sin^2\theta_2)}$$
$$= \frac{r_1}{r_2}\frac{[(\cos\theta_1\cos\theta_2 + \sin\theta_1\sin\theta_2) + j(\sin\theta_1\cos\theta_2 - \cos\theta_1\sin\theta_2)]}{1}$$
$$= \frac{r_1}{r_2}[\cos(\theta_1 - \theta_2) + j\sin(\theta_1 - \theta_2)]$$

So, for division, the rule is ...........

**18**

divide the $r$'s and subtract the angles

That is correct.

e.g. $\dfrac{6(\cos 72^\circ + j\sin 72^\circ)}{2(\cos 41^\circ + j\sin 41^\circ)} = 3(\cos 31^\circ + j\sin 31^\circ)$

So we now have two important rules:

If $z_1 = r_1(\cos\theta_1 + j\sin\theta_1)$ and $z_2 = r_2(\cos\theta_2 + j\sin\theta_2)$

then (a) $z_1z_2 = r_1r_2[\cos(\theta_1 + \theta_2) + j\sin(\theta_1 + \theta_2)]$

and (b) $\dfrac{z_1}{z_2} = \dfrac{r_1}{r_2}[\cos(\theta_1 - \theta_2) + j\sin(\theta_1 - \theta_2)]$

The results are still, of course, in polar form.

Now here is one for you to think about.

If $z_1 = 8(\cos 65^\circ + j\sin 65^\circ)$ and $z_2 = 4(\cos 23^\circ + j\sin 23^\circ)$ then

(a) $z_1z_2 = \ldots\ldots\ldots\ldots$

and (b) $\dfrac{z_1}{z_2} = \ldots\ldots\ldots\ldots$

**19**

(a) $z_1z_2 = 32(\cos 88^\circ + j\sin 88^\circ)$

(b) $\dfrac{z_1}{z_2} = 2(\cos 42^\circ + j\sin 42^\circ)$

Of course, we can combine the rules in a single example:

e.g. $\dfrac{5(\cos 60^\circ + j\sin 60^\circ) \times 4(\cos 30^\circ + j\sin 30^\circ)}{2(\cos 50^\circ + j\sin 50^\circ)}$

$= \dfrac{20(\cos 90^\circ + j\sin 90^\circ)}{2(\cos 50^\circ + j\sin 50^\circ)}$

$= 10(\cos 40^\circ + j\sin 40^\circ)$

What does the following product become?

$4(\cos 20^\circ + j\sin 20^\circ) \times 3(\cos 30^\circ + j\sin 30^\circ) \times 2(\cos 40^\circ + j\sin 40^\circ)$

*Result in next frame*

**20**

$24(\cos 90^\circ + j\sin 90^\circ)$

i.e. $(4 \times 3 \times 2)[\cos(20^\circ + 30^\circ + 40^\circ) + j\sin(20^\circ + 30^\circ + 40^\circ)]$

$= 24(\cos 90^\circ + j\sin 90^\circ)$

Now what about a few revision examples on the work we have done so far?

*Move to the next frame*

## Review exercise

**21**

Work all these questions and then turn on to Frame 22 and check your results.

1 Express in polar form, $z = -4 + j2$.

2 Express in true polar form, $z = 5(\cos 55^\circ - j\sin 55^\circ)$.

3 Simplify the following, giving the results in polar form:

(a) $3(\cos 143^\circ + j\sin 143^\circ) \times 4(\cos 57^\circ + j\sin 57^\circ)$

(b) $\dfrac{10(\cos 126^\circ + j\sin 126^\circ)}{2(\cos 72^\circ + j\sin 72^\circ)}$

4 Express in the form $a + jb$:

(a) $2(\cos 30^\circ + j\sin 30^\circ)$

(b) $5(\cos 57^\circ - j\sin 57^\circ)$

*Solutions are in Frame 22. Move on and see how you have fared*

**22**

1

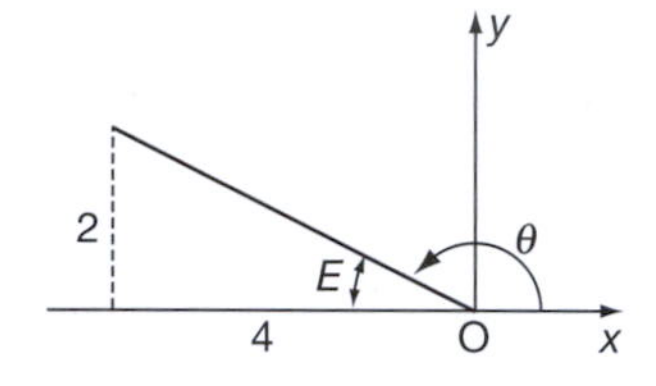


$$r^2 = 2^2 + 4^2 = 4 + 16 = 20$$
$$\therefore\ r = 4{\cdot}472$$
$$\tan E = 0{\cdot}5 \quad \therefore\ E = 26^\circ 34'$$
$$\therefore\ \theta = 153^\circ 26'$$

$$z = -4 + j2 = 4{\cdot}472(\cos 153^\circ 26' + j\sin 153^\circ 26')$$

2 $z = 5(\cos 55^\circ - j\sin 55^\circ) = 5[\cos(-55^\circ) + j\sin(-55^\circ)]$
$= 5(\cos 305^\circ + j\sin 305^\circ)$

3 (a) $3(\cos 143^\circ + j\sin 143^\circ) \times 4(\cos 57^\circ + j\sin 57^\circ)$
$= 3 \times 4[\cos(143^\circ + 57^\circ) + j\sin(143^\circ + 57^\circ)]$
$= 12(\cos 200^\circ + j\sin 200^\circ)$

(b) $\dfrac{10(\cos 126^\circ + j\sin 126^\circ)}{2(\cos 72^\circ + j\sin 72^\circ)}$
$= \dfrac{10}{2}[\cos(126^\circ - 72^\circ) + j\sin(126^\circ - 72^\circ)]$
$= 5(\cos 54^\circ + j\sin 54^\circ)$

4 (a) $2(\cos 30^\circ + j\sin 30^\circ)$
$= 2(0{\cdot}866 + j0{\cdot}5) = 1{\cdot}732 + j$

(b) $5(\cos 57^\circ - j\sin 57^\circ)$
$= 5(0{\cdot}5446 - j0{\cdot}8387)$
$= 2{\cdot}723 - j4{\cdot}193$

*Now continue the Programme in Frame 23*

## Roots of a complex number

**23**

Now we are ready to go on to a very important section which follows from our work on multiplication of complex numbers in polar form.

We have already established that:

if $z_1 = r_1(\cos\theta_1 + j\sin\theta_1)$ and $z_2 = r_2(\cos\theta_2 + j\sin\theta_2)$ then

$$z_1 z_2 = r_1 r_2[\cos(\theta_1 + \theta_2) + j\sin(\theta_1 + \theta_2)]$$

So if $z_3 = r_3(\cos\theta_3 + j\sin\theta_3)$ then we have

$$z_1 z_2 z_3 = r_1 r_2[\cos(\theta_1 + \theta_2) + j\sin(\theta_1 + \theta_2)]r_3(\cos\theta_3 + j\sin\theta_3)$$
$$= \dots\dots\dots\dots$$

**24**

$$z_1 z_2 z_3 = r_1 r_2 r_3[\cos(\theta_1 + \theta_2 + \theta_3) + j\sin(\theta_1 + \theta_2 + \theta_3)]$$

because in multiplication, we multiply the moduli and add the arguments

Now suppose that $z_1$, $z_2$, $z_3$ are all alike and that each is equal to $z = r(\cos\theta + j\sin\theta)$. Then the result above becomes:

$$z_1 z_2 z_3 = z^3 = r.r.r[\cos(\theta + \theta + \theta) + j\sin(\theta + \theta + \theta)]$$
$$= r^3(\cos 3\theta + j\sin 3\theta)$$

or $$z^3 = [r(\cos\theta + j\sin\theta)]^3 = r^3(\cos\theta + j\sin\theta)^3$$
$$= r^3(\cos 3\theta + j\sin 3\theta)$$

That is, if we wish to cube a complex number in polar form, we just cube the modulus ($r$ value) and multiply the argument ($\theta$) by 3.

Similarly, to square a complex number in polar form, we square the modulus ($r$ value) and multiply the argument ($\theta$) by $\dots\dots\dots\dots$

**25**

2 i.e. $[r(\cos\theta + j\sin\theta)]^2 = r^2(\cos 2\theta + j\sin 2\theta)$

Let us take another look at these results:

$$[r(\cos\theta + j\sin\theta)]^2 = r^2(\cos 2\theta + j\sin 2\theta)$$
$$[r(\cos\theta + j\sin\theta)]^3 = r^3(\cos 3\theta + j\sin 3\theta)$$

Similarly:

$$[r(\cos\theta + j\sin\theta)]^4 = r^4(\cos 4\theta + j\sin 4\theta)$$
$$[r(\cos\theta + j\sin\theta)]^5 = r^5(\cos 5\theta + j\sin 5\theta) \quad \text{and so on}$$

In general, then, we can say:

$$[r(\cos\theta + j\sin\theta)]^n = \dots\dots\dots\dots$$

**26**

$$[r(\cos\theta + j\sin\theta)]^n = \boxed{r^n(\cos n\theta + j\sin n\theta)}$$

This general result is very important and is called *DeMoivre's theorem*. It says that to raise a complex number in polar form to any power $n$, we raise the $r$ to the power $n$ and multiply the angle by $n$:

e.g. $[4(\cos 50^\circ + j\sin 50^\circ]^2 = 4^2[\cos(2\times 50^\circ) + j\sin(2\times 50^\circ)]$

$= 16(\cos 100^\circ + j\sin 100^\circ)$

and $[3(\cos 110^\circ + j\sin 110^\circ)]^3 = 27(\cos 330^\circ + j\sin 330^\circ)$

and in the same way:

$[2(\cos 37^\circ + j\sin 37^\circ)]^4 = \dots\dots\dots\dots$

**27**

$$\boxed{16(\cos 148^\circ + j\sin 148^\circ)}$$

This is where the polar form really comes into its own! For DeMoivre's theorem also applies when we are raising the complex number to a fractional power, i.e. when we are finding the roots of a complex number.

e.g. To find the square root of $z = 4(\cos 70^\circ + j\sin 70^\circ)$

We have $\sqrt{z} = z^{\frac{1}{2}} = [4(\cos 70^\circ + j\sin 70^\circ)]^{\frac{1}{2}}$ i.e. $n = \dfrac{1}{2}$

$$= 4^{\frac{1}{2}}\left(\cos\frac{70^\circ}{2} + j\sin\frac{70^\circ}{2}\right)$$

$$= 2(\cos 35^\circ + j\sin 35^\circ)$$

It works every time, no matter whether the power is positive, negative, whole number or fraction. In fact, DeMoivre's theorem is so important, let us write it down again. Here goes:

If $z = r(\cos\theta + j\sin\theta)$, then $z^n = \dots\dots\dots\dots$

**28**

$$z = r(\cos\theta + j\sin\theta),\ \text{then}\ \boxed{z^n = r^n(\cos n\theta + j\sin n\theta)}$$

for any value of $n$.

Look again at finding a root of a complex number. Let us find the cube root of $z = 8(\cos 120^\circ + j\sin 120^\circ)$. Here is the given complex number shown on an Argand diagram:

$z = 8\underline{|120^\circ}$

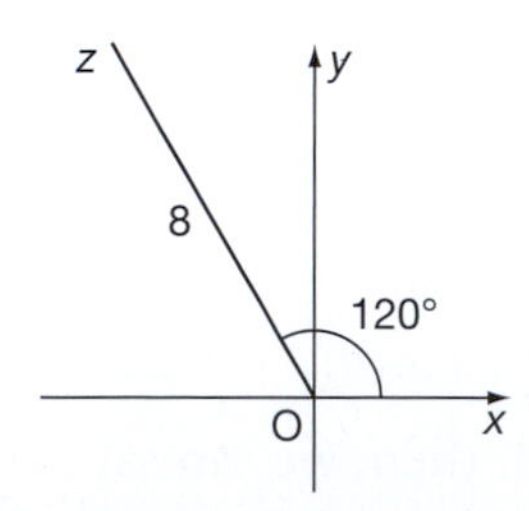


Of course, we could say that $\theta$ was '1 revolution + 120°': the vector would still be in the same position, or, for that matter (2 revs + 120°), (3 revs + 120°) etc. i.e. $z = 8\underline{|120^\circ}$ or $8\underline{|480^\circ}$ or $8\underline{|840^\circ}$ or $8\underline{|1200^\circ}$ etc. and if now apply DeMoivre's theorem to each of these, we get:

$$z^{\frac{1}{3}} = 8^{\frac{1}{3}}\underline{\left|\frac{120^\circ}{3}\right.}\ \text{or}\ 8^{\frac{1}{3}}\underline{\left|\frac{480^\circ}{3}\right.}\ \text{or}\ \dots\dots\dots\dots\ \text{or}\ \dots\dots\dots\dots\ \text{etc.}$$

**29**

$$z^{\frac{1}{3}} = 8^{\frac{1}{3}}\underline{\left|\frac{120^\circ}{3}\right.} \quad \text{or} \quad 8^{\frac{1}{3}}\underline{\left|\frac{480^\circ}{3}\right.} \quad \text{or} \quad 8^{\frac{1}{3}}\underline{\left|\frac{840^\circ}{3}\right.} \quad \text{or} \quad 8^{\frac{1}{3}}\underline{\left|\frac{1200^\circ}{3}\right.}$$

If we simplify these, we get:

$z^{\frac{1}{3}} = 2\underline{|40^\circ}$ or $2\underline{|160^\circ}$ or $2\underline{|280^\circ}$ or $2\underline{|400^\circ}$etc.

If we put each of these on an Argand diagram, as follows:

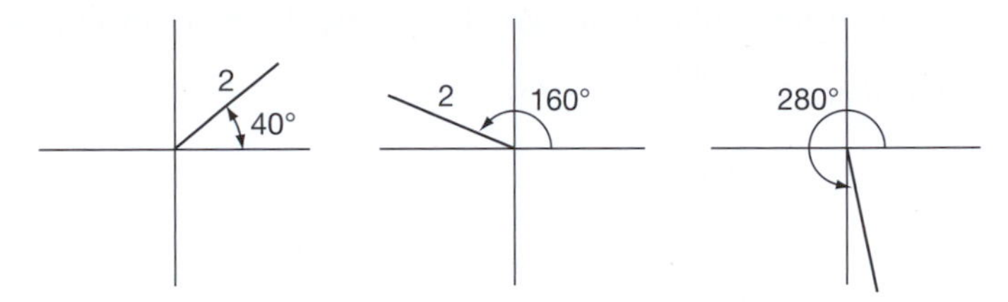


we see we have three quite different results for the cube root of $z$ and also that the fourth diagram would be a repetition of the first. Any subsequent calculations merely repeat these three positions.

*Make a sketch of the first three vectors on a single Argand diagram*

**30**

Here they are. The cube roots of $z = 8(\cos 120^\circ + j \sin 120^\circ)$:

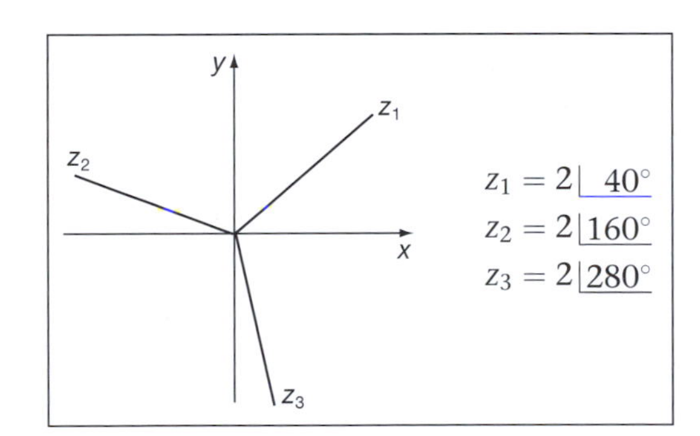


We see, therefore, that there are 3 cube roots of a complex number. Also, if you consider the angles, you see that the 3 roots are equally spaced round the diagram, any two adjacent complex numbers being separated by ........... degrees.

**31**

$120^\circ$

That is right. Therefore all we need to do in practice is to find the first of the roots and simply add $120^\circ$ on to get the next – and so on.

Notice that the three cube roots of a complex number are equal in modulus (or length) and equally spaced at intervals of $\frac{360^\circ}{3}$ i.e. $120^\circ$.

*Now let us take another example. On to the next frame*

**32**

To find the three cube roots of $z = 5(\cos 225^\circ + j\sin 225^\circ)$
The first root is given by

$$z_1 = z^{\frac{1}{3}} = 5^{\frac{1}{3}}\left(\cos\frac{225^\circ}{3} + j\sin\frac{225^\circ}{3}\right)$$
$$= 1{\cdot}71(\cos 75^\circ + j\sin 75^\circ)$$
$$z_1 = 1{\cdot}71\underline{|75^\circ}$$

We know that the other cube roots are the same length (modulus), i.e. 1·71, and separated at intervals of $\frac{360^\circ}{3}$, i.e. 120°.

So the three cube roots are:

$$z_1 = 1{\cdot}71\underline{|75^\circ} \qquad z_2 = 1{\cdot}71\underline{|195^\circ} \qquad z_3 = 1{\cdot}71\underline{|315^\circ}$$

It helps to see them on an Argand diagram, so sketch them on a combined diagram.

**33**

Here they are:

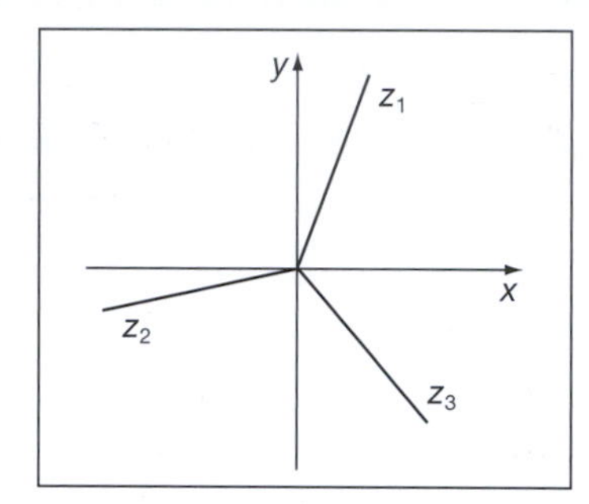


We find any roots of a complex number in the same way:

(a) Apply DeMoivre's theorem to find the first of the $n$ roots.

(b) The other roots will then be distributed round the diagram at regular intervals of $\frac{360^\circ}{n}$.

A complex number, therefore, has:

2 square roots, separated by $\frac{360^\circ}{2}$ i.e. 180°

3 cube roots, separated by $\frac{360^\circ}{3}$ i.e. 120°

4 fourth roots, separated by $\frac{360^\circ}{4}$ i.e. 90°

5 fifth roots, separated by ........... etc.

**34**

$$\boxed{\frac{360^\circ}{5} \text{ i.e. } 72^\circ}$$

And now: To find the 5 fifth roots of $12\underline{|300^\circ}$

$$z = 12\underline{|300^\circ} \quad \therefore\ z_1 = 12^{\frac{1}{5}}\underline{\left|\frac{300^\circ}{5}\right.} = 12^{\frac{1}{5}}\underline{|60^\circ}$$

Now, $12^{\frac{1}{5}} = 1{\cdot}644$ to 3 dp and so the first of the 5 fifth roots is therefore $z_1 = 1{\cdot}644\underline{|60^\circ}$

▶

The others will be of the same modulus, i.e. 1·644, and equally separated at intervals of $\frac{360^\circ}{5}$ i.e. 72°.

So the required 5 fifth roots of $12\underline{|300^\circ}$ are:

$z_1 = 1{\cdot}644\underline{|60^\circ}$

$z_2 = 1{\cdot}644\underline{|132^\circ}$

$z_3 = 1{\cdot}644\underline{|204^\circ}$

$z_4 = 1{\cdot}644\underline{|276^\circ}$

$z_5 = 1{\cdot}644\underline{|348^\circ}$

Sketch them on an Argand diagram, as before.

---

**35**

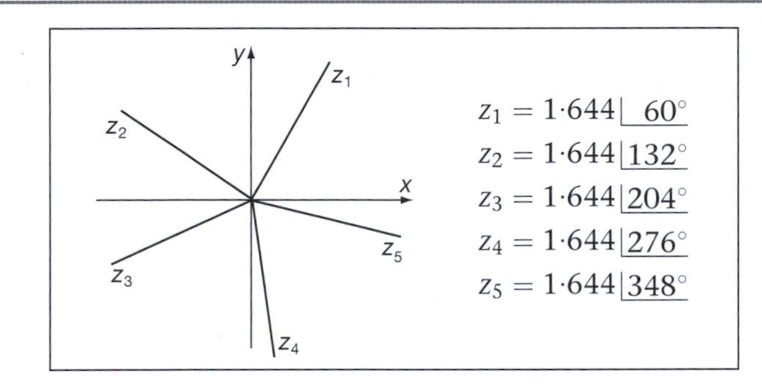


$z_1 = 1{\cdot}644\underline{|60^\circ}$
$z_2 = 1{\cdot}644\underline{|132^\circ}$
$z_3 = 1{\cdot}644\underline{|204^\circ}$
$z_4 = 1{\cdot}644\underline{|276^\circ}$
$z_5 = 1{\cdot}644\underline{|348^\circ}$

Although there are 5 fifth roots of a complex number, we are sometimes asked to find the *principal root*. This is always the root which is nearest to the positive $x$-axis.

In some cases, it may be the first root. In others, it may be the last root. The only test is to see which root is nearest to the positive $x$-axis. If the first and last root are equidistant from the positive $x$-axis, the principal root is taken to be the first root.

In the example above, the *principal root* is therefore ...........

---

**36**

$z_5 = 1{\cdot}644\underline{|348^\circ}$

Good. Now here is another example worked in detail. Follow it.

We have to find the 4 fourth roots of $z = 7(\cos 80^\circ + j\sin 80^\circ)$

The first root $z_1 = 7^{\frac{1}{4}}\underline{\left|\frac{80^\circ}{4}\right.} = 7^{\frac{1}{4}}\underline{|20^\circ}$

Since $7^{\frac{1}{4}} = 1{\cdot}627$ to 3 dp

$z_1 = 1{\cdot}627\underline{|20^\circ}$

The other roots will be separated by intervals of $\frac{360^\circ}{4} = 90^\circ$

Therefore the 4 fourth roots are:

$z_1 = 1{\cdot}627\underline{|20^\circ}$

$z_2 = 1{\cdot}627\underline{|110^\circ}$

$z_3 = 1{\cdot}627\underline{|200^\circ}$

$z_4 = 1{\cdot}627\underline{|290^\circ}$

And once again, draw an Argand diagram to illustrate these roots.

**37**

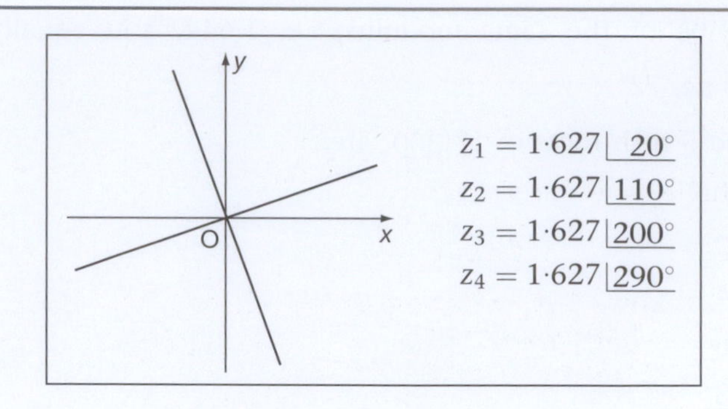


And in this example the principal fourth root is ...........

**38**

$$z_1 = 1{\cdot}627\underline{|20^\circ}$$

since it is the root nearest to the positive $x$-axis.

Now you can do one entirely on your own. Here it is.

Find the three cube roots of $6(\cos 240^\circ + j\sin 240^\circ)$. Represent them on an Argand diagram and indicate which is the principal cube root.

*When you have finished it, move on to Frame 39 and check your results*

**39**

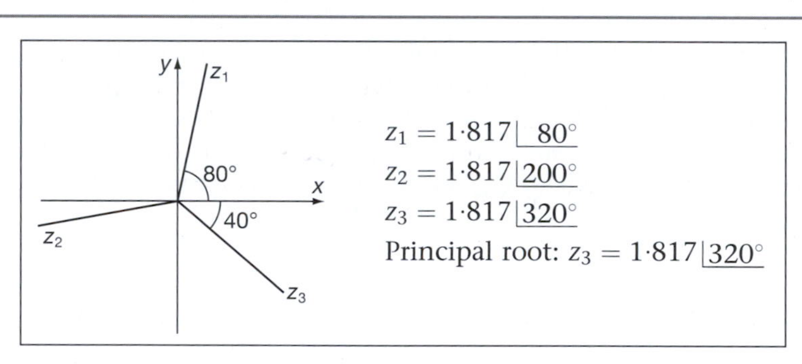


Here is the working:

$$z = 6\underline{|240^\circ} \qquad z_1 = 6^{\frac{1}{3}}\underline{\left|\frac{240^\circ}{3}\right.} = 1{\cdot}817\underline{|80^\circ}$$

$$\text{Interval between roots} = \frac{360^\circ}{3} = 120^\circ$$

Therefore the roots are:

$z_1 = 1{\cdot}817\underline{|80^\circ}$

$z_2 = 1{\cdot}817\underline{|200^\circ}$

$z_3 = 1{\cdot}817\underline{|320^\circ}$

The principal root is the root nearest to the positive $x$-axis. In this case, then, the principal root is $z_3 = 1{\cdot}817\underline{|320^\circ}$

*On to the next frame*

# Expansions

## Expansions of sin $n\theta$ and cos $n\theta$

40

By DeMoivre's theorem, we know that:

$$\cos n\theta + j\sin n\theta = (\cos\theta + j\sin\theta)^n$$

The method is simply to expand the right-hand side as a binomial series, after which we can equate real and imaginary parts.

An example will soon show you how it is done:

To find expansions for $\cos 3\theta$ and $\sin 3\theta$.

We have:

$$\cos 3\theta + j\sin 3\theta = (\cos\theta + j\sin\theta)^3$$
$$= (c + js)^3 \qquad \text{where } c \equiv \cos\theta \text{ and } s \equiv \sin\theta$$

Now expand this by the binomial series – like $(a+b)^3$ so that

$$\cos 3\theta + j\sin 3\theta = \dots\dots\dots\dots$$

41

$$c^3 + j3c^2s - 3cs^2 - js^3$$

Because

$$\cos 3\theta + j\sin 3\theta = c^3 + 3c^2(js) + 3c(js)^2 + (js)^3$$
$$= c^3 + j3c^2s - 3cs^2 - js^3 \qquad \text{since } j^2 = -1$$
$$= (c^3 - 3cs^2) + j(3c^2s - s^3) \qquad j^3 = -j$$

Now, equating real parts and imaginary parts, we get

$$\cos 3\theta = \dots\dots\dots\dots$$
and
$$\sin 3\theta = \dots\dots\dots\dots$$

42

$$\cos 3\theta = \cos^3\theta - 3\cos\theta\sin^2\theta$$
$$\sin 3\theta = 3\cos^2\theta\sin\theta - \sin^3\theta$$

If we wish, we can replace $\sin^2\theta$ by $(1 - \cos^2\theta)$
and $\cos^2\theta$ by $(1 - \sin^2\theta)$

so that we could write the results above as:

$\cos 3\theta = \dots\dots\dots$ (all in terms of $\cos\theta$)
$\sin 3\theta = \dots\dots\dots$ (all in terms of $\sin\theta$)

43

$$\cos 3\theta = 4\cos^3\theta - 3\cos\theta$$
$$\sin 3\theta = 3\sin\theta - 4\sin^3\theta$$

Because

$$\cos 3\theta = \cos^3\theta - 3\cos\theta(1 - \cos^2\theta)$$
$$= \cos^3\theta - 3\cos\theta + 3\cos^3\theta$$
$$= 4\cos^3\theta - 3\cos\theta$$

▶

and

$$\sin 3\theta = 3(1 - \sin^2\theta)\sin\theta - \sin^3\theta$$
$$= 3\sin\theta - 3\sin^3\theta - \sin^3\theta$$
$$= 3\sin\theta - 4\sin^3\theta$$

While these results are useful, it is really the method that counts.

So now do this one in just the same way:

Obtain an expression for $\cos 4\theta$ in terms of $\cos\theta$.

*When you have finished, check your result with the next frame*

**44**

$$\boxed{\cos 4\theta = 8\cos^4\theta - 8\cos^2\theta + 1}$$

Working: $\cos 4\theta + j\sin 4\theta = (\cos\theta + j\sin\theta)^4$

$$= (c + js)^4$$
$$= c^4 + 4c^3(js) + 6c^2(js)^2 + 4c(js)^3 + (js)^4$$
$$= c^4 + j4c^3s - 6c^2s^2 - j4cs^3 + s^4$$
$$= (c^4 - 6c^2s^2 + s^4) + j(4c^3s - 4cs^3)$$

Equating real parts: $\cos 4\theta = c^4 - 6c^2s^2 + s^4$

$$= c^4 - 6c^2(1 - c^2) + (1 - c^2)^2$$
$$= c^4 - 6c^2 + 6c^4 + 1 - 2c^2 + c^4$$
$$= 8c^4 - 8c^2 + 1$$
$$= 8\cos^4\theta - 8\cos^2\theta + 1$$

Now for a different problem.

*On to the next frame*

**45**

## Expansions for $\cos^n\theta$ and $\sin^n\theta$

Let $z = \cos\theta + j\sin\theta$

then $\dfrac{1}{z} = z^{-1} = \cos\theta - j\sin\theta$

$$\therefore\ z + \frac{1}{z} = 2\cos\theta \text{ and } z - \frac{1}{z} = j2\sin\theta$$

Also, by DeMoivre's theorem:

$$z^n = \cos n\theta + j\sin n\theta$$

and $\dfrac{1}{z^n} = z^{-n} = \cos n\theta - j\sin n\theta$

$$\therefore\ z^n + \frac{1}{z^n} = 2\cos n\theta \text{ and } z^n - \frac{1}{z^n} = j2\sin n\theta$$

Let us collect these four results together: $z = \cos\theta + j\sin\theta$

| | |
|---|---|
| $z + \dfrac{1}{z} = 2\cos\theta$ | $z - \dfrac{1}{z} = j2\sin\theta$ |
| $z^n + \dfrac{1}{z^n} = 2\cos n\theta$ | $z^n - \dfrac{1}{z^n} = j2\sin n\theta$ |

*Make a note of these results in your record book. Then move on and we will see how we use them*

46

We shall expand $\cos^3\theta$ as an example.

From our results: $z+\dfrac{1}{z}=2\cos\theta$

$$\therefore\ (2\cos\theta)^3=\left(z+\frac{1}{z}\right)^3$$
$$=z^3+3z^2\left(\frac{1}{z}\right)+3z\left(\frac{1}{z^2}\right)+\frac{1}{z^3}$$
$$=z^3+3z+3\frac{1}{z}+\frac{1}{z^3}$$

Now here is the trick: we rewrite this, collecting the terms up in pairs from the two extreme ends, thus:

$$(2\cos\theta)^3=\left(z^3+\frac{1}{z^3}\right)+3\left(z+\frac{1}{z}\right)$$

And, from the four results that we noted:

$$z+\frac{1}{z}=\ldots\ldots\ldots\ldots$$

and $$z^3+\frac{1}{z^3}=\ldots\ldots\ldots\ldots$$

47

$$z+\frac{1}{z}=2\cos\theta;\ \ z^3+\frac{1}{z^3}=2\cos 3\theta$$

$$\therefore\ (2\cos\theta)^3=2\cos 3\theta+3\times 2\cos\theta$$
$$8\cos^3\theta=2\cos 3\theta+6\cos\theta$$
$$4\cos^3\theta=\cos 3\theta+3\cos\theta$$
$$\cos^3\theta=\frac{1}{4}(\cos 3\theta+3\cos\theta)$$

Now one for you:

Find an expression for $\sin^4\theta$.

Work in the same way, but, this time, remember that

$z-\dfrac{1}{z}=j2\sin\theta$ and $z^n-\dfrac{1}{z^n}=j2\sin n\theta$.

*When you have obtained a result, check it with the next frame*

48

$$\sin^4\theta=\frac{1}{8}[\cos 4\theta-4\cos 2\theta+3]$$

Because we have:

$$z-\frac{1}{z}=j2\sin\theta;\ \ z^n-\frac{1}{z^n}=j2\sin n\theta$$

$$\therefore\ (j2\sin\theta)^4=\left(z-\frac{1}{z}\right)^4$$
$$=z^4-4z^3\left(\frac{1}{z}\right)+6z^2\left(\frac{1}{z^2}\right)-4z\left(\frac{1}{z^3}\right)+\frac{1}{z^4}$$
$$=\left(z^4+\frac{1}{z^4}\right)-4\left(z^2+\frac{1}{z^2}\right)+6$$

▶

Now: $z^n + \dfrac{1}{z^n} = 2\cos n\theta$

$$\therefore\ 16\sin^4\theta = 2\cos 4\theta - 4 \times 2\cos 2\theta + 6$$

$$\therefore\ \sin^4\theta = \frac{1}{8}[\cos 4\theta - 4\cos 2\theta + 3]$$

They are all done in the same way: once you know the trick, the rest is easy.

*Now let us move on to something new*

## Loci problems

**49**

We are sometimes required to find the locus of a point which moves in the Argand diagram according to some stated condition. Before we work through one or two examples of this kind, let us just revise a couple of useful points.

You will remember that when we were representing a complex number in polar form, i.e. $z = a + jb = r(\cos\theta + j\sin\theta)$, we said that:

(a) $r$ is called the *modulus* of z and is written 'mod z' or $|z|$ and

(b) $\theta$ is called the *argument* of z and is written 'arg z'.

Also $r = \sqrt{a^2 + b^2}$ and $\theta = \tan^{-1}\left\{\dfrac{b}{a}\right\}$

so that $|z| = \sqrt{a^2 + b^2}$ and $\arg z = \tan^{-1}\left\{\dfrac{b}{a}\right\}$

Similarly, if $z = x + jy$ then $|z| = \ldots\ldots\ldots\ldots$

and $\arg z = \ldots\ldots\ldots\ldots$

**50**

$$|z| = \sqrt{x^2 + y^2} \text{ and } \arg z = \tan^{-1}\left\{\frac{y}{x}\right\}$$

Keep those in mind and we are now ready to tackle some examples.

**Example 1**

If $z = x + jy$, find the locus defined as $|z| = 5$.

Now we know that in this case, $|z| = \sqrt{x^2 + y^2}$

The locus is defined as $\sqrt{x^2 + y^2} = 5$

$$\therefore\ x^2 + y^2 = 25$$

This is a circle, with centre at the origin and with radius 5.

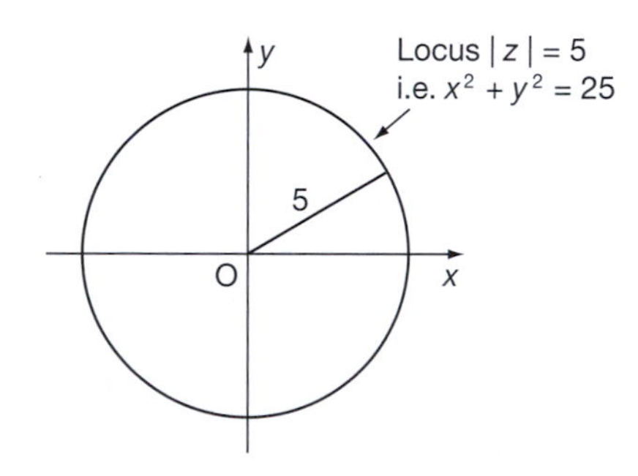


*That was easy enough. Move on for Example 2*

51

**Example 2**

If $z = x + jy$, find the locus defined as $\arg z = \dfrac{\pi}{4}$.

In this case: $\arg z = \tan^{-1}\left\{\dfrac{y}{x}\right\}$ $\therefore$ $\tan^{-1}\left\{\dfrac{y}{x}\right\} = \dfrac{\pi}{4}$

$$\therefore \frac{y}{x} = \tan\frac{\pi}{4} = \tan 45^\circ = 1 \quad \therefore \frac{y}{x} = 1 \;\therefore y = x$$

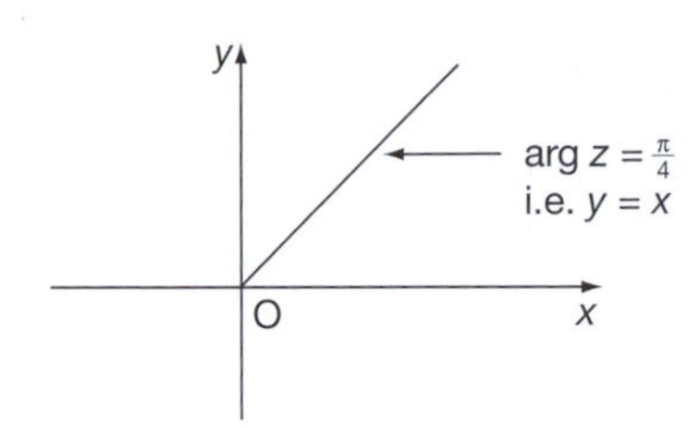


So the locus $\arg z = \dfrac{\pi}{4}$ is therefore the straight line $y = x$ and $y > 0$.

All locus problems at this stage are fundamentally of one of these kinds. Of course, the given condition may look a trifle more involved, but the approach is always the same.

*Let us look at a more complicated one. Next frame*

52

**Example 3**

If $z = x + jy$, find the equation of the locus $\left|\dfrac{z+1}{z-1}\right| = 2$.

Since $z = x + jy$:

$$z + 1 = x + jy + 1 = (x+1) + jy = r_1\angle\theta_1 = z_1$$
$$z - 1 = x + jy - 1 = (x-1) + jy = r_2\angle\theta_2 = z_2$$

$$\therefore \frac{z+1}{z-1} = \frac{r_1\angle\theta_1}{r_2\angle\theta_2} = \frac{r_1}{r_2}\angle\theta_1 - \theta_2$$

$$\therefore \left|\frac{z+1}{z-1}\right| = \frac{r_1}{r_2} = \frac{|z_1|}{|z_2|} = \frac{\sqrt{(x+1)^2 + y^2}}{\sqrt{(x-1)^2 + y^2}}$$

$$\therefore \frac{\sqrt{(x+1)^2 + y^2}}{\sqrt{(x-1)^2 + y^2}} = 2$$

$$\therefore \frac{(x+1)^2 + y^2}{(x-1)^2 + y^2} = 4$$

All that now remains is to multiply across by the denominator and tidy up the result. So finish it off in its simplest form.

*The answer is in the next frame*

**53**

$$3x^2 - 10x + 3 + 3y^2 = 0$$

We had $\dfrac{(x+1)^2 + y^2}{(x-1)^2 + y^2} = 4$

So therefore $\quad (x+1)^2 + y^2 = 4\left\{(x-1)^2 + y^2\right\}$

$$x^2 + 2x + 1 + y^2 = 4(x^2 - 2x + 1 + y^2)$$
$$= 4x^2 - 8x + 4 + 4y^2$$
$$\therefore\ 3x^2 - 10x + 3 + 3y^2 = 0$$

This is the equation of the given locus.

Although this takes longer to write out than either of the first two examples, the basic principle is the same. The given condition must be a function of either the modulus or the argument.

*Move on now to Frame 54 for Example 4*

**54**

**Example 4**

If $z = x + jy$, find the equation of the locus $\arg(z^2) = -\dfrac{\pi}{4}$.

$z = x + jy = r\underline{|\theta} \quad \therefore\ \arg z = \theta = \tan^{-1}\left\{\dfrac{y}{x}\right\}$

$\therefore\ \tan\theta = \dfrac{y}{x}$

$\therefore$ By DeMoivre's theorem, $z^2 = r^2\underline{|2\theta}$

$\therefore\ \arg(z^2) = 2\theta = -\dfrac{\pi}{4}$

$\therefore\ \tan 2\theta = \tan\left(-\dfrac{\pi}{4}\right) = -1$

$\therefore\ \dfrac{2\tan\theta}{1 - \tan^2\theta} = -1$

$\therefore\ 2\tan\theta = \tan^2\theta - 1$

But $\quad \tan\theta = \dfrac{y}{x} \quad \therefore\ \dfrac{2y}{x} = \dfrac{y^2}{x^2} - 1$

$$2xy = y^2 - x^2 \quad \therefore\ y^2 = x^2 + 2xy$$

In that example, the given condition was a function of the argument. Here is one for you to do:

If $z = x + jy$, find the equation of the locus $\arg(z + 1) = \dfrac{\pi}{3}$.

*Do it carefully; then check with the next frame*

**55**

$$y = \sqrt{3}(x + 1) \text{ for } y > 0$$

Here is the solution set out in detail.

If $z = x + jy$, find the locus $\arg(z + 1) = \dfrac{\pi}{3}$.

$z = x + jy \quad \therefore\ z + 1 = x + jy + 1 = (x + 1) + jy$

$\arg(z + 1) = \tan^{-1}\left\{\dfrac{y}{x+1}\right\} = \dfrac{\pi}{3} \quad \therefore\ \dfrac{y}{x+1} = \tan\dfrac{\pi}{3} = \sqrt{3}$

$$y = \sqrt{3}(x + 1) \text{ for } y > 0$$

And that is all there is to that.

▶

Now do this one. You will have no trouble with it.

If $z = x + jy$, find the equation of the locus $|z - 1| = 5$

*When you have finished it, move on to Frame 56*

56

$$x^2 - 2x + y^2 = 24$$

Here it is: $z = x + jy$, given locus $|z - 1| = 5$

$$z - 1 = x + jy - 1 = (x - 1) + jy$$

$$\therefore |z - 1| = \sqrt{(x-1)^2 + y^2} = 5 \qquad \therefore (x-1)^2 + y^2 = 25$$

$$\therefore x^2 - 2x + 1 + y^2 = 25 \qquad \therefore x^2 - 2x + y^2 = 24$$

Every one is very much the same.

This brings us to the end of this Programme, except for the final **Can You?** checklist and **Test exercise**. Before you work through them, read down the **Review summary** (Frame 57), just to refresh your memory of what we have covered in this Programme.

*So on now to Frame 57*

## Review summary

57

**1**

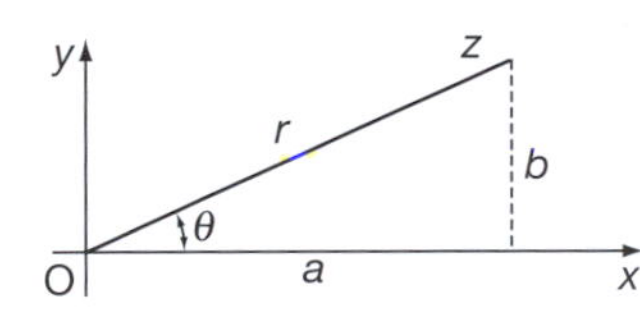


*Polar form of a complex number*

$$z = a + jb = r(\cos\theta + j\sin\theta) = r\underline{|\theta}$$

$$r = \text{mod } z = |z| = \sqrt{a^2 + b^2}$$

$$\theta = \arg z = \tan^{-1}\left\{\frac{b}{a}\right\}$$

**2**

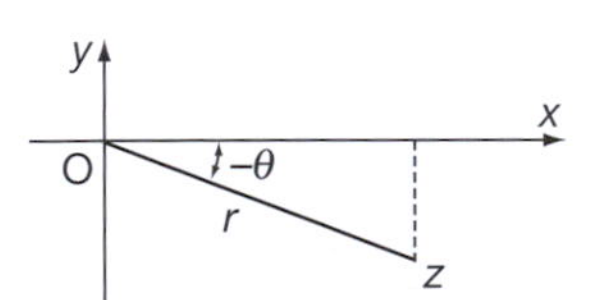


*Negative angles*

$$z = r(\cos[-\theta] + j\sin[-\theta])$$

$$\cos[-\theta] = \cos\theta$$

$$\sin[-\theta] = -\sin\theta$$

$$\therefore z = r(\cos\theta - j\sin\theta) = r\overline{|\theta}$$

**3** *Multiplication and division in polar form*

If $\quad z_1 = r_1\underline{|\theta_1}; \quad z_2 = r_2\underline{|\theta_2}$

then $\quad z_1 z_2 = r_1 r_2\underline{|\theta_1 + \theta_2}$

$$\frac{z_1}{z_2} = \frac{r_1}{r_2}\underline{|\theta_1 - \theta_2}$$

**4** *DeMoivre's theorem*

If $z = r(\cos\theta + j\sin\theta)$, then $z^n = r^n(\cos n\theta + j\sin n\theta)$

**5** *Loci problems*

If $\quad z = x + jy, \quad |z| = \sqrt{x^2 + y^2}$

$$\arg z = \tan^{-1}\left\{\frac{y}{x}\right\}$$

*That's it! Now you are ready for the **Can You?** checklist in Frame 58 and the **Test exercise** in Frame 59*

## Can You?

**58**

### Checklist 2

Check this list before and after you try the end of Programme test.

On a scale of 1 to 5 how confident are you that you can: Frames

- Use the shorthand form for a complex number in polar form? [1] to [6]
  *Yes* ☐ ☐ ☐ ☐ ☐ *No*
- Write complex numbers in polar form using negative angles? [7] to [12]
  *Yes* ☐ ☐ ☐ ☐ ☐ *No*
- Multiply and divide complex numbers in polar form? [13] to [20]
  *Yes* ☐ ☐ ☐ ☐ ☐ *No*
- Use DeMoivre's theorem? [23] to [26]
  *Yes* ☐ ☐ ☐ ☐ ☐ *No*
- Find the roots of a complex number? [27] to [39]
  *Yes* ☐ ☐ ☐ ☐ ☐ *No*
- Demonstrate trigonometric identities of multiple angles using complex numbers? [40] to [48]
  *Yes* ☐ ☐ ☐ ☐ ☐ *No*
- Solve loci problems using complex numbers? [49] to [56]
  *Yes* ☐ ☐ ☐ ☐ ☐ *No*

## Test exercise 2

**59**

1 Express in polar form, $z = -5 - j3$.

2 Express in the form $a + jb$:

(a) $2\underline{|156^{\circ}}$

(b) $5\overline{|37^{\circ}}$

3 If $z_1 = 12(\cos 125^{\circ} + j \sin 125^{\circ})$ and $z_2 = 3(\cos 72^{\circ} + j \sin 72^{\circ})$, find (a) $z_1 z_2$ and (b) $\dfrac{z_1}{z_2}$ giving the results in polar form.

4 If $z = 2(\cos 25^{\circ} + j \sin 25^{\circ})$, find $z^3$ in polar form.

5 Find the three cube roots of $8(\cos 264^{\circ} + j \sin 264^{\circ})$ and state which of them is the principal cube root. Show all three roots on an Argand diagram.

6 Expand $\sin 4\theta$ in powers of $\sin\theta$ and $\cos\theta$.

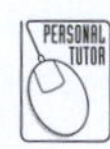


7 Express $\cos^4\theta$ in terms of cosines of multiples of $\theta$.

8 If $z = x + jy$, find the equations of the two loci defined by:

(a) $|z - 4| = 3$ (b) $\arg(z + 2) = \dfrac{\pi}{6}$

## Further problems 2

60

1 If $z = x + jy$, where $x$ and $y$ are real, find the values of $x$ and $y$ when $\dfrac{3z}{1-j} + \dfrac{3z}{j} = \dfrac{4}{3-j}$.

2 In the Argand diagram, the origin is the centre of an equilateral triangle and one vertex of the triangle is the point $3 + j\sqrt{3}$. Find the complex numbers representing the other vertices.

3 Express $2 + j3$ and $1 - j2$ in polar form and apply DeMoivre's theorem to evaluate $\dfrac{(2+j3)^4}{1-j2}$. Express the result in the form $a + jb$ and in exponential form.

4 Find the fifth roots of $-3 + j3$ in polar form and in exponential form.

5 Express $5 + j12$ in polar form and hence evaluate the principal value of $\sqrt[3]{(5+j12)}$, giving the results in the form $a + jb$ and in the form $re^{j\theta}$.

6 Determine the fourth roots of $-16$, giving the results in the form $a + jb$.

7 Find the fifth roots of $-1$, giving the results in polar form. Express the principal root in the form $re^{j\theta}$.

8 Determine the roots of the equation $x^3 + 64 = 0$ in the form $a + jb$, where $a$ and $b$ are real.

9 Determine the three cube roots of $\dfrac{2-j}{2+j}$ giving the result in modulus/argument form. Express the principal root in the form $a + jb$.

10 Show that the equation $z^3 = 1$ has one real root and two other roots which are not real, and that, if one of the non-real roots is denoted by $\omega$, the other is then $\omega^2$. Mark on the Argand diagram the points which represent the three roots and show that they are the vertices of an equilateral triangle.

11 Determine the fifth roots of $(2 - j5)$, giving the results in modulus/argument form. Express the principal root in the form $a + jb$ and in the form $re^{j\theta}$.

12 Solve the equation $z^2 + 2(1 + j)z + 2 = 0$, giving each result in the form $a + jb$, with $a$ and $b$ correct to 2 places of decimals.

13 Express $e^{1-j\pi/2}$ in the form $a + jb$.

14 Obtain the expansion of $\sin 7\theta$ in powers of $\sin\theta$.

15 Express $\sin^6 x$ as a series of terms which are cosines of angles that are multiples of $x$.

16 If $z = x + jy$, where $x$ and $y$ are real, show that the locus $\left|\dfrac{z-2}{z+2}\right| = 2$ is a circle and determine its centre and radius.

17 If $z = x + jy$, show that the locus $\arg\left\{\dfrac{z-1}{z-j}\right\} = \dfrac{\pi}{6}$ is a circle. Find its centre and radius.

▶

18 If $z = x + jy$, determine the Cartesian equation of the locus of the point $z$ which moves in the Argand diagram so that

$|z + j2|^2 + |z - j2|^2 = 40$.

19 If $z = x + jy$, determine the equations of the two loci:

(a) $\left|\dfrac{z+2}{z}\right| = 3$ and (b) $\arg\left\{\dfrac{z+2}{z}\right\} = \dfrac{\pi}{4}$

20 If $z = x + jy$, determine the equations of the loci in the Argand diagram, defined by:

(a) $\left|\dfrac{z+2}{z-1}\right| = 2$ and (b) $\arg\left\{\dfrac{z-1}{z+2}\right\} = \dfrac{\pi}{2}$

21 Prove that:

(a) if $|z_1 + z_2| = |z_1 - z_2|$, the difference of the arguments of $z_1$ and $z_2$ is $\dfrac{\pi}{2}$

(b) if $\arg\left\{\dfrac{z_1 + z_2}{z_1 - z_2}\right\} = \dfrac{\pi}{2}$, then $|z_1| = |z_2|$

22 If $z = x + jy$, determine the loci in the Argand diagram, defined by:

(a) $|z + j2|^2 - |z - j2|^2 = 24$

(b) $|z + jk|^2 + |z - jk|^2 = 10k^2 \quad (k > 0)$

**61**

Now visit the companion website for this book at www.palgrave.com/stroud for more questions applying this mathematics to science and engineering.

# Hyperbolic functions

## Learning outcomes

When you have completed this Programme you will be able to:

- ☐ Define the hyperbolic functions in terms of the exponential function
- ☐ Express the hyperbolic functions as power series
- ☐ Recognize the graphs of the hyperbolic functions
- ☐ Evaluate hyperbolic functions and their inverses
- ☐ Determine the logarithmic form of the inverse hyperbolic functions
- ☐ Prove hyperbolic trigonometric identities
- ☐ Understand the relationship between the circular and the hyperbolic trigonometric functions

# Introduction

**1**

The cosine of an angle was first defined as the ratio of two sides of a right-angled triangle - adjacent over hypotenuse. In Programme F.11 of Part I you learnt how to extend the definition of a sine and a cosine to *any* angle, positive or negative. You might just check that out to refresh your memory by re-reading Frames 2 to 4 of Programme F.11.

Now, in Frames 56 to 59 of Programme 1 in Part II you learnt how a complex number of unit length could be written in either polar or exponential form, giving rise to the equations:

$$\cos\theta + j\sin\theta = e^{j\theta}$$
$$\cos\theta - j\sin\theta = e^{-j\theta}$$

If these two equations are added you find that:

$$2\cos\theta = e^{j\theta} + e^{-j\theta} \text{ so that } \cos\theta = \frac{e^{j\theta} + e^{-j\theta}}{2}$$

If $\theta$ is replaced by $jx$ in this last equation you find that:

$$\cos jx = \frac{e^{jjx} + e^{-jjx}}{2} = \frac{e^{-x} + e^{x}}{2}$$

where the right-hand side is *entirely real*. In fact, you have seen this before in Frame 50 of Programme F.11, it is the even part of the exponential function which is called the *hyperbolic cosine*:

$$\cosh x = \frac{e^{x} + e^{-x}}{2} \text{ so that } \cos jx = \cosh x$$

The graph of $y = \cosh x$ is called a *catenary* from the Latin word *catena* meaning chain because the shape of the graph is the shape of a hanging chain.

*Move on to Frame 2 and start the Programme*

**2**

You may remember that of the many functions that can be expressed as a series of powers of $x$, a common one is $e^x$:

$$e^x = 1 + x + \frac{x^2}{2!} + \frac{x^3}{3!} + \frac{x^4}{4!} + \dots$$

If we replace $x$ by $-x$, we get:

$$e^{-x} = 1 - x + \frac{x^2}{2!} - \frac{x^3}{3!} + \frac{x^4}{4!} - \dots$$

and these two functions $e^x$ and $e^{-x}$ are the foundations of the definitions we are going to use.

(a) If we take the value of $e^x$, subtract $e^{-x}$, and divide by 2, we form what is defined as the hyperbolic sine of $x$:

$$\frac{e^x - e^{-x}}{2} = \textit{hyperbolic sine} \text{ of } x$$

This is a lot to write every time we wish to refer to it, so we shorten it to sinh $x$, the $h$ indicating its connection with the hyperbola. We pronounce it 'shine $x$'.

$$\frac{e^x - e^{-x}}{2} = \sinh x$$

So, in the same way, $\dfrac{e^y - e^{-y}}{2}$ would be written as ...........

**3**

$\boxed{\sinh y}$

In much the same way, we have two other definitions:

(b) $\dfrac{e^x + e^{-x}}{2} =$ *hyperbolic cosine* of $x$

$= \cosh x$ [pronounced 'cosh $x$']

(c) $\dfrac{e^x - e^{-x}}{e^x + e^{-x}} =$ *hyperbolic tangent* of $x$

$= \tanh x$ [pronounced 'than $x$']

We must start off by learning these definitions, for all the subsequent developments depend on them.

So now then; what was the definition of sinh $x$?

$\sinh x = \ldots\ldots\ldots\ldots$

**4**

$$\boxed{\sinh x = \frac{e^x - e^{-x}}{2}}$$

Here they are together so that you can compare them:

$$\sinh x = \frac{e^x - e^{-x}}{2}$$
$$\cosh x = \frac{e^x + e^{-x}}{2}$$
$$\tanh x = \frac{e^x - e^{-x}}{e^x + e^{-x}}$$

Make a copy of these in your record book for future reference when necessary.

**5**

$$\boxed{\sinh x = \frac{e^x - e^{-x}}{2};\quad \cosh x = \frac{e^x + e^{-x}}{2};\quad \tanh x = \frac{e^x - e^{-x}}{e^x + e^{-x}}}$$

We started the programme by referring to $e^x$ and $e^{-x}$ as series of powers of $x$. It should not be difficult therefore to find series at least for $\sinh x$ and for $\cosh x$. Let us try.

(a) *Series for sinh x*

$$e^x = 1 + x + \frac{x^2}{2!} + \frac{x^3}{3!} + \frac{x^4}{4!} + \ldots$$
$$e^{-x} = 1 - x + \frac{x^2}{2!} - \frac{x^3}{3!} + \frac{x^4}{4!} - \ldots$$

If we subtract, we get:

$$e^x - e^{-x} = 2x + \frac{2x^3}{3!} + \frac{2x^5}{5!} \ldots$$

Divide by 2:

$$\frac{e^x - e^{-x}}{2} = \sinh x = x + \frac{x^3}{3!} + \frac{x^5}{5!} + \ldots$$

(b) If we add the series for $e^x$ and $e^{-x}$, we get a similar result. What is it?

*When you have decided, move on to Frame 6*

**6**

$$\cosh x = 1 + \frac{x^2}{2!} + \frac{x^4}{4!} + \frac{x^6}{6!} + \dots$$

Because we have:

$$e^x = 1 + x + \frac{x^2}{2!} + \frac{x^3}{3!} + \frac{x^4}{4!} + \dots$$

$$e^{-x} = 1 - x + \frac{x^2}{2!} - \frac{x^3}{3!} + \frac{x^4}{4!} - \dots$$

$$\therefore \; e^x + e^{-x} = 2 + \frac{2x^2}{2!} + \frac{2x^4}{4!} + \dots$$

$$\therefore \; \frac{e^x + e^{-x}}{2} = \cosh x = 1 + \frac{x^2}{2!} + \frac{x^4}{4!} + \dots$$

*Move on to Frame 7*

**7**

So we have:

$$\sinh x = x + \frac{x^3}{3!} + \frac{x^5}{5!} + \frac{x^7}{7!} + \dots$$

$$\cosh x = 1 + \frac{x^2}{2!} + \frac{x^4}{4!} + \frac{x^6}{6!} + \dots$$

*Note*: All terms positive: $\sinh x$ has all the odd powers
$\cosh x$ has all the even powers

We cannot easily get a series for $\tanh x$ by this process, so we will leave that one to some other time.

Make a note of these two series in your record book. Then, cover up what you have done so far and see if you can write down the definitions of:

(a) $\sinh x = \dots\dots\dots\dots$

(b) $\cosh x = \dots\dots\dots\dots$

(c) $\tanh x = \dots\dots\dots\dots$ No looking!

**8**

$$\sinh x = \frac{e^x - e^{-x}}{2}; \quad \cosh x = \frac{e^x + e^{-x}}{2}; \quad \tanh x = \frac{e^x - e^{-x}}{e^x + e^{-x}}$$

All correct? Right.

## Graphs of hyperbolic functions

**9**

We shall get to know quite a lot about these hyperbolic functions if we sketch the graphs of these functions. Since they depend on the values of $e^x$ and $e^{-x}$, we had better just refresh our memories of what these graphs look like.

▶

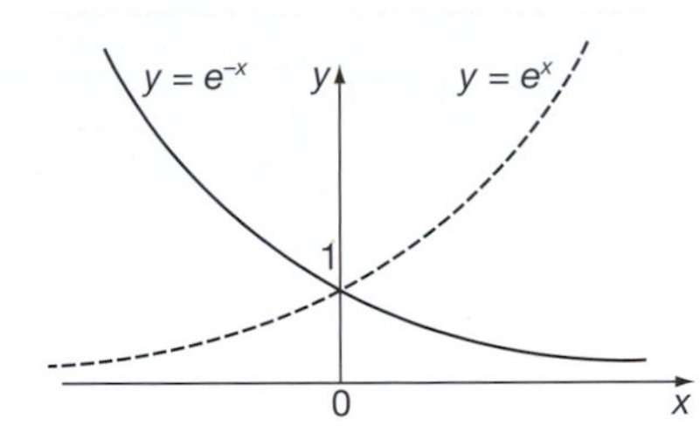


$y = e^x$ and $y = e^{-x}$ cross the $y$-axis at the point $y = 1$ $(e^0 = 1)$. Each graph then approaches the $x$-axis as an asymptote, getting nearer and nearer to it as it goes away to infinity in each direction, without actually crossing it,.

So, for what range of values of $x$ are $e^x$ and $e^{-x}$ positive?

**10**

$e^x$ and $e^{-x}$ are positive for all values of $x$

Correct, since the graphs are always above the $x$-axis.

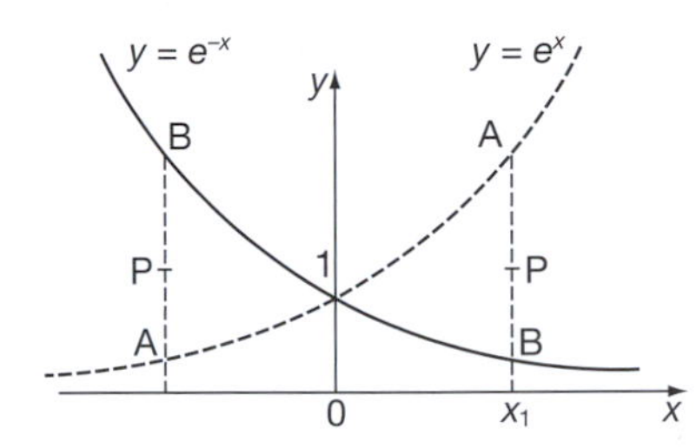


At any value of $x$, e.g. $x = x_1$,

$$\cosh x = \frac{e^x + e^{-x}}{2},$$

i.e. the value of $\cosh x$ is the average of the values of $e^x$ and $e^{-x}$ at that value of $x$. This is given by P, the mid-point of AB.

If we can imagine a number of ordinates (or verticals) like AB and we plot their mid-points, we shall obtain the graph of $y = \cosh x$.

Can you sketch in what the graph will look like?

**11**

Here it is:

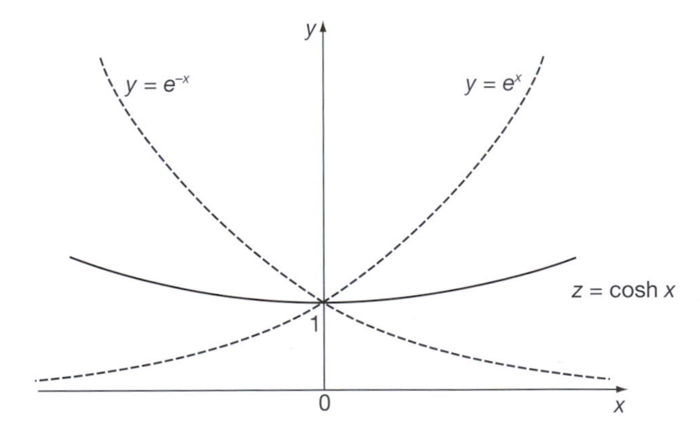


We see from the graph of $y = \cosh x$ that:

(a) $\cosh 0 = 1$

(b) the value of $\cosh x$ is never less than 1

(c) the curve is symmetrical about the $y$-axis,

i.e. $\cosh(-x) = \cosh x$

(d) for any given value of $\cosh x$, there are two values of $x$, equally spaced about the origin, i.e. $x = \pm a$.

Now let us see about the graph of $y = \sinh x$ in the same sort of way.

**12**

$$\sinh x = \frac{e^x - e^{-x}}{2}$$

On the diagram:

$CA = e^x$
$CB = e^{-x}$
$BA = e^x - e^{-x}$
$BP = \dfrac{e^x - e^{-x}}{2}$

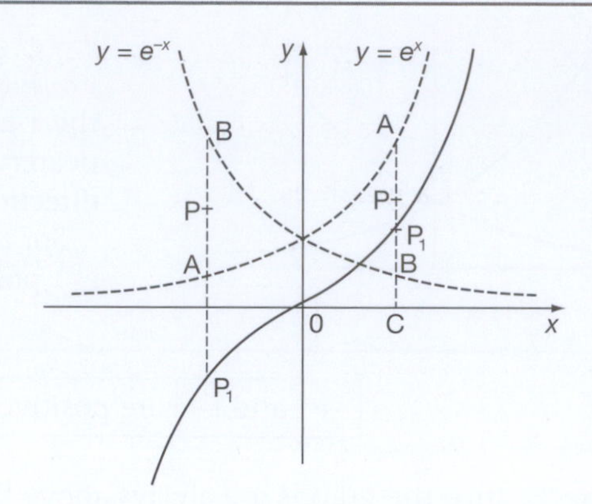


The corresponding point on the graph of $y = \sinh x$ is thus obtained by standing the ordinate BP on the $x$-axis at C, i.e. $P_1$.

Note that on the left of the origin, BP is negative and is therefore placed below the $x$-axis.

So what can we say about $y = \sinh x$?

**13**

From the graph of $y = \sinh x$, we see:

(a) $\sinh 0 = 0$

(b) $\sinh x$ can have all values from $-\infty$ to $+\infty$

(c) the curve is symmetrical about the origin, i.e.

$$\sinh(-x) = -\sinh x$$

(d) for a given value of $\sinh x$, there is only one real value of $x$.

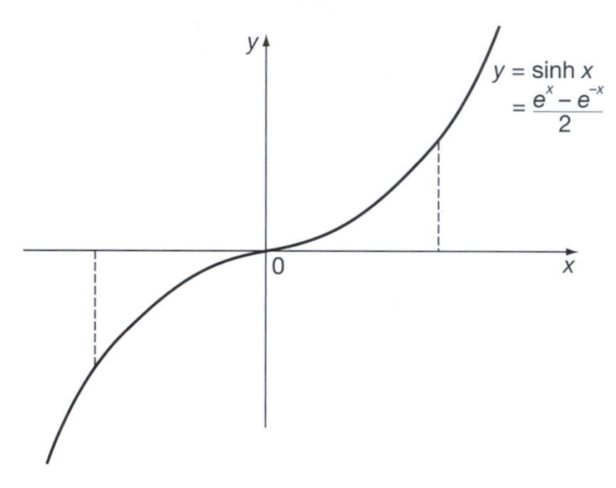


If we draw $y = \sinh x$ and $y = \cosh x$ on the same graph, what do we get?

**14**

Note that $y = \sinh x$ is always outside $y = \cosh x$, but gets nearer to it as $x$ increases:

i.e. as $x \to \infty$, $\sinh x \to \cosh x$

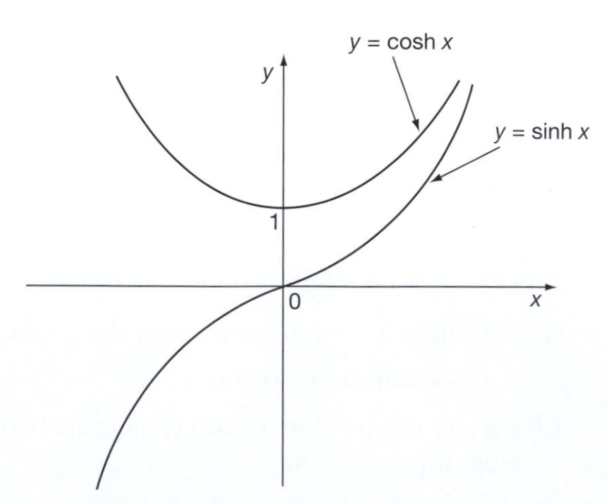


And now let us consider the graph of $y = \tanh x$.

*Move on*

**15**

It is not easy to build $y = \tanh x$ directly from the graphs of $y = e^x$ and $y = e^{-x}$.

If, however, we take values of $e^x$ and $e^{-x}$ and then calculate $y = \dfrac{e^x - e^{-x}}{e^x + e^{-x}}$ and plot points, we get a graph as shown:

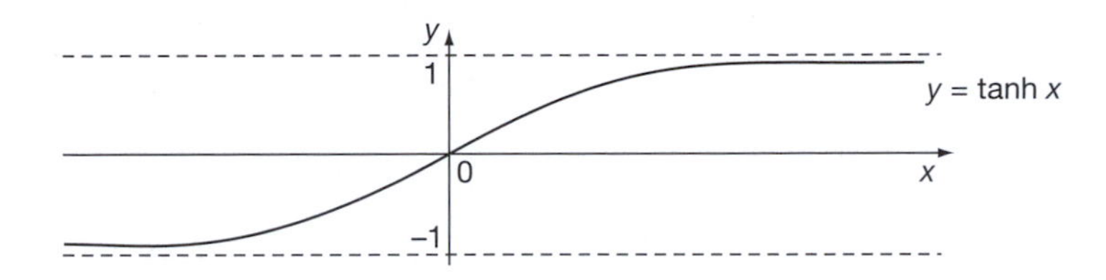


We see:

(a) $\tanh 0 = 0$

(b) $\tanh x$ always lies between $y = -1$ and $y = 1$

(c) $\tanh(-x) = -\tanh x$

(d) as $x \to \infty$, $\tanh x \to 1$

as $x \to -\infty$, $\tanh x \to -1$

Finally, let us now sketch all three graphs on one diagram so that we can compare them and distinguish between them.

**16**

Here they are:

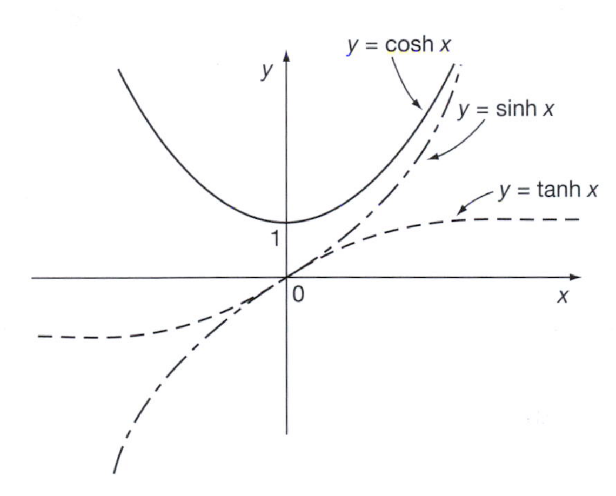


One further point to note:

At the origin, $y = \sinh x$ and $y = \tanh x$ coincide and have the same slope.

It is worthwhile to remember this combined diagram: sketch it in your record book for reference.

## Review exercise

**17** Fill in the following:

(a) $\dfrac{e^x + e^{-x}}{2} = \ldots\ldots\ldots\ldots$

(b) $\dfrac{e^x - e^{-x}}{e^x + e^{-x}} = \ldots\ldots\ldots\ldots$

(c) $\dfrac{e^x - e^{-x}}{2} = \ldots\ldots\ldots\ldots$

(d)

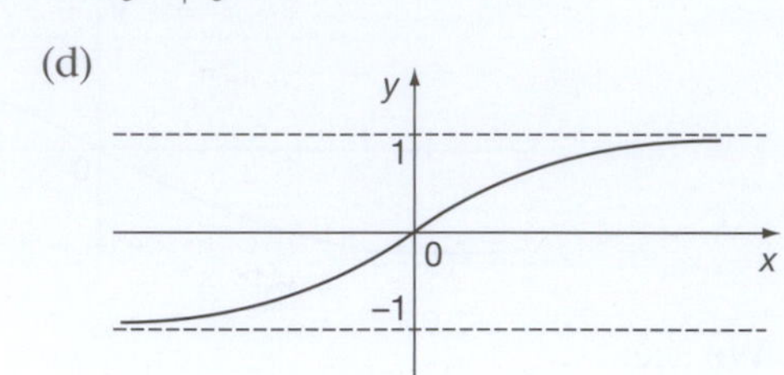


$y = \ldots\ldots\ldots\ldots$

(e)

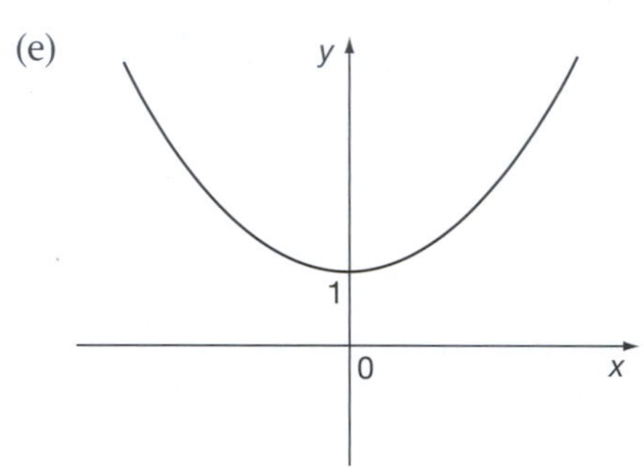


$y = \ldots\ldots\ldots\ldots$

(f)

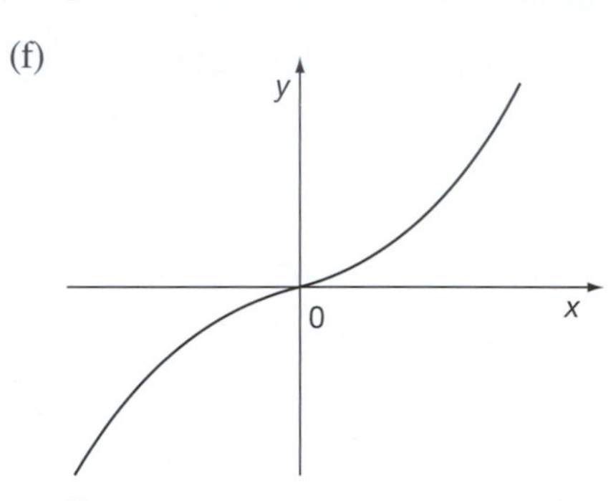


$y = \ldots\ldots\ldots\ldots$

*Results in the next frame. Check your answers carefully*

**18** Here are the results: check yours.

(a) $\dfrac{e^x + e^{-x}}{2} = \cosh x$

(b) $\dfrac{e^x - e^{-x}}{e^x + e^{-x}} = \tanh x$

(c) $\dfrac{e^x - e^{-x}}{2} = \sinh x$

(d)

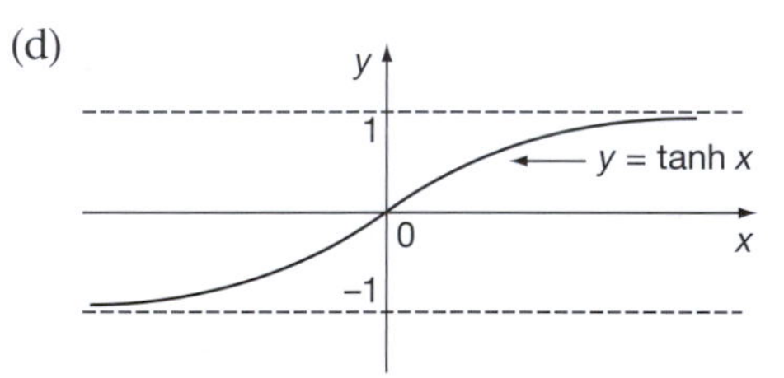


$y = \tanh x$

(e)

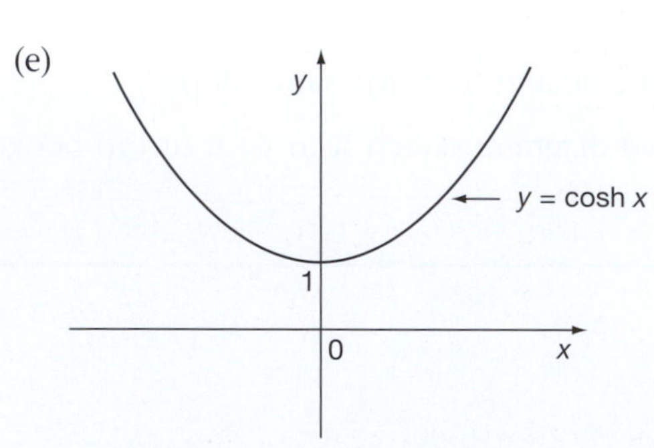


$y = \cosh x$

(f)

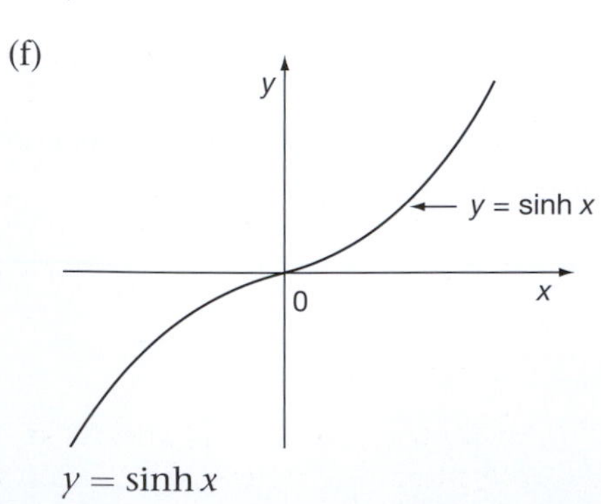


$y = \sinh x$

*Now we can continue with the next piece of work*

## Evaluation of hyperbolic functions

**19**

The values of $\sinh x$, $\cosh x$ and $\tanh x$ can be found using a calculator in just the same manner as the values of the circular trigonometric expressions were found. However, if your calculator does not possess the facility to work out hyperbolic expressions then their values can still be found by using the exponential key instead.

**Example 1**

To evaluate sinh 1·275

$$\text{Now } \sinh x = \frac{1}{2}(e^{x} - e^{-x}) \quad \therefore \quad \sinh 1{\cdot}275 = \frac{1}{2}(e^{1{\cdot}275} - e^{-1{\cdot}275}).$$

We now have to evaluate $e^{1{\cdot}275}$ and $e^{-1{\cdot}275}$.

Using your calculator, you will find that:

$$e^{1{\cdot}275} = 3{\cdot}579 \text{ and } e^{-1{\cdot}275} = \frac{1}{3{\cdot}579} = 0{\cdot}2794$$

$$\therefore \quad \sinh 1{\cdot}275 = \frac{1}{2}(3{\cdot}579 - 0{\cdot}279)$$

$$= \frac{1}{2}(3{\cdot}300) = 1{\cdot}65$$

$$\therefore \quad \sinh 1{\cdot}275 = 1{\cdot}65$$

In the same way, you now find the value of cosh 2·156.

*When you have finished, move on to Frame 20*

**20**

$$\boxed{\cosh 2{\cdot}156 = 4{\cdot}38}$$

Here is the working:

**Example 2**

$$\cosh 2{\cdot}156 = \frac{1}{2}(e^{2{\cdot}156} + e^{-2{\cdot}156})$$

$$\therefore \quad \cosh 2{\cdot}156 = \frac{1}{2}(8{\cdot}637 + 0{\cdot}116)$$

$$= \frac{1}{2}(8{\cdot}753) = 4{\cdot}377$$

$$\therefore \quad \cosh 2{\cdot}156 = 4{\cdot}38$$

Right, one more. Find the value of tanh 1·27.

*When you have finished, move on to Frame 21*

**21**

$$\boxed{\tanh 1{\cdot}27 = 0{\cdot}8538}$$

Here is the working.

**Example 3**

$$\tanh 1{\cdot}27 = \frac{e^{1{\cdot}27} - e^{-1{\cdot}27}}{e^{1{\cdot}27} + e^{-1{\cdot}27}}$$

$$\therefore \quad \tanh 1{\cdot}27 = \frac{3{\cdot}561 - 0{\cdot}281}{3{\cdot}561 + 0{\cdot}281} = \frac{3{\cdot}280}{3{\cdot}842}$$

$$\tanh 1{\cdot}27 = 0{\cdot}8538$$

So, evaluating sinh, cosh and tanh is easy enough and depends mainly on being able to evaluate $e^{k}$, where $k$ is a given number.

*And now let us look at the reverse process. So on to Frame 22*

## Inverse hyperbolic functions

**22**

**Example 1**

Find $\sinh^{-1} 1{\cdot}475$, i.e. find the value of $x$ such that $\sinh x = 1{\cdot}475$.

Here it is:

$$\sinh x = 1{\cdot}475 \quad \therefore \quad \frac{1}{2}(e^x - e^{-x}) = 1{\cdot}475$$

$$\therefore \; e^x - \frac{1}{e^x} = 2{\cdot}950$$

Multiplying both sides by $e^x$ : $(e^x)^2 - 1 = 2{\cdot}95(e^x)$

$$(e^x)^2 - 2{\cdot}95(e^x) - 1 = 0$$

This is a quadratic equation and can be solved as usual, giving:

$$e^x = \frac{2{\cdot}95 \pm \sqrt{2{\cdot}95^2 + 4}}{2} = \frac{2{\cdot}95 \pm \sqrt{8{\cdot}703 + 4}}{2}$$

$$= \frac{2{\cdot}95 \pm \sqrt{12{\cdot}703}}{2} = \frac{2{\cdot}95 \pm 3{\cdot}564}{2}$$

$$= \frac{6{\cdot}514}{2} \text{ or } -\frac{0{\cdot}614}{2} = 3{\cdot}257 \text{ or } -0{\cdot}307$$

But $e^x$ is always positive for real values of $x$. Therefore the only real solution is given by $e^x = 3{\cdot}257$.

$$\therefore \; x = \ln 3{\cdot}257 = 1{\cdot}1808$$

$$\therefore \; x = 1{\cdot}1808$$

**Example 2**

Now you find $\cosh^{-1} 2{\cdot}364$ in the same way.

**23**

$$\boxed{\cosh^{-1} 2{\cdot}364 = \pm 1{\cdot}5054 \text{ to 4 dp}}$$

Because

To evaluate $\cosh^{-1} 2{\cdot}364$, let $x = \cosh^{-1} 2{\cdot}364$

$$\therefore \; \cosh x = 2{\cdot}364 \quad \therefore \quad \frac{e^x + e^{-x}}{2} = 2{\cdot}364 \quad \therefore \; e^x + \frac{1}{e^x} = 4{\cdot}728$$

$$(e^x)^2 - 4{\cdot}728(e^x) + 1 = 0$$

$$e^x = \frac{4{\cdot}728 \pm \sqrt{4{\cdot}728^2 - 4}}{2}$$

$$= \frac{1}{2}(4{\cdot}728 \pm 4{\cdot}284\ldots)$$

$$= 4{\cdot}5060\ldots \text{ or } 0{\cdot}2219\ldots$$

Therefore:

$$x = \ln 4{\cdot}5060\ldots \text{ or } \ln 0{\cdot}2219\ldots$$

$$= \pm 1{\cdot}5054 \text{ to 4 dp}$$

Before we do the next one, do you remember the exponential definition of $\tanh x$? Well, what is it?

**24**

$$\tanh x = \frac{e^x - e^{-x}}{e^x + e^{-x}}$$

That being so, we can now evaluate $\tanh^{-1} 0{\cdot}623$.

Let $x = \tanh^{-1} 0{\cdot}623 \qquad \therefore \quad \tanh x = 0{\cdot}623$

$$\therefore \quad \frac{e^x - e^{-x}}{e^x + e^{-x}} = 0{\cdot}623$$

$$\therefore \quad e^x - e^{-x} = 0{\cdot}623(e^x + e^{-x})$$

$$\therefore \quad (1 - 0{\cdot}623)e^x = (1 + 0{\cdot}623)e^{-x}$$

$$0{\cdot}377e^x = 1{\cdot}623e^{-x}$$

$$= \frac{1{\cdot}623}{e^x}$$

$$\therefore \quad (e^x)^2 = \frac{1{\cdot}623}{0{\cdot}377}$$

$$\therefore \quad e^x = 2{\cdot}075$$

$$\therefore \quad x = \ln 2{\cdot}075 = 0{\cdot}7299$$

$$\therefore \quad \tanh^{-1} 0{\cdot}623 = 0{\cdot}730$$

Now one for you to do on your own. Evaluate $\sinh^{-1} 0{\cdot}5$.

**25**

$$\sinh^{-1} 0{\cdot}5 = 0{\cdot}4812$$

Check your working.

Let $x = \sinh^{-1} 0{\cdot}5 \qquad \therefore \quad \sinh x = 0{\cdot}5$

$$\therefore \quad \frac{e^x - e^{-x}}{2} = 0{\cdot}5 \qquad \therefore \quad e^x - \frac{1}{e^x} = 1$$

$$\therefore \quad (e^x)^2 - 1 = e^x$$

$$\therefore \quad (e^x)^2 - (e^x) - 1 = 0$$

$$e^x = \frac{1 \pm \sqrt{1+4}}{2} = \frac{1 \pm \sqrt{5}}{2}$$

$$= \frac{3{\cdot}2361}{2} \text{ or } \frac{-1{\cdot}2361}{2}$$

$$= 1{\cdot}6180 \text{ or } -0{\cdot}6180 \qquad e^x = -0{\cdot}6180 \text{ gives no real value of } x$$

$$\therefore \quad x = \ln 1{\cdot}6180 = 0{\cdot}4812$$

$$\sinh^{-1} 0{\cdot}5 = 0{\cdot}4812$$

And just one more! Evaluate $\tanh^{-1} 0{\cdot}75$.

**26**

$$\boxed{\tanh^{-1} 0{\cdot}75 = 0{\cdot}9730}$$

Let $x = \tanh^{-1} 0{\cdot}75 \quad \therefore \quad \tanh x = 0{\cdot}75$

$$\therefore \quad \frac{e^x - e^{-x}}{e^x + e^{-x}} = 0{\cdot}75$$

$$e^x - e^{-x} = 0{\cdot}75(e^x + e^{-x})$$

$$(1 - 0{\cdot}75)e^x = (1 + 0{\cdot}75)e^{-x}$$

$$0{\cdot}25e^x = 1{\cdot}75e^{-x}$$

$$(e^x)^2 = \frac{1{\cdot}75}{0{\cdot}25} = 7$$

$$e^x = \pm\sqrt{7} = \pm 2{\cdot}6458$$

But remember that $e^x$ cannot be negative for real values of $x$.

Therefore $e^x = 2{\cdot}6458$ is the only real solution.

$$\therefore \quad x = \ln 2{\cdot}6458 = 0{\cdot}9730$$

$$\tanh^{-1} 0{\cdot}75 = 0{\cdot}9730$$

## Log form of the inverse hyperbolic functions

**27**

Let us do the same thing in a general way.

To find $\tanh^{-1} x$ in log form.

As usual, we start off with:

Let $y = \tanh^{-1} x \quad \therefore \quad x = \tanh y$

$$\therefore \quad \frac{e^y - e^{-y}}{e^y + e^{-y}} = x \quad \therefore \quad e^y - e^{-y} = x(e^y + e^{-y})$$

$$e^y(1 - x) = e^{-y}(1 + x) = \frac{1}{e^y}(1 + x)$$

$$e^{2y} = \frac{1 + x}{1 - x}$$

$$\therefore \quad 2y = \ln\left\{\frac{1 + x}{1 - x}\right\}$$

$$\therefore \quad y = \tanh^{-1} x = \frac{1}{2}\ln\left\{\frac{1 + x}{1 - x}\right\}$$

So that: $\tanh^{-1} 0{\cdot}5 = \frac{1}{2}\ln\left\{\frac{1{\cdot}5}{0{\cdot}5}\right\}$

$$= \frac{1}{2}\ln 3 = \frac{1}{2}(1{\cdot}0986) = 0{\cdot}5493$$

And similarly: $\tanh^{-1}(-0{\cdot}6) = \ldots\ldots\ldots\ldots$

**28**

$$\boxed{\tanh^{-1}(-0{\cdot}6) = -0{\cdot}6931}$$

Because

$$\tanh^{-1} x = \frac{1}{2}\ln\left\{\frac{1+x}{1-x}\right\}$$

$$\therefore\ \tanh^{-1}(-0{\cdot}6) = \frac{1+(-0{\cdot}6)}{1-(-0{\cdot}6)}$$

$$= \frac{1}{2}\ln\left\{\frac{1-0{\cdot}6}{1+0{\cdot}6}\right\}$$

$$= \frac{1}{2}\ln\left\{\frac{0{\cdot}4}{1{\cdot}6}\right\}$$

$$= \frac{1}{2}\ln 0{\cdot}25$$

$$= \frac{1}{2}(-1{\cdot}3863)$$

$$= -0{\cdot}6931$$

Now, in the same way, find an expression for $\sinh^{-1} x$.

Start off by saying: Let $y = \sinh^{-1} x \quad \therefore\ x = \sinh y$

$$\therefore\ \frac{e^y - e^{-y}}{2} = x \quad \therefore\ e^y - e^{-y} = 2x \quad \therefore\ e^y - \frac{1}{e^y} = 2x$$

$$(e^y)^2 - 2x(e^y) - 1 = 0$$

*Now finish it off – result in Frame 29*

**29**

$$\boxed{\sinh^{-1} x = \ln\left\{x + \sqrt{x^2+1}\right\}}$$

Because

$$(e^y)^2 - 2x(e^y) - 1 = 0$$

$$e^y = \frac{2x \pm \sqrt{4x^2+4}}{2}$$

$$= \frac{2x \pm 2\sqrt{x^2+1}}{2}$$

$$= x \pm \sqrt{x^2+1}$$

$$e^y = x + \sqrt{x^2+1} \text{ or } e^y = x - \sqrt{x^2+1}$$

At first sight, there appear to be two results, but notice this:

In the second result $\quad \sqrt{x^2+1} > x$

$\therefore\ e^y = x -$ (something $> x$), i.e. negative

Therefore we can discard the second result as far as we are concerned since powers of $e$ are always positive. (Remember the graph of $e^x$.)

The only real solution then is given by $e^y = x + \sqrt{x^2+1}$

$$y = \sinh^{-1} x = \ln\left\{x + \sqrt{x^2+1}\right\}$$

**30**

Finally, let us find the general expression for $\cosh^{-1} x$.

$$\text{Let} \quad y = \cosh^{-1} x \qquad \therefore \; x = \cosh y = \frac{e^y + e^{-y}}{2}$$

$$\therefore \; e^y + \frac{1}{e^y} = 2x \qquad \therefore \; (e^y)^2 - 2x(e^y) + 1 = 0$$

$$\therefore \; e^y = \frac{2x \pm \sqrt{4x^2 - 4}}{2} = x \pm \sqrt{x^2 - 1}$$

$$\therefore \; e^y = x + \sqrt{x^2 - 1} \quad \text{and} \quad e^y = x - \sqrt{x^2 - 1}$$

Both these results are positive, since $\sqrt{x^2 - 1} < x$.

$$\text{However} \quad \frac{1}{x + \sqrt{x^2 - 1}}$$

$$= \frac{1}{x + \sqrt{x^2 - 1}} \cdot \frac{x - \sqrt{x^2 - 1}}{x - \sqrt{x^2 - 1}}$$

$$= \frac{x - \sqrt{x^2 - 1}}{x^2 - (x^2 - 1)}$$

$$= x - \sqrt{x^2 - 1}$$

So our results can be written:

$$e^y = x + \sqrt{x^2 - 1} \quad \text{and} \quad e^y = \frac{1}{x + \sqrt{x^2 - 1}}$$

$$e^y = x + \sqrt{x^2 - 1} \text{ or } \left\{x + \sqrt{x^2 - 1}\right\}^{-1}$$

$$\therefore \; y = \ln\left\{x + \sqrt{x^2 - 1}\right\} \text{ or } -\ln\left\{x + \sqrt{x^2 - 1}\right\}$$

$$\therefore \qquad \cosh^{-1} x = \pm \ln\left\{x + \sqrt{x^2 - 1}\right\}$$

Notice that the plus and minus signs give two results which are symmetrical about the $y$-axis (agreeing with the graph of $y = \cosh x$).

**31**

Here are the three general results collected together:

$$\sinh^{-1} x = \ln\left\{x + \sqrt{x^2 + 1}\right\}$$

$$\cosh^{-1} x = \pm \ln\left\{x + \sqrt{x^2 - 1}\right\}$$

$$\tanh^{-1} x = \frac{1}{2} \ln\left\{\frac{1 + x}{1 - x}\right\}$$

Add these to your list in your record book. They will be useful. Compare the first two carefully, for they are very nearly alike. Note also that:

(a) $\sinh^{-1} x$ has only one value

(b) $\cosh^{-1} x$ has two values.

*So what comes next? We shall see in Frame 32*

## Hyperbolic identities

32

There is no need to recoil in horror. You will see before long that we have an easy way of doing these. First of all, let us consider one or two relationships based on the basic definitions.

(1) The first set are really definitions themselves. Like the trig ratios, we have reciprocal hyperbolic functions:

(a) $\coth x$ (i.e. hyperbolic cotangent) $= \dfrac{1}{\tanh x}$

(b) $\operatorname{sech} x$ (i.e. hyperbolic secant) $= \dfrac{1}{\cosh x}$

(c) $\operatorname{cosech} x$ (i.e. hyperbolic cosecant) $= \dfrac{1}{\sinh x}$

These, by the way, are pronounced (a) coth, (b) sheck and (c) co-sheck respectively.

These remind us, once again, how like trig functions these hyperbolic functions are.

*Make a list of these three definitions: then move on to Frame 33*

33

(2) Let us consider $\dfrac{\sinh x}{\cosh x} = \dfrac{e^x - e^{-x}}{2} \div \dfrac{e^x + e^{-x}}{2}$

$$= \frac{e^x - e^{-x}}{e^x + e^{-x}} = \tanh x$$

$$\therefore \tanh x = \frac{\sinh x}{\cosh x} \qquad \left\{ \begin{array}{l} \text{Very much like} \\ \tan\theta = \dfrac{\sin\theta}{\cos\theta} \end{array} \right\}$$

(3) $\cosh x = \frac{1}{2}(e^x + e^{-x})$; $\quad \sinh x = \frac{1}{2}(e^x - e^{-x})$

Add these results: $\cosh x + \sinh x = e^x$

Subtract: $\cosh x - \sinh x = e^{-x}$

Multiply these two expressions together:

$(\cosh x + \sinh x)(\cosh x - \sinh x) = e^x . e^{-x}$

$\therefore \cosh^2 x - \sinh^2 x = 1$

$\left\{ \begin{array}{l} \text{In trig, we have } \cos^2\theta + \sin^2\theta = 1, \\ \text{so there is a difference in sign here.} \end{array} \right\}$

*On to Frame 34*

34

(4) We have just established that $\cosh^2 x - \sinh^2 x = 1$.

Divide by $\cosh^2 x$: $\quad 1 - \dfrac{\sinh^2 x}{\cosh^2 x} = \dfrac{1}{\cosh^2 x}$

$$\therefore 1 - \tanh^2 x = \operatorname{sech}^2 x$$

$$\therefore \operatorname{sech}^2 x = 1 - \tanh^2 x$$

Something like $\sec^2\theta = 1 + \tan^2\theta$, isn't it?

▶

(5) If we start again with $\cosh^2 x - \sinh^2 x = 1$ and divide this time by $\sinh^2 x$, we get:

$$\frac{\cosh^2 x}{\sinh^2 x} - 1 = \frac{1}{\sinh^2 x}$$

$$\therefore \coth^2 x - 1 = \operatorname{cosech}^2 x$$

$$\therefore \operatorname{cosech}^2 x = \coth^2 x - 1$$

{In trig, we have $\operatorname{cosec}^2\theta = 1 + \cot^2\theta$, so there is a sign difference here too.}

*Move on to Frame 35*

**35**

(6) We have already used the fact that

$$\cosh x + \sinh x = e^x \text{ and } \cosh x - \sinh x = e^{-x}$$

If we square each of these statements, we obtain

(a) ...........

(b) ...........

**36**

$$\text{(a)} \quad \cosh^2 x + 2\sinh x \cosh x + \sinh^2 x = e^{2x}$$
$$\text{(b)} \quad \cosh^2 x - 2\sinh x \cosh x + \sinh^2 x = e^{-2x}$$

So if we subtract as they stand, we get:

$$4\sinh x \cosh x = e^{2x} - e^{-2x}$$

$$\therefore 2\sinh x \cosh x = \frac{e^{2x} - e^{-2x}}{2} = \sinh 2x$$

$$\therefore \sinh 2x = 2\sinh x \cosh x$$

If however we add the two lines together, we get ...........

**37**

$$2(\cosh^2 x + \sinh^2 x) = e^{2x} + e^{-2x}$$

$$\therefore \cosh^2 x + \sinh^2 x = \frac{e^{2x} + e^{-2x}}{2} = \cosh 2x$$

$$\therefore \cosh 2x = \cosh^2 x + \sinh^2 x$$

We already know that $\cosh^2 x - \sinh^2 x = 1$

$$\therefore \cosh^2 x = 1 + \sinh^2 x$$

Substituting this in our previous result, we have:

$$\cosh 2x = 1 + \sinh^2 x + \sinh^2 x$$

$$\therefore \cosh 2x = 1 + 2\sinh^2 x$$

Or we could say $\cosh^2 x - 1 = \sinh^2 x$

$$\therefore \cosh 2x = \cosh^2 x + (\cosh^2 x - 1)$$

$$\therefore \cosh 2x = 2\cosh^2 x - 1$$

Now we will collect all these hyperbolic identities together and compare them with the corresponding trig identities.

*These are all listed in the next frame, so move on*

38

| | *Trig identities* | *Hyperbolic identities* |
|---|---|---|
| (1) | $\cot x = 1/\tan x$ | $\coth x = 1/\tanh x$ |
| | $\sec x = 1/\cos x$ | $\text{sech}\, x = 1/\cosh x$ |
| | $\text{cosec}\, x = 1/\sin x$ | $\text{cosech}\, x = 1/\sinh x$ |
| (2) | $\cos^2 x + \sin^2 x = 1$ | $\cosh^2 x - \sinh^2 x = 1$ |
| | $\sec^2 x = 1 + \tan^2 x$ | $\text{sech}^2 x = 1 - \tanh^2 x$ |
| | $\text{cosec}^2 x = 1 + \cot^2 x$ | $\text{cosech}^2 x = \coth^2 x - 1$ |
| (3) | $\sin 2x = 2\sin x \cos x$ | $\sinh 2x = 2\sinh x \cosh x$ |
| | $\cos 2x = \cos^2 x - \sin^2 x$ | $\cosh 2x = \cosh^2 x + \sinh^2 x$ |
| | $= 1 - 2\sin^2 x$ | $= 1 + 2\sinh^2 x$ |
| | $= 2\cos^2 x - 1$ | $= 2\cosh^2 x - 1$ |

If we look at these results, we find that some of the hyperbolic identities follow exactly the trig identities: others have a difference in sign. This change of sign occurs whenever $\sin^2 x$ in the trig results is being converted into $\sinh^2 x$ to form the corresponding hyperbolic identities. This sign change also occurs when $\sin^2 x$ is involved without actually being written as such. For example, $\tan^2 x$ involves $\sin^2 x$ since $\tan^2 x$ could be written as $\dfrac{\sin^2 x}{\cos^2 x}$. The change of sign therefore occurs with:

$\tan^2 x$ when it is being converted into $\tanh^2 x$

$\cot^2 x$ when it is being converted into $\coth^2 x$

$\text{cosec}^2 x$ when it is being converted into $\text{cosech}^2 x$

The sign change also occurs when we have a product of two sinh terms,

e.g. the trig identity $\cos(A + B) = \cos A \cos B - \sin A \sin B$ gives
the hyperbolic identity $\cosh(A + B) = \cosh A \cosh B + \sinh A \sinh B$.

Apart from this one change, the hyperbolic identities can be written down from the trig identities which you already know.

For example:

$$\tan 2x = \frac{2\tan x}{1 - \tan^2 x} \text{ becomes } \tanh 2x = \frac{2\tanh x}{1 + \tanh^2 x}$$

So provided you know your trig identities, you can apply the rule to form the corresponding hyperbolic identities.

## Relationship between trigonometric and hyperbolic functions

39

From our previous work on complex numbers, we know that:

$$e^{j\theta} = \cos\theta + j\sin\theta$$

and $$e^{-j\theta} = \cos\theta - j\sin\theta$$

Adding these two results together, we have:

$e^{j\theta} + e^{-j\theta} = \ldots\ldots\ldots\ldots$

**40**

$2\cos\theta$

So that: $\cos\theta = \dfrac{e^{j\theta} + e^{-j\theta}}{2}$

which is of the form $\dfrac{e^x + e^{-x}}{2}$, with $x$ replaced by $(j\theta)$

$\therefore \cos\theta = \ldots\ldots\ldots\ldots$

**41**

$\cosh j\theta$

Here, then, is our first relationship:

$\cos\theta = \cosh j\theta$

*Make a note of that for the moment: then on to Frame 42*

**42**

If we return to our two original statements:

$$e^{j\theta} = \cos\theta + j\sin\theta$$
$$e^{-j\theta} = \cos\theta - j\sin\theta$$

and this time subtract, we get a similar kind of result

$e^{j\theta} - e^{-j\theta} = \ldots\ldots\ldots\ldots$

**43**

$2j\sin\theta$

So that: $j\sin\theta = \dfrac{e^{j\theta} - e^{-j\theta}}{2}$

$= \ldots\ldots\ldots\ldots$

**44**

$\sinh j\theta$

So: $\sinh j\theta = j\sin\theta$

*Make a note of that also*

**45**

So far, we have two important results:

(a) $\cosh j\theta = \cos\theta$
(b) $\sinh j\theta = j\sin\theta$

Now if we substitute $\theta = jx$ in the first of these results, we have:

$\cos jx = \cosh(j^2x) = \cosh(-x) \quad \therefore \quad \cos jx = \cosh x \quad [\text{since } \cosh(-x) = \cosh x]$

Writing this in reverse order, gives:

$\cosh x = \cos jx$ Another result to note

Now do exactly the same with the second result above, i.e. put $\theta = jx$ in the relationship $j\sin\theta = \sinh j\theta$ and simplify the result. What do you get?

**46**

$$j \sinh x = \sin jx$$

Because we have: $j \sin \theta = \sinh j\theta$

$$j \sin jx = \sinh(j^2 x)$$
$$= \sinh(-x)$$
$$= -\sinh x \quad [\text{since } \sinh(-x) = -\sinh x]$$

Finally, divide both sides by $j$, and we have

$\sin jx = j \sinh x$

*On to the next frame*

**47**

Now let us collect together the results we have established. They are so nearly alike, that we must distinguish between them:

| $\sin jx = j \sinh x$ | $\sinh jx = j \sin x$ |
|---|---|
| $\cos jx = \cosh x$ | $\cosh jx = \cos x$ |

and, by division, we can also obtain:

| $\tan jx = j \tanh x$ | $\tanh jx = j \tan x$ |
|---|---|

Copy the complete table into your record book for future use.

**48**

Here is an example of an application of these results:

Find an expansion for $\sin(x + jy)$.
Now we know that:

$$\sin(A + B) = \sin A \cos B + \cos A \sin B$$
$$\therefore \quad \sin(x + jy) = \sin x \cos jy + \cos x \sin jy$$

so using the results we have listed, we can replace

$\cos jy$ by ...........

and $\sin jy$ by ...........

**49**

| $\cos jy = \cosh y$ | $\sin jy = j \sinh y$ |
|---|---|

So that:

$$\sin(x + jy) = \sin x \cos jy + \cos x \sin jy$$

becomes $\quad \sin(x + jy) = \sin x \cosh y + j \cos x \sinh y$

*Note:* $\sin(x + jy)$ is a function of the angle $(x + jy)$, which is, of course, a complex quantity. In this case, $(x + jy)$ is referred to as a *complex variable* and you will most likely deal with this topic at a later stage of your course.

Meanwhile, here is just one example for you to work through:

Find an expansion for $\cos(x - jy)$.

*Then check with Frame 50*

**50**

$$\boxed{\cos(x - jy) = \cos x \cosh y + j \sin x \sinh y}$$

Here is the working:

$$\cos(A - B) = \cos A \cos B + \sin A \sin B$$

$$\therefore \quad \cos(x - jy) = \cos x \cos jy + \sin x \sin jy$$

$$\text{But} \quad \cos jy = \cosh y$$

$$\text{and} \quad \sin jy = j \sinh y$$

$$\therefore \quad \cos(x - jy) = \cos x \cosh y + j \sin x \sinh y$$

**51**

All that now remains is the **Can You?** checklist and the **Test exercise**, but before working through them, look through your notes, or revise any parts of the Programme on which you are not perfectly clear.

*When you are ready, move on to the next frame*

## Can You?

**52**

### Checklist 3

Check this list before and after you try the end of Programme test.

On a scale of 1 to 5 how confident are you that you can: Frames

- Define the hyperbolic functions in terms of the exponential function? — Frames 1 to 4
  *Yes* ☐ ☐ ☐ ☐ ☐ *No*
- Express the hyperbolic functions as power series? — Frames 5 to 8
  *Yes* ☐ ☐ ☐ ☐ ☐ *No*
- Recognize the graphs of the hyperbolic functions? — Frames 9 to 16
  *Yes* ☐ ☐ ☐ ☐ ☐ *No*
- Evaluate hyperbolic functions and their inverses? — Frames 19 to 26
  *Yes* ☐ ☐ ☐ ☐ ☐ *No*
- Determine the logarithmic form of the inverse hyperbolic functions? — Frames 27 to 31
  *Yes* ☐ ☐ ☐ ☐ ☐ *No*
- Prove hyperbolic trigonometric identities? — Frames 32 to 38
  *Yes* ☐ ☐ ☐ ☐ ☐ *No*
- Understand the relationship between the circular and the hyperbolic trigonometric functions? — Frames 39 to 50
  *Yes* ☐ ☐ ☐ ☐ ☐ *No*

## Test exercise 3

53

1 On the same axes, draw sketch graphs of (a) $y = \sinh x$, (b) $y = \cosh x$, (c) $y = \tanh x$.

2 If $L = 2C \sinh \dfrac{H}{2C}$, find $L$ when $H = 63$ and $C = 50$.

3 If $v^2 = 1{\cdot}8L \tanh \dfrac{6{\cdot}3d}{L}$, find $v$ when $d = 40$ and $L = 315$.

4 Calculate from first principles, the value of:

(a) $\sinh^{-1} 1{\cdot}532$ (b) $\cosh^{-1} 1{\cdot}25$

5 If $\tanh x = \dfrac{1}{3}$, find $e^{2x}$ and hence evaluate $x$.

6 The curve assumed by a heavy chain or cable is:

$y = C \cosh \dfrac{x}{C}$

If $C = 50$, calculate : (a) the value of $y$ when $x = 109$

(b) the value of $x$ when $y = 75$.

7 Simplify $\dfrac{1 + \sinh 2A + \cosh 2A}{1 - \sinh 2A - \cosh 2A}$.

8 Obtain the expansion of $\sin(x - jy)$ in terms of the trigonometric and hyperbolic functions of $x$ and $y$.

## Further problems 3

54

1 Prove that $\cosh 2x = 1 + 2 \sinh^2 x$.

2 Express $\cosh 2x$ and $\sinh 2x$ in exponential form and hence solve for real values of $x$, the equation:

$2 \cosh 2x - \sinh 2x = 2$

3 If $\sinh x = \tan y$, show that $x = \ln(\tan y \pm \sec y)$.

4 If $a = c \cosh x$ and $b = c \sinh x$, prove that

$(a + b)^2 e^{-2x} = a^2 - b^2$

5 Evaluate: (a) $\tanh^{-1} 0{\cdot}75$ and (b) $\cosh^{-1} 2$.

6 Prove that $\tanh^{-1} \left\{ \dfrac{x^2 - 1}{x^2 + 1} \right\} = \ln x$.

7 Express (a) $\cosh \dfrac{1 + j}{2}$ and (b) $\sinh \dfrac{1 + j}{2}$ in the form $a + jb$, giving $a$ and $b$ to 4 significant figures.

▶

8 Prove that:

(a) $\sinh(x+y) = \sinh x \cosh y + \cosh x \sinh y$

(b) $\cosh(x+y) = \cosh x \cosh y + \sinh x \sinh y$

Hence prove that:

$$\tanh(x+y) = \frac{\tanh x + \tanh y}{1 + \tanh x \tanh y}$$

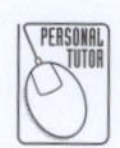

9 Show that the coordinates of any point on the hyperbola $\dfrac{x^2}{a^2} - \dfrac{y^2}{b^2} = 1$ can be represented in the form $x = a\cosh u$, $y = b\sinh u$.

10 Solve for real values of $x$:
$3\cosh 2x = 3 + \sinh 2x$

11 Prove that: $\dfrac{1 + \tanh x}{1 - \tanh x} = e^{2x}$.

12 If $t = \tanh\dfrac{x}{2}$, prove that $\sinh x = \dfrac{2t}{1-t^2}$ and $\cosh x = \dfrac{1+t^2}{1-t^2}$.

Hence solve the equation:
$7\sinh x + 20\cosh x = 24$

13 If $x = \ln\tan\left\{\dfrac{\pi}{4} + \dfrac{\theta}{2}\right\}$, find $e^x$ and $e^{-x}$, and hence show that $\sinh x = \tan\theta$.

14 Given that $\sinh^{-1} x = \ln\left\{x + \sqrt{x^2+1}\right\}$, determine $\sinh^{-1}(2+j)$ in the form $a + jb$.

15 If $\tan\dfrac{x}{2} = \tan A \tanh B$, prove that: $\tan x = \dfrac{\sin 2A \sinh 2B}{1 + \cos 2A \cosh 2B}$

16 Prove that $\sinh 3\theta = 3\sinh\theta + 4\sinh^3\theta$.

17 If $\lambda = \dfrac{at}{2}\left\{\dfrac{\sinh at + \sin at}{\cosh at - \cos at}\right\}$, calculate $\lambda$ when $a = 0{\cdot}215$ and $t = 5$.

18 Prove that $\tanh^{-1}\left\{\dfrac{x^2 - a^2}{x^2 + a^2}\right\} = \ln\dfrac{x}{a}$.

**55** Now visit the companion website for this book at www.palgrave.com/stroud for more questions applying this mathematics to science and engineering.

# Determinants

### Learning outcomes

When you have completed this Programme you will be able to:

- ☐ Expand a $2 \times 2$ determinant
- ☐ Solve pairs of simultaneous linear equations in two variables using $2 \times 2$ determinants
- ☐ Expand a $3 \times 3$ determinant
- ☐ Solve three simultaneous linear equations in three variables using $3 \times 3$ determinants
- ☐ Determine the consistency of sets of simultaneous linear equations
- ☐ Use the properties of determinants to solve equations written in determinant form

# Determinants

**1**

You are quite familiar with the method of solving a pair of simultaneous equations by elimination.

e.g. To solve $2x + 3y + 2 = 0$ (a)
$3x + 4y + 6 = 0$ (b)

we could first find the value of $x$ by eliminating $y$. To do this, of course, we should multiply (a) by 4 and (b) by 3 to make the coefficient of $y$ the same in each equation.

So $8x + 12y + 8 = 0$
$9x + 12y + 18 = 0$

Then by subtraction, we get $x + 10 = 0$, i.e. $x = -10$. By substituting back in either equation, we then obtain $y = 6$.

So, finally, $x = -10$, $y = 6$

That was trivial. You have done similar ones many times before. In just the same way, if:

$a_1x + b_1y + d_1 = 0$ (a)
$a_2x + b_2y + d_2 = 0$ (b)

then to eliminate $y$ we make the coefficients of $y$ in the two equations identical by multiplying

(a) by ........... and (b) by ...........

**2**

(a) by $b_2$ and (b) by $b_1$

Correct, of course. So the equations:

$$a_1x + b_1y + d_1 = 0$$
$$a_2x + b_2y + d_2 = 0$$

become

$$a_1b_2x + b_1b_2y + b_2d_1 = 0$$
$$a_2b_1x + b_1b_2y + b_1d_2 = 0$$

Subtracting, we get:

$$(a_1b_2 - a_2b_1)x + b_2d_1 - b_1d_2 = 0$$

so that $(a_1b_2 - a_2b_1)x = b_1d_2 - b_2d_1$

Then $x =$ ...........

**3**

$$x = \frac{b_1d_2 - b_2d_1}{a_1b_2 - a_2b_1}$$

In practice, this result can give a finite value for $x$ only if the denominator is not zero. That is, the equations:

$a_1x + b_1y + d_1 = 0$
$a_2x + b_2y + d_2 = 0$

give a finite value for $x$ provided that $(a_1b_2 - a_2b_1) \neq 0$.

▶

Consider these equations:

$$3x + 2y - 5 = 0$$
$$4x + 3y - 7 = 0$$

In this case, $a_1 = 3$, $b_1 = 2$, $a_2 = 4$, $b_2 = 3$

$$a_1b_2 - a_2b_1 = 3 \times 3 - 4 \times 2$$
$$= 9 - 8 = 1$$

This is not zero, so there $\left\{ \begin{matrix} \text{will} \\ \text{will not} \end{matrix} \right\}$ be a finite value of $x$.

---

**4**

will

The expression $a_1b_2 - a_2b_1$ is therefore an important one in the solution of simultaneous equations. We have a shorthand notation for this:

$$a_1b_2 - a_2b_1 = \begin{vmatrix} a_1 & b_1 \\ a_2 & b_2 \end{vmatrix}$$

For $\begin{vmatrix} a_1 & b_1 \\ a_2 & b_2 \end{vmatrix}$ to represent $a_1b_2 - a_2b_1$ then we must multiply the terms diagonally to form the product terms in the expansion: we multiply $\begin{vmatrix} a_1 & \\ & b_2 \end{vmatrix}$ and then subtract the product $\begin{vmatrix} & b_1 \\ a_2 & \end{vmatrix}$ i.e. $+ \searrow$ and $- \nearrow$

e.g. $\begin{vmatrix} 3 & 7 \\ 5 & 2 \end{vmatrix} = \begin{vmatrix} 3 & \\ & 2 \end{vmatrix} - \begin{vmatrix} & 7 \\ 5 & \end{vmatrix} = 3 \times 2 - 5 \times 7 = 6 - 35 = -29$

So $\begin{vmatrix} 6 & 5 \\ 1 & 2 \end{vmatrix} = \begin{vmatrix} 6 & \\ & 2 \end{vmatrix} - \begin{vmatrix} & 5 \\ 1 & \end{vmatrix} = \ldots\ldots\ldots\ldots$

---

**5**

$\begin{vmatrix} 6 & 5 \\ 1 & 2 \end{vmatrix} = 12 - 5 = 7$

$\begin{vmatrix} a_1 & b_1 \\ a_2 & b_2 \end{vmatrix}$ is called a *determinant* of the second order (since it has two rows and two columns) and represents $a_1b_2 - a_2b_1$. You can easily remember this as $+ \searrow - \nearrow$.

Just for practice, evaluate the following determinants:

(a) $\begin{vmatrix} 4 & 2 \\ 5 & 3 \end{vmatrix}$, (b) $\begin{vmatrix} 7 & 4 \\ 6 & 3 \end{vmatrix}$, (c) $\begin{vmatrix} 2 & 1 \\ 4 & -3 \end{vmatrix}$

*Finish all three: then move on to Frame 6*

---

**6**

(a) $\begin{vmatrix} 4 & 2 \\ 5 & 3 \end{vmatrix} = 4 \times 3 - 5 \times 2 = 12 - 10 = \boxed{2}$

(b) $\begin{vmatrix} 7 & 4 \\ 6 & 3 \end{vmatrix} = 7 \times 3 - 6 \times 4 = 21 - 24 = \boxed{-3}$

(c) $\begin{vmatrix} 2 & 1 \\ 4 & -3 \end{vmatrix} = 2(-3) - 4 \times 1 = -6 - 4 = \boxed{-10}$

▶

Now, in solving the equations $\begin{cases} a_1x + b_1y + d_1 = 0 \\ a_2x + b_2y + d_2 = 0 \end{cases}$

we found that $x = \dfrac{b_1d_2 - b_2d_1}{a_1b_2 - a_2b_1}$ and the numerator and the denominator can each be written as a determinant.

$b_1d_2 - b_2d_1 = \ldots\ldots\ldots\ldots;\quad a_1b_2 - a_2b_1 = \ldots\ldots\ldots\ldots$

**7**

$$\begin{vmatrix} b_1 & d_1 \\ b_2 & d_2 \end{vmatrix};\quad \begin{vmatrix} a_1 & b_1 \\ a_2 & b_2 \end{vmatrix}$$

If we eliminate $x$ from the original equations and find an expression for $y$, we obtain $y = -\left\{\dfrac{a_1d_2 - a_2d_1}{a_1b_2 - a_2b_1}\right\}$

So, for any pair of simultaneous equations:

$$a_1x + b_1y + d_1 = 0$$
$$a_2x + b_2y + d_2 = 0$$

we have $\quad x = \dfrac{b_1d_2 - b_2d_1}{a_1b_2 - a_2b_1}$ and $y = -\left\{\dfrac{a_1d_2 - a_2d_1}{a_1b_2 - a_2b_1}\right\}$

Each of these numerators and denominators can be expressed as a determinant.

So, $\quad x = \ldots\ldots\ldots\ldots$ and $y = \ldots\ldots\ldots\ldots$

**8**

$$x = \frac{\begin{vmatrix} b_1 & d_1 \\ b_2 & d_2 \end{vmatrix}}{\begin{vmatrix} a_1 & b_1 \\ a_2 & b_2 \end{vmatrix}} \text{ and } y = -\frac{\begin{vmatrix} a_1 & d_1 \\ a_2 & d_2 \end{vmatrix}}{\begin{vmatrix} a_1 & b_1 \\ a_2 & b_2 \end{vmatrix}}$$

$$\therefore \frac{x}{\begin{vmatrix} b_1 & d_1 \\ b_2 & d_2 \end{vmatrix}} = \frac{1}{\begin{vmatrix} a_1 & b_1 \\ a_2 & b_2 \end{vmatrix}} \text{ and } \frac{y}{\begin{vmatrix} a_1 & d_1 \\ a_2 & d_2 \end{vmatrix}} = \frac{-1}{\begin{vmatrix} a_1 & b_1 \\ a_2 & b_2 \end{vmatrix}}$$

We can combine these results, thus:

$$\frac{x}{\begin{vmatrix} b_1 & d_1 \\ b_2 & d_2 \end{vmatrix}} = \frac{-y}{\begin{vmatrix} a_1 & d_1 \\ a_2 & d_2 \end{vmatrix}} = \frac{1}{\begin{vmatrix} a_1 & b_1 \\ a_2 & b_2 \end{vmatrix}}$$

*Make a note of these results and then move on to the next frame*

**9**

So if $\quad \begin{cases} a_1x + b_1y + d_1 = 0 \\ a_2x + b_2y + d_2 = 0 \end{cases}$

Then $\quad \dfrac{x}{\begin{vmatrix} b_1 & d_1 \\ b_2 & d_2 \end{vmatrix}} = \dfrac{-y}{\begin{vmatrix} a_1 & d_1 \\ a_2 & d_2 \end{vmatrix}} = \dfrac{1}{\begin{vmatrix} a_1 & b_1 \\ a_2 & b_2 \end{vmatrix}}$

Each variable is divided by a determinant. Let us see how we can get them from the original equations.

▶

(a) Consider $\dfrac{x}{\begin{vmatrix} b_1 & d_1 \\ b_2 & d_2 \end{vmatrix}}$. Let us denote the determinant in the denominator by $\triangle_1$, i.e. $\triangle_1 = \begin{vmatrix} b_1 & d_1 \\ b_2 & d_2 \end{vmatrix}$

To form $\triangle_1$ from the given equations, omit the $x$-terms and write down the coefficients and constant terms in the order in which they stand:

$$\begin{cases} a_1x + b_1y + d_1 = 0 \\ a_2x + b_2y + d_2 = 0 \end{cases} \text{gives} \begin{vmatrix} b_1 & d_1 \\ b_2 & d_2 \end{vmatrix}$$

(b) Similarly for $\dfrac{-y}{\begin{vmatrix} a_1 & d_1 \\ a_2 & d_2 \end{vmatrix}}$, let $\triangle_2 = \begin{vmatrix} a_1 & d_1 \\ a_2 & d_2 \end{vmatrix}$

To form $\triangle_2$ from the given equations, omit the $y$-terms and write down the coefficients and constant terms in the order in which they stand:

$$\begin{cases} a_1x + b_1y + d_1 = 0 \\ a_2x + b_2y + d_2 = 0 \end{cases} \text{gives } \triangle_2 = \begin{vmatrix} a_1 & d_1 \\ a_2 & d_2 \end{vmatrix}$$

(c) For the expression $\dfrac{1}{\begin{vmatrix} a_1 & b_1 \\ a_2 & b_2 \end{vmatrix}}$, denote the determinant by $\triangle_0$.

To form $\triangle_0$ from the given equations, omit the constant terms and write down the coefficients in the order in which they stand:

$$\begin{cases} a_1x + b_1y + d_1 = 0 \\ a_2x + b_2y + d_2 = 0 \end{cases} \text{gives} \begin{vmatrix} a_1 & b_1 \\ a_2 & b_2 \end{vmatrix}$$

Note finally that $\dfrac{x}{\triangle_1} = -\dfrac{y}{\triangle_2} = \dfrac{1}{\triangle_0}$

*Now let us do some examples, so on to Frame 10*

---

**10**

**Example 1**

To solve the equations $\begin{cases} 5x + 2y + 19 = 0 \\ 3x + 4y + 17 = 0 \end{cases}$

The key to the method is:

$$\frac{x}{\triangle_1} = \frac{-y}{\triangle_2} = \frac{1}{\triangle_0}$$

To find $\triangle_0$, omit the constant terms:

$$\therefore \triangle_0 = \begin{vmatrix} 5 & 2 \\ 3 & 4 \end{vmatrix} = 5 \times 4 - 3 \times 2 = 20 - 6 = 14$$

$$\therefore \triangle_0 = 14 \quad \text{(a)}$$

Now, to find $\triangle_1$, omit the $x$-terms:

$\therefore \triangle_1 = \dots\dots\dots\dots$

**11**

$$\boxed{\triangle_1 = -42}$$

Because $\triangle_1 = \begin{vmatrix} 2 & 19 \\ 4 & 17 \end{vmatrix} = 34 - 76 = -42 \quad \text{(b)}$

Similarly, to find $\triangle_2$, omit the $y$-terms:

$$\triangle_2 = \begin{vmatrix} 5 & 19 \\ 3 & 17 \end{vmatrix} = 85 - 57 = 28 \quad \text{(c)}$$

Substituting the values of $\triangle_1$, $\triangle_2$, $\triangle_0$ in the key, we get

$$\frac{x}{-42} = \frac{-y}{28} = \frac{1}{14}$$

from which $x = \dots\dots\dots\dots$ and $y = \dots\dots\dots\dots$

**12**

$$\boxed{x = \frac{-42}{14} = -3; \quad -y = \frac{28}{14}, \quad y = -2}$$

Now for another example.

**Example 2**

Solve by determinants $\begin{cases} 2x + 3y - 14 = 0 \\ 3x - 2y + 5 = 0 \end{cases}$

First of all, write down the key:

$$\frac{x}{\triangle_1} = \frac{-y}{\triangle_2} = \frac{1}{\triangle_0}$$

(Note that the terms are alternately positive and negative.)

Then $\quad \triangle_0 = \begin{vmatrix} 2 & 3 \\ 3 & -2 \end{vmatrix} = -4 - 9 = -13 \quad \text{(a)}$

Now you find $\triangle_1$ and $\triangle_2$ in the same way.

**13**

$$\boxed{\triangle_1 = -13; \quad \triangle_2 = 52}$$

Because we have $\begin{cases} 2x + 3y - 14 = 0 \\ 3x - 2y + 5 = 0 \end{cases}$

$$\therefore \triangle_1 = \begin{vmatrix} 3 & -14 \\ -2 & 5 \end{vmatrix} = \begin{vmatrix} 3 & \\ & 5 \end{vmatrix} - \begin{vmatrix} & -14 \\ -2 & \end{vmatrix}$$

$$= 15 - 28$$

$$= -13 \qquad \therefore \triangle_1 = -13$$

$$\triangle_2 = \begin{vmatrix} 2 & -14 \\ 3 & 5 \end{vmatrix} = \begin{vmatrix} 2 & \\ & 5 \end{vmatrix} - \begin{vmatrix} & -14 \\ 3 & \end{vmatrix}$$

$$= 10 - (-42)$$

$$= 52 \qquad \therefore \triangle_2 = 52$$

▶

So that $\frac{x}{\triangle_1} = \frac{-y}{\triangle_2} = \frac{1}{\triangle_0}$

and $\triangle_1 = -13; \quad \triangle_2 = 52; \quad \triangle_0 = -13$

$$\therefore x = \frac{\triangle_1}{\triangle_0} = \frac{-13}{-13} = 1 \qquad \therefore x = 1$$

$$-y = \frac{\triangle_2}{\triangle_0} = \frac{52}{-13} = -4 \qquad \therefore y = 4$$

Do not forget the key:

$$\frac{x}{\triangle_1} = \frac{-y}{\triangle_2} = \frac{1}{\triangle_0}$$

with alternate plus and minus signs.

*Make a note of this in your record book*

---

**14**

Here is another one: do it on your own.

**Example 3**

Solve by determinants: $\begin{cases} 4x - 3y + 20 = 0 \\ 3x + 2y - 2 = 0 \end{cases}$

First of all, write down the key.

Then off you go: find $\triangle_0$, $\triangle_1$ and $\triangle_2$ and hence determine the values of $x$ and $y$.

*When you have finished, move on to Frame 15*

---

**15**

$$\boxed{x = -2; \quad y = 4}$$

Here is the working in detail:

$$\begin{cases} 4x - 3y + 20 = 0 \\ 3x + 2y - 2 = 0 \end{cases} \qquad \frac{x}{\triangle_1} = \frac{-y}{\triangle_2} = \frac{1}{\triangle_0}$$

$$\triangle_0 = \begin{vmatrix} 4 & -3 \\ 3 & 2 \end{vmatrix} = 8 - (-9) = 8 + 9 = 17$$

$$\triangle_1 = \begin{vmatrix} -3 & 20 \\ 2 & -2 \end{vmatrix} = 6 - 40 = -34$$

$$\triangle_2 = \begin{vmatrix} 4 & 20 \\ 3 & -2 \end{vmatrix} = -8 - 60 = -68$$

$$x = \frac{\triangle_1}{\triangle_0} = \frac{-34}{17} = -2 \qquad \therefore x = -2$$

$$-y = \frac{\triangle_2}{\triangle_0} = \frac{-68}{17} = -4 \qquad \therefore y = 4$$

Now, by way of revision, complete the following:

(a) $\begin{vmatrix} 5 & 6 \\ 7 & 4 \end{vmatrix} = \dots\dots\dots\dots$

(b) $\begin{vmatrix} 5 & -2 \\ -3 & -4 \end{vmatrix} = \dots\dots\dots\dots$

(c) $\begin{vmatrix} a & d \\ b & c \end{vmatrix} = \dots\dots\dots\dots$

(d) $\begin{vmatrix} p & q \\ r & s \end{vmatrix} = \dots\dots\dots\dots$

---

**16**

Here are the results. You must have got them correct.

(a) $20 - 42 = -22$
(b) $-20 - 6 = -26$
(c) $ac - bd$
(d) $ps - rq$

*For the next section of the work, move on to Frame 17*

## Determinants of the third order

**17**

A determinant of the third order will contain 3 rows and 3 columns, thus:

$$\begin{vmatrix} a_1 & b_1 & c_1 \\ a_2 & b_2 & c_2 \\ a_3 & b_3 & c_3 \end{vmatrix}$$

Each element in the determinant is associated with its *minor*, which is found by omitting the row and column containing the element concerned.

e.g. the minor of $a_1$ is $\begin{vmatrix} b_2 & c_2 \\ b_3 & c_3 \end{vmatrix}$ obtained $\begin{vmatrix} a_1 & b_1 & c_1 \\ a_2 & b_2 & c_2 \\ a_3 & b_3 & c_3 \end{vmatrix}$

the minor of $b_1$ is $\begin{vmatrix} a_2 & c_2 \\ a_3 & c_3 \end{vmatrix}$ obtained $\begin{vmatrix} a_1 & b_1 & c_1 \\ a_2 & b_2 & c_2 \\ a_3 & b_3 & c_3 \end{vmatrix}$

the minor of $c_1$ is $\begin{vmatrix} a_2 & b_2 \\ a_3 & b_3 \end{vmatrix}$ obtained $\begin{vmatrix} a_1 & b_1 & c_1 \\ a_2 & b_2 & c_2 \\ a_3 & b_3 & c_3 \end{vmatrix}$

So, in the same way, the minor of $a_2$ is ...........

**18**

Minor of $a_2$ is $\begin{vmatrix} b_1 & c_1 \\ b_3 & c_3 \end{vmatrix}$

Because, to find the minor of $a_2$, we simply ignore the row and column containing $a_2$, i.e.

$$\begin{vmatrix} a_1 & b_1 & c_1 \\ a_2 & b_2 & c_2 \\ a_3 & b_3 & c_3 \end{vmatrix}$$

Similarly, the minor of $b_3$ is ...........

19

$$\text{Minor of } b_3 \text{ is } \begin{vmatrix} a_1 & c_1 \\ a_2 & c_2 \end{vmatrix}$$

i.e. omit the row and column containing $b_3$.

$$\begin{matrix} a_1 & b_1 & c_1 \\ a_2 & b_2 & c_2 \\ a_3 & b_3 & c_3 \end{matrix}$$

*Now on to Frame 20*

20

## Evaluation of a third-order determinant

To expand a determinant of the third order, we can write down each element along the top row, multiply it by its minor and give the terms a plus or minus sign alternately:

$$\begin{vmatrix} a_1 & b_1 & c_1 \\ a_2 & b_2 & c_2 \\ a_3 & b_3 & c_3 \end{vmatrix} = a_1\begin{vmatrix} b_2 & c_2 \\ b_3 & c_3 \end{vmatrix} - b_1\begin{vmatrix} a_2 & c_2 \\ a_3 & c_3 \end{vmatrix} + c_1\begin{vmatrix} a_2 & b_2 \\ a_3 & b_3 \end{vmatrix}$$

Then, of course, we already know how to expand a determinant of the second order by multiplying diagonally, $+ \searrow - \nearrow$

**Example 1**

$$\begin{vmatrix} 1 & 3 & 2 \\ 4 & 5 & 7 \\ 2 & 4 & 8 \end{vmatrix} = 1\begin{vmatrix} 5 & 7 \\ 4 & 8 \end{vmatrix} - 3\begin{vmatrix} 4 & 7 \\ 2 & 8 \end{vmatrix} + 2\begin{vmatrix} 4 & 5 \\ 2 & 4 \end{vmatrix}$$

$$= 1(5 \times 8 - 4 \times 7) - 3(4 \times 8 - 2 \times 7) + 2(4 \times 4 - 2 \times 5)$$
$$= 1(40 - 28) - 3(32 - 14) + 2(16 - 10)$$
$$= 1(12) - 3(18) + 2(6)$$
$$= 12 - 54 + 12 = -30$$

21

Here is another.

**Example 2**

$$\begin{vmatrix} 3 & 2 & 5 \\ 4 & 6 & 7 \\ 2 & 9 & 2 \end{vmatrix} = 3\begin{vmatrix} 6 & 7 \\ 9 & 2 \end{vmatrix} - 2\begin{vmatrix} 4 & 7 \\ 2 & 2 \end{vmatrix} + 5\begin{vmatrix} 4 & 6 \\ 2 & 9 \end{vmatrix}$$

$$= 3(12 - 63) - 2(8 - 14) + 5(36 - 12)$$
$$= 3(-51) - 2(-6) + 5(24)$$
$$= -153 + 12 + 120 = -21$$

Now here is one for you to do.

**Example 3**

Evaluate $\begin{vmatrix} 2 & 7 & 5 \\ 4 & 6 & 3 \\ 8 & 9 & 1 \end{vmatrix}$

Expand along the top row, multiply each element by its minor, and assign alternate + and − signs to the products.

*When you are ready, move on to Frame 22 for the result*

**22**

38

Because
$$\begin{vmatrix} 2 & 7 & 5 \\ 4 & 6 & 3 \\ 8 & 9 & 1 \end{vmatrix} = 2\begin{vmatrix} 6 & 3 \\ 9 & 1 \end{vmatrix} - 7\begin{vmatrix} 4 & 3 \\ 8 & 1 \end{vmatrix} + 5\begin{vmatrix} 4 & 6 \\ 8 & 9 \end{vmatrix}$$
$$= 2(6 - 27) - 7(4 - 24) + 5(36 - 48)$$
$$= 2(-21) - 7(-20) + 5(-12)$$
$$= -42 + 140 - 60$$
$$= 38$$

We obtained the result above by expanding along the top row of the given determinant. If we expand down the first column in the same way, still assigning alternate + and – signs to the products, we get:

$$\begin{vmatrix} 2 & 7 & 5 \\ 4 & 6 & 3 \\ 8 & 9 & 1 \end{vmatrix} = 2\begin{vmatrix} 6 & 3 \\ 9 & 1 \end{vmatrix} - 4\begin{vmatrix} 7 & 5 \\ 9 & 1 \end{vmatrix} + 8\begin{vmatrix} 7 & 5 \\ 6 & 3 \end{vmatrix}$$
$$= 2(6 - 27) - 4(7 - 45) + 8(21 - 30)$$
$$= 2(-21) - 4(-38) + 8(-9)$$
$$= -42 + 152 - 72$$
$$= 38$$

which is the same result as that which we obtained before.

**23**

We can, if we wish, expand along any row or column in the same way, multiplying each element by its minor, so long as we assign to each product the appropriate + or – sign. The appropriate 'place signs' are given by

$$\begin{matrix} + & - & + & - & + & \dots & \dots \\ - & + & - & + & - & \dots & \dots \\ + & - & + & - & + & \dots & \dots \\ - & + & - & + & - & \dots & \dots \end{matrix}$$

etc., etc.

The key element (in the top left-hand corner) is always +. The others are then alternately + or –, as you proceed along any row or down any column.
So in the determinant:

$$\begin{vmatrix} 1 & 3 & 7 \\ 5 & 6 & 9 \\ 4 & 2 & 8 \end{vmatrix}$$ the 'place sign' of the element 9 is ...........

**24**

–

Because in a third-order determinant, the 'place signs' are:

$$\begin{vmatrix} + & - & + \\ - & + & - \\ + & - & + \end{vmatrix}$$

Remember that the top left-hand element always has a + place sign. The others follow from it.

▶

Now consider this one: $\begin{vmatrix} 3 & 7 & 2 \\ 6 & 8 & 4 \\ 1 & 9 & 5 \end{vmatrix}$

If we expand down the middle column, we get

$$\begin{vmatrix} 3 & 7 & 2 \\ 6 & 8 & 4 \\ 1 & 9 & 5 \end{vmatrix} = -7\begin{vmatrix} 6 & 4 \\ 1 & 5 \end{vmatrix} + 8\begin{vmatrix} 3 & 2 \\ 1 & 5 \end{vmatrix} - 9\begin{vmatrix} 3 & 2 \\ 6 & 4 \end{vmatrix} = \ldots\ldots\ldots\ldots$$

*Finish it off. Then move on for the result*

---

**25**

$-78$

Because $-7\begin{vmatrix} 6 & 4 \\ 1 & 5 \end{vmatrix} + 8\begin{vmatrix} 3 & 2 \\ 1 & 5 \end{vmatrix} - 9\begin{vmatrix} 3 & 2 \\ 6 & 4 \end{vmatrix}$

$$= -7(30 - 4) + 8(15 - 2) - 9(12 - 12)$$
$$= -7(26) + 8(13) - 9(0) = -182 + 104 = -78$$

So now you do this one:

Evaluate $\begin{vmatrix} 2 & 3 & 4 \\ 6 & 1 & 3 \\ 5 & 7 & 2 \end{vmatrix}$ by expanding along the *bottom* row.

*When you have done it, move to Frame 26 for the answer*

---

**26**

$119$

We have $\begin{vmatrix} 2 & 3 & 4 \\ 6 & 1 & 3 \\ 5 & 7 & 2 \end{vmatrix}$ and remember $\begin{vmatrix} + & - & + \\ - & + & - \\ + & - & + \end{vmatrix}$

$$= 5\begin{vmatrix} 3 & 4 \\ 1 & 3 \end{vmatrix} - 7\begin{vmatrix} 2 & 4 \\ 6 & 3 \end{vmatrix} + 2\begin{vmatrix} 2 & 3 \\ 6 & 1 \end{vmatrix} = 5(9 - 4) - 7(6 - 24) + 2(2 - 18)$$
$$= 5(5) - 7(-18) + 2(-16) = 25 + 126 - 32 = 119$$

One more: Evaluate $\begin{vmatrix} 1 & 2 & 8 \\ 7 & 3 & 1 \\ 4 & 6 & 9 \end{vmatrix}$ by expanding along the *middle* row.

*Result in Frame 27*

---

**27**

$143$

Because $\begin{vmatrix} 1 & 2 & 8 \\ 7 & 3 & 1 \\ 4 & 6 & 9 \end{vmatrix} = -7\begin{vmatrix} 2 & 8 \\ 6 & 9 \end{vmatrix} + 3\begin{vmatrix} 1 & 8 \\ 4 & 9 \end{vmatrix} - 1\begin{vmatrix} 1 & 2 \\ 4 & 6 \end{vmatrix}$

$$= -7(18 - 48) + 3(9 - 32) - 1(6 - 8)$$
$$= -7(-30) + 3(-23) - 1(-2)$$
$$= 210 - 69 + 2 = 143$$

We have seen how we can use second-order determinants to solve simultaneous equations in 2 unknowns.

We can now extend the method to solve simultaneous equations in 3 unknowns.

*So move on to Frame 28*

## Simultaneous equations in three unknowns

**28**

Consider the equations:

$$\begin{cases} a_1x + b_1y + c_1z + d_1 = 0 \\ a_2x + b_2y + c_2z + d_2 = 0 \\ a_3x + b_3y + c_3z + d_3 = 0 \end{cases}$$

If we find $x$, $y$ and $z$ by the elimination method, we obtain results that can be expressed in determinant form thus:

$$\frac{x}{\begin{vmatrix} b_1 & c_1 & d_1 \\ b_2 & c_2 & d_2 \\ b_3 & c_3 & d_3 \end{vmatrix}} = \frac{-y}{\begin{vmatrix} a_1 & c_1 & d_1 \\ a_2 & c_2 & d_2 \\ a_3 & c_3 & d_3 \end{vmatrix}} = \frac{z}{\begin{vmatrix} a_1 & b_1 & d_1 \\ a_2 & b_2 & d_2 \\ a_3 & b_3 & d_3 \end{vmatrix}} = \frac{-1}{\begin{vmatrix} a_1 & b_1 & c_1 \\ a_2 & b_2 & c_2 \\ a_3 & b_3 & c_3 \end{vmatrix}}$$

We can remember this more easily in this form:

$$\frac{x}{\triangle_1} = \frac{-y}{\triangle_2} = \frac{z}{\triangle_3} = \frac{-1}{\triangle_0}$$

where $\triangle_1 =$ the det of the coefficients omitting the $x$-terms

$\triangle_2 =$ the det of the coefficients omitting the $y$-terms

$\triangle_3 =$ the det of the coefficients omitting the $z$-terms

$\triangle_0 =$ the det of the coefficients omitting the constant terms.

Notice that the signs are alternately plus and minus.

Let us work through a numerical example.

**Example 1**

Find the value of $x$ from the equations:

$$\begin{cases} 2x + 3y - z - 4 = 0 \\ 3x + y + 2z - 13 = 0 \\ x + 2y - 5z + 11 = 0 \end{cases}$$

First the key: $\dfrac{x}{\triangle_1} = \dfrac{-y}{\triangle_2} = \dfrac{z}{\triangle_3} = \dfrac{-1}{\triangle_0}$

To find the value of $x$, we use $\dfrac{x}{\triangle_1} = \dfrac{-1}{\triangle_0}$, i.e. we must find $\triangle_1$ and $\triangle_0$.

(a) To find $\triangle_0$, omit the constant terms.

$$\therefore \triangle_0 = \begin{vmatrix} 2 & 3 & -1 \\ 3 & 1 & 2 \\ 1 & 2 & -5 \end{vmatrix} = 2\begin{vmatrix} 1 & 2 \\ 2 & -5 \end{vmatrix} - 3\begin{vmatrix} 3 & 2 \\ 1 & -5 \end{vmatrix} - 1\begin{vmatrix} 3 & 1 \\ 1 & 2 \end{vmatrix}$$

$$= -18 + 51 - 5 = 28$$

(b) Now you find $\triangle_1$, in the same way.

**29**

$$\boxed{\triangle_1 = -56}$$

Because $\triangle_1 = \begin{vmatrix} 3 & -1 & -4 \\ 1 & 2 & -13 \\ 2 & -5 & 11 \end{vmatrix} = 3(22 - 65) + 1(11 + 26) - 4(-5 - 4)$

$$= 3(-43) + 1(37) - 4(-9)$$
$$= -129 + 37 + 36$$
$$= -129 + 73 = -56$$

But $\dfrac{x}{\triangle_1} = \dfrac{-1}{\triangle_0} \quad \therefore \dfrac{x}{-56} = \dfrac{-1}{28}$

$\therefore x = \dfrac{56}{28} = 2 \quad \therefore x = 2$

Note that by this method we can evaluate any one of the variables, without necessarily finding the others. Let us do another example.

**Example 2**

Find $y$, given that:

$$\begin{cases} 2x + y - 5z + 11 = 0 \\ x - y + z - 6 = 0 \\ 4x + 2y - 3z + 8 = 0 \end{cases}$$

First, the key, which is ...........

**30**

$$\boxed{\frac{x}{\triangle_1} = \frac{-y}{\triangle_2} = \frac{z}{\triangle_3} = \frac{-1}{\triangle_0}}$$

To find $y$, we use $\dfrac{-y}{\triangle_2} = \dfrac{-1}{\triangle_0}$

Therefore, we must find $\triangle_2$ and $\triangle_0$.

The equations are $\begin{cases} 2x + y - 5z + 11 = 0 \\ x - y + z - 6 = 0 \\ 4x + 2y - 3z + 8 = 0 \end{cases}$

To find $\triangle_2$, omit the $y$-terms.

$$\therefore \triangle_2 = \begin{vmatrix} 2 & -5 & 11 \\ 1 & 1 & -6 \\ 4 & -3 & 8 \end{vmatrix} = 2\begin{vmatrix} 1 & -6 \\ -3 & 8 \end{vmatrix} + 5\begin{vmatrix} 1 & -6 \\ 4 & 8 \end{vmatrix} + 11\begin{vmatrix} 1 & 1 \\ 4 & -3 \end{vmatrix}$$
$$= 2(8 - 18) + 5(8 + 24) + 11(-3 - 4)$$
$$= -20 + 160 - 77 = 63$$

To find $\triangle_0$, omit the constant terms

$\therefore \triangle_0 =$ ...........

**31**

$$\boxed{\triangle_0 = -21}$$

Because

$$\triangle_0 = \begin{vmatrix} 2 & 1 & -5 \\ 1 & -1 & 1 \\ 4 & 2 & -3 \end{vmatrix} = 2\begin{vmatrix} -1 & 1 \\ 2 & -3 \end{vmatrix} - 1\begin{vmatrix} 1 & 1 \\ 4 & -3 \end{vmatrix} - 5\begin{vmatrix} 1 & -1 \\ 4 & 2 \end{vmatrix}$$

$$= 2(3-2) - 1(-3-4) - 5(2+4)$$

$$= 2 + 7 - 30 = -21$$

So we have $\dfrac{-y}{\triangle_2} = \dfrac{-1}{\triangle_0}$ $\quad \therefore y = \dfrac{\triangle_2}{\triangle_0} = \dfrac{63}{-21}$

$$\therefore y = -3$$

The important things to remember are:

(a) The key: $\dfrac{x}{\triangle_1} = \dfrac{-y}{\triangle_2} = \dfrac{z}{\triangle_3} = \dfrac{-1}{\triangle_0}$

with alternate + and – signs.

(b) To find $\triangle_1$, which is associated with $x$ in this case, omit the $x$-terms and form a determinant with the remaining coefficients and constant terms. Similarly for $\triangle_2$, $\triangle_3$, $\triangle_0$.

*Next frame*

## Review exercise

**32**

Here is a short review exercise on the work so far.

Find the following by the use of determinants:

1 $\left\{\begin{array}{c} x + 2y - 3z - 3 = 0 \\ 2x - y - z - 11 = 0 \\ 3x + 2y + z + 5 = 0 \end{array}\right\}$ Find $y$.

2 $\left\{\begin{array}{c} 3x - 4y + 2z + 8 = 0 \\ x + 5y - 3z + 2 = 0 \\ 5x + 3y - z + 6 = 0 \end{array}\right\}$ Find $x$ and $z$.

3 $\left\{\begin{array}{c} 2x - 2y - z - 3 = 0 \\ 4x + 5y - 2z + 3 = 0 \\ 3x + 4y - 3z + 7 = 0 \end{array}\right\}$ Find $x$, $y$ and $z$.

*When you have finished them all, check your answers with those given in the next frame*

**33**

Here are the answers:

1 $y = -4$

2 $x = -2$; $\quad z = 5$

3 $x = 2$; $\quad y = -1$; $\quad z = 3$

If you have them *all* correct, turn straight on to *Frame 52*.

▶

If you have not got them all correct, it is well worth spending a few minutes seeing where you may have gone astray, for one of the main applications of determinants is in the solution of simultaneous equations.

*If you made any slips, move to Frame 34*

**34**

The answer to question **1** in the review test was $\boxed{y = -4}$

Did you get that one right? If so, move on straight away to Frame 41. If you did not manage to get it right, let us work through it in detail.

The equations were $\begin{cases} x + 2y - 3z - 3 = 0 \\ 2x - y - z - 11 = 0 \\ 3x + 2y + z + 5 = 0 \end{cases}$

Copy them down on your paper so that we can refer to them as we go along.

The first thing, always, is to write down the key to the solutions. In this case:

$$\frac{x}{\Delta_1} = \dots = \dots = \dots$$

To fill in the missing terms, take each variable in turn, divide it by the associated determinant, and include the appropriate sign.

So what do we get?

*On to Frame 35*

**35**

$$\boxed{\frac{x}{\Delta_1} = \frac{-y}{\Delta_2} = \frac{z}{\Delta_3} = \frac{-1}{\Delta_0}}$$

The signs go alternately + and −.

In this question, we have to find $y$, so we use the second and last terms in the key.

i.e. $\frac{-y}{\Delta_2} = \frac{-1}{\Delta_0} \quad \therefore y = \frac{\Delta_2}{\Delta_0}$.

So we have to find $\Delta_2$ and $\Delta_0$.

To find $\Delta_2$, we ...........

**36**

form a determinant of the coefficients omitting those of the $y$-terms

So $$\Delta_2 = \begin{vmatrix} 1 & -3 & -3 \\ 2 & -1 & -11 \\ 3 & 1 & 5 \end{vmatrix}$$

Expanding along the top row, this gives:

$$\Delta_2 = 1\begin{vmatrix} -1 & -11 \\ 1 & 5 \end{vmatrix} - (-3)\begin{vmatrix} 2 & -11 \\ 3 & 5 \end{vmatrix} + (-3)\begin{vmatrix} 2 & -1 \\ 3 & 1 \end{vmatrix}$$

We now evaluate each of these second-order determinants by the usual process of multiplying diagonally, remembering the sign convention that + ↘ and − ↗

So we get $\Delta_2 = \dots\dots\dots$

**37**

$$\boxed{\triangle_2 = 120}$$

Because

$$\begin{aligned}\triangle_2 &= 1(-5+11) + 3(10+33) - 3(2+3)\\ &= 6 + 3(43) - 3(5)\\ &= 6 + 129 - 15\\ &= 135 - 15 = 120\end{aligned}$$

$\therefore \triangle_2 = 120$

We also have to find $\triangle_0$, i.e. the determinant of the coefficients omitting the constant terms.

So $\quad \triangle_0 = \begin{vmatrix} \dots & \dots & \dots \\ \dots & \dots & \dots \\ \dots & \dots & \dots \end{vmatrix}$

**38**

$$\boxed{\triangle_0 = \begin{vmatrix} 1 & 2 & -3 \\ 2 & -1 & -1 \\ 3 & 2 & 1 \end{vmatrix}}$$

If we expand this along the top row, we get

$\triangle_0 = \dots\dots\dots\dots$

**39**

$$\boxed{\triangle_0 = 1\begin{vmatrix} -1 & -1 \\ 2 & 1 \end{vmatrix} - 2\begin{vmatrix} 2 & -1 \\ 3 & 1 \end{vmatrix} - 3\begin{vmatrix} 2 & -1 \\ 3 & 2 \end{vmatrix}}$$

Now, evaluating the second-order determinants in the usual way gives

$\triangle_0 = \dots\dots\dots\dots$

**40**

$$\boxed{\triangle_0 = -30}$$

Because

$$\begin{aligned}\triangle_0 &= 1(-1+2) - 2(2+3) - 3(4+3)\\ &= 1(1) - 2(5) - 3(7)\\ &= 1 - 10 - 21 = -30\end{aligned}$$

So $\triangle_0 = -30$

So we have

$$y = \frac{\triangle_2}{\triangle_0} = \frac{120}{-30} = -4$$

$\therefore y = -4$

Every one is done in the same way.

Did you get review exercise **2** correct?

If so, turn straight on to *Frame 50*.

If not, have another go at it, now that we have worked through **1** in detail.

*When you have finished, move on to Frame 41*

**41**

The answers to review exercise **2** were $\boxed{\begin{array}{l} x = -2 \\ z = 5 \end{array}}$

Did you get those correct? If so, turn on right away to *Frame 50*. If not, follow through the working. Here it is:

The equations were:

$$\begin{cases} 3x - 4y + 2z + 8 = 0 \\ x + 5y - 3z + 2 = 0 \\ 5x + 3y - z + 6 = 0 \end{cases}$$

Copy them down on to your paper.

The key to the solutions is:

$$\frac{x}{\triangle_1} = \ldots\ldots\ldots\ldots = \ldots\ldots\ldots\ldots = \ldots\ldots\ldots\ldots$$

*Fill in the missing terms and then move on to Frame 42*

**42**

$$\boxed{\frac{x}{\triangle_1} = \frac{-y}{\triangle_2} = \frac{z}{\triangle_3} = \frac{-1}{\triangle_0}}$$

We have to find $x$ and $z$. $\therefore$ We shall use:

$$\frac{x}{\triangle_1} = -\frac{1}{\triangle_0} \quad \text{i.e. } x = -\frac{\triangle_1}{\triangle_0}$$

and $$\frac{z}{\triangle_3} = \frac{-1}{\triangle_0} \quad \text{i.e. } z = -\frac{\triangle_3}{\triangle_0}$$

So we must find $\triangle_1$, $\triangle_3$ and $\triangle_0$.

(a) To find $\triangle_1$, form the determinant of coefficients omitting those of the $x$-terms.

$$\therefore \triangle_1 = \begin{vmatrix} \ldots & \ldots & \ldots \\ \ldots & \ldots & \ldots \\ \ldots & \ldots & \ldots \end{vmatrix}$$

**43**

$$\boxed{\triangle_1 = \begin{vmatrix} -4 & 2 & 8 \\ 5 & -3 & 2 \\ 3 & -1 & 6 \end{vmatrix}}$$

Now expand along the top row.

$$\triangle_1 = -4\begin{vmatrix} -3 & 2 \\ -1 & 6 \end{vmatrix} - 2\begin{vmatrix} 5 & 2 \\ 3 & 6 \end{vmatrix} + 8\begin{vmatrix} 5 & -3 \\ 3 & -1 \end{vmatrix}$$

$$= \ldots\ldots\ldots$$

*Finish it off: then on to Frame 44*

**44**

$$\boxed{\triangle_1 = 48}$$

Because $$\begin{aligned} \triangle_1 &= -4(-18 + 2) - 2(30 - 6) + 8(-5 + 9) \\ &= -4(-16) - 2(24) + 8(4) \\ &= 64 - 48 + 32 = 96 - 48 = 48 \end{aligned}$$

$$\therefore \triangle_1 = 48$$

▶

(b) To find $\triangle_3$, form the determinant of coefficients omitting the $z$-terms.

$$\therefore \triangle_3 = \begin{vmatrix} \dots & \dots & \dots \\ \dots & \dots & \dots \\ \dots & \dots & \dots \end{vmatrix}$$

**45**

$$\triangle_3 = \begin{vmatrix} 3 & -4 & 8 \\ 1 & 5 & 2 \\ 5 & 3 & 6 \end{vmatrix}$$

Expanding this along the top row gives

$\triangle_3 = \dots\dots\dots\dots$

**46**

$$\triangle_3 = 3\begin{vmatrix} 5 & 2 \\ 3 & 6 \end{vmatrix} + 4\begin{vmatrix} 1 & 2 \\ 5 & 6 \end{vmatrix} + 8\begin{vmatrix} 1 & 5 \\ 5 & 3 \end{vmatrix}$$

Now evaluate the second-order determinants and finish it off. So that

$\triangle_3 = \dots\dots\dots$

*On to Frame 47*

**47**

$$\triangle_3 = -120$$

Because
$$\begin{aligned} \triangle_3 &= 3(30-6) + 4(6-10) + 8(3-25) \\ &= 3(24) + 4(-4) + 8(-22) \\ &= 72 - 16 - 176 \\ &= 72 - 192 = -120 \end{aligned}$$
$$\therefore \triangle_3 = -120$$

(c) Now we want to find $\triangle_0$.

$$\triangle_0 = \begin{vmatrix} \dots & \dots & \dots \\ \dots & \dots & \dots \\ \dots & \dots & \dots \end{vmatrix}$$

**48**

$$\triangle_0 = \begin{vmatrix} 3 & -4 & 2 \\ 1 & 5 & -3 \\ 5 & 3 & -1 \end{vmatrix}$$

Now expand this along the top row as we have done before. Then evaluate the second-order determinants which will appear and so find the value of $\triangle_0$.
Work it right through: so that

$\triangle_0 = \dots\dots\dots$

**49**

$$\boxed{\triangle_0 = 24}$$

Because $\triangle_0 = 3\begin{vmatrix} 5 & -3 \\ 3 & -1 \end{vmatrix} + 4\begin{vmatrix} 1 & -3 \\ 5 & -1 \end{vmatrix} + 2\begin{vmatrix} 1 & 5 \\ 5 & 3 \end{vmatrix}$

$$= 3(-5+9) + 4(-1+15) + 2(3-25)$$
$$= 3(4) + 4(14) + 2(-22)$$
$$= 12 + 56 - 44$$
$$= 68 - 44 = 24$$
$$\therefore \triangle_0 = 24$$

So we have: $\triangle_1 = 48, \quad \triangle_3 = -120, \quad \triangle_0 = 24$

Also we know that:

$$x = -\frac{\triangle_1}{\triangle_0} \text{ and } z = -\frac{\triangle_3}{\triangle_0}$$

So that $x = \ldots\ldots\ldots$ and $z = \ldots\ldots\ldots$

**50**

$$\boxed{\begin{aligned} x &= -\frac{48}{24} = -2 \\ z &= -\frac{(-120)}{24} = 5 \end{aligned}}$$

Well, there you are. The method is the same every time – but take care not to make a slip with the signs.

Now what about review exercise **3**. Did you get that right? If so, move on straight away to *Frame 52*.

If not, have another go at it. Here are the equations again: copy them down and then find $x$, $y$ and $z$.

$$2x - 2y - z - 3 = 0$$
$$4x + 5y - 2z + 3 = 0$$
$$3x + 4y - 3z + 7 = 0$$

*When you have finished this one, move on to the next frame and check your results*

**51**

$$\boxed{x = 2, \quad y = -1, \quad z = 3}$$

Here are the main steps, so that you can check your own working:

$$\frac{x}{\triangle_1} = \frac{-y}{\triangle_2} = \frac{z}{\triangle_3} = \frac{-1}{\triangle_0}$$

$$\triangle_1 = \begin{vmatrix} -2 & -1 & -3 \\ 5 & -2 & 3 \\ 4 & -3 & 7 \end{vmatrix} = 54 \qquad \triangle_2 = \begin{vmatrix} 2 & -1 & -3 \\ 4 & -2 & 3 \\ 3 & -3 & 7 \end{vmatrix} = 27$$

$$\triangle_3 = \begin{vmatrix} 2 & -2 & -3 \\ 4 & 5 & 3 \\ 3 & 4 & 7 \end{vmatrix} = 81 \qquad \triangle_0 = \begin{vmatrix} 2 & -2 & -1 \\ 4 & 5 & -2 \\ 3 & 4 & -3 \end{vmatrix} = -27$$

▶

$$\frac{x}{\Delta_1} = -\frac{1}{\Delta_0} \qquad \therefore x = -\frac{\Delta_1}{\Delta_0} = -\frac{54}{-27} = 2 \qquad \therefore x = 2$$

$$\frac{-y}{\Delta_2} = -\frac{1}{\Delta_0} \qquad \therefore y = \frac{\Delta_2}{\Delta_0} = \frac{27}{-27} = -1 \qquad \therefore y = -1$$

$$\frac{z}{\Delta_3} = -\frac{1}{\Delta_0} \qquad \therefore z = -\frac{\Delta_3}{\Delta_0} = -\frac{81}{-27} = +3 \qquad \therefore z = +3$$

All correct now?

*On to Frame 52 then for the next section of the work*

## Consistency of a set of equations

**52**

Let us consider the following three equations in two unknowns.

$$3x - y - 4 = 0 \quad \text{(a)}$$
$$2x + 3y - 8 = 0 \quad \text{(b)}$$
$$x - 2y + 3 = 0 \quad \text{(c)}$$

If we solve equations (b) and (c) in the usual way, we find that $x = 1$ and $y = 2$.

If we now substitute these values in the left-hand side of (a), we obtain $3x - y - 4 = 3 - 2 - 4 = -3$ (and not 0 as the equation states).

The solutions of (b) and (c) do not satisfy (a) and the three given equations do not have a common solution. They are thus not *consistent*. There are no values of $x$ and $y$ which satisfy all three equations.

If equations are consistent, they have a ....................... 

**53**

common solution

Let us now consider the three equations:

$$3x + y - 5 = 0 \quad \text{(a)}$$
$$2x + 3y - 8 = 0 \quad \text{(b)}$$
$$x - 2y + 3 = 0 \quad \text{(c)}$$

The solutions of (b) and (c) are, as before, $x = 1$ and $y = 2$. Substituting these in (a) gives:

$$3x + y - 5 = 3 + 2 - 5 = 0$$

i.e. all three equations have the common solution $x = 1$, $y = 2$ and the equations are said to be ............

consistent

54

Now we will take the general case:

$$a_1x + b_1y + d_1 = 0 \qquad \text{(a)}$$
$$a_2x + b_2y + d_2 = 0 \qquad \text{(b)}$$
$$a_3x + b_3y + d_3 = 0 \qquad \text{(c)}$$

If we solve equations (b) and (c):

i.e. $\begin{cases} a_2x + b_2y + d_2 = 0 \\ a_3x + b_3y + d_3 = 0 \end{cases}$ we get $\dfrac{x}{\triangle_1} = \dfrac{-y}{\triangle_2} = \dfrac{1}{\triangle_0}$

where $\triangle_1 = \begin{vmatrix} b_2 & d_2 \\ b_3 & d_3 \end{vmatrix}, \quad \triangle_2 = \begin{vmatrix} a_2 & d_2 \\ a_3 & d_3 \end{vmatrix}, \quad \triangle_0 = \begin{vmatrix} a_2 & b_2 \\ a_3 & b_3 \end{vmatrix}$

so that $x = \dfrac{\triangle_1}{\triangle_0}$ and $y = -\dfrac{\triangle_2}{\triangle_0}$

If these results also satisfy equation (a), then $a_1 \cdot \dfrac{\triangle_1}{\triangle_0} + b_1 \cdot \dfrac{-\triangle_2}{\triangle_0} + d_1 = 0$

i.e. $a_1.\triangle_1 - b_1.\triangle_2 + d_1.\triangle_0 = 0$

i.e. $a_1\begin{vmatrix} b_2 & d_2 \\ b_3 & d_3 \end{vmatrix} - b_1\begin{vmatrix} a_2 & d_2 \\ a_3 & d_3 \end{vmatrix} + d_1\begin{vmatrix} a_2 & b_2 \\ a_3 & b_3 \end{vmatrix} = 0$

i.e. $\begin{vmatrix} a_1 & b_1 & d_1 \\ a_2 & b_2 & d_2 \\ a_3 & b_3 & d_3 \end{vmatrix} = 0$

which is therefore the condition that the three given equations are *consistent*.

So three simultaneous equations in two unknowns are consistent if the determinant of coefficients is ...........

zero

55

**Example 1**

Test for consistency:

$$\begin{cases} 2x + y - 5 = 0 \\ x + 4y + 1 = 0 \\ 3x - y - 10 = 0 \end{cases}$$

For the equations to be consistent $\begin{vmatrix} 2 & 1 & -5 \\ 1 & 4 & 1 \\ 3 & -1 & -10 \end{vmatrix}$ must be zero

$$\begin{vmatrix} 2 & 1 & -5 \\ 1 & 4 & 1 \\ 3 & -1 & -10 \end{vmatrix} = 2\begin{vmatrix} 4 & 1 \\ -1 & -10 \end{vmatrix} - 1\begin{vmatrix} 1 & 1 \\ 3 & -10 \end{vmatrix} - 5\begin{vmatrix} 1 & 4 \\ 3 & -1 \end{vmatrix}$$
$$= 2(-40 + 1) - 1(-10 - 3) - 5(-1 - 12)$$
$$= 2(-39) - (-13) - 5(-13)$$
$$= -78 + 13 + 65 = -78 + 78 = 0$$

The given equations therefore ........... consistent.
(are/are not)

**56**

are

**Example 2**

Find the value of $k$ for which the following equations are consistent:

$$\begin{cases} 3x+y+2=0 \\ 4x+2y-k=0 \\ 2x-y+3k=0 \end{cases} \qquad \text{For consistency: } \begin{vmatrix} 3 & 1 & 2 \\ 4 & 2 & -k \\ 2 & -1 & 3k \end{vmatrix} = 0$$

$$\therefore 3\begin{vmatrix} 2 & -k \\ -1 & 3k \end{vmatrix} - 1\begin{vmatrix} 4 & -k \\ 2 & 3k \end{vmatrix} + 2\begin{vmatrix} 4 & 2 \\ 2 & -1 \end{vmatrix} = 0$$

$$3(6k-k) - 1(12k+2k) + 2(-4-4) = 0$$

$$\therefore 15k - 14k - 16 = 0 \quad \therefore k - 16 = 0 \quad \therefore k = 16$$

Now one for you, done in just the same way.

**Example 3**

$$\text{Given } \begin{cases} x+(k+1)y+1=0 \\ 2kx+5y-3=0 \\ 3x+7y+1=0 \end{cases}$$

find the values of $k$ for which the equations are consistent.

**57**

$k = 2$ or $-\dfrac{1}{2}$

The condition for consistency is that:

$$\begin{vmatrix} 1 & k+1 & 1 \\ 2k & 5 & -3 \\ 3 & 7 & 1 \end{vmatrix} = 0$$

$$\therefore 1\begin{vmatrix} 5 & -3 \\ 7 & 1 \end{vmatrix} - (k+1)\begin{vmatrix} 2k & -3 \\ 3 & 1 \end{vmatrix} + 1\begin{vmatrix} 2k & 5 \\ 3 & 7 \end{vmatrix} = 0$$

$$(5+21) - (k+1)(2k+9) + (14k-15) = 0$$

$$26 - 2k^2 - 11k - 9 + 14k - 15 = 0$$

$$-2k^2 + 3k + 2 = 0$$

$$\therefore 2k^2 - 3k - 2 = 0 \quad \therefore (2k+1)(k-2) = 0$$

$$\therefore k = 2 \text{ or } k = -\frac{1}{2}$$

Finally, one more for you to do.

**Example 4**

$$\text{Find the values of } k \text{ for consistency when } \begin{cases} x+y-k=0 \\ kx-3y+11=0 \\ 2x+4y-8=0 \end{cases}$$

58

$$\boxed{k = 1 \text{ or } -\frac{1}{2}}$$

Because

$$\begin{vmatrix} 1 & 1 & -k \\ k & -3 & 11 \\ 2 & 4 & -8 \end{vmatrix} = 0$$

$$1\begin{vmatrix} -3 & 11 \\ 4 & -8 \end{vmatrix} - 1\begin{vmatrix} k & 11 \\ 2 & -8 \end{vmatrix} - k\begin{vmatrix} k & -3 \\ 2 & 4 \end{vmatrix} = 0$$

$$\therefore (24 - 44) - (-8k - 22) - k(4k + 6) = 0$$

$$\therefore -20 + 8k + 22 - 4k^2 - 6k = 0$$

$$-4k^2 + 2k + 2 = 0$$

$$\therefore 2k^2 - k - 1 = 0 \quad \therefore (2k + 1)(k - 1) = 0$$

$$\therefore k = 1 \text{ or } k = -\frac{1}{2}$$

## Properties of determinants

59

Expanding a determinant in which the elements are large numbers can be a very tedious affair. It is possible, however, by knowing something of the properties of determinants, to simplify the working. So here are some of the main properties. Make a note of them in your record book for future reference.

1 The value of a determinant remains unchanged if rows are changed to columns and columns to rows.

$$\begin{vmatrix} a_1 & a_2 \\ b_1 & b_2 \end{vmatrix} = \begin{vmatrix} a_1 & b_1 \\ a_2 & b_2 \end{vmatrix}$$

2 If two rows (or two columns) are interchanged, the sign of the determinant is changed.

$$\begin{vmatrix} a_2 & b_2 \\ a_1 & b_1 \end{vmatrix} = -\begin{vmatrix} a_1 & b_1 \\ a_2 & b_2 \end{vmatrix}$$

3 If two rows (or two columns) are identical, the value of the determinant is zero.

$$\begin{vmatrix} a_1 & a_1 \\ a_2 & a_2 \end{vmatrix} = 0$$

4 If the elements of any one row (or column) are all multiplied by a common factor, the determinant is multiplied by that factor.

$$\begin{vmatrix} ka_1 & kb_2 \\ a_2 & b_2 \end{vmatrix} = k\begin{vmatrix} a_1 & b_1 \\ a_2 & b_2 \end{vmatrix}$$

5 If the elements of any row (or column) are increased (or decreased) by equal multiples of the corresponding elements of any other row (or column), the value of the determinant is unchanged.

$$\begin{vmatrix} a_1 + kb_1 & b_1 \\ a_2 + kb_2 & b_2 \end{vmatrix} = \begin{vmatrix} a_1 & b_1 \\ a_2 & b_2 \end{vmatrix} \text{ and } \begin{vmatrix} a_1 & b_1 \\ a_2 + ka_1 & b_2 + kb_1 \end{vmatrix} = \begin{vmatrix} a_1 & b_1 \\ a_2 & b_2 \end{vmatrix}$$

*Note*: The properties stated above are general and apply not only to second-order determinants but to determinants of any order.

*Move on for some examples*

**60**

**Example 1**

Evaluate $\begin{vmatrix} 427 & 429 \\ 369 & 371 \end{vmatrix}$

Of course we could evaluate this by the usual method

$(427)(371) - (369)(429)$

which is rather deadly! On the other hand, we could apply our knowledge of the properties of determinants, thus:

$$\begin{vmatrix} 427 & 429 \\ 369 & 371 \end{vmatrix} = \begin{vmatrix} 427 & 429-427 \\ 369 & 371-369 \end{vmatrix}$$ (Rule 5) Subtract column 1 from column 2}

$$= \begin{vmatrix} 427 & 2 \\ 369 & 2 \end{vmatrix}$$

$$= \begin{vmatrix} 58 & 0 \\ 369 & 2 \end{vmatrix}$$ (Rule 5) Subtract row 2 from row 1

$$= (58)(2) - (0) = 116$$

Naturally, the more zero elements we can arrange, the better.

*For another example, move to Frame 61*

**61**

**Example 2**

Evaluate

$$\begin{vmatrix} 1 & 2 & 2 \\ 4 & 3 & 5 \\ 4 & 2 & 7 \end{vmatrix}$$ Column 2 minus column 3 will give us one zero

$$= \begin{vmatrix} 1 & 0 & 2 \\ 4 & -2 & 5 \\ 4 & -5 & 7 \end{vmatrix}$$ Column 3 minus twice (column 1) will give another zero

$$= \begin{vmatrix} 1 & 0 & 0 \\ 4 & -2 & -3 \\ 4 & -5 & -1 \end{vmatrix}$$ Now expand along the top row

$$= \begin{vmatrix} -2 & -3 \\ -5 & -1 \end{vmatrix}$$ We could take a factor $(-1)$ from the top row and another factor $(-1)$ from the bottom row (rule 4)

$$= (-1)(-1)\begin{vmatrix} 2 & 3 \\ 5 & 1 \end{vmatrix}$$

$$= 1(2 - 15) = -13$$

*Next frame*

**62**

**Example 3**

Evaluate $\begin{vmatrix} 4 & 2 & 2 \\ 2 & 4 & 2 \\ 2 & 2 & 4 \end{vmatrix}$

You do that one, but by way of practice, apply as many of the listed properties as possible. It is quite fun.

*When you have finished it, move on to Frame 63*

**63**

The answer is $\boxed{32}$, but what we are more interested in is the method of applying the properties, so follow it through. This is one way of doing it; not the only way by any means.

$$\begin{vmatrix} 4 & 2 & 2 \\ 2 & 4 & 2 \\ 2 & 2 & 4 \end{vmatrix}$$

We can take out a factor 2 from each row, giving a factor $2^3$, i.e. 8 outside the determinant

$$= 8\begin{vmatrix} 2 & 1 & 1 \\ 1 & 2 & 1 \\ 1 & 1 & 2 \end{vmatrix}$$

Column 2 minus column 3 will give one zero in the top row

$$= 8\begin{vmatrix} 2 & 0 & 1 \\ 1 & 1 & 1 \\ 1 & -1 & 2 \end{vmatrix}$$

Column 1 minus twice(column 3) will give another zero in the same row

$$= 8\begin{vmatrix} 0 & 0 & 1 \\ -1 & 1 & 1 \\ -3 & -1 & 2 \end{vmatrix}$$

Expanding along the top row will now reduce this to a second-order determinant

$$= 8\begin{vmatrix} -1 & 1 \\ -3 & -1 \end{vmatrix}$$

Now row 2 + row 1

$$= 8\begin{vmatrix} -1 & 1 \\ -4 & 0 \end{vmatrix}$$

$$= -8\begin{vmatrix} 1 & 1 \\ 4 & 0 \end{vmatrix} = -8(-4) = 32$$

**64**

Here is another type of problem.

**Example 4**

Solve the equation $\begin{vmatrix} x & 5 & 3 \\ 5 & x+1 & 1 \\ -3 & -4 & x-2 \end{vmatrix} = 0$

In this type of question, we try to establish common factors wherever possible.

For example, if we add row 2 and row 3 to row 1, we get:

$$\begin{vmatrix} (x+2) & (x+2) & (x+2) \\ 5 & x+1 & 1 \\ -3 & -4 & x-2 \end{vmatrix} = 0$$

Taking out the common factor $(x+2)$ gives:

$$(x+2)\begin{vmatrix} 1 & 1 & 1 \\ 5 & x+1 & 1 \\ -3 & -4 & x-2 \end{vmatrix} = 0$$

Now if we take column 1 from column 2 and also from column 3, what do we get?

*When you have done it, move on to the next frame*

**65**

We now have $(x+2)\begin{vmatrix} 1 & 0 & 0 \\ 5 & x-4 & -4 \\ -3 & -1 & x+1 \end{vmatrix} = 0$

Expanding along the top row reduces this to a second-order determinant:

$(x+2)\begin{vmatrix} x-4 & -4 \\ -1 & x+1 \end{vmatrix} = 0$

If we now multiply out the determinant, we get

$$(x+2)[(x-4)(x+1)-4] = 0$$
$$\therefore (x+2)(x^2-3x-8) = 0$$
$$\therefore x+2 = 0 \text{ or } x^2-3x-8 = 0$$

which finally gives $x = -2$ or $x = \dfrac{3 \pm \sqrt{41}}{2}$

Finally, here is one for you to do on your own.

**Example 5**

Solve the equation:

$$\begin{vmatrix} 5 & x & 3 \\ x+2 & 2 & 1 \\ -3 & 2 & x \end{vmatrix} = 0$$

*Check your working with that given in the next frame*

**66**

$$\boxed{x = -4 \text{ or } 1 \pm \sqrt{6}}$$

Here is one way of doing the problem:

$$\begin{vmatrix} 5 & x & 3 \\ x+2 & 2 & 1 \\ -3 & 2 & x \end{vmatrix} = 0$$ Adding row 2 and row 3 to row 1, gives

$$\begin{vmatrix} x+4 & x+4 & x+4 \\ x+2 & 2 & 1 \\ -3 & 2 & x \end{vmatrix} = 0$$ Taking out the common factor $(x+4)$

$$(x+4)\begin{vmatrix} 1 & 1 & 1 \\ x+2 & 2 & 1 \\ -3 & 2 & x \end{vmatrix} = 0$$ Take column 3 from column 1 and from column 2

$$(x+4)\begin{vmatrix} 0 & 0 & 1 \\ x+1 & 1 & 1 \\ -x-3 & 2-x & x \end{vmatrix} = 0$$ This now reduces to second order

$$(x+4)\begin{vmatrix} x+1 & 1 \\ -x-3 & 2-x \end{vmatrix} = 0$$ Subtract column 2 from column 1

$$(x+4)\begin{vmatrix} x & 1 \\ -5 & 2-x \end{vmatrix} = 0$$ We now finish it off

$$\therefore (x+4)(2x-x^2+5) = 0$$
$$\therefore x+4 = 0 \text{ or } x^2-2x-5 = 0$$

which gives $x = -4$ or $x = 1 \pm \sqrt{6}$

▶

You have now reached the end of this Programme on determinants except for the **Can You?** checklist and **Test exercise** which follow. Before you work through them, brush up on any parts of the work about which you are at all uncertain.

## Can You?

### Checklist 4

67

Check this list before and after you try the end of Programme test.

On a scale of 1 to 5 how confident you are that you can:

| | Frames |
|---|---|
| • Expand a $2 \times 2$ determinant?<br>*Yes* ☐ ☐ ☐ ☐ ☐ *No* | 1 to 6 |
| • Solve pairs of simultaneous linear equations in two variables using $2 \times 2$ determinants?<br>*Yes* ☐ ☐ ☐ ☐ ☐ *No* | 7 to 16 |
| • Expand a $3 \times 3$ determinant?<br>*Yes* ☐ ☐ ☐ ☐ ☐ *No* | 17 to 27 |
| • Solve three simultaneous linear equations in three variables using $3 \times 3$ determinants?<br>*Yes* ☐ ☐ ☐ ☐ ☐ *No* | 28 to 51 |
| • Determine the consistency of sets of simultaneous linear equations?<br>*Yes* ☐ ☐ ☐ ☐ ☐ *No* | 52 to 58 |
| • Use the properties of determinants to solve equations written in determinant form?<br>*Yes* ☐ ☐ ☐ ☐ ☐ *No* | 59 to 66 |

## Test exercise 4

68

If you have worked steadily through the Programme, you should have no difficulty with this exercise. Take your time and work carefully. There is no extra credit for speed. Off you go then. They are all quite straightforward.

**1** Evaluate

(a) $\begin{vmatrix} 1 & 1 & 2 \\ 2 & 1 & 1 \\ 1 & 2 & 1 \end{vmatrix}$ (b) $\begin{vmatrix} 1 & 2 & 3 \\ 3 & 1 & 2 \\ 2 & 3 & 1 \end{vmatrix}$

**2** By determinants, find the value of $x$, given:

$$\begin{cases} 2x + 3y - z + 3 = 0 \\ x - 4y + 2z - 14 = 0 \\ 4x + 2y - 3z + 6 = 0 \end{cases}$$

**3** Use determinants to solve completely:

$$\begin{cases} x - 3y + 4z - 5 = 0 \\ 2x + y + z - 3 = 0 \\ 4x + 3y + 5z - 1 = 0 \end{cases}$$

▶

4 Find the values of $k$ for which the following equations are consistent:

$$\begin{cases} 3x + 5y + k = 0 \\ 2x + y - 5 = 0 \\ (k+1)x + 2y - 10 = 0 \end{cases}$$

5 Solve the equation:

$$\begin{vmatrix} x+1 & -5 & -6 \\ -1 & x & 2 \\ -3 & 2 & x+1 \end{vmatrix} = 0$$

*Now you can continue with the next Programme*

## Further problems 4

**69**

1 Evaluate: (a) $\begin{vmatrix} 3 & 5 & 7 \\ 11 & 9 & 13 \\ 15 & 17 & 19 \end{vmatrix}$ (b) $\begin{vmatrix} 1 & 428 & 861 \\ 2 & 535 & 984 \\ 3 & 642 & 1107 \end{vmatrix}$

2 Evaluate: (a) $\begin{vmatrix} 25 & 3 & 35 \\ 16 & 10 & -18 \\ 34 & 6 & 38 \end{vmatrix}$ (b) $\begin{vmatrix} 155 & 226 & 81 \\ 77 & 112 & 39 \\ 74 & 111 & 37 \end{vmatrix}$

3 Solve by determinants:

$$4x - 5y + 7z = -14$$
$$9x + 2y + 3z = 47$$
$$x - y - 5z = 11$$

4 Use determinants to solve the equations:

$$4x - 3y + 2z = -7$$
$$6x + 2y - 3z = 33$$
$$2x - 4y - z = -3$$

5 Solve by determinants:

$$3x + 2y - 2z = 16$$
$$4x + 3y + 3z = 2$$
$$2x - y + z = 0$$

6 Find the values of $\lambda$ for which the following equations are consistent:

$$5x + (\lambda + 1)y - 5 = 0$$
$$(\lambda - 1)x + 7y + 5 = 0$$
$$3x + 5y + 1 = 0$$

7 Determine the values of $k$ for which the following equations have solutions:

$$4x - (k-2)y - 5 = 0$$
$$2x + y - 10 = 0$$
$$(k+1)x - 4y - 9 = 0$$

**8** (a) Find the values of $k$ which satisfy the equation:

$$\begin{vmatrix} k & 1 & 0 \\ 1 & k & 1 \\ 0 & 1 & k \end{vmatrix} = 0$$

(b) Factorize

$$\begin{vmatrix} 1 & 1 & 1 \\ a & b & c \\ a^3 & b^3 & c^3 \end{vmatrix}$$

**9** Solve the equation:

$$\begin{vmatrix} x & 2 & 3 \\ 2 & x+3 & 6 \\ 3 & 4 & x+6 \end{vmatrix} = 0$$

**10** Find the values of $x$ that satisfy the equation:

$$\begin{vmatrix} x & 3+x & 2+x \\ 3 & -3 & -1 \\ 2 & -2 & -2 \end{vmatrix} = 0$$

**11** Express

$$\begin{vmatrix} 1 & 1 & 1 \\ a^2 & b^2 & c^2 \\ (b+c)^2 & (c+a)^2 & (a+b)^2 \end{vmatrix}$$

as a product of linear factors.

**12** A resistive network gives the following equations:

$$2(i_3 - i_2) + 5(i_3 - i_1) = 24$$
$$(i_2 - i_3) + 2i_2 + (i_2 - i_1) = 0$$
$$5(i_1 - i_3) + 2(i_1 - i_2) + i_1 = 6$$

Simplify the equations and use determinants to find the value of $i_2$ correct to two significant figures.

**13** Show that $(a+b+c)$ is a factor of the determinant

$$\begin{vmatrix} b+c & a & a^3 \\ c+a & b & b^3 \\ a+b & c & c^3 \end{vmatrix}$$

and express the determinant as a product of five factors.

**14** Find values of $k$ for which the following equations are consistent:

$$x + (1+k)y + 1 = 0$$
$$(2+k)x + 5y - 10 = 0$$
$$x + 7y + 9 = 0$$

**15** Express

$$\begin{vmatrix} 1+x^2 & yz & 1 \\ 1+y^2 & zx & 1 \\ 1+z^2 & xy & 1 \end{vmatrix}$$

as a product of four linear factors.

**16** Solve the equation:

$$\begin{vmatrix} x+1 & x+2 & 3 \\ 2 & x+3 & x+1 \\ x+3 & 1 & x+2 \end{vmatrix} = 0$$

▶

**17** If $x$, $y$, $z$ satisfy the equations:

$$\left(\frac{1}{2}M_1 + M_2\right)x - M_2y = W$$

$$-M_2x + 2M_2y + (M_1 - M_2)z = 0$$

$$-M_2y + \left(\frac{1}{2}M_1 + M_2\right)z = 0$$

evaluate $x$ in terms of $W$, $M_1$ and $M_2$.

**18** Three currents, $i_1$, $i_2$, $i_3$, in a network are related by the following equations:

$$2i_1 + 3i_2 + 8i_3 = 30$$

$$6i_1 - i_2 + 2i_3 = 4$$

$$3i_1 - 12i_2 + 8i_3 = 0$$

By the use of determinants, find the value of $i_1$ and hence solve completely the three equations.

**19** If

$$k(x - a) + 2x - z = 0$$

$$k(y - a) + 2y - z = 0$$

$$k(z - a) - x - y + 2z = 0$$

show that $x = \dfrac{ak(k+3)}{k^2 + 4k + 2}$.

**20** Find the angles between $\theta = 0$ and $\theta = \pi$ that satisfy the equation:

$$\begin{vmatrix} 1 + \sin^2\theta & \cos^2\theta & 4\sin 2\theta \\ \sin^2\theta & 1 + \cos^2\theta & 4\sin 2\theta \\ \sin^2\theta & \cos^2\theta & 1 + 4\sin 2\theta \end{vmatrix} = 0$$

**70**

Now visit the companion website for this book at www.palgrave.com/stroud for more questions applying this mathematics to science and engineering.

Programme 5 Frames 1 to 59

# Matrices

## Learning outcomes

When you have completed this Programme you will be able to:

- ☐ Define a matrix
- ☐ Understand what is meant by the equality of two matrices
- ☐ Add and subtract two matrices
- ☐ Multiply a matrix by a scalar and multiply two matrices together
- ☐ Obtain the transpose of a matrix
- ☐ Recognize special types of matrix
- ☐ Obtain the determinant, cofactors and adjoint of a square matrix
- ☐ Obtain the inverse of a non-singular matrix
- ☐ Use matrices to solve a set of linear equations using inverse matrices
- ☐ Use the Gaussian elimination method to solve a set of linear equations
- ☐ Evaluate eigenvalues and eigenvectors

# Matrices – definitions

**1**

A *matrix* is a set of real or complex numbers (or *elements*) arranged in rows and columns to form a rectangular array.

A matrix having $m$ rows and $n$ columns is called an $m \times n$ (i.e. '$m$ by $n$') matrix and is referred to as having *order* $m \times n$.

A matrix is indicated by writing the array within brackets

e.g. $\begin{pmatrix} 5 & 7 & 2 \\ 6 & 3 & 8 \end{pmatrix}$ is a $2 \times 3$ matrix, i.e. a '2 by 3' matrix, where

5, 7, 2, 6, 3, 8 are the elements of the matrix.

Note that, in describing the matrix, the number of rows is stated first and the number of columns second.

$\begin{pmatrix} 5 & 6 & 4 \\ 2 & -3 & 2 \\ 7 & 8 & 7 \\ 6 & 7 & 5 \end{pmatrix}$ is a matrix of order $4 \times 3$, i.e. 4 rows and 3 columns.

So the matrix $\begin{pmatrix} 6 & 4 \\ 0 & 1 \\ 2 & 3 \end{pmatrix}$ is of order ...........

and the matrix $\begin{pmatrix} 2 & 5 & 3 & 4 \\ 6 & 7 & 4 & 9 \end{pmatrix}$ is of order ...........

**2**

$3 \times 2; \quad 2 \times 4$

A matrix is simply an array of numbers: there is no arithmetical connection between the elements and it therefore differs from a determinant in that the elements cannot be multiplied together in any way to find a numerical value of the matrix. A matrix has no numerical value. Also, in general, rows and columns cannot be interchanged as was the case with determinants.

*Row matrix*: A row matrix consists of 1 row only.

e.g. $(4 \quad 3 \quad 7 \quad 2)$ is a row matrix of order $1 \times 4$.

*Column matrix*: A column matrix consists of 1 column only.

e.g. $\begin{pmatrix} 6 \\ 3 \\ 8 \end{pmatrix}$ is a column matrix of order $3 \times 1$.

To conserve space in printing, a column matrix is sometimes written on one line but with 'curly' brackets, e.g. $\{\,6\ 3\ 8\,\}$ is the same column matrix of order $3 \times 1$.

*Move on to the next frame*

**3**

So, from what we have already said:

(a) $\begin{pmatrix} 5 \\ 2 \end{pmatrix}$ is a ........... matrix of order ...........

(b) $(4 \quad 0 \quad 7 \quad 3)$ is a ........... matrix of order ...........

(c) $\{2 \quad 6 \quad 9\}$ is a ........... matrix of order ...........

**4**

(a) column, $2 \times 1$ (b) row, $1 \times 4$ (c) column, $3 \times 1$

We use a simple row matrix in stating the $x$- and $y$-coordinates of a point relative to the $x$- and $y$-axes. For example, if P is the point (3, 5) then the 3 is the $x$-coordinate and the 5 the $y$-coordinate. In matrices generally, however, no commas are used to separate the elements.

*Single element matrix*: A single number may be regarded as a $1 \times 1$ matrix, i.e. having 1 row and 1 column.

*Double suffix notation*: Each element in a matrix has its own particular 'address' or location which can be defined by a system of double suffixes, the first indicating the row, the second the column, thus:

$$\begin{pmatrix} a_{11} & a_{12} & a_{13} & a_{14} \\ a_{21} & a_{22} & a_{23} & a_{24} \\ a_{31} & a_{32} & a_{33} & a_{34} \end{pmatrix}$$

$\therefore$ $a_{23}$ indicates the element in the second row and third column.

Therefore, in the matrix

$$\begin{pmatrix} 6 & -5 & 1 & -3 \\ 2 & -4 & 8 & 3 \\ 4 & -7 & -6 & 5 \\ -2 & 9 & 7 & -1 \end{pmatrix}$$

the location of (a) the element 3 can be stated as ...........
(b) the element $-1$ can be stated as ...........
(c) the element 9 can be stated as ...........

**5**

(a) $a_{24}$ (b) $a_{44}$ (c) $a_{42}$

*Move on*

## Matrix notation

**6**

Where there is no ambiguity, a whole matrix can be denoted by a single general element enclosed in brackets, or by a single letter printed in bold type. This is a very neat shorthand and saves much space and writing. For example:

$$\begin{pmatrix} a_{11} & a_{12} & a_{13} & a_{14} \\ a_{21} & a_{22} & a_{23} & a_{24} \\ a_{31} & a_{32} & a_{33} & a_{34} \end{pmatrix} \text{ can be denoted by } (a_{ij}) \text{ or } (a) \text{ or by } \mathbf{A}.$$

$$\text{Similarly } \begin{pmatrix} x_1 \\ x_2 \\ x_3 \end{pmatrix} \text{ can be denoted by } (x_i) \text{ or } (x) \text{ or simply by } \mathbf{x}.$$

For an $(m \times n)$ matrix, we use a bold capital letter, e.g. **A**. For a row or column matrix, we use a lower-case bold letter, e.g. **x**. (In handwritten work, we can indicate bold-face type by a wavy line placed under the letter, e.g. $\underset{\sim}{A}$ or $\underset{\sim}{x}$.)

So, if **B** represents a $2 \times 3$ matrix, write out the elements $b_{ij}$ in the matrix, using the double suffix notation. This gives ...........

7

$$\mathbf{B} = \begin{pmatrix} b_{11} & b_{12} & b_{13} \\ b_{21} & b_{22} & b_{23} \end{pmatrix}$$

*Next frame*

## Equal matrices

8

By definition, two matrices are said to be equal if corresponding elements throughout are equal. Thus, the two matrices must also be of the same order.

So, if $\begin{pmatrix} a_{11} & a_{12} & a_{13} \\ a_{21} & a_{22} & a_{23} \end{pmatrix} = \begin{pmatrix} 4 & 6 & 5 \\ 2 & 3 & 7 \end{pmatrix}$

then $a_{11} = 4$; $a_{12} = 6$; $a_{13} = 5$; $a_{21} = 2$; etc.

Therefore, if $(a_{ij}) = (x_{ij})$ then $a_{ij} = x_{ij}$ for all values of $i$ and $j$.

So, if $\begin{pmatrix} a & b & c \\ d & e & f \\ g & h & k \end{pmatrix} = \begin{pmatrix} 5 & -7 & 3 \\ 1 & 2 & 6 \\ 0 & 4 & 8 \end{pmatrix}$

then $d = \dots\dots\dots\dots$; $b = \dots\dots\dots\dots$; $a - k = \dots\dots\dots\dots$

9

$$d = 1; \quad b = -7; \quad a - k = -3$$

## Addition and subtraction of matrices

10

To be added or subtracted, two matrices must be of the *same order*. The sum or difference is then determined by adding or subtracting corresponding elements.

e.g. $\begin{pmatrix} 4 & 2 & 3 \\ 5 & 7 & 6 \end{pmatrix} + \begin{pmatrix} 1 & 8 & 9 \\ 3 & 5 & 4 \end{pmatrix} = \begin{pmatrix} 4+1 & 2+8 & 3+9 \\ 5+3 & 7+5 & 6+4 \end{pmatrix}$

$= \begin{pmatrix} 5 & 10 & 12 \\ 8 & 12 & 10 \end{pmatrix}$

and $\begin{pmatrix} 6 & 5 & 12 \\ 9 & 4 & 8 \end{pmatrix} - \begin{pmatrix} 3 & 7 & 1 \\ 2 & 10 & -5 \end{pmatrix} = \begin{pmatrix} 6-3 & 5-7 & 12-1 \\ 9-2 & 4-10 & 8+5 \end{pmatrix}$

$= \begin{pmatrix} 3 & -2 & 11 \\ 7 & -6 & 13 \end{pmatrix}$

So, (a) $\begin{pmatrix} 6 & 5 & 4 & 1 \\ 2 & 3 & -7 & 8 \end{pmatrix} + \begin{pmatrix} 1 & 4 & 2 & 3 \\ 6 & -1 & 0 & 5 \end{pmatrix} = \dots\dots\dots\dots$

(b) $\begin{pmatrix} 8 & 3 & 6 \\ 5 & 2 & 7 \\ 1 & 0 & 4 \end{pmatrix} - \begin{pmatrix} 1 & 2 & 3 \\ 4 & 5 & 6 \\ 7 & 8 & 9 \end{pmatrix} = \dots\dots\dots\dots$

11

(a) $\begin{pmatrix} 7 & 9 & 6 & 4 \\ 8 & 2 & -7 & 13 \end{pmatrix}$ (b) $\begin{pmatrix} 7 & 1 & 3 \\ 1 & -3 & 1 \\ -6 & -8 & -5 \end{pmatrix}$

# Multiplication of matrices

12

## Scalar multiplication

To multiply a matrix by a single number (i.e. a scalar), each individual element of the matrix is multiplied by that factor:

$$\text{e.g. } 4 \times \begin{pmatrix} 3 & 2 & 5 \\ 6 & 1 & 7 \end{pmatrix} = \begin{pmatrix} 12 & 8 & 20 \\ 24 & 4 & 28 \end{pmatrix}$$

i.e. in general, $k(a_{ij}) = (ka_{ij})$.

It also means that, in reverse, we can take a common factor out of every element – not just one row or one column as in determinants.

Therefore, $\begin{pmatrix} 10 & 25 & 45 \\ 35 & 15 & 50 \end{pmatrix}$ can be written . . . . . . . . . . .

13

$$5 \times \begin{pmatrix} 2 & 5 & 9 \\ 7 & 3 & 10 \end{pmatrix}$$

## Multiplication of two matrices

Two matrices can be multiplied together only when the number of columns in the first is equal to the number of rows in the second.

$$\text{e.g. if } \mathbf{A} = (a_{ij}) = \begin{pmatrix} a_{11} & a_{12} & a_{13} \\ a_{21} & a_{22} & a_{23} \end{pmatrix} \text{ and } \mathbf{b} = (b_i) = \begin{pmatrix} b_1 \\ b_2 \\ b_3 \end{pmatrix}$$

$$\text{then } \mathbf{A}.\mathbf{b} = \begin{pmatrix} a_{11} & a_{12} & a_{13} \\ a_{21} & a_{22} & a_{23} \end{pmatrix} \cdot \begin{pmatrix} b_1 \\ b_2 \\ b_3 \end{pmatrix}$$

$$= \begin{pmatrix} a_{11}b_1 + a_{12}b_2 + a_{13}b_3 \\ a_{21}b_1 + a_{22}b_2 + a_{23}b_3 \end{pmatrix}$$

i.e. each element in the top row of **A** is multiplied by the corresponding element in the first column of **b** and the products added. Similarly, the second row of the product is found by multiplying each element in the second row of **A** by the corresponding element in the first column of **b**.

▶

**Example 1**

$$\begin{pmatrix} 4 & 7 & 6 \\ 2 & 3 & 1 \end{pmatrix} \cdot \begin{pmatrix} 8 \\ 5 \\ 9 \end{pmatrix} = \begin{pmatrix} 4 \times 8 + 7 \times 5 + 6 \times 9 \\ 2 \times 8 + 3 \times 5 + 1 \times 9 \end{pmatrix} = \begin{pmatrix} 32 + 35 + 54 \\ 16 + 15 + 9 \end{pmatrix} = \begin{pmatrix} 121 \\ 40 \end{pmatrix}$$

Similarly $\begin{pmatrix} 2 & 3 & 5 & 1 \\ 4 & 6 & 0 & 7 \end{pmatrix} \cdot \begin{pmatrix} 3 \\ 4 \\ 2 \\ 9 \end{pmatrix} = \ldots\ldots\ldots\ldots$

## 14

$$\boxed{\begin{pmatrix} 6 + 12 + 10 + 9 \\ 12 + 24 + 0 + 63 \end{pmatrix} = \begin{pmatrix} 37 \\ 99 \end{pmatrix}}$$

In just the same way, if $\mathbf{A} = \begin{pmatrix} 3 & 6 & 8 \\ 1 & 0 & 2 \end{pmatrix}$ and $\mathbf{b} = \begin{pmatrix} 7 \\ 4 \\ 5 \end{pmatrix}$ then $\mathbf{A}.\mathbf{b} = \ldots\ldots\ldots\ldots$

## 15

$$\boxed{\begin{pmatrix} 85 \\ 17 \end{pmatrix}}$$

The same process is carried out for each row and column.

**Example 2**

If $\mathbf{A} = (a_{ij}) = \begin{pmatrix} 1 & 5 \\ 2 & 7 \\ 3 & 4 \end{pmatrix}$ and $\mathbf{B} = (b_{ij}) = \begin{pmatrix} 8 & 4 & 3 & 1 \\ 2 & 5 & 8 & 6 \end{pmatrix}$

$$\text{then} \quad \mathbf{A}.\mathbf{B} = \begin{pmatrix} 1 & 5 \\ 2 & 7 \\ 3 & 4 \end{pmatrix} \cdot \begin{pmatrix} 8 & 4 & 3 & 1 \\ 2 & 5 & 8 & 6 \end{pmatrix}$$

$$= \begin{pmatrix} 1 \times 8 + 5 \times 2 & 1 \times 4 + 5 \times 5 & 1 \times 3 + 5 \times 8 & 1 \times 1 + 5 \times 6 \\ 2 \times 8 + 7 \times 2 & 2 \times 4 + 7 \times 5 & 2 \times 3 + 7 \times 8 & 2 \times 1 + 7 \times 6 \\ 3 \times 8 + 4 \times 2 & 3 \times 4 + 4 \times 5 & 3 \times 3 + 4 \times 8 & 3 \times 1 + 4 \times 6 \end{pmatrix}$$

$$= \begin{pmatrix} 8 + 10 & 4 + 25 & 3 + 40 & 1 + 30 \\ 16 + 14 & 8 + 35 & 6 + 56 & 2 + 42 \\ 24 + 8 & 12 + 20 & 9 + 32 & 3 + 24 \end{pmatrix}$$

$$= \begin{pmatrix} 18 & 29 & 43 & 31 \\ 30 & 43 & 62 & 44 \\ 32 & 32 & 41 & 27 \end{pmatrix}$$

Note that multiplying a $(3 \times 2)$ matrix and a $(2 \times 4)$ matrix gives a product matrix of order $(3 \times 4)$

i.e. order $(3 \times 2) \times$ order $(2 \times 4) \rightarrow$ order $(3 \times 4)$.

(same number 2)

In general then, the product of an $(l \times m)$ matrix and an $(m \times n)$ matrix has order $(l \times n)$.

If $\mathbf{A} = \begin{pmatrix} 2 & 4 & 6 \\ 3 & 9 & 5 \end{pmatrix}$ and $\mathbf{B} = \begin{pmatrix} 7 & 1 \\ -2 & 9 \\ 4 & 3 \end{pmatrix}$ then $\mathbf{A}.\mathbf{B} = \ldots\ldots\ldots\ldots$

16

$$\begin{pmatrix} 30 & 56 \\ 23 & 99 \end{pmatrix}$$

$$\text{since } \mathbf{A.B} = \begin{pmatrix} 2 & 4 & 6 \\ 3 & 9 & 5 \end{pmatrix} \cdot \begin{pmatrix} 7 & 1 \\ -2 & 9 \\ 4 & 3 \end{pmatrix}$$

$$= \begin{pmatrix} 14 - 8 + 24 & 2 + 36 + 18 \\ 21 - 18 + 20 & 3 + 81 + 15 \end{pmatrix} = \begin{pmatrix} 30 & 56 \\ 23 & 99 \end{pmatrix}$$

**Example 3**

It follows that a matrix can be squared only if it is itself a square matrix, i.e. the number of rows equals the number of columns.

$$\text{If } \mathbf{A} = \begin{pmatrix} 4 & 7 \\ 5 & 2 \end{pmatrix}$$

$$\mathbf{A}^2 = \begin{pmatrix} 4 & 7 \\ 5 & 2 \end{pmatrix} \cdot \begin{pmatrix} 4 & 7 \\ 5 & 2 \end{pmatrix}$$

$$= \begin{pmatrix} 16 + 35 & 28 + 14 \\ 20 + 10 & 35 + 4 \end{pmatrix} = \begin{pmatrix} 51 & 42 \\ 30 & 39 \end{pmatrix}$$

Remember that multiplication of matrices is defined only when ...........

17

the number of columns in the first
= the number of rows in the second

That is correct. $\begin{pmatrix} 1 & 5 & 6 \\ 4 & 9 & 7 \end{pmatrix} \cdot \begin{pmatrix} 2 & 3 & 5 \\ 8 & 7 & 1 \end{pmatrix}$ has no meaning.

$$\left.\begin{array}{r} \text{If } \mathbf{A} \text{ is an } (m \times n) \text{ matrix} \\ \text{and } \mathbf{B} \text{ is an } (n \times m) \text{ matrix} \end{array}\right\} \text{ then products } \mathbf{A.B} \text{ and } \mathbf{B.A} \text{ are possible.}$$

**Example**

$$\text{If } \mathbf{A} = \begin{pmatrix} 1 & 2 & 3 \\ 4 & 5 & 6 \end{pmatrix} \text{ and } \mathbf{B} = \begin{pmatrix} 7 & 10 \\ 8 & 11 \\ 9 & 12 \end{pmatrix}$$

$$\text{then } \mathbf{A.B} = \begin{pmatrix} 1 & 2 & 3 \\ 4 & 5 & 6 \end{pmatrix} \cdot \begin{pmatrix} 7 & 10 \\ 8 & 11 \\ 9 & 12 \end{pmatrix}$$

$$= \begin{pmatrix} 7 + 16 + 27 & 10 + 22 + 36 \\ 28 + 40 + 54 & 40 + 55 + 72 \end{pmatrix} = \begin{pmatrix} 50 & 68 \\ 122 & 167 \end{pmatrix}$$

$$\text{and } \mathbf{B.A} = \begin{pmatrix} 7 & 10 \\ 8 & 11 \\ 9 & 12 \end{pmatrix} \cdot \begin{pmatrix} 1 & 2 & 3 \\ 4 & 5 & 6 \end{pmatrix}$$

$$= \begin{pmatrix} 7 + 40 & 14 + 50 & 21 + 60 \\ 8 + 44 & 16 + 55 & 24 + 66 \\ 9 + 48 & 18 + 60 & 27 + 72 \end{pmatrix} = \begin{pmatrix} 47 & 64 & 81 \\ 52 & 71 & 90 \\ 57 & 78 & 99 \end{pmatrix}$$

▶

Note that, in matrix multiplication, $\mathbf{A.B} \neq \mathbf{B.A}$, i.e. multiplication is not commutative. The order of the factors is important!

In the product **A.B**, **B** is *pre-multiplied* by **A**
and **A** is *post-multiplied* by **B**.

So, if $\mathbf{A} = \begin{pmatrix} 5 & 2 \\ 7 & 4 \\ 3 & 1 \end{pmatrix}$ and $\mathbf{B} = \begin{pmatrix} 9 & 2 & 4 \\ -2 & 3 & 6 \end{pmatrix}$

then $\mathbf{A.B} = \ldots\ldots\ldots\ldots$

and $\mathbf{B.A} = \ldots\ldots\ldots\ldots$

**18**

$$\mathbf{A.B} = \begin{pmatrix} 41 & 16 & 32 \\ 55 & 26 & 52 \\ 25 & 9 & 18 \end{pmatrix}; \quad \mathbf{B.A} = \begin{pmatrix} 71 & 30 \\ 29 & 14 \end{pmatrix}$$

## Transpose of a matrix

**19**

If the rows and columns of a matrix are interchanged:

i.e. the first row becomes the first column,
the second row becomes the second column,
the third row becomes the third column, etc.,

then the new matrix so formed is called the *transpose* of the original matrix. If **A** is the original matrix, its transpose is denoted by $\tilde{\mathbf{A}}$ or $\mathbf{A}^{\mathrm{T}}$. We shall use the latter.

$$\therefore \text{ If } \mathbf{A} = \begin{pmatrix} 4 & 6 \\ 7 & 9 \\ 2 & 5 \end{pmatrix}, \text{ then } \mathbf{A}^{\mathrm{T}} = \begin{pmatrix} 4 & 7 & 2 \\ 6 & 9 & 5 \end{pmatrix}$$

Therefore, given that

$$\mathbf{A} = \begin{pmatrix} 2 & 7 & 6 \\ 3 & 1 & 5 \end{pmatrix} \text{ and } \mathbf{B} = \begin{pmatrix} 4 & 0 \\ 3 & 7 \\ 1 & 5 \end{pmatrix}$$

then $\mathbf{A.B} = \ldots\ldots\ldots\ldots$ and $(\mathbf{A.B})^{\mathrm{T}} = \ldots\ldots\ldots\ldots$

**20**

$$\mathbf{A.B} = \begin{pmatrix} 35 & 79 \\ 20 & 32 \end{pmatrix}; \quad (\mathbf{A.B})^{\mathrm{T}} = \begin{pmatrix} 35 & 20 \\ 79 & 32 \end{pmatrix}$$

## Special matrices

21

(a) *Square matrix* is a matrix of order $m \times m$.

e.g. $\begin{pmatrix} 1 & 2 & 5 \\ 6 & 8 & 9 \\ 1 & 7 & 4 \end{pmatrix}$ is a $3 \times 3$ matrix

A square matrix $(a_{ij})$ is *symmetric* if $a_{ij} = a_{ji}$, e.g. $\begin{pmatrix} 1 & 2 & 5 \\ 2 & 8 & 9 \\ 5 & 9 & 4 \end{pmatrix}$

i.e. it is symmetrical about the leading diagonal.

Note that $\mathbf{A} = \mathbf{A}^{\mathrm{T}}$.

A square matrix $(a_{ij})$ is skew-symmetric if $a_{ij} = -a_{ji}$ e.g. $\begin{pmatrix} 0 & 2 & 5 \\ -2 & 0 & 9 \\ -5 & -9 & 0 \end{pmatrix}$

In that case, $\mathbf{A} = -\mathbf{A}^{\mathrm{T}}$.

(b) *Diagonal matrix* is a square matrix with all elements zero except those on the leading diagonal, thus $\begin{pmatrix} 5 & 0 & 0 \\ 0 & 2 & 0 \\ 0 & 0 & 7 \end{pmatrix}$

(c) *Unit matrix* is a diagonal matrix in which the elements on the leading diagonal are all unity, i.e. $\begin{pmatrix} 1 & 0 & 0 \\ 0 & 1 & 0 \\ 0 & 0 & 1 \end{pmatrix}$

The unit matrix is denoted by **I**.

If $\mathbf{A} = \begin{pmatrix} 5 & 2 & 4 \\ 1 & 3 & 8 \\ 7 & 9 & 6 \end{pmatrix}$ and $\mathbf{I} = \begin{pmatrix} 1 & 0 & 0 \\ 0 & 1 & 0 \\ 0 & 0 & 1 \end{pmatrix}$ then $\mathbf{A.I} = \dots\dots\dots\dots$

22

$$\begin{pmatrix} 5 & 2 & 4 \\ 1 & 3 & 8 \\ 7 & 9 & 6 \end{pmatrix} \text{ i.e. } \mathbf{A.I} = \mathbf{A}$$

Similarly, if we form the product **I.A** we obtain:

$$\mathbf{I.A} = \begin{pmatrix} 1 & 0 & 0 \\ 0 & 1 & 0 \\ 0 & 0 & 1 \end{pmatrix} \cdot \begin{pmatrix} 5 & 2 & 4 \\ 1 & 3 & 8 \\ 7 & 9 & 6 \end{pmatrix}$$

$$= \begin{pmatrix} 5+0+0 & 2+0+0 & 4+0+0 \\ 0+1+0 & 0+3+0 & 0+8+0 \\ 0+0+7 & 0+0+9 & 0+0+6 \end{pmatrix} = \begin{pmatrix} 5 & 2 & 4 \\ 1 & 3 & 8 \\ 7 & 9 & 6 \end{pmatrix} = \mathbf{A}$$

$\mathbf{A.I} = \mathbf{I.A} = \mathbf{A}$

Therefore, the unit matrix **I** behaves very much like the unit factor in ordinary algebra and arithmetic.

▶

(d) *Null matrix*: A null matrix is one whose elements are all zero.

i.e. $\begin{pmatrix} 0 & 0 & 0 \\ 0 & 0 & 0 \\ 0 & 0 & 0 \end{pmatrix}$ and is denoted by **0**.

If $\mathbf{A.B} = \mathbf{0}$, we cannot say that therefore $\mathbf{A} = \mathbf{0}$ or $\mathbf{B} = \mathbf{0}$

for if $\mathbf{A} = \begin{pmatrix} 2 & 1 & -3 \\ 6 & 3 & -9 \end{pmatrix}$ and $\mathbf{B} = \begin{pmatrix} 1 & 9 \\ 4 & -6 \\ 2 & 4 \end{pmatrix}$

$$\text{then}\quad \mathbf{A.B} = \begin{pmatrix} 2 & 1 & -3 \\ 6 & 3 & -9 \end{pmatrix} \cdot \begin{pmatrix} 1 & 9 \\ 4 & -6 \\ 2 & 4 \end{pmatrix}$$

$$= \begin{pmatrix} 2+4-6 & 18-6-12 \\ 6+12-18 & 54-18-36 \end{pmatrix} = \begin{pmatrix} 0 & 0 \\ 0 & 0 \end{pmatrix}$$

That is, $\mathbf{A.B} = \mathbf{0}$, but clearly $\mathbf{A} \neq \mathbf{0}$ and $\mathbf{B} \neq \mathbf{0}$.

*Now a short revision exercise.* Do these without looking back.

**1** If $\mathbf{A} = \begin{pmatrix} 4 & 6 & 5 & 7 \\ 3 & 1 & 9 & 4 \end{pmatrix}$ and $\mathbf{B} = \begin{pmatrix} 2 & 8 & 3 & -1 \\ 5 & 2 & -4 & 6 \end{pmatrix}$

determine (a) $\mathbf{A} + \mathbf{B}$ and (b) $\mathbf{A} - \mathbf{B}$.

**2** If $\mathbf{A} = \begin{pmatrix} 4 & 3 \\ 2 & 7 \\ 6 & 1 \end{pmatrix}$ and $\mathbf{B} = \begin{pmatrix} 5 & 9 & 2 \\ 4 & 0 & 8 \end{pmatrix}$

determine (a) $5\mathbf{A}$; (b) $\mathbf{A.B}$; (c) $\mathbf{B.A}$.

**3** If $\mathbf{A} = \begin{pmatrix} 2 & 6 \\ 5 & 7 \\ 4 & 1 \end{pmatrix}$ and $\mathbf{B} = \begin{pmatrix} 3 & 2 \\ 0 & 7 \\ 2 & 3 \end{pmatrix}$ then $\mathbf{A.B} = \ldots\ldots\ldots\ldots$

**4** Given that $\mathbf{A} = \begin{pmatrix} 4 & 2 & 6 \\ 1 & 8 & 7 \end{pmatrix}$ determine (a) $\mathbf{A}^{\mathrm{T}}$ and (b) $\mathbf{A.A}^{\mathrm{T}}$.

*When you have completed them, check your results with the next frame*

**23**

Here are the solutions. Check your results.

**1** (a) $\mathbf{A} + \mathbf{B} = \begin{pmatrix} 6 & 14 & 8 & 6 \\ 8 & 3 & 5 & 10 \end{pmatrix}$; (b) $\mathbf{A} - \mathbf{B} = \begin{pmatrix} 2 & -2 & 2 & 8 \\ -2 & -1 & 13 & -2 \end{pmatrix}$

**2** (a) $5\mathbf{A} = \begin{pmatrix} 20 & 15 \\ 10 & 35 \\ 30 & 5 \end{pmatrix}$ (b) $\mathbf{A.B} = \begin{pmatrix} 32 & 36 & 32 \\ 38 & 18 & 60 \\ 34 & 54 & 20 \end{pmatrix}$ (c) $\mathbf{B.A} = \begin{pmatrix} 50 & 80 \\ 64 & 20 \end{pmatrix}$

**3** $\mathbf{A.B} = \begin{pmatrix} 2 & 6 \\ 5 & 7 \\ 4 & 1 \end{pmatrix} \cdot \begin{pmatrix} 3 & 2 \\ 0 & 7 \\ 2 & 3 \end{pmatrix}$ is not possible since the number of columns in the first must be equal to the number of rows in the second.

▶

4 $A = \begin{pmatrix} 4 & 2 & 6 \\ 1 & 8 & 7 \end{pmatrix} \quad \therefore \quad A^T = \begin{pmatrix} 4 & 1 \\ 2 & 8 \\ 6 & 7 \end{pmatrix}$

$$A.A^T = \begin{pmatrix} 4 & 2 & 6 \\ 1 & 8 & 7 \end{pmatrix} \cdot \begin{pmatrix} 4 & 1 \\ 2 & 8 \\ 6 & 7 \end{pmatrix} = \begin{pmatrix} 16+4+36 & 4+16+42 \\ 4+16+42 & 1+64+49 \end{pmatrix}$$

$$= \begin{pmatrix} 56 & 62 \\ 62 & 114 \end{pmatrix}$$

*Now move on to the next frame*

---

# Determinant of a square matrix

**24**

The determinant of a square matrix is the determinant having the same elements as those of the matrix. For example:

the determinant of $\begin{pmatrix} 5 & 2 & 1 \\ 0 & 6 & 3 \\ 8 & 4 & 7 \end{pmatrix}$ is $\begin{vmatrix} 5 & 2 & 1 \\ 0 & 6 & 3 \\ 8 & 4 & 7 \end{vmatrix}$ and the value of this

determinant is $5(42-12) - 2(0-24) + 1(0-48)$
$= 5(30) - 2(-24) + 1(-48)$
$= 150 + 48 - 48 = 150$

Note that the transpose of the matrix is $\begin{pmatrix} 5 & 0 & 8 \\ 2 & 6 & 4 \\ 1 & 3 & 7 \end{pmatrix}$ and the

determinant of the transpose is $\begin{vmatrix} 5 & 0 & 8 \\ 2 & 6 & 4 \\ 1 & 3 & 7 \end{vmatrix}$ the value of which is

$5(42-12) - 0(14-4) + 8(6-6) = 5(30) = 150$.

That is, the determinant of a square matrix has the same value as that of the determinant of the transposed matrix.

A matrix whose determinant is zero is called a *singular* matrix.

The determinant of the matrix $\begin{pmatrix} 3 & 2 & 5 \\ 4 & 7 & 9 \\ 1 & 8 & 6 \end{pmatrix}$ has the value ...........

and the determinant value of the diagonal matrix $\begin{pmatrix} 2 & 0 & 0 \\ 0 & 5 & 0 \\ 0 & 0 & 4 \end{pmatrix}$ has the

value ...........

---

**25**

$$\begin{vmatrix} 3 & 2 & 5 \\ 4 & 7 & 9 \\ 1 & 8 & 6 \end{vmatrix} = 3(-30) - 2(15) + 5(25) = 5$$

$$\begin{vmatrix} 2 & 0 & 0 \\ 0 & 5 & 0 \\ 0 & 0 & 4 \end{vmatrix} = 2(20) + 0 + 0 = 40$$

▶

## Cofactors

If $\mathbf{A} = (a_{ij})$ is a square matrix, we can form a determinant of its elements:

$$\begin{vmatrix} a_{11} & a_{12} & a_{13} & \dots & a_{1n} \\ a_{21} & a_{22} & a_{23} & \dots & a_{2n} \\ a_{31} & a_{32} & a_{33} & \dots & a_{3n} \\ \vdots & & \vdots & & \vdots \\ a_{n1} & a_{n2} & a_{n3} & \dots & a_{nn} \end{vmatrix}$$

Each element gives rise to a *cofactor*, which is simply the minor of the element in the determinant together with its 'place sign', which was described in detail in the previous programme.

For example, the determinant of the matrix $\mathbf{A} = \begin{pmatrix} 2 & 3 & 5 \\ 4 & 1 & 6 \\ 1 & 4 & 0 \end{pmatrix}$ is

$\det \mathbf{A} = |\mathbf{A}| = \begin{vmatrix} 2 & 3 & 5 \\ 4 & 1 & 6 \\ 1 & 4 & 0 \end{vmatrix}$ which has a value of 45.

The minor of the element 2 is $\begin{vmatrix} 1 & 6 \\ 4 & 0 \end{vmatrix} = 0 - 24 = -24$.

The place sign is $+$. Therefore the cofactor of the element 2 is $+(-24)$ i.e. $-24$.

Similarly, the minor of the element 3 is $\begin{pmatrix} 4 & 6 \\ 1 & 0 \end{pmatrix} = 0 - 6 = -6$.

The place sign is $-$. Therefore the cofactor of the element 3 is $-(-6) = 6$.

In each case the minor is found by striking out the line and column containing the element in question and forming a determinant of the remaining elements. The appropriate place signs are given by:

$$\begin{pmatrix} + & - & + & - & \dots \\ - & + & - & + & \\ + & - & + & & \\ \vdots & & & & \end{pmatrix}$$

alternate plus and minus from the top left-hand corner which carries a $+$

Therefore, in the example above, the minor of the element 6 is $\begin{vmatrix} 2 & 3 \\ 1 & 4 \end{vmatrix}$ i.e. $8 - 3 = 5$. The place sign is $-$. Therefore the cofactor of the element 6 is $-5$.

So, for the matrix $\begin{pmatrix} 7 & 1 & -2 \\ 6 & 5 & 4 \\ 3 & 8 & 9 \end{pmatrix}$, the cofactor of the element 3 is ...........

and that of the element 4 is ...........

**26**

Cofactor of 3 is $4 - (-10) = 14$
Cofactor of 4 is $-(56 - 3) = -53$

## Adjoint of a square matrix

If we start afresh with $\mathbf{A} = \begin{pmatrix} 2 & 3 & 5 \\ 4 & 1 & 6 \\ 1 & 4 & 0 \end{pmatrix}$, its determinant

$\det \mathbf{A} = |\mathbf{A}| = \begin{vmatrix} 2 & 3 & 5 \\ 4 & 1 & 6 \\ 1 & 4 & 0 \end{vmatrix}$ from which we can form a new matrix $\mathbf{C}$ of the cofactors.

▶

$$\mathbf{C} = \begin{pmatrix} A_{11} & A_{12} & A_{13} \\ A_{21} & A_{22} & A_{23} \\ A_{31} & A_{32} & A_{33} \end{pmatrix} \quad \text{where} \quad \begin{array}{l} A_{11} \text{ is the cofactor of } a_{11} \\ A_{ij} \text{ is the cofactor of } a_{ij} \text{ etc.} \end{array}$$

$$A_{11} = +\begin{vmatrix} 1 & 6 \\ 4 & 0 \end{vmatrix} = +(0-24) = -24 \qquad A_{12} = -\begin{vmatrix} 4 & 6 \\ 1 & 0 \end{vmatrix} = -(0-6) = 6 \qquad A_{13} = +\begin{vmatrix} 4 & 1 \\ 1 & 4 \end{vmatrix} = +(16-1) = 15$$

$$A_{21} = -\begin{vmatrix} 3 & 5 \\ 4 & 0 \end{vmatrix} = -(0-20) = 20 \qquad A_{22} = +\begin{vmatrix} 2 & 5 \\ 1 & 0 \end{vmatrix} = +(0-5) = -5 \qquad A_{23} = -\begin{vmatrix} 2 & 3 \\ 1 & 4 \end{vmatrix} = -(8-3) = -5$$

$$A_{31} = +\begin{vmatrix} 3 & 5 \\ 1 & 6 \end{vmatrix} = +(18-5) = 13 \qquad A_{32} = -\begin{vmatrix} 2 & 5 \\ 4 & 6 \end{vmatrix} = -(12-20) = 8 \qquad A_{33} = +\begin{vmatrix} 2 & 3 \\ 4 & 1 \end{vmatrix} = +(2-12) = -10$$

$$\therefore \quad \text{The matrix of cofactors is } \mathbf{C} = \begin{pmatrix} -24 & 6 & 15 \\ 20 & -5 & -5 \\ 13 & 8 & -10 \end{pmatrix}$$

$$\text{and the transpose of } \mathbf{C}, \text{ i.e. } \mathbf{C}^{\mathrm{T}} = \begin{pmatrix} -24 & 20 & 13 \\ 6 & -5 & 8 \\ 15 & -5 & -10 \end{pmatrix}$$

This is called the *adjoint* of the original matrix **A** and is written adj **A**.

Therefore, to find the adjoint of a square matrix **A**:

(a) we form the matrix **C** of cofactors,

(b) we write the transpose of **C**, i.e. $\mathbf{C}^{\mathrm{T}}$.

Hence the adjoint of $\begin{pmatrix} 5 & 2 & 1 \\ 3 & 1 & 4 \\ 4 & 6 & 3 \end{pmatrix}$ is ...........

27

$$\text{adj } \mathbf{A} = \mathbf{C}^{\mathrm{T}} = \begin{pmatrix} -21 & 0 & 7 \\ 7 & 11 & -17 \\ 14 & -22 & -1 \end{pmatrix}$$

## Inverse of a square matrix

28

The adjoint of a square matrix is important, since it enables us to form the inverse of the matrix. If each element of the adjoint of **A** is divided by the value of the determinant of **A**, i.e. $|\mathbf{A}|$, (provided $|\mathbf{A}| \neq 0$), the resulting matrix is called the *inverse* of **A** and is denoted by $\mathbf{A}^{-1}$.

▶

For the matrix which we used in the last frame, $\mathbf{A} = \begin{pmatrix} 2 & 3 & 5 \\ 4 & 1 & 6 \\ 1 & 4 & 0 \end{pmatrix}$

$$\det \mathbf{A} = |\mathbf{A}| = \begin{vmatrix} 2 & 3 & 5 \\ 4 & 1 & 6 \\ 1 & 4 & 0 \end{vmatrix} = 2(0 - 24) - 3(0 - 6) + 5(16 - 1) = 45,$$

$$\text{the matrix of cofactors } \mathbf{C} = \begin{pmatrix} -24 & 6 & 15 \\ 20 & -5 & -5 \\ 13 & 8 & -10 \end{pmatrix}$$

$$\text{and the adjoint of } \mathbf{A}, \text{ i.e. } \mathbf{C}^{\mathrm{T}} = \begin{pmatrix} -24 & 20 & 13 \\ 6 & -5 & 8 \\ 15 & -5 & -10 \end{pmatrix}$$

Then the inverse of **A** is given by

$$\mathbf{A}^{-1} = \begin{pmatrix} -\frac{24}{45} & \frac{20}{45} & \frac{13}{45} \\ \frac{6}{45} & -\frac{5}{45} & \frac{8}{45} \\ \frac{15}{45} & -\frac{5}{45} & -\frac{10}{45} \end{pmatrix} = \frac{1}{45}\begin{pmatrix} -24 & 20 & 13 \\ 6 & -5 & 8 \\ 15 & -5 & -10 \end{pmatrix}$$

*Therefore, to form the inverse of a square matrix* **A**:

(a) Evaluate the determinant of **A**, i.e. $|\mathbf{A}|$

(b) Form a matrix **C** of the cofactors of the elements of $|\mathbf{A}|$

(c) Write the transpose of **C**, i.e. $\mathbf{C}^{\mathrm{T}}$, to obtain the adjoint of **A**

(d) Divide each element of $\mathbf{C}^{\mathrm{T}}$ by $|\mathbf{A}|$

(e) The resulting matrix is the inverse $\mathbf{A}^{-1}$ of the original matrix **A**.

Let us work through an example in detail:

$$\text{To find the inverse of } \mathbf{A} = \begin{pmatrix} 1 & 2 & 3 \\ 4 & 1 & 5 \\ 6 & 0 & 2 \end{pmatrix}$$

(a) Evaluate the determinant of **A**, i.e. $|\mathbf{A}|$, $|\mathbf{A}| = \ldots\ldots\ldots\ldots$

## 29

$$\boxed{|\mathbf{A}| = 28}$$

Because

$$|\mathbf{A}| = \begin{pmatrix} 1 & 2 & 3 \\ 4 & 1 & 5 \\ 6 & 0 & 2 \end{pmatrix} = 1(2 - 0) - 2(8 - 30) + 3(0 - 6) = 28$$

(b) Now form the matrix of the cofactors. $\mathbf{C} = \ldots\ldots\ldots\ldots$

## 30

$$\boxed{\mathbf{C} = \begin{pmatrix} 2 & 22 & -6 \\ -4 & -16 & 12 \\ 7 & 7 & -7 \end{pmatrix}}$$

Because

$$A_{11} = +(2 - 0) = 2; \quad A_{12} = -(8 - 30) = 22; \quad A_{13} = +(0 - 6) = -6$$
$$A_{21} = -(4 - 0) = -4; \quad A_{22} = +(2 - 18) = -16; \quad A_{23} = -(0 - 12) = 12$$
$$A_{31} = +(10 - 3) = 7; \quad A_{32} = -(5 - 12) = 7; \quad A_{33} = +(1 - 8) = -7$$

▶

(c) Next we have to write down the transpose of **C** to obtain the adjoint of **A**.

adj $\mathbf{A} = \mathbf{C}^{\mathrm{T}} = \ldots\ldots\ldots\ldots$

**31**

$$\text{adj } \mathbf{A} = \mathbf{C}^{\mathrm{T}} = \begin{pmatrix} 2 & -4 & 7 \\ 22 & -16 & 7 \\ -6 & 12 & -7 \end{pmatrix}$$

(d) Finally, we divide the elements of adj **A** by the value of $|\mathbf{A}|$, i.e. 28, to arrive at $\mathbf{A}^{-1}$, the inverse of **A**.

$\therefore\ \mathbf{A}^{-1} = \ldots\ldots\ldots\ldots$

**32**

$$\mathbf{A}^{-1} = \begin{pmatrix} \frac{2}{28} & -\frac{4}{28} & \frac{7}{28} \\ \frac{22}{28} & -\frac{16}{28} & \frac{7}{28} \\ -\frac{6}{28} & \frac{12}{28} & -\frac{7}{28} \end{pmatrix} = \frac{1}{28}\begin{pmatrix} 2 & -4 & 7 \\ 22 & -16 & 7 \\ -6 & 12 & -7 \end{pmatrix}$$

Every one is done in the same way. Work the next one right through on your own.

Determine the inverse of the matrix $\mathbf{A} = \begin{pmatrix} 2 & 7 & 4 \\ 3 & 1 & 6 \\ 5 & 0 & 8 \end{pmatrix}$

$\mathbf{A}^{-1} = \ldots\ldots\ldots\ldots$

**33**

$$\mathbf{A}^{-1} = \frac{1}{38}\begin{pmatrix} 8 & -56 & 38 \\ 6 & -4 & 0 \\ -5 & 35 & -19 \end{pmatrix}$$

Here are the details:

$$\det \mathbf{A} = |\mathbf{A}| = \begin{vmatrix} 2 & 7 & 4 \\ 3 & 1 & 6 \\ 5 & 0 & 8 \end{vmatrix} = 2(8) - 7(-6) + 4(-5) = 38$$

Cofactors:

$A_{11} = +(8 - 0) = 8$; $\quad A_{12} = -(24 - 30) = 6$; $\quad A_{13} = +(0 - 5) = -5$

$A_{21} = -(56 - 0) = -56$; $\quad A_{22} = +(16 - 20) = -4$; $\quad A_{23} = -(0 - 35) = 35$

$A_{31} = +(42 - 4) = 38$; $\quad A_{32} = -(12 - 12) = 0$; $\quad A_{33} = +(2 - 21) = -19$

$$\therefore\ \mathbf{C} = \begin{pmatrix} 8 & 6 & -5 \\ -56 & -4 & 35 \\ 38 & 0 & -19 \end{pmatrix} \quad \therefore\ \mathbf{C}^{\mathrm{T}} = \begin{pmatrix} 8 & -56 & 38 \\ 6 & -4 & 0 \\ -5 & 35 & -19 \end{pmatrix}$$

$$\text{then } \mathbf{A}^{-1} = \frac{1}{38}\begin{pmatrix} 8 & -56 & 38 \\ 6 & -4 & 0 \\ -5 & 35 & -19 \end{pmatrix}$$

Now let us find some uses for the inverse.

## Product of a square matrix and its inverse

From a previous example, we have seen that when $\mathbf{A} = \begin{pmatrix} 1 & 2 & 3 \\ 4 & 1 & 5 \\ 6 & 0 & 2 \end{pmatrix}$

$$\mathbf{A}^{-1} = \frac{1}{28}\begin{pmatrix} 2 & -4 & 7 \\ 22 & -16 & 7 \\ -6 & 12 & -7 \end{pmatrix}$$

$$\text{Then } \mathbf{A}^{-1}.\mathbf{A} = \frac{1}{28}\begin{pmatrix} 2 & -4 & 7 \\ 22 & -16 & 7 \\ -6 & 12 & -7 \end{pmatrix}.\begin{pmatrix} 1 & 2 & 3 \\ 4 & 1 & 5 \\ 6 & 0 & 2 \end{pmatrix}$$

$$= \frac{1}{28}\begin{pmatrix} 2-16+42 & 4-4+0 & 6-20+14 \\ 22-64+42 & 44-16+0 & 66-80+14 \\ -6+48-42 & -12+12+0 & -18+60-14 \end{pmatrix}$$

$$= \frac{1}{28}\begin{pmatrix} 28 & 0 & 0 \\ 0 & 28 & 0 \\ 0 & 0 & 28 \end{pmatrix} = \begin{pmatrix} 1 & 0 & 0 \\ 0 & 1 & 0 \\ 0 & 0 & 1 \end{pmatrix} = \mathbf{I} \qquad \therefore\ \mathbf{A}^{-1}.\mathbf{A} = \mathbf{I}$$

$$\text{Also } \mathbf{A}.\mathbf{A}^{-1} = \begin{pmatrix} 1 & 2 & 3 \\ 4 & 1 & 5 \\ 6 & 0 & 2 \end{pmatrix} \times \frac{1}{28}\begin{pmatrix} 2 & -4 & 7 \\ 22 & -16 & 7 \\ -6 & 12 & -7 \end{pmatrix}$$

$$= \frac{1}{28}\begin{pmatrix} 1 & 2 & 3 \\ 4 & 1 & 5 \\ 6 & 0 & 2 \end{pmatrix}.\begin{pmatrix} 2 & -4 & 7 \\ 22 & -16 & 7 \\ -6 & 12 & -7 \end{pmatrix} = \dots\dots\dots\dots$$

*Finish it off*

**34**

$$\mathbf{A}.\mathbf{A}^{-1} = \frac{1}{28}\begin{pmatrix} 28 & 0 & 0 \\ 0 & 28 & 0 \\ 0 & 0 & 28 \end{pmatrix} = \begin{pmatrix} 1 & 0 & 0 \\ 0 & 1 & 0 \\ 0 & 0 & 1 \end{pmatrix} = \mathbf{I}$$

$\therefore\ \mathbf{A}.\mathbf{A}^{-1} = \mathbf{A}^{-1}.\mathbf{A} = \mathbf{I}$

That is, the product of a square matrix and its inverse, in whatever order the factors are written, is the unit matrix of the same matrix order.

# Solution of a set of linear equations

**35**

Consider the set of linear equations:

$$\begin{aligned} a_{11}x_1 + a_{12}x_2 + a_{13}x_3 + \dots\dots\dots + a_{1n}x_n &= b_1 \\ a_{21}x_1 + a_{22}x_2 + a_{23}x_3 + \dots\dots\dots + a_{2n}x_n &= b_2 \\ \vdots \qquad \vdots \qquad\qquad\qquad \vdots \;&\;\; \vdots \\ a_{n1}x_1 + a_{n2}x_2 + a_{n3}x_3 + \dots\dots\dots + a_{nn}x_n &= b_n \end{aligned}$$

From our knowledge of matrix multiplication, this can be written in matrix form:

$$\begin{pmatrix} a_{11} & a_{12} & a_{13} & \dots & a_{1n} \\ a_{21} & a_{22} & a_{23} & \dots & a_{2n} \\ \vdots & \vdots & \vdots & & \vdots \\ a_{n1} & a_{n2} & a_{n3} & \dots & a_{nn} \end{pmatrix} \cdot \begin{pmatrix} x_1 \\ x_2 \\ \vdots \\ x_n \end{pmatrix} = \begin{pmatrix} b_1 \\ b_2 \\ \vdots \\ b_n \end{pmatrix} \quad \text{i.e. } \mathbf{A.x} = \mathbf{b}$$

$$\text{where } \mathbf{A} = \begin{pmatrix} a_{11} & a_{12} & \dots & a_{1n} \\ a_{21} & a_{22} & \dots & a_{2n} \\ \vdots & \vdots & & \vdots \\ a_{n1} & a_{n2} & \dots & a_{nn} \end{pmatrix}; \quad \mathbf{x} = \begin{pmatrix} x_1 \\ x_2 \\ \vdots \\ x_n \end{pmatrix}; \quad \text{and } \mathbf{b} = \begin{pmatrix} b_1 \\ b_2 \\ \vdots \\ b_n \end{pmatrix}$$

If we multiply both sides of the matrix equation by the inverse of **A**, we have:

$$\mathbf{A}^{-1}.\mathbf{A}.\mathbf{x} = \mathbf{A}^{-1}.\mathbf{b}$$

But $\mathbf{A}^{-1}.\mathbf{A} = \mathbf{I} \quad \therefore \quad \mathbf{I}.\mathbf{x} = \mathbf{A}^{-1}.\mathbf{b} \quad \text{i.e. } \mathbf{x} = \mathbf{A}^{-1}.\mathbf{b}$

Therefore, if we form the inverse of the matrix of coefficients and pre-multiply matrix **b** by it, we shall determine the matrix of the solutions of **x**.

**Example**

To solve the set of equations:

$$\begin{aligned} x_1 + 2x_2 + x_3 &= 4 \\ 3x_1 - 4x_2 - 2x_3 &= 2 \\ 5x_1 + 3x_2 + 5x_3 &= -1 \end{aligned}$$

First write the set of equations in matrix form, which gives ...........

---

**36**

$$\begin{pmatrix} 1 & 2 & 1 \\ 3 & -4 & -2 \\ 5 & 3 & 5 \end{pmatrix} \cdot \begin{pmatrix} x_1 \\ x_2 \\ x_3 \end{pmatrix} = \begin{pmatrix} 4 \\ 2 \\ -1 \end{pmatrix}$$

i.e. $\mathbf{A.x} = \mathbf{b} \qquad \therefore \quad \mathbf{x} = \mathbf{A}^{-1}.\mathbf{b}$

So the next step is to find the inverse of **A** where **A** is the matrix of the coefficients of **x**. We have already seen how to determine the inverse of a matrix, so in this case

$\mathbf{A}^{-1} = $ ...........

---

**37**

$$\mathbf{A}^{-1} = -\frac{1}{35}\begin{pmatrix} -14 & -7 & 0 \\ -25 & 0 & 5 \\ 29 & 7 & -10 \end{pmatrix}$$

Because

$$|\mathbf{A}| = \begin{vmatrix} 1 & 2 & 1 \\ 3 & -4 & -2 \\ 5 & 3 & 5 \end{vmatrix} = -14 - 50 + 29 = 29 - 64 \quad \therefore \quad |\mathbf{A}| = -35$$

Cofactors:

$$\begin{aligned} &A_{11} = +(-20 + 6) = -14; \quad A_{12} = -(15 + 10) = -25; \quad A_{13} = +(9 + 20) = 29 \\ &A_{21} = -(10 - 3) = -7; \quad A_{22} = +(5 - 5) = 0; \quad A_{23} = -(3 - 10) = 7 \\ &A_{31} = +(-4 + 4) = 0; \quad A_{32} = -(-2 - 3) = 5; \quad A_{33} = +(-4 - 6) = -10 \end{aligned}$$

▶

$$\therefore\ \mathbf{C}=\begin{pmatrix}-14 & -25 & 29\\ -7 & 0 & 7\\ 0 & 5 & -10\end{pmatrix}\quad \therefore\ \text{adj}\,\mathbf{A}=\mathbf{C}^{\mathrm{T}}=\begin{pmatrix}-14 & -7 & 0\\ -25 & 0 & 5\\ 29 & 7 & -10\end{pmatrix}$$

$$\text{Now } |\mathbf{A}|=-35\quad \therefore\ \mathbf{A}^{-1}=\frac{\text{adj}\,\mathbf{A}}{|\mathbf{A}|}=-\frac{1}{35}\begin{pmatrix}-14 & -7 & 0\\ -25 & 0 & 5\\ 29 & 7 & -10\end{pmatrix}$$

$$\therefore\ \mathbf{x}=\mathbf{A}^{-1}.\mathbf{b}=-\frac{1}{35}\begin{pmatrix}-14 & -7 & 0\\ -25 & 0 & 5\\ 29 & 7 & -10\end{pmatrix}\cdot\begin{pmatrix}4\\ 2\\ -1\end{pmatrix}=\dots\dots\dots\dots$$

*Multiply it out*

## 38

$$\boxed{\mathbf{x}=-\frac{1}{35}\begin{pmatrix}-70\\ -105\\ 140\end{pmatrix}=\begin{pmatrix}2\\ 3\\ -4\end{pmatrix}}$$

$$\text{So finally } \mathbf{x}=\begin{pmatrix}x_1\\ x_2\\ x_3\end{pmatrix}=\begin{pmatrix}2\\ 3\\ -4\end{pmatrix}\quad \therefore\ x_1=2;\ x_2=3;\ x_3=-4$$

Once you have found the inverse, the rest is simply $\mathbf{x}=\mathbf{A}^{-1}.\mathbf{b}$. Here is another example to solve in the same way:

$$\begin{aligned}\text{If}\quad 2x_1 \quad -x_2 \quad +3x_3 &= 2\\ x_1 \quad +3x_2 \quad -x_3 &= 11\\ 2x_1 \quad -2x_2 \quad +5x_3 &= 3\end{aligned}$$

then $x_1=\dots\dots\dots\dots;\ x_2=\dots\dots\dots\dots;\ x_3=\dots\dots\dots\dots$

## 39

$$\boxed{x_1=-1;\ x_2=5;\ x_3=3}$$

The essential intermediate results are as follows:

$$\begin{pmatrix}2 & -1 & 3\\ 1 & 3 & -1\\ 2 & -2 & 5\end{pmatrix}.\begin{pmatrix}x_1\\ x_2\\ x_3\end{pmatrix}=\begin{pmatrix}2\\ 11\\ 3\end{pmatrix}\quad \text{i.e. } \mathbf{A}.\mathbf{x}=\mathbf{b}\quad \therefore\ \mathbf{x}=\mathbf{A}^{-1}.\mathbf{b}$$

$$\det \mathbf{A}=|\mathbf{A}|=9$$

$$\mathbf{C}=\begin{pmatrix}13 & -7 & -8\\ -1 & 4 & 2\\ -8 & 5 & 7\end{pmatrix}\quad \therefore\ \text{adj}\,\mathbf{A}=\mathbf{C}^{\mathrm{T}}=\begin{pmatrix}13 & -1 & -8\\ -7 & 4 & 5\\ -8 & 2 & 7\end{pmatrix}$$

$$\mathbf{A}^{-1}=\frac{\mathbf{C}^{\mathrm{T}}}{|\mathbf{A}|}=\frac{1}{9}\begin{pmatrix}13 & -1 & -8\\ -7 & 4 & 5\\ -8 & 2 & 7\end{pmatrix}$$

$$\mathbf{x}=\mathbf{A}^{-1}.\mathbf{b}=\frac{1}{9}\begin{pmatrix}13 & -1 & -8\\ -7 & 4 & 5\\ -8 & 2 & 7\end{pmatrix}\cdot\begin{pmatrix}2\\ 11\\ 3\end{pmatrix}=\frac{1}{9}\begin{pmatrix}-9\\ 45\\ 27\end{pmatrix}=\begin{pmatrix}-1\\ 5\\ 3\end{pmatrix}$$

$$\therefore\ \mathbf{x}=\begin{pmatrix}x_1\\ x_2\\ x_3\end{pmatrix}=\begin{pmatrix}-1\\ 5\\ 3\end{pmatrix}$$

$$\therefore\ x_1=-1;\ x_2=5;\ x_3=3$$

▶

## Gaussian elimination method for solving a set of linear equations

$$\begin{pmatrix} a_{11} & a_{12} & a_{13} & \dots & a_{1n} \\ a_{21} & a_{22} & a_{23} & \dots & a_{2n} \\ \vdots & \vdots & \vdots & & \vdots \\ a_{n1} & a_{n2} & a_{n3} & \dots & a_{nn} \end{pmatrix} \cdot \begin{pmatrix} x_1 \\ x_2 \\ \vdots \\ x_n \end{pmatrix} = \begin{pmatrix} b_1 \\ b_2 \\ \vdots \\ b_n \end{pmatrix} \quad \text{i.e. } \mathbf{A.x} = \mathbf{b}$$

All the information for solving the set of equations is provided by the matrix of coefficients **A** and the column matrix **b**. If we write the elements of **b** within the matrix **A**, we obtain the *augmented matrix* **B** of the given set of equations.

$$\text{i.e. } \mathbf{B} = \begin{pmatrix} a_{11} & a_{12} & a_{13} & \dots & a_{1n} & b_1 \\ a_{21} & a_{22} & a_{23} & \dots & a_{2n} & b_2 \\ \vdots & \vdots & & & \vdots & \vdots \\ a_{n1} & a_{n2} & a_{n3} & \dots & a_{nn} & b_n \end{pmatrix}$$

(a) We then eliminate the elements other than $a_{11}$ from the first column by subtracting $a_{21}/a_{11}$ times the first row from the second row and $a_{31}/a_{11}$ times the first row from the third row, etc.

(b) This gives a new matrix of the form:

$$\begin{pmatrix} a_{11} & a_{12} & a_{13} & \dots & a_{1n} & b_1 \\ 0 & c_{22} & c_{23} & \dots & c_{2n} & d_2 \\ \vdots & \vdots & \vdots & & \vdots & \vdots \\ 0 & c_{n2} & c_{n3} & \dots & c_{nn} & d_n \end{pmatrix}$$

The process is then repeated to eliminate $c_{i2}$ from the third and subsequent rows.

*A specific example will explain the method, so move on to the next frame*

**40**

To solve
$$x_1 + 2x_2 - 3x_3 = 3$$
$$2x_1 - x_2 - x_3 = 11$$
$$3x_1 + 2x_2 + x_3 = -5$$

This can be written $\begin{pmatrix} 1 & 2 & -3 \\ 2 & -1 & -1 \\ 3 & 2 & 1 \end{pmatrix} \cdot \begin{pmatrix} x_1 \\ x_2 \\ x_3 \end{pmatrix} = \begin{pmatrix} 3 \\ 11 \\ -5 \end{pmatrix}$

The augmented matrix becomes $\begin{pmatrix} 1 & 2 & -3 & 3 \\ 2 & -1 & -1 & 11 \\ 3 & 2 & 1 & -5 \end{pmatrix}$

Now subtract $\dfrac{2}{1}$ times the first row from the second row

and $\dfrac{3}{1}$ times the first row from the third row.

This gives $\begin{pmatrix} 1 & 2 & -3 & 3 \\ 0 & -5 & 5 & 5 \\ 0 & -4 & 10 & -14 \end{pmatrix}$

Now subtract $\dfrac{-4}{-5}$, i.e. $\dfrac{4}{5}$, times the second row from the third row.

The matrix becomes $\begin{pmatrix} 1 & 2 & -3 & 3 \\ 0 & -5 & 5 & 5 \\ 0 & 0 & 6 & -18 \end{pmatrix}$

Note that as a result of these steps, the matrix of coefficients of $x$ has been reduced to a triangular matrix.

▶

Finally, we detach the right-hand column back to its original position:

$$\begin{pmatrix} 1 & 2 & -3 \\ 0 & -5 & 5 \\ 0 & 0 & 6 \end{pmatrix} \cdot \begin{pmatrix} x_1 \\ x_2 \\ x_3 \end{pmatrix} = \begin{pmatrix} 3 \\ 5 \\ -18 \end{pmatrix}$$

Then, by 'back-substitution', starting from the bottom row we get:

$$6x_3 = -18 \;\therefore\; x_3 = -3$$
$$-5x_2 + 5x_3 = 5 \;\therefore\; -5x_2 = 5 + 15 = 20 \;\therefore\; x_2 = -4$$
$$x_1 + 2x_2 - 3x_3 = 3 \;\therefore\; x_1 - 8 + 9 = 3 \;\therefore\; x_1 = 2$$
$$\therefore\; x_1 = 2;\; x_2 = -4;\; x_3 = -3$$

Note that when dealing with the augmented matrix, we may, if we wish:

(a) interchange two rows
(b) multiply any row by a non-zero factor
(c) add (or subtract) a constant multiple of any one row to (or from) another.

These operations are permissible since we are really dealing with the coefficients of both sides of the equations.

*Now for another example: move on to the next frame*

**41**

Solve the following set of equations:

$$x_1 - 4x_2 - 2x_3 = 21$$
$$2x_1 + x_2 + 2x_3 = 3$$
$$3x_1 + 2x_2 - x_3 = -2$$

First write the equations in matrix form, which is ...........

**42**

$$\begin{pmatrix} 1 & -4 & -2 \\ 2 & 1 & 2 \\ 3 & 2 & -1 \end{pmatrix} \cdot \begin{pmatrix} x_1 \\ x_2 \\ x_3 \end{pmatrix} = \begin{pmatrix} 21 \\ 3 \\ -2 \end{pmatrix}$$

The augmented matrix is then ...........

**43**

$$\begin{pmatrix} 1 & -4 & -2 & 21 \\ 2 & 1 & 2 & 3 \\ 3 & 2 & -1 & -2 \end{pmatrix}$$

We can now eliminate the $x_1$ coefficients from the second and third rows by ........... and ...........

**44**

subtracting 2 times the first row from the second row
and 3 times the first row from the third row.

So the matrix now becomes $\begin{pmatrix} 1 & -4 & -2 & 21 \\ 0 & 9 & 6 & -39 \\ 0 & 14 & 5 & -65 \end{pmatrix}$

and the next stage is to subtract from the third row ........... times the second row.

45

$$\boxed{\frac{14}{9}}$$

If we do this, the matrix becomes $\begin{pmatrix} 1 & -4 & -2 & | & 21 \\ 0 & 9 & 6 & | & -39 \\ 0 & 0 & -4{\cdot}33 & | & -4{\cdot}33 \end{pmatrix}$

Re-forming the matrix equation $\begin{pmatrix} 1 & -4 & -2 \\ 0 & 9 & 6 \\ 0 & 0 & -4{\cdot}33 \end{pmatrix} \cdot \begin{pmatrix} x_1 \\ x_2 \\ x_3 \end{pmatrix} = \begin{pmatrix} 21 \\ -39 \\ -4{\cdot}33 \end{pmatrix}$

Now, starting from the bottom row, you can finish it off.

$x_1 = \dots\dots\dots\dots$; $x_2 = \dots\dots\dots\dots$; $x_3 = \dots\dots\dots\dots$

46

$$\boxed{x_1 = 3;\ x_2 = -5;\ x_3 = 1}$$

Now for something rather different.

# Eigenvalues and eigenvectors

47

## Eigenvalues

In many applications of matrices to technological problems involving coupled oscillations and vibrations, equations of the form

$$\mathbf{A}.\mathbf{x} = \lambda\mathbf{x}$$

occur, where $\mathbf{A} = [a_{ij}]$ is a square matrix and $\lambda$ is a number (scalar). Clearly, $\mathbf{x} = \mathbf{0}$ is a solution for any value of $\lambda$ and is not normally useful. For non-trivial solutions, i.e. $\mathbf{x} \neq \mathbf{0}$, the values of $\lambda$ are called the *eigenvalues*, *characteristic values* or *latent roots* of the matrix $\mathbf{A}$ and the corresponding solutions of the given equations $\mathbf{A}.\mathbf{x} = \lambda\mathbf{x}$ are called the *eigenvectors* or characteristic vectors of $\mathbf{A}$.

Expressed as a set of separate equations, we have:

$$\begin{pmatrix} a_{11} & a_{12} & \dots & a_{1n} \\ a_{21} & a_{22} & \dots & a_{2n} \\ \vdots & \vdots & & \vdots \\ a_{n1} & a_{n2} & \dots & a_{nn} \end{pmatrix} \cdot \begin{pmatrix} x_1 \\ x_2 \\ \vdots \\ x_n \end{pmatrix} = \lambda \begin{pmatrix} x_1 \\ x_2 \\ \vdots \\ x_n \end{pmatrix}$$

i.e.
$$\begin{aligned} a_{11}x_1 + a_{12}x_2 + \dots + a_{1n}x_n &= \lambda x_1 \\ a_{21}x_1 + a_{22}x_2 + \dots + a_{2n}x_n &= \lambda x_2 \\ \vdots \quad\quad \vdots \quad\quad\quad\quad \vdots \quad &\quad \vdots \\ a_{n1}x_1 + a_{n2}x_2 + \dots + a_{nn}x_n &= \lambda x_n \end{aligned}$$

Bringing the right-hand-side terms to the left-hand side, this simplifies to:

$$\begin{aligned} (a_{11} - \lambda)x_1 + a_{12}x_2 + \dots + a_{1n}x_n &= 0 \\ a_{21}x_1 + (a_{22} - \lambda)x_2 + \dots + a_{2n}x_n &= 0 \\ \vdots \quad\quad\quad \vdots \quad\quad\quad\quad \vdots \quad &\quad \vdots \\ a_{n1}x_1 + a_{n2}x_2 + \dots + (a_{nn} - \lambda)x_n &= 0 \end{aligned}$$

▶

$$\text{i.e.} \quad \begin{pmatrix} (a_{11}-\lambda) & a_{12} & \dots & a_{1n} \\ a_{21} & (a_{22}-\lambda) & \dots & a_{2n} \\ \vdots & \vdots & & \vdots \\ a_{n1} & a_{n2} & \dots & (a_{nn}-\lambda) \end{pmatrix} \cdot \begin{pmatrix} x_1 \\ x_2 \\ \vdots \\ x_n \end{pmatrix} = \begin{pmatrix} 0 \\ 0 \\ \vdots \\ 0 \end{pmatrix}$$

$\mathbf{A.x} = \lambda\mathbf{x}$ becomes $\mathbf{A.x} - \lambda\mathbf{x} = \mathbf{0}$

and then $(\mathbf{A} - \lambda\mathbf{I})\mathbf{x} = \mathbf{0}$

Note that the unit matrix is introduced since we can subtract only a matrix from another matrix.

For this set of homogeneous linear equations (i.e. right-hand constants all zero) to have a non-trivial solution, $|\mathbf{A} - \lambda\mathbf{I}|$ must be zero (see Programme 4, Frame 54).

$$|\mathbf{A} - \lambda\mathbf{I}| = \begin{vmatrix} (a_{11}-\lambda) & a_{12} & \dots & a_{1n} \\ a_{21} & (a_{22}-\lambda) & \dots & a_{2n} \\ \vdots & \vdots & & \vdots \\ a_{n1} & a_{n2} & \dots & (a_{nn}-\lambda) \end{vmatrix} = 0$$

$|\mathbf{A} - \lambda\mathbf{I}|$ is called the *characteristic determinant* of $\mathbf{A}$ and $|\mathbf{A} - \lambda\mathbf{I}| = 0$ is the *characteristic equation*. On expanding the determinant, this gives a polynomial of degree $n$ and the solution of the characteristic equation gives the values of $\lambda$, i.e. the eigenvalues of $\mathbf{A}$.

**Example 1**

To find the eigenvalues of the matrix $\mathbf{A} = \begin{pmatrix} 4 & -1 \\ 2 & 1 \end{pmatrix}$.

$\mathbf{A.x} = \lambda\mathbf{x}$ i.e. $(\mathbf{A} - \lambda\mathbf{I})\mathbf{x} = 0$

Characteristic determinant: $|\mathbf{A} - \lambda\mathbf{I}| = \begin{vmatrix} (4-\lambda) & -1 \\ 2 & (1-\lambda) \end{vmatrix}$

Characteristic equation: $|\mathbf{A} - \lambda\mathbf{I}| = 0$

$\therefore\ (4-\lambda)(1-\lambda) + 2 = 0 \quad \therefore\ 4 - 5\lambda + \lambda^2 + 2 = 0$

$\therefore\ \lambda^2 - 5\lambda + 6 = 0 \ \therefore\ (\lambda - 2)(\lambda - 3) = 0$

$\therefore\ \lambda = 2 \text{ or } 3 \ \therefore\ \lambda_1 = 2;\ \lambda_2 = 3.$

**Example 2**

To find the eigenvalues of the matrix $\mathbf{A} = \begin{pmatrix} 2 & 3 & -2 \\ 1 & 4 & -2 \\ 2 & 10 & -5 \end{pmatrix}$

The characteristic determinant is ...........

**48**

$$\begin{vmatrix} (2-\lambda) & 3 & -2 \\ 1 & (4-\lambda) & -2 \\ 2 & 10 & (-5-\lambda) \end{vmatrix}$$

Expanding this, we get:

$$(2-\lambda)\{(4-\lambda)(-5-\lambda) + 20\} - 3\{(-5-\lambda) + 4\} - 2\{10 - 2(4-\lambda)\}$$
$$= (2-\lambda)\{-20 + \lambda + \lambda^2 + 20\} + 3(1+\lambda) - 2(2+2\lambda)$$
$$= (2-\lambda)\{\lambda^2 + \lambda\} + 3(1+\lambda) - 4(1+\lambda)$$
$$= (2-\lambda)\lambda(\lambda+1) - (1+\lambda) = (1+\lambda)(2\lambda - \lambda^2 - 1) = -(1+\lambda)(1-\lambda)^2$$

$\therefore$ Characteristic equation: $(1+\lambda)(1-\lambda)^2 = 0 \ \therefore\ \lambda = -1, 1, 1$

$\therefore\ \lambda_1 = -1;\ \lambda_2 = 1;\ \lambda_3 = 1$

▶

Now one for you to do. Find the eigenvalues of the matrix

$$\mathbf{A} = \begin{pmatrix} 1 & -1 & 0 \\ 1 & 2 & 1 \\ -2 & 1 & -1 \end{pmatrix}$$

Work through the steps in the same manner.

$\lambda = \ldots\ldots\ldots\ldots$

---

49

$$\boxed{\lambda_1 = -1; \quad \lambda_2 = 1; \quad \lambda_3 = 2}$$

Here is the working:

Characteristic equation: $\begin{vmatrix} (1-\lambda) & -1 & 0 \\ 1 & (2-\lambda) & 1 \\ -2 & 1 & (-1-\lambda) \end{vmatrix} = 0$

$$\therefore \ (1-\lambda)\{(2-\lambda)(-1-\lambda) - 1\} + 1(-1-\lambda+2) + 0 = 0$$
$$(1-\lambda)\{\lambda^2 - \lambda - 3\} + 1 - \lambda = 0$$
$$\therefore \ 1 - \lambda = 0 \text{ or } \lambda^2 - \lambda - 2 = 0$$
$$\therefore \ \lambda = 1 \text{ or } (\lambda+1)(\lambda-2) = 0 \ \text{ i.e. } \lambda = -1 \text{ or } 2$$
$$\therefore \ \lambda_1 = -1; \ \lambda_2 = 1; \ \lambda_3 = 2$$

## Eigenvectors

Each eigenvalue obtained has corresponding to it a solution of **x** called an *eigenvector*. In matrices, the term 'vector' indicates a row matrix or column matrix.

### Example 1

Consider the equation $\mathbf{A}.\mathbf{x} = \lambda\mathbf{x}$ where $\mathbf{A} = \begin{pmatrix} 4 & 1 \\ 3 & 2 \end{pmatrix}$

The characteristic equation is $\begin{vmatrix} (4-\lambda) & 1 \\ 3 & (2-\lambda) \end{vmatrix} = 0$

$$\therefore \ (4-\lambda)(2-\lambda) - 3 = 0 \ \therefore \ \lambda^2 - 6\lambda + 5 = 0$$
$$\therefore \ (\lambda-1)(\lambda-5) = 0 \ \therefore \ \lambda = 1 \text{ or } 5$$
$$\lambda_1 = 1; \ \lambda_2 = 5$$

For $\lambda_1 = 1$, the equation $\mathbf{A}.\mathbf{x} = \lambda\mathbf{x}$ becomes:

$$\begin{pmatrix} 4 & 1 \\ 3 & 2 \end{pmatrix} . \begin{pmatrix} x_1 \\ x_2 \end{pmatrix} = 1. \begin{pmatrix} x_1 \\ x_2 \end{pmatrix}$$

$$\left.\begin{aligned} 4x_1 + x_2 &= x_1 \\ 3x_1 + 2x_2 &= x_2 \end{aligned}\right\} \text{ Either of these gives } x_2 = -3x_1$$

This result merely tells us that whatever value $x_1$ has, the value of $x_2$ is $-3$ times it. Therefore, the eigenvector $x_1 = \begin{pmatrix} k \\ -3k \end{pmatrix}$ is the general form of an infinite number of such eigenvectors. The simplest eigenvector is therefore $\mathbf{x}_1 = \begin{pmatrix} 1 \\ -3 \end{pmatrix}$.

For $\lambda_2 = 5$, a similar result can be obtained. Determine the eigenvector in the same way.

$\mathbf{x}_2 = \ldots\ldots\ldots\ldots$

---

**50**

$$\mathbf{x}_2 = \begin{pmatrix} k \\ k \end{pmatrix} \text{ is the general solution; } \mathbf{x}_2 = \begin{pmatrix} 1 \\ 1 \end{pmatrix} \text{ is a solution}$$

Because, when $\lambda_2 = 5$, $\begin{pmatrix} 4 & 1 \\ 3 & 2 \end{pmatrix} \cdot \begin{pmatrix} x_1 \\ x_2 \end{pmatrix} = 5\begin{pmatrix} x_1 \\ x_2 \end{pmatrix} = \begin{pmatrix} 5x_1 \\ 5x_2 \end{pmatrix}$

$\therefore\ 4x_1 + x_2 = 5x_1 \ \therefore\ x_1 = x_2 \ \therefore\ x_2 = \begin{pmatrix} 1 \\ 1 \end{pmatrix}$ is a solution

Therefore, $\mathbf{x}_1 = \begin{pmatrix} 1 \\ -3 \end{pmatrix}$ is an eigenvector corresponding to $\lambda_1 = 1$

and $\mathbf{x}_2 = \begin{pmatrix} 1 \\ 1 \end{pmatrix}$ is an eigenvector corresponding to $\lambda_2 = 5$.

**Example 2**

Determine the eigenvalues and eigenvectors for the equation

$$\mathbf{A.x} = \lambda\mathbf{x} \text{ where } \mathbf{A} = \begin{pmatrix} 2 & 0 & 1 \\ -1 & 4 & -1 \\ -1 & 2 & 0 \end{pmatrix}$$

The characteristic equation is $\begin{vmatrix} (2-\lambda) & 0 & 1 \\ -1 & (4-\lambda) & -1 \\ -1 & 2 & -\lambda \end{vmatrix} = 0$

$$\therefore\ (2-\lambda)\{-\lambda(4-\lambda)+2\} + 1\{-2+(4-\lambda)\} = 0$$
$$\therefore\ (2-\lambda)\{\lambda^2 - 4\lambda + 2\} + (2-\lambda) = 0$$
$$\therefore\ (2-\lambda)\{\lambda^2 - 4\lambda + 3\} = 0 \qquad \therefore\ \lambda = \ldots\ldots\ldots\ldots$$

**51**

$$\lambda = 1,\ 2,\ 3$$

For $\lambda_1 = 1$: $\begin{pmatrix} 1 & 0 & 1 \\ -1 & 3 & -1 \\ -1 & 2 & -1 \end{pmatrix} \cdot \begin{pmatrix} x_1 \\ x_2 \\ x_3 \end{pmatrix} = \begin{pmatrix} 0 \\ 0 \\ 0 \end{pmatrix}$ using $(\mathbf{A} - \lambda\mathbf{I}).\mathbf{x} = \mathbf{0}$

$$x_1 + x_3 = 0 \ \therefore\ x_3 = -x_1$$
$$-x_1 + 2x_2 - x_3 = 0 \ \therefore\ -x_1 + 2x_2 + x_1 = 0 \ \therefore\ x_2 = 0$$

$\therefore\ \mathbf{x}_1 = \begin{pmatrix} 1 \\ 0 \\ -1 \end{pmatrix}$ is an eigenvector corresponding to $\lambda_1 = 1$.

For $\lambda_2 = 2$: $\begin{pmatrix} 0 & 0 & 1 \\ -1 & 2 & -1 \\ -1 & 2 & -2 \end{pmatrix} \cdot \begin{pmatrix} x_1 \\ x_2 \\ x_3 \end{pmatrix} = \begin{pmatrix} 0 \\ 0 \\ 0 \end{pmatrix}$

Therefore, an eigenvector corresponding to $\lambda_2 = 2$ is

$\mathbf{x}_2 = \ldots\ldots\ldots\ldots$

**52**

$$\mathbf{x}_2 = \begin{pmatrix} 2 \\ 1 \\ 0 \end{pmatrix}$$

Because $x_3 = 0$ and $-x_1 + 2x_2 - x_3 = 0 \ \therefore\ x_1 = 2x_2$.

For $\lambda_3 = 3$, we can find an eigenvector in the same way. This gives

$\mathbf{x}_3 = \ldots\ldots\ldots\ldots$

53

$$\mathbf{x}_3 = \begin{pmatrix} 1 \\ 2 \\ 1 \end{pmatrix}$$

Because with $\lambda_3 = 3$: $\begin{pmatrix} -1 & 0 & 1 \\ -1 & 1 & -1 \\ -1 & 2 & -3 \end{pmatrix} \cdot \begin{pmatrix} x_1 \\ x_2 \\ x_3 \end{pmatrix} = \begin{pmatrix} 0 \\ 0 \\ 0 \end{pmatrix}$

$\therefore \quad -x_1 + x_3 = 0 \ \therefore \ x_3 = x_1$

$-x_1 + x_2 - x_3 = 0 \ \therefore \ -2x_1 + x_2 = 0 \ \therefore \ x_2 = 2x_1.$

So, collecting our results together, we have:

$\mathbf{x}_1 = \begin{pmatrix} 1 \\ 0 \\ -1 \end{pmatrix}$ is an eigenvector corresponding to $\lambda_1 = 1$

$\mathbf{x}_2 = \begin{pmatrix} 2 \\ 1 \\ 0 \end{pmatrix}$ is an eigenvector corresponding to $\lambda_2 = 2$

$\mathbf{x}_3 = \begin{pmatrix} 1 \\ 2 \\ 1 \end{pmatrix}$ is an eigenvector corresponding to $\lambda_3 = 3$

Here is one for you to do on your own. The method is the same as before.

If $\mathbf{A.x} = \lambda\mathbf{x}$ where $\mathbf{A} = \begin{pmatrix} 1 & -1 & 0 \\ 1 & 2 & 1 \\ -2 & 1 & -1 \end{pmatrix}$ and the eigenvalues are known to be $\lambda_1 = -1$, $\lambda_2 = 1$ and $\lambda_3 = 2$, determine corresponding eigenvectors.

$\mathbf{x}_1 = \ldots\ldots\ldots\ldots$

$\mathbf{x}_2 = \ldots\ldots\ldots\ldots$

$\mathbf{x}_3 = \ldots\ldots\ldots\ldots$

54

$$\mathbf{x}_1 = \begin{pmatrix} 1 \\ 2 \\ -7 \end{pmatrix}; \quad \mathbf{x}_2 = \begin{pmatrix} 1 \\ 0 \\ -1 \end{pmatrix}; \quad \mathbf{x}_3 = \begin{pmatrix} 1 \\ -1 \\ -1 \end{pmatrix}$$

Using $\begin{pmatrix} (1-\lambda) & -1 & 0 \\ 1 & (2-\lambda) & 1 \\ -2 & 1 & (-1-\lambda) \end{pmatrix} \cdot \begin{pmatrix} x_1 \\ x_2 \\ x_3 \end{pmatrix} = \begin{pmatrix} 0 \\ 0 \\ 0 \end{pmatrix}$

simple substitution of the values of $\lambda$ in turn and the knowledge of how to multiply matrices together give the results indicated.

As we have seen, a basic knowledge of matrices provides a neat and concise way of dealing with sets of linear equations. In practice, the numerical coefficients are not always simple numbers, neither is the number of equations in the set limited to three. In more extensive problems, recourse to computing facilities is a great help, but the underlying methods are still the same.

All that now remains is to check down the **Review summary** and the **Can You?** checklist. Then you can work through the **Test exercise**.

## Review summary

55

1 *Matrix* – a rectangular array of numbers (elements).

2 *Order* – a matrix order of $(m \times n)$ denotes $m$ rows and $n$ columns.

3 *Row matrix* – one row only.

4 *Column matrix* – one column only.

5 *Double suffix notation* – $a_{34}$ denotes element in 3rd row and 4th column.

6 *Equal matrices* – corresponding elements equal.

7 *Addition and subtraction of matrices* – add or subtract corresponding elements. Therefore, for addition or subtraction, matrices must be of the same order.

8 *Multiplication of matrices*

(a) *Scalar multiplier* – every element multiplied by the same constant, i.e. $k[a_{ij}] = [ka_{ij}]$.

(b) *Matrix multiplier* – product **A**.**B** possible only if the number of columns in **A** equals the number of rows in **B**.

$$\overrightarrow{\begin{pmatrix} a & b & c \\ d & e & f \end{pmatrix}} \cdot \begin{pmatrix} g & j \\ h & k \\ i & l \end{pmatrix}\Bigg\downarrow = \begin{pmatrix} ag + bh + ci & aj + bk + cl \\ dg + eh + fi & dj + ek + fl \end{pmatrix}$$

9 *Square matrix* – of order $(m \times m)$

(a) *Symmetric* if $a_{ij} = a_{ji}$, e.g. $\begin{pmatrix} 3 & 2 & 4 \\ 2 & 6 & 1 \\ 4 & 1 & 5 \end{pmatrix}$

(b) *Skew symmetric* if $a_{ij} = -a_{ji}$, e.g. $\begin{pmatrix} 0 & 2 & 4 \\ -2 & 0 & 1 \\ -4 & -1 & 0 \end{pmatrix}$

10 *Diagonal matrix* – all elements zero except those on the leading diagonal.

11 *Unit matrix* – a diagonal matrix with elements on the leading diagonal all unity, i.e. $\begin{pmatrix} 1 & 0 & 0 \\ 0 & 1 & 0 \\ 0 & 0 & 1 \end{pmatrix}$ denoted by **I**.

12 *Null matrix* – all elements zero.

13 *Transpose of a matrix* – rows and columns interchanged. Transpose of **A** is $\mathbf{A}^{\mathrm{T}}$.

14 *Cofactors* – minors of the elements of $|\mathbf{A}|$ together with the respective 'place signs' of the elements.

15 *Adjoint of a square matrix* **A** – form matrix **C** of the cofactors of the elements of $|\mathbf{A}|$, then the adjoint of **A** is $\mathbf{C}^{\mathrm{T}}$, i.e. the transpose of **C**. $\therefore$ adj $\mathbf{A} = \mathbf{C}^{\mathrm{T}}$.

16 *Inverse of a square matrix* **A**

$$\mathbf{A}^{-1} = \frac{\text{adj } \mathbf{A}}{|\mathbf{A}|} = \frac{\mathbf{C}^{\mathrm{T}}}{|\mathbf{A}|}$$

17 Product of a square matrix and its inverse

$$\mathbf{A}.\mathbf{A}^{-1} = \mathbf{A}^{-1}.\mathbf{A} = \mathbf{I}$$

18 *Solution of a set of linear equations*

$$\mathbf{A}.\mathbf{x} = \mathbf{b} \quad \mathbf{x} = \mathbf{A}^{-1}.\mathbf{b}$$

19 *Gaussian elimination method* – reduce augmented matrix to triangular form, then use 'back substitution'.

▶

20 *Eigenvalues* – values of $\lambda$ for which $\mathbf{A.x} = \lambda\mathbf{x}$.

21 *Eigenvectors* – solutions of $\mathbf{x}$ corresponding to particular values of $\lambda$.

*Now for the* **Can You?** *checklist*

## Can You?

### Checklist 5

56

Check this list before and after you try the end of Programme test.

On a scale of 1 to 5 how confident are you that you can:

| | | | | | | | Frames |
|---|---|---|---|---|---|---|---|
| • Define a matrix? | | | | | | | 1 to 7 |
| *Yes* | ☐ | ☐ | ☐ | ☐ | ☐ | *No* | |
| • Understand what is meant by the equality of two matrices? | | | | | | | 8 to 9 |
| *Yes* | ☐ | ☐ | ☐ | ☐ | ☐ | *No* | |
| • Add and subtract two matrices? | | | | | | | 10 to 11 |
| *Yes* | ☐ | ☐ | ☐ | ☐ | ☐ | *No* | |
| • Multiply a matrix by a scalar and multiply two matrices together? | | | | | | | 12 to 18 |
| *Yes* | ☐ | ☐ | ☐ | ☐ | ☐ | *No* | |
| • Obtain the transpose of a matrix? | | | | | | | 19 to 20 |
| *Yes* | ☐ | ☐ | ☐ | ☐ | ☐ | *No* | |
| • Recognize special types of matrix? | | | | | | | 21 to 23 |
| *Yes* | ☐ | ☐ | ☐ | ☐ | ☐ | *No* | |
| • Obtain the determinant, cofactors and adjoint of a square matrix? | | | | | | | 24 to 27 |
| *Yes* | ☐ | ☐ | ☐ | ☐ | ☐ | *No* | |
| • Obtain the inverse of a non-singular matrix? | | | | | | | 28 to 34 |
| *Yes* | ☐ | ☐ | ☐ | ☐ | ☐ | *No* | |
| • Use matrices to solve a set of linear equations using inverse matrices? | | | | | | | 35 to 39 |
| *Yes* | ☐ | ☐ | ☐ | ☐ | ☐ | *No* | |
| • Use the Gaussian elimination method to solve a set of linear equations? | | | | | | | 39 to 46 |
| *Yes* | ☐ | ☐ | ☐ | ☐ | ☐ | *No* | |
| • Evaluate eigenvalues and eigenvectors? | | | | | | | 47 to 54 |
| *Yes* | ☐ | ☐ | ☐ | ☐ | ☐ | *No* | |

## Test exercise 5

57

The questions are all straightforward and based on the work covered. You will have no trouble.

1 If $\mathbf{A} = \begin{pmatrix} 2 & 4 & 6 & 3 \\ 1 & 7 & 0 & 4 \end{pmatrix}$ and $\mathbf{B} = \begin{pmatrix} 3 & 5 & 2 & 7 \\ 9 & 1 & 6 & 3 \end{pmatrix}$ determine

(a) $\mathbf{A} + \mathbf{B}$ and (b) $\mathbf{A} - \mathbf{B}$.

2 Given that $\mathbf{A} = \begin{pmatrix} 6 & 0 & 4 \\ 1 & 5 & -3 \end{pmatrix}$ and $\mathbf{B} = \begin{pmatrix} 2 & 9 \\ 8 & 0 \\ -4 & 7 \end{pmatrix}$ determine

(a) $3\mathbf{A}$, (b) $\mathbf{A.B}$, (c) $\mathbf{B.A}$.

▶

3 If $A = \begin{pmatrix} 2 & 3 & 5 \\ 1 & 7 & 4 \\ 8 & 0 & 6 \end{pmatrix}$, form the transpose $A^T$ and determine the matrix product $A^T.I$.

4 Show that the square matrix $A = \begin{pmatrix} 3 & 2 & 4 \\ 1 & 5 & 3 \\ -1 & 8 & 2 \end{pmatrix}$ is a singular matrix.

5 If $A = \begin{pmatrix} 1 & 4 & 3 \\ 6 & 2 & 5 \\ 1 & 7 & 0 \end{pmatrix}$, determine (a) $|A|$ and (b) adj **A**.

6 Find the inverse of the matrix $A = \begin{pmatrix} 2 & 1 & 4 \\ 3 & 5 & 1 \\ 2 & 0 & 6 \end{pmatrix}$

7 Express the following set of linear equations in matrix form:

$$2x_1 + 4x_2 - 5x_3 = -7$$
$$x_1 - 3x_2 + x_3 = 10$$
$$3x_1 + 5x_2 + 3x_3 = 2$$

8 Solve the following set of linear equations by the matrix method:

$$x_1 + 3x_2 + 2x_3 = 3$$
$$2x_1 - x_2 - 3x_3 = -8$$
$$5x_1 + 2x_2 + x_3 = 9$$

9 For the following set of simultaneous equations:

(a) form the augmented coefficient matrix

(b) solve the equations by Gaussian elimination.

$$x_1 + 2x_2 + 3x_3 = 5$$
$$3x_1 - x_2 + 2x_3 = 8$$
$$4x_1 - 6x_2 - 4x_3 = -2$$

10 If $A.x = \lambda x$, where $A = \begin{pmatrix} 2 & 2 & -2 \\ 1 & 3 & 1 \\ 1 & 2 & 2 \end{pmatrix}$ determine the eigenvalues of the matrix **A** and an eigenvector corresponding to each eigenvalue.

---

## Further problems 5

58

1 If $A = \begin{pmatrix} 7 & 2 \\ 3 & 1 \end{pmatrix}$ and $B = \begin{pmatrix} 4 & 6 \\ 5 & 8 \end{pmatrix}$, determine:

(a) **A** + **B**, (b) **A** − **B**, (c) **A.B**, (d) **B.A**

2 If $A = \begin{pmatrix} j & 0 \\ 0 & -j \end{pmatrix}$ $B = \begin{pmatrix} 0 & 1 \\ -1 & 0 \end{pmatrix}$ $C = \begin{pmatrix} 0 & j \\ j & 0 \end{pmatrix}$ $I = \begin{pmatrix} 1 & 0 \\ 0 & 1 \end{pmatrix}$,

where $j = \sqrt{-1}$, express (a) **A.B**, (b) **B.C**, (c) **C.A** and (d) $A^2$ in terms of other matrices.

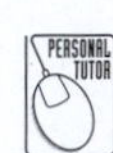

3 If $A = \begin{pmatrix} 1 & 0{\cdot}5 \\ 0{\cdot}5 & 0{\cdot}1 \end{pmatrix}$ and $B = \begin{pmatrix} 1 & 2 \\ 2 & 3 \end{pmatrix}$, determine:

(a) $B^{-1}$, (b) **A.B**, (c) $B^{-1}.A$

▶

4 Determine the value of $k$ for which the following set of homogeneous equations has non-trivial solutions:

$$4x_1 + 3x_2 - x_3 = 0$$
$$7x_1 - x_2 - 3x_3 = 0$$
$$3x_1 - 4x_2 + kx_3 = 0$$

5 Express the following sets of simultaneous equations in matrix form:

(a) $2x_1 - 3x_2 - x_3 = 2$
$x_1 + 4x_2 + 2x_3 = 3$
$x_1 - x_2 + x_3 = 5$

(b) $x_1 - 2x_2 - x_3 + 3x_4 = 10$
$2x_1 + 3x_2 + x_4 = 8$
$x_1 - 4x_3 - 2x_4 = 3$
$-x_2 + 3x_3 + x_4 = -7$

In **6** to **10** solve, where possible, the sets of equations by a matrix method:

6 $2i_1 + i_2 + i_3 = 8$
$5i_1 - 3i_2 + 2i_3 = 3$
$7i_1 + i_2 + 3i_3 = 20$

7 $3x + 2y + 4z = 3$
$x + y + z = 2$
$2x - y + 3z = -3$

8 $4i_1 - 5i_2 + 6i_3 = 3$
$8i_1 - 7i_2 - 3i_3 = 9$
$7i_1 - 8i_2 + 9i_3 = 6$

9 $3x + 2y + 5z = 1$
$x - y + z = 4$
$6x + 4y + 10z = 7$

10 $3x_1 + 2x_2 - 2x_3 = 16$
$4x_1 + 3x_2 + 3x_3 = 2$
$-2x_1 + x_2 - x_3 = 1$

In **11** to **13** form the augmented matrix and solve the sets of equations by Gaussian elimination:

11 $5i_1 - i_2 + 2i_3 = 3$
$2i_1 + 4i_2 + i_3 = 8$
$i_1 + 3i_2 - 3i_3 = 2$

12 $i_1 + 2i_2 + 3i_3 = -4$
$2i_1 + 6i_2 - 3i_3 = 33$
$4i_1 - 2i_2 + i_3 = 3$

13 $7i_1 - 4i_2 = 12$
$-4i_1 + 12i_2 - 6i_3 = 0$
$-6i_2 + 14i_3 = 0$

14 In a star-connected circuit, currents $i_1$, $i_2$, $i_3$ flowing through impedances $Z_1$, $Z_2$, $Z_3$, are given by:

$$i_1 + i_2 + i_3 = 0$$
$$Z_1 i_1 - Z_2 i_2 = e_1 - e_2$$
$$Z_2 i_1 - Z_3 i_3 = e_2 - e_3$$

If $Z_1 = 10$; $Z_2 = 8$; $Z_3 = 3$; $e_1 - e_2 = 65$; $e_2 - e_3 = 160$; apply matrix methods to determine the values of $i_1$, $i_2$, $i_3$.

▶

**15** Currents of $i_1$, $i_2$, $i_3$ in a network are related by the following equations:

$$Z_1 i_1 + Z_3 i_3 = V$$
$$Z_2 i_2 - Z_3 i_3 = 0$$
$$i_1 - i_2 - i_3 = 0$$

Determine expressions for $i_1$, $i_2$, $i_3$, in terms of $Z_1$, $Z_2$, $Z_3$ and $V$.

**16** to **20** refer to the vector equation $\mathbf{A.x} = \lambda\mathbf{x}$. For the coefficient matrix **A** given in each case, determine the eigenvalues and an eigenvector corresponding to each eigenvalue:

**16** $\mathbf{A} = \begin{pmatrix} 2 & 1 & 1 \\ 1 & 3 & 2 \\ -1 & 1 & 2 \end{pmatrix}$

**17** $\mathbf{A} = \begin{pmatrix} 1 & 2 & 2 \\ 1 & 3 & 1 \\ 2 & 2 & 1 \end{pmatrix}$

**18** $\mathbf{A} = \begin{pmatrix} 2 & 0 & 1 \\ -1 & 4 & -1 \\ -1 & 2 & 0 \end{pmatrix}$

**19** $\mathbf{A} = \begin{pmatrix} 1 & -4 & -2 \\ 0 & 3 & 1 \\ 1 & 2 & 4 \end{pmatrix}$

**20** $\mathbf{A} = \begin{pmatrix} 3 & 0 & 3 \\ 0 & 3 & 3 \\ 2 & 3 & 1 \end{pmatrix}$

Now visit the companion website for this book at www.palgrave.com/stroud for more questions applying this mathematics to science and engineering.

# Vectors

### Learning outcomes

When you have completed this Programme you will be able to:

- ☐ Define a vector
- ☐ Represent a vector by a directed straight line
- ☐ Add vectors
- ☐ Write a vector in terms of component vectors
- ☐ Write a vector in terms of component unit vectors
- ☐ Set up a coordinate system for representing vectors
- ☐ Obtain the direction cosines of a vector
- ☐ Calculate the scalar product of two vectors
- ☐ Calculate the vector product of two vectors
- ☐ Determine the angle between two vectors
- ☐ Evaluate the direction ratios of a vector

# Introduction: scalar and vector quantities

**1**

Physical quantities can be divided into two main groups, scalar quantities and vector quantities.

(a) A *scalar quantity* is one that is defined completely by a single number with appropriate units, e.g. length, area, volume, mass, time, etc. Once the units are stated, the quantity is denoted entirely by its size or *magnitude*.

(b) A *vector quantity* is defined completely when we know not only its magnitude (with units) but also the direction in which it operates, e.g. force, velocity, acceleration. A vector quantity necessarily involves *direction* as well as magnitude.

So (a) a speed of 10 km/h is a scalar quantity, but
(b) a velocity of '10 km/h due north' is a ........... quantity.

**2**

vector

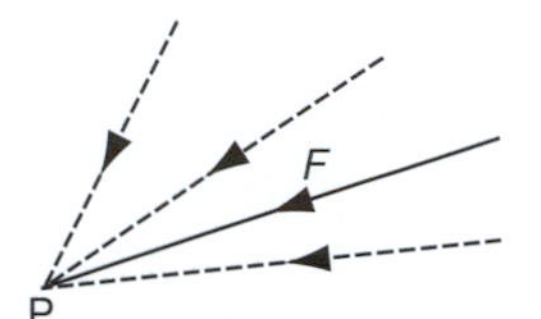


A force $F$ acting at a point P is a vector quantity, since to define it completely we must give:

(a) its magnitude, and also
(b) its ...........

**3**

direction

So that:

(a) A temperature of 100°C is a ........... quantity.
(b) An acceleration of 9·8 m/s$^2$ vertically downwards is a ........... quantity.
(c) The weight of a 7 kg mass is a ........... quantity.
(d) The sum of £500 is a ........... quantity.
(e) A north-easterly wind of 20 knots is a ........... quantity.

**4**

(a) scalar
(b) vector
(c) vector
(d) scalar
(e) vector

Since, in (b), (c) and (e) the complete description of the quantity includes not only its ..........., but also its ...........

**5**

magnitude
direction

*Move on to Frame 6*

# Vector representation

**6**

A vector quantity can be represented graphically by a line, drawn so that:

(a) the *length* of the line denotes the magnitude of the quantity, according to some stated vector scale

(b) the *direction* of the line denotes the direction in which the vector quantity acts. The sense of the direction is indicated by an arrowhead.

e.g. A horizontal force of 35 N acting to the right, would be indicated by a line and if the chosen vector scale were 1 cm ≡ 10 N, the line would be ........... cm long.

**7**

The vector quantity AB is referred to as

$\overline{AB}$ or **a**

The magnitude of the vector quantity is written $|\overline{AB}|$, or $|\mathbf{a}|$, or simply AB or $a$.

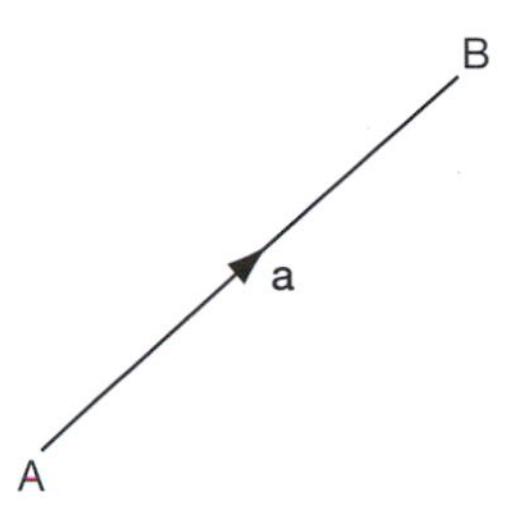


Note that $\overline{BA}$ would represent a vector quantity of the same magnitude but with opposite sense.

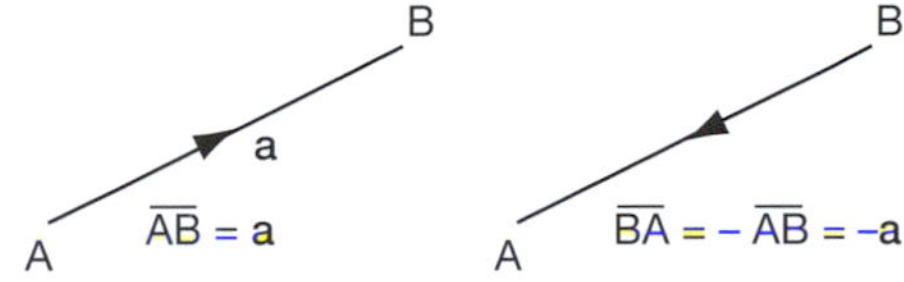


*On to Frame 8*

## Two equal vectors

**8**

If two vectors, **a** and **b**, are said to be equal, they have the same magnitude and the same direction. If **a** = **b**, then

(a) $a = b$ (magnitudes equal)

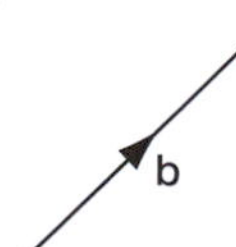


(b) the direction of **a** = direction of **b**, i.e. the two vectors are parallel and in the same sense.

Similarly, if two vectors **a** and **b** are such that **b** = −**a**, what can we say about:

(a) their magnitudes,

(b) their directions?

**9**

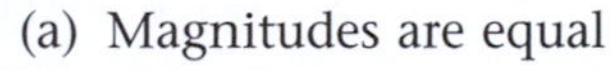

(a) Magnitudes are equal

(b) The vectors are parallel but opposite in sense

i.e. if **b** = −**a**, then

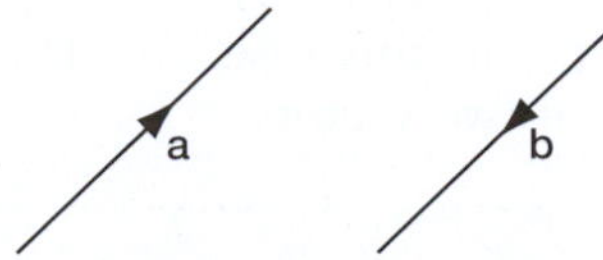

10

## Types of vector

(a) A *position vector* represents the position of a point in space relative to a predefined set of coordinates.

(b) A *free vector* is not restricted in any way. It is completely defined by its magnitude and direction and can be drawn as any one of a set of equal-length parallel lines.

Most of the vectors we shall consider will be free vectors

*So on now to Frame 11*

11

## Addition of vectors

The sum of two vectors, $\overline{AB}$ and $\overline{BC}$, is defined as the single or equivalent or resultant vector $\overline{AC}$

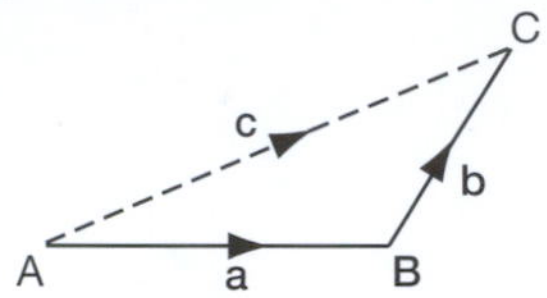


i.e. $\overline{AB} + \overline{BC} = \overline{AC}$

or $\mathbf{a} + \mathbf{b} = \mathbf{c}$

To find the sum of two vectors **a** and **b** then, we draw them as a chain, starting the second where the first ends: the sum **c** is given by the single vector joining the start of the first to the end of the second.

e.g. if $\mathbf{p} \equiv$ a force of 40 N, acting in the direction due east
$\mathbf{q} \equiv$ a force of 30 N, acting in the direction due north

then the magnitude of the vector sum $r$ of these two forces will be ...........

12

$$r = 50 \text{ N}$$

Because

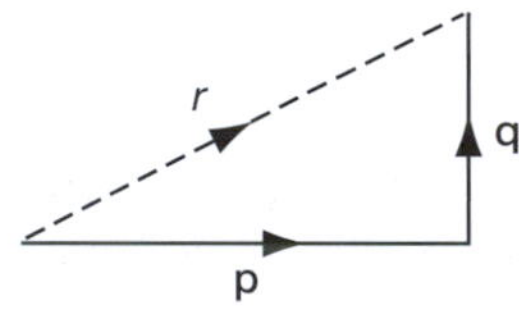


$$r^2 = p^2 + q^2$$
$$= 1600 + 900 = 2500$$
$$r = \sqrt{2500} = 50 \text{ N}$$

## The sum of a number of a vectors a + b + c + d + ...

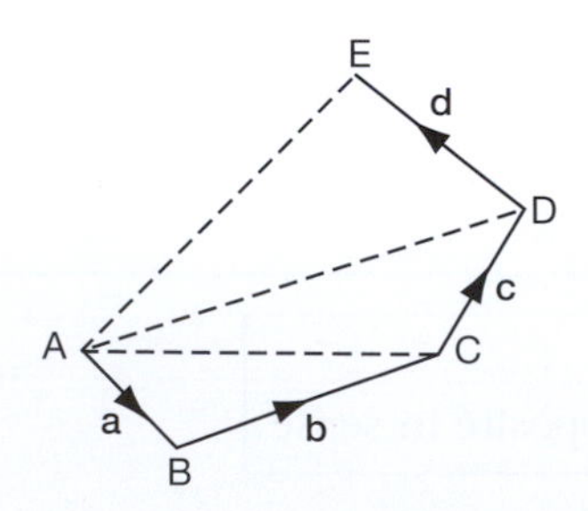


(a) Draw the vectors as a chain.

(b) Then:

$$\mathbf{a} + \mathbf{b} = \overline{AC}$$
$$\overline{AC} + \mathbf{c} = \overline{AD}$$
$$\therefore \quad \mathbf{a} + \mathbf{b} + \mathbf{c} = \overline{AD}$$
$$\overline{AD} + \mathbf{d} = \overline{AE}$$
$$\therefore \quad \mathbf{a} + \mathbf{b} + \mathbf{c} + \mathbf{d} = \overline{AE}$$

i.e. the sum of all vectors, **a**, **b**, **c**, **d**, is given by the single vector joining the start of the first to the end of the last – in this case, $\overline{AE}$. This follows directly from our previous definition of the sum of two vectors.

Similarly:

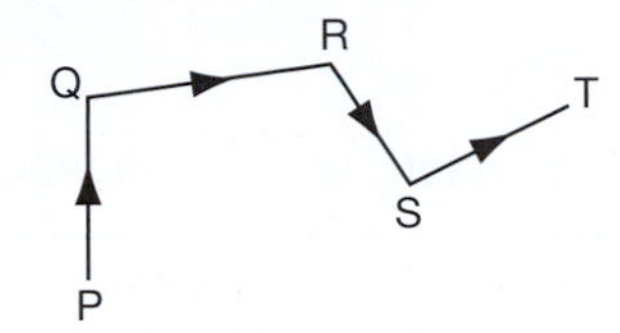


$\overline{PQ} + \overline{QR} + \overline{RS} + \overline{ST} = \ldots\ldots\ldots\ldots$

**13**

$\overline{PT}$

Now suppose that in another case, we draw the vector diagram to find the sum of **a**, **b**, **c**, **d**, **e**, and discover that the resulting diagram is, in fact, a closed figure.

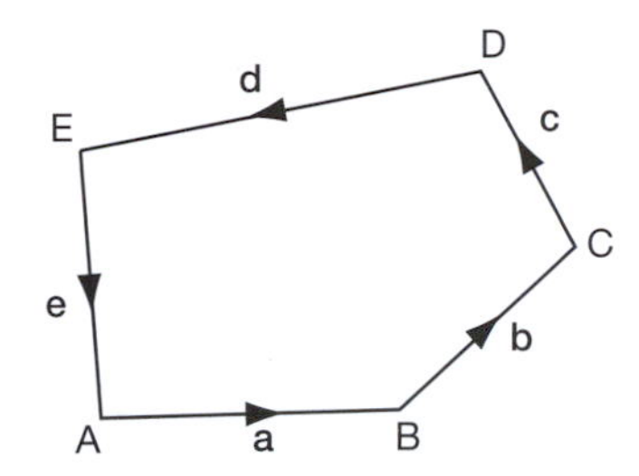


What is the sum of the vectors **a**, **b**, **c**, **d**, **e** in this case?

*Think carefully and when you have decided, move on to Frame 14*

**14**

Sum of the vectors = **0**

Because we said in the previous case, that the vector sum was given by the single equivalent vector joining the beginning of the first vector to the end of the last.

But, if the vector diagram is a closed figure, the end of the last vector coincides with the beginning of the first, so that the resultant sum is a vector with no magnitude.

Now for some examples:

Find the vector sum $\overline{AB} + \overline{BC} + \overline{CD} + \overline{DE} + \overline{EF}$.

Without drawing a diagram, we can see that the vectors are arranged in a chain, each beginning where the previous one left off. The sum is therefore given by the vector joining the beginning of the first vector to the end of the last.

$\therefore$ Sum $= \overline{AF}$

In the same way:

$\overline{AK} + \overline{KL} + \overline{LP} + \overline{PQ} = \ldots\ldots\ldots\ldots$

**15**

$\overline{AQ}$

Right. Now what about this one?

Find the sum of $\overline{AB} - \overline{CB} + \overline{CD} - \overline{ED}$

We must beware of the negative vectors. Remember that $-\overline{CB} = \overline{BC}$, i.e. the same magnitude and direction but in the opposite sense.
Also $-\overline{ED} = \overline{DE}$

$$\begin{aligned}\therefore \overline{AB} - \overline{CB} + \overline{CD} - \overline{ED} &= \overline{AB} + \overline{BC} + \overline{CD} + \overline{DE}\\ &= \overline{AE}\end{aligned}$$

▶

Now you do this one:

Find the vector sum $\overline{AB}+\overline{BC}-\overline{DC}-\overline{AD}$

*When you have the result, move on to Frame 16*

**16**

$\boxed{0}$

Because

$\overline{AB}+\overline{BC}-\overline{DC}-\overline{AD}=\overline{AB}+\overline{BC}+\overline{CD}+\overline{DA}$

and the lettering indicates that the end of the last vector coincides with the beginning of the first. The vector diagram is thus a closed figure and therefore the sum of the vectors is **0**.

Now here are some for you to do:

(a) $\overline{PQ}+\overline{QR}+\overline{RS}+\overline{ST}=\dots\dots\dots\dots$

(b) $\overline{AC}+\overline{CL}-\overline{ML}=\dots\dots\dots\dots$

(c) $\overline{GH}+\overline{HJ}+\overline{JK}+\overline{KL}+\overline{LG}=\dots\dots\dots\dots$

(d) $\overline{AB}+\overline{BC}+\overline{CD}+\overline{DB}=\dots\dots\dots\dots$

*When you have finished all four, check with the results in the next frame*

**17**

Here are the results:

(a) $\overline{PQ}+\overline{QR}+\overline{RS}+\overline{ST}=\overline{PT}$

(b) $\overline{AC}+\overline{CL}-\overline{ML}=\overline{AC}+\overline{CL}+\overline{LM}=\overline{AM}$

(c) $\overline{GH}+\overline{HJ}+\overline{JK}+\overline{KL}+\overline{LG}=\mathbf{0}$

[Since the end of the last vector coincides with the beginning of the first.]

(d) $\overline{AB}+\overline{BC}+\overline{CD}+\overline{DB}=\overline{AB}$

The last three vectors form a closed figure and therefore the sum of these three vectors is zero, leaving only $\overline{AB}$ to be considered.

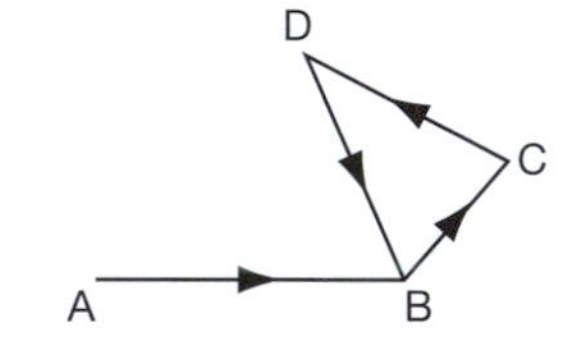


*Now on to Frame 18*

## Components of a given vector

**18**

Just as $\overline{AB}+\overline{BC}+\overline{CD}+\overline{DE}$ can be replaced by $\overline{AE}$, so any single vector $\overline{PT}$ can be replaced by any number of component vectors so long as they form a chain in the vector diagram, beginning at P and ending at T.

e.g.

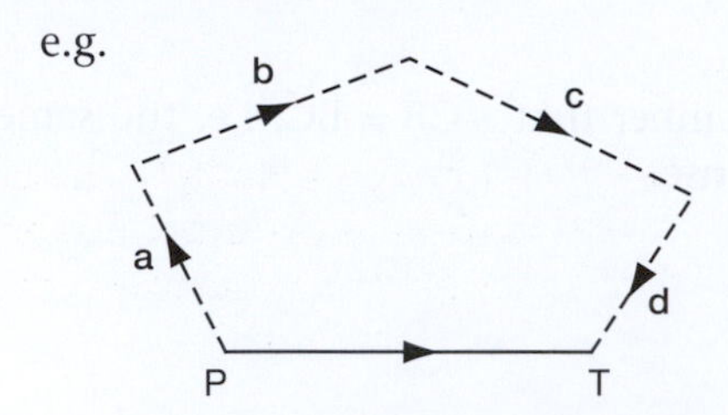


$\overline{PT}=\mathbf{a}+\mathbf{b}+\mathbf{c}+\mathbf{d}$

▶

**Example 1**

ABCD is a quadrilateral, with G and H the mid-points of DA and BC respectively. Show that $\overline{AB} + \overline{DC} = 2\overline{GH}$.

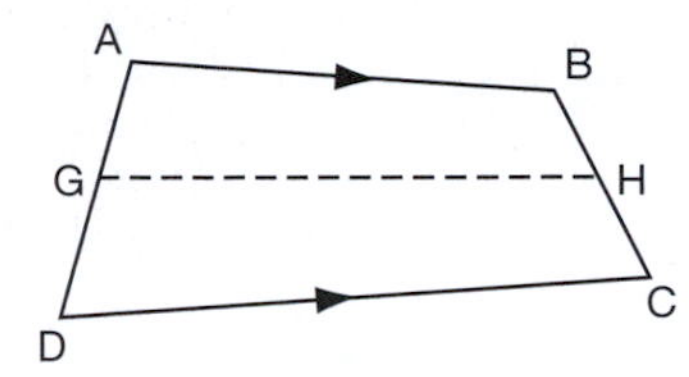


We can replace vector $\overline{AB}$ by any chain of vectors so long as they start at A and end at B, e.g. we could say

$$\overline{AB} = \overline{AG} + \overline{GH} + \overline{HB}$$

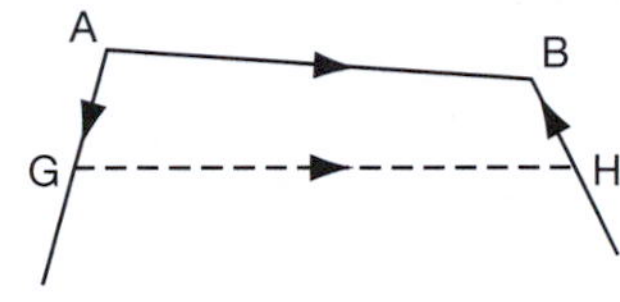


Similarly, we could say $\overline{DC} = \ldots\ldots\ldots\ldots$

---

**19**

$$\boxed{\overline{DC} = \overline{DG} + \overline{GH} + \overline{HC}}$$

So we have:

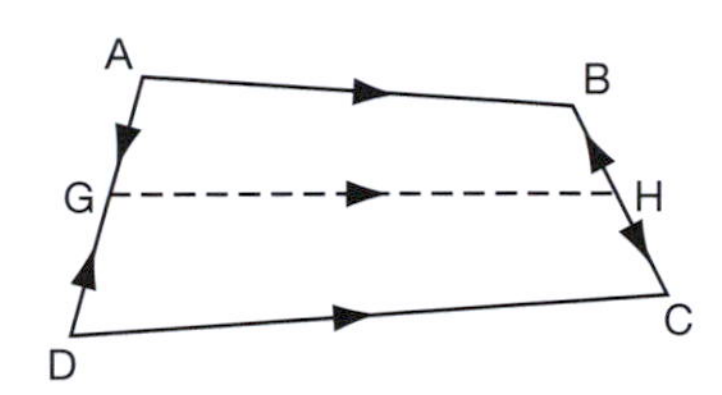


$$\overline{AB} = \overline{AG} + \overline{GH} + \overline{HB}$$
$$\overline{DC} = \overline{DG} + \overline{GH} + \overline{HC}$$

$$\therefore\ \overline{AB} + \overline{DC} = \overline{AG} + \overline{GH} + \overline{HB} + \overline{DG} + \overline{GH} + \overline{HC}$$
$$= 2\overline{GH} + (\overline{AG} + \overline{DG}) + (\overline{HB} + \overline{HC})$$

Now, G is the mid-point of AD. Therefore, vectors $\overline{AG}$ and $\overline{DG}$ are equal in length but opposite in sense.

$$\therefore\ \overline{DG} = -\overline{AG}$$
$$\text{Similarly}\ \ \overline{HC} = -\overline{HB}$$
$$\therefore\ \overline{AB} + \overline{DC} = 2\overline{GH} + (\overline{AG} - \overline{AG}) + (\overline{HB} - \overline{HB})$$
$$= 2\overline{GH}$$

*Next frame*

---

**20**

**Example 2**

Points L, M, N are mid-points of the sides AB, BC, CA of the triangle ABC. Show that:

(a) $\overline{AB} + \overline{BC} + \overline{CA} = \mathbf{0}$

(b) $2\overline{AB} + 3\overline{BC} + \overline{CA} = 2\overline{LC}$

(c) $\overline{AM} + \overline{BN} + \overline{CL} = \mathbf{0}$

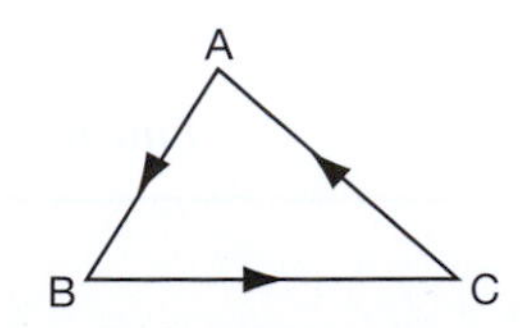


(a) We can dispose of the first part straight away without any trouble. We can see from the vector diagram that $\overline{AB} + \overline{BC} + \overline{CA} = \mathbf{0}$ since these three vectors form a $\ldots\ldots\ldots\ldots$

## 21

closed figure

Now for part (b):

To show that $2\overline{AB} + 3\overline{BC} + \overline{CA} = 2\overline{LC}$

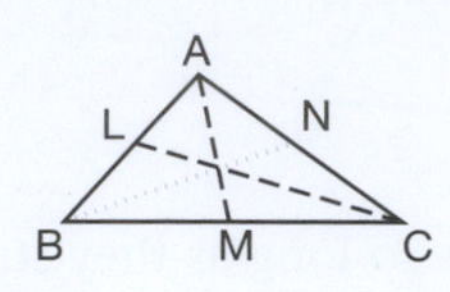


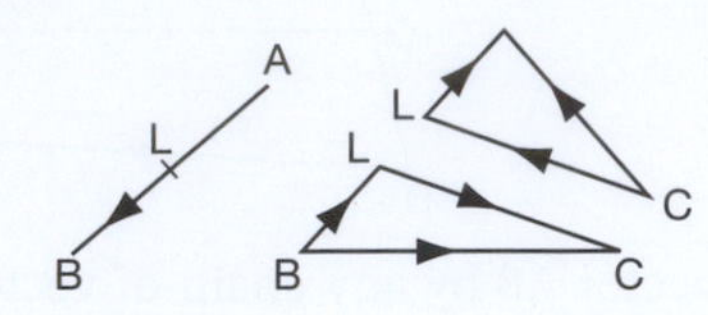


From the figure:

$\overline{AB} = 2\overline{AL}$; $\overline{BC} = \overline{BL} + \overline{LC}$; $\overline{CA} = \overline{CL} + \overline{LA}$

$\therefore\ 2\overline{AB} + 3\overline{BC} + \overline{CA} = 4\overline{AL} + 3\overline{BL} + 3\overline{LC} + \overline{CL} + \overline{LA}$

Now $\overline{BL} = -\overline{AL}$; $\overline{CL} = -\overline{LC}$; $\overline{LA} = -\overline{AL}$

Substituting these in the previous line, gives

$2\overline{AB} + 3\overline{BC} + \overline{CA} = \ldots\ldots\ldots\ldots$

## 22

$2\overline{LC}$

Because

$$\begin{aligned}2\overline{AB} + 3\overline{BC} + \overline{CA} &= 4\overline{AL} + 3\overline{BL} + 3\overline{LC} + \overline{CL} + \overline{LA}\\ &= 4\overline{AL} - 3\overline{AL} + 3\overline{LC} - \overline{LC} - \overline{AL}\\ &= 4\overline{AL} - 4\overline{AL} + 3\overline{LC} - \overline{LC}\\ &= 2\overline{LC}\end{aligned}$$

Now part (c):

To prove that $\overline{AM} + \overline{BN} + \overline{CL} = \mathbf{0}$

From the figure in Frame 21, we can say:

$\overline{AM} = \overline{AB} + \overline{BM}$

$\overline{BN} = \overline{BC} + \overline{CN}$

Similarly

$\overline{CL} = \ldots\ldots\ldots\ldots$

## 23

$\overline{CL} = \overline{CA} + \overline{AL}$

$$\begin{aligned}\text{So }\ \overline{AM} + \overline{BN} + \overline{CL} &= \overline{AB} + \overline{BM} + \overline{BC} + \overline{CN} + \overline{CA} + \overline{AL}\\ &= (\overline{AB} + \overline{BC} + \overline{CA}) + (\overline{BM} + \overline{CN} + \overline{AL})\\ &= (\overline{AB} + \overline{BC} + \overline{CA}) + \frac{1}{2}(\overline{BC} + \overline{CA} + \overline{AB})\\ &= \ldots\ldots\ldots\ldots\end{aligned}$$

*Finish it off*

**24**

$$\boxed{\overline{AM} + \overline{BN} + \overline{CL} = \mathbf{0}}$$

Because $\overline{AM} + \overline{BN} + \overline{CL} = (\overline{AB} + \overline{BC} + \overline{CA}) + \frac{1}{2}(\overline{BC} + \overline{CA} + \overline{AB})$

Now $\overline{AB} + \overline{BC} + \overline{CA}$ is a closed figure $\quad\therefore$ Vector sum $= \mathbf{0}$

and $\overline{BC} + \overline{CA} + \overline{AB}$ is a closed figure $\quad\therefore$ Vector sum $= \mathbf{0}$

$\therefore \overline{AM} + \overline{BN} + \overline{CL} = \mathbf{0}$

Here is another.

**Example 3**

ABCD is a quadrilateral in which P and Q are the mid-points of the diagonals AC and BD respectively.

Show that $\quad \overline{AB} + \overline{AD} + \overline{CB} + \overline{CD} = 4\overline{PQ}$

First, just draw the figure.

*Then move on to Frame 25*

**25**

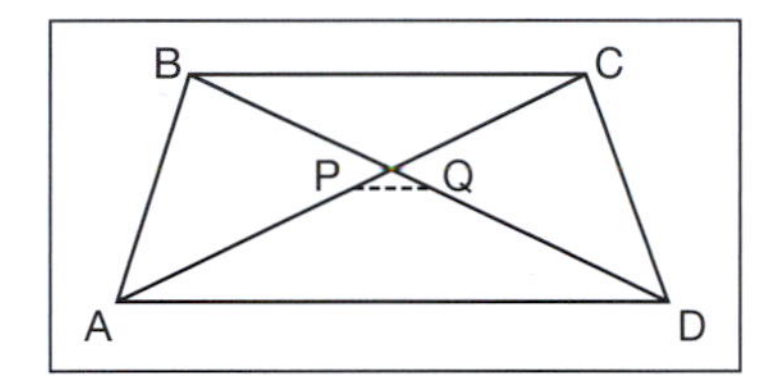


To prove that $\overline{AB} + \overline{AD} + \overline{CB} + \overline{CD} = 4\overline{PQ}$

Taking the vectors on the left-hand side, one at a time, we can write:

$\overline{AB} = \overline{AP} + \overline{PQ} + \overline{QB}$

$\overline{AD} = \overline{AP} + \overline{PQ} + \overline{QD}$

$\overline{CB} = \dots\dots\dots\dots$

$\overline{CD} = \dots\dots\dots\dots$

**26**

$$\boxed{\overline{CB} = \overline{CP} + \overline{PQ} + \overline{QB};\quad \overline{CD} = \overline{CP} + \overline{PQ} + \overline{QD}}$$

Adding all four lines together, we have:

$$\begin{aligned}\overline{AB} + \overline{AD} + \overline{CB} + \overline{CD} &= 4\overline{PQ} + 2\overline{AP} + 2\overline{CP} + 2\overline{QB} + 2\overline{QD}\\ &= 4\overline{PQ} + 2(\overline{AP} + \overline{CP}) + 2(\overline{QB} + \overline{QD})\end{aligned}$$

Now what can we say about $(\overline{AP} + \overline{CP})$?

**27**

$$\boxed{\overline{AP} + \overline{CP} = \mathbf{0}}$$

Because P is the mid-point of AC. $\therefore$ AP = PC

$\therefore \overline{CP} = -\overline{PC} = -\overline{AP}$

$\therefore \overline{AP} + \overline{CP} = \overline{AP} - \overline{AP} = \mathbf{0}.$

In the same way, $(\overline{QB} + \overline{QD}) = \dots\dots\dots\dots$

**28**

$$\boxed{\overline{QB}+\overline{QD}=\mathbf{0}}$$

Since Q is the mid-point of BD. $\therefore\ \overline{QD}=-\overline{QB}$

$$\therefore\ \overline{QB}+\overline{QD}=\overline{QB}-\overline{QB}=\mathbf{0}$$

$$\therefore\ \overline{AB}+\overline{AD}+\overline{CB}+\overline{CD}=4\overline{PQ}+\mathbf{0}+\mathbf{0}=4\overline{PQ}$$

**29**

Here is one more.

**Example 4**

Prove by vectors that the line joining the mid-points of two sides of a triangle is parallel to the third side and half its length.

Let D and E be the mid-points of AB and AC respectively.

We have $\quad \overline{DE}=\overline{DA}+\overline{AE}$

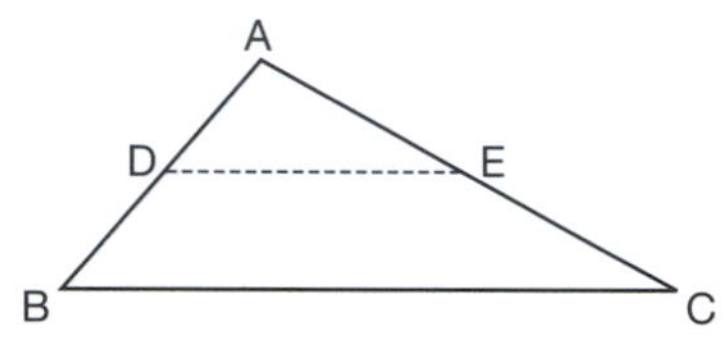


Now express $\overline{DA}$ and $\overline{AE}$ in terms of $\overline{BA}$ and $\overline{AC}$ respectively and see if you can get the required results.

*Then on to Frame 30*

**30**

Here is the working. Check through it.

$$\begin{aligned}\overline{DE}&=\overline{DA}+\overline{AE}\\&=\frac{1}{2}\overline{BA}+\frac{1}{2}\overline{AC}\\&=\frac{1}{2}(\overline{BA}+\overline{AC})\\\overline{DE}&=\frac{1}{2}\overline{BC}\end{aligned}$$

$\therefore\ \overline{DE}$ is half the magnitude (length) of $\overline{BC}$ and acts in the same direction.

i.e. DE and BC are parallel.

*Now for the next section of the work: move on to Frame 31*

**31**

## Components of a vector in terms of unit vectors

The vector $\overline{OP}$ is defined by its magnitude ($r$) and its direction ($\theta$). It could also be defined by its two components in the O$x$ and O$y$ directions.

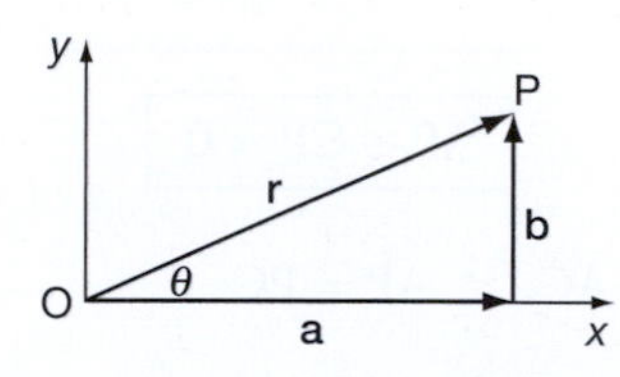


i.e. $\overline{OP}$ is equivalent to a vector **a** in the O$x$ direction + a vector **b** in the O$y$ direction.

i.e. $\overline{OP}=\mathbf{a}$ (along O$x$) + $\mathbf{b}$ (along O$y$)

▶

If we now define $\mathbf{i}$ to be a *unit vector* in the O$x$ direction,

then $\mathbf{a} = a\,\mathbf{i}$

Similarly, if we define $\mathbf{j}$ to be a *unit vector* in the O$y$ direction,

then $\mathbf{b} = b\,\mathbf{j}$

So that the vector OP can be written as:

$\mathbf{r} = a\mathbf{i} + b\,\mathbf{j}$

where $\mathbf{i}$ and $\mathbf{j}$ are unit vectors in the O$x$ and O$y$ directions.

---

**32**

Let $\mathbf{z}_1 = 2\mathbf{i} + 4\mathbf{j}$ and $\mathbf{z}_2 = 5\mathbf{i} + 2\mathbf{j}$

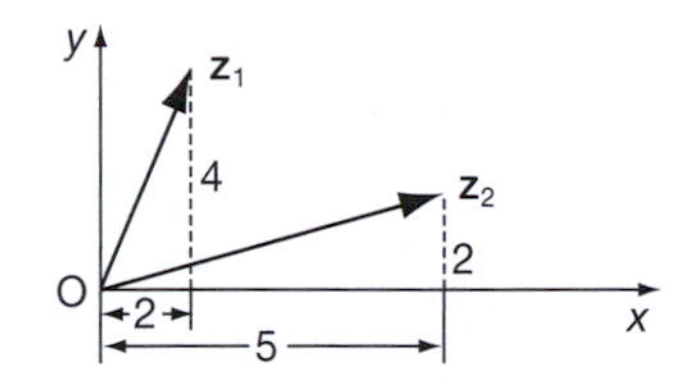


To find $\mathbf{z}_1 + \mathbf{z}_2$, draw the two vectors in a chain.

$\mathbf{z}_1 + \mathbf{z}_2 = \overline{\text{OB}} = (2+5)\mathbf{i} + (4+2)\mathbf{j} = 7\mathbf{i} + 6\mathbf{j}$

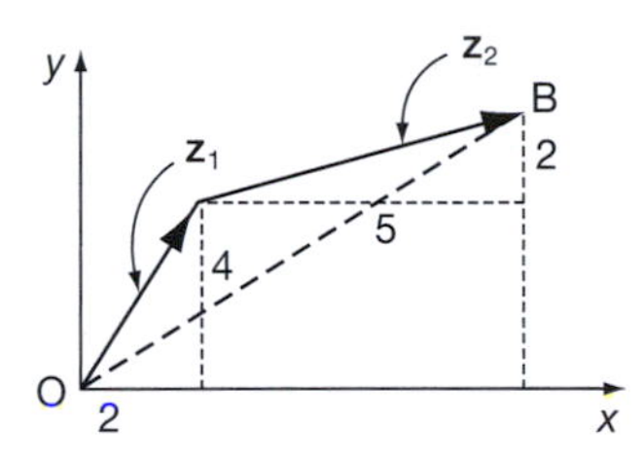


i.e. total up the vector components along O$x$,
and total up the vector components along O$y$

Of course, we can do this without a diagram:

If $\mathbf{z}_1 = 3\mathbf{i} + 2\mathbf{j}$ and $\mathbf{z}_2 = 4\mathbf{i} + 3\mathbf{j}$

$$\begin{aligned}\mathbf{z}_1 + \mathbf{z}_2 &= 3\mathbf{i} + 2\mathbf{j} + 4\mathbf{i} + 3\mathbf{j}\\ &= 7\mathbf{i} + 5\mathbf{j}\end{aligned}$$

And in much the same way, $\mathbf{z}_2 - \mathbf{z}_1 = \dots\dots\dots\dots$

---

**33**

$$\boxed{\mathbf{z}_2 - \mathbf{z}_1 = \mathbf{i} + \mathbf{j}}$$

Because

$$\begin{aligned}\mathbf{z}_2 - \mathbf{z}_1 &= (4\mathbf{i} + 3\mathbf{j}) - (3\mathbf{i} + 2\mathbf{j})\\ &= 4\mathbf{i} + 3\mathbf{j} - 3\mathbf{i} - 2\mathbf{j}\\ &= 1\mathbf{i} + 1\mathbf{j}\\ &= \mathbf{i} + \mathbf{j}\end{aligned}$$

Similarly, if $\mathbf{z}_1 = 5\mathbf{i} - 2\mathbf{j}$; $\mathbf{z}_2 = 3\mathbf{i} + 3\mathbf{j}$; $\mathbf{z}_3 = 4\mathbf{i} - 1\mathbf{j}$

then (a) $\mathbf{z}_1 + \mathbf{z}_2 + \mathbf{z}_3 = \dots\dots\dots\dots$

and (b) $\mathbf{z}_1 - \mathbf{z}_2 - \mathbf{z}_3 = \dots\dots\dots\dots$

*When you have the results, move on to Frame 34*

---

**34**

(a) $12\mathbf{i}$ (b) $-2\mathbf{i} - 4\mathbf{j}$

Here is the working:

(a) $\mathbf{z}_1 + \mathbf{z}_2 + \mathbf{z}_3 = 5\mathbf{i} - 2\mathbf{j} + 3\mathbf{i} + 3\mathbf{j} + 4\mathbf{i} - 1\mathbf{j}$
$= (5 + 3 + 4)\mathbf{i} + (3 - 2 - 1)\mathbf{j} = 12\mathbf{i}$

(b) $\mathbf{z}_1 - \mathbf{z}_2 - \mathbf{z}_3 = (5\mathbf{i} - 2\mathbf{j}) - (3\mathbf{i} + 3\mathbf{j}) - (4\mathbf{i} - 1\mathbf{j})$
$= (5 - 3 - 4)\mathbf{i} + (-2 - 3 + 1)\mathbf{j} = -2\mathbf{i} - 4\mathbf{j}$

Now this one.

If $\overline{OA} = 3\mathbf{i} + 5\mathbf{j}$ and $\overline{OB} = 5\mathbf{i} - 2\mathbf{j}$, find $\overline{AB}$.

As usual, a diagram will help. Here it is:

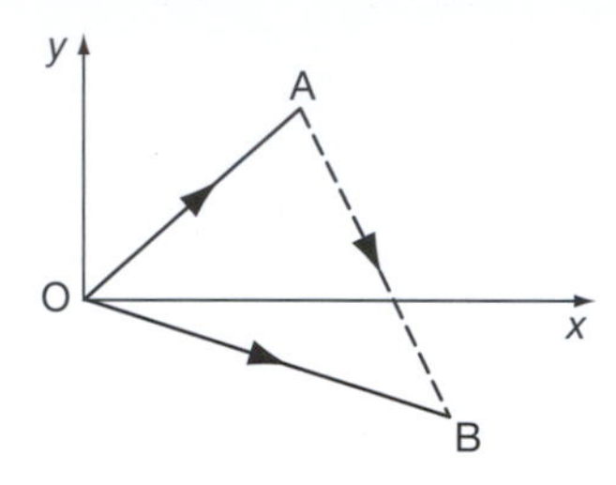


First of all, from the diagram, write down a relationship between the vectors. Then express them in terms of the unit vectors.

$\overline{AB} = \ldots\ldots\ldots\ldots$

**35**

$\overline{AB} = 2\mathbf{i} - 7\mathbf{j}$

Because we have

$\overline{OA} + \overline{AB} = \overline{OB}$ (from diagram)

$\therefore \overline{AB} = \overline{OB} - \overline{OA}$

$= (5\mathbf{i} - 2\mathbf{j}) - (3\mathbf{i} + 5\mathbf{j}) = 2\mathbf{i} - 7\mathbf{j}$

*On to Frame 36*

## Vectors in space

**36**

So far we have only considered vectors that lie in the $x$–$y$ plane. These vectors are described using the mutually perpendicular unit vectors **i** and **j** that lie in the $x$- and $y$-directions respectively. To consider vectors in space we need a third direction with an axis that is perpendicular to both the $x$ and $y$ axes. This is provided by the $z$-axis.The $x$-, $y$- and $z$-axes define three mutually perpendicular directions for the axes of reference which obey the 'right-hand' rule.

O$x$, O$y$, O$z$ form a right-handed set if rotation from O$x$ to O$y$ takes a right-handed corkscrew action along the positive direction of O$z$.

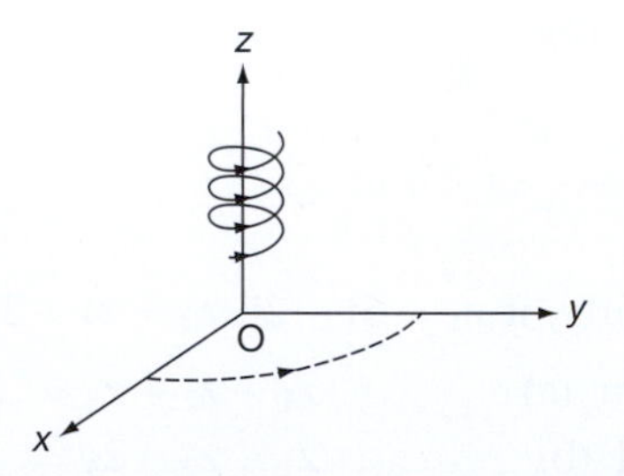


Similarly, rotation from O$y$ to O$z$ gives right-hand corkscrew action along the positive direction of ...........

37

Ox

Vector $\overline{OP}$ is defined by its components

$a$ along O$x$
$b$ along O$y$
$c$ along O$z$

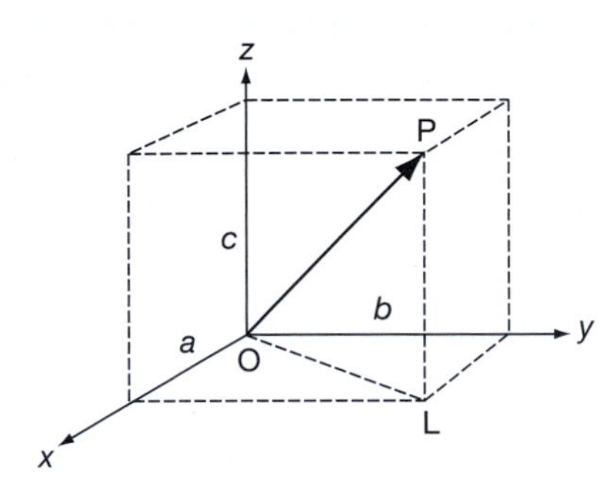


Let $\mathbf{i}$ = unit vector in O$x$ direction
$\mathbf{j}$ = unit vector in O$y$ direction
$\mathbf{k}$ = unit vector in O$z$ direction

Then

$$\overline{OP} = a\mathbf{i} + b\mathbf{j} + c\mathbf{k}$$

Also

$$OL^2 = a^2 + b^2$$

and

$$OP^2 = OL^2 + c^2$$
$$OP^2 = a^2 + b^2 + c^2$$

So, if $\mathbf{r} = a\mathbf{i} + b\mathbf{j} + c\mathbf{k}$, then

$$r = \sqrt{a^2 + b^2 + c^2}$$

This gives us an easy way of finding the magnitude of a vector expressed in terms of the unit vectors.

Now you can do this one:

If $\overline{PQ} = 4\mathbf{i} + 3\mathbf{j} + 2\mathbf{k}$, then

$|\overline{PQ}|$ = ...........

38

$|\overline{PQ}| = \sqrt{29} = 5{\cdot}385$

Because

$$\overline{PQ} = 4\mathbf{i} + 3\mathbf{j} + 2\mathbf{k}$$
$$|\overline{PQ}| = \sqrt{4^2 + 3^2 + 2^2}$$
$$= \sqrt{16 + 9 + 4}$$
$$= \sqrt{29}$$
$$= 5{\cdot}385$$

*Now move on to Frame 39*

# Direction cosines

**39**

The direction of a vector in three dimensions is determined by the angles which the vector makes with the three axes of reference.

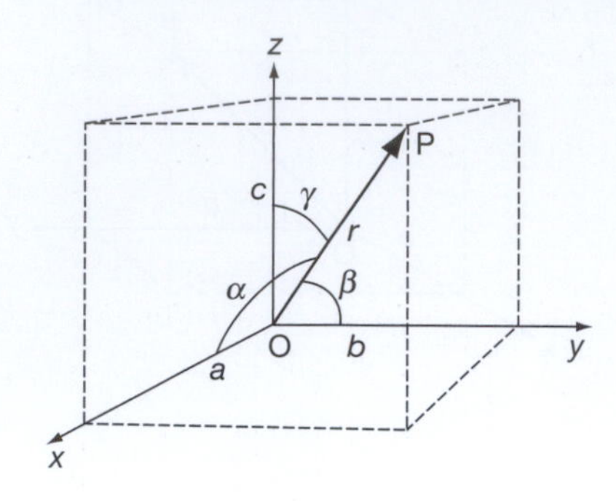


Let $\overline{OP} = \mathbf{r} = a\mathbf{i} + b\mathbf{j} + c\mathbf{k}$

Then

$$\frac{a}{r} = \cos\alpha \qquad \therefore\ a = r\cos\alpha$$

$$\frac{b}{r} = \cos\beta \qquad b = r\cos\beta$$

$$\frac{c}{r} = \cos\gamma \qquad c = r\cos\gamma$$

Also

$$a^2 + b^2 + c^2 = r^2$$

$$\therefore\ r^2\cos^2\alpha + r^2\cos^2\beta + r^2\cos^2\gamma = r^2$$

$$\therefore\ \cos^2\alpha + \cos^2\beta + \cos^2\gamma = 1$$

If $l = \cos\alpha$

$m = \cos\beta$

$n = \cos\gamma$ then $l^2 + m^2 + n^2 = 1$

*Note*: $[l, m, n]$ written in square brackets are called the *direction cosines* of the vector $\overline{OP}$ and are the values of the cosines of the angles which the vector makes with the three axes of reference.

So for the vector $\mathbf{r} = a\mathbf{i} + b\mathbf{j} + c\mathbf{k}$

$$l = \frac{a}{r};\quad m = \frac{b}{r};\quad n = \frac{c}{r};$$

and, of course

$$r = \sqrt{a^2 + b^2 + c^2}$$

So, with that in mind, find the direction cosines $[l, m, n]$ of the vector

$\mathbf{r} = 3\mathbf{i} - 2\mathbf{j} + 6\mathbf{k}$

*Then to Frame 40*

**40**

$\mathbf{r} = 3\mathbf{i} - 2\mathbf{j} + 6\mathbf{k}$

$$\therefore\ a = 3,\ b = -2,\ c = 6,\ r = \sqrt{9 + 4 + 36}$$

$$\therefore\ r = \sqrt{49} = 7$$

$$\therefore\ l = \frac{3}{7};\quad m = -\frac{2}{7};\quad n = \frac{6}{7}$$

Just as easy as that!

*On to the next frame*

## Scalar product of two vectors

**41**

If **a** and **b** are two vectors, the *scalar product* of **a** and **b** is defined as the *scalar* (number) $ab\cos\theta$ where $a$ and $b$ are the magnitudes of the vectors **a** and **b** and $\theta$ is the angle between them.

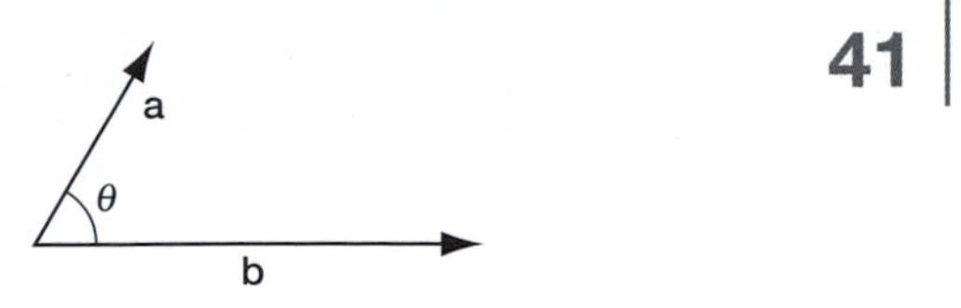


The scalar product is denoted by **a**.**b** (often called the 'dot product' for obvious reasons).

$\therefore\ \mathbf{a}.\mathbf{b} = ab\cos\theta$

$= a \times$ projection of **b** on **a**

$= b \times$ projection of **a** on **b**

In both cases the result is a *scalar* quantity.

For example:

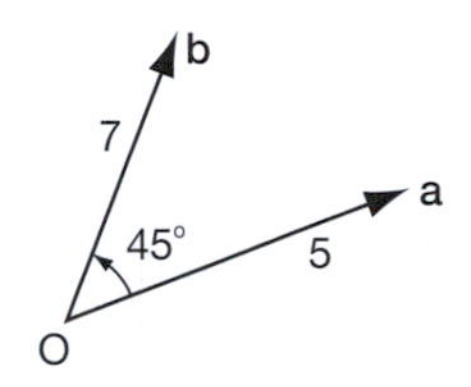


$\mathbf{a}.\mathbf{b} = \dots\dots\dots\dots$

**42**

$$\mathbf{a}.\mathbf{b} = \frac{35\sqrt{2}}{2}$$

Because we have:

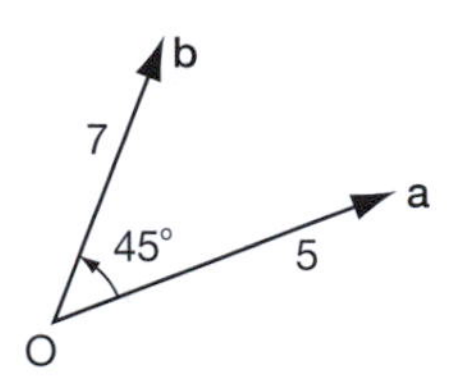


$$\mathbf{a}.\mathbf{b} = a.b\cos\theta$$
$$= 5.7.\cos 45^\circ$$
$$= 35.\frac{1}{\sqrt{2}} = \frac{35\sqrt{2}}{2}$$

Now what about this case:

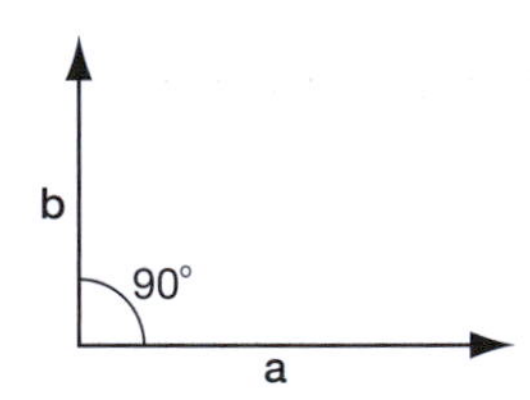


The scalar product of **a** and **b** is

$\mathbf{a}.\mathbf{b} = \dots\dots\dots\dots$

**43**

$$\boxed{0}$$

Because in this case $\mathbf{a}.\mathbf{b} = ab\cos 90^\circ = ab0 = 0$. So the scalar product of any two vectors at right-angles to each other is always zero.

And in this case now, with two vectors in the same direction, $\theta = 0^\circ$

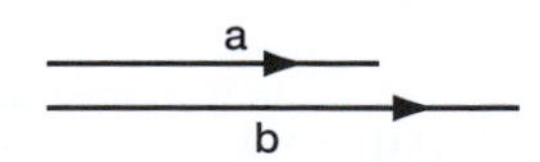


so $\mathbf{a}.\mathbf{b} = \dots\dots\dots\dots$

**44**

$ab$

Because $\mathbf{a}.\mathbf{b} = ab\cos 0^\circ = ab.1 = ab$

Now suppose our two vectors are expressed in terms of the unit vectors **i**, **j** and **k**.

Let $\quad \mathbf{a} = a_1\mathbf{i} + a_2\mathbf{j} + a_3\mathbf{k}$

and $\quad \mathbf{b} = b_1\mathbf{i} + b_2\mathbf{j} + b_3\mathbf{k}$

Then
$$\mathbf{a}.\mathbf{b} = (a_1\mathbf{i} + a_2\mathbf{j} + a_3\mathbf{k}).(b_1\mathbf{i} + b_2\mathbf{j} + b_3\mathbf{k})$$
$$= a_1b_1\mathbf{i}.\mathbf{i} + a_1b_2\mathbf{i}.\mathbf{j} + a_1b_3\mathbf{i}.\mathbf{k} + a_2b_1\mathbf{j}.\mathbf{i} + a_2b_2\mathbf{j}.\mathbf{j} + a_2b_3\mathbf{j}.\mathbf{k}$$
$$+ a_3b_1\mathbf{k}.\mathbf{i} + a_3b_2\mathbf{k}.\mathbf{j} + a_3b_3\mathbf{k}.\mathbf{k}$$

This can now be simplified.

Because $\mathbf{i}.\mathbf{i} = (1)(1)(\cos 0^\circ) = 1$

$\therefore\ \mathbf{i}.\mathbf{i} = 1;\ \ \mathbf{j}.\mathbf{j} = 1;\ \ \mathbf{k}.\mathbf{k} = 1 \qquad$ (a)

Also $\mathbf{i}.\mathbf{j} = (1)(1)(\cos 90^\circ) = 0$

$\therefore\ \mathbf{i}.\mathbf{j} = 0;\ \ \mathbf{j}.\mathbf{k} = 0;\ \ \mathbf{k}.\mathbf{i} = 0 \qquad$ (b)

So, using the results (a) and (b), we get:

$\mathbf{a}.\mathbf{b} = \ldots\ldots\ldots\ldots$

**45**

$\mathbf{a}.\mathbf{b} = a_1b_1 + a_2b_2 + a_3b_3$

Because

$$\mathbf{a}.\mathbf{b} = a_1b_1.1 + a_1b_2.0 + a_1b_3.0 + a_2b_1.0 + a_2b_2.1 + a_2b_3.0$$
$$+ a_3b_1.0 + a_3b_2.0 + a_3b_3.1$$
$$\therefore\ \mathbf{a}.\mathbf{b} = a_1b_1 + a_2b_2 + a_3b_3$$

i.e. we just sum the products of the coefficients of the unit vectors along the corresponding axes.

For example:

If $\mathbf{a} = 2\mathbf{i} + 3\mathbf{j} + 5\mathbf{k}$ and $\mathbf{b} = 4\mathbf{i} + 1\mathbf{j} + 6\mathbf{k}$

then $\mathbf{a}.\mathbf{b} = 2 \times 4 + 3 \times 1 + 5 \times 6$
$= 8 + 3 + 30$
$= 41 \qquad \therefore\ \mathbf{a}.\mathbf{b} = 41$

One for you: If $\mathbf{p} = 3\mathbf{i} - 2\mathbf{j} + 1\mathbf{k};\ \ \mathbf{q} = 2\mathbf{i} + 3\mathbf{j} - 4\mathbf{k}$

then $\mathbf{p}.\mathbf{q} = \ldots\ldots\ldots\ldots$

**46**

$-4$

Because

$\mathbf{p}.\mathbf{q} = 3 \times 2 + (-2) \times 3 + 1 \times (-4)$
$= 6 - 6 - 4$
$= -4 \qquad \therefore\ \mathbf{p}.\mathbf{q} = -4$

*Now on to Frame 47*

## Vector product of two vectors

**47**

The vector product of **a** and **b** is written $\mathbf{a} \times \mathbf{b}$ (often called the 'cross product') and is defined as a *vector* having magnitude $ab \sin \theta$ where $\theta$ is the angle between the two given vectors. The product vector acts in a direction perpendicular to both **a** and **b** in such a sense that **a**, **b** and $\mathbf{a} \times \mathbf{b}$ form a right-handed set – in that order.

$$|\mathbf{a} \times \mathbf{b}| = ab \sin \theta$$

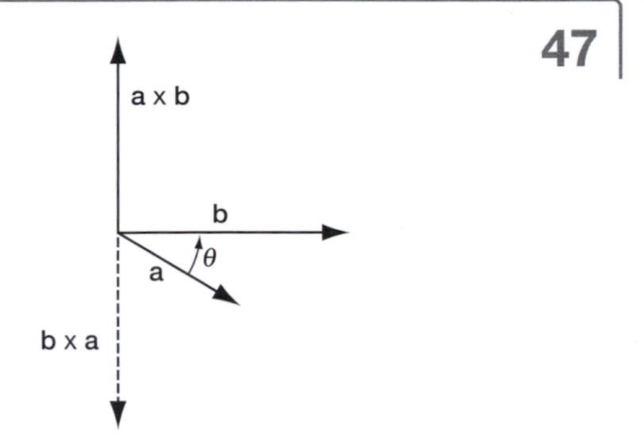


Note that $\mathbf{b} \times \mathbf{a}$ reverses the direction of rotation and the product vector would now act downwards, i.e.

$$\mathbf{b} \times \mathbf{a} = -(\mathbf{a} \times \mathbf{b})$$

If $\theta = 0°$, then $|\mathbf{a} \times \mathbf{b}| = \ldots\ldots\ldots\ldots$

and if $\theta = 90°$, then $|\mathbf{a} \times \mathbf{b}| = \ldots\ldots\ldots\ldots$

---

**48**

$$\theta = 0°, \quad |\mathbf{a} \times \mathbf{b}| = 0$$
$$\theta = 90°, \quad |\mathbf{a} \times \mathbf{b}| = ab$$

If **a** and **b** are given in terms of the unit vectors **i**, **j** and **k**:

$$\mathbf{a} = a_1\mathbf{i} + a_2\mathbf{j} + a_3\mathbf{k} \text{ and } \mathbf{b} = b_1\mathbf{i} + b_2\mathbf{j} + b_3\mathbf{k}$$

Then:

$$\mathbf{a} \times \mathbf{b} = a_1b_1\mathbf{i} \times \mathbf{i} + a_1b_2\mathbf{i} \times \mathbf{j} + a_1b_3\mathbf{i} \times \mathbf{k} + a_2b_1\mathbf{j} \times \mathbf{i} + a_2b_2\mathbf{j} \times \mathbf{j}$$
$$+ a_2b_3\mathbf{j} \times \mathbf{k} + a_3b_1\mathbf{k} \times \mathbf{i} + a_3b_2\mathbf{k} \times \mathbf{j} + a_3b_3\mathbf{k} \times \mathbf{k}$$

But $|\mathbf{i} \times \mathbf{i}| = (1)(1)(\sin 0°) = 0 \quad \therefore \mathbf{i} \times \mathbf{i} = \mathbf{j} \times \mathbf{j} = \mathbf{k} \times \mathbf{k} = 0$ (a)

Also $|\mathbf{i} \times \mathbf{j}| = (1)(1)(\sin 90°) = 1$ and $\mathbf{i} \times \mathbf{j}$ is in the direction of **k**, i.e. $\mathbf{i} \times \mathbf{j} = \mathbf{k}$ (same magnitude and same direction). Therefore:

$$\mathbf{i} \times \mathbf{j} = \mathbf{k}$$
$$\mathbf{j} \times \mathbf{k} = \mathbf{i}$$
$$\mathbf{k} \times \mathbf{i} = \mathbf{j} \qquad \text{(b)}$$

And remember too that therefore:

$$\mathbf{i} \times \mathbf{j} = -(\mathbf{j} \times \mathbf{i})$$
$$\mathbf{j} \times \mathbf{k} = -(\mathbf{k} \times \mathbf{j})$$
$$\mathbf{k} \times \mathbf{i} = -(\mathbf{i} \times \mathbf{k})$$ since the sense of rotation is reversed

Now with the results of (a) and (b), and this last reminder, you can simplify the expression for $\mathbf{a} \times \mathbf{b}$.

Remove the zero terms and tidy up what is left.

..................................................

*Then on to Frame 49*

**49**

$$\mathbf{a}\times\mathbf{b} = (a_2b_3 - a_3b_2)\mathbf{i} - (a_1b_3 - a_3b_1)\mathbf{j} + (a_1b_2 - a_2b_1)\mathbf{k}$$

Because

$$\mathbf{a}\times\mathbf{b} = a_1b_1\mathbf{0} + a_1b_2\mathbf{k} + a_1b_3(-\mathbf{j}) + a_2b_1(-\mathbf{k}) + a_2b_2\mathbf{0} + a_2b_3\mathbf{i} + a_3b_1\mathbf{j} + a_3b_2(-\mathbf{i}) + a_3b_3\mathbf{0}$$

$$\mathbf{a}\times\mathbf{b} = (a_2b_3 - a_3b_2)\mathbf{i} - (a_1b_3 - a_3b_1)\mathbf{j} + (a_1b_2 - a_2b_1)\mathbf{k}$$

and you may recognize this as the pattern of a determinant where the first row is made up of the vectors **i**, **j** and **k**.

So now we have that:

If $\mathbf{a} = a_1\mathbf{i} + a_2\mathbf{j} + a_3\mathbf{k}$ and $\mathbf{b} = b_1\mathbf{i} + b_2\mathbf{j} + b_3\mathbf{k}$ then:

$$\mathbf{a}\times\mathbf{b} = \begin{vmatrix} \mathbf{i} & \mathbf{j} & \mathbf{k} \\ a_1 & a_2 & a_3 \\ b_1 & b_2 & b_3 \end{vmatrix} = (a_2b_3 - a_3b_2)\mathbf{i} - (a_1b_3 - a_3b_1)\mathbf{j} + (a_1b_2 - a_2b_1)\mathbf{k}$$

and that is the easiest way to write out the vector product of two vectors.

*Notes*:

(a) The top row consists of the unit vectors in order **i**, **j**, **k**.

(b) The second row consists of the coefficients of **a**.

(c) The third row consists of the coefficients of **b**.

For example, if $\mathbf{p} = 2\mathbf{i} + 4\mathbf{j} + 3\mathbf{k}$ and $\mathbf{q} = \mathbf{i} + 5\mathbf{j} - 2\mathbf{k}$, first write down the determinant that represents the vector product $\mathbf{p}\times\mathbf{q}$.

**50**

$$\mathbf{p}\times\mathbf{q} = \begin{vmatrix} \mathbf{i} & \mathbf{j} & \mathbf{k} \\ 2 & 4 & 3 \\ 1 & 5 & -2 \end{vmatrix} \begin{matrix} \text{Unit vectors} \\ \text{Coefficients of } \mathbf{p} \\ \text{Coefficients of } \mathbf{q} \end{matrix}$$

And now, expanding the determinant, we get:

$\mathbf{p}\times\mathbf{q} = \ldots\ldots\ldots\ldots$

**51**

$$\mathbf{p}\times\mathbf{q} = -23\mathbf{i} + 7\mathbf{j} + 6\mathbf{k}$$

Because

$$\mathbf{p}\times\mathbf{q} = \begin{vmatrix} \mathbf{i} & \mathbf{j} & \mathbf{k} \\ 2 & 4 & 3 \\ 1 & 5 & -2 \end{vmatrix} = \mathbf{i}\begin{vmatrix} 4 & 3 \\ 5 & -2 \end{vmatrix} - \mathbf{j}\begin{vmatrix} 2 & 3 \\ 1 & -2 \end{vmatrix} + \mathbf{k}\begin{vmatrix} 2 & 4 \\ 1 & 5 \end{vmatrix}$$

$$= \mathbf{i}(-8 - 15) - \mathbf{j}(-4 - 3) + \mathbf{k}(10 - 4)$$

$$= -23\mathbf{i} + 7\mathbf{j} + 6\mathbf{k}$$

▶

So, by way of review:

(a) *Scalar product* ('dot product')

$\mathbf{a}.\mathbf{b} = ab\cos\theta$ a scalar quantity

(b) *Vector product* ('cross product')

$\mathbf{a} \times \mathbf{b}$ = vector of magnitude $ab\sin\theta$, acting in a direction to make **a**, **b** and $\mathbf{a} \times \mathbf{b}$ a right-handed set. Also:

$$\mathbf{a} \times \mathbf{b} = \begin{vmatrix} \mathbf{i} & \mathbf{j} & \mathbf{k} \\ a_1 & a_2 & a_3 \\ b_1 & b_2 & b_3 \end{vmatrix} = (a_2b_3 - a_3b_2)\mathbf{i} - (a_1b_3 - a_3b_1)\mathbf{j} + (a_1b_2 - a_2b_1)\mathbf{k}$$

And here is one final example on this point.

Find the vector product of **p** and **q** where:

$\mathbf{p} = 3\mathbf{i} - 4\mathbf{j} + 2\mathbf{k}$ and $\mathbf{q} = 2\mathbf{i} + 5\mathbf{j} - \mathbf{k}$

---

**52**

$$\boxed{\mathbf{p} \times \mathbf{q} = -6\mathbf{i} + 7\mathbf{j} + 23\mathbf{k}}$$

Because

$$\mathbf{p} \times \mathbf{q} = \begin{vmatrix} \mathbf{i} & \mathbf{j} & \mathbf{k} \\ 3 & -4 & 2 \\ 2 & 5 & -1 \end{vmatrix}$$
$$= \mathbf{i}\begin{vmatrix} -4 & 2 \\ 5 & -1 \end{vmatrix} - \mathbf{j}\begin{vmatrix} 3 & 2 \\ 2 & -1 \end{vmatrix} + \mathbf{k}\begin{vmatrix} 3 & -4 \\ 2 & 5 \end{vmatrix}$$
$$= \mathbf{i}(4 - 10) - \mathbf{j}(-3 - 4) + \mathbf{k}(15 + 8)$$
$$= -6\mathbf{i} + 7\mathbf{j} + 23\mathbf{k}$$

Remember that the order in which the vectors appear in the vector product is important. It is a simple matter to verify that:

$\mathbf{q} \times \mathbf{p} = 6\mathbf{i} - 7\mathbf{j} - 23\mathbf{k} = -(\mathbf{p} \times \mathbf{q})$

*On to Frame 53*

---

## Angle between two vectors

**53**

Let **a** be one vector with direction cosines $[l, m, n]$

Let **b** be the other vector with direction cosines $[l', m', n']$

We have to find the angle between these two vectors.

Let $\overline{\mathrm{OP}}$ and $\overline{\mathrm{OP}}'$ be *unit* vectors parallel to **a** and **b** respectively. Then P has coordinates $(l, m, n)$ and P′ has coordinates $(l', m', n')$.

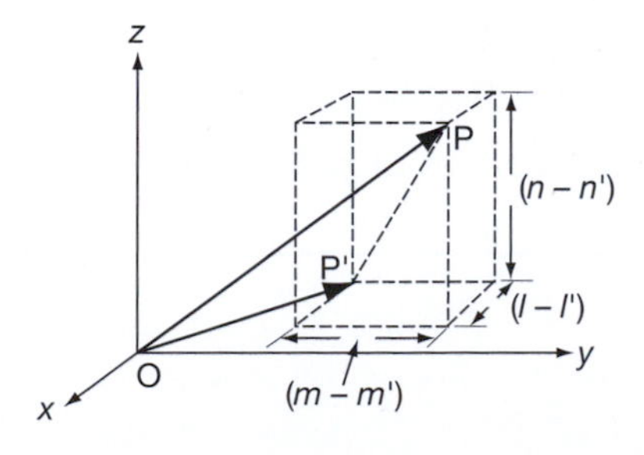


►

Then

$$(\text{PP}')^2 = (l - l')^2 + (m - m')^2 + (n - n')^2$$
$$= l^2 - 2.l.l' + l'^2 + m^2 - 2.m.m' + m'^2 + n^2 - 2n.n' + n'^2$$
$$= (l^2 + m^2 + n^2) + (l'^2 + m'^2 + n'^2) - 2(ll' + mm' + nn')$$

But $(l^2 + m^2 + n^2) = 1$ and $(l'^2 + m'^2 + n'^2) = 1$ as was proved earlier.

$$\therefore\ (\text{PP}')^2 = 2 - 2(ll' + mm' + nn') \qquad \text{(a)}$$

Also, by the cosine rule:

$$(\text{PP}')^2 = \text{OP}^2 + \text{OP}'^2 - 2.\text{OP}.\text{OP}'.\cos\theta$$
$$= 1 + 1 - 2.1.1.\cos\theta$$
$$= 2 - 2\cos\theta \qquad \text{(b)}$$

$\left\{ \overline{\text{OP}} \text{ and } \overline{\text{OP}'} \text{ are unit vectors} \right\}$

So from (a) and (b), we have:

$$(\text{PP}')^2 = 2 - 2(ll' + mm' + nn')$$

and $(\text{PP}')^2 = 2 - 2\cos\theta$

$\therefore \quad \cos\theta = \ldots\ldots\ldots\ldots$

---

**54**

$$\boxed{\cos\theta = ll' + mm' + nn'}$$

That is, just sum the products of the corresponding direction cosines of the two given vectors.

So, if $[l, m, n] = [0{\cdot}54, 0{\cdot}83, -0{\cdot}14]$
and $[l', m', n'] = [0{\cdot}25, 0{\cdot}60, 0{\cdot}76]$

the angle between the vectors is $\theta = \ldots\ldots\ldots\ldots$

---

**55**

$$\boxed{\theta = 58^\circ 13'}$$

Because, we have:

$$\cos\theta = ll' + mm' + nn'$$
$$= (0{\cdot}54)(0{\cdot}25) + (0{\cdot}83)(0{\cdot}60) + (-0{\cdot}14)(0{\cdot}76)$$
$$= 0{\cdot}1350 + 0{\cdot}4980 - 0{\cdot}1064$$
$$= 0{\cdot}6330 - 0{\cdot}1064 \qquad = 0{\cdot}5266$$

$\theta = 58^\circ 13'$

*Note*: For *parallel vectors*, $\theta = 0^\circ \ \therefore\ ll' + mm' + nn' = 1$

For *perpendicular vectors*, $\theta = 90^\circ,\ \therefore\ ll' + mm' + nn' = 0$

Now an example for you to work:

Find the angle between the vectors

$\mathbf{p} = 2\mathbf{i} + 3\mathbf{j} + 4\mathbf{k}$ and $\mathbf{q} = 4\mathbf{i} - 3\mathbf{j} + 2\mathbf{k}$

First of all, find the direction cosines of $\mathbf{p}$. So do that.

**56**

$$l = \frac{2}{\sqrt{29}} \qquad m = \frac{3}{\sqrt{29}} \qquad n = \frac{4}{\sqrt{29}}$$

Because

$$p = |\mathbf{p}| = \sqrt{2^2 + 3^2 + 4^2} = \sqrt{4 + 9 + 16} = \sqrt{29}$$

$$\therefore\ l = \frac{a}{p} = \frac{2}{\sqrt{29}}$$

$$m = \frac{b}{p} = \frac{3}{\sqrt{29}}$$

$$n = \frac{c}{p} = \frac{4}{\sqrt{29}}$$

$$\therefore\ [l, m, n] = \left[\frac{2}{\sqrt{29}}, \frac{3}{\sqrt{29}}, \frac{4}{\sqrt{29}}\right]$$

Now find the direction cosines $[l', m', n']$ of $\mathbf{q}$ in just the same way.

*When you have done that move on to the next frame*

**57**

$$l' = \frac{4}{\sqrt{29}} \qquad m' = \frac{-3}{\sqrt{29}} \qquad n' = \frac{2}{\sqrt{29}}$$

Because

$$q = |\mathbf{q}| = \sqrt{4^2 + 3^2 + 2^2}$$

$$= \sqrt{16 + 9 + 4}$$

$$= \sqrt{29}$$

$$\therefore\ [l', m', n'] = \left[\frac{4}{\sqrt{29}}, \frac{-3}{\sqrt{29}}, \frac{2}{\sqrt{29}}\right]$$

We already know that, for $\mathbf{p}$:

$$[l, m, n] = \left[\frac{2}{\sqrt{29}}, \frac{3}{\sqrt{29}}, \frac{4}{\sqrt{29}}\right]$$

So, using $\cos\theta = ll' + mm' + nn'$, you can finish it off and find the angle $\theta$. Off you go.

**58**

$$\theta = 76°2'$$

Because

$$\cos\theta = \frac{2}{\sqrt{29}}\cdot\frac{4}{\sqrt{29}} + \frac{3}{\sqrt{29}}\cdot\frac{(-3)}{\sqrt{29}} + \frac{4}{\sqrt{29}}\cdot\frac{2}{\sqrt{29}}$$

$$= \frac{8}{29} - \frac{9}{29} + \frac{8}{29}$$

$$= \frac{7}{29} = 0{\cdot}2414$$

$$\therefore\ \theta = 76°2'$$

*Now on to Frame 59*

## Direction ratios

**59**

If $\overline{\text{OP}} = a\mathbf{i} + b\mathbf{j} + c\mathbf{k}$, we know that:

$$|\overline{\text{OP}}| = r = \sqrt{a^2 + b^2 + c^2}$$

and that the direction cosines of $\overline{\text{OP}}$ are given by:

$$l = \frac{a}{r}, \quad m = \frac{b}{r}, \quad n = \frac{c}{r}$$

We can see that the components, $a$, $b$, $c$, are proportional to the direction cosines, $l$, $m$, $n$, respectively and they are sometimes referred to as the *direction ratios* of the vector $\overline{\text{OP}}$.

*Note*: The direction ratios can be converted into the direction cosines by dividing each of them by $r$ (the magnitude of the vector).

Now read through the summary of the work we have covered in this Programme.

*So move on to Frame 60*

## Review summary

**60**

1 A *scalar* quantity has magnitude only; a *vector* quantity has both magnitude and direction.

2 The axes of reference, O$x$, O$y$, O$z$, are chosen so that they form a right-handed set. The symbols $\mathbf{i}$, $\mathbf{j}$, $\mathbf{k}$ denote *unit vectors* in the directions O$x$, O$y$, O$z$, respectively.

If $\overline{\text{OP}} = a\mathbf{i} + b\mathbf{j} + c\mathbf{k}$, then $|\overline{\text{OP}}| = r = \sqrt{a^2 + b^2 + c^2}$

3 The *direction cosines* $[l, m, n]$ are the cosines of the angles between the vector and the axes O$x$, O$y$, O$z$ respectively.

For any vector: $l = \frac{a}{r}$, $m = \frac{b}{r}$, $n = \frac{c}{r}$; and $l^2 + m^2 + n^2 = 1$

4 *Scalar product* ('dot product')

$\mathbf{a}.\mathbf{b} = ab\cos\theta$ where $\theta$ is the angle between $\mathbf{a}$ and $\mathbf{b}$.

If $\mathbf{a} = a_1\mathbf{i} + a_2\mathbf{j} + a_3\mathbf{k}$ and $\mathbf{b} = b_1\mathbf{i} + b_2\mathbf{j} + b_3\mathbf{k}$

then $\mathbf{a}.\mathbf{b} = a_1b_1 + a_2b_2 + a_3b_3$

5 *Vector product* ('cross product')

$\mathbf{a} \times \mathbf{b} = (ab\sin\theta)$ in direction perpendicular to $\mathbf{a}$ and $\mathbf{b}$, so that $\mathbf{a}$, $\mathbf{b}$ and $(\mathbf{a} \times \mathbf{b})$ form a right-handed set.

Also $\mathbf{a} \times \mathbf{b} = \begin{vmatrix} \mathbf{i} & \mathbf{j} & \mathbf{k} \\ a_1 & a_2 & a_3 \\ b_1 & b_2 & b_3 \end{vmatrix}$

6 *Angle between two vectors*

$\cos\theta = ll' + mm' + nn'$

For perpendicular vectors, $ll' + mm' + nn' = 0$

Now you are ready for the **Can You?** checklist and **Test exercise**.

*So off you go*

## Can You?

### Checklist 6

61

Check this list before and after you try the end of Programme test.

On a scale of 1 to 5 how confident are you that you can:

| | Frames |
|---|---|
| • Define a vector? *Yes* ☐ ☐ ☐ ☐ ☐ *No* | 1 to 5 |
| • Represent a vector by a directed straight line? *Yes* ☐ ☐ ☐ ☐ ☐ *No* | 6 to 10 |
| • Add vectors? *Yes* ☐ ☐ ☐ ☐ ☐ *No* | 11 to 17 |
| • Write a vector in terms of component vectors? *Yes* ☐ ☐ ☐ ☐ ☐ *No* | 18 to 30 |
| • Write a vector in terms of component unit vectors? *Yes* ☐ ☐ ☐ ☐ ☐ *No* | 31 to 35 |
| • Set up a coordinate system for representing vectors? *Yes* ☐ ☐ ☐ ☐ ☐ *No* | 36 to 38 |
| • Obtain the direction cosines of a vector? *Yes* ☐ ☐ ☐ ☐ ☐ *No* | 39 to 40 |
| • Calculate the scalar product of two vectors? *Yes* ☐ ☐ ☐ ☐ ☐ *No* | 41 to 46 |
| • Calculate the vector product of two vectors? *Yes* ☐ ☐ ☐ ☐ ☐ *No* | 47 to 52 |
| • Determine the angle between two vectors? *Yes* ☐ ☐ ☐ ☐ ☐ *No* | 53 to 58 |
| • Evaluate the direction ratios of a vector? *Yes* ☐ ☐ ☐ ☐ ☐ *No* | 59 |

## Test exercise 6

62

Take your time: the problems are all straightforward so avoid careless slips. Diagrams often help where appropriate.

1 If $\overline{OA} = 4\mathbf{i} + 3\mathbf{j}$, $\overline{OB} = 6\mathbf{i} - 2\mathbf{j}$, $\overline{OC} = 2\mathbf{i} - \mathbf{j}$, find $\overline{AB}$, $\overline{BC}$ and $\overline{CA}$, and deduce the lengths of the sides of the triangle ABC.

2 Find the direction cosines of the vector joining the two points (4, 2, 2) and (7, 6, 14).

3 If $\mathbf{a} = 2\mathbf{i} + 2\mathbf{j} - \mathbf{k}$ and $\mathbf{b} = 3\mathbf{i} - 6\mathbf{j} + 2\mathbf{k}$, find (a) $\mathbf{a}.\mathbf{b}$ and (b) $\mathbf{a} \times \mathbf{b}$.

4 If $\mathbf{a} = 5\mathbf{i} + 4\mathbf{j} + 2\mathbf{k}$, $\mathbf{b} = 4\mathbf{i} - 5\mathbf{j} + 3\mathbf{k}$ and $\mathbf{c} = 2\mathbf{i} - \mathbf{j} - 2\mathbf{k}$, where $\mathbf{i}$, $\mathbf{j}$, $\mathbf{k}$ are the unit vectors, determine:

(a) the value of $\mathbf{a}.\mathbf{b}$ and the angle between the vectors $\mathbf{a}$ and $\mathbf{b}$

(b) the magnitude and the direction cosines of the product vector $\mathbf{a} \times \mathbf{b}$ and also the angle which this product vector makes with the vector $\mathbf{c}$.

# Further problems 6

**63**

1 The centroid of the triangle OAB is denoted by G. If O is the origin and $\overline{OA} = 4\mathbf{i} + 3\mathbf{j}$, $\overline{OB} = 6\mathbf{i} - \mathbf{j}$, find $\overline{OG}$ in terms of the unit vectors, **i** and **j**.

2 Find the direction cosines of the vectors whose direction ratios are (3, 4, 5) and (1, 2, −3). Hence find the angle between the two vectors.

3 Find the modulus and the direction cosines of each of the vectors $3\mathbf{i} + 7\mathbf{j} - 4\mathbf{k}$, $\mathbf{i} - 5\mathbf{j} - 8\mathbf{k}$ and $6\mathbf{i} - 2\mathbf{j} + 12\mathbf{k}$. Find also the modulus and the direction cosines of their sum.

4 If $\mathbf{a} = 2\mathbf{i} + 4\mathbf{j} - 3\mathbf{k}$ and $\mathbf{b} = \mathbf{i} + 3\mathbf{j} + 2\mathbf{k}$, determine the scalar and vector products, and the angle between the two given vectors.

5 If $\overline{OA} = 2\mathbf{i} + 3\mathbf{j} - \mathbf{k}$ and $\overline{OB} = \mathbf{i} - 2\mathbf{j} + 3\mathbf{k}$, determine:

(a) the value of $\overline{OA}.\overline{OB}$

(b) the product $\overline{OA} \times \overline{OB}$ in terms of the unit vectors

(c) the cosine of the angle between $\overline{OA}$ and $\overline{OB}$

6 Find the cosine of the angle between the vectors $2\mathbf{i} + 3\mathbf{j} - \mathbf{k}$ and $3\mathbf{i} - 5\mathbf{j} + 2\mathbf{k}$.

7 Find the scalar product $\mathbf{a}.\mathbf{b}$ and the vector product $\mathbf{a} \times \mathbf{b}$, when

(a) $\mathbf{a} = \mathbf{i} + 2\mathbf{j} - \mathbf{k}$, $\mathbf{b} = 2\mathbf{i} + 3\mathbf{j} + \mathbf{k}$

(b) $\mathbf{a} = 2\mathbf{i} + 3\mathbf{j} + 4\mathbf{k}$, $\mathbf{b} = 5\mathbf{i} - 2\mathbf{j} + \mathbf{k}$

8 Find the unit vector perpendicular to each of the vectors $2\mathbf{i} - \mathbf{j} + \mathbf{k}$ and $3\mathbf{i} + 4\mathbf{j} - \mathbf{k}$, where **i**, **j**, **k** are the mutually perpendicular unit vectors. Calculate the sine of the angle between the two vectors.

9 If A is the point (1, −1, 2), B is the point (−1, 2, 2) and C is the point (4, 3, 0), find the direction cosines of $\overline{BA}$ and $\overline{BC}$, and hence show that the angle ABC = 69°14′.

10 If $\mathbf{a} = 3\mathbf{i} - \mathbf{j} + 2\mathbf{k}$, $\mathbf{b} = \mathbf{i} + 3\mathbf{j} - 2\mathbf{k}$, determine the magnitude and direction cosines of the product vector $(\mathbf{a} \times \mathbf{b})$ and show that it is perpendicular to a vector $\mathbf{c} = 9\mathbf{i} + 2\mathbf{j} + 2\mathbf{k}$.

11 **a** and **b** are vectors defined by $\mathbf{a} = 8\mathbf{i} + 2\mathbf{j} - 3\mathbf{k}$ and $\mathbf{b} = 3\mathbf{i} - 6\mathbf{j} + 4\mathbf{k}$, where **i**, **j**, **k** are mutually perpendicular unit vectors.

(a) Calculate $\mathbf{a}.\mathbf{b}$ and show that **a** and **b** are perpendicular to each other.

(b) Find the magnitude and the direction cosines of the product vector $\mathbf{a} \times \mathbf{b}$.

12 If the position vectors of P and Q are $\mathbf{i} + 3\mathbf{j} - 7\mathbf{k}$ and $5\mathbf{i} - 2\mathbf{j} + 4\mathbf{k}$ respectively, find $\overline{PQ}$ and determine its direction cosines.

13 If position vectors, $\overline{OA}$, $\overline{OB}$, $\overline{OC}$, are defined by $\overline{OA} = 2\mathbf{i} - \mathbf{j} + 3\mathbf{k}$, $\overline{OB} = 3\mathbf{i} + 2\mathbf{j} - 4\mathbf{k}$, $\overline{OC} = -\mathbf{i} + 3\mathbf{j} - 2\mathbf{k}$, determine:

(a) the vector $\overline{AB}$

(b) the vector $\overline{BC}$

(c) the vector product $\overline{AB} \times \overline{BC}$

(d) the unit vector perpendicular to the plane ABC.

**14** Given that $\mathbf{a} = 2\mathbf{i} - 3\mathbf{j} + \mathbf{k}$, $\mathbf{b} = 4\mathbf{i} + \mathbf{j} - 3\mathbf{k}$ and $\mathbf{c} = -\mathbf{i} + \mathbf{j} - 3\mathbf{k}$ find:

(a) $(\mathbf{a} \cdot \mathbf{b})\mathbf{c}$

(b) $\mathbf{a} \cdot (\mathbf{b} \times \mathbf{c})$ (called the scalar triple product)

(c) $\mathbf{a} \times (\mathbf{b} \times \mathbf{c})$ (called the vector triple product).

**15** Given that $\mathbf{a} = 5\mathbf{i} + 2\mathbf{j} - \mathbf{k}$ and $\mathbf{b} = \mathbf{i} - 3\mathbf{j} + \mathbf{k}$ find $(\mathbf{a} + \mathbf{b}) \times (\mathbf{a} - \mathbf{b})$.

**16** Prove that the area of the parallelogram with adjacent sides $\mathbf{a}$ and $\mathbf{b}$ is given as $|\mathbf{a} \times \mathbf{b}|$.

**17** Show that if $\mathbf{a}$, $\mathbf{b}$ and $\mathbf{c}$ lie in the same plane then $\mathbf{a} \cdot (\mathbf{b} \times \mathbf{c}) = 0$.

**18** Show that the volume $V$ of the parallelepiped with adjacent edges $\mathbf{a}$, $\mathbf{b}$ and $\mathbf{c}$ is given as the magnitude of the scalar triple product. That is: $V = |\mathbf{a} \cdot (\mathbf{b} \times \mathbf{c})| = |\mathbf{b} \cdot (\mathbf{c} \times \mathbf{a})| = |\mathbf{c} \cdot (\mathbf{a} \times \mathbf{b})|$.

**19** Given the triangle of vectors $\mathbf{a}$, $\mathbf{b}$ and $\mathbf{c}$ where $\mathbf{c} = \mathbf{b} - \mathbf{a}$ derive the cosine law for plane triangles $c^2 = a^2 + b^2 - 2ab\cos\theta$ where $\mathbf{a} \cdot \mathbf{b} = ab\cos\theta$.

**20** The point (1, 2, 4) lies in plane $P$ which is perpendicular to the vector $\mathbf{a} = 2\mathbf{i} - 3\mathbf{j} + \mathbf{k}$. Given $\mathbf{r} = x\mathbf{i} + y\mathbf{j} + z\mathbf{k}$ is the position vector of another point in the plane find the equation of the plane.

Now visit the companion website for this book at www.palgrave.com/stroud for more questions applying this mathematics to science and engineering.

**64**

# Differentiation

**Learning outcomes**

When you have completed this Programme you will be able to:

- ☐ Differentiate by using a list of standard derivatives
- ☐ Apply the chain rule
- ☐ Apply the product and quotient rules
- ☐ Perform logarithmic differentiation
- ☐ Differentiate implicit functions
- ☐ Differentiate parametric equations

# Standard derivatives

**1**

Here is a revision list of the standard derivatives which you have no doubt used many times before. Copy out the list into your record book and memorize those with which you are less familiar – possibly **4**, **6**, **10**, **11** and **12**. Here they are:

| | $y = f(x)$ | $\frac{dy}{dx}$ |
|---|---|---|
| 1 | $x^n$ | $nx^{n-1}$ |
| 2 | $e^x$ | $e^x$ |
| 3 | $e^{kx}$ | $ke^{kx}$ |
| 4 | $a^x$ | $a^x . \ln a$ |
| 5 | $\ln x$ | $\frac{1}{x}$ |
| 6 | $\log_a x$ | $\frac{1}{x . \ln a}$ |
| 7 | $\sin x$ | $\cos x$ |
| 8 | $\cos x$ | $-\sin x$ |
| 9 | $\tan x$ | $\sec^2 x$ |
| 10 | $\cot x$ | $-\text{cosec}^2 x$ |
| 11 | $\sec x$ | $\sec x . \tan x$ |
| 12 | $\text{cosec}\ x$ | $-\text{cosec}\ x . \cot x$ |
| 13 | $\sinh x$ | $\cosh x$ |
| 14 | $\cosh x$ | $\sinh x$ |

*The last two are proved in Frame 2, so move on*

**2**

The derivatives of $\sinh x$ and $\cosh x$ are easily obtained by remembering the exponential definitions, and also that:

$$\frac{d}{dx}\{e^x\} = e^x \text{ and } \frac{d}{dx}\{e^{-x}\} = -e^{-x}$$

(a) $y = \sinh x \qquad y = \frac{e^x - e^{-x}}{2}$

$$\therefore \frac{dy}{dx} = \frac{e^x - (-e^{-x})}{2} = \frac{e^x + e^{-x}}{2} = \cosh x$$

$$\therefore \frac{d}{dx}(\sinh x) = \cosh x$$

(b) $y = \cosh x \qquad y = \frac{e^x + e^{-x}}{2}$

$$\therefore \frac{dy}{dx} = \frac{e^x + (-e^{-x})}{2} = \frac{e^x - e^{-x}}{2} = \sinh x$$

$$\therefore \frac{d}{dx}(\cosh x) = \sinh x$$

Note that there is no minus sign involved as there is when differentiating the trig function $\cos x$.

We will find the derivative of $\tanh x$ later on.

*Move on to Frame 3*

**3**

Let us see if you really do know those basic derivatives. First of all cover up the list you have copied and then write down the derivatives of the following. All very easy.

| | | | | | |
|---|---|---|---|---|---|
| 1 | $x^5$ | 8 | $\cosh x$ | 15 | $\cot x$ |
| 2 | $\sin x$ | 9 | $\log_{10} x$ | 16 | $a^x$ |
| 3 | $e^{3x}$ | 10 | $e^x$ | 17 | $x^{-4}$ |
| 4 | $\ln x$ | 11 | $\cos x$ | 18 | $\log_a x$ |
| 5 | $\tan x$ | 12 | $\sinh x$ | 19 | $\sqrt{x}$ |
| 6 | $2^x$ | 13 | $\operatorname{cosec} x$ | 20 | $e^{x/2}$ |
| 7 | $\sec x$ | 14 | $a^3$ | | |

*When you have finished them all, move on to the next frame to check your results*

**4**

Here are the results. Check yours carefully and make a special note of any where you may have slipped up.

| | | | | | |
|---|---|---|---|---|---|
| 1 | $5x^4$ | 8 | $\sinh x$ | 15 | $-\operatorname{cosec}^2 x$ |
| 2 | $\cos x$ | 9 | $1/(x \ln 10)$ | 16 | $a^x \ln a$ |
| 3 | $3e^{3x}$ | 10 | $e^x$ | 17 | $-4x^{-5}$ |
| 4 | $1/x$ | 11 | $-\sin x$ | 18 | $1/(x \ln a)$ |
| 5 | $\sec^2 x$ | 12 | $\cosh x$ | 19 | $1/2x^{-\frac{1}{2}} = 1/(2\sqrt{x})$ |
| 6 | $2^x \ln 2$ | 13 | $-\operatorname{cosec} x . \cot x$ | 20 | $1/2e^{x/2}$ |
| 7 | $\sec x . \tan x$ | 14 | $0$ | | |

If by chance you have not got them all correct, it is well worthwhile returning to Frame 1, or to the list you copied, and brushing up where necessary. These are the tools for all that follows.

*When you are sure you know the basic results, move on*

## Functions of a function

**5**

$\sin x$ is a function of $x$ since the value of $\sin x$ depends on the value of the angle $x$. Similarly, $\sin(2x + 5)$ is a function of the angle $(2x + 5)$ since the value of the sine depends on the value of this angle.

i.e. $\sin(2x + 5)$ is a function of $(2x + 5)$

But $(2x + 5)$ is itself a function of $x$, since its value depends on $x$.

i.e. $(2x + 5)$ is a function of $x$

$\sin(2x + 5)$ is therefore a function of a function of $x$ and such expressions are referred to generally as *functions of a function*.

So $e^{\sin y}$ is a function of a function of ...........

6

$\boxed{y}$

Because $e^{\sin y}$ depends on the value of the index $\sin y$ and $\sin y$ depends on $y$. Therefore $e^{\sin y}$ is a function of a function of $y$.

We very often need to find the derivatives of such functions of a function. We could do them from first principles.

As a first example, differentiate with respect to $x$, $y = \cos(5x - 4)$.

Let $u = (5x - 4)$ $\therefore$ $y = \cos u$ $\therefore$ $\dfrac{dy}{du} = -\sin u = -\sin(5x - 4)$.

But this gives us $\dfrac{dy}{du}$, not $\dfrac{dy}{dx}$. To convert our result into the required derivative, we use $\dfrac{dy}{dx} = \dfrac{dy}{du} \cdot \dfrac{du}{dx}$, i.e. we multiply $\dfrac{dy}{du}$ (which we have) by $\dfrac{du}{dx}$ to obtain $\dfrac{dy}{dx}$ (which we want); $\dfrac{du}{dx}$ is found from the substitution $u = (5x - 4)$, i.e. $\dfrac{du}{dx} = 5$.

$$\therefore \frac{d}{dx}\{\cos(5x - 4)\} = -\sin(5x - 4) \times 5 = -5\sin(5x - 4)$$

So now find from first principles the derivative of $y = e^{\sin x}$. (As before, put $u = \sin x$.)

7

$$\boxed{\frac{d}{dx}\{e^{\sin x}\} = \cos x . e^{\sin x}}$$

Because $y = e^{\sin x}$, put $u = \sin x$ $\therefore$ $y = e^u$ $\therefore$ $\dfrac{dy}{du} = e^u$

But $\dfrac{dy}{dx} = \dfrac{dy}{du} \cdot \dfrac{du}{dx}$ and $\dfrac{du}{dx} = \cos x$

$$\therefore \frac{d}{dx}\{e^{\sin x}\} = e^{\sin x} . \cos x$$

This is quite general.

If $y = f(u)$ and $u = F(x)$, then $\dfrac{dy}{dx} = \dfrac{dy}{du} \cdot \dfrac{du}{dx}$, i.e. if $y = \ln F$, where $F$ is a function of $x$, then:

$$\frac{dy}{dx} = \frac{dy}{dF} \cdot \frac{dF}{dx} = \frac{1}{F} \cdot \frac{dF}{dx}$$

So, if $y = \ln \sin x$ $\quad \dfrac{dy}{dx} = \dfrac{1}{\sin x} . \cos x = \cot x$

It is of utmost importance not to forget this factor $\dfrac{dF}{dx}$, so beware!

8

Just two more examples:

(a) $y = \tan(5x - 4)$ $\qquad$ Basic standard form is $y = \tan x$, $\dfrac{dy}{dx} = \sec^2 x$

In this case $(5x - 4)$ replaces the single $x$

$$\therefore \frac{dy}{dx} = \sec^2(5x - 4) \times \text{the derivative of the function } (5x - 4)$$

$$= \sec^2(5x - 4) \times 5 = 5\sec^2(5x - 4)$$

▶

(b) $y = (4x - 3)^5$ Basic standard form is $y = x^5$, $\dfrac{dy}{dx} = 5x^4$

Here $(4x - 3)$ replaces the single $x$

$$\therefore \frac{dy}{dx} = 5(4x - 3)^4 \times \text{the derivative of the function } (4x - 3)$$

$$= 5(4x - 3)^4 \times 4 = 20(4x - 3)^4$$

So what about this one?

If $y = \cos(7x + 2)$, then $\dfrac{dy}{dx} = \ldots\ldots\ldots\ldots$

**9**

$$\boxed{\frac{dy}{dx} = -7\sin(7x + 2)}$$

Right, now you differentiate these:

1 $y = (4x - 5)^6$
2 $y = e^{3-x}$
3 $y = \sin 2x$
4 $y = \cos(x^2)$
5 $y = \ln(3 - 4\cos x)$

*The results are in Frame 10. Check to see that yours are correct*

**10**

1 $y = (4x - 5)^6$ $\quad \dfrac{dy}{dx} = 6(4x - 5)^5.4 = 24(4x - 5)^5$

2 $y = e^{3-x}$ $\quad \dfrac{dy}{dx} = e^{3-x}(-1) = -e^{3-x}$

3 $y = \sin 2x$ $\quad \dfrac{dy}{dx} = \cos 2x.2 = 2\cos 2x$

4 $y = \cos(x^2)$ $\quad \dfrac{dy}{dx} = -\sin(x^2).2x = -2x\sin(x^2)$

5 $y = \ln(3 - 4\cos x)$ $\quad \dfrac{dy}{dx} = \dfrac{1}{3 - 4\cos x}.(4\sin x) = \dfrac{4\sin x}{3 - 4\cos x}$

Now do these

6 $y = e^{\sin 2x}$
7 $y = \sin^2 x$
8 $y = \ln\cos 3x$
9 $y = \cos^3(3x)$
10 $y = \log_{10}(2x - 1)$

Take your time to do them.

*When you are satisfied with your results, check them against the results in Frame 11*

11

6 $y = e^{\sin 2x}$ $\quad \frac{dy}{dx} = e^{\sin 2x}.2\cos 2x = 2\cos 2x.e^{\sin 2x}$

7 $y = \sin^2 x$ $\quad \frac{dy}{dx} = 2\sin x \cos x = \sin 2x$

8 $y = \ln \cos 3x$ $\quad \frac{dy}{dx} = \frac{1}{\cos 3x}(-3\sin 3x) = -3\tan 3x$

9 $y = \cos^3(3x)$ $\quad \frac{dy}{dx} = 3\cos^2(3x).(-3\sin 3x) = -9\sin 3x \cos^2 3x$

10 $y = \log_{10}(2x - 1)$ $\quad \frac{dy}{dx} = \frac{1}{(2x-1)\ln 10}.2 = \frac{2}{(2x-1)\ln 10}$

*All correct? Now on with the Programme. Next frame*

12

## Products

Of course, we may need to differentiate functions which are products or quotients of two of the functions.

If $y = uv$, where $u$ and $v$ are functions, of $x$, then you already know that:

$$\frac{dy}{dx} = u\frac{dv}{dx} + v\frac{du}{dx}$$

e.g. If $y = x^3.\sin 3x$

$$\text{then} \quad \frac{dy}{dx} = x^3.3\cos 3x + 3x^2 \sin 3x$$
$$= 3x^2(x\cos 3x + \sin 3x)$$

Every one is done the same way. To differentiate a product:

(a) put down the first, differentiate the second; plus

(b) put down the second, differentiate the first.

So what is the derivative of $e^{2x}\ln 5x$?

13

$$\boxed{\frac{dy}{dx} = e^{2x}\left(\frac{1}{x} + 2\ln 5x\right)}$$

Because $y = e^{2x}\ln 5x$, i.e. $u = e^{2x}$, $v = \ln 5x$

$$\frac{dy}{dx} = e^{2x}\frac{1}{5x}.5 + 2e^{2x}\ln 5x$$
$$= e^{2x}\left(\frac{1}{x} + 2\ln 5x\right)$$

Now here is a short test for you to do. Find $\frac{dy}{dx}$ when:

1 $y = x^2 \tan x$

2 $y = e^{5x}(3x + 1)$

3 $y = x\cos 2x$

4 $y = x^3 \sin 5x$

5 $y = x^2 \ln \sinh x$

*When you have completed all five, move on to Frame 14*

**14**

1 $y = x^2 \tan x \quad \therefore \dfrac{dy}{dx} = x^2 \sec^2 x + 2x \tan x$

$$= x(x \sec^2 x + 2 \tan x)$$

2 $y = e^{5x}(3x+1) \quad \therefore \dfrac{dy}{dx} = e^{5x}.3 + 5e^{5x}(3x+1)$

$$= e^{5x}(3 + 15x + 5) = e^{5x}(8 + 15x)$$

3 $y = x \cos 2x \quad \therefore \dfrac{dy}{dx} = x(-2 \sin 2x) + 1. \cos 2x$

$$= \cos 2x - 2x \sin 2x$$

4 $y = x^3 \sin 5x \quad \therefore \dfrac{dy}{dx} = x^3.5 \cos 5x + 3x^2 \sin 5x$

$$= x^2(5x \cos 5x + 3 \sin 5x)$$

5 $y = x^2 \ln \sinh x \quad \therefore \dfrac{dy}{dx} = x^2 \dfrac{1}{\sinh x} \cosh x + 2x \ln \sinh x$

$$= x(x \coth x + 2 \ln \sinh x)$$

So much for the product. What about the quotient?

*Next frame*

**15**

## Quotients

In the case of the quotient, if $u$ and $v$ are functions of $x$, and $y = \dfrac{u}{v}$

then $\dfrac{dy}{dx} = \dfrac{v\dfrac{du}{dx} - u\dfrac{dv}{dx}}{v^2}$

Here are two examples:

If $y = \dfrac{\sin 3x}{x+1}$, $\dfrac{dy}{dx} = \dfrac{(x+1)3 \cos 3x - \sin 3x.1}{(x+1)^2}$

If $y = \dfrac{\ln x}{e^{2x}}$, $\dfrac{dy}{dx} = \dfrac{e^{2x}\dfrac{1}{x} - \ln x.2e^{2x}}{e^{4x}}$

$$= \frac{e^{2x}\left(\dfrac{1}{x} - 2 \ln x\right)}{e^{4x}}$$

$$= \frac{\dfrac{1}{x} - 2 \ln x}{e^{2x}}$$

If you can differentiate the separate functions, the rest is easy.

You do this one. If $y = \dfrac{\cos 2x}{x^2}$, $\dfrac{dy}{dx} = \dots\dots\dots\dots$

**16**

$$\frac{d}{dx}\left\{\frac{\cos 2x}{x^2}\right\} = \frac{-2(x\sin 2x + \cos 2x)}{x^3}$$

Because
$$\frac{d}{dx}\left\{\frac{\cos 2x}{x^2}\right\} = \frac{x^2(-2\sin 2x) - \cos 2x.2x}{x^4}$$
$$= \frac{-2x(x\sin 2x + \cos 2x)}{x^4}$$
$$= \frac{-2(x\sin 2x + \cos 2x)}{x^3}$$

So: for $y = uv$,
$$\frac{dy}{dx} = u\frac{dv}{dx} + v\frac{du}{dx} \quad \text{(a)}$$

for $y = \dfrac{u}{v}$
$$\frac{dy}{dx} = \frac{v\dfrac{du}{dx} - u\dfrac{dv}{dx}}{v^2} \quad \text{(b)}$$

Be sure that you remember these.

You can prove the derivative of $\tan x$ by the quotient method, for if $y = \tan x$, $y = \dfrac{\sin x}{\cos x}$.

Then by the quotient rule, $\dfrac{dy}{dx} = \ldots\ldots\ldots\ldots$ (work it through in detail)

**17**

$$\frac{dy}{dx} = \sec^2 x$$

Because $y = \dfrac{\sin x}{\cos x}$ $\therefore \dfrac{dy}{dx} = \dfrac{\cos x.\cos x + \sin x.\sin x}{\cos^2 x} = \dfrac{1}{\cos^2 x} = \sec^2 x$

In the same way we can obtain the derivative of $\tanh x$

$$y = \tanh x = \frac{\sinh x}{\cosh x} \quad \therefore \frac{dy}{dx} = \frac{\cosh x.\cosh x - \sinh x.\sinh x}{\cosh^2 x}$$
$$= \frac{\cosh^2 x - \sinh^2 x}{\cosh^2 x} = \frac{1}{\cosh^2 x} = \text{sech}^2 x \quad \therefore \frac{d}{dx}(\tanh x) = \text{sech}^2 x$$

Add this last result to your list of derivatives in your record book.

So what is the derivative of $\tanh(5x + 2)$?

**18**

$$\frac{d}{dx}\left\{\tanh(5x+2)\right\} = 5\text{sech}^2(5x+2)$$

Because we have:

If $\dfrac{d}{dx}\left\{\tanh x\right\} = \text{sech}^2 x$ then

$$\frac{d}{dx}\left\{\tanh(5x+2)\right\} = \text{sech}^2(5x+2) \times \text{derivative of } (5x+2)$$
$$= \text{sech}^2(5x+2) \times 5 = 5\text{sech}^2(5x+2)$$

*Fine. Now move on to Frame 19 for the next part of the Programme*

## Logarithmic differentiation

**19**

The rules for differentiating a product or a quotient that we have revised are used when there are just two-factor functions, i.e. $uv$ or $\dfrac{u}{v}$. Where there are more than two functions in any arrangement top or bottom, the derivative is best found by what is known as 'logarithmic differentiation'.

It all depends on the basic fact that $\dfrac{\mathrm{d}}{\mathrm{d}x}\left\{\ln x\right\} = \dfrac{1}{x}$ and that if $x$ is replaced by a function $F$ then $\dfrac{\mathrm{d}}{\mathrm{d}x}\left\{\ln F\right\} = \dfrac{1}{F}\cdot\dfrac{\mathrm{d}F}{\mathrm{d}x}$. Bearing that in mind, let us consider the case where $y = \dfrac{uv}{w}$, where $u$, $v$ and w - and also $y$ - are functions of $x$.

First take logs to the base $e$.

$$\ln y = \ln u + \ln v - \ln w$$

Now differentiate each side with respect to $x$, remembering that $u$, $v$, $w$ and $y$ are all functions of $x$. What do we get?

**20**

$$\boxed{\frac{1}{y}\cdot\frac{\mathrm{d}y}{\mathrm{d}x} = \frac{1}{u}\cdot\frac{\mathrm{d}u}{\mathrm{d}x} + \frac{1}{v}\cdot\frac{\mathrm{d}v}{\mathrm{d}x} - \frac{1}{w}\cdot\frac{\mathrm{d}w}{\mathrm{d}x}}$$

So to get $\dfrac{\mathrm{d}y}{\mathrm{d}x}$ by itself, we merely have to multiply across by $y$. Note that when we do this, we put the grand function that $y$ represents:

$$\frac{\mathrm{d}y}{\mathrm{d}x} = \frac{uv}{w}\left\{\frac{1}{u}\cdot\frac{\mathrm{d}u}{\mathrm{d}x} + \frac{1}{v}\cdot\frac{\mathrm{d}v}{\mathrm{d}x} - \frac{1}{w}\cdot\frac{\mathrm{d}w}{\mathrm{d}x}\right\}$$

This is not a formula to memorize, but a *method* of working, since the actual terms on the right-hand side will depend on the functions you start with.

Let us do an example to make it quite clear.

If $y = \dfrac{x^2 \sin x}{\cos 2x}$, find $\dfrac{\mathrm{d}y}{\mathrm{d}x}$

The first step in the process is ...........

**21**

$$\boxed{\text{To take logs of both sides}}$$

$$y = \frac{x^2 \sin x}{\cos 2x} \quad \therefore\ \ln y = \ln(x^2) + \ln(\sin x) - \ln(\cos 2x)$$

Now differentiate both sides with respect to $x$, remembering that $\dfrac{\mathrm{d}}{\mathrm{d}x}(\ln F) = \dfrac{1}{F}\cdot\dfrac{\mathrm{d}F}{\mathrm{d}x}$

$$\frac{1}{y}\cdot\frac{\mathrm{d}y}{\mathrm{d}x} = \frac{1}{x^2}.2x + \frac{1}{\sin x}.\cos x - \frac{1}{\cos 2x}.(-2\sin 2x)$$

$$= \frac{2}{x} + \cot x + 2\tan 2x$$

$$\therefore\ \frac{\mathrm{d}y}{\mathrm{d}x} = \frac{x^2 \sin x}{\cos 2x}\left\{\frac{2}{x} + \cot x + 2\tan 2x\right\}$$

This is a pretty complicated result, but the original function was also somewhat involved!

▶

Do this one on your own:

If $y = x^4 e^{3x} \tan x$, then $\dfrac{dy}{dx} = \ldots\ldots\ldots\ldots$

**22**

$$\frac{dy}{dx} = x^4 e^{3x} \tan x \left\{ \frac{4}{x} + 3 + \frac{\sec^2 x}{\tan x} \right\}$$

Here is the working. Follow it through.

$$y = x^4 e^{3x} \tan x \quad \therefore \ \ln y = \ln(x^4) + \ln(e^{3x}) + \ln(\tan x)$$

$$\frac{1}{y} \cdot \frac{dy}{dx} = \frac{1}{x^4}.4x^3 + \frac{1}{e^{3x}}.3e^{3x} + \frac{1}{\tan x}.\sec^2 x$$

$$= \frac{4}{x} + 3 + \frac{\sec^2 x}{\tan x}$$

$$\therefore \ \frac{dy}{dx} = x^4 e^{3x} \tan x \left\{ \frac{4}{x} + 3 + \frac{\sec^2 x}{\tan x} \right\}$$

There it is.

Always use the logarithmic differentiation method where there are more than two functions involved in a product or quotient (or both).

Here is just one more for you to do.

Find $\dfrac{dy}{dx}$, given that $y = \dfrac{e^{4x}}{x^3 \cosh 2x}$

**23**

$$\frac{dy}{dx} = \frac{e^{4x}}{x^3 \cosh 2x} \left\{ 4 - \frac{3}{x} - 2 \tanh 2x \right\}$$

Here is the working. Check yours.

$$y = \frac{e^{4x}}{x^3 \cosh 2x}$$

$$\therefore \ \ln y = \ln(e^{4x}) - \ln(x^3) - \ln(\cosh 2x)$$

$$\therefore \ \frac{1}{y} \frac{dy}{dx} = \frac{1}{e^{4x}}.4e^{4x} - \frac{1}{x^3}.3x^2 - \frac{1}{\cosh 2x}.2 \sinh 2x$$

$$= 4 - \frac{3}{x} - 2 \tanh 2x$$

$$\therefore \ \frac{dy}{dx} = \frac{e^{4x}}{x^3 \cosh 2x} \left\{ 4 - \frac{3}{x} - 2 \tanh 2x \right\}$$

Well now, before continuing with the rest of the Programme, here is a revision exercise on the work so far for you to deal with.

*Move on for details*

## Review exercise

**24**

Differentiate with respect to $x$:

1 (a) $\ln 4x$ (b) $\ln(\sin 3x)$

2 $e^{3x}\sin 4x$

3 $\dfrac{\sin 2x}{2x+5}$

4 $\dfrac{(3x+1)\cos 2x}{e^{2x}}$

5 $x^5\sin 2x\cos 4x$

*When you have finished them all (and not before) move on to Frame 25 to check your results*

**25**

1 (a) $y=\ln 4x \quad \therefore \dfrac{dy}{dx}=\dfrac{1}{4x}.4=\dfrac{1}{x}$

(b) $y=\ln\sin 3x \quad \therefore \dfrac{dy}{dx}=\dfrac{1}{\sin 3x}.3\cos 3x$

$$=3\cot 3x$$

2 $y=e^{3x}\sin 4x \quad \therefore \dfrac{dy}{dx}=e^{3x}4\cos 4x+3e^{3x}\sin 4x$

$$=e^{3x}(4\cos 4x+3\sin 4x)$$

3 $y=\dfrac{\sin 2x}{2x+5} \quad \therefore \dfrac{dy}{dx}=\dfrac{(2x+5)2\cos 2x-2\sin 2x}{(2x+5)^2}$

4 $y=\dfrac{(3x+1)\cos 2x}{e^{2x}}$

$$\therefore \ln y=\ln(3x+1)+\ln(\cos 2x)-\ln(e^{2x})$$

$$\therefore \frac{1}{y}\frac{dy}{dx}=\frac{1}{3x+1}.3+\frac{1}{\cos 2x}.(-2\sin 2x)-\frac{1}{e^{2x}}.2e^{2x}$$

$$=\frac{3}{3x+1}-2\tan 2x-2$$

$$\frac{dy}{dx}=\frac{(3x+1)\cos 2x}{e^{2x}}\left\{\frac{3}{3x+1}-2\tan 2x-2\right\}$$

5 $y=x^5\sin 2x\cos 4x$

$$\therefore \ln y=\ln(x^5)+\ln(\sin 2x)+\ln(\cos 4x)$$

$$\therefore \frac{1}{y}\frac{dy}{dx}=\frac{1}{x^5}.5x^4+\frac{2\cos 2x}{\sin 2x}+\frac{1}{\cos 4x}(-4\sin 4x)$$

$$=\frac{5}{x}+2\cot 2x-4\tan 4x$$

$$\frac{dy}{dx}=x^5\sin 2x\cos 4x\left\{\frac{5}{x}+2\cot 2x-4\tan 4x\right\}$$

*So far so good. Now on to the next part of the Programme in Frame 26*

## Implicit functions

**26**

If $y = x^2 - 4x + 2$, $y$ is completely defined in terms of $x$, and $y$ is called an *explicit function* of $x$.

When the relationship between $x$ and $y$ is more involved, it may not be possible (or desirable) to separate $y$ completely on the left-hand side, e.g. $xy + \sin y = 2$. In such a case as this, $y$ is called an *implicit function* of $x$, because a relationship of the form $y = f(x)$ is implied in the given equation.

It may still be necessary to determine the derivatives of $y$ with respect to $x$ and in fact this is not at all difficult. All we have to remember is that $y$ is a function of $x$, even if it is difficult to see what it is. In fact, this is really an extension of our 'function of a function' routine.

$x^2 + y^2 = 25$, as it stands is an example of an ........... function.

**27**

implicit

Once again, all we have to remember is that $y$ is a function of $x$. So, if $x^2 + y^2 = 25$, let us find $\frac{dy}{dx}$

If we differentiate as it stands with respect to $x$, we get

$$2x + 2y\frac{dy}{dx} = 0$$

Note that we differentiate $y^2$ as a function squared, giving 'twice times the function, times the derivative of the function'. The rest is easy.

$$2x + 2y\frac{dy}{dx} = 0$$

$$\therefore y\frac{dy}{dx} = -x \qquad \therefore \frac{dy}{dx} = -\frac{x}{y}$$

As you will have noticed, with an implicit function the derivative may contain (and usually does) both $x$ and ...........

**28**

$y$

Let us look at some examples.

If $x^2 + y^2 - 2x - 6y + 5 = 0$, find $\frac{dy}{dx}$ and $\frac{d^2y}{dx^2}$ at $x = 3$, $y = 2$.

Differentiate as it stands with respect to $x$.

$$2x + 2y\frac{dy}{dx} - 2 - 6\frac{dy}{dx} = 0$$

$$\therefore (2y - 6)\frac{dy}{dx} = 2 - 2x$$

$$\therefore \frac{dy}{dx} = \frac{2 - 2x}{2y - 6} = \frac{1 - x}{y - 3}$$

$$\therefore \text{at } (3, 2) \qquad \frac{dy}{dx} = \frac{1 - 3}{2 - 3} = \frac{-2}{-1} = 2$$

▶

Then $\quad \dfrac{d^2y}{dx^2} = \dfrac{d}{dx}\left\{\dfrac{1-x}{y-3}\right\} = \dfrac{(y-3)(-1)-(1-x)\dfrac{dy}{dx}}{(y-3)^2}$

$$= \frac{(3-y)-(1-x)\dfrac{dy}{dx}}{(y-3)^2}$$

at $(3, 2)$ $\quad \dfrac{d^2y}{dx^2} = \dfrac{(3-2)-(1-3)2}{(2-3)^2} = \dfrac{1-(-4)}{1} = 5$

$$\therefore \text{ At } (3, 2) \quad \frac{dy}{dx} = 2, \ \frac{d^2y}{dx^2} = 5$$

Now this one. If $x^2 + 2xy + 3y^2 = 4$, find $\dfrac{dy}{dx}$

Away you go, but beware of the product term. When you come to $2xy$ treat this as $(2x)(y)$.

---

**29**

Here is the working:

$$x^2 + 2xy + 3y^2 = 4$$

$$2x + 2x\frac{dy}{dx} + 2y + 6y\frac{dy}{dx} = 0$$

$$\therefore (2x + 6y)\frac{dy}{dx} = -(2x + 2y)$$

$$\therefore \frac{dy}{dx} = -\frac{(2x + 2y)}{(2x + 6y)} = -\frac{(x + y)}{(x + 3y)}$$

And now, just one more:

If $x^3 + y^3 + 3xy^2 = 8$, find $\dfrac{dy}{dx}$

*Turn to Frame 30 for the solution*

---

**30**

Solution in detail:

$$x^3 + y^3 + 3xy^2 = 8$$

$$3x^2 + 3y^2\frac{dy}{dx} + 3x.2y\frac{dy}{dx} + 3y^2 = 0$$

$$\therefore (y^2 + 2xy)\frac{dy}{dx} = -(x^2 + y^2)$$

$$\therefore \frac{dy}{dx} = -\frac{(x^2 + y^2)}{(y^2 + 2xy)}$$

That is really all there is to it. All examples are tackled the same way. The key to it is simply that '$y$ is a function of $x$' and then apply the 'function of a function' routine.

*Now on to the last section of this particular Programme, which starts in Frame 31*

## Parametric equations

31

In some cases, it is more convenient to represent a function by expressing $x$ and $y$ separately in terms of a third independent variable, e.g. $y = \cos 2t$, $x = \sin t$. In this case, any value we give to $t$ will produce a pair of values for $x$ and $y$, which could if necessary be plotted and provide one point of the curve of $y = f(x)$.

The third variable, e.g. $t$, is called a *parameter*, and the two expressions for $x$ and $y$ *parametric equations*. We may still need to find the derivatives of the function with respect to $x$, so how do we go about it?

Let us take the case already quoted above. The parametric equations of a function are given as $y = \cos 2t$, $x = \sin t$. We are required to find expressions for $\frac{dy}{dx}$ and $\frac{d^2y}{dx^2}$.

*Move to the next frame to see how we go about it*

32

$y = \cos 2t$, $x = \sin t$. Find $\frac{dy}{dx}$ and $\frac{d^2y}{dx^2}$

From $y = \cos 2t$, we can get $\frac{dy}{dt} = -2\sin 2t$

From $x = \sin t$, we can get $\frac{dx}{dt} = \cos t$

We can now use the fact that $\frac{dy}{dx} = \frac{dy}{dt} \cdot \frac{dt}{dx}$

so that
$$\frac{dy}{dx} = -2\sin 2t.\frac{1}{\cos t} \qquad \left(\text{noting that } \frac{dt}{dx} = \frac{1}{dx/dt}\right)$$
$$= -4\sin t\cos t.\frac{1}{\cos t}$$
$$\therefore \frac{dy}{dx} = -4\sin t$$

That was easy enough. Now how do we find the second derivative? We *cannot* get it by finding $\frac{d^2y}{dt^2}$ and $\frac{d^2x}{dt^2}$ from the parametric equations and joining them together as we did for the first derivative. So what do we do?

*On to the next frame and all will be revealed!*

33

To find the second derivative, we must go back to the very meaning of $\frac{d^2y}{dx^2}$

i.e. $\frac{d^2y}{dx^2} = \frac{d}{dx}\left(\frac{dy}{dx}\right) = \frac{d}{dx}\left(-4\sin t\right)$

But we cannot differentiate a function of $t$ directly with respect to $x$.

Therefore we say $\frac{d}{dx}\left(-4\sin t\right) = \frac{d}{dt}\left(-4\sin t\right).\frac{dt}{dx}$.

$$\therefore \frac{d^2y}{dx^2} = -4\cos t.\frac{1}{\cos t} = -4$$
$$\therefore \frac{d^2y}{dx^2} = -4$$

▶

Let us work through another one. What about this? The parametric equations of a function are given as:

$y = 3\sin\theta - \sin^3\theta,\ x = \cos^3\theta$

Find $\dfrac{dy}{dx}$ and $\dfrac{d^2y}{dx^2}$

*Move on to Frame 34*

**34**

$$y = 3\sin\theta - \sin^3\theta \quad \therefore \frac{dy}{d\theta} = 3\cos\theta - 3\sin^2\theta\cos\theta = 3\cos\theta(1 - \sin^2\theta)$$

$$x = \cos^3\theta \quad \therefore \frac{dx}{d\theta} = 3\cos^2\theta(-\sin\theta) = -3\cos^2\theta\sin\theta$$

$$\frac{dy}{dx} = \frac{dy}{d\theta}\cdot\frac{d\theta}{dx}$$

$$= 3\cos\theta(1 - \sin^2\theta).\frac{1}{-3\cos^2\theta\sin\theta} \qquad \left[\text{Remember: } \frac{d\theta}{dx} = \frac{1}{dx/d\theta}\right]$$

$$= \frac{3\cos^3\theta}{-3\cos^2\theta\sin\theta} \qquad \therefore \frac{dy}{dx} = -\cot\theta$$

Also $\dfrac{d^2y}{dx^2} = \dfrac{d}{dx}(-\cot\theta) = \dfrac{d}{d\theta}(-\cot\theta)\dfrac{d\theta}{dx}$

$$= -(-\text{cosec}^2\theta)\frac{1}{-3\cos^2\theta\sin\theta}$$

$$\therefore \frac{d^2y}{dx^2} = \frac{-1}{3\cos^2\theta\sin^3\theta}$$

Now here is one for you to do in just the same way.

If $x = \dfrac{2 - 3t}{1 + t},\ y = \dfrac{3 + 2t}{1 + t},$ find $\dfrac{dy}{dx}$

*When you have done it, move on to Frame 35*

**35**

$$\boxed{\frac{dy}{dx} = \frac{1}{5}}$$

Because

$$x = \frac{2 - 3t}{1 + t} \quad \therefore \frac{dx}{dt} = \frac{(1 + t)(-3) - (2 - 3t)}{(1 + t)^2}$$

$$y = \frac{3 + 2t}{1 + t} \quad \therefore \frac{dy}{dt} = \frac{(1 + t)(2) - (3 + 2t)}{(1 + t)^2}$$

$$\frac{dx}{dt} = \frac{-3 - 3t - 2 + 3t}{(1 + t)^2} = \frac{-5}{(1 + t)^2}$$

$$\frac{dy}{dt} = \frac{2 + 2t - 3 - 2t}{(1 + t)^2} = \frac{-1}{(1 + t)^2}$$

$$\frac{dy}{dx} = \frac{dy}{dt}\cdot\frac{dt}{dx} = \frac{-1}{(1 + t)^2}\cdot\frac{(1 + t)^2}{-5} = \frac{1}{5} \qquad \therefore \frac{dy}{dx} = \frac{1}{5}$$

And now here is one more for you to do to finish up this part of the work. It is done in just the same way as the others.

If $x = a(\cos\theta + \theta\sin\theta)$ and $y = a(\sin\theta - \theta\cos\theta)$

find $\dfrac{dy}{dx}$ and $\dfrac{d^2y}{dx^2}$

36

Here it is, set out like the previous examples.

$$x = a(\cos\theta + \theta\sin\theta)$$

$$\therefore \frac{dx}{d\theta} = a(-\sin\theta + \theta\cos\theta + \sin\theta) = a\theta\cos\theta$$

$$y = a(\sin\theta - \theta\cos\theta)$$

$$\therefore \frac{dy}{d\theta} = a(\cos\theta + \theta\sin\theta - \cos\theta) = a\theta\sin\theta$$

$$\frac{dy}{dx} = \frac{dy}{d\theta}.\frac{d\theta}{dx} = a\theta\sin\theta.\frac{1}{a\theta\cos\theta} = \tan\theta$$

$$\frac{d^2y}{dx^2} = \frac{d}{dx}(\tan\theta) = \frac{d}{d\theta}(\tan\theta).\frac{d\theta}{dx}$$

$$= \sec^2\theta.\frac{1}{a\theta\cos\theta}$$

$$\therefore \frac{d^2y}{dx^2} = \frac{1}{a\theta\cos^3\theta}$$

You have now reached the end of this Programme on differentiation, much of which has been useful revision of what you have done before. This brings you now to the **Can You?** checklist and **Test exercise**, so move on to them and work through them carefully.

*Next frame*

## Can You?

### Checklist 7

37

Check this list before and after you try the end of Programme test.

On a scale of 1 to 5 how confident are you that you can:

| | Frames |
|---|---|
| • Differentiate by using a list of standard derivatives?<br>*Yes* ☐ ☐ ☐ ☐ ☐ *No* | 1 to 4 |
| • Apply the chain rule?<br>*Yes* ☐ ☐ ☐ ☐ ☐ *No* | 5 to 11 |
| • Apply the product and quotient rules?<br>*Yes* ☐ ☐ ☐ ☐ ☐ *No* | 12 to 18 |
| • Perform logarithmic differentiation?<br>*Yes* ☐ ☐ ☐ ☐ ☐ *No* | 19 to 23 |
| • Differentiate implicit functions?<br>*Yes* ☐ ☐ ☐ ☐ ☐ *No* | 26 to 30 |
| • Differentiate parametric equations?<br>*Yes* ☐ ☐ ☐ ☐ ☐ *No* | 31 to 36 |

## Test exercise 7

**38**

Write out the solutions carefully. They are all quite straightforward.

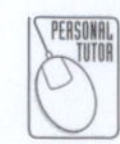

1 Differentiate the following with respect to $x$:

(a) $\tan 2x$ (b) $(5x+3)^6$ (c) $\cosh^2 x$

(d) $\log_{10}(x^2-3x-1)$ (e) $\ln\cos 3x$ (f) $\sin^3 4x$

(g) $e^{2x}\sin 3x$ (h) $\dfrac{x^4}{(x+1)^2}$ (i) $\dfrac{e^{4x}\sin x}{x\cos 2x}$

2 If $x^2+y^2-2x+2y=23$, find $\dfrac{dy}{dx}$ and $\dfrac{d^2y}{dx^2}$ at the point where $x=-2$, $y=3$.

3 Find an expression for $\dfrac{dy}{dx}$ when $x^3+y^3+4xy^2=5$.

4 If $x=3(1-\cos\theta)$ and $y=3(\theta-\sin\theta)$ find $\dfrac{dy}{dx}$ and $\dfrac{d^2y}{dx^2}$ in their simplest forms.

## Further problems 7

**39**

1 Differentiate with respect to $x$:

(a) $\ln\left\{\dfrac{\cos x+\sin x}{\cos x-\sin x}\right\}$ (b) $\ln(\sec x+\tan x)$

(c) $\sin^4 x\cos^3 x$

2 Find $\dfrac{dy}{dx}$ when:

(a) $y=\dfrac{x\sin x}{1+\cos x}$ (b) $y=\ln\left\{\dfrac{1-x^2}{1+x^2}\right\}$

3 If $y$ is a function of $x$, and $x=\dfrac{e^t}{e^t+1}$ show that $\dfrac{dy}{dt}=x(1-x)\dfrac{dy}{dx}$.

4 Find $\dfrac{dy}{dx}$ when $x^3+y^3-3xy^2=8$.

5 Differentiate:

(a) $y=e^{\sin^2 5x}$ (b) $y=\ln\left\{\dfrac{\cosh x-1}{\cosh x+1}\right\}$ (c) $y=\ln\left\{e^x\left(\dfrac{x-2}{x+2}\right)^{3/4}\right\}$

6 Differentiate:

(a) $y=x^2\cos^2 x$ (b) $y=\ln\left\{x^2\sqrt{1-x^2}\right\}$ (c) $y=\dfrac{e^{2x}\ln x}{(x-1)^3}$

7 If $(x-y)^3=A(x+y)$, prove that $(2x+y)\dfrac{dy}{dx}=x+2y$.

8 If $x^2-xy+y^2=7$, find $\dfrac{dy}{dx}$ and $\dfrac{d^2y}{dx^2}$ at $x=3$, $y=2$.

9 If $x^2+2xy+3y^2=1$, prove that $(x+3y)^3\dfrac{d^2y}{dx^2}+2(x^2+2xy+3y^2)=0$.

**10** If $x = \ln \tan \dfrac{\theta}{2}$ and $y = \tan\theta - \theta$, prove that

$$\frac{d^2y}{dx^2} = \tan^2\theta \sin\theta(\cos\theta + 2\sec\theta).$$

**11** If $y = 3e^{2x}\cos(2x - 3)$, verify that $\dfrac{d^2y}{dx^2} - 4\dfrac{dy}{dx} + 8y = 0$.

**12** The parametric equations of a curve are $x = \cos 2\theta$, $y = 1 + \sin 2\theta$. Find $\dfrac{dy}{dx}$ and $\dfrac{d^2y}{dx^2}$ at $\theta = \pi/6$. Find also the equation of the curve as a relationship between $x$ and $y$.

**13** If $y = \left\{x + \sqrt{1 + x^2}\right\}^{3/2}$, show that $4(1 + x^2)\dfrac{d^2y}{dx^2} + 4x\dfrac{dy}{dx} - 9y = 0$.

**14** Find $\dfrac{dy}{dx}$ and $\dfrac{d^2y}{dx^2}$ if $x = a\cos^3\theta$, $y = a\sin^3\theta$.

**15** If $x = 3\cos\theta - \cos^3\theta$, $y = 3\sin\theta - \sin^3\theta$, express $\dfrac{dy}{dx}$ and $\dfrac{d^2y}{dx^2}$ in terms of $\theta$.

**16** Show that $y = e^{-2mx}\sin 4mx$ is a solution of the equation

$$\frac{d^2y}{dx^2} + 4m\frac{dy}{dx} + 20m^2y = 0.$$

**17** If $y = \sec x$, prove that $y\dfrac{d^2y}{dx^2} = \left(\dfrac{dy}{dx}\right)^2 + y^4$.

**18** Prove that $x = Ae^{-kt}\sin pt$, satisfies the equation

$$\frac{d^2x}{dt^2} + 2k\frac{dx}{dt} + (p^2 + k^2)x = 0.$$

**19** If $y = e^{-kt}(A\cosh qt + B\sinh qt)$ where $A$, $B$, $q$ and $k$ are constants, show that $\dfrac{d^2y}{dt^2} + 2k\dfrac{dy}{dt} + (k^2 - q^2)y = 0$.

**20** If $\sinh y = \dfrac{4\sinh x - 3}{4 + 3\sinh x}$, show that $\dfrac{dy}{dx} = \dfrac{-5}{4 + 3\sinh x}$.

Now visit the companion website for this book at www.palgrave.com/stroud for more questions applying this mathematics to science and engineering.

**40**

# Differentiation applications

## Learning outcomes

When you have completed this Programme you will be able to:

- ☐ Differentiate the inverse trigonometric functions
- ☐ Differentiate the inverse hyperbolic functions
- ☐ Identify and locate a maximum and a minimum
- ☐ Identify and locate a point of inflexion

# Differentiation of inverse trigonometric functions

**1**

$\sin^{-1}x$, $\cos^{-1}x$, $\tan^{-1}x$ depend, of course, on the values assigned to $x$. They are therefore functions of $x$ and we may well be required to find their derivatives. So let us deal with them in turn.

(1) Let $y = \sin^{-1}x$. We have to find $\dfrac{dy}{dx}$

First of all, write this inverse statement as a direct statement:

$$y = \sin^{-1}x \qquad \therefore\ x = \sin y$$

Now we can differentiate this with respect to $y$ and obtain $\dfrac{dx}{dy}$

$$\frac{dx}{dy} = \cos y \qquad \therefore\ \frac{dy}{dx} = \ldots\ldots\ldots\ldots$$

**2**

$$\boxed{\frac{dy}{dx} = \frac{1}{\cos y}}$$

Now we express $\cos y$ in terms of $x$, thus:

We know that $\cos^2 y + \sin^2 y = 1$

$$\therefore\ \cos^2 y = 1 - \sin^2 y = 1 - x^2 \qquad (\text{since } x = \sin y)$$

$$\therefore\ \cos y = \sqrt{1 - x^2}$$

$$\therefore\ \frac{dy}{dx} = \frac{1}{\sqrt{1 - x^2}}$$

$$\frac{d}{dx}\left\{\sin^{-1}x\right\} = \frac{1}{\sqrt{1 - x^2}}$$

(2) Now you can determine $\dfrac{d}{dx}\{\cos^{-1}x\}$ in exactly the same way.

*Go through the same steps and finally check your result with that in Frame 3*

**3**

$$\boxed{\frac{d}{dx}\left\{\cos^{-1}x\right\} = \frac{-1}{\sqrt{1 - x^2}}}$$

Here is the working:

$$\text{Let}\quad y = \cos^{-1}x \qquad \therefore\ x = \cos y$$

$$\therefore\ \frac{dx}{dy} = -\sin y \qquad \therefore\ \frac{dy}{dx} = \frac{-1}{\sin y}$$

$$\cos^2 y + \sin^2 y = 1 \qquad \therefore\ \sin^2 y = 1 - \cos^2 y = 1 - x^2$$

$$\sin y = \sqrt{1 - x^2}$$

$$\therefore\ \frac{dy}{dx} = \frac{-1}{\sqrt{1 - x^2}} \qquad \therefore\ \frac{d}{dx}\left\{\cos^{-1}x\right\} = \frac{-1}{\sqrt{1 - x^2}}$$

So we have two very similar results:

$$\left.\begin{aligned} &(1)\quad \frac{d}{dx}\{\sin^{-1}x\} = \frac{1}{\sqrt{1 - x^2}} \\ &(2)\quad \frac{d}{dx}\{\cos^{-1}x\} = \frac{-1}{\sqrt{1 - x^2}} \end{aligned}\right\}\ \text{Different only in sign}$$

▶

(3) Now you find the derivative of $\tan^{-1} x$. The working is slightly different, but the general method the same. See what you get and then move to Frame 4 where the detailed working is set out.

**4**

$$\boxed{\frac{d}{dx}\left\{\tan^{-1} x\right\} = \frac{1}{1+x^2}}$$

Working: Let $y = \tan^{-1} x \quad \therefore x = \tan y$

$$\frac{dx}{dy} = \sec^2 y = 1 + \tan^2 y = 1 + x^2$$

$$\frac{dx}{dy} = 1 + x^2 \quad \therefore \frac{dy}{dx} = \frac{1}{1+x^2}$$

$$\frac{d}{dx}\left\{\tan^{-1} x\right\} = \frac{1}{1+x^2}$$

Let us collect these three results together. Here they are:

$$\frac{d}{dx}\left\{\sin^{-1} x\right\} = \frac{1}{\sqrt{1-x^2}} \qquad (1)$$

$$\frac{d}{dx}\left\{\cos^{-1} x\right\} = \frac{-1}{\sqrt{1-x^2}} \qquad (2)$$

$$\frac{d}{dx}\left\{\tan^{-1} x\right\} = \frac{1}{1+x^2} \qquad (3)$$

Copy these results into your record book. You will need to remember them.

*On to the next frame*

**5**

Of course, these derivatives can occur in all the usual combinations, e.g. products, quotients, etc.

**Example 1**

Find $\frac{dy}{dx}$, given that $y = (1 - x^2)\sin^{-1} x$

Here we have a product

$$\therefore \frac{dy}{dx} = (1 - x^2)\frac{1}{\sqrt{1-x^2}} + \sin^{-1} x.(-2x)$$

$$= \sqrt{1-x^2} - 2x.\sin^{-1} x$$

**Example 2**

If $y = \tan^{-1}(2x - 1)$, find $\frac{dy}{dx}$

This time, it is a function of a function

$$\frac{dy}{dx} = \frac{1}{1 + (2x-1)^2}.2 = \frac{2}{1 + 4x^2 - 4x + 1}$$

$$= \frac{2}{2 + 4x^2 - 4x} = \frac{1}{2x^2 - 2x + 1}$$

and so on.

There you are. Now here is a short exercise. Do all the questions.

##  Review exercise

6

Differentiate with respect to $x$:

1 $y = \sin^{-1} 5x$

2 $y = \cos^{-1} 3x$

3 $y = \tan^{-1} 2x$

4 $y = \sin^{-1}(x^2)$

5 $y = x^2.\sin^{-1}\left(\frac{x}{2}\right)$

*When you have finished them all, check your results with those in Frame 7*

7

1 $y = \sin^{-1} 5x \quad \therefore \frac{dy}{dx} = \frac{1}{\sqrt{1-(5x)^2}}.5 = \frac{5}{\sqrt{1-25x^2}}$

2 $y = \cos^{-1} 3x \quad \therefore \frac{dy}{dx} = \frac{-1}{\sqrt{1-(3x)^2}}.3 = \frac{-3}{\sqrt{1-9x^2}}$

3 $y = \tan^{-1} 2x \quad \therefore \frac{dy}{dx} = \frac{1}{1+(2x)^2}.2 = \frac{2}{1+4x^2}$

4 $y = \sin^{-1}(x^2) \quad \therefore \frac{dy}{dx} = \frac{1}{\sqrt{1-(x^2)^2}}.2x = \frac{2x}{\sqrt{1-x^4}}$

5 $y = x^2.\sin^{-1}\left(\frac{x}{2}\right) \quad \therefore \frac{dy}{dx} = x^2\frac{1}{\sqrt{\left\{1-\left(\frac{x}{2}\right)^2\right\}}}.\frac{1}{2} + 2x.\sin^{-1}\left(\frac{x}{2}\right)$

$$= \frac{x^2}{2\sqrt{\left\{1-\frac{x^2}{4}\right\}}} + 2x.\sin\left(\frac{x}{2}\right)$$

$$= \frac{x^2}{\sqrt{4-x^2}} + 2x.\sin^{-1}\left(\frac{x}{2}\right)$$

*Right, now on to the next frame*

## Derivatives of inverse hyperbolic functions

8

In just the same way that we have inverse trig functions, so we have inverse hyperbolic functions and we would not be unduly surprised if their derivatives bore some resemblance to those of the inverse trig functions.

Anyway, let us see what we get. The method is very much as before.

(4) $y = \sinh^{-1} x$ To find $\dfrac{dy}{dx}$

First express the inverse statement as a direct statement:

$$y = \sinh^{-1} x \quad \therefore x = \sinh y \quad \therefore \frac{dx}{dy} = \cosh y \quad \therefore \frac{dy}{dx} = \frac{1}{\cosh y}$$

We now need to express $\cosh y$ in terms of $x$.

We know that $\cosh^2 y - \sinh^2 y = 1 \quad \therefore \cosh^2 y = \sinh^2 y + 1 = x^2 + 1$

$$\cosh y = \sqrt{x^2 + 1}$$

$$\frac{dy}{dx} = \frac{1}{\sqrt{x^2 + 1}} \quad \therefore \frac{d}{dx}\left\{\sinh^{-1} x\right\} = \frac{1}{\sqrt{x^2 + 1}}$$

Let us obtain similar results for $\cosh^{-1} x$ and $\tanh^{-1} x$ and then we will take a look at them.

*So on to the next frame*

**9**

We have just established $\dfrac{d}{dx}\left\{\sinh^{-1} x\right\} = \dfrac{1}{\sqrt{x^2 + 1}}$

(5) $y = \cosh^{-1} x \quad \therefore x = \cosh y$

$$\therefore \frac{dx}{dy} = \sinh y \quad \therefore \frac{dy}{dx} = \frac{1}{\sinh y}$$

Now $\cosh^2 y - \sinh^2 y = 1 \quad \therefore \sinh^2 y = \cosh^2 y - 1 = x^2 - 1$

$$\therefore \sinh y = \sqrt{x^2 - 1}$$

$$\therefore \frac{dy}{dx} = \frac{1}{\sqrt{x^2 - 1}} \quad \therefore \frac{d}{dx}\left\{\cosh^{-1} x\right\} = \frac{1}{\sqrt{x^2 - 1}}$$

Now you can deal with the remaining one.

(6) If $y = \tanh^{-1} x$, $\dfrac{dy}{dx} = \dots\dots\dots\dots$

Tackle it in much the same way as we did for $\tan^{-1} x$, remembering this time, however, that $\text{sech}^2 x = 1 - \tanh^2 x$. You will find that useful.

*When you have finished, move to Frame 10*

**10**

$$\boxed{\frac{dy}{dx} = \frac{1}{1 - x^2}}$$

Because

$$y = \tanh^{-1} x \quad \therefore x = \tanh y$$

$$\therefore \frac{dx}{dy} = \text{sech}^2 y = 1 - \tanh^2 y = 1 - x^2 \quad \therefore \frac{dy}{dx} = \frac{1}{1 - x^2}$$

$$\frac{d}{dx}\left\{\tanh^{-1} x\right\} = \frac{1}{1 - x^2}$$

▶

Now here are the results, all together, so that we can compare them:

$$\frac{d}{dx}\left\{\sinh^{-1} x\right\} = \frac{1}{\sqrt{x^2+1}} \qquad (4)$$

$$\frac{d}{dx}\left\{\cosh^{-1} x\right\} = \frac{1}{\sqrt{x^2-1}} \qquad (5)$$

$$\frac{d}{dx}\left\{\tanh^{-1} x\right\} = \frac{1}{1-x^2} \qquad (6)$$

Make a note of these in your record book. You will need to remember these results.

*Now on to Frame 11*

## 11

Here are some examples, using the previous results

**Example 1**

$y = \cosh^{-1}\left\{3-2x\right\}$

$$\therefore \frac{dy}{dx} = \frac{1}{\sqrt{(3-2x)^2-1}}.(-2) = \frac{-2}{\sqrt{9-12x+4x^2-1}}$$

$$= \frac{-2}{\sqrt{8-12x+4x^2}} = \frac{-2}{2\sqrt{x^2-3x+2}} = \frac{-1}{\sqrt{x^2-3x+2}}$$

**Example 2**

$y = \tanh^{-1}\left(\frac{3x}{4}\right)$

$$\therefore \frac{dy}{dx} = \frac{1}{1-\left(\frac{3x}{4}\right)^2}\cdot\frac{3}{4} = \frac{1}{1-\frac{9x^2}{16}}\cdot\frac{3}{4}$$

$$= \frac{16}{16-9x^2}\cdot\frac{3}{4} = \frac{12}{16-9x^2}$$

**Example 3**

$y = \sinh^{-1}\{\tan x\}$

$$\therefore \frac{dy}{dx} = \frac{1}{\sqrt{\tan^2 x+1}}.\sec^2 x = \frac{\sec^2 x}{\sqrt{\sec^2 x}} = \sec x$$

## 12

Now here are a few exercises for you to do.

Differentiate:

1. $y = \sinh^{-1} 3x$
2. $y = \cosh^{-1}\left(\frac{5x}{2}\right)$
3. $y = \tanh^{-1}(\tan x)$
4. $y = \sinh^{-1}\left\{\sqrt{x^2-1}\right\}$
5. $y = \cosh^{-1}(e^{2x})$

*Finish them all. Then move on to Frame 13 for the results*

**13**

1 $y = \sinh^{-1} 3x \quad \therefore \dfrac{dy}{dx} = \dfrac{1}{\sqrt{(3x)^2 + 1}}.3 = \dfrac{3}{\sqrt{9x^2 + 1}}$

2 $y = \cosh^{-1}\left(\dfrac{5x}{2}\right) \quad \therefore \dfrac{dy}{dx} = \dfrac{1}{\sqrt{\left(\dfrac{5x}{2}\right)^2 - 1}}.\dfrac{5}{2} = \dfrac{5}{2\sqrt{\dfrac{25x^2}{1} - 4}}$

$$= \frac{5}{2\sqrt{\dfrac{25x^2 - 4}{4}}} = \frac{5}{\sqrt{25x^2 - 4}}$$

3 $y = \tanh^{-1}(\tan x) \quad \therefore \dfrac{dy}{dx} = \dfrac{1}{1 - \tan^2 x}.\sec^2 x = \dfrac{\sec^2 x}{1 - \tan^2 x}$

4 $y = \sinh^{-1}\left\{\sqrt{x^2 - 1}\right\}$

$$\therefore \frac{dy}{dx} = \frac{1}{\sqrt{x^2 - 1 + 1}}.\frac{1}{2}(x^2 - 1)^{-\frac{1}{2}}(2x) = \frac{1}{\sqrt{x^2 - 1}}$$

5 $y = \cosh^{-1}(e^{2x}) \quad \therefore \dfrac{dy}{dx} = \dfrac{1}{\sqrt{(e^{2x})^2 - 1}}.2e^{2x} = \dfrac{2e^{2x}}{\sqrt{e^{4x} - 1}}$

All correct?

*On then to Frame 14*

**14**

Before we leave these inverse trig and hyperbolic functions, let us look at them all together.

| *Inverse trig functions* | | *Inverse hyperbolic functions* | |
|---|---|---|---|
| $y$ | $\dfrac{dy}{dx}$ | $y$ | $\dfrac{dy}{dx}$ |
| $\sin^{-1} x$ | $\dfrac{1}{\sqrt{1 - x^2}}$ | $\sinh^{-1} x$ | $\dfrac{1}{\sqrt{x^2 + 1}}$ |
| $\cos^{-1} x$ | $\dfrac{-1}{\sqrt{1 - x^2}}$ | $\cosh^{-1} x$ | $\dfrac{1}{\sqrt{x^2 - 1}}$ |
| $\tan^{-1} x$ | $\dfrac{1}{1 + x^2}$ | $\tanh^{-1} x$ | $\dfrac{1}{1 - x^2}$ |

It would be a good idea to copy down this combined table, so that you compare and use the results. Do that: it will help you to remember them and to distinguish clearly between them.

**15**

Before you do a revision exercise, cover up the table you have just copied and see if you can complete the following correctly:

1 If $y = \sin^{-1} x$ $\quad \dfrac{dy}{dx} = \ldots\ldots\ldots\ldots$

2 If $y = \cos^{-1} x$ $\quad \dfrac{dy}{dx} = \ldots\ldots\ldots\ldots$

3 If $y = \tan^{-1} x$ $\quad \dfrac{dy}{dx} = \ldots\ldots\ldots\ldots$

4 If $y = \sinh^{-1} x$ $\quad \dfrac{dy}{dx} = \ldots\ldots\ldots\ldots$

5 If $y = \cosh^{-1} x$ $\quad \dfrac{dy}{dx} = \ldots\ldots\ldots\ldots$

6 If $y = \tanh^{-1} x$ $\quad \dfrac{dy}{dx} = \ldots\ldots\ldots\ldots$

Now check your results with your table and make a special point of brushing up any of which you are not really sure.

## Review exercise

**16**

Differentiate the following with respect to $x$:

1 $\tan^{-1}(\sinh x)$

2 $\sinh^{-1}(\tan x)$

3 $\cosh^{-1}(\sec x)$

4 $\tanh^{-1}(\sin x)$

5 $\sin^{-1}\left(\dfrac{x}{a}\right)$

Take care with these; we have mixed them up to some extent.

*When you have finished them all – and you are sure you have done what was required – check your results with those in Frame 17*

**17**

1 $y = \tan^{-1}(\sinh x)$ $\quad \dfrac{d}{dx}\left\{\tan^{-1} x\right\} = \dfrac{1}{1+x^2}$

$$\therefore \frac{dy}{dx} = \frac{1}{1+\sinh^2 x}.\cosh x = \frac{\cosh x}{\cosh^2 x} = \text{sech } x$$

2 $y = \sinh^{-1}(\tan x)$ $\quad \dfrac{d}{dx}\left\{\sinh^{-1} x\right\} = \dfrac{1}{\sqrt{x^2+1}}$

$$\therefore \frac{dy}{dx} = \frac{1}{\sqrt{\tan^2 x + 1}}.\sec^2 x = \frac{\sec^2 x}{\sqrt{\sec^2 x}} = \sec x$$

3 $y = \cosh^{-1}(\sec x)$ $\quad \dfrac{d}{dx}\left\{\cosh^{-1} x\right\} = \dfrac{1}{\sqrt{x^2-1}}$

$$\therefore \frac{dy}{dx} = \frac{1}{\sqrt{\sec^2 x - 1}}.\sec x.\tan x = \frac{\sec x.\tan x}{\sqrt{\tan^2 x}} = \sec x$$

▶

4 $y = \tanh^{-1}(\sin x)$  $\dfrac{d}{dx}\left\{\tanh^{-1} x\right\} = \dfrac{1}{1-x^2}$

$$\therefore \frac{dy}{dx} = \frac{1}{1-\sin^2 x}.\cos x = \frac{\cos x}{\cos^2 x} = \sec x$$

5 $y = \sin^{-1}\left\{\dfrac{x}{a}\right\}$  $\dfrac{d}{dx}\left\{\sin^{-1} x\right\} = \dfrac{1}{\sqrt{1-x^2}}$

$$\therefore \frac{dy}{dx} = \frac{1}{\sqrt{1-\left(\frac{x}{a}\right)^2}}.\frac{1}{a} = \frac{1}{a}.\frac{1}{\sqrt{1-\frac{x^2}{a^2}}}$$

$$= \frac{1}{a}.\frac{1}{\sqrt{\frac{a^2-x^2}{a^2}}} = \frac{1}{\sqrt{a^2-x^2}}$$

If you have got those all correct, or nearly all correct, you now know quite a lot about the derivatives of inverse trig and hyperbolic functions.

*You are now ready to move on to the next topic of this Programme, so off you go to Frame 18*

## Maximum and minimum values

**18**

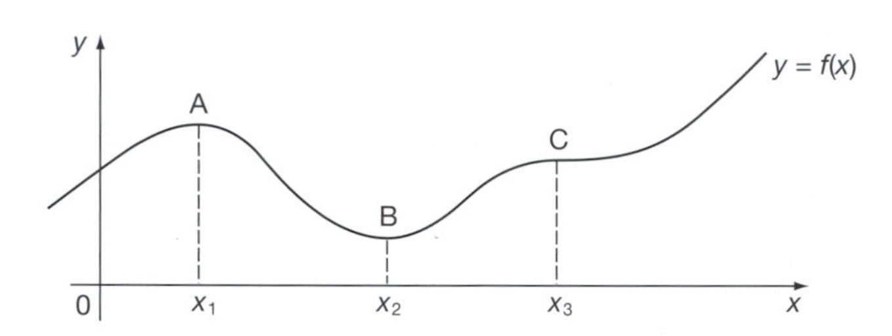


You are already familiar with the basic techniques for finding maximum and minimum values of a function. You have done this kind of operation many times in the past, but just to refresh your memory, let us consider some function, $y = f(x)$ whose graph is shown above.

At the point A, i.e. at $x = x_1$, a maximum value of $y$ occurs since at A, the $y$-value is greater than the $y$-values on either side of it and close to it.

Similarly, at B, $y$ is a ..........., since the $y$-value at the point B is less than the $y$-values on either side of it and close to it.

19

At B, $y$ is a *minimum* value

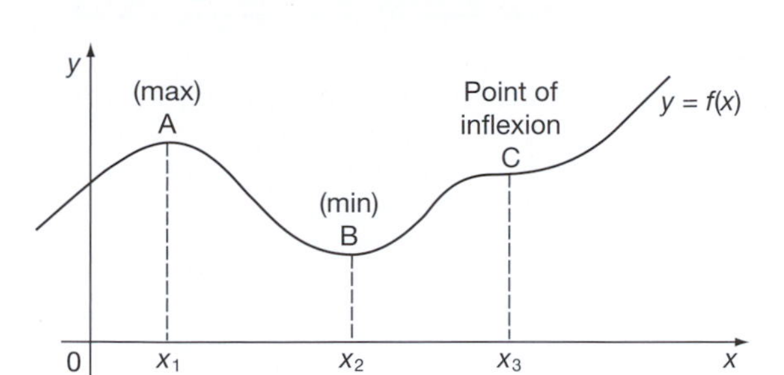


The point C is worth a second consideration. It looks like 'half a max and half a min'. The curve flattens out at C, but instead of dipping down, it then goes on with an increasingly positive gradient. Such a point is an example of a *point of inflexion*, i.e. it is essentially a form of S-bend.

Points A, B and C are called *stationary points* on the graph, or *stationary values of y*, and while you know how to find the positions of A and B, you may know considerably less about points of inflexion. We shall be taking a special look at these.

*On to Frame 20*

If we consider the gradient of the graph as we travel left to right, we can draw a graph to show how this gradient varies. We have no actual values for the gradient, but we can see whether it is positive or negative, more or less steep. The graph we obtain is the first derived curve of the function and we are really plotting the values of $\frac{dy}{dx}$ against values of $x$.

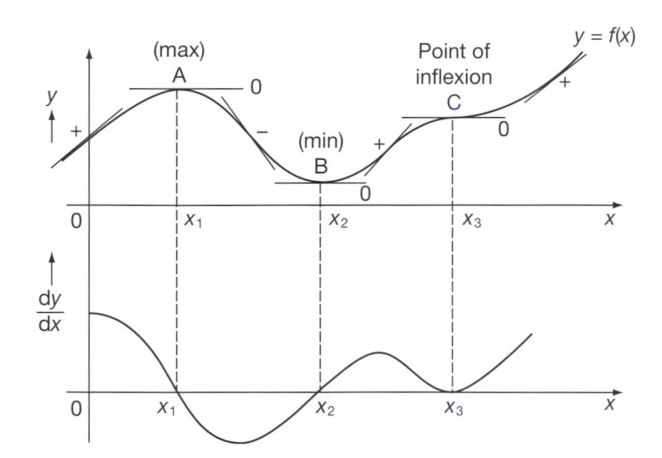


We see that at $x = x_1, x_2, x_3$ (corresponding to our three stationary points), the graph of $\frac{dy}{dx}$ is at the $x$-axis and at no other points.

Therefore, we obtain the first rule, which is that for stationary points,

$$\frac{dy}{dx} = \ldots\ldots\ldots\ldots$$

*Move on to Frame 21*

**21**

For stationary points, A, B, C: $\dfrac{dy}{dx} = 0$

If we now trace the gradient of the *first derived curve* and plot this against $x$, we obtain the *second derived curve*, which shows values of $\dfrac{d^2y}{dx^2}$ against $x$.

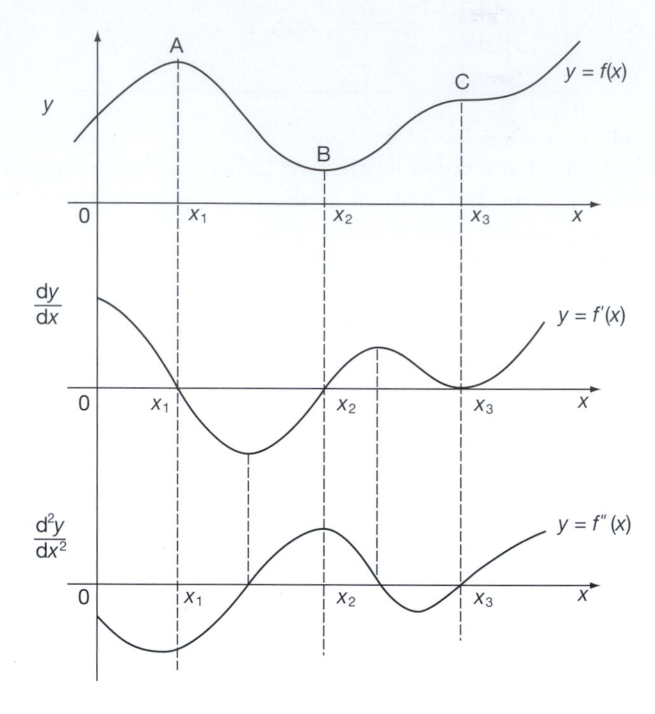


From the first derived curve, we see that for stationary points:

$$\frac{dy}{dx} = 0$$

For the second derived curve, we see that:

| | |
|---|---|
| for maximum $y$, | $\dfrac{d^2y}{dx^2}$ is negative |
| for minimum $y$, | $\dfrac{d^2y}{dx^2}$ is positive |
| for point of inflexion | $\dfrac{d^2y}{dx^2}$ is zero |

Copy the diagram into your record book. It summarizes all the facts on max and min values so far.

**22**

From the results we have just established, we can now determine:

(a) the values of $x$ at which stationary points occur, by differentiating the function and then solving the equation $\frac{dy}{dx} = 0$

(b) the corresponding values of $y$ at these points by merely substituting the $x$-values found, in $y = f(x)$

(c) the type of each stationary point (max, min, or point of inflexion) by testing in the expression for $\frac{d^2y}{dx^2}$

With this information, we can go a long way towards drawing a sketch of the curve. So let us apply these results to a straightforward example in the next frame.

**23**

Find the stationary points on the graph of the function $y = \frac{x^3}{3} - \frac{x^2}{2} - 2x + 5$.

Distinguish between them and sketch the graph of the function.

There are, of course, two stages:

(a) Stationary points are given by $\frac{dy}{dx} = 0$

(b) The type of each stationary point is determined by substituting the roots of the equation $\frac{dy}{dx} = 0$ in the expression for $\frac{d^2y}{dx^2}$

If $\frac{d^2y}{dx^2}$ is negative, then $y$ is a maximum.

If $\frac{d^2y}{dx^2}$ is positive, then $y$ is a minimum.

If $\frac{d^2y}{dx^2}$ is zero, then $y$ may be a point of inflexion.

We shall need both the first and second derivatives, so make sure you are ready.

If $y = \frac{x^3}{3} - \frac{x^2}{2} - 2x + 5$, then $\frac{dy}{dx} = \ldots\ldots\ldots\ldots$ and $\frac{d^2y}{dx^2} = \ldots\ldots\ldots\ldots$

**24**

$$\frac{dy}{dx} = x^2 - x - 2; \qquad \frac{d^2y}{dx^2} = 2x - 1$$

(a) Stationary points occur at $\frac{dy}{dx} = 0$

$\therefore x^2 - x - 2 = 0 \qquad \therefore (x - 2)(x + 1) = 0 \qquad \therefore x = 2$ and $x = -1$

i.e. stationary points occur at $x = 2$ and $x = -1$.

▶

(b) To determine the type of each stationary point, substitute $x = 2$ and then $x = -1$ in the expression for $\dfrac{d^2y}{dx^2}$

At $x = 2$, $\dfrac{d^2y}{dx^2} = 4 - 1 = 3$, i.e. positive $\quad \therefore x = 2$ gives $y_{min}$.

At $x = -1$, $\dfrac{d^2y}{dx^2} = -2 - 1$, i.e. negative $\quad \therefore x = -1$ gives $y_{max}$

Substituting in $y = f(x)$ gives $x = 2$, $y_{min} = 1\frac{2}{3}$ and $x = -1$, $y_{max} = 6\frac{1}{6}$.

Also, we can see at a glance from the function, that when $x = 0$, $y = 5$.

*You can now sketch the graph of the function. Do it*

---

**25**

We know that:

(a) at $x = -1$, $y_{max} = 6\frac{1}{6}$

(b) at $x = 2$, $y_{min} = 1\frac{2}{3}$

(c) at $x = 0$, $y = 5$

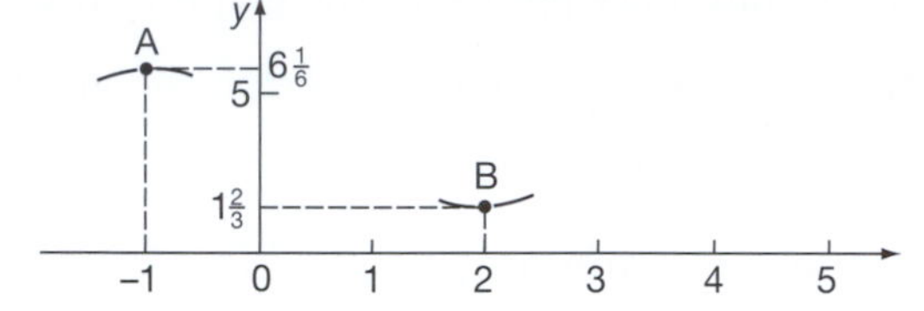


Joining up with a smooth curve gives:

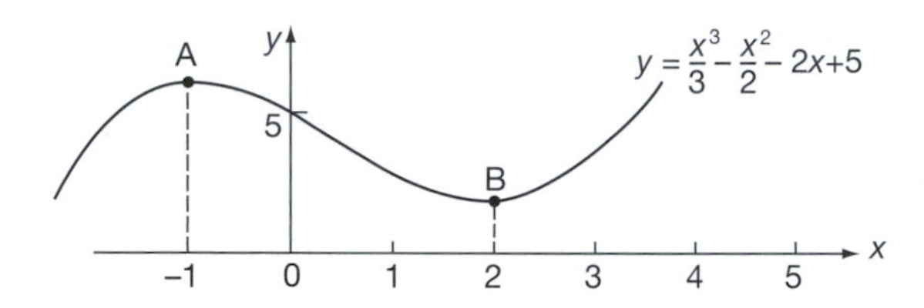


There is no point of inflexion, like the point C, on this particular graph.

All that was just by way of refreshing your memory on work you have done before. Now let us take a wider look at points of inflexion.

*Move on*

---

## Points of inflexion

**26**

The point C that we considered on our first diagram was rather a special kind of point of inflexion. In general, it is not necessary for the curve at a point of inflexion to have zero gradient.

A *point of inflexion* (P-of-I) is defined simply as a point on a curve at which the *direction of bending* changes, i.e. from a right-hand bend (R.H.) to a left-hand bend (L.H.), or from a left-hand bend to a right-hand bend.

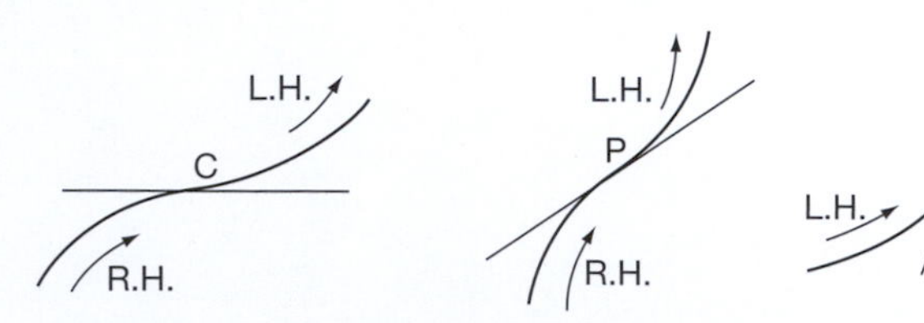

The point C we considered is, of course, a P-of-I, but it is not *essential* at a P-of-I for the gradient to be zero. Points P and Q are perfectly good points of inflexion and in fact in these cases the gradient is:

$$\left\{\begin{matrix} \text{positive} \\ \text{negative} \\ \text{zero} \end{matrix}\right\} \quad \text{Which?}$$

---

**27**

At the points of inflexion, P and Q, the gradient is in fact

positive

Correct. The gradient can of course be positive, negative or zero in any one case, but there is no restriction on its sign.

A point of inflexion, then, is simply a point on a curve at which there is a change in the ........... of ...........

---

**28**

Point of inflexion: a point at which there is a change in the

direction of bending

If the gradient at a P-of-I is not zero, it will not appear in our usual max and min routine, for $\frac{dy}{dx}$ will not be zero. How, then, are we going to find where such points of inflexion occur?

Let us sketch the graphs of the gradients as we did before:

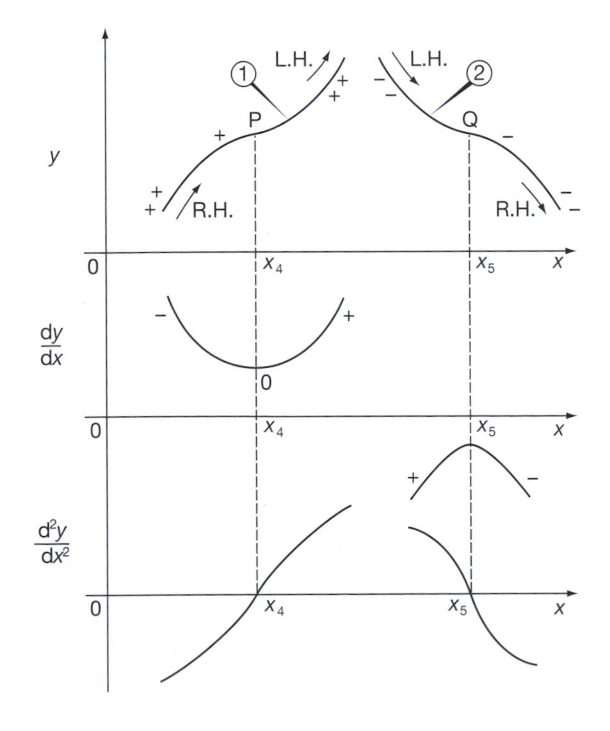

P and Q are points of inflexion.

In curve 1, the gradient is always positive, ++ indicating a greater positive gradient than +.

Similarly in curve 2, the gradient is always negative.

In curve 1, $\frac{dy}{dx}$ reaches a minimum value but not zero.

In curve 2, $\frac{dy}{dx}$ reaches a maximum value but not zero.

For both points of inflexion, i.e. at $x = x_4$ and $x = x_5$, $\frac{d^2y}{dx^2} = 0$

We see that where points of inflexion occur, $\frac{d^2y}{dx^2} = 0$

So, is this the clue we have been seeking? If so, it simply means that to find the points of inflexion we differentiate the function of the curve twice and solve the equation $\frac{d^2y}{dx^2} = 0$

*That sounds easy enough! But move on to the next frame to see what is involved*

**29**

We have just found that

$$\text{where points of inflexion occur, } \frac{d^2y}{dx^2} = 0$$

This is perfectly true. Unfortunately, this is not the whole of the story, for it is also possible for $\frac{d^2y}{dx^2}$ to be zero at points other than points of inflexion!

So if we solve $\frac{d^2y}{dx^2} = 0$, we cannot as yet be sure whether the solution $x = a$ gives a point of inflexion or not. How can we decide?

Let us consider just one more set of graphs. This should clear the matter up.

30

Let S be a true point of inflexion and T a point on $y = f(x)$ as shown. Clearly, T is not a point of inflexion.

The first derived curves could well look like this.

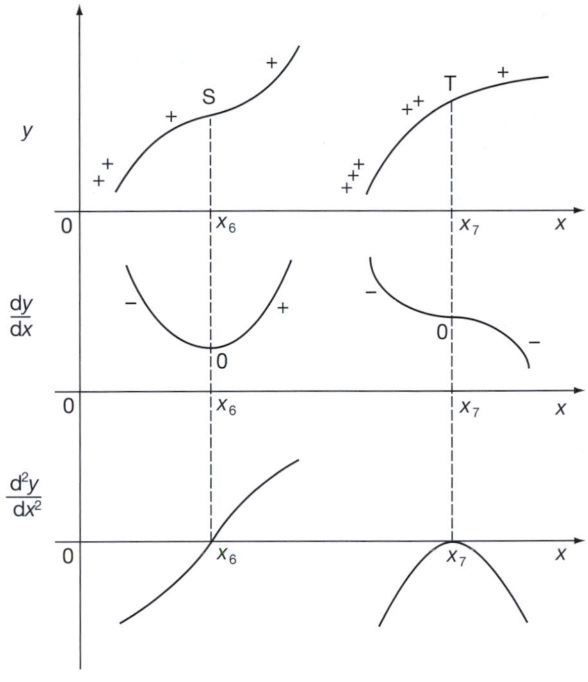


*Notice the difference* between the two second derived curves.

Although $\frac{d^2y}{dx^2}$ is zero for each (at $x = x_6$ and $x = x_7$), how do they differ?

*When you have discovered the difference, move on to Frame 31*

31

In the case of the real P-of-I, the graph of $\frac{d^2y}{dx^2}$ crosses the $x$-axis. In the case of no P-of-I, the graph of $\frac{d^2y}{dx^2}$ only touches the $x$-axis and $\frac{d^2y}{dx^2}$ does not change sign.

This is the clue we have been after, and gives us our final rule:

*For a point of inflexion,* $\frac{d^2y}{dx^2} = 0$ *and there is a change of sign of* $\frac{d^2y}{dx^2}$ *as we go through the point.*

(In the phoney case, there is no change of sign.)

So, to find where points of inflexion occur:

(a) we differentiate $y = f(x)$ twice to get $\frac{d^2y}{dx^2}$

(b) we solve the equation $\frac{d^2y}{dx^2} = 0$

(c) we test to see whether or not a change of sign occurs in $\frac{d^2y}{dx^2}$ as we go through this value of $x$.

For points of inflexion, then, $\frac{d^2y}{dx^2} = 0$, with ........... of ...........

**32**

For a P-of-I, $\dfrac{d^2y}{dx^2} = 0$ with $\boxed{\text{change of sign}}$

This last phrase is all-important.

**Example 1**

Find the points of inflexion, if any, on the graph of the function:

$$y = \frac{x^3}{3} - \frac{x^2}{2} - 2x + 5$$

(a) *Differentiate twice.* $\dfrac{dy}{dx} = x^2 - x - 2$, $\dfrac{d^2y}{dx^2} = 2x - 1$

For P-of-I, $\dfrac{d^2y}{dx^2} = 0$, with change of sign. $\therefore\ 2x - 1 = 0 \quad \therefore\ x = \dfrac{1}{2}$

If there is a P-of-I, it occurs at $x = \dfrac{1}{2}$.

(b) *Test for change of sign.* We take a point just before $x = \dfrac{1}{2}$, i.e. $x = \dfrac{1}{2} - a$, and a point just after $x = \dfrac{1}{2}$, i.e. $x = \dfrac{1}{2} + a$, where $a$ is a small positive quantity, and investigate the sign of $\dfrac{d^2y}{dx^2}$ at these two values of $x$.

*Move on*

**33**

$$\frac{d^2y}{dx^2} = 2x - 1$$

(a) At $x = \dfrac{1}{2} - a$, $\dfrac{d^2y}{dx^2} = 2\left(\dfrac{1}{2} - a\right) - 1 = 1 - 2a - 1$

$$= -2a \quad \text{(negative)}$$

(b) At $x = \dfrac{1}{2} + a$, $\dfrac{d^2y}{dx^2} = 2\left(\dfrac{1}{2} + a\right) - 1 = 1 + 2a - 1$

$$= 2a \quad \text{(positive)}$$

There *is* a change in sign of $\dfrac{d^2y}{dx^2}$ as we go through $x = \dfrac{1}{2}$

$\therefore$ There *is* a point of inflexion at $x = \dfrac{1}{2}$

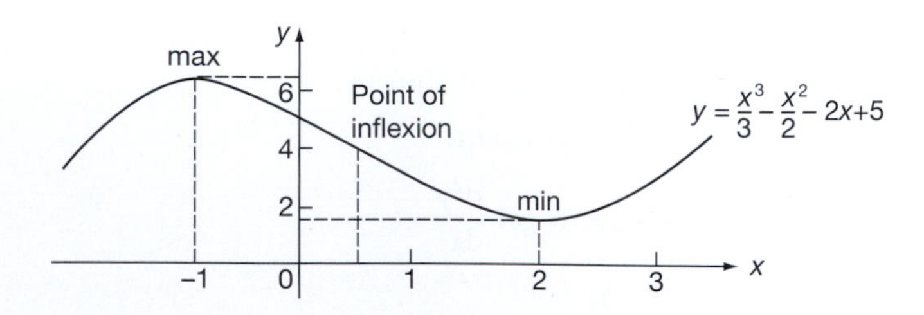


If you look at the sketch graph of this function which you have already drawn, you will see the point of inflexion where the right-hand curve changes to the left-hand curve.

**34**

**Example 2**

Find the points of inflexion on the graph of the function:

$y = 3x^5 - 5x^4 + x + 4$

First, differentiate twice and solve the equation $\frac{d^2y}{dx^2} = 0$. This will give the values of $x$ at which there are possibly points of inflexion. We cannot be sure until we have then tested for a change of sign in $\frac{d^2y}{dx^2}$. We will do that in due course.

So start off by finding an expression for $\frac{d^2y}{dx^2}$ and solving the equation $\frac{d^2y}{dx^2} = 0$.

*When you have done that, move on to the next frame*

**35**

We have $\quad y = 3x^5 - 5x^4 + x + 4$

$$\therefore \frac{dy}{dx} = 15x^4 - 20x^3 + 1$$

$$\therefore \frac{d^2y}{dx^2} = 60x^3 - 60x^2 = 60x^2(x - 1)$$

For P-of-I, $\frac{d^2y}{dx^2} = 0$, with change of sign.

$\therefore 60x^2(x - 1) = 0 \qquad \therefore x = 0$ or $x = 1$

If there is a point of inflexion, it occurs at $x = 0$, $x = 1$, or both. Now comes the test for a change of sign. For each of the two values of $x$ we have found, i.e. $x = 0$ and $x = 1$, take points on either side of it, differing from it by a very small amount $a$, where $0 < a < 1$.

(a) For $x = 0$

$$\left.\begin{aligned} \text{At } x = -a, \quad \frac{d^2y}{dx^2} &= 60(-a)^2(-a - 1) \\ &= (+)(+)(-) = \text{negative} \\ \text{At } x = +a, \quad \frac{d^2y}{dx^2} &= 60(+a)^2(a - 1) \\ &= (+)(+)(-) = \text{negative} \end{aligned}\right\} \begin{array}{l}\text{No sign change.} \\ \text{No P-of-I.}\end{array}$$

(b) For $x = 1$

$$\left.\begin{aligned} \text{At } x = 1 - a, \quad \frac{d^2y}{dx^2} &= 60(1 - a)^2(1 - a - 1) \\ &= (+)(+)(-) = \text{negative} \\ \text{At } x = 1 + a, \quad \frac{d^2y}{dx^2} &= 60(1 + a)^2(1 + a - 1) \\ &= (+)(+)(+) = \text{positive} \end{aligned}\right\} \begin{array}{l}\text{Change in sign.} \\ \therefore \text{ P-of-I.}\end{array}$$

Therefore, the only point of inflexion occurs when $x = 1$, i.e. at the point

$x = 1, y = 3$

That is just about all there is to it. The functions with which we have to deal differ, of course, from problem to problem, but the method remains the same.

*Now go on to the next frame and complete the* **Can You?** *checklist and* **Test exercise**

## Can You?

**36**

### Checklist 8

Check this list before and after you try the end of Programme test.

On a scale of 1 to 5 how confident are you that you can: — Frames

- Differentiate the inverse trigonometric functions? — 1 to 7
  *Yes* ☐ ☐ ☐ ☐ ☐ *No*
- Differentiate the inverse hyperbolic functions? — 8 to 17
  *Yes* ☐ ☐ ☐ ☐ ☐ *No*
- Identify and locate a maximum and a minimum? — 18 to 25
  *Yes* ☐ ☐ ☐ ☐ ☐ *No*
- Identify and locate a point of inflexion? — 26 to 35
  *Yes* ☐ ☐ ☐ ☐ ☐ *No*

## Test exercise 8

**37**

The questions are all very straightforward and should not cause you any anxiety.

1 Evaluate:

(a) $\cos^{-1}(-0{\cdot}6428)$ (b) $\tan^{-1}(-0{\cdot}7536)$

2 Differentiate with respect to $x$:

(a) $y = \sin^{-1}(3x + 2)$ (d) $y = \cosh^{-1}(1 - 3x)$

(b) $y = \dfrac{\cos^{-1} x}{x}$ (e) $y = \sinh^{-1}(\cos x)$

(c) $y = x^2 \tan^{-1}\left(\dfrac{x}{2}\right)$ (f) $y = \tanh^{-1} 5x$

3 Find the stationary values of $y$ and the points of inflexion on the graph of each of the following functions, and in each case, draw a sketch graph of the function:

(a) $y = x^3 - 6x^2 + 9x + 6$ (b) $y = x + \dfrac{1}{x}$ (c) $y = xe^{-x}$

*Well done. You are now ready for the next Programme*

## Further problems 8

**38**

1 Differentiate:

(a) $\tan^{-1}\left\{\dfrac{1 + \tan x}{1 - \tan x}\right\}$ (b) $x\sqrt{1 - x^2} - \sin^{-1}\sqrt{1 - x^2}$

2 If $y = \dfrac{\sin^{-1} x}{\sqrt{1 - x^2}}$, prove that for $0 \leq x < 1$:

(a) $(1 - x^2)\dfrac{dy}{dx} = xy + 1$ (b) $(1 - x^2)\dfrac{d^2y}{dx^2} - 3x\dfrac{dy}{dx} = y$

3 Find $\frac{dy}{dx}$ when:

(a) $y = \tan^{-1}\left\{\frac{4\sqrt{x}}{1-4x}\right\}$ (b) $y = \tanh^{-1}\left\{\frac{2x}{1+x^2}\right\}$

4 Find the coordinates of the point of inflexion on the curves:

(a) $y = (x-2)^2(x-7)$
(b) $y = 4x^3 + 3x^2 - 18x - 9$

5 Find the values of $x$ for which the function $y = f(x)$, defined by $y(3x-2) = (3x-1)^2$ has maximum and minimum values and distinguish between them. Sketch the graph of the function.

6 Find the values of $x$ at which maximum and minimum values of $y$ and points of inflexion occur on the curve $y = 12\ln x + x^2 - 10x$.

7 If $4x^2 + 8xy + 9y^2 - 8x - 24y + 4 = 0$, show that when $\frac{dy}{dx} = 0$, $x + y = 1$ and $\frac{d^2y}{dx^2} = \frac{4}{8-5y}$. Hence find the maximum and minimum values of $y$.

8 Determine the smallest positive value of $x$ at which a point of inflexion occurs on the graph of $y = 3e^{2x}\cos(2x-3)$.

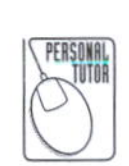

9 If $y^3 = 6xy - x^3 - 1$, prove that $\frac{dy}{dx} = \frac{2y - x^2}{y^2 - 2x}$ and that the maximum value of $y$ occurs where $x^3 = 8 + 2\sqrt{14}$ and the minimum value where $x^3 = 8 - 2\sqrt{14}$.

10 For the curve $y = e^{-x}\sin x$, express $\frac{dy}{dx}$ in the form $Ae^{-x}\cos(x + a)$ and show that the points of inflexion occur at $x = \frac{\pi}{2} + k\pi$ for any integral value of $k$.

11 Find the stationary points and points of inflexion on the following curves, and, in each case, sketch the graph:

(a) $y = 2x^3 - 5x^2 + 4x - 1$ (b) $y = \frac{x(x-1)}{x-2}$

(c) $y = x + \sin x$ (take $x$ and $y$ scales as multiples of $\pi$)

12 Find the values of $x$ at which points of inflexion occur on the following curves:

(a) $y = e^{-x^2}$ (b) $y = e^{-2x}(2x^2 + 2x + 1)$

(c) $y = x^4 - 10x^2 + 7x + 4$

13 The signalling range $(x)$ of a submarine cable is proportional to $r^2\ln\left(\frac{1}{r}\right)$, where $r$ is the ratio of the radii of the conductor and cable. Find the value of $r$ for maximum range.

14 The power transmitted by a belt drive is proportional to $Tv - \frac{wv^3}{g}$, where $v$ = speed of the belt, $T$ = tension on the driving side and $\omega$ = weight per unit length of belt. Find the speed at which the transmitted power is a maximum.

15 A right circular cone has a given curved surface $A$. Show that, when its volume is a maximum, the ratio of the height to the base radius is $\sqrt{2} : 1$.

16 The motion of a particle performing damped vibrations is given by $y = e^{-t}\sin 2t$, $y$ being the displacement from its mean position at time $t$. Show that $y$ is a maximum when $t = \frac{1}{2}\tan^{-1}(2)$ and determine this maximum displacement to three significant figures.

17 The cross-section of an open channel is a trapezium with base 6 cm and sloping sides each 10 cm wide. Calculate the width across the open top so that the cross-sectional area of the channel shall be a maximum.

18 The velocity ($v$) of a piston is related to the angular velocity ($\omega$) of the crank by the relationship $v = \omega r\left\{\sin\theta + \frac{r}{2\ell}\sin 2\theta\right\}$ where $r$ = length of crank and $\ell$ = length of connecting rod. Find the first positive value of $\theta$ for which $v$ is a maximum, for the case when $\ell = 4r$.

19 A right circular cone of base radius $r$ has a total surface area $S$ and volume $V$. Prove that $9V^2 = r^2(S^2 - 2\pi r^2 S)$. If $S$ is constant, prove that the vertical angle ($\theta$) of the cone for maximum volume is given by $\theta = 2\sin^{-1}\left(\frac{1}{3}\right)$.

20 Show that the equation $4\frac{d^2x}{dt^2} + 4\mu\frac{dx}{dt} + \mu^2 x = 0$ is satisfied by

$x = (At + B)e^{-\mu t/2}$, where $A$ and $B$ are arbitrary constants.

If $x = 0$ and $\frac{dx}{dt} = C$ when $t = 0$, find $A$ and $B$ and show that the maximum value of $x$ is $\frac{2C}{\mu e}$ and that this occurs when $t = \frac{2}{\mu}$.

39 Now visit the companion website for this book at www.palgrave.com/stroud for more questions applying this mathematics to science and engineering.

# Tangents, normals and curvature

## Learning outcomes

When you have completed this Programme you will be able to:

- ☐ Evaluate the gradient of a straight line
- ☐ Recognize the relationship satisfied by two mutually perpendicular straight lines
- ☐ Derive the equations of a tangent and a normal to a curve
- ☐ Evaluate the curvature and radius of curvature at a point on a curve
- ☐ Locate the centre of curvature for a point on a curve

# Equation of a straight line

**1**

The basic equation of a straight line is $y = mx + c$ (refer to Programme F.12),

$$\text{where } m = \text{gradient} = \frac{dy}{dx}$$
$$c = \text{intercept on real } y\text{-axis}$$

Note that if the scales of $x$ and $y$ are identical, $\frac{dy}{dx} = \tan\theta$

e.g. To find the equation of the straight line passing through P (3, 2) and Q (−2, 1), we could argue thus:

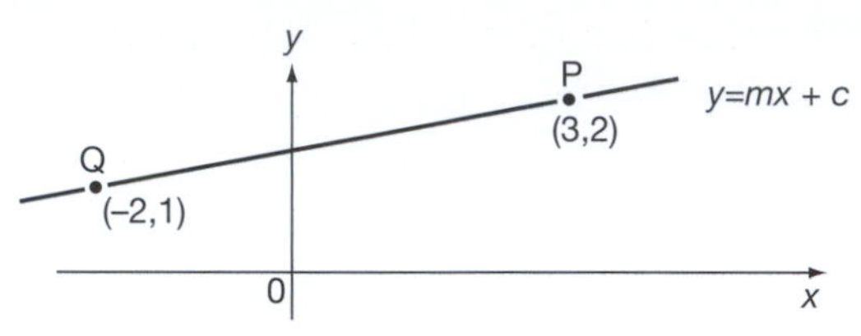


Line passes through P, i.e. when $x = 3$, $y = 2$ $\therefore 2 = m.3 + c$.

Line passes through Q, i.e. when $x = -2$, $y = 1$ $\therefore 1 = m(-2) + c$.

So we obtain a pair of simultaneous equations from which the values of $m$ and $c$ can be found. Therefore the equation is ...........

**2**

We find $m = 1/5$ and $c = 7/5$. Therefore the equation of the line is
$$y = \frac{x}{5} + \frac{7}{5}, \quad \text{i.e. } 5y = x + 7$$

Sometimes we are given the gradient, $m$, of a straight line passing through a given point $(x_1, y_1)$ and we are required to find its equation. In that case, it is more convenient to use the form:

$$y - y_1 = m(x - x_1)$$

For example, the equation of the line passing through the point (5, 3) with gradient 2 is simply ........... which simplifies to ...........

*Move on to the next frame*

**3**

$$y - 3 = 2(x - 5)$$
$$\text{i.e. } y - 3 = 2x - 10 \qquad \therefore y = 2x - 7$$

Similarly, the equation of the line through the point (−2, −1) and having a gradient $\frac{1}{2}$ is

$$y - (-1) = \frac{1}{2}\left\{x - (-2)\right\} \quad \therefore y + 1 = \frac{1}{2}(x + 2)$$

$$2y + 2 = x + 2 \qquad \therefore y = \frac{x}{2}$$

So, in the same way, the line passing through (2, −3) and having gradient (−2) is ...........

**4**

$$\boxed{y = 1 - 2x}$$

Because

$$y - (-3) = -2(x - 2)$$
$$\therefore y + 3 = -2x + 4 \qquad \therefore y = 1 - 2x$$

Right. So in general terms, the equation of the line passing through the point $(x_1, y_1)$ with gradient $m$ is ...........

*Move on to Frame 5*

**5**

$$\boxed{y - y_1 = m(x - x_1)}$$ It is well worth remembering.

So for one last time:

If a point P has coordinates (4, 3) and the gradient $m$ of a straight line through P is 2, then the equation of the line is thus:

$$y - 3 = 2(x - 4)$$
$$= 2x - 8$$
$$\therefore y = 2x - 5$$

The equation of the line through P, perpendicular to the line we have just considered, will have a gradient $m_1$, such that $mm_1 = -1$

i.e. $m_1 = -\frac{1}{m}$. And since $m = 2$, then $m_1 = -\frac{1}{2}$. This line passes through (4, 3) and its equation is therefore:

$$y - 3 = -\frac{1}{2}(x - 4)$$
$$= -x/2 + 2$$
$$y = -\frac{x}{2} + 5, \quad 2y = 10 - x$$

**6**

If $m$ and $m_1$ represent the gradients of two lines perpendicular to each other, then $mm_1 = -1$ or $m_1 = -\frac{1}{m}$

Consider the two straight lines:

$2y = 4x - 5$ and $6y = 2 - 3x$

If we convert each of these to the form $y = mx + c$, we get:

(a) $y = 2x - \frac{5}{2}$ and (b) $y = -\frac{1}{2}x + \frac{1}{3}$

So in (a) the gradient $m = 2$ and in (b) the gradient $m_1 = -\frac{1}{2}$

We notice that, in this case, $m_1 = -\frac{1}{m}$ or that $mm_1 = -1$

Therefore we know that the two given lines are at right-angles to each other.

Which of these represents a pair of lines perpendicular to each other:

(a) $y = 3x - 5$ and $3y = x + 2$
(b) $2y = x - 5$ and $y = 6 - x$
(c) $y - 3x - 2 = 0$ and $3y + x + 9 = 0$
(d) $5y - x = 4$ and $2y + 10x + 3 = 0$

**7**

(c) and (d)

Because

If we convert each to the form $y = mx + c$, we get:

(a) $y = 3x - 5$ and $y = \frac{x}{3} + \frac{2}{3}$

$m = 3;\ m_1 = \frac{1}{3} \qquad \therefore mm_1 \neq -1 \qquad$ Not perpendicular.

(b) $y = \frac{x}{2} - \frac{5}{2}$ and $y = -x + 6$

$m = \frac{1}{2};\ m_1 = -1 \qquad \therefore mm_1 \neq -1 \qquad$ Not perpendicular.

(c) $y = 3x + 2$ and $y = -\frac{x}{3} - 3$

$m = 3;\ m_1 = -\frac{1}{3} \qquad \therefore mm_1 = -1 \qquad$ Perpendicular.

(d) $y = \frac{x}{5} + \frac{4}{5}$ and $y = -5x - \frac{3}{2}$

$m = \frac{1}{5};\ m_1 = -5 \qquad \therefore mm_1 = -1 \qquad$ Perpendicular.

Do you agree with these?

**8**

Remember that if $y = mx + c$ and $y = m_1x + c_1$ are perpendicular to each other, then:

$mm_1 = -1$, i.e. $m_1 = -\frac{1}{m}$

Here is one further example:

A line AB passes through the point P (3, −2) with gradient $-\frac{1}{2}$. Find its equation and also the equation of the line CD through P perpendicular to AB.

*When you have finished, check your results with those in Frame 9*

**9**

Equation of AB:

$$y - (-2) = -\frac{1}{2}(x - 3)$$

$$\therefore y + 2 = -\frac{x}{2} + \frac{3}{2}$$

$$\therefore y = -\frac{x}{2} - \frac{1}{2}$$

$$\therefore 2y + x + 1 = 0 \text{ or } y = -\frac{x}{2} - \frac{1}{2}$$

Equation of CD:

$$\text{gradient } m_1 = -\frac{1}{m} = -\frac{1}{-\frac{1}{2}} = 2$$

$$y - (-2) = 2(x - 3)$$

$$y + 2 = 2x - 6$$

$$y = 2x - 8$$

▶

So we have:

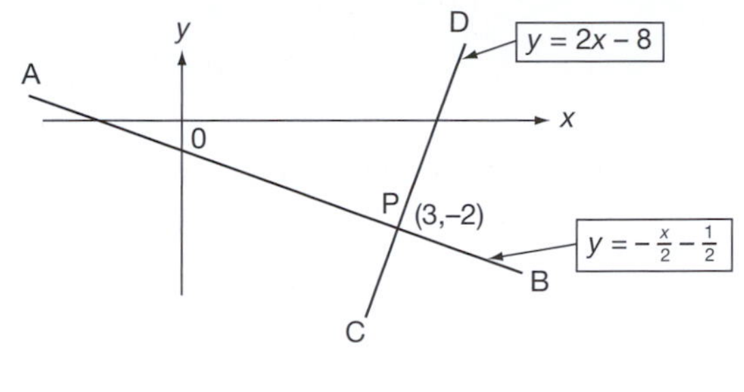


$$mm_1 = \ldots\ldots\ldots\ldots$$

---

**10**

$$\boxed{mm_1 = -1}$$

And now, just one more to do on your own.

The point P (3, 4) is a point on the line $y = 5x - 11$.

Find the equation of the line through P which is perpendicular to the given line.

*That should not take long. When you have finished it, move on to the next frame*

---

**11**

$$\boxed{5y + x = 23}$$

Because

gradient of the given line, $y = 5x - 11$ is 5

gradient of required line $= -\dfrac{1}{5}$

The line passes through P, i.e. when $x = 3$, $y = 4$.

$$y - 4 = -\frac{1}{5}(x - 3)$$

$$5y - 20 = -x + 3 \qquad \therefore \; 5y + x = 23$$

---

## Tangents and normals to a curve at a given point

**12**

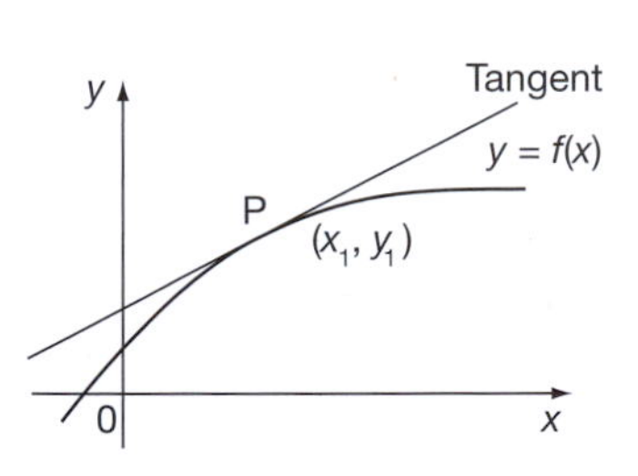


The gradient of a curve, $y = f(x)$, at a point P on the curve is given by the gradient of the tangent at P. It is also given by the value of $\dfrac{dy}{dx}$ at the point P, which we can calculate, knowing the equation of the curve. Thus we can calculate the gradient of the tangent to the curve at any point P.

What else do we know about the tangent which will help us to determine its equation?

---

**13**

> We know that the tangent passes through P, i.e. when $x = x_1$, $y = y_1$

Correct. This is sufficient information for us to find the equation of the tangent. Let us do an example.

▶

Find the equation of the tangent to the curve $y = 2x^3 + 3x^2 - 2x - 3$ at the point P, $x = 1$, $y = 0$.

$$\frac{dy}{dx} = 6x^2 + 6x - 2$$

Gradient of tangent $= \left\{\frac{dy}{dx}\right\}_{x=1} = 6 + 6 - 2 = 10$, i.e. $m = 10$

Passes through P, i.e. $x = 1$, $y = 0$.

$y - y_1 = m(x - x_1)$ gives $y - 0 = 10(x - 1)$

Therefore the tangent is $y = 10x - 10$

We could also, if required, find the equation of the normal at P which is defined as the line through P perpendicular to the tangent at P. We know, for example, that the gradient of the normal is ............

## 14

$$\text{Gradient of normal} = \frac{-1}{\text{Gradient of tangent}} = -\frac{1}{10}$$

The normal also passes through P, i.e. when $x = 1$, $y = 0$.

$\therefore$ Equation of normal is $y - 0 = -\frac{1}{10}(x - 1)$

$10y = -x + 1 \qquad \therefore\ 10y + x = 1$

That was very easy. Do this one just to get your hand in:

Find the equations of the tangent and normal to the curve

$$y = x^3 - 2x^2 + 3x - 1$$

at the point (2, 5).

Off you go. Do it in just the same way.

*When you have got the results, move on to Frame 15*

## 15

Tangent: $y = 7x - 9$ Normal: $7y + x = 37$

Here are the details:

$y = x^3 - 2x^2 + 3x - 1$

$\therefore\ \frac{dy}{dx} = 3x^2 - 4x + 3 \qquad \therefore$ At P (2, 5), $\frac{dy}{dx} = 12 - 8 + 3 = 7$

Tangent passes through (2, 5), i.e. $x = 2$, $y = 5$

$y - 5 = 7(x - 2) \qquad$ Tangent is $y = 7x - 9$

For normal, gradient $= \frac{-1}{\text{gradient of tangent}} = -\frac{1}{7}$

Normal passes through P (2, 5)

$\therefore\ y - 5 = -\frac{1}{7}(x - 2)$

$7y - 35 = -x + 2$

Normal is $7y + x = 37$

You will perhaps remember doing all this long ago.

*Anyway, on to Frame 16*

**16**

The equation of the curve may, of course, be presented as an implicit function or as a pair of parametric equations. But this will not worry you for you already know how to differentiate functions in these two forms. Let us have an example or two.

Find the equations of the tangent and normal to the curve

$x^2 + y^2 + 3xy - 11 = 0$ at the point $x = 1$, $y = 2$.

First of all we must find $\frac{dy}{dx}$ at (1, 2). So differentiate right away.

$$2x + 2y\frac{dy}{dx} + 3x\frac{dy}{dx} + 3y = 0$$

$$(2y + 3x)\frac{dy}{dx} = -(2x + 3y)$$

$$\frac{dy}{dx} = -\frac{2x + 3y}{2y + 3x}$$

Therefore, at $x = 1$, $y = 2$,

$$\frac{dy}{dx} = \dots\dots\dots\dots$$

**17**

$$\boxed{\frac{dy}{dx} = -\frac{8}{7}}$$

Because

$$\frac{dy}{dx} = -\frac{2+6}{4+3} = -\frac{8}{7}$$

Now we proceed as for the previous cases.

Tangent passes through (1, 2) $\therefore\ y - 2 = -\frac{8}{7}(x - 1)$

$$7y - 14 = -8x + 8$$

$\therefore$ Tangent is $7y + 8x = 22$

Now to find the equation of the normal.

$$\text{Gradient} = \frac{-1}{\text{Gradient of tangent}} = \frac{7}{8}$$

Normal passes through (1, 2) $\therefore\ y - 2 = \frac{7}{8}(x - 1)$

$$8y - 16 = 7x - 7$$

$\therefore$ Normal is $8y = 7x + 9$ That's that!

Now try this one:

Find the equations of the tangent and normal to the curve

$$x^3 + x^2y + y^3 - 7 = 0$$

at the point $x = 2$, $y = 3$.

**18**

Tangent: $31y + 24x = 141$ Normal: $24y = 31x + 10$

Here is the working:

$$x^3 + x^2y + y^3 - 7 = 0$$

$$3x^2 + x^2\frac{dy}{dx} + 2xy + 3y^2\frac{dy}{dx} = 0$$

$$(x^2 + 3y^2)\frac{dy}{dx} = -(3x^2 + 2xy) \quad \therefore \frac{dy}{dx} = -\frac{3x^2 + 2xy}{x^2 + 3y^2}$$

$$\therefore \text{At } (2, 3) \quad \frac{dy}{dx} = -\frac{12 + 12}{4 + 27} = -\frac{24}{31}$$

(a) Tangent passes through (2, 3) $\therefore y - 3 = -\frac{24}{31}(x - 2)$

$$31y - 93 = -24x + 48 \qquad \therefore 31y + 24x = 141$$

(b) Normal gradient $= \frac{31}{24}$. Passes through (2, 3) $\therefore y - 3 = \frac{31}{24}(x - 2)$

$$24y - 72 = 31x - 62 \qquad \therefore 24y = 31x + 10$$

*Now on to the next frame for another example*

**19**

Now what about this one?

The parametric equations of a curve are $x = \frac{3t}{1+t}$, $y = \frac{t^2}{1+t}$

Find the equations of the tangent and normal at the point for which $t = 2$.

First find the value of $\frac{dy}{dx}$ when $t = 2$.

$$x = \frac{3t}{1+t} \quad \therefore \frac{dx}{dt} = \frac{(1+t)3 - 3t}{(1+t)^2} = \frac{3 + 3t - 3t}{(1+t)^2} = \frac{3}{(1+t)^2}$$

$$y = \frac{t^2}{1+t} \quad \therefore \frac{dy}{dt} = \frac{(1+t)2t - t^2}{(1+t)^2} = \frac{2t + 2t^2 - t^2}{(1+t)^2} = \frac{2t + t^2}{(1+t)^2}$$

$$\frac{dy}{dx} = \frac{dy}{dt}.\frac{dt}{dx}$$

$$= \frac{2t + t^2}{(1+t)^2}.\frac{(1+t)^2}{3}$$

$$= \frac{2t + t^2}{3}$$

$$\therefore \text{At } t = 2, \frac{dy}{dx} = \frac{8}{3}$$

To get the equation of the tangent, we must know the $x$- and $y$-values of a point through which it passes.

At P:

$$x = \frac{3t}{1+t} = \frac{6}{1+2} = \frac{6}{3} = 2,$$

$$y = \frac{t^2}{1+t} = \frac{4}{3}$$

*Continued in Frame 20*

**20**

So the tangent has a gradient of $\frac{8}{3}$ and passes through $\left(2, \frac{4}{3}\right)$

$\therefore$ Its equation is $y - \frac{4}{3} = \frac{8}{3}(x-2)$

$3y - 4 = 8x - 16 \quad \therefore\ 3y = 8x - 12$ (Tangent)

For the normal, gradient $= \dfrac{-1}{\text{gradient of tangent}} = -\frac{3}{8}$

Also passes through $\left(2, \frac{4}{3}\right) \qquad \therefore\ y - \frac{4}{3} = -\frac{3}{8}(x-2)$

$24y - 32 = -9x + 18 \qquad \therefore\ 24y + 9x = 50$ (Normal)

Now you do this one. When you are satisfied with your result, check it with the results in Frame 21. Here it is:

If $y = \cos 2t$ and $x = \sin t$, find the equations of the tangent and normal to the curve at $t = \frac{\pi}{6}$.

**21**

| Tangent: $2y + 4x = 3$     Normal: $4y = 2x + 1$ |
|---|

Working:

$y = \cos 2t \qquad \therefore\ \dfrac{dy}{dt} = -2\sin 2t = -4\sin t \cos t$

$x = \sin t \qquad \therefore\ \dfrac{dx}{dt} = \cos t$

$\dfrac{dy}{dx} = \dfrac{dy}{dt} \cdot \dfrac{dt}{dx} = \dfrac{-4\sin t \cos t}{\cos t} = -4\sin t$

At $t = \frac{\pi}{6}$, $\quad \dfrac{dy}{dx} = -4\sin\frac{\pi}{6} = -4\left(\frac{1}{2}\right) = -2$

$\therefore$ gradient of tangent $= -2$

Passes through $\quad x = \sin\frac{\pi}{6} = 0{\cdot}5;\ y = \cos\frac{\pi}{3} = 0{\cdot}5$

$\therefore$ Tangent is $y - \frac{1}{2} = -2\left(x - \frac{1}{2}\right) \qquad \therefore\ 2y - 1 = -4x + 2$

$\therefore\ 2y + 4x = 3$ (Tangent)

Gradient of normal $= \frac{1}{2}$. Line passes through $(0{\cdot}5, 0{\cdot}5)$

Equation is $\quad y - \frac{1}{2} = \frac{1}{2}\left(x - \frac{1}{2}\right)$

$\therefore\ 4y - 2 = 2x - 1$

$\therefore\ 4y = 2x + 1$ (Normal)

**22**

Before we leave this part of the Programme, let us revise the fact that we can easily find the angle between two intersecting curves.

Since the gradient of a curve at $(x_1, y_1)$ is given by the value of $\frac{dy}{dx}$ at that point, and $\frac{dy}{dx} = \tan\theta$, where $\theta$ is the angle of slope, then we can use these facts to determine the angle between the curves at their point of intersection. One example will be sufficient.

▶

Find the angle between $y^2 = 8x$ and $x^2 + y^2 = 16$ at their point of intersection for which $y$ is positive.

First find the point of intersection:

i.e. solve $y^2 = 8x$ and $x^2 + y^2 = 16$

We have

$$x^2 + 8x = 16 \qquad \therefore \; x^2 + 8x - 16 = 0$$

$$x = \frac{-8 \pm \sqrt{64 + 64}}{2} = \frac{-8 \pm \sqrt{128}}{2}$$

$$= \frac{-8 \pm 11{\cdot}314}{2} = \frac{3{\cdot}314}{2} \text{ or } \frac{-19{\cdot}314}{2}$$

$x = 1{\cdot}657$ or $[-9{\cdot}657]$ Not a real point of intersection

When $x = 1{\cdot}657$, $y^2 = 8(1{\cdot}657) = 13{\cdot}256$, $y = 3{\cdot}641$.

Coordinates of P are $x = 1{\cdot}657$, $y = 3{\cdot}641$.

Now we have to find $\dfrac{dy}{dx}$ for each of the two curves. Do that.

---

**23**

(a) $y^2 = 8x \qquad \therefore \; 2y\dfrac{dy}{dx} = 8 \qquad \therefore \; \dfrac{dy}{dx} = \dfrac{4}{y} = \dfrac{4}{3{\cdot}641} = \dfrac{1}{0{\cdot}910} = 1{\cdot}099$

$\tan\theta_1 = 1{\cdot}099 \qquad \therefore \; \theta_1 = 47^\circ 42'$

(b) Similarly for $x^2 + y^2 = 16$:

$2x + 2y\dfrac{dy}{dx} = 0 \qquad \therefore \; \dfrac{dy}{dx} = -\dfrac{x}{y} = -\dfrac{1{\cdot}657}{3{\cdot}641} = -0{\cdot}4551$

$\tan\theta_2 = -0{\cdot}4551 \qquad \therefore \; \theta_2 = -24^\circ 28'$

Finally

$$\theta = \theta_1 - \theta_2 = 47^\circ 42' - (-24^\circ 28')$$
$$= 47^\circ 42' + 24^\circ 28'$$
$$= 72^\circ 10'$$

That just about covers all there is to know about finding tangents and normals to a curve. We now look at another application of differentiation.

---

## Curvature

**24**

The value of $\dfrac{dy}{dx}$ at any point on a curve denotes the gradient of the curve at that point. Curvature is concerned with how quickly the curve is changing direction in the neighbourhood of that point.

Let us see in the next few frames what it is all about.

**25**

Let us first consider the change in direction of a curve $y = f(x)$ between the points P and Q as shown. The direction of a curve is measured by the gradient of the tangent.

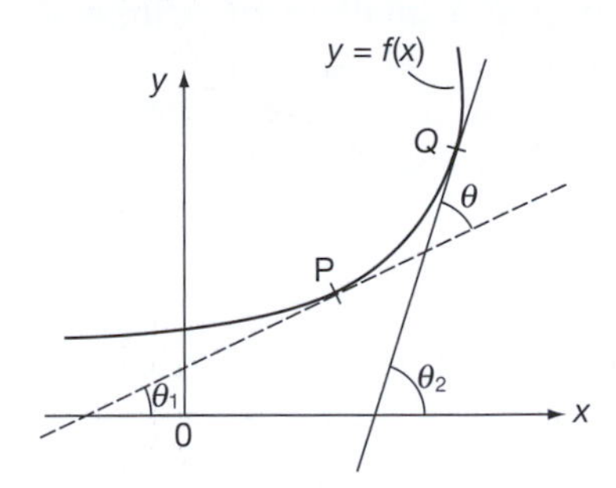


$$\text{Gradient at P} = \tan\theta_1 = \left\{\frac{dy}{dx}\right\}_P$$

$$\text{Gradient at Q} = \tan\theta_2 = \left\{\frac{dy}{dx}\right\}_Q$$

These can be calculated, knowing the equation of the curve.

From the values of $\tan\theta_1$ and $\tan\theta_2$, the angles $\theta_1$ and $\theta_2$ can be found using a calculator. Then from the diagram, $\theta = \theta_2 - \theta_1$.

If we are concerned with how fast the curve is bending, we must consider not only the change in direction from P to Q, but also the length of ........... which provides this change in direction.

**26**

The arc PQ

That is, we must know not only the change of direction, but also how far along the curve we must go to obtain this change in direction.

Now let us consider the two points, P and Q, near to each other, so that PQ is a small arc ($= \delta s$). The change in direction will not be great, so that if $\theta$ is the slope at P, then the angle of slope at Q can be put as $\theta + \delta\theta$.

The change in direction from P to Q is therefore $\delta\theta$.

The length of arc from P to Q is $\delta s$.

The average rate of change of direction with arc from P to Q is

$$\frac{\text{the change in direction from P to Q}}{\text{the length of arc from P to Q}} = \frac{\delta\theta}{\delta s}$$

This could be called the average curvature from P to Q. If Q now moves down towards P, i.e. $\delta s \to 0$, we finally get $\dfrac{d\theta}{ds}$, which is the *curvature* at P. It tells us how quickly the curve is bending in the immediate neighbourhood of P.

**27**

In practice, it is difficult to find $\dfrac{d\theta}{ds}$ since we should need a relationship between $\theta$ and $s$, and usually all we have is the equation of the curve, $y = f(x)$ and the coordinates of P. So we must find some other way round it.

▶

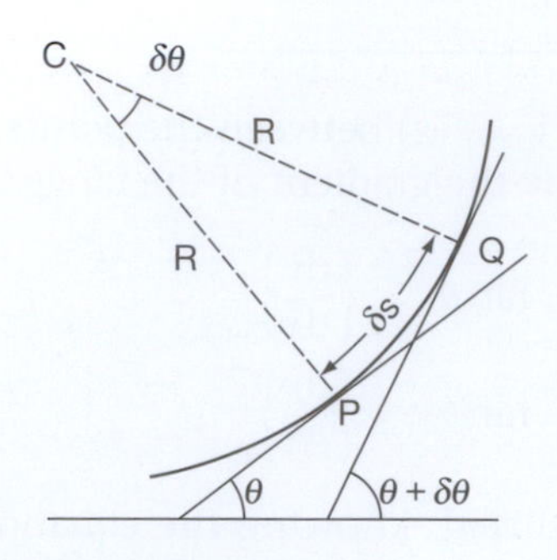


Let the normals at P and Q meet at C. Since P and Q are close, CP ≈ QC (= $R$ say) and the arc PQ can be thought of as a small arc of a circle of radius $R$. Note that PCQ = $\delta\theta$ (for if the tangent turns through $\delta\theta$, the radius at right-angles to it will also turn through the same angle).

You remember that the arc of a circle of radius $r$ which subtends an angle $\theta$ radians at the centre is given by arc = $r\theta$. So, in the diagram above,

arc PQ= $\delta s$ =...........

**28**

arc PQ= $\delta s = R\delta\theta$

$$\delta s = R\delta\theta \qquad \therefore \frac{\delta\theta}{\delta s} = \frac{1}{R}$$

If $\delta s \to 0$, this becomes $\dfrac{d\theta}{ds} = \dfrac{1}{R}$ which is the curvature at P.

That is, we can state the curvature at a point, in terms of the radius $R$ of the circle we have considered. This is called the *radius of curvature*, and the point C the *centre of curvature*.

So we have now found that we can obtain the curvature $\dfrac{d\theta}{ds}$ if we have some way of finding the radius of curvature $R$.

If $R$ is large, is the curvature large or small?

If you think 'large', move on to Frame 29.

If you think 'small' go on to Frame 30.

**29**

Your answer was: 'If $R$ is large, the curvature is large'.

This is not so. For the curvature $= \dfrac{d\theta}{ds}$ and we have just shown that $\dfrac{d\theta}{ds} = \dfrac{1}{R}$. $R$ is the denominator, so that a large value for $R$ gives a small value for the fraction $\dfrac{1}{R}$ and hence a small value for the curvature.

You can see it this way. If you walk round a circle with a large radius $R$, then the curve is relatively a gentle one, i.e. small value of curvature, but if $R$ is small, the curve is more abrupt.

So once again, if $R$ is large, the curvature is ...........

**30**

If $R$ is large, the curvature is small

Correct, since the curvature $\dfrac{d\theta}{ds} = \dfrac{1}{R}$

In practice, we often indicate the curvature in terms of the radius of curvature $R$, since this is something we can appreciate.

▶

Let us consider our two points P and Q again. Since $\delta s$ is very small, there is little difference between the arc PQ and the chord PQ, or between the direction of the chord and that of the tangent. So, when

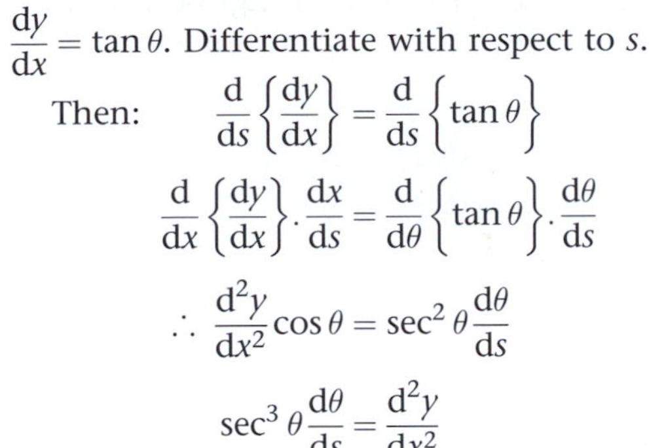


$$\delta s \to 0, \quad \frac{dy}{dx} = \tan\theta \quad \text{and} \quad \frac{dx}{ds} = \cos\theta$$

$\frac{dy}{dx} = \tan\theta$. Differentiate with respect to $s$.

Then:
$$\frac{d}{ds}\left\{\frac{dy}{dx}\right\} = \frac{d}{ds}\left\{\tan\theta\right\}$$

$$\frac{d}{dx}\left\{\frac{dy}{dx}\right\}\cdot\frac{dx}{ds} = \frac{d}{d\theta}\left\{\tan\theta\right\}\cdot\frac{d\theta}{ds}$$

$$\therefore \quad \frac{d^2y}{dx^2}\cos\theta = \sec^2\theta\frac{d\theta}{ds}$$

$$\sec^3\theta\frac{d\theta}{ds} = \frac{d^2y}{dx^2}$$

Now: $\sec^3\theta = (\sec^2\theta)^{3/2} = (1+\tan^2\theta)^{3/2} = \left\{1+\left(\frac{dy}{dx}\right)^2\right\}^{3/2}$

$$\therefore \frac{d\theta}{ds} = \frac{1}{R} = \frac{\dfrac{d^2y}{dx^2}}{\left\{1+\left(\dfrac{dy}{dx}\right)^2\right\}^{3/2}} \qquad \therefore R = \frac{\left\{1+\left(\dfrac{dy}{dx}\right)^2\right\}^{3/2}}{\dfrac{d^2y}{dx^2}}$$

Now we have got somewhere. For knowing the equation $y = f(x)$ of the curve, we can calculate the first and second derivatives at the point P and substitute these values in the formula for $R$.

This is an important result. Copy it down and remember it. You may never be asked to prove it, but you will certainly be expected to know it and to apply it.

*So now for some examples. Move on to Frame 31*

---

**31**

**Example 1**

Find the radius of curvature for the hyperbola $xy = 4$ at the point $x = 2$, $y = 2$.

$$R = \frac{\left\{1+\left(\dfrac{dy}{dx}\right)^2\right\}^{3/2}}{\dfrac{d^2y}{dx^2}}$$

So all we need to find are $\frac{dy}{dx}$ and $\frac{d^2y}{dx^2}$ at $(2, 2)$

$xy = 4 \qquad \therefore y = \frac{4}{x} = 4x^{-1} \qquad \therefore \frac{dy}{dx} = -4x^{-2} = \frac{-4}{x^2}$

and $\qquad \frac{d^2y}{dx^2} = 8x^{-3} = \frac{8}{x^3}$

At $(2, 2) \qquad \frac{dy}{dx} = -\frac{4}{4} = -1; \qquad \frac{d^2y}{dx^2} = \frac{8}{8} = 1$

$$\therefore R = \frac{\left\{1+(-1)^2\right\}^{3/2}}{1} = \frac{\{1+1\}^{3/2}}{1} = (2)^{3/2} = 2\sqrt{2}$$

$\therefore R = 2\sqrt{2} = 2{\cdot}828$ units

*There we are. Another example ion Frame 32*

**32**

**Example 2**

If $y = x + 3x^2 - x^3$, find $R$ at $x = 0$.

$$R = \frac{\left\{1 + \left(\frac{dy}{dx}\right)^2\right\}^{3/2}}{\frac{d^2y}{dx^2}}$$

$$\frac{dy}{dx} = 1 + 6x - 3x^2 \qquad \therefore \text{ At } x = 0, \quad \frac{dy}{dx} = 1 \qquad \therefore \left(\frac{dy}{dx}\right)^2 = 1$$

$$\frac{d^2y}{dx^2} = 6 - 6x \qquad \therefore \text{ At } x = 0, \quad \frac{d^2y}{dx^2} = 6$$

$$R = \frac{\{1+1\}^{3/2}}{6} = \frac{2^{3/2}}{6} = \frac{2\sqrt{2}}{6} = \frac{\sqrt{2}}{3}$$

$\therefore R = 0{\cdot}471$ units

Now you do this one:

Find the radius of curvature of the curve $y^2 = \frac{x^3}{4}$ at the point $\left(1, \frac{1}{2}\right)$

*When you have finished, check with the solution in Frame 33*

**33**

$$\boxed{R = 5{\cdot}21 \text{ units}}$$

Here is the solution in full:

$$y^2 = \frac{x^3}{4} \qquad \therefore 2y\frac{dy}{dx} = \frac{3x^2}{4} \qquad \therefore \frac{dy}{dx} = \frac{3x^2}{8y}$$

$$\therefore \text{ At } \left(1, \frac{1}{2}\right), \frac{dy}{dx} = \frac{3}{4} \qquad \therefore \left(\frac{dy}{dx}\right)^2 = \frac{9}{16}$$

$$\frac{dy}{dx} = \frac{3x^2}{8y} \qquad \therefore \frac{d^2y}{dx^2} = \frac{8y(6x) - 3x^2.8\frac{dy}{dx}}{64y^2}$$

$$\therefore \text{ At } \left(1, \frac{1}{2}\right), \frac{d^2y}{dx^2} = \frac{24 - 24.\frac{3}{4}}{16} = \frac{24 - 18}{16} = \frac{3}{8}$$

$$R = \frac{\left\{1 + \left(\frac{dy}{dx}\right)^2\right\}^{3/2}}{\frac{d^2y}{dx^2}} = \frac{\left\{1 + \frac{9}{16}\right\}^{3/2}}{\frac{3}{8}} = \frac{\left\{\frac{25}{16}\right\}^{3/2}}{\frac{3}{8}} = \frac{8}{3}.\frac{125}{64} = \frac{125}{24} = 5\frac{5}{24}$$

$\therefore R = 5{\cdot}21$ units

**34**

Of course, the equation of the curve could be an implicit function, as in the previous example, or a pair of parametric equations.

e.g. If $x = \theta - \sin\theta$ and $y = 1 - \cos\theta$, find $R$ when $\theta = 60^\circ = \frac{\pi}{3}$ radians.

$$\left.\begin{aligned} x &= \theta - \sin\theta \quad \therefore \frac{dx}{d\theta} = 1 - \cos\theta \\ y &= 1 - \cos\theta \quad \therefore \frac{dy}{d\theta} = \sin\theta \end{aligned}\right\} \quad \frac{dy}{dx} = \frac{dy}{d\theta}.\frac{d\theta}{dx}$$

$$\therefore \frac{dy}{dx} = \sin\theta.\frac{1}{1 - \cos\theta} = \frac{\sin\theta}{1 - \cos\theta}$$

▶

At $\theta = 60^\circ$, $\sin\theta = \dfrac{\sqrt{3}}{2}$, $\cos\theta = \dfrac{1}{2}$, $\dfrac{dy}{dx} = \dfrac{\sqrt{3}}{1}$

$$\frac{d^2y}{dx^2} = \frac{d}{dx}\left\{\frac{\sin\theta}{1-\cos\theta}\right\} = \frac{d}{d\theta}\left\{\frac{\sin\theta}{1-\cos\theta}\right\}.\frac{d\theta}{dx}$$

$$= \frac{(1-\cos\theta)\cos\theta - \sin\theta.\sin\theta}{(1-\cos\theta)^2}.\frac{1}{1-\cos\theta}$$

$$= \frac{\cos\theta - \cos^2\theta - \sin^2\theta}{(1-\cos\theta)^3} = \frac{\cos\theta - 1}{(1-\cos\theta)^3} = \frac{-1}{(1-\cos\theta)^2}$$

$\therefore$ At $\theta = 60^\circ$, $\dfrac{d^2y}{dx^2} = \dfrac{-1}{(1-\frac{1}{2})^2} = \dfrac{-1}{\frac{1}{4}} = -4$

$\therefore R = \dfrac{\{1+3\}^{3/2}}{-4} = \dfrac{2^3}{-4} = \dfrac{8}{-4} = -2$ $\quad\therefore R = -2$ units

---

**35**

You notice in this last example that the value of $R$ is negative. This merely indicates which way the curve is bending. Since $R$ is a physical length, then for all practical purposes, $R$ is taken as 2 units long.

If the value of $R$ is to be used in further calculations however, it is usually necessary to maintain the negative sign. You will see an example of this later.

Here is one for you to do in just the same way as before:

Find the radius of curvature of the curve $x = 2\cos^3\theta$, $y = 2\sin^3\theta$, at the point for which $\theta = \dfrac{\pi}{4} = 45^\circ$.

*Work through it and then go to Frame 36 to check your work*

---

**36**

$R = 3$ units

Because

$x = 2\cos^3\theta \quad \therefore \dfrac{dx}{d\theta} = 6\cos^2\theta(-\sin\theta) = -6\sin\theta\cos^2\theta$

$y = 2\sin^3\theta \quad \therefore \dfrac{dy}{d\theta} = 6\sin^2\theta\cos\theta$

$$\frac{dy}{dx} = \frac{dy}{d\theta}.\frac{d\theta}{dx} = \frac{6\sin^2\theta\cos\theta}{-6\sin\theta\cos^2\theta} = -\frac{\sin\theta}{\cos\theta} = -\tan\theta$$

At $\theta = 45^\circ$, $\dfrac{dy}{dx} = -1 \quad \therefore \left(\dfrac{dy}{dx}\right)^2 = 1$

Also $\quad \dfrac{d^2y}{dx^2} = \dfrac{d}{dx}\left\{-\tan\theta\right\} = \dfrac{d}{d\theta}\left\{-\tan\theta\right\}\dfrac{d\theta}{dx} = \dfrac{-\sec^2\theta}{-6\sin\theta\cos^2\theta}$

$$= \frac{1}{6\sin\theta\cos^4\theta}$$

$\therefore$ At $\theta = 45^\circ$, $\dfrac{d^2y}{dx^2} = \dfrac{1}{6\left(\frac{1}{\sqrt{2}}\right)\left(\frac{1}{4}\right)} = \dfrac{4\sqrt{2}}{6} = \dfrac{2\sqrt{2}}{3}$

$$R = \frac{\left\{1+\left(\frac{dy}{dx}\right)^2\right\}^{3/2}}{\frac{d^2y}{dx^2}} = \frac{\left\{1+1\right\}^{3/2}}{\frac{2\sqrt{2}}{3}} = \frac{3}{2\sqrt{2}}2^{3/2}$$

$$= \frac{3\times 2\sqrt{2}}{2\sqrt{2}} = 3$$

$R = 3$ units

37

## Centre of curvature

To get a complete picture, we need to know also the position of the centre of the circle of curvature for the point P $(x_1, y_1)$.

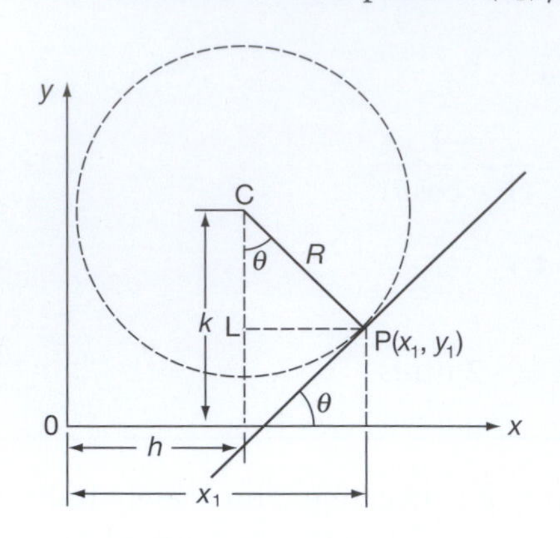


If the centre C is the point $(h, k)$, we can see from the diagram that:

$$h = x_1 - \text{LP} = x_1 - R\sin\theta$$
$$k = y_1 + \text{LC} = y_1 + R\cos\theta$$

$$\text{That is, } \begin{Bmatrix} h = x_1 - R\sin\theta \\ k = y_1 + R\cos\theta \end{Bmatrix}$$

where $x_1$ and $y_1$ are the coordinates of P, $R$ is the radius of curvature at P and $\theta$ is the angle of slope at P,

i.e. $\tan\theta = \left\{\frac{dy}{dx}\right\}_P$

38

As an example, find the radius of curvature and the coordinates of the centre of curvature of the curve $y = \frac{11-4x}{3-x}$ at the point (2, 3).

$$\frac{dy}{dx} = \frac{(3-x)(-4)-(11-4x)(-1)}{(3-x)^2} = \frac{-12+4x+11-4x}{(3-x)^2} = \frac{-1}{(3-x)^2}$$

$$\therefore \text{ At } x = 2, \ \frac{dy}{dx} = \frac{-1}{1} = -1 \qquad \therefore \left(\frac{dy}{dx}\right)^2 = 1$$

$$\frac{d^2y}{dx^2} = \frac{d}{dx}\left\{-(3-x)^{-2}\right\} = 2(3-x)^{-3}(-1) = \frac{-2}{(3-x)^3}$$

$$\therefore \text{ At } x = 2, \ \frac{d^2y}{dx^2} = \frac{-2}{1} = -2$$

$$R = \frac{\left\{1 + \left(\frac{dy}{dx}\right)^2\right\}^{3/2}}{\frac{d^2y}{dx^2}} = \frac{\{1+1\}^{3/2}}{-2} = \frac{2\sqrt{2}}{-2} = -\sqrt{2}$$

$$R = -\sqrt{2}$$

Now before we find the centre of curvature $(h, k)$ we must find the angle of slope $\theta$ from the fact that $\tan\theta = \frac{dy}{dx}$ at P.

i.e. $\tan\theta = -1 \qquad \therefore \theta = -45^\circ \qquad (\theta$ measured between $\pm 90^\circ)$

$\therefore \sin\theta = \ldots\ldots\ldots\ldots$ and $\cos\theta = \ldots\ldots\ldots\ldots$

**39**

$$\sin\theta = -\frac{1}{\sqrt{2}} \qquad \cos\theta = \frac{1}{\sqrt{2}}$$

So we have:

$x_1 = 2, \quad y_1 = 3$

$R = -\sqrt{2}$

$\sin\theta = -\frac{1}{\sqrt{2}}, \quad \cos\theta = \frac{1}{\sqrt{2}}$

$$\therefore\ h = x_1 - R\sin\theta = 2 - \left(-\sqrt{2}\right)\left(-\frac{1}{\sqrt{2}}\right) = 2 - 1 = 1, \ \ h = 1$$

$$k = y_1 + R\cos\theta = 3 + \left(-\sqrt{2}\right)\left(\frac{1}{\sqrt{2}}\right) = 3 - 1 = 2, \ \ k = 2$$

$\therefore$ centre of curvature C is the point $(1, 2)$

*Note*: If, by chance, the calculated value of $R$ is negative, the minus sign must be included when we substitute for $R$ in the expressions for $h$ and $k$.

*Next frame for a final example*

**40**

Find the radius of curvature and the centre of curvature for the curve $y = \sin^2\theta$, $x = 2\cos\theta$, at the point for which $\theta = \frac{\pi}{3}$.

Before we rush off and deal with this one, let us heed an important *warning*. You will remember that the centre of curvature $(h, k)$ is given by:

$$\left.\begin{aligned} h &= x_1 - R\sin\theta \\ k &= y_1 + R\cos\theta \end{aligned}\right\} \text{ and in these expressions}$$

$\theta$ is the angle of slope of the curve at the point being considered

i.e. $\tan\theta = \left\{\frac{dy}{dx}\right\}_P$

Now, in the problem stated above, $\theta$ is a parameter and not the angle of slope at any particular point. In fact, if we proceed with our usual notation, we shall be using $\theta$ to stand for two completely different things and that can be troublesome, to say the least.

So the safest thing to do is this. *Where you have to find the centre of curvature of a curve given in parametric equations involving $\theta$, change the symbol of the parameter to something other than $\theta$*. Then you will be safe. The trouble occurs only when we find C, not when we are finding $R$ only.

**41**

So, in this case, we will rewrite the problem thus:

Find the radius of curvature and the centre of curvature for the curve $y = \sin^2 t$, $x = 2\cos t$, at the point for which $t = \frac{\pi}{3}$.

Start off by finding the radius of curvature only. Then check your result so far with the solution given in the next frame before setting out to find the centre of curvature.

## 42

$R = -2{\cdot}795$, i.e. $2{\cdot}795$ units

Here is the working:

$$y = \sin^2 t \qquad \therefore \frac{dy}{dt} = 2\sin t\cos t$$

$$x = 2\cos t \qquad \therefore \frac{dx}{dt} = -2\sin t$$

$$\frac{dy}{dx} = \frac{dy}{dt}\cdot\frac{dt}{dx} = \frac{2\sin t\cos t}{-2\sin t} = -\cos t$$

$$\text{At } t = 60^\circ, \quad \frac{dy}{dx} = -\cos 60^\circ = -\frac{1}{2} \qquad \therefore \frac{dy}{dx} = -\frac{1}{2}$$

$$\text{Also } \frac{d^2y}{dx^2} = \frac{d}{dx}\left\{-\cos t\right\} = \frac{d}{dt}\left\{-\cos t\right\}.\frac{dt}{dx} = \frac{\sin t}{-2\sin t} = -\frac{1}{2}$$

$$\therefore \frac{d^2y}{dx^2} = -\frac{1}{2}$$

$$R = \frac{\left\{1 + \left(\frac{dy}{dx}\right)^2\right\}^{3/2}}{\frac{d^2y}{dx^2}} = \frac{\left\{1 + \frac{1}{4}\right\}^{3/2}}{-\frac{1}{2}} = -2\left\{\frac{5}{4}\right\}^{3/2}$$

$$= \frac{-10\sqrt{5}}{8} = \frac{-5\sqrt{5}}{4} = \frac{-5}{4}(2{\cdot}2361)$$

$$= \frac{-11{\cdot}1805}{4} = -2{\cdot}7951$$

$$R = -2{\cdot}795$$

*All correct so far? Move on to the next frame then*

## 43

Now to find the centre of curvature $(h, k)$:

$$h = x_1 - R\sin\theta$$
$$k = y_1 + R\cos\theta$$

where $\quad \tan\theta = \frac{dy}{dx} = -\frac{1}{2} \qquad \therefore \theta = -26^\circ 34' \quad (\theta \text{ between } \pm 90^\circ)$

$\therefore \sin(-26^\circ 34') = -0{\cdot}4472; \quad \cos(-26^\circ 34') = 0{\cdot}8944$

Also $\quad x_1 = 2\cos 60^\circ = 2.\frac{1}{2} = 1$

$$y_1 = \sin^2 60^\circ = \left\{\frac{\sqrt{3}}{2}\right\}^2 = \frac{3}{4}$$

and you have already proved that $R = -2{\cdot}795$.

What then are the coordinates of the centre of curvature?

*Calculate them and when you have finished, move on to the next frame*

44

$$h = -0{\cdot}25; \quad k = -1{\cdot}75$$

Because

$$h = 1 - (-2{\cdot}795)(-0{\cdot}4472)$$
$$= 1 - 1{\cdot}250$$
$$\therefore h = -0{\cdot}25$$

and

$$k = 0{\cdot}75 + (-2{\cdot}795)(0{\cdot}8944)$$
$$= 0{\cdot}75 - 2{\cdot}50$$
$$\therefore k = -1{\cdot}75$$

Therefore, the centre of curvature is the point $(-0{\cdot}25, -1{\cdot}75)$

This brings us to the end of this particular Programme. If you have followed it carefully and carried out the exercises set, you must know quite a lot about the topics we have covered. So move on now and look at the **Can You?** checklist before you work through the **Test exercise**.

## Can You?

### Checklist 9

45

Check this list before and after you try the end of Programme test.

On a scale of 1 to 5 how confident are you that you can:

| | Frames |
|---|---|
| • Evaluate the gradient of a straight line?<br>*Yes* ☐ ☐ ☐ ☐ ☐ *No* | 1 to 5 |
| • Recognize the relationship satisfied by two mutually perpendicular straight lines?<br>*Yes* ☐ ☐ ☐ ☐ ☐ *No* | 6 to 11 |
| • Derive the equations of a tangent and a normal to a curve?<br>*Yes* ☐ ☐ ☐ ☐ ☐ *No* | 12 to 23 |
| • Evaluate the curvature and radius of curvature at a point on a curve?<br>*Yes* ☐ ☐ ☐ ☐ ☐ *No* | 24 to 36 |
| • Locate the centre of curvature for a point on a curve?<br>*Yes* ☐ ☐ ☐ ☐ ☐ *No* | 37 to 44 |

# Test exercise 9

**46**

The questions are all straightforward.

1 Find the angle between the curves $x^2 + y^2 = 4$ and $5x^2 + y^2 = 5$ at their point of intersection for which $x$ and $y$ are positive.

2 Find the equations of the tangent and normal to the curve $y^2 = 11 - \dfrac{10}{4 - x}$ at the point (6, 4).

3 The parametric equations of a function are $x = 2\cos^3\theta$, $y = 2\sin^3\theta$. Find the equation of the normal at the point for which $\theta = \dfrac{\pi}{4} = 45^\circ$.

4 If $x = 1 + \sin 2\theta$, $y = 1 + \cos\theta + \cos 2\theta$, find the equation of the tangent at $\theta = 60^\circ$.

5 Find the radius of curvature and the coordinates of the centre of curvature at the point $x = 4$ on the curve whose equation is $y = x^2 + 5\ln x - 24$.

6 Given that $x = 1 + \sin\theta$, $y = \sin\theta - \frac{1}{2}\cos 2\theta$, show that $\dfrac{d^2y}{dx^2} = 2$.

Find the radius of curvature and the centre of curvature for the point on this curve where $\theta = 30^\circ$.

*Now you are ready for the next Programme*

# Further problems 9

**47**

1 Find the equation of the normal to the curve $y = \dfrac{2x}{x^2 + 1}$ at the point (3, 0·6) and the equation of the tangent at the origin.

2 Find the equations of the tangent and normal to the curve $4x^3 + 4xy + y^2 = 4$ at (0, 2), and find the coordinates of a further point of intersection of the tangent and the curve.

3 Obtain the equations of the tangent and normal to the ellipse $\dfrac{x^2}{169} + \dfrac{y^2}{25} = 1$ at the point $(13\cos\theta, 5\sin\theta)$. If the tangent and normal meet the $x$-axis at the points T and N respectively, show that ON.OT is constant, O being the origin of coordinates.

4 If $x^2y + xy^2 - x^3 - y^3 + 16 = 0$, find $\dfrac{dy}{dx}$ in its simplest form. Hence find the equation of the normal to the curve at the point (1, 3).

5 Find the radius of curvature of the catenary $y = c\cosh\left(\dfrac{x}{c}\right)$ at the point $(x_1, y_1)$.

6 If $2x^2 + y^2 - 6y - 9x = 0$, determine the equation of the normal to the curve at the point (1, 7).

7 Show that the equation of the tangent to the curve $x = 2a\cos^3 t$, $y = a\sin^3 t$, at any point P $\left(0 \le t \le \dfrac{\pi}{2}\right)$ is $x\sin t + 2y\cos t - 2a\sin t\cos t = 0$.

If the tangent at P cuts the $y$-axis at Q, determine the area of the triangle POQ.

**8** Find the equation of the normal at the point $x = a\cos\theta,\ y = b\sin\theta$, of the ellipse $\dfrac{x^2}{a^2} + \dfrac{y^2}{b^2} = 1$. The normal at P on the ellipse meets the major axis of the ellipse at N. Show that the locus of the mid-point of PN is an ellipse and state the lengths of its principal axes.

**9** For the point where the curve $y = \dfrac{x - x^2}{1 + x^2}$ passes through the origin, determine:

(a) the equations of the tangent and normal to the curve
(b) the radius of curvature
(c) the coordinates of the centre of curvature.

**10** In each of the following cases, find the radius of curvature and the coordinates of the centre of curvature for the point stated.

(a) $\dfrac{x^2}{25} + \dfrac{y^2}{16} = 1$ at (0, 4)
(b) $y^2 = 4x - x^2 - 3$ at $x = 2{\cdot}5$
(c) $y = 2\tan\theta,\ x = 3\sec\theta$ at $\theta = 45°$.

**11** Find the radius of curvature at the point (1, 1) on the curve $x^3 - 2xy + y^3 = 0$.

**12** If $3ay^2 = x(x - a)^2$ with $a > 0$, prove that the radius of curvature at the point $(3a, 2a)$ is $\dfrac{50a}{3}$.

**13** If $x = 2\theta - \sin 2\theta$ and $y = 1 - \cos 2\theta$, show that $\dfrac{dy}{dx} = \cot\theta$ and that $\dfrac{d^2y}{dx^2} = \dfrac{-1}{4\sin^4\theta}$. If $\rho$ is the radius of curvature at any point on the curve, show that $\rho^2 = 8y$.

**14** Find the radius of curvature of the curve $2x^2 + y^2 - 6y - 9x = 0$ at the point $(1, 7)$.

**15** Prove that the centre of curvature $(h, k)$ at the point P $(at^2, 2at)$ on the parabola $y^2 = 4ax$ has coordinates $h = 2a + 3at^2$, $k = -2at^3$.

**16** If $\rho$ is the radius of curvature at any point P on the parabola $x^2 = 4ay$, and S is the point $(0, a)$, show that $\rho = 2\sqrt{(\text{SP})^3/\text{SO}}$, where O is the origin of coordinates.

**17** The parametric equations of a curve are $x = \cos t + t\sin t$, $y = \sin t - t\cos t$. Determine an expression for the radius of curvature ($\rho$) and for the coordinates $(h, k)$ of the centre of curvature in terms of $t$.

**18** Find the radius of curvature and the coordinates of the centre of curvature of the curve $y = 3\ln x$, at the point where it meets the $x$-axis.

**19** Show that the numerical value of the radius of curvature at the point $(x_1,\ y_1)$ on the parabola $y^2 = 4ax$ is $\dfrac{2(a + x_1)^{3/2}}{a^{1/2}}$. If C is the centre of curvature at the origin O and S is the point $(a,\ 0)$, show that OC = 2(OS).

▶

20 The equation of a curve is $4y^2 = x^2(2 - x^2)$:

(a) Determine the equations of the tangents at the origin.

(b) Show that the angle between these tangents is $\tan^{-1}(2\sqrt{2})$.

(c) Find the radius of curvature at the point (1, 1/2).

48

Now visit the companion website for this book at www.palgrave.com/stroud for more questions applying this mathematics to science and engineering.

# Sequences

## Learning outcomes

When you have completed this Programme you will be able to:

- ☐ Recognize a sequence as a function
- ☐ Plot the graphs of sequences
- ☐ Recognize specific types of sequence
- ☐ Recognize arithmetic, geometric and harmonic sequences
- ☐ Convert the descriptive prescription of the output from arithmetic and geometric sequences into a recursive description and recognize the importance of initial terms
- ☐ Generate the recursive prescription of a sequence from a given sequence of numbers
- ☐ Determine the order and generate the terms of a difference equation
- ☐ Obtain the solution to an homogeneous, linear difference equation
- ☐ Derive the limit of a sequence using the rules of limits

# Functions with integer input

1

## Sequences

Any function $f$ whose input is restricted to positive or negative integer values $n$ has an output in the form of a sequence of numbers. Accordingly, such a function is called a **sequence**. For example the function defined by the prescription

$f(n) = 5n - 2$ where $n$ is a positive integer $\geq 1$

is a sequence. The first three output values corresponding to the successive input values $n = 1$, 2 and 3 are:

$$f(1) = 5 \times 1 - 2 = 3$$
$$f(2) = 5 \times 2 - 2 = 8$$
$$f(3) = 5 \times 3 - 2 = 13$$

Each output value of the sequence is called a **term** of the sequence, so 3, 8 and 13 are the first three terms of the sequence.

So, given the sequence $f(n) = 6 - 3n$ where $n$ is a positive integer $\geq 1$, the first three terms are:

............

*The answer is in the next frame*

2

3, 0, −3

Because:

$$f(1) = 6 - 3 \times 1 = 3$$
$$f(2) = 6 - 3 \times 2 = 0$$
$$f(3) = 6 - 3 \times 3 = -3$$

*Next frame*

3

## Graphs of sequences

Since the output of a sequence consists of a sequence of discrete numbers (the terms of the sequence) the graph of a sequence will take the form of a collection of isolated points in the Cartesian plane. For example, consider the sequence defined by the prescription

$$f(n) = \left(\frac{1}{2}\right)^n \quad n = 0, 1, 2, 3, \ldots$$

[Notice that here the first term corresponds to $n = 0$. The first term of a sequence need not necessarily correspond to $n = 1$, it can correspond to any integer value depending upon the purpose to which the sequence is to be put.]

▶

This sequence has the following successive terms:

$$f(0) = \left(\frac{1}{2}\right)^0 = 1$$
$$f(1) = \left(\frac{1}{2}\right)^1 = \frac{1}{2}$$
$$f(2) = \left(\frac{1}{2}\right)^2 = \frac{1}{4}$$
$$f(3) = \left(\frac{1}{2}\right)^3 = \frac{1}{8}$$
$$f(4) = \left(\frac{1}{2}\right)^4 = \frac{1}{16}$$
$$\ldots$$

and the following graph:

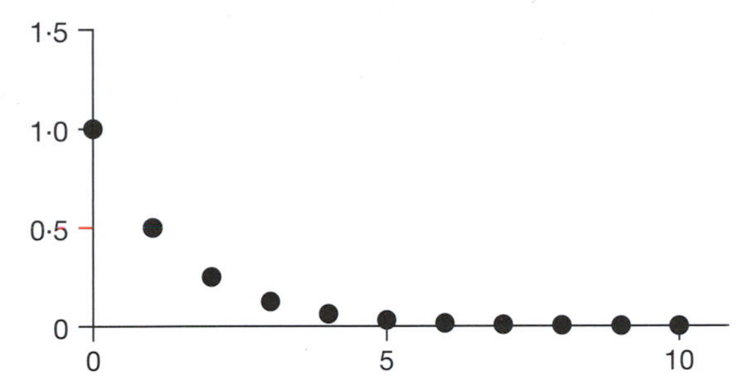


So the sequence defined by the prescription

$f(n) = n^2 \quad n = 0, 1, 2, \ldots$

has the following graph: ............

*The answer is in the next frame*

---

**4**

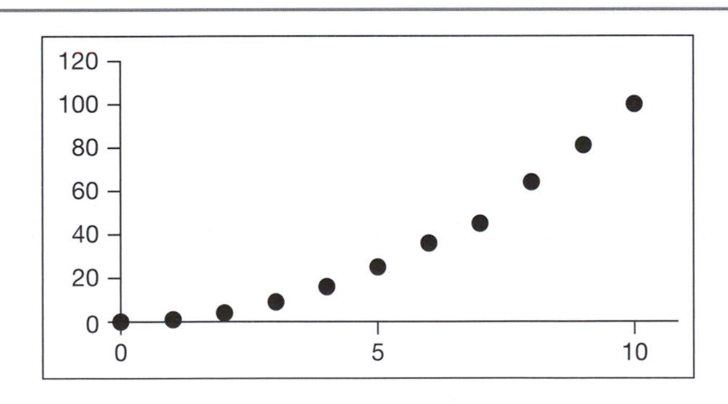


Because

$$f(0) = 0^2 = 0$$
$$f(1) = 1^2 = 1$$
$$f(2) = 2^2 = 4$$
$$f(3) = 3^2 = 9$$
$$f(4) = 4^2 = 16$$
$$\ldots$$

There are many different forms of sequence so let's start with some forms that you will have already seen.

*Next frame*

5

## Arithmetic sequence

Consider the sequence 2, 4, 6, 8, 10, ... where each term is obtained from the previous term by *adding* 2 to it. Such a sequence is called an **arithmetic sequence**. In its general form an arithmetic sequence is given by a prescription of the form:

$$f(n) = a + nd \quad n = 0, 1, 2, 3, \ldots$$

That is:

$$f(0) = a$$
$$f(1) = a + d$$
$$f(2) = a + 2d$$
$$f(3) = a + 3d$$
$$\ldots$$

The number $a$ is the *first term* because it is the output from the function when the input is $n = 0$. The number $d$ is called the *common difference* because it is the difference between any pair of successive terms and, despite being called a difference it is in fact the number that is *added* to a term to find the next term in the sequence. For example, consider the sequence

$$f(n) = 5 + 4n \quad n = 0, 1, 2, 3, \ldots$$

Comparing this prescription with that of the prescription for a general arithmetic sequences

$$f(n) = a + nd \quad n = 0, 1, 2, 3, \ldots$$

we can see that the first term is $a = 5$, the common difference is $d = 4$ and the first four terms are:

$$f(0) = 5 + 0 \times 4 = 5$$
$$f(1) = 5 + 1 \times 4 = 5 + 4 = 9$$
$$f(2) = 5 + 2 \times 4 = 5 + 8 = 13$$
$$f(3) = 5 + 3 \times 4 = 5 + 12 = 17$$

So the first term $a$, common difference $d$ and first four term of the arithmetic sequence defined by

$$f(n) = 3 + 2n \quad n = 0, 1, 2, 3, \ldots$$

are

$a =$ ..........., $d =$ ........... and ...........

*The answer is in the next frame*

6

$a = 3$, $d = 2$ and 3, 5, 7, 9

Comparing $f(n) = 3 + 2n \quad n = 0, 1, 2, 3, \ldots$ with the general form of an arithmetic sequence $f(n) = a + nd \quad n = 0, 1, 2, 3, \ldots$ we can see that $a = 3$, $d = 2$ and:

$$f(0) = 3 + 0 \times 2 = 3$$
$$f(1) = 3 + 1 \times 2 = 3 + 2 = 5$$
$$f(2) = 3 + 2 \times 2 = 3 + 4 = 7$$
$$f(3) = 3 + 3 \times 2 = 3 + 6 = 9$$

▶

Also, given the sequence of terms for example:

$-3, 4, 11, 18, \ldots$

we observe that each term is obtained by adding 7 to the previous term. This means that it is an arithmetic sequence with a common difference 7 and first term −3. The general term of this sequence is then obtained by substituting $a = -3$ and $d = 7$ into the prescription for the general arithmetic sequence $f(n) = a + nd$ giving:

$f(n) = -3 + 7n \quad n = 0, 1, 2, 3, \ldots$

So the general term of the sequence of terms 8, 5, 2, − 1, ... is ...........

*Next frame*

**7**

$$f(n) = 8 - 3n \quad n = 0, 1, 2, 3, \ldots$$

Because

The first term is $a = 8$ and each term is obtained from its predecessor by subtracting 3 giving the common difference as $d = -3$. The general term of the sequence is then obtained by substituting $a = 8$ and $d = -3$ into the prescription for the general arithmetic sequence $f(n) = a + nd$ giving:

$f(n) = 8 - 3n \quad n = 0, 1, 2, 3, \ldots$

## Geometric sequence

Consider the sequence 2, 4, 8, 16, 32, ... where each term is obtained from the previous term by *multiplying it* by 2. Such a sequence is called a **geometric sequence**. In its general form a geometric sequence is given by a prescription of the form:

$f(n) = Ar^n \quad n = 0, 1, 2, 3, \ldots$

That is

$f(0) = A$
$f(1) = Ar$
$f(2) = Ar^2$
$f(3) = Ar^3$
$\ldots$

The number $A$ is the *first term* because it is the output from the function when the input is $n = 0$. The number $r$ is called the *common ratio* because it is the ratio between any pair of successive terms – it is the number that a term is multiplied by to find the next term in the sequence. For example, consider the sequence:

$f(n) = 2(3)^n \quad n = 0, 1, 2, 3, \ldots$

Comparing this prescription with that of the prescription for a general geometric sequence

$f(n) = Ar^n \quad n = 0, 1, 2, 3, \ldots$

we can see that the first term is $A = 2$, the common ratio is $r = 3$ and the first four terms are:

$f(0) = 2$
$f(1) = 2 \times 3 = 6$
$f(2) = 2 \times 3^2 = 18$
$f(3) = 2 \times 3^3 = 54$

▶

So the first term $A$, the common ratio $r$ and first four terms of the geometric sequence defined by

$$f(n) = 5(2)^n \quad n = 0, 1, 2, 3, \ldots$$

are

$A = \ldots\ldots\ldots\ldots$, $r = \ldots\ldots\ldots\ldots$ and $\ldots\ldots\ldots\ldots$

*Next frame for the answer*

**8**

$A = 5$, $r = 2$ and 5, 10, 20, 40

Because

Comparing $f(n) = 5(2)^n \quad n = 0, 1, 2, 3, \ldots$ with the general form of a geometric sequence $f(n) = Ar^n \quad n = 0, 1, 2, 3, \ldots$ we see that $A = 5$, $r = 2$ and:

$$f(0) = 5$$
$$f(1) = 5 \times 2 = 10$$
$$f(2) = 5 \times 2^2 = 20$$
$$f(3) = 5 \times 2^3 = 40$$

Also, given the sequence of terms, for example:

$6, \ -12, \ 24, \ -48, \ 96, \ \ldots$

we observe that each term is obtained by multiplying the previous term by $-2$. This means that it is a geometric sequence with a common ratio $-2$ and first term 6. The general term of this sequence is then obtained by substituting $A = 6$ and $r = -2$ into the prescription for the general arithmetic sequence $f(n) = Ar^n$ giving

$$f(n) = 6(-2)^n \quad n = 0, 1, 2, 3, \ldots$$

So the general term of the sequence of terms 4, 12, 36, 108, ... is $\ldots\ldots\ldots\ldots$

*Next frame*

**9**

$f(n) = 4(3)^n \quad n = 0, 1, 2, 3, \ldots$

Because

The first term is $A = 4$ and each term is obtained from its predecessor by multiplying it by 3 giving the common ratio as $r = 3$. The general term of the sequence is then obtained by substituting $A = 4$ and $r = 3$ into the prescription for the general arithmetic sequence $f(n) = Ar^n$ giving:

$$f(n) = 4(3)^n \quad n = 0, 1, 2, 3, \ldots$$

## Harmonic sequence

A sequence of terms is said to be an **harmonic sequence** if the reciprocals of its terms form an arithmetic sequence. Accordingly, the sequence defined by the prescription

$$f(n) = \frac{1}{n} \quad n = 1, 2, 3, \ldots$$

is an *harmonic sequence*. The sequence of output values is then

$$1, \frac{1}{2}, \frac{1}{3}, \frac{1}{4}, \ldots$$

▶

and their reciprocals are 1, 2, 3, 4, ... which form an arithmetic sequence with first term 1 and common difference 1.

Notice that in the geometric and arithmetic sequences the first term is given for $n = 0$ whereas in this harmonic sequence the first term is given for $n = 1$. There is no hard and fast rule for the value of $n$ to start off a sequence – it is all a matter of circumstance.

*And now for a different way to generate the terms of a sequence in the next frame*

## Recursive prescriptions

10

So far we have defined a sequence by giving a prescription for the output $f(n)$ in terms of the input $n$. For example, in Frame 1 we considered the sequence defined by the prescription

$$f(n) = 5n - 2 \quad \text{where } n \text{ is a positive integer} \geq 1$$

The first three output values corresponding to the successive input values $n = 1$, 2 and 3 are:

$$f(1) = 5 \times 1 - 2 = 3$$
$$f(2) = 5 \times 2 - 2 = 8$$
$$f(3) = 5 \times 3 - 2 = 13$$

An alternative way of describing the terms of this sequence can be found from the following consideration:

$$f(1) = 5 \times 1 - 2 = 3$$
$$f(2) = 5 \times 2 - 2 = 8 = 3 + 5 \quad \text{that is } f(2) = f(1) + 5$$
$$f(3) = 5 \times 3 - 2 = 13 = 8 + 5 \text{ that is } f(3) = f(2) + 5$$
$$\ldots$$

Continuing in this way we see that adding 5 to the $n$th term gives the $(n + 1)$th term. That is:

$$f(n) + 5 = f(n + 1) \text{ or } f(n + 1) = f(n) + 5$$

The value of any term is the value of the previous term plus 5 and provided we know that the first term is 3 we can compute any other term.

$$f(1) = 3$$
$$f(2) = 3 + 5 = 8$$
$$f(3) = 8 + 5 = 13$$
$$f(4) = 13 + 5 = 18$$
$$\ldots$$

A process such as this one that repeatedly refers back to known values to compute an unknown value is called a **recursive process**. To obtain this recursive form in a systematic way we start with the general term of the sequence that is given as $f(n) = 5n - 2$. Increasing the input from $n$ to $n + 1$ we see that:

$$\begin{aligned} f(n + 1) &= 5(n + 1) - 2 \\ &= 5n + 5 - 2 \\ &= 5n - 2 + 5 \\ &= f(n) + 5 \end{aligned}$$

That is:

$$f(n + 1) = f(n) + 5 \quad \text{where } f(1) = 3$$

▶

This description where each term of the sequence is seen to depend upon another term of the same sequence is called a *recursive prescription* and can make the computing of the terms of the sequence more efficient and very amenable to a spreadsheet implementation. In this particular example we simply start with 3 and just add 5 to each preceding term to get:

3, 8, 13, 18, 23, 28, 33, 38, ...

Notice that without the initial term $f(1) = 3$ the recursive prescription would be of little worth because we would not know how to start the sequence.

Try one yourself. The recursive prescription of the sequence:

$f(n) = 7n + 4$ where $n$ is an integer $\geq 1$

is ...........

*Next frame*

**11**

$$f(n+1) = f(n) + 7 \text{ where } f(1) = 11$$

Because:

The general term of the sequence is given as $f(n) = 7n + 4$ so increasing the input from $n$ to $n + 1$ we see that:

$$\begin{aligned} f(n+1) &= 7(n+1) + 4 \\ &= 7n + 7 + 4 \\ &= 7n + 4 + 7 \\ &= f(n) + 7 \end{aligned}$$

That is:

$f(n+1) = f(n) + 7$ where $f(1) = 7 + 4 = 11$

From this recursive prescription the first five terms of the sequence are then:

...........

*Next frame*

**12**

11, 18, 25, 32, 39

Because:

Since $f(n+1) = f(n) + 7$ where $f(1) = 11$ then:

$$\begin{aligned} f(2) &= f(1) + 7 = 11 + 7 = 18 \\ f(3) &= f(2) + 7 = 18 + 7 = 25 \\ f(4) &= f(3) + 7 = 25 + 7 = 32 \\ f(5) &= f(4) + 7 = 32 + 7 = 39 \end{aligned}$$

Try another one just to make sure. The recursive prescription and first five terms of the sequence:

$f(n) = 8 - 2n$ where $n$ is an integer $\geq 0$

are ...........

*The answer is in the next frame*

13

$$f(n+1) = f(n) - 2 \text{ where } f(0) = 8$$
First five terms 8, 6, 4, 2, 0

Because

$$\begin{aligned} f(n+1) &= 8 - 2(n+1) \\ &= 8 - 2n - 2 \\ &= f(n) - 2 \end{aligned}$$

and $f(0) = 8 - 0 = 8$ so that

$$\begin{aligned} f(1) &= f(0) - 2 = 8 - 2 = 6 \\ f(2) &= f(1) - 2 = 6 - 2 = 4 \\ f(3) &= f(2) - 2 = 4 - 2 = 2 \\ f(4) &= f(3) - 2 = 2 - 2 = 0 \end{aligned}$$

We can create a recursive prescription for a geometric sequence as well. Given the general term of a geometric sequence as:

$$f(n) = Ar^n$$

then the next term in the sequence is:

$$\begin{aligned} f(n+1) &= Ar^{n+1} \\ &= Ar^n \times r \\ &= f(n) \times r \end{aligned}$$

That is

$$f(n+1) = rf(n)$$

For example, to find the recursive prescription for the geometric sequence with general term

$$f(n) = 5(4)^n \quad \text{where } n = 0, 1, 2, 3, \ldots$$

we note that:

$$\begin{aligned} f(n+1) &= 5(4)^{n+1} \\ &= 5(4)^n \times 4 \\ &= f(n) \times 4 \end{aligned}$$

So that $f(n+1) = 4f(n)$ where $f(0) = 5$ and

$$\begin{aligned} f(1) &= 4f(0) = 20 \\ f(2) &= 4f(1) = 80 \\ f(3) &= 4f(2) = 320 \\ f(4) &= 4f(3) = 1280 \\ &\ldots \end{aligned}$$

You try one. The recursive prescription and first five terms of the sequence:

$$f(n) = 6\left(\frac{1}{3}\right)^n \text{ where } n \geq 0 \text{ is } \ldots\ldots\ldots\ldots$$

*Next frame for the answer*

**14**

$$f(n+1) = \frac{f(n)}{3} \text{ where } f(0) = 6$$

First five terms $6,\ 2,\ \frac{2}{3},\ \frac{2}{9},\ \frac{2}{27}$

Because

$$\begin{aligned} f(n+1) &= 6\left(\frac{1}{3}\right)^{n+1} \\ &= 6\left(\frac{1}{3}\right)^{n}\left(\frac{1}{3}\right) \\ &= f(n) \times \left(\frac{1}{3}\right) \end{aligned}$$

so $f(n+1) = \frac{f(n)}{3}$ and $f(0) = 6$ so that

$$\begin{aligned} f(1) &= \frac{f(0)}{3} = \frac{6}{3} = 2, \\ f(2) &= \frac{f(1)}{3} = \frac{2}{3}, \\ f(3) &= \frac{f(2)}{3} = \frac{2/3}{3} = \frac{2}{9}, \\ f(4) &= \frac{f(3)}{3} = \frac{2/9}{3} = \frac{2}{27} \end{aligned}$$

and finally, the recursive prescription and first five terms of the sequence:

$f(n) = 5^n$ where $n$ is an integer $\geq -3$

are ............

*The answers are in the following frame*

**15**

$$f(n+1) = 5f(n) \text{ where } f(-3) = \frac{1}{125}$$

First five terms $\frac{1}{125},\ \frac{1}{25},\ \frac{1}{5},\ 1,\ 5$

Because:

$$\begin{aligned} f(n+1) &= 5^{n+1} \\ &= 5^n \times 5 \\ &= f(n) \times 5 \end{aligned}$$

So $f(n+1) = 5f(n)$ and the first term is $f(-3) = 5^{-3} = \frac{1}{125}$ so that

$$\begin{aligned} f(-2) &= 5 \times f(-3) = 5 \times \frac{1}{125} = \frac{1}{25}, \\ f(-1) &= 5 \times f(-2) = 5 \times \frac{1}{25} = \frac{1}{5}, \\ f(0) &= 5 \times f(-1) = 5 \times \frac{1}{5} = 1, \\ f(1) &= 5 \times f(0) = 5 \times 1 = 5 \end{aligned}$$

The first five terms are then $\frac{1}{125},\ \frac{1}{25},\ \frac{1}{5},\ 1,\ 5$

**16**

## Other sequences

Arithmetic, geometric and harmonic sequences are just three specific types of sequence, each of whose general term conforms to a given pattern. For example, every arithmetic sequence has a general term that can be written in the form $f(n) = a + nd$. This is not the case for a myriad of other sequences each of which stand alone and do not conform to any particular type. Take, for example, the sequence of numbers:

$$0, 1, 1, 2, 3, 5, 8, 13, 21, \ldots$$

By examining this sequence of numbers closely you will see that each number from the third number onwards is the sum of the two previous numbers:

$$1 = 0 + 1$$
$$2 = 1 + 1$$
$$3 = 1 + 2$$
$$5 = 2 + 3$$
$$\ldots$$

If we represent the general term of this sequence by $f(n)$ then the fact that any term of the sequence is found by adding the two previous terms can be expressed as:

$$f(n) = f(n-1) + f(n-2) \text{ where } f(0) = 0 \text{ and } f(1) = 1$$

or as

$$f(n+2) = f(n+1) + f(n) \text{ where } f(0) = 0 \text{ and } f(1) = 1$$

It does not matter which equation we use to describe the recursive sequence as they are entirely equivalent. Notice that this recursive prescription requires **two** initial terms for it to get started. This particular sequence is called the *Fibonacci sequence* but most such sequences are not honoured with a name. See how you get on with each of the following.

By close inspection of the terms of each of the following sequences from the third term onwards find the general term of each sequence:

(a) $1, 2, 1, -1, -2, -1, 1, \ldots\ldots\ldots\ldots$

(b) $0, 1, 1, 3, 5, 11, 21, 43, \ldots\ldots\ldots\ldots$

(c) $1, 2, 2, 4, 8, 32, 256, 8192, \ldots\ldots\ldots\ldots$

*The answers are in the following frame*

**17**

(a) $f(n+2) = f(n+1) - f(n)$
(b) $f(n+2) = f(n+1) + 2f(n)$
(c) $f(n+2) = f(n+1)f(n)$

▶

Because

(a) $1, 2, 1, -1, -2, -1, 1, \ldots$
Here each term from the third term onwards is obtained by subtracting from the previous term the next previous term:

$$\begin{aligned}1 &= 2 - 1\\ -1 &= 1 - 2\\ -2 &= -1 - 1\\ -1 &= -2 - (-1)\\ 1 &= -1 - (-2)\\ &\ldots\end{aligned}$$

The general term is then:

$f(n) = f(n-1) - f(n-2)$ where $f(0) = 1$ and $f(1) = 2$

or

$f(n+2) = f(n+1) - f(n)$ where $f(0) = 1$ and $f(1) = 2$

(b) $0, 1, 1, 3, 5, 11, 21, 43, \ldots$
Here each term from the third term onwards is obtained by adding the previous term to twice the next previous term:

$$\begin{aligned}1 &= 1 + 2 \times 0\\ 3 &= 1 + 2 \times 1\\ 5 &= 3 + 2 \times 1\\ 11 &= 5 + 2 \times 3\\ 21 &= 11 + 2 \times 5\\ 43 &= 21 + 2 \times 11\\ &\ldots\end{aligned}$$

The general term is then:

$f(n) = f(n-1) + 2f(n-2)$ where $f(0) = 0$ and $f(1) = 1$

or

$f(n+2) = f(n+1) + 2f(n)$ where $f(0) = 0$ and $f(1) = 1$

(c) $1, 2, 2, 4, 8, 32, 256, 8192, \ldots$
Here each term from the third term onwards is obtained by multiplying the previous term by the next previous term:

$$\begin{aligned}2 &= 2 \times 1\\ 4 &= 2 \times 2\\ 8 &= 4 \times 2\\ 32 &= 8 \times 4\\ 256 &= 32 \times 8\\ 8192 &= 256 \times 32\\ &\ldots\end{aligned}$$

The general term is then:

$f(n) = f(n-1)f(n-2)$ where $f(0) = 1$ and $f(1) = 2$

or

$f(n+2) = f(n+1)f(n)$ where $f(0) = 1$ and $f(1) = 2$

***Now let us pause for a brief summary and exercise covering what we have just studied on specific types of sequence.***

##  Review summary

18

1 *Sequences*
Any function $f$ whose input is restricted to positive or negative integer values $n$ has an output in the form of a sequence of numbers. Accordingly, such a function is called a **sequence**.

2 *Graphs of sequences*
Since the range of a sequence consists of a sequence of discrete numbers – the terms of the sequence – the graph of a sequence will take the form of a collection of isolated points in the Cartesian plane.

3 *Arithmetic sequence*
Any sequence defined by a prescription of the form
$f(n) = a + nd \quad n + 0, 1, 2, \ldots$
is called an arithmetic sequence.

4 *Geometric sequence*
Any sequence defined by a prescription of the form
$f(n) = Ar^n \quad n = 0, 1, 2, \ldots$
is called a geometric sequence.

5 *Harmonic sequence*
The sequence defined by the prescription
$$f(n) = \frac{1}{g(n)}$$
where $g(n)$ is a term of an arithmetic sequence is called an **harmonic sequence**.

6 *Recursive prescriptions*
A prescription where each term of a sequence is seen to depend upon another term of the same sequence is called a *recursive prescription* and can make the computing of the terms of the sequence more efficient and very amenable to a spreadsheet implementation. For example:
$f(n+1) = f(n) + 5$ where $f(1) = 3$
which tells us that each term of a sequence is obtained by adding 5 to the previous term and that the first term of the sequences is 3.

##  Review exercise

19

1 Plot the graph of:
$f(n) = n - 3$ where $n \geq 0$

2 Find the next two terms and form of the general term for each of the following sequences:
(a) 7, 14, 21, 28, 35, ...
(b) 6, 12, 24, 48, ...
(c) 1, 2, 3, 5, 8, ...
(d) 1, 2, 0, −4, −4, 4 ...

3 Find the recursive description of the prescription:
$f(n) = 12 - 4n$ where $n$ is an integer $\geq 1$

**20**

1 The graph of $f(n) = n - 3$ where $n \geq 0$

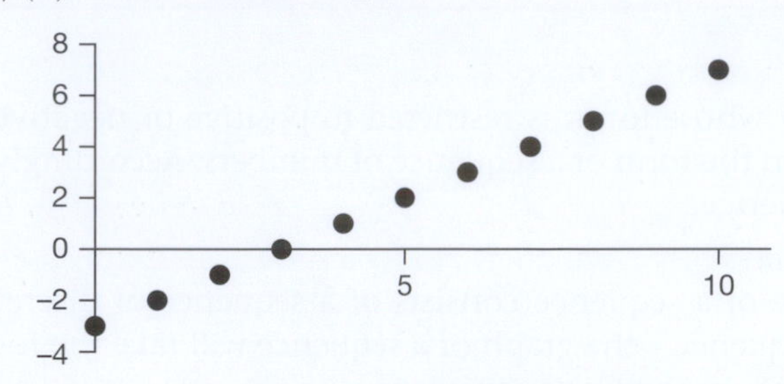


2 (a) 7, 14, 21, 28, 35, ... The next two terms are 42 and 49 because each term is the same as the previous one plus 7. The general term is $f(n) = 7 + 7n$ where $n \geq 0$.

(b) 6, 12, 24, 48, ... The next two terms are 96 and 192 because each term is double the previous one. The general term is $f(n) = 6(2^n)$ where $n \geq 0$.

(c) 1, 2, 3, 5, 8, ... The next two terms are 13 and 21 because each term is the sum of the previous two terms. The general term is

$$f(n+2) = f(n+1) + f(n) \text{ where } n \geq 0,\ f(0) = 1 \text{ and } f(1) = 2$$

(d) 1, 2, 0, −4, −4, 4 ... The next two terms are 12 and −20 because each term is the difference between the previous term and twice the next previous term. The general term is

$$f(n+2) = f(n+1) - 2f(n) \text{ where } n \geq 0,\ f(0) = 1 \text{ and } f(1) = 2$$

3 Each term of the sequence:

$f(n) = 12 - 4n$ where $n$ is an integer $\geq 1$

is obtained by subtracting 4 from the previous term where the first term is 8. To obtain the recursive prescription we find that:

$$\begin{aligned} f(n+1) &= 12 - 4(n+1) \\ &= 12 - 4n - 4 \\ &= f(n) - 4 \end{aligned}$$

Therefore:

$f(n+1) = f(n) - 4$ where $f(1) = 8$

*Now let us move on. Next frame*

## Difference equations

**21**

The prescription for a sequence $f(n) = 5n - 2$ can be written as the recursive equation $f(n+1) = f(n) + 5$. It can also be written as an equation with all the unknowns on the left-hand side as:

$$f(n+1) - f(n) = 5$$

In this form it is an example of a *first-order, constant coefficient, linear* ***difference*** *equation* also referred to as a *linear* ***recurrence*** *relation*. It is linear because there are no products of terms such as $f(n+1)f(n)$, it is first-order because $f(n+1)$ is just one term away from $f(n)$, and it has constant coefficients (the numbers multiplying the $f(n)$ and $f(n+1)$ are constants and do not involve $n$). The order of a difference equation is taken from the maximum number of terms between any pair of terms so that, for example:

(a) $f(n+2) + 2f(n) = 3n^4 + 2$ is a second-order difference equation because $f(n+2)$ is two terms away from $f(n)$.

▶

(b) $-3f(n+3) - f(n+2) + 5f(n-1) + 4f(n-2) = -6n^2$ is a fifth-order difference equation because $f(n+3)$ is five terms away from $f(n-2)$.

So the order of the difference equation:

$$89f(n-3) + 17f(n+1) - 3f(n+5) = 13n^2 - 2n^4 \text{ is } \ldots\ldots\ldots\ldots$$

*The answer is in the following frame*

---

22

$$\boxed{8}$$

Because:

$f(n+5)$ is 8 terms away from $f(n-3)$

In order to generate the terms of the sequence from the recursive description it is necessary to have as many initial terms as the order of the difference equation. For example if we are given the second-order difference equation with a single initial term:

$f(n+2) + 2f(n) = 0$ where $f(1) = 1$

Then we shall only be able to generate the values of those terms that correspond to odd values of $n$. Let's see why:

$n = 1$: $f(1+2) + 2f(1) = 0$ that is $f(3) + 2f(1) = 0$ so $f(3) = -2$
$n = 2$: $f(2+2) + 2f(2) = 0$ that is $f(4) + 2f(2) = 0$ so $f(4) = -2f(2)$

and since we do not know the value of $f(2)$ we cannot find the value of $f(4)$. Similarly, since $f(5) = -2f(3)$ then $f(5) = 4$ but we cannot find the value of $f(6)$ because $f(6) = -2f(4)$ and we do not know the value of $f(4)$. If, however, we had been given a second initial term, namely $f(2)$ then we would have been able to find the values of all those terms corresponding to an even value of $n$.

So, to emphasize the point we repeat that in order to generate the terms of the sequence from the recursive description it is necessary to have as many initial terms as the order of the difference equation.

*Next frame*

---

## Solving difference equations

23

We have seen how the prescription for a sequence such as:

$f(n) = 5n - 2$ where $n$ is an integer $\geq 1$

can be manipulated to create the difference equation:

$f(n+1) - f(n) = 5$ where $f(1) = 3$

What we wish to be able to do now is to reverse this process. That is, given the difference equation we wish to find the prescription for the sequence which is the *solution to the difference equation*. To begin this topic we shall look at first-order, homogeneous difference equations of the form:

$af(n+1) + bf(n) = 0$

where the $a$ and $b$ are known constants. An equation of this form with zero on the right hand side of this equation is called an homogeneous equation. We shall consider a specific example which will typify the method of obtaining the solution.

*Next frame*

**24**

Consider the difference equation:

$f(n+1) + 9f(n) = 0$ for $n \geq 0$ where $f(0) = 6$

To find the form of the general term $f(n)$ that satisfies this equation we first assume a solution of the form:

$f(n) = Kw^n$ where $K$ and $w$ are non-zero real numbers and $n$ is an integer.

If we substitute this form for $f(n)$ into the difference equation we find that:

$$Kw^{n+1} + 9Kw^n = 0$$

Factorizing out $Kw^n$ gives us the *characteristic equation*:

$$\begin{aligned} Kw^{n+1} + 9Kw^n &= Kw^n(w+9) \\ &= 0 \end{aligned}$$

This means that $w + 9 = 0$ and so $w = -9$. We shall, therefore, write the solution as:

$$f(n) = K(-9)^n$$

If we now consider the initial term $f(0) = 6$ we see that:

$$f(0) = K \times (-9)^0 = K = 6$$

So that the solution to the difference equation is given as:

$$f(n) = 6(-9)^n$$

As a check that the working was correct we substitute this solution back into the difference equation to obtain:

$$\begin{aligned} f(n+1) + 9f(n) &= 6(-9)^{n+1} + 9 \times 6(-9)^n \\ &= (-9)^n(6(-9) + 9 \times 6) \\ &= 0 \end{aligned}$$

so confirming the validity of the solution. Try one yourself.

The solution to the difference equation:

$f(n+1) - 12f(n) = 0$ for $n \geq 1$ where $f(1) = -3$ is ...........

*Next frame*

**25**

$$\boxed{f(n) = -\frac{12^n}{4}}$$

Because:

Given $f(n+1) - 12f(n) = 0$ for $n \geq 1$ where $f(1) = -3$

To find the form of the general term $f(n)$ that satisfies this equation we first assume a solution of the form:

$f(n) = Kw^n$ where $K$ and $w$ are non-zero real numbers

If we substitute this form for $f(n)$ into the difference equation we find that:

$$Kw^{n+1} - 12Kw^n = 0$$

Factorizing out $Kw^n$ gives us the characteristic equation:

$$\begin{aligned} Kw^{n+1} - 12Kw^n &= Kw^n(w-12) \\ &= 0 \end{aligned}$$

▶

This means that $w = 12$. We shall, therefore, write the solution as:

$$f(n) = K \times 12^n$$

If we now consider the initial term we see that:

$$f(1) = K \times 12^1 = 12K = -3 \text{ so } K = -\frac{1}{4}$$

So that the solution to the difference equation is given as:

$$f(n) = -\frac{12^n}{4}$$

As a check that the working was correct we substitute this solution back into the difference equation to obtain:

$$\begin{aligned} f(n+1) - 12f(n) &= -\frac{12^{n+1}}{4} - 12 \times \left(-\frac{12^n}{4}\right) \\ &= -\frac{12^{n+1}}{4} + \frac{12^{n+1}}{4} \\ &= 0 \end{aligned}$$

so confirming the validity of the solution.

*Next frame*

## Second-order, homogeneous equations

26

The solution to a second-order, homogeneous difference equation follows the same procedure with the exception that the characteristic equation has two roots and so a linear combination of two solutions is required. Again, we proceed by example.

Consider the difference equation:

$$f(n+2) - 7f(n+1) + 12f(n) = 0 \text{ for } n \geq 0 \text{ where } f(0) = 0 \text{ and } f(1) = 1$$

This is a second-order difference equation because $f(n+2)$ is two terms away from $f(n)$. To find the form of the general term $f(n)$ that satisfies this equation we first assume as before a solution of the form:

$f(n) = Kw^n$ where $K$ and $w$ are non-zero real numbers and $n$ is an integer.

If we substitute this form for $f(n)$ into the difference equation we find that:

$$Kw^{n+2} - 7Kw^{n+1} + 12Kw^n = 0$$

Factorizing out $Kw^n$ gives us the characteristic equation:

$$\begin{aligned} Kw^{n+2} - 7Kw^{n+1} + 12Kw^n &= Kw^n(w^2 - 7w + 12) \\ &= Kw^n(w-3)(w-4) \\ &= 0 \end{aligned}$$

This means that $w - 3 = 0$ or $w - 4 = 0$ giving $w = 3$ or $4$ giving $f(n) = A \times 3^n$ and $f(n) = B \times 4^n$ as two distinct solutions where $A$ and $B$ are two constants. These two solutions can be combined as:

$$f(n) = A \times 3^n + B \times 4^n$$

which, by substitution into the difference equation, can be seen to ...........

*Next frame*

**27**

satisfy the difference equation

Because:

$$\begin{aligned} &f(n+2) - 7f(n+1) + 12f(n) \\ &= A \times 3^{n+2} + B \times 4^{n+2} - 7(A \times 3^{n+1} + B \times 4^{n+1}) + 12(A \times 3^n + B \times 4^n) \\ &= 3^n[A \times 3^2 - 7A \times 3^1 + 12A] + 4^n[B \times 4^2 - 7B \times 4^1 + 12B] \\ &= 3^n[9A - 21A + 12A] + 4^n[16B - 28B + 12B] \\ &= 0 \end{aligned}$$

We shall, therefore, write the general solution as:

$$f(n) = A \times 3^n + B \times 4^n$$

If we now consider the two given terms we see that:

$$f(0) = A \times 3^0 + B \times 4^0 = A + B = 0$$

$$f(1) = A \times 3^1 + B \times 4^1 = 3A + 4B = 1$$

Solving these two simultaneous equations gives $A = -1$ and $B = 1$ so that the solution to the difference equation is given as:

$$\begin{aligned} f(n) &= (-1) \times 3^n + (1) \times 4^n \\ &= 4^n - 3^n \end{aligned}$$

As a check that the working was correct we substitute this solution back into the difference equation to obtain:

$$\begin{aligned} f(n+2) - 7f(n+1) + 12f(n) &= 4^{n+2} - 3^{n+2} - 7(4^{n+1} - 3^{n+1}) + 12(4^n - 3^n) \\ &= 4^n(4^2 - 7 \times 4 + 12) - 3^n(3^2 - 7 \times 3 + 12) \\ &= 4^n(16 - 28 + 12) - 3^n(9 - 21 + 12) \\ &= 0 \end{aligned}$$

so confirming the validity of the solution.

Now you try one. The solution to the difference equation:

$$f(n+2) - 5f(n+1) + 6f(n) = 0 \text{ for } n \geq 0 \text{ where } f(0) = 1 \text{ and } f(1) = -1$$

is . . . . . . . . . . .

*Next frame*

**28**

$$f(n) = 2^{n+2} - 3^{n+1}$$

Because:

Given $f(n+2) - 5f(n+1) + 6f(n) = 0$ for $n \geq 0$ where $f(0) = 1$ and $f(1) = -1$

To find the form of the general term $f(n)$ that satisfies this equation we first assume a solution of the form:

$f(n) = Kw^n$ where $K$ and $w$ are non-zero real numbers

If we substitute this form for $f(n)$ into the difference equation we find that:

$$Kw^{n+2} - 5Kw^{n+1} + 6Kw^n = 0$$

Factorizing out $Kw^n$ gives us the characteristic equation:

$$\begin{aligned} Kw^{n+2} - 5Kw^{n+1} + 6Kw^n &= Kw^n(w^2 - 5w + 6) \\ &= Kw^n(w-2)(w-3) \\ &= 0 \end{aligned}$$

▶

This means that $w = 2$ or 3. We shall, therefore, permit all possibilities by writing:

$$f(n) = A \times 2^n + B \times 3^n$$

If we now consider the two given terms we see that:

$$f(0) = A \times 2^0 + B \times 3^0 = A + B = 1$$
$$f(1) = A \times 2^1 + B \times 3^1 = 2A + 3B = -1$$

Solving these two simultaneous equations gives $A = 4$ and $B = -3$ so that the solution to the difference equation is given as:

$$\begin{aligned} f(n) &= 4 \times 2^n - 3 \times 3^n \\ &= 2^{n+2} - 3^{n+1} \end{aligned}$$

As a check that the working was correct we substitute this solution back into the difference equation to obtain:

$$\begin{aligned} f(n+2) - 5f(n+1) + 6f(n) &= 2^{n+4} - 3^{n+3} - 5(2^{n+3} - 3^{n+2}) + 6(2^{n+2} - 3^{n+1}) \\ &= 2^{n+2}(2^2 - 5 \times 2 + 6) - 3^{n+1}(3^2 - 5 \times 3 + 6) \\ &= 2^{n+2}(4 - 10 + 6) - 3^{n+1}(9 - 15 + 6) \\ &= 0 \end{aligned}$$

so confirming the validity of the solution.

*Next frame*

## Equal roots of the characteristic equation

**29**

If the characteristic equation associated with a second-order, homogeneous difference equation has two equal roots then the solution to the difference equation takes on a different form. For example, to obtain the solution to the difference equation:

$$f(n+2) - 4f(n+1) + 4f(n) = 0 \text{ for } n \geq 0 \text{ where } f(0) = 0 \text{ and } f(1) = 1$$

we proceed as usual and assume that the general term $f(n)$ that satisfies this equation takes the form:

$f(n) = Kw^n$ where $K$ and $w$ are non-zero real numbers

If we substitute this form for $f(n)$ into the difference equation we find that:

$$Kw^{n+2} - 4Kw^{n+1} + 4Kw^n = 0$$

Factorizing out $Kw^n$ gives us the characteristic equation:

$$\begin{aligned} Kw^{n+2} - 4Kw^{n+1} + 4Kw^n &= Kw^n(w^2 - 4w + 4) \\ &= Kw^n(w - 2)^2 \\ &= 0 \end{aligned}$$

This means that the characteristic equation has two equal roots, $w = 2$ which unfortunately only provides one solution of the usual form $f(n) = A \times 2^n$. The difference equation is of second order and requires two distinct solutions. By substituting $f(n) = Bn \times 2^n$ into the difference equation we find that:

$$\begin{aligned} f(n+2) - 4f(n+1) + 4f(n) &= B(n+2)2^{n+2} - 4B(n+1)2^{n+1} + 4Bn2^n \\ &= B2^n\{(n+2)2^2 - 4(n+1)2^1 + 4n\} \\ &= B2^n\{4n + 8 - 8n - 8 + 4n\} \\ &= 0 \end{aligned}$$

Therefore $f(n) = Bn \times 2^n$ is also a solution to the difference equation giving the general solution as:

$$f(n) = (A + Bn) \times 2^n$$

▶

Again, applying the two given terms $f(0) = 0$ and $f(1) = 1$ we find the solution to be:

$f(n) = \ldots\ldots\ldots\ldots$

*Next frame*

**30**

$$\boxed{f(n) = n2^{n-1}}$$

Because:

Applying the two given terms $f(0) = 0$ and $f(1) = 1$ to $f(n) = (A + Bn) \times 2^n$ we find:

$$f(0) = (A + 0) \times 2^0 = A = 0$$
$$f(1) = (A + B) \times 2^1 = 2B = 1 \text{ so } B = 1/2$$

So that the solution to the difference equation is given as:

$$\begin{aligned} f(n) &= \left(\frac{1}{2}\right) n2^n \\ &= \frac{n2^n}{2} \\ &= n2^{n-1} \end{aligned}$$

As a check of the working, substitute this solution back into the difference equation to obtain:

$$\begin{aligned} f(n+2) - 4f(n+1) + 4f(n) &= (n+2)2^{n+1} - 4(n+1)2^n + 4n2^{n-1} \\ &= 2^{n-1}\left((n+2)2^2 - 4(n+1) \times 2 + 4n\right) \\ &= 2^{n-1}(4n + 8 - 8n - 8 + 4n) \\ &= 0 \end{aligned}$$

so confirming the validity of the solution.

Now you try one from the beginning. The solution to the difference equation:

$f(n+2) + 2f(n+1) + f(n) = 0$ for $n \geq 1$ where $f(1) = 0$ and $f(2) = 2$

is $\ldots\ldots\ldots\ldots$

*Next frame*

**31**

$$\boxed{f(n) = (2n - 2)(-1)^n}$$

Because:

Given $f(n+2) + 2f(n+1) + f(n) = 0$ for $n \geq 1$ where $f(1) = 0$ and $f(2) = 2$

We proceed as usual and assume that the general term $f(n)$ that satisfies this equation takes the form:

$f(n) = Kw^n$ where $K$ and $w$ are non-zero real numbers

If we substitute this form for $f(n)$ into the difference equation we find that:

$$Kw^{n+2} + 2Kw^{n+1} + Kw^n = 0$$

Factorizing out $Kw^n$ gives us the characteristic equation:

$$\begin{aligned} Kw^{n+2} + 2Kw^{n+1} + Kw^n &= Kw^n(w^2 + 2w + 1) \\ &= Kw^n(w+1)^2 \\ &= 0 \end{aligned}$$

▶

This means that the characteristic equation has two equal roots, $w = -1$. In this case the solution is given as:

$$f(n) = (A + Bn) \times (-1)^n$$

If we now consider the two given terms we see that:

$$f(1) = (A + B) \times (-1)^1 = -A - B = 0$$

$$f(2) = (A + 2B) \times (-1)^2 = A + 2B = 2 \text{ with solution } A = -B = -2$$

So that the solution to the difference equation is given as:

$$f(n) = (2n - 2)(-1)^n$$

As a check of the working substitute this solution back into the difference equation to obtain:

$$\begin{aligned}
&f(n+2) + 2f(n+1) + f(n) \\
&= (2[n+2] - 2)(-1)^{n+2} + 2(2[n+1] - 2)(-1)^{n+1} + (2n - 2)(-1)^n \\
&= (-1)^n\left[(2n + 4 - 2)(-1)^2 + 2(2n + 2 - 2)(-1)^1 + 2n - 2\right] \\
&= (-1)^n[2n + 2 - 4n + 2n - 2] \\
&= 0
\end{aligned}$$

so confirming the validity of the solution.

***We now pause for a summary of difference equations and their solutions***

---

##  Review summary

32

1 *Difference equations*
Any equation of the form:

$$a_{n+m}f(n+m) + a_{n+m-1}f(n+m-1) + a_{n+m-2}f(n+m-2) + \ldots + a_n f(n) = 0$$

where the $a$'s are constants is an $m$th-order, homogeneous, constant coefficient, linear difference equation also referred to as a linear recurrence relation.

2 *Solving first-order difference equations*
Any first-order difference equation $af(n+1) + bf(n) = 0$ will have a solution of the form $f(n) = Kw^n$ where the value of $n$ can be found from the characteristic equation obtained by substituting the solution $f(n) = Kw^n$ into the difference equation. The value of $K$ is found by applying the initial term.

3 *Solving second-order difference equations*
Any second-order difference equation $af(n+2) + bf(n+1) + cf(n) = 0$ will have a solution of the form $f(n) = Kw^n$ where the value of $n$ can be found from the characteristic equation obtained by substituting the solution $f(n) = Kw^n$ into the difference equation. Since the characteristic equation is a quadratic there will be two values of $w$ yielding the solution:

$$f(n) = \begin{cases} A{w_1}^n + B{w_2}^n & \text{if } w_1 \neq w_2 \text{ and} \\ (A + Bn)w^n & \text{if } w_1 = w_2 = w \end{cases}$$

The values of $A$ and $B$ are found by applying two given terms.

---

## Review exercise

**33**

1 Solve each of the following difference equations:

(a) $f(n+1)+6f(n)=0$ where $f(0)=-2$

(b) $f(n+2)-4f(n)=0$ where $f(0)=-1$ and $f(1)=1$

(c) $f(n+2)-6f(n+1)+9f(n)=0$ where $f(0)=1$ and $f(1)=3$

**34**

1 (a) $f(n+1)+6f(n)=0$ where $f(0)=-2$

Assuming a solution of the form $f(n)=Kw^n$ gives the characteristic equation:

$$Kw^n(w+6)=0 \text{ therefore } w=-6$$

This gives the solution as $f(n)=K(-6)^n$. Applying the initial term $f(0)=-2$ yields $-2=K(-6)^0=K$ so that $f(n)=-2(-6)^n$.

(b) $f(n+2)-4f(n)=0$ where $f(0)=-1$ and $f(1)=1$

Assuming a solution of the form $f(n)=Kw^n$ gives the characteristic equation:

$$Kw^n(w^2-4)=0 \text{ therefore } w=\pm 2$$

This gives the solution as $f(n)=A(2)^n+B(-2)^n$. Applying the two given terms yields:

$$f(0)=A+B=-1$$
$$f(1)=2A-2B=1$$

so that $A=-\frac{1}{4}$ and $B=-\frac{3}{4}$ giving

$$f(n)=-\frac{1}{4}(2)^n-\frac{3}{4}(-2)^n=-2^{n-2}-3(-1)^n 2^{n-2}$$

(c) $f(n+2)-6f(n+1)+9f(n)=0$ where $f(0)=1$ and $f(1)=3$

Assuming a solution of the form $f(n)=Kw^n$ gives the characteristic equation:

$$Kw^n(w-3)^2=0 \text{ therefore } w=3 \text{ (coincident roots)}$$

This gives the solution as $f(n)=(A+Bn)(3)^n$. Applying the two given terms yields:

$$f(0)=(A+0)(3)^0=A=1$$
$$f(1)=(A+B)3=3$$

so that $A=1$ and $B=0$ giving $f(n)=3^n$

*And now for something different. Next frame*

## Limits of sequences

**35**

Each successive term of a sequence $f(n)$ is obtained by increasing the input integer $n$ by unity. As a consequence it is natural to ask how the output behaves as the input integer takes on larger and larger values. For example, consider the sequence with output

$$f(n)=\frac{n+1}{n} \text{ where } n=1, 2, 3, \ldots$$

The respective outputs for each of the inputs $n=1$, 10, 100, 1000 and 10 000 are

. . . . . . . . . . . .

*Next frame*

36

2, 1·1, 1·01, 1·001, 1·0001

Because

$$f(1) = \frac{1+1}{1} = 2, \quad f(10) = \frac{10+1}{10} = 1{\cdot}1, \quad f(100) = \frac{100+1}{100} = 1{\cdot}01,$$
$$f(1000) = \frac{1000+1}{1000} = 1{\cdot}001, \quad f(10\,000) = \frac{10\,000+1}{10\,000} = 1{\cdot}0001$$

So as $n$ becomes larger $f(n) = \dfrac{n+1}{n}$ becomes closer to ...........

*Next frame*

37

1

Because

$$\begin{aligned} f(10) &= 1 + 0{\cdot}1 \\ f(100) &= 1 + 0{\cdot}01 \\ f(1000) &= 1 + 0{\cdot}001 \\ f(10\,000) &= 1 + 0{\cdot}0001 \end{aligned}$$

So, as $n$ increases the number added to unity gets smaller and smaller and closer to zero and the output becomes closer and closer to 1. However, the output never attains the number 1 because no matter how large the input the output will always be a shade greater than 1. We say that:

as $n$ becomes arbitrarily large so $f(n) = \dfrac{n+1}{n}$ becomes close to 1

## Infinity

There is no largest integer; this fact is embodied in the statement that *the integers increase without bound* – no matter how large an integer you can think of you can always add 1 to it to obtain an even larger integer. An alternative description of this idea is to say that *the integers increase to infinity* where infinity is represented by the symbol $\infty$ (negative infinity is represented by $-\infty$). Unfortunately, because we have a symbol for it there is a temptation to give infinity some numerical aspect that it does not possess. It must be clearly understood that although infinity is a well-established concept it cannot be defined numerically and so it cannot be used as a number in any arithmetic calculations. So the statement:

as $n$ becomes arbitrarily large so $f(n) = \dfrac{n+1}{n}$ becomes close to 1

is written:

as $n$ *approaches infinity* so $f(n) = \dfrac{n+1}{n}$ *approaches* 1

So as $n$ approaches infinity $f(n) = \dfrac{1}{n}$ approaches ...........

*Next frame*

**38**

$\boxed{0}$

Because

$$f(1) = \frac{1}{1} = 1$$

$$f(100) = \frac{1}{100} = 0{\cdot}01$$

$$f(10\,000) = \frac{1}{10\,000} = 0{\cdot}0001$$

So as $n$ approaches infinity $f(n) = \frac{1}{n}$ approaches 0

Symbolically we use an arrow to indicate *approaching* and write

as $n \to \infty$ so $f(n) = \frac{1}{n} \to 0$

## Limits

The number that the output of a sequence approaches as the input increases without bound is called the **limit** of the sequence. For example, no matter how large $n$ becomes $f(n) = \frac{1}{n}$ never attains the value of 0 but it can become as close to 0 as we wish by choosing $n$ to be sufficiently large. We call 0 the

limit of $f(n) = \frac{1}{n}$ as $n$ approaches infinity

and write

$$\underset{n\to\infty}{Lim}\left(\frac{1}{n}\right) = 0$$

If a sequence has a finite limit we say that it **converges** to the limit and so $f(n) = \frac{1}{n}$ converges to 0. This means that:

$$\underset{n\to\infty}{Lim}\left(\frac{n+1}{n}\right) = \ldots\ldots\ldots\ldots$$

*Next frame*

**39**

$\boxed{1}$

Because

As $n$ approaches infinity so $f(n) = \frac{n+1}{n}$ approaches 1, that is $\underset{n\to\infty}{Lim}\left(\frac{n+1}{n}\right) = 1$

Try another.

$$\underset{n\to\infty}{Lim}\,(0.1)^n = \ldots\ldots\ldots\ldots$$

*Next frame*

40

$\boxed{0}$

Because

$(0{\cdot}1)^1 = 0{\cdot}1$

$(0{\cdot}1)^{10} = 0{\cdot}0000000001$

As $n$ increases so $(0{\cdot}1)^n$ rapidly decreases towards zero. Therefore, $\underset{n\to\infty}{Lim}\,(0{\cdot}1)^n = 0$

Indeed, $\underset{n\to\infty}{Lim}\,\alpha^n = 0$ provided $-1 < \alpha < 1$. This is an important fact so make a note of it in your record book or highlight it here.

## Infinite limits

Sometimes as $n$ becomes large so does $f(n)$. For example, the output from the sequence $f(n) = n^3$ becomes large even faster than $n$. In this case we write the limit as:

$$\underset{n\to\infty}{Lim}\,(n^3) = \infty$$

*Be aware*. This notation can be misleading if it is not correctly understood. It does not mean what it appears to mean, namely that the limit is *equal* to infinity. It cannot be equal to infinity because infinity is not numerically defined so nothing can be said to be equal to it. What it does mean is that as $n$ becomes large without bound then so does $n^3$. The notation is unfortunate but it is common usage so we have to accept it. If a sequence does not have a finite limit it is said to **diverge** thus the sequence with general term $f(n) = n^3$ diverges.

So $\underset{n\to\infty}{Lim}\left(\dfrac{n^2+1}{n}\right) = \ldots\ldots\ldots\ldots$

*Next frame*

41

$\boxed{\infty}$

Because

$$\begin{aligned}\underset{n\to\infty}{Lim}\left(\frac{n^2+1}{n}\right) &= \underset{n\to\infty}{Lim}\left(\frac{n^2}{n}+\frac{1}{n}\right)\\ &= \underset{n\to\infty}{Lim}\left(n+\frac{1}{n}\right)\\ &= \infty\end{aligned}$$

*Next frame*

42

## Rules of limits

Limits can be manipulated algebraically according to the following rules: assume

$\underset{n\to\infty}{Lim}\,f(n) = F$ and $\underset{n\to\infty}{Lim}\,g(n) = G$

### Multiplication by a constant

The limit of an expression multiplied by a constant is the constant multiplying the limit of the expression:

$\underset{n\to\infty}{Lim}\,kf(n) = k\underset{n\to\infty}{Lim}\,f(n) = kF$ where $k$ is a constant.

▶

In particular,

if $f(n) = 1$ then $\lim_{n\to\infty} k = k$

So that, for example since $\lim_{n\to\infty} \frac{n+1}{n} = 1$ then $\lim_{n\to\infty} 3\frac{n+1}{n} = 3\lim_{n\to\infty} \frac{n+1}{n} = 3$ and

$$\lim_{n\to\infty} \frac{2n+2}{3n} = \ldots\ldots\ldots$$

*Next frame*

**43**

$$\boxed{\frac{2}{3}}$$

Because

$$\begin{aligned}\lim_{n\to\infty} \frac{2n+2}{3n} &= \lim_{n\to\infty} \frac{2}{3}\left(\frac{n+1}{n}\right)\\ &= \frac{2}{3}\lim_{n\to\infty}\left(\frac{n+1}{n}\right)\\ &= \frac{2}{3}\end{aligned}$$

### Sums and differences

The limit of a sum (or difference) is the sum (or difference) of the limits:

$$\lim_{n\to\infty} \{f(n) + g(n)\} = \lim_{n\to\infty} f(n) + \lim_{n\to\infty} g(n) = F + G$$

and

$$\lim_{n\to\infty} \{f(n) - g(n)\} = \lim_{n\to\infty} f(n) - \lim_{n\to\infty} g(n) = F - G$$

So that, for example since:

$$\lim_{n\to\infty} \frac{n+1}{n} = 1 \text{ and } \lim_{n\to\infty} \frac{1}{n} = 0$$

then

$$\begin{aligned}\lim_{n\to\infty}\left(\frac{n+1}{5n} - \frac{6}{n}\right) &= \lim_{n\to\infty}\left(\frac{n+1}{5n}\right) - \lim_{n\to\infty}\left(\frac{6}{n}\right)\\ &= \frac{1}{5}\lim_{n\to\infty}\left(\frac{n+1}{n}\right) - 6\lim_{n\to\infty}\left(\frac{1}{n}\right)\\ &= \frac{1}{5}\times 1 - 6\times 0\\ &= \frac{1}{5}\end{aligned}$$

$$\text{So } \lim_{n\to\infty}\left(\frac{3}{7n} + \frac{4(n+1)}{5n}\right) = \ldots\ldots\ldots$$

*Next frame*

44

$$\boxed{\frac{4}{5}}$$

Because

$$\begin{aligned}\lim_{n\to\infty}\left(\frac{3}{7n}+\frac{4(n+1)}{5n}\right)&=\lim_{n\to\infty}\left(\frac{3}{7n}\right)+\lim_{n\to\infty}\left(\frac{4(n+1)}{5n}\right)\\&=\frac{3}{7}\lim_{n\to\infty}\left(\frac{1}{n}\right)+\frac{4}{5}\lim_{n\to\infty}\left(\frac{n+1}{n}\right)\\&=\frac{3}{7}\times 0+\frac{4}{5}\times 1\\&=\frac{4}{5}\end{aligned}$$

**Products and quotients**

The limit of a product is the product of the limits:

$$\lim_{n\to\infty}\{f(n)\times g(n)\}=\lim_{n\to\infty}f(n)\times\lim_{n\to\infty}g(n)=FG$$

and the limit of a quotient is the quotient of the limits:

$$\lim_{n\to\infty}\left\{\frac{f(n)}{g(n)}\right\}=\frac{\lim_{n\to\infty}f(n)}{\lim_{n\to\infty}g(n)}=\frac{F}{G}\text{ provided } G\neq 0$$

So, for example

$$\begin{aligned}\lim_{n\to\infty}\left(\frac{n+1}{n^2}\right)&=\lim_{n\to\infty}\left[\left(\frac{n+1}{n}\right)\times\frac{1}{n}\right]\\&=\lim_{n\to\infty}\frac{n+1}{n}\times\lim_{n\to\infty}\frac{1}{n}\\&=1\times 0\\&=0\end{aligned}$$

and $$\lim_{n\to\infty}\left\{\frac{3-\frac{4}{n}}{\frac{2}{n}+5}\right\}=\frac{\lim_{n\to\infty}\left\{3-\frac{4}{n}\right\}}{\lim_{n\to\infty}\left\{\frac{2}{n}+5\right\}}=\frac{3}{5}$$

So $\lim_{n\to\infty}\left\{\left(\frac{7}{n}-4\right)\times\left(6+\frac{1}{n}\right)\right\}=$ ........... and $\lim_{n\to\infty}\left\{\frac{\frac{7}{n}-4}{6+\frac{1}{n}}\right\}=$ ...........

*Next frame*

**45**

$$-24 \text{ and } -\frac{2}{3}$$

Because

$$\lim_{n\to\infty}\left\{\left(\frac{7}{n}\times 4\right)\times\left(6+\frac{1}{n}\right)\right\} = \lim_{n\to\infty}\left(\frac{7}{n}-4\right)\times\lim_{n\to\infty}\left(6+\frac{1}{n}\right)$$
$$= (-4)\times 6$$
$$= -24$$

and

$$\lim_{n\to\infty}\left\{\frac{\dfrac{7}{n}-4}{6+\dfrac{1}{n}}\right\} = \frac{\lim_{n\to\infty}\left\{\dfrac{7}{n}-4\right\}}{\lim_{n\to\infty}\left\{6+\dfrac{1}{n}\right\}}$$
$$= \frac{-4}{6} = -\frac{2}{3}$$

Finally, $\lim_{n\to\infty} k^n = \begin{cases} 0 \text{ if } -1 < k < 1 \\ 1 \text{ if } k = 1 \\ \infty \text{ if } k > 1 \\ \textit{undefined} \text{ if } k \leq -1 \end{cases}$

so $\lim_{n\to\infty}\left(\frac{1}{3}\right)^n = 0$ whereas $\lim_{n\to\infty}(3)^n = \infty$ and $\lim_{n\to\infty}(-3)^n$ is undefined.

The latter is undefined because as $n$ increases and repeatedly switches between being even and being odd the output $(-3)^n$ also increases and repeatedly switches between being positive and being negative. Because it makes no sense to say whether infinity is even or odd it is not possible to decide whether the limit is $\infty$ or $-\infty$ so the limit is undefined. *Also since such a sequence does not have a finite limit it is said to diverge.*

So, determine whether each of the following sequences converge or diverge:

(a) $f(n) = 0{\cdot}01^n$

(b) $f(n) = \left(\frac{1}{0{\cdot}01}\right)^n$

(c) $f(n) = \left(-\frac{1}{0{\cdot}01}\right)^n$

*Next frame*

**46**

(a) converges to 0
(b) diverges with limit $\infty$
(c) diverges with limit undefined

Because

(a) $0{\cdot}01 < 1$ and so $\lim_{n\to\infty}(0{\cdot}01)^n = 0$. The sequence converges to zero

(b) $\frac{1}{0{\cdot}01} = 10 > 1$ and so $\lim_{n\to\infty}\left(\frac{1}{0{\cdot}01}\right)^n = \infty$. The sequence diverges

(c) $-10 < -1$ and so $\lim_{n\to\infty}\left(-\frac{1}{0{\cdot}01}\right)^n = \textit{undefined}$. The sequence diverges

*Next frame*

## Indeterminate limits

47

Sometimes when trying to determine the limit of a quotient the limits of both the numerator and the denominator are infinite. Such a limit is called an **indeterminate** limit and cannot be found without some manipulation on the quotient. For example, the limit

$$\lim_{n\to\infty}\left(\frac{n^2+3}{4n^2+6}\right)$$

is indeterminate because both numerator and denominator have infinite limits. To attack this problem we first divide both numerator and denominator by $n^2$ to give the result:

$$\lim_{n\to\infty}\left(\frac{1+\dfrac{3}{n^2}}{4+\dfrac{6}{n^2}}\right)=\frac{\lim\limits_{n\to\infty}\left(1+\dfrac{3}{n^2}\right)}{\lim\limits_{n\to\infty}\left(4+\dfrac{6}{n^2}\right)}$$

$$=\frac{1}{4}\quad\text{since } \lim_{n\to\infty}\frac{3}{n^2}=0 \text{ and } \lim_{n\to\infty}\frac{6}{n^2}=0$$

So $\lim\limits_{n\to\infty}\left(\dfrac{7n^2-2}{5n^2+8}\right)=$ ...........

*Next frame*

48

$$\boxed{\frac{7}{5}}$$

Because

This is an indeterminate limit so dividing numerator and denominator by $n^2$ gives:

$$\lim_{n\to\infty}\left(\frac{7-\dfrac{2}{n^2}}{5+\dfrac{8}{n^2}}\right)=\frac{\lim\limits_{n\to\infty}\left(7-\dfrac{2}{n^2}\right)}{\lim\limits_{n\to\infty}\left(5+\dfrac{8}{n^2}\right)}$$

$$=\frac{7}{5}\quad\text{since } \lim_{n\to\infty}\frac{2}{n^2}=0 \text{ and } \lim_{n\to\infty}\frac{8}{n^2}=0$$

***We shall now summarize this last aspect of the Programme and follow the summary with a brief exercise.***

## Review summary

49

1 *Limits*

The limit of a sequence is the number that the output approaches as the input increases to infinity. A sequence with a finite limit is said to be convergent and to converge to that limit. A sequence without a finite limit is said to diverge.

▶

2 *Rules of limits*

Limits can be manipulated algebraically according to the following rules: assume

$\lim\limits_{n\to\infty} f(n) = F$ and $\lim\limits_{n\to\infty} g(n) = G$

**Multiplication by a constant**

$\lim\limits_{n\to\infty} kf(n) = k\lim\limits_{n\to\infty} f(n) = kF$ where $k$ is a constant

**Sums and differences**

The limit of a sum (or difference) is the sum (or difference) of the limits:

$$\lim_{n\to\infty}\{f(n) \pm g(n)\} = \lim_{n\to\infty}\{f(n)\} \pm \lim_{n\to\infty}\{g(n)\}$$

**Products and quotients**

The limit of a product is the product of the limits:

$$\lim_{n\to\infty}\{f(n) \times g(n)\} = \lim_{n\to\infty} f(n) \times \lim_{n\to\infty} g(n) = FG$$

and the limit of a quotient is the quotient of the limits:

$$\lim_{n\to\infty}\left\{\frac{f(n)}{g(n)}\right\} = \frac{\lim\limits_{n\to\infty} f(n)}{\lim\limits_{n\to\infty} g(n)} = \frac{F}{G} \text{ provided } G \neq 0$$

If the limits of both the numerator and the denominator in a quotient are infinite the limit is called an **indeterminate** limit and cannot be found without some manipulation on the quotient.

Finally:

$$\lim_{n\to\infty} k^n = \begin{cases} 0 \text{ if } -1 < k < 1 \\ 1 \text{ if } k = 1 \\ \infty \text{ if } k > 1 \\ \textit{undefined} \text{ if } k \leq -1 \end{cases}$$

---

## Review exercise

**50**

Determine each of the following limits

1 (a) $\lim\limits_{n\to\infty}\left(12 - \dfrac{11}{n^2}\right)$

(b) $\lim\limits_{n\to\infty}(-10)^{-n}$

(c) $\lim\limits_{n\to\infty}\left(\dfrac{3n^2 - 2n + 4}{-6 - 2n - 7n^2}\right)$

**51**

1 (a) $\lim\limits_{n\to\infty}\left(12 - \dfrac{11}{n^2}\right) = \lim\limits_{n\to\infty}(12) - \lim\limits_{n\to\infty}\left(\dfrac{11}{n^2}\right)$

$= 12 - 0$

$= 12$

(b) $\lim\limits_{n\to\infty}(-10)^{-n} = \lim\limits_{n\to\infty}\left(-\dfrac{1}{10}\right)^n$

$= 0$ since $-1 < -0{\cdot}1 < 1$

▶

(c) $$\lim_{n\to\infty}\left(\frac{3n^2-2n+4}{-6-2n-7n^2}\right)=\lim_{n\to\infty}\left(\frac{3-\frac{2}{n}+\frac{4}{n^2}}{-\frac{6}{n^2}-\frac{2}{n}-7}\right)$$
$$=\frac{\lim_{n\to\infty}\left(3-\frac{2}{n}+\frac{4}{n^2}\right)}{\lim_{n\to\infty}\left(-\frac{6}{n^2}-\frac{2}{n}-7\right)}$$
$$=\frac{3}{-7}$$
$$=-\frac{3}{7}$$

Now go on to the next frame and complete the **Can You?** checklist and **Test exercise**. After that, a set of **Further problems** provides additional valuable practice.

---

## Can you?

### Checklist 10

52

Check this list before and after you try the end of Programme test.

On a scale of 1 to 5 how confident are you that you can: Frames

- Recognize a sequence as a function — Frames 1 to 2
  *Yes* ☐ ☐ ☐ ☐ ☐ *No*
- Plot the graphs of sequences — Frames 3 to 4
  *Yes* ☐ ☐ ☐ ☐ ☐ *No*
- Recognize arithmetic, geometric and harmonic sequences — Frames 5 to 9
  *Yes* ☐ ☐ ☐ ☐ ☐ *No*
- Convert the descriptive prescription of the output from arithmetic and geometric sequences into a recursive description and recognize the importance of initial terms — Frames 10 to 14
  *Yes* ☐ ☐ ☐ ☐ ☐ *No*
- Generate the recursive prescription of a sequence from a given sequence of numbers — Frames 15 to 16
  *Yes* ☐ ☐ ☐ ☐ ☐ *No*
- Determine the order and generate the terms of a difference equation — Frames 20 to 21
  *Yes* ☐ ☐ ☐ ☐ ☐ *No*
- Obtain the solution to an homogeneous, linear difference equation — Frames 22 to 28
  *Yes* ☐ ☐ ☐ ☐ ☐ *No*
- Derive the limit of a sequence using the rules of limits — Frames 29 to 45
  *Yes* ☐ ☐ ☐ ☐ ☐ *No*

## Test exercise 10

53

1 Find the next two terms and form of the general term for each of the following sequences:

(a) 1, 3, 5, 7, ...

(b) −4, −1, 2, 5, ...

(c) 16, 8, 4, 2, ...

(d) −3, 6, −12, 24, ...

(e) 1, 1, 1, 3, 5, 9, ...

(f) 1, 1, 4, 10, 28, 76, ...

2 Plot the graph of each of the following sequences:

(a) $f(n) = 3 - 2n \quad n \geq 0$

(b) $f(n) = 2(-1)^n \quad n \geq 0$

(c) $f(n) = \dfrac{(-1)^n}{n} \quad n \geq 1$

3 Find a recursive description corresponding to each of the following prescriptions for the output of a sequence:

(a) $f(n) = 5n - 9$ where $n$ is an integer $\geq 1$

(b) $f(n) = 23 - 4n$ where $n$ is an integer $\geq 0$

(c) $f(n) = 3^{-n}$ where $n$ is an integer $\geq -2$

4 Find the general term of each of the following sequences:

(a) $f(n+1) - 25f(n-1) = 0$ where $f(0) = 1$ and $f(1) = 5$

(b) $f(n+2) + f(n+1) - 12f(n) = 0$ where $f(-1) = 0$ and $f(0) = 1$

(c) $f(n+2) - 16f(n+1) + 64f(n) = 0$ where $f(0) = 0$ and $f(1) = 1$

5 Evaluate the following limits:

(a) $\underset{n\to\infty}{Lim}\left(1 - \dfrac{n}{100}\right)$

(b) $\underset{n\to\infty}{Lim}\, 5^n$

(c) $\underset{n\to\infty}{Lim}\, 1000(0{\cdot}999)^n$

(d) $\underset{n\to\infty}{Lim}\, 25(-1)^n$

(e) $\underset{n\to\infty}{Lim}\dfrac{n}{n+1}$

(f) $\underset{n\to\infty}{Lim}\dfrac{n^2+n+1}{n^2-n-1}$

6 Which of the following sequences converge and which diverge?

(a) $f(n) = 3 - 3n$

(b) $f(n) = (0{\cdot}99)^{99n}$

(c) $f(n+1) = \dfrac{f(n)+2}{5}$, $f(0) = 1$

# Further problems 10

**54**

1 Find the next two terms and form of the general term for each of the following sequences:

(a) 1, 6, 11, 16, . . .

(b) −101, −99, −97, −95, . . .

(c) 10, 1, 0·1, 0·01, . . .

(d) 1234·5, 123·45, 12·345, . . .

(e) 2, 4, −2, 6, −8, . . .

(f) 5, 8, 13, 21, . . .

2 Plot the graph of each of the following sequences:

(a) $f(n) = -5 + 3n \quad n \geq 0$

(b) $f(n) = -7(-2)^n \quad n \geq 0$

(c) $f(n) = \dfrac{(-1)^n}{n^2} \quad n \geq 1$

3 Solve each of the following difference equations:

(a) $f(n+2) + 5f(n+1) + 6f(n) = 0$ where $f(0) = 0$ and $f(1) = 1$

(b) $3f(n+2) - 7f(n+1) + 2f(n) = 0$ where $f(0) = 1$ and $f(1) = 0$

(c) $f(n+2) - 49f(n) = 0$ where $f(0) = 1$ and $f(1) = 1$

4 Evaluate the following limits:

(a) $\underset{n\to\infty}{Lim}\,(5 + 9n)$

(b) $\underset{n\to\infty}{Lim}\left(\dfrac{1}{4}\right)^{-3n}$

(c) $\underset{n\to\infty}{Lim}\,0{\cdot}999(-1)^n$

(d) $\underset{n\to\infty}{Lim}\,152^{-2n}$

(e) $\underset{n\to\infty}{Lim}\,\dfrac{n+7}{n-7}$

(f) $\underset{n\to\infty}{Lim}\,\dfrac{1-n-n^2}{5n^2-2n-3}$

5 Which of the following sequences converge and which diverge?

(a) $f(n) = 6 + n$

(b) $f(n) = (n+1)^2$

(c) $f(n) = \dfrac{16f(n-1) - 2}{32}$, $f(0) = 1$

(d) $f(n) = [f(n-1)]^2$, $-1 < f(n) < 1$

(e) $f(n) = \dfrac{f(n-1)}{0{\cdot}1}$, $f(0) = 1$

6 The fourth and the eighth terms of an arithmetic sequence are 14 and 22 respectively. Find the tenth term.

7 The first, third and sixth terms of an arithmetic sequence form three successive terms of a geometric sequence. If the first term of both the arithmetic and geometric sequence is 8, find the second, third and fourth terms and the general term of the geometric sequence.

▶

8 Five numbers are in a geometric sequence. The first is 10 and the fifth is 160, what are the other three numbers?

9 For what value of $n \geq 0$ does $\dfrac{2n+3}{n+4}$ differ from 2 by 0·02 to 2 dp?

10 Find $\lim\limits_{n\to\infty} f(n)$ where:

(a) $f(n) = \dfrac{1}{n}\{n + (-1)^n\}$

(b) $f(n) = \dfrac{3^n}{1 - 3^n}$

(c) $f(n) = \dfrac{(0{\cdot}3)^n}{1 - (0{\cdot}3)^n}$

(d) $f(n) = \dfrac{1}{n}\{5n(-1)^n - 3\}$

11 If three numbers in geometric sequence are respectively subtracted from three numbers also in geometric sequence leaving remainders that are also in geometric sequence show that all three geometric sequences have the same common ratio.

12 The seventh and eleventh terms of an arithmetic sequence are $7b + 5c$ and $11b + 9c$. Find the first term and the common difference.

13 If $x$, $5x$, $6x + 9$ form three successive terms of an arithmetic sequence, find the next four terms.

14 Why do the terms of the sequence:

$$f(n) = \frac{1}{2n-1} : n \geq 1$$

form an harmonic sequence?

15 The twelfth and nineteenth terms of an harmonic sequence are respectively $\dfrac{1}{5}$ and $\dfrac{3}{22}$. Find the fourth term.

16 If $m$ and $n$ are the $(p+q)$th and $(p-q)$th terms respectively of an arithmetic sequence show that the $p$th term is $\dfrac{m+n}{2}$.

17 If $\dfrac{1}{b+c}, \dfrac{1}{c+a}, \dfrac{1}{a+b}$ form three successive terms of an arithmetic sequence show that $a^2$, $b^2$, $c^2$ also form three successive terms of another arithmetic sequence.

18 If $m$ and $n$ are the $(p+q)$th and $(p-q)$th terms respectively of a geometric sequence, show that the $p$th term is $\sqrt{mn}$.

19 Show that the difference equation $g(n+2) - g(n+1) - 6g(n) = 0$ can be derived from the coupled difference equations:

$$f(n+1) = g(n)$$
$$g(n+1) = g(n) + 6f(n)$$

Find $f(n)$ and $g(n)$ given that $f(1) = 0$ and $g(1) = 1$.

20 Solve the equation for the Fibonacci sequence:

$f(n+2) = f(n+1) + f(n)$ where $f(0) = 0$, $f(1) = 1$ and $n \geq 0$.

**55**

Now visit the companion website for this book at www.palgrave.com/stroud for more questions applying this mathematics to science and engineering.

# Series 1

### Learning outcomes

When you have completed this Programme you will be able to:

- ☐ Manipulate arithmetic and geometric series
- ☐ Manipulate series of powers of the natural numbers
- ☐ Determine the limiting values of arithmetic and geometric series
- ☐ Determine the limiting values of simple indeterminate forms
- ☐ Apply various convergence tests to infinite series
- ☐ Distinguish between absolute and conditional convergence

# Series

**1**

If the terms of the sequence, $f(1), f(2), f(3), \ldots, f(r), \ldots$ are added together to form the sum:

$$\sum_{r=1}^{n} f(r) = f(1) + f(2) + f(3) + \ldots + f(n)$$

then what is formed is called a **series** – a sum of terms of a sequence. Just as there are many types of sequence so there are many types of series and the first one we shall look at is the arithmetic series that is formed from the arithmetic sequence.

*Move to the next frame*

**2**

## Arithmetic series

As an example of an arithmetic series consider:

$2 + 5 + 8 + 11 + 14 + \ldots$ which is derived from the sequence 2, 5, 8, 11, 14, ...

You will note that each term of the sequence can be written from the previous terms by simply adding on a constant 3. This regular increment is called the *common difference* and is found by selecting any term and subtracting from it the previous term

e.g. $11 - 8 = 3$; $5 - 2 = 3$; etc.

*Move on to the next frame*

**3**

The *general arithmetic series* containing the first $n$ terms of the arithmetic sequence

$f(r) = a + rd$

can be written as:

$$\sum_{r=0}^{n-1}(a + rd) = a + (a + d) + (a + 2d) + (a + 3d) + \ldots + (a + [n - 1]d)$$

This can also be written as:

$$\sum_{r=0}^{n-1}(a + rd) = (a + [n - 1]d) + (a + [n - 2]d) + (a + [n - 3]d) + \ldots + a$$

Adding these together gives:

$$2\sum_{r=0}^{n-1}(a + rd)$$
$$= (2a + [n - 1]d) + (2a + [n - 1]d) + (2a + [n - 1]d) + \ldots + (2a + [n - 1]d)$$
$$= n(2a + [n - 1]d)$$

and so:

$$\sum_{r=0}^{n-1}(a + rd) = \frac{n}{2}(2a + [n - 1]d)$$

By way of warming up, find the sum of the first 20 terms of the series:

$10 + 6 + 2 - 2 - 6 \ldots$ etc.

*Then move to Frame 4*

**4**

$$\sum_{r=0}^{19}(10-4r)=-560$$

Because

For the series $10+6+2-2-6\ldots$ etc, $a=10$ and $d=6-10=-4$
Therefore, for $n=20$

$$\begin{aligned}\sum_{r=0}^{n-1}(a+rd)&=\sum_{r=0}^{20-1}(10-4r)\\&=\frac{n}{2}(2a+[n-1]d)\\&=\frac{20}{2}(20-[20-1]4)\\&=-560\end{aligned}$$

Here is another example.
If the 7th term of an arithmetic series is 22 and the 12th term is 37, find the series.

We know 7th term $=22$ $\therefore a+6d=22$
and 12th term $=37$ $\therefore a+11d=37$
$5d=15$ $\therefore d=3$
$\therefore a=4$

So the series is $4+7+10+13+16+\ldots$ etc.

Here is one for you to do:

The 6th term of an arithmetic series is $-5$ and the 10th term is $-21$. Find the sum of the first 30 terms.

**5**

$$\sum_{r=0}^{29}(15-4r)=-1290$$

Because

6th term $=-5$ $\therefore a+5d=-5$
10th term $=-21$ $\therefore a+9d=-21$
$4d=-16$ $\therefore d=-4$ so $a=15$

$\therefore a=15,\ d=-4,\ n=30$ and $\sum_{r=0}^{n-1}(a+rd)=\frac{n}{2}(2a+[n-1]d)$

Therefore $\sum_{r=0}^{29}(15-4r)=\frac{30}{2}(30+29[-4])=15(30-116)=15(-86)=-1290$

## Arithmetic mean

We are sometimes required to find the arithmetic mean of two numbers, $P$ and $Q$. This means that we have to insert a number $A$ between $P$ and $Q$, so that $P+A+Q$ forms an arithmetic series.

$A-P=d$ and $Q-A=d$

$\therefore A-P=Q-A \qquad 2A=P+Q \qquad \therefore A=\frac{P+Q}{2}$

The arithmetic mean of two numbers, then, is simply their average.

Therefore, the arithmetic mean of 23 and 58 is ...........

**6**

40·5

If we are required to insert 3 arithmetic means between two given numbers, $P$ and $Q$, it means that we have to supply three numbers, $A$, $B$, $C$ between $P$ and $Q$, so that $P + A + B + C + Q$ forms an arithmetic series.

For example: Insert 3 arithmetic means between 8 and 18.

Let the means be denoted by $A$, $B$, $C$.

Then $8 + A + B + C + 18$ forms an arithmetic series.

First term, $a = 8$; fifth term $= a + 4d = 18$

$$\left.\begin{aligned} \therefore a &= 8 \\ a + 4d &= 18 \end{aligned}\right\} 4d = 10 \qquad \therefore d = 2{\cdot}5$$

$$\left.\begin{aligned} A &= 8 + 2{\cdot}5 = 10{\cdot}5 \\ B &= 8 + 5 = 13 \\ C &= 8 + 7{\cdot}5 = 15{\cdot}5 \end{aligned}\right\} \text{Required arithmetic means are } 10{\cdot}5,\ 13,\ 15{\cdot}5$$

Now you find five arithmetic means between 12 and 21·6.

*Then move to Frame 7*

**7**

13·6, 15·2, 16·8, 18·4, 20

Here is the working:

Let the 5 arithmetic means be $A$, $B$, $C$, $D$, $E$.

Then $12 + A + B + C + D + E + 21{\cdot}6$ forms an arithmetic series.

$\therefore a = 12; \quad a + 6d = 21{\cdot}6$

$\therefore 6d = 9{\cdot}6 \quad \therefore d = 1{\cdot}6$

Then $A = 12 + 1{\cdot}6 = 13{\cdot}6 \qquad A = 13{\cdot}6$

$B = 12 + 3{\cdot}2 = 15{\cdot}2 \qquad B = 15{\cdot}2$

$C = 12 + 4{\cdot}8 = 16{\cdot}8 \qquad C = 16{\cdot}8$

$D = 12 + 6{\cdot}4 = 18{\cdot}4 \qquad D = 18{\cdot}4$

$E = 12 + 8{\cdot}0 = 20{\cdot}0 \qquad E = 20$

So that is it! Once you have done one, the others are just like it.

Now we will see how much you remember about *geometric series*.

*So, on to Frame 8*

**8**

## Geometric series

An example of a geometric series is the series:

$1 + 3 + 9 + 27 + 81 + \ldots$ etc.

Here you can see that any term can be written from the previous term by multiplying it by a constant factor 3. This constant factor is called the *common ratio* and is found by selecting any term and dividing it by the previous one:

e.g. $27 \div 9 = 3; \quad 9 \div 3 = 3$; etc.

A geometric series therefore has the form:

$a + ar + ar^2 + ar^3 + \ldots$ etc.

where $a =$ first term, $r =$ common ratio.

So in the geometric series $5 - 10 + 20 - 40 + \ldots$ etc. the common ratio, $r$, is ...........

9

$$r = \frac{20}{-10} = -2$$

The general *geometric series* containing the first $n$ terms of the geometric sequence

$$f(k) = ar^k$$

can be written as:

$$\sum_{k=0}^{n-1}(ar^k) = a + ar + ar^2 + ar^3 + \ldots + ar^{n-1}$$

Notice that we use $k$ as the symbol for the counting number rather than $r$, which has already been used for the common ratio. Multiplying this equation throughout by $r$ gives:

$$r\sum_{k=0}^{n-1}(ar^k) = ar + ar^2 + ar^3 + \ldots + ar^n$$

Subtracting the second from the first term by term gives:

$$\sum_{k=0}^{n-1}(ar^k) - r\sum_{k=0}^{n-1}(ar^k) = a - ar^n = a(1 - r^n)$$

and so:

$$\sum_{k=0}^{n-1}(ar^k) = \frac{a(1 - r^n)}{1 - r}$$

By way of warming up, find the sum of the first 8 terms of the series:

$$8 + 4 + 2 + 1 + \frac{1}{2} + \ldots \text{ etc}$$

*Then move to Frame 10*

10

$$\sum_{k=0}^{7} 8\left(\frac{1}{2}\right)^k = 15\frac{15}{16}$$

Because

For the series $8 + 4 + 2 + 1 + \frac{1}{2} + \ldots etc., a = 8$ and $r = \frac{2}{4} = \frac{1}{2}$ and $n = 8$.

Therefore

$$\begin{aligned}\sum_{k=0}^{n-1} ar^k &= \sum_{k=0}^{8-1} 8\left(\frac{1}{2}\right)^k \\ &= \frac{a(1 - r^n)}{1 - r} \\ &= \frac{8\left(1 - \left(\frac{1}{2}\right)^8\right)}{1 - \frac{1}{2}} \\ &= 8 \div \frac{1}{2}\left(1 - \frac{1}{256}\right) = 16\left(\frac{255}{256}\right) = \frac{255}{16} = 15\frac{15}{16}\end{aligned}$$

Now here is another example.

▶

If the 5th term of a geometric series is 162 and the 8th term is 4374, find the series.

We have 5th term = 162 $\therefore ar^4 = 162$

8th term = 4374 $\therefore ar^7 = 4374$

$$\frac{ar^7}{ar^4} = \frac{4374}{162}$$

$$\therefore r^3 = 27$$

$$\therefore r = 3$$

$$\therefore a = \ldots\ldots\ldots\ldots$$

11

$a = 2$

Because

$ar^4 = 162;\ ar^7 = 4374$ and $r = 3$

$\therefore a.3^4 = 162 \quad \therefore a = \dfrac{162}{81} \quad \therefore a = 2$

$\therefore$ The series is: $2 + 6 + 18 + 54 + \ldots$ etc.

Of course, now that we know the values of $a$ and $r$, we could calculate the value of any term or the sum of a given number of terms.

For this same series, find

(a) the 10th term

(b) the sum of the first 10 terms.

*When you have finished, move to Frame 12*

12

39 366, 59 048

Because $a = 2;\ r = 3$

(a) 10th term $= ar^9 = 2 \times 3^9 = 2(19\,683) = 39\,366$

(b) $\displaystyle\sum_{k=0}^{9} 2 \times 3^k = \frac{a(1 - r^{10})}{1 - r} = \frac{2(1 - 3^{10})}{1 - 3}$

$$= \frac{2(1 - 59\,049)}{-2} = 59\,048$$

## Geometric mean

The geometric mean of two given numbers $P$ and $Q$ is a number $A$ such that $P + A + Q$ form a geometric series.

$$\frac{A}{P} = r \text{ and } \frac{Q}{A} = r$$

$$\therefore \frac{A}{P} = \frac{Q}{A} \quad \therefore A^2 = PQ \quad A = \sqrt{PQ}$$

So the geometric mean of 2 numbers is the square root of their product.

Therefore, the geometric mean of 4 and 25 is ...........

13

$$A = \sqrt{4 \times 25} = \sqrt{100} = 10$$

To insert 3 GMs between two given numbers, $P$ and $Q$, means to insert 3 numbers, $A$, $B$, $C$, such that $P + A + B + C + Q$ form a geometric series.

For example, insert 4 geometric means between 5 and 1215.

Let the means be $A$, $B$, $C$, $D$. Then $5 + A + B + C + D + 1215$ form a geometric series.

i.e. $a = 5$ and $ar^5 = 1215$

$\therefore r^5 = \dfrac{1215}{5} = 243 \quad \therefore r = 3$

$\therefore A = 5 \times 3 = 15$

$B = 5 \times 9 = 45$

$C = 5 \times 27 = 135$

$D = 5 \times 81 = 405$

The required geometric means are: 15, 45, 135, 405

Now here is one for you to do: Insert two geometric means between 5 and 8·64.

*Then on to Frame 14*

14

Required geometric means are 6·0, 7·2

Because

Let the means be $A$ and $B$.

Then $5 + A + B + 8{\cdot}64$ form a geometric series.

$\therefore a = 5; \quad \therefore ar^3 = 8{\cdot}64; \quad \therefore r^3 = 1{\cdot}728; \quad \therefore r = 1{\cdot}2$

$$\left.\begin{aligned} A &= 5 \times 1{\cdot}2 = 6{\cdot}0 \\ B &= 5 \times 1{\cdot}44 = 7{\cdot}2 \end{aligned}\right\} \text{ Required means are } 6{\cdot}0 \text{ and } 7{\cdot}2$$

Arithmetic and geometric series are, of course, special kinds of series. There are other special series that are worth knowing. These consist of the series of the powers of the natural numbers. So let us look at these in the next frame.

# Series of powers of the natural numbers

15

## Sum of natural numbers

The series $1 + 2 + 3 + 4 + 5 + \ldots + n = \sum_{r=1}^{n} r$.

This series, you will see, is an example of an arithmetic series, where $a = 1$ and $d = 1$. The sum of the first $n$ terms is given by:

$$\sum_{r=1}^{n} r = 1 + 2 + 3 + 4 + 5 = \ldots + n$$

$$= \frac{n}{2}(2a + [n-1]d) = \frac{n}{2}(2 + n - 1) = \frac{n(n+1)}{2}$$

$$\sum_{r=1}^{n} r = \frac{n(n+1)}{2}$$

So, the sum of the first 100 natural numbers is ...........

*Then on to Frame 16*

16

$$\sum_{r=1}^{100} r = 5050$$

Because

$$\sum_{r=1}^{n} r = \frac{100(101)}{2} = 50(101) = 5050$$

## Sum of squares

That was easy enough. Now let us look at this one: To establish the result for the sum of $n$ terms of the series $1^2 + 2^2 + 3^2 + 4^2 + 5^2 + \ldots + n^2$, we make use of the identity:

$$(n+1)^3 = n^3 + 3n^2 + 3n + 1$$

We write this as:

$$(n+1)^3 - n^3 = 3n^2 + 3n + 1$$

Replacing $n$ by $n-1$, we get

$$n^3 - (n-1)^3 = 3(n-1)^2 + 3(n-1) + 1$$

and again $(n-1)^3 - (n-2)^3 = 3(n-2)^2 + 3(n-2) + 1$

and $(n-2)^3 - (n-3)^3 = 3(n-3)^2 + 3(n-3) + 1$

Continuing like this, we should eventually arrive at:

$$3^3 - 2^3 = 3 \times 2^2 + 3 \times 2 + 1$$
$$2^3 - 1^3 = 3 \times 1^2 + 3 \times 1 + 1$$

If we now add all these results together, we find on the left-hand side that all the terms disappear except the first and the last.

$$(n+1)^3 - 1^3 = 3\left\{n^2 + (n-1)^2 + (n-2)^2 + \ldots + 2^2 + 1^2\right\}$$
$$+ 3\left\{n + (n-1) + (n-2) + \ldots + 2 + 1\right\} + n(1)$$
$$= 3.\sum_{r=1}^{n} r^2 + 3\sum_{r=1}^{n} r + n$$

$$\therefore\ n^3 + 3n^2 + 3n + 1 - 1 = 3\sum_{r=1}^{n} r^2 + 3\sum_{r=1}^{n} r + n = 3\sum_{r=1}^{n} r^2 + 3\frac{n(n+1)}{2} + n$$

$$\therefore\ n^3 + 3n^2 + 2n = 3\sum_{r=1}^{n} r^2 + \frac{3}{2}(n^2 + n)$$

$$\therefore\ 2n^3 + 6n^2 + 4n = 6\sum_{r=1}^{n} r^2 + 3n^2 + 3n$$

$$6\sum_{r=1}^{n} r^2 = 2n^3 + 3n^2 + n$$

$$\therefore\ \sum_{r=1}^{n} r^2 = \frac{n(n+1)(2n+1)}{6}$$

So, the sum of the first 12 terms of the series $1^2 + 2^2 + 3^2 + \ldots$ is ...........

17

650

Because $\sum_{r=1}^{n} r^2 = \frac{n(n+1)(2n+1)}{6}$

so $\sum_{r=1}^{12} r^2 = \frac{12(13)(25)}{6} = 26(25) = 650$

## Sum of cubes

The sum of the cubes of the natural numbers is found in much the same way. This time, we use the identity:

$$(n+1)^4 = n^4 + 4n^3 + 6n^2 + 4n + 1$$

We rewrite it as before:

$$(n+1)^4 - n^4 = 4n^3 + 6n^2 + 4n + 1$$

If we now do the same trick as before and replace $n$ by $(n-1)$ over and over again, and finally total up the results we get the result:

$$\sum_{r=1}^{n} r^3 = \left\{\frac{n(n+1)}{2}\right\}^2$$

Note in passing that $\sum_{r=1}^{n} r^3 = \left\{\sum_{r=1}^{n} r\right\}^2$

18

Let us collect together these last three results. Here they are:

1 $\sum_{r=1}^{n} r = \frac{n(n+1)}{2}$

2 $\sum_{r=1}^{n} r^2 = \frac{n(n+1)(2n+1)}{6}$

3 $\sum_{r=1}^{n} r^3 = \left\{\frac{n(n+1)}{2}\right\}^2$

These are handy results, so copy them into your record book.

*Now move on to Frame 19 and we can see an example of the use of these results*

19

Find the sum of the series $\sum_{r=1}^{5} r(3+2r)$

$$\begin{aligned}\sum_{r=1}^{5} r(3+2r) &= \sum_{r=1}^{5}(3r+2r^2)\\ &= \sum_{r=1}^{5} 3r + \sum_{r=1}^{5} 2r^2 = 3\sum_{r=1}^{5} r + 2\sum_{r=1}^{5} r^2\\ &= 3.\frac{5.6}{2} + 2.\frac{5.6.11}{6} = 45 + 110 = 155\end{aligned}$$

It is just a question of using the established results. Here is one for you to do in the same manner.

Find the sum of the series $\sum_{r=1}^{4}(2r + r^3)$

*Working in Frame 20*

**20**

$\boxed{120}$

Because $\sum_{r=1}^{4}(2r+r^3) = 2\sum_{r=1}^{4} r + \sum_{r=1}^{4} r^3$

$$= \frac{2.4.5}{2} + \left\{\frac{4.5}{2}\right\}^2 = 20 + 100 = 120$$

Remember:

$$\text{Sum of the first } n \text{ natural numbers} = \frac{n(n+1)}{2}$$

$$\text{Sum of squares of the first } n \text{ natural numbers} = \frac{n(n+1)(2n+1)}{6}$$

$$\text{Sum of cubes of the first } n \text{ natural numbers} = \left\{\frac{n(n+1)}{2}\right\}^2$$

## Infinite series

**21**

So far, we have been concerned with a finite number of terms of a given series. When we are dealing with the sum of an infinite number of terms of a series, we must be careful about the steps we take.

For example, consider the infinite series $1 + \frac{1}{2} + \frac{1}{4} + \frac{1}{8} + \ldots$ ad inf, where the 'ad inf' is an abbreviation for the Latin *ad infinitum* which means *and so on for ever without end.*

This we recognize as an *infinite* geometric series in which $a = 1$ and $r = \frac{1}{2}$. The sum of the first $n$ terms is therefore given by [refer to Frame 9]:

$$\sum_{k=0}^{n-1}\left(\frac{1}{2}\right)^k = \frac{1\left(1 - \left[\frac{1}{2}\right]^n\right)}{1 - \frac{1}{2}}$$

$$= 2\left(1 - \frac{1}{2^n}\right)$$

Now as $n$ increases so $2^n$ increases and $\frac{1}{2^n}$ decreases towards zero [refer to Frame 45, Programme 10]. That is, as

$$\text{as } n \to \infty \text{ so } \frac{1}{2^n} \to 0$$

Hence:

$$\underset{n\to\infty}{Lim}\sum_{k=0}^{n-1}\left(\frac{1}{2}\right)^k = \underset{n\to\infty}{Lim}\left[2\left(1 - \frac{1}{2^n}\right)\right]$$

$$= 2\left(1 - \underset{n\to\infty}{Lim}\left[\frac{1}{2^n}\right]\right)$$

$$= 2(1 - 0)$$

$$= 2$$

▶

We have a shorthand notation for the limit on the left, namely:

$$\lim_{n\to\infty}\sum_{k=0}^{n-1}\left(\frac{1}{2}\right)^k = \sum_{k=0}^{\infty}\left(\frac{1}{2}\right)^k$$

so that

$$\sum_{k=0}^{\infty}\left(\frac{1}{2}\right)^k = 2$$

Beware of this notation – it appears to suggest that we can add up an infinite number of terms, and that is something we just cannot do despite the fact that we do often refer to *the sum of an infinite number of terms*. As usual we must bow to historical precedence and accept the description but make sure that you do understand that it means the *limit* – that we can take the sum of the series as near to the value 2 as we please by taking a sufficiently large number of terms.

*Next frame*

---

**22**

This is not always possible with an infinite series, for in the case of an arithmetic series things are very different.

Consider the infinite series $1 + 3 + 5 + 7 + \ldots$

This is an arithmetic series in which $a = 1$ and $d = 2$.

$$\text{Then } \sum_{r=0}^{n-1}(1+2r) = \frac{n}{2}(2a + [n-1]d)$$

$$= \frac{n}{2}(2 + 2n - 2)$$

$$= n^2$$

Of course, in this case, if $n$ is large then the value of $n^2$ is very large. In fact, if $n \to \infty$, then $n^2 \to \infty$, which is not a definite numerical value and therefore of little use to us.

This always happens with an arithmetic series: if we try to find the 'sum to infinity', we always obtain $+\infty$ or $-\infty$ as the result, depending on the actual series.

*Move on now to Frame 23*

---

**23**

In the previous two frames, we made two important points:

(a) We cannot evaluate the sum of an infinite number of terms of an arithmetic series because the result is always infinite.

(b) We can sometimes evaluate the sum of an infinite number of terms of a geometric series since, for such a series,

$$\sum_{r=0}^{n-1} ar^n = \frac{a(1-r^n)}{1-r} \text{ and } provided\ |r| < 1, \text{ then as } n \to \infty,\ r^n \to 0.$$

In that case

$$\sum_{k=1}^{\infty} ar^n = \frac{a(1-0)}{1-r} = \frac{a}{1-r}$$

So, find the 'sum to infinity' of the series

$20 + 4 + 0{\cdot}8 + 0{\cdot}16 + 0{\cdot}032 + \ldots\ldots\ldots\ldots$

---

**24**

$$\boxed{25}$$

Because

For the series $20 + 4 + 0{\cdot}8 + 0{\cdot}16 + 0{\cdot}032 + \ldots$

$$a = 20; \quad r = \frac{0{\cdot}8}{4} = 0{\cdot}2 = \frac{1}{5} \quad \therefore \sum_{k=1}^{n-1} 20\left(\frac{1}{5}\right)^n = \frac{a}{1-r} = \frac{20}{1-\frac{1}{5}} = 20\left(\frac{5}{4}\right) = 25$$

## Limiting values

**25**

In this Programme, we have already seen that we have sometimes to determine the limiting value of a sum of $n$ terms as $n \to \infty$. Before we leave this topic, let us look a little further into the process of finding limiting values. A few examples will suffice.

*So move on to Frame 26*

**26**

**Example 1**

To find the limiting value of $\dfrac{5n+3}{2n-7}$ as $n \to \infty$.

We cannot just substitute $n = \infty$ in the expression and simplify the result, since $\infty$ is not an ordinary number and does not obey the normal rules. So we do it this way:

$$\frac{5n+3}{2n-7} = \frac{5+3/n}{2-7/n} \quad \text{(dividing top and bottom by } n\text{)}$$

$$\underset{n\to\infty}{Lim}\left\{\frac{5n+3}{2n-7}\right\} = \underset{n\to\infty}{Lim}\frac{5+3/n}{2-7/n}$$

Now when $n \to \infty$, $3/n \to 0$ and $7/n \to 0$

$$\therefore \underset{n\to\infty}{Lim}\frac{5n+3}{2n-7} = \underset{n\to\infty}{Lim}\frac{5+3/n}{2-7/n} = \frac{5+0}{2-0} = \frac{5}{2}$$

We can always deal with fractions of the form $\dfrac{c}{n}, \dfrac{c}{n^2}, \dfrac{c}{n^3}$, etc., because when $n \to \infty$, each of these tends to zero, which is a precise value.

Let us try another example.

*On to the next frame then*

**27**

**Example 2**

To find the limiting value of $\dfrac{2n^2+4n-3}{5n^2-6n+1}$ as $n \to \infty$.

First of all, we divide top and bottom by the highest power of $n$ which is involved, in this case $n^2$.

$$\frac{2n^2+4n-3}{5n^2-6n+1} = \frac{2+4/n-3/n^2}{5-6/n+1/n^2}$$

$$\therefore \underset{n\to\infty}{Lim}\frac{2n^2+4n-3}{5n^2-6n+1} = \underset{n\to\infty}{Lim}\frac{2+4/n-3/n^2}{5-6/n+1/n^2}$$

$$= \frac{2+0-0}{5-0+0} = \frac{2}{5}$$

▶

**Example 3**

To find $\lim_{n\to\infty} \dfrac{n^3 - 2}{2n^3 + 3n - 4}$

In this case, the first thing is to ...........

*Move on to Frame 28*

28

Divide top and bottom by $n^3$

Right. So we get:

$$\frac{n^3 - 2}{2n^3 + 3n - 4} = \frac{1 - 2/n^3}{2 + 3/n^2 - 4/n^3}$$

$$\therefore \lim_{n\to\infty} \frac{n^3 - 2}{2n^3 + 3n - 4} = \ldots\ldots\ldots\ldots$$

*Finish it off. Then move on to Frame 29*

29

$$\frac{1}{2}$$

*Next frame*

30

## Convergent and divergent series

A series in which the sum of $n$ terms of the series tends to a definite value, as $n \to \infty$, is called a *convergent* series. If the sum does not tend to a definite value as $n \to \infty$, the series is said to be *divergent*.

For example, consider the geometric series: $1 + \dfrac{1}{3} + \dfrac{1}{9} + \dfrac{1}{27} + \dfrac{1}{81} + \ldots$

We know that for a geometric series, $\displaystyle\sum_{k=0}^{n-1} ar^k = \frac{a(1 - r^n)}{1 - r}$, so in this case, since $a = 1$ and $r = \dfrac{1}{3}$, we have:

$$\sum_{k=0}^{n-1} \left(\frac{1}{3}\right)^k = \frac{1\left(1 - \dfrac{1}{3^n}\right)}{1 - \dfrac{1}{3}} = \frac{1 - \dfrac{1}{3^n}}{\dfrac{2}{3}} = \frac{3}{2}\left(1 - \frac{1}{3^n}\right)$$

$$\therefore \text{ As } n \to \infty, \ \frac{1}{3^n} \to 0 \quad \therefore \lim_{n\to\infty} \sum_{k=0}^{n-1} \left(\frac{1}{3}\right)^k = \frac{3}{2}$$

The sum of $n$ terms of this series tends to the definite value of $\dfrac{3}{2}$ as $n \to \infty$. It is therefore a ........................ series.
(convergent/divergent)

**31**

convergent

If $\sum_{k=0}^{n-1} f(k)$ tends to a definite value as $n \to \infty$, the series is *convergent*.

If $\sum_{k=0}^{n-1} f(k)$ does not tend to a definite value as $n \to \infty$, the series is *divergent*.

Here is another series. Let us investigate this one.

$1 + 3 + 9 + 27 + 81 + \dots$

This is also a geometric series with $a = 1$ and $r = 3$.

$$\therefore \sum_{k=0}^{n-1} 3^k = \frac{a(1-r^n)}{1-r} = \frac{1(1-3^n)}{1-3} = \frac{1-3^n}{-2}$$

$$= \frac{3^n - 1}{2}$$

Of course, when $n \to \infty$, $3^n \to \infty$ also.

$$\therefore \sum_{k=0}^{n-1} 3^k = \infty \text{ (which is not a definite numerical value)}$$

So in this case, the series is ...........

**32**

divergent

We can make use of infinite series only when they are convergent and it is necessary, therefore, to have some means of testing whether or not a given series is, in fact, convergent.

Of course, we could determine the limiting value of a sum of $n$ terms as $n \to \infty$, as we did in the examples a moment ago, and this would tell us directly whether the series in question tended to a definite value (i.e. was convergent) or not.

That is the fundamental test, but unfortunately, it is not always easy to find a formula for a sum of $n$ terms and we have therefore to find a test for convergence which uses the terms themselves.

Remember the notation for series in general. We shall denote the terms by $u_1 + u_2 + u_3 + u_4 + \dots$

*So now move on to Frame 33*

**33**

## Tests for convergence

**Test 1. A series cannot be convergent unless its terms ultimately tend to zero, i.e. unless $\lim_{n\to\infty} u_n = 0$**

If $\lim_{n\to\infty} u_n \neq 0$, the series is divergent.

This is almost just common sense, for if the sum is to approach some definite value as the value of $n$ increases, the numerical value of the individual terms must diminish. For example, we have already seen that:

(a) the series $1 + \frac{1}{3} + \frac{1}{9} + \frac{1}{27} + \frac{1}{81} + \dots$ converges

while (b) the series $1 + 3 + 9 + 27 + 81 + \dots$ diverges.

So what would you say about the series

$$\sum_{r=1}^{\infty} \frac{1}{r} = 1 + \frac{1}{2} + \frac{1}{3} + \frac{1}{4} + \frac{1}{5} + \frac{1}{6} + \ldots ?$$

Just by looking at it, do you think this series converges or diverges?

---

**34**

Most likely you said that the series converges since it was clear that the numerical value of the terms decreases as $n$ increases. If so, I am afraid you were wrong, for we shall show later that, in fact, the series

$$\sum_{r=1}^{\infty} \frac{1}{r} = 1 + \frac{1}{2} + \frac{1}{3} + \frac{1}{4} + \frac{1}{5} + \ldots \text{ diverges.}$$

It was rather a trick question, but be very clear about what the rule states. It says:

*A series cannot be convergent unless its terms ultimately tend to zero*, i.e. $\lim_{n \to \infty} u_n = 0$. It does not say that if the terms tend to zero, then the series is convergent. In fact, it is quite possible for the terms to tend to zero without the series converging – as in the example stated.

In practice, then, we use the rule in the following form:

If $\lim_{n \to \infty} u_n = 0$, the series *may* converge or diverge and we must test further.

If $\lim_{n \to \infty} u_n \neq 0$, we can be sure that the series diverges.

*Make a note of these two statements*

---

**35**

Before we leave the series

$$\sum_{r=1}^{\infty} \frac{1}{r} = 1 + \frac{1}{2} + \frac{1}{3} + \frac{1}{4} + \frac{1}{5} + \frac{1}{6} + \ldots + \frac{1}{n} + \ldots$$

here is the proof that, although $\lim_{n \to \infty} u_n = 0$, the series does, in fact, diverge.

We can, of course, if we wish, group the terms as follows:

$$\sum_{r=1}^{\infty} \frac{1}{r} = 1 + \frac{1}{2} + \left\{\frac{1}{3} + \frac{1}{4}\right\} + \left\{\frac{1}{5} + \frac{1}{6} + \frac{1}{7} + \frac{1}{8}\right\} + \ldots$$

Now $\left\{\frac{1}{3} + \frac{1}{4}\right\} > \left\{\frac{1}{4} + \frac{1}{4}\right\} = \frac{1}{2}$

and $\left\{\frac{1}{5} + \frac{1}{6} + \frac{1}{7} + \frac{1}{8}\right\} > \left\{\frac{1}{8} + \frac{1}{8} + \frac{1}{8} + \frac{1}{8}\right\} = \frac{1}{2}$ etc.

So that $\sum_{r=1}^{\infty} \frac{1}{r} > 1 + \frac{1}{2} + \frac{1}{2} + \frac{1}{2} + \frac{1}{2} + \frac{1}{2} + \ldots$

$$\therefore \sum_{r=1}^{\infty} \frac{1}{r} = \infty$$

This is not a definite numerical value, so the series is ...........

**36**

divergent

The best we can get from Test 1 is that a series *may* converge. We must therefore apply a further test.

**Test 2. The comparison test**

*A series of positive terms is convergent if its terms are less than the corresponding terms of a positive series which is known to be convergent. Similarly, the series is divergent if its terms are greater than the corresponding terms of a series which is known to be divergent.*

A couple of examples will show how we apply this particular test.

*So move on to the next frame*

**37**

To test the series

$$1+\frac{1}{2^2}+\frac{1}{3^3}+\frac{1}{4^4}+\frac{1}{5^5}+\frac{1}{6^6}+\dots+\frac{1}{n^n}+\dots$$

we can compare it with the series

$$1+\frac{1}{2^2}+\frac{1}{2^3}+\frac{1}{2^4}+\frac{1}{2^5}+\frac{1}{2^6}+\dots+\dots$$

which is known to converge.

If we compare corresponding terms after the first two terms, we see that $\frac{1}{3^3}<\frac{1}{2^3}$; $\frac{1}{4^4}<\frac{1}{2^4}$; and so on for all the further terms, so that, after the first two terms, the terms of the first series are each less than the corresponding terms of the series known to converge.

The first series also, therefore, ...........

**38**

converges

The difficulty with the comparison test is knowing which convergent series to use as a standard. A useful series for this purpose is this one:

$$\frac{1}{1^p}+\frac{1}{2^p}+\frac{1}{3^p}+\frac{1}{4^p}+\frac{1}{5^p}+\dots+\frac{1}{n^p}+\dots=\sum_{n=1}^{\infty}\frac{1}{n^p}$$

It can be shown that:

(a) if $p>1$, the series converges

(b) if $p\leq 1$, the series diverges.

So what about the series $\sum_{n=1}^{\infty}\frac{1}{n^2}$?

Does it converge or diverge?

**39**

Converge

Because the series $\sum\frac{1}{n^2}$ is the series $\sum\frac{1}{n^p}$ with $p>1$.

Let us look at another example:

To test the series $\frac{1}{1 \times 2} + \frac{1}{2 \times 3} + \frac{1}{3 \times 4} + \frac{1}{4 \times 5} + \dots$

If we take our standard series

$$\frac{1}{1^p} + \frac{1}{2^p} + \frac{1}{3^p} + \frac{1}{4^p} + \frac{1}{5^p} + \frac{1}{6^p} + \dots$$

when $p = 2$, we get

$$\frac{1}{1^2} + \frac{1}{2^2} + \frac{1}{3^2} + \frac{1}{4^2} + \frac{1}{5^2} + \frac{1}{6^2} + \dots$$

which we know to converge.

But $\frac{1}{1 \times 2} < \frac{1}{1^2}$; $\frac{1}{2 \times 3} < \frac{1}{2^2}$; $\frac{1}{3 \times 4} < \frac{1}{3^2}$; etc.

Each term of the given series is less than the corresponding term in the series known to converge.

Therefore ............

---

**40**

The given series converges

It is not always easy to devise a suitable comparison series, so we look for yet another test to apply, and here it is:

**Test 3: D'Alembert's ratio test for positive terms**

Let $u_1 + u_2 + u_3 + u_4 + \dots u_n + \dots$ be a series of *positive terms*. Find expressions for $u_n$ and $u_{n+1}$, i.e. the $n$th term and the $(n+1)$th term, and form the ratio $\frac{u_{n+1}}{u_n}$. Determine the limiting value of this ratio as $n \to \infty$.

If $\lim_{n \to \infty} \frac{u_{n+1}}{u_n} < 1$, the series converges

If $\lim_{n \to \infty} \frac{u_{n+1}}{u_n} > 1$, the series diverges

If $\lim_{n \to \infty} \frac{u_{n+1}}{u_n} = 1$, the series may converge or diverge and the test gives us no definite information.

*Copy out D'Alembert's ratio test into your record book. Then on to Frame 41*

---

**41**

Here it is again:

*D'Alembert's ratio test* for positive terms:

If $\lim_{n \to \infty} \frac{u_{n+1}}{u_n} < 1$, the series *converges*

If $\lim_{n \to \infty} \frac{u_{n+1}}{u_n} > 1$, the series *diverges*

If $\lim_{n \to \infty} \frac{u_{n+1}}{u_n} = 1$, the result is inconclusive.

▶

For example: To test the series $\frac{1}{1}+\frac{3}{2}+\frac{5}{2^2}+\frac{7}{2^3}+\ldots$

We first of all decide on the pattern of the terms and hence write down the $n$th term. In this case $u_n=\frac{2n-1}{2^{n-1}}$. The $(n+1)$th term will then be the same with $n$ replaced by $(n+1)$

$$\text{i.e. } u_{n+1}=\frac{2n+1}{2^n}$$

$$\therefore \frac{u_{n+1}}{u_n}=\frac{2n+1}{2^n}\cdot\frac{2^{n-1}}{2n-1}=\frac{1}{2}\cdot\frac{2n+1}{2n-1}$$

We now have to find the limiting value of this ratio as $n\to\infty$. From our prevous work on limiting values, we know that the next step, then, is to divide top and bottom by ...........

**42**

Divide top and bottom by $n$

$$\text{So } \underset{n\to\infty}{Lim}\frac{u_{n+1}}{u_n}=\underset{n\to\infty}{Lim}\frac{1}{2}\cdot\frac{2n+1}{2n-1}=\underset{n\to\infty}{Lim}\frac{1}{2}\cdot\frac{2+1/n}{2-1/n}=\frac{1}{2}\cdot\frac{2+0}{2-0}=\frac{1}{2}$$

Because in this case, $\underset{n\to\infty}{Lim}\frac{u_{n+1}}{u_n}<1$, we know that the given series is *convergent.*

Let us do another one in the same way:

Apply D'Alembert's ratio test to the series

$$\frac{1}{2}+\frac{2}{3}+\frac{3}{4}+\frac{4}{5}+\frac{5}{6}+\ldots$$

First of all, we must find an expression for $u_n$.

In this series, $u_n=$ ...........

**43**

$$\frac{1}{2}+\frac{2}{3}+\frac{3}{4}+\frac{4}{5}+\ldots \quad \boxed{u_n=\frac{n}{n+1}}$$

Then $u_{n+1}$ is found by simply replacing $n$ by $(n+1)$.

$$\therefore u_{n+1}=\frac{n+1}{n+2}$$

$$\text{So that } \frac{u_{n+1}}{u_n}=\frac{n+1}{n+2}\cdot\frac{n+1}{n}=\frac{n^2+2n+1}{n^2+2n}$$

We now have to find $\underset{n\to\infty}{Lim}\frac{u_{n+1}}{u_n}$ and in order to do that we must divide top and bottom, in this case, by ...........

**44**

$\boxed{n^2}$

$$\therefore \underset{n\to\infty}{Lim}\frac{u_{n+1}}{u_n}=\underset{n\to\infty}{Lim}\frac{n^2+2n+1}{n^2+2n}=\underset{n\to\infty}{Lim}\frac{1+2/n+1/n^2}{1+2/n}$$

$$=\frac{1+0+0}{1+0}=1$$

$\therefore \underset{n\to\infty}{Lim}\frac{u_{n+1}}{u_n}=1$, which is inconclusive and merely tells us that the series may be convergent or divergent. So where do we go from there?

▶

We have of course, forgotten about Test 1, which states that:

(a) if $\lim_{n\to\infty} u_n = 0$, the series *may* be convergent

(b) if $\lim_{n\to\infty} u_n \neq 0$, the series is certainly *divergent*.

In our present series: $u_n = \dfrac{n}{n+1}$

$$\therefore \lim_{n\to\infty} u_n = \lim_{n\to\infty} \frac{n}{n+1} = \lim_{n\to\infty} \frac{1}{1+1/n} = 1$$

This is *not* zero. Therefore the series is *divergent*.

Now you do this one entirely on your own:

Test the series $\dfrac{1}{5} + \dfrac{2}{6} + \dfrac{2^2}{7} + \dfrac{2^3}{8} + \dfrac{2^4}{9} + \dots$

*When you have finished, check your result with that in Frame 45*

---

**45**

Here is the solution in detail: see if you agree with it.

$$\frac{1}{5} + \frac{2}{6} + \frac{2^2}{7} + \frac{2^3}{8} + \frac{2^4}{9} + \dots$$

$$u_n = \frac{2^{n-1}}{4+n}; \quad u_{n+1} = \frac{2^n}{5+n}$$

$$\therefore \frac{u_{n+1}}{u_n} = \frac{2^n}{5+n} \cdot \frac{4+n}{2^{n-1}}$$

The power $2^{n-1}$ cancels with the power $2^n$ to leave a single factor 2.

$$\therefore \frac{u_{n+1}}{u_n} = \frac{2(4+n)}{5+n}$$

$$\therefore \lim_{n\to\infty} \frac{u_{n+1}}{u_n} = \lim_{n\to\infty} \frac{2(4+n)}{5+n} = \lim_{n\to\infty} \frac{2(4/n+1)}{5/n+1} = \frac{2(0+1)}{0+1} = 2$$

$$\therefore \lim_{n\to\infty} \frac{u_{n+1}}{u_n} = 2$$

And since the limiting value is >1, we know the series is ...........

---

**46**

divergent

*Next frame*

---

**47**

## Absolute convergence

So far we have considered series with positive terms only. Some series consist of alternate positive and negative terms.

For example, the series $1 - \dfrac{1}{2} + \dfrac{1}{3} - \dfrac{1}{4} + \dots$ is in fact convergent

while the series $1 + \dfrac{1}{2} + \dfrac{1}{3} + \dfrac{1}{4} + \dots$ is divergent.

If $u_n$ denotes the $n$th term of a series in general, it may well be positive or negative. But $|u_n|$, or 'mod $u_n$' denotes the numerical value of $u_n$, so that if $u_1 + u_2 + u_3 + u_4 + \dots$ is a series of mixed terms, i.e. some positive, some negative, then the series $|u_1| + |u_2| + |u_3| + |u_4| + \dots$ will be a series of positive terms.

So if $\sum u_n = 1 - 3 + 5 - 7 + 9 - \dots$

Then $\sum |u_n| = \dots\dots\dots\dots$

**48**

$$\sum |u_n| = 1 + 3 + 5 + 7 + 9 + \ldots$$

*Note*: If a series $\sum u_n$ is convergent, then the series $\sum |u_n|$ may very well not be convergent, as in the example stated in the previous frame. But if $\sum |u_n|$ is found to be convergent, we can be sure that $\sum u_n$ is convergent.

If $\sum |u_n|$ converges, the series $\sum u_n$ is said to be *absolutely convergent*.

If $\sum |u_n|$ is not convergent, but $\sum u_n$ does converge, then $\sum u_n$ is said to be *conditionally convergent*.

So, if $\sum u_n = 1 - \frac{1}{2} + \frac{1}{3} - \frac{1}{4} + \frac{1}{5} - \ldots$ converges

and $\sum |u_n| = 1 + \frac{1}{2} + \frac{1}{3} + \frac{1}{4} + \frac{1}{5} + \ldots$ diverges

then $\sum u_n$ is ...................... convergent.

(absolutely or conditionally)

**49**

conditionally

As an example, find the range of values of $x$ for which the following series is absolutely convergent:

$$\frac{x}{2 \times 5} - \frac{x^2}{3 \times 5^2} + \frac{x^3}{4 \times 5^3} - \frac{x^4}{5 \times 5^4} + \frac{x^5}{6 \times 5^5} - \ldots$$

$$|u_n| = \frac{x^n}{(n+1)5^n};$$

$$|u_{n+1}| = \frac{x^{n+1}}{(n+2)5^{n+1}}$$

$$\therefore \left|\frac{u_{n+1}}{u_n}\right| = \frac{x^{n+1}}{(n+2)5^{n+1}} \cdot \frac{(n+1)5^n}{x^n}$$

$$= \frac{x(n+1)}{5(n+2)}$$

$$= \frac{x(1+1/n)}{5(1+2/n)}$$

$$\therefore \lim_{n\to\infty} \left|\frac{u_{n+1}}{u_n}\right| = \frac{x}{5}$$

For absolute convergence $\lim_{n\to\infty} \left|\frac{u_{n+1}}{u_n}\right| < 1$.

$\therefore$ Series convergent when $\left|\frac{x}{5}\right| < 1$, i.e. for $|x| < 5$.

You have now reached the end of this Programme, except for the **Can You?** checklist and the **Test exercise** which follow. Before you work through them, here is a summary of the topics we have covered. Read through it carefully: it will refresh your memory on what we have been doing.

*On to Frame 50*

##  Review summary

50

1 *Arithmetic series*: $a + (a + d) + (a + 2d) + (a + 3d) + \dots$

$$u_n = a + (n-1)d \qquad \sum_{r=0}^{n-1}(a + rd) = \frac{n}{2}(2a + [n-1]d)$$

2 *Geometric series*: $a + ar + ar^2 + ar^3 + \dots$

$$u_n = ar^{n-1} \qquad \sum_{k=0}^{n-1} ar^k = \frac{a(1-r^n)}{1-r}$$

If $|r| < 1$, $\displaystyle\sum_{k=0}^{\infty} ar^k = \frac{a}{1-r}$

3 *Powers of natural numbers*

$$\sum_{r=1}^{n} r = \frac{n(n+1)}{2} \qquad \sum_{r=1}^{n} r^2 = \frac{n(n+1)(2n+1)}{6}$$

$$\sum_{r=1}^{n} r^3 = \left\{\frac{n(n+1)}{2}\right\}^2$$

4 *Infinite series*: $\displaystyle\sum_{k=1}^{\infty} u_k = u_1 + u_2 + u_3 + u_4 + \dots + u_n + \dots$

If $\displaystyle\lim_{n\to\infty} \sum_{k=1}^{\infty} u_k$ is a definite value, series is convergent

If $\displaystyle\lim_{n\to\infty} \sum_{k=1}^{\infty} u_k$ is not a definite value, series is divergent.

5 *Tests for convergence*

(1) If $\displaystyle\lim_{n\to\infty} u_n = 0$, the series may be convergent

If $\displaystyle\lim_{n\to\infty} u_n \neq 0$ the series is certainly divergent.

(2) *Comparison test* – Useful standard series

$$\frac{1}{1^p} + \frac{1}{2^p} + \frac{1}{3^p} + \frac{1}{4^p} + \frac{1}{5^p} + \dots + \frac{1}{n^p} \dots$$

For $p > 1$, series converges: for $p \leq 1$, series diverges.

(3) *D'Alembert's ratio test for positive terms*

If $\displaystyle\lim_{n\to\infty} \frac{u_{n+1}}{u_n} < 1$, series converges

If $\displaystyle\lim_{n\to\infty} \frac{u_{n+1}}{u_n} > 1$, series diverges

If $\displaystyle\lim_{n\to\infty} \frac{u_{n+1}}{u_n} = 1$, inconclusive.

(4) *For general series*

(a) If $\sum |u_n|$ converges, $\sum u_n$ is absolutely convergent.

(b) If $\sum |u_n|$ diverges, but $\sum u_n$ converges, then $\sum u_n$ is conditionally convergent.

Now you are ready for the **Can You?** checklist and the **Test exercise**.

*So move on to Frame 51*

## Can You?

**51**

### Checklist 11

Check this list before and after you try the end of Programme test.

On a scale of 1 to 5 how confident are you that you can: **Frames**

- Manipulate arithmetic and geometric series? 1 to 14
  *Yes* ☐ ☐ ☐ ☐ ☐ *No*
- Manipulate series of powers of the natural numbers? 15 to 20
  *Yes* ☐ ☐ ☐ ☐ ☐ *No*
- Determine the limiting values of arithmetic and geometric series? 21 to 24
  *Yes* ☐ ☐ ☐ ☐ ☐ *No*
- Determine the limiting values of simple indeterminate forms? 25 to 29
  *Yes* ☐ ☐ ☐ ☐ ☐ *No*
- Apply various convergence tests to infinite series? 30 to 46
  *Yes* ☐ ☐ ☐ ☐ ☐ *No*
- Distinguish between absolute and conditional convergence? 47 to 49
  *Yes* ☐ ☐ ☐ ☐ ☐ *No*

## Test exercise 11

**52**

Take your time and work carefully.

1 The 3rd term of an arithmetic series is 34 and the 17th term is $-8$. Find the sum of the first 20 terms

2 For the series $1 + 1{\cdot}2 + 1{\cdot}44 + \ldots$find the 6th term and the sum of the first 10 terms.

3 Evaluate $\sum_{r=1}^{8} r(3 + 2r + r^2)$

4 Determine whether each of the following series is convergent:

(a) $\dfrac{2}{2 \times 3} + \dfrac{2}{3 \times 4} + \dfrac{2}{4 \times 5} + \dfrac{2}{5 \times 6} + \ldots$

(b) $\dfrac{2}{1^2} + \dfrac{2^2}{2^2} + \dfrac{2^3}{3^2} + \dfrac{2^4}{4^2} + \ldots + \dfrac{2^n}{n^2} + \ldots$

(c) $u_n = \dfrac{1 + 2n^2}{1 + n^2}$

(d) $u_n = \dfrac{1}{n!}$

5 Find the range of values of $x$ for which each of the following series is convergent or divergent:

(a) $1 + x + \dfrac{x^2}{2!} + \dfrac{x^3}{3!} + \dfrac{x^4}{4!} + \ldots$

(b) $\dfrac{x}{1 \times 2} + \dfrac{x^2}{2 \times 3} + \dfrac{x^3}{3 \times 4} + \dfrac{x^4}{4 \times 5} \ldots$

(c) $\sum_{n=1}^{\infty} \dfrac{(n + 1)}{n^3} x^n$

# Further problems 11

1 Find the sum of $n$ terms of the series

$$1^2 + 3^2 + 5^2 + \ldots + (2n-1)^2$$

2 Find the sum to $n$ terms of

$$\frac{1}{1.2.3} + \frac{3}{2.3.4} + \frac{5}{3.4.5} + \frac{7}{4.5.6} + \ldots$$

3 Sum to $n$ terms, the series

$$1.3.5 + 2.4.6 + 3.5.7 + \ldots$$

4 Evaluate the following:

(a) $\displaystyle\sum_{r=1}^{n} r(r+3)$ (b) $\displaystyle\sum_{r=1}^{n} (r+1)^3$

5 Find the sum to infinity of the series

$$1 + \frac{4}{3!} + \frac{6}{4!} + \frac{8}{5!} + \ldots$$

6 For the series $5 - \frac{5}{2} + \frac{5}{4} - \frac{5}{8} + \ldots + \frac{(-1)^{n-1}5}{2^{n-1}}$

find an expression for the sum of the first $n$ terms. Also if the series converges, find the sum to infinity.

7 Find the limiting values of:

(a) $\dfrac{3x^2 + 5x - 4}{5x^2 - x + 7}$ as $x \to \infty$

(b) $\dfrac{x^2 + 5x - 4}{2x^2 - 3x + 1}$ as $x \to \infty$

8 Determine whether each of the following series converges or diverges:

(a) $\displaystyle\sum_{n=1}^{\infty} \frac{n}{n+2}$ (b) $\displaystyle\sum_{n=1}^{\infty} \frac{n}{n^2+1}$

(c) $\displaystyle\sum_{n=1}^{\infty} \frac{1}{n^2+1}$ (d) $\displaystyle\sum_{n=0}^{\infty} \frac{1}{(2n+1)!}$

9 Find the range of values of $x$ for which the series

$$\frac{x}{27} + \frac{x^2}{125} + \ldots + \frac{x^n}{(2n+1)^3} + \ldots$$

is absolutely convergent.

10 Show that the series

$$1 + \frac{x}{1 \times 2} + \frac{x^2}{2 \times 3} + \frac{x^3}{3 \times 4} + \ldots$$

is absolutely convergent when $-1 < x < +1$.

11 Determine the range of values of $x$ for which the following series is convergent:

$$\frac{x}{1.2.3} + \frac{x^2}{2.3.4} + \frac{x^3}{3.4.5} + \frac{x^4}{4.5.6} + \ldots$$

12 Find the range of values of $x$ for convergence for the series

$$x + \frac{2^4x^2}{2!} + \frac{3^4x^3}{3!} + \frac{4^4x^4}{4!} + \ldots$$

▶

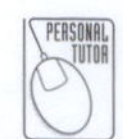

**13** Investigate the convergence of the series

$$\frac{1}{1\times 2}+\frac{x}{2\times 3}+\frac{x^2}{3\times 4}+\frac{x^3}{4\times 5}+\dots \text{ for } x>0.$$

**14** Show that the following series is convergent: $2+\frac{3}{2}\cdot\frac{1}{4}+\frac{4}{3}\cdot\frac{1}{4^2}+\frac{5}{4}\cdot\frac{1}{4^3}+\dots$

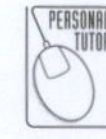

**15** Prove that

$$\frac{1}{\sqrt{1}}+\frac{1}{\sqrt{2}}+\frac{1}{\sqrt{3}}+\frac{1}{\sqrt{4}}+\dots \text{ is divergent}$$

and that

$$\frac{1}{1^2}+\frac{1}{2^2}+\frac{1}{3^2}+\frac{1}{4^2}+\dots \text{ is convergent.}$$

**16** Determine whether each of the following series is convergent or divergent:

(a) $\sum_{n=1}^{\infty}\frac{1}{2n(2n+1)}$ (b) $\sum_{n=1}^{\infty}\frac{1+3n^2}{1+n^2}$

(c) $\sum_{n=1}^{\infty}\frac{n}{\sqrt{4n^2+1}}$ (d) $\sum_{n=1}^{\infty}\frac{3n+1}{3n^2-2}$

**17** Show that the series

$$1+\frac{2x}{5}+\frac{3x^2}{25}+\frac{4x^3}{125}+\dots \text{ is convergent}$$

if $-5<x<5$ and for no other values of $x$.

**18** Investigate the convergence of:

(a) $1+\frac{3}{2\times 4}+\frac{7}{4\times 9}+\frac{15}{8\times 16}+\frac{31}{16\times 25}+\dots$

(b) $\frac{1}{1\times 2}+\frac{1}{2\times 2^2}+\frac{1}{3\times 2^3}+\frac{1}{4\times 2^4}+\dots$

**19** Find the range of values of $x$ for which the following series is convergent:

$$\frac{(x-2)}{1}+\frac{(x-2)^2}{2}+\frac{(x-2)^3}{3}+\dots+\frac{(x-2)^n}{n}+\dots$$

**20** If $u_r=r(2r+1)+2^{r+1}$, find the value of $\sum_{r=1}^{n}u_r$.

Now visit the companion website for this book at www.palgrave.com/stroud for more questions applying this mathematics to science and engineering.

# Series 2

## Learning outcomes

When you have completed this Programme you will be able to:

- ☐ Derive the power series for $\sin x$
- ☐ Use Maclaurin's series to derive series of common functions
- ☐ Use Maclaurin's series to derive the binomial series
- ☐ Derive power series expansions of miscellaneous functions using known expansions of common functions
- ☐ Use power series expansions in numerical approximations
- ☐ Extend Maclaurin's series to Taylor's series
- ☐ Use l'Hôpital's rule to evaluate limits of indeterminate forms

# Power series

1

## Introduction

In the first Programme on series, we saw how important it is to know something of the convergence properties of any infinite series we may wish to use and to appreciate the conditions in which the series is valid.

This is very important, since it is often convenient to represent a function as a series of ascending powers of the variable. This, in fact, is just how a computer finds the value of the sine of a given angle. Instead of storing the whole of the mathematical tables, it sums up the terms of a series representing the sine of an angle.

That is just one example. There are many occasions when we have need to express a function of $x$ as an infinite series of powers of $x$. It is not at all difficult to express a function in this way, as you will soon see in this Programme.

*So make a start and move on to Frame 2*

2

Suppose we wish to express $\sin x$ as a series of ascending powers of $x$. The series will be of the form

$$\sin x \equiv a + bx + cx^2 + dx^3 + ex^4 + \dots$$

where $a$, $b$, $c$, etc. are constant coefficients, i.e. numerical factors of some kind. Notice that we have used the 'equivalent' sign and not the usual 'equals' sign. The statement is not an equation: it is an identity. The right-hand side does not *equal* the left-hand side: the RHS *is* the LHS expressed in a different form and the expression is therefore true for any value of $x$ that we like to substitute.

Can you pick out an identity from these?

$$(x+4)^2 = 3x^2 - 2x + 1$$
$$(2x+1)^2 = 4x^2 + 4x - 3$$
$$(x+2)^2 = x^2 + 4x + 4$$

*When you have decided, move on to Frame 3*

3

$$\boxed{(x+2)^2 \equiv x^2 + 4x + 4}$$

Correct. This is the only identity of the three, since it is the only one in which the RHS is the LHS written in a different form. Right. Now back to our series:

$$\sin x \equiv a + bx + cx^2 + dx^3 + ex^4 + \dots$$

To establish the series, we have to find the values of the constant coefficients $a$, $b$, $c$, $d$, etc.

Suppose we substitute $x = 0$ on both sides.

Then $\quad \sin 0 \equiv a + 0 + 0 + 0 + 0 + \dots$

and since $\sin 0 = 0$, we immediately get the value of $a$.

$a = \dots\dots\dots\dots$

4

$$\boxed{a = 0}$$

Now can we substitute some value for $x$, which will make all the terms disappear except the second? If we could, we should then find the value of $b$. Unfortunately, we cannot find any such substitution, so what is the next step?

▶

Here is the series once again:

$\sin x \equiv a + bx + cx^2 + dx^3 + ex^4 + \ldots$

and so far we know that $a = 0$.

The key to the whole business is simply this:

*Differentiate both sides with respect to x.*

On the left, we get $\cos x$.

On the right the terms are simply powers of $x$, so we get

$\cos x \equiv \ldots\ldots\ldots\ldots$

---

**5**

$$\cos x \equiv b + c.2x + d.3x^2 + c.4x^3 + \ldots$$

This is still an identity, so we can substitute in it any value for $x$ we like.

Notice that the $a$ has now disappeared from the scene and that the constant term at the beginning of the expression is now $b$.

So what do you suggest that we substitute in the identity as it now stands, in order that all the terms except the first shall vanish?

We substitute $x = \ldots\ldots\ldots\ldots$ again.

---

**6**

Substitute $x = 0$ again

Correct: because then all the terms will disappear except the first and we shall be able to find $b$.

$\cos x \equiv b + c.2x + d.3x^2 + e.4x^3 + \ldots$

Put $x = 0$:

$\therefore\ \cos 0 = 1 = b + 0 + 0 + 0 + 0 + \ldots \qquad \therefore b = 1$

So far, so good. We have found the values of $a$ and $b$. To find $c$ and $d$ and all the rest, we merely repeat the process over and over again at each successive stage.

i.e. *Differentiate both sides with respect to x and substitute* $\ldots\ldots\ldots\ldots$

---

**7**

Substitute $x = 0$

So we now get this, from the beginning:

$\sin x \equiv a + bx + cx^2 + dx^3 + ex^4 + fx^5 + \ldots$

Put $x = 0$. $\therefore\ \sin 0 = 0 = a + 0 + 0 + 0 + \ldots \qquad \therefore a = 0$

Differentiate. $\cos x \equiv b + c.2x + d.3x^2 + e.4x^3 + f.5x^4 + \ldots$
Put $x = 0$. $\therefore\ \cos 0 = 1 + b + 0 + 0 + 0 + \ldots \qquad \therefore b = 1$

Differentiate. $-\sin x \equiv c.2 + d.3.2x + e.4.3x^2 + f.5.4x^3 + \ldots$
Put $x = 0$. $\therefore\ -\sin 0 = 0 = c.2 + 0 + 0 + \ldots \qquad \therefore c = 0$

Differentiate. $-\cos x \equiv d.3.2.1 + e.4.3.2x + f.5.4.3x^2 + \ldots$
Put $x = 0$. $\therefore\ -\cos 0 = -1 = d.3! + 0 + 0 + \ldots \qquad \therefore d = -\dfrac{1}{3!}$

And again: $\sin x \equiv e.4.3.2.1 + f.5.4.3.2x + \ldots$
Put $x = 0$. $\therefore\ \sin 0 = 0 = e.4! + 0 + 0 + \ldots \qquad \therefore e = 0$

Once more: $\cos x \equiv f.5.4.3.2.1 + \ldots$
Put $x = 0$. $\therefore\ \cos 0 = 1 = f.5! + 0 + \ldots \qquad \therefore f = \dfrac{1}{5!}$

etc.

▶

All that now remains is to put these values for the constant coefficients back into the original series.

$$\sin x \equiv 0 + 1.x + 0.x^2 + -\frac{1}{3!}x^3 + 0.x^4 + \frac{1}{5!}x^5 + \dots$$

$$\text{i.e. } \sin x \equiv x - \frac{x^3}{3!} + \frac{x^5}{5!} - \dots$$

Now we have obtained the first few terms of an infinite series representing the function $\sin x$, and you can see how the terms are likely to proceed.

Write down the first six terms of the series for $\sin x$.

*When you have done so, move on to Frame 8*

**8**

$$\sin x \equiv x - \frac{x^3}{3!} + \frac{x^5}{5!} - \frac{x^7}{7!} + \frac{x^9}{9!} - \frac{x^{11}}{11!} + \dots$$

Provided we can differentiate a given function over and over again, and find the values of the derivatives when we put $x = 0$, then this method would enable us to express any function as a series of ascending powers of $x$.

However, it entails a considerable amount of writing, so we now establish a general form of such a series, which can be applied to most functions with very much less effort. This general series is known as *Maclaurin's series*.

*So move on to Frame 9 and we will find out all about it*

**9**

## Maclaurin's series

To establish the series, we repeat the process of the previous example, but work with a general function, $f(x)$, instead of $\sin x$. The first derivative of $f(x)$ will be denoted by $f'(x)$; the second by $f''(x)$; the third by $f'''(x)$; and so on. Here it is then:

Let $f(x) = a + bx + cx^2 + dx^3 + ex^4 + fx^5 + \dots$

Put $x = 0$. Then $f(0) = a + 0 + 0 + 0 + \dots$ $\quad\therefore a = f(0)$

i.e. $a$ = the value of the function with $x$ put equal to 0.

Differentiate. $f'(x) = b + c.2x + d.3x^2 + e.4x^3 + f.5x^4 + \dots$

Put $x = 0$. $\therefore f'(0) = b + 0 + 0 + \dots$ $\quad\therefore b = f'(0)$

Differentiate. $f''(x) = c.2.1 + d.3.2x + e.4.3x^2 + f.5.4x^3 \dots$

Put $x = 0$. $\therefore f''(0) = c.2! + 0 + 0 + \dots$ $\quad\therefore c = \dfrac{f''(0)}{2!}$

Now go on and find $d$ and $e$, remembering that we denote

$$\frac{d}{dx}\left\{f''(x)\right\} \text{ by } f'''(x) \text{ and } \frac{d}{dx}\left\{f'''(x)\right\} \text{ by } f^{iv}(x)$$

So, $d = \dots\dots\dots\dots$ and $e = \dots\dots\dots\dots$

10

$$d = \frac{f'''(0)}{3!}; \quad e = \frac{f^{iv}(0)}{4!}$$

Here it is. We had:

$f''(x) = c.2.1 + d.3.2x + e.4.3x^2 + f.5.4x^3 + \dots$

Differentiate. $\therefore f'''(x) = d.3.2.1 + e.4.3.2x + f.5.4.3x^2 + \dots$
Put $x = 0$. $\therefore f'''(0) = d.3! + 0 + 0 + \dots$ $\quad \therefore d = \frac{f'''(0)}{3!}$

Differentiate. $\therefore f^{iv}(x) = e.4.3.2.1 + f.5.4.3.2x + \dots$
Put $x = 0$. $\therefore f^{iv}(0) = e.4! + 0 + 0 + \dots$ $\quad \therefore e = \frac{f^{iv}(0)}{4!}$

etc. So $a = f(0)$; $b = f'(0)$; $c = \frac{f''(0)}{2!}$; $d = \frac{f'''(0)}{3!}$; $e = \frac{f^{iv}(0)}{4!}$; ...

Now, in just the same way as we did with our series for $\sin x$, we put the expressions for $a$, $b$, $c$, ... etc. back into the original series and get:

$f(x) = \dots\dots\dots$

11

$$f(x) = f(0) + f'(0).x + \frac{f''(0)}{2!}.x^2 + \frac{f'''(0)}{3!}.x^3 + \dots$$

and this is usually written as

$$f(x) = f(0) + x.f'(0) + \frac{x^2}{2!}.f''(0) + \frac{x^3}{3!}.f'''(0) + \dots$$

I

This is *Maclaurin's series* and important!

Notice how tidy each term is.

The term in $x^2$ is divided by 2! and multiplied by $f''(0)$
The term in $x^3$ is divided by 3! and multiplied by $f'''(0)$
The term in $x^4$ is divided by 4! and multiplied by $f^{iv}(0)$

Copy the series into your record book for future reference.

*Move on to Frame 12*

12

*Maclaurin's series*

$$f(x) = f(0) + x.f'(0) + \frac{x^2}{2!}.f''(0) + \frac{x^3}{3!}.f'''(0) + \dots$$

Now we will use Maclaurin's series to find a series for $\sinh x$. We have to find the successive derivatives of $\sinh x$ and put $x = 0$ in each. Here goes, then:

$f(x) = \sinh x$ $\quad f(0) = \sinh 0 = 0$
$f'(x) = \cosh x$ $\quad f'(0) = \cosh 0 = 1$
$f''(x) = \sinh x$ $\quad f''(0) = \sinh 0 = 0$
$f'''(x) = \cosh x$ $\quad f'''(0) = \cosh 0 = 1$
$f^{iv}(x) = \sinh x$ $\quad f^{iv}(0) = \sinh 0 = 0$
$f^{v}(x) = \cosh x$ $\quad f^{v}(0) = \cosh 0 = 1$ etc.

$$\therefore \sinh x = 0 + x.1 + \frac{x^2}{2!}.(0) + \frac{x^3}{3!}.(1) + \frac{x^4}{4!}.(0) + \frac{x^5}{5!}.(1) + \dots$$

$$\therefore \sinh x = x + \frac{x^3}{3!} + \frac{x^5}{5!} + \frac{x^7}{7!} + \dots$$

*Move on to Frame 13*

**13**

Now let us find a series for $\ln(1+x)$ in just the same way:

$$f(x) = \ln(1+x) \qquad \therefore f(0) = \ldots\ldots\ldots\ldots$$

$$f'(x) = \frac{1}{1+x} = (1+x)^{-1} \qquad \therefore f'(0) = \ldots\ldots\ldots\ldots$$

$$f''(x) = -(1+x)^{-2} = \frac{-1}{(1+x)^2} \qquad \therefore f''(0) = \ldots\ldots\ldots\ldots$$

$$f'''(x) = 2(1+x)^{-3} = \frac{2}{(1+x)^3} \qquad \therefore f'''(0) = \ldots\ldots\ldots\ldots$$

$$f^{iv}(x) = -3.2(1+x)^{-4} = -\frac{3.2}{(1+x)^4} \qquad \therefore f^{iv}(0) = \ldots\ldots\ldots\ldots$$

$$f^{v}(x) = 4.3.2(1+x)^{-5} = \frac{4!}{(1+x)^5} \qquad \therefore f^{v}(0) = \ldots\ldots\ldots\ldots$$

Evaluate the derivatives when $x = 0$, remembering that $\ln 1 = 0$, and substitute back into Maclaurin's series to obtain the series for $\ln(1+x)$.

So, $\ln(1+x) = \ldots\ldots\ldots\ldots$

**14**

$$\boxed{\ln(1+x) = x - \frac{x^2}{2} + \frac{x^3}{3} - \frac{x^4}{4} + \frac{x^5}{5} - \ldots}$$

$f(0) = \ln 1 = 0;\quad f'(0) = 1;\quad f''(0) = -1;\quad f'''(0) = 2;$
$f^{iv}(0) = -3!;\quad f^{v}(0) = 4!;\quad \ldots$

Also

$$f(x) = f(0) + x.f'(0) + \frac{x^2}{2!}f''(0) + \frac{x^3}{3!}f'''(0) + \ldots$$

$$\ln(1+x) = 0 + x.1 + \frac{x^2}{2!}(-1) + \frac{x^3}{3!}(2) + \frac{x^4}{4!}(-3!) + \frac{x^5}{5!}(4!) + \ldots$$

$$\ln(1+x) = x - \frac{x^2}{2} + \frac{x^3}{3} - \frac{x^4}{4} + \frac{x^5}{5} - \ldots$$

**NB** In this series, the denominators are the natural numbers, not factorials!

*Another example in Frame 15*

**15**

Expand $\sin^2 x$ as a series of ascending powers of $x$.

Maclaurin's series:

$$f(x) = f(0) + x.f'(0) + \frac{x^2}{2!}.f''(0) + \frac{x^3}{3!}.f'''(0) + \ldots$$

$$\therefore f(x) = \sin^2 x \qquad f(0) = \ldots\ldots\ldots\ldots$$

$$f'(x) = 2\sin x \cos x = \sin 2x \qquad f'(0) = \ldots\ldots\ldots\ldots$$

$$f''(x) = 2\cos 2x \qquad f''(0) = \ldots\ldots\ldots\ldots$$

$$f'''(x) = -4\sin 2x \qquad f'''(0) = \ldots\ldots\ldots\ldots$$

$$f^{iv}(x) = \ldots\ldots\ldots\ldots \qquad f^{iv}(0) = \ldots\ldots\ldots\ldots$$

Finish it off: find the first three non-vanishing terms of the series.

*Then move on to Frame 16*

16

$$\sin^2 x = x^2 - \frac{x^4}{3} + \frac{2x^6}{45} \ldots$$

Because

$f(x) = \sin^2 x \qquad \therefore f(0) = 0$

$f'(x) = 2\sin x \cos x = \sin 2x \qquad \therefore f'(0) = 0$

$f''(x) = 2\cos 2x \qquad \therefore f''(0) = 2$

$f'''(x) = -4\sin 2x \qquad \therefore f'''(0) = 0$

$f^{iv}(x) = -8\cos 2x \qquad \therefore f^{iv}(0) = -8$

$f^{v}(x) = 16\sin 2x \qquad \therefore f^{v}(0) = 0$

$f^{vi}(x) = 32\cos 2x \qquad \therefore f^{vi}(0) = 32 \quad$ etc.

$$f(x) = f(0) + x.f'(0) + \frac{x^2}{2!}.f''(0) + \frac{x^3}{3!}.f'''(0) + \ldots$$

$$\therefore \sin^2 x = 0 + x(0) + \frac{x^2}{2!}(2) + \frac{x^3}{3!}(0) + \frac{x^4}{4!}(-8) + \frac{x^5}{5!}(0) + \frac{x^6}{6!}(32) + \ldots$$

$$\therefore \sin^2 x = x^2 - \frac{x^4}{3} + \frac{2x^6}{45} + \ldots$$

Next we will find the series for $\tan x$. This is a little heavier but the method is always the same.

*Move to Frame 17*

17

**Series for tan** $x$

$f(x) = \tan x \qquad \therefore f(0) = 0$

$\therefore f'(x) = \sec^2 x \qquad \therefore f'(0) = 1$

$\therefore f''(x) = 2\sec^2 x \tan x \qquad \therefore f''(0) = 0$

$$\begin{aligned}\therefore f'''(x) &= 2\sec^4 x + 4\sec^2 x \tan^2 x\\ &= 2\sec^4 x + 4(1 + \tan^2 x)\tan^2 x\\ &= 2\sec^4 x + 4\tan^2 x + 4\tan^4 x \qquad \therefore f'''(0) = 2\end{aligned}$$

$$\begin{aligned}\therefore f^{iv}(x) &= 8\sec^4 x \tan x + 8\tan x \sec^2 x + 16\tan^3 x \sec^2 x\\ &= 8(1 + t^2)^2 t + 8t(1 + t^2) + 16t^3(1 + t^2) \qquad [\text{putting } t \equiv \tan x]\\ &= 8(1 + 2t^2 + t^4)t + 8t + 8t^3 + 16t^3 + 16t^5\\ &= 16t + 40t^3 + 24t^5 \qquad \therefore f^{iv}(0) = 0\end{aligned}$$

$\therefore f^{v}(x) = 16\sec^2 x + 120t^2.\sec^2 x + 120t^4.\sec^2 x \qquad \therefore f^{v}(0) = 16$

$\therefore \tan x = \ldots\ldots\ldots\ldots$

18

$$\therefore \tan x = x + \frac{x^3}{3} + \frac{2x^5}{15} + \ldots$$

*Move on to Frame 19*

## 19 Standard series

Using Maclaurin's series we can build up a list of series representing many of the common functions – we have already found series for $\sin x$, $\sinh x$ and $\ln(1+x)$.

To find a series for $\cos x$, we could apply the same technique all over again. However, let us be crafty about it. Suppose we take the series for $\sin x$ and differentiate both sides with respect to $x$ just once, we get:

$$\sin x = x - \frac{x^3}{3!} + \frac{x^5}{5!} - \frac{x^7}{7!} + \dots$$

Differentiate:

$$\cos x = 1 - \frac{3x^2}{3!} + \frac{5x^4}{5!} - \frac{7x^6}{7!} + \dots$$

$$\therefore \cos x = 1 - \frac{x^2}{2!} + \frac{x^4}{4!} - \frac{x^6}{6!} + \dots$$

In the same way, we can obtain the series for $\cosh x$. We already know that

$$\sinh x = x + \frac{x^3}{3!} + \frac{x^5}{5!} + \frac{x^7}{7!} + \dots$$

so if we differentiate both sides we shall establish a series for $\cosh x$.

What do we get?

## 20

We get:

$$\sinh x = x + \frac{x^3}{3!} + \frac{x^5}{5!} + \frac{x^7}{7!} + \frac{x^9}{9!} + \dots$$

Differentiate:

$$\cosh x = 1 + \frac{3x^2}{3!} + \frac{5x^4}{5!} + \frac{7x^6}{7!} + \frac{9x^8}{9!} + \dots$$

giving:

$$\cosh x = 1 + \frac{x^2}{2!} + \frac{x^4}{4!} + \frac{x^6}{6!} + \frac{x^8}{8!} + \dots$$

*Let us pause at this point and take stock of the series we have obtained.*
*We will make a list of them, so move on to Frame 21*

## 21 Summary

Here are the standard series that we have established so far:

$$\sin x = x - \frac{x^3}{3!} + \frac{x^5}{5!} - \frac{x^7}{7!} + \frac{x^9}{9!} + \dots \qquad \text{II}$$

$$\cos x = 1 - \frac{x^2}{2!} + \frac{x^4}{4!} - \frac{x^6}{6!} + \frac{x^8}{8!} + \dots \qquad \text{III}$$

$$\tan x = x + \frac{x^3}{3} + \frac{2x^5}{15} + \dots \qquad \text{IV}$$

$$\sinh x = x + \frac{x^3}{3!} + \frac{x^5}{5!} + \frac{x^7}{7!} + \dots \qquad \text{V}$$

$$\cosh x = 1 + \frac{x^2}{2!} + \frac{x^4}{4!} + \frac{x^6}{6!} + \frac{x^8}{8!} + \dots \qquad \text{VI}$$

$$\ln(1+x) = x - \frac{x^2}{2} + \frac{x^3}{3} - \frac{x^4}{4} + \frac{x^5}{5} + \dots \qquad \text{VII}$$

Make a note of these six series in your record book.

*Then move on to Frame 22*

22

## The binomial series

By the same method, we can apply Maclaurin's series to obtain the binomial series for $(1+x)^n$. Here it is:

$$\begin{aligned}
f(x) &= (1+x)^n & f(0) &= 1\\
f'(x) &= n.(1+x)^{n-1} & f'(0) &= n\\
f''(x) &= n(n-1).(1+x)^{n-2} & f''(0) &= n(n-1)\\
f'''(x) &= n(n-1)(n-2).(1+x)^{n-3} & f'''(0) &= n(n-1)(n-2)\\
f^{iv}(x) &= n(n-1)(n-2)(n-3).(1+x)^{n-4} & f^{iv}(0) &= n(n-1)(n-2)(n-3)\\
&\text{etc.} & &\text{etc.}
\end{aligned}$$

General Maclaurin's series:

$$f(x) = f(0) + x.f'(0) + \frac{x^2}{2!}f''(0) + \frac{x^3}{3!}f'''(0) + \dots$$

Therefore, in this case:

$$(1+x)^n = 1 + xn + \frac{x^2}{2!}n(n-1) + \frac{x^3}{3!}n(n-1)(n-2) + \dots$$

$$(1+x)^n = 1 + nx + \frac{n(n-1)}{2!}x^2 + \frac{n(n-1)(n-2)}{3!}x^3 + \dots \qquad \text{VIII}$$

By replacing $x$ wherever it occurs by $(-x)$, determine the series for $(1-x)^n$.

*When finished, move to Frame 23*

23

$$(1-x)^n = 1 - nx + \frac{n(n-1)}{2!}x^2 - \frac{n(n-1)(n-2)}{3!}x^3 + \dots$$

Now we will work through another example. Here it is:

To find a series for $\tan^{-1} x$.

As before, we need to know the successive derivatives in order to insert them in Maclaurin's series.

$$f(x) = \tan^{-1} x \text{ and } f'(x) = \frac{1}{1+x^2}$$

If we differentiate again, we get $f''(x) = -\frac{2x}{(1+x^2)^2}$, after which the working becomes rather heavy, so let us be crafty and see if we can avoid unnecessary work.

We have $f(x) = \tan^{-1} x$ and $f'(x) = \frac{1}{1+x^2} = (1+x^2)^{-1}$. If we now expand $(1+x^2)^{-1}$ as a binomial series, we shall have a series of powers of $x$ from which we can easily find the higher derivatives.

*So see how it works out in the next frame*

## 24

*To find a series for* $\tan^{-1}x$

$$f(x)=\tan^{-1}x \qquad \therefore f(0)=0$$

$$\therefore f'(x)=\frac{1}{1+x^2}=(1+x^2)^{-1}$$

$$=1-x^2+\frac{(-1)(-2)}{1.2}x^4+\frac{(-1)(-2)(-3)}{1.2.3}x^6+\dots$$

$$=1-x^2+x^4-x^6+x^8-\dots \qquad f'(0)=1$$

$$\therefore f''(x)=-2x+4x^3-6x^5+8x^7-\dots \qquad f''(0)=0$$

$$\therefore f'''(x)=-2+12x^2-30x^4+56x^6-\dots \qquad f'''(0)=-2$$

$$\therefore f^{iv}(x)=24x-120x^3+336x^5-\dots \qquad f^{iv}(0)=0$$

$$\therefore f^{v}(x)=24-360x^2+1680x^4-\dots \qquad f^{v}(0)=24 \text{ etc.}$$

$$\therefore \tan^{-1}x=f(0)+x.f'(0)+\frac{x^2}{2!}f''(0)+\frac{x^3}{3!}f'''(0)+\dots \qquad \text{IX}$$

Substituting the values for the derivatives, gives us that $\tan^{-1}x=$ ...........

*Then on to Frame 25*

## 25

$$\tan^{-1}x=0+x(1)+\frac{x^2}{2!}(0)+\frac{x^3}{3!}(-2)+\frac{x^4}{4!}(0)+\frac{x^5}{5!}(24)+\dots$$

$$\boxed{\tan^{-1}x=x-\frac{x^3}{3}+\frac{x^5}{5}-\frac{x^7}{7}+\dots} \qquad \text{X}$$

This is also a useful series, so make a note of it.

Another series which you already know quite well is the series for $e^x$. Do you remember how it goes? Here it is anyway:

$$e^x=1+x+\frac{x^2}{2!}+\frac{x^3}{3!}+\frac{x^4}{4!}+\dots \qquad \text{XI}$$

and if we simply replace $x$ by $(-x)$, we obtain the series for $e^{-x}$

$$e^{-x}=1-x+\frac{x^2}{2!}-\frac{x^3}{3!}+\frac{x^4}{4!}-\dots \qquad \text{XII}$$

So now we have quite a few. Add the last two to your list.

*And then on to the next frame*

## 26

Once we have established these standard series, we can of course combine them as necessary, as a couple of examples will show.

**Example 1**

Find the first three terms of the series for $e^x.\ln(1+x)$.

We know that $e^x=1+x+\dfrac{x^2}{2!}+\dfrac{x^3}{3!}+\dfrac{x^4}{4!}+\dots$

and that $\ln(1+x)=x-\dfrac{x^2}{2}+\dfrac{x^3}{3}-\dfrac{x^4}{4}+\dots$

So $e^x.\ln(1+x)=\left\{1+x+\dfrac{x^2}{2!}+\dfrac{x^3}{3!}+\dfrac{x^4}{4!}+\dots\right\}\left\{x-\dfrac{x^2}{2}+\dfrac{x^3}{3}-\dots\right\}$

▶

Now we have to multiply these series together. There is no constant term in the second series, so the lowest power of $x$ in the product will be $x$ itself. This can only be formed by multiplying the 1 in the first series by the $x$ in the second.

The $x^2$ term is found by multiplying $1 \times \left(-\frac{x^2}{2}\right)$ and $x \times x$ $\Bigg\}$ $x^2 - \frac{x^2}{2} = \frac{x^2}{2}$

The $x^3$ term is found by multiplying $1 \times \frac{x^3}{3}$ and $x \times \left(-\frac{x^2}{2}\right)$ and $\frac{x^2}{2} \times x$ $\Bigg\}$ $\frac{x^3}{3} - \frac{x^2}{2} + \frac{x^3}{2} = \frac{x^3}{3}$

and so on.

---

**27**

$$\therefore e^x.\ln(1+x) = x + \frac{x^2}{2} + \frac{x^3}{3} + \dots$$

It is not at all difficult, provided you are careful to avoid missing any of the products of the terms.

Here is one for you to do in the same way:

**Example 2**

Find the first four terms of the series for $e^x \sinh x$.

*Take your time over it: then check your working with that in Frame 28*

---

**28**

Here is the solution. Look through it carefully to see if you agree with the result.

$$e^x = 1 + x + \frac{x^2}{2!} + \frac{x^3}{3!} + \frac{x^4}{4!} + \dots$$

$$\sinh x = x + \frac{x^3}{3!} + \frac{x^5}{5!} + \frac{x^7}{7!} + \dots$$

$$e^x.\sinh x = \left\{1 + x + \frac{x^2}{2!} + \frac{x^3}{3!} + \dots\right\}\left\{x + \frac{x^3}{3!} + \frac{x^5}{5!} + \dots\right\}$$

Lowest power is $x$

Term in $x = 1.x = x$

Term in $x^2 = x.x = x^2$

Term in $x^3 = 1.\frac{x^3}{3!} + \frac{x^2}{2!}.x = x^3\left(\frac{1}{6} + \frac{1}{2}\right) = \frac{2x^3}{3}$

Term in $x^4 = x.\frac{x^3}{3!} + \frac{x^3}{3!}.x = x^4\left(\frac{1}{6} + \frac{1}{6}\right) = \frac{x^4}{3}$

$$\therefore e^x.\sinh x = x + x^2 + \frac{2x^3}{3} + \frac{x^4}{3} + \dots$$

*There we are. Now move on to Frame 29*

## Approximate values

**29**

This is a very obvious application of series and you will surely have done some examples on this topic at some time in the past. Here is just an example or two to refresh your memory.

**Example 1**

Evaluate $\sqrt{1{\cdot}02}$ correct to 5 decimal places.

$$1{\cdot}02 = 1 + 0{\cdot}02$$

$$\sqrt{1{\cdot}02} = (1 + 0{\cdot}02)^{\frac{1}{2}}$$

$$= 1 + \frac{1}{2}(0{\cdot}02) + \frac{\frac{1}{2}\left(-\frac{1}{2}\right)}{1.2}(0{\cdot}02)^2 + \frac{\frac{1}{2}\left(-\frac{1}{2}\right)\left(-\frac{3}{2}\right)}{1.2.3}(0{\cdot}02)^3 + \ldots$$

$$= 1 + 0{\cdot}01 - \frac{1}{8}(0{\cdot}0004) + \frac{1}{16}(0{\cdot}000008) - \ldots$$

$$= 1 + 0{\cdot}01 - 0{\cdot}00005 + 0{\cdot}0000005 - \ldots$$

$$= 1{\cdot}010001 - 0{\cdot}000050$$

$$= 1{\cdot}009951$$

$$\therefore \sqrt{1{\cdot}02} = 1{\cdot}00995$$

*Note*: Whenever we substitute a value for $x$ in any one of the standard series, we must be satisfied that the substitution value for $x$ is within the range of values of $x$ for which the series is valid.

The present series for $(1 + x)^n$ is valid for $|x| < 1$, so we are safe enough on this occasion.

Here is one for you to do.

**Example 2**

Evaluate $\tan^{-1} 0{\cdot}1$ correct to 4 decimal places.

*Complete the working and then check with the next frame*

**30**

$$\boxed{\tan^{-1} 0{\cdot}1 = 0{\cdot}0997}$$

$$\tan^{-1} x = x - \frac{x^3}{3} + \frac{x^5}{5} - \frac{x^7}{7} + \ldots$$

$$\therefore \tan^{-1}(0{\cdot}1) = 0{\cdot}1 - \frac{0{\cdot}001}{3} + \frac{0{\cdot}00001}{5} - \frac{0{\cdot}0000001}{7} + \ldots$$

$$= 0{\cdot}1 - 0{\cdot}00033 + 0{\cdot}000002 - \ldots$$

$$= 0{\cdot}0997$$ [*Note*: $x$ is measured in radians.]

## Taylor's series

31

Maclaurin's series $f(x) = f(0) + x.f'(0) + \frac{x^2}{2!}f''(0) + \ldots$ expresses a function in terms of its derivatives at $x = 0$, i.e. at the point K.

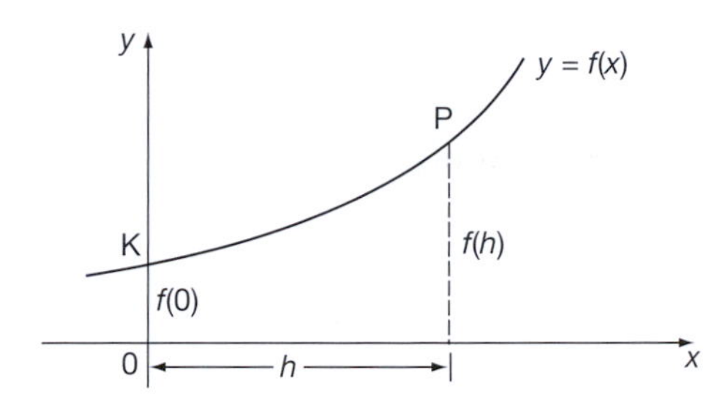


At P, $f(h) = f(0) + h.f'(0) + \frac{h^2}{2!}f''(0) + \frac{h^3}{3!}f'''(0) \ldots$

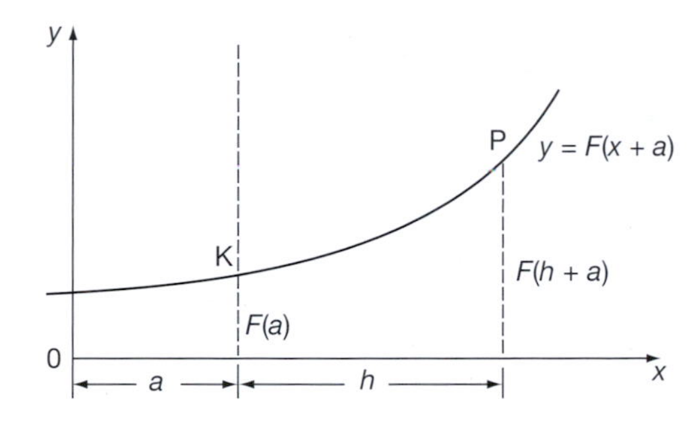


If we now move the $y$-axis $a$ units to the left, the equation of the curve relative to the new axes now becomes $y = F(a + x)$ and the value at K is now $F(a)$.

At P, $F(a + h) = F(a) + h.F'(a) + \frac{h^2}{2!}F''(a) + \frac{h^3}{3!}F'''(a) + \ldots$

This is, in fact, a general series and holds good when $a$ and $h$ are both variables. If we write $a = x$ in this result, we obtain:

$$f(x + h) = f(x) + h.f'(x) + \frac{h^2}{2!}f''(x) + \frac{h^3}{3!}f'''(x) + \ldots$$

which is the usual form of **Taylor's series**.

32

Maclaurin's series and Taylor's series are very much alike in some respects. In fact, Maclaurin's series is really a special case of Taylor's.

*Maclaurin's series:* $f(x) = f(0) + x.f'(0) + \frac{x^2}{2!}f''(0) + \frac{x^3}{3!}f'''(0) + \ldots$

*Taylor's series:* $f(x + h) = f(x) + h.f'(x) + \frac{h^2}{2!}f''(x) + \frac{h^3}{3!}f'''(x) + \ldots$

Copy the two series down together: it will help you remember them.

33

**Example 1**

Show that, if $h$ is small, then

$$\tan^{-1}(x+h) = \tan^{-1}x + \frac{h}{1+x^2} - \frac{xh^2}{(1+x^2)^2} \text{ approximately.}$$

Taylor's series states:

$$f(x+h) = f(x) + h.f'(x) + \frac{h^2}{2!}f''(x) + \frac{h^3}{3!}f'''(x) + \dots$$

where $f(x)$ is the function obtained by putting $h = 0$ in the function $f(x+h)$.

In this case then, $f(x) = \tan^{-1}x$

$$\therefore f'(x) = \frac{1}{1+x^2} \text{ and } f''(x) = -\frac{2x}{(1+x^2)^2}$$

Putting these expressions back into the series, we have:

$$\tan^{-1}(x+h) = \tan^{-1}x + h.\frac{1}{1+x^2} - \frac{h^2}{2!}.\frac{2x}{(1+x^2)^2} + \dots$$

$$= \tan^{-1}x + \frac{h}{1+x^2} - \frac{xh^2}{(1+x^2)^2} \quad \text{approximately}$$

Why are we justified in omitting the terms that follow?

34

The following terms contain higher powers of $h$ which, by definition, are small. These terms will therefore be very small

**Example 2**

Express $\sin(x+h)$ as a series of powers of $h$ and evaluate $\sin 44^\circ$ correct to 5 decimal places.

$$\sin(x+h) = f(x) + h.f'(x) + \frac{h^2}{2!}f''(x) + \frac{h^3}{3!}f'''(x) + \dots$$

$$f(x) = \sin x; \quad f'(x) = \cos x; \quad f''(x) = -\sin x$$

$$f'''(x) = -\cos x; \quad f^{iv}(x) = \sin x; \text{ etc.}$$

$$\therefore \sin(x+h) = \sin x + h\cos x - \frac{h^2}{2!}\sin x - \frac{h^3}{3!}\cos x + \dots$$

$$\sin 44^\circ = \sin(45^\circ - 1^\circ) = \sin\left(\frac{\pi}{4} - 0{\cdot}01745\right) \text{ and } \sin\frac{\pi}{4} = \cos\frac{\pi}{4} = \frac{1}{\sqrt{2}}$$

$$\therefore \sin 44^\circ = \frac{1}{\sqrt{2}}\left\{1 + h - \frac{h^2}{2} - \frac{h^3}{6} + \dots\right\} \qquad h = -0{\cdot}01745$$

$$= \frac{1}{\sqrt{2}}\left\{1 - 0{\cdot}01745 - \frac{0{\cdot}0003046}{2} + \frac{0{\cdot}0000053}{6} + \dots\right\}$$

$$= \frac{1}{\sqrt{2}}\{1 - 0{\cdot}01745 - 0{\cdot}0001523 + 0{\cdot}0000009 - \dots\}$$

$$= 0{\cdot}7071(0{\cdot}982395)$$

$$= 0{\cdot}69466$$

*Remember* that in all the trigonometric expansions, the angle *must* be in radians.

# Limiting values – indeterminate forms

**35**

In the first Programme on series, we had occasion to find the limiting value of $\dfrac{u_{n+1}}{u_n}$ as $n \to \infty$. Sometimes, we have to find the limiting value of a function of $x$ when $x \to 0$, or perhaps when $x \to a$.

$$\text{e.g. } \underset{x\to 0}{Lim}\left\{\frac{x^2+5x-14}{x^2-5x+8}\right\} = \frac{0+0-14}{0-0+8} = -\frac{14}{8} = -\frac{7}{4}$$

That is easy enough, but suppose we have to find

$$\underset{x\to 2}{Lim}\left\{\frac{x^2+5x-14}{x^2-5x+6}\right\}$$

Putting $x = 2$ in the function gives $\dfrac{4+10-14}{4-10+6} = \dfrac{0}{0}$ and what is the value of $\dfrac{0}{0}$?

Is it zero? Is it 1? Is it indeterminate?

*When you have decided, move on to Frame 36*

**36**

$\dfrac{0}{0}$, as it stands, is | indeterminate |

We can sometimes, however, use our knowledge of series to help us out of the difficulty. Let us consider an example or two.

**Example 1**

Find $\underset{x\to 0}{Lim}\left\{\dfrac{\tan x - x}{x^3}\right\}$

If we substitute $x = 0$ in the function, we get the result $\dfrac{0}{0}$ which is indeterminate. So how do we proceed?

Well, we already know that $\tan x = x + \dfrac{x^3}{3} + \dfrac{2x^5}{15} + \dots$ So if we replace $\tan x$ by its series in the given function, we get:

$$\underset{x\to 0}{Lim}\left\{\frac{\tan x - x}{x^3}\right\} = \underset{x\to 0}{Lim}\left\{\frac{\left(x+\dfrac{x^3}{3}+\dfrac{2x^5}{15}+\dots\right)-x}{x^3}\right\}$$

$$= \underset{x\to 0}{Lim}\left\{\frac{1}{3}+\frac{2x^2}{15}+\dots\right\} = \frac{1}{3}$$

$\therefore \underset{x\to 0}{Lim}\left\{\dfrac{\tan x - x}{x^3}\right\} = \dfrac{1}{3}$ and the job is done!

*Move on to Frame 37 for another example*

**37**

**Example 2**

To find $\lim_{x\to 0}\left\{\frac{\sinh x}{x}\right\}$

Direct substitution of $x=0$ gives $\frac{\sin 0}{0}$ which is $\frac{0}{0}$ again. So we will express $\sinh x$ by its series, which is

$\sinh x = \ldots\ldots\ldots\ldots$

(If you do not remember, you will find it in your list of standard series which you have been compiling. Look it up.)

*Then on to Frame 38*

**38**

$$\boxed{\sinh x = x + \frac{x^3}{3!} + \frac{x^5}{5!} + \frac{x^7}{7!} + \ldots}$$

$$\text{So } \lim_{x\to 0}\left\{\frac{\sinh x}{x}\right\} = \lim_{x\to 0}\left\{\frac{x + \frac{x^3}{3!} + \frac{x^5}{5!} + \frac{x^7}{7!} + \ldots}{x}\right\}$$

$$= \lim_{x\to 0}\left\{1 + \frac{x^2}{3!} + \frac{x^4}{5!} + \ldots\right\}$$

$$= 1 + 0 + 0 + \ldots = 1$$

$$\therefore \lim_{x\to 0}\left\{\frac{\sinh x}{x}\right\} = 1$$

Now, in very much the same way, you find $\lim_{x\to 0}\left\{\frac{\sin^2 x}{x^2}\right\}$

*Work it through: then check your result with that in the next frame*

**39**

$$\boxed{\lim_{x\to 0}\left\{\frac{\sin^2 x}{x^2}\right\} = 1}$$

Here is the working:

$$\lim_{x\to 0}\left\{\frac{\sin^2 x}{x^2}\right\} = \lim_{x\to 0}\left\{\frac{x^2 - \frac{x^4}{3} + \frac{2x^6}{45} - \ldots}{x^2}\right\} \qquad \text{(see Frame 16)}$$

$$= \lim_{x\to 0}\left\{1 - \frac{x^2}{3} + \frac{2x^4}{45} - \ldots\right\} = 1$$

$$\therefore \lim_{x\to 0}\left\{\frac{\sin^2 x}{x^2}\right\} = 1$$

Here is one more for you to do in like manner.

Find $\lim_{x\to 0}\left\{\frac{\sinh x - x}{x^3}\right\}$

*Then on to Frame 40*

40

$$\underset{x\to 0}{Lim}\left\{\frac{\sinh x - x}{x^3}\right\} = \frac{1}{6}$$

Here is the working in detail:

$$\sinh x = x + \frac{x^3}{3!} + \frac{x^5}{5!} + \frac{x^7}{7!} + \dots$$

$$\therefore \quad \frac{\sinh x - x}{x^3} = \frac{x + \frac{x^3}{3!} + \frac{x^5}{5!} + \frac{x^7}{7!} + \dots - x}{x^3}$$

$$= \frac{1}{3!} + \frac{x^2}{5!} + \frac{x^4}{7!} + \dots$$

$$\therefore \underset{x\to 0}{Lim}\left\{\frac{\sinh x - x}{x^3}\right\} = \underset{x\to 0}{Lim}\left\{\frac{1}{3!} + \frac{x^2}{5!} + \frac{x^4}{7!} + \dots\right\}$$

$$= \frac{1}{3!} = \frac{1}{6}$$

$$\therefore \underset{x\to 0}{Lim}\left\{\frac{\sinh x - x}{x^3}\right\} = \frac{1}{6}$$

So there you are: they are all done the same way:

(a) Express the given function in terms of power series.

(b) Simplify the function as far as possible.

(c) Then determine the limiting value – which should now be possible.

Of course, there may well be occasions when direct substitution gives the indeterminate form $\frac{0}{0}$ and we do not know the series expansion of the function concerned. What are we going to do then?

All is not lost! – for we do in fact have another method of finding limiting values which, in many cases, is quicker than the series method. It all depends upon the application of a rule which we must first establish, so move to the next frame for details.

*Next frame*

41

## L'Hôpital's rule for finding limiting values

Suppose we have to find the limiting value of a function $F(x) = \frac{f(x)}{g(x)}$ at $x = a$, when direct substitution of $x = a$ gives the indeterminate form $\frac{0}{0}$, i.e. at $x = a$, $f(x) = 0$ and $g(x) = 0$.

If we represent the circumstances graphically, the diagram would look like this:

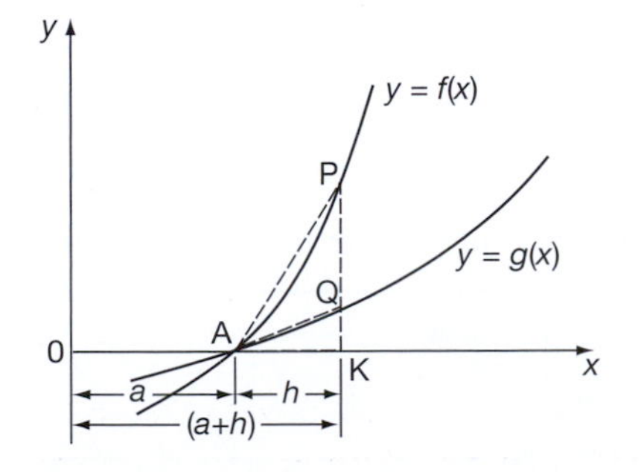


Note that at $x = a$, both of the graphs $y = f(x)$ and $y = g(x)$ cross the $x$-axis, so that at $x = a$, $f(x) = 0$ and $g(x) = 0$.

▶

At a point K, i.e. $x = (a+h)$, $\text{KP} = f(a+h)$ and $\text{KQ} = g(a+h)$

$$\frac{f(a+h)}{g(a+h)} = \frac{\text{KP}}{\text{KQ}}$$

Now divide top and bottom by AK:

$$\frac{f(a+h)}{g(a+h)} = \frac{\text{KP/AK}}{\text{KQ/AK}} = \frac{\tan \text{PAK}}{\tan \text{QAK}}$$

Now $\quad \underset{x \to a}{Lim} \dfrac{f(x)}{g(x)} = \underset{h \to 0}{Lim} \dfrac{f(a+h)}{g(a+h)} = \underset{h \to 0}{Lim} \dfrac{\tan \text{PAK}}{\tan \text{QAK}} = \dfrac{f'(a)}{g'(a)}$

i.e. the limiting value of $\dfrac{f(x)}{g(x)}$ as $x \to a$ (at which the function value by direct substitution gives $\dfrac{0}{0}$) is given by the ratio of the derivatives of numerator and denominator at $x = a$ (provided, of course, that $f'(a)$ and $g'(a)$ are not both zero themselves!). That is:

$$\underset{x \to a}{Lim} \left\{\frac{f(x)}{g(x)}\right\} = \frac{f'(a)}{g'(a)} = \underset{x \to a}{Lim} \left\{\frac{f'(x)}{g'(x)}\right\}$$

$$\therefore \underset{x \to a}{Lim} \left\{\frac{f(x)}{g(x)}\right\} = \underset{x \to a}{Lim} \left\{\frac{f'(x)}{g'(x)}\right\}$$

This is known as *l'Hôpital's rule* and is extremely useful for finding limiting values when the derivatives of the numerator and denominator can easily be found.

*Copy the rule into your record book. Now we will use it*

## 42

$$\boxed{\underset{x \to a}{Lim} \left\{\frac{f(x)}{g(x)}\right\} = \underset{x \to a}{Lim} \left\{\frac{f'(x)}{g'(x)}\right\}}$$

**Example 1**

To find $\underset{x \to 1}{Lim} \left\{\dfrac{x^3 + x^2 - x - 1}{x^2 + 2x - 3}\right\}$

Note first that if we substitute $x = 1$, we get the indeterminate form $\dfrac{0}{0}$. Therefore we apply l'Hôpital's rule.

We therefore differentiate numerator and denominator separately (*not* as a quotient):

$$\underset{x \to 1}{Lim} \left\{\frac{x^3 + x^2 - x - 1}{x^2 + 2x - 3}\right\} = \underset{x \to 1}{Lim} \left\{\frac{3x^2 + 2x - 1}{2x + 2}\right\}$$

$$= \frac{3 + 2 - 1}{2 + 2} = \frac{4}{4} = 1$$

$$\therefore \underset{x \to 1}{Lim} \left\{\frac{x^3 + x^2 - x - 1}{x^2 + 2x - 3}\right\} = 1$$

and that is all there is to it!

*Let us do another example, so, on to the next frame*

43

**Example 2**

Determine $\underset{x \to 0}{Lim}\left\{\dfrac{\cosh x - e^x}{x}\right\}$

We first of all try direct substitution, but we find that this leads us to the result $\dfrac{1-1}{0}$, i.e. $\dfrac{0}{0}$ which is indeterminate. Therefore, apply l'Hôpital's rule:

$$\underset{x \to a}{Lim}\left\{\frac{f(x)}{g(x)}\right\} = \underset{x \to a}{Lim}\left\{\frac{f'(x)}{g'(x)}\right\}$$

i.e. differentiate top and bottom separately and substitute the given value of $x$ in the derivatives:

$$\therefore \underset{x \to 0}{Lim}\left\{\frac{\cosh x - e^x}{x}\right\} = \underset{x \to 0}{Lim}\left\{\frac{\sinh x - e^x}{1}\right\} = \frac{0-1}{1} = -1$$

$$\therefore \underset{x \to 0}{Lim}\left\{\frac{\cosh x - e^x}{x}\right\} = -1$$

Now you can do this one:

Determine $\underset{x \to 0}{Lim}\left\{\dfrac{x^2 - \sin 3x}{x^2 + 4x}\right\}$

44

$$\boxed{\underset{x \to 0}{Lim}\left\{\frac{x^2 - \sin 3x}{x^2 + 4x}\right\} = -\frac{3}{4}}$$

The working is simply this:

Direct substitution gives $\dfrac{0}{0}$, so we apply l'Hôpital's rule which gives

$$\underset{x \to 0}{Lim}\left\{\frac{x^2 - \sin 3x}{x^2 + 4x}\right\} = \underset{x \to 0}{Lim}\left\{\frac{2x - 3\cos 3x}{2x + 4}\right\}$$
$$= \frac{0-3}{0+4}$$
$$= -\frac{3}{4}$$

*Warning*: l'Hôpital's rule applies only when the indeterminate form arises. If the limiting value can be found by direct substitution, the rule will not work. An example will soon show this.

Consider $\underset{x \to 2}{Lim}\left\{\dfrac{x^2 + 4x - 3}{5 - 2x}\right\}$

By direct substitution, the limiting value $= \dfrac{4+8-3}{5-4} = 9$. By l'Hôpital's rule $\underset{x \to 2}{Lim}\left\{\dfrac{f'(x)}{g'(x)}\right\} = \underset{x \to 2}{Lim}\left\{\dfrac{2x+4}{-2}\right\} = -4$. As you will see, these results do not agree.

*Before using l'Hôpital's rule, therefore, you must satisfy yourself that direct substitution gives the indeterminate form* $\dfrac{0}{0}$. If it does, you may use the rule, but not otherwise.

**45**

Let us look at another example.

**Example 3**

Determine $\underset{x\to 0}{Lim}\left\{\dfrac{x-\sin x}{x^2}\right\}$

By direct substitution, limiting value $=\dfrac{0-0}{0}=\dfrac{0}{0}$.

Apply l'Hôpital's rule:

$$\underset{x\to 0}{Lim}\left\{\frac{x-\sin x}{x^2}\right\}=\underset{x\to 0}{Lim}\left\{\frac{1-\cos x}{2x}\right\}$$

We now find, with some horror, that substituting $x=0$ in the derivatives again produces the indeterminate form $\dfrac{0}{0}$. So what do you suggest we do now to find $\underset{x\to 0}{Lim}\left\{\dfrac{1-\cos x}{2x}\right\}$ (without bringing in the use of series)? Any ideas?

We ...........

**46**

We apply the rule a second time

Correct, for our immediate problem now is to find $\underset{x\to 0}{Lim}\left\{\dfrac{1-\cos x}{2x}\right\}$.

If we do that, we get:

$$\underset{x\to 0}{Lim}\left\{\frac{x-\sin x}{x^2}\right\}=\underbrace{\underset{x\to 0}{Lim}\left\{\frac{1-\cos x}{2x}\right\}}_{\text{First stage}}=\underbrace{\underset{x\to 0}{Lim}\left\{\frac{\sin x}{2}\right\}}_{\text{Second stage}}=\frac{0}{2}=0$$

$$\therefore \underset{x\to 0}{Lim}\left\{\frac{x-\sin x}{x^2}\right\}=0$$

So now we have the rule complete:

For limiting values when the indeterminate form $\left(\text{i.e. } \dfrac{0}{0}\right)$ exists, apply l'Hôpital's rule:

$$\underset{x\to a}{Lim}\left\{\frac{f(x)}{g(x)}\right\}=\underset{x\to a}{Lim}\left\{\frac{f'(x)}{g'(x)}\right\}$$

and continue to do so until a stage is reached where the limits of the numerator and denominator are not simultaneously zero.

*Next frame*

47

Just one more example to illustrate the point

**Example 4**

Determine $\underset{x\to 0}{Lim}\left\{\dfrac{\sinh x - \sin x}{x^3}\right\}$

Direct substitution gives $\dfrac{0-0}{0}$, i.e. $\dfrac{0}{0}$. (indeterminate)

$$\begin{aligned}\underset{x\to 0}{Lim}\left\{\frac{\sinh x - \sin x}{x^3}\right\} &= \underset{x\to 0}{Lim}\left\{\frac{\cosh x - \cos x}{3x^2}\right\}, \text{ giving } \frac{1-1}{0}=\frac{0}{0}\\ &= \underset{x\to 0}{Lim}\left\{\frac{\sinh x + \sin x}{6x}\right\}, \text{ giving } \frac{0+0}{0}=\frac{0}{0}\\ &= \underset{x\to 0}{Lim}\left\{\frac{\cosh x + \cos x}{6}\right\} = \frac{1+1}{6}=\frac{1}{3}\end{aligned}$$

$$\therefore\ \underset{x\to 0}{Lim}\left\{\frac{\sinh x - \sin x}{x^3}\right\} = \frac{1}{3}$$

*Note*: We apply l'Hôpital's rule again and again until we reach the stage where the limits of the numerator and the denominator are *not* simultaneously zero. We shall then arrive at a definite limiting value of the function.

*Move on to Frame 48*

48

Here are three *Review Examples* for you to do. Work through all of them and then check your working with the results set out in the next frame. They are all straightforward and easy, so do not peep at the official solutions before you have done them all.

Determine

(a) $\underset{x\to 1}{Lim}\left\{\dfrac{x^3-2x^2+4x-3}{4x^2-5x+1}\right\}$

(b) $\underset{x\to 0}{Lim}\left\{\dfrac{\tan x - x}{\sin x - x}\right\}$

(c) $\underset{x\to 0}{Lim}\left\{\dfrac{x\cos x - \sin x}{x^3}\right\}$

*Solutions in Frame 49*

49

(a) $\underset{x\to 1}{Lim}\left\{\dfrac{x^3-2x^2+4x-3}{4x^2-5x+1}\right\}$ $\left(\text{Substitution gives } \dfrac{0}{0}\right)$

$$= \underset{x\to 1}{Lim}\left\{\frac{3x^2-4x+4}{8x-5}\right\} = \frac{3}{3} = 1$$

$$\therefore\ \underset{x\to 1}{Lim}\left\{\frac{x^3-2x^2+4x-3}{4x^2-5x+1}\right\} = 1$$

▶

(b) $\underset{x \to 0}{Lim}\left\{\dfrac{\tan x - x}{\sin x - x}\right\}$ $\left(\text{Substitution gives } \dfrac{0}{0}\right)$

$= \underset{x \to 0}{Lim}\left\{\dfrac{\sec^2 x - 1}{\cos x - 1}\right\}$ $\left(\text{still gives } \dfrac{0}{0}\right)$

$= \underset{x \to 0}{Lim}\left\{\dfrac{2\sec^2 x \tan x}{-\sin x}\right\}$ (and again!)

$= \underset{x \to 0}{Lim}\left\{\dfrac{2\sec^2 x \sec^2 x + 4\sec^2 x \tan^2 x}{-\cos x}\right\} = \dfrac{2+0}{-1} = -2$

$\therefore \underset{x \to 0}{Lim}\left\{\dfrac{\tan x - x}{\sin x - x}\right\} = -2$

(c) $\underset{x \to 0}{Lim}\left\{\dfrac{x\cos x - \sin x}{x^3}\right\}$ $\left(\text{Substitution gives } \dfrac{0}{0}\right)$

$= \underset{x \to 0}{Lim}\left\{\dfrac{-x\sin x + \cos x - \cos x}{3x^2}\right\}$

$= \underset{x \to 0}{Lim}\left\{\dfrac{-\sin x}{3x}\right\} = \underset{x \to 0}{Lim}\left\{\dfrac{-\cos x}{3}\right\} = -\dfrac{1}{3}$

$\therefore \underset{x \to 0}{Lim}\left\{\dfrac{x\cos x - \sin x}{x^3}\right\} = -\dfrac{1}{3}$

You have now reached the end of the Programme, except for the **Can You?** checklist and the **Test exercise** which follow.

*Now move on to Frames 50 and 51*

---

## Can You?

**50**

### Checklist 12

Check this list before and after you try the end of Programme test.

On a scale of 1 to 5 how confident are you that you can: **Frames**

- Derive the power series for $\sin x$? 1 to 8
  *Yes* ☐ ☐ ☐ ☐ ☐ *No*
- Use Maclaurin's series to derive series of common functions? 9 to 21
  *Yes* ☐ ☐ ☐ ☐ ☐ *No*
- Use Maclaurin's series to derive the binomial series? 22 to 23
  *Yes* ☐ ☐ ☐ ☐ ☐ *No*
- Derive power series expansions of miscellaneous functions using known expansions of common functions? 24 to 28
  *Yes* ☐ ☐ ☐ ☐ ☐ *No*
- Use power series expansions in numerical approximations? 29 to 30
  *Yes* ☐ ☐ ☐ ☐ ☐ *No*
- Extend Maclaurin's series to Taylor's series? 31 to 34
  *Yes* ☐ ☐ ☐ ☐ ☐ *No*
- Use l'Hôpital's rule to evaluate limits of indeterminate forms? 35 to 49
  *Yes* ☐ ☐ ☐ ☐ ☐ *No*

## Test exercise 12

51

The questions are all straightforward and you will have no trouble with them. Work through at your own speed. There is no need to hurry.

1 State Maclaurin's series.

2 Find the first 4 non-zero terms in the expansion of $\cos^2 x$.

3 Find the first 3 non-zero terms in the series for $\sec x$.

4 Show that $\tan^{-1} x = x - \dfrac{x^3}{3} + \dfrac{x^5}{5} - \dfrac{x^7}{7} + \dots$

5 Assuming the series for $e^x$ and $\tan x$, determine the series for $e^x . \tan x$ up to and including the term in $x^4$.

6 Evaluate $\sqrt{1{\cdot}05}$ correct to 5 significant figures.

7 Find:

(a) $\underset{x \to 0}{Lim} \left\{ \dfrac{1 - 2\sin^2 x - \cos^3 x}{5x^2} \right\}$

(b) $\underset{x \to 0}{Lim} \left\{ \dfrac{\tan x . \tan^{-1} x - x^2}{x^6} \right\}$

(c) $\underset{x \to 0}{Lim} \left\{ \dfrac{x - \sin x}{x - \tan x} \right\}$

8 Expand $\cos(x + h)$ as a series of powers of $h$ and hence evaluate $\cos 31^\circ$ correct to 5 decimal places.

## Further problems 12

52

1 Prove that $\cos x = 1 - \dfrac{x^2}{2!} + \dfrac{x^4}{4!} - \dfrac{x^6}{6!} + \dots$ and that the series is valid for all values of $x$. Deduce the power series for $\sin^2 x$ and show that, if $x$ is small, $\dfrac{\sin^2 x - x^2 \cos x}{x^4} = \dfrac{1}{6} + \dfrac{x^2}{360}$ approximately.

2 Apply Maclaurin's series to establish a series for $\ln(1 + x)$. If $1 + x = \dfrac{b}{a}$, show that $(b^2 - a^2)/2ab = x - \dfrac{x^2}{2} + \dfrac{x^3}{2} - \dots$

Hence show that, if $b$ is nearly equal to $a$, then $(b^2 - a^2)/2ab$ exceeds $\ln\left(\dfrac{b}{a}\right)$ by approximately $(b - a)^3/6a^3$.

3 Evaluate:

(a) $\underset{x \to 0}{Lim} \left\{ \dfrac{\sin x - \cos x}{x^3} \right\}$

(b) $\underset{x \to 0}{Lim} \left\{ \dfrac{\tan x - \sin x}{x^3} \right\}$

(c) $\underset{x \to 0}{Lim} \left\{ \dfrac{\sin x - x}{x^3} \right\}$

(d) $\underset{x \to 0}{Lim} \left\{ \dfrac{\tan x - x}{x - \sin x} \right\}$

(e) $\underset{x \to 0}{Lim} \left\{ \dfrac{1 - 2\sin^2 x - \cos^3 x}{5x^2} \right\}$

▶

4 Write down the expansions of:

(a) $\cos x$ and (b) $\dfrac{1}{1+x}$, and hence show that

$$\frac{\cos x}{1+x} = 1 - x + \frac{x^2}{2} - \frac{x^3}{2} + \frac{13x^4}{24} - \ldots$$

5 State the series for $\ln(1+x)$ and the range of values of $x$ for which it is valid. Assuming the series for $\sin x$ and for $\cos x$, find the series for $\ln\left(\dfrac{\sin x}{x}\right)$ and $\ln(\cos x)$ as far as the term in $x^4$. Hence show that, if $x$ is small, $\tan x$ is approximately equal to $x.e^{\frac{x^2}{3}}$.

6 Use Maclaurin's series to obtain the expansion of $e^x$ and of $\cos x$ in ascending powers of $x$ and hence determine

$$\underset{x \to 0}{Lim}\left\{\frac{e^x + e^{-x} - 2}{2\cos 2x - 2}\right\}.$$

7 Find the first four terms in the expansion of $\dfrac{x-3}{(1-x)^2(2+x^2)}$ in ascending powers of $x$.

8 Write down the series for $\ln(1+x)$ in ascending powers of $x$ and state the conditions for convergence.
If $a$ and $b$ are small compared with $x$, show that

$$\ln(x+a) - \ln x = \frac{a}{b}\left(1 + \frac{b-a}{2x}\right)\{\ln(x+b) - \ln x\}.$$

9 Find the value of $k$ for which the expansion of $(1+kx)\left(1+\dfrac{x}{6}\right)^{-1}\ln(1+x)$ contains no term in $x^2$.

10 Evaluate

(a) $\underset{x \to 0}{Lim}\left\{\dfrac{\sinh x - \tanh x}{x^3}\right\}$

(b) $\underset{x \to 1}{Lim}\left\{\dfrac{\ln x}{x^2 - 1}\right\}$

(c) $\underset{x \to 0}{Lim}\left\{\dfrac{x + \sin x}{x^2 + x}\right\}$

11 If $u_r$ and $u_{r-1}$ indicate the $r$th term and the $(r-1)$th term respectively of the expansion of $(1+x)^n$, determine an expression, in its simplest form, for the ratio $\dfrac{u_r}{u_{r-1}}$. Hence show that in the binomial expansion of $(1 + 0{\cdot}03)^{12}$ the $r$th term is less than one-tenth of the $(r-1)$th term if $r > 4$. Use the expansion to evaluate $(1{\cdot}03)^{12}$ correct to three places of decimals.

12 By the use of Maclaurin's series, show that

$$\sin^{-1} x = x + \frac{x^3}{6} + \frac{3x^5}{40} + \ldots$$

Assuming the series for $e^x$, obtain the expansion of $e^x \sin^{-1} x$, up to and including the term in $x^4$. Hence show that, when $x$ is small, the graph of $y = e^x \sin^{-1} x$ approximates to the parabola $y = x^2 + x$.

▶

**13** By application of Maclaurin's series, determine the first two non-vanishing terms of a series for $\ln\cos x$. Express $(1+\cos\theta)$ in terms of $\cos\frac{\theta}{2}$ and show that, if $\theta$ is small, $\ln(1+\cos\theta)=\ln 2-\frac{\theta^2}{4}-\frac{\theta^4}{96}$ approximately.

**14** If $x$ is small, show that:

(a) $\sqrt{\frac{1+x}{1-x}}\approx 1+x+\frac{x^2}{2}$

(b) $\frac{\sqrt{(1+3x^2)e^x}}{1-x}\approx 1+\frac{3x}{2}+\frac{25x^2}{8}$

**15** Prove that:

(a) $\frac{x}{e^x-1}=1-\frac{x}{2}+\frac{x^2}{12}-\frac{x^4}{720}+\dots$

(b) $\frac{x}{e^x+1}=\frac{x}{2}-\frac{x^2}{4}+\frac{x^4}{48}-\dots$

**16** Find:

(a) $\underset{x\to 0}{Lim}\left\{\frac{\sinh^{-1}x-x}{x^3}\right\}$

(b) $\underset{x\to 0}{Lim}\left\{\frac{e^{\sin x}-1-x}{x^2}\right\}$

**17** Find the first three terms in the expansion of $\frac{\sinh x.\ln(1+x)}{x^2(1+x)^3}$.

**18** The field strength of a magnet $(H)$ at a point on the axis, distance $x$ from its centre, is given by

$$H=\frac{M}{2l}\left\{\frac{1}{(x-l)^2}-\frac{1}{(x+l)^2}\right\}$$

where $2l$ = length of magnet and $M$ = moment. Show that if $l$ is very small compared with $x$, then $H\approx\frac{2M}{x^3}$.

**19** Expand $[\ln(1+x)]^2$ in powers of $x$ up to and including the term in $x^4$. Hence determine whether $\cos 2x-[\ln(1+x)]^2$ has a maximum value, minimum value, or point of inflexion at $x=0$.

**20** If $l$ is the length of a circular arc, $a$ is the length of the chord of the whole arc, and $b$ is the length of the chord of half the arc, show that:

(a) $a=2r\sin\frac{l}{2r}$ and (b) $b=2r\sin\frac{l}{4r}$, where $r$ is the radius of the circle.

By expanding $\sin\frac{l}{2r}$ and $\sin\frac{l}{4r}$ as series, show that $l=\frac{8b-a}{3}$ approximately.

Now visit the companion website for this book at www.palgrave.com/stroud for more questions applying this mathematics to science and engineering.

53

# Curves and curve fitting

## Learning outcomes

When you have completed this Programme you will be able to:

- ☐ Draw sketch graphs of standard curves
- ☐ Determine the equations of asymptotes parallel to the $x$- and $y$-axes
- ☐ Sketch the graphs of curves with asymptotes, stationary points and other features
- ☐ Fit graphs to data using the 'straight-line' forms
- ☐ Fit graphs to data using the method of least squares
- ☐ Understand what is meant by the correlation of two variables
- ☐ Calculate the Pearson product-moment coefficient
- ☐ Calculate Spearman's rank correlation coefficient

# Introduction

1

The purpose of this Programme is eventually to devise a reliable method for establishing the relationship between two variables, corresponding values of which have been obtained as a result of tests or experimentation. These results in practice are highly likely to include some errors, however small, owing to the imperfect materials used, the limitations of the measuring devices and the shortcomings of the operator conducting the test and recording the results.

There are methods by which we can minimize any further errors or guesswork in processing the results and, indeed, eradicate some of the errors already inherent in the recorded results, but before we consider this important section of the work, some revision of the shape of standard curves and the systematic sketching of curves from their equations would be an advantage.

# Standard curves

2

## Straight line

The equation is a first-degree relationship and can always be expressed in the form $y = mx + c$, where

$m$ denotes the gradient, i.e. $\dfrac{dy}{dx}$

$c$ denotes the intercept on the $y$-axis.

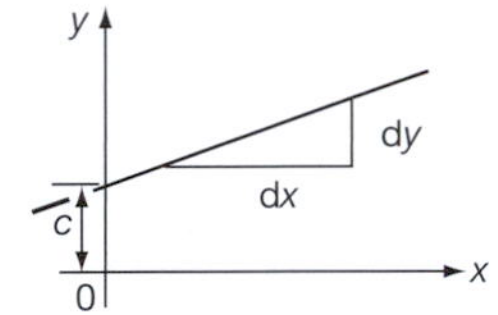


Any first-degree equation gives a straight-line graph.

To find where the line crosses the $x$-axis, put $y = 0$.

To find where the line crosses the $y$-axis, put $x = 0$.

Therefore, the line $2y + 3x = 6$ crosses the axes at ...........

3

(2, 0) and (0, 3)

Because when $x = 0,\quad 2y = 6\quad \therefore y = 3$

and when $y = 0,\quad 3x = 6\quad \therefore x = 2$

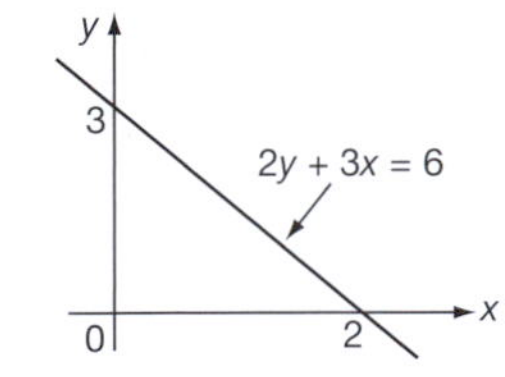


We can establish the equation of a given straight line by substituting in $y = mx + c$ the $x$- and $y$-coordinates of any three points on the line. Of course, two points are sufficient to determine values of $m$ and $c$, but the third point is taken as a check.

So, if (1, 6), (3, 2) and (5, −2) lie on a straight line, its equation is ...........

## 4

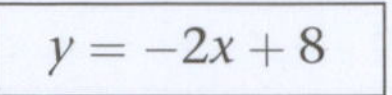


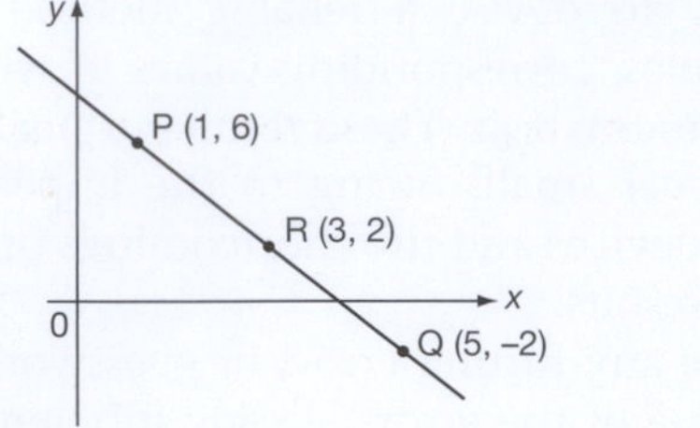


$$(1, 6) \qquad 6 = 1m + c$$
$$(5, -2) \qquad -2 = 5m + c$$
$$\therefore\ -8 = 4m \quad \therefore\ m = -2$$
$$y = -2x + c$$

When $x = 1$, $y = 6$ $\therefore\ 6 = -2 + c$ $\therefore\ c = 8$ $\therefore\ y = -2x + 8$

Check: When $x = 3$, $y = -6 + 8 = 2$ which agrees with the third point.

### Second-degree curves

The basic second-degree curve is $y = x^2$, a parabola symmetrical about the $y$-axis and existing only for $y \geq 0$.

$y = ax^2$ gives a thinner parabola if $a > 1$ and a flatter parabola if $a < 1$.

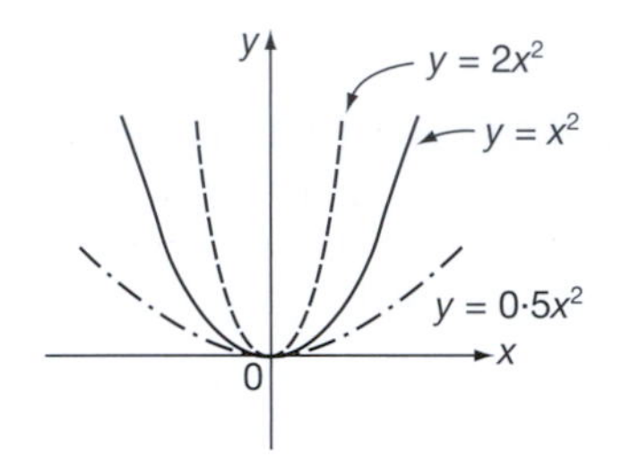


The general second-degree curve is $y = ax^2 + bx + c$, where the three coefficients, $a$, $b$ and $c$, determine the position of the vertex and the 'width' of the parabola.

*Change of vertex:* If the parabola $y = x^2$ is moved parallel to itself to a vertex position at (2, 3), its equation relative to the new axes is $Y = X^2$.

Considering a sample point P, we see that:

$Y = y - 3$ and $X = x - 2$

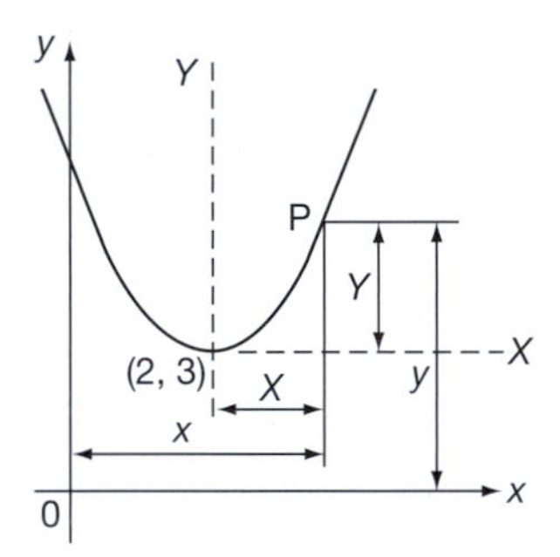


So, in terms of the original variables, $x$ and $y$, the equation of the new parabola is ............

## 5

$$\boxed{y = x^2 - 4x + 7}$$

Because $Y = X^2$ becomes $y - 3 = (x - 2)^2$ i.e. $y - 3 = x^2 - 4x + 4$ which simplifies to $y = x^2 - 4x + 7$.

*Note*: If the coefficient of $x^2$ is negative, the parabola is inverted e.g. $y = -2x^2 + 6x + 5$.

The vertex is at (1·5, 9·5).

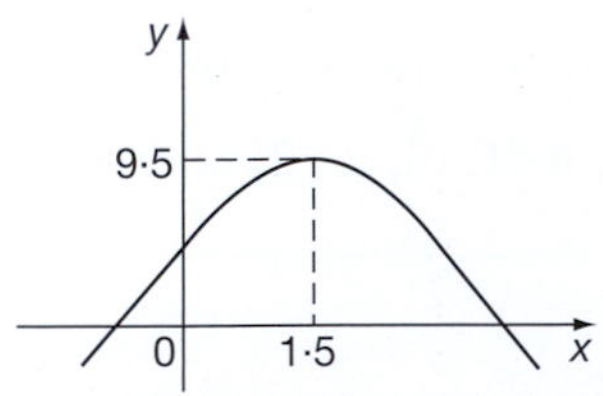


The parabola cuts the $y$-axis at $y =$ ........... and the $x$-axis at $x =$ ........... and $x =$ ...........

6

$y = 5$; $x = -0{\cdot}68$ and $x = 3{\cdot}68$

## Third-degree curves

The basic third-degree curve is $y = x^3$ which passes through the origin. For $x$ positive, $y$ is positive and for $x$ negative, $y$ is negative.

Writing $y = -x^3$ turns the curve upside down:

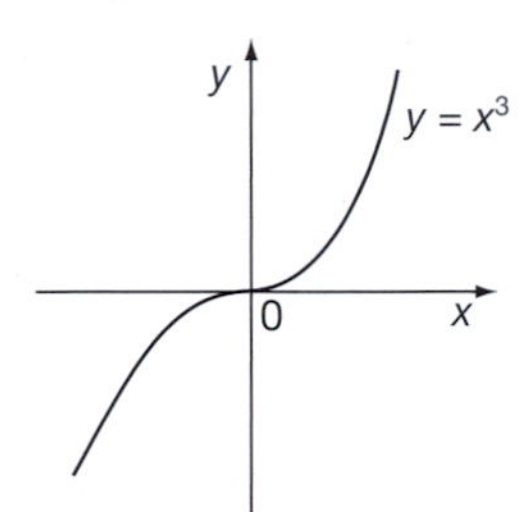


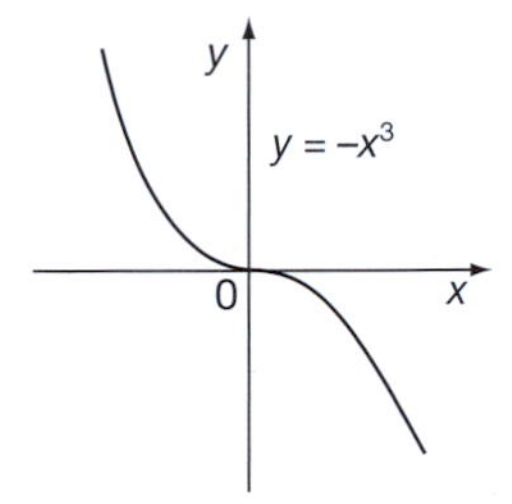


In general, a third-degree curve has a more accentuated double bend and cuts the $x$-axis in three points which may have (a) three real and different values, (b) two values the same and one different, or (c) one real value and two complex values.

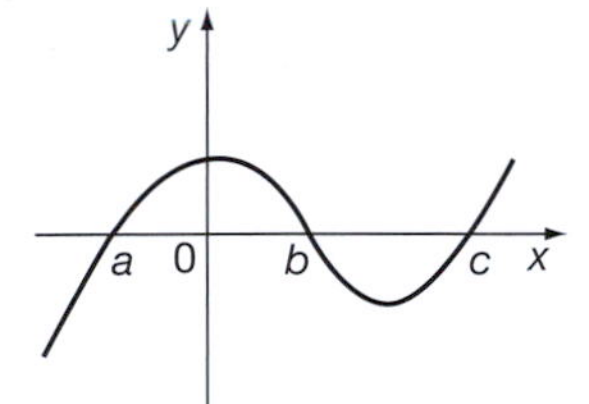


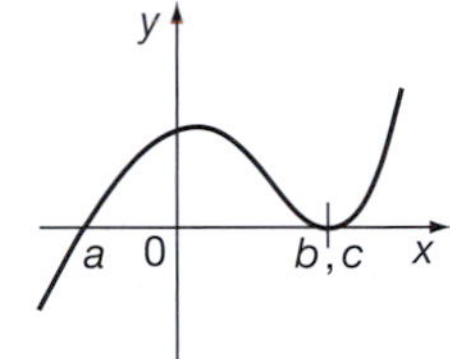


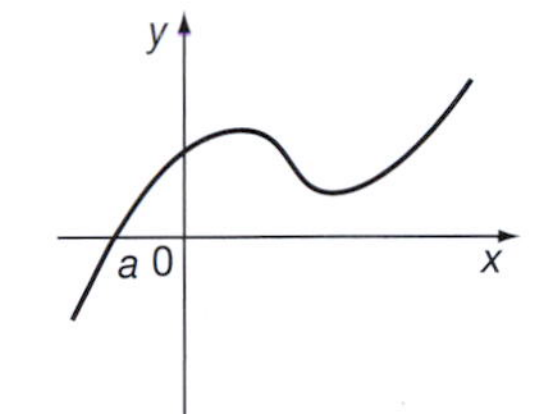


*Now let us collect our ideas so far by working through a short exercise.*
*Move on to the next frame*

7

As an exercise, sketch the graphs of the following, indicating relevant information. Do not plot the graphs in detail.

1 $y = 2x - 5$
2 $y = \dfrac{x}{3} + 7$
3 $y = -2x + 4$
4 $2y + 5x - 6 = 0$
5 $y = x^2 + 4$
6 $y = (x - 3)^2$
7 $y = (x + 2)^2 - 4$
8 $y = x - x^2$
9 $y = x^3 - 4$
10 $y = 2 - (x + 3)^3$

*When you have completed the whole set, check your results*
*with those in the next frame*

8

Here are the results:

1

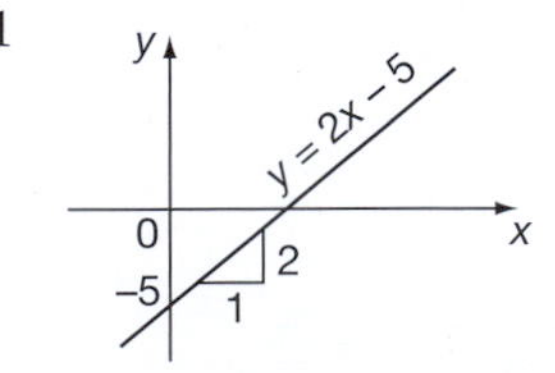


2

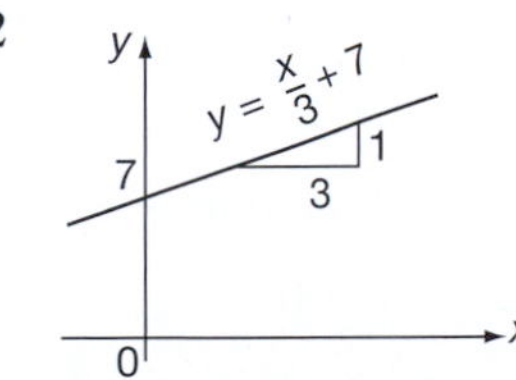

3

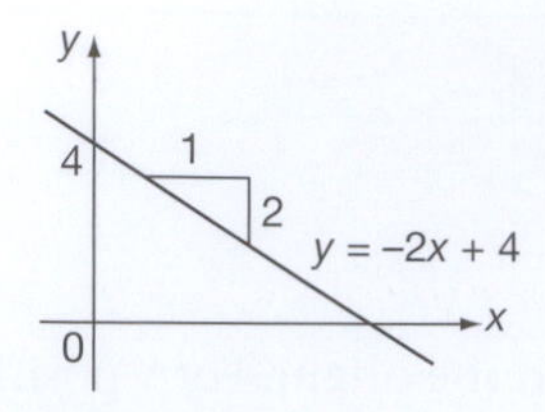


4

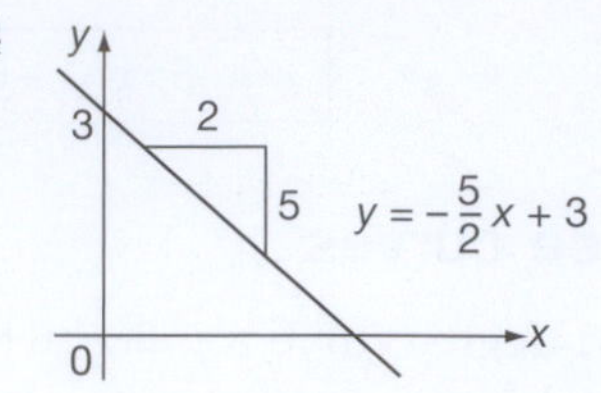


5

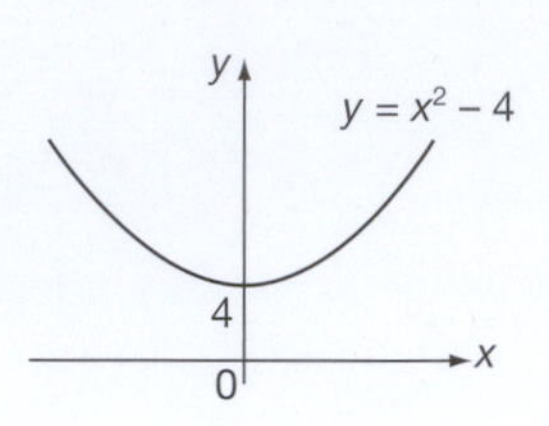


6

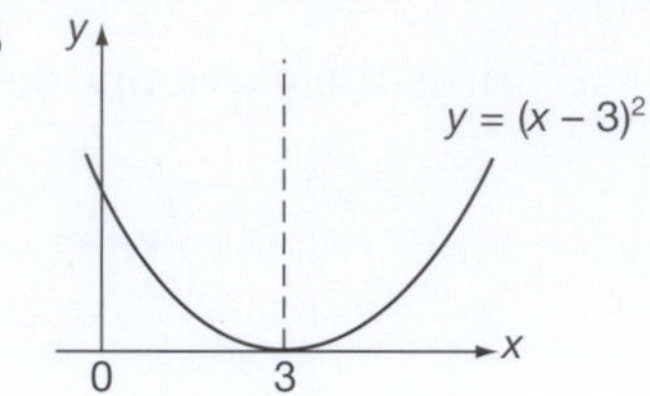


7

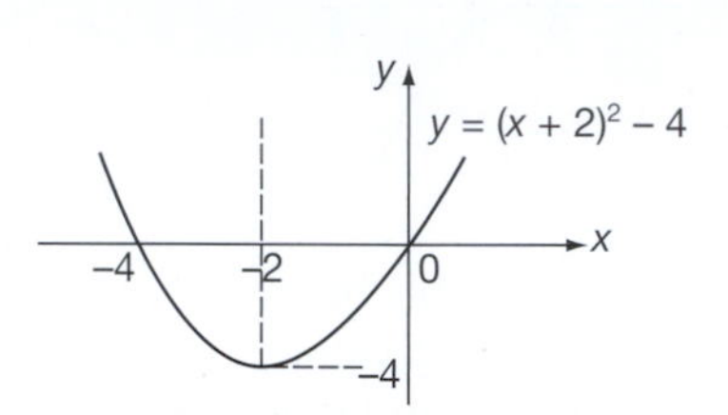


8

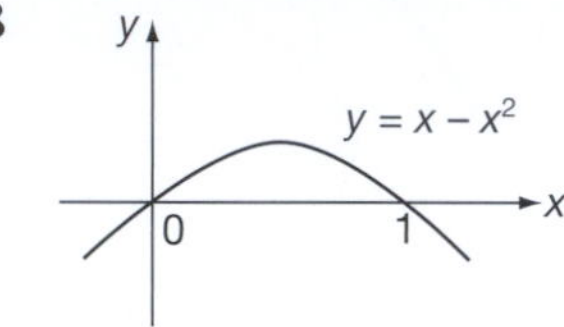


9

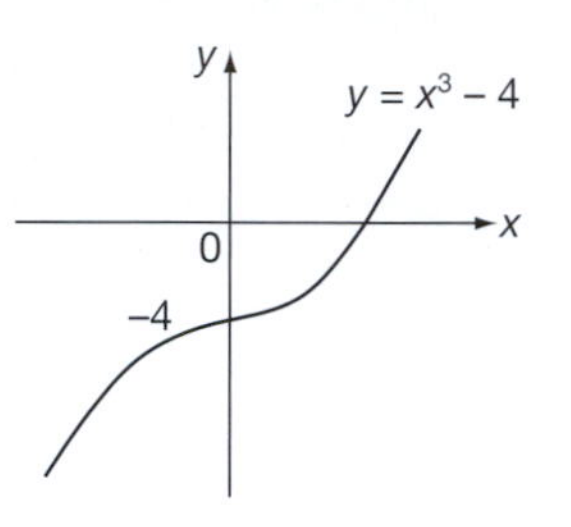


10

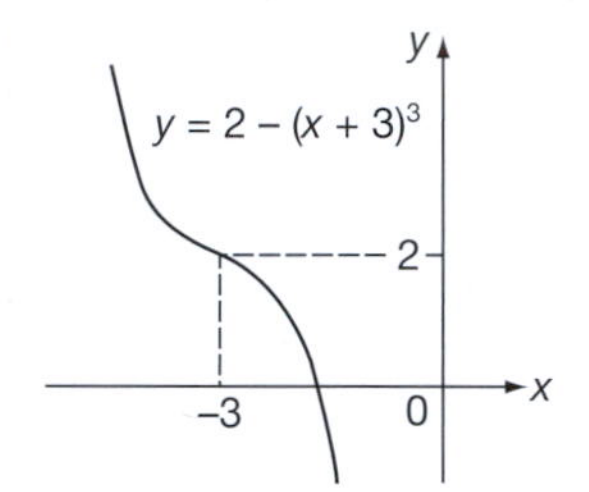


*Now we will revise a further set of curves. Next frame*

**9**

## Circle

The simplest case of the circle is with centre at the origin and radius $r$:

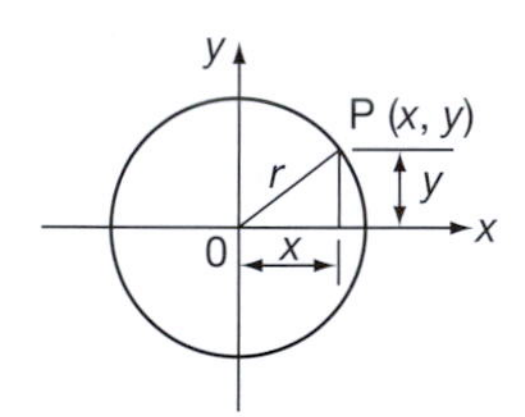


The equation is then $x^2 + y^2 = r^2$

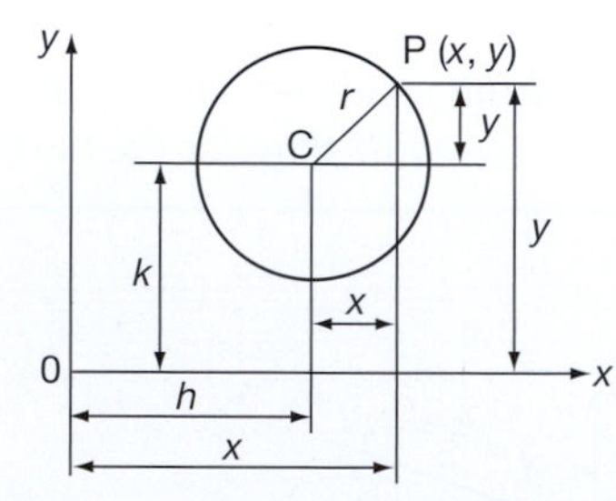


Moving the centre to a new point $(h, k)$ gives $X^2 + Y^2 = r^2$

where $Y = y - k$

and $X = x - h$

$\therefore\ (x - h)^2 + (y - k)^2 = r^2$

The general equation of a circle is

$$x^2 + y^2 + 2gx + 2fy + c = 0$$

where the centre is the point $(-g, -f)$ and radius $= \sqrt{g^2 + f^2 - c}$.

▶

Note that, for a second-degree equation to represent a circle:

(a) the coefficients of $x^2$ and $y^2$ are identical

(b) there is no product term in $xy$.

So the equation $x^2 + y^2 + 2x - 6y - 15 = 0$ represents a circle with centre ........... and radius ...........

**10**

centre (−1, 3); radius 5

Because

$$\left.\begin{array}{ll} 2g = 2 & \therefore g = 1 \\ 2f = -6 & \therefore f = -3 \end{array}\right\} \quad \therefore \text{ centre } (-g, -f) = (-1, 3)$$

also $c = -15 \quad \therefore$ radius $= \sqrt{g^2 + f^2 - c} = \sqrt{1 + 9 + 15} = \sqrt{25} = 5$.

## Ellipse

The equation of an ellipse is $\dfrac{x^2}{a^2} + \dfrac{y^2}{b^2} = 1$

where $a$ = semi-major axis $\quad \therefore y = 0,\ x = \pm a$

and $b$ = semi-minor axis $\quad \therefore x = 0,\ y = \pm b$

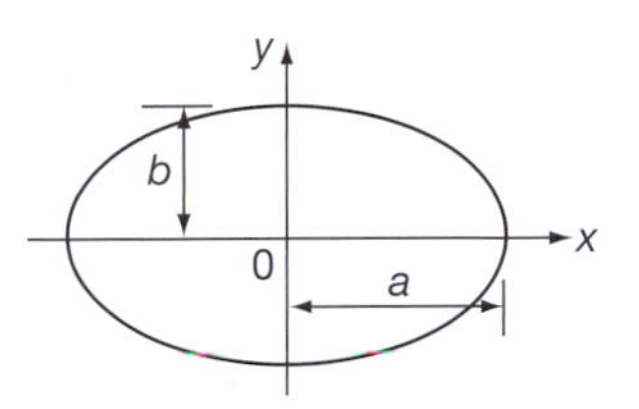


Of course, when $a^2$ and $b^2$ are equal (say $r^2$) we obtain ...........

**11**

the equation of a circle, i.e. $x^2 + y^2 = r^2$

## Hyperbola

The equation of a hyperbola is

$$\frac{x^2}{a^2} - \frac{y^2}{b^2} = 1$$

When $y = 0,\ x = \pm a$.

When $x = 0,\ y^2 = -b^2 \quad \therefore$ the curve does not cross the $y$-axis.

Note that the opposite arms of the hyperbola gradually approach two straight lines (asymptotes).

*Rectangular hyperbola*

If the asymptotes are at right-angles to each other, the curve is then a *rectangular hyperbola*. A more usual form of the rectangular hyperbola is to rotate the figure through 45 and to make the asymptotes the axes of $x$ and $y$. The equation of the curve relative to the new axes then becomes

$$xy = c \qquad \text{i.e. } y = \frac{c}{x}$$

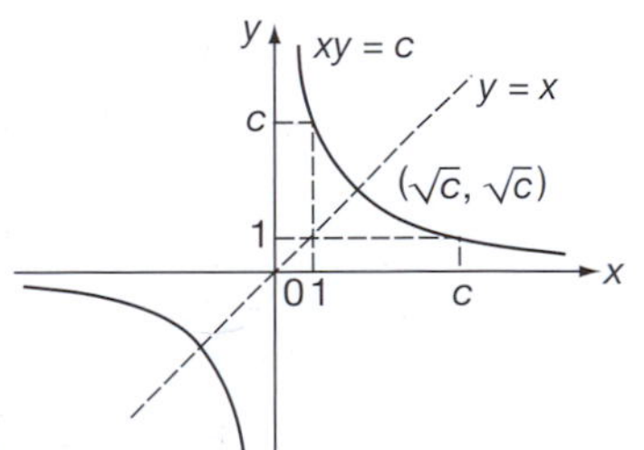


Three points are easily located.

If $xy = c$:

(a) when $x = 1,\ y = c$

(b) when $y = 1,\ x = c$

(c) the line $y = x$ cuts $xy = c$ at the point $(\pm\sqrt{c},\ \pm\sqrt{c})$.

These three points are a great help in sketching a rectangular hyperbola. Rectangular hyperbolas frequently occur in practical considerations.

*Now for another short exercise, so move on to the next frame*

**12**

As an exercise, sketch the graphs of the following, showing relevant facts:

| | |
|---|---|
| **1** $x^2 + y^2 = 12{\cdot}25$ | **2** $x^2 + 4y^2 = 100$ |
| **3** $x^2 + y^2 - 4x + 6y - 3 = 0$ | **4** $2x^2 - 3x + 4y + 2y^2 = 0$ |
| **5** $\dfrac{x^2}{36} - \dfrac{y^2}{49} = 1$ | **6** $xy = 5$ |

*When you have sketched all six, compare your results with those in the next frame*

**13**

The sketch graphs are as follows:

**1**

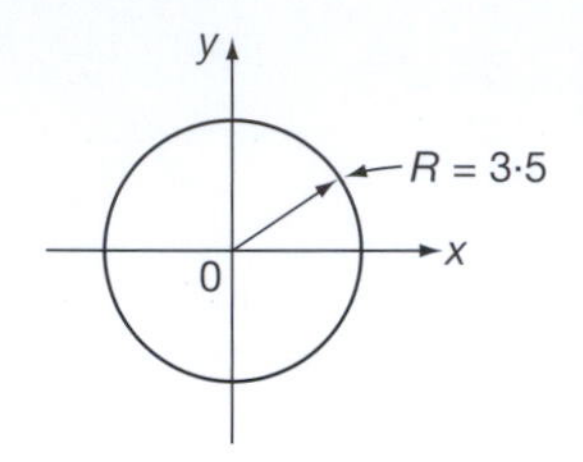


**2**

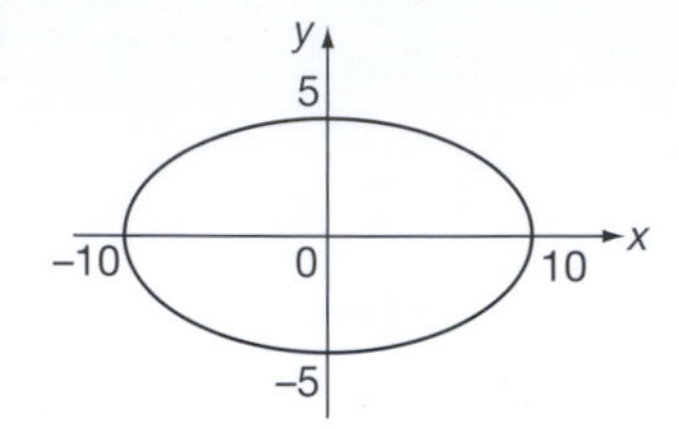


**3**

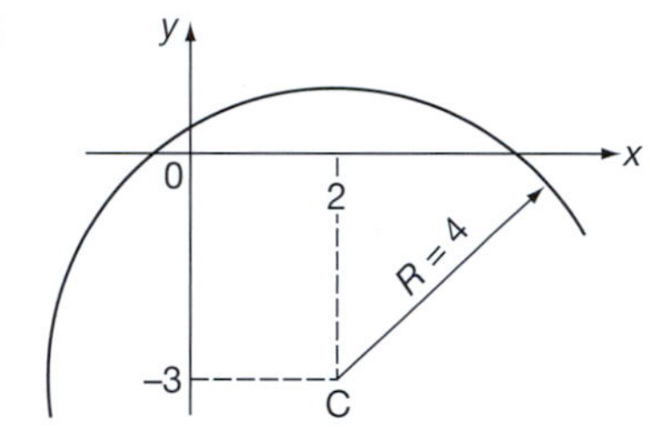


**4**

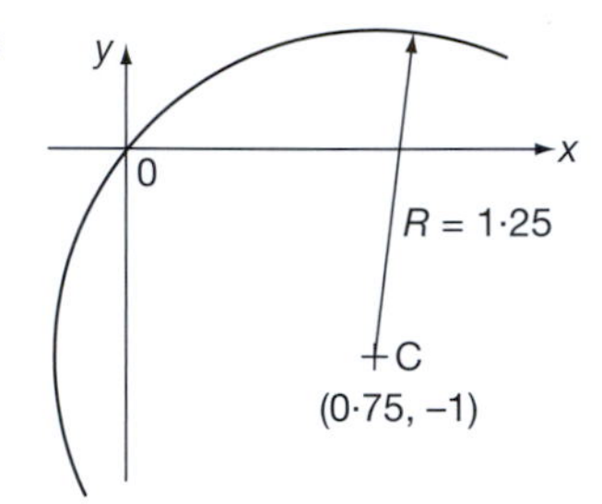


**5**

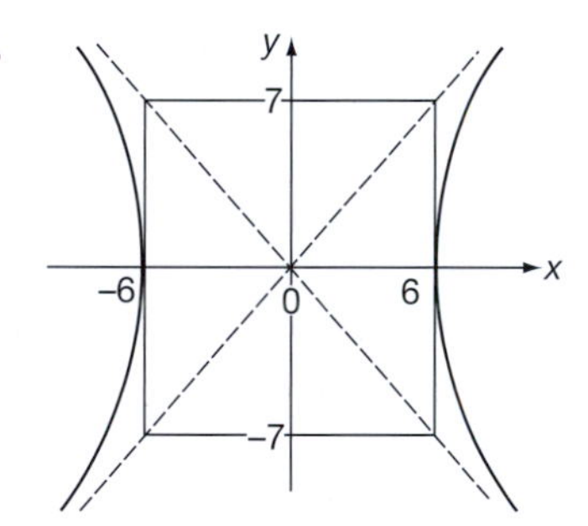


**6**

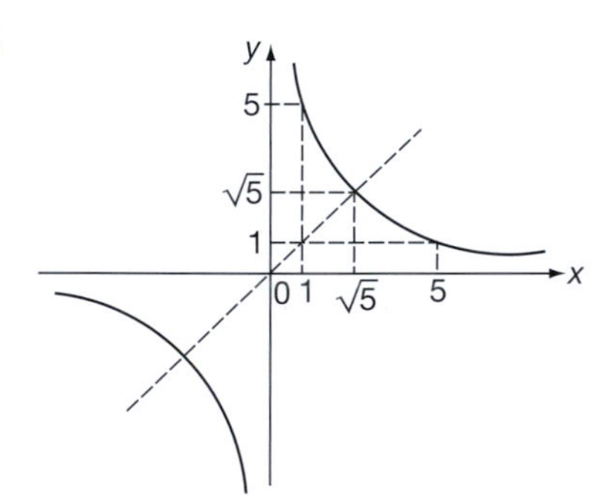


*On to Frame 14 for a third set of curves frequently occurring*

14

## Logarithmic curves

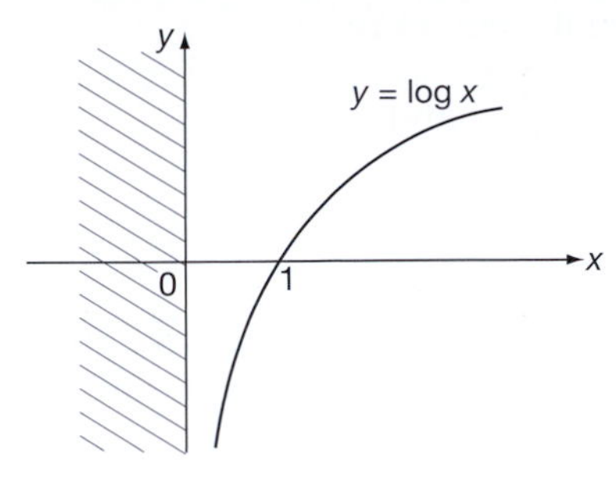


If $y = \log x$, then when $x = 1$, $y = \log 1 = 0$ i.e. the curve crosses the $x$-axis at $x = 1$.

Also, $\log x$ does not exist for $x < 0$.
$y = \log x$ flattens out as $x \to \infty$, but continues to increase at an ever-decreasing rate.

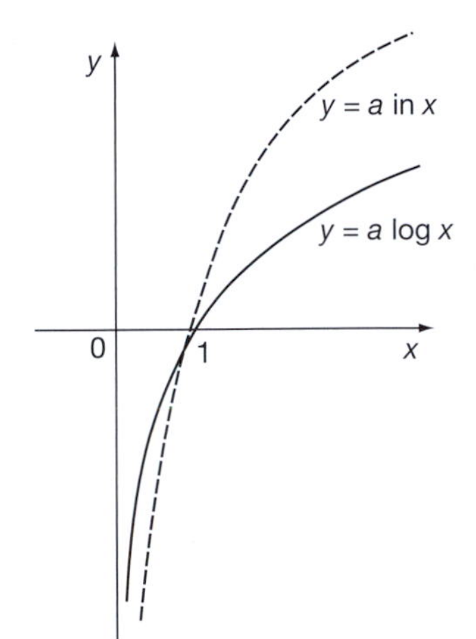


The graph of $y = \ln x$ also has the same shape and crosses the $x$-axis at $x = 1$, but the function values are different.

The graphs of $y = a \log x$ and $y = a \ln x$ are similar, but with all ordinates multiplied by the constant factor $a$.

*Continued in the next frame*

15

## Exponential curves

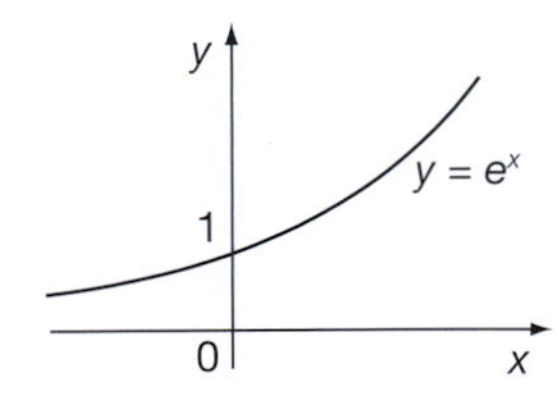


$y = e^x$ crosses the $y$-axis at $x = 0$

i.e. $y = e^0 = 1$

As $x \to \infty$, $y \to \infty$

as $x \to -\infty$, $y \to 0$

Sometimes known as the *growth curve*.

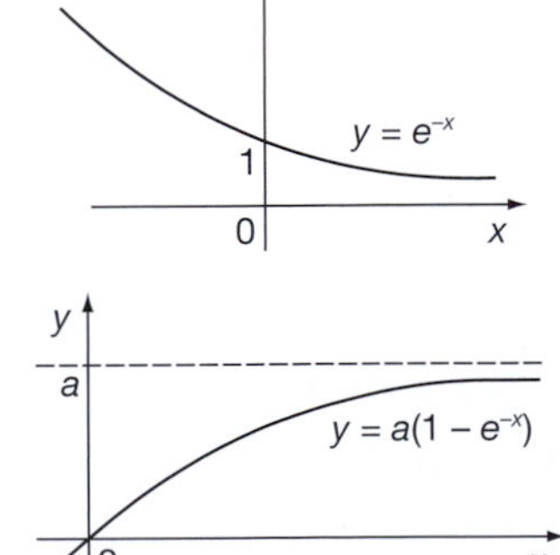


$y = e^{-x}$ also crosses the $y$-axis at $y = 1$.

As $x \to \infty$, $y \to 0$

as $x \to -\infty$, $y \to \infty$

Sometimes known as the *decay curve*.

In electrical work, we also frequently have curves of the form $y = a(1 - e^{-x})$. This is an inverted exponential curve, passing through the origin and tending to $y = a$ as asymptote as $x \to \infty$ (since $e^{-x} \to 0$ as $x \to \infty$).

▶

## Hyperbolic curves

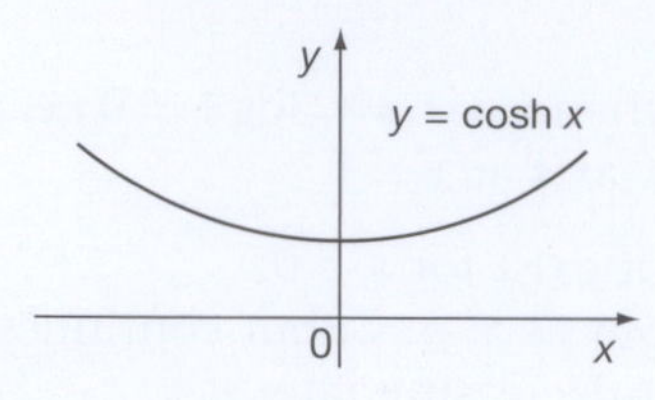


Combination of the curves for $y = e^x$ and $y = e^{-x}$ gives the hyperbolic curves of

$$y = \cosh x = \frac{e^x + e^{-x}}{2} \quad \text{and}$$

$$y = \sinh x = \frac{e^x - e^{-x}}{2}$$

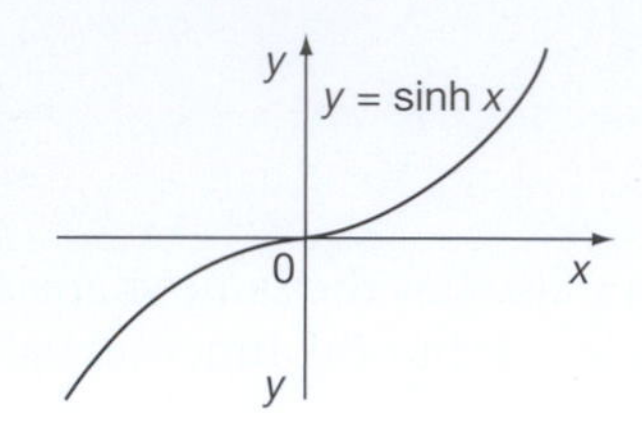


We have already dealt with these functions in detail in Programme 3, so refer back if you need to revise them further.

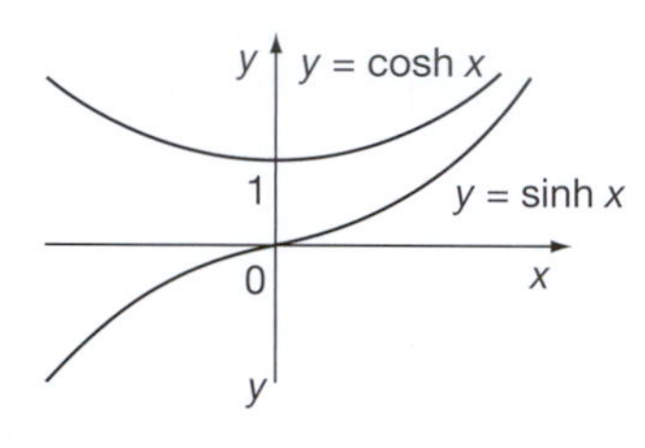


If we draw the two graphs on the same axes, we see that $y = \sinh x$ is always outside $y = \cosh x$, i.e. for any particular value of $x$, $\cosh x > \sinh x$.

*We have one more type of curve to list, so move on to the next frame*

## 16 Trigonometrical curves

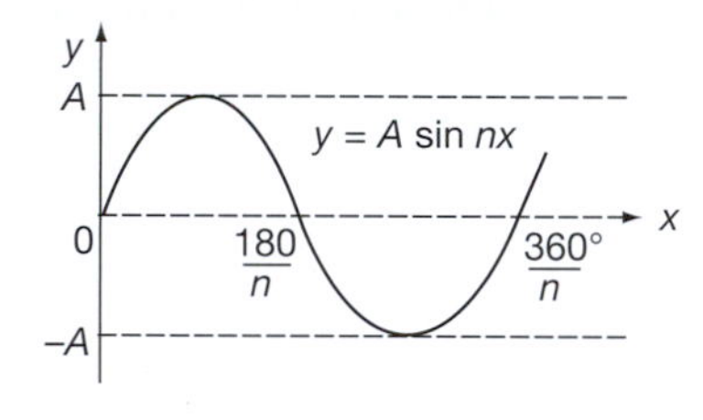


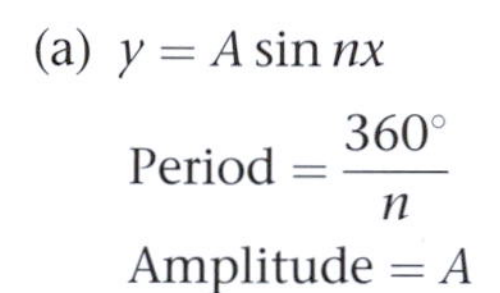

The most commonly occurring in practice is the sine curve.

(a) $y = A \sin nx$

$$\text{Period} = \frac{360^\circ}{n}$$

$$\text{Amplitude} = A$$

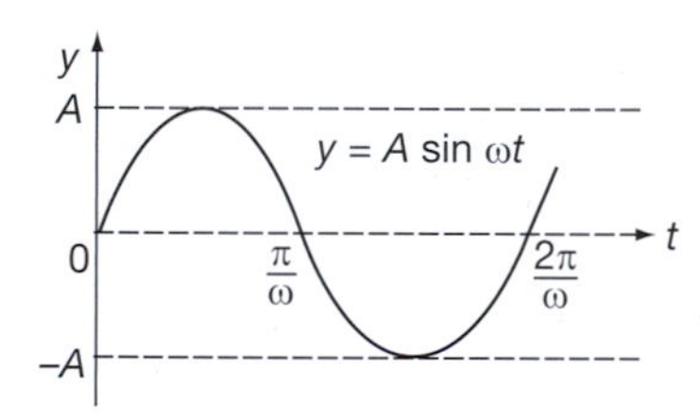


(b) $y = A \sin \omega t$

$$\text{Period} = \frac{2\pi}{\omega}$$

$$\text{Amplitude} = A$$

Now, as an exercise, we can sketch a further selection of curves.

Sketch the following pairs of curves on the same axes. Label each graph clearly and show relevant information.

1 $y = \cosh x$ and $y = 2 \cosh x$

2 $y = \sinh x$ and $y = \sinh 2x$

3 $y = e^x$ and $y = e^{3x}$

4 $y = e^{-x}$ and $y = 2e^{-x}$

5 $y = 5 \sin x$ and $y = 3 \sin 2x$

6 $y = 4 \sin \omega t$ and $y = 2 \sin 3\omega t$

17

Here they are:

1

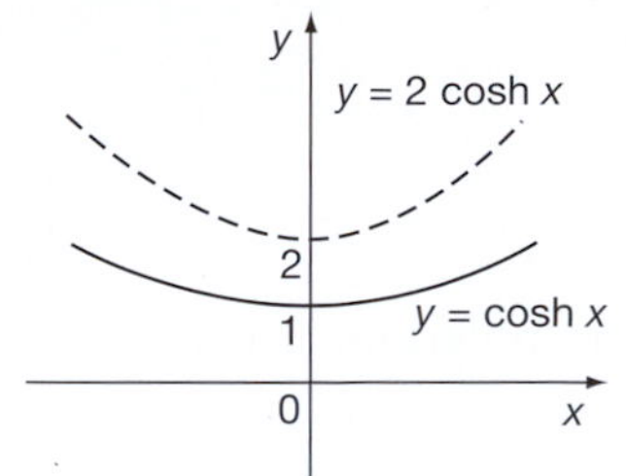


2

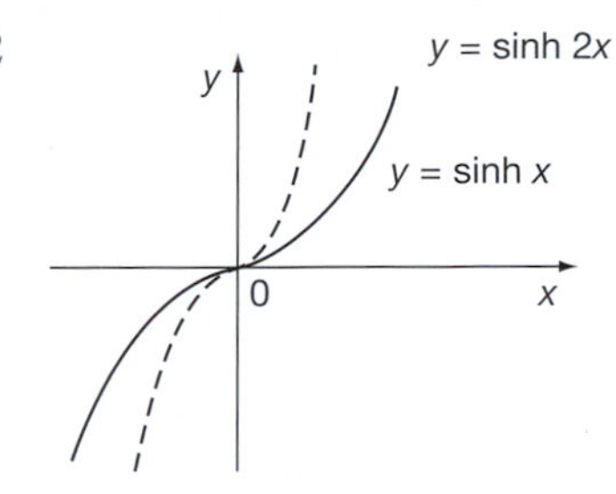


3

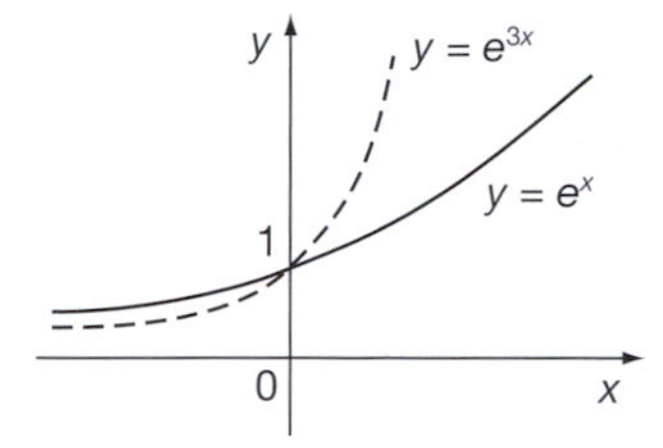


4

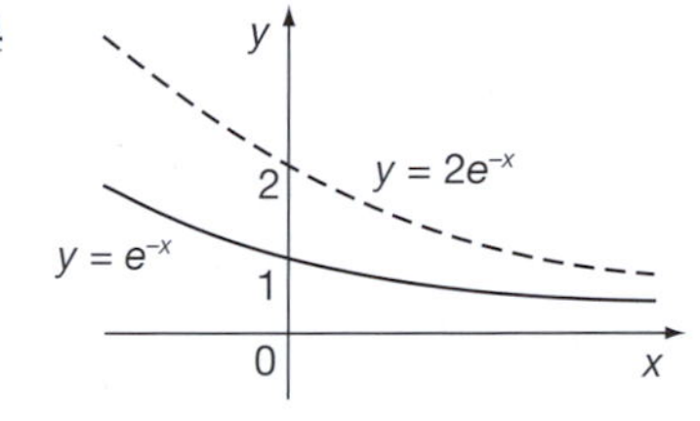


5

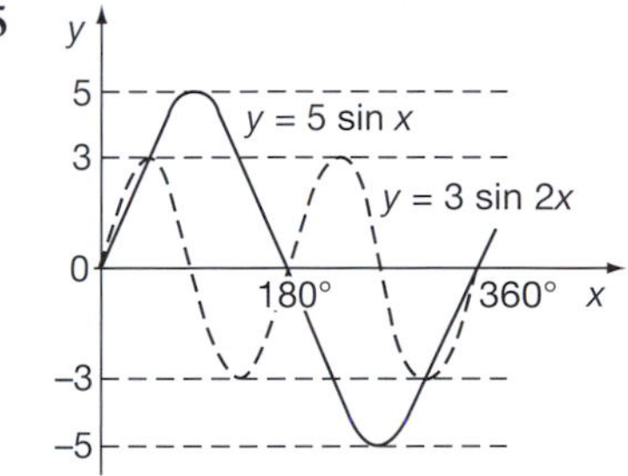


6

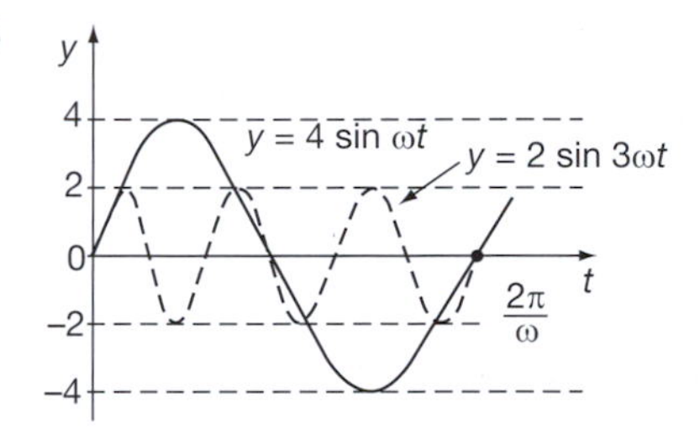


## Asymptotes

18

We have already made references to asymptotes in the previous work and it is always helpful to know where asymptotes occur when sketching curves of functions.

*Definition*: An *asymptote* to a curve is a straight line to which the curve approaches as the distance from the origin increases. It can also be thought of as a tangent to the curve at infinity – that is, the curve touches the asymptote at two coincident points at infinity.

*Condition for roots at infinity*: Consider the polynomial equation

$$a_0x^n + a_1x^{n-1} + \ldots + a_{n-2}x^2 + a_{n-1}x + a_n = 0$$

Because we cannot handle the behaviour of this equation when $x$ approaches infinity we change the variable to

$p = \dfrac{1}{x}$ and consider the behaviour of the resulting equation as $p$ approaches zero

▶

Substituting $p = \frac{1}{x}$ we find that:

$$a_0 \frac{1}{p^n} + a_1 \frac{1}{p^{n-1}} + \ldots + a_{n-2} \frac{1}{p^2} + a_{n-1} \frac{1}{p} + a_n = 0$$

Multiplying through by $p^n$ gives:

$$a_0 + a_1 p + a_2 p^2 \ldots + a_{n-1} p^{n-1} + a_n p^n = 0$$

If we now impose the conditions $a_0 = a_1 = 0$ then the polynomial equation in $p$ becomes:

$$p^2 (a_2 + a_3 p + \ldots + a_{n-1} p^{n-3} + a_n p^{n-2}) = 0$$

That is:

$$p^2 = 0 \text{ or } a_2 + a_3 p + \ldots + a_{n-1} p^{n-3} + a_n p^{n-2} = 0$$

Taking $p^2 = 0$ yields two coincident roots at the point $p = 0$. Put another way, the conditions for two coincident roots at $p = 0$ are

$$a_0 = 0 \text{ and } a_1 = 0$$

Because $p = \frac{1}{x}$ we can say that the conditions for two coincident roots at infinity of the polynomial equation in $x$ are that the coefficients of the two highest powers of $x$ are zero, that is:

$$a_0 = 0 \text{ and } a_1 = 0$$

*Therefore, the original equation*

$$a_0 x^n + a_1 x^{n-1} + \ldots + a_{n-1} x + a_n = 0$$

*will have two infinite roots if* $a_0 = 0$ and $a_1 = 0$.

## Determination of an asymptote

From the result we have just established, to find an asymptote to $y = f(x)$:

(a) substitute $y = mx + c$ in the given equation and simplify

(b) equate to zero ...........

**19**

the coefficients of the two highest powers of $x$ and so determine the values of $m$ and $c$

Let us work through an example to see how it develops.

**20**

To find the asymptote to the curve $x^2 y - 5y - x^3 = 0$.

Substitute $y = mx + c$ in the equation:

$$x^2(mx + c) - 5(mx + c) - x^3 = 0$$

$$mx^3 + cx^2 - 5mx - 5c - x^3 = 0$$

$$(m - 1)x^3 + cx^2 - 5mx - 5c = 0 \quad \text{this is our polynomial equation in } x$$

Equating to zero the coefficients of the two highest powers of $x$:

$$\left.\begin{array}{ll} m - 1 = 0 & \therefore m = 1 \\ c = 0 & c = 0 \end{array}\right\} \quad \therefore \text{ Asymptote is } y = x$$

▶

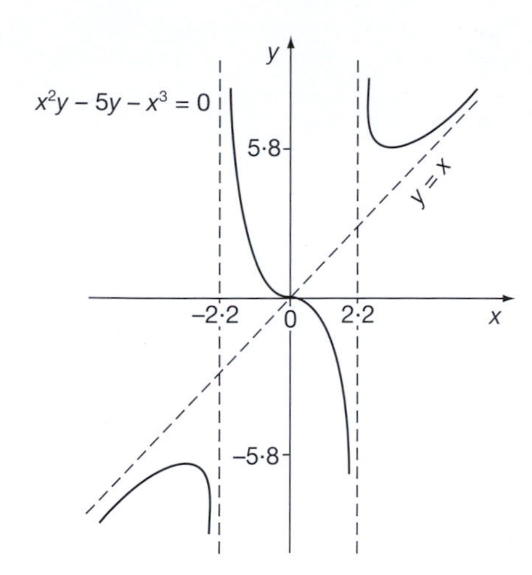


In fact, the graph of $x^2y - 5y - x^3 = 0$ is as shown on the left and we can see that the curve approaches $y = x$ as $x \to \infty$ and as $x \to -\infty$.

From the graph, however, it appears that there are two further asymptotes which are the lines ........... and ...........

**21**

$x = -2{\cdot}2$ and $x = 2{\cdot}2$

These are two lines parallel to the $y$-axis.

## Asymptotes parallel to the $x$- and $y$-axes

These can be found by a simple rule.

*For the curve $y = f(x)$, the asymptotes parallel to the x-axis can be found by equating the coefficient of the highest power of x to zero. Similarly, the asymptotes parallel to the y-axis can be found by equating the coefficient of the highest power of y to zero.*

**Example 1**

Find the asymptotes, if any, of $y = \dfrac{x-2}{2x+3}$.

First get everything on one line by multiplying by the denominator, $(2x + 3)$:

$y(2x + 3) = x - 2 \quad \therefore\ 2xy + 3y - x + 2 = 0$

(a) Asymptote parallel to $x$-axis: Equate the coefficient of the highest power of $x$ to zero.

$(2y - 1)x + 3y + 2 = 0$
$\therefore\ 2y - 1 = 0 \quad \therefore\ 2y = 1 \quad \therefore\ y = 0{\cdot}5$
$\therefore\ y = 0{\cdot}5$ is an asymptote.

(b) Asymptote parallel to $y$-axis: Equate the coefficient of the highest power of $y$ to zero.

$\therefore$ ........... is also an asymptote.

**22**

$x = -1{\cdot}5$

Because rearranging the equation to obtain the highest power of $y$:

$(2x + 3)y - x + 2 = 0$

gives

$2x + 3 = 0 \quad \therefore\ x = -1{\cdot}5$
$\therefore\ x = -1{\cdot}5$ is an asymptote.

▶

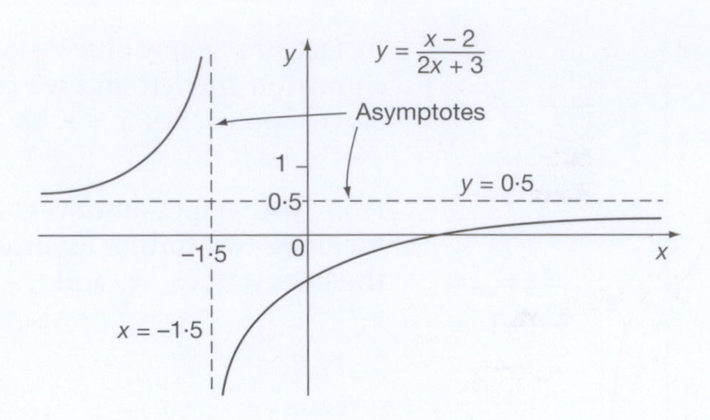


In fact, the graph is positioned as shown above. The only asymptotes are $y = 0{\cdot}5$ (parallel to the $x$-axis) and $x = -1{\cdot}5$ (parallel to the $y$-axis).

Let us do another.

**Example 2**

Find the asymptotes of the curve $x^2(x^2 + 2) = y^3(x + 5)$.

(a) Parallel to the $x$-axis: $x^4 + 2x^2 = (x + 5)y^3$.
Equate the coefficient of the highest power of $x$ to zero.
Highest power of $x$ is $x^4$. Its coefficient is 1, which gives $1 = 0$. This not the equation of a line. Therefore, there is no asymptote parallel to the $x$-axis.

(b) Parallel to the $y$-axis: This gives ...........

## 23

$x = -5$ is an asymptote

Because for an asymptote parallel to the $y$-axis, we equate the highest power of $y$ to zero. $\therefore\ x + 5 = 0 \quad \therefore\ x = -5$ Therefore, $x = -5$ is an asymptote parallel to the $y$-axis.

Now, to find a general asymptote, i.e. not necessarily parallel to either axis, we carry out the method described earlier, which was to ...........

## 24

substitute $y = mx + c$ in the equation and equate the coefficients of the two highest powers of $x$ to zero

If we do that with the equation of this example, we get ...........

*Work it right through to the end*

## 25

$y = x - \dfrac{5}{3}$ is also an asymptote

Because substituting $y = mx + c$ in $x^2(x^2 + 2) = y^3(x + 5)$ we have

$$x^4 + 2x^2 = (m^3x^3 + 3m^2x^2c + 3mxc^2 + c^3)(x + 5)$$

$$= m^3x^4 + 3m^2x^3c + 3mx^2c^2 + c^3x$$

$$+ 5m^3x^3 + 15m^2x^2c + 15mxc^2 + 5c^3$$

$$\therefore\ (m^3 - 1)x^4 + (5m^3 + 3m^2c)x^3 + (15m^2c + 3mc^2 - 2)x^2$$

$$+ (15mc^2 + c^3)x + 5c^3 = 0$$

Equating to zero the coefficients of the two highest powers of $x$:

$m^3 - 1 = 0 \qquad \therefore m^3 = 1 \qquad \therefore m = 1$

$5m^3 + 3m^2c = 0 \qquad \therefore 5 + 3c = 0 \qquad \therefore c = \dfrac{-5}{3}$

$$\therefore y = x - \frac{5}{3} \text{ is an asymptote}$$

There are, then, two asymptotes:

$x = -5$

and $y = x - \dfrac{5}{3}$

In fact, the graph is shown on the right.

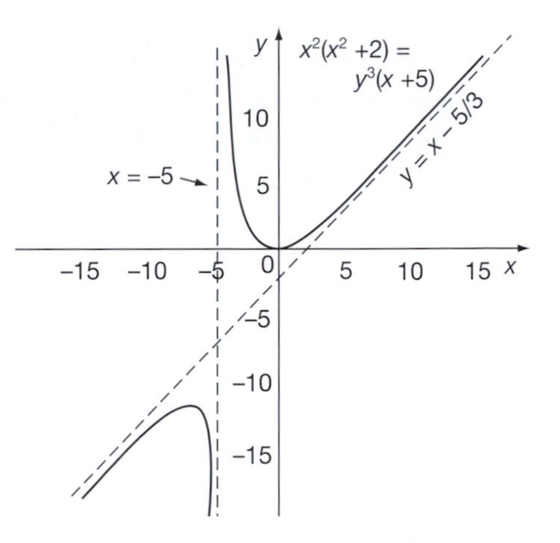


Here are two similar problems for you to do as an exercise on your own. Just apply the rules. There are no tricks.

Determine the asymptotes, if any, for the following curves:

**1** $x^4 - 2x^3y + 10x^2 - 7y^2 = 0$

**2** $x^3 - xy^2 + 4y^2 - 5 = 0$

---

**26**

| |
|---|
| **1** $y = \dfrac{x}{2}$ is the only asymptote |
| **2** $y = x + 2;\quad y = -x - 2;\quad x = 4$ |

**1** There are no asymptotes parallel to the axes.

Substituting $y = mx + c$ and collecting up terms gives:

$$(1 - 2m)x^4 - 2cx^3 + (10 - 7m^2)x^2 - 14mcx - 7c^2 = 0$$

from which $\left.\begin{array}{l} 1 - 2m = 0 \quad \therefore m = \dfrac{1}{2} \\ c = 0 \end{array}\right\} \quad \therefore y = \dfrac{x}{2}$ is the only asymptote.

**2** $x = 4$ is an asymptote parallel to the $y$-axis.

There is no asymptote parallel to the $x$-axis.

Putting $y = mx + c$ and simplifying produces:

$$(1 - m^2)x^3 + (4m^2 - 2mc)x^2 + (8mc - c^2)x + 4c^2 - 5 = 0$$

so that $\quad 1 - m^2 = 0 \quad \therefore m^2 = 1 \quad \therefore m = \pm 1$

and $\quad 4m^2 - 2mc = 0 \quad \therefore 2m = c.\;$ When $m = 1$, $c = 2$ and

when $m = -1$, $c = -2 \quad \therefore y = x + 2$ and $y = -x - 2$ are asymptotes.

*Now let us apply these methods in a wider context starting in the next frame*

---

# Systematic curve sketching, given the equation of the curve

**27**

If, for $y = f(x)$, the function $f(x)$ is known, the graph of the function can be plotted by calculating $x$- and $y$-coordinates of a number of selected points. This can be a tedious occupation and result in some significant features being overlooked – a problem that can occur even when an electronic spreadsheet is used. However, considerable information about the shape and positioning of the curve can be obtained by a systematic analysis of the given equation. There is a list of steps we can take.

## Symmetry

Inspect the equation for symmetry:

(a) If only even powers of $y$ occur, the curve is symmetrical about the $x$-axis.

(b) If only even powers of $x$ occur, the curve is symmetrical about the $y$-axis.

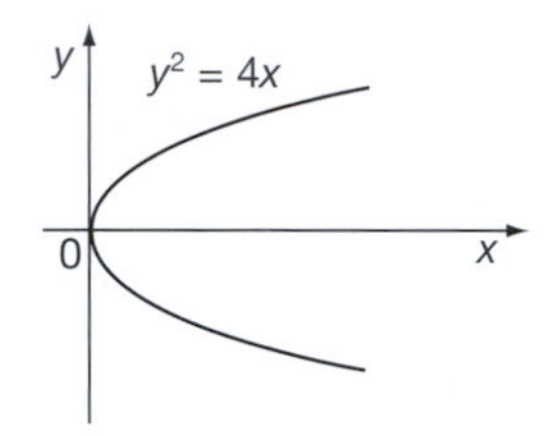


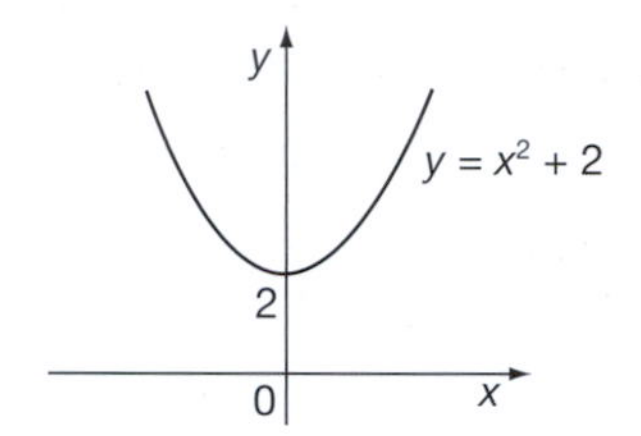


(c) If only even powers of $y$ and also only even powers of $x$ occur, then ...........

**28**

the curve is symmetrical about both axes

For example:

$25x^2 + 16y^2 = 400$ is symmetrical about both axes.

$y^2 + 3y - 2 = (x^2 + 7)^2$ is symmetrical about the $y$-axis, but not about the $x$-axis since ...........

**29**

both odd and even powers of $y$ occur

## Intersection with the axes

Points at which the curve crosses the $x$- and $y$-axes.

Crosses the $x$-axis: Put $y = 0$ and solve for $x$.

Crosses the $y$-axis: Put $x = 0$ and solve for $y$.

So, the curve $y^2 + 3y - 2 = x + 8$ crosses the $x$- and $y$-axes at ...........

30

$x$-axis at $x = -10$
$y$-axis at $y = 2$ and $y = -5$

In fact, the curve is:

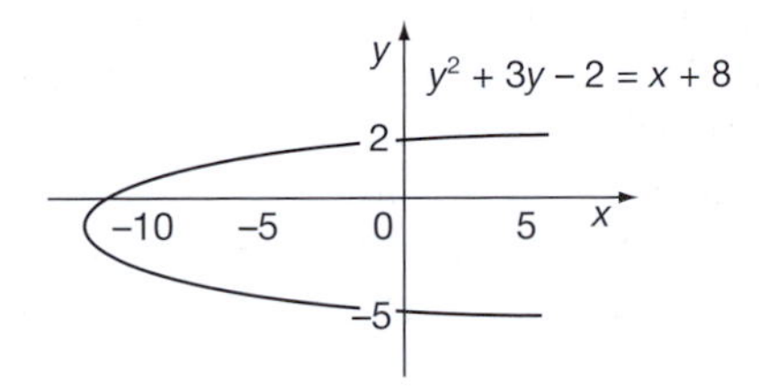


## Change of origin

Look for a possible change of origin to simplify the equation. For example, for the curve $4(y + 3) = (x - 4)^2$, if we change the origin by putting $Y = y + 3$ and $X = x - 4$, the equation becomes $4Y = X^2$ which is a parabola symmetrical about the axis of $Y$.

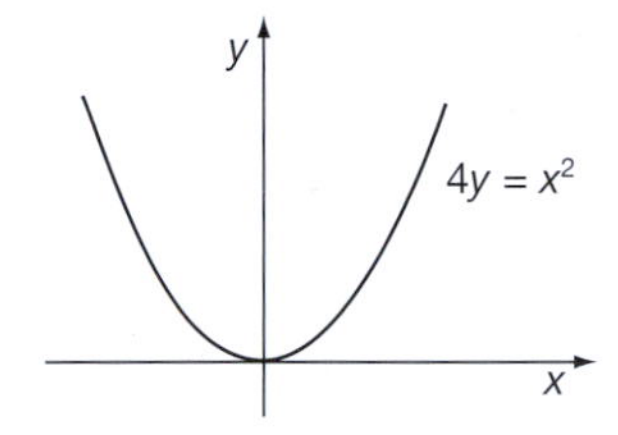


So the curve relative to the original $x$- and $y$-axes is positioned thus:

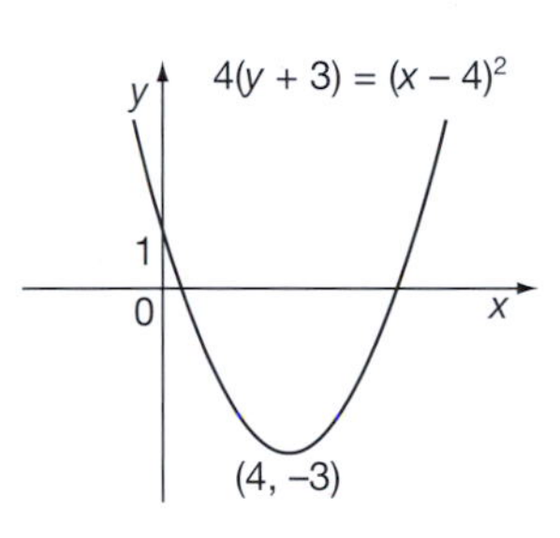


31

## Asymptotes

We have already dealt with asymptotes in some detail. We investigate (a) asymptotes parallel to the axes, and (b) those of a more general nature.

(a) *Parallel to the axes*

(i) Express the equation 'on one line', i.e. remove algebraic fractions.

(ii) Equate to zero the coefficient of the highest power of $y$ to find the asymptote parallel to the $y$-axis.

(iii) Equate to zero the coefficient of the highest power of $x$ to find the asymptote parallel to the $x$-axis.

As an example, find the asymptotes parallel to the axes for the curve $y = \dfrac{(x-1)(x+6)}{(x+3)(x-4)}$.

32

$x = -3; \quad x = 4; \quad y = 1$

Because $y(x + 3)(x - 4) = (x - 1)(x + 6)$

Asymptotes parallel to the $y$-axis: $(x + 3)(x - 4) = 0 \quad \therefore x = -3$ and $x = 4$.

Rearranging the equation gives $(y - 1)x^2 - (y + 5)x - 12y + 6 = 0$

Asymptote parallel to the $x$-axis: $y - 1 = 0 \quad \therefore y = 1$

▶

The graph of the function is as shown to the right:

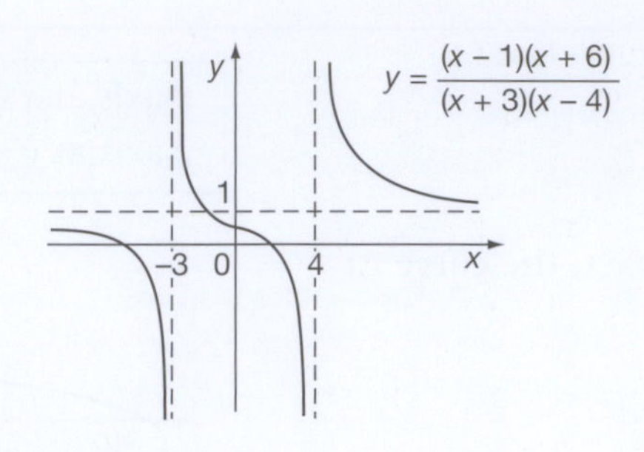


(b) *General asymptotes*

Substitute $y = mx + c$ and equate the coefficient of the two highest powers of $x$ to zero to determine $m$ and $c$.

Thus, for the curve $x^4 - 2x^3y + 5x^2 - 4y^2 = 0$, the asymptote is ...........

## 33

$$\boxed{y = \frac{x}{2}}$$

Because substituting $y = mx + c$ and simplifying the left-hand side, gives:

$$(1 - 2m)x^4 - 2cx^3 + (5 - 4m^2)x^2 - 8mcx - 4c^2 = 0$$

$$\left.\begin{array}{ll} \therefore 1 - 2m = 0 & \therefore m = \frac{1}{2} \\ 2c = 0 & \therefore c = 0 \end{array}\right\} \quad \therefore y = \frac{x}{2} \text{ is an asymptote.}$$

### Large and small values of *x* and *y*

If $x$ or $y$ is *small*, higher powers of $x$ or $y$ become negligible and hence only lower powers of $x$ or $y$ appearing in the equation provide an approximate simpler form.

Similarly, if $x$ or $y$ is *large*, the higher powers have predominance and lower powers can be neglected, i.e. when $x$ is large

$$y^2 = 2x^2 - 7x + 4$$

approximates to

$$y^2 = 2x^2, \text{ i.e. } y = \pm x\sqrt{2}.$$

### Stationary points

Maximum and minimum values; points of inflexion. We have dealt at length in a previous Programme with this whole topic. We will just summarize the results at this stage:

$\frac{dy}{dx} = 0$ and $\frac{d^2y}{dx^2}$ is negative (curve concave downwards) then the point is a *maximum*

$\frac{dy}{dx} = 0$ and $\frac{d^2y}{dx^2}$ is positive (curve concave upwards) then the point is a *minimum*

$\frac{dy}{dx} = 0$ and $\frac{d^2y}{dx^2} = 0$ with a change of sign through the stationary point then the point is a *point of inflexion*.

▶

## Limitations

Restrictions on the possible range of values that $x$ or $y$ may have. For example, consider:

$$y^2 = \frac{(x+1)(x-3)}{x+4}$$

For $x < -4$, $y^2$ is negative $\therefore$ no real values of $y$.
For $-4 < x < -1$, $y^2$ is positive $\therefore$ real values of $y$ exist.
For $-1 < x < 3$, $y^2$ is negative $\therefore$ no real values of $y$.
For $3 < x$, $y^2$ is positive $\therefore$ real values of $y$ exist.

The curve finally looks like this:

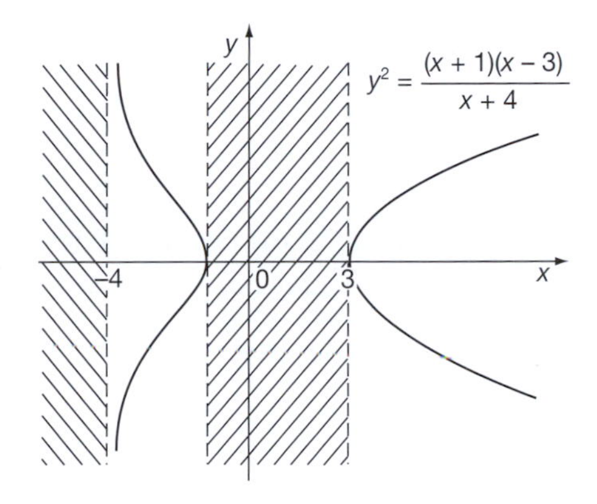


**34**

In practice, not all of these considerations are applicable in any one particular case. Let us work in detail through one such example.

Sketch the curve whose equation is $y = \dfrac{(x+2)(x-3)}{x+1}$.

(a) *Symmetry*: First write the equation 'on the line'.

$$y(x+1) = (x+2)(x-3) = x^2 - x - 6$$

Both odd and even powers of $x$ occur, $\therefore$ no symmetry about $y$-axis.
Only odd powers of $y$ occur, $\therefore$ no symmetry about $x$-axis.

(b) *Crossing the axes*: This is done simply by putting $y = 0$ and $x = 0$, so, in this example, the curve crosses the axes at ...........

**35**

$x$-axis at $x = -2$ and at $x = 3$
$y$-axis at $y = -6$

(c) *Stationary points*: The first essential is to find the values of $x$ at which $\dfrac{dy}{dx} = 0$.

Obtain an expression for $\dfrac{dy}{dx}$ and solve $\dfrac{dy}{dx} = 0$ for values of $x$.

This gives $\dfrac{dy}{dx} = 0$ at $x =$ ...........

**36**

no real values of $x$

Because if $y = \dfrac{x^2 - x - 6}{x + 1}$, $\dfrac{dy}{dx} = \dfrac{(x+1)(2x-1) - (x^2 - x - 6)}{(x+1)^2}$

$$= \frac{x^2 + 2x + 5}{(x+1)^2}$$

For stationary points, $\dfrac{dy}{dx} = 0 \quad \therefore x^2 + 2x + 5 = 0 \quad \therefore x = \dfrac{-2 \pm \sqrt{-16}}{2}$

i.e. $x$ is complex. Therefore, there are no stationary points on the graph.

(d) *When x is very small*: $y \approx -\dfrac{x+6}{x+1}$ i.e. $y \approx -6$.

*When x is very large*: $y \approx \dfrac{x^2}{x}$ i.e. $x \quad \therefore y \approx x$.

(e) *Asymptotes*:

(i) First find any asymptotes parallel to the axes. These are ...........

**37**

parallel to the $y$-axis: $x = -1$

Because $y(x+1) - x^2 + x + 6 = 0 \qquad \therefore x + 1 = 0 \qquad \therefore x = -1$.

(ii) Now investigate the general asymptote, if any.
This gives ...........

**38**

$y = x - 2$

This is obtained, as usual, by putting $y = mx + c$ in the equation:

$$(mx + c)(x + 1) = (x + 2)(x - 3)$$

$$\therefore mx^2 + mx + cx + c = x^2 - x - 6$$

$$(m - 1)x^2 + (m + c + 1)x + c + 6 = 0$$

Equating the coefficients of the two highest powers of $x$ to zero:

$m - 1 = 0 \quad \therefore m = 1$

and $m + c + 1 = 0 \quad \therefore c + 2 = 0 \quad \therefore c = -2$

$\therefore y = x - 2$ is an asymptote.

So, collecting our findings together, we have

(a) No symmetry about the $x$- or $y$-axis.
(b) Curve crosses the $x$-axis at $x = -2$ and at $x = 3$.
(c) Curve crosses the $y$-axis at $y = -6$.
(d) There are no stationary points on the curve.
(e) Near $x = 0$, the curve approximates to $y = -6$.
(f) For numerically large values of $x$, the curve approximates to $y = x$, i.e. when $x$ is large and positive, $y$ is large and positive and when $x$ is large and negative, $y$ is large and negative.
(g) The only asymptotes are $x = -1$ (parallel to the $y$-axis) and $y = x - 2$.

*With these facts before us, we can now sketch the curve.*
*Do that and then check your result with that shown in the next frame*

Here is the graph as it should appear. You can, of course, always plot an odd point or two in critical positions if extra help is needed.

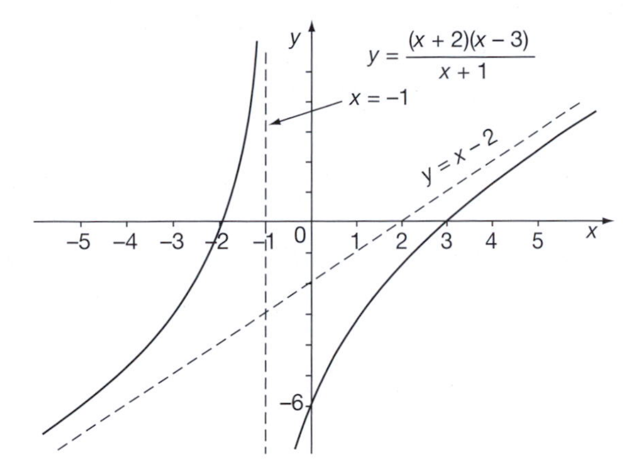


*Now let us move on to something rather different*

# Curve fitting

40

Readings recorded from a test or experiment normally include errors of various kinds and therefore the points plotted from these data are scattered about the positions they should ideally occupy. Unless very few readings are taken, it can be assumed that the inherent errors will be of a random nature, resulting in some of the values being slightly too high and some slightly too low. Having plotted the points, we then draw as the graph the middle line of this narrow band of points. It may well be that the line drawn does not pass through any of the actual plotted points, but from now on it is this line which is used to determine the relationship between the two variables.

## Straight-line law

For example, values of $V$ and $h$ are recorded in a test:

| $h$ | 6·0 | 10 | 14 | 18 | 21 | 25 |
|---|---|---|---|---|---|---|
| $V$ | 5·5 | 7·0 | 9·5 | 12·5 | 13·5 | 16·5 |

If the law relating $V$ and $h$ is $V = ah + b$, where $a$ and $b$ are constants,

(a) plot the graph of $V$ against $h$

(b) determine the values of $a$ and $b$.

(a) Plotting the points is quite straightforward. Do it carefully on squared paper. You get …………

## 41

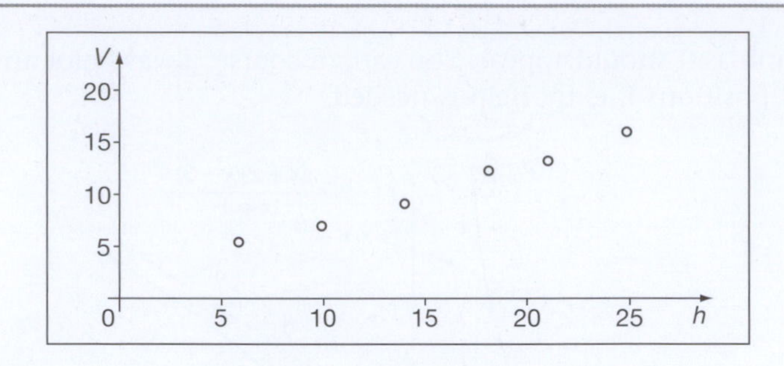


We now estimate by eye the straight position for a straight-line graph drawn down the middle of this band of points. Draw the line on your graph.

## 42

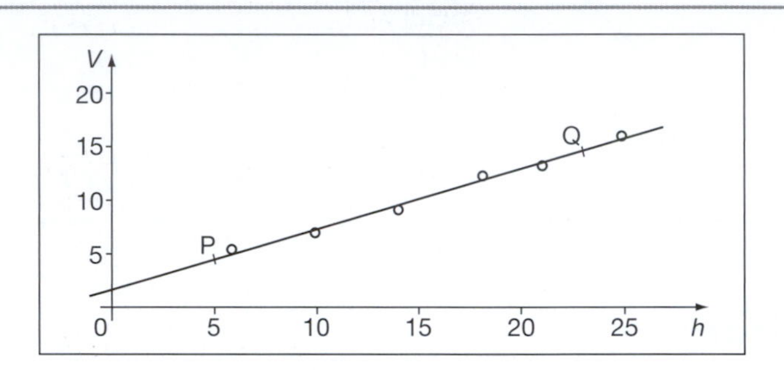


Henceforth, we shall use this line as representing the equation and ignore the actual points we plotted.

(b) The law is first degree, i.e. $V = ah + b$

Compare this with $y = mx + c$.

Therefore, if we find the values of $m$ and $c$ in the normal way, these values will be those of $a$ and $b$.

We now select two convenient points *on the line* and read off their $x$- and $y$-coordinates. For instance, P (5, 4·5) and Q (23, 15). Substituting these values in $y = mx + c$ gives two equations from which we can find the values of $m$ and $c$, which are ............

## 43

$$m = 0{\cdot}583; \quad c = 1{\cdot}58$$

Because we have $\left.\begin{aligned} 4{\cdot}5 &= 5m + c \\ 15{\cdot}0 &= 23m + c \end{aligned}\right\}$ $\therefore 10{\cdot}5 = 18m \quad \therefore m = \dfrac{10{\cdot}5}{18} = 0{\cdot}583$

and then $4{\cdot}5 = 5(0{\cdot}583) + c \quad \therefore c = 4{\cdot}5 - 2{\cdot}917 \quad \therefore c = 1{\cdot}583$

The equation of the line is $y = 0{\cdot}583x + 1{\cdot}58$

and the law relating $V$ and $h$ is $V = 0{\cdot}583h + 1{\cdot}58$

**Provided we can express the law in straight-line form, they are all tackled in the same manner.**

## Graphs of the form $y = ax^n$, where $a$ and $n$ are constants

To convert this into 'straight-line form', we take logarithms of both sides. The equation then becomes ...........

**44**

$$\log y = n \log x + \log a$$

If we compare this result with $Y = mX + c$ we see we have to plot ........... along the $Y$-axis and ........... along the $X$-axis to obtain a straight-line graph.

**45**

$\log y$ along the $Y$-axis
$\log x$ along the $X$-axis

$$\left.\begin{aligned}\log y &= n \log x + \log a\\ Y &= mX + c\end{aligned}\right\}$$

If we then find $m$ and $c$ from the straight line as before, then $m = n$
and $c = \log a \quad \therefore\ a = 10^c$.

Let us work through an example.
Values of $x$ and $y$ are related by the equation $y = ax^n$:

| $x$ | 2 | 5 | 12 | 25 | 32 | 40 |
|---|---|---|---|---|---|---|
| $y$ | 5·62 | 13·8 | 52·5 | 112 | 160 | 200 |

Determine the values of the constants $a$ and $n$.

$$y = ax^n \quad \therefore\ \log y = n \log x + \log a$$
$$Y = mX + c$$

We must first compile a table showing the corresponding values of $\log x$ and $\log y$.

*Do that and check with the results shown before moving on*

**46**

| $\log x$ | 0·3010 | 0·6990 | 1·079 | 1·398 | 1·505 | 1·602 |
|---|---|---|---|---|---|---|
| $\log y$ | 0·750 | 1·14 | 1·72 | 2·05 | 2·20 | 2·30 |

Now plot these points on graph paper, $\log x$ along the $x$-axis
$\log y$ along the $y$-axis
and draw the estimated best straight-line graph.

**47**

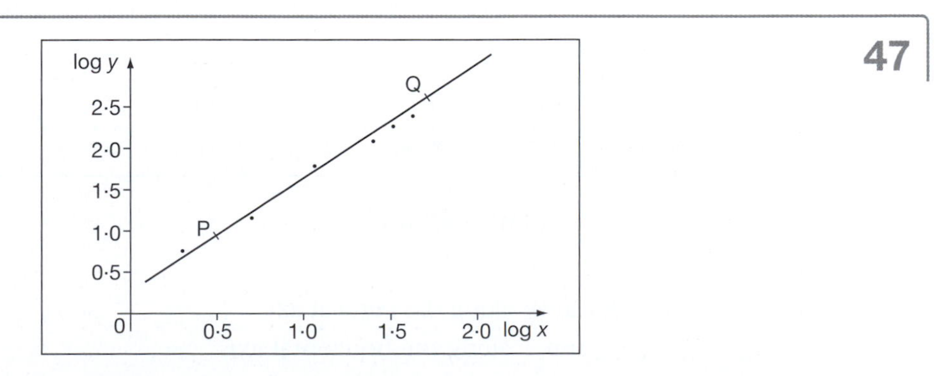


If we select from the graph two points, P (0·50, 0·94) and Q (1·70, 2·45) we can calculate the values of $m$ and $c$ which are $m =$ ........... and $c =$ ...........

**48**

$$m = 1{\cdot}258; \quad c = 0{\cdot}311$$

Because $Y = mX + c \quad \therefore \quad 2{\cdot}45 = 1{\cdot}70m + c$

$$0{\cdot}94 = 0{\cdot}50m + c$$

$$\therefore \quad 1{\cdot}51 = 1{\cdot}20m \qquad \therefore m = 1{\cdot}258; \text{ then } c = 0{\cdot}311.$$

Therefore $Y = 1{\cdot}258X + 0{\cdot}311$

$$\log y = n \log x + \log a$$

$$\therefore n = 1{\cdot}258 \text{ and } \log a = 0{\cdot}311 \quad \therefore a = 2{\cdot}05$$

Therefore, the equation is $y = 2{\cdot}05x^{1{\cdot}26}$.

## Graphs of the form $y = ae^{nx}$

Exponential relationships occur frequently in technical situations. As before, our first step is to convert the equation to 'straight-line form' by taking logs of both sides. We could use common logarithms as we did previously, but the work involved is less if we use natural logarithms.

So, taking natural logarithms of both sides, we express the equation in the form ............

**49**

$$\ln y = nx + \ln a$$

If we compare this with the straight-line equation, we have:

$$\ln y = nx + \ln a$$

$$Y = mX + c$$

which shows that if we plot values of $\ln y$ along the $Y$-axis and values of just $x$ along the $X$-axis, the value of $m$ will give the value of $n$, and the value of $c$ will be $\ln a$, hence $a = e^c$.

Let us do an example.

The following values of $W$ and $T$ are related by the law $W = ae^{nT}$ where $a$ and $n$ are constants:

| $T$ | 3·0 | 10 | 15 | 30 | 50 | 90 |
|---|---|---|---|---|---|---|
| $W$ | 3·857 | 1·974 | 1·733 | 0·4966 | 0·1738 | 0·0091 |

We need values of $\ln W$, so compile a table showing values of $T$ and $\ln W$.

**50**

| $T$ | 3·0 | 10 | 15 | 30 | 50 | 90 |
|---|---|---|---|---|---|---|
| $\ln W$ | 1·35 | 0·68 | 0·55 | −0·70 | −1·75 | −4·70 |

$W = ae^{nT} \qquad \therefore \ln W = nT + \ln a$

$$Y = mX + c$$

Therefore, we plot $\ln W$ along the vertical axis

and $T$ along the horizontal axis

to obtain a straight-line graph, from which $m = n$

and $c = \ln a \quad \therefore a = e^c$.

So, plot the points; draw the best straight-line graph; and from it determine the values of $n$ and $a$. The required law is therefore

$W = \dots\dots\dots\dots$

---

**51**

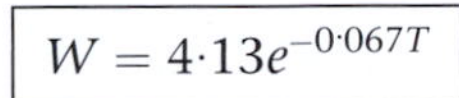


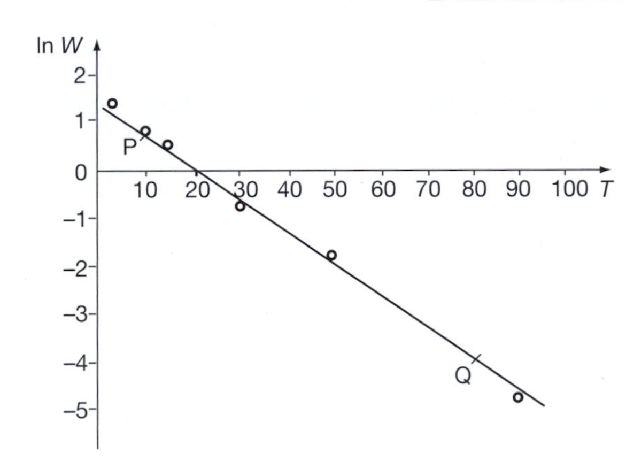


$P = (10, 0{\cdot}75)$; $Q = (80, \; -3{\cdot}93)$

$$-3{\cdot}93 = 80m + c$$
$$0{\cdot}75 = 10m + c$$
$$\therefore \; -4{\cdot}68 = 70m$$
$$\therefore m = -0{\cdot}0668$$
$$\therefore c = 0{\cdot}75 + 0{\cdot}668$$
$$= 1{\cdot}418$$
$$m = n = -0{\cdot}0668$$
$$c = \ln a = 1{\cdot}418$$
$$\therefore a = 4{\cdot}13$$

$\therefore$ The required law is

$$W = 4{\cdot}13e^{-0{\cdot}067T}$$

If we can express the equation or law in straight-line form, the same method can be applied.

How about these? How could you arrange for each of the following to give a straight-line graph?

**1** $y = ax^2 + b$

**3** $y = \dfrac{a}{x} + b$

**2** $y = ax + b$

**4** $y = ax^2 + bx$

*Check your suggestions with the next frame*

---

**52**

Here they are:

| *Equation* | *Straight-line form* | *Plot* |
|---|---|---|
| $y = ax^2 + b$ | $y = ax^2 + b$ | $y$ against $x^2$ |
| $y = ax + b$ | $y = ax + b$ | $y$ against $x$ |
| $y = \dfrac{a}{x} + b$ | $y = a\dfrac{1}{x} + b$ | $y$ against $\dfrac{1}{x}$ |
| $y = ax^2 + bx$ | $\dfrac{y}{x} = ax + b$ | $\dfrac{y}{x}$ against $x$ |

And now just one more. If we convert $y = \dfrac{a}{x + b}$ to straight-line form, it becomes

$\dots\dots\dots\dots$

53

$$y = -\frac{1}{b}xy + \frac{a}{b}$$

Because

$$\text{if } y = \frac{a}{x+b}, \text{ then } xy + by = a \quad \therefore by = -xy + a$$

$$\therefore y = -\frac{1}{b}xy + \frac{a}{b}$$

That is, if we plot values of $y$ against values of the product $xy$, we shall obtain a straight-line graph from which $m = -\frac{1}{b}$ and $c = \frac{a}{b}$.

From these, $a$ and $b$ can be easily found.

*Finally, to what is perhaps the most important part of this Programme.*
*Let us start with a new frame*

## Method of least squares

54

All the methods which involve drawing the best straight line by eye are approximate only and depend upon the judgement of the operator. Quite considerable variation can result from various individuals' efforts to draw the 'best straight line'.

The *method of least squares* determines the best straight line entirely by calculation, using the set of recorded results. The form of the equation has to be chosen and this is where the previous revision will be useful.

*Let us start with the case of a linear relationship*

55

### Fitting a straight-line graph

We have to fit a straight line $y = a + bx$ to a set of plotted points $(x_1, y_1)$, $(x_2, y_2)\ldots$ $(x_n, y_n)$ so that the sum of the squares of the distances to this straight line from the given set of points is a minimum. The distance of any point from the line is measured along an ordinate, i.e. in the $y$-direction.

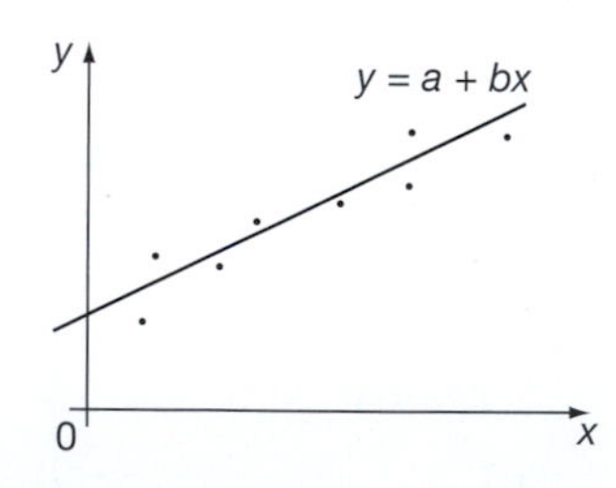


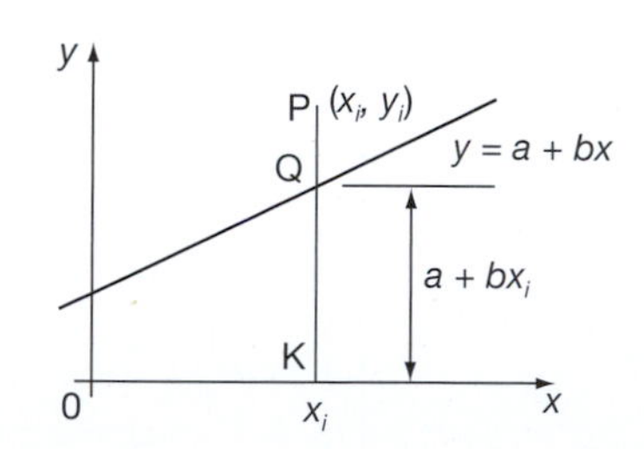


If we take a sample point P $(x_i, y_i)$:

QK is the value of $y = a + bx$ at $x = x_i$, i.e. $a + bx_i$.

PQ is the difference between PK and QK, i.e. $y_i - a - bx_i$.

$$\therefore PQ^2 = (y_i - a - bx_i)^2$$

Therefore, the sum $S$ of the squares of these differences for all $n$ such points is given by $S = \sum_{i=1}^{n}(y_i - a - bx_i)^2$.

We have to determine the values $a$ and $b$ so that $S$ shall be a minimum. The right-hand side contains two unknowns $a$ and $b$. Therefore, for the sum of the squares to be a minimum:

$$\frac{\partial S}{\partial a} = 0 \text{ and } \frac{\partial S}{\partial b} = 0$$

$$\frac{\partial S}{\partial a} = -2\sum_{i=1}^{n}(y_i - a - bx_i) = 0; \quad \frac{\partial S}{\partial b} = -2\sum_{i=1}^{n}x_i(y_i - a - bx_i) = 0$$

The first gives $\sum_{i=1}^{n} y_i - na - b\sum_{i=1}^{n} x_i = 0$ $\left[\textit{Note}: \sum_{i=1}^{n} a = na.\right]$

i.e. $an + b\sum_{i=1}^{n} x_i = \sum_{i=1}^{n} y_i$ (a)

The second gives $\sum_{i=1}^{n} x_i y_i - a\sum_{i=1}^{n} x_i - b\sum_{i=1}^{n} x_i^2 = 0$

i.e. $a\sum_{i=1}^{n} x_i + b\sum_{i=1}^{n} x_i^2 = \sum_{i=1}^{n} x_i y_i$ (b)

Equations (a) and (b) are called the *normal equations* of the problem and we will now write them without the suffixes, remembering that the $x$- and $y$-values are those of the recorded values.

$$\left.\begin{aligned} an + b\Sigma x &= \Sigma y \\ a\Sigma x + b\Sigma x^2 &= \Sigma xy \end{aligned}\right\} \text{ for all the } n \text{ given pairs of values.}$$

From these *normal equations*, the specific values of $a$ and $b$ can be determined.

*We will now work through some examples*

56

**Example 1**

Apply the method of least squares to fit a straight-line relationship for the following points:

| $x$ | −2·4 | −0·8 | 0·3 | 1·9 | 3·2 |
|---|---|---|---|---|---|
| $y$ | −5·0 | −1·5 | 2·5 | 6·4 | 11·0 |

For this set, $n = 5$ and the normal equations are

$$\left.\begin{aligned} an + b\sum x &= \sum y \\ a\sum x + b\sum x^2 &= \sum xy \end{aligned}\right\} \text{ where } y = a + bx.$$

Therefore, we need to sum the values of $x$, $y$, $x^2$ and $xy$. This is best done in table form:

| | $x$ | $y$ | $x^2$ | $xy$ |
|---|---|---|---|---|
| | −2·4 | −5·0 | 5·76 | 12·0 |
| | −0·8 | −1·5 | 0·64 | 1·2 |
| | 0·3 | 2·5 | 0·09 | 0·75 |
| | 1·9 | 6·4 | 3·61 | 12·16 |
| | 3·2 | 11·0 | 10·24 | 35·2 |
| $\Sigma$ | 2·2 | 13·4 | 20·34 | 61·31 |

The normal equations now become ........... and ...........

**57**

$$5a + 2{\cdot}2b = 13{\cdot}4$$
$$2{\cdot}2a + 20{\cdot}34b = 61{\cdot}31$$

Dividing through each equation by the coefficient of $a$ gives:

$$a + 0{\cdot}440b = 2{\cdot}68$$
$$a + 9{\cdot}245b = 27{\cdot}87$$
$$\therefore\ 8{\cdot}805b = 25{\cdot}19 \quad \therefore\ b = 2{\cdot}861$$
$$\therefore\ a = 2{\cdot}68 - 1{\cdot}2588 \quad \therefore\ a = 1{\cdot}421$$

Therefore, the best straight line for the given values is

$$y = 1{\cdot}42 + 2{\cdot}86x$$

## Using a spreadsheet

If you have access to the Microsoft Excel™ spreadsheet then there is a function that will calculate the least-square parameters for you. For example, if in this problem the $x$-values are ranged in cells **A1** to **A5** and the corresponding $y$-values in cells **B1** to **B5** then highlight the two cells **A7** and **B7** and type:

**=LINEST(B1:B5,A1:A5)**

Then, with cells **A7** and **B7** still highlighted press **Ctrl-SHIFT-Enter** (hold down the **Ctrl** and **SHIFT** keys together and press **Enter**) and the numbers 2·86052... and 1·42137... will appear in cells **A7** and **B7** respectively.

The number 2·86052... in cell **A7** is the *gradient* (2·86 to 2dp) and the number 1·42137... in cell **B7** is the *vertical intercept* (1·42 to 2 dp) of the straight line.

To see how well the method works, plot the set of values for $x$ and $y$ and also the straight line whose equation we have just found on the same axes.

**58**

The result should look like this:

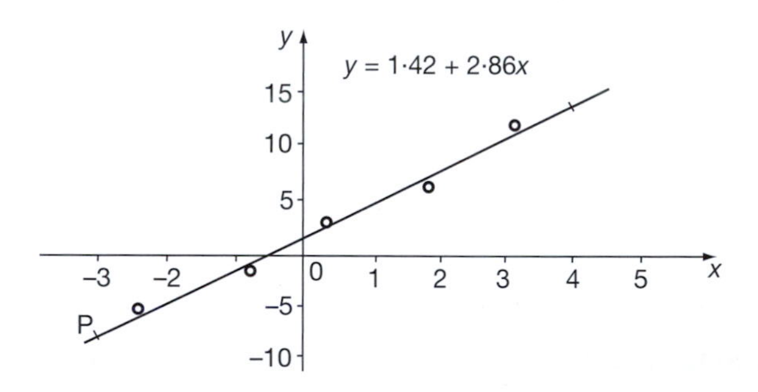


Any relationship that can be expressed in straight-line form can be dealt with in the same way.

### Example 2

It is required to fit the best rectangular hyperbola $xy = c$ to the set of values given below:

| $x$ | 0·5 | 1·0 | 2·0 | 3·0 | 4·0 | 5·0 |
|---|---|---|---|---|---|---|
| $y$ | 62 | 28 | 17 | 9·0 | 7·0 | 5·0 |

▶

In this case, $n = 6$. Also $y = c.\dfrac{1}{x}$

Ordinate difference between a point and the curve $= y_i - \dfrac{c}{x_i}$

The sum of the squares $S = \sum_{i=1}^{n}\left(y_i - \dfrac{c}{x_i}\right)^2$ and for $S$ to be a minimum,

$$\frac{\partial S}{\partial c} = 0 \quad \frac{\partial S}{\partial c} = -2\sum_{i=1}^{n}\frac{1}{x_i}\left(y_i - \frac{c}{x_i}\right) = 0$$

$$\therefore \sum_{i=1}^{n}\frac{y_i}{x_i} - c\sum_{i=1}^{n}\frac{1}{x_i{}^2} = 0$$

So this time we need values of $x$, $y$, $\dfrac{1}{x}$, $\dfrac{y}{x}$ and $\dfrac{1}{x^2}$:

| $x$ | $y$ | $\dfrac{1}{x}$ | $\dfrac{y}{x}$ | $\dfrac{1}{x^2}$ |
|---|---|---|---|---|
| 0·5 | 62 | 2·0 | 124 | 4·0 |
| 1·0 | 28 | 1·0 | 28 | 1·0 |
| 2·0 | 17 | 0·5 | 8·5 | 0·25 |
| 3·0 | 9 | 0·333 | 3 | 0·111 |
| 4·0 | 7 | 0·25 | 1·75 | 0·0625 |
| 5·0 | 5 | 0·2 | 1 | 0·04 |
| | | $\Sigma$ | 166·25 | 5·4635 |

From this we can find $c$ and the equation of the required hyperbola which is therefore ............

---

**59**

$xy = 30{\cdot}4$

Because

$$\sum_{i=1}^{n}\frac{y}{x} - c\sum_{i=1}^{n}\frac{1}{x^2} = 0 \quad \therefore 166{\cdot}25 = 5{\cdot}4635c \quad \therefore c = 30{\cdot}4$$

Here is the graph showing the plotted values and also the rectangular hyperbola we have just obtained:

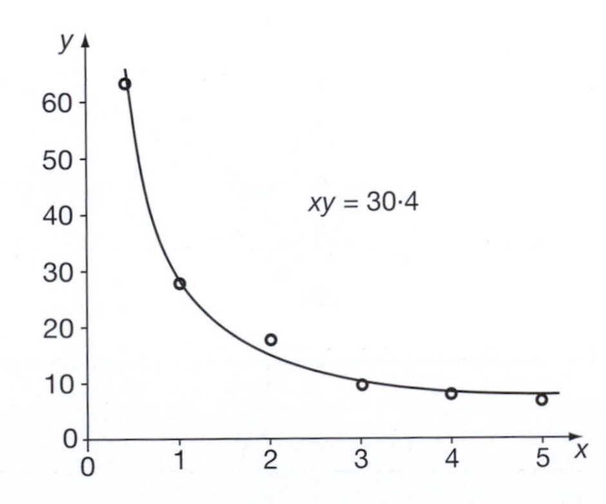

## 60

**Example 3**

Values of $x$ and $y$ are related by the law $y = a\ln x$. Determine the value of $a$ to provide the best fit for the following set of values:

| $x$ | 0·3 | 0·9 | 1·7 | 2·5 | 4·0 | 10·0 |
|---|---|---|---|---|---|---|
| $y$ | −7·54 | −0·672 | 3·63 | 4·41 | 8·22 | 12·2 |

$y = a\ln x$ will give a straight line if we plot $y$ against $\ln x$. Therefore, let $u = \ln x$.

Ordinate difference $= y_i - a\ln x_i = y_i - au_i \quad \therefore S = \sum_{i=1}^{n}(y_i - au_i)^2$

For minimum $S$, $\dfrac{\partial S}{\partial a} = 0. \qquad \therefore -2\sum_{i=1}^{n} u_i(y_i - au_i) = 0$

$$\therefore \sum_{i=1}^{n} u_i y_i - a\sum_{i=1}^{n} u_i^2 = 0$$

Therefore we need values of $x$, $\ln x$ (i.e. $u$), $y$, $uy$ and $u^2$:

| $x$ | $u\ (=\ln x)$ | $y$ | $uy$ | $u^2$ |
|---|---|---|---|---|
| 0·3 | −1·204 | −7·54 | 9·078 | 1·450 |
| 0·9 | −0·105 | −0·672 | 0·0708 | 0·0111 |
| 1·7 | 0·531 | 3·63 | 1·9262 | 0·2816 |
| 2·5 | 0·916 | 4·41 | 4·0408 | 0·8396 |
| 4·0 | 1·3863 | 8·22 | 11·3953 | 1·9218 |
| 10·0 | 2·3026 | 12·20 | 28·092 | 5·302 |

Now total up the appropriate columns and finish the problem off. The equation is finally …………

## 61

$$\boxed{y = 5{\cdot}57\ln x}$$

Because

$\sum uy = 54{\cdot}603$ and $\sum u^2 = 9{\cdot}806$

and since $\sum uy - a\sum u^2 = 0$,

then $a = \dfrac{\sum uy}{\sum u^2} = 5{\cdot}568$

$\therefore y = 5{\cdot}57\ln x$

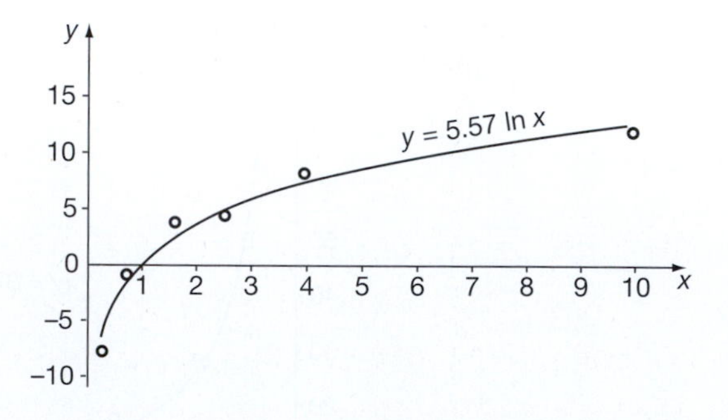

62

**Example 4**

A test on the length of tool life at different cutting speeds gave the following results:

| Speed $v$ (m/min) | 120 | 130 | 170 | 200 |
|---|---|---|---|---|
| Life $T$ (min) | 62 | 25 | 7·2 | 2·8 |

If the law relating $v$ and $T$ is $v = aT^k$ determine the constants $a$ and $k$ to give the best fit.

$$v = aT^k \qquad \therefore \log v = k \log T + \log a$$
$$Y = mX + c$$

| $X = \log T$ | 1·792 | 1·398 | 0·8573 | 0·4472 |
|---|---|---|---|---|
| $Y = \log v$ | 2·079 | 2·114 | 2·230 | 2·301 |

Arguing as before, $S = \sum_{i=1}^{n} (Y_i - mX_i - c)^2$

$$\frac{\partial S}{\partial m} = -2\sum_{i=1}^{n} X_i(Y_i - mX_i - c) = 0 \qquad \therefore \sum_{i=1}^{n} X_iY_i - m\sum_{i=1}^{n} X_i^2 - c\sum_{i=1}^{n} X_i = 0$$

$$\frac{\partial S}{\partial c} = -2\sum_{i=1}^{n} (Y_i - mX_i - c) = 0 \qquad \therefore \sum_{i=1}^{n} Y_i - m\sum_{i=1}^{n} X_i - nc = 0$$

So we now need columns for $X$, $Y$, $XY$ and $X^2$.

Compile the appropriate table and finish off the problem, so finding the required law ...........

63

$$v = 237T^{-0{\cdot}173}$$

Here is the working:

| $X$ | $Y$ | $XY$ | $X^2$ |
|---|---|---|---|
| 1·792 | 2·079 | 3·726 | 3·211 |
| 1·398 | 2·114 | 2·955 | 1·954 |
| 0·8573 | 2·230 | 1·912 | 0·735 |
| 0·4472 | 2·301 | 1·029 | 0·200 |
| $\sum$ 4·4945 | 8·724 | 9·622 | 6·100 |

The normal equations give $9{\cdot}622 - m6{\cdot}100 - c4{\cdot}495 = 0$

and $8{\cdot}724 - m4{\cdot}495 - c4 = 0.$

Dividing through each equation by the coefficient of $m$, we have:

$$m + 0{\cdot}7369c = 1{\cdot}5774$$
$$m + 0{\cdot}8899c = 1{\cdot}9408$$
$$\therefore 0{\cdot}1530c = 0{\cdot}3634 \qquad \therefore c = 2{\cdot}375$$

▶

Then $m = 1{\cdot}5774 - 0{\cdot}7369(2{\cdot}375) = 1{\cdot}5774 - 1{\cdot}7503 = -0{\cdot}1729$

$\therefore\ Y = -0{\cdot}173X + 2{\cdot}375$

$m = n$

$\therefore\ n = -0{\cdot}173$

$c = \log a$

$\therefore\ \log a = 2{\cdot}3752 \quad \therefore\ a = 237{\cdot}25$

The law is

$v = 237T^{-0{\cdot}173}$

The same principles can be applied to fit curves of higher degree to sets of values but, in general, fitting a second-degree curve involves three normal equations in three unknowns and curves of higher degree involve an increasing number of simultaneous equations, so that the working tends to become unwieldy without some form of computing facility.

# Correlation

## 64 Correlation

Calculating the least squares fit as we have just done is called *regression analysis* because it finds the line to which the plotted data points fall back – that is, regress to. In such an analysis there is an implicit assumption that there is a linear relationship between the two variables and that the value of one variable can be found from the value of the other variable by using the equation of the straight line so found.

If two variables vary together they are said to be **correlated**. For example, the fuel consumed by a car is correlated to the number of miles travelled. The exact relationship is not easily found because so many other factors are involved such as speed of travel, condition of the engine, tyre pressures and so on. What is important is that as the mileage increases so does the fuel used – there is a correlation.

It is, however, most important to note that whilst cause-and-effect does display correlation between the cause and the effect as in the case of mileage travelled (cause) and petrol consumed (effect) the converse does not; the correlation of variables $x$ and $y$ does not imply cause-and-effect between $x$ and $y$. For example, if it were noted that during the month of July in a particular year the temperature rose week by week whereas the average share price fell week by week we would say that there was a correlation between the temperature and the share price but clearly there was not cause-and-effect; the rising temperature did not cause the share price to fall and the falling share price did not cause the temperature to rise!

You may think that these are trite observations but it is surprising how many statistical surveys do imply a cause-and-effect when in fact only a correlation is observed. Indeed, there is a clear distinction between the regression analysis of the least squares method and correlation analysis. Regression analysis does assume a linear relationship where the value of one variable is directly related to the value of the other variable whereas in correlation analysis neither variable is assumed to influence the other; we are simply looking at two sets of numbers to find out if when the numbers in one set change the numbers in the other set change in unison or not.

So how do we measure correlation?

*Move to the next frame*

## Measures of correlation

65

The strength of the correlation between two variables is given by a **correlation coefficient** – a number whose value ranges from $-1$ to $+1$. The strongest *positive* correlation is when the correlation coefficient is $+1$. In this case the two variables increase and decrease together in perfect unison. For example, the distance $s$ travelled by a car moving at a constant velocity $v$ is given by the equation:

$$s = vt$$

where $t$ is the time taken to cover the distance. A plot of $s$ values against the corresponding $t$ values gives a straight line of gradient $v$ passing through the origin. If distances $s_i$ were measured and the corresponding times $t_i$ recorded then they would exhibit perfect correlation with a correlation coefficient of 1.

The strongest *negative* correlation is when the correlation coefficient is $-1$. In this case one variable will increase or decrease as the other decreases or increases respectively, again in perfect unison. An example of such a perfect negative correlation would be the lengths of adjacent sides $a$ and $b$ of rectangles of a given constant area $A$. In this case:

$$ab = A \text{ so that } a = \frac{A}{b}$$

so as one side increased or decreased the adjacent side would decrease or increase in perfect unison. If adjacent lengths $a_i$ and $b_i$ were recorded then they would exhibit perfect correlation with a correlation coefficient of $-1$.

When the correlation coefficient is zero then there is no correlation between the two variables whatsoever.

*Move to the next frame*

## The Pearson product-moment correlation coefficient

66

The *Pearson product-moment correlation coefficient* $r$ gives the strength of a linear relationship between the $n$ values of two variables $x_i$ and $y_i$ for $i = 1 \ldots n$, where $r$ is given by the rather daunting equation:

$$r = \frac{n\sum_{i=1}^{n} x_i y_i - \left(\sum_{i=1}^{n} x_i\right)\left(\sum_{i=1}^{n} y_i\right)}{\sqrt{\left[n\sum_{i=1}^{n} x_i^2 - \left(\sum_{i=1}^{n} x_i\right)^2\right]\left[n\sum_{i=1}^{n} y_i^2 - \left(\sum_{i=1}^{n} y_i\right)^2\right]}}$$

However, by taking things one step at a time it is not as bad as it first appears. For example, a factory produces engine components and finds that its profit profile can be given as:

| Units sold $x$ | 100 | 200 | 300 | 400 | 500 | 600 |
|---|---|---|---|---|---|---|
| Profit $y$ (£1000s) | 1·4 | 3·1 | 4·8 | 5·5 | 6·4 | 8·1 |

▶

To find the Pearson product-moment correlation coefficient $r$ from this data we must first construct the following table:

| $i$ | $x_i$ | $y_i$ | $x_iy_i$ | $x_i^2$ | $y_i^2$ |
|---|---|---|---|---|---|
| 1 | 100 | 1·4 | 140 | 10 000 | 1·96 |
| 2 | 200 | 3·1 | ........... | ........... | ........... |
| 3 | 300 | 4·8 | ........... | ........... | ........... |
| 4 | 400 | 5·5 | ........... | ........... | ........... |
| 5 | 500 | 6·4 | ........... | ........... | ........... |
| 6 | 600 | 8·1 | ........... | ........... | ........... |
| | | | | | |
| | $\sum_{i=1}^{6} x_i = \ldots$ | $\sum_{i=1}^{6} y_i = \ldots$ | $\sum_{i=1}^{6} x_iy_i = \ldots$ | $\sum_{i=1}^{6} x_i^2 = \ldots$ | $\sum_{i=1}^{6} y_i^2 = \ldots$ |

*Complete the table.*
*The answer is in the next frame*

**67**

| $i$ | $x_i$ | $y_i$ | $x_iy_i$ | $x_i^2$ | $y_i^2$ |
|---|---|---|---|---|---|
| 1 | 100 | 1·4 | 140 | 10 000 | 1·96 |
| 2 | 200 | 3·1 | 620 | 40 000 | 9·61 |
| 3 | 300 | 4·8 | 1440 | 90 000 | 23·04 |
| 4 | 400 | 5·5 | 2200 | 160 000 | 30·25 |
| 5 | 500 | 6·4 | 3200 | 250 000 | 40·96 |
| 6 | 600 | 8·1 | 4860 | 360 000 | 65·61 |
| | | | | | |
| | $\sum_{i=1}^{6} x_i = 2100$ | $\sum_{i=1}^{6} y_i = 29{\cdot}3$ | $\sum_{i=1}^{6} x_iy_i = 12\,460$ | $\sum_{i=1}^{6} x_i^2 = 910\,000$ | $\sum_{i=1}^{6} y_i^2 = 171.43$ |

It is a straightforward matter to substitute these numbers into:

$$r = \frac{6\sum_{i=1}^{6} x_iy_i - \left(\sum_{i=1}^{6} x_i\right)\left(\sum_{i=1}^{6} y_i\right)}{\sqrt{\left[6\sum_{i=1}^{6} x_i^2 - \left(\sum_{i=1}^{6} x_i\right)^2\right]\left[6\sum_{i=1}^{6} y_i^2 - \left(\sum_{i=1}^{6} y_i\right)^2\right]}} = \ldots\ldots\ldots\ldots$$

*The answer is in the next frame*

**68**

$$\boxed{r = 0{\cdot}99}$$

Because

$$r = \frac{6(12\,460) - (2100)(29{\cdot}3)}{\sqrt{\left[6(910\,000) - (2100)^2\right]\left[6(171{\cdot}43) - (29{\cdot}3)^2\right]}}$$
$$= \frac{13\,230}{\sqrt{[1050\,000][170{\cdot}09]}}$$
$$= \frac{13\,230}{13\,363{\cdot}9}$$
$$= 0{\cdot}99$$

A strong correlation between the profit and the number of units sold, as one would hope to be the case.

As another example, consider the following table recording the unit output and the size of the workforce involved in production from seven different manufacturing plants.

| Number of staff $x$ | 6 | 11 | 12 | 9 | 11 | 7 | 9 |
|---|---|---|---|---|---|---|---|
| Units of production $y$ | 21 | 18 | 33 | 39 | 25 | 27 | 21 |

To calculate the Pearson product-moment correlation coefficient the following table is constructed:

| $i$ | $x_i$ | $y_i$ | $x_i y_i$ | $x_i^2$ | $y_i^2$ |
|---|---|---|---|---|---|
| 1 | 6 | 21 | 126 | 36 | 441 |
| 2 | 11 | 18 | ........... | ........... | ........... |
| 3 | 12 | 33 | ........... | ........... | ........... |
| 4 | 9 | 39 | ........... | ........... | ........... |
| 5 | 11 | 25 | ........... | ........... | ........... |
| 6 | 7 | 27 | ........... | ........... | ........... |
| 7 | 9 | 21 | ........... | ........... | ........... |
| | | | | | |
| | $\sum_{i=1}^{7} x_i = \ldots$ | $\sum_{i=1}^{7} y_i = \ldots$ | $\sum_{i=1}^{7} x_i y_i = \ldots$ | $\sum_{i=1}^{7} x_i^2 = \ldots$ | $\sum_{i=1}^{7} y_i^2 = \ldots$ |

*Complete the table.*
*The answer is in the next frame*

## 69

| $i$ | $x_i$ | $y_i$ | $x_iy_i$ | $x_i^2$ | $y_i^2$ |
|---|---|---|---|---|---|
| 1 | 6 | 21 | 126 | 36 | 441 |
| 2 | 11 | 18 | 198 | 121 | 324 |
| 3 | 12 | 33 | 396 | 144 | 1089 |
| 4 | 9 | 39 | 351 | 81 | 1521 |
| 5 | 11 | 25 | 275 | 121 | 625 |
| 6 | 7 | 27 | 189 | 49 | 729 |
| 7 | 9 | 21 | 189 | 81 | 441 |
| | | | | | |
| | $\sum_{i=1}^{7} x_i = 65$ | $\sum_{i=1}^{7} y_i = 184$ | $\sum_{i=1}^{7} x_iy_i = 1724$ | $\sum_{i=1}^{7} x_i^2 = 633$ | $\sum_{i=1}^{7} y_i^2 = 5170$ |

It is a straightforward matter to substitute these numbers into:

$$r = \frac{n\sum_{i=1}^{n} x_iy_i - \left(\sum_{i=1}^{n} x_i\right)\left(\sum_{i=1}^{n} y_i\right)}{\sqrt{\left[n\sum_{i=1}^{n} x_i^2 - \left(\sum_{i=1}^{n} x_i\right)^2\right]\left[n\sum_{i=1}^{n} y_i^2 - \left(\sum_{i=1}^{n} y_i\right)^2\right]}}$$

$$= \ldots\ldots\ldots\ldots$$

*Insert the correct value for n.*
*The answer is in the next frame*

## 70

$$\boxed{r = 0{\cdot}16}$$

Because:

The number of pairs of data is $n = 7$, so that:

$$r = \frac{7\sum_{i=1}^{7} x_iy_i - \left(\sum_{i=1}^{7} x_i\right)\left(\sum_{i=1}^{7} y_i\right)}{\sqrt{\left[7\sum_{i=1}^{7} x_i^2 - \left(\sum_{i=1}^{7} x_i\right)^2\right]\left[7\sum_{i=1}^{7} y_i^2 - \left(\sum_{i=1}^{7} y_i\right)^2\right]}}$$

Furthermore:

$$r = \frac{7(1724) - (65)(184)}{\sqrt{\left[7(633) - (65)^2\right]\left[7(5170) - (184)^2\right]}}$$

$$= \frac{108}{\sqrt{[206][2334]}}$$

$$= \frac{108}{693{\cdot}4}$$

$$= 0{\cdot}16$$

▶

So a weak correlation between the size of the workforce at any given site and the number of units manufactured is exhibited. This could be taken to imply that productivity was more dependent upon location than upon the size of the workforce. One also needs to be aware that there may be a reasonable correlation but that it is not linear.

*Move to the next frame*

71

## Spearman's rank correlation coefficient

Another method of measuring correlation that does not use the actual values of the data but rather the rankings of the data values is *Spearman's rank correlation coefficient*. For example, five colleges had their mathematics courses ranked according to their quality by two outside academics. Their rankings of the quality scores is as follows where 1 represents the highest quality and 5 the lowest:

| College | | 1 | 2 | 3 | 4 | 5 |
|---|---|---|---|---|---|---|
| Academic | A | 2 | 3 | 1 | 4 | 5 |
| | B | 1 | 4 | 2 | 3 | 5 |

In the table that now follows are tabulated the differences in the rankings and their squares:

| College | Academic | | | |
|---|---|---|---|---|
| | A | B | $d_i$ | $d_i^2$ |
| 1 | 2 | 1 | 1 | 1 |
| 2 | 3 | 4 | −1 | 1 |
| 3 | 1 | 2 | −1 | 1 |
| 4 | 4 | 3 | 1 | 1 |
| 5 | 5 | 5 | 0 | 0 |
| | | | | $\sum_{i=1}^{5} d_i^2 = 4$ |

The Spearman's rank correlation coefficient is then given as:

$$r_S = 1 - \frac{6\sum_{i=1}^{n} d_i^2}{n(n^2 - 1)}$$

so that:

$r_S = \dots\dots\dots\dots$

*Next frame*

**72**

$r_S = 0{\cdot}8$

Because:

$$r_S = 1 - \frac{6\sum_{i=1}^{5} d_i^2}{5(5^2 - 1)}$$
$$= 1 - \frac{6 \times 4}{5 \times 24}$$
$$= 1 - 0.2$$
$$= 0.8$$

There is a reasonably high correlation between the two academics and this indicates that their judgements have a good measure of agreement.

Now you try another one. The following table records the sales of two employees of an office supply company:

| Office supply sales | Value of sales | |
|---|---|---|
| | A | B |
| A4 paper | 1200 | 400 |
| Ballpoint pens | 600 | 1000 |
| Folders | 800 | 700 |
| Printer ink units | 2100 | 1800 |
| Miscellaneous item packs | 750 | 1200 |

By ranking this data the Spearman rank correlation coefficient is found to be:

$r_S = \ldots\ldots\ldots\ldots$

*Next frame*

**73**

$r_S = 0{\cdot}1$

Because:

In the table that now follows, the sales values are ranked according to size where 1 represents the highest sale, 2 the next highest and so on. Also tabulated are the differences in the rankings and their squares:

| Office supply sales | Value of sales | | | |
|---|---|---|---|---|
| | A | B | $d_i$ | $d_i^2$ |
| A4 paper | 2 | 5 | −3 | 9 |
| Ballpoint pens | 5 | 3 | 2 | 4 |
| Folders | 3 | 4 | −1 | 1 |
| Printer ink units | 1 | 1 | 0 | 0 |
| Miscellaneous item packs | 4 | 2 | 2 | 4 |
| | | | | $\sum_{i=1}^{5} d_i^2 = 18$ |

The Spearman's rank correlation coefficient is then given as:

$$r_S = 1 - \frac{6\sum_{i=1}^{n} d_i^2}{n(n^2-1)} \quad \text{so that:}$$

$$r_S = 1 - \frac{6\sum_{i=1}^{5} d_i^2}{5(5^2-1)}$$

$$= 1 - \frac{6 \times 18}{5 \times 24}$$

$$= 1 - 0{\cdot}9 = 0{\cdot}1$$

There is very little correlation between the two employees' sales – just because A sells a lot of one particular item does not mean that B will also.

*Let's now pause and summarize the work on correlation*

##  Review summary

74

1 *Standard curves* Refer back to Frames 2 to 17.

2 *Asymptotes* Rewrite the given equation, if necessary, on one line.

(a) *Asymptotes parallel to the axes*
Parallel to the $x$-axis: equate to zero the coefficient of the highest power of $x$.
Parallel to the $y$-axis: equate to zero the coefficient of the highest power of $y$.

(b) *General asymptotes*
Substitute $y = mx + c$ in the equation of the curve and equate to zero the coefficients of the two highest powers of $x$.

3 *Systematic curve sketching.*

(a) *Symmetry*
Even powers only of $x$: curve symmetrical about $y$-axis.
Even powers only of $y$: curve symmetrical about $x$-axis.
Even powers only of both $x$ and $y$: curves symmetrical about both axes.

(b) *Intersection with the axes* Put $x = 0$ and $y = 0$.

(c) *Change of origin* to simplify analysis.

(d) *Asymptotes*

(e) *Large and small values* of $x$ or $y$.

(f) *Stationary points*

(g) *Limitations* on possible range of values of $x$ or $y$.

4 *Curve fitting.*

(a) Express the law or equation in straight-line form.

(b) Plot values and draw the best straight line as the middle line of the band of points.

(c) Determine $m$ and $c$ from the $x$- and $y$-coordinates of two points *on the line*. Check with the coordinates of a third point.

(d) Line of 'best fit' can be calculated by the *method of least squares*.
Refer back to Frames 54 to 63.

(e) The Microsoft Excel™ spreadsheet has the formula =LINEST(y-values, x-values) that automatically calculates the least-square parameters $a$ and $b$ for the straight line $y = b + ax$.

▶

5 *Correlation*: If two variables vary together they are said to be correlated. Cause and effect may be the reason for the correlation but it may also be a coincidence.

6 *Measures of correlation*: The strength of the correlation between two variables is given by a correlation coefficient whose value ranges from $-1$ for perfect negative correlation to $+1$ for perfect positive correlation.

(a) *Pearson product-moment correlation coefficient* $r$: This gives the strength of a linear relationship between the $n$ values of the two variables $x_i$ and $y_i$ for $i = 1 \ldots n$, where $r$ is given by the rather daunting equation:

$$r = \frac{n\sum_{i=1}^{n} x_i y_i - \left(\sum_{i=1}^{n} x_i\right)\left(\sum_{i=1}^{n} y_i\right)}{\sqrt{\left[n\sum_{i=1}^{n} x_i^2 - \left(\sum_{i=1}^{n} x_i\right)^2\right]\left[n\sum_{i=1}^{n} y_i^2 - \left(\sum_{i=1}^{n} y_i\right)^2\right]}}$$

(b) *Spearman's rank correlation coefficient* $r_S$: By taking the ranking of the values of two variables the Spearman rank correlation coefficient measures the correlation of the ranking. The coefficient is given as:

$$r_S = 1 - \frac{6\sum_{i=1}^{n} d_i^2}{n(n^2 - 1)}$$

where the $n$ values of the two variables are ranked and $d_i$ is the difference in ranking of the $i$th pair.

---

## Can You?

**75**

### Checklist 13

Check this list before and after you try the end of Programme test.

On a scale of 1 to 5 how confident are you that you can:

| | | | | | | | Frames |
|---|---|---|---|---|---|---|---|
| • Draw sketch graphs of standard curves? | | | | | | | |
| *Yes* | ☐ | ☐ | ☐ | ☐ | ☐ | *No* | 1 to 17 |
| • Determine the equations of asymptotes parallel to the $x$- and $y$-axes? | | | | | | | |
| *Yes* | ☐ | ☐ | ☐ | ☐ | ☐ | *No* | 18 to 26 |
| • Sketch the graphs of curves with asymptotes, stationary points and other features? | | | | | | | |
| *Yes* | ☐ | ☐ | ☐ | ☐ | ☐ | *No* | 27 to 39 |
| • Fit graphs to data using the 'straight-line' forms? | | | | | | | |
| *Yes* | ☐ | ☐ | ☐ | ☐ | ☐ | *No* | 40 to 53 |
| • Fit graphs to data using the method of least squares? | | | | | | | |
| *Yes* | ☐ | ☐ | ☐ | ☐ | ☐ | *No* | 54 to 63 |
| • Understand what is meant by the correlation of two variables | | | | | | | |
| *Yes* | ☐ | ☐ | ☐ | ☐ | ☐ | *No* | 64 to 65 |
| • Calculate the Pearson product-moment correlation coefficient | | | | | | | |
| *Yes* | ☐ | ☐ | ☐ | ☐ | ☐ | *No* | 66 to 70 |
| • Calculate Spearman's rank correlation coefficient | | | | | | | |
| *Yes* | ☐ | ☐ | ☐ | ☐ | ☐ | *No* | 71 to 73 |

## Test exercise 13

76

1 Without detailed plotting of points, sketch the graphs of the following showing relevant information on the graphs:

(a) $y = (x-3)^2 + 5$
(b) $y = 4x - x^2$
(c) $4y^2 + 24y - 14 - 16x + 4x^2 = 0$
(d) $5xy = 40$
(e) $y = 6 - e^{-2x}$

2 Determine the asymptotes of the following curves:

(a) $x^2y - 9y + x^3 = 0$
(b) $y^2 = \dfrac{x(x-2)(x+4)}{x-4}$

3 Analyze and sketch the graph of the function $y = \dfrac{(x-3)(x+5)}{x+2}$

4 Express the following in 'straight-line' form and state the variables to be plotted on the $x$- and $y$-axes to give a straight line:

(a) $y = Ax + Bx^2$
(b) $y = x + Ae^{kx}$
(c) $y = \dfrac{A}{B+x}$
(d) $x^2(y^2 - 1) = k$

5 The force, $P$ newtons, required to keep an object moving at a speed, $V$ metres per second, was recorded.

| $P$ | 126 | 178 | 263 | 398 | 525 | 724 |
|---|---|---|---|---|---|---|
| $V$ | 1·86 | 2·34 | 2·75 | 3·63 | 4·17 | 4·79 |

If the law connecting $P$ and $V$ is of the form $V = aP^k$, where $a$ and $k$ are constants, apply the method of least squares to obtain the values of $a$ and $k$ that give the best fit to the given set of values.

6 Find the Pearson product-moment correlation coefficient for the following data to find the strength of the linear relationship between the two variables $x$ and $y$:

| $x$ | 25 | 56 | 144 | 225 | 625 |
|---|---|---|---|---|---|
| $y$ | 13·1 | 28·1 | 70·2 | 109 | 301 |

7 Five different manufacturers produce their own brand of pain-killer. Two individuals were then asked to rank the effectiveness of each product with the following results.

| **Make** | **1** | **2** | **3** | **4** | **5** |
|---|---|---|---|---|---|
| *Trial 1* | 5 | 3 | 4 | 1 | 2 |
| *Trial 2* | 2 | 3 | 5 | 4 | 1 |

Calculate Spearman's rank correlation coefficient for this data.

# Further problems 13

77

1 For each of the following curves, determine the asymptotes parallel to the $x$- and $y$-axes:

(a) $xy^2 + x^2 - 1 = 0$

(b) $x^2y^2 = 4(x^2 + y^2)$

(c) $y = \dfrac{x^2 - 3x + 5}{x - 3}$

(d) $y = \dfrac{x(x+4)}{(x+3)(x+2)}$

(e) $x^2y^2 - x^2 = y^2 + 1$

(f) $y^2 = \dfrac{x}{x-2}$

2 Determine all the asymptotes of each of the following curves:

(a) $x^3 - xy^2 + 4x - 16 = 0$

(b) $xy^3 - x^2y + 3x^3 - 4y^3 - 1 = 0$

(c) $y^3 + 2y^2 - x^2y + y - x + 4 = 0$

3 Analyze and sketch the graphs of the following functions:

(a) $y = x + \dfrac{1}{x}$

(b) $y = \dfrac{1}{x^2 + 1}$

(c) $y^2 = \dfrac{x}{x-2}$

(d) $y = \dfrac{(x-1)(x+4)}{(x-2)(x-3)}$

(e) $y(x+2) = (x+3)(x-4)$

(f) $x^2(y^2 - 25) = y$

(g) $xy^2 - x^2y + x + y = 2$

4 Variables $x$ and $y$ are thought to be related by the law $y = a + bx^2$. Determine the values of $a$ and $b$ that best fit the set of values given.

| $x$ | 5·0 | 7·5 | 12 | 15 | 25 |
|---|---|---|---|---|---|
| $y$ | 13·1 | 28·1 | 70·2 | 109 | 301 |

5 By plotting a suitable graph, show that $P$ and $W$ are related by a law of the form $P = a\sqrt{W} + b$, where $a$ and $b$ are constants, and determine the values of $a$ and $b$.

| $W$ | 7·0 | 10 | 15 | 24 | 40 | 60 |
|---|---|---|---|---|---|---|
| $P$ | 9·76 | 11·0 | 12·8 | 15·4 | 18·9 | 22·4 |

6 If $R = a + \dfrac{b}{d^2}$, find the best values for $a$ and $b$ from the set of corresponding values given below:

| $d$ | 0·1 | 0·2 | 0·3 | 0·5 | 0·8 | 1·5 |
|---|---|---|---|---|---|---|
| $R$ | 5·78 | 2·26 | 1·60 | 1·27 | 1·53 | 1·10 |

7 Two quantities, $x$ and $y$, are related by the law $y = \dfrac{a}{1 - bx^2}$, where $a$ and $b$ are constants. Using the values given below, draw a suitable graph and hence determine the best values of $a$ and $b$.

| $x$ | 4 | 6 | 8 | 10 | 11 | 12 |
|---|---|---|---|---|---|---|
| $y$ | 4·89 | 5·49 | 6·62 | 9·00 | 11·4 | 16·1 |

▶

8 The pressure $p$ and volume $v$ of a mass of gas in a container are related by the law $pv^n = c$, where $n$ and $c$ are constants. From the values given below, plot a suitable graph and hence determine the values of $n$ and $c$.

| $v$ | 4·60 | 7·20 | 10·1 | 15·3 | 20·4 | 30·0 |
|---|---|---|---|---|---|---|
| $p$ | 14·2 | 7·59 | 4·74 | 2·66 | 1·78 | 1·04 |

9 The current, $I$ milliamperes, in a circuit is measured for various values of applied voltage $V$ volts. If the law connecting $I$ and $V$ is $I = aV^n$, where $a$ and $n$ are constants, draw a suitable graph and determine the values of $a$ and $n$ that best fit the set of recorded values.

| $V$ | 8 | 12 | 15 | 20 | 28 | 36 |
|---|---|---|---|---|---|---|
| $I$ | 41·1 | 55·6 | 65·8 | 81·6 | 105 | 127 |

10 Values of $x$ and $y$ are thought to be related by a law of the form $y = ax + b\ln x$, where $a$ and $b$ are constants. By drawing a suitable graph, test whether this is so and determine the values of $a$ and $b$.

| $x$ | 10·4 | 32·0 | 62·8 | 95·7 | 136 | 186 |
|---|---|---|---|---|---|---|
| $y$ | 8·14 | 12·8 | 16·3 | 19·2 | 22·1 | 25·3 |

11 The following pairs of values of $x$ and $y$ are thought to satisfy the law $y = ax^2 + \frac{b}{x}$. Draw a suitable graph to confirm that this is so and determine the values of the constants $a$ and $b$.

| $x$ | 1 | 3 | 4 | 5 | 6 | 7 |
|---|---|---|---|---|---|---|
| $y$ | 5·18 | 15·9 | 27·0 | 41·5 | 59·3 | 80·4 |

12 In a test on breakdown voltages, $V$ kilovolts, for insulation of different thicknesses, $t$ millimetres, the following results were obtained:

| $t$ | 2·0 | 3·0 | 5·0 | 10 | 14 | 18 |
|---|---|---|---|---|---|---|
| $V$ | 153 | 200 | 282 | 449 | 563 | 666 |

If the law connecting $V$ and $t$ is $V = at^n$, draw a suitable graph and determine the values of the constants $a$ and $n$.

13 The torque, $T$ newton metres, required to rotate shafts of different diameters, $D$ millimetres, on a machine is shown below. If the law is $T = aD^n$, where $a$ and $n$ are constants, draw a suitable graph and hence determine the values of $a$ and $n$.

| $D$ | 7·0 | 10 | 18 | 25 | 40 |
|---|---|---|---|---|---|
| $T$ | 0·974 | 1·71 | 4·33 | 7·28 | 15·3 |

▶

14 The price of a particular commodity was reduced by different percentages in different branches of a store to determine the effect of price change on sales. The percentage changes in subsequent sales were recorded in the following table.

| Price cut (%) | 5 | 7 | 10 | 14 | 20 |
|---|---|---|---|---|---|
| Sales change (%) | 3 | 4 | 8 | 8 | 10 |

Calculate the Pearson product-moment correlation coefficient and comment on the result.

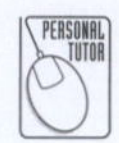


15 Ten candidates for a teaching post were ranked by two members of the interviewing panel. Their ranking is as follows:

| Candidate | | 1 | 2 | 3 | 4 | 5 | 6 | 7 | 8 | 9 | 10 |
|---|---|---|---|---|---|---|---|---|---|---|---|
| Panel member | 1 | 8 | 10 | 3 | 9 | 6 | 5 | 1 | 7 | 2 | 4 |
| | 2 | 10 | 8 | 7 | 6 | 9 | 4 | 1 | 5 | 2 | 3 |

Calculate Spearman's rank correlation coefficient and comment on the measure of agreement between the two panel members.

78

Now visit the companion website for this book at www.palgrave.com/stroud for more questions applying this mathematics to science and engineering.

# Partial differentiation 1

## Learning outcomes

When you have completed this Programme you will be able to:

- ☐ Find the first partial derivatives of a function of two real variables
- ☐ Find second-order partial derivatives of a function of two real variables
- ☐ Calculate errors using partial differentiation

# Partial differentiation

**1**

The volume $V$ of a cylinder of radius $r$ and height $h$ is given by

$V = \pi r^2 h$

i.e. $V$ depends on two quantities, the values of $r$ and $h$.

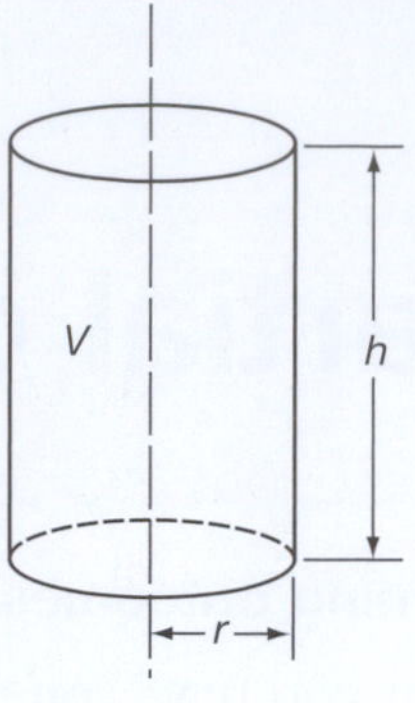


If we keep $r$ constant and increase the height $h$, the volume $V$ will increase. In these circumstances, we can consider the derivative of $V$ with respect to $h$ – but only if $r$ is kept constant.

i.e. $\left[\dfrac{dV}{dh}\right]_{r\text{ constant}}$ is written $\dfrac{\partial V}{\partial h}$

Notice the new type of 'delta'. We already know the meaning of $\dfrac{\delta y}{\delta x}$ and $\dfrac{dy}{dx}$. Now we have a new one, $\dfrac{\partial V}{\partial h}$. $\dfrac{\partial V}{\partial h}$ is called the *partial derivative* of $V$ with respect to $h$ and implies that for our present purpose, the value of $r$ is considered as being kept

...........

**2**

constant

$V = \pi r^2 h$. To find $\dfrac{\partial V}{\partial h}$, we differentiate the given expression, taking all symbols except $V$ and $h$ as being constant $\therefore \dfrac{\partial V}{\partial h} = \pi r^2.1 = \pi r^2$

Of course, we could have considered $h$ as being kept constant, in which case, a change in $r$ would also produce a change in $V$. We can therefore talk about $\dfrac{\partial V}{\partial r}$ which simply means that we now differentiate $V = \pi r^2 h$ with respect to $r$, taking all symbols except $V$ and $r$ as being constant for the time being.

$\therefore \dfrac{\partial V}{\partial r} = \pi 2rh = 2\pi rh$

In the statement $V = \pi r^2 h$, $V$ is expressed as a function of two variables, $r$ and $h$. It therefore has two partial derivatives, one with respect to ........... and one with respect to ...........

**3**

One with respect to $r$; one with respect to $h$

Another example:

Let us consider the area of the curved surface of the cylinder $A = 2\pi rh$

$A$ is a function of $r$ and $h$, so we can find $\dfrac{\partial A}{\partial r}$ and $\dfrac{\partial A}{\partial h}$

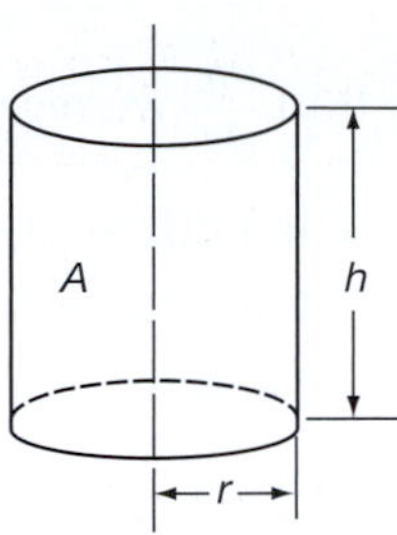


To find $\dfrac{\partial A}{\partial r}$ we differentiate the expression for $A$ with respect to $r$, keeping all other symbols constant.

To find $\dfrac{\partial A}{\partial h}$ we differentiate the expression for $A$ with respect to $h$, keeping all other symbols constant.

So, if $A = 2\pi rh$, then $\dfrac{\partial A}{\partial r} =$ ...........and $\dfrac{\partial A}{\partial h} =$ ...........

---

**4**

$$\frac{\partial A}{\partial r} = 2\pi h \text{ and } \frac{\partial A}{\partial h} = 2\pi r$$

Of course, we are not restricted to the mensuration of the cylinder. The same will happen with any function which is a function of two independent variables. For example, consider $z = x^2y^3$.

Here $z$ is a function of $x$ and $y$. We can therefore find $\dfrac{\partial z}{\partial x}$ and $\dfrac{\partial z}{\partial y}$.

(a) To find $\dfrac{\partial z}{\partial x}$, differentiate with respect to $x$, regarding $y$ as a constant.

$$\therefore \frac{\partial z}{\partial x} = 2xy^3$$

(b) To find $\dfrac{\partial z}{\partial y}$, differentiate with respect to $y$, regarding $x$ as a constant.

$$\frac{\partial z}{\partial y} = x^2 3y^2 = 3x^2y^2$$

Partial differentiation is easy! For we regard every independent variable, except the one with respect to which we are differentiating, as being for the time being ...........

---

**5**

constant

Here are some examples. 'With respect to' is abbreviated to w.r.t.

**Example 1**

$u = x^2 + xy + y^2$

(a) To find $\dfrac{\partial u}{\partial x}$, we regard $y$ as being constant.

Partial diff w.r.t. $x$ of $x^2 = 2x$

Partial diff w.r.t. $x$ of $xy = y$ ($y$ is a constant factor)

Partial diff w.r.t. $x$ of $y^2 = 0$ ($y^2$ is a constant term)

$$\frac{\partial u}{\partial x} = 2x + y$$

▶

(b) To find $\dfrac{\partial u}{\partial y}$, we regard $x$ as being constant.

Partial diff w.r.t. $y$ of $x^2 = 0$ ($x^2$ is a constant term)
Partial diff w.r.t. $y$ of $xy = x$ ($x$ is a constant factor)
Partial diff w.r.t. $y$ of $y^2 = 2y$

$$\frac{\partial u}{\partial y} = x + 2y$$

*Another example in Frame 6*

6

**Example 2**

$$z = x^3 + y^3 - 2x^2y$$

$$\frac{\partial z}{\partial x} = 3x^2 + 0 - 4xy = 3x^2 - 4xy$$

$$\frac{\partial z}{\partial y} = 0 + 3y^2 - 2x^2 = 3y^2 - 2x^2$$

And it is all just as easy as that.

**Example 3**

$$z = (2x - y)(x + 3y)$$

This is a product, and the usual product rule applies except that we keep $y$ constant when finding $\dfrac{\partial z}{\partial x}$, and $x$ constant when finding $\dfrac{\partial z}{\partial y}$.

$$\frac{\partial z}{\partial x} = (2x - y)(1 + 0) + (x + 3y)(2 - 0) = 2x - y + 2x + 6y = 4x + 5y$$

$$\frac{\partial z}{\partial y} = (2x - y)(0 + 3) + (x + 3y)(0 - 1) = 6x - 3y - x - 3y = 5x - 6y$$

Here is one for you to do.

If $z = (4x - 2y)(3x + 5y)$, find $\dfrac{\partial z}{\partial x}$ and $\dfrac{\partial z}{\partial y}$

*Find the results and then move on to Frame 7*

7

$$\frac{\partial z}{\partial x} = 24x + 14y; \qquad \frac{\partial z}{\partial y} = 14x - 20y$$

Because $z = (4x - 2y)(3x + 5y)$, i.e. product

$$\therefore \frac{\partial z}{\partial x} = (4x - 2y)(3 + 0) + (3x + 5y)(4 - 0)$$
$$= 12x - 6y + 12x + 20y = 24x + 14y$$

$$\frac{\partial z}{\partial y} = (4x - 2y)(0 + 5) + (3x + 5y)(0 - 2)$$
$$= 20x - 10y - 6x - 10y = 14x - 20y$$

There we are. Now what about this one?

▶

**Example 4**

If $z = \dfrac{2x - y}{x + y}$, find $\dfrac{\partial z}{\partial x}$ and $\dfrac{\partial z}{\partial y}$

Applying the quotient rule, we have:

$$\frac{\partial z}{\partial x} = \frac{(x+y)(2-0) - (2x-y)(1+0)}{(x+y)^2} = \frac{3y}{(x+y)^2}$$

and $$\frac{\partial z}{\partial y} = \frac{(x+y)(0-1) - (2x-y)(0+1)}{(x+y)^2} = \frac{-3x}{(x+y)^2}$$

That was not difficult. Now you do this one:

If $z = \dfrac{5x + y}{x - 2y}$, find $\dfrac{\partial z}{\partial x}$ and $\dfrac{\partial z}{\partial y}$

*When you have finished, on to the next frame*

---

**8**

$$\frac{\partial z}{\partial x} = \frac{-11y}{(x-2y)^2}; \quad \frac{\partial z}{\partial y} = \frac{11x}{(x-2y)^2}$$

Here is the working:

(a) To find $\dfrac{\partial z}{\partial x}$, we regard $y$ as being constant.

$$\therefore \frac{\partial z}{\partial x} = \frac{(x-2y)(5+0) - (5x+y)(1-0)}{(x-2y)^2}$$
$$= \frac{5x - 10y - 5x - y}{(x-2y)^2} = \frac{-11y}{(x-2y)^2}$$

(b) To find $\dfrac{\partial z}{\partial y}$, we regard $x$ as being constant.

$$\therefore \frac{\partial z}{\partial y} = \frac{(x-2y)(0+1) - (5x+y)(0-2)}{(x-2y)^2}$$
$$= \frac{x - 2y + 10x + 2y}{(x-2y)^2} = \frac{11y}{(x-2y)^2}$$

In practice, we do not write down the zeros that occur in the working, but this is how we think.

*Let us do one more example, so move on to the next frame*

---

**9**

**Example 5**

If $z = \sin(3x + 2y)$ find $\dfrac{\partial z}{\partial x}$ and $\dfrac{\partial z}{\partial y}$

Here we have what is clearly a 'function of a function'. So we apply the usual procedure, except to remember that when we are finding:

(a) $\dfrac{\partial z}{\partial x}$, we treat $y$ as constant, and

(b) $\dfrac{\partial z}{\partial y}$, we treat $x$ as constant.

▶

Here goes then.

$$\frac{\partial z}{\partial x} = \cos(3x + 2y) \times \frac{\partial}{\partial x}(3x + 2y) = \cos(3x + 2y) \times 3 = 3\cos(3x + 2y)$$
$$\frac{\partial z}{\partial y} = \cos(3x + 2y) \times \frac{\partial}{\partial y}(3x + 2y) = \cos(3x + 2y) \times 2 = 2\cos(3x + 2y)$$

There it is. So in partial differentiation, we can apply all the ordinary rules of normal differentiation, except that we regard the independent variables, other than the one we are using, for the time being as …………

**10**

constant

Fine. Now here is a short exercise for you to do by way of revision.

In each of the following cases, find $\frac{\partial z}{\partial x}$ and $\frac{\partial z}{\partial y}$:

1 $z = 4x^2 + 3xy + 5y^2$

2 $z = (3x + 2y)(4x - 5y)$

3 $z = \tan(3x + 4y)$

4 $z = \dfrac{\sin(3x + 2y)}{xy}$

*Finish them all, then move on to Frame 11 for the results*

**11**

Here are the answers:

1 $z = 4x^2 + 3xy + 5y^2$ $\quad \frac{\partial z}{\partial x} = 8x + 3y$ $\quad \frac{\partial z}{\partial y} = 3x + 10y$

2 $z = (3x + 2y)(4x - 5y)$ $\quad \frac{\partial z}{\partial x} = 24x - 7y$ $\quad \frac{\partial z}{\partial y} = -7x - 20y$

3 $z = \tan(3x + 4y)$ $\quad \frac{\partial z}{\partial x} = 3\sec^2(3x + 4y)$ $\quad \frac{\partial z}{\partial y} = 4\sec^2(3x + 4y)$

4 $z = \dfrac{\sin(3x + 2y)}{xy}$

$$\frac{\partial z}{\partial x} = \frac{3x\cos(3x + 2y) - \sin(3x + 2y)}{x^2y} \qquad \frac{\partial z}{\partial y} = \frac{2y\cos(3x + 2y) - \sin(3x + 2y)}{xy^2}$$

If you have got *all* the answers correct, turn straight on to Frame 15. If you have not got all these answers, or are at all uncertain, move to Frame 12.

**12**

Let us work through these examples in detail.

1 $z = 4x^2 + 3xy + 5y^2$

To find $\frac{\partial z}{\partial x}$, regard $y$ as a constant:

$$\therefore \frac{\partial z}{\partial x} = 8x + 3y + 0, \text{ i.e. } 8x + 3y \qquad \therefore \frac{\partial z}{\partial x} = 8x + 3y$$

Similarly, regarding $x$ as constant:

$$\frac{\partial z}{\partial y} = 0 + 3x + 10y, \text{ i.e. } 3x + 10y \qquad \therefore \frac{\partial z}{\partial y} = 3x + 10y$$

▶

2 $z = (3x + 2y)(4x - 5y)$ Product rule

$$\frac{\partial z}{\partial x} = (3x + 2y)(4) + (4x - 5y)(3)$$

$$= 12x + 8y + 12x - 15y = 24x - 7y$$

$$\frac{\partial z}{\partial y} = (3x + 2y)(-5) + (4x - 5y)(2)$$

$$= -15x - 10y + 8x - 10y = -7x - 20y$$

*Move on for the solutions to **3** and **4***

13

3 $z = \tan(3x + 4y)$

$$\frac{\partial z}{\partial x} = \sec^2(3x + 4y)(3) = 3\sec^2(3x + 4y)$$

$$\frac{\partial z}{\partial y} = \sec^2(3x + 4y)(4) = 4\sec^2(3x + 4y)$$

4 $z = \dfrac{\sin(3x + 2y)}{xy}$

$$\frac{\partial z}{\partial x} = \frac{xy\cos(3x + 2y)(3) - \sin(3x + 2y)(y)}{x^2y^2}$$

$$= \frac{3x\cos(3x + 2y) - \sin(3x + 2y)}{x^2y}$$

Now have another go at finding $\dfrac{\partial z}{\partial y}$ in the same way.

*Then check it with Frame 14*

14

Here it is:

$$z = \frac{\sin(3x + 2y)}{xy}$$

$$\therefore \frac{\partial z}{\partial y} = \frac{xy\cos(3x + 2y).(2) - \sin(3x + 2y).(x)}{x^2y^2}$$

$$= \frac{2y\cos(3x + 2y) - \sin(3x + 2y)}{xy^2}$$

That should have cleared up any troubles. This business of partial differentiation is perfectly straightforward. All you have to remember is that for the time being, all the independent variables except the one you are using are kept constant – and behave like constant factors or constant terms according to their positions.

*On you go now to Frame 15 and continue the Programme*

15

Right. Now let us move on a step.

Consider $z = 3x^2 + 4xy - 5y^2$

Then $\dfrac{\partial z}{\partial x} = 6x + 4y$ and $\dfrac{\partial z}{\partial y} = 4x - 10y$

The expression $\dfrac{\partial z}{\partial x} = 6x + 4y$ is itself a function of $x$ and $y$. We could therefore find its partial derivatives with respect to $x$ or to $y$.

▶

(a) If we differentiate it partially w.r.t. $x$, we get:

$$\frac{\partial}{\partial x}\left\{\frac{\partial z}{\partial x}\right\} \text{ and this is written } \frac{\partial^2 z}{\partial x^2}$$ (much like an ordinary second derivative, but with the partial $\partial$)

$$\therefore \frac{\partial^2 z}{\partial x^2} = \frac{\partial}{\partial x}(6x + 4y) = 6$$

This is called the second partial derivative of $z$ with respect to $x$.

(b) If we differentiate partially w.r.t. $y$, we get:

$$\frac{\partial}{\partial y}\left\{\frac{\partial z}{\partial x}\right\} \text{ and this is written } \frac{\partial^2 z}{\partial y.\partial x}$$

Note that the operation now being performed is given by the left-hand of the two symbols in the denominator.

$$\frac{\partial^2 z}{\partial y.\partial x} = \frac{\partial}{\partial y}\left\{\frac{\partial z}{\partial x}\right\} = \frac{\partial}{\partial y}\{6x + 4y\} = 4$$

**16**

So we have this:

$$z = 3x^2 + 4xy - 5y^2$$

$$\frac{\partial z}{\partial x} = 6x + 4y \qquad \frac{\partial z}{\partial y} = 4x - 10y$$

$$\frac{\partial^2 z}{\partial x^2} = 6 \qquad \frac{\partial^2 z}{\partial y.\partial x} = 4$$

Of course, we could carry out similar steps with the expression for $\frac{\partial z}{\partial y}$ on the right. This would give us:

$$\frac{\partial^2 z}{\partial y^2} = -10 \qquad \frac{\partial^2 z}{\partial x.\partial y} = 4$$

Note that $\frac{\partial^2 z}{\partial y.\partial x}$ means $\frac{\partial}{\partial y}\left\{\frac{\partial z}{\partial x}\right\}$ so $\frac{\partial^2 z}{\partial x.\partial y}$ means ...........

**17**

$$\boxed{\frac{\partial^2 z}{\partial x.\partial y} \text{ means } \frac{\partial}{\partial x}\left\{\frac{\partial z}{\partial y}\right\}}$$

Collecting our previous results together then, we have:

$$z = 3x^2 + 4xy - 5y^2$$

$$\frac{\partial z}{\partial x} = 6x + 4y \qquad \frac{\partial z}{\partial y} = 4x - 10y$$

$$\frac{\partial^2 z}{\partial x^2} = 6 \qquad \frac{\partial^2 z}{\partial y^2} = -10$$

$$\frac{\partial^2 z}{\partial y.\partial x} = 4 \qquad \frac{\partial^2 z}{\partial x.\partial y} = 4$$

We see in this case that $\frac{\partial^2 z}{\partial y.\partial x} = \frac{\partial^2 z}{\partial x.\partial y}$. There are then, *two* first derivatives and *four* second derivatives, though the last two seem to have the same value.

▶

Here is one for you to do.

If $z = 5x^3 + 3x^2y + 4y^3$, find $\frac{\partial z}{\partial x}, \frac{\partial z}{\partial y}, \frac{\partial^2 z}{\partial x^2}, \frac{\partial^2 z}{\partial y^2}, \frac{\partial^2 z}{\partial x.\partial y}$ and $\frac{\partial^2 z}{\partial y.\partial x}$

*When you have completed all that, move to Frame 18*

---

**18**

Here are the results:

$$z = 5x^3 + 3x^2y + 4y^3$$

$$\frac{\partial z}{\partial x} = 15x^2 + 6xy \qquad \frac{\partial z}{\partial y} = 3x^2 + 12y^2$$

$$\frac{\partial^2 z}{\partial x^2} = 30x + 6y \qquad \frac{\partial^2 z}{\partial y^2} = 24y$$

$$\frac{\partial^2 z}{\partial y.\partial x} = 6x \qquad \frac{\partial^2 z}{\partial x.\partial y} = 6x$$

Again in this example also, we see that $\frac{\partial^2 z}{\partial y.\partial x} = \frac{\partial^2 z}{\partial x.\partial y}$. Now do this one.

It looks more complicated, but it is done in just the same way. Do not rush at it; take your time and all will be well. Here it is. Find all the first and second partial derivatives of $z = x\cos y - y\cos x$.

*Then to Frame 19*

---

**19**

Check your results with these.

$$z = x\cos y - y\cos x$$

When differentiating w.r.t. $x$, $y$ is constant (and therefore $\cos y$ also).
When differentiating w.r.t. $y$, $x$ is constant (and therefore $\cos x$ also).

So we get:

$$\frac{\partial z}{\partial x} = \cos y + y.\sin x \qquad \frac{\partial z}{\partial y} = -x.\sin y - \cos x$$

$$\frac{\partial^2 z}{\partial x^2} = y.\cos x \qquad \frac{\partial^2 z}{\partial y^2} = -x.\cos y$$

$$\frac{\partial^2 z}{\partial y.\partial x} = -\sin y + \sin x \qquad \frac{\partial^2 z}{\partial x.\partial y} = -\sin y + \sin x$$

And again, $\frac{\partial^2 z}{\partial y.\partial x} = \frac{\partial^2 z}{\partial x.\partial y}$

In fact this will always be so for the functions you are likely to meet, so that there are really *three* different second partial derivatives (and not four). In practice, if you have found $\frac{\partial^2 z}{\partial y.\partial x}$ it is a useful check to find $\frac{\partial^2 z}{\partial x.\partial y}$ separately. They should give the same result, of course.

---

**20**

What about this one?

If $V = \ln(x^2 + y^2)$, prove that $\dfrac{\partial^2 V}{\partial x^2} + \dfrac{\partial^2 V}{\partial y^2} = 0$

This merely entails finding the two second partial derivatives and substituting them in the left-hand side of the statement. So here goes:

$$V = \ln(x^2 + y^2)$$
$$\frac{\partial V}{\partial x} = \frac{1}{(x^2 + y^2)} 2x$$
$$= \frac{2x}{x^2 + y^2}$$
$$\frac{\partial^2 V}{\partial x^2} = \frac{(x^2 + y^2)2 - 2x.2x}{(x^2 + y^2)^2}$$
$$= \frac{2x^2 + 2y^2 - 4x^2}{(x^2 + y^2)^2} = \frac{2y^2 - 2x^2}{(x^2 + y^2)^2} \qquad \text{(a)}$$

Now you find $\dfrac{\partial^2 V}{\partial y^2}$ in the same way and hence prove the given identity.

*When you are ready, move on to Frame 21*

**21**

We had found that $\dfrac{\partial^2 V}{\partial x^2} = \dfrac{2y^2 - 2x^2}{(x^2 + y^2)^2}$

So making a fresh start from $V = \ln(x^2 + y^2)$, we get:

$$\frac{\partial V}{\partial y} = \frac{1}{(x^2 + y^2)}.2y = \frac{2y}{x^2 + y^2}$$
$$\frac{\partial^2 V}{\partial y^2} = \frac{(x^2 + y^2)2 - 2y.2y}{(x^2 + y^2)^2}$$
$$= \frac{2x^2 + 2y^2 - 4y^2}{(x^2 + y^2)^2} = \frac{2x^2 - 2y^2}{(x^2 + y^2)^2} \qquad \text{(b)}$$

Substituting now the two results in the identity, gives:

$$\frac{\partial^2 V}{\partial x^2} + \frac{\partial^2 V}{\partial y^2} = \frac{2y^2 - 2x^2}{(x^2 + y^2)^2} + \frac{2x^2 - 2y^2}{(x^2 + y^2)^2}$$
$$= \frac{2y^2 - 2x^2 + 2x^2 - 2y^2}{(x^2 + y^2)^2} = 0$$

*Now on to Frame 22*

**22**

Here is another kind of example that you should see.

**Example 1**

If $V = f(x^2 + y^2)$, show that $x\dfrac{\partial V}{\partial y} - y\dfrac{\partial V}{\partial x} = 0$

Here we are told that $V$ is a function of $(x^2 + y^2)$ but the precise nature of the function is not given. However, we can treat this as a 'function of a function' and write $f'(x^2 + y^2)$ to represent the derivative of the function w.r.t. its own combined variable $(x^2 + y^2)$.

▶

$$\therefore \frac{\partial V}{\partial x} = f'(x^2+y^2) \times \frac{\partial}{\partial x}(x^2+y^2) = f'(x^2+y^2).2x$$

$$\frac{\partial V}{\partial y} = f'(x^2+y^2).\frac{\partial}{\partial y}(x^2+y^2) = f'(x^2+y^2).2y$$

$$\therefore x\frac{\partial V}{\partial y} - y\frac{\partial V}{\partial x} = x.f'(x^2+y^2).2y - y.f'(x^2+y^2).2x$$

$$= 2xy.f'(x^2+y^2) - 2xy.f'(x^2+y^2)$$

$$= 0$$

*Let us have another one of that kind in the next frame*

**23**

**Example 2**

If $z = f\left\{\frac{y}{x}\right\}$, show that $x\frac{\partial z}{\partial x} + y\frac{\partial z}{\partial y} = 0$

Much the same as before:

$$\frac{\partial z}{\partial x} = f'\left\{\frac{y}{x}\right\}.\frac{\partial}{\partial x}\left\{\frac{y}{x}\right\} = f'\left\{\frac{y}{x}\right\}\left(-\frac{y}{x^2}\right) = -\frac{y}{x^2}f'\left\{\frac{y}{x}\right\}$$

$$\frac{\partial z}{\partial y} = f'\left\{\frac{y}{x}\right\}.\frac{\partial}{\partial y}\left\{\frac{y}{x}\right\} = f'\left\{\frac{y}{x}\right\}.\frac{1}{x} = \frac{1}{x}f'\left\{\frac{y}{x}\right\}$$

$$\therefore x\frac{\partial z}{\partial x} + y\frac{\partial z}{\partial y} = x\left(-\frac{y}{x^2}\right)f'\left\{\frac{y}{x}\right\} + y\frac{1}{x}f'\left\{\frac{y}{x}\right\}$$

$$= -\frac{y}{x}f'\left\{\frac{y}{x}\right\} + \frac{y}{x}f'\left\{\frac{y}{x}\right\}$$

$$= 0$$

And one for you, just to get your hand in:

If $V = f(ax+by)$, show that $b\frac{\partial V}{\partial x} - a\frac{\partial V}{\partial y} = 0$

*When you have done it, check your working against that in Frame 24*

**24**

Here is the working; this is how it goes.

$$V = f(ax+by)$$

$$\therefore \frac{\partial V}{\partial x} = f'(ax+by).\frac{\partial}{\partial x}(ax+by)$$

$$= f'(ax+by).a = a.f'(ax+by) \qquad \text{(a)}$$

$$\frac{\partial z}{\partial y} = f'(ax+by).\frac{\partial}{\partial y}(ax+by)$$

$$= f'(ax+by).b = b.f'(ax+by) \qquad \text{(b)}$$

$$\therefore b\frac{\partial V}{\partial x} - a\frac{\partial V}{\partial y} = ab.f'(ax+by) - ab.f'(ax+by)$$

$$= 0$$

*Move on to Frame 25*

**25**

So to sum up so far.

Partial differentiation is easy, no matter how complicated the expression to be differentiated may seem.

To differentiate partially w.r.t. $x$, all independent variables other than $x$ are constant for the time being.

To differentiate partially w.r.t. $y$, all independent variables other than $y$ are constant for the time being.

So that, if $z$ is a function of $x$ and $y$, i.e. if $z = f(x, y)$, we can find:

$$\frac{\partial z}{\partial x} \quad \frac{\partial z}{\partial y}$$

$$\frac{\partial^2 z}{\partial x^2} \quad \frac{\partial^2 z}{\partial y^2}$$

$$\frac{\partial^2 z}{\partial y.\partial x} \quad \frac{\partial^2 z}{\partial x.\partial y}$$

And also: $\dfrac{\partial^2 z}{\partial y.\partial x} = \dfrac{\partial^2 z}{\partial x.\partial y}$

*Now for a revision exercise*

## Review exercise

**26**

1 Find all first and second partial derivatives for each of the following functions:

(a) $z = 3x^2 + 2xy + 4y^2$

(b) $z = \sin xy$

(c) $z = \dfrac{x+y}{x-y}$

2 If $z = \ln(e^x + e^y)$, show that $\dfrac{\partial z}{\partial x} + \dfrac{\partial z}{\partial y} = 1$.

3 If $z = x.f(xy)$, express $x\dfrac{\partial z}{\partial x} - y\dfrac{\partial z}{\partial y}$ in its simplest form.

*When you have finished, check with the solutions in Frame 27*

**27**

1 (a) $z = 3x^2 + 2xy + 4y^2$

$$\frac{\partial z}{\partial x} = 6x + 2y \qquad \frac{\partial z}{\partial y} = 2x + 8y$$

$$\frac{\partial^2 z}{\partial x^2} = 6 \qquad \frac{\partial^2 z}{\partial y^2} = 8$$

$$\frac{\partial^2 z}{\partial y.\partial x} = 2 \qquad \frac{\partial^2 z}{\partial x.\partial y} = 2$$

(b) $z = \sin xy$

$$\frac{\partial z}{\partial x} = y\cos xy \qquad \frac{\partial z}{\partial y} = x\cos xy$$

$$\frac{\partial^2 z}{\partial x^2} = -y^2\sin xy \qquad \frac{\partial^2 z}{\partial y^2} = -x^2\sin xy$$

$$\frac{\partial^2 z}{\partial y.\partial x} = y(-x\sin xy) + \cos xy \qquad \frac{\partial^2 z}{\partial x.\partial y} = x(-y\sin xy) + \cos xy$$

$$= \cos xy - xy\sin xy \qquad = \cos xy - xy\sin xy$$

▶

(c) $z = \dfrac{x+y}{x-y}$

$$\frac{\partial z}{\partial x} = \frac{(x-y)1-(x+y)1}{(x-y)^2} = \frac{-2y}{(x-y)^2}$$

$$\frac{\partial z}{\partial y} = \frac{(x-y)1-(x+y)(-1)}{(x-y)^2} = \frac{2x}{(x-y)^2}$$

$$\frac{\partial^2 z}{\partial x^2} = (-2y)\frac{(-2)}{(x-y)^3} = \frac{4y}{(x-y)^3}$$

$$\frac{\partial^2 z}{\partial y^2} = 2x\frac{(-2)}{(x-y)^3}(-1) = \frac{4x}{(x-y)^3}$$

$$\frac{\partial^2 z}{\partial y.\partial x} = \frac{(x-y)^2(-2)-(-2y)2(x-y)(-1)}{(x-y)^4}$$

$$= \frac{-2(x-y)^2-4y(x-y)}{(x-y)^4}$$

$$= \frac{-2}{(x-y)^2} - \frac{4y}{(x-y)^3}$$

$$= \frac{-2x+2y-4y}{(x-y)^3} = \frac{-2x-2y}{(x-y)^3}$$

$$\frac{\partial^2 z}{\partial x.\partial y} = \frac{(x-y)^2(2)-2x.2(x-y)1}{(x-y)^4}$$

$$= \frac{2(x-y)^2-4x(x-y)}{(x-y)^4}$$

$$= \frac{2}{(x-y)^2} - \frac{4x}{(x-y)^3}$$

$$= \frac{2x-2y-4x}{(x-y)^3} = \frac{-2x-2y}{(x-y)^3}$$

**2** $z = \ln(e^x + e^y)$

$$\frac{\partial z}{\partial x} = \frac{1}{e^x+e^y}.e^x \qquad \frac{\partial z}{\partial y} = \frac{1}{e^x+e^y}.e^y$$

$$\frac{\partial z}{\partial x} + \frac{\partial z}{\partial y} = \frac{e^x}{e^x+e^y} + \frac{e^y}{e^x+e^y}$$

$$= \frac{e^x+e^y}{e^x+e^y} = 1$$

$$\frac{\partial z}{\partial x} + \frac{\partial z}{\partial y} = 1$$

**3** $z = x.f(xy)$

$$\frac{\partial z}{\partial x} = x.f'(xy).y + f(xy)$$

$$\frac{\partial z}{\partial y} = x.f'(xy).x$$

$$x\frac{\partial z}{\partial x} - y\frac{\partial z}{\partial y} = x^2yf'(xy) + xf(xy) - x^2yf'(xy)$$

$$x\frac{\partial z}{\partial x} - y\frac{\partial z}{\partial y} = xf(xy) = z$$

▶

That was a pretty good revision test. Do not be unduly worried if you made a slip or two in your working. Try to avoid doing so, of course, but you are doing fine. Now on to the next part of the Programme.

So far we have been concerned with the technique of partial differentiation. Now let us look at one of its applications.

*So move on to Frame 28*

## Small increments

**28**

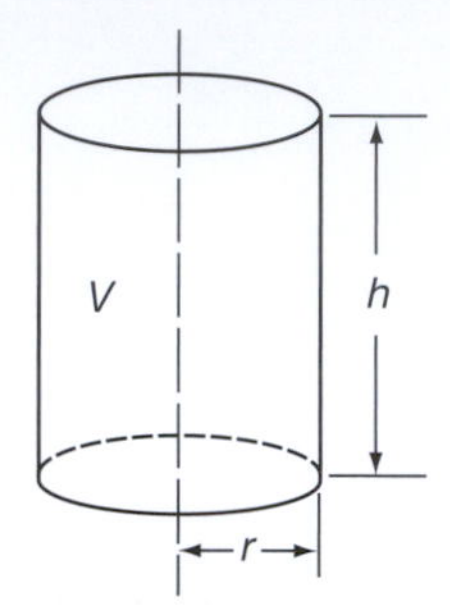


If we return to the volume of the cylinder with which we started this Programme, we have once again that $V = \pi r^2 h$. We have seen that we can find $\dfrac{\partial V}{\partial r}$ with $h$ constant, and $\dfrac{\partial V}{\partial h}$ with $r$ constant.

$$\frac{\partial V}{\partial r} = 2\pi rh; \quad \frac{\partial V}{\partial h} = \pi r^2$$

Now let us see what we get if $r$ and $h$ both change simultaneously.

If $r$ becomes $r + \delta r$, and $h$ becomes $h + \delta h$, let $V$ become $V + \delta V$. Then the new volume is given by:

$$\begin{aligned} V + \delta V &= \pi(r + \delta r)^2(h + \delta h) \\ &= \pi(r^2 + 2r\delta r + [\delta r]^2)(h + \delta h) \\ &= \pi(r^2 h + 2rh\delta r + h[\delta r]^2 + r^2\delta h + 2r\delta r\delta h + [\delta r]^2\delta h) \end{aligned}$$

Subtract $V = \pi r^2 h$ from each side, giving:

$$\delta V = \pi(2rh\delta r + h[\delta r]^2 + r^2\delta h + 2r\delta r\delta h + [\delta r]^2\delta h)$$

$\approx \pi(2rh\delta r + r^2\delta h)$ since $\delta r$ and $\delta h$ are small and all the remaining terms are of a higher degree of smallness.

Therefore

$$\delta V \approx 2\pi rh\delta r + \pi r^2\delta h,$$

that is:

$$\delta V \approx \frac{\partial V}{\partial r}\delta r + \frac{\partial V}{\partial h}\delta h$$

Let us now do a numerical example to see how it all works out.

*On to Frame 29*

29

A cylinder has dimensions $r = 5$ cm, $h = 10$ cm. Find the approximate increase in volume when $r$ increases by 0·2 cm and $h$ decreases by 0·1 cm.
Well now

$$V = \pi r^2 h \text{ so } \frac{\partial V}{\partial r} = 2\pi rh \qquad \frac{\partial V}{\partial h} = \pi r^2$$

In this case, when $r = 5$ cm, $h = 10$ cm so

$$\frac{\partial V}{\partial r} = 2\pi 5.10 = 100\pi \qquad \frac{\partial V}{\partial h} = \pi r^2 = \pi 5^2 = 25\pi$$

$\delta r = 0{\cdot}2$ and $\delta h = -0{\cdot}1$ (minus because $h$ is decreasing)

$$\therefore \delta V \approx \frac{\partial V}{\partial r}.\delta r + \frac{\partial V}{\partial h}.\delta h$$

$$\delta V = 100\pi(0{\cdot}2) + 25\pi(-0{\cdot}1)$$

$$= 20\pi - 2{\cdot}5\pi = 17{\cdot}5\pi$$

$$\therefore \delta V \approx 54{\cdot}98 \text{ cm}^3$$

i.e. the volume increases by 54·98 $\text{cm}^3$

Just like that!

30

This kind of result applies not only to the volume of the cylinder, but to any function of two independent variables. Here is an example:

If $z$ is a function of $x$ and $y$, i.e. $z = f(x, y)$ and if $x$ and $y$ increase by small amounts $\delta x$ and $\delta y$, the increase $\delta z$ will also be relatively small. If we expand $\delta z$ in powers of $\delta x$ and $\delta y$, we get:

$$\delta z = A\delta x + B\delta y + \text{higher powers of } \delta x \text{ and } \delta y,$$

where $A$ and $B$ are functions of $x$ and $y$.

If $y$ remains constant, so that $\delta y = 0$, then:

$$\delta z = A\delta x + \text{higher powers of } \delta x$$

$$\therefore \frac{\delta z}{\delta x} = A. \text{ So that if } \delta x \to 0, \text{ this becomes } A = \frac{\partial z}{\partial x}$$

Similarly, if $x$ remains constant, making $\delta y \to 0$ gives $B = \dfrac{\partial z}{\partial y}$

$$\therefore \delta z = \frac{\partial z}{\partial x}\delta x + \frac{\partial z}{\partial y}\delta y + \text{higher powers of very small quantities which can be ignored}$$

$$\delta z = \frac{\partial z}{\partial x}\delta x + \frac{\partial z}{\partial y}\delta y$$

31

So, if $z = f(x, y)$

$$\delta z = \frac{\partial z}{\partial x}\delta x + \frac{\partial z}{\partial y}\delta y$$

This is the key to all the forthcoming applications and will be quoted over and over again.

The result is quite general and a similar result applies for a function of three independent variables. For example:

If $z = f(x, y, w)$

$$\text{then } \delta z = \frac{\partial z}{\partial x}\delta x + \frac{\partial z}{\partial y}\delta y + \frac{\partial z}{\partial w}\delta w$$

▶

If we remember the rule for a function of two independent variables, we can easily extend it when necessary.

Here it is once again:

$$\text{If } z = f(x, y) \text{ then } \delta z = \frac{\partial z}{\partial x}\delta x + \frac{\partial z}{\partial y}\delta y$$

Copy this result into your record book in a prominent position, such as it deserves!

## 32

Now for a couple of examples

**Example 1**

If $I = \frac{V}{R}$, and $V = 250$ volts and $R = 50$ ohms, find the change in $I$ resulting from an increase of 1 volt in $V$ and an increase of 0·5 ohm in $R$.

$$I = f(V, R) \qquad \therefore \delta I = \frac{\partial I}{\partial V}\delta V + \frac{\partial I}{\partial R}\delta R$$

$$\frac{\partial I}{\partial V} = \frac{1}{R} \text{ and } \frac{\partial I}{\partial R} = -\frac{V}{R^2}$$

$$\therefore \delta I = \frac{1}{R}\delta V - \frac{V}{R^2}\delta R$$

So when $R = 50$, $V = 250$, $\delta V = 1$ and $\delta R = 0{\cdot}5$:

$$\begin{aligned}\delta I &= \frac{1}{50}(1) - \frac{250}{2500}(0{\cdot}5)\\ &= \frac{1}{50} - \frac{1}{20}\\ &= 0{\cdot}02 - 0{\cdot}05 = -0{\cdot}03\end{aligned}$$

i.e. $I$ decreases by 0·03 amperes

## 33

Here is another example.

**Example 2**

If $y = \frac{ws^3}{d^4}$, find the percentage increase in $y$ when $w$ increases by 2 per cent, $s$ decreases by 3 per cent and $d$ increases by 1 per cent.

Notice that, in this case, $y$ is a function of three variables, $w$, $s$ and $d$. The formula therefore becomes:

$$\delta y = \frac{\partial y}{\partial w}\delta w + \frac{\partial y}{\partial s}\delta s + \frac{\partial y}{\partial d}\delta d$$

We have

$$\frac{\partial y}{\partial w} = \frac{s^3}{d^4}; \quad \frac{\partial y}{\partial s} = \frac{3ws^2}{d^4}; \quad \frac{\partial y}{\partial d} = -\frac{4ws^3}{d^5}$$

$$\therefore \delta y = \frac{s^3}{d^4}\delta w + \frac{3ws^2}{d^4}\delta s + \frac{-4ws^3}{d^5}\delta d$$

Now then, what are the values of $\delta w, \delta s$ and $\delta d$?

Is it true to say that $\delta w = \frac{2}{100}$; $\delta s = \frac{-3}{100}$; $\delta d = \frac{1}{100}$?

If not, why not?

*Next frame*

34

No. It is not correct

Because $\delta w$ is not $\dfrac{2}{100}$ of a unit, but 2 per cent of $w$, i.e. $\delta w = \dfrac{2}{100}$ of $w = \dfrac{2w}{100}$.

Similarly, $\delta s = \dfrac{-3}{100}$ of $s = \dfrac{-3s}{100}$ and $\delta d = \dfrac{d}{100}$. Now that we have cleared that point up, we can continue with the problem.

$$\begin{aligned}\delta y &= \frac{s^3}{d^4}\left(\frac{2w}{100}\right) + \frac{3ws^2}{d^4}\left(\frac{-3s}{100}\right) - \frac{4ws^3}{d^5}\left(\frac{d}{100}\right)\\ &= \frac{ws^3}{d^4}\left(\frac{2}{100}\right) - \frac{ws^3}{d^4}\left(\frac{9}{100}\right) - \frac{ws^3}{d^4}\left(\frac{4}{100}\right)\\ &= \frac{ws^3}{d^4}\left\{\frac{2}{100} - \frac{9}{100} - \frac{4}{100}\right\}\\ &= y\left\{-\frac{11}{100}\right\} = -11 \text{ per cent of } y\end{aligned}$$

i.e. $y$ decreases by 11 per cent

Remember that where the increment of $w$ is given as 2 per cent, it is *not* $\dfrac{2}{100}$ of a unit, but $\dfrac{2}{100}$ of $w$, and the symbol $w$ must be included.

*Move on to Frame 35*

35

Now here is an exercise for you to do.

$P = w^2hd$. If errors of up to 1 per cent (plus or minus) are possible in the measured values of $w$, $h$ and $d$, find the maximum possible percentage error in the calculated values of $P$.

This is very much like the previous example, so you will be able to deal with it without any trouble. Work it right through and then go on to Frame 36 and check your result.

36

$P = w^2hd$. $\therefore \delta P = \dfrac{\partial P}{\partial w}.\delta w + \dfrac{\partial P}{\partial h}.\delta h + \dfrac{\partial P}{\partial d}.\delta d$

$\dfrac{\partial P}{\partial w} = 2whd$; $\dfrac{\partial P}{\partial h} = w^2d$; $\dfrac{\partial P}{\partial d} = w^2h$

$\delta P = 2whd.\delta w + w^2d.\delta h + w^2h.\delta d$

Now $\quad \delta w = \pm\dfrac{w}{100}$; $\delta h = \pm\dfrac{h}{100}$, $\delta d = \pm\dfrac{d}{100}$

$$\begin{aligned}\delta P &= 2whd\left(\pm\frac{w}{100}\right) + w^2d\left(\pm\frac{h}{100}\right) + w^2h\left(\pm\frac{d}{100}\right)\\ &= \pm\frac{2w^2hd}{100} \pm \frac{w^2dh}{100} \pm \frac{w^2hd}{100}\end{aligned}$$

The greatest possible error in $P$ will occur when the signs are chosen so that they are all of the same kind, i.e. all plus or minus. If they were mixed, they would tend to cancel each other out.

$$\therefore \delta P = \pm w^2hd\left\{\frac{2}{100} + \frac{1}{100} + \frac{1}{100}\right\} = \pm P\left(\frac{4}{100}\right)$$

$\therefore$ Maximum possible error in $P$ is 4 per cent of $P$

▶

Finally, here is one last example for you to do. Work right through it and then check your results with those in Frame 37.

The two sides forming the right-angle of a right-angled triangle are denoted by $a$ and $b$. The hypotenuse is $h$. If there are possible errors of $\pm 0{\cdot}5$ per cent in measuring $a$ and $b$, find the maximum possible error in calculating (a) the area of the triangle and (b) the length of $h$.

---

**37**

(a) $\delta A = 1$ per cent of $A$
(b) $\delta h = 0{\cdot}5$ per cent of $h$

Here is the working in detail:

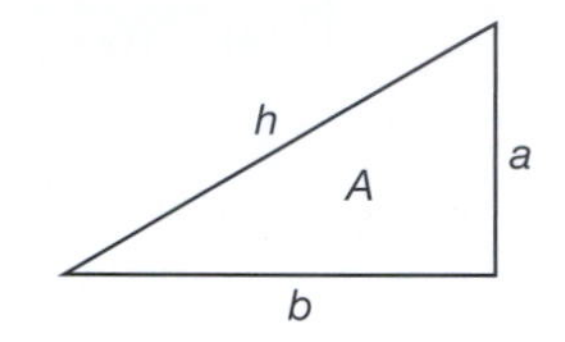


(a) $A = \dfrac{a.b}{2}$ $\qquad \delta A = \dfrac{\partial A}{\partial a}.\delta a + \dfrac{\partial A}{\partial b}.\delta b$

$$\frac{\partial A}{\partial a} = \frac{b}{2}; \quad \frac{\partial A}{\partial b} = \frac{a}{2}; \quad \delta a = \pm\frac{a}{200}; \quad \delta b = \pm\frac{b}{200}$$

$$\delta A = \frac{b}{2}\left(\pm\frac{a}{200}\right) + \frac{a}{2}\left(\pm\frac{b}{200}\right)$$

$$= \pm\frac{a.b}{2}\left[\frac{1}{200} + \frac{1}{200}\right] = \pm A.\frac{1}{100}$$

$\therefore\ \delta A = 1$ per cent of $A$

(b) $h = \sqrt{a^2 + b^2} = (a^2 + b^2)^{\frac{1}{2}}$

$$\delta h = \frac{\partial h}{\partial a}\delta a + \frac{\partial h}{\partial b}\delta b$$

$$\frac{\partial h}{\partial a} = \frac{1}{2}(a^2 + b^2)^{-\frac{1}{2}}(2a) = \frac{a}{\sqrt{a^2 + b^2}}$$

$$\frac{\partial h}{\partial b} = \frac{1}{2}(a^2 + b^2)^{-\frac{1}{2}}(2b) = \frac{b}{\sqrt{a^2 + b^2}}$$

Also $\qquad \delta a = \pm\dfrac{a}{200}; \quad \delta b = \pm\dfrac{b}{200}$

$$\therefore\ \delta h = \frac{a}{\sqrt{a^2 + b^2}}\left(\pm\frac{a}{200}\right) + \frac{b}{\sqrt{a^2 + b^2}}\left(\pm\frac{b}{200}\right)$$

$$= \pm\frac{1}{200}\frac{a^2 + b^2}{\sqrt{a^2 + b^2}}$$

$$= \pm\frac{1}{200}\sqrt{a^2 + b^2} = \pm\frac{1}{200}(h)$$

$\therefore\ \delta h = 0{\cdot}5$ per cent of $h$

That brings us to the end of this particular Programme. We shall meet partial differentiation again in the next Programme when we shall consider some more of its applications. But for the time being, there remain only the **Can You?** checklist and the **Test exercise**.

*So on now to Frames 38 and 39*

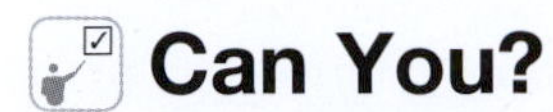

# Can You?

**Checklist 14** **38**

Check this list before and after you try the end of Programme test.

On a scale of 1 to 5 how confident you are that you can: Frames

- Find the first partial derivatives of a function of two real variables? 1 to 14
  *Yes* ☐ ☐ ☐ ☐ ☐ *No*
- Find second-order partial derivatives of a function of two real variables? 15 to 25
  *Yes* ☐ ☐ ☐ ☐ ☐ *No*
- Calculate errors using partial differentiation? 28 to 37
  *Yes* ☐ ☐ ☐ ☐ ☐ *No*

# Test exercise 14

**39**

Take your time over the questions; do them carefully.

1 Find all first and second partial derivatives of the following:

(a) $z = 4x^3 - 5xy^2 + 3y^3$

(b) $z = \cos(2x + 3y)$

(c) $z = e^{x^2 - y^2}$

(d) $z = x^2 \sin(2x + 3y)$

2 (a) If $V = x^2 + y^2 + z^2$, express in its simplest form

$$x\frac{\partial V}{\partial x} + y\frac{\partial V}{\partial y} + z\frac{\partial V}{\partial z}.$$

(b) If $z = f(x + ay) + F(x - ay)$, find $\frac{\partial^2 z}{\partial x^2}$ and $\frac{\partial^2 z}{\partial y^2}$ and hence prove that

$$\frac{\partial^2 z}{\partial y^2} = a^2 \cdot \frac{\partial^2 z}{\partial x^2}.$$

3 The power $P$ dissipated in a resistor is given by $P = \frac{E^2}{R}$.

If $E = 200$ volts and $R = 8$ ohms, find the change in $P$ resulting from a drop of 5 volts in $E$ and an increase of 0·2 ohm in $R$.

4 If $\theta = kHLV^{-\frac{1}{2}}$, where $k$ is a constant, and there are possible errors of $\pm 1$ per cent in measuring $H$, $L$ and $V$, find the maximum possible error in the calculated value of $\theta$.

*That's it*

## Further problems 14

**40**

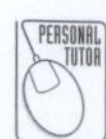

1 If $z = \dfrac{1}{x^2 + y^2 - 1}$, show that $x\dfrac{\partial z}{\partial x} + y\dfrac{\partial z}{\partial y} = -2z(1+z)$.

2 Prove that, if $V = \ln(x^2 + y^2)$, then $\dfrac{\partial^2 V}{\partial x^2} + \dfrac{\partial^2 V}{\partial y^2} = 0$.

3 If $z = \sin(3x + 2y)$, verify that $3\dfrac{\partial^2 z}{\partial y^2} - 2\dfrac{\partial^2 z}{\partial x^2} = 6z$.

4 If $u = \dfrac{x + y + z}{(x^2 + y^2 + z^2)^{\frac{1}{2}}}$, show that $x\dfrac{\partial u}{\partial x} + y\dfrac{\partial u}{\partial y} + z\dfrac{\partial u}{\partial z} = 0$.

5 Show that the equation $\dfrac{\partial^2 z}{\partial x^2} + \dfrac{\partial^2 z}{\partial y^2} = 0$, is satisfied by

$$z = \ln\sqrt{x^2 + y^2} + \frac{1}{2}\tan^{-1}\left(\frac{y}{x}\right)$$

6 If $z = e^x(x\cos y - y\sin y)$, show that $\dfrac{\partial^2 z}{\partial x^2} + \dfrac{\partial^2 z}{\partial y^2} = 0$.

7 If $u = (1 + x)\sin(5x - 2y)$, verify that $4\dfrac{\partial^2 u}{\partial x^2} + 20\dfrac{\partial^2 u}{\partial x.\partial y} + 25\dfrac{\partial^2 u}{\partial y^2} = 0$.

8 If $z = f\left(\dfrac{y}{x}\right)$, show that $x^2\dfrac{\partial^2 z}{\partial x^2} + 2xy\dfrac{\partial^2 z}{\partial x.\partial y} + y^2\dfrac{\partial^2 z}{\partial y^2} = 0$.

9 If $z = (x + y).f\left(\dfrac{y}{x}\right)$, where $f$ is an arbitrary function, show that

$$x\frac{\partial z}{\partial x} + y\frac{\partial z}{\partial y} = z.$$

10 In the formula $D = \dfrac{Eh^3}{12(1 - v^2)}$, $h$ is given as $0{\cdot}1 \pm 0{\cdot}002$ and $v$ as $0{\cdot}3 \pm 0{\cdot}02$. Express the approximate maximum error in $D$ in terms of $E$.

11 The formula $z = \dfrac{a^2}{x^2 + y^2 - a^2}$ is used to calculate $z$ from observed values of $x$ and $y$. If $x$ and $y$ have the same percentage error $p$, show that the percentage error in $z$ is approximately $-2p(1 + z)$.

12 In a balanced bridge circuit, $R_1 = R_2R_3/R_4$. If $R_2, R_3, R_4$ have known tolerances of $\pm x$ per cent, $\pm y$ per cent, $\pm z$ per cent respectively, determine the maximum percentage error in $R_1$, expressed in terms of $x$, $y$ and $z$.

13 The deflection $y$ at the centre of a circular plate suspended at the edge and uniformly loaded is given by $y = \dfrac{kwd^4}{t^3}$, where $w$ = total load, $d$ = diameter of plate, $t$ = thickness and $k$ is a constant.

Calculate the approximate percentage change in $y$ if $w$ is increased by 3 per cent, $d$ is decreased by $2\frac{1}{2}$ per cent and $t$ is increased by 4 per cent.

14 The coefficient of rigidity $(n)$ of a wire of length $(L)$ and uniform diameter $(d)$ is given by $n = \dfrac{AL}{d^4}$, where $A$ is a constant. If errors of $\pm 0{\cdot}25$ per cent and $\pm 1$ per cent are possible in measuring $L$ and $d$ respectively, determine the maximum percentage error in the calculated value of $n$.

▶

**15** If $k/k_0 = (T/T_0)^n.p/760$, show that the change in $k$ due to small changes of $a$ per cent in $T$ and $b$ per cent in $p$ is approximately $(na + b)$ per cent.

**16** The deflection $y$ at the centre of a rod is known to be given by $y = \dfrac{kwl^3}{d^4}$, where $k$ is a constant. If $w$ increases by 2 per cent, $l$ by 3 per cent, and $d$ decreases by 2 per cent, find the percentage increase in $y$.

**17** The displacement $y$ of a point on a vibrating stretched string, at a distance $x$ from one end, at time $t$, is given by

$$\frac{\partial^2 y}{\partial t^2} = c^2.\frac{\partial^2 y}{\partial x^2}$$

Show that one solution of this equation is $y = A\sin\dfrac{px}{c}.\sin(pt + a)$, where $A$, $p$, $c$ and $a$ are constants.

**18** If $y = A\sin(px + a)\cos(qt + b)$, find the error in $y$ due to small errors $\delta x$ and $\delta t$ in $x$ and $t$ respectively.

**19** Show that $\phi = Ae^{-kt/2}\sin pt\cos qx$, satisfies the equation

$$\frac{\partial^2\phi}{\partial x^2} = \frac{1}{c^2}\left\{\frac{\partial^2\phi}{\partial t^2} + k\frac{\partial\phi}{\partial t}\right\}, \text{ provided that } p^2 = c^2q^2 - \frac{k^2}{4}.$$

**20** Show that (a) the equation $\dfrac{\partial^2 V}{\partial x^2} + \dfrac{\partial^2 V}{\partial y^2} + \dfrac{\partial^2 V}{\partial z^2} = 0$ is satisfied by $V = \dfrac{1}{\sqrt{x^2 + y^2 + z^2}}$, and that (b) the equation $\dfrac{\partial^2 V}{\partial x^2} + + \dfrac{\partial^2 V}{\partial y^2} = 0$ is satisfied by $V = \tan^{-1}\left(\dfrac{y}{x}\right)$.

Now visit the companion website for this book at www.palgrave.com/stroud for more questions applying this mathematics to science and engineering. **41**

# Partial differentiation 2

## Learning outcomes

When you have completed this Programme you will be able to:

- ☐ Derive the first- and second-order partial derivatives of a function of two real variables
- ☐ Apply partial differentiation to rate-of-change problems
- ☐ Apply partial differentiation to change-of-variable problems

# Partial differentiation

**1**

In the first part of the Programme on partial differentiation, we established a result which, we said, would be the foundation of most of the applications of partial differentiation to follow.

You surely remember it: it went like this:

If $z$ is a function of two independent variables, $x$ and $y$, i.e. if $z = f(x, y)$, then

$$\delta z = \frac{\partial z}{\partial x}\delta x + \frac{\partial z}{\partial y}\delta y$$

We were able to use it, just as it stands, to work out certain problems on small increments, errors and tolerances. It is also the key to much of the work of this Programme, so copy it down into your record book, thus:

If $z = f(x, y)$ then $\delta z = \dfrac{\partial z}{\partial x}\delta x + \dfrac{\partial z}{\partial y}\delta y$

**2**

If $z = f(x, y)$, then $\delta z = \dfrac{\partial z}{\partial x}\delta x + \dfrac{\partial z}{\partial y}\delta y$

In this expression, $\dfrac{\partial z}{\partial x}$ and $\dfrac{\partial z}{\partial y}$ are the partial derivatives of $z$ with respect to $x$ and $y$ respectively, and you will remember that to find:

(a) $\dfrac{\partial z}{\partial x}$, we differentiate the function $z$, with respect to $x$, keeping all independent variables other than $x$, for the time being, . . . . . . . . . . .

(b) $\dfrac{\partial z}{\partial y}$, we differentiate the function $z$ with respect to $y$, keeping all independent variables other than $y$, for the time being, . . . . . . . . . . .

**3**

constant (in both cases)

An example, just to remind you:

If $z = x^3 + 4x^2y - 3y^3$

then $\dfrac{\partial z}{\partial x} = 3x^2 + 8xy - 0$ ($y$ is constant)

and $\dfrac{\partial z}{\partial y} = 0 + 4x^2 - 9y^2$ ($x$ is constant)

In practice, of course, we do not write down the zero terms.

Before we tackle any further applications, we must be expert at finding partial derivatives, so with the reminder above, have a go at this one.

(1) If $z = \tan(x^2 - y^2)$, find $\dfrac{\partial z}{\partial x}$ and $\dfrac{\partial z}{\partial y}$

*When you have finished it, check with the next frame*

**4**

$$\frac{\partial z}{\partial x} = 2x \sec^2(x^2 - y^2); \quad \frac{\partial z}{\partial y} = -2y \sec^2(x^2 - y^2)$$

Because

$$\begin{aligned} z &= \tan(x^2 - y^2) \\ \therefore \frac{\partial z}{\partial x} &= \sec^2(x^2 - y^2) \times \frac{\partial}{\partial x}(x^2 - y^2) \\ &= \sec^2(x^2 - y^2)(2x) \\ &= 2x \sec^2(x^2 - y^2) \end{aligned}$$

and

$$\begin{aligned} \frac{\partial z}{\partial y} &= \sec^2(x^2 - y^2) \times \frac{\partial}{\partial y}(x^2 - y^2) \\ &= \sec^2(x^2 - y^2)(-2y) \\ &= -2y \sec^2(x^2 - y^2) \end{aligned}$$

That was easy enough. Now do this one:

(2) If $z = e^{2x-3y}$, find $\dfrac{\partial^2 z}{\partial x^2}, \dfrac{\partial^2 z}{\partial y^2}, \dfrac{\partial^2 z}{\partial x.\partial y}$

*Finish them all. Then move on to Frame 5 and check your results*

**5**

Here are the results in detail:

$$z = e^{2x-3y} \qquad \therefore \frac{\partial z}{\partial x} = e^{2x-3y}.2 = 2.e^{2x-3y}$$

$$\frac{\partial z}{\partial y} = e^{2x-3y}(-3) = -3.e^{2x-3y}$$

$$\frac{\partial^2 z}{\partial x^2} = 2.e^{2x-3y}.2 = 4.e^{2x-3y}$$

$$\frac{\partial^2 z}{\partial y^2} = -3.e^{2x-3y}(-3) = 9.e^{2x-3y}$$

$$\frac{\partial^2 z}{\partial x.\partial y} = -3.e^{2x-3y}.2 = -6.e^{2x-3y}$$

All correct?

You remember, too, that in the 'mixed' second partial derivative, the order of differentiating does not matter. So in this case, since

$$\frac{\partial^2 z}{\partial x.\partial y} = -6.e^{2x-3y}, \text{ then } \frac{\partial^2 z}{\partial y.\partial x} = \dots\dots\dots\dots$$

**6**

$$\frac{\partial^2 z}{\partial x.\partial y} = \frac{\partial^2 z}{\partial y.\partial x} = -6.e^{2x-3y}$$

Well now, before we move on to new work, see what you make of these.

Find all the first and second partial derivatives of the following:

(a) $z = x \sin y$

(b) $z = (x + y) \ln(xy)$

*When you have found all the derivatives, check your work with the solutions in the next frame*

7

Here they are. Check your results carefully.

(a) $z = x\sin y$

$$\therefore \frac{\partial z}{\partial x} = \sin y \qquad \frac{\partial z}{\partial y} = x\cos y$$

$$\frac{\partial^2 z}{\partial x^2} = 0 \qquad \frac{\partial^2 z}{\partial y^2} = -x\sin y$$

$$\frac{\partial^2 z}{\partial y.\partial x} = \cos y \qquad \frac{\partial^2 z}{\partial x.\partial y} = \cos y$$

(b) $z = (x+y)\ln(xy)$

$$\therefore \frac{\partial z}{\partial x} = (x+y)\frac{1}{xy}.y + \ln(xy) = \frac{(x+y)}{x} + \ln(xy)$$

$$\frac{\partial z}{\partial y} = (x+y)\frac{1}{xy}.x + \ln(xy) = \frac{(x+y)}{y} + \ln(xy)$$

$$\therefore \frac{\partial^2 z}{\partial x^2} = \frac{x-(x+y)}{x^2} + \frac{1}{xy}.y = \frac{x-x-y}{x^2} + \frac{1}{x}$$

$$= \frac{x-y}{x^2}$$

$$\frac{\partial^2 z}{\partial y^2} = \frac{y-(x+y)}{y^2} + \frac{1}{xy}.x = \frac{y-x-y}{y^2} + \frac{1}{y}$$

$$= \frac{y-x}{y^2}$$

$$\frac{\partial^2 z}{\partial y.\partial x} = \frac{1}{x} + \frac{1}{xy}.x = \frac{1}{x} + \frac{1}{y}$$

$$= \frac{y+x}{xy}$$

$$\frac{\partial^2 z}{\partial x.\partial y} = \frac{1}{y} + \frac{1}{xy}.y = \frac{1}{y} + \frac{1}{x}$$

$$= \frac{x+y}{xy}$$

Well now, that was just by way of warming up with work you have done before. Let us now move on to the next section of this Programme.

## Rate-of-change problems

8

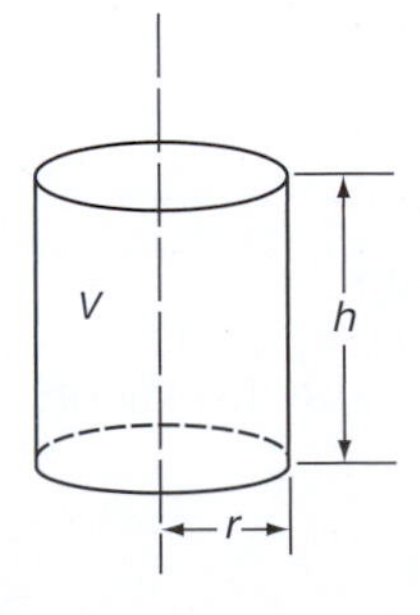


Let us consider a cylinder of radius $r$ and height $h$ as before. Then the volume is given by

$$V = \pi r^2 h$$

$$\therefore \frac{\partial V}{\partial r} = 2\pi rh \quad \text{and} \quad \frac{\partial V}{\partial h} = \pi r^2$$

▶

Since $V$ is a function of $r$ and $h$, we also know that

$$\delta V = \frac{\partial V}{\partial r}\delta r + \frac{\partial V}{\partial h}\delta h$$

(Here it is, popping up again!)

Now divide both sides by $\delta t$: $\dfrac{\delta V}{\delta t} = \dfrac{\partial V}{\partial r}\cdot\dfrac{\delta r}{\delta t} + \dfrac{\partial V}{\partial h}\cdot\dfrac{\delta h}{\delta t}$

Then if $\delta t \to 0$, $\dfrac{\delta V}{\delta t} \to \dfrac{dV}{dt}$, $\dfrac{\delta r}{\delta t} \to \dfrac{dr}{dt}$, $\dfrac{\delta h}{\delta t} \to \dfrac{dh}{dt}$, but the partial derivatives, which do not contain $\delta t$, will remain unchanged.

So our result now becomes $\dfrac{dV}{dt} = \ldots\ldots\ldots\ldots$

**9**

$$\boxed{\frac{dV}{dt} = \frac{\partial V}{\partial r}\cdot\frac{dr}{dt} + \frac{\partial V}{\partial h}\cdot\frac{dh}{dt}}$$

This result is really the key to problems of the kind we are about to consider. If we know the rate at which $r$ and $h$ are changing, we can now find the corresponding rate of change of $V$. Like this:

**Example 1**

The radius of a cylinder increases at the rate of 0·2 cm/s while the height decreases at the rate of 0·5 cm/s. Find the rate at which the volume is changing at the instant when $r = 8$ cm and $h = 12$ cm.

*Warning*: The first inclination is to draw a diagram and to put in the given values for its dimensions, i.e. $r = 8$ cm, $h = 12$ cm. This we *must NOT do*, for the radius and height are changing and the given values are instantaneous values only. Therefore on the diagram we keep the symbols $r$ and $h$ to indicate that they are variables.

**10**

Here it is then:

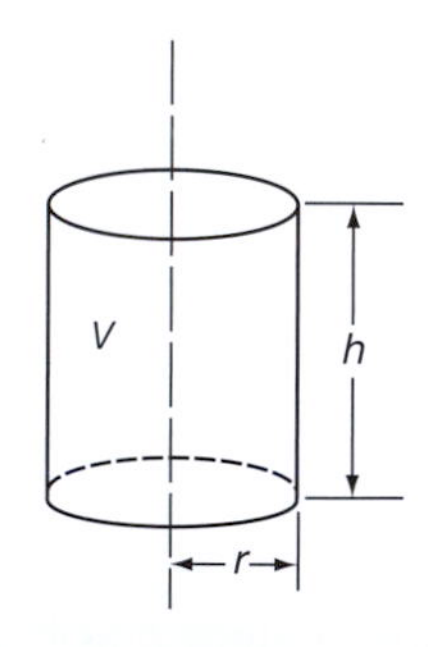


$$V = \pi r^2 h$$

$$\delta V = \frac{\partial V}{\partial r}\delta r + \frac{\partial V}{\partial h}\delta h$$

$$\therefore \frac{dV}{dt} = \frac{\partial V}{\partial r}\cdot\frac{dr}{dt} + \frac{\partial V}{\partial h}\cdot\frac{dh}{dt}$$

$$\frac{\partial V}{\partial r} = 2\pi rh; \quad \frac{\partial V}{\partial h} = \pi r^2$$

$$\therefore \frac{dV}{dt} = 2\pi rh\frac{dr}{dt} + \pi r^2\frac{dh}{dt}$$

Now at the instant we are considering:

$r = 8$, $h = 12$, $\dfrac{dr}{dt} = 0{\cdot}2$, $\dfrac{dh}{dt} = -0{\cdot}5$ (minus since $h$ is decreasing)

So you can now substitute these values in the last statement and finish off the calculation, giving:

$\dfrac{dV}{dt} = \ldots\ldots\ldots\ldots$

11

$$\frac{dV}{dt} = 20{\cdot}1 \text{ cm}^3/\text{s}$$

Because

$$\begin{aligned}\frac{dV}{dt} &= 2\pi rh.\frac{dr}{dt} + \pi r^2\frac{dh}{dt}\\ &= 2\pi 8.12.(0{\cdot}2) + \pi 64(-0{\cdot}5)\\ &= 38{\cdot}4\pi - 32\pi\\ &= 6{\cdot}4\pi = 20{\cdot}1 \text{ cm}^3/\text{s}\end{aligned}$$

Now another one.

**Example 2**

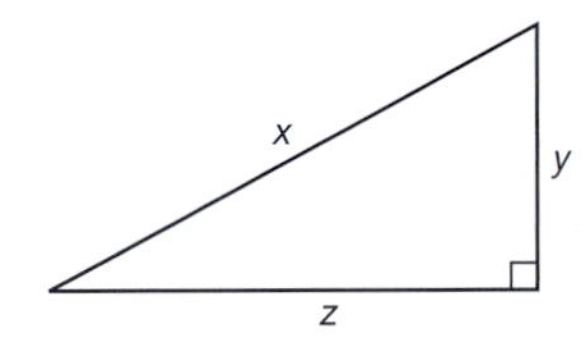


In the right-angled triangle shown, $x$ is increasing at 2 cm/s while $y$ is decreasing at 3 cm/s. Calculate the rate at which $z$ is changing when $x = 5$ cm and $y = 3$ cm.

The first thing to do, of course, is to express $z$ in terms of $x$ and $y$. That is not difficult.

$z = \dots\dots\dots\dots$

12

$$z = \sqrt{x^2 - y^2}$$

$$z = \sqrt{x^2 - y^2} = (x^2 - y^2)^{\frac{1}{2}}$$

$$\therefore\ \delta z = \frac{\partial z}{\partial x}\delta x + \frac{\partial z}{\partial y}\delta y \qquad \text{(The key to the whole business)}$$

$$\therefore\ \frac{dz}{dt} = \frac{\partial z}{\partial x}.\frac{dx}{dt} + \frac{\partial z}{\partial y}.\frac{dy}{dt}$$

In this case $\dfrac{\partial z}{\partial x} = \dfrac{1}{2}(x^2 - y^2)^{-\frac{1}{2}}(2x) = \dfrac{x}{\sqrt{x^2 - y^2}}$

$$\frac{\partial z}{\partial y} = \frac{1}{2}(x^2 - y^2)^{-\frac{1}{2}}(-2y) = \frac{-y}{\sqrt{x^2 - y^2}}$$

$$\frac{dz}{dt} = \frac{x}{\sqrt{x^2 - y^2}}.\frac{dx}{dt} - \frac{y}{\sqrt{x^2 - y^2}}.\frac{dy}{dt}$$

So far so good. Now for the numerical values:

$$x = 5,\ y = 3,\ \frac{dx}{dt} = 2,\ \frac{dy}{dt} = -3$$

$$\frac{dz}{dt} = \dots\dots\dots\dots$$

*Finish it off, then move to Frame 13*

**13**

$$\boxed{\frac{dz}{dt} = 4{\cdot}75 \text{ cm/s}}$$

Because we have $\dfrac{dz}{dt} = \dfrac{5}{\sqrt{5^2 - 3^2}}(2) - \dfrac{3}{\sqrt{5^2 - 3^2}}(-3)$

$$= \frac{5(2)}{4} + \frac{3(3)}{4} = \frac{10}{4} + \frac{9}{4} = \frac{19}{4} = 4{\cdot}75 \text{ cm/s}$$

$\therefore$ Side $z$ increases at the rate of 4· 75 cm/s

Now here is

**Example 3**

The total surface area $S$ of a cone of base radius $r$ and perpendicular height $h$ is given by

$$S = \pi r^2 + \pi r\sqrt{r^2 + h^2}$$

If $r$ and $h$ are each increasing at the rate of 0·25 cm/s, find the rate at which $S$ is increasing at the instant when $r = 3$ cm and $h = 4$ cm.

Do that one entirely on your own. Take your time: there is no need to hurry. Be quite sure that each step you write down is correct.

*Then move to Frame 14 and check your result*

**14**

Here is the solution in detail:

$$S = \pi r^2 + \pi r\sqrt{r^2 + h^2} = \pi r^2 + \pi r(r^2 + h^2)^{\frac{1}{2}}$$

$$\delta S = \frac{\partial S}{\partial r}.\delta r + \frac{\partial S}{\partial h}.\delta h \quad \therefore \quad \frac{dS}{dt} = \frac{\partial S}{\partial r}.\frac{dr}{dt} + \frac{\partial S}{\partial h}.\frac{dh}{dt}$$

(1) $$\frac{\partial S}{\partial r} = 2\pi r + \pi r.\frac{1}{2}(r^2 + h^2)^{-\frac{1}{2}}(2r) + \pi(r^2 + h^2)^{\frac{1}{2}}$$

$$= 2\pi r + \frac{\pi r^2}{\sqrt{r^2 + h^2}} + \pi\sqrt{r^2 + h^2}$$

When $r = 3$ and $h = 4$:

$$\frac{\partial S}{\partial r} = 2\pi 3 + \frac{\pi 9}{5} + \pi 5 = 11\pi + \frac{9\pi}{5} = \frac{64\pi}{5}$$

(2) $$\frac{\partial S}{\partial h} = \pi r\frac{1}{2}(r^2 + h^2)^{-\frac{1}{2}}(2h) = \frac{\pi rh}{\sqrt{r^2 + h^2}}$$

$$= \frac{\pi 3.4}{5} = \frac{12\pi}{5}$$

Also we are given that $\dfrac{dr}{dt} = 0{\cdot}25$ and $\dfrac{dh}{dt} = 0{\cdot}25$

$$\therefore \frac{dS}{dt} = \frac{64\pi}{5}.\frac{1}{4} + \frac{12\pi}{5}.\frac{1}{4}$$

$$= \frac{16\pi}{5} + \frac{3\pi}{5} = \frac{19\pi}{5}$$

$$= 3{\cdot}8\pi = 11{\cdot}94 \text{ cm}^2\text{/s}$$

15

So there we are. Rate-of-change problems are all very much the same. What you must remember is simply this:

(a) The basic statement

$$\text{If } z = f(x, y) \text{ then } \delta z = \frac{\partial z}{\partial x}.\delta x + \frac{\partial z}{\partial y}.\delta y \qquad \text{(a)}$$

(b) Divide this result by $\delta t$ and make $\delta t \to 0$. This converts the result into the form for rate-of-change problems:

$$\frac{dz}{dt} = \frac{\partial z}{\partial x}\cdot\frac{dx}{dt} + \frac{\partial z}{\partial y}\cdot\frac{dy}{dt} \qquad \text{(b)}$$

The second result follows directly from the first. Make a note of both of these in your record book for future reference.

*Then for the next part of the work, move on to Frame 16*

16

Partial differentiation can also be used with advantage in finding *derivatives of implicit functions*.

For example, suppose we are required to find an expression for $\frac{dy}{dx}$ when we are given that $x^2 + 2xy + y^3 = 0$.

We can set about it in this way:

Let $z$ stand for the function of $x$ and $y$, i.e. $z = x^2 + 2xy + y^3$. Again we use the basic relationship $\delta z = \frac{\partial z}{\partial x}\delta x + \frac{\partial z}{\partial y}\delta y$.

If we divide both sides by $\delta x$, we get:

$$\frac{\delta z}{\delta x} = \frac{\partial z}{\partial x} + \frac{\partial z}{\partial y}\cdot\frac{\delta y}{\delta x}$$

$$\text{Now, if } \delta x \to 0, \quad \frac{dz}{dx} = \frac{\partial z}{\partial x} + \frac{\partial z}{\partial y}\cdot\frac{dy}{dx}$$

If we now find expressions for $\frac{\partial z}{\partial x}$ and $\frac{\partial z}{\partial y}$, we shall be quite a way towards finding $\frac{dy}{dx}$ (which you see at the end of the expression).

In this particular example, where $z = x^2 + 2xy + y^3$,

$\frac{\partial z}{\partial x} = \ldots\ldots\ldots\ldots$ and $\frac{\partial z}{\partial y} = \ldots\ldots\ldots\ldots$

## 17

$$\frac{\partial z}{\partial x} = 2x + 2y; \quad \frac{\partial z}{\partial y} = 2x + 3y^2$$

Substituting these in our previous result gives us:

$$\frac{dz}{dx} = (2x + 2y) + (2x + 3y^2)\frac{dy}{dx}$$

If only we knew $\frac{dz}{dx}$, we could rearrange this result and obtain an expression for $\frac{dy}{dx}$.

So where can we find out something about $\frac{dz}{dx}$?

Refer back to the beginning of the problem. We have used $z$ to stand for $x^2 + 2xy + y^3$ and we were told initially that $x^2 + 2xy + y^3 = 0$. Therefore $z = 0$, i.e. $z$ is a constant (in this case zero) and hence $\frac{dz}{dx} = 0$.

$$\therefore 0 = (2x + 2y) + (2x + 3y^2)\frac{dy}{dx}$$

From this we can find $\frac{dy}{dx}$. So finish it off.

$$\frac{dy}{dx} = \dots\dots\dots\dots$$

*On to Frame 18*

## 18

$$\frac{dy}{dx} = -\frac{2x + 2y}{2x + 3y^2}$$

This is almost a routine that always works. In general, we have:

If $f(x, y) = 0$, find $\frac{dy}{dx}$

Let $z = f(x, y)$ then $\delta z = \frac{\partial z}{\partial x}\delta x + \frac{\partial z}{\partial y}\delta y$. Divide by $\delta x$ and make $\delta x \to 0$, in which case:

$$\frac{dz}{dx} = \frac{\partial z}{\partial x} + \frac{\partial z}{\partial y}.\frac{dy}{dx}$$

But $z = 0$ (constant) $\therefore \frac{dz}{dx} = 0 \quad \therefore 0 = \frac{\partial z}{\partial x} + \frac{\partial z}{\partial y}.\frac{dy}{dx}$

giving $\frac{dy}{dx} = -\frac{\partial z}{\partial x}\bigg/\frac{\partial z}{\partial y}$

The easiest form to remember is the one that comes direct from the basic result:

$$\delta z = \frac{\partial z}{\partial x}\delta x + \frac{\partial z}{\partial y}\delta y$$

Divide by $\delta x$, etc.

$$\frac{dz}{dx} = \frac{\partial z}{\partial x} + \frac{\partial z}{\partial y}.\frac{dy}{dx} \quad \left\{\frac{dz}{dx} = 0\right\}$$

Make a note of this result.

**19**

Now for some examples.

**Example 1**

If $e^{xy}+x+y=1$, evaluate $\dfrac{dy}{dx}$ at (0, 0), The function can be written $e^{xy}+x+y-1=0$.

$$\text{Let } z=e^{xy}+x+y-1 \qquad \delta z=\frac{\partial z}{\partial x}.\delta x+\frac{\partial z}{\partial y}.\delta y \qquad \therefore\ \frac{dz}{dx}=\frac{\partial z}{\partial x}+\frac{\partial z}{\partial y}\cdot\frac{dy}{dx}$$

$$\frac{\partial z}{\partial x}=e^{xy}.y+1; \quad \frac{\partial z}{\partial y}=e^{xy}.x+1 \qquad \therefore\ \frac{dz}{dx}=(y.e^{xy}+1)+(x.e^{xy}+1)\frac{dy}{dx}$$

$$\text{But } z=0 \quad \therefore\ \frac{dz}{dx}=0 \quad \therefore\ \frac{dy}{dx}=-\left\{\frac{y.e^{xy}+1}{x.e^{xy}+1}\right\}$$

$$\text{At } x=0,\ y=0,\ \frac{dy}{dx}=-\frac{1}{1}=-1 \qquad \therefore\ \frac{dy}{dx}=-1$$

All very easy so long as you can find partial derivatives correctly.

*On to Frame 20*

**20**

Now here is:

**Example 2**

If $xy+\sin y=2$, find $\dfrac{dy}{dx}$

Let $z=xy+\sin y-2=0$

$$\delta z=\frac{\partial z}{\partial x}\delta x+\frac{\partial z}{\partial y}\delta y$$

$$\frac{dz}{dx}=\frac{\partial z}{\partial x}+\frac{\partial z}{\partial y}\cdot\frac{dy}{dx}$$

$$\frac{\partial z}{\partial x}=y; \quad \frac{\partial z}{\partial y}=x+\cos y$$

$$\therefore\ \frac{dz}{dx}=y+(x+\cos y)\frac{dy}{dx}$$

$$\text{But } z=0 \quad \therefore\ \frac{dz}{dx}=0$$

$$\therefore\ \frac{dy}{dx}=\frac{-y}{x+\cos y}$$

Here is one for you to do:

**Example 3**

Find an expression for $\dfrac{dy}{dx}$ when $x\tan y=y\sin x$. Do it all on your own.

*Then check your working with that in Frame 21*

**21**

$$\boxed{\frac{dy}{dx}=-\frac{\tan y-y\cos x}{x\sec^2 y-\sin x}}$$

Did you get that? If so, go straight on to Frame 22. If not, here is the working below. Follow it through and see where you have gone astray!

▶

$$x\tan y = y\sin x \qquad \therefore\ x\tan y - y\sin x = 0$$

Let $z = x\tan y - y\sin x = 0$

$$\delta z = \frac{\partial z}{\partial x}\delta x + \frac{\partial z}{\partial y}\delta y$$

$$\frac{dz}{dx} = \frac{\partial z}{\partial x} + \frac{\partial z}{\partial y}\cdot\frac{dy}{dx}$$

$$\frac{\partial z}{\partial x} = \tan y - y\cos x; \qquad \frac{\partial z}{\partial y} = x\sec^2 y - \sin x$$

$$\therefore\ \frac{dz}{dx} = (\tan y - y\cos x) + (x\sec^2 y - \sin x)\frac{dy}{dx}$$

But $z = 0 \quad \therefore\ \dfrac{dz}{dx} = 0$

$$\frac{dy}{dx} = -\frac{\tan y - y\cos x}{x\sec^2 y - \sin x}$$

*On now to Frame 22*

---

**22**

Right. Now here is just one more for you to do. They are really very much the same.

**Example 4**

If $e^{x+y} = x^2y^2$, find an expression for $\dfrac{dy}{dx}$

$$e^{x+y} - x^2y^2 = 0. \quad \text{Let } z = e^{x+y} - x^2y^2 = 0$$

$$\delta z = \frac{\partial z}{\partial x}\delta x + \frac{\partial z}{\partial y}\delta y$$

$$\frac{dz}{dx} = \frac{\partial z}{\partial x} + \frac{\partial z}{\partial y}\cdot\frac{dy}{dx}$$

So continue with the good work and finish it off, finally getting that

$$\frac{dy}{dx} = \dots\dots\dots\dots$$

*Then move to Frame 23*

---

**23**

$$\boxed{\frac{dy}{dx} = \frac{2xy^2 - e^{x+y}}{e^{x+y} - 2x^2y}}$$

Because $z = e^{x+y} - x^2y^2 = 0$

$$\frac{\partial z}{\partial x} = e^{x+y} - 2xy^2; \qquad \frac{\partial z}{\partial y} = e^{x+y} - 2x^2y$$

$$\therefore\ \frac{dz}{dx} = (e^{x+y} - 2xy^2) + (e^{x+y} - 2x^2y)\frac{dy}{dx}$$

But $z = 0 \quad \therefore\ \dfrac{dz}{dx} = 0$

$$\therefore\ \frac{dy}{dx} = -\frac{(e^{x+y} - 2xy^2)}{(e^{x+y} - 2x^2y)}$$

$$\therefore\ \frac{dy}{dx} = \frac{2xy^2 - e^{x+y}}{(e^{x+y} - 2x^2y)}$$

That is how they are all done. But now there is one more process that you must know how to tackle.

*So on to Frame 24*

## Change of variables

24

If $z$ is a function of $x$ and $y$, i.e. $z = f(x, y)$, and $x$ and $y$ are themselves functions of two other variables $u$ and $v$, then $z$ is also a function of $u$ and $v$. We may therefore need to find $\frac{\partial z}{\partial u}$ and $\frac{\partial z}{\partial v}$. How do we go about it?

$$z = f(x, y) \qquad \therefore \delta z = \frac{\partial z}{\partial x}\delta x + \frac{\partial z}{\partial y}\delta y$$

Divide both sides by $\delta u$:

$$\frac{\delta z}{\delta u} = \frac{\partial z}{\partial x}\cdot\frac{\delta x}{\delta u} + \frac{\partial z}{\partial y}\cdot\frac{\delta y}{\delta u}$$

If $v$ is kept constant for the time being, then $\frac{\delta x}{\delta u}$ when $\delta u \to 0$ becomes $\frac{\partial x}{\partial u}$ and $\frac{\delta y}{\delta u}$ becomes $\frac{\partial y}{\partial u}$.

$$\left.\begin{aligned} \therefore \quad \frac{\partial z}{\partial u} &= \frac{\partial z}{\partial x}\cdot\frac{\partial x}{\partial u} + \frac{\partial z}{\partial y}\cdot\frac{\partial y}{\partial u} \\ \text{and} \quad \frac{\partial z}{\partial v} &= \frac{\partial z}{\partial x}\cdot\frac{\partial x}{\partial v} + \frac{\partial z}{\partial y}\cdot\frac{\partial y}{\partial v} \end{aligned}\right\} \text{ Note these}$$

*Next frame*

25

Here is an example of this work.

If $z = x^2 + y^2$, where $x = r\cos\theta$ and $y = r\sin 2\theta$, find $\frac{\partial z}{\partial r}$ and $\frac{\partial z}{\partial \theta}$

$$\frac{\partial z}{\partial r} = \frac{\partial z}{\partial x}\cdot\frac{\partial x}{\partial r} + \frac{\partial z}{\partial y}\cdot\frac{\partial y}{\partial r}$$

$$\text{and} \quad \frac{\partial z}{\partial \theta} = \frac{\partial z}{\partial x}\cdot\frac{\partial x}{\partial \theta} + \frac{\partial z}{\partial y}\cdot\frac{\partial y}{\partial \theta}$$

$$\text{Now,} \quad \frac{\partial z}{\partial x} = 2x \qquad \left[\frac{\partial z}{\partial y} = 2y\right]$$

$$\frac{\partial x}{\partial r} = \cos\theta \qquad \left[\frac{\partial y}{\partial r} = \sin 2\theta\right]$$

$$\therefore \quad \frac{\partial z}{\partial r} = 2x\cos\theta + 2y\sin 2\theta$$

$$\text{and} \quad \frac{\partial x}{\partial \theta} = -r\sin\theta \quad \text{and} \quad \frac{\partial y}{\partial \theta} = 2r\cos 2\theta$$

$$\therefore \quad \frac{\partial z}{\partial \theta} = 2x(-r\sin\theta) + 2y(2r\cos 2\theta)$$

$$\frac{\partial z}{\partial \theta} = 4yr\cos 2\theta - 2xr\sin\theta$$

And in these two results, the symbols $x$ and $y$ can be replaced by $r\cos\theta$ and $r\sin 2\theta$ respectively.

**26**

One more example:

If $z = e^{xy}$ where $x = \ln(u+v)$ and $y = \sin(u-v)$, find $\dfrac{\partial z}{\partial u}$ and $\dfrac{\partial z}{\partial v}$.

$$\text{We have } \frac{\partial z}{\partial u} = \frac{\partial z}{\partial x}.\frac{\partial x}{\partial u} + \frac{\partial z}{\partial y}.\frac{\partial y}{\partial u} = y.e^{xy}.\frac{1}{u+v} + x.e^{xy}.\cos(u-v)$$

$$= e^{xy}\left\{\frac{y}{u+v} + x.\cos(u-v)\right\}$$

$$\text{and } \frac{\partial z}{\partial v} = \frac{\partial z}{\partial x}.\frac{\partial x}{\partial v} + \frac{\partial z}{\partial y}.\frac{\partial y}{\partial v} = y.e^{xy}.\frac{1}{u+v} + x.e^{xy}.\{-\cos(u-v)\}$$

$$= e^{xy}\left\{\frac{y}{u+v} - x.\cos(u-v)\right\}$$

*Now move on to Frame 27*

**27**

Here is one for you to do on your own. All that it entails is to find the various partial derivatives and to substitute them in the established results:

$$\frac{\partial z}{\partial u} = \frac{\partial z}{\partial x}.\frac{\partial x}{\partial u} + \frac{\partial z}{\partial y}.\frac{\partial y}{\partial u} \quad \text{and} \quad \frac{\partial z}{\partial v} = \frac{\partial z}{\partial x}.\frac{\partial x}{\partial v} + \frac{\partial z}{\partial y}.\frac{\partial y}{\partial v}$$

So you do this one:

If $z = \sin(x+y)$, where $x = u^2 + v^2$ and $y = 2uv$, find $\dfrac{\partial z}{\partial u}$ and $\dfrac{\partial z}{\partial v}$

The method is the same as before.

*When you have completed the work, check with the result in Frame 28*

**28**

$z = \sin(x+y); \quad x = u^2 + v^2; \quad y = 2uv$

$$\frac{\partial z}{\partial x} = \cos(x+y); \qquad \frac{\partial z}{\partial y} = \cos(x+y)$$

$$\frac{\partial x}{\partial u} = 2u \qquad \frac{\partial y}{\partial u} = 2v$$

$$\frac{\partial z}{\partial u} = \frac{\partial z}{\partial x}.\frac{\partial x}{\partial u} + \frac{\partial z}{\partial y}.\frac{\partial y}{\partial u}$$

$$= \cos(x+y).2u + \cos(x+y).2v$$

$$= 2(u+v)\cos(x+y)$$

$$\text{Also } \frac{\partial z}{\partial v} = \frac{\partial z}{\partial x}.\frac{\partial x}{\partial v} + \frac{\partial z}{\partial y}.\frac{\partial y}{\partial v}$$

$$\frac{\partial x}{\partial v} = 2v; \qquad \frac{\partial y}{\partial v} = 2u$$

$$\frac{\partial z}{\partial v} = \cos(x+y).2v + \cos(x+y).2u$$

$$= 2(u+v)\cos(x+y)$$

29

You have now reached the end of this Programme and know quite a bit about partial differentiation. We have established some important results during the work, so let us list them once more.

1 *Small increments*

$$z = f(x, y) \quad \delta z = \frac{\partial z}{\partial x}\delta x + \frac{\partial z}{\partial y}\delta y \qquad \text{(a)}$$

2 *Rates of change*

$$\frac{dz}{dt} = \frac{\partial z}{\partial x}\cdot\frac{dx}{dt} + \frac{\partial z}{\partial y}\cdot\frac{dy}{dt} \qquad \text{(b)}$$

3 *Implicit functions*

$$\frac{dz}{dx} = \frac{\partial z}{\partial x} + \frac{\partial z}{\partial y}\cdot\frac{dy}{dx} \qquad \text{(c)}$$

4 *Change of variables*

$$\frac{\partial z}{\partial u} = \frac{\partial z}{\partial x}\cdot\frac{\partial x}{\partial u} + \frac{\partial z}{\partial y}\cdot\frac{\partial y}{\partial u}$$
$$\frac{\partial z}{\partial v} = \frac{\partial z}{\partial x}\cdot\frac{\partial x}{\partial v} + \frac{\partial z}{\partial y}\cdot\frac{\partial y}{\partial v} \qquad \text{(d)}$$

All that now remains is the **Can You?** checklist and the **Test exercise**, so move to Frames 30 and 31 and work through them carefully at your own speed.

## Can You?

### Checklist 15

30

Check this list before and after you try the end of Programme test.

On a scale of 1 to 5 how confident are you that you can: Frames

- Derive the first- and second-order partial derivatives of a function of two real variables? 1 to 7
  *Yes* ☐ ☐ ☐ ☐ ☐ *No*
- Apply partial differentiation to rate-of-change problems? 8 to 23
  *Yes* ☐ ☐ ☐ ☐ ☐ *No*
- Apply partial differentiation to change-of-variable problems? 24 to 29
  *Yes* ☐ ☐ ☐ ☐ ☐ *No*

## Test exercise 15

31

Take your time and work carefully. The questions are just like those you have been doing quite successfully.

1 Use partial differentiation to determine expressions for $\frac{dy}{dx}$ in the following cases:

(a) $x^3 + y^3 - 2x^2y = 0$

(b) $e^x \cos y = e^y \sin x$

(c) $\sin^2 x - 5\sin x \cos y + \tan y = 0$

2 The base radius of a cone, $r$, is decreasing at the rate of 0·1 cm/s while the perpendicular height, $h$, is increasing at the rate of 0·2 cm/s. Find the rate at which the volume, $V$, is changing when $r = 2$ cm and $h = 3$ cm.

3 If $z = 2xy - 3x^2y$ and $x$ is increasing at 2 cm/s, determine at what rate $y$ must be changing in order that $z$ shall be neither increasing nor decreasing at the instant when $x = 3$ cm and $y = 1$ cm.

4 If $z = x^4 + 2x^2y + y^3$ and $x = r\cos\theta$ and $y = r\sin\theta$, find $\dfrac{\partial z}{\partial r}$ and $\dfrac{\partial z}{\partial \theta}$ in their simplest forms.

## Further problems 15

32

1 If $F = f(x, y)$ where $x = e^u \cos v$ and $y = e^u \sin v$, show that
$\dfrac{\partial F}{\partial u} = x\dfrac{\partial F}{\partial x} + y\dfrac{\partial F}{\partial y}$ and $\dfrac{\partial F}{\partial v} = -y\dfrac{\partial F}{\partial x} + x\dfrac{\partial F}{\partial y}$.

2 Given that $z = x^3 + y^3$ and $x^2 + y^2 = 1$, determine an expression for $\dfrac{dz}{dx}$ in terms of $x$ and $y$.

3 If $z = f(x, y) = 0$, show that $\dfrac{dy}{dx} = -\dfrac{\partial z}{\partial x} \Big/ \dfrac{\partial z}{\partial y}$. The curves $2y^2 + 3x - 8 = 0$ and $x^3 + 2xy^3 + 3y - 1 = 0$ intersect at the point (2, −1). Find the tangent of the angle between the tangents to the curves at this point.

4 If $u = (x^2 - y^2)f(t)$ where $t = xy$ and $f$ denotes an arbitrary function, prove that $\dfrac{\partial^2 u}{\partial x.\partial y} = (x^2 - y^2)\{t.f''(t) + 3f'(t)\}$. [*Note*: $f''(t)$ is the second derivative of $f(t)$ w.r.t. $t$.]

5 If $V = xy/(x^2 + y^2)^2$ and $x = r\cos\theta$, $y = r\sin\theta$, show that
$$\frac{\partial^2 V}{\partial r^2} + \frac{1}{r}\frac{\partial V}{\partial r} + \frac{1}{r^2}\frac{\partial^2 V}{\partial \theta^2} = 0.$$

6 If $u = f(x, y)$ where $x = r^2 - s^2$ and $y = 2rs$, prove that
$$r\frac{\partial u}{\partial r} - s\frac{\partial u}{\partial s} = 2(r^2 + s^2)\frac{\partial u}{\partial x}.$$

7 If $f = F(x, y)$ and $x = re^{\theta}$ and $y = re^{-\theta}$, prove that
$2x\dfrac{\partial f}{\partial x} = r\dfrac{\partial f}{\partial r} + \dfrac{\partial f}{\partial \theta}$ and $2y\dfrac{\partial f}{\partial y} = r\dfrac{\partial f}{\partial r} - \dfrac{\partial f}{\partial \theta}$.

8 If $z = x\ln(x^2 + y^2) - 2y\tan^{-1}\left(\dfrac{y}{x}\right)$ verify that $x\dfrac{\partial z}{\partial x} + y\dfrac{\partial z}{\partial y} = z + 2x$.

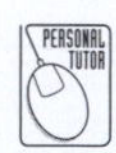

9 By means of partial differentiation, determine $\dfrac{dy}{dx}$ in each of the following cases:

(a) $xy + 2y - x = 4$

(b) $x^3y^2 - 2x^2y + 3xy^2 - 8xy = 5$

(c) $\dfrac{4y}{x} + \dfrac{2x}{y} = 3$

**10** If $z = 3xy - y^3 + (y^2 - 2x)^{3/2}$, verify that:

(a) $\dfrac{\partial^2 z}{\partial x.\partial y} = \dfrac{\partial^2 z}{\partial y.\partial x}$ and (b) $\dfrac{\partial^2 z}{\partial x^2}\cdot\dfrac{\partial^2 z}{\partial y^2} = \left(\dfrac{\partial^2 z}{\partial x.\partial y}\right)^2$

**11** If $f = \dfrac{1}{\sqrt{1 - 2xy + y^2}}$, show that $y\dfrac{\partial f}{\partial y} = (x - y)\dfrac{\partial f}{\partial x}$.

**12** If $z = x.f\left(\dfrac{y}{x}\right) + F\left(\dfrac{y}{x}\right)$, prove that:

(a) $x\dfrac{\partial z}{\partial x} + y\dfrac{\partial z}{\partial y} = z - F\left(\dfrac{y}{x}\right)$ (b) $x^2\dfrac{\partial^2 z}{\partial x^2} + 2xy\dfrac{\partial^2 z}{\partial x.\partial y} + y^2\dfrac{\partial^2 z}{\partial y^2} = 0$

**13** If $z = e^{k(r-x)}$, where $k$ is a constant, and $r^2 = x^2 + y^2$, prove:

(a) $\left(\dfrac{\partial z}{\partial x}\right)^2 + \left(\dfrac{\partial z}{\partial y}\right)^2 + 2zk\dfrac{\partial z}{\partial x} = 0$ (b) $\dfrac{\partial^2 z}{\partial x^2} + \dfrac{\partial^2 z}{\partial y^2} + 2k\dfrac{\partial z}{\partial x} = \dfrac{kz}{r}$

**14** If $z = f(x - 2y) + F(3x + y)$, where $f$ and $F$ are arbitrary functions, and if $\dfrac{\partial^2 z}{\partial x^2} + a\dfrac{\partial^2 z}{\partial x.\partial y} + b\dfrac{\partial^2 z}{\partial y^2} = 0$, find the values of $a$ and $b$.

**15** If $z = xy/(x^2 + y^2)^2$, verify that $\dfrac{\partial^2 z}{\partial x^2} + \dfrac{\partial^2 z}{\partial y^2} = 0$.

**16** If $\sin^2 x - 5\sin x\cos y + \tan y = 0$, find $\dfrac{dy}{dx}$ by using partial differentiation.

**17** Find $\dfrac{dy}{dx}$ by partial differentiation, when $x\tan y = y\sin x$.

**18** If $V = \tan^{-1}\left\{\dfrac{2xy}{x^2 - y^2}\right\}$, prove that:

(a) $x\dfrac{\partial V}{\partial x} + y\dfrac{\partial V}{\partial y} = 0$ (b) $\dfrac{\partial^2 V}{\partial x^2} + \dfrac{\partial^2 V}{\partial y^2} = 0$

**19** Prove that, if $z = 2xy + x.f\left(\dfrac{y}{x}\right)$ then $x\dfrac{\partial z}{\partial x} + y\dfrac{\partial z}{\partial y} = z + 2xy$.

**20** (a) Find $\dfrac{dy}{dx}$ given that $x^2y + \sin xy = 0$.

(b) Find $\dfrac{dy}{dx}$ given that $x\sin xy = 1$.

Now visit the companion website for this book at www.palgrave.com/stroud for more questions applying this mathematics to science and engineering.

# Integration 1

## Learning outcomes

When you have completed this Programme you will be able to:

- ☐ Integrate standard expressions using a table of standard forms
- ☐ Integrate functions of a linear form
- ☐ Evaluate integrals with integrands of the form $f'(x)/f(x)$ and $f(x).f'(x)$
- ☐ Integrate by parts
- ☐ Integrate by partial fractions
- ☐ Integrate trigonometric functions

# Introduction

**1**

You are already familiar with the basic principles of integration and have had plenty of practice at some time in the past. However, that was some time ago, so let us first of all brush up our ideas of the fundamentals.

Integration is the reverse of differentiation. In differentiation, we start with a function and proceed to find its derivative. In integration, we start with the derivative and have to work back to find the function from which it has been derived.

e.g. $\frac{\mathrm{d}}{\mathrm{d}x}(x^3+5)=3x^2$. Therefore it is true, in this case, to say that the integral of $3x^2$, with respect to $x$, is the function from which it came, i.e. $\int 3x^2\,\mathrm{d}x = x^3+5$. However, if we had to find $\int 3x^2\,\mathrm{d}x$, without knowing the past history of the function, we should have no indication of the size of the constant term involved, since all trace of it is lost in the derivative. All we can do is to indicate the constant term by a symbol, e.g. $C$.

So, in general, $\int 3x^2\,\mathrm{d}x = x^3 + C$

Although we cannot determine the value of this *constant of integration* without extra information about the function, it is vitally important that we should always include it in our results. There are just one or two occasions when we are permitted to leave it out, not because it is not there, but because in some prescribed situation, it will cancel out in subsequent working. Such occasions, however, are very rare and, in general, the *constant of integration must be included in the result.*

If you omit the constant of integration, your work will be slovenly and, furthermore, it will be completely wrong! So, *do not forget the constant of integration.*

## Standard integrals

Every derivative, when written in reverse, gives us an integral:

e.g. $\frac{\mathrm{d}}{\mathrm{d}x}(\sin x)=\cos x \qquad \therefore \int \cos x\,\mathrm{d}x = \sin x + C$

It follows then that our list of standard derivatives will form the basis of a list of standard integrals – sometimes slightly modified to give a neater expression.

**2**

Here is a list of basic derivatives and the basic integrals that go with them:

1 $\dfrac{d}{dx}(x^n) = nx^{n-1}$ $\quad\therefore \displaystyle\int x^n\,dx = \frac{x^{n+1}}{n+1} + C \quad \left\{\begin{matrix}\text{provided}\\ n \neq -1\end{matrix}\right\}$

2 $\dfrac{d}{dx}(\ln x) = \dfrac{1}{x}$ $\quad\therefore \displaystyle\int \frac{1}{x}\,dx = \ln x + C$

3 $\dfrac{d}{dx}(e^x) = e^x$ $\quad\therefore \displaystyle\int e^x\,dx = e^x + C$

4 $\dfrac{d}{dx}(e^{kx}) = ke^{kx}$ $\quad\therefore \displaystyle\int e^{kx}\,dx = \frac{e^{kx}}{k} + C$

5 $\dfrac{d}{dx}(a^x) = a^x \ln a$ $\quad\therefore \displaystyle\int a^x\,dx = \frac{a^x}{\ln a} + C$

6 $\dfrac{d}{dx}(\cos x) = -\sin x$ $\quad\therefore \displaystyle\int \sin x\,dx = -\cos x + C$

7 $\dfrac{d}{dx}(\sin x) = \cos x$ $\quad\therefore \displaystyle\int \cos x\,dx = \sin x + C$

8 $\dfrac{d}{dx}(\tan x) = \sec^2 x$ $\quad\therefore \displaystyle\int \sec^2 x\,dx = \tan x + C$

9 $\dfrac{d}{dx}(\cosh x) = \sinh x$ $\quad\therefore \displaystyle\int \sinh x\,dx = \cosh x + C$

10 $\dfrac{d}{dx}(\sinh x) = \cosh x$ $\quad\therefore \displaystyle\int \cosh x\,dx = \sinh x + C$

11 $\dfrac{d}{dx}(\sin^{-1} x) = \dfrac{1}{\sqrt{1-x^2}}$ $\quad\therefore \displaystyle\int \frac{1}{\sqrt{1-x^2}}\,dx = \sin^{-1} x + C$

12 $\dfrac{d}{dx}(\cos^{-1} x) = \dfrac{-1}{\sqrt{1-x^2}}$ $\quad\therefore \displaystyle\int \frac{-1}{\sqrt{1-x^2}}\,dx = \cos^{-1} x + C$

13 $\dfrac{d}{dx}(\tan^{-1} x) = \dfrac{1}{1+x^2}$ $\quad\therefore \displaystyle\int \frac{1}{1+x^2}\,dx = \tan^{-1} x + C$

14 $\dfrac{d}{dx}(\sinh^{-1} x) = \dfrac{1}{\sqrt{x^2+1}}$ $\quad\therefore \displaystyle\int \frac{1}{\sqrt{x^2+1}}\,dx = \sinh^{-1} x + C$

15 $\dfrac{d}{dx}(\cosh^{-1}) = \dfrac{1}{\sqrt{x^2-1}}$ $\quad\therefore \displaystyle\int \frac{1}{\sqrt{x^2-1}}\,dx = \cosh^{-1} x + C$

16 $\dfrac{d}{dx}(\tanh^{-1} x) = \dfrac{1}{1-x^2}$ $\quad\therefore \displaystyle\int \frac{1}{1-x^2}\,dx = \tanh^{-1} x + C$

Spend a little time copying this list carefully into your record book as a reference list.

**3**

Here is a second look at the last six results, which are less familiar to you than the others:

$$\int \frac{1}{\sqrt{1-x^2}}\,dx = \sin^{-1} x + C \qquad \int \frac{1}{\sqrt{x^2+1}}\,dx = \sinh^{-1} x + C$$

$$\int \frac{-1}{\sqrt{1-x^2}}\,dx = \cos^{-1} x + C \qquad \int \frac{1}{\sqrt{x^2-1}}\,dx = \cosh^{-1} x + C$$

$$\int \frac{1}{1+x^2}\,dx = \tan^{-1} x + C \qquad \int \frac{1}{1-x^2}\,dx = \tanh^{-1} x + C$$

*Notice*: (a) How alike the two sets are in shape.
(b) Where the small, but all important, differences occur.

*On to Frame 4*

**4**

Now cover up the lists you have just copied down and complete the following:

(a) $\int e^{5x}\,dx = \ldots\ldots\ldots$

(b) $\int x^7\,dx = \ldots\ldots\ldots$

(c) $\int \sqrt{x}\,dx = \ldots\ldots\ldots$

(d) $\int \sin x\,dx = \ldots\ldots\ldots$

(e) $\int 2\sinh x\,dx = \ldots\ldots\ldots$

(f) $\int \frac{5}{x}\,dx = \ldots\ldots\ldots$

(g) $\int \frac{1}{\sqrt{1-x^2}}\,dx = \ldots\ldots\ldots$

(h) $\int 5^x\,dx = \ldots\ldots\ldots$

(i) $\int \frac{1}{\sqrt{x^2-1}}\,dx = \ldots\ldots\ldots$

(j) $\int \frac{1}{1+x^2}\,dx = \ldots\ldots\ldots$

*When you have finished them all, check your results with those given in the next frame*

**5**

Here they are

(a) $\int e^{5x}\,dx = \frac{e^{5x}}{5} + C$

(b) $\int x^7\,dx = \frac{x^8}{8} + C$

(c) $\int \sqrt{x}\,dx = \int x^{1/2}\,dx = 2\frac{x^{3/2}}{3} + C$

(d) $\int \sin x\,dx = -\cos x + C$

(e) $\int 2\sinh x\,dx = 2\cosh x + C$

(f) $\int \frac{5}{x}\,dx = 5\ln x + C$

(g) $\int \frac{1}{\sqrt{1-x^2}}\,dx = \sin^{-1} x + C$

(h) $\int 5^x\,dx = \frac{5^x}{\ln 5} + C$

(i) $\int \frac{1}{\sqrt{x^2-1}}\,dx = \cosh^{-1} x + C$

(j) $\int \frac{1}{1+x^2}\,dx = \tan^{-1} x + C$

All correct or nearly so? At the moment, these are fresh in your mind, but have a look at your list of standard integrals whenever you have a few minutes to spare. It will help you to remember them.

*Now move on to Frame 6*

## Functions of a linear function of $x$

**6**

We are very often required to integrate functions like those in the standard list, but where $x$ is replaced by a linear function of $x$, e.g. $\int (5x-4)^6\,dx$, which is very much like $\int x^6\,dx$ except that $x$ is replaced by $(5x-4)$. If we put $z$ to stand for $(5x-4)$, the integral becomes $\int z^6\,dx$ and before we can complete the operation, we must change the variable, thus:

$$\int z^6\,dx = \int z^6 \frac{dx}{dz}\,dz$$

Now $\frac{dx}{dz}$ can be found from the substitution $z = 5x - 4$ for $\frac{dz}{dx} = 5$, therefore $\frac{dx}{dz} = \frac{1}{5}$ and the integral becomes:

$$\int z^6\,dx = \int z^6 \frac{dx}{dz}\,dz = \int z^6\left(\frac{1}{5}\right)\,dz = \frac{1}{5}\int z^6 dz = \frac{1}{5}\cdot\frac{z^7}{7} + C$$

Finally, we must express $z$ in terms of the original variable, $x$, so that:

$$\int (5x-4)^6\,dx = \ldots\ldots\ldots\ldots$$

**7**

$$\int (5x-4)^6\,dx = \frac{(5x-4)^7}{5.7} + C = \boxed{\frac{(5x-4)^7}{35} + C}$$

The corresponding standard integral is $\int x^6\,dx = \frac{x^7}{7} + C$. We see, therefore, that when $x$ is replaced by $(5x-4)$, the 'power' rule still applies, i.e. $(5x-4)$ replaces the single $x$ in the result, so long as we also *divide by the coefficient of* $x$, in this case 5:

$$\int x^6\,dx = \frac{x^7}{7} + C \qquad \therefore \quad \int (5x-4)^6\,dx = \frac{(5x-4)^7}{35} + C$$

This will always happen when we integrate functions of a *linear* function of $x$.

e.g. $\int e^x\,dx = e^x + C \quad \therefore \quad \int e^{3x+4}\,dx = \frac{e^{3x+4}}{3} + C$

i.e. $(3x+4)$ replaces $x$ in the integral

then $(3x+4)$ replaces $x$ in the result, provided we also divide by the coefficient of $x$.

Similarly, since $\int \cos x\,dx = \sin x + C$

then $\int \cos(2x+5)\,dx = \ldots\ldots\ldots\ldots$

8

$$\int \cos(2x+5)\,dx = \frac{\sin(2x+5)}{2} + C$$

Similarly:

$$\int \sec^2 x\,dx = \tan x + C \qquad \therefore \int \sec^2 4x\,dx = \frac{\tan 4x}{4} + C$$

$$\int \frac{1}{x}\,dx = \ln x + C \qquad \therefore \int \frac{1}{2x+3}\,dx = \frac{\ln(2x+3)}{2} + C$$

$$\int \sinh x\,dx = \cosh x + C \qquad \therefore \int \sinh(3-4x)\,dx = \frac{\cosh(3-4x)}{-4}$$

$$= -\frac{\cosh(3-4x)}{4} + C$$

$$\int \sin x\,dx = -\cos x + C \qquad \therefore \int \sin 3x\,dx = -\frac{\cos 3x}{3} + C$$

$$\int e^x\,dx = e^x + C \qquad \therefore \int e^{4x}\,dx = \frac{e^{4x}}{4} + C$$

So if a *linear* function of $x$ replaces the single $x$ in the standard integral, the same linear function of $x$ replaces the single $x$ in the result, so long as we also remember to ............

9

divide by the coefficient of $x$

Now you can do these quite happily and do not forget the constants of integration!

1. $\int (2x-7)^3\,dx$
2. $\int \cos(7x+2)\,dx$
3. $\int e^{5x+4}\,dx$
4. $\int \sinh 7x\,dx$
5. $\int \frac{1}{4x+3}\,dx$
6. $\int \frac{1}{1+(2x)^2}\,dx$
7. $\int \sec^2(3x+1)\,dx$
8. $\int \sin(2x-5)\,dx$
9. $\int \cosh(1+4x)\,dx$
10. $\int 3^{5x}\,dx$

*Finish them all, then move on to Frame 10 and check your results*

10

Here are the results:

1. $\int (2x-7)^3\,dx = \frac{(2x-7)^4}{2.4} + C = \frac{(2x-7)^4}{8} + C$
2. $\int \cos(7x+2)\,dx = \frac{\sin(7x+2)}{7} + C$
3. $\int e^{5x+4}\,dx = \frac{e^{5x+4}}{5} + C$
4. $\int \sinh 7x\,dx = \frac{\cosh 7x}{7} + C$

▶

5 $\int \frac{1}{4x+3}\,dx = \frac{\ln(4x+3)}{4} + C$

6 $\int \frac{1}{1+(2x)^2}\,dx = \frac{\tan^{-1}(2x)}{2} + C$

7 $\int \sec^2(3x+1)\,dx = \frac{\tan(3x+1)}{3} + C$

8 $\int \sin(2x-5)\,dx = -\frac{\cos(2x-5)}{2} + C$

9 $\int \cosh(1+4x)\,dx = \frac{\sinh(1+4x)}{4} + C$

10 $\int 3^{5x}\,dx = \frac{3^{5x}}{5\ln 3} + C$

*Now we can start the next section of the Programme. So move on to Frame 11*

---

## Integrals of the forms $\frac{f'(x)}{f(x)}$ dx and $f(x).f'(x)$ dx

**11**

Consider the integral $\int \frac{2x+3}{x^2+3x-5}\,dx$. This is not one of our standard integrals, so how shall we tackle it? This is an example of a type of integral which is very easy to deal with but which depends largely on how keen your wits are.

You will notice that if we differentiate the denominator, we obtain the expression in the numerator. So, let $z$ stand for the denominator, i.e. $z = x^2 + 3x - 5$.

$$\therefore \quad \frac{dz}{dx} = 2x+3 \qquad \therefore\ dz \equiv (2x+3)\,dx$$

The given integral can then be written in terms of $z$:

$$\int \frac{(2x+3)}{x^2+3x-5}\,dx = \int \frac{dz}{z} \quad \text{and we know that} \quad \int \frac{1}{z}\,dz = \ln z + C$$

$$= \ln z + C$$

If we now put back what $z$ stands for in terms of $x$, we get

$$\int \frac{(2x+3)}{x^2+3x-5}\,dx = \ldots\ldots\ldots\ldots$$

**12**

$$\boxed{\int \frac{(2x+3)}{x^2+3x-5}\,dx = \ln(x^2+3x-5) + C}$$

Any integral in which the numerator is the derivative of the denominator will be of the kind $\int \frac{f'(x)}{f(x)}\,dx = \ln\{f(x)\} + C$

e.g. $\int \frac{3x^2}{x^3-4}\,dx$ is of the form $\int \frac{dz}{z}$ since $\frac{d}{dx}(x^3-4) = 3x^2$, i.e. the derivative of the denominator appears as the numerator. Therefore we can say at once, without any further working:

$$\int \frac{3x^2}{x^3-4}\,dx = \ln(x^3-4) + C$$

▶

Similarly $\displaystyle\int \frac{6x^2}{x^3-4}\,dx = 2\int \frac{3x^2}{x^3-4}\,dx = 2\ln(x^3-4)+C$

and $\displaystyle\int \frac{2x^2}{x^3-4}\,dx = \frac{2}{3}\int \frac{3x^2}{x^3-4}\,dx = \frac{2}{3}\ln(x^3-4)+C$

and $\displaystyle\int \frac{x^2}{x^3-4}\,dx = \ldots\ldots\ldots\ldots$

**13**

$$\boxed{\int \frac{x^2}{x^3-4}\,dx = \frac{1}{3}\cdot\int \frac{3x^2}{x^3-4}\,dx = \frac{1}{3}\ln(x^3-4)+C}$$

We shall always get this log form of the result, then, whenever the numerator is the derivative of the denominator, or is a multiple or sub-multiple of it. Here is an example:

$$\int \cot x\,dx = \int \frac{\cos x}{\sin x}\,dx$$

and since we know that $\cos x$ is the derivative of $\sin x$, then:

$$\int \cot x\,dx = \int \frac{\cos x}{\sin x}\,dx = \ln\sin x + C$$

In the same way:

$$\int \tan x\,dx = \int \frac{\sin x}{\cos x}\,dx$$
$$= \ldots\ldots\ldots\ldots$$

**14**

$$\int \tan x\,dx = \int \frac{\sin x}{\cos x}\,dx = -\int \frac{(-\sin x)}{\cos x}\,dx$$
$$= \boxed{-\ln\cos x + C}$$

Whenever we are confronted by an integral in the form of a quotient, our first reaction is to see whether the numerator is the derivative of the denominator. If so, the result is simply the log of the denominator:

e.g. $\displaystyle\int \frac{4x-8}{x^2-4x+5}\,dx = \ldots\ldots\ldots\ldots$

**15**

$$\int \frac{4x-8}{x^2-4x+5}\,dx = 2\int \frac{2x-4}{x^2-4x+5}\,dx = \boxed{2\ln(x^2-4x+5)+C}$$

Here you are: complete the following:

**1** $\displaystyle\int \frac{\sec^2 x}{\tan x}\,dx = \ldots\ldots\ldots\ldots$

**2** $\displaystyle\int \frac{2x+4}{x^2+4x-1}\,dx = \ldots\ldots\ldots\ldots$

**3** $\displaystyle\int \frac{\sinh x}{\cosh x}\,dx = \ldots\ldots\ldots\ldots$

**4** $\displaystyle\int \frac{x-3}{x^2-6x+2}\,dx = \ldots\ldots\ldots\ldots$

**16**

Here are the results: check yours.

1 $\displaystyle\int \frac{\sec^2 x}{\tan x}\,dx = \ln \tan x + C$

2 $\displaystyle\int \frac{2x+4}{x^2+4x-1}\,dx = \ln(x^2+4x-1) + C$

3 $\displaystyle\int \frac{\sinh x}{\cosh x}\,dx = \ln \cosh x + C$

4 $\displaystyle\int \frac{x-3}{x^2-6x+2}\,dx = \frac{1}{2}\ln(x^2-6x+2) + C$

*Now move on to Frame 17*

**17**

In very much the same way, we sometimes have integrals such as

$$\int \tan x.\sec^2 x\,dx$$

This, of course, is not a quotient but a product. Nevertheless we notice that one function ($\sec^2 x$) of the product is the derivative of the other function ($\tan x$).

If we put $z = \tan x$, then $dz \equiv \sec^2 x\,dx$ and the integral can then be written $\int z\,dz$ which gives $\frac{z^2}{2} + C$

$$\therefore \int \tan x.\sec^2 x\,dx = \frac{\tan^2 x}{2} + C$$

Here, then, we have a product where one factor is the derivative of the other. We could write it as:

$$\int \tan x.d(\tan x)$$

This is just like $\int z\,dz$ which gives $\frac{z^2}{2} + C$.

$$\therefore \int \tan x.\sec^2 x\,dx = \int \tan x.d(\tan x) = \frac{\tan^2 x}{2} + C$$

*On to the next frame*

**18**

Here is another example of the same kind:

$$\int \sin x.\cos x\,dx = \int \sin x.d(\sin x), \text{ i.e. like } \int z\,dz = \frac{\sin^2 x}{2} + C$$

The only thing you have to spot is that one factor of the product is the derivative of the other, or is some multiple of it.

**Example 1**

$$\begin{aligned}\int \frac{\ln x}{x}\,dx &= \int \ln x.\frac{1}{x}\,dx \\ &= \int \ln x.d(\ln x) \\ &= \frac{(\ln x)^2}{2} + C\end{aligned}$$

▶

**Example 2**

$$\int \frac{\sin^{-1} x}{\sqrt{1-x^2}}\,dx = \int \sin^{-1} x.\frac{1}{\sqrt{1-x^2}}\,dx$$
$$= \int \sin^{-1} x.d(\sin^{-1} x)$$
$$= \frac{(\sin^{-1} x)^2}{2} + C$$

**Example 3**

$$\int \sinh x.\cosh x\,dx = \ldots\ldots\ldots\ldots$$

19

$$\int \sinh x.\cosh x\,dx = \int \sinh x.d(\sinh x)$$
$$= \boxed{\frac{\sinh^2 x}{2} + C}$$

Now here is a short revision exercise for you to do. Finish all four and then check your results with those in the next frame.

1 $\displaystyle\int \frac{2x+3}{x^2+3x-7}\,dx$ 2 $\displaystyle\int \frac{\cos x}{1+\sin x}\,dx$

3 $\displaystyle\int (x^2+7x-4)(2x+7)\,dx$ 4 $\displaystyle\int \frac{4x^2}{x^3-7}\,dx$

*Results in Frame 20*

20

1 $\displaystyle\int \frac{2x+3}{x^2+3x-7}\,dx$

Notice that the top is exactly the derivative of the bottom, i.e. $\displaystyle\int \frac{dz}{z}$

$$\therefore \int \frac{2x+3}{x^2+3x-7}\,dx = \int \frac{d(x^2+3x-7)}{x^2+3x-7}$$
$$= \ln(x^2+3x-7) + C$$

2 $$\int \frac{\cos x}{1+\sin x}\,dx = \int \frac{d(1+\sin x)}{1+\sin x}$$
$$= \ln(1+\sin x) + C$$

3 $$\int (x^2+7x-4)(2x+7)\,dx = \int (x^2+7x-4).d(x^2+7x-4)$$
$$= \frac{(x^2+7x-4)^2}{2} + C$$

4 $$\int \frac{4x^2}{x^3-7}\,dx = \frac{4}{3}\int \frac{3x^2}{x^3-7}\,dx$$
$$= \frac{4}{3}\ln(x^3-7) + C$$

Always be prepared for these types of integrals. They are often missed, but very easy if you spot them.

*Now on to the next part of the work that starts in Frame 21*

## Integration of products – integration by parts

**21**

We often need to integrate a product where either function is *not* the derivative of the other. For example, in the case of

$$\int x^2.\ln x\,dx$$

$\ln x$ is not the derivative of $x^2$

$x^2$ is not the derivative of $\ln x$

so in situations like this, we have to find some other method of dealing with the integral. Let us establish the rule for such cases.

If $u$ and $v$ are functions of $x$, then we know that

$$\frac{d}{dx}(uv) = u\frac{dv}{dx} + v\frac{du}{dx}$$

Now integrate both sides with respect to $x$. On the left, we get back to the function from which we started:

$$uv = \int u\frac{dv}{dx}\,dx + \int v\frac{du}{dx}\,dx$$

and rearranging the terms, we have:

$$\int u\frac{dv}{dx}\,dx = uv - \int v\frac{du}{dx}\,dx$$

On the left-hand side, we have a product of two factors to integrate. One factor is chosen as the function $u$, the other is thought of as being the derivative of some function $v$. To find $v$, of course, we must integrate this particular factor separately. Then, knowing $u$ and $v$ we can substitute in the right-hand side and so complete the routine.

You will notice that we finish up with another product to integrate on the end of the line, but, unless we are very unfortunate, this product will be easier to tackle than the original one.

This then is the key to the routine:

$$\int u\frac{dv}{dx}\,dx = uv - \int v\frac{du}{dx}\,dx$$

For convenience, this can be memorized as:

$$\int u\,dv = uv - \int v\,du$$

In this form it is easier to remember, but the previous line gives its meaning in detail. This method is called *integration by parts*.

**22**

So $\displaystyle\int u\frac{dv}{dx}\,dx = uv - \int v\frac{du}{dx}\,dx$

i.e. $\displaystyle\int u\,dv = uv - \int v\,du$

Copy these results into your record book. You will soon learn them. Now for some examples involving integration by parts.

▶

**Example 1**

$$\int x^2 . \ln x \, dx$$

The two factors are $x^2$ and $\ln\ x$, and we have to decide which to take as $u$ and which to link to $dv$. If we choose $x^2$ to be $u$ and $\ln x\, dx$ to be $dv$, then we shall have to integrate $\ln x$ in order to find $v$. Unfortunately, $\int \ln x dx$ is not in our basic list of standard integrals, therefore we must allocate $u$ and $dv$ the other way round, i.e. $u = \ln x$ so that $du = \frac{1}{x}\, dx$ and $dv = x^2\, dx$ so that $v = \frac{x^3}{3}$. (We omit the integration constant here because we are in the middle of evaluating the integral. The integration constant will come out eventually.)

$$\begin{aligned}\therefore \int x^2 \ln x\, dx &= \int u\, dv \\ &= uv - \int v du \\ &= \ln x \left(\frac{x^3}{3}\right) - \frac{1}{3}\int x^3 \frac{1}{x} dx\end{aligned}$$

Notice that we can tidy up the writing of the second integral by writing the constant factors involved, outside the integral:

$$\begin{aligned}\therefore \int x^2 \ln x\, dx &= \ln x \left(\frac{x^3}{3}\right) - \frac{1}{3}\int x^3 . \frac{1}{x}\, dx = \frac{x^3}{3}\ln x - \frac{1}{3}\int x^2\, dx \\ &= \frac{x^3}{3}.\ln x - \frac{1}{3}.\frac{x^3}{3} + C = \frac{x^3}{3}\left\{\ln x - \frac{1}{3}\right\} + C\end{aligned}$$

Note that if one of the factors of the product to be integrated is a log term, this must be chosen as ............ ($u$ or $dv$)

---

**23**

$u$

**Example 2**

$\int x^2 e^{3x} dx$ Let $u = x^2$ so that $du = 2x\, dx$ and $dv = e^{3x}\, dx$ so that $v = \frac{e^{3x}}{3}$

Then $\int x^2 e^{3x}\, dx = uv - \int v du$

$$= x^2\left(\frac{e^{3x}}{3}\right) - \frac{2}{3}\int e^{3x} x\, dx$$

The integral $\int e^{3x} x\, dx$ will also have to be integrated by parts. So that:

$$\begin{aligned}&= \frac{x^2 e^{3x}}{3} - \frac{2}{3}\left\{x\left(\frac{e^{3x}}{3}\right) - \frac{1}{3}\int e^{3x}\, dx\right\} \\ &= \frac{x^2 e^{3x}}{3} - \frac{2xe^{3x}}{9} + \frac{2}{9}\frac{e^{3x}}{3} + C \\ &= \frac{e^{3x}}{3}\left\{x^2 - \frac{2x}{3} + \frac{2}{9}\right\} + C\end{aligned}$$

*On to Frame 24*

**24**

In Example 1 we saw that if one of the factors is a log function, that log function *must* be taken as $u$.

In Example 2 we saw that, provided there is no log term present, the power of $x$ is taken as $u$. (By the way, this method holds good only for positive whole-number powers of $x$. For other powers, a different method must be applied.)

So which of the two factors should we choose to be $u$ in each of the following cases?

(a) $\int x.\ln x\,dx$ (b) $\int x^3.\sin x\,dx$

**25**

(a) In $\int x.\ln x\,dx,\ u = \ln x$

(b) In $\int x^3 \sin x\,dx,\ u = x^3$

Right. Now for a third example.

**Example 3**

$\int e^{3x}\sin x\,dx$. Here we have neither a log factor nor a power of $x$. Let us try putting $u = e^{3x}$ and $dv = \sin x\,dx$ so $v = -\cos x$.

$$\therefore \int e^{3x}\sin x\,dx = e^{3x}(-\cos x) + 3\int \cos x.e^{3x}\,dx$$

$$= -e^{3x}\cos x + 3\int e^{3x}\cos x\,dx \qquad \text{(and again by parts)}$$

$$= -e^{3x}\cos x + 3\left\{e^{3x}(\sin x) - 3\int \sin x.e^{3x}\,dx\right\}$$

and it looks as though we are back where we started.

However, let us write $I$ for the integral $\int e^{3x}\sin x\,dx$:

$$I = -e^{3x}\cos x + 3e^{3x}\sin x - 9I$$

Then, treating this as a simple equation, we get:

$$10I = e^{3x}(3\sin x - \cos x) + C_1$$

$$I = \frac{e^{3x}}{10}(3\sin x - \cos x) + C$$

Whenever we integrate functions of the form $e^{kx}\sin x$ or $e^{kx}\cos x$, we get similar types of results after applying the rule twice.

*Move on to Frame 26*

**26**

The three examples we have considered enable us to form a priority order for $u$:

(a) $\ln x$ (b) $x^n$ (c) $e^{kx}$

i.e. If one factor is a log function, that must be taken as '$u$'.

If there is no log function but a power of $x$, that becomes '$u$'.

If there is neither a log function nor a power of $x$, then the exponential function is taken as '$u$'.

Remembering the priority order will save a lot of false starts.

▶

So which would you choose as '$u$' in the following cases?

(a) $\int x^4 \cos 2x \, dx$ $\quad u = \ldots\ldots\ldots\ldots$

(b) $\int x^4 e^{3x} \, dx$ $\quad u = \ldots\ldots\ldots\ldots$

(c) $\int x^3 \ln(x+4) \, dx$ $\quad u = \ldots\ldots\ldots\ldots$

(d) $\int e^{2x} \cos 4x \, dx$ $\quad u = \ldots\ldots\ldots\ldots$

---

**27**

(a) $\int x^4 \cos 2x \, dx$ $\quad u = x^4$

(b) $\int x^4 e^{3x} \, dx$ $\quad u = x^4$

(c) $\int x^3 \ln(x+4) \, dx$ $\quad u = \ln(x+4)$

(d) $\int e^{2x} \cos 4x \, dx$ $\quad u = e^{2x}$

Right. Now look at this one:

$$\int e^{5x} \sin 3x \, dx$$

Following our rule for priority for $u$, in this case, we should put $u = \ldots\ldots\ldots\ldots$

---

**28**

$$\int e^{5x} \sin 3x \, dx \qquad \boxed{\therefore u = e^{5x}}$$

Correct. Make a note of that priority list for $u$ in your record book.
Then go ahead and determine the integral given above.

*When you have finished, check your working with that set out in the next frame*

---

**29**

$$\boxed{\int e^{5x} \sin 3x \, dx = \frac{3e^{5x}}{34}\left\{\frac{5}{3}\sin 3x - \cos 3x\right\} + C}$$

Here is the working. Follow it through.

$$\int e^{5x} \sin 3x \, dx = e^{5x}\left(-\frac{\cos 3x}{3}\right) + \frac{5}{3}\int \cos 3x . e^{5x} \, dx$$

$$= -\frac{e^{5x}\cos 3x}{3} + \frac{5}{3}\left\{e^{5x}\left(\frac{\sin 3x}{3}\right) - \frac{5}{3}\int \sin 3x . e^{5x} \, dx\right\}$$

$$\therefore I = -\frac{e^{5x}\cos 3x}{3} + \frac{5}{9}e^{5x}\sin 3x - \frac{25}{9}I$$

$$\frac{34}{9}I = \frac{e^{5x}}{3}\left\{\frac{5}{3}\sin 3x - \cos 3x\right\} + C_1$$

$$I = \frac{3e^{5x}}{34}\left\{\frac{5}{3}\sin 3x - \cos 3x\right\} + C$$

There you are. Now do these in much the same way. Finish them both before moving on to the next frame.

(a) $\int x \ln x \, dx$ $\qquad$ (b) $\int x^3 e^{2x} \, dx$

*Solutions in Frame 30*

**30**

(a) $\int x \ln x \, dx = \ln x \left(\frac{x^2}{2}\right) - \frac{1}{2}\int x^2 . \frac{1}{x} \, dx$

$= \frac{x^2 \ln x}{2} - \frac{1}{2}\int x \, dx$

$= \frac{x^2 \ln x}{2} - \frac{1}{2} . \frac{x^2}{2} + C$

$= \frac{x^2}{2}\left\{\ln x - \frac{1}{2}\right\} + C$

(b) $\int x^3 e^{2x} \, dx = x^3 \left(\frac{e^{2x}}{2}\right) - \frac{3}{2}\int e^{2x} x^2 \, dx$

$= \frac{x^3 e^{2x}}{2} - \frac{3}{2}\left\{x^2\left(\frac{e^{2x}}{2}\right) - \frac{2}{2}\int e^{2x} x \, dx\right\}$

$= \frac{x^3 e^{2x}}{2} - \frac{3x^2 e^{2x}}{4} + \frac{3}{2}\left\{x\left(\frac{e^{2x}}{2}\right) - \frac{1}{2}\int e^{2x} \, dx\right\}$

$= \frac{x^3 e^{2x}}{2} - \frac{3x^2 e^{2x}}{4} + \frac{3xe^{2x}}{4} - \frac{3}{4}\frac{e^{2x}}{2} + C$

$= \frac{e^{2x}}{2}\left\{x^3 - \frac{3x^2}{2} + \frac{3x}{2} - \frac{3}{4}\right\} + C$

That is it. You can now deal with the integration of products.

*The next section of the Programme begins in Frame 31, so move on now and continue the good work*

## Integration by partial fractions

**31**

Suppose we have $\int \frac{x+1}{x^2 - 3x + 2} \, dx$. Clearly this is not one of our standard types, and the numerator is *not* the derivative of the denominator. So how do we go about this one?

In such a case as this, we first of all express the rather cumbersome algebraic fraction in terms of its *partial fractions*, i.e. a number of simpler algebraic fractions which we shall most likely be able to integrate separately without difficulty. $\frac{x+1}{x^2 - 3x + 2}$ can, in fact, be expressed as $\frac{3}{x-2} - \frac{2}{x-1}$.

$$\therefore \int \frac{x+1}{x^2 - 3x + 2} \, dx = \int \frac{3}{x-2} \, dx - \int \frac{2}{x-1} \, dx$$

$$= \ldots\ldots\ldots\ldots$$

**32**

$$\boxed{3\ln(x-2) - 2\ln(x-1) + C}$$

The method, of course, hinges on one's ability to express the given function in terms of its partial fractions.

The rules of *partial fractions* are as follows:

(a) The numerator of the given function must be of lower degree than that of the denominator. If it is not, then first of all divide out by long division.

(b) Factorize the denominator into its prime factors. This is important, since the factors obtained determine the shape of the partial fractions.

(c) A linear factor $(ax + b)$ gives a partial fraction of the form $\dfrac{A}{ax + b}$

(d) Factors $(ax + b)^2$ give partial fractions $\dfrac{A}{ax + b} + \dfrac{B}{(ax + b)^2}$

(e) Factors $(ax + b)^3$ give partial fractions $\dfrac{A}{ax + b} + \dfrac{B}{(ax + b)^2} + \dfrac{C}{(ax + b)^3}$

(f) A quadratic factor $(ax^2 + bx + c)$ gives a partial fraction $\dfrac{Ax + B}{ax^2 + bx + c}$

Copy down this list of rules into your record book for reference. It will be well worth it.

*Then on to the next frame*

**33**

Now for some examples.

**Example 1**

$$\int \frac{x + 1}{x^2 - 3x + 2}\,dx$$

$$\frac{x + 1}{x^2 - 3x + 2} = \frac{x + 1}{(x - 1)(x - 2)} = \frac{A}{x - 1} + \frac{B}{x - 2}$$

Multiply both sides by the denominator $(x - 1)(x - 2)$:

$$x + 1 = A(x - 2) + B(x - 1)$$

This is an identity and true for any value of $x$ we like to substitute. Where possible, choose a value of $x$ which will make one of the brackets zero.

Let $(x - 1) = 0$, i.e. substitute $x = 1$

$$\therefore\ 2 = A(-1) + B(0) \qquad \therefore\ A = -2$$

Let $(x - 2) = 0$, i.e. substitute $x = 2$

$$\therefore\ 3 = A(0) + B(1) \qquad \therefore\ B = 3$$

So the integral can now be written . . . . . . . . . . .

**34**

$$\int \frac{x+1}{x^2-3x+2}\,dx = \int \frac{3}{x-2}\,dx - \int \frac{2}{x-1}\,dx$$

Now the rest is easy.

$$\int \frac{x+1}{x^2-3x+2}\,dx = 3\int \frac{1}{x-2}\,dx - 2\int \frac{1}{x-1}\,dx$$

$$= 3\ln(x-2) - 2\ln(x-1) + C$$ (Do not forget the constant of integration!)

And now another one.

**Example 2**

To determine $\int \frac{x^2}{(x+1)(x-1)^2}\,dx$

Numerator = 2nd degree; denominator = 3rd degree. Rule (a) is satisfied.

Denominator already factorized into its prime factors. Rule (b) is satisfied.

$$\frac{x^2}{(x+1)(x-1)^2} = \frac{A}{x+1} + \frac{B}{x-1} + \frac{C}{(x-1)^2}$$

Clear the denominators: $x^2 = A(x-1)^2 + B(x+1)(x-1) + C(x+1)$

Put $(x-1) = 0$, i.e. $x = 1$ $\quad \therefore 1 = A(0) + B(0) + C(2) \quad \therefore C = \frac{1}{2}$

Put $(x+1) = 0$, i.e. $x = -1$ $\quad \therefore 1 = A(4) + B(0) + C(0) \quad \therefore A = \frac{1}{4}$

When the crafty substitution has come to an end, we can find the remaining constants (in this case, just $B$) by equating coefficients. Choose the highest power involved, i.e. $x^2$ in this example:

$$[x^2] \quad \therefore 1 = A + B \quad \therefore B = 1 - A = 1 - \frac{1}{4} \quad \therefore B = \frac{3}{4}$$

$$\therefore \frac{x^2}{(x+1)(x-1)^2} = \frac{1}{4}\cdot\frac{1}{x+1} + \frac{3}{4}\cdot\frac{1}{x-1} + \frac{1}{2}\cdot\frac{1}{(x-1)^2}$$

$$\therefore \int \frac{x^2}{(x+1)(x-1)^2}\,dx = \frac{1}{4}\int \frac{1}{x+1}\,dx + \frac{3}{4}\int \frac{1}{x-1}\,dx + \frac{1}{2}\int (x-1)^{-2}\,dx$$

$$= \dots\dots\dots\dots$$

**35**

$$\int \frac{x^2}{(x+1)(x-1)^2}\,dx = \frac{1}{4}\ln(x+1) + \frac{3}{4}\ln(x-1) - \frac{1}{2(x-1)} + C$$

**Example 3**

To determine $\int \frac{x^2+1}{(x+2)^3}\,dx$

Rules (a) and (b) of partial fractions are satisfied. The next stage is to write down the form of the partial fractions.

$$\frac{x^2+1}{(x+2)^3} = \dots\dots\dots\dots$$

**36**

$$\frac{x^2+1}{(x+2)^3}=\frac{A}{x+2}+\frac{B}{(x+2)^2}+\frac{C}{(x+2)^3}$$

Now clear the denominators by multiplying both sides by $(x+2)^3$. So we get:

$x^2+1=\ldots\ldots\ldots\ldots$

**37**

$$x^2+1=A(x+2)^2+B(x+2)+C$$

We now put $(x+2)=0$, i.e. $x=-2$

$\therefore\ 4+1=A(0)+B(0)+C \qquad \therefore\ C=5$

There are no other brackets in this identity so we now equate coefficients, starting with the highest power involved, i.e. $x^2$. What does that give us?

**38**

$$A=1$$

Because

$$x^2+1=A(x+2)^2+B(x+2)+C \qquad C=5$$
$$[x^2] \qquad \therefore\ 1=A$$

We now go to the other extreme and equate the lowest power involved, i.e. the constant terms (or absolute terms) on each side:

$[\text{CT}] \qquad \therefore\ 1=4A+2B+C$

$\therefore\ 1=4+2B+5 \qquad \therefore\ 2B=-8 \qquad \therefore\ B=-4$

$$\therefore\ \frac{x^2+1}{(x+2)^3}=\frac{1}{x+2}-\frac{4}{(x+2)^2}+\frac{5}{(x+2)^3}$$

$$\therefore\ \int\frac{x^2+1}{(x+2)^3}\,dx=\ldots\ldots\ldots\ldots$$

**39**

$$\int\frac{x^2+1}{(x+2)^3}\,dx=\ln(x+2)-4\frac{(x+2)^{-1}}{-1}+5\frac{(x+2)^{-2}}{-2}+C$$
$$=\ln(x+2)+\frac{4}{x+2}-\frac{5}{2(x+2)^2}+C$$

*Now for another example, move on to Frame 40*

**40**

**Example 4**

To find $\displaystyle\int\frac{x^2}{(x-2)(x^2+1)}\,dx$

In this example, we have a quadratic factor which will not factorize any further.

$$\therefore\ \frac{x^2}{(x-2)(x^2+1)}=\frac{A}{x-2}+\frac{Bx+C}{x^2+1}$$
$$\therefore\ x^2=A(x^2+1)+(x-2)(Bx+C)$$

Put $(x-2)=0$, i.e. $x=2$

$\therefore\ 4=A(5)+0 \qquad \therefore\ A=\dfrac{4}{5}$

▶

Equate coefficients:

$$[x^2] \quad 1 = A + B \quad \therefore B = 1 - A = 1 - \frac{4}{5} \quad \therefore B = \frac{1}{5}$$

$$[CT] \quad 0 = A - 2C \quad \therefore C = A/2 \quad \therefore C = \frac{2}{5}$$

$$\therefore \frac{x^2}{(x-2)(x^2+1)} = \frac{4}{5}.\frac{1}{x-2} + \frac{\frac{1}{5}x + \frac{2}{5}}{x^2+1}$$

$$= \frac{4}{5}.\frac{1}{x-2} + \frac{1}{5}.\frac{x}{x^2+1} + \frac{2}{5}.\frac{1}{x^2+1}$$

$$\therefore \int \frac{x^2}{(x-2)(x^2+1)}\,dx = \ldots\ldots\ldots\ldots$$

**41**

$$\int \frac{x^2}{(x-2)(x^2+1)}\,dx = \frac{4}{5}\ln(x-2) + \frac{1}{10}\ln(x^2+1) + \frac{2}{5}\tan^{-1}x + C$$

Here is one for you to do on your own.

**Example 5**

Determine $\displaystyle\int \frac{4x^2+1}{x(2x-1)^2}\,dx$

Rules (a) and (b) are satisfied, and the form of the partial fractions will be

$$\frac{4x^2+1}{x(2x-1)^2} = \frac{A}{x} + \frac{B}{2x-1} + \frac{C}{(2x-1)^2}$$

*Off you go then. When you have finished it completely, move on to Frame 42*

**42**

$$\int \frac{4x^2+1}{x(2x-1)^2}\,dx = \ln x - \frac{2}{2x-1} + C$$

Check through your working in detail.

$$\frac{4x^2+1}{x(2x-1)^2} = \frac{A}{x} + \frac{B}{2x-1} + \frac{C}{(2x-1)^2}$$

$$\therefore 4x^2 + 1 = A(2x-1)^2 + Bx(2x-1) + Cx$$

Put $(2x-1) = 0$, i.e. $x = 1/2$

$$\therefore 2 = A(0) + B(0) + \frac{C}{2} \quad \therefore C = 4$$

$$\left.\begin{array}{lll}[x^2] & 4 = 4A + 2B & \therefore 2A + B = 2 \\ [CT] & 1 = A & \end{array}\right\} \quad \begin{array}{l} A = 1 \\ B = 0 \end{array}$$

$$\therefore \frac{4x^2+1}{x(2x-1)^2} = \frac{1}{x} + \frac{4}{(2x-1)^2}$$

$$\therefore \int \frac{4x^2+1}{x(2x-1)^2}\,dx = \int \frac{1}{x}\,dx + 4\int (2x-1)^{-2}\,dx$$

$$= \ln x + \frac{4.(2x-1)^{-1}}{-1.2} + C$$

$$= \ln x - \frac{2}{2x-1} + C$$

*Move on to Frame 43*

43

We have done quite a number of integrals of one type or another in our work so far. We have covered:

1 the basic standard integrals
2 functions of a linear function $x$
3 integrals in which one part is the derivative of the other part
4 integration by parts, i.e. integration of products
5 integration by partial fractions.

Before we finish this Programme on integration, let us look particularly at some types of integrals involving trig functions.

*So, on we go to Frame 44*

## Integration of trigonometric functions

44

(a) *Powers of* $\sin x$ *and of* $\cos x$

(i) We already know that

$$\int \sin x \, dx = -\cos x + C$$

$$\int \cos x \, dx = \sin x + C$$

(ii) To integrate $\sin^2 x$ and $\cos^2 x$, we express the function in terms of the cosine of the double angle:

$$\cos 2x = 1 - 2\sin^2 x \text{ and } \cos 2x = 2\cos^2 x - 1$$

$$\therefore \sin^2 x = \frac{1}{2}(1 - \cos 2x) \text{ and } \cos^2 x = \frac{1}{2}(1 + \cos 2x)$$

$$\therefore \int \sin^2 x \, dx = \frac{1}{2}\int (1 - \cos 2x) \, dx = \frac{x}{2} - \frac{\sin 2x}{4} + C$$

$$\therefore \int \cos^2 x \, dx = \frac{1}{2}\int (1 + \cos 2x) dx = \frac{x}{2} + \frac{\sin 2x}{4} + C$$

Notice how nearly alike these two results are. One must be careful to distinguish between them, so make a note of them in your record book for future reference.

*Then move on to Frame 45*

45

(iii) To integrate $\sin^3 x$ and $\cos^3 x$.

To integrate $\sin^3 x$, we release one of the factors, $\sin x$, from the power and convert the remaining $\sin^2 x$ into $(1 - \cos^2 x)$, thus:

$$\int \sin^3 x \, dx = \int \sin^2 x . \sin x \, dx$$

$$= \int (1 - \cos^2 x) \sin x \, dx$$

$$= \int \sin x \, dx - \int \cos^2 x . \sin x \, dx$$

$$= -\cos x + \frac{\cos^3 x}{3} + C$$

▶

We do not normally remember this as a standard result, but we certainly do remember the method by which we can find

$\int \sin^3 x \, dx$ when necessary.

So, in a similar way, you can now find $\int \cos^3 x \, dx$.

*When you have done it, move on to Frame 46*

**46**

$$\boxed{\int \cos^3 x \, dx = \sin x - \frac{\sin^3 x}{3} + C}$$

Because

$$\int \cos^3 x \, dx = \int \cos^2 x . \cos x \, dx = \int (1 - \sin^2 x) \cos x \, dx$$
$$= \int \cos x \, dx - \int \sin^2 x . \cos x \, dx = \sin x - \frac{\sin^3 x}{3} + C$$

Now what about this one?

(iv) To integrate $\sin^4 x$ and $\cos^4 x$:

$$\int \sin^4 x \, dx = \int (\sin^2 x)^2 \, dx = \int \frac{(1 - \cos 2x)^2}{4} \, dx$$
$$= \int \frac{1 - 2\cos 2x + \cos^2 2x}{4} \, dx \qquad Note \left\{ \begin{array}{l} \cos^2 x = \frac{1}{2}(1 + \cos 2x) \\ \cos^2 2x = \frac{1}{2}(1 + \cos 4x) \end{array} \right.$$
$$= \frac{1}{4} \int \left( 1 - 2\cos 2x + \frac{1}{2} + \frac{1}{2} . \cos 4x \right) dx$$
$$= \frac{1}{4} \int \left( \frac{3}{2} - 2\cos 2x + \frac{1}{2} \cos 4x \right) dx$$
$$= \frac{1}{4} \left\{ \frac{3x}{2} - \frac{2\sin 2x}{2} + \frac{1}{2} . \frac{\sin 4x}{4} \right\} + C = \frac{3x}{8} - \frac{\sin 2x}{4} + \frac{\sin 4x}{32} + C$$

Remember not this result, but the *method*.

Now you find $\int \cos^4 x \, dx$ in much the same way.

**47**

$$\boxed{\int \cos^4 x \, dx = \frac{3x}{8} + \frac{\sin 2x}{4} + \frac{\sin 4x}{32} + C}$$

The working is very much like that of the previous example:

$$\int \cos^4 x \, dx = \int (\cos^2 x)^2 \, dx = \int \frac{(1 + \cos 2x)^2}{4} \, dx$$
$$= \int \frac{(1 + 2\cos 2x + \cos^2 2x)}{4} \, dx = \frac{1}{4} \int \left( 1 + 2\cos 2x + \frac{1}{2} + \frac{1}{2} . \cos 4x \right) dx$$
$$= \frac{1}{4} \int \left( \frac{3}{2} + 2\cos 2x + \frac{1}{2} . \cos 4x \right) dx = \frac{1}{4} \left\{ \frac{3x}{2} + \sin 2x + \frac{\sin 4x}{8} \right\} + C$$
$$= \frac{3x}{8} + \frac{\sin 2x}{4} + \frac{\sin 4x}{32} + C$$

*On to the next frame*

48

(v) To integrate $\sin^5 x$ and $\cos^5 x$.

We can integrate $\sin^5 x$ in very much the same way as we found the integral of $\sin^3 x$:

$$\int \sin^5 x\,dx = \int \sin^4 x.\sin x\,dx = \int (1-\cos^2 x)^2 \sin x\,dx$$

$$= \int (1 - 2\cos^2 x + \cos^4 x)\sin x\,dx$$

$$= \int \sin x\,dx - 2\int \cos^2 x.\sin x\,dx + \int \cos^4 x.\sin x\,dx$$

$$= -\cos x + \frac{2\cos^3 x}{3} - \frac{\cos^5 x}{5} + C$$

Similarly:

$$\int \cos^5 x\,dx = \int \cos^4 x.\cos x\,dx = \int (1-\sin^2 x)^2 \cos x\,dx$$

$$= \int (1 - 2\sin^2 x + \sin^4 x)\cos x\,dx$$

$$= \int \cos x\,dx - 2\int \sin^2 x.\cos x\,dx + \int \sin^4 x.\cos x\,dx$$

$$= \sin x - \frac{2\sin^3 x}{3} + \frac{\sin^5 x}{5} + C$$

Note the method, but do not try to memorize these results. Sometimes we need to integrate higher powers of $\sin x$ and $\cos x$ than those we have considered. In those cases, we make use of a different approach which we shall deal with in due course.

49

(b) *Products of sines and cosines*

Finally, while we are dealing with the integrals of trig functions, let us consider one further type. Here is an example:

$$\int \sin 4x.\cos 2x\,dx$$

To determine this, we make use of the identity

$$2\sin A\cos B = \sin(A+B) + \sin(A-B)$$

$$\therefore\ \sin 4x.\cos 2x = \frac{1}{2}(2\sin 4x\cos 2x)$$

$$= \frac{1}{2}\left\{\sin(4x+2x) + \sin(4x-2x)\right\}$$

$$= \frac{1}{2}\left\{\sin 6x + \sin 2x\right\}$$

$$\therefore\ \int \sin 4x\cos 2x\,dx = \frac{1}{2}\int(\sin 6x + \sin 2x)\,dx$$

$$= -\frac{\cos 6x}{12} - \frac{\cos 2x}{4} + C$$

**50**

This type of integral means, of course, that you must know your trig identities. Do they need polishing up? Now is the chance to revise some of them, anyway.

There are four identities very like the one we have just used:

$$2\sin A\cos B = \sin(A+B) + \sin(A-B)$$
$$2\cos A\sin B = \sin(A+B) - \sin(A-B)$$
$$2\cos A\cos B = \cos(A+B) + \cos(A-B)$$
$$2\sin A\sin B = \cos(A-B) - \cos(A+B)$$

Remember that the compound angles are interchanged in the last line. These are important and very useful, so copy them down into your record book and learn them.

*Now move to Frame 51*

**51**

Now another example of the same kind:

$$\int \cos 5x\sin 3x\,dx = \frac{1}{2}\int (2\cos 5x\sin 3x)\,dx$$
$$= \frac{1}{2}\int\left\{\sin(5x+3x) - \sin(5x-3x)\right\}dx = \frac{1}{2}\int\left\{\sin 8x - \sin 2x\right\}dx$$
$$= \frac{1}{2}\left\{-\frac{\cos 8x}{8} + \frac{\cos 2x}{2}\right\} + C$$
$$= \frac{\cos 2x}{4} - \frac{\cos 8x}{16} + C$$

And now here is one for you to do: $\int \cos 6x\cos 4x\,dx = \dots\dots\dots\dots$

*Off you go. Finish it, then move on to Frame 52*

**52**

$$\boxed{\int \cos 6x\cos 4x\,dx = \frac{\sin 10x}{20} + \frac{\sin 2x}{4} + C}$$

Because

$$\int \cos 6x\cos 4x\,dx = \frac{1}{2}\int 2\cos 6x\cos 4x\,dx$$
$$= \frac{1}{2}\int\left\{\cos 10x + \cos 2x\right\}dx = \frac{1}{2}\left\{\frac{\sin 10x}{10} + \frac{\sin 2x}{2}\right\} + C$$
$$= \frac{\sin 10x}{20} + \frac{\sin 2x}{4} + C$$

Well, there you are. They are all done in the same basic way. Here is one last one for you to do. Take care!

$$\int \sin 5x\sin x\,dx = \dots\dots\dots\dots$$

This will use the last of our four trig identities, the one in which the compound angles are interchanged, so do not get caught.

*When you have finished, move on to Frame 53*

**53**

Well, here it is, worked out in detail. Check your result.

$$\begin{aligned}\int \sin 5x \sin x \, dx &= \frac{1}{2}\int 2 \sin 5x \sin x \, dx \\ &= \frac{1}{2}\int\left\{\cos(5x - x) - \cos(5x + x)\right\} dx \\ &= \frac{1}{2}\int\left\{\cos 4x - \cos 6x\right\} dx \\ &= \frac{1}{2}\left\{\frac{\sin 4x}{4} - \frac{\sin 6x}{6}\right\} + C \\ &= \frac{\sin 4x}{8} - \frac{\sin 6x}{12} + C\end{aligned}$$

This bring us to the end of the first of the Programmes on integration. Now for the **Can You?** checklist and the **Test exercise** on the work you have been doing in this Programme.

*So when you are ready, move on*

## Can You?

### Checklist 16

**54**

Check this list before and after you try the end of Programme test.

On a scale of 1 to 5 how confident are you that you can:

| | Frames |
|---|---|
| • Integrate standard expressions using a table of standard forms?<br>*Yes* ☐ ☐ ☐ ☐ ☐ *No* | 1 to 5 |
| • Integrate functions of a linear form?<br>*Yes* ☐ ☐ ☐ ☐ ☐ *No* | 6 to 10 |
| • Evaluate integrals with integrands of the form $f'(x)/f(x)$ and $f(x).f'(x)$?<br>*Yes* ☐ ☐ ☐ ☐ ☐ *No* | 11 to 20 |
| • Integrate by parts?<br>*Yes* ☐ ☐ ☐ ☐ ☐ *No* | 21 to 30 |
| • Integrate by partial fractions?<br>*Yes* ☐ ☐ ☐ ☐ ☐ *No* | 31 to 42 |
| • Integrate trigonometric functions?<br>*Yes* ☐ ☐ ☐ ☐ ☐ *No* | 44 to 53 |

## Test exercise 16

**55**

The integrals are all quite straightforward, so you will have no trouble with them. Take your time: there is no need to hurry and no extra marks for speed! Look back through the notes you have made in your record book and brush up any points on which you are not perfectly clear.

Determine the following integrals:

1 $\displaystyle\int e^{\cos x}\sin x\,\mathrm{d}x$ 2 $\displaystyle\int\frac{\ln x}{\sqrt{x}}\,\mathrm{d}x$

3 $\displaystyle\int\tan^2 x\,\mathrm{d}x$ 4 $\displaystyle\int x^2\sin 2x\,\mathrm{d}x$

5 $\displaystyle\int e^{-3x}\cos 2x\,\mathrm{d}x$ 6 $\displaystyle\int\sin^5 x\,\mathrm{d}x$

7 $\displaystyle\int\cos^4 x\,\mathrm{d}x$ 8 $\displaystyle\int\frac{4x+2}{x^2+x+5}\,\mathrm{d}x$

9 $\displaystyle\int x\sqrt{(1+x^2)}\,\mathrm{d}x$ 10 $\displaystyle\int\frac{2x-1}{x^2-8x+15}\,\mathrm{d}x$

11 $\displaystyle\int\frac{2x^2+x+1}{(x-1)(x^2+1)}\,\mathrm{d}x$ 12 $\displaystyle\int\sin 5x\cos 3x\,\mathrm{d}x$

*You are now ready to start the next Programme on integration*

## Further problems 16

**56**

Determine the following:

1 $\displaystyle\int\frac{3x^2}{(x-1)(x^2+x+1)}\,\mathrm{d}x$ 2 $\displaystyle\int_0^{\pi/2}\sin 7x\cos 5x\,\mathrm{d}x$

3 $\displaystyle\int\frac{\sin 2x}{1+\cos^2 x}\,\mathrm{d}x$ 4 $\displaystyle\int_0^{a/2}x^2(a^2-x^2)^{-3/2}\,\mathrm{d}x$

5 $\displaystyle\int_0^{\pi}x\sin^2 x\,\mathrm{d}x$ 6 $\displaystyle\int\frac{2x+1}{(x^2+x+1)^{3/2}}\,\mathrm{d}x$

7 $\displaystyle\int\frac{x+1}{(x-1)(x^2+x+1)}\,\mathrm{d}x$ 8 $\displaystyle\int\frac{x^2}{x+1}\,\mathrm{d}x$

9 $\displaystyle\int\frac{2x^2+x+1}{(x+1)(x^2+1)}\,\mathrm{d}x$ 10 $\displaystyle\int_0^{\pi}(\pi-x)\cos x\,\mathrm{d}x$

11 $\displaystyle\int_0^{n}x^2(n-x)^p\,\mathrm{d}x$, for $p>0$ 12 $\displaystyle\int\frac{4x^2-7x+13}{(x-2)(x^2+1)}\,\mathrm{d}x$

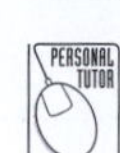

13 $\displaystyle\int_0^{\pi/2}\sin 5x\cos 3x\,\mathrm{d}x$ 14 $\displaystyle\int\frac{\sin^{-1}x}{\sqrt{1-x^2)}}\,\mathrm{d}x$

 **15** $\displaystyle\int_0^1 \frac{x^2 - 2x}{(2x+1)(x^2+1)}\,dx$

**16** $\displaystyle\int_0^{\pi} x^2 \sin x\,dx$

 **17** $\displaystyle\int_0^{\pi} x^2 \sin^2 x\,dx$

**18** $\displaystyle\int_0^1 x \tan^{-1} x\,dx$

**19** $\displaystyle\int \frac{dx}{x^2(1+x^2)}$

**20** $\displaystyle\int x\sqrt{1+x^2}\,dx$

**21** $\displaystyle\int \frac{8-x}{(x-2)^2(x+1)}\,dx$

**22** $\displaystyle\int_0^{\pi} e^{2x} \cos 4x\,dx$

**23** $\displaystyle\int_0^{\pi/2} \sin^5 x \cos^3 x\,dx$

**24** $\displaystyle\int_0^{\pi/6} e^{2\theta} \cos 3\theta\,d\theta$

 **25** $\displaystyle\int_0^{\pi/\omega} \sin \omega t \cos 2\omega t\,dt$

**26** $\displaystyle\int \tan^2 x \sec^2 x\,dx$

 **27** $\displaystyle\int \frac{2x+3}{(x-4)(5x+2)}\,dx$

**28** $\displaystyle\int \frac{dx}{\sqrt{x^2+4x+4}}$

 **29** $\displaystyle\int \frac{5x^2+11x-2}{(x+5)(x^2+9)}\,dx$

**30** $\displaystyle\int \frac{x-1}{9x^2-18x+17}\,dx$

 **31** $\displaystyle\int \frac{4x^5}{x^4-1}\,dx$

**32** $\displaystyle\int x^2 \ln(1+x^2)\,dx$

 **33** $\displaystyle\int \frac{\cos\theta - \sin\theta}{\cos\theta + \sin\theta}\,d\theta$

**34** $\displaystyle\int \frac{1-\sin\theta}{\cos^2\theta}\,d\theta$

 **35** $\displaystyle\int \frac{2x-3}{(x-1)(x-2)(x+3)}\,dx$

**36** $\displaystyle\int_0^{\pi/3} \frac{\sin x}{(1+\cos x)^2}\,dx$

 **37** $\displaystyle\int_1^2 (x-1)^2 \ln x\,dx$

**38** $\displaystyle\int \frac{4x^2-x+12}{x(x^2+4)}\,dx$

 **39** $\displaystyle\int \frac{x^3+x+1}{x^4+x^2}\,dx$

**40** If $L\dfrac{di}{dt} + Ri = E$, where $L$, $R$ and $E$ are constants, and it is known that $i = 0$ at $t = 0$, show that

$$\int_0^t (Ei - Ri^2)dt = \frac{Li^2}{2}.$$

*Note*: Some of the integrals above are *definite integrals*, so here is a reminder.

In $\displaystyle\int_a^b f(x)\,dx$, the values of $a$ and $b$ are called the *limits* of the integral.

If $\displaystyle\int f(x)\,dx = F(x) + C$

then $\displaystyle\int_a^b f(x)\,dx = [F(x)]_{x=b} - [F(x)]_{x=a}$

Now visit the companion website for this book at www.palgrave.com/stroud for more questions applying this mathematics to science and engineering.

# Integration 2

## Learning outcomes

When you have completed this Programme you will be able to:

- ☐ Evaluate integrals with integrands of the form $\pm 1/(Z^2 - A^2)$
- ☐ Evaluate integrals with integrands of the form $1/(Z^2 + A^2)$
- ☐ Evaluate integrals with integrands of the form $1/\sqrt{A^2 - Z^2}$
- ☐ Evaluate integrals with integrands of the form $1/\sqrt{Z^2 + A^2}$
- ☐ Evaluate integrals with integrands of the form $1/\sqrt{Z^2 - A^2}$
- ☐ Evaluate integrals with integrands of the form $\sqrt{A^2 - Z^2}$, $\sqrt{Z^2 + A^2}$ and $\sqrt{Z^2 - A^2}$
- ☐ Evaluate integrals with integrands of the form $1/(a + b\sin^2 x + c\cos^2 x)$
- ☐ Evaluate integrals with integrands of the form $1/(a + b\sin x + c\cos x)$

1

**1 Consider the integral $\int \frac{dZ}{Z^2 - A^2}$**

From our work on integration, you will recognize that the denominator can be factorized and that the function can therefore be expressed in its *partial fractions*:

$$\frac{1}{Z^2 - A^2} = \frac{1}{(Z-A)(Z+A)} = \frac{P}{Z-A} + \frac{Q}{Z+A}$$

where $P$ and $Q$ are constants.

$$\therefore \quad 1 = P(Z+A) + Q(Z-A)$$

$$\text{Put } Z = A \quad \therefore \quad 1 = P(2A) + Q(0) \quad \therefore \quad P = \frac{1}{2A}$$

$$\text{Put } Z = -A \quad \therefore \quad 1 = P(0) + Q(-2A) \quad \therefore \quad Q = -\frac{1}{2A}$$

$$\therefore \quad \frac{1}{Z^2 - A^2} = \frac{1}{2A} \cdot \frac{1}{Z-A} - \frac{1}{2A} \cdot \frac{1}{Z+A}$$

$$\therefore \quad \int \frac{1}{Z^2 - A^2}\, dZ = \frac{1}{2A} \int \frac{1}{Z-A}\, dZ - \frac{1}{2A} \int \frac{1}{Z+A}\, dZ$$

$$= \ldots\ldots\ldots\ldots$$

2

$$\int \frac{1}{Z^2 - A^2}\, dZ = \frac{1}{2A} \cdot \ln(Z-A) - \frac{1}{2A} \cdot \ln(Z+A) + C$$

$$= \boxed{\frac{1}{2A} \cdot \ln\left\{\frac{Z-A}{Z+A}\right\} + C}$$

This is the first of nine standard results which we are going to establish in this Programme. They are useful to remember since the standard results will remove the need to work each example in detail, as you will see.

$$\text{We have} \quad \int \frac{1}{Z^2 - A^2}\, dZ = \frac{1}{2A} \ln\left\{\frac{Z-A}{Z+A}\right\} + C$$

$$\therefore \quad \int \frac{1}{Z^2 - 16}\, dZ = \int \frac{1}{Z^2 - 4^2}\, dZ = \frac{1}{8} \ln\left\{\frac{Z-4}{Z+4}\right\} + C$$

$$\text{and} \quad \int \frac{1}{x^2 - 5}\, dx = \int \frac{1}{x^2 - (\sqrt{5})^2}\, dx = \frac{1}{2\sqrt{5}} \ln\left\{\frac{x-\sqrt{5}}{x+\sqrt{5}}\right\} + C$$

(Note that 5 can be written as the square of its own square root.)

$$\text{So} \quad \int \frac{1}{Z^2 - A^2}\, dZ = \frac{1}{2A} \ln\left\{\frac{Z-A}{Z+A}\right\} + C \qquad (1)$$

*Copy this result into your record book and move on to Frame 3*

3

We had:

$$\boxed{\int \frac{dZ}{Z^2 - A^2} = \frac{1}{2A} \ln\left\{\frac{Z-A}{Z+A}\right\} + C}$$

So therefore:

$$\int \frac{dZ}{Z^2 - 25} = \ldots\ldots\ldots\ldots$$

$$\int \frac{dZ}{Z^2 - 7} = \ldots\ldots\ldots\ldots$$

**4**

$$\int \frac{dZ}{Z^2 - 25} = \int \frac{dZ}{Z^2 - 5^2} = \boxed{\frac{1}{10} \cdot \ln\left\{\frac{Z-5}{Z+5}\right\} + C}$$

$$\int \frac{dZ}{Z^2 - 7} = \int \frac{dZ}{Z^2 - (\sqrt{7})^2} = \boxed{\frac{1}{2\sqrt{7}} \cdot \ln\left\{\frac{Z-\sqrt{7}}{Z+\sqrt{7}}\right\} + C}$$

Now what about this one?

$$\int \frac{1}{x^2 + 4x + 2} dx$$

At first sight, this seems to have little to do with the standard result, or to the examples we have done so far. However, let us rewrite the denominator, thus:

$$x^2 + 4x + 2 = x^2 + 4x \qquad + 2 \quad \text{(Nobody would argue with that!)}$$

Now we complete the square with the first two terms, by adding on the square of half the coefficient of $x$:

$$x^2 + 4x + 2 = x^2 + 4x + 2^2 + 2 - 2^2$$

and of course we must subtract an equal amount, i.e. 4, to keep the identity true.

$$\therefore \quad x^2 + 4x + 2 = \underbrace{x^2 + 4x + 2^2}\, + 2 - 4$$

$$= (x+2)^2 - 2$$

So $\displaystyle\int \frac{1}{x^2 + 4x + 2} dx$ can be written $\displaystyle\int \frac{1}{\dots\dots\dots\dots} dx$

*Move on to Frame 5*

**5**

$$\int \frac{1}{x^2 + 4x + 2} dx = \boxed{\int \frac{1}{(x+2)^2 - 2} dx}$$

Then we can express the constant 2 as the square of its own square root.

$$\therefore \quad \int \frac{1}{x^2 + 4x + 2} dx = \int \frac{1}{(x+2)^2 - (\sqrt{2})^2} dx$$

You will see that we have rewritten the given integral in the form $\displaystyle\int \frac{1}{Z^2 - A^2} dZ$ where, in this case, $Z = (x+2)$ and $A = \sqrt{2}$. Now the standard result was:

$$\int \frac{1}{Z^2 - A^2} dZ = \frac{1}{2A} \ln\left\{\frac{Z-A}{Z+A}\right\} + C$$

Substituting our expressions for $Z$ and $A$ in this result, gives:

$$\int \frac{1}{x^2 + 4x + 2} dx = \int \frac{1}{(x+2)^2 - (\sqrt{2})^2} dx$$

$$= \frac{1}{2\sqrt{2}} \ln\left\{\frac{x + 2 - \sqrt{2}}{x + 2 + \sqrt{2}}\right\} + C$$

Once we have found our particular expressions for $Z$ and $A$, all that remains is to substitute these expressions in the standard result.

*On now to Frame 6*

6

Here is another example:

$$\int \frac{1}{x^2+6x+4}\,dx$$

First complete the square with the first two terms of the given denominator and subtract an equal amount:

$$\begin{aligned} x^2+6x+4 &= x^2+6x \qquad +4 \\ &= \underbrace{x^2+6x+3^2}\,+4-9 \\ &= (x+3)^2-5 \\ &= (x+3)^2-(\sqrt{5})^2 \end{aligned}$$

So $\displaystyle\int \frac{1}{x^2+6x+4}\,dx = \int \frac{1}{(x+3)^2-(\sqrt{5})^2}\,dx$

$= \ldots\ldots\ldots\ldots$

7

$$\boxed{\int \frac{1}{x^2+6x+4}\,dx = \frac{1}{2\sqrt{5}}\ln\frac{x+3-\sqrt{5}}{x+3+\sqrt{5}}+C}$$

And another on your own:

Find $\displaystyle\int \frac{1}{x^2+10x+18}\,dx$

*When you have finished, move on to Frame 8*

8

$$\int \frac{1}{x^2-10x+18}\,dx = \boxed{\frac{1}{2\sqrt{7}}\ln\left\{\frac{x-5-\sqrt{7}}{x-5+\sqrt{7}}\right\}+C}$$

Because

$$\begin{aligned} x^2-10x+18 &= x^2-10x \qquad +18 \\ &= x^2-10x+5^2+18-25 \\ &= (x-5)^2-7 \\ &= (x-5)^2-(\sqrt{7})^2 \end{aligned}$$

$$\therefore \quad \int \frac{1}{x^2-10x+18}\,dx = \frac{1}{2\sqrt{7}}\ln\left\{\frac{x-5-\sqrt{7}}{x-5+\sqrt{7}}\right\}+C$$

*Now on to Frame 9*

**9**

Now what about this one? $\int \frac{1}{5x^2 - 2x - 4}\,dx$

In order to complete the square, as we have done before, the coefficient of $x^2$ must be 1. Therefore, in the denominator, we must first of all take out a factor 5 to reduce the second-degree term to a single $x^2$.

$$\therefore \quad \int \frac{1}{5x^2 - 2x - 4}\,dx = \frac{1}{5}\int \frac{1}{x^2 - \frac{2}{5}x - \frac{4}{5}}\,dx$$

Now we can proceed as in the previous examples:

$$\begin{aligned} x^2 - \frac{2}{5}x - \frac{4}{5} &= x^2 - \frac{2}{5}x \qquad\quad - \frac{4}{5} \\ &= x^2 - \frac{2}{5}x + \left(\frac{1}{5}\right)^2 - \frac{4}{5} - \frac{1}{25} \\ &= \left(x - \frac{1}{5}\right)^2 - \frac{21}{25} \\ &= \left(x - \frac{1}{5}\right)^2 - \left(\frac{\sqrt{21}}{5}\right)^2 \end{aligned}$$

$$\therefore \quad \int \frac{1}{5x^2 - 2x - 4}\,dx = \ldots\ldots\ldots\ldots$$

(Remember the factor 1/5 in the front!)

**10**

$$\boxed{\int \frac{1}{5x^2 - 2x - 4}\,dx = \frac{1}{2\sqrt{21}}\ln\left\{\frac{5x - 1 - \sqrt{21}}{5x - 1 + \sqrt{21}}\right\} + C}$$

Here is the working: follow it through.

$$\begin{aligned} \int \frac{1}{5x^2 - 2x - 4}\,dx &= \frac{1}{5}\int \frac{1}{\left(x - \frac{1}{5}\right)^2 - \left(\frac{\sqrt{21}}{5}\right)^2}\,dx \\ &= \frac{1}{5}\cdot\frac{5}{2\sqrt{21}}\ln\left\{\frac{x - 1/5 - \sqrt{21}/5}{x - 1/5 + \sqrt{21}/5}\right\} + C \\ &= \frac{1}{2\sqrt{21}}\ln\left\{\frac{5x - 1 - \sqrt{21}}{5x - 1 + \sqrt{21}}\right\} + C \end{aligned}$$

**2 Now, in very much the same way, let us establish the second standard result by considering $\int \frac{dZ}{A^2 - Z^2}$**

This looks rather like the last one and can be determined again by partial fractions. Work through it on your own and determine the general result.

*Then move on to Frame 11 and check your working*

11

$$\int \frac{dZ}{A^2 - Z^2} = \frac{1}{2A} \ln\left\{\frac{A+Z}{A-Z}\right\} + C$$

Because

$$\frac{1}{A^2 - Z^2} = \frac{1}{(A-Z)(A+Z)} = \frac{P}{A-Z} + \frac{Q}{A+Z}$$

$$\therefore \quad 1 = P(A+Z) + Q(A-Z)$$

$$\text{Put } Z = A: \quad \therefore \quad 1 = P(2A) + Q(0) \quad \therefore \quad P = \frac{1}{2A}$$

$$\text{Put } Z = -A: \quad \therefore \quad 1 = P(0) + Q(2A) \quad \therefore \quad Q = \frac{1}{2A}$$

$$\therefore \quad \int \frac{1}{A^2 - Z^2}\, dZ = \frac{1}{2A} \cdot \int \frac{1}{A+Z}\, dZ + \frac{1}{2A} \int \frac{1}{A-Z}\, dZ$$

$$= \frac{1}{2A} \cdot \ln(A+Z) - \frac{1}{2A} \cdot \ln(A-Z) + C$$

$$\therefore \quad \int \frac{1}{A^2 - Z^2}\, dZ = \frac{1}{2A} \ln\left\{\frac{A+Z}{A-Z}\right\} + C \qquad (2)$$

Copy this second standard form into your record book and compare it with the first result. They are very much alike.

*Move to Frame 12*

12

So we have:

$$\int \frac{dZ}{Z^2 - A^2} = \frac{1}{2A} \ln\left\{\frac{Z-A}{Z+A}\right\} + C$$

$$\int \frac{dZ}{A^2 - Z^2} = \frac{1}{2A} \ln\left\{\frac{A+Z}{A-Z}\right\} + C$$

Note how nearly alike these two results are.

Now for some examples on the second standard form.

**Example 1**

$$\int \frac{1}{9 - x^2}\, dx = \int \frac{1}{3^2 - x^2}\, dx = \frac{1}{6} \ln\left\{\frac{3+x}{3-x}\right\} + C$$

**Example 2**

$$\int \frac{1}{5 - x^2}\, dx = \int \frac{1}{(\sqrt{5})^2 - x^2}\, dx = \frac{1}{2\sqrt{5}} \ln\left\{\frac{\sqrt{5}+x}{\sqrt{5}-x}\right\} + C$$

**Example 3**

$$\int \frac{1}{3 - x^2}\, dx = \ldots\ldots\ldots\ldots$$

**13**

$$\frac{1}{2\sqrt{3}}\ln\left\{\frac{\sqrt{3}+x}{\sqrt{3}-x}\right\}+C$$

**Example 4**

$$\int\frac{1}{3+6x-x^2}\,dx$$

We complete the square in the denominator as before, but we must be careful of the signs – and do not forget, the coefficient of $x^2$ must be 1.

So we do it like this:

$$3+6x-x^2=3-(x^2-6x\qquad)$$

Note that we put the $x^2$ term and the $x$ term inside brackets with a minus sign outside. Naturally, the $6x$ becomes $-6x$ inside the brackets. Now we can complete the square inside the brackets and *add* on a similar amount outside the brackets (since everything inside the brackets is negative).

$$\begin{aligned}\text{So}\quad 3+6x-x^2&=3-(x^2-6x+3^2)+9\\&=12-(x-3)^2\\&=(2\sqrt{3})^2-(x-3)^2\end{aligned}$$

In this case, then, $A=2\sqrt{3}$ and $Z=(x-3)$

$$\begin{aligned}\therefore\quad\int\frac{1}{3+6x-x^2}\,dx&=\int\frac{1}{(2\sqrt{3})^2-(x-3)^2}\,dx\\&=\ldots\ldots\ldots\ldots\end{aligned}$$

*Finish it off*

**14**

$$\frac{1}{4\sqrt{3}}\ln\left\{\frac{2\sqrt{3}+x-3}{2\sqrt{3}-x+3}\right\}+C$$

Here is another example of the same type:

**Example 5**

$$\int\frac{1}{9-4x-x^2}\,dx$$

First of all we carry out the 'completing the square' routine:

$$\begin{aligned}9-4x-x^2&=9-(x^2+4x\qquad)\\&=9-(x^2+4x+2^2)+4\\&=13-(x+2)^2\\&=(\sqrt{13})^2-(x+2)^2\end{aligned}$$

In this case, $A=\sqrt{13}$ and $Z=(x+2)$.

Now we know that $\displaystyle\int\frac{dZ}{A^2-Z^2}=\frac{1}{2A}\ln\left\{\frac{A+Z}{A-Z}\right\}+C$

So that, in this example $\displaystyle\int\frac{1}{9-4x-x^2}\,dx=\ldots\ldots\ldots\ldots$

15

$$\boxed{\frac{1}{2\sqrt{13}} \ln\left\{\frac{\sqrt{13}+x+2}{\sqrt{13}-x-2}\right\} + C}$$

**Example 6**

$$\int \frac{1}{5+4x-2x^2}\,dx$$

Remember that we must first remove the factor 2 from the denominator to reduce the coefficient of $x^2$ to 1.

$$\therefore \quad \int \frac{1}{5+4x-2x^2}\,dx = \frac{1}{2}\int \frac{1}{\frac{5}{2}+2x-x^2}\,dx$$

Now we proceed as before:

$$\begin{aligned}
\frac{5}{2}+2x-x^2 &= \frac{5}{2}-(x^2-2x \qquad )\\
&= \frac{5}{2}-(x^2-2x+1^2)+1\\
&= \frac{7}{2}-(x-1)^2\\
&= (\sqrt{3{\cdot}5})^2-(x-1)^2
\end{aligned}$$

$$\therefore \quad \int \frac{1}{5+4x-2x^2}\,dx = \ldots\ldots\ldots\ldots$$

(Do not forget the factor 2 we took out of the denominator.)

---

16

$$\boxed{\frac{1}{4\sqrt{3{\cdot}5}} \ln\left\{\frac{\sqrt{3{\cdot}5}+x-1}{\sqrt{3{\cdot}5}-x+1}\right\} + C}$$

Right. Now just one more.

**Example 7**

Determine $\displaystyle\int \frac{1}{6-6x-5x^2}\,dx$

What is the first thing to do?

---

17

> Reduce the coefficient of $x^2$ to 1,
> i.e. take out a factor 5 from the denominator

Correct. Let us do it then.

$$\int \frac{1}{6-6x-5x^2}\,dx = \frac{1}{5}\int \frac{1}{\frac{6}{5}-\frac{6}{5}x-x^2}\,dx$$

Now can you complete the square as usual and finish it off.

*Then move to Frame 18*

**18**

$$\int \frac{1}{6-6x-5x^2}\,dx = \frac{1}{2\sqrt{39}}\ln\left\{\frac{\sqrt{39}+5x+3}{\sqrt{39}-5x-3}\right\}+C$$

Because

$$\int \frac{1}{6-6x-5x^2}\,dx = \frac{1}{5}\int \frac{1}{\frac{6}{5}-\frac{6}{5}x-x^2}\,dx$$

$$\frac{6}{5}-\frac{6}{5}x-x^2 = \frac{6}{5}-\left(x^2+\frac{6}{5}x\qquad\right)$$

$$= \frac{6}{5}-\left\{x^2+\frac{6}{5}x+\left(\frac{3}{5}\right)^2\right\}+\frac{9}{25}$$

$$= \frac{39}{25}-\left(x+\frac{3}{5}\right)^2 = \left(\frac{\sqrt{39}}{5}\right)^2-\left(x+\frac{3}{5}\right)^2$$

So that $\quad A = \frac{\sqrt{39}}{5}$ and $Z = \left(x+\frac{3}{5}\right)$

Now $\quad \int \frac{1}{A^2-Z^2}\,dZ = \frac{1}{2A}\ln\left\{\frac{A+Z}{A-Z}\right\}+C$

$$\therefore \quad \int \frac{1}{6-6x-x^2} = \frac{1}{5}\cdot\frac{5}{2\sqrt{39}}\ln\left\{\frac{\sqrt{39}/5+x+3/5}{\sqrt{39}/5-x-3/5}\right\}+C$$

$$= \frac{1}{2\sqrt{39}}\ln\left\{\frac{\sqrt{39}+5x+3}{\sqrt{39}-5x-3}\right\}+C$$

*Now move to Frame 19*

**19**

By way of revision, cover up your notes and complete the following. Do not work out the integrals in detail; just quote the results.

(a) $\int \frac{dZ}{Z^2-A^2} = \ldots\ldots\ldots\ldots$

(b) $\int \frac{dZ}{A^2-Z^2} = \ldots\ldots\ldots\ldots$

*Check your results with Frame 20*

**20**

$$\int \frac{dZ}{Z^2-A^2} = \frac{1}{2A}\ln\left\{\frac{Z-A}{Z+A}\right\}+C$$
$$\int \frac{dZ}{A^2-Z^2} = \frac{1}{2A}\ln\left\{\frac{A+Z}{A-Z}\right\}+C$$

### 3 Now for the third standard form

Consider $\int \frac{dZ}{Z^2+A^2}$

Here the denominator will not factorize, so we cannot apply the rules of partial fractions. We now turn to substitution, i.e. we try to find a substitution for $Z$ which will enable us to write the integral in a form which we already know how to tackle.

▶

Suppose we put $Z = A\tan\theta$.

Then $Z^2 + A^2 = A^2\tan^2\theta + A^2 = A^2(1+\tan^2\theta) = A^2\sec^2\theta$

Also $\dfrac{dZ}{d\theta} = A\sec^2\theta$ i.e. $dZ = A\sec^2\theta\, d\theta$

The integral now becomes:

$$\int \frac{1}{Z^2 + A^2}\, dZ = \int \frac{1}{A^2\sec^2\theta} \cdot A\sec^2\theta\, d\theta = \int \frac{1}{A}\, d\theta = \frac{1}{A}\cdot\theta + C$$

This is a nice simple result, but we cannot leave it like that, for $\theta$ is a variable we introduced in the working. We must express $\theta$ in terms of the original variable $Z$:

$$Z = A\tan\theta \quad \therefore \quad \frac{Z}{A} = \tan\theta \quad \therefore \quad \theta = \tan^{-1}\frac{Z}{A}$$

$$\therefore \int \frac{1}{Z^2 + A^2}\, dZ = \frac{1}{A}\tan^{-1}\left\{\frac{Z}{A}\right\} + C \qquad (3)$$

*Add this one to your growing list of standard forms*

---

**21**

$$\int \frac{1}{Z^2 + A^2}\, dZ = \frac{1}{A}\tan^{-1}\left\{\frac{Z}{A}\right\} + C$$

**Example 1**

$$\int \frac{1}{x^2 + 16}\, dx = \int \frac{1}{x^2 + 4^2}\, dx = \frac{1}{4}\tan^{-1}\left\{\frac{x}{4}\right\} + C$$

**Example 2**

$$\int \frac{1}{x^2 + 10x + 30}\, dx$$

As usual, we complete the square in the denominator:

$$\begin{aligned} x^2 + 10x + 30 &= x^2 + 10x \qquad + 30 \\ &= x^2 + 10x + 5^2 + 30 - 25 \\ &= (x+5)^2 + 5 = (x+5)^2 + (\sqrt{5})^2 \end{aligned}$$

$$\begin{aligned} \therefore \int \frac{1}{x^2 + 10x + 30}\, dx &= \int \frac{1}{(x+5)^2 + (\sqrt{5})^2}\, dx \\ &= \ldots\ldots\ldots\ldots \end{aligned}$$

---

**22**

$$\frac{1}{\sqrt{5}} \cdot \tan^{-1}\left\{\frac{x+5}{\sqrt{5}}\right\} + C$$

Once you know the standard form, you can find the expressions for Z and A in any example and then substitute these in the result. Here you are; do this one on your own.

**Example 3**

Determine $\displaystyle\int \frac{1}{2x^2 + 12x + 32}\, dx$

Take your time over it. Remember all the rules we have used and then you cannot go wrong.

*When you have completed it, move to Frame 23 and check your working*

---

**23**

$$\int \frac{1}{2x^2 + 12x + 32}\,dx = \frac{1}{2\sqrt{7}}\tan^{-1}\left\{\frac{x+3}{\sqrt{7}}\right\} + C$$

Check your working.

$$\int \frac{1}{2x^2 + 12x + 32}\,dx = \frac{1}{2}\int \frac{1}{x^2 + 6x + 16}\,dx$$

$$\begin{aligned} x^2 + 6x + 16 &= x^2 + 6x \qquad + 16 \\ &= \underbrace{x^2 + 6x + 3^2} + 16 - 9 \\ &= (x+3)^2 + 7 \\ &= (x+3)^2 + (\sqrt{7})^2 \end{aligned}$$

So $Z = (x+3)$ and $A = \sqrt{7}$

$$\int \frac{1}{Z^2 + A^2}\,dZ = \frac{1}{A}\tan^{-1}\left\{\frac{Z}{A}\right\} + C$$

$$\therefore \quad \int \frac{1}{2x^2 + 12x + 32}\,dx = \frac{1}{2}\cdot\frac{1}{\sqrt{7}}\tan^{-1}\left\{\frac{x+3}{\sqrt{7}}\right\} + C$$

*Now move to Frame 24*

**24**

**4 Let us now consider a different integral,** $\int \frac{1}{\sqrt{A^2 - Z^2}}\,dZ$

We clearly cannot employ partial fractions, because of the root sign. So we must find a suitable substitution.

Put $\qquad Z = A\sin\theta$

Then $\qquad A^2 - Z^2 = A^2 - A^2\sin^2\theta$

$$= A^2(1 - \sin^2\theta) = A^2\cos^2\theta$$

$$\sqrt{A^2 - Z^2} = A\cos\theta$$

Also $\frac{dZ}{d\theta} = A\cos\theta \quad \therefore \quad dZ = A\cos\theta.d\theta$

So the integral becomes:

$$\int \frac{1}{\sqrt{A^2 - Z^2}}\,dZ = \int \frac{1}{A\cos\theta}.A\cos\theta.d\theta$$

$$= \int d\theta = \theta + C$$

Expressing $\theta$ in terms of the original variable:

$$Z = A\sin\theta \qquad \therefore \quad \sin\theta = \frac{Z}{A} \qquad \therefore \quad \theta = \sin^{-1}\frac{Z}{A}$$

$$\therefore \quad \int \frac{1}{\sqrt{A^2 - Z^2}}\,dZ = \sin^{-1}\left\{\frac{Z}{A}\right\} + C \qquad (4)$$

*This is our next standard form, so add it to your record book.*
*Then move on to Frame 25*

**25**

$$\int \frac{1}{\sqrt{A^2 - Z^2}}\,dZ = \sin^{-1}\left\{\frac{Z}{A}\right\} + C$$

**Example 1**

$$\int \frac{1}{\sqrt{25 - x^2}}\,dx = \int \frac{1}{\sqrt{5^2 - x^2}}\,dx$$
$$= \sin^{-1}\left\{\frac{x}{5}\right\} + C$$

**Example 2**

$$\int \frac{1}{\sqrt{3 - 2x - x^2}}\,dx$$

As usual $3 - 2x - x^2 = 3 - (x^2 + 2x \qquad)$

$$= 3 - (x^2 + 2x + 1^2) + 1$$
$$= 4 - (x+1)^2 = 2^2 - (x+1)^2$$

So in this case: $A = 2$ and $Z = (x + 1)$

$$\int \frac{1}{\sqrt{3 - 2x - x^2}}\,dx = \int \frac{1}{\sqrt{2^2 - (x+1)^2}}\,dx$$
$$= \sin^{-1}\left\{\frac{x+1}{2}\right\} + C$$

Similarly:

**Example 3**

$$\int \frac{1}{\sqrt{5 - 4x - x^2}}\,dx = \ldots\ldots\ldots\ldots$$

**26**

$$\int \frac{1}{\sqrt{5 - 4x - x^2}}\,dx = \sin^{-1}\left\{\frac{x+2}{3}\right\} + C$$

Because

$$5 - 4x - x^2 = 5 - (x^2 + 4x \qquad)$$
$$= 5 - (x^2 + 4x + 2^2) + 4$$
$$= 9 - (x+2)^2 = 3^2 - (x+2)^2$$
$$\therefore \int \frac{1}{\sqrt{5 - 4x - x^2}}\,dx = \sin^{-1}\left\{\frac{x+2}{3}\right\} + C$$

Now this one:

**Example 4**

Determine $\displaystyle\int \frac{1}{\sqrt{14 - 12x - 2x^2}}\,dx$

Before we can complete the square, we must reduce the coefficient of $x^2$ to 1, i.e. we must divide the expression $14 - 12x - 2x^2$ by 2, but note that this becomes $\sqrt{2}$ when brought outside the root sign.

$$\int \frac{1}{\sqrt{14 - 12x - 2x^2}}\,dx = \frac{1}{\sqrt{2}}\int \frac{1}{\sqrt{7 - 6 - x^2}}\,dx$$

*Now finish that as in the previous example*

**27**

$$\int \frac{1}{\sqrt{14-12x-2x^2}}\,dx = \frac{1}{\sqrt{2}}\sin^{-1}\left\{\frac{x+3}{4}\right\} + C$$

Because

$$\int \frac{1}{\sqrt{14-12x-2x^2}}\,dx = \frac{1}{\sqrt{2}}\int \frac{1}{\sqrt{7-6x-x^2}}\,dx$$

$$\begin{aligned}7-6x-x^2 &= 7-(x^2+6x \qquad )\\ &= 7-(x^2+6x+3^2)+9\\ &= 16-(x+3)^2\\ &= 4^2-(x+3)^2\end{aligned}$$

So $A=4$ and $Z=(x+3)$

$$\int \frac{1}{\sqrt{A^2-Z^2}}\,dZ = \sin^{-1}\left\{\frac{Z}{A}\right\} + C$$

$$\therefore \quad \int \frac{1}{\sqrt{14-12x-2x^2}}\,dx = \frac{1}{\sqrt{2}}\sin^{-1}\left\{\frac{x+3}{4}\right\} + C$$

**28**

**5 Let us now look at the next standard integral in the same way**

To determine $\int \frac{dZ}{\sqrt{Z^2+A^2}}$. Again we try to find a convenient substitution for $Z$, but no trig substitution converts the function into a form that we can manage.

We therefore have to turn to the *hyperbolic identities* and put $Z = a\sinh\theta$.

Then $\quad Z^2+A^2 = A^2\sinh^2\theta + A^2 = A^2(\sinh^2\theta+1)$

Remember $\quad \cosh^2\theta - \sinh^2\theta = 1 \quad \therefore \quad \cosh^2\theta = \sinh^2\theta + 1$

$$\therefore \quad Z^2+A^2 = A^2\cosh^2\theta \quad \therefore \quad \sqrt{Z^2+A^2} = A\cosh\theta$$

Also $\quad \frac{dZ}{d\theta} = A\cosh\theta \quad \therefore \quad dZ = A\cosh\theta.d\theta$

So $\quad \int \frac{dZ}{\sqrt{Z^2+A^2}} = \int \frac{1}{A\cosh\theta}\cdot A\cosh\theta\ d\theta = \int d\theta = \theta + C$

But $\quad Z = A\sinh\theta \quad \therefore \quad \sinh\theta = \frac{Z}{A} \quad \therefore \theta = \sinh^{-1}\left\{\frac{Z}{A}\right\}$

$$\therefore \quad \int \frac{dZ}{\sqrt{Z^2+A^2}} = \sinh^{-1}\left\{\frac{Z}{A}\right\} + C \qquad (5)$$

Copy this result into your record book for future reference.

Then $\int \frac{1}{\sqrt{x^2+4}}\,dx = \ldots\ldots\ldots\ldots$

**29**

$$\int \frac{1}{\sqrt{x^2+4}}\,dx = \sinh^{-1}\left\{\frac{x}{2}\right\} + C$$

Once again, all we have to do is to find the expressions for $Z$ and $A$ in any particular example and substitute in the standard form.

Now you can do this one all on your own.

Determine $\int \frac{1}{\sqrt{x^2+5x+12}}\,dx$

*Complete the working: then check with Frame 30*

**30**

$$\int \frac{1}{\sqrt{x^2+5x+12}}\,dx = \sinh^{-1}\left\{\frac{2x+5}{\sqrt{23}}\right\} + C$$

Here is the working set out in detail:

$$\begin{aligned} x^2+5x+12 &= x^2+5x \qquad\quad +12 \\ &= x^2+5x+\left(\frac{5}{2}\right)^2+12-\frac{25}{4} \\ &= \left(x+\frac{5}{2}\right)^2+\frac{23}{4} \\ &= \left(x+\frac{5}{2}\right)^2+\left(\frac{\sqrt{23}}{2}\right)^2 \end{aligned}$$

So that $Z = x+\dfrac{5}{2}$ and $A = \dfrac{\sqrt{23}}{2}$

$$\begin{aligned} \therefore \quad \int \frac{1}{\sqrt{x^2+5x+12}}\,dx &= \sinh^{-1}\left\{\frac{x+\dfrac{5}{2}}{\sqrt{23}/2}\right\} + C \\ &= \sinh^{-1}\left\{\frac{2x+5}{\sqrt{23}}\right\} + C \end{aligned}$$

Now do one more:

$$\int \frac{1}{\sqrt{2x^2+8x+15}}\,dx = \ldots\ldots\ldots\ldots$$

**31**

$$\frac{1}{\sqrt{2}}\sinh^{-1}\left\{\frac{(x+2)\sqrt{2}}{\sqrt{7}}\right\} + C$$

Here is the working:

$$\int \frac{1}{\sqrt{2x^2+8x+15}}\,dx = \frac{1}{\sqrt{2}}\int \frac{1}{\sqrt{x^2+4x+\dfrac{15}{2}}}\,dx$$

$$\begin{aligned} x^2+4x+\frac{15}{2} &= x^2+4x \qquad +\frac{15}{2} \\ &= x^2+4x+2^2+\frac{15}{2}-4 \\ &= (x+2)^2+\frac{7}{2} \\ &= (x+2)^2+\left(\sqrt{\frac{7}{2}}\right)^2 \end{aligned}$$

So that $Z = (x+2)$ and $A = \sqrt{\dfrac{7}{2}}$

$$\begin{aligned} \therefore \quad \int \frac{1}{\sqrt{2x^2+8x+15}}\,dx &= \frac{1}{\sqrt{2}}\sinh^{-1}\left\{\frac{x+2}{\sqrt{\dfrac{7}{2}}}\right\} + C \\ &= \frac{1}{\sqrt{2}}\sinh^{-1}\frac{(x+2)\sqrt{2}}{\sqrt{7}} + C \end{aligned}$$

*Fine. Now on to Frame 32*

**32**

Now we will establish another standard result.

**6 Consider** $\displaystyle\int \frac{dZ}{\sqrt{A^2 - Z^2}}$

The substitution here is to put $Z = A\cosh\theta$.

$$Z^2 - A^2 = A^2\cosh^2\theta - A^2 = A^2(\cosh^2\theta - 1) = A^2\sinh^2\theta$$

$$\therefore \quad \sqrt{Z^2 - A^2} = A\sinh\theta$$

Also $\quad Z = A\cosh\theta \;\therefore\; dZ = A\sinh\theta\, d\theta$

$$\therefore \quad \int \frac{dZ}{\sqrt{Z^2 - A^2}} = \int \frac{1}{A\sinh\theta} \cdot A\sinh\theta\, d\theta = \int d\theta = \theta + C$$

$$Z = A\cosh\theta \;\therefore\; \cosh\theta = \frac{Z}{A} \;\therefore\; \theta = \cosh^{-1}\left\{\frac{Z}{A}\right\} + C$$

$$\therefore \quad \int \frac{dZ}{\sqrt{Z^2 - A^2}} = \cosh^{-1}\left\{\frac{Z}{A}\right\} + C \qquad (6)$$

This makes the sixth standard result we have established. Add it to your list.

*Then move on to Frame 33*

**33**

$$\boxed{\int \frac{dZ}{\sqrt{Z^2 - A^2}} = \cosh^{-1}\left\{\frac{Z}{A}\right\} + C}$$

**Example 1**

$$\int \frac{1}{\sqrt{x^2 - 9}}\, dx = \cosh^{-1}\left\{\frac{x}{3}\right\} + C$$

**Example 2**

$$\int \frac{1}{\sqrt{x^2 + 6x + 1}}\, dx = \ldots\ldots\ldots\ldots$$

You can do that one on your own. The method is the same as before: just complete the square and find out what $Z$ and $A$ are in this case and then substitute in the standard result.

**34**

$$\boxed{\int \frac{1}{\sqrt{x^2 + 6x + 1}}\, dx = \cosh^{-1}\left\{\frac{x+3}{2\sqrt{2}}\right\} + C}$$

Here it is:

$$\begin{aligned} x^2 + 6x + 1 &= x^2 + 6x \qquad + 1 \\ &= x^2 + 6x + 3^2 + 1 - 9 \\ &= (x+3)^2 - 8 \\ &= (x+3)^2 - (2\sqrt{2})^2 \end{aligned}$$

So that $Z = (x+3)$ and $A = 2\sqrt{2}$

$$\begin{aligned} \therefore \quad \int \frac{1}{\sqrt{x^2 + 6x + 1}}\, dx &= \int \frac{1}{\sqrt{(x+3)^2 - (2\sqrt{2})^2}}\, dx \\ &= \cosh^{-1}\left\{\frac{x+3}{2\sqrt{2}}\right\} + C \end{aligned}$$

▶

Let us now collect together the results we have established so far so that we can compare them.

*So move on to Frame 35*

**35**

Here are our standard forms so far, with the method indicated in each case:

| | | |
|---|---|---|
| 1 | $\displaystyle\int\frac{\mathrm{d}Z}{Z^2-A^2}=\frac{1}{2A}\ln\left\{\frac{Z-A}{Z+A}\right\}+C$ | Partial fractions |
| 2 | $\displaystyle\int\frac{\mathrm{d}Z}{A^2-Z^2}=\frac{1}{2A}\ln\left\{\frac{A+Z}{A-Z}\right\}+C$ | Partial fractions |
| 3 | $\displaystyle\int\frac{\mathrm{d}Z}{Z^2+A^2}=\frac{1}{A}\tan^{-1}\left\{\frac{Z}{A}\right\}+C$ | Put $Z=A\tan\theta$ |
| 4 | $\displaystyle\int\frac{\mathrm{d}Z}{\sqrt{A^2-Z^2}}=\sin^{-1}\left\{\frac{Z}{A}\right\}+C$ | Put $Z=A\sin\theta$ |
| 5 | $\displaystyle\int\frac{\mathrm{d}Z}{\sqrt{Z^2+A^2}}=\sinh^{-1}\left\{\frac{Z}{A}\right\}+C$ | Put $Z=A\sinh\theta$ |
| 6 | $\displaystyle\int\frac{\mathrm{d}Z}{\sqrt{Z^2-A^2}}=\cosh^{-1}\left\{\frac{Z}{A}\right\}+C$ | Put $Z=A\cosh\theta$ |

Note that the first three make one group (without square roots).

Note that the second three make a group with the square roots in the denominators.

You should make an effort to memorize these six results, for you will be expected to know them and to be able to quote them and use them in various examples.

**36**

You will remember that in the Programme on hyperbolic functions, we obtained the result $\sinh^{-1}x=\ln\{x+\sqrt{x^2+1}\}$

$$\therefore\quad \sinh^{-1}\left\{\frac{Z}{A}\right\}=\ln\left\{\frac{Z}{A}+\sqrt{\frac{Z^2}{A^2}+1}\right\}$$

$$=\ln\left\{\frac{Z}{A}+\sqrt{\frac{Z^2+A^2}{A^2}}\right\}$$

$$=\ln\left\{\frac{Z}{A}+\frac{\sqrt{Z^2+A^2}}{A}\right\}$$

$$\sinh^{-1}\left\{\frac{Z}{A}\right\}=\ln\left\{\frac{Z+\sqrt{Z^2+A^2}}{A}\right\}$$

Similarly:

$$\cosh^{-1}\left\{\frac{Z}{A}\right\}=\ln\left\{\frac{Z+\sqrt{Z^2-A^2}}{A}\right\}$$

This means that the results of standard integrals **5** and **6** can be expressed either as inverse hyperbolic functions or in log form according to the needs of the exercise.

*Move on now to Frame 37*

**37**

The remaining three standard integrals in our list are:

7 $\int \sqrt{A^2 - Z^2}.dZ$

8 $\int \sqrt{Z^2 + A^2}.dZ$

9 $\int \sqrt{Z^2 - A^2}.dZ$

In each case, the appropriate substitution is the same as with the corresponding integral in which the same expression occurred in the denominator.

i.e. for $\int \sqrt{A^2 - Z^2}.dZ$ put $Z = A\sin\theta$

$\int \sqrt{Z^2 + A^2}.dZ$ put $Z = A\sinh\theta$

$\int \sqrt{Z^2 - A^2}.dZ$ put $Z = A\cosh\theta$

Making these substitutions, gives the following results:

$$\int \sqrt{A^2 - Z^2}.dZ = \frac{A^2}{2}\left\{\sin^{-1}\left(\frac{Z}{A}\right) + \frac{Z\sqrt{A^2 - Z^2}}{A^2}\right\} + C \qquad (7)$$

$$\int \sqrt{Z^2 + A^2}.dZ = \frac{A^2}{2}\left\{\sinh^{-1}\left(\frac{Z}{A}\right) + \frac{Z\sqrt{Z^2 + A^2}}{A^2}\right\} + C \qquad (8)$$

$$\int \sqrt{Z^2 - A^2}.dZ = \frac{A^2}{2}\left\{\frac{Z\sqrt{Z^2 - A^2}}{A^2} - \cosh^{-1}\left(\frac{Z}{A}\right)\right\} + C \qquad (9)$$

These results are more complicated and difficult to remember but the method of using them is much the same as before. Copy them down.

**38**

Let us see how the first of these results is obtained:

$\int \sqrt{A^2 - Z^2}.dZ$ Put $Z = A\sin\theta$

$\therefore\ A^2 - Z^2 = A^2 - A^2\sin^2\theta = A^2(1 - \sin^2\theta) = A^2\cos^2\theta$

$\therefore\ \sqrt{A^2 - Z^2} = A\cos\theta$ Also $dZ = A\cos\theta\, d\theta$

$$\int \sqrt{A^2 - Z^2}.dZ = \int A\cos\theta.A\cos\theta\, d\theta = A^2\int \cos^2\theta\, d\theta$$

$$= A^2\left[\frac{\theta}{2} + \frac{\sin 2\theta}{4}\right] + C = \frac{A^2}{2}\left\{\theta + \frac{2\sin\theta\cos\theta}{2}\right\} + C$$

Now $\sin\theta = \dfrac{Z}{A}$ and $\cos^2\theta = 1 - \dfrac{Z^2}{A^2} = \dfrac{A^2 - Z^2}{A^2}$ $\therefore$ $\cos\theta = \dfrac{\sqrt{A^2 - Z^2}}{A}$

$$\therefore\ \int \sqrt{A^2 - Z^2}.dZ = \frac{A^2}{2}\left\{\sin^{-1}\left(\frac{Z}{A}\right) + \frac{Z}{A}\cdot\frac{\sqrt{A^2 - Z^2}}{A}\right\}$$

$$= \frac{A^2}{2}\left\{\sin^{-1}\left(\frac{Z}{A}\right) + \frac{Z\sqrt{A^2 - Z^2}}{A^2}\right\} + C$$

The other two are proved in a similar manner.

*Now on to Frame 39*

**39**

Here is an example:

$$\int \sqrt{x^2 + 4x + 13}.dx$$

First of all complete the square and find $Z$ and $A$ as before. Right. Do that.

**40**

$$\boxed{x^2 + 4x + 13 = (x+2)^2 + 3^2}$$

So that, in this case $\boxed{Z = x + 2}$ and $\boxed{A = 3}$

$$\therefore \quad \int \sqrt{x^2 + 4x + 13}.dx = \int \sqrt{(x+2)^2 + 3^2}.dx$$

This is of the form:

$$\int \sqrt{Z^2 + A^2}.dZ = \frac{A^2}{2}\left\{\sinh^{-1}\left(\frac{Z}{A}\right) + \frac{Z\sqrt{Z^2 + A^2}}{A^2}\right\} + C$$

So, substituting our expressions for $Z$ and $A$, we get:

$$\int \sqrt{x^2 + 4x + 13}.dx = \ldots\ldots\ldots\ldots$$

**41**

$$\boxed{\int \sqrt{x^2 + 4x + 13}.dx = \frac{9}{2}\left\{\sinh^{-1}\left(\frac{x+2}{3}\right) + \frac{(x+2)\sqrt{x^2 + 4x + 13}}{9}\right\} + C}$$

We see then that the use of any of these standard forms merely involves completing the square as we have done on many occasions, finding the expressions for $Z$ and $A$, and substituting these in the appropriate result. This means that you can now tackle a wide range of integrals which were beyond your ability before you worked through this Programme.

Now, by way of review, *without looking at your notes*, complete the following:

(a) $\displaystyle\int \frac{dZ}{Z^2 - A^2} = \ldots\ldots\ldots\ldots$ (b) $\displaystyle\int \frac{dZ}{A^2 - Z^2} = \ldots\ldots\ldots\ldots$ (c) $\displaystyle\int \frac{dZ}{Z^2 + A^2} = \ldots\ldots\ldots\ldots$

**42**

$$\int \frac{dZ}{Z^2 - A^2} = \frac{1}{2A}\cdot \ln\left\{\frac{Z - A}{Z + A}\right\} + C$$

$$\int \frac{dZ}{A^2 - Z^2} = \frac{1}{2A}\cdot \ln\left\{\frac{A + Z}{A - Z}\right\} + C$$

$$\int \frac{dZ}{Z^2 + A^2} = \frac{1}{A}\cdot \tan^{-1}\left\{\frac{Z}{A}\right\} + C$$

And now the second group:

$\displaystyle\int \frac{dZ}{\sqrt{A^2 - Z^2}} = \ldots\ldots\ldots\ldots$ $\displaystyle\int \frac{dZ}{\sqrt{Z^2 + A^2}} = \ldots\ldots\ldots\ldots$ $\displaystyle\int \frac{dZ}{\sqrt{Z^2 - A^2}} = \ldots\ldots\ldots\ldots$

43

$$\int \frac{dZ}{\sqrt{A^2 - Z^2}} = \sin^{-1}\left\{\frac{Z}{A}\right\} + C$$
$$\int \frac{dZ}{\sqrt{Z^2 + A^2}} = \sinh^{-1}\left\{\frac{Z}{A}\right\} + C$$
$$\int \frac{dZ}{\sqrt{Z^2 - A^2}} = \cosh^{-1}\left\{\frac{Z}{A}\right\} + C$$

You will not have remembered the third group, but here they are again. Take another look at them:

$$\int \sqrt{A^2 - Z^2}.dZ = \frac{A^2}{2}\left\{\sin^{-1}\left(\frac{Z}{A}\right) + \frac{Z\sqrt{A^2 - Z^2}}{A^2}\right\} + C$$

$$\int \sqrt{Z^2 + A^2}.dZ = \frac{A^2}{2}\left\{\sinh^{-1}\left(\frac{Z}{A}\right) + \frac{Z\sqrt{Z^2 + A^2}}{A^2}\right\} + C$$

$$\int \sqrt{Z^2 - A^2}.dZ = \frac{A^2}{2}\left\{\frac{Z\sqrt{Z^2 - A^2}}{A^2} - \cosh^{-1}\left(\frac{Z}{A}\right)\right\} + C$$

Notice that the square root in the result is the same root as that in the integral in each case.

That ends that particular section of the Programme, but there are other integrals that require substitution of some kind, so we will now deal with one or two of these.

*Move on to Frame 44*

44

**Integrals of the form $\int \frac{1}{a + b\sin^2 x + c\cos^2 x}\,dx$**

**Example 1**

Consider $\int \frac{1}{3 + \cos^2 x}\,dx$, which is different from any we have had before. It is certainly not one of the standard forms.

The key to the method is to substitute $t = \tan x$ in the integral. Of course, $\tan x$ is not mentioned in the integral, but if $\tan x = t$, we can soon find corresponding expressions for $\sin x$ and $\cos x$. Draw a sketch diagram, thus:

$\tan x = t$

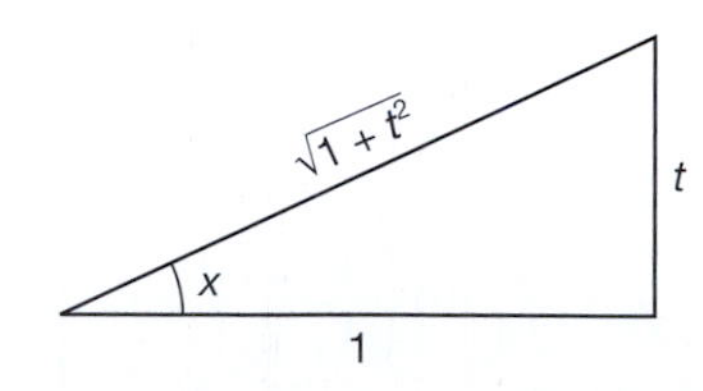


$$\therefore \quad \sin x = \frac{t}{\sqrt{1+t^2}}$$
$$\therefore \quad \cos x = \frac{1}{\sqrt{1+t^2}}$$

Also, since $t = \tan x$, $\frac{dt}{dx} = \sec^2 x = 1 + \tan^2 x = 1 + t^2$

$$\therefore \quad \frac{dx}{dt} = \frac{1}{1+t^2} \quad \therefore \quad dx = \frac{dt}{1+t^2}$$

Then $3 + \cos^2 x = 3 + \frac{1}{1+t^2} = \frac{3 + 3t^2 + 1}{1+t^2} = \frac{4 + 3t^2}{1+t^2}$

▶

So the integral now becomes: $\displaystyle\int \frac{1}{3+\cos^2 x}.dx = \int \frac{1+t^2}{4+3t^2}\cdot\frac{dt}{1+t^2}$

$$= \int \frac{1}{4+3t^2}\,dt$$

$$= \frac{1}{3}\int \frac{1}{\frac{4}{3}+t^2}\,dt$$

and from what we have done in the earlier part of this Programme, this is ...........

**45**

$$\frac{1}{3}\int \frac{1}{\frac{4}{3}+t^2}\,dt = \frac{1}{3}\cdot\frac{\sqrt{3}}{2}\tan^{-1}\left\{\frac{t}{2/\sqrt{3}}\right\}+C$$

$$= \frac{1}{3}\frac{\sqrt{3}}{2}\tan^{-1}\left\{\frac{t\sqrt{3}}{2}\right\}+C$$

Finally, since $t = \tan x$, we can return to the original variable and obtain:

$$\int \frac{1}{3+\cos^2 x}\,dx = \frac{1}{2\sqrt{3}}\tan^{-1}\left\{\frac{\sqrt{3}.\tan x}{2}\right\}+C$$

*Move to Frame 46*

**46**

The method is the same for all integrals of the type

$$\int \frac{1}{a+b\sin^2 x+c\cos^2 x}\,dx$$

In practice, some of the coefficients may be zero and those terms missing from the function. But the routine remains the same.

Use the substitution $t = \tan x$. That is all there is to it.

From the diagram:

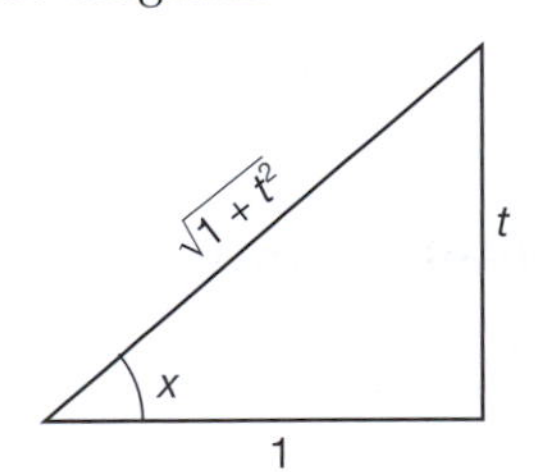


we get

$\sin x = \ldots\ldots\ldots\ldots$

$\cos x = \ldots\ldots\ldots\ldots$

**47**

$$\sin x = \frac{t}{\sqrt{1+t^2}}; \qquad \cos x = \frac{1}{\sqrt{1+t^2}}$$

We also have to change the variable.

$$t = \tan x \quad \therefore \quad \frac{dt}{dx} = \sec^2 x = 1+\tan^2 x = 1+t^2$$

$$\therefore \quad \frac{dx}{dt} = \frac{1}{1+t^2};$$

$$dx = \ldots\ldots\ldots\ldots$$

**48**

$$dx = \frac{dt}{1+t^2}$$

Armed with these substitutions we can deal with any integral of the present type. This does not give us a standard result, but provides us with a standard method.

We will work through another example in the next frame, but first of all, what were those substitutions?

$\sin x = \dots\dots\dots\dots \qquad \cos x = \dots\dots\dots\dots$

**49**

$$\sin x = \frac{t}{\sqrt{1+t^2}} \qquad \cos x = \frac{1}{\sqrt{1+t^2}}$$

Right. Now for an example.

**Example 2**

Determine $\displaystyle\int \frac{1}{2\sin^2 x + 4\cos^2 x}\,dx$

Using the substitution above, and that $dx = \dfrac{dt}{1+t^2}$, we have:

$$2\sin^2 x + 4\cos^2 x = \frac{2t^2}{1+t^2} + \frac{4}{1+t^2} = \frac{2t^2+4}{1+t^2}$$

$$\therefore \quad \int \frac{1}{2\sin^2 x + 4\cos^2 x}\,dx = \int \frac{1+t^2}{2t^2+4} \cdot \frac{dt}{1+t^2}$$

$$= \frac{1}{2}\int \frac{1}{t^2+2}\,dt$$

$$= \dots\dots\dots\dots$$

**50**

$$\frac{1}{2\sqrt{2}}\tan^{-1}\left\{\frac{t}{\sqrt{2}}\right\} + C$$

and since $t = \tan x$, we can return to the original variable, so that:

$$\int \frac{1}{2\sin^2 x + 4\cos^2 x} \cdot dx = \frac{1}{2\sqrt{2}}\tan^{-1}\left\{\frac{\tan x}{\sqrt{2}}\right\} + C$$

Now here is one for you to do on your own.

Remember the substitutions:

$$t = \tan x \qquad \sin x = \frac{t}{\sqrt{1+t^2}}$$

$$\cos x = \frac{1}{\sqrt{1+t^2}}$$

$$dx = \frac{dt}{(1+t^2)}$$

Right, then here it is:

**Example 3**

$$\int \frac{1}{2\cos^2 x + 1}\,dx = \dots\dots\dots\dots$$

*Work it right through to the end and then check your result and your working with that in the next frame*

51

$$\int \frac{1}{2\cos^2 x + 1}\,dx = \frac{1}{\sqrt{3}}\tan^{-1}\left\{\frac{\tan x}{\sqrt{3}}\right\} + C$$

Here is the working:

$$2\cos^2 x + 1 = \frac{2}{1+t^2} + 1 = \frac{2+1+t^2}{1+t^2} = \frac{3+t^2}{1+t^2}$$

$$\therefore \int \frac{1}{2\cos^2 x + 1}\,dx = \int \frac{1+t^2}{3+t^2}\cdot\frac{dt}{1+t^2}$$

$$= \int \frac{1}{3+t^2}\,dt = \frac{1}{\sqrt{3}}\tan^{-1}\left(\frac{t}{\sqrt{3}}\right) + C$$

$$= \frac{1}{\sqrt{3}}\tan^{-1}\left\{\frac{\tan x}{\sqrt{3}}\right\} + C$$

So whenever we have an integral of this type, with $\sin^2 x$ and/or $\cos^2 x$ in the denominator, the key to the whole business is to make the substitution

$t = \ldots\ldots\ldots\ldots$

52

$$t = \tan x$$

Let us now consider the integral $\int \frac{1}{5 + 4\cos x}\,dx$

This is clearly not one of the last type, for the trig function in the denominator is $\cos x$ and not $\cos^2 x$.

In fact, this is an example of a further group of integrals that we are going to cover in this Programme. In general they are of the form $\int \frac{1}{a + b\sin x + c\cos x}\,dx$, i.e. sines and cosines in the denominator but not squared.

*So move on to Frame 53 and we will start to find out something about these integrals*

53

**Integrals of the type $\int \frac{1}{a + b\sin x + c\cos x}\,dx$**

The key this time is to substitute $t = \tan\frac{x}{2}$

From this, we can find corresponding expressions for $\sin\frac{x}{2}$ and $\cos\frac{x}{2}$ from a simple diagram as before, but it also means that we must express $\sin x$ and $\cos x$ in terms of the trig ratios of the half angle – so it will entail a little more work, but only a little, so do not give up. It is a lot easier than it sounds.

First of all let us establish the substitutions in detail.

$t = \tan\frac{x}{2}$

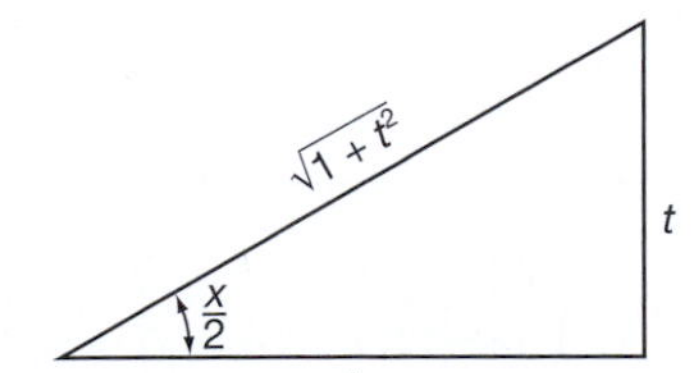


$$\therefore \quad \sin\frac{x}{2} = \frac{t}{\sqrt{1+t^2}}$$

$$\therefore \quad \cos\frac{x}{2} = \frac{1}{\sqrt{1+t^2}}$$

$$\sin x = 2\sin\frac{x}{2}\cos\frac{x}{2} = 2.\frac{t}{\sqrt{1+t^2}}.\frac{1}{\sqrt{1+t^2}} = \frac{2t}{1+t^2}$$

$$\cos x = \cos^2\frac{x}{2} - \sin^2\frac{x}{2} = \frac{1}{1+t^2} - \frac{t^2}{1+t^2} = \frac{1-t^2}{1+t^2}$$

Also, since $t = \tan\frac{x}{2}$, $\frac{dt}{dx} = \frac{1}{2}\sec^2\frac{x}{2} = \frac{1}{2}\left(1+\tan^2\frac{x}{2}\right)$

$$= \frac{1+t^2}{2}$$

$$\frac{dx}{dt} = \frac{2}{1+t^2}$$

$$dx = \frac{2dt}{1+t^2}$$

So we have:

If $t = \tan\frac{x}{2}$ $\quad \sin x = \frac{2t}{1+t^2}$

$$\cos x = \frac{1-t^2}{1+t^2}$$

$$dx = \frac{2dt}{1+t^2}$$

It is worth remembering these substitutions for use in examples. So copy them down into your record book for future reference. Then we shall be ready to use them.

*On to Frame 54*

54

**Example 1**

$$\int\frac{dx}{5+4\cos x}$$

Using the substitution $t = \tan\frac{x}{2}$, we have:

$$5+4\cos x = 5+4\frac{(1-t^2)}{(1+t^2)}$$

$$= \frac{5+5t^2+4-4t^2}{1+t^2}$$

$$= \frac{9+t^2}{1+t^2}$$

$$\therefore \int\frac{dx}{5+4\cos x} = \int\frac{1+t^2}{9+t^2}.\frac{2dt}{1+t^2}$$

$$= 2\int\frac{dt}{9+t^2}$$

$$= \dots\dots\dots\dots$$

55

$$\frac{2}{3}\tan^{-1}\left\{\frac{t}{3}\right\} + C$$

$$= \frac{2}{3}\tan^{-1}\left\{\frac{\tan x/2}{3}\right\} + C$$

Here is another.

**Example 2**

$$\int \frac{dx}{3\sin x + 4\cos x}$$

Using the substitution $t = \tan\frac{x}{2}$

$$3\sin x + 4\cos x = \frac{6t}{1+t^2} + \frac{4(1-t^2)}{1+t^2}$$

$$= \frac{4+6t-4t^2}{1+t^2}$$

$$\therefore \quad \int \frac{dx}{3\sin x + 4\cos x} = \int \frac{1+t^2}{4+6t-4t^2} \cdot \frac{2dt}{1+t^2}$$

$$= \int \frac{1}{2+3t-2t^2}\,dt$$

$$= \frac{1}{2}\int \frac{1}{1+\frac{3}{2}t-t^2}\,dt$$

Now complete the square in the denominator as we are doing earlier in the Programme and finish it off.

*Then on to Frame 56*

---

**56**

$$\boxed{\frac{1}{5}\cdot\ln\left\{\frac{1+2\tan x/2}{4-2\tan x/2}\right\}+C}$$

Because

$$1+\frac{3}{2}t-t^2 = 1-\left(t^2-\frac{3}{2}t \qquad\right)$$

$$= 1-\left(t^2-\frac{3}{2}t+\left[\frac{3}{4}\right]^2\right)+\frac{9}{16}$$

$$= \frac{25}{16}-\left(t-\frac{3}{4}\right)^2$$

$$= \left(\frac{5}{4}\right)^2-\left(t-\frac{3}{4}\right)^2$$

$$\text{Integral, } I = \frac{1}{2}\int \frac{1}{\left(\frac{5}{4}\right)^2-\left(t-\frac{3}{4}\right)^2}\,dt$$

$$= \frac{1}{2.\frac{5}{2}}\ln\left\{\frac{5/4+t-3/4}{5/4-t+3/4}\right\}+C$$

$$= \frac{1}{5}\ln\left\{\frac{1/2+t}{2-t}\right\}+C$$

$$= \frac{1}{5}\ln\left\{\frac{1+2t}{4-2t}\right\}+C$$

$$= \frac{1}{5}\ln\left\{\frac{1+2\tan x/2}{4-2\tan x/2}\right\}+C$$

▶

And here is one more for you, all on your own. Finish it: then check your working with that in the next frame. Here it is.

**Example 3**

$$\int \frac{1}{1+\sin x-\cos x}\,dx = \ldots\ldots\ldots\ldots$$

**57**

$$\boxed{\ln\left\{\frac{\tan x/2}{1+\tan x/2}\right\}+C}$$

Here is the working:

$$1+\sin x-\cos x = 1+\frac{2t}{1+t^2}-\frac{1-t^2}{1+t^2}$$

$$= \frac{1+t^2+2t-1+t^2}{1+t^2}$$

$$= \frac{2(t^2+t)}{1+t^2}$$

$$I = \int \frac{1+t^2}{2(t^2+t)}\cdot\frac{2dt}{1+t^2}$$

$$= \int \frac{1}{t^2+t}dt$$

$$= \int \left(\frac{1}{t}-\frac{1}{1+t}\right)dt$$

$$\ln\left\{\frac{t}{1+t}\right\}+C$$

$$= \ln\left\{\frac{\tan x/2}{1+\tan x/2}\right\}+C$$

You have now reached the end of this Programme except for the **Can You?** checklist and the **Test exercise** which follow.

*Move on to Frames 58 and 59*

## Can You?

### Checklist 17

58

Check this list before and after you try the end of Programme test.

On a scale of 1 to 5 how confident are you that you can:

| | Frames |
|---|---|
| • Evaluate integrals with integrands of the form $\pm 1/(Z^2 - A^2)$? <br> *Yes* ☐ ☐ ☐ ☐ ☐ *No* | 1 to 19 |
| • Evaluate integrals with integrands of the form $1/(Z^2 + A^2)$? <br> *Yes* ☐ ☐ ☐ ☐ ☐ *No* | 20 to 23 |
| • Evaluate integrals with integrands of the form $1/\sqrt{A^2 - Z^2}$? <br> *Yes* ☐ ☐ ☐ ☐ ☐ *No* | 24 to 27 |
| • Evaluate integrals with integrands of the form $1/\sqrt{Z^2 + A^2}$? <br> *Yes* ☐ ☐ ☐ ☐ ☐ *No* | 28 to 31 |
| • Evaluate integrals with integrands of the form $1/\sqrt{Z^2 - A^2}$? <br> *Yes* ☐ ☐ ☐ ☐ ☐ *No* | 32 to 34 |
| • Evaluate integrals with integrands of the form $\sqrt{A^2 - Z^2}$, $\sqrt{Z^2 + A^2}$ and $\sqrt{Z^2 - A^2}$? <br> *Yes* ☐ ☐ ☐ ☐ ☐ *No* | 37 to 43 |
| • Evaluate integrals with integrands of the form $1/(a + b\sin^2 x + c\cos^2 x)$? <br> *Yes* ☐ ☐ ☐ ☐ ☐ *No* | 44 to 52 |
| • Evaluate integrals with integrands of the form $1/(a + b\sin x + c\cos x)$? <br> *Yes* ☐ ☐ ☐ ☐ ☐ *No* | 53 to 57 |

## Test exercise 17

59

Before you work through the questions, brush up any parts of the Programme about which you are not perfectly clear. Look back through the Programme if you want to do so. There is no hurry. Your success is all that matters. The integrals here are just like those we have been doing in the Programme, so you will find them quite straightforward.

Determine the following:

1 $\displaystyle\int \frac{1}{\sqrt{49 - x^2}}\,dx$ 2 $\displaystyle\int \frac{dx}{x^2 + 3x - 5}$

3 $\displaystyle\int \frac{dx}{2x^2 + 8x + 9}$ 4 $\displaystyle\int \frac{1}{\sqrt{3x^2 + 16}}\,dx$

5 $\displaystyle\int \frac{dx}{9 - 8x - x^2}$ 6 $\displaystyle\int \sqrt{1 - x - x^2}\,dx$

7 $\displaystyle\int \frac{1}{\sqrt{5x^2 + 10x - 16}}$ 8 $\displaystyle\int \frac{dx}{1 + 2\sin^2 x}$

9 $\displaystyle\int \frac{dx}{2\cos x + 3\sin x}$ 10 $\displaystyle\int \sec x\,dx$

*You are now ready for the next Programme. Well done!*

# Further problems 17

**60** Determine the following:

1. $\displaystyle\int \frac{dx}{x^2+12x+15}$
2. $\displaystyle\int \frac{dx}{8-12x-x^2}$
3. $\displaystyle\int \frac{dx}{x^2+14x+60}$
4. $\displaystyle\int \frac{x-8}{x^2+4x+16}\,dx$
5. $\displaystyle\int \frac{dx}{\sqrt{x^2+12x+48}}$
6. $\displaystyle\int \frac{dx}{\sqrt{17-14x-x^2}}$
7. $\displaystyle\int \frac{dx}{\sqrt{x^2+16x+36}}$
8. $\displaystyle\int \frac{6x-5}{\sqrt{x^2-12x+52}}\,dx$
9. $\displaystyle\int \frac{dx}{2+\cos x}$
10. $\displaystyle\int_0^{\pi/2} \frac{dx}{4\sin^2 x+5\cos^2 x}$

11. $\displaystyle\int \frac{dx}{x^2+5x+5}$
12. $\displaystyle\int \frac{3x^3-4x^2+3x}{x^2+1}\,dx$
13. $\displaystyle\int \sqrt{3-2x-x^2}\,dx$
14. $\displaystyle\int_2^4 \frac{dx}{\sqrt{6x-8-x^2}}$

15. $\displaystyle\int \frac{dx}{\sqrt{x^2+4x-21}}$
16. $\displaystyle\int \frac{dx}{4\sin^2 x+9\cos^2 x}$

17. $\displaystyle\int \frac{dx}{3\sin x-4\cos x}$
18. $\displaystyle\int_0^1 \sqrt{\frac{x}{2-x}}\,dx$ (Put $x=2\sin^2\theta$)

19. $\displaystyle\int \frac{x+3}{\sqrt{1-x^2}}\,dx$
20. $\displaystyle\int \frac{\cos x}{2-\cos x}\,dx$

21. $\displaystyle\int \frac{x^2-x+14}{(x+2)(x^2+4)}\,dx$
22. $\displaystyle\int \frac{dx}{5+4\cos^2 x}$

23. $\displaystyle\int \frac{x+2}{\sqrt{x^2+9}}\,dx$
24. $\displaystyle\int \frac{dx}{\sqrt{2x^2-7x+5}}$

25. $\displaystyle\int_1^4 \frac{dx}{\sqrt{(x+2)(4-x)}}$
26. $\displaystyle\int \frac{d\theta}{2\sin^2\theta-\cos^2\theta}$

27. $\displaystyle\int \frac{x+3}{\sqrt{x^2+2x+10}}\,dx$
28. $\displaystyle\int \sqrt{15-2x-x^2}\,dx$

29. $\displaystyle\int_0^a \frac{dx}{(a^2+x^2)^2}$ (Put $x=a\tan\theta$)
30. $\displaystyle\int \frac{a^2dx}{(x+a)(x^2+2a^2)}$

**61** Now visit the companion website for this book at www.palgrave.com/stroud for more questions applying this mathematics to science and engineering.

# Reduction formulas

## Learning outcomes

When you have completed this Programme you will be able to:

- ☐ Integrate by parts and generate a reduction formula
- ☐ Integrate by parts using a reduction formula
- ☐ Evaluate integrals with integrands of the form $\sin^n x$ and $\cos^n x$ using reduction formulas

**1**

In an earlier Programme on integration, we dealt with the method of *integration by parts*, and you have had plenty of practice in that since that time. You remember that it can be stated thus:

$$\int u\,dv = uv - \int v\,du$$

So just to refresh your memory, do this one to start with:

$$\int x^2 e^x\,dx = \ldots\ldots\ldots\ldots$$

*When you have finished, move on to Frame 2*

**2**

$$\boxed{\int x^2 e^x\,dx = e^x[x^2 - 2x + 2] + C}$$

Here is the working, so that you can check your solution:

$$\int x^2 e^x\,dx = x^2(e^x) - 2\int e^x x\,dx$$

$$= x^2 e^x - 2[x(e^x) - \int e^x\,dx]$$

$$= x^2 e^x - 2xe^x + 2e^x + C$$

$$= e^x[x^2 - 2x + 2] + C$$

*On to Frame 3*

**3**

Now let us try the same thing with this one:

$$\int x^n e^x\,dx = x^n(e^x) - n\int e^x x^{n-1}\,dx$$

$$= x^n e^x - n\int e^x x^{n-1}\,dx$$

Now you will see that the integral on the right, i.e. $\int e^x x^{n-1}\,dx$, is of exactly the same form as the one we started with, i.e. $\int e^x x^n\,dx$, except for the fact that $n$ has now been replaced by $(n-1)$.

Then, if we denote $\int x^n e^x\,dx$ by $I_n$

we can denote $\int x^{n-1} e^x\,dx$ by $I_{n-1}$

So our result

$$\int x^n e^x\,dx = x^n e^x - n\int e^x x^{n-1}\,dx$$

can be written

$$I_n = x^n e^x - \ldots\ldots\ldots\ldots$$

*Then on to Frame 4*

4

$$\boxed{I_n = x^n e^x - n.I_{n-1}}$$

This relationship is called a *reduction formula* since it expresses an integral in $n$ in terms of the same integral in $(n-1)$. Here it is again.

If $I_n = \int x^n e^x \, dx$

then $I_n = x^n e^x - n.I_{n-1}$

Make a note of this result in your record book, since we shall be using it in the examples that follow.

*Then to the example in Frame 5*

5

Consider $\int x^2 e^x \, dx$

This is, of course, the case of $I_n = \int x^n e^x \, dx$ in which $n = 2$.

We know that $I_n = x^n e^x - n.I_{n-1}$ applies to this integral, so putting $n = 2$, we get

$$I_2 = x^2 e^x - 2.I_1$$

and then $I_1 = x^1 e^x - 1.I_0$

Now we can easily evaluate $I_0$ the normal manner:

$$I_0 = \int x^0 e^x \, dx + \int 1 e^x \, dx = \int e^x \, dx = e^x + C$$

$$\text{So } \; I_2 = x^2 e^x - 2.I_1$$

$$\text{and } \; I_1 = xe^x - e^x + C_1$$

$$\therefore I_2 = x^2 e^x - 2xe^x + 2e^x + C$$

$$= e^x[x^2 - 2x + 2] + C$$

And that is it. Once you have established the reduction formula for a particular type of integral, its use is very simple.

In just the same way, using the same reduction formula, determine the integral $\int x^3 e^x \, dx$.

*Then check with the next frame*

6

$$\boxed{\int x^3 e^x \, dx = e^x[x^3 - 3x^2 + 6x - 6] + C}$$

Here is the working. Check yours. $I_n = x^n e^x - n.I_{n-1}$

$n = 3$ $\quad I_3 = x^3 e^x - 3.I_2$

$n = 2$ $\quad I_2 = x^2 e^x - 2.I_1$

$n = 1$ $\quad I_1 = xe^x - 1.I_0$

and $\quad I_0 = \int x^0 e^x \, dx = \int e^x \, dx = e^x + C$

$$\therefore I_3 = x^3 e^x - 3.I_2$$

$$= x^3 e^x - 3[x^2 e^x + 2.I_1]$$

$$= x^3 e^x - 3[x^2 e^x - 2(xe^x - 1.I_0)]$$

$$= x^3 e^x - 3[x^2 e^x - 2(xe^x - [e^x + C_1])]$$

$$= x^3 e^x - 3x^2 e^x + 6xe^x - 6e^x - 6C_1$$

$$= e^x[x^3 - 3x^2 + 6x - 6] + C$$

*Now move on to Frame 7*

**7**

Let us now find a reduction formula for the integral $\int x^n \cos x \, dx$.

$$I_n = \int x^n \cos x \, dx$$
$$= x^n(\sin x) - n\int \sin x \, x^{n-1} \, dx$$
$$= x^n \sin x - n\int x^{n-1} \sin x \, dx$$

Note that this is *not* a reduction formula yet, since the integral on the right is *not* of the same form as that of the original integral. So let us apply the integration-by-parts routine again:

$$I_n = x^n \sin x - n\int x^{n-1} \sin x \, dx$$
$$= x^n \sin x - n \ldots\ldots\ldots\ldots$$

**8**

$$\boxed{I_n = x^n \sin x + nx^{n-1} \cos x - n(n-1)\int x^{n-2} \cos x \, dx}$$

Now you will see that the integral $\int x^{n-2} \cos x \, dx$ is the same as the integral $\int x^n \cos x \, dx$, with $n$ replaced by ...........

**9**

$$\boxed{n-2}$$

i.e. $I_n = x^n \sin x + nx^{n-1} \cos x - n(n-1)I_{n-2}$

So this is the reduction formula for $I_n = \int x^n \cos x \, dx$

Copy the result down in your record book and then use it to find $\int x^2 \cos x \, dx$. First of all, put $n = 2$ in the result, which then gives ...........

**10**

$$\boxed{I_2 = x^2 \sin x + 2x \cos x - 2.1.I_0}$$

Now $I_0 = \int x^0 \cos x \, dx = \int \cos x \, dx = \sin x + C_1$

And so $I_2 = x^2 \sin x + 2x \cos x - 2 \sin x + C$

Now you know what it is all about, how about this one?

Find a reduction formula for $\int x^n \sin x \, dx$.

Apply the integration-by-parts routine: it is very much like the previous one.

*When you have finished, move on to Frame 11*

11

$$I_n = -x^n \cos x + nx^{n-1} \sin x - n(n-1)I_{n-2}$$

Because

$$I_n = \int x^n \sin x \, dx$$
$$= x^n(-\cos x) + n \int \cos x \, x^{n-1} \, dx$$
$$= -x^n \cos x + n\left\{x^{n-1}(\sin x) - (n-1)\int \sin x \, x^{n-2} \, dx\right\}$$
$$\therefore I_n = -x^n \cos x + nx^{n-1} \sin x - n(n-1)I_{n-2}$$

Make a note of the result, and then let us find $\int x^3 \sin x \, dx$.

Putting $n = 3$, $I_3 = -x^3 \cos x + 3x^2 \sin x - 3.2.I_1$

and then $I_1 = \int x \sin x \, dx$

$= \dots\dots\dots\dots$

*Find this and then finish the result – then on to Frame 12*

12

$$I_1 = -x \cos x + \sin x + C_1$$

So that $I_3 = -x^3 \cos x + 3x^2 \sin x - 6.I_1$

$\therefore I_3 = -x^3 \cos x + 3x^2 \sin x + 6x \cos x - 6 \sin x + C$

Note that a reduction formula can be repeated until the value of $n$ decreases to $n = 1$ or $n = 0$, when the final integral is determined by normal methods.

*Now move on to Frame 13 for the next section of the work*

13

Let us now see, in the following example, what complications there are when the integral has limits.

To determine $\int_0^{\pi} x^n \cos x \, dx$.

Now we have already established that, if $I_n = \int x^n \cos x \, dx$, then

$I_n = x^n \sin x + nx^{n-1} \cos x - n(n-1)I_{n-2}$

If we now define $I_n = \int_0^{\pi} x^n \cos x \, dx$, all we have to do is to apply the limits to the calculated terms on the right-hand side of our result:

$$I_n = \Big[x^n \sin x + nx^{n-1} \cos x\Big]_0^{\pi} - n(n-1)I_{n-2}$$
$$= [0 + n\pi^{n-1}(-1)] - [0 + 0] - n(n-1)I_{n-2}$$
$$\therefore I_n = -n\pi^{n-1} - n(n-1)I_{n-2}$$

This, of course, often simplifies the reduction formula and is much quicker than obtaining the complete general result and then having to substitute the limits.

Use the result above to evaluate $\int_0^{\pi} x^4 \cos x \, dx$.

First put $n = 4$, giving $I_4 = \dots\dots\dots\dots$

**14**

$$I_4 = -4\pi^3 - 4.3.I_2$$

Now put $n = 2$ to find $I_2$, which is $I_2 = \ldots\ldots\ldots\ldots$

**15**

$$I_2 = -2.\pi - 2.1.I_0$$

$$\text{and } I_0 = \int_0^{\pi} x^0 \cos x \, dx = \int_0^{\pi} \cos x \, dx = \Big[\sin x\Big]_0^{\pi} = 0$$

$$\text{So we have } I_4 = -4\pi^3 - 12I_2$$

$$I_2 = -2\pi$$

$$\text{and } \therefore I_4 = \ldots\ldots\ldots\ldots$$

**16**

$$\int_0^{\pi} x^4 \cos x \, dx = I_4 = -4\pi^3 + 24\pi$$

Now here is one for you to do in very much the same way.

Evaluate $\displaystyle\int_0^{\pi} x^5 \cos x \, dx$.

*Work it right through and then check your working with Frame 17*

**17**

$$I_5 = -5\pi^4 + 60\pi^2 - 240$$

Working:

$$I_n = -n\pi^{n-1} - n(n-1)I_{n-2}$$

$$\therefore I_5 = -5\pi^4 - 5.4.I_3$$

$$I_3 = -3\pi^2 - 3.2.I_1$$

$$\text{and } I_1 = \int_0^{\pi} x \cos x \, dx = \Big[x(\sin x)\Big]_0^{\pi} - \int_0^{\pi} \sin x \, dx$$

$$= [0 - 0] - \Big[-\cos x\Big]_0^{\pi}$$

$$= \Big[\cos x\Big]_0^{\pi} = (-1) - (1) = -2$$

$$\therefore I_5 = -5\pi^4 - 20I_3$$

$$I_3 = -3\pi^2 - 6(-2)$$

$$\therefore I_5 = -5\pi^4 + 60\pi^2 - 240$$

*Move on to Frame 18*

18

*Reduction formulas for* (a) $\int \sin^n x \, dx$ and (b) $\int \cos^n x \, dx$.

(a) $\int \sin^n x \, dx$

Let $I_n = \int \sin^n x \, dx = \int \sin^{n-1} x . \sin x \, dx = \int \sin^{n-1} x . d(-\cos x)$

Then, integration by parts, gives:

$$I_n = \sin^{n-1} x.(-\cos x) + (n-1) \int \cos x . \sin^{n-2} x . \cos x \, dx$$

$$= -\sin^{n-1} x . \cos x + (n-1) \int \cos^2 x . \sin^{n-2} x \, dx$$

$$= -\sin^{n-1} x . \cos x + (n-1) \int (1 - \sin^2 x) \sin^{n-2} x \, dx$$

$$= -\sin^{n-1} x . \cos x + (n-1) \left\{ \int \sin^{n-2} x \, dx - \int \sin^n x \, dx \right\}$$

$$\therefore I_n = -\sin^{n-1} x . \cos x + (n-1) I_{n-2} - (n-1) I_n$$

Now bring the last term over to the left-hand side, and we have:

$$n.I_n = -\sin^{n-1} x . \cos x + (n-1) I_{n-2}$$

So finally, if $I_n = \int \sin^n x \, dx$, $I_n = -\frac{1}{n} \sin^{n-1} x . \cos x + \frac{n-1}{n} I_{n-2}$

Make a note of this result, and then use it to find $\int \sin^6 x \, dx$.

---

19

$$I_6 = -\frac{1}{6} \sin^5 x . \cos x - \frac{5}{24} \sin^3 x . \cos x - \frac{5}{16} \sin x . \cos x + \frac{5x}{16} + C$$

Because

$$I_6 = -\frac{1}{6} \sin^5 x . \cos x + \frac{5}{6} . I_4$$

$$I_4 = -\frac{1}{4} \sin^3 x . \cos x + \frac{3}{4} . I_2$$

$$I_2 = -\frac{1}{2} \sin x . \cos x + \frac{1}{2} . I_0 \text{ where } I_0 = \int dx = x + C_0$$

$$\therefore I_6 = -\frac{1}{6} \sin^5 x . \cos x + \frac{5}{6} \left[ -\frac{1}{4} \sin^3 x . \cos x + \frac{3}{4} . I_2 \right]$$

$$= -\frac{1}{6} \sin^5 x . \cos x - \frac{5}{24} \sin^3 x . \cos x + \frac{5}{8} \left[ -\frac{1}{2} \sin x . \cos x + \frac{x}{2} \right] + C$$

$$= -\frac{1}{6} \sin^5 x . \cos x - \frac{5}{24} \sin^3 x . \cos x - \frac{5}{16} \sin x . \cos x + \frac{5x}{16} + C$$

**20**

(b) $\int \cos^n x \, dx$

$$\text{Let } I_n = \int \cos^n x \, dx = \int \cos^{n-1} x . \cos x \, dx = \int \cos^{n-1} x \, d(\sin x)$$

$$= \cos^{n-1} x . \sin x - (n-1) \int \sin x . \cos^{n-2} x . (-\sin x) \, dx$$

$$= \cos^{n-1} x . \sin x + (n-1) \int \sin^2 x . \cos^{n-2} x \, dx$$

$$= \cos^{n-1} x . \sin x + (n-1) \int (1 - \cos^2 x) . \cos^{n-2} x \, dx$$

$$= \cos^{n-1} x . \sin x + (n-1) \left\{ \int \cos^{n-2} x \, dx - \int \cos^n x \, dx \right\}$$

Now finish it off, so that $I_n = \ldots\ldots\ldots\ldots$

**21**

$$\boxed{I_n = \frac{1}{n} \cos^{n-1} x . \sin x + \frac{n-1}{n} . I_{n-2}}$$

Because

$$I_n = \cos^{n-1} x . \sin x + (n-1) I_{n-2} - (n-1) I_n$$

$$n . I_n = \cos^{n-1} x . \sin x + (n-1) I_{n-2}$$

$$\therefore I_n = \frac{1}{n} \cos^{n-1} x . \sin x + \frac{n-1}{n} I_{n-2}$$

Add this result to your list and then apply it to find $\int \cos^5 x \, dx$.

*When you have finished it, move to Frame 22*

**22**

$$\boxed{\int \cos^5 x \, dx = \frac{1}{5} \cos^4 x . \sin x + \frac{4}{15} \cos^2 x . \sin x + \frac{8}{15} \sin x + C}$$

Here it is:

$$I_5 = \frac{1}{5} \cos^4 x . \sin x + \frac{4}{5} I_3$$

$$I_3 = \frac{1}{3} \cos^2 x . \sin x + \frac{2}{3} I_1$$

$$\text{And } I_1 = \int \cos x \, dx = \sin x + C_1$$

$$\therefore I_5 = \frac{1}{5} \cos^4 x . \sin x + \frac{4}{5} \left[ \frac{1}{3} \cos^2 x . \sin x + \frac{2}{3} \sin x \right] + C$$

$$= \frac{1}{5} \cos^4 x . \sin x + \frac{4}{15} \cos^2 x . \sin x + \frac{8}{15} \sin x + C$$

*On to Frame 23*

**23**

The integrals $\int \sin^n x \, dx$ and $\int \cos^n x \, dx$ with limits $x = 0$ and $x = \pi/2$, give some interesting and useful results.

We already know the reduction formula:

$$\int \sin^n x \, dx = I_n = -\frac{1}{n} \sin^{n-1} x . \cos x + \frac{n-1}{n} I_{n-2}$$

Inserting the limits:

$$I_n = \left[ -\frac{1}{n} \sin^{n-1} x . \cos x \right]_0^{\pi/2} + \frac{n-1}{n} I_{n-2}$$

$$= [0 - 0] + \frac{n-1}{n} I_{n-2}$$

$$\therefore I_n = \frac{n-1}{n} I_{n-2}$$

And if you do the same with the reduction formula for $\int \cos^n x \, dx$, you get exactly the same result.

So for $\int_0^{\pi/2} \sin^n x \, dx$ and $\int_0^{\pi/2} \cos^n x \, dx$, we have:

$$I_n = \frac{n-1}{n} I_{n-2}$$

Also:

(a) If $n$ is even, the formula eventually reduces to $I_0$

i.e. $\int_0^{\pi/2} 1 \, dx = [x]_0^{\pi/2} = \pi/2$

$\therefore I_0 = \pi/2$

(b) If $n$ is odd, the formula eventually reduces to $I_1$

i.e $\int_0^{\pi/2} \sin x \, dx = [-\cos x]_0^{\pi/2} = -(-1)$

$\therefore I_1 = 1$

So now, all on your own, evaluate $\int_0^{\pi/2} \sin^5 x \, dx$. What do you get?

**24**

$$I_5 = \frac{4}{5} \cdot \frac{2}{3} . 1 = \frac{8}{15}$$

Because

$$I_5 = \frac{4}{5} . I_3$$

$I_3 = \frac{2}{3} . I_1$ and we know that $I_1 = 1$

$$\therefore I_5 = \frac{4}{5} \cdot \frac{2}{3} . 1 = \frac{8}{15}$$

In the same way, find $\int_0^{\pi/2} \cos^6 x \, dx$.

*Then to Frame 25*

**25**

$$I_6 = \frac{5\pi}{32}$$

Because

$$I_6 = \frac{5}{6}.I_4$$
$$I_4 = \frac{3}{4}.I_2$$
$$I_2 = \frac{1}{2}.I_0$$
$$\text{and } I_0 = \frac{\pi}{2}$$
$$\therefore I_6 = \frac{5}{6}.\frac{3}{4}.\frac{1}{2}.\frac{\pi}{2} = \frac{5\pi}{32}$$

*Note*: All the natural numbers from $n$ down to 1 appear alternately on the bottom or top of the expression. In fact, if we start writing the numbers with the value of $n$ on the bottom, we can obtain the result with very little working:

$$\frac{(n-1) \qquad (n-3) \qquad (n-5)\ldots}{n \qquad (n-2) \qquad (n-4) \qquad \ldots} \quad \text{etc.}$$

If $n$ is odd, the factors end with 1 on the bottom

e.g. $\dfrac{6.4.2}{7.5.3.1}$ and that is all there is to it.

If $n$ is even, the factor 1 comes on top and then we add the factor $\pi/2$

e.g. $\dfrac{7.5.3.1}{8.6.4.2}.\dfrac{\pi}{2}$

So (a) $\displaystyle\int_0^{\pi/2} \sin^4 x\,dx = \ldots\ldots\ldots\ldots$ and (b) $\displaystyle\int_0^{\pi/2} \cos^5 x\,dx = \ldots\ldots\ldots\ldots$

**26**

$$\int_0^{\pi/2} \sin^4 x\,dx = \frac{3\pi}{16}, \quad \int_0^{\pi/2} \cos^5 x\,dx = \frac{8}{15}$$

This result for evaluating $\int \sin^n x\,dx$ or $\int \cos^n x\,dx$ between the limits $x = 0$ and $x = \pi/2$, is known as *Wallis's formula*. It is well worth remembering, so make a few notes on it.

*Then on to Frame 27 for a further example*

**27** Here is another example on the same theme.

Evaluate $\displaystyle\int_0^{\pi/2} \sin^5 x.\cos^2 x\,dx.$

We can write:

$$\int_0^{\pi/2} \sin^5 x.\cos^2 x\,dx = \int_0^{\pi/2} \sin^5 x(1-\sin^2 x)\,dx$$
$$= \int_0^{\pi/2} (\sin^5 x - \sin^7 x)\,dx$$
$$= I_5 - I_7 = \ldots\ldots\ldots\ldots$$

*Finish it off*

28

$$\boxed{\frac{8}{105}}$$

$$I_5 = \frac{4.2}{5.3.1} = \frac{8}{15};$$

$$I_7 = \frac{6.4.2}{7.5.3.1} = \frac{16}{35}$$

$$\therefore I_5 - I_7 = \frac{8}{15} - \frac{16}{35} = \frac{8}{105}$$

All that remains now is the **Can You?** checklist and the **Test exercise.**

*On then to Frames 29 and 30*

## Can You?

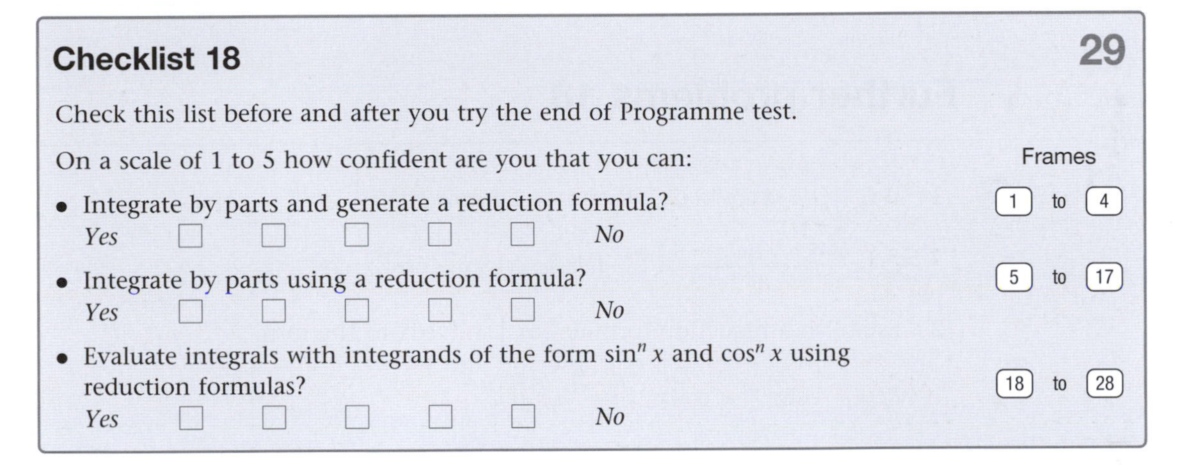

### Checklist 18

29

Check this list before and after you try the end of Programme test.

On a scale of 1 to 5 how confident are you that you can:

| | Frames |
|---|---|
| • Integrate by parts and generate a reduction formula? *Yes* ☐ ☐ ☐ ☐ ☐ *No* | 1 to 4 |
| • Integrate by parts using a reduction formula? *Yes* ☐ ☐ ☐ ☐ ☐ *No* | 5 to 17 |
| • Evaluate integrals with integrands of the form $\sin^n x$ and $\cos^n x$ using reduction formulas? *Yes* ☐ ☐ ☐ ☐ ☐ *No* | 18 to 28 |

## Test exercise 18

30

Before you work the exercise, look back through your notes and revise any points on which you are not absolutely certain; there should not be many. Take your time over the exercise: there are no prizes for speed! The questions are all straightforward and should cause no difficulty. Here they are then.

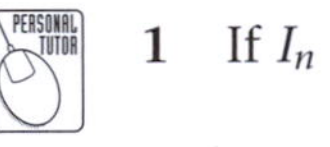


1 If $I_n = \int x^n e^{2x}\,dx$, show that

$$I_n = \frac{x^n e^{2x}}{2} - \frac{n}{2}.I_{n-1}$$

and hence evaluate $\int x^3 e^{2x}\,dx$.

2 Evaluate:

(a) $\int_0^{\pi/2} \sin^2 x.\cos^6 x\,dx$

(b) $\int_0^{\pi/2} \sin^4 x.\cos^5 x\,dx$

▶

3 By the substitution $x = a\sin\theta$, determine
$$\int_0^a x^3(a^2 - x^2)^{3/2}\,dx.$$

4 By writing $\tan^n x$ as $\tan^{n-2} x.(\sec^2 x - 1)$, obtain a reduction formula for
$$\int \tan^n x\,dx.$$

Hence show that $I_n = \int_0^{\pi/4} \tan^n x\,dx = \dfrac{1}{n-1} - I_{n-2}$.

5 By the substitution $x = \sin^2\theta$, determine a reduction formula for the integral
$$\int x^{5/2}(1 - x)^{3/2}\,dx.$$

Hence evaluate
$$\int_0^1 x^{5/2}(1 - x)^{3/2}\,dx.$$

---

## Further problems 18

31

1 If $I_n = \int_0^{\pi/2} x\cos^n x\,dx$, when $n > 1$, show that
$$I_n = \frac{n(n-1)I_{n-2} - 1}{n^2}. \qquad [\textit{Hint}:\ \cos^n x = \cos^2 x.\cos^{n-2} x]$$

2 Establish a reduction formula for $\int \sin^n x\,dx$ in the form
$$I_n = -\frac{1}{n}\sin^{n-1} x.\cos x + \frac{n-1}{n}I_{n-2}$$
and hence determine $\int \sin^7 x\,dx$.

3 If $I_n = \int_0^{\infty} x^n e^{-ax}\,dx$, show that $I_n = \dfrac{n}{a}I_{n-1}$. Hence evaluate $\int_0^{\infty} x^9 e^{-2x}\,dx$.

4 If $I_n = \int_0^{\pi} e^{-x}\sin^n x\,dx$, show that $I_n = \dfrac{n(n-1)}{n^2+1}I_{n-2}$.

5 If $I_n = \int_0^{\pi/2} x^n \sin x\,dx$, prove that, for $n \geq 2$,
$$I_n = n\left(\frac{\pi}{2}\right)^{n-1} - n(n-1)I_{n-2}.$$
Hence evaluate $I_3$ and $I_4$.

6 If $I_n = \int x^n e^x\,dx$, obtain a reduction formula for $I_n$ in terms of $I_{n-1}$ and hence determine $\int x^4 e^x\,dx$.

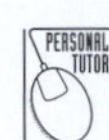

**7** If $I_n = \int \sec^n x \, dx$, prove that

$$I_n = \frac{1}{n-1} \tan x \sec^{n-2} x + \frac{n-2}{n-1} I_{n-2} \quad (n \geq 2).$$

Hence evaluate $\int_0^{\pi/6} \sec^8 x \, dx$.

**8** If $I_n = \int_0^{\pi/2} e^{-x} \cos^n x \, dx$, where $n \geq 2$, prove that:

(a) $I_n = 1 - n \int_0^{\pi/2} e^{-x} \sin x . \cos^{n-1} x \, dx$

(b) $(n^2 + 1)I_n = 1 + n(n-1)I_{n-2}$

Show that $I_6 = \dfrac{263 - 144e^{-\pi/2}}{629}$.

**9** If $I_n = \int (x^2 + a^2)^n \, dx$, show that

$$I_n = \frac{1}{2n+1} [x(x^2 + a^2)^n + 2na^2 I_{n-1}].$$

**10** If $I_n = \int \cot^n x \, dx$, $(n > 1)$, show that

$$I_n = -\frac{\cot^{n-1} x}{(n-1)} - I_{n-2}.$$

Hence determine $I_6$.

**11** If $I_n = \int (\ln x)^n \, dx$, show that

$I_n = x(\ln x)^n - n.I_{n-1}$.

Hence find $\int (\ln x)^3 \, dx$.

**12** If $I_n = \int \cosh^n x \, dx$, prove that

$$I_n = \frac{1}{n} \cosh^{n-1} x . \sinh x + \frac{n-1}{n} I_{n-2}.$$

Hence evaluate $\int_0^a \cosh^3 x \, dx$, where $a = \cosh^{-1}(\sqrt{2})$.

Now visit the companion website for this book at www.palgrave.com/stroud for more questions applying this mathematics to science and engineering.

# Integration applications 1

## Learning outcomes

When you have completed this Programme you will be able to:

- ☐ Evaluate the area beneath a curve
- ☐ Evaluate the area beneath a curve given in parametric form
- ☐ Determine the mean value of a function between two points
- ☐ Evaluate the root mean square (rms) value of a function

# Basic applications

**1**

We now look at some of the applications to which integration can be put. Some you already know from earlier work: others will be new to you. So let us start with the one you first met long ago.

## Areas under curves

*To find the area bounded by the curve $y = f(x)$, the x axis and the ordinates at $x = a$ and $x = b$.*

There is, of course, no mensuration formula for this, since its shape depends on the function $f(x)$. Do you remember how you established the method for finding this area?

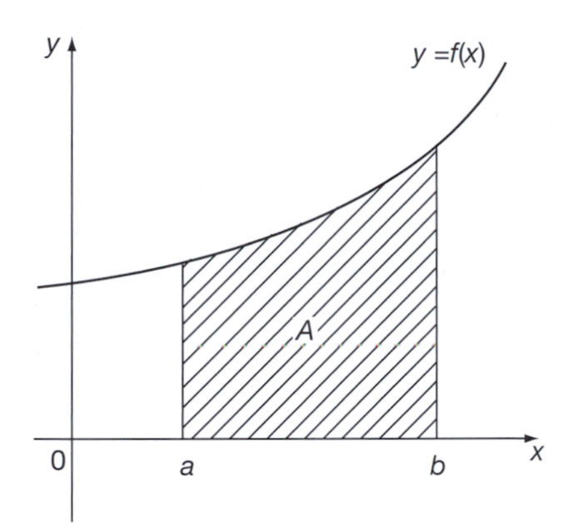


Move on to Frame 2

**2**

Let us revise this, for the same principles are applied in many other cases.

Let $P(x, y)$ be a point on the curve $y = f(x)$ and let $A_x$ denote the area under the curve measured from some point away to the left of the diagram.

The point Q, near to P, will have coordinates $(x + \delta x, y + \delta y)$ and the area is increased by the extent of the shaded strip. Denote this by $\delta A_x$.

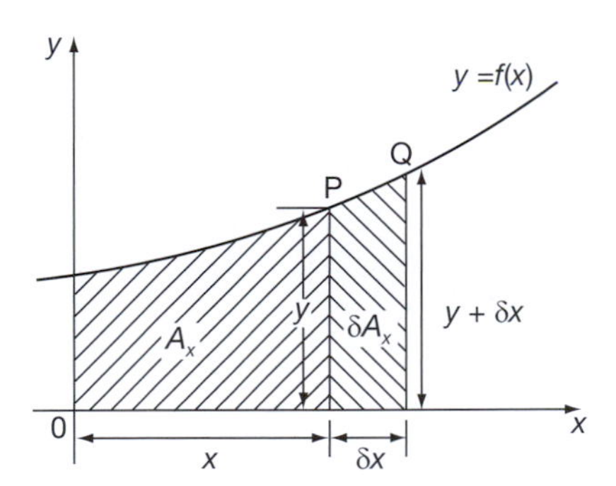


If we 'square off' the strip at the level of P, then we can say that the area of the strip is approximately equal to that of the rectangle (omitting PQR).

i.e. area of strip $\delta A_x \approx$ ...........

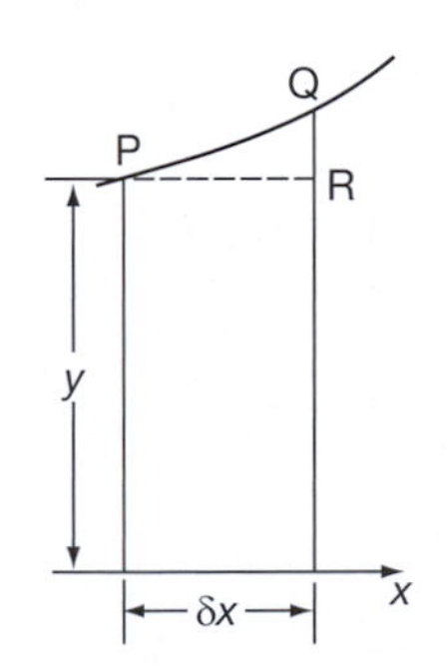


Move to Frame 3

**3**

$$\boxed{\delta A_x \approx y\delta x}$$

Therefore, $\dfrac{\delta A_x}{\delta x} \approx y$

i.e. the total area of the strip divided by the width, $\delta x$, of the strip gives approximately the value $y$.

The area above the rectangle represents the error in our stated approximation, but if we reduce the width of the strips, the total error is very much reduced.

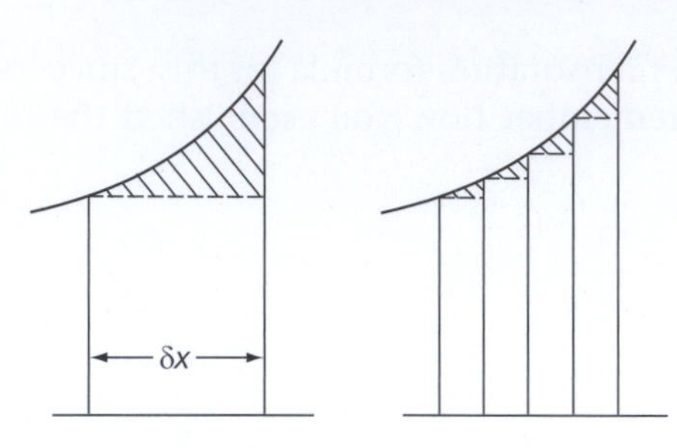


If we continue this process and make $\delta x \to 0$, then in the end the error will vanish, and, at the same time, $\dfrac{\delta A_x}{\delta x} \to$ ...........

**4**

$$\boxed{\frac{\delta A_x}{\delta x} \to \frac{dA_x}{dx}}$$

Correct. So we have $\dfrac{dA_x}{dx} = y$ (no longer an approximation)

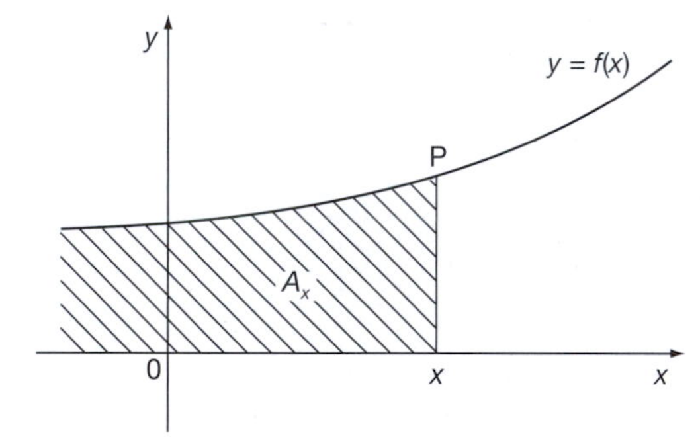


$$A_x = \int y\,dx$$
$$= \int f(x)\,dx$$
$$A_x = F(x) + C$$

and this represents the area under the curve up to the point P.

*Note*: As it stands, this result would not give us a numerical value for the area, because we do not know from what point the measurement of the area began (somewhere off to the left of the figure). Nevertheless, we can make good use of the result.

*So move on now to Frame 5*

5

$A_x = \int y\,dx$ gives the area up to the point P $(x, y)$.

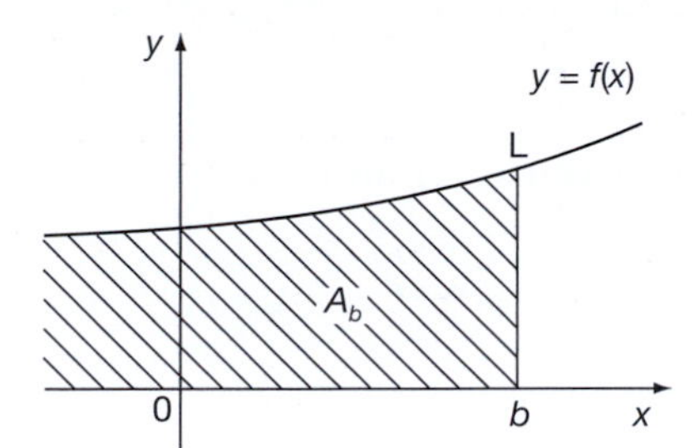


So:

(a) If we substitute $x = b$, we have the area up to the point L

i.e. $A_b = \int y\,dx$ with $x = b$.

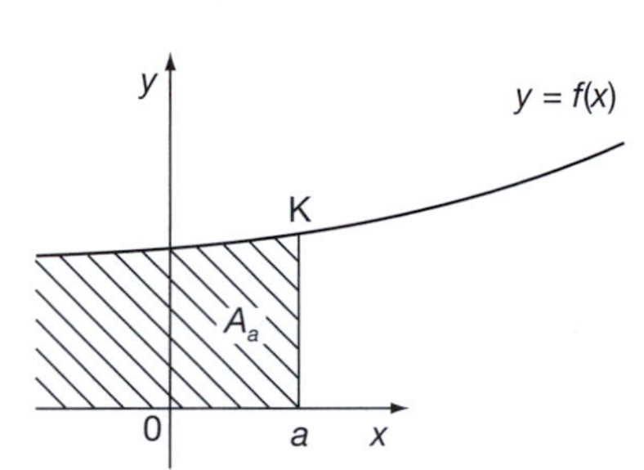


(b) If we substitute $x = a$, we have the area up to point K

i.e. $A_a = \int y\,dx$ with $x = a$.

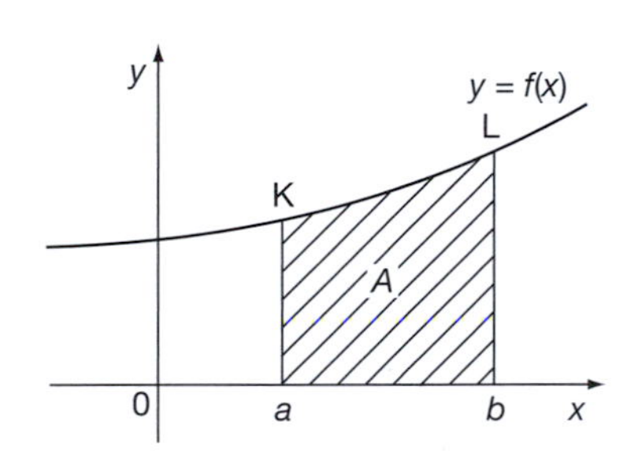


If we now subtract the second result from the first, we have the area under the curve between the ordinates at $x = a$ and $x = b$.

i.e. $A = \int y\,dx_{(x=b)} - \int y\,dx_{(x=a)}$

This is written

$$A = \int_a^b y\,dx$$

and the boundary values $a$ and $b$ are called the *limits* of the integral.

Remember: the higher limit goes at the top.
the lower limit goes at the bottom. } That seems logical.

So, the area under the curve $y = f(x)$ between $x = 1$ and $x = 5$ is written

$$A = \int_1^5 y\,dx$$

Similarly, the area under the curve $y = f(x)$ between $x = -5$ and $x = -1$ is written

$A = \ldots\ldots\ldots\ldots$

*On to Frame 6*

6

$$A = \int_{-5}^{-1} y\,dx$$

Let us do a simple example:

Find the area under the curve $y = x^2 + 2x + 1$ between $x = 1$ and $x = 2$.

$$A = \int_1^2 y\,dx = \int_1^2 (x^2 + 2x + 1)\,dx$$

$$= \left[\frac{x^3}{3} + x^2 + x + C\right]_1^2$$

$$= \left(\frac{8}{3} + 4 + 2 + C\right) - \left(\frac{1}{3} + 1 + 1 + C\right)$$

(putting $x = 2$) (putting $x = 1$)

$$= \left(8\frac{2}{3} + C\right) - \left(2\frac{1}{3} + C\right)$$

$$= 6\frac{1}{3}\text{units}^2$$

*Note*: When we have limits to substitute, the constant of integration appears in each bracket and will therefore always disappear. In practice therefore, we may leave out the constant of integration when we have limits, since we know it will always vanish in the next line of working. Now you do this one:

Find the area under the curve $y = 3x^2 + 4x - 5$ between $x = 1$ and $x = 3$.

*Then move on to Frame 7*

7

$$A = 32\text{ units}^2$$

Because

$$A = \int_1^3 (3x^2 + 4x - 5)\,dx = \left[x^3 + 2x^2 - 5x\right]_1^3$$

$$= \left(27 + 18 - 15\right) - \left(1 + 2 - 5\right)$$

$$= \left(30\right) - \left(-2\right)$$

$$= 32\text{ units}^2$$

## Definite integrals

An integral with limits is called a *definite integral*.

With a definite integral, the constant of integration may be omitted, not because it is not there, but because …………

*On to Frame 8*

8

... it occurs in both brackets and disappears in subsequent working

So, to evaluate a definite integral:

(a) Integrate the function (omitting the constant of integration) and enclose within square brackets with the limits at the right-hand end.
(b) Substitute the upper limit.
(c) Substitute the lower limit.
(d) Subtract the second result from the first result.

$$\int_a^b f(x)\,dx = [F(x)]_a^b = F(b) - F(a) \quad \text{where } f(x) = F'(x)$$

Now, you evaluate this one:

$$\int_{-1}^{\frac{1}{2}} 4e^{2x}\,dx = \ldots\ldots\ldots\ldots$$

9

5·166

Here it is:

$$\int_{-1}^{\frac{1}{2}} 4e^{2x}\,dx = 4\left[\frac{e^{2x}}{2}\right]_{-1}^{\frac{1}{2}} = 2\left[e^{2x}\right]_{-1}^{\frac{1}{2}}$$
$$= 2(e - e^{-2})$$
$$= 2\left(e - \frac{1}{e^2}\right)$$
$$= 5{\cdot}166$$

Now, what about this one: $\int_0^{\pi/2} x\cos x\,dx$.

First of all, forget about the limits.

$$\int x\cos x\,dx = \ldots\ldots\ldots\ldots$$

*When you have done that part, move on to Frame 10*

10

$$\int x\cos x\,dx = x(\sin x) - \int \sin x\,dx$$
$$= x\sin x + \cos x + C$$

$$\therefore \int_0^{\pi/2} x\cos x\,dx = \left[x\sin x + \cos x\right]_0^{\pi/2}$$
$$= \ldots\ldots\ldots\ldots \qquad \text{You finish it off.}$$

**11**

$$\boxed{\frac{\pi}{2}-1}$$

Because

$$\begin{aligned}\int_0^{\pi/2} x\cos x\,dx &= \left[x\sin x+\cos x\right]_0^{\pi/2}\\ &= \left(\frac{\pi}{2}+0\right)-(0+1)\\ &= \frac{\pi}{2}-1\end{aligned}$$

If you can integrate the given function, the rest is easy.

*So move to the next frame and work a few exercises on your own*

**12** Evaluate:

1 $\displaystyle\int_1^2 (2x-3)^4\,dx$ 2 $\displaystyle\int_0^5 \frac{1}{x+5}\,dx$ 3 $\displaystyle\int_{-3}^3 \frac{dx}{x^2+9}$ 4 $\displaystyle\int_1^e x^2\ln x\,dx$

*When you have finished them all, check your results with the solutions given in the next frame*

**13**

1 $$\int_1^2 (2x-3)^4\,dx = \left[\frac{(2x-3)^5}{10}\right]_1^2 = \frac{1}{10}\left\{(1)^5-(-1)^5\right\}$$

$$= \frac{1}{10}\{(1)-(-1)\} = \frac{2}{10} = \frac{1}{5}$$

2 $$\int_0^5 \frac{1}{x+5}\,dx = \Big[\ln(x+5)\Big]_0^5$$

$$= \ln 10 - \ln 5 = \ln\frac{10}{5} = \ln 2$$

3 $$\int_{-3}^3 \frac{dx}{x^2+9} = \left[\frac{1}{3}\tan^{-1}\frac{x}{3}\right]_{-3}^3$$

$$= \frac{1}{3}\{(\tan^{-1}1)-(\tan^{-1}[-1])\}$$

$$= \frac{1}{3}\left\{\frac{\pi}{4}-\left(-\frac{\pi}{4}\right)\right\} = \frac{\pi}{6}$$

4 $$\int x^2\ln x\,dx = \ln x\left(\frac{x^3}{3}\right)-\frac{1}{3}\int x^3.\frac{1}{x}\,dx$$

$$= \frac{x^3\ln x}{3}-\frac{x^3}{9}+C$$

$$\therefore \int_1^e x^2\ln x\,dx = \left[\frac{x^3\ln x}{3}-\frac{x^3}{9}\right]_1^e$$

$$= \left(\frac{e^3}{3}-\frac{e^3}{9}\right)-\left(0-\frac{1}{9}\right)$$

$$= \frac{2e^3}{9}+\frac{1}{9} = \frac{1}{9}(2e^3+1)$$

*On to Frame 14*

**14**

In very many practical applications we shall be using definite integrals, so let us practise a few more.

Do these:

5 $\int_0^{\pi/2} \frac{\sin 2x}{1+\cos^2 x}\,dx$

6 $\int_1^2 xe^x\,dx$

7 $\int_0^{\pi} x^2 \sin x\,dx$

*Finish them off and then check the solutions in the next frame*

**15**

5 $$\int_0^{\pi/2} \frac{\sin 2x}{1+\cos^2 x}\,dx = \left[-\ln(1+\cos^2 x)\right]_0^{\pi/2}$$
$$= (-\ln(1+0)) - (-\ln(1+1))$$
$$= (-\ln 1 + \ln 2)$$
$$= \ln 2$$

6 $$\int xe^x\,dx = x(e^x) - \int e^x\,dx$$
$$= xe^x - e^x + C$$
$$\therefore \int_1^2 xe^x\,dx = \left[e^x(x-1)\right]_1^2$$
$$= e^2 - 0$$
$$= e^2$$

7 $$\int x^2 \sin x\,dx = x^2(-\cos x) + 2\int x\cos x\,dx$$
$$= -x^2\cos x + 2\left\{x(\sin x) - \int \sin x\,dx\right\}$$
$$= -x^2\cos x + 2x\sin x + 2\cos x + C$$
$$\therefore \int_0^{\pi} x^2 \sin x\,dx = \left[(2-x^2)\cos x + 2x\sin x\right]_0^{\pi}$$
$$= ((2-\pi^2)(-1)+0) - (2+0)$$
$$= \pi^2 - 2 - 2$$
$$= \pi^2 - 4$$

*Now move on to Frame 16*

**16**

Before we move on to the next piece of work, here is just one more example for you to do on areas.

Find the area bounded by the curve $y = x^2 - 6x + 5$, the $x$-axis, and the ordinates at $x = 1$ and $x = 3$.

*Work it through and then move on to Frame 17*

## 17

$$I = -5\frac{1}{3} \text{ units so } A = 5\frac{1}{3} \text{ units}$$

Here is the working:

$$I = \int_1^3 y\,dx = \int_1^3 (x^2 - 6x + 5)\,dx$$

$$= \left[\frac{x^3}{3} - 3x^2 + 5x\right]_1^3$$

$$= \left(9 - 27 + 15\right) - \left(\frac{1}{3} - 3 + 5\right)$$

$$= \left(-3\right) - \left(2\frac{1}{3}\right)$$

$$= -5\frac{1}{3} \text{ units}^2$$

The area is then $A = 5\frac{1}{3}$ units because there is no such thing as a negative area. If you are concerned about the negative sign of the result, let us sketch the graph of the function. Here it is:

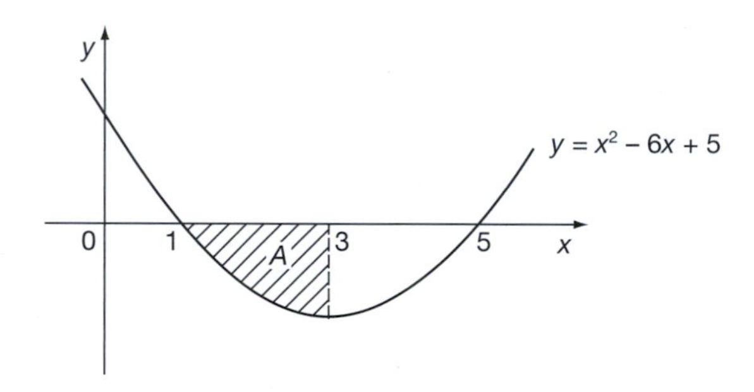


We find that between the limits we are given, the area lies below the $x$-axis. For such an area, $y$ is negative as is the integral

$$\therefore\ y\delta x \text{ is negative} \qquad \therefore\ I \text{ is negative.}$$

So remember:

> Whilst area is always positive, if the area is below the $x$-axis the corresponding integral is negative.

*Next frame*

## 18

The danger comes when we are integrating between limits and part of the area is above the $x$-axis and part below it. In this case, that part of the integral corresponding to the area above the $x$-axis will partly or wholly cancel out that part of the integral corresponding to the area below the $x$-axis. If this is likely to happen, sketch the curve and perform the integration in two parts.

*Now move to Frame 19*

## Parametric equations

19

**Example 1**

A curve has parametric equations $x = at^2$, $y = 2at$. Find the area bounded by the curve, the $x$-axis and the ordinates at $t = 1$ and $t = 2$.

We know that $A = \int_\alpha^\beta y\,dx$ where $\alpha$ and $\beta$ are the limits or boundary values of the variable. Replacing $y$ by $2at$, gives

$$A = \int_\alpha^\beta 2at\,dx$$

but we cannot integrate a function of $t$ with respect to $x$ directly. We therefore have to change the variable of the integral and we do it thus:

We are given $x = at^2 \quad \therefore \frac{dx}{dt} = 2at \quad \therefore dx = 2at\,dt$

We now have $A = \int_1^2 2at.2at\,dt = \int_1^2 4a^2t^2\,dt$

$= \ldots\ldots\ldots\ldots$ Finish it off.

20

$$A = \int_1^2 4a^2t^2dt = 4a^2\left[\frac{t^3}{3}\right]_1^2 = 4a^2\left(\frac{8}{3} - \frac{1}{3}\right) = \frac{28a^2}{3}$$

The method is always the same:

(a) express $x$ and $y$ in terms of the parameter
(b) change the variable
(c) insert limits of the parameter.

**Example 2**

If $x = a\sin\theta$, $y = b\cos\theta$, find the area under the curve between $\theta = 0$ and $\theta = \pi$.

$$A = \int_\alpha^\beta y\,dx = \int_0^\pi b\cos\theta.a\cos\theta.d\theta \qquad \begin{aligned} x &= a\sin\theta \\ dx &= a\cos\theta\,d\theta \end{aligned}$$

$$= ab\int_0^\pi \cos^2\theta\,d\theta$$

$= \ldots\ldots\ldots\ldots$

21

$$A = \frac{\pi ab}{2}$$

Because

$$A = ab\int_0^\pi \cos^2\theta\,d\theta = ab\left[\frac{\theta}{2} + \frac{\sin 2\theta}{4}\right]_0^\pi = ab\left(\frac{\pi}{2}\right) = \frac{\pi ab}{2}$$

Now do this one on your own.

▶

**Example 3**

If $x = \theta - \sin\theta,\ y = 1 - \cos\theta$, find the area under the curve between $\theta = 0$ and $\theta = \pi$.

*When you have finished it, move on to Frame 22*

**22**

$$A = \frac{3\pi}{2}\text{units}^2$$

Working:

$$A = \int_{\alpha}^{\beta} y\,dx$$

$$= \int_0^{\pi}(1-\cos\theta)(1-\cos\theta)d\theta \qquad \begin{aligned} y &= (1-\cos\theta) \\ x &= (\theta - \sin\theta) \\ dx &= (1-\cos\theta)d\theta \end{aligned}$$

$$= \int_0^{\pi}(1 - 2\cos\theta + \cos^2\theta)d\theta$$

$$= \left[\theta - 2\sin\theta + \frac{\theta}{2} + \frac{\sin 2\theta}{4}\right]_0^{\pi}$$

$$= \left(\frac{3\pi}{2}\right) - \left(0\right)$$

$$= \frac{3\pi}{2}\text{units}^2$$

## Mean values

**23**

To find the mean height of the students in a class, we could measure their individual heights, total the results and divide by the number of students. That is, in such cases, the *mean value* is simply the *average* of the separate values we were considering.

To find the mean value of a continuous function, however, requires further consideration.

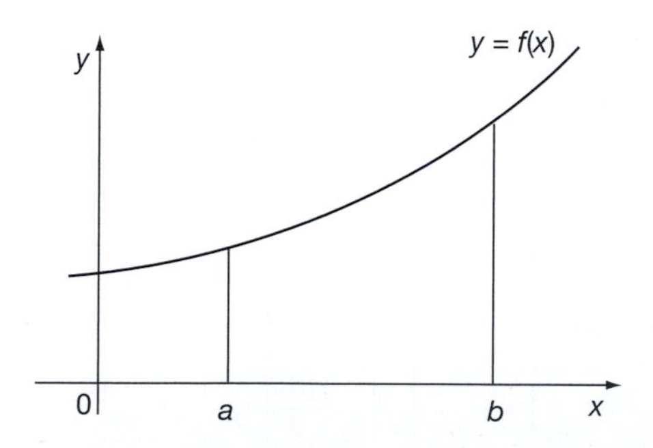


When we set out to find the mean of the function $y = f(x)$ between $x = a$ and $x = b$, we are no longer talking about separate items but a quantity which is continuously changing from $x = a$ to $x = b$. If we estimate the mean height of the figure in the diagram, over the given range, we are selecting a value $M$ such that the part of the figure cut off would fill in the space below.

In other words, the area of the figure between $x = a$ and $x = b$ is shared out equally along the base line of the figure to produce the rectangle.

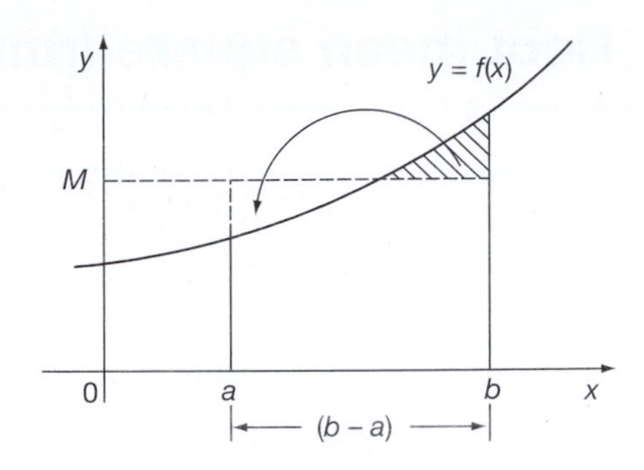


$$\therefore M = \frac{\text{Area}}{\text{Base line}}$$

$$= \frac{A}{b-a}$$

$$\therefore M = \frac{1}{b-a}\int_a^b y\,dx$$

So, to find the mean value of a function between two limits, find the area under the curve between those limits and divide by ...........

*On to Frame 24*

---

**24**

length of the base line

So it is really an application of areas.

**Example 1**

To find the mean value of $y = 3x^2 + 4x + 1$ between $x = -1$ and $x = 2$.

$$M = \frac{1}{b-a}\int_a^b y\,dx = \frac{1}{2-(-1)}\int_{-1}^{2}(3x^2 + 4x + 1)\,dx$$

$$= \frac{1}{3}\left[x^3 + 2x^2 + x\right]_{-1}^{2} = \frac{1}{3}\left\{(8 + 8 + 2) - (-1 + 2 - 1)\right\}$$

$$= \frac{1}{3}\left\{18\right\} = 6$$

$$\therefore M = 6$$

Here is one for you:

**Example 2**

Find the mean value of $y = 3\sin 5t + 2\cos 3t$ between $t = 0$ and $t = \pi$.

*Check your result with Frame 25*

---

**25**

Here is the working in full:

$$M = \frac{1}{\pi - 0}\int_0^{\pi}(3\sin 5t + 2\cos 3t)\,dt$$

$$= \frac{1}{\pi}\left[\frac{-3\cos 5t}{5} + \frac{2\sin 3t}{3}\right]_0^{\pi}$$

$$= \frac{1}{\pi}\left\{\left(\frac{-3\cos 5\pi}{5} + \frac{2\sin 3\pi}{3}\right) - \left(-\frac{3}{5} + 0\right)\right\}$$

$$= \frac{1}{\pi}\left\{\frac{3}{5} + \frac{3}{5}\right\}$$

$$M = \frac{6}{5\pi}$$

---

## Root mean square (rms) value

**26**

The phrase 'rms value of $y$' stands for 'the square *r*oot of the *m*ean value of the *s*quares of $y$' between some stated limits.

For example, if we are asked to find the rms value of $y = x^2 + 3$ between $x = 1$ and $x = 3$, we have:

$$\text{rms} = \sqrt{\text{Mean value of } y^2 \text{ between } x = 1 \text{ and } x = 3}$$

$$\therefore (\text{rms})^2 = \text{Mean value of } y^2 \text{ between } x = 1 \text{ and } x = 3$$

$$= \frac{1}{3-1}\int_1^3 y^2\,dx$$

$$= \ldots\ldots\ldots\ldots$$

*Next frame*

**27**

$$(\text{rms})^2 = \frac{1}{2}\int_1^3 (x^4 + 6x^2 + 9)\,dx$$

$$= \frac{1}{2}\left[\frac{x^5}{5} + 2x^3 + 9x\right]_1^3$$

$$= \frac{1}{2}\left\{\left(\frac{243}{5} + 54 + 27\right) - \left(\frac{1}{5} + 2 + 9\right)\right\}$$

$$= \frac{1}{2}\left\{48{\cdot}6 + 81 - 11{\cdot}2\right\} = \frac{1}{2}\left\{129{\cdot}6 - 11{\cdot}2\right\}$$

$$= \frac{1}{2}\left\{118{\cdot}4\right\} = 59{\cdot}2$$

$$\text{rms} = \sqrt{59{\cdot}2} = 7{\cdot}694$$

$$\therefore \text{rms} = 7{\cdot}69$$

So, *in words*, the rms value of $y$ between $x = a$ and $x = b$ means

........................

(Write it out)

*Then to the next frame*

**28**

'. . . the square root of the mean value of the squares of $y$ between $x = a$ and $x = b$'

There are three distinct steps:

(1) Square the given function.

(2) Find the mean value of the result over the interval given.

(3) Take the square root of the mean value.

So here is an example for you to do:

Find the rms value of $y = 400 \sin 200\pi t$ between $t = 0$ and $t = \dfrac{1}{100}$.

*When you have the result, move on to Frame 29*

**29**

See if you agree with this:

$$y^2 = 160\,000\sin^2 200\pi t$$
$$= 160\,000.\frac{1}{2}(1 - \cos 400\pi t)$$
$$= 80\,000(1 - \cos 400\pi t)$$
$$\therefore\ (\text{rms})^2 = \frac{1}{\frac{1}{100} - 0}\int_0^{1/100} 80\,000(1 - \cos 400\pi t)\,\text{d}t$$
$$= 10 \times 80\,000\left[t - \frac{\sin 400\pi t}{400\pi}\right]_0^{1/100}$$
$$= 8 \times 10^6\left(\frac{1}{100} - 0\right) = 8 \times 10^4$$
$$\therefore\ \text{rms} = \sqrt{8 \times 10^4} = 200\sqrt{2} = 282{\cdot}8$$

*Now on to Frame 30*

**30**

Before we come to the end of this particular Programme, let us think back yet again to the beginning of the work. We were, of course, considering the area bounded by the curve $y = f(x)$, the $x$-axis and the ordinates at $x = a$ and $x = b$.

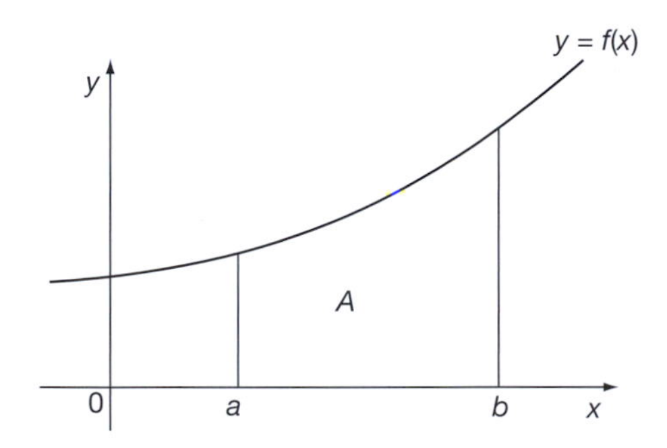


We found that

$$A = \int_a^b y\,\text{d}x$$

Let us look at the figure again.

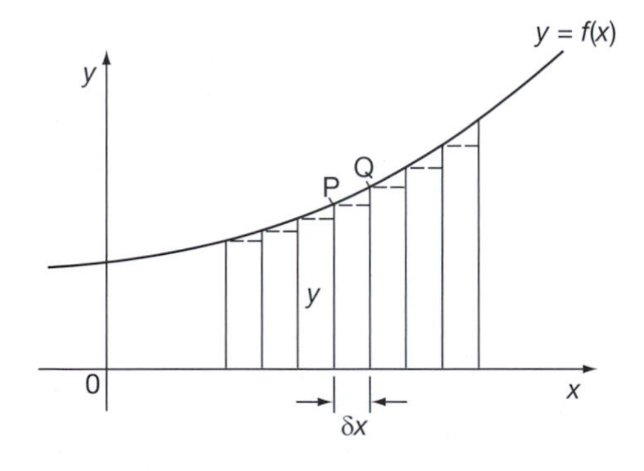


If P is the point $(x, y)$ then the area of the strip $\delta A$ is given by

$$\delta A \approx y.\delta x$$

If we divide the complete figure up into a series of such strips, then the total area is given approximately by the sum of the areas of these strips.

i.e. $A$ = sum of the strips between $x = a$ and $x = b$

i.e. $A \approx \sum_{x=a}^{x=b} y.\delta x$

$\sum$ = 'the sum of all terms like...'

▶

The error in our approximation is caused by ignoring the area over each rectangle. But if the strips are made narrower, this error progressively decreases and, at the same time, the number of strips required to cover the figure increases. Finally, when $\delta x \to 0$:

$A =$ sum of an infinite number of minutely thin rectangles

$$\therefore A = \int_a^b y\,\mathrm{d}x = \sum_{x=a}^{x=b} y.\delta x \text{ when } \delta x \to 0$$

It is sometimes convenient, therefore, to regard integration as a summing process on an infinite number of minutely small quantities, each of which is too small to exist alone.

We shall make use of this idea at a later date.

*Next frame*

## Review summary

**31**

1 *Areas under curves*

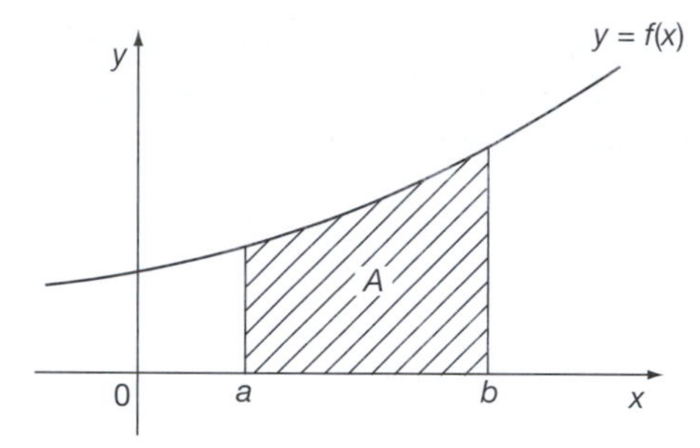


$$A = \int_b^a y\,\mathrm{d}x$$

If the area is below the $x$-axis the integral will be negative.

2 *Definite integrals*

A definite integral is an integral with limits:

$$\int_a^b y\,\mathrm{d}x = [F(x)]_a^b = F(b) - F(a) \text{ where } y = f(x) = F'(x)$$

3 *Parametric equations*

$x = f(t),\; y = F(t)$

$$\int_{x_1}^{x_2} y\,\mathrm{d}x = \int_{x_1}^{x_2} F(t)\,\mathrm{d}x = \int_{t_1}^{t_2} F(t).\frac{\mathrm{d}x}{\mathrm{d}t}\,\mathrm{d}t$$

4 *Mean values*

$$M = \frac{1}{b-a}\int_a^b y\,\mathrm{d}x$$

5 *rms values*

$$(\text{rms})^2 = \frac{1}{b-a}\int_a^b y^2\,\mathrm{d}x$$

6 *Integration as a summing process*

$$\text{When } \delta x \to 0,\quad \sum_{x=a}^{x=b} y.\delta x = \int_a^b y\,\mathrm{d}x$$

All now now remains is the **Can You?** checklist and the **Test exercise**. Before you work through them, make sure there is nothing that you wish to brush up.

*On then to Frames 32 and 33*

## Can You?

### Checklist 19

32

Check this list before and after you try the end of Programme test.

On a scale of 1 to 5 how confident are you that you can: Frames

- Evaluate the area beneath a curve? 1 to 18
  *Yes* ☐ ☐ ☐ ☐ ☐ *No*
- Evaluate the area beneath a curve given in parametric form? 19 to 22
  *Yes* ☐ ☐ ☐ ☐ ☐ *No*
- Determine the mean value of a function between two points? 23 to 25
  *Yes* ☐ ☐ ☐ ☐ ☐ *No*
- Evaluate the root mean square (rms) value of a function? 26 to 30
  *Yes* ☐ ☐ ☐ ☐ ☐ *No*

## Test exercise 19

33

The questions are all very straightforward, but take your time.

1 Find the area bounded by the curves $y = 3e^{2x}$ and $y = 3e^{-x}$ and the ordinates at $x = 1$ and $x = 2$.

2 The parametric equations of a curve are
$y = 2\sin\dfrac{\pi}{10}t,\ \ x = 2 + 2t - 2\cos\dfrac{\pi}{10}t.$
Find the area under the curve between $t = 0$ and $t = 10$.

3 Find the mean value of $y = \dfrac{5}{2 - x - 3x^2}$ between $x = -\dfrac{1}{3}$ and $x = +\dfrac{1}{3}$.

4 Calculate the rms value of $i = 20 + 100\sin 100\pi t$ between $t = 0$ and $t = 1/50$.

5 If $i = I\sin\omega t$ and $v = L\dfrac{\mathrm{d}i}{\mathrm{d}t} + Ri$, find the mean value of the product $vi$ between $t = 0$ and $t = \dfrac{2\pi}{\omega}$.

6 If $i = 300\sin 100\pi t + I$, and the rms value of $i$ between $t = 0$ and $t = 0{\cdot}02$ is 250, determine the value of $I$.

## Further problems 19

34

1 Find the mean height of the curve $y = 3x^2 + 5x - 7$ above the $x$-axis between $x = -2$ and $x = 3$.

2 Find the rms value of $i = \cos x + \sin x$ over the range $x = 0$ to $\dfrac{3\pi}{4}$.

3 Determine the area of one arch of the cycloid $x = \theta - \sin\theta$, $y = 1 - \cos\theta$, i.e. find the area of the plane figure bounded by the curve and the $x$-axis between $\theta = 0$ and $\theta = 2\pi$.

▶

4 Find the area enclosed by the curves $y = \sin x$ and $y = \sin 2x$, between $x = 0$ and $x = \frac{\pi}{3}$.

5 If $i = 0{\cdot}2 \sin 10\pi t + 0{\cdot}01 \sin 30\pi t$, find the mean value of $i$ between $t = 0$ and $t = 0{\cdot}2$.

6 If $i = i_1 \sin pt + i_2 \sin 2pt$, where $p$ is a constant, show that the mean value of $i^2$ over a period is $\frac{1}{2}(i_1^2 + i_2^2)$.

7 Sketch the curves $y = 4e^x$ and $y = 9 \sinh x$, and show that they intersect when $x = \ln 3$. Find the area bounded by the two curves and the $y$-axis.

8 If $v = v_0 \sin \omega t$ and $i = i_0 \sin(\omega t - a)$, find the mean value of $vi$ between $t = 0$ and $t = \frac{2\pi}{\omega}$.

9 If $i = \frac{E}{R} + I \sin \omega t$, where $E$, $R$, $I$ and $\omega$ are constants, find the rms value of $i$ over the range $t = 0$ to $t = \frac{2\pi}{\omega}$.

10 The parametric equations of a curve are
$x = a\cos^2 t \sin t, \; y = a \cos t \sin^2 t$.
Show that the area enclosed by the curve between $t = 0$ and $t = \frac{\pi}{2}$ is $\frac{\pi a^2}{32}$ units$^2$.

11 Find the area bounded by the curve $(1 - x^2)y = (x - 2)(x - 3)$, the $x$-axis and the ordinates at $x = 2$ and $x = 3$.

12 Find the area enclosed by the curve $a(a - x)y = x^3$, the $x$-axis and the line $2x = a$.

13 Prove that the area bounded by the curve $y = \tanh x$ and the straight line $y = 1$ between $x = 0$ and $x = \infty$, is $\ln 2$.

14 Prove that the curve defined by $x = \cos^3 t, \; y = 2 \sin^3 t$ encloses an area of $\frac{3\pi}{4}$ units$^2$ between $t = 0$ and $t = 2\pi$.

15 Find the mean value of $y = xe^{-x/a}$ between $x = 0$ and $x = a$.

16 A plane figure is bounded by the curves $2y = x^2$ and $x^3 y = 16$, the $x$-axis and the ordinate at $x = 4$. Calculate the area enclosed.

17 Find the area of the loop of the curve $y^2 = x^4(4 + x)$.

18 If $i = I_1 \sin(\omega t + \alpha) + I_2 \sin(2\omega t + \beta)$, where $I_1$, $I_2$, $\omega$, $\alpha$, and $\beta$ are constants, find the rms value of $i$ over a period, i.e. from $t = 0$ to $t = \frac{2\pi}{\omega}$.

19 Show that the area enclosed by the curve $x = a(2t - \sin 2t), \; y = 2a \sin^2 t$ and the $x$-axis between $t = 0$ and $t = \pi$ is $3\pi a^2$ units$^2$.

20 A plane figure is bounded by the curves $y = 1/x^2, \; y = e^{x/2} - 3$ and the lines $x = 1$ and $x = 2$. Determine the extent of the area of the figure.

**35**

Now visit the companion website for this book at www.palgrave.com/stroud for more questions applying this mathematics to science and engineering.

# Integration applications 2

## Learning outcomes

When you have completed this Programme you will be able to:

- ☐ Calculate volumes of revolution
- ☐ Locate the centroid of a plane figure
- ☐ Locate the centre of gravity of a solid of revolution
- ☐ Determine the lengths of curves
- ☐ Determine the lengths of curves given by parametric equations
- ☐ Calculate surfaces of revolution
- ☐ Calculate surfaces of revolution using parametric equations
- ☐ Use the two rules of Pappus

## Introduction

**1**

In the previous Programme, we saw how integration could be used

(a) to calculate areas under plane curves

(b) to find mean values of functions,

(c) to find rms values of functions.

We are now going to deal with a few more applications of integration: with some of these you will already be familiar and the work will serve as revision; others may be new to you. Anyway let us make a start, so move on to Frame 2.

## Volume of a solid of revolution

**2**

If the plane figure bounded by the curve $y = f(x)$, the $x$-axis and the ordinates at $x = a$ and $x = b$, rotates through a complete revolution about the $x$-axis, it will generate a solid symmetrical about the $x$-axis.

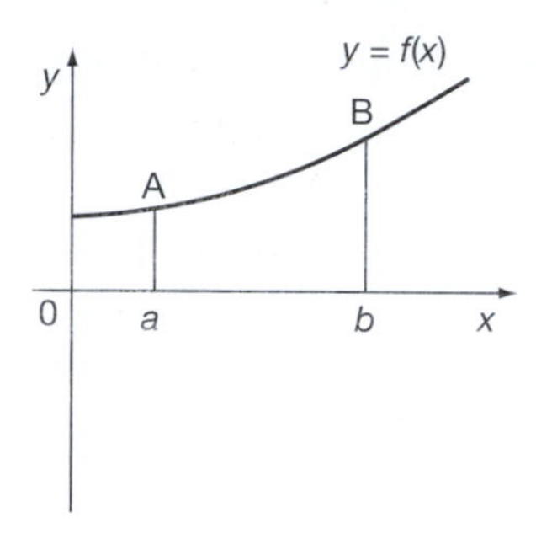


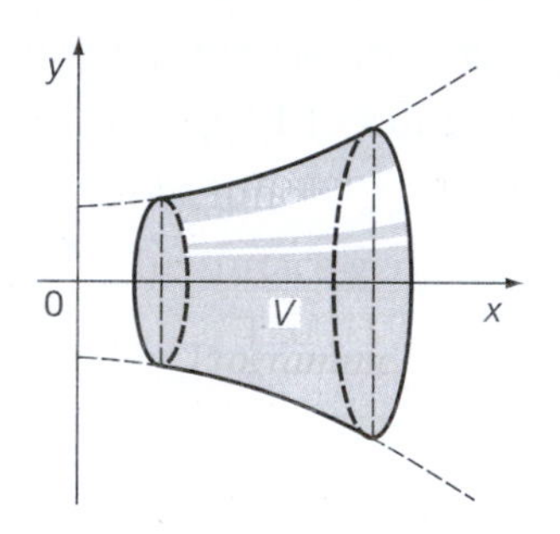


Let V be the volume of the solid generated.

To find $V$, let us first consider a thin strip of the original plane figure.

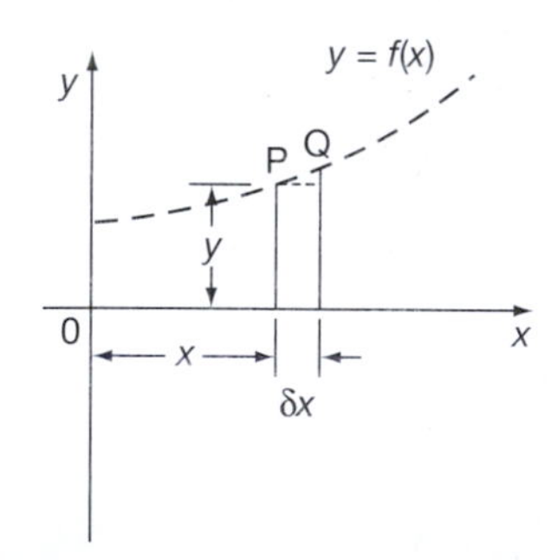


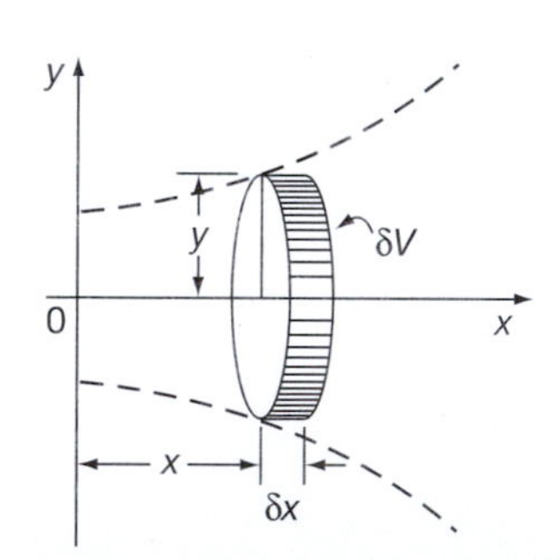


The volume generated by the strip is approximately equal to the volume generated by the rectangle.

i.e. $\delta V \approx$ ...........

**3**

$$\boxed{\delta V \approx y^2.\delta x}$$

Correct, since the solid generated is a flat cylinder.

▶

If we divide the whole plane figure up into a number of such strips, each will contribute its own flat disc with volume $\pi y^2.\delta x$.

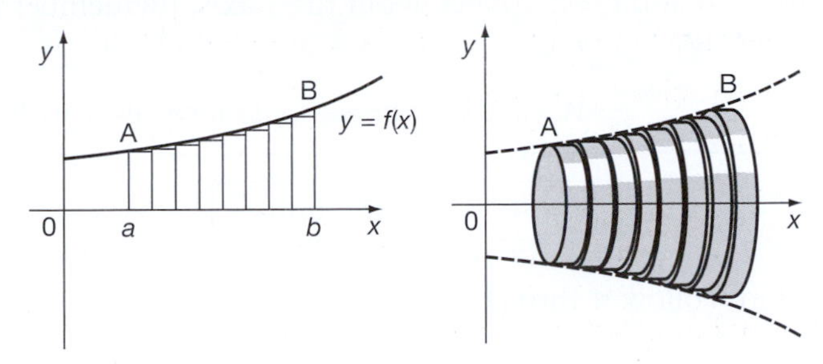


$\therefore$ Total volume, $V \approx \sum_{x=a}^{x=b} \pi y^2.\delta x$

The error in the approximation is due to the areas above the rectangles, which cause the step formation in the solid. However, if $\delta x \to 0$, the error disappears, so finally

$V = \ldots\ldots\ldots\ldots$

**4**

$$V = \int_a^b \pi y^2.\mathrm{d}x$$

This is a standard result, which you have doubtless seen many times before, so make a note of it in your record book and move on to Frame 5.

**5**

Here is an example:

Find the volume generated when the plane figure bounded by $y = 5\cos 2x$, the $x$-axis and ordinates at $x = 0$ and $x = \frac{\pi}{4}$, rotates about the $x$-axis through a complete revolution.

We have: $V = \int_0^{\pi/4} \pi y^2.\,\mathrm{d}x = 25\pi \int_0^{\pi/4} \cos^2 2x\,\mathrm{d}x$

Express this in terms of the double angle (i.e. $4x$) and finish it off.

*Then move on to Frame 6*

**6**

$$V = \frac{25\pi^2}{8} \text{ units}^3$$

Because

$$V = \pi \int_0^{\pi/4} y^2\,\mathrm{d}x = 25\pi \int_0^{\pi/4} \cos^2 2x\,\mathrm{d}x$$

$$= \frac{25\pi}{2} \int_0^{\pi/4} (1 + \cos 4x)\,\mathrm{d}x$$

$$\cos 2\theta = 2\cos^2\theta - 1$$

$$\cos^2\theta = \frac{1}{2}(1 + \cos 2\theta)$$

$$= \frac{25\pi}{2}\left[x + \frac{\sin 4x}{4}\right]_0^{\pi/4}$$

$$= \frac{25\pi}{2}\left(\left\{\frac{\pi}{4} + 0\right\} - \left\{0 + 0\right\}\right) = \frac{25\pi^2}{8} \text{ units}^3$$

Now what about this example?

The parametric equations of a curve are $x = 3t^2$, $y = 3t - t^2$. Find the volume generated when the plane figure bounded by the curve, the $x$-axis and the ordinates corresponding to $t = 0$ and $t = 2$, rotates about the $x$-axis. [Remember to change the variable of the integral!]

*Work it right through and then check with the next frame*

**7**

$$\boxed{V = 49{\cdot}62\pi = 156 \text{ units}^3}$$

Here is the solution. Follow it through.

$$V = \int_a^b \pi y^2 \,\mathrm{d}x \qquad y = 3t - t^2$$

$$V = \int_{t=0}^{t=2} \pi(3t - t^2)^2 \,\mathrm{d}x \qquad x = 3t^2, \quad \mathrm{d}x = 6t\mathrm{d}t$$

$$= \pi \int_0^2 (9t^2 - 6t^3 + t^4)6t\mathrm{d}t$$

$$= 6\pi \int_0^2 (9t^3 - 6t^4 + t^5)\mathrm{d}t$$

$$= 6\pi \left[ \frac{9t^4}{4} - \frac{6t^5}{5} + \frac{t^6}{6} \right]_0^2$$

$$= 6\pi(36 - 38{\cdot}4 + 10{\cdot}67)$$

$$= 6\pi(46{\cdot}67 - 38{\cdot}4)$$

$$= 6\pi(8{\cdot}27)$$

$$= 49{\cdot}6\pi$$

$$= 156 \text{ units}^3$$

So they are all done in very much the same way.

*Move on now to Frame 8*

**8**

Here is a slightly different example.

Find the volume generated when the plane figure bounded by the curve $y = x^2 + 5$, the $x$-axis and the ordinates $x = 1$ and $x = 3$, rotates about the $y$-axis through a complete revolution.

Note that this time the figure rotates about the axis of $y$.

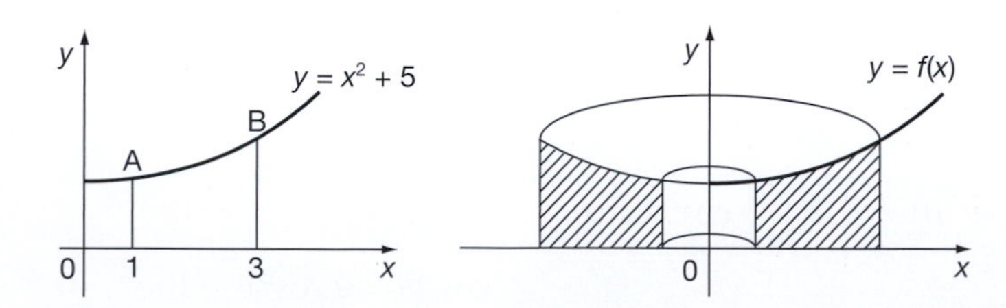


Half of the solid formed is shown in the right-hand diagram. We have no standard formula for this case. [$V = \int_a^b \pi y^2 \,\mathrm{d}x$ refers to rotation about the $x$-axis.] In such cases, we build up the integral from first principles.

*To see how we go about this, move on to Frame 9*

**9**

Here it is: note the general method.

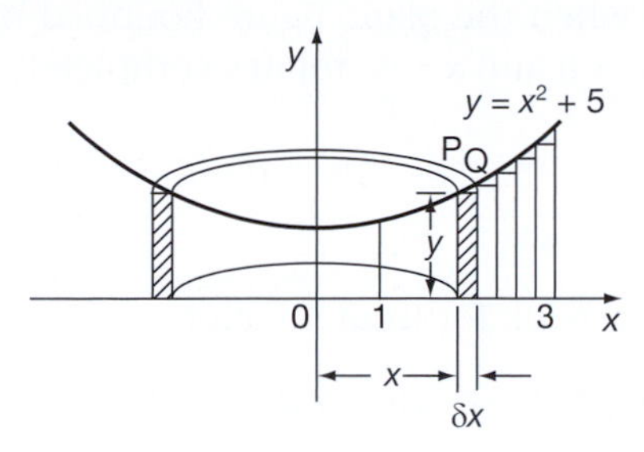


If we rotate an elementary strip PQ, we can say:

Volume generated by the strip $\approx$ volume generated by rectangle (i.e. hollow thin cylinder)

$\therefore \; \delta V \approx$ area of cross-section $\times$ circumference

$\delta V \approx y\delta x.2\pi x \approx 2\pi xy\delta x$

For all such strips between $x = 1$ and $x = 3$:

$$V \approx \sum \delta V \approx \sum_{x=1}^{x=3} 2\pi xy.\delta x$$

As usual, if $\delta x \to 0$, the error disappears and we finally obtain:

$$V = 2\int_1^3 \pi xy\mathrm{d}x$$

Since $y = x^2 + 5$, we can now substitute for $y$ and finish the calculation.

*Do that, and then on to the next frame*

**10**

$$\boxed{V = 80\pi \text{ units}^3}$$

Here is the working: check yours.

$$\begin{aligned} V &= \int_1^3 2\pi xy\mathrm{d}x \\ &= 2\pi\int_1^3 x(x^2+5)\,\mathrm{d}x \\ &= 2\pi\int_1^3 (x^3+5x)\,\mathrm{d}x \\ &= 2\pi\left[\frac{x^4}{4}+\frac{5x^2}{2}\right]_1^3 \\ &= 2\pi\left(\left\{\frac{81}{4}+\frac{45}{2}\right\}-\left\{\frac{1}{4}+\frac{5}{2}\right\}\right) \\ &= 2\pi\left(\frac{80}{4}+\frac{40}{2}\right) \\ &= 2\pi\left(20+20\right) = 80\pi \text{ units}^3 \end{aligned}$$

Whenever we have a problem not covered by our standard results, we build up the integral from first principles.

**11**

The last result is often required, so let us write it out again.

The volume generated when the plane figure bounded by the curve $y = f(x)$, the $x$-axis and the ordinates $x = a$ and $x = b$, rotates completely about the $y$-axis is given by:

$$V = 2\pi\int_a^b xy\,dx$$

Copy this into your record book for future reference.

*Then on to Frame 12, where we will deal with another application of integration*

## Centroid of a plane figure

**12**

The position of the centroid of a plane figure depends not only on the extent of the area but also on how the area is distributed. It is very much like the idea of the centre of gravity of a thin plate, but we cannot call it a centre of gravity, since a plane figure has no mass.

We can find its position, however, by taking an elementary strip and then taking moments (a) about the $y$-axis to find $\bar{x}$, and (b) about the $x$-axis to find $\bar{y}$. No doubt, you remember the results. Here they are:

$$A\bar{x} \approx \sum_{x=a}^{x=b} x.y\delta x$$

$$A\bar{y} \approx \sum_{x=a}^{x=b} \frac{y}{2}.y\delta x$$

Which give $\bar{x} = \dfrac{\int_a^b xy\,dx}{\int_a^b y\,dx}$, $\quad \bar{y} = \dfrac{\frac{1}{2}\int_a^b y^2\,dx}{\int_a^b y\,dx}$

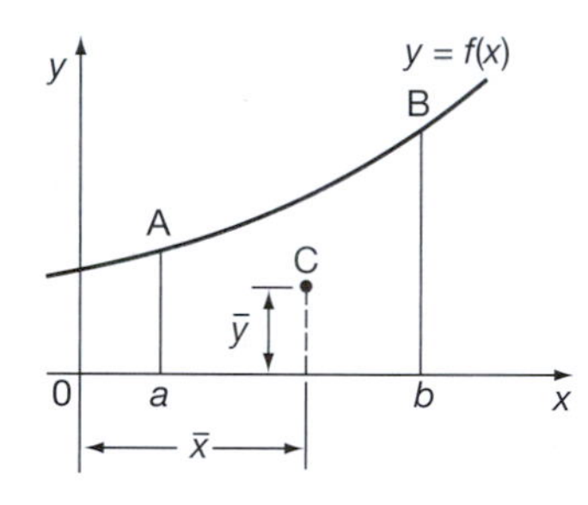


*Add these to your list of results.*

**13**

Now let us do one example. Here goes.

Find the position of the centroid of the figure bounded by $y = e^{2x}$, the $x$-axis, the $y$-axis and the ordinate at $x = 2$.

First, to find $\bar{x}$:

$$\bar{x} = \frac{\int_0^2 xy\,dx}{\int_0^2 y\,dx}$$

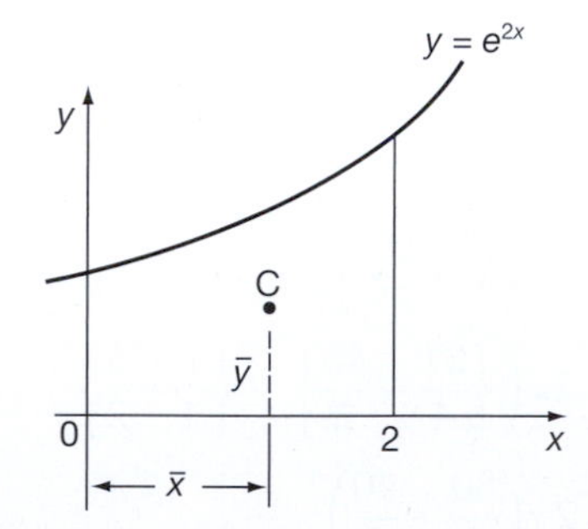


We evaluate the two integrals quite separately, so let $\bar{x} = \dfrac{I_1}{I_2}$.

Then $I_1 = \int_0^2 xe^{2x}\,dx = \ldots\ldots\ldots\ldots$

**14**

$$I_1 = \frac{3e^4 + 1}{4}$$

Because

$$\begin{aligned} I_1 &= \int_0^2 xe^{2x}\,dx \\ &= \left[x\left(\frac{e^{2x}}{2}\right) - \frac{1}{2}\int e^{2x}\,dx\right]_0^2 \\ &= \left[\frac{xe^{2x}}{2} - \frac{e^{2x}}{4}\right]_0^2 \\ &= \left(e^4 - \frac{e^4}{4}\right) - \left(-\frac{1}{4}\right) \\ &= \frac{3e^4}{4} + \frac{1}{4} = \frac{3e^4 + 1}{4} \end{aligned}$$

Similarly, $I_2 = \int_0^2 e^{2x}\,dx$ which gives $I_2 = \ldots\ldots\ldots\ldots$

**15**

$$I_2 = \frac{e^4 - 1}{2}$$

Because

$$\begin{aligned} I_2 &= \int_0^2 e^{2x}\,dx \\ &= \left[\frac{e^{2x}}{2}\right]_0^2 = \frac{e^4}{2} - \frac{1}{2} = \frac{e^4 - 1}{2} \end{aligned}$$

So, therefore, $\bar{x} = \dfrac{I_1}{I_2} = \dfrac{3e^4 + 1}{4} \times \dfrac{2}{e^4 - 1}$

$= \ldots\ldots\ldots\ldots$

**16**

$$\bar{x} = 1{\cdot}537$$

$$\bar{x} = \frac{3e^4 + 1}{2(e^4 - 1)} = \frac{3(54{\cdot}60) + 1}{2(54{\cdot}60 - 1)} = \frac{163{\cdot}8 + 1}{109{\cdot}2 - 2} = \frac{164{\cdot}8}{107{\cdot}2}$$

$\therefore\ \bar{x} = 1{\cdot}537$

Now we have to find $\bar{y}$:

$$\bar{y} = \frac{\int_0^2 \frac{1}{2} y^2\,dx}{\int_0^2 y\,dx} = \frac{I_3}{I_2}$$

Note that the denominator is the same as before.

$$I_3 = \frac{1}{2}\int_0^2 y^2\,dx = \ldots\ldots\ldots\ldots$$

**17**

$$I_3 = \frac{1}{8}(e^8 - 1) \;\therefore\; \bar{y} = \frac{1}{4}(e^4 + 1)$$

$$I_3 = \frac{1}{2}\int_0^2 y^2\,dx = \frac{1}{2}\int_0^2 e^{4x}\,dx$$

$$= \frac{1}{2}\left[\frac{e^{4x}}{4}\right]_0^2 = \frac{1}{8}(e^8 - 1)$$

$$\therefore \bar{y} = \frac{I_3}{I_2}$$

$$= \frac{\frac{1}{8}(e^8 - 1)}{\frac{1}{2}(e^4 - 1)} = \frac{1}{4}(e^4 + 1) = \frac{1}{4}(54{\cdot}60 + 1)$$

$$= \frac{55{\cdot}60}{4} = 13{\cdot}90$$

So the results are:

$\bar{x} = 1{\cdot}537;\ \bar{y} = 13{\cdot}90$

Now do this example on your own in just the same way.

Find the position of the centroid of the figure bounded by the curve $y = 5\sin 2x$, the $x$-axis and the ordinates at $x = 0$ and $x = \frac{\pi}{6}$.

(First of all find $\bar{x}$ and check your result before going on to find $\bar{y}$.)

**18**

$$\bar{x} = 0{\cdot}3424$$

$$I_1 = \int_0^{\pi/6} xy\,dx = 5\int_0^{\pi/6} x\sin 2x\,dx$$

$$= 5\left[x\frac{(-\cos 2x)}{2} + \frac{1}{2}\int_0^{\pi/6}\cos 2x\,dx\right]$$

$$= 5\left[-\frac{x\cos 2x}{2} + \frac{\sin 2x}{4}\right]_0^{\pi/6}$$

$$= 5\left(-\frac{\pi}{6}\cdot\frac{1}{2}\cdot\frac{1}{2} + \frac{\sqrt{3}}{8}\right)$$

$$= 5\left(\frac{\sqrt{3}}{8} - \frac{\pi}{24}\right) = \frac{5}{4}\left(\frac{\sqrt{3}}{2} - \frac{\pi}{6}\right)$$

$$\text{Also } I_2 = \int_0^{\pi/6} 5\sin 2x\,dx = 5\left[-\frac{\cos 2x}{2}\right]_0^{\pi/6} = -\frac{5}{2}\left(\frac{1}{2} - 1\right) = \frac{5}{4}$$

$$\therefore \bar{x} = \frac{5}{4}\left(\frac{\sqrt{3}}{2} - \frac{\pi}{6}\right)\cdot\frac{4}{5} = \left(\frac{\sqrt{3}}{2} - \frac{\pi}{6}\right)$$

$$= 0{\cdot}8660 - 0{\cdot}5236 \qquad \therefore \bar{x} = 0{\cdot}3424$$

Do you agree with that? If so, push on and find $\bar{y}$.

*When you have finished, move on to Frame 19*

19

$$\bar{y} = 1{\cdot}535$$

Here is the working in detail.

$$I_3 = \frac{1}{2}\int_0^{\pi/6} 25\sin^2 2x\,dx$$

$$= \frac{25}{2}\int_0^{\pi/6} \frac{1}{2}(1 - \cos 4x)\,dx$$

$$= \frac{25}{4}\left[x - \frac{\sin 4x}{4}\right]_0^{\pi/6}$$

$$= \frac{25}{4}\left(\frac{\pi}{6} - \frac{\sin(2\pi/3)}{4}\right) \qquad \sin\frac{2\pi}{3} = \sin\frac{\pi}{3} = \frac{\sqrt{3}}{2}$$

$$= \frac{25}{4}\left(\frac{\pi}{6} - \frac{\sqrt{3}}{8}\right)$$

$$= \frac{25}{4}\left(0{\cdot}5236 - 0{\cdot}2165\right)$$

$$= \frac{25}{4}\left(0{\cdot}3071\right) = 25(0{\cdot}07678) = 1{\cdot}919$$

Therefore:

$$y = \frac{I_3}{I_2} = \frac{1{\cdot}919}{5/4} = \frac{(1{\cdot}919)4}{5} = 1{\cdot}535$$

So the final results are $\bar{x} = 0{\cdot}342$, $\bar{y} = 1{\cdot}535$

*Now to Frame 20*

## Centre of gravity of a solid of revolution

20

Here is another application of integration not very different from the last.

To find the position of the centre of gravity of the solid formed when the plane figure bounded by the curve $y = f(x)$, the $x$-axis and the ordinates at $x = a$ and $x = b$, rotates about the $x$-axis.

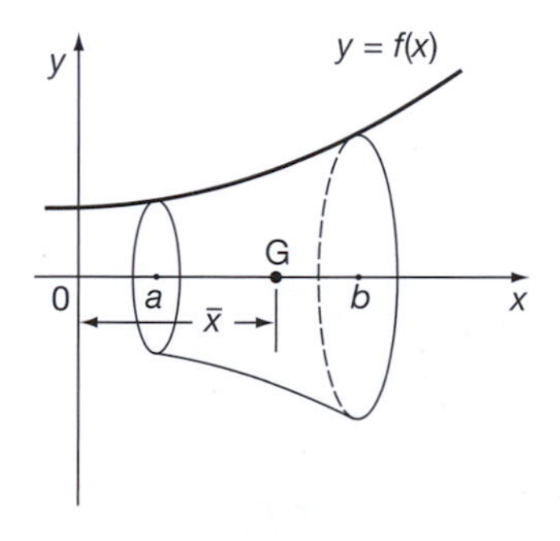


If we take elementary discs and sum the moments of volume (or mass) about the $y$-axis, we can calculate $\bar{x}$.

This gives $\bar{x} = \dfrac{\int_a^b xy^2\,dx}{\int_a^b y^2\,dx}$

What about $\bar{y}$? Clearly, $\bar{y} = \ldots\ldots\ldots\ldots$

**21**

$\boxed{\bar{y} = 0}$

Correct, since the solid generated is symmetrical about the $x$-axis and therefore the centre of gravity lies on this axis, i.e. $\bar{y} = 0$.

So we have to find only $\bar{x}$, using

$$\bar{x} = \frac{\displaystyle\int_a^b xy^2\,dx}{\displaystyle\int_a^b y^2\,dx} = \frac{I_1}{I_2}$$

and we proceed in much the same way as did for centroids.

Now do this example, all on your own.

Find the position of the centre of gravity of the solid formed when the plane figure bounded by the curve $x^2 + y^2 = 16$, the $x$-axis and the ordinates $x = 1$ and $x = 3$, rotates about the $x$-axis.

*When you have finished, move to Frame 22*

**22**

$\boxed{\bar{x} = 1{\cdot}89,\ \bar{y} = 0}$

Check your working.

$$I_1 = \int_1^3 x(16 - x^2)dx = \int_1^3 (16x - x^3)dx = \left[8x^2 - \frac{x^4}{4}\right]_1^3$$

$$= \left(72 - \frac{81}{4}\right) - \left(8 - \frac{1}{4}\right) = 64 - 20 = 44 \qquad \therefore I_1 = 44$$

$$I_2 = \int_1^3 (16 - x^2)dx = \left[16x - \frac{x^3}{3}\right]_1^3 = (48 - 9) - \left(16 - \frac{1}{3}\right)$$

$$= 23\frac{1}{3} \qquad \therefore I_2 = 23\frac{1}{3} \qquad \therefore \bar{x} = \frac{I_1}{I_2} = \frac{44}{1}\cdot\frac{3}{70} = \frac{132}{70} = 1{\cdot}89$$

So $\bar{x} = 1{\cdot}89,\ \bar{y} = 0$

They are all done in the same manner.

Now for something that may be new to you

*Move on to Frame 23*

## Length of a curve

**23**

To find the length of the arc of the curve $y = f(x)$ between $x = a$ and $x = b$.

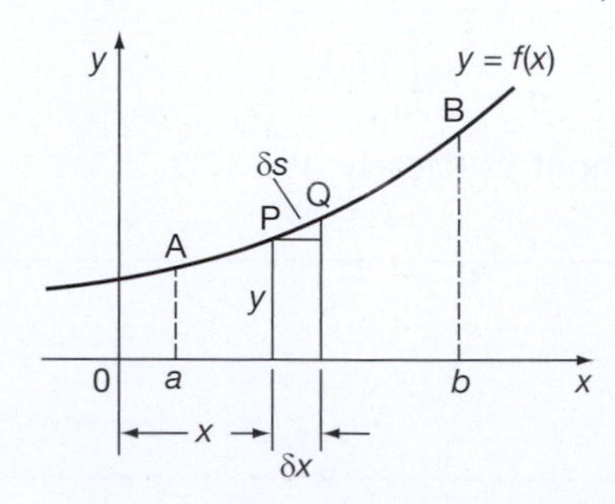


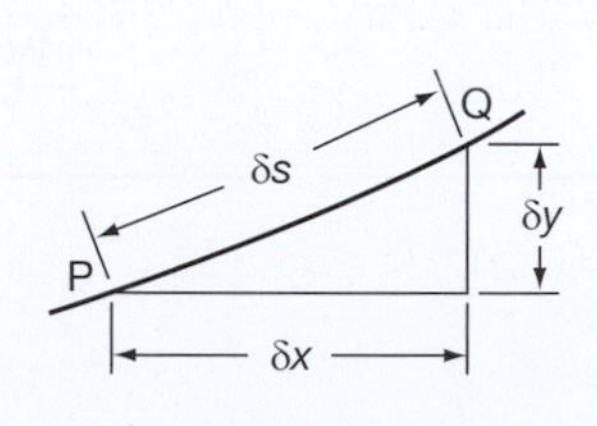

Let P be the point $(x, y)$ and Q a point on the curve near to P. Let $\delta s$ = length of the small arc PQ.

Then $\quad (\delta s)^2 \approx (\delta x)^2 + (\delta y)^2 \qquad \therefore \dfrac{(\delta s)^2}{(\delta x)^2} \approx 1 + \dfrac{(\delta y)^2}{(\delta x)^2}$

$$\left(\frac{\delta s}{\delta x}\right)^2 \approx 1 + \left(\frac{\delta y}{\delta x}\right)^2 \qquad \therefore \frac{\delta s}{\delta x} \approx \sqrt{1 + \left(\frac{\delta y}{\delta x}\right)^2}$$

If $\delta x \to 0 \quad \dfrac{ds}{dx} = \sqrt{1 + \log\left(\dfrac{dy}{dx}\right)^2} \qquad \therefore s = \displaystyle\int_a^b \sqrt{1 + \left(\frac{dy}{dx}\right)^2}.dx$

Make a note of this result

*Then on to the next frame*

---

**24**

**Example 1**

Find the length of the curve $y^2 = x^3$ between $x = 0$ and $x = 4$.

$$y^2 = x^3 \quad \therefore y = x^{3/2} \quad \therefore \frac{dy}{dx} = \frac{3}{2}x^{\frac{1}{2}} \quad \therefore 1 + \left(\frac{dy}{dx}\right)^2 = 1 + \frac{9x}{4}$$

$$\therefore s = \int_0^4 \left(1 + \frac{9x}{4}\right)^{\frac{1}{2}} dx = \left[\frac{2}{3}\cdot\frac{4}{9}\left(1 + \frac{9x}{4}\right)^{3/2}\right]_0^4$$

$$= \frac{8}{27}\left(10\sqrt{10} - 1\right) = \frac{8}{27}(31{\cdot}62 - 1)$$

$$= \frac{8}{27}(30{\cdot}62) = 9{\cdot}07 \text{ units}$$

That is all there is to it. Now here is one for you:

**Example 2**

Find the length of the curve $y = 10\cosh\dfrac{x}{10}$ between $x = -1$ and $x = 2$.

*Finish it, then move to Frame 25*

---

**25**

$s = 3{\cdot}015$ units

Here is the working set out.

$$y = 10\cosh\frac{x}{10} \qquad s = \int_{-1}^{2} \sqrt{1 + \left(\frac{dy}{dx}\right)^2}.dx$$

$$\frac{dy}{dx} = \sinh\frac{x}{10} \quad \therefore 1 + \left(\frac{dy}{dx}\right)^2 = 1 + \sinh^2\frac{x}{10} = \cosh^2\frac{x}{10}$$

$$\therefore s = \int_{-1}^{2} \sqrt{\cosh^2\frac{x}{10}}.dx = \int_{-1}^{2} \cosh\frac{x}{10}dx = \left[10\sinh\frac{x}{10}\right]_{-1}^{2}$$

$$= 10\left(\sinh 0{\cdot}2 - \sinh(-0{\cdot}1)\right) \qquad \sinh(-x) = -\sinh x$$

$$= 10\left(\sinh 0{\cdot}2 + \sinh 0{\cdot}1\right) = 10\left(0{\cdot}2013 + 0{\cdot}1002\right)$$

$$= 10\left(0{\cdot}3015\right) = 3{\cdot}015 \text{ units}$$

*Now to Frame 26*

## 26 Parametric equations

Instead of changing the variable of the integral as we have done before when the curve is defined in terms of parametric equations, we establish a special form of the result which saves a deal of working when we use it.

Here it is:

Let $y = f(t), \quad x = F(t)$

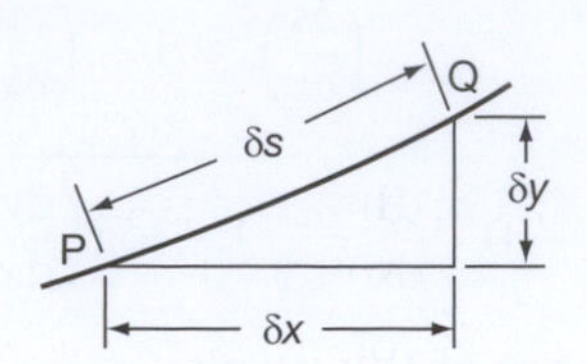


As before:

$$(\delta s)^2 \approx (\delta x)^2 + (\delta y)^2$$

Divide by $(\delta t)^2$:

$$\therefore \left(\frac{\delta s}{\delta t}\right)^2 \approx \left(\frac{\delta x}{\delta t}\right)^2 + \left(\frac{\delta y}{\delta t}\right)^2$$

If $\delta t \to 0$, this becomes:

$$\left(\frac{ds}{dt}\right)^2 = \left(\frac{dx}{dt}\right)^2 + \left(\frac{dy}{dt}\right)^2$$

$$\frac{ds}{dt} = \sqrt{\left(\frac{dx}{dt}\right)^2 + \left(\frac{dy}{dt}\right)^2}$$

$$\therefore s = \int_{t=t_1}^{t=t_2} \sqrt{\left(\frac{dx}{dt}\right)^2 + \left(\frac{dy}{dt}\right)^2}.dt$$

*This is a very useful result. Make a note of it in your record book and then move on to the next frame*

## 27

**Example 1**

Find the length of the curve $x = 2\cos^3\theta, \quad y = 2\sin^3\theta$ between the points corresponding to $\theta = 0$ and $\theta = \pi/2$.

Remember $$s = \int_0^{\pi/2} \sqrt{\left(\frac{dx}{d\theta}\right)^2 + \left(\frac{dy}{d\theta}\right)^2}.d\theta$$

We have $$\frac{dx}{d\theta} = 6\cos^2\theta(-\sin\theta) = -6\cos^2\theta\sin\theta$$

$$\frac{dy}{d\theta} = 6\sin^2\theta\cos\theta$$

$$\begin{aligned}\therefore \left(\frac{dx}{d\theta}\right)^2 + \left(\frac{dy}{d\theta}\right)^2 &= 36\cos^4\theta\sin^2\theta + 36\sin^4\theta\cos^2\theta \\ &= 36\sin^2\theta\cos^2\theta(\cos^2\theta + \sin^2\theta) \\ &= 36\sin^2\theta\cos^2\theta\end{aligned}$$

$$\therefore \sqrt{\left(\frac{dx}{d\theta}\right)^2 + \left(\frac{dy}{d\theta}\right)^2} = 6\sin\theta\cos\theta = 3\sin 2\theta$$

$$\therefore s = \int_0^{\pi/2} 3\sin 2\theta \, d\theta$$

$= \ldots\ldots\ldots\ldots$ Finish it off.

28

$s = 3$ units

Because we had $s = \int_0^{\pi/2} 3\sin 2\theta \, d\theta = 3\left[-\frac{\cos 2\theta}{2}\right]_0^{\pi/2}$

$$= 3\left\{\left(\frac{1}{2}\right) - \left(-\frac{1}{2}\right)\right\} = 3 \text{ units}$$

It is all very straightforward and not at all difficult. Just take care not to make any silly slips that would wreck the results.

Here is one for you to do in much the same way.

**Example 2**

Find the length of the curve $x = 5(2t - \sin 2t)$, $y = 10\sin^2 t$ between $t = 0$ and $t = \pi$.

*When you have completed it, move on to Frame 29*

29

$s = 40$ units

Because

$$x = 5(2t - \sin 2t), \quad y = 10\sin^2 t$$

$$\therefore \frac{dx}{dt} = 5(2 - 2\cos 2t) = 10(1 - \cos 2t)$$

$$\frac{dy}{dt} = 20\sin t \cos t = 10\sin 2t$$

$$\left(\frac{dx}{dt}\right)^2 + \left(\frac{dy}{dt}\right)^2 = 100(1 - 2\cos 2t + \cos^2 2t) + 100\sin^2 2t$$

$$= 100(1 - 2\cos 2t + \cos^2 2t + \sin^2 2t)$$

$$= 200(1 - \cos 2t) \qquad \text{but } \cos 2t = 1 - 2\sin^2 t$$

$$= 400\sin^2 t$$

$$\therefore \sqrt{\left(\frac{dx}{dt}\right)^2 + \left(\frac{dy}{dt}\right)^2} = 20\sin t$$

$$\therefore s = \int_0^{\pi} 20\sin t \, dt$$

$$= 20\left[-\cos t\right]_0^{\pi}$$

$$= 20\{(1) - (-1)\}$$

$$= 40 \text{ units}$$

*Next frame*

30

So, for the lengths of curves, there are two forms:

(a) $s = \int_{x_1}^{x_2} \sqrt{1 + \left(\frac{dy}{dx}\right)^2} . dx$ when $y = F(x)$

(b) $s = \int_{\theta_1}^{\theta_2} \sqrt{\left(\frac{dx}{d\theta}\right)^2 + \left(\frac{dy}{d\theta}\right)^2} . d\theta$ for parametric equations.

Just check that you have made a note of these in your record book.

*Now move on to Frame 31 and we will consider a further application of integration. This will be the last for this Programme*

## Surface of revolution

**31**

If an arc of a curve rotates about an axis, it will generate a surface. Let us take the general case.

Find the area of the surface generated when the arc of the curve $y = f(x)$ between $x = x_1$ and $x = x_2$, rotates about the $x$-axis through a complete revolution.

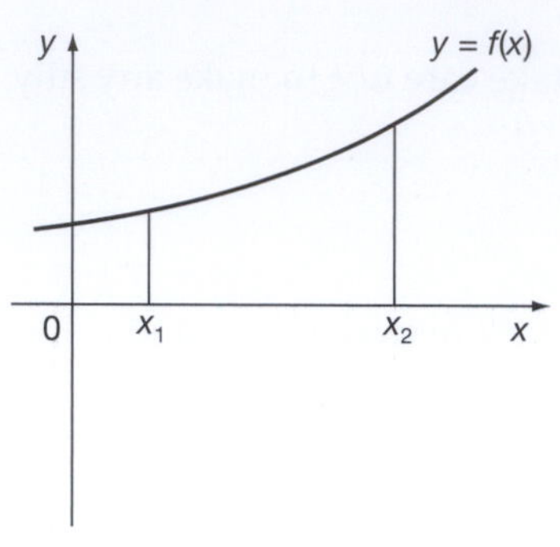


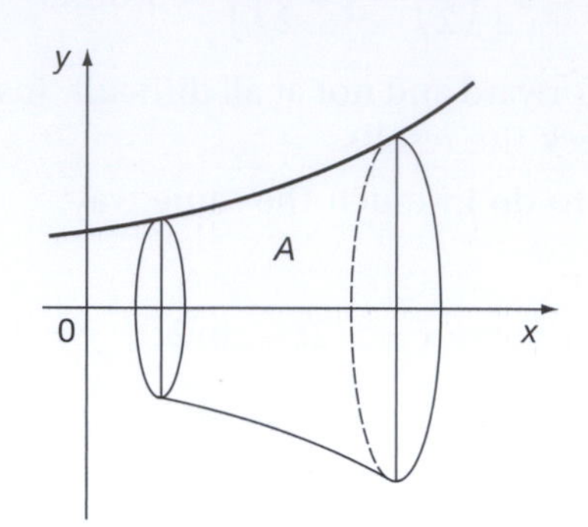


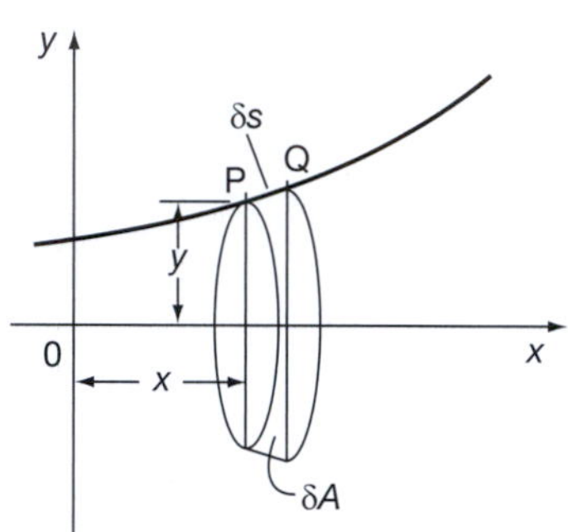


If we rotate a small element of arc $\delta s$ units long, it will generate a thin band of area $\delta A$.

Then $\delta A \approx 2\pi y.\delta s$.

Dividing by $\delta x$, gives $\dfrac{\delta A}{\delta x} \approx 2\pi y \dfrac{\delta s}{\delta x}$

and if $\delta x \to 0$ $\dfrac{dA}{dx} = 2\pi y \dfrac{ds}{dx}$

Now we have previously seen that $\dfrac{ds}{dx} = \sqrt{1 + \left(\dfrac{dy}{dx}\right)^2}$

$$\therefore \ \frac{dA}{dx} = 2\pi y \sqrt{1 + \left(\frac{dy}{dx}\right)^2}$$

So that $A =$ ...........

**32**

$$\boxed{A = \int_{x_1}^{x_2} 2\pi y \sqrt{1 + \left(\frac{dy}{dx}\right)^2}.dx}$$

This is another standard result, so copy it down into your record book.

*Then on to the next frame*

33

Here is another example requiring the previous result.

Find the area generated when the arc of the parabola $y^2 = 8x$ between $x = 0$ and $x = 2$, rotates about the $x$-axis.

We have

$$A = \int_0^2 2\pi y \sqrt{1 + \left(\frac{dy}{dx}\right)^2}.dx$$

$$y^2 = 8x \quad \therefore y = 2\sqrt{2}\, x^{\frac{1}{2}} \quad \therefore \frac{dy}{dx} = \sqrt{2}x^{-\frac{1}{2}} \quad \therefore \left(\frac{dy}{dx}\right)^2 = \frac{2}{x}$$

$$\therefore 1 + \left(\frac{dy}{dx}\right)^2 = 1 + \frac{2}{x} = \frac{x+2}{x}$$

$$\therefore A = \int_0^2 2\pi 2\sqrt{2}\, x^{\frac{1}{2}} \sqrt{\frac{x+2}{x}}.dx$$

$$= \int_0^2 4\sqrt{2}.\pi.x^{\frac{1}{2}} \frac{(x+2)^{\frac{1}{2}}}{x^{\frac{1}{2}}} dx$$

$$= 4\sqrt{2}.\pi \int_0^2 (x+2)^{\frac{1}{2}} dx$$

$= \ldots\ldots\ldots\ldots$ Finish it off; then move on.

34

$$\boxed{A = 19{\cdot}5\pi = 61{\cdot}3 \text{ units}^2}$$

Because we had

$$A = 4\sqrt{2}.\pi \int_0^2 (x+2)^{\frac{1}{2}} dx$$

$$= 4\sqrt{2}.\pi \left[\frac{(x+2)^{3/2}}{3/2}\right]_0^2$$

$$= \frac{8\sqrt{2}\pi}{3}\left((8) - (2\sqrt{2})\right)$$

$$= \frac{8\pi}{3}\left(8\sqrt{2} - 4\right) = \frac{8\pi}{3}(7{\cdot}314)$$

$$= 19{\cdot}5\pi = 61{\cdot}3 \text{ units}^2$$

*Now continue the good work by moving on to Frame 35*

35

## Parametric equations

We have seen that if we rotate a small arc $\delta s$, the area $\delta A$ of the thin band generated is given by

$$\delta A \approx 2\pi y.\delta s$$

If we divide by $\delta\theta$, we get

$$\frac{\delta A}{\delta\theta} \approx 2\pi y.\frac{\delta s}{\delta\theta}$$

and if $\delta\theta \to 0$, this becomes

$$\frac{dA}{d\theta} = 2\pi y.\frac{ds}{d\theta}$$

▶

We already have established in our work on lengths of curves that

$$\frac{ds}{d\theta}=\sqrt{\left(\frac{dx}{d\theta}\right)^2+\left(\frac{dy}{d\theta}\right)^2}$$

$$\therefore\ \frac{dA}{d\theta}=2\pi y\sqrt{\left(\frac{dx}{d\theta}\right)^2+\left(\frac{dy}{d\theta}\right)^2}$$

$$\therefore\ A=\int_{\theta_1}^{\theta_2}2\pi y\sqrt{\left(\frac{dx}{d\theta}\right)^2+\left(\frac{dy}{d\theta}\right)^2}.d\theta$$

This is a special form of the result for use when the curve is defined as a pair of parametric equations.

*On to Frame 36*

## 36

**Example 1**

Find the area generated when the curve $x=a(\theta-\sin\theta)$, $y=a(1-\cos\theta)$ between $\theta=0$ and $\theta=\pi$, rotates about the $x$-axis through a complete revolution.

Here $\frac{dx}{d\theta}=a(1-\cos\theta)\quad\therefore\ \left(\frac{dx}{d\theta}\right)^2=a^2(1-2\cos\theta+\cos^2\theta)$

$\frac{dy}{d\theta}=a\sin\theta\quad\therefore\ \left(\frac{dy}{d\theta}\right)^2=a^2\sin^2\theta$

$$\therefore\ \left(\frac{dx}{d\theta}\right)^2+\left(\frac{dy}{d\theta}\right)^2=a^2(1-2\cos\theta+\cos^2\theta+\sin^2\theta)$$

$$=2a^2(1-\cos\theta)\qquad\text{but }\cos\theta=1-2\sin^2\frac{\theta}{2}$$

$$=4a^2\sin^2\frac{\theta}{2}$$

$$\therefore\ \sqrt{\left(\frac{dx}{d\theta}\right)^2+\left(\frac{dy}{d\theta}\right)^2}=\dots\dots\dots\dots$$

Finish the integral and so find the area of the surface generated.

## 37

$$\boxed{\sqrt{\left(\frac{dx}{d\theta}\right)^2+\left(\frac{dy}{d\theta}\right)^2}=2a\sin\frac{\theta}{2}}$$

$$A=\int_0^{\pi}2\pi y\sqrt{\left(\frac{dx}{d\theta}\right)^2+\left(\frac{dy}{d\theta}\right)^2}.d\theta$$

$$=2\pi\int_0^{\pi}a(1-\cos\theta).2a\sin\frac{\theta}{2}.d\theta=2\pi\int_0^{\pi}a\left(2\sin^2\frac{\theta}{2}\right).2a\sin\frac{\theta}{2}.d\theta$$

$$=8\pi a^2\int_0^{\pi}\left(1-\cos^2\frac{\theta}{2}\right)\sin\frac{\theta}{2}.d\theta$$

$$=8\pi a^2\int_0^{\pi}\left(\sin\frac{\theta}{2}-\cos^2\frac{\theta}{2}\sin\frac{\theta}{2}\right)d\theta$$

$$=8\pi a^2\left[-2\cos\frac{\theta}{2}+\frac{2\cos^3\theta/2}{3}\right]_0^{\pi}$$

$$=8\pi a^2\big((0)-(-2+2/3)\big)$$

$$=8\pi a^2\big(4/3\big)=\frac{32\pi a^2}{3}\ \text{units}^2$$

▶

Here is one final one for you to do.

**Example 2**

Find the surface area generated when the arc of the curve $y = 3t^2$, $x = 3t - t^3$ between $t = 0$ and $t = 1$, rotates about the $x$-axis through $2\pi$ radians.

*When you have finished – next frame.*

**38**

Here it is in full.

$$y = 3t^2 \quad \therefore \frac{dy}{dt} = 6t \quad \therefore \left(\frac{dy}{dt}\right)^2 = 36t^2$$

$$x = 3t - t^3 \quad \therefore \frac{dx}{dt} = 3 - 3t^2 = 3(1 - t^2) \quad \therefore \left(\frac{dx}{dt}\right)^2 = 9(1 - 2t^2 + t^4)$$

$$\left(\frac{dx}{dt}\right)^2 + \left(\frac{dy}{dt}\right)^2 = 9 - 18t^2 + 9t^4 + 36t^2$$

$$= 9 + 18t^2 + 9t^4 = 9(1 + t^2)^2$$

$$\therefore A = \int_0^1 2\pi 3t^2 \sqrt{9(1 + t^2)^2}.dt$$

$$= 18\pi \int_0^1 t^2(1 + t^2)dt = 18\pi \int_0^1 (t^2 + t^4)dt$$

$$= 18\pi \left[\frac{t^3}{3} + \frac{t^5}{5}\right]_0^1 = 18\pi\left(\frac{1}{3} + \frac{1}{5}\right) = 18\pi\frac{8}{15}$$

$$= \frac{48\pi}{5} \text{ units}^2$$

# Rules of Pappus

**39**

There are two useful rules worth knowing which can well be included with this stage of the work. In fact we have used them already in our work just by common sense. Here they are:

1. If an arc of a plane curve rotates about an axis in its plane, the area of the surface generated is equal to the length of the line multiplied by the distance travelled by its centroid.
2. If a plane figure rotates about an axis in its plane, the volume generated is equal to the area of the figure multiplied by the distance travelled by its centroid.

You can see how much alike they are.

By the way, there is just one proviso in using the rules of Pappus: the axis of rotation must not cut the rotating arc or plane figure.

So copy the rules down into your record book. You may need to refer to them at some future time.

*Now on to Frame 40*

## Review summary

40

**1** *Volume of a solid of revolution*

(a) *About x-axis*

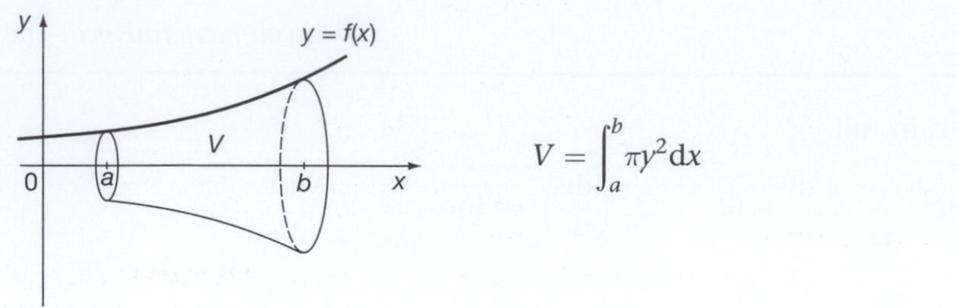


$$V=\int_a^b \pi y^2\,\mathrm{d}x \qquad (1)$$

Parametric equations $V=\int_{\theta_1}^{\theta_2}\pi y^2.\frac{\mathrm{d}x}{\mathrm{d}\theta}.\mathrm{d}\theta$ (2)

(b) *About y-axis*

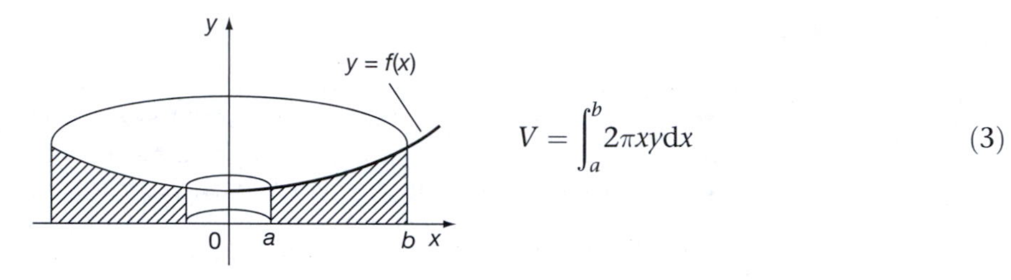


$$V=\int_a^b 2\pi xy\,\mathrm{d}x \qquad (3)$$

**2** *Centroid of a plane figure*

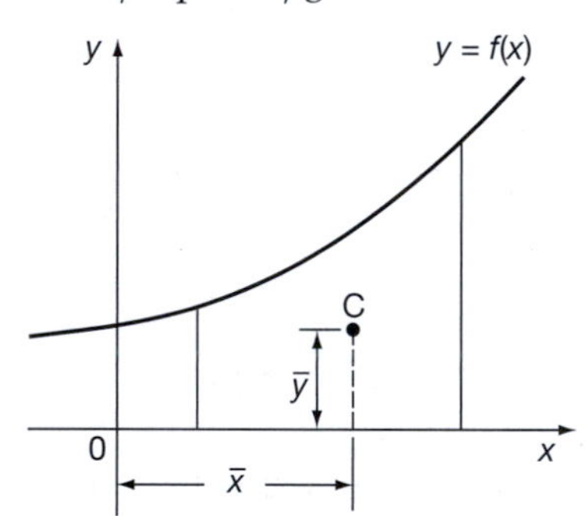


$$\bar{x}=\frac{\int_a^b xy\,\mathrm{d}x}{\int_a^b y\,\mathrm{d}x} \qquad (4)$$

$$\bar{y}=\frac{\int_a^b \frac{1}{2}y^2\mathrm{d}x}{\int_a^b y\,\mathrm{d}x} \qquad (5)$$

**3** *Centre of gravity of a solid of revolution*

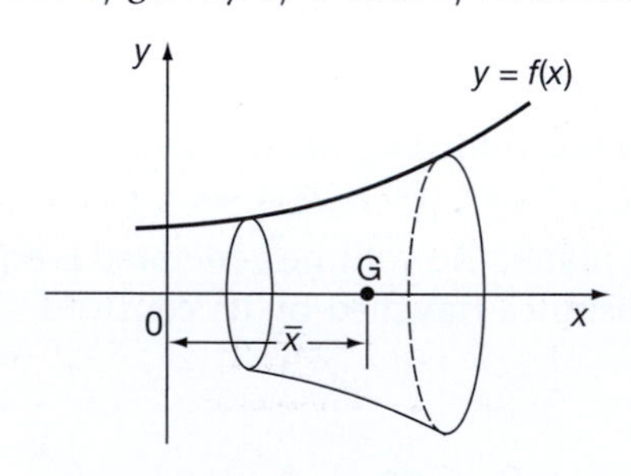


$$\bar{x}=\frac{\int_a^b xy^2\mathrm{d}x}{\int_a^b y^2\mathrm{d}x} \qquad (6)$$

$$\bar{y}=0$$

▶

4 *Length of a curve*

$y = f(x)$ $$s = \int_{x_1}^{x_2} \sqrt{1 + \left(\frac{dy}{dx}\right)^2}.dx \quad (7)$$

Parametric equation $$s = \int_{\theta_1}^{\theta_2} \sqrt{\left(\frac{dx}{d\theta}\right)^2 + \left(\frac{dy}{d\theta}\right)^2}.d\theta \quad (8)$$

5 *Surface of revolution*

$y = f(x)$ $$A = \int_{x_1}^{x_2} 2\pi y \sqrt{1 + \left(\frac{dy}{dx}\right)^2}.dx \quad (9)$$

Parametric equation $$A = \int_{\theta_1}^{\theta_2} 2\pi y \sqrt{\left(\frac{dx}{d\theta}\right)^2 + \left(\frac{dy}{d\theta}\right)^2}.d\theta \quad (10)$$

*All that remains now is the* **Can You?** *checklist and the* **Test exercise,** *so when you are ready, move on and work through them*

---

# Can You?

## Checklist 20

**41**

Check this list before you try the end of Programme test.

On a scale of 1 to 5 how confident are you that you can:

| | | | | | | | Frames |
|---|---|---|---|---|---|---|---|
| • Calculate volumes of revolution? | | | | | | | 1 to 11 |
| *Yes* | ☐ | ☐ | ☐ | ☐ | ☐ | *No* | |
| • Locate the centroid of a plane figure? | | | | | | | 12 to 19 |
| *Yes* | ☐ | ☐ | ☐ | ☐ | ☐ | *No* | |
| • Locate the centre of gravity of a solid of revolution? | | | | | | | 20 to 22 |
| *Yes* | ☐ | ☐ | ☐ | ☐ | ☐ | *No* | |
| • Determine the lengths of curves? | | | | | | | 23 to 25 |
| *Yes* | ☐ | ☐ | ☐ | ☐ | ☐ | *No* | |
| • Determine the lengths of curves given by parametric equations? | | | | | | | 26 to 30 |
| *Yes* | ☐ | ☐ | ☐ | ☐ | ☐ | *No* | |
| • Calculate surfaces of revolution? | | | | | | | 31 to 34 |
| *Yes* | ☐ | ☐ | ☐ | ☐ | ☐ | *No* | |
| • Calculate surfaces of revolution using parametric equations? | | | | | | | 35 to 38 |
| *Yes* | ☐ | ☐ | ☐ | ☐ | ☐ | *No* | |
| • Use the two rules of Pappus? | | | | | | | 39 |
| *Yes* | ☐ | ☐ | ☐ | ☐ | ☐ | *No* | |

## Test exercise 20

**42**

The problems are all straightforward so you should have no trouble with them. Work steadily: take your time. Off you go.

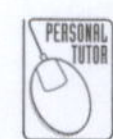

1. Find the position of the centroid of the plane figure bounded by the curve $y = 4 - x^2$ and the two axes of reference.

2. The curve $y^2 = x(1-x)^2$ between $x = 0$ and $x = 1$, rotates about the $x$-axis through $2\pi$ radians. Find the position of the centre of gravity of the solid so formed.

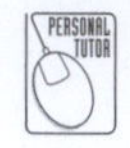

3. If $x = a(\theta - \sin\theta)$, $y = a(1 - \cos\theta)$, find the volume generated when the plane figure bounded by the curve, the $x$-axis and the ordinates at $\theta = 0$ and $\theta = 2\pi$, rotates about the $x$-axis through a complete revolution.

4. Find the length of the curve $8(y + \ln x) = x^2$ between $x = 1$ and $x = e$.

5. The arc of the catenary $y = 5\cosh\dfrac{x}{5}$ between $x = 0$ and $x = 5$, rotates about the $x$-axis. Find the area of the surface so generated.

6. Find the length of the curve $x = 5(\cos\theta + \theta\sin\theta)$, $y = 5(\sin\theta - \theta\cos\theta)$ between $\theta = 0$ and $\theta = \dfrac{\pi}{2}$.

7. The parametric equations of a curve are $x = e^t\sin t$, $y = e^t\cos t$. If the arc of this curve between $t = 0$ and $t = \dfrac{\pi}{2}$ rotates through a complete revolution about the $x$-axis, calculate the area of the surface generated.

*Now you are all ready for the next Programme. Well done, keep it up!*

## Further problems 20

**43**

1. Find the length of the curve
$$y = \frac{x}{2} - \frac{x^2}{4} + \frac{1}{2}\ln(1-x)$$
between $x = 0$ and $x = \dfrac{1}{2}$.

2. For the catenary $y = 5\cosh\dfrac{x}{5}$, calculate:

   (a) the length of arc of the curve between $x = 0$ and $x = 2$

   (b) the surface area generated when this arc rotates about the $x$-axis through a complete revolution.

3. The plane figure bounded by the parabola $y^2 = 4ax$, the $x$-axis and the ordinate at $x = a$, is rotated a complete revolution about the line $x = -a$. Find the volume of the solid generated.

4. A plane figure is enclosed by the parabola $y^2 = 4x$ and the line $y = 2x$. Determine (a) the position of the centroid of the figure, and (b) the centre of gravity of the solid formed when the plane figure rotates completely about the $x$-axis.

▶

5 The area bounded by $y^2x = 4a^2(2a - x)$, the $x$-axis and the ordinates $x = a$, $x = 2a$, is rotated through a complete revolution about the $x$-axis. Show that the volume generated is $4\pi a^3(2\ln 2 - 1)$.

6 Find the length of the curve $x^{2/3} + y^{2/3} = 4$ between $x = 0$ and $x = 8$.

7 Find the length of the arc of the curve $6xy = x^4 + 3$, between $x = 1$ and $x = 2$.

8 A solid is formed by the rotation about the $y$-axis of the area bounded by the $y$-axis, the lines $y = -5$ and $y = 4$, and an arc of the curve $2x^2 - y^2 = 8$. Given that the volume of the solid is $\dfrac{135\pi}{2}$, find the distance of the centre of gravity from the $x$-axis.

9 The line $y = x - 1$ is a tangent to the curve $y = x^3 - 5x^2 + 8x - 4$ at $x = 1$ and cuts the curve again at $x = 3$. Find the $x$-coordinate of the centroid of the plane figure so formed.

10 Find by integration, the area of the minor segment of the circle $x^2 + y^2 = 4$ cut off by the line $y = 1$. If this plane figure rotates about the $x$-axis through $2\pi$ radians, calculate the volume of the solid generated and hence obtain the distance of the centroid of the minor segment from the $x$-axis.

11 If the parametric equations of a curve are $x = 3a\cos\theta - a\cos 3\theta$, $y = 3a\sin\theta - a\sin 3\theta$, show that the length of arc between points corresponding to $\theta = 0$ and $\theta = \phi$ is $6a(1 - \cos\phi)$.

12 A curve is defined by the parametric equations $x = \theta - \sin\theta$, $y = 1 - \cos\theta$:

(a) Determine the length of the curve between $\theta = 0$ and $\theta = 2\pi$.

(b) If the arc in (a) rotates through a complete revolution about the $x$-axis, determine the area of the surface generated.

(c) Deduce the distance of the centroid of the arc from the $x$-axis.

13 Find the length of the curve $y = \cosh x$ between $x = 0$ and $x = 1$. Show that the area of the surface of revolution obtained by rotating the arc through four right-angles about the $y$-axis is $\dfrac{2\pi(e-1)}{e}$ units.

14 A parabolic reflector is formed by revolving the arc of the parabola $y^2 = 4ax$ from $x = 0$ to $x = h$ about the $x$-axis. If the diameter of the reflector is $2l$, show that the area of the reflecting surface is

$$\frac{\pi l}{6h^2}\left\{(l^2 + 4h^2)^{3/2} - l^3\right\}.$$

15 A segment of a sphere has a base radius $r$ and maximum height $h$. Prove that its volume is $\dfrac{\pi h}{6}\{h^2 + 3r^2\}$.

16 A groove, semi-circular in section and 1 cm deep, is turned in a solid cylindrical shaft of diameter 6 cm. Find the volume of material removed and the surface area of the groove.

17 Prove that the length of arc of the parabola $y^2 = 4ax$, between the points where $y = 0$ and $y = 2a$, is $a\left\{\sqrt{2} + \ln(1 + \sqrt{2})\right\}$. This arc is rotated about the $x$-axis through $2\pi$ radians. Find the area of the surface generated. Hence find the distance of the centroid of the arc from the line $y = 0$.

▶

18 A cylindrical hole of length $2a$ is bored centrally through a sphere. Prove that the volume of material remaining is $\dfrac{4\pi a^3}{3}$.

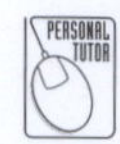


19 Prove that the centre of gravity of the zone of a thin uniform spherical shell, cut off by two parallel planes, is halfway between the centres of the two circular end sections.

20 Sketch the curve $3ay^2 = x(x-a)^2$, when $a > 0$. Show that $\dfrac{dy}{dx} = \pm\dfrac{3x-a}{2\sqrt{3ax}}$ and hence prove that the perimeter of the loop is $4a/\sqrt{3}$ units.

**44**

Now visit the companion website for this book at www.palgrave.com/stroud for more questions applying this mathematics to science and engineering.

# Integration applications 3

## Learning outcomes

When you have completed this Programme you will be able to:

- ☐ Determine moments of inertia
- ☐ Determine the radius of gyration
- ☐ Use the parallel axes theorem
- ☐ Use the perpendicular axes theorem for thin plates
- ☐ Determine moments of inertia using standard results
- ☐ Determine second moments of area
- ☐ Determine centres of pressure

# Moments of inertia

**1**

The amount of work that an object of mass $m$, moving with velocity $v$, will do against a resistance before coming to rest, depends on the values of these two quantities: its mass and its velocity.

The store of energy possessed by the object, due to its movement, is called its *kinetic energy* (KE) and it can be shown experimentally that the kinetic energy of a moving object is proportional

(a) to its mass, and

(b) to the square of its ...........

**2**

velocity

That is:

$\text{KE} \propto mv^2 \quad \therefore \text{KE} = kmv^2$

and if standard units of mass and velocity are used, the value of the constant $k$ is $\frac{1}{2}$.

$\therefore \text{KE} = \frac{1}{2}mv^2$

No doubt, you have met and used that result elsewhere.

It is important, so make a note of it.

**3**

$$\text{KE} = \frac{1}{2}mv^2$$

In many applications in engineering, we are concerned with objects that are rotating – wheels, cams, shafts, armatures, etc. – and we often refer to their movement in terms of 'revolutions per second'. Each particle of the rotating object, however, has a linear velocity, and so has its own store of KE – and it is the KE of rotating objects that we are concerned with in this part of the Programme.

*So move on to Frame 4*

**4**

Let us consider a single particle P of mass $m$ rotating about an axis X with constant angular velocity $\omega$ radians per second.

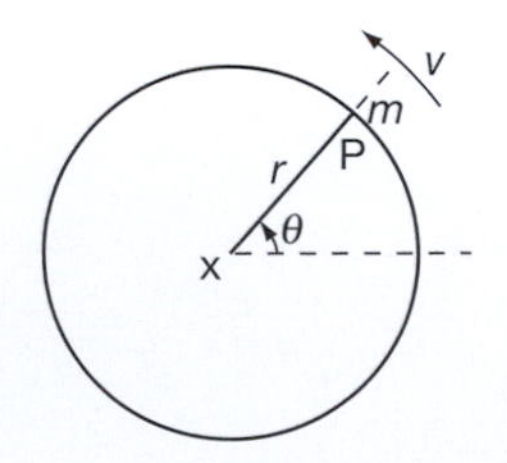


This means that the angle $\theta$ at the centre is increasing at the rate of $\omega$ radians/per second.

Of course, the linear velocity of P, $v$ cm/s, depends upon two quantities:

(1) the angular velocity ($\omega$ rad/s)

and also (2) ...........

how far P is from the centre

**5**

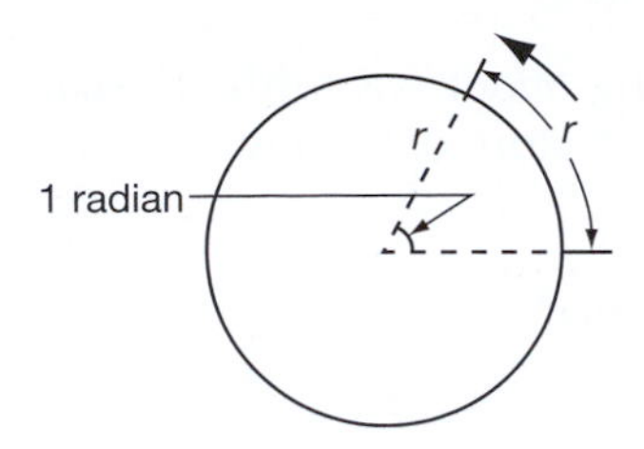


To generate an angle of 1 radian in a second, P must move round the circle a distance equal to 1 radius length, i.e. $r$ (cm).

If $\theta$ is increasing at 1 rad/s, P is moving at $r$ cm/s
If $\theta$ is increasing at 2 rad/s, P is moving at $2r$ cm/s
If $\theta$ is increasing at 3 rad/s, P is moving at $3r$ cm/s, etc.

So, in general:

if $\theta$ is increasing at $\omega$ rad/s, P is moving at $\omega r$ cm/s.

Therefore, if the angular velocity of P is $\omega$ rad/s, the linear velocity, $v$, of P is ...........

---

$v = \omega r$

**6**

We have already established that the kinetic energy of an object of mass $m$ moving with velocity $v$ is given by

KE = ...........

---

$\text{KE} = \frac{1}{2}mv^2$

**7**

So, for our rotating particle, we have:

$$\begin{aligned}\text{KE} &= \frac{1}{2}mv^2 \\ &= \frac{1}{2}m(\omega r)^2 \\ &= \frac{1}{2}m\omega^2 r^2\end{aligned}$$

and changing the order of the factors we can write:

$$\text{KE} = \frac{1}{2}\omega^2.mr^2$$

where $\omega$ = the angular velocity of the particle P about the axis (rad/s)
$m$ = mass of P
$r$ = distance of P from the axis of rotation.

*Make a note of that result: we shall certainly need it again*

## 8

$$\boxed{KE = \frac{1}{2}\omega^2.mr^2}$$

If we now have a whole system of particles, all rotating about XX with the same angular velocity $\omega$ rad/s, each particle contributes its own store of energy:

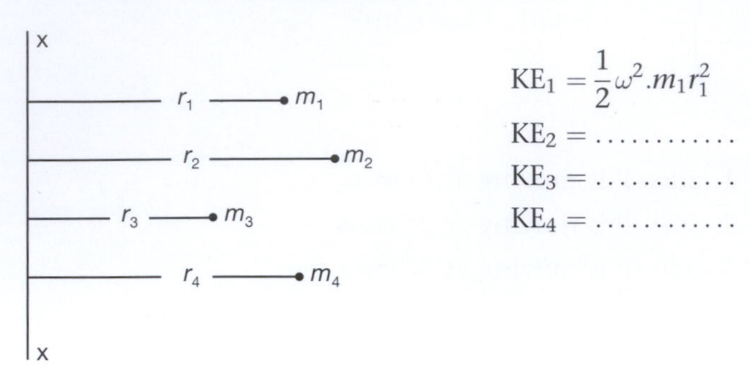


$KE_1 = \frac{1}{2}\omega^2.m_1r_1^2$

$KE_2 = \ldots\ldots\ldots\ldots$

$KE_3 = \ldots\ldots\ldots\ldots$

$KE_4 = \ldots\ldots\ldots\ldots$

## 9

$$\boxed{\begin{array}{ll} KE_1 = \frac{1}{2}\omega^2.m_1r_1^2 & KE_2 = \frac{1}{2}\omega^2.m_2r_2^2 \\ KE_3 = \frac{1}{2}\omega^2.m_3r_3^2 & KE_4 = \frac{1}{2}\omega^4.m_4r_4^2 \end{array}}$$

So that, the total energy of the system (or solid object) is given by:

$$KE = KE_1 + KE_2 + KE_3 + KE_4 + \ldots$$

$$= \frac{1}{2}\omega^2.m_1r_1^2 + \frac{1}{2}\omega^2.m_2r_2^2 + \frac{1}{2}\omega^2.m_3r_3^2 + \ldots$$

$$KE = \sum \frac{1}{2}\omega^2.mr^2 \quad \text{(summing over all particles)}$$

$$KE = \frac{1}{2}\omega^2.\sum mr^2 \quad \text{(since } \omega \text{ is a constant)}$$

This is another result to note.

## 10

$$\boxed{KE = \frac{1}{2}\omega^2.\sum mr^2}$$

This result is the product of two distinct factors.

(a) $\frac{1}{2}\omega^2$ can be varied by speeding up or slowing down the rate of rotation,

but (b) $\sum mr^2$ is a property of the rotating object. It depends on the total mass but also on where that mass is distributed in relation to the axis XX. It is a physical property of the object and is called its *second moment of mass*, or its *moment of inertia* (denoted by the symbol I).

$\therefore$ I $= \sum mr^2$ (for all the particles)

▶

As an example, for the system of particles shown, find its moment of inertia about the axis XX.

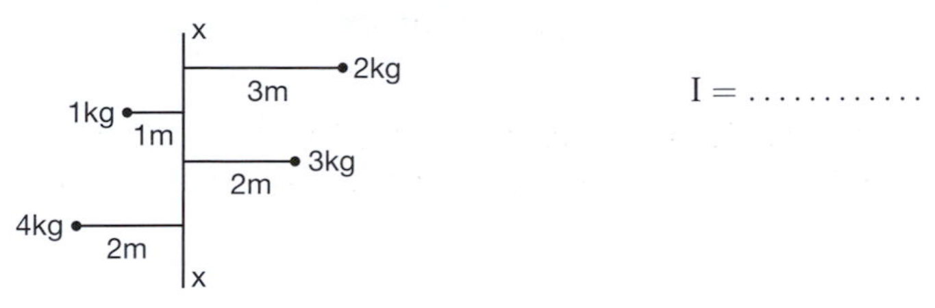


I = ...........

**11**

I = 47 kg m²

Because

$$I = \sum mr^2$$
$$= 2.3^2 + 1.1^2 + 3.2^2 + 4.2^2$$
$$= 18 + 1 + 12 + 16$$
$$= 47 \text{ kg m}^2$$

*Move on to Frame 12*

**12**

## Radius of gyration

If we imagine the total mass $M$ of the system arranged at a distance $k$ from the axis, so that the KE of $M$ would be the same as the total KE of the distributed particles,

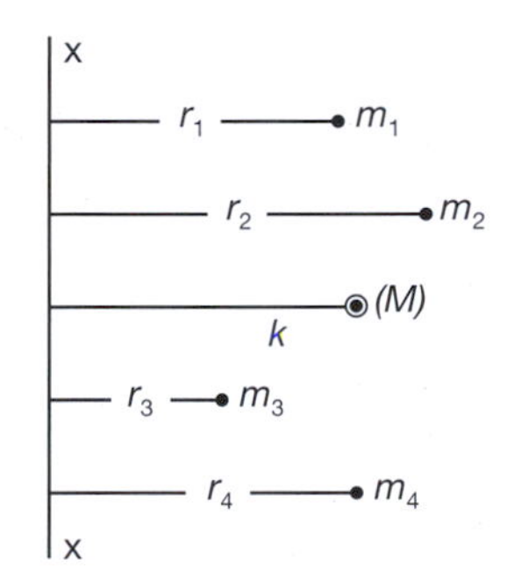


then $\frac{1}{2}\omega^2.Mk^2 = \frac{1}{2}\omega^2.\sum mr^2$

$\therefore\ Mk^2 = \sum mr^2$

and $k$ is called the *radius of gyration* of the object about the particular axis of rotation.

So, we have $I = \sum mr^2$; $Mk^2 = I$, where $M = \sum m$

I = moment of inertia (or second moment of mass)

$k$ = radius of gyration about the given axis.

*Now let us apply some of these results, so on you go to Frame 13*

**13**

### Example 1

To find the moment of inertia (I) and the radius of gyration ($k$) of a uniform thin rod about an axis through one end perpendicular to the length of the rod.

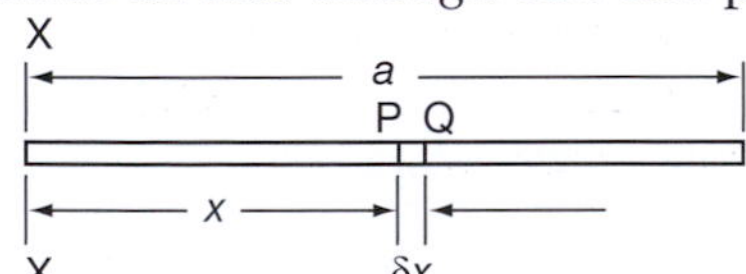


Let $\rho$ = mass per unit length of rod

Mass of element PQ $= \rho.\delta x$.

$\therefore$ Second moment of mass of PQ about XX = mass × (distance)²

$$= \rho.\delta x.x^2 = \rho x^2.\delta x$$

$\therefore$ Total second moment for all such elements can be written

I ≈ ...........

**14**

$$I \approx \sum_{x=0}^{a} \rho x^2.\delta x$$

The approximation sign is included since $x$ is the distance up to the left-hand side of the element PQ. But, if $\delta x \to 0$, this becomes:

$$I = \int_0^a \rho x^2.dx$$

$$= \rho\left[\frac{x^3}{3}\right]_0^a$$

$$= \frac{\rho a^3}{3}$$

$$\therefore I = \frac{\rho a^3}{3}$$

Now, to find $k$, we shall use $Mk^2 = I$, so we must first determine the total mass $M$.

Since $\rho$ = mass per unit length of rod, and the rod is $a$ units long, the total mass,

$M = \ldots\ldots\ldots\ldots$

**15**

$$M = a\rho$$

$$Mk^2 = I \qquad \therefore a\rho.k^2 = \frac{\rho a^3}{3}$$

$$\therefore k^2 = \frac{a^2}{3} \qquad \therefore k = \frac{a}{\sqrt{3}}$$

$$\therefore I = \frac{\rho a^3}{3} \quad \text{and } k = \frac{a}{\sqrt{3}}$$

Now for another:

**Example 2**

Find I for a rectangular plate about an axis through its centre of gravity parallel to one side, as shown.

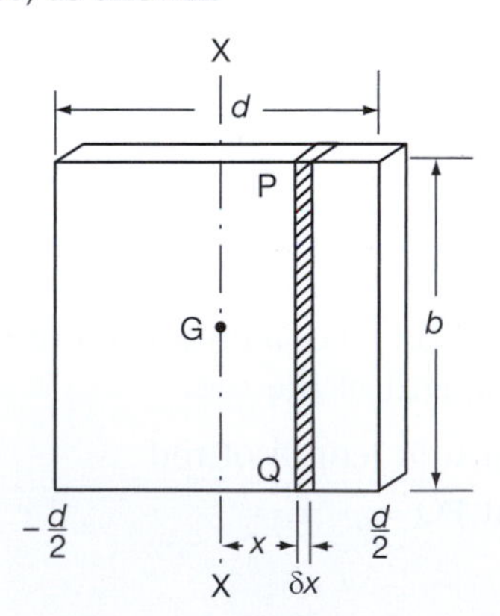


Let $\rho$ = mass per unit area of plate.

Mass of strip PQ $= b.\delta x.\rho$

Second moment of mass of strip about XX

$$\approx b\delta x\rho.x^2$$

(i.e. mass × distance$^2$)

$\therefore$ Total second moment for all strips covering the figure

$$I \approx \sum_{x=}^{x=} \ldots\ldots\ldots\ldots$$

16

$$I \approx \sum_{x=-d/2}^{x=d/2} b\rho x^2.\delta x$$

Did you remember the limits?

So now, if $\delta x \to 0$:

$$I = \int_{-d/2}^{d/2} b\rho x^2.dx$$

$$= b\rho \left[\frac{x^3}{3}\right]_{-d/2}^{d/2}$$

$$= b\rho \left\{ \left(\frac{d^3}{24}\right) - \left(-\frac{d^3}{24}\right) \right\} = \frac{b\rho d^3}{12}$$

$$\therefore I = \frac{bd^3\rho}{12}$$

and since the total mass $M = bd\rho$, $I = \dfrac{Md^2}{12}$.

$$\therefore I = \frac{bd^3\rho}{12} = \frac{Md^2}{12}$$

$$\text{and } k = \frac{d}{\sqrt{12}} = \frac{d}{2\sqrt{3}}$$

*This is a useful standard result for a rectangular plate, so make a note of it for future use*

17

Here is an example, very much like the last, for you to do.

**Example 3**

Find I for a rectangular plate, 20 cm $\times$10 cm, of mass 2 kg, about an axis 5 cm from one 20-cm side as shown.

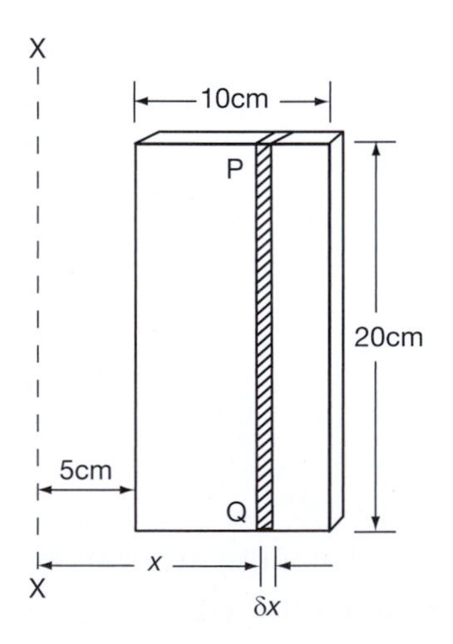


Take a strip parallel to the axis and argue as above. *Note* that, in this case,

$$\rho = \frac{2}{10.20} = \frac{2}{200} = 0{\cdot}01$$

$$\text{i.e. } \rho = 0{\cdot}01 \text{ kg/cm}^2$$

*Finish it off and then move to the next frame*

**18**

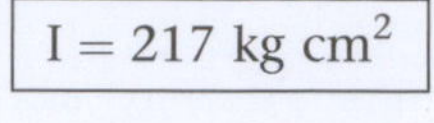

$\mathrm{I} = 217 \text{ kg cm}^2$

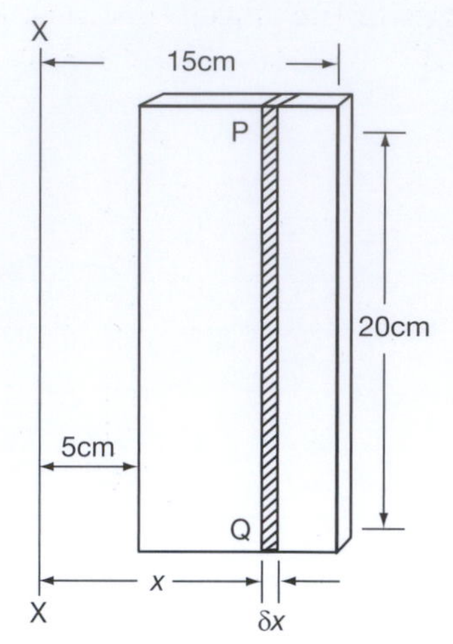


Here is the working in full:

$\rho = 0{\cdot}01 \text{ kg cm}^2$

Area of strip $= 20.\delta x$

$\therefore$ Mass of strip $= 20.\delta x.\rho$

$\therefore$ 2nd moment of mass of strip about XX $\approx 20.\delta x.\rho.x^2$

$$\therefore \text{ Total 2nd moment of mass} = \mathrm{I} \approx \sum_{x=5}^{x=15} 20\rho x^2.\delta x.$$

$$\text{If } \delta x \to 0, \quad \mathrm{I} = \int_5^{15} 20\rho x^2.\mathrm{d}x$$

$$= 20\rho \left[\frac{x^3}{3}\right]_5^{15}$$

$$= \frac{20\rho}{3}\left\{3375 - 125\right\}$$

$$= \frac{20}{3}\left\{3250\right\}\frac{1}{100}$$

$$= \frac{650}{3}$$

$$= 217 \text{ kg cm}^2$$

Now, for the same problem, find the value of $k$.

**19**

$k = 10{\cdot}4 \text{ cm}$

Because

$Mk^2 = \mathrm{I}$ and $M = 2$ kg

$\therefore 2k^2 = 217 \quad \therefore k^2 = 108{\cdot}5$

$\therefore k = \sqrt{108{\cdot}5} = 10{\cdot}4$ cm

Normally, then, we find I this way:

(a) Take an elementary strip parallel to the axis of rotation at a distance $x$ from it.

(b) Form an expression for its second moment of mass about the axis.

(c) Sum for all such strips.

(d) Convert to integral form and evaluate.

It is just as easy as that!

## Parallel axes theorem

**20**

If I is known about an axis through the centre of gravity of the object, we can easily write down the value of I about any other axis parallel to the first and a known distance from it.

Let G be the centre of gravity of the object

Let $m$ = mass of the strip PQ

Then $\quad I_G = \sum mx^2$

and $\quad I_{AB} = \sum m(x+l)^2$

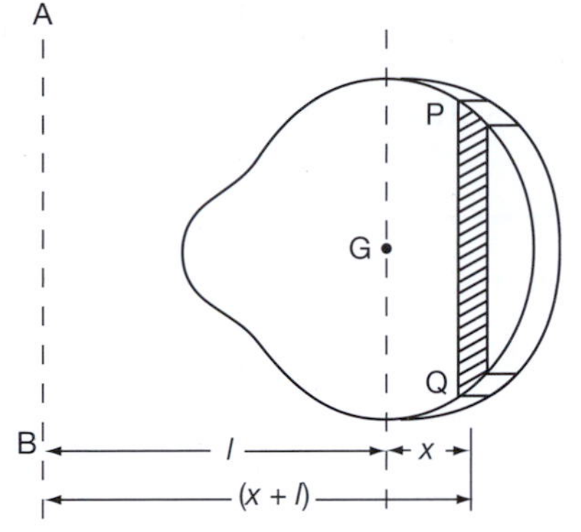


$$\therefore I_{AB} = \sum m(x^2 + 2lx + l^2)$$

$$= \sum mx^2 + \sum 2mxl + \sum ml^2$$

$$= \sum mx^2 + 2l\sum mx + l^2\sum m \quad \text{(since } l \text{ is a constant)}$$

Now $\sum mx^2 = \ldots\ldots\ldots\ldots$

and $\sum m = \ldots\ldots\ldots\ldots$

**21**

$$\boxed{\sum mx^2 = I_G; \sum m = M}$$

Right. In the middle term we have $\sum mx$. This equals 0, since the axis XX by definition passes through the centre of gravity of the solid.

In our previous result, then:

$$\sum mx^2 = I_G; \quad \sum mx = 0; \quad \sum m = M$$

and substituting these in, we get:

$$I_{AB} = I_G + Ml^2$$

Thus, if we know $I_G$, we can obtain $I_{AB}$ by simply adding on the product of the total mass × square of the distance of transfer.

*This result is important: make a note of it in your record book*

**22**

**Example 1**

To find I about the axis AB for the rectangular plate shown below.

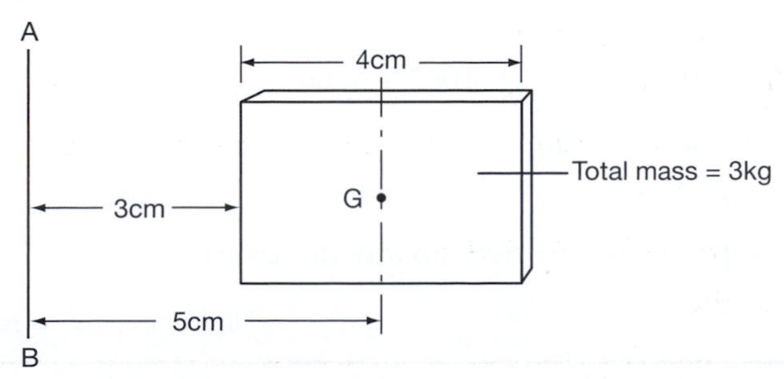

We have: $$I_G = \frac{Md^2}{12} = \frac{3.16}{12} = 4 \text{ kg cm}^2$$
$$\therefore I_{AB} = I_G + Ml^2$$
$$= 4 + 3.25$$
$$= 4 + 75 = 79 \text{ kg cm}^2$$
$$\therefore I_{AB} = 79 \text{ kg cm}^2$$

As easy as that!

*Next frame*

**23**

You do this one:

**Example 2**

A metal door, 40 cm × 60 cm, has a mass of 8 kg and is hinged along one 60-cm side.

Here is the figure

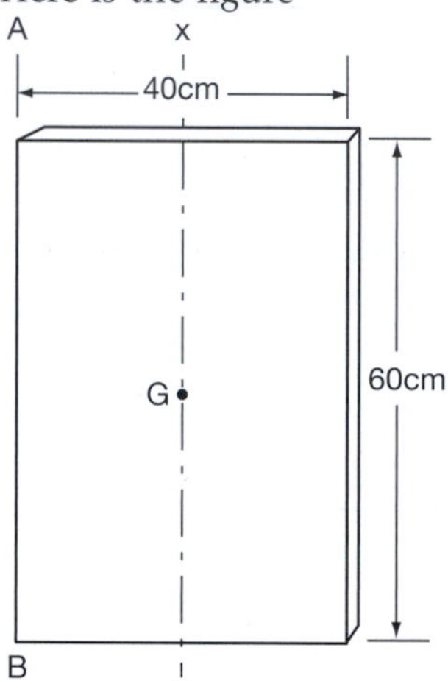


Calculate:

(a) I about XX, the axis through the centre of gravity.

(b) I about the line of hinge, AB.

(c) $k$ about AB.

*Find all three results: then move on to Frame 24 and check your working*

**24**

$$I_{XX} = 1067 \text{ kg cm}^2; \quad I_{AB} = 4267 \text{ kg cm}^2; \quad k_{AB} = 23{\cdot}1 \text{ cm}$$

Here is the working:

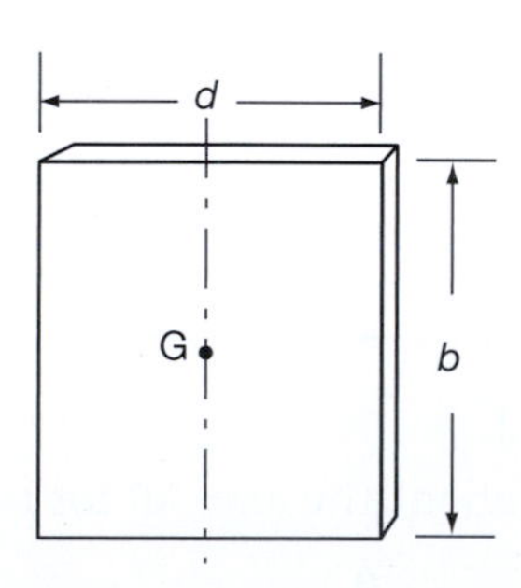


(a) $I_G = \frac{Md^2}{12}$

$= \frac{8.40^2}{12} = \frac{8.1600}{12}$

$= \frac{3200}{3} = 1067 \text{ kg cm}^2$

(b) $I_{AB} = I_G + Ml^2 = 1067 + 8.20^2 = 1067 + 3200$

$= 4267 \text{ kg cm}^2$

(c) $Mk^2 = I_{AB} \quad \therefore 8k^2 = 4267 \quad \therefore k^2 = 533{\cdot}4 \quad \therefore k = 23{\cdot}1 \text{ cm}$

If you make any slips, be sure to clear up any difficulties.

*Then move on to the next example*

**25**

Let us now consider wheels, cams, etc. – basically rotating discs.

To find the moment of inertia of a circular plate about an axis through its centre, perpendicular to the plane of the plate.

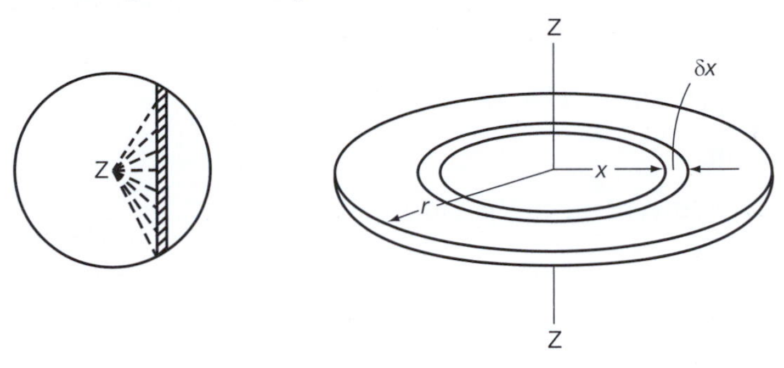


If we take a slice across the disc as an elementary strip, we are faced with the difficulty that all points in the strip are not at the same distance from the axis. We therefore take a circular strip as shown.

Mass of strip $\approx 2\pi x.\delta x.\rho$ ($\rho =$ mass per unit area of plate)

$\therefore$ 2nd moment of strip about ZZ $\approx$ ...........

---

**26**

$$2\pi x.\delta x.\rho.x^2$$

$\therefore$ 2nd moment of strip about ZZ $= 2\pi\rho x^3.\delta x$

$\therefore$ Total 2nd moment for all such circular strips about ZZ, is given by:

$$\mathrm{I_Z} \approx \sum_{x=0}^{x=r} 2\pi\rho x^3.\delta x$$

$$\text{If } \delta x \to 0: \quad \mathrm{I_Z} = \int_0^r 2\pi\rho x^3.\mathrm{d}x = 2\pi\rho\left[\frac{x^4}{4}\right]_0^r$$

$$= \frac{2\pi\rho r^4}{4} = \frac{\pi r^4\rho}{2}$$

Total mass, $M = \pi r^2\rho$

$$\therefore \mathrm{I_Z} = \frac{\pi r^4\rho}{2} = \frac{M.r^2}{2} \text{ and } k = \frac{r}{\sqrt{2}}$$

This is another standard result, so note it down.

*Next frame*

---

**27**

$$\mathrm{I_Z} = \frac{\pi r^4\rho}{2} = \frac{M.r^2}{2}$$

As an example, find the radius of gyration of a metal disc of radius 6 cm and total mass 0·5 kg.

We know that, for a circular disc:

$$\mathrm{I_Z} = \frac{M.r^2}{2} \text{ and, of course, } Mk^2 = \mathrm{I}.$$

So off you go and find the value of $k$.

## 28

$$k = 4{\cdot}24 \text{ cm}$$

$$I_Z = \frac{M.r^2}{2} = \frac{0{\cdot}5.36}{2} = 9 \text{ kg cm}^2$$

$$Mk^2 = I \quad \therefore \frac{1}{2}k^2 = 9 \quad \therefore k^2 = 18$$

$$\therefore k = 4{\cdot}24 \text{ cm.}$$

They are all done in very much the same way.

*Move on to Frame 29*

## 29

### Perpendicular axes theorem (for thin plates)

Let $\delta m$ be a small mass at P.

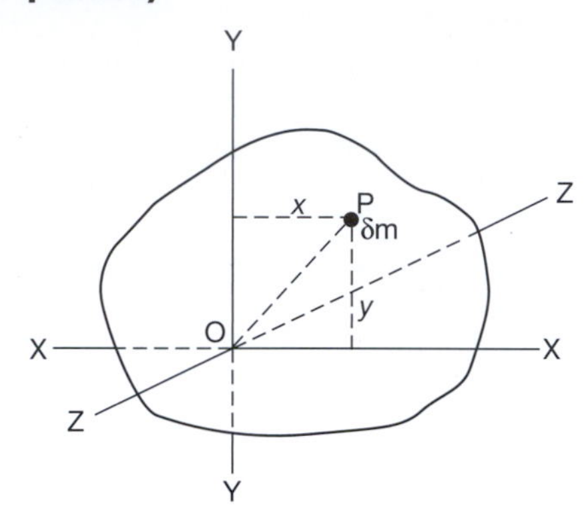


$$\text{Then } I_X \approx \sum \delta m.y^2$$

$$\text{and } I_Y \approx \sum \delta m.x^2.$$

Let ZZ be the axis perpendicular to both XX and YY.

$$\text{Then } I_Z = \sum \delta m.(\text{OP})^2$$

$$= \sum \delta m.(x^2 + y^2)$$

$$= \sum \delta m.y^2 + \sum \delta m.x^2$$

$$\therefore I_Z = I_X + I_Y$$

Therefore if we know the second moment about two perpendicular axes in the plane of the plate, the second moment about a third axis, perpendicular to both (through the point of intersection) is given by

$$I_Z = I_X + I_Y$$

And that is another result to note.

## 30

To find I for a circular disc about a diameter as axis.

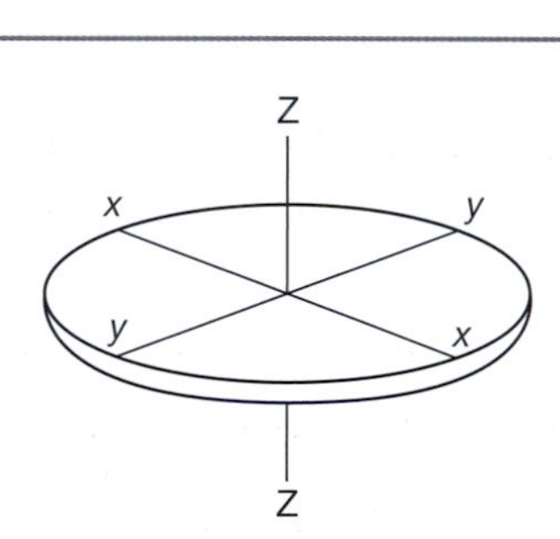


We have already established that

$$I_Z = \frac{\pi r^4 \rho}{2} = \frac{M.r^2}{2}$$

Let XX and YY be two diameters perpendicular to each other.

$$\text{Then we know } I_X + I_Y = I_Z = \frac{M.r^2}{2}$$

But all diameters are identical

$$\therefore I_X = I_Y \quad \therefore 2I_X = \frac{M.r^2}{2} \quad \therefore I_X = \frac{M.r^2}{4}$$

$\therefore$ For a circular disc:

$$I_Z = \frac{\pi r^4 \rho}{2} = \frac{M.r^2}{2} \quad \text{and} \quad I_X = \frac{\pi r^4 \rho}{4} = \frac{M.r^2}{4}$$

Make a note of these too.

**31**

As an example, find I for a circular disc, 40 cm diameter, and of mass 12 kg:

(a) about the normal axis (Z-axis)

(b) about a diameter as axis

(c) about a tangent as axis.

Work it through on your own. When you have obtained (b) you can find (c) by applying the parallel axes theorem.

*Then check with the next frame*

**32**

$$I_Z = 2400 \text{ kg cm}^2; \quad I_X = 1200 \text{ kg cm}^2; \quad I_T = 6000 \text{ kg cm}^2$$

Because

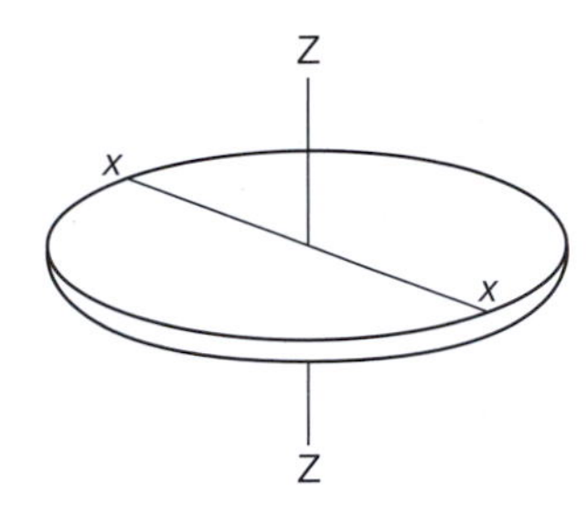


(a) $I_Z = \dfrac{M.r^2}{2} = \dfrac{12.20^2}{2}$

$= 2400 \text{ kg cm}^2$

(b) $I_X = \dfrac{M.r^2}{4} = \dfrac{1}{2} I_Z = 1200 \text{ kg cm}^2$

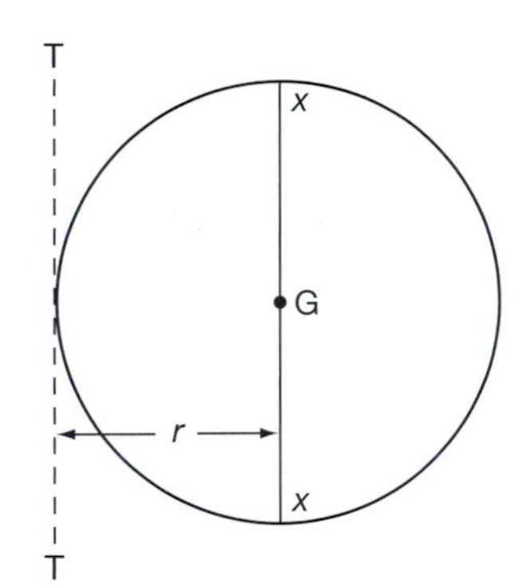


(c) $I_X = 1200 \text{ kg cm}^2$

By the parallel axes theorem

$I_T = I_X + Ml^2$

$= 1200 + 12.20^2$

$= 1200 + 4800$

$= 6000 \text{ kg cm}^2$

In the course of our work, we have established a number of important results, so, at this point, let us collect them together, so that we can see them as a whole.

*On then to the next frame*

## Useful standard results

**33**

**1** $I = \sum mr^2; \quad Mk^2 = I$

**2** *Rectangular plate* ($\rho =$ mass/unit area)

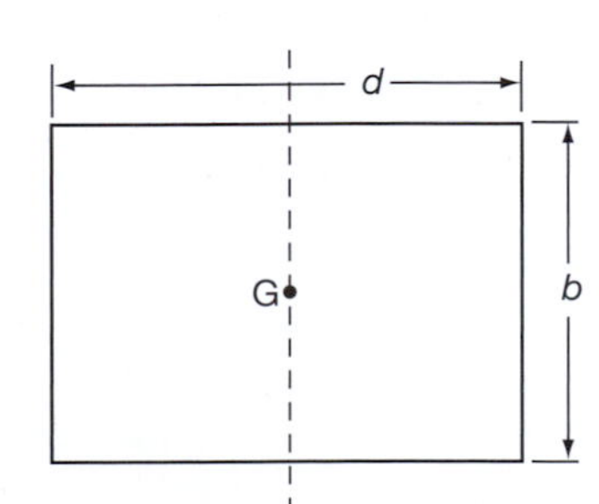


$$I_G = \frac{bd^3\rho}{12} = \frac{M.d^2}{12}$$

▶

**3** *Circular disc*

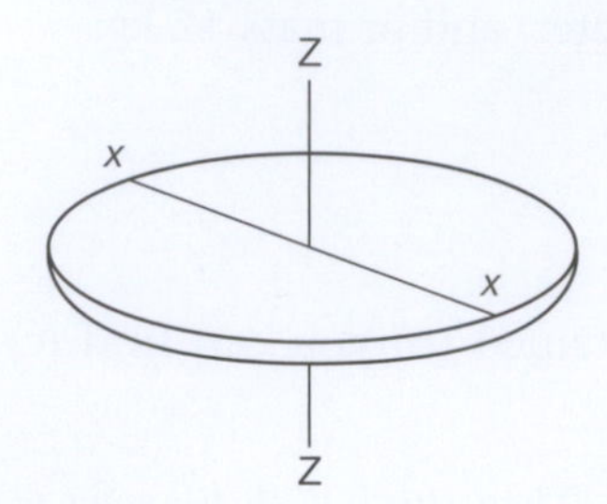


$$I_Z = \frac{\pi r^4 \rho}{2} = \frac{M.r^2}{2}$$
$$I_X = \frac{\pi r^4 \rho}{4} = \frac{M.r^2}{4}$$

**4** *Parallel axes theorem*

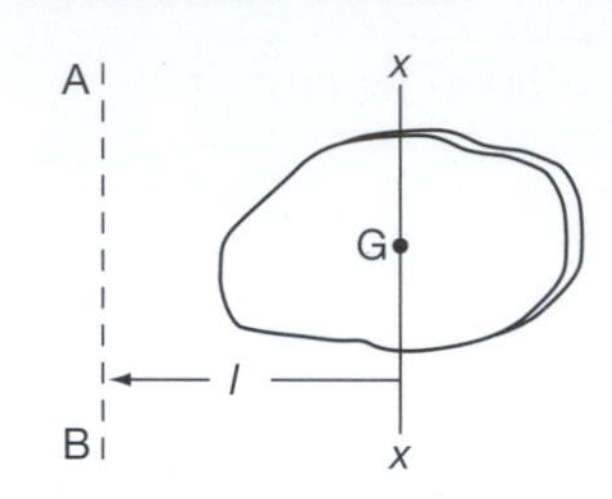


$$I_{AB} = I_G + Ml^2$$

**5** *Perpendicular axes theorem*

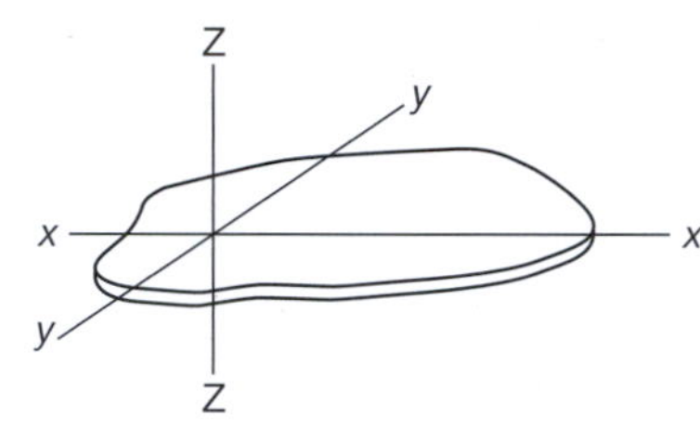


$$I_Z = I_X + I_Y$$

These standard results cover a large number of problems, but sometimes it is better to build up expressions in particular cases from first principles.

*Let us see an example using that method*

**34**

**Example 1**

Find I for the hollow shaft shown, about its natural axis.
Density of material = 0·008 kg cm$^3$.

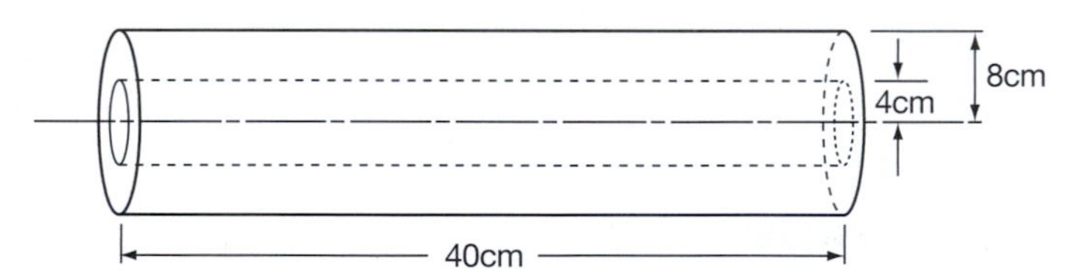


First consider a thin shell, distance $x$ from the axis.

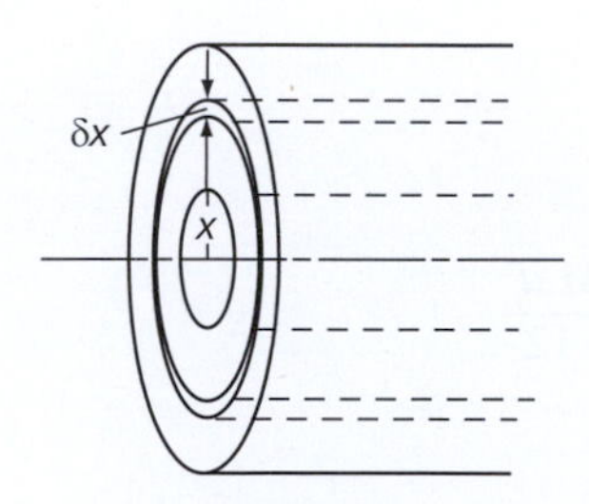


Mass of shell $\approx 2\pi x.\delta x.40\rho$ (kg)

$\therefore$ 2nd moment about XX $\approx 2\pi x.\delta x.40\rho.x^2$

$\approx 80\pi\rho x^3.\delta x$

$\therefore$ Total 2nd moment $= \sum_{x=4}^{x=8} 80\pi\rho x^3.\delta x$

Now, if $\delta x \to 0$, I = ...........

*Finish it off, then check with the next frame*

35

$$I = 1930 \text{ kg cm}^2$$

Because

$$I = 80\pi\rho\int_4^8 x^3 dx = 80\pi\rho\left[\frac{x^4}{4}\right]_4^8$$
$$= \frac{80\pi\rho}{4}(64^2 - 16^2)$$
$$= 20\pi\rho.48.80 = 20\pi.48.80.0{\cdot}008$$
$$= 614{\cdot}4\pi = 1930 \text{ kg cm}^2$$

Here is another:

**Example 2**

Find I and $k$ for the solid cone shown, about its natural axis of symmetry.

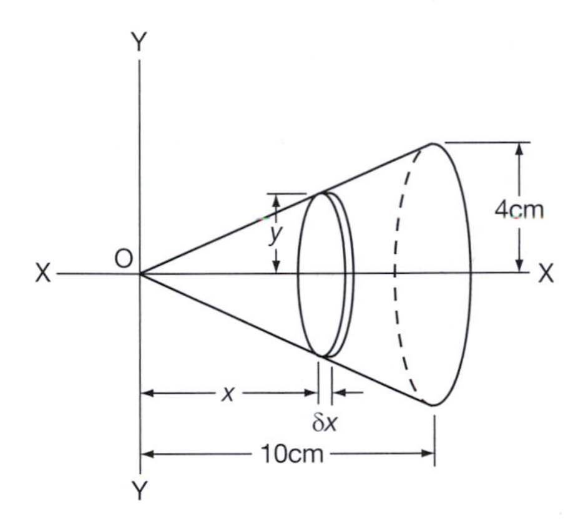


First take an elementary disc at distance $x$ from the origin. For this disc, OX is the normal axis, so

$I_X$ . . . . . . . . . . .

Then sum for all the discs, etc.

*Finish it off*

36

$$I_X = 256\pi\rho;\ k = 2{\cdot}19 \text{ cm}$$

Here is the working.

For elementary disc: $I_X = \dfrac{\pi y^4.\delta x\rho}{2}$

$$\therefore \text{ Total } \quad I_X \approx \sum_{x=0}^{x=10} \frac{\pi y^4.\delta x\rho}{2}$$

If $\delta x \to 0$: $\quad I_X = \displaystyle\int_0^{10} \frac{\pi\rho y^4}{2} dx = \frac{\pi\rho}{2}\int_0^{10} y^4 dx$

Now, from the figure, the slope of the generating line is 4/10.

$$\therefore y = \frac{4x}{10}$$

$$\therefore I_X = \frac{\pi\rho}{2}\int_0^{10}\left(\frac{4x}{10}\right)^4 dx$$
$$= \frac{\pi\rho}{2} 0{\cdot}16^2\left[\frac{x^5}{5}\right]_0^{10}$$
$$= \frac{\pi\rho 0{\cdot}0256}{2}\left[\frac{10^5}{5}\right]$$
$$= \pi\rho 0{\cdot}0256 \times 10^4 = 256\pi\rho$$

▶

Now we proceed to find $k$.

$$\text{Total mass} = M = \frac{1}{3}\pi 4^2 10\rho = \frac{160\pi\rho}{3}$$

$$Mk^2 = \text{I}$$

$$\therefore \frac{160\pi\rho}{3}k^2 = 256\pi\rho$$

$$\therefore k^2 = \frac{3.256.\pi\rho}{160.\pi\rho}$$

$$= \frac{3.64}{40} = 4{\cdot}8$$

$$\therefore k = \sqrt{4{\cdot}8} = 2{\cdot}19 \text{ cm}$$

*Move now to Frame 37*

## Second moment of area

**37**

In the theory of bending of beams, the expression $\sum ar^2$, relating to the cross-section of the beam, has to be evaluated. This expression is called the *second moment of area* of the section and although it has nothing to do with kinetic energy of rotation, the mathematics involved is clearly very much akin to that for moments of inertia, i.e. $\sum mr^2$.

Indeed, all the results we have obtained for thin plates could apply to plane figures, provided always that 'mass' is replaced by 'area'. In fact, the mathematical processes are so nearly alike that the same symbol (I) is used in practice both for *moment of inertia* and for *second moment of area.*

**38**

| **Moments of inertia** | **Second moments of area** |
|---|---|
| $\text{I} = \sum mr^2$ | $\text{I} = \sum ar^2$ |
| $Mk^2 = \text{I}$ | $Ak^2 = \text{I}$ |
| *Rectangular plate* | *Rectangle* |
| $\text{I}_G = \frac{bd^3\rho}{12}$ | $\text{I}_C = \frac{bd^3}{12}$ |
| $= \frac{M.d^2}{12}$ | $= \frac{A.d^2}{12}$ |
| *Circular plate* | *Circle* |
| $\text{I}_Z = \frac{\pi r^4\rho}{2}$ | $\text{I}_Z = \frac{\pi r^4}{2}$ |
| $= \frac{Mr^2}{2}$ | $= \frac{Ar^2}{2}$ |
| $\text{I}_X = \frac{\pi r^4\rho}{4}$ | $\text{I}_X = \frac{\pi r^4}{4}$ |
| $= \frac{M.r^2}{4}$ | $= \frac{A.r^2}{4}$ |

*Parallel axes theorem* – applies to both:

$$\text{I}_{AB} = \text{I}_G + Ml^2 \qquad \text{I}_{AB} = \text{I}_C + Al^2$$

*Perpendicular axes theorem* – applies to thin plates and plane figures only:

$$\text{I}_Z = \text{I}_X + \text{I}_Y$$

*Move on*

39

There is really nothing new about this: all we do is replace 'mass' by 'area'.

**Example 1**

Find the second moment of area of a rectangle about an axis through one corner perpendicular to the plane of the figure.

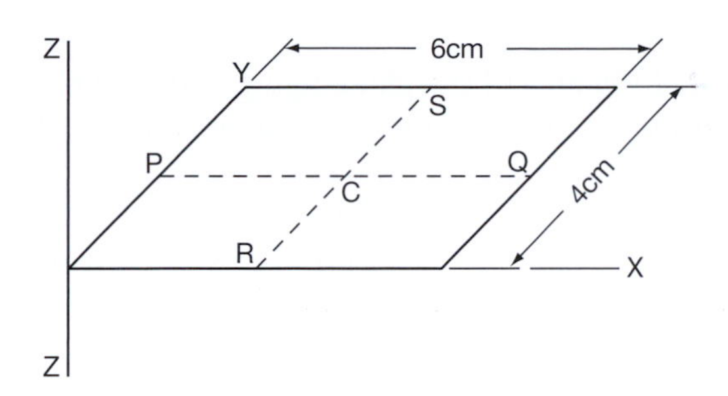


$$I_{PQ} = \frac{bd^3}{12} = \frac{6.4^3}{12} = 32 \text{ cm}^4$$

By the parallel axes theorem, $I_X = \ldots\ldots\ldots\ldots$

40

$$\boxed{I_X = 128 \text{ cm}^4}$$

Because

$$I_X = 32 + 24.2^2 = 32 + 24.4$$
$$= 32 + 96 = 128 \text{ cm}^4$$

Also $I_{RS} = \dfrac{bd^3}{12} = \ldots\ldots\ldots\ldots$

41

$$\boxed{I_{RS} = 72 \text{ cm}^4}$$

Because

$$I_{RS} = \frac{4.6^3}{12} = 72 \text{ cm}^4$$

$\therefore I_Y = \ldots\ldots\ldots\ldots$

42

$$\boxed{I_Y = 288 \text{ cm}^4}$$

Because, again by the parallel axes theorem:

$I_Y = 72 + 24.3^2 = 72 + 216 = 288 \text{ cm}^4$

So we have therefore: $I_X = 128 \text{ cm}^4$

and $I_Y = 288 \text{ cm}^4$

$\therefore I_Z$ (which is perpendicular to both $I_X$ and $I_Y$) $= \ldots\ldots\ldots\ldots$

**43**

$$I_Z = 416\text{ cm}^4$$

When the plane figure is bounded by an analytical curve, we proceed in much the same way.

**Example 2**

Find the second moment of area of the plane figure bounded by the curve $y = x^2 + 3$, the $x$-axis and the ordinates at $x = 1$ and $x = 3$, about the $y$-axis.

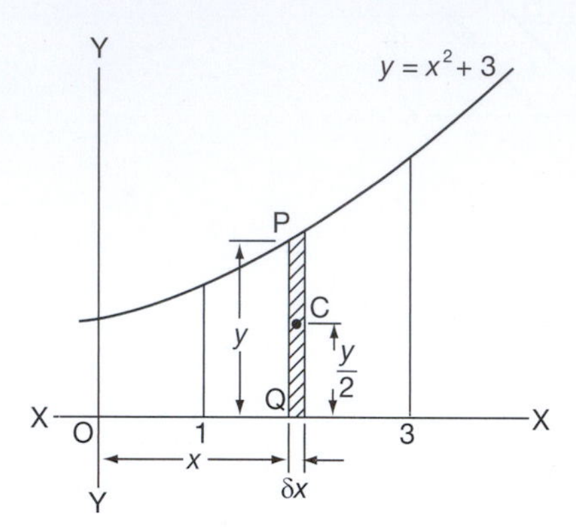


Area of strip PQ $= y.\delta x$

$\therefore$ 2nd moment of strip about OY $= y.\delta x.x^2$

$= x^2.y.\delta x$

$$\therefore I_Y \approx \sum_{x=1}^{x=3} x^2 y \delta x$$

If $\delta x \to 0$, $\quad I_Y = \int_1^3 x^2 y dx = \ldots\ldots\ldots\ldots$

*Finish it off*

**44**

$$I_Y = 74{\cdot}4\text{ units}^4$$

Because

$$I_Y = \int_1^3 x^2(x^2+3)dx = \int_1^3 (x^4 + 3x^2)dx$$

$$= \left[\frac{x^5}{5} + x^3\right]_1^3$$

$$= \left(\frac{243}{5} + 27\right) - \left(\frac{1}{5} + 1\right)$$

$$= \frac{242}{5} + 26 = 48{\cdot}4 + 26$$

$$= 74{\cdot}4\text{ units}^4$$

*Note*: Had we been asked to find $I_X$, we should take second moment of the strip about OX, i.e. $\frac{y^3}{3}\delta x$; sum for all strips $\sum_{x=1}^{x=3} \frac{y^3}{3}\delta x$; and then evaluate the integral.

*Now for one further example, so move on to the next frame*

45

**Example 3**

For the triangle PQR shown, find the second moment of area and $k$ about an axis AB through the vertex and parallel to the base.

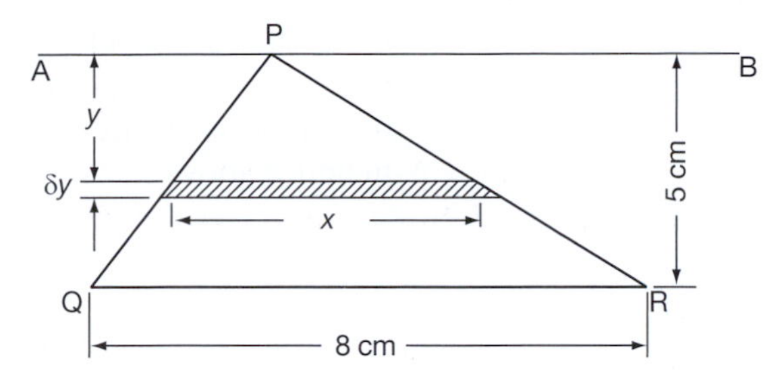


First consider an elementary strip. Area of strip $= x.\delta y$

$\therefore$ 2nd moment of strip about AB $= x.\delta y.y^2 = xy^2.\delta y$

$\therefore$ Total 2nd moment about AB for all such strips $\approx \sum_{y=0}^{y=5} xy^2.\delta y$

If $\delta y \to 0$ : $\quad I_{AB} = \int_0^5 xy^2 dy$

We must now write $x$ in terms of $y$ – and we can obtain this from the figure by similar triangles.

Finish the work, so that $I_{AB} = \ldots\ldots\ldots\ldots$

46

$$\boxed{I_{AB} = 250 \text{ cm}^4; \quad k = 3{\cdot}536 \text{ cm}}$$

Because we have $\dfrac{x}{y} = \dfrac{8}{5} \quad \therefore x = \dfrac{8y}{5}$

$$\therefore I_{AB} = \int_0^5 xy^2 dy = \frac{8}{5}\int_0^5 y^3 dy = \frac{8}{5}\left[\frac{y^4}{4}\right]_0^5$$

$$= \frac{8}{20}(5^4 - 0) = \frac{8}{20}(625) = 250 \text{ cm}^4$$

Also, total area, $A = \dfrac{5.8}{2} = 20 \text{ cm}^2$

$$\therefore Ak^2 = I \quad \therefore 20k^2 = 250$$
$$\therefore k^2 = 12{\cdot}5$$
$$\therefore k = 3{\cdot}536 \text{ cm}$$

*Next frame*

47

## Composite figures

If a figure is made up of a number of standard figures whose individual second moments about a given axis are $I_1, I_2, I_3$, etc., then the second moment of the composite figure about the same axis is simply the sum of $I_1, I_2, I_3$, etc.

Similarly, if a figure whose second moment about a given axis is $I_2$ is removed from a larger figure with second moment $I_1$ about the same axis, the second moment of the remaining figure is $I = I_1 - I_2$.

Now for something new.

# Centre of pressure

48

## Pressure at a point P, depth $z$ below the surface of a liquid

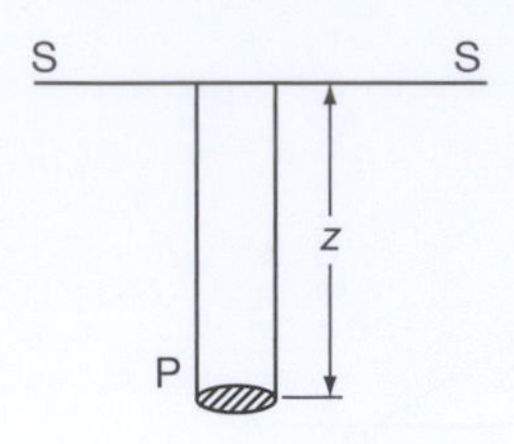


If we have a perfect liquid, the pressure at P, i.e. the thrust on unit area at P, is due to the weight of the column of liquid of height $z$ above it.

Pressure at P is $p = wz$ where $w =$ weight of unit volume of the liquid. Also, the pressure at P operates equally in all directions.

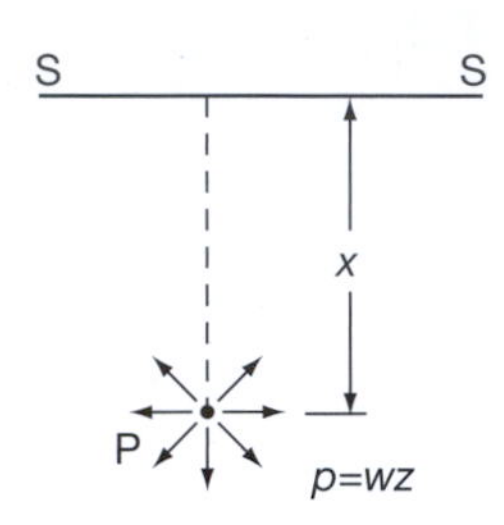


Note that, in our considerations, we shall ignore the atmospheric pressure which is also acting on the surface of the liquid.

The pressure, then, at any point in a liquid is proportional to the ........... of the point below the surface.

49

depth

## Total thrust on a vertical plate immersed in liquid

Consider a thin strip at a depth $z$ below the surface of the liquid.

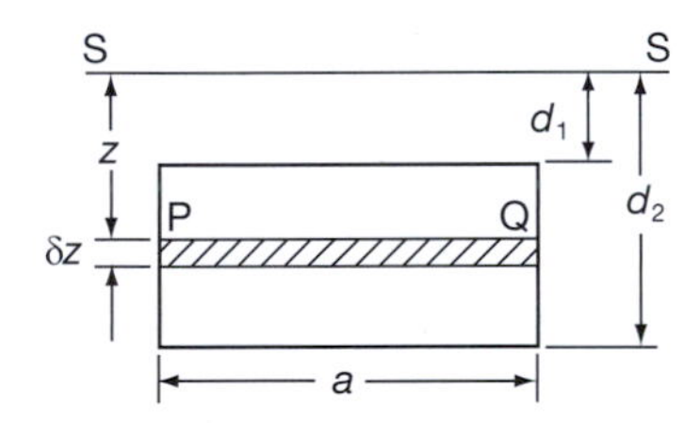


Pressure at P $= wz$.

$\therefore$ Thrust on strip PQ $\approx wz$(area of strip)

$$\approx w.z.a.\delta z$$

Then the total thrust on the whole plate

$$\approx \sum_{z=d_1}^{z=d_2} awz\delta z$$

If $\delta z \to 0$, total thrust $= \int_{d_1}^{d_2} awz\,\mathrm{d}z = \ldots\ldots\ldots\ldots$

**50**

$$\boxed{\frac{aw}{2}\left(d_2^2 - d_1^2\right)}$$

Because

$$\text{total thrust} = aw\left[\frac{z^2}{2}\right]_{d_1}^{d_2} = \frac{aw}{2}\left(d_2^2 - d_1^2\right)$$

This can be written:

$$\text{total thrust} = \frac{aw}{2}(d_2 - d_1)(d_2 + d_1)$$
$$= wa(d_2 - d_1)\left(\frac{d_2 + d_1}{2}\right)$$

Now, $\left(\frac{d_2 + d_1}{2}\right)$ is the depth halfway down the plate, i.e. it indicates the depth of the centre of gravity of the plate. Denote this by $\bar{z}$.

Then, total thrust $= wa(d_2 - d_1)\bar{z} = a(d_2 - d_1)w\bar{z}$.

Also, $a(d_2 - d_1)$ is the total area of the plate.

So we finally obtain the fact that:

total thrust = area of plate × pressure at the centre of gravity of the plate.

In fact, this result applies whatever the shape of the plate, so copy the result down for future use.

*On to the next frame*

**51**

Total thrust = area of plate × pressure at the centre of gravity of plate

So, if $w$ is the weight per unit volume of liquid, determine the total thrust on the following plates, immersed as shown:

(a)

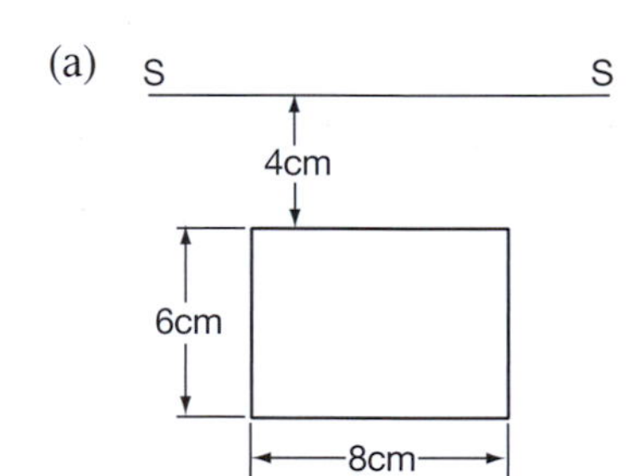


(b)

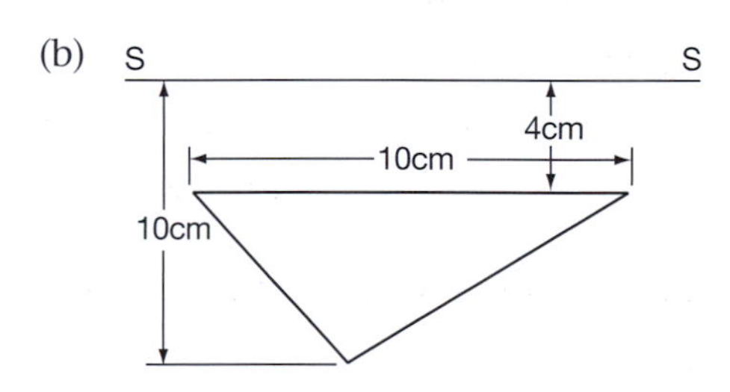


So, thrust (a) = ........... and thrust (b) = ...........

## 52

$$\text{thrust (a)} = 336w; \quad \text{thrust (b)} = 180w$$

Because in each case:

total thrust = area of surface × pressure at the centre of gravity

(a)

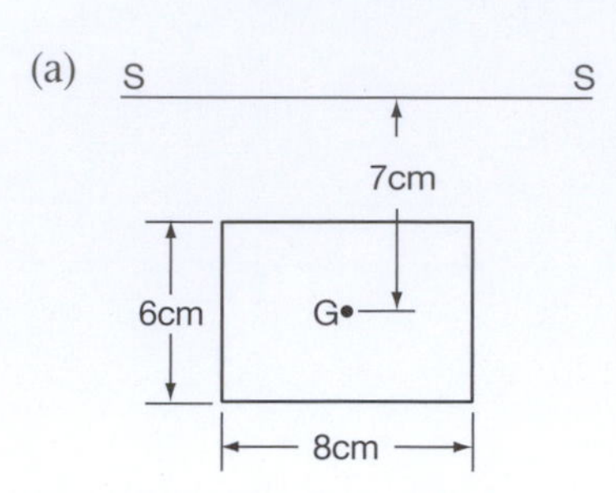


Area $= 6 \times 8 = 48\ \text{cm}^2$

Pressure at G $= 7w$

$\therefore$ Total thrust $= 48.7w$

$= 336w$

(b)

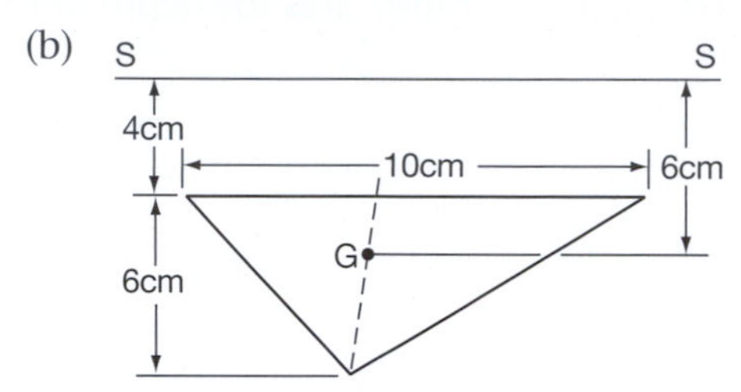


Area $= \dfrac{10 \times 6}{2} = 30\ \text{cm}^2$

Pressure at G $= 6w$

$\therefore$ Total thrust $= 30.6w$

$= 180w$

*On to the next frame*

## 53

If the plate is not vertical, but inclined at an angle $\theta$ to the horizontal, the rule still holds good.

For example:

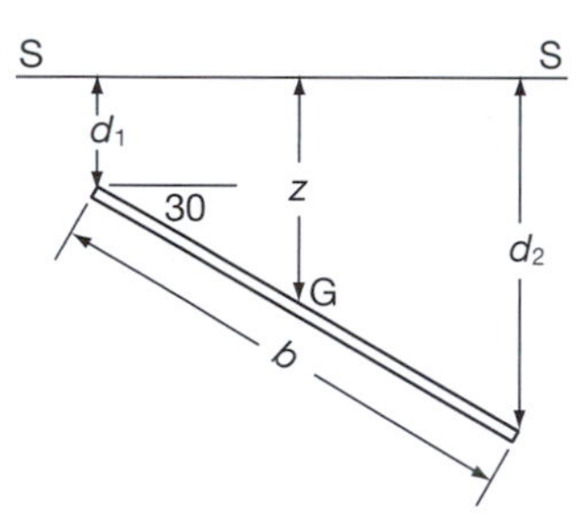


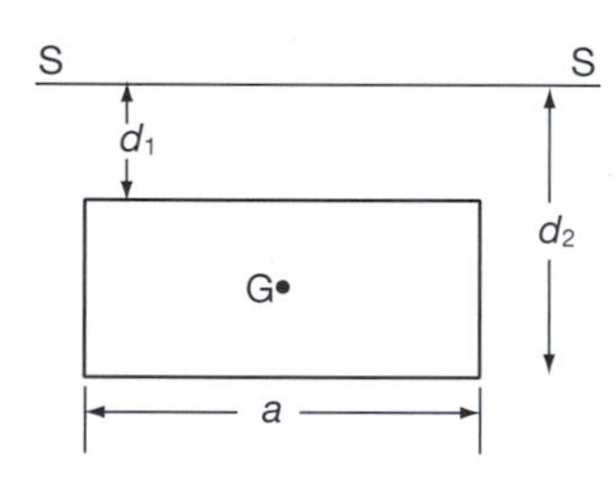


$$\text{Depth of G} = d_1 + \frac{b}{2}\sin 30^\circ = d_1 + \frac{b}{4}$$

$$\text{Pressure at G} = \left(d_1 + \frac{b}{4}\right)w$$

Total area $= ab$

$\therefore$ Total thrust $= \dots\dots\dots\dots$

## 54

$$ab\left(d_1 + \frac{b}{4}\right)w$$

Remember this general rule enables us to calculate the total thrust on an immersed surface in almost any set of circumstances.

So make a note of it:

total thrust = area of surface × pressure at the centre of gravity

*Then on to Frame 55*

## Depth of the centre of pressure

55

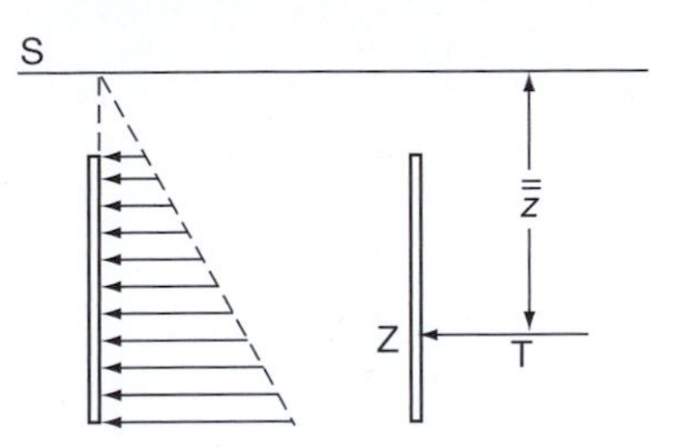


The pressure on an immersed plate increases with depth and we have seen how to find the total thrust $T$ on the plate.

The resultant of these forces is a single force equal to the total thrust, $T$, in magnitude and acting at a point $Z$ called the *centre of pressure* of the plate. Let $\bar{\bar{z}}$ denote the depth of the centre of pressure.

To find $\bar{\bar{z}}$ we take moments of forces about the axis where the plane of the plate cuts the surface of the liquid. Let us consider our same rectangular plate again.

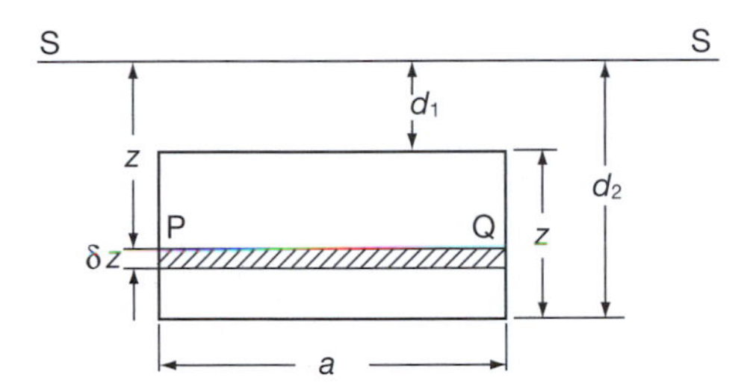


The area of the strip PQ = ...........

56

$a.\delta z$

The pressure at the level of PQ= ...........

57

$zw$

So the thrust on the strip PQ = ...........

58

$a.\delta z.w.z$ i.e. $awz\delta z$

The moment of this thrust about the axis in the surface is therefore

$= awz\delta z.z$

$= awz^2.\delta z$

So that the sum of the moments of thrusts on all such strips

= ...........

**59**

$$\sum_{d_1}^{d_2} awz^2 \delta z$$

Now, if $\delta z \to 0$:

the sum of the moments of thrusts $= \int_{d_1}^{d_2} awz^2\,dz$

Also, the total thrust on the whole plate = ...........

**60**

$$\int_{d_1}^{d_2} awz\,dz$$

Right. Now the total thrust $\times\, \bar{\bar{z}} =$ sum of moments of all individual thrusts

$$\therefore \int_{d_1}^{d_2} awz\,dz \times \bar{\bar{z}} = \int_{d_1}^{d_2} awz^2\,dz$$

$$\therefore \text{Total thrust} \times \bar{\bar{z}} = w\int_{d_1}^{d_2} az^2\,dz$$

$$= w\mathrm{I}$$

Therefore, we have:

$$\bar{\bar{z}} = \frac{w\mathrm{I}}{\text{total thrust}} = \frac{wAk^2}{Aw\bar{z}}$$

$$\therefore \bar{\bar{z}} = \frac{k^2}{\bar{z}}$$

*Make a note of that and then move on*

**61** So we have these two important results:

(a) The total thrust on a submerged surface = total area of face $\times$ pressure at its centroid (depth $\bar{z}$).

(b) The resultant thrust acts at the centre of pressure, the depth of which, $\bar{\bar{z}}$, is given by $\bar{\bar{z}} = \dfrac{k^2}{\bar{z}}$.

*Now for an example on this*

**62** **Example 1**

For a vertical rectangular dam, 40 m $\times$ 20 m, the top edge of the dam coincides with the surface level. Find the depth of the centre of pressure.

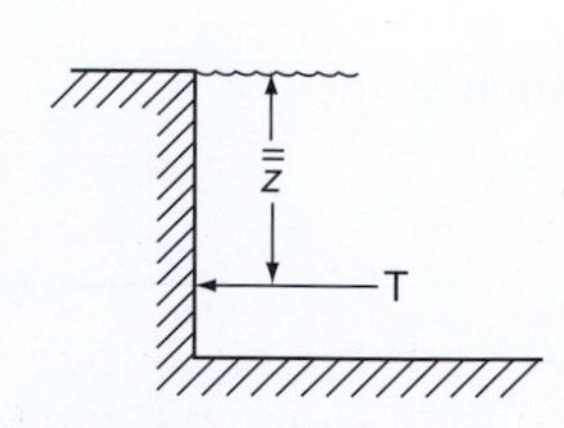


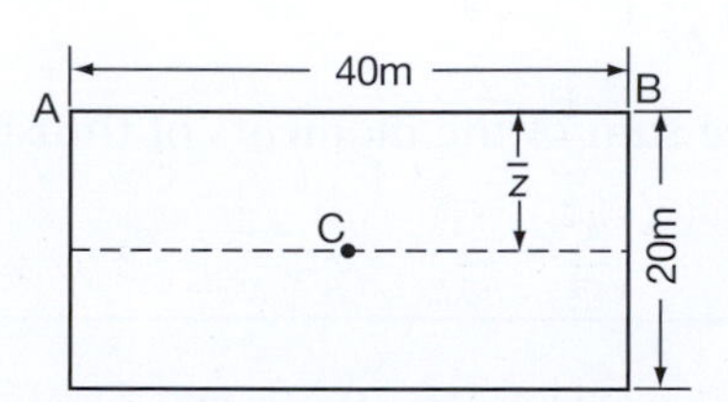


In this case, $\bar{z} = 10$ m.

▶

To find $k^2$ about AB:

$$I_C = \frac{Ad^2}{12} = \frac{40.20.400}{12} = \frac{80\,000}{3}\ m^4$$

$$I_{AB} = I_C + Al^2 = \frac{80\,000}{3} + 800.100$$

$$= \frac{4}{3}.(80\,000)$$

$$Ak^2 = I \quad \therefore\ k^2 = \frac{4}{3}.\frac{80\,000}{800} = \frac{400}{3}$$

$$\therefore\ \bar{\bar{z}} = \frac{k^2}{\bar{z}} = \frac{400}{3.10} = \frac{40}{3} = 13{\cdot}33\ m$$

*Note that, in this case*:

(a) the centroid is halfway down the rectangle

but (b) the centre of pressure is two-thirds of the way down the rectangle.

63

Here is one for you.

**Example 2**

An outlet from a storage tank is closed by a circular cover hung vertically. The diameter of the cover = 1 m and the top of the cover is 2·5 m below the surface of the liquid. Determine the depth of the centre of pressure of the cover.

*Work completely through it: then check your working with the next frame*

64

$$\bar{\bar{z}} = 3{\cdot}02\ m$$

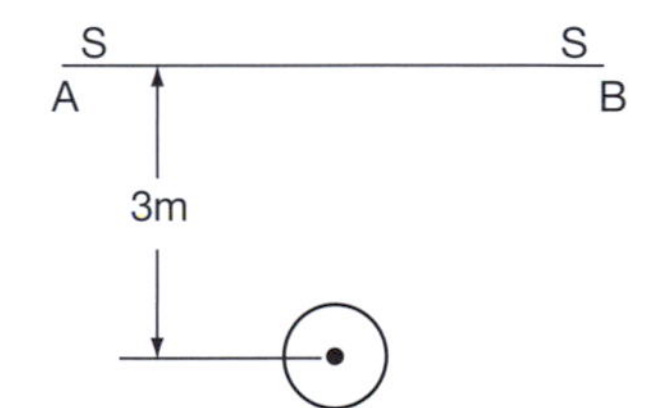


We have:

(a) Depth of centroid = $\bar{z}$ = 3 m.

(b) To find $k^2$ about AB:

$$I_C = \frac{Ar^2}{4} = \frac{\pi\left(\frac{1}{2}\right)^2.\left(\frac{1}{2}\right)^2}{4} = \frac{\pi}{64}$$

$$I_{AB} = \frac{\pi}{64} + A.3^2$$

$$= \frac{\pi}{64} + \pi\left(\frac{1}{2}\right)^2.9$$

$$= \frac{\pi}{64} + \frac{9\pi}{4} = \frac{145\pi}{64}$$

For AB: $k^2 = \frac{I_{AB}}{A} = \frac{145\pi}{64}.\frac{4}{\pi} = \frac{145}{16}$

$$\bar{\bar{z}} = \frac{k^2}{\bar{z}} = \frac{145}{16}.\frac{1}{3} = \frac{145}{48} = 3{\cdot}02\ m$$

And that brings us to the end of this piece of work. Before you work through the **Can You?** checklist and **Test exercise**, check down the revision summary that follows in Frame 65 and brush up any part of the Programme about which you may not be absolutely clear.

*When you are ready, move on*

## Review summary

65

1 SECOND MOMENTS

| *Moments of Inertia* | *2nd Moments of Area* |
|---|---|
| (a) $I = \sum mr^2$ | (a) $I = \sum ar^2$ |
| $Mk^2 = I$ | $Ak^2 = I$ |
| (b) Rectangular plate: | (b) Rectangle: |
| $I_G = \frac{bd^3\rho}{12} = \frac{M.d^2}{12}$ | $I_C = \frac{b.d}{12} = \frac{A.d^2}{12}$ |
| (c) Circular disc: | (c) Circle: |
| $I_Z = \frac{\pi r^4 \rho}{2} = \frac{M.r^2}{2}$ | $I_Z = \frac{\pi r^4}{2} = \frac{Ar^2}{2}$ |
| $I_X = \frac{\pi r^4 \rho}{4} = \frac{M.r^2}{4}$ | $I_Z = \frac{\pi r^4}{4} = \frac{Ar^2}{4}$ |
| (d) Parallel axes theorem: | |
| $I_{AB} = I_G + Ml^2$ | $I_{AB} = I_C + Al^2$ |

(e) Perpendicular axes theorem (thin plates and plane figure only):

$$I_Z = I_X + I_Y$$

2 CENTRES OF PRESSURE

(a) Pressure at depth $z = wz$ ($w$ = weight of unit volume of liquid).

(b) Total thrust on plane surface
= area of surface × pressure at the centroid.

(c) Depth of centre of pressure ($\bar{\bar{z}}$)

Total thrust × $\bar{\bar{z}}$ = sum of moments of distributed thrust

$$\bar{\bar{z}} = \frac{k^2}{\bar{z}}$$

where $k$ = radius of gyration of figure about axis in surface of liquid
$\bar{z}$ = depth of centroid.

*Note*: The magnitude of the total thrust = (area × pressure at the centroid) but it acts through the centre of pressure.

*Now for the* **Can You?** *checklist in the next frame*

## Can You?

### Checklist 21

66

Check this list before and after you try the end of Programme test.

On a scale of 1 to 5 how confident are you that you can: Frames

- Determine moments of inertia? 1 to 11
  *Yes* ☐ ☐ ☐ ☐ ☐ *No*
- Determine the radius of gyration? 12 to 19
  *Yes* ☐ ☐ ☐ ☐ ☐ *No*
- Use the parallel axes theorem? 20 to 28
  *Yes* ☐ ☐ ☐ ☐ ☐ *No*
- Use the perpendicular axes theorem for thin plates? 29 to 32
  *Yes* ☐ ☐ ☐ ☐ ☐ *No*
- Determine moments of inertia using standard results? 33 to 36
  *Yes* ☐ ☐ ☐ ☐ ☐ *No*
- Determine second moments of area? 37 to 47
  *Yes* ☐ ☐ ☐ ☐ ☐ *No*
- Determine centres of pressure? 48 to 64
  *Yes* ☐ ☐ ☐ ☐ ☐ *No*

## Test exercise 21

Work through all the questions. They are very much like those we have just been doing, so will cause you no difficulty: there are no tricks. Take your time and work carefully.

1 (a) Find the moment of inertia of a rectangular plate, of sides $a$ and $b$, about an axis through the mid-point of the plate and perpendicular to the plane of the plate. (b) Hence find also the moment of inertia about an axis parallel to the first axis and passing through one corner of the plate. (c) Find the radius of gyration about the second axis.

2 Show that the radius of gyration of a thin rod of length $l$ about an axis through its centre and perpendicular to the rod is $\dfrac{l}{2\sqrt{3}}$.
An equilateral triangle ABC is made of three identical thin rods each of length $l$. Find the radius of gyration of the triangle about an axis through A, perpendicular to the plane of ABC.

3 A plane figure is bounded by the curve $xy = 4$, the $x$-axis and the ordinates at $x = 2$ and $x = 4$. Calculate the square of the radius of gyration of the figure: (a) about the $x$-axis, and (b) about the $y$-axis.

4 Prove that the radius of gyration of a uniform solid cone with base radius $r$ about its natural axis is $\sqrt{\dfrac{3r^2}{10}}$.

5 An equilateral triangular plate is immersed in water vertically with one edge in the surface. If the length of each side is $a$, find the total thrust on the plate and the depth of the centre of pressure.

## Further problems 21

68

1 A plane figure is enclosed by the curve $y = a \sin x$ and the $x$-axis between $x = 0$ and $x = \pi$. Show that the radius of gyration of the figure about the $x$-axis is $\dfrac{a\sqrt{2}}{3}$.

2 A length of thin uniform wire of mass $M$ is made into a circle of radius $a$. Find the moment of inertia of the wire about a diameter as axis.

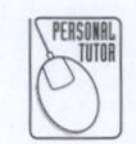

3 A solid cylinder of mass $M$ has a length $l$ and radius $r$, Show that its moment of inertia about a diameter of the base is $M\left[\dfrac{r^2}{4} + \dfrac{l^2}{3}\right]$.

4 Show that the moment of inertia of a solid sphere of radius $r$ and mass $M$, about a diameter as axis, is $\dfrac{2}{5}Mr^2$.

5 Prove that, if $k$ is the radius of gyration of an object about an axis through its centre of gravity, and $k_1$ is the radius of gyration about another axis parallel to the first and at a distance $l$ from it, then $k_1 = \sqrt{k^2 + l^2}$.

6 A plane figure is bounded by the parabola $y^2 = 4ax$, the $x$-axis and the ordinate $x = c$. Find the radius of gyration of the figure: (a) about the $x$-axis, and (b) about the $y$-axis.

7 Prove that the moment of inertia of a hollow cylinder of length $l$, with inner and outer radii $r$ and $R$ respectively, and total mass $M$, about its natural axis, is given by $\mathrm{I} = \dfrac{1}{2}M(R^2 + r^2)$.

8 Show that the depth of the centre of pressure of a vertical triangle with one side in the surface is $\dfrac{1}{2}h$, if $h$ is the perpendicular height of the triangle.

9 Calculate the second moment of area of a square of side $a$ about a diagonal as axis.

10 Find the moment of inertia of a solid cone of mass $M$ and base radius $r$ and height $h$, about a diameter of the base as axis. Find also the radius of gyration.

11 A thin plate in the form of a trapezium with parallel sides of length $a$ and $b$, distance $d$ apart, is immersed vertically in water with the side of length $a$ in the surface. Prove that the depth of the centre of pressure $(\bar{\bar{z}})$ is given by $\bar{\bar{z}} = \dfrac{d(a + 3b)}{2(a + 2b)}$.

12 Find the second moment of area of an ellipse about its major axis.

13 A square plate of side $a$ is immersed vertically in water with its upper side horizontal and at a depth $d$ below the surface. Prove that the centre of pressure is at a distance $\dfrac{a^2}{6(a + 2d)}$ below the centre of the square.

14 Find the total thrust and the depth of the centre of pressure when a semicircle of radius $a$ is immersed vertically in liquid with its diameter in the surface.

**15** A plane figure is bounded by the curve $y = e^x$, the $x$-axis, the $y$-axis and the ordinate $x = 1$. Calculate the radius of gyration of the figure: (a) about the $x$-axis, and (b) about the $y$-axis.

**16** A vertical dam is a parabolic segment of width 12 m and maximum depth 4 m at the centre. If the water reaches the top of the dam, find the total thrust on the face.

**17** A circle of diameter 6 cm is removed from the centre of a rectangle measuring 10 cm by 16 cm. For the figure that remains, calculate the radius of gyration about one 10-cm side as axis.

**18** Prove that the moment of inertia of a thin hollow spherical shell of mass $M$ and radius $r$, about a diameter as axis is $\frac{2}{3}Mr^2$.

**19** A semicircular plate of radius $a$ is immersed vertically in water, with its diameter horizontal and the centre of the arc just touching the surface. Find the depth of the centre of pressure.

**20** A thin plate of uniform thickness and total mass $M$ is bounded by the curve $y = c\cosh\frac{x}{c}$, the $x$-axis, the $y$-axis and the ordinate $x = a$. Show that the moment of inertia of the plate about the $y$-axis is

$$M\left\{a^2 - 2ca\coth^{-1}\left(\frac{a}{c}\right) + 2c^2\right\}.$$

Now visit the companion website for this book at www.palgrave.com/stroud for more questions applying this mathematics to science and engineering.

**69**

# Approximate integration

## Learning outcomes

When you have completed this Programme you will be able to:

- ☐ Recognize when an integral cannot be evaluated directly
- ☐ Approximate integrals using series expansions
- ☐ Use Simpson's rule to approximate the area beneath a curve

# Introduction

**1**

In previous Programmes, we have seen how to deal with various types of integral, but there are still some integrals that look simple enough, but which cannot be determined by any of the standard methods we have studied.

For instance $\int_0^{\frac{1}{2}} xe^x\,dx$ can be evaluated by the method of integration by parts.

$$\int_0^{\frac{1}{2}} xe^x\,dx = \ldots\ldots\ldots\ldots$$

What do you get?

**2**

$$\boxed{1 - \frac{1}{2}\sqrt{e}}$$

Because

$$\int_0^{\frac{1}{2}} xe^x\,dx = \left[x(e^x)\right]_0^{\frac{1}{2}} - \int_0^{\frac{1}{2}} e^x\,dx$$

$$= \left[xe^x - e^x\right]_0^{\frac{1}{2}} = \left[e^x(x-1)\right]_0^{\frac{1}{2}}$$

$$= e^{\frac{1}{2}}\left(-\frac{1}{2}\right) - e^0(-1) = 1 - \frac{1}{2}\sqrt{e}$$

That was easy enough, and this method depends, of course, on the fact that on each application of the routine, the power of $x$ decreases by 1, until it disappears, leaving $\int e^x\,dx$ to be completed without difficulty.

But suppose we try to evaluate $\int_0^{\frac{1}{2}} x^{\frac{1}{2}}e^x\,dx$ by the same method. The process now breaks down. Work through it and see if you can decide why.

*When you have come to a conclusion, move on to the next frame*

**3**

Reducing the power of $x$ by 1 at each application of the method will never give $x^0$, i.e. the power of $x$ will never disappear and so the resulting integral will always be a product

Because we get:

$$\int_0^{\frac{1}{2}} x^{\frac{1}{2}}e^x\,dx = \left[x^{\frac{1}{2}}(e^x)\right]_0^{\frac{1}{2}} - \frac{1}{2}\int_0^{\frac{1}{2}} e^x x^{-\frac{1}{2}}\,dx$$

and in the process, we have hopped over $x^0$.

So here is a complication. The present Programme will show you how to deal with this and similar integrals that do not fit into our normal patterns.

*So on, then, to Frame 4*

## Approximate integration

**4**

First of all, the results we shall get will be approximate in value, but like many other 'approximate' methods in mathematics, this does not imply that they are 'rough and ready' and of little significance.

The word 'approximate' in this context simply means that the numerical value cannot be completely defined, but that we can state the value to as many decimal places as we wish.

e.g. To say $x = \sqrt{3}$ is exact, but to say $x = 1{\cdot}732$ is an approximate result since, in fact, $\sqrt{3}$ has a value $1{\cdot}7320\ldots$ with an infinite number of decimal places.

Let us not be worried, then, by approximate values: we use them whenever we quote a result correct to a stated number of decimal places, or significant figures:

$$\pi = 3\tfrac{1}{7} \quad \pi = 3{\cdot}142 \quad \pi = 3{\cdot}14159$$

are all ........... values.

**5**

approximate

We note, of course, that an approximate value can be made nearer and nearer to the real value by taking a larger number of decimal places – and that usually means more work!

Evaluation of definite integrals is often required in science and engineering problems: a numerical approximation of the result is then quite satisfactory.

Let us see two methods that we can apply when the standard routines fail.

*On to Frame 6*

**6**

### Method 1: *By series*

Consider the integral $\int_0^{\frac{1}{2}} x^{\frac{1}{2}} e^x \, dx$, which we have already seen cannot be evaluated by the normal means. We have to convert this into some other form that we can deal with.

Now we know that:

$$e^x = 1 + x + \frac{x^2}{2!} + \frac{x^3}{3!} + \frac{x^4}{4!} + \ldots$$

$$\therefore \; x^{\frac{1}{2}} e^x = x^{\frac{1}{2}} \left\{ 1 + x + \frac{x^2}{2!} + \frac{x^3}{3!} + \frac{x^4}{4!} + \ldots \right\}$$

$$\therefore \; \int_0^{\frac{1}{2}} x^{\frac{1}{2}} e^x \, dx = \int_0^{\frac{1}{2}} x^{\frac{1}{2}} \left\{ 1 + x + \frac{x^2}{2!} + \frac{x^3}{3!} + \frac{x^4}{4!} + \ldots \right\} dx$$

$$= \int_0^{\frac{1}{2}} \left\{ x^{1/2} + x^{3/2} + \frac{x^{5/2}}{2!} + \frac{x^{7/2}}{3!} + \frac{x^{9/2}}{4!} + \ldots \right\} dx$$

Now these are simply powers of $x$, so on the next line, we have

$$= \ldots\ldots\ldots\ldots$$

**7**

$$I = \left[\frac{2x^{3/2}}{3} + \frac{2x^{5/2}}{5} + \frac{2x^{7/2}}{7.2} + \frac{2x^{9/2}}{9.6} + \frac{2x^{11/7}}{11.24} + \dots\right]_0^{\frac{1}{2}}$$
$$= \left[\frac{2x^{3/2}}{3} + \frac{2x^{5/2}}{5} + \frac{x^{7/2}}{7} + \frac{x^{9/2}}{27} + \frac{x^{11/2}}{132} + \dots\right]_0^{\frac{1}{2}}$$

To ease the calculation, take out the factor $x^{\frac{1}{2}}$:

$$I = \left[x^{\frac{1}{2}}\left\{\frac{2x}{3} + \frac{2x^2}{5} + \frac{x^3}{7} + \frac{x^4}{27} + \frac{x^5}{132} + \dots\right\}\right]_0^{\frac{1}{2}}$$
$$= \frac{1}{\sqrt{2}}\left\{\frac{1}{3} + \frac{2}{4.5} + \frac{1}{8.7} + \frac{1}{16.27} + \frac{1}{32.132} + \dots\right\}$$
$$= \frac{\sqrt{2}}{2}\left\{0{\cdot}3333 + 0{\cdot}1000 + 0{\cdot}0179 + 0{\cdot}0023 + 0{\cdot}0002 + \dots\right\}$$
$$= \frac{\sqrt{2}}{2}\left\{0{\cdot}4537\right\}$$
$$= 1{\cdot}414(0{\cdot}2269)$$
$$= 0{\cdot}3208$$

All we do is to express the function as a series and integrate the powers of $x$ one at a time.

*Let us see another example, so move on to Frame 8*

**8**

Here is another.

To evaluate $\displaystyle\int_0^{\frac{1}{2}} \frac{\ln(1+x)}{\sqrt{x}}\,dx$

First we expand $\ln(1+x)$ as a power series. Do you remember what it is?

$\ln(1+x) = \dots\dots\dots\dots$

**9**

$$\ln(1+x) \equiv x - \frac{x^2}{2} + \frac{x^3}{3} - \frac{x^4}{4} + \frac{x^5}{5} - \dots$$

$$\therefore\ \frac{\ln(1+x)}{\sqrt{x}} = x^{-\frac{1}{2}}\left\{x - \frac{x^2}{2} + \frac{x^3}{3} - \frac{x^4}{4} + \frac{x^5}{5} - \dots\right\}$$
$$= x^{\frac{1}{2}} - \frac{x^{3/2}}{2} + \frac{x^{5/2}}{3} - \frac{x^{7/2}}{4} + \frac{x^{9/2}}{5} - \dots$$

$$\therefore\ \int \frac{\ln(1+x)}{\sqrt{x}}\,dx = \dots$$

**10**

$$\int \frac{\ln(1+x)}{\sqrt{x}}\,\mathrm{d}x = \frac{2}{3}x^{3/2} - \frac{x^{5/2}}{5} + \frac{2x^{7/2}}{21} - \frac{x^{9/2}}{18} + \ldots$$

So that, applying the limits, we get:

$$\int_0^{\frac{1}{2}} \frac{\ln(1+x)}{\sqrt{x}}\,\mathrm{d}x = \left[x^{\frac{1}{2}}\left\{\frac{2x}{3} - \frac{x^2}{5} + \frac{2x^3}{21} - \frac{x^4}{18} + \ldots\right\}\right]_0^{\frac{1}{2}}$$

$$= \frac{1}{\sqrt{2}}\left\{\frac{1}{3} - \frac{1}{20} + \frac{1}{84} - \frac{1}{288} + \ldots\right\}$$

$$= 0{\cdot}7071\left\{0{\cdot}3333 - 0{\cdot}0500 + 0{\cdot}0119 - 0{\cdot}0035 + \ldots\right\}$$

$$= 0{\cdot}7071(0{\cdot}2917) = 0{\cdot}2063$$

Here is one for you to do in very much the same way.

Evaluate $\int_0^1 \sqrt{x}.\cos x\,\mathrm{d}x$

*Complete the working and then check your result with that given in the next frame*

**11**

0·531to 3 decimal places

Because

$$\cos x = 1 - \frac{x^2}{2!} + \frac{x^4}{4!} - \frac{x^6}{6!} + \frac{x^8}{8!} - \ldots$$

$$\therefore\ \sqrt{x}\cos x = x^{\frac{1}{2}} - \frac{x^{5/2}}{2} + \frac{x^{9/2}}{24} - \frac{x^{13/2}}{720} + \ldots$$

$$\therefore\ \int_0^1 \sqrt{x}\cos x\,\mathrm{d}x = \left[\frac{2x^{3/2}}{3} - \frac{x^{7/2}}{7} + \frac{x^{11/2}}{132} - \frac{x^{15/2}}{5400} + \ldots\right]_0^1$$

$$= \left\{\frac{2}{3} - \frac{1}{7} + \frac{1}{132} - \frac{1}{5400} + \ldots\right\}$$

$$= 0{\cdot}6667 - 0{\cdot}1429 + 0{\cdot}007576 - 0{\cdot}000185 + \ldots$$

$$= 0{\cdot}531 \text{ to 3 dp}$$

*Check carefully if you made a slip. Then on to Frame 12*

**12**

The method, then, is really very simple, provided the function can readily be expressed in the form of a series.

But we must use this method with caution. Remember that we are dealing with infinite series which are valid only for values of $x$ for which the series converges. In many cases, if the limits are less than 1 we are safe, but with limits greater than 1 we must be extra careful. For instance, the integral $\int_2^4 \frac{1}{1+x^3}\,\mathrm{d}x$ would give a divergent series when the limits were substituted. So what tricks can we employ in a case such as this?

*On to the next frame, and we will find out*

13

To evaluate $\int_2^4 \frac{1}{1+x^3}\,dx$

We first of all take out the factor $x^3$ from the denominator:

$$\frac{1}{1+x^3} \equiv \frac{1}{x^3}\left\{\frac{1}{\frac{1}{x^3}+1}\right\} = \frac{1}{x^3}\left\{1+\frac{1}{x^3}\right\}^{-1}$$

This is better, for if $x^3$ is going to be greater than 1 when we substitute the limits, $\frac{1}{x^3}$ will be ...........

14

less than 1

Right. So in this form we can expand without further trouble.

$$I = \int_2^4 x^{-3}\left\{1 - \frac{1}{x^3} + \frac{1}{x^6} - \frac{1}{x^9} + \ldots\right\} dx$$
$$= \int_2^4 x^{-3}\left\{1 - x^{-3} + x^{-6} - x^{-9} + \ldots\right\} dx$$
$$= \int_2^4 \left\{x^{-3} - x^{-6} + x^{-9} - x^{-12} + \ldots\right\} dx$$
$$= \ldots\ldots\ldots\ldots$$ Now finish it off.

15

0·088 to 3 decimal places

Because

$$I = \int_2^4 \left\{x^{-3} - x^{-6} + x^{-9} - x^{-12} + \ldots\right\} dx$$
$$= \left[-\frac{x^{-2}}{2} + \frac{x^{-5}}{5} - \frac{x^{-8}}{8} + \frac{x^{-11}}{11} - \ldots\right]_2^4$$
$$= \left[-\frac{1}{2x^2} + \frac{1}{5x^5} - \frac{1}{8x^8} + \frac{1}{11x^11} - \ldots\right]_2^4$$
$$= \left\{-\frac{1}{32} + \frac{1}{5120} - \frac{1}{524\,288} + \ldots\right\} - \left\{-\frac{1}{8} + \frac{1}{160} - \frac{1}{2048} + \ldots\right\}$$
$$= -0{\cdot}03125 + 0{\cdot}00020 - 0{\cdot}00000 + 0{\cdot}12500 - 0{\cdot}00625 + 0{\cdot}00049$$
$$= 0{\cdot}12569 - 0{\cdot}03750$$
$$= 0{\cdot}08819$$
$$= 0{\cdot}088 \text{ to 3 dp}$$

## 16 Method 2: *By Simpson's rule*

Integration by series is rather tedious and cannot always be applied, so let us start afresh and try to discover some other method of obtaining the approximate value of a definite integral.

We know, of course, that integration can be used to calculate the area under a curve $y = f(x)$ between two given points $x = a$ and $x = b$.

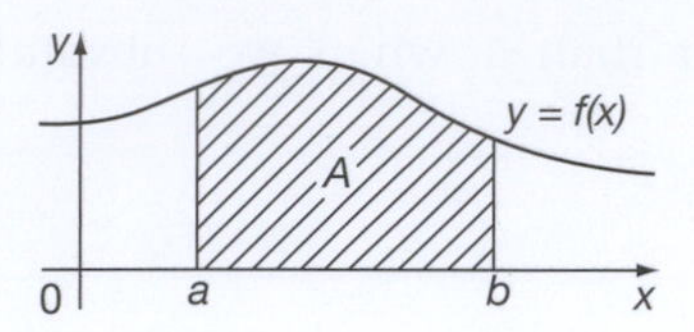


$$A = \int_a^b y\,dx = \int_a^b f(x)\,dx$$

Sometimes, however, it is not possible to evaluate the integral so, if only we could find the area $A$ by some other means, this would give us an approximate numerical value of the integral we have to evaluate. There are various practical ways of doing this and the one we shall choose is to apply Simpson's rule.

*So on to Frame 17*

## 17

To find the area under the curve $y = f(x)$ between $x = a$ and $x = b$:

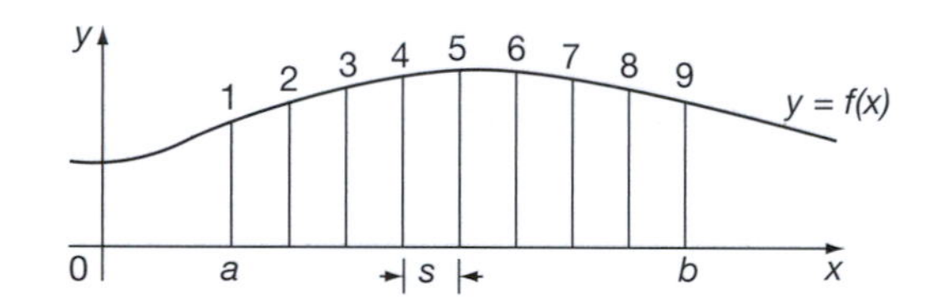


(a) Divide the figure into any even number ($n$) of equal-width strips (width $s$).

(b) Number and measure each ordinate: $y_1, y_2, y_3, \ldots, y_{n+1}$.
The number of ordinates will be one more than the number of strips.

(c) The area $A$ of the figure is then given by *Simpson's rule*:

$$A \approx \frac{s}{3}\left[(F + L) + 4E + 2R\right]$$

where $s$ = width of each strip

$F + L$ = sum of the first and last ordinates

$4E = 4 \times$ the sum of the even-numbered ordinates

$2R = 2 \times$ the sum of the remaining odd-numbered ordinates.

*Note*: Each ordinate is used once – and only once.

*Make a note of this result in your record book for future reference*

**18**

$$A \approx \frac{s}{3}\left[(F+L)+4E+2R\right]$$

The symbols themselves remind you of what they represent.

We shall now evaluate $\int_2^6 y\,dx$ for the function $y = f(x)$, the graph of which is shown:

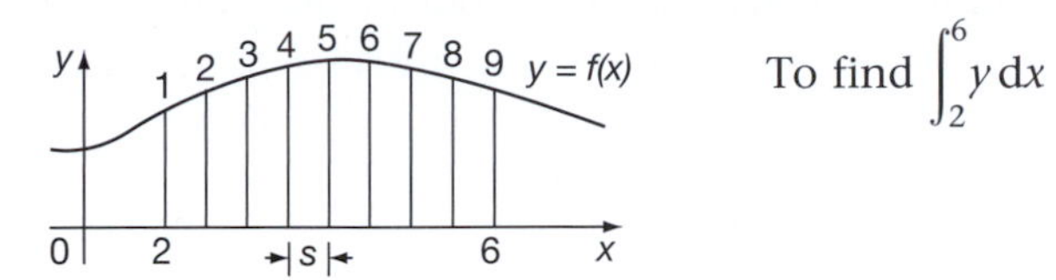


To find $\int_2^6 y\,dx$

If we take 8 strips, then $s = \frac{6-2}{8} = \frac{4}{8} = \frac{1}{2}$. $\quad s = \frac{1}{2}$

Suppose we find lengths of the ordinates to be as follows:

| Ordinate no. | 1 | 2 | 3 | 4 | 5 | 6 | 7 | 8 | 9 |
|---|---|---|---|---|---|---|---|---|---|
| Length | 7·5 | 8·2 | 10·3 | 11·5 | 12·4 | 12·8 | 12·3 | 11·7 | 11·5 |

Then we have:

$$F + L = 7{\cdot}5 + 11{\cdot}5 = 19$$
$$4E = 4(8{\cdot}2 + 11{\cdot}5 + 12{\cdot}8 + 11{\cdot}7) = 4(44{\cdot}2) = 176{\cdot}8$$
$$2R = 2(10{\cdot}3 + 12{\cdot}4 + 12{\cdot}3) = 2(35) = 70$$

So that:

$$A \approx \frac{1/2}{3}\left[19 + 176{\cdot}8 + 70\right]$$
$$= \frac{1}{6}\left[265{\cdot}8\right] = 44{\cdot}3 \quad \therefore A = 44{\cdot}3 \text{ units}^2$$

$$\therefore \int_2^6 f(x)\,dx \approx 44{\cdot}3$$

The accuracy of the result depends on the number of strips into which we divide the figure. A larger number of thinner strips gives a more accurate result.

Simpson's rule is important: it is well worth remembering.

Here it is again: write it out, but replace the query marks with the appropriate coefficients:

$$A \approx \frac{s}{?}\left[(F+L) + ?E + ?R\right]$$

**19**

$$A \approx \frac{s}{3}\left[(F+L)+4E+2R\right]$$

In practice, we do not have to plot the curve in order to measure the ordinates. We calculate them at regular intervals. Here is an example.

▶

**Example 1**

To evaluate $\int_0^{\pi/3} \sqrt{\sin x}\, dx$, using six intervals.

(a) Find the value of $s$:

$$s = \frac{\pi/3 - 0}{6} = \frac{\pi}{18} \quad (= 10^\circ \text{ intervals})$$

(b) Calculate the values of $y$ (i.e. $\sqrt{\sin x}$) at intervals of $\pi/18$ between $x = 0$ (lower limit) and $x = \pi/3$ (upper limit), and set your work out in the form of the table below:

| $x$ | | $\sin x$ | $\sqrt{\sin x}$ |
|---|---|---|---|
| 0 | (0°) | 0·0000 | 0·0000 |
| $\pi/18$ | (10°) | 0·1736 | 0·4167 |
| $\pi/9$ | (20°) | 0·3420 | ...... |
| $\pi/6$ | (30°) | 0·5000 | ...... |
| $2\pi/9$ | (40°) | ...... | ...... |
| $5\pi/18$ | (50°) | ...... | ...... |
| $\pi/3$ | (60°) | ...... | ...... |

Leave the right-hand side of your page blank for the moment.

Copy and complete the table as shown on the left-hand side above.

## 20

Here it is: check your results so far.

| $x$ | | $\sin x$ | $\sqrt{\sin x}$ | (1) $F + L$ | (2) $E$ | (3) $R$ |
|---|---|---|---|---|---|---|
| 0 | (0°) | 0·0000 | 0·0000 | | | |
| $\pi/18$ | (10°) | 0·1736 | 0·4167 | | | |
| $\pi/9$ | (20°) | 0·3420 | 0·5848 | | | |
| $\pi/6$ | (30°) | 0·5000 | 0·7071 | | | |
| $2\pi/9$ | (40°) | 0·6428 | 0·8017 | | | |
| $5\pi/18$ | (50°) | 0·7660 | 0·8752 | | | |
| $\pi/3$ | (60°) | 0·8660 | 0·9306 | | | |

Now form three more columns on the right-hand side, headed as shown, and transfer the final results across as indicated. This will automatically sort out the ordinates into the correct groups.

*Then on to Frame 21*

## 21

*Note that*:

(a) You start in column 1

(b) You then zig-zag down the two right-hand columns.

(c) You finish back in column 1.

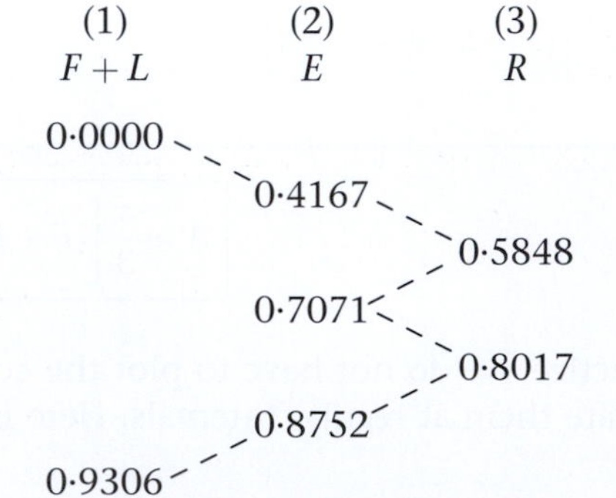


*Now total up each of the three columns*

**22**

Your results should be:

| (1) | (2) | (3) |
|---|---|---|
| $F+L$ | $E$ | $R$ |
| 0·9306 | 1·999 | 1·3865 |

Now: (a) Multiply column (2) by 4 so as to give $4E$.

(b) Multiply column (3) by 2 so as to give $2R$.

(c) Transfer the result in columns (2) and (3) to column (1) and total column (1) to obtain $(F+L)+4E+2R$.

*Now do that.*

**23**

This gives:

| | $F+L$ | $E$ | $R$ |
|---|---|---|---|
| $F+L \longrightarrow$ | 0·9306 | 1·999 | 1·3865 |
| $4E \longrightarrow$ | 7·996 | 4 | 2 |
| $2R \longrightarrow$ | 2·773 | 7·996 | 2·773 |
| $(F+L)+4E+2R \longrightarrow$ | 11·6996 | | |

The formula is $A \approx \frac{s}{3}\left[(F+L)+4E+2R\right]$ so to find $A$ we simply need to multiply our last result by $\frac{s}{3}$. Remember $s=\pi/18$.

So now you can finish it off.

$$\int_0^{\pi/3} \sqrt{\sin x}\,\mathrm{d}x = \dots\dots\dots\dots$$

**24**

0·681

Because

$$\begin{aligned}A &\approx \frac{s}{3}\left[(F+L)+4E+2R\right]\\ &\approx \frac{\pi/18}{3}\left[11{\cdot}6996\right]\\ &\approx (\pi/54)\left[11{\cdot}6996\right]\\ &\approx 0{\cdot}6807\end{aligned}$$

$$\therefore \int_0^{\pi/3} \sqrt{\sin x}\,\mathrm{d}x \approx 0{\cdot}681$$

Before we do another example, let us see the last solution complete.

▶

To evaluate $\int_0^{\pi/3} \sqrt{\sin x}\,dx$ by Simpson's rule, using 6 intervals.

$$s = \frac{\pi/3 - 0}{6} = \pi/18 \quad (= 10^\circ \text{ intervals})$$

| $x$ | | $\sin x$ | $\sqrt{\sin x}$ | $F+L$ | $E$ | $R$ |
|---|---|---|---|---|---|---|
| 0 | (0°) | 0·0000 | 0·0000 | 0·0000 | | |
| $\pi/18$ | (10°) | 0·1736 | 0·4167 | | 0·4167 | |
| $\pi/9$ | (20°) | 0·3420 | 0·5848 | | | 0·5848 |
| $\pi/6$ | (30°) | 0·5000 | 0·7071 | | 0·7071 | |
| $2\pi/9$ | (40°) | 0·6428 | 0·8017 | | | 0·8017 |
| $5\pi/18$ | (50°) | 0·7660 | 0·8752 | | 0·8752 | |
| $\pi/3$ | (60°) | 0·8660 | 0·9306 | 0·9306 | | |
| | | $F+L$ | $\longrightarrow$ | 0·9306 | 1·999 | 1·3865 |
| | | $4E$ | $\longrightarrow$ | 7·996 | 4 | 2 |
| | | $2R$ | $\longrightarrow$ | 2·773 | 7·996 | 2·773 |
| | | $(F+L)+4E+2R$ | $\longrightarrow$ | 11·6996 | | |

$$I \approx \frac{s}{3}\Big[(F+L) + 4E + 2R\Big]$$
$$\approx \frac{\pi}{54}\big[11{\cdot}6996\big]$$
$$\approx 0{\cdot}6807$$
$$\therefore \int_0^{\pi/3} \sqrt{\sin x}\,dx \approx 0{\cdot}681$$

Now we tackle another example and set it out in much the same way.

*Move to Frame 25*

**25**

**Example 2**

To evaluate $\int_{0\cdot2}^{1\cdot0} \sqrt{1 + x^3}\,dx$, using 8 intervals.

First of all, find the value of $s$ in this case.

$s = \ldots\ldots\ldots\ldots$

**26**

| 0·1 |
|---|

Because

$$s = \frac{1{\cdot}0 - 0{\cdot}2}{8} = \frac{0{\cdot}8}{8} = 0{\cdot}1$$
$$s = 0{\cdot}1$$

Now write the column headings required to build up the function values. What will they be on this occasion?

27

| $x$ | $x^3$ | $1+x^3$ | $\sqrt{1+x^3}$ | $F+L$ | $E$ | $R$ |
|---|---|---|---|---|---|---|

Right. So your table will look like this, with $x$ ranging from 0·2 to 1·0:

| $x$ | $x^3$ | $1+x^3$ | $\sqrt{1+x^3}$ | $F+L$ | $E$ | $R$ |
|---|---|---|---|---|---|---|
| 0·2 | 0·008 | 1·008 | 1·0040 | | | |
| 0·3 | 0·027 | 1·027 | 1·0134 | | | |
| 0·4 | 0·064 | | | | | |
| 0·5 | 0·125 | | | | | |
| 0·6 | 0·216 | | | | | |
| 0·7 | 0·343 | | | | | |
| 0·8 | | | | | | |
| 0·9 | | | | | | |
| 1·0 | | | | | | |
| | | | $F+L\longrightarrow$ | | | |
| | | | $4E\longrightarrow$ | | 4 | 2 |
| | | | $2R\longrightarrow$ | | | |
| | | | $(F+L)+4E+2R\longrightarrow$ | | | |

Copy down and complete the table above and finish off the working to evaluate $\int_{0\cdot 2}^{1\cdot 0}\sqrt{1+x^3}\,dx$.

*Check with the next frame*

28

$$\int_{0\cdot 2}^{1\cdot 0}\sqrt{1+x^3}\,dx = 0\cdot 911$$

| $x$ | $x^3$ | $1+x^3$ | $\sqrt{1+x^3}$ | $F+L$ | $E$ | $R$ |
|---|---|---|---|---|---|---|
| 0·2 | 0·008 | 1·008 | 1·0040 | 1·0040 | | |
| 0·3 | 0·027 | 1·027 | 1·0134 | | 1·0134 | |
| 0·4 | 0·064 | 1·064 | 1·0315 | | | 1·0315 |
| 0·5 | 0·125 | 1·125 | 1·0607 | | 1·0607 | |
| 0·6 | 0·216 | 1·216 | 1·1027 | | | 1·1027 |
| 0·7 | 0·343 | 1·343 | 1·1589 | | 1·1589 | |
| 0·8 | 0·512 | 1·512 | 1·2296 | | | 1·2296 |
| 0·9 | 0·729 | 1·729 | 1·3149 | | 1·3149 | |
| 1·0 | 1·000 | 2·000 | 1·4142 | 1·4142 | | |
| | | | $F+L\longrightarrow$ | 2·4182 | 4·5479 | 3·3638 |
| | | | $4E\longrightarrow$ | 18·1916 | 4 | 2 |
| | | | $2R\longrightarrow$ | 6·7276 | 18·1916 | 6·7276 |
| | | | $(F+L)+4E+2R\longrightarrow$ | 27·3374 | | |

▶

$$I = \frac{s}{3}\left[(F + L) + 4E + 2R\right]$$
$$= \frac{0{\cdot}1}{3}\left[27{\cdot}3374\right] = \frac{1}{3}\left[2{\cdot}73374\right] = 0{\cdot}9112$$
$$\therefore \int_{0{\cdot}2}^{1{\cdot}0} \sqrt{1 + x^3}\,dx \approx 0{\cdot}911$$

*There it is. Next frame*

## 29

Here is another one: let us work through it together.

**Example 3**

Using Simpson's rule with 8 intervals, evaluate $\int_1^3 y\,dx$, where the values of $y$ at regular intervals of $x$ are given.

| $x$ | 1·0 | 1·25 | 1·50 | 1·75 | 2·00 | 2·25 | 2·50 | 2·75 | 3·00 |
|---|---|---|---|---|---|---|---|---|---|
| $y$ | 2·45 | 2·80 | 3·44 | 4·20 | 4·33 | 3·97 | 3·12 | 2·38 | 1·80 |

If these function values are to be used as they stand, they must satisfy the requirements for Simpson's rule, which are:

(a) the function values must be spaced at ........... intervals of $x$, and

(b) there must be an ........... number of strips and therefore an ........... number of ordinates.

## 30

regular; even; odd

These conditions are satisfied in each case, so we can go ahead and evaluate the integral. In fact, the working will be a good deal easier for we are told the function value and there is no need to build them up as we had to do before.

In this example, $s =$ ...........

## 31

$s = 0{\cdot}25$

Because

$$s = \frac{3 - 1}{8} = \frac{2}{8} = 0{\cdot}25$$

Off you go, then. Set out your table and evaluate the integral defined by the values given in Frame 29. When you have finished, move on to Frame 32 to check your working.

32

$$\boxed{6{\cdot}62}$$

| $x$ | $y$ | $F+L$ | $E$ | $R$ |
|---|---|---|---|---|
| 1·0 | 2·45 | 2·45 | | |
| 1·25 | 2·80 | | 2·80 | |
| 1·50 | 3·44 | | | 3·44 |
| 1·75 | 4·20 | | 4·20 | |
| 2·00 | 4·33 | | | 4·33 |
| 2·25 | 3·97 | | 3·97 | |
| 2·50 | 3·12 | | | 3·12 |
| 2·75 | 2·38 | | 2·38 | |
| 3·00 | 1·80 | 1·80 | | |
| | $F+L\longrightarrow$ | 4·25 | 13·35 | 10·89 |
| | $4E\longrightarrow$ | 53·40 | 4 | 2 |
| | $2R\longrightarrow$ | 21·78 | 53·40 | 21·78 |
| | $(F+L)+4E+2R\longrightarrow$ | 79·43 | | |

$$I=\frac{s}{3}\left[(F+L)+4E+2R\right]=\frac{0{\cdot}25}{3}\left[79{\cdot}43\right]$$

$$=\frac{1}{12}\left[79{\cdot}43\right]=6{\cdot}62$$

$$\therefore \int_1^3 y\,\mathrm{d}x\approx 6{\cdot}62$$

33

Here is one further example.

**Example 4**

A pin moves along a straight guide so that its velocity $v$ (cm/s) when it is distance $x$ (cm) from the beginning of the guide at time $t$ (s) is as given in the table below:

| $t$ (s) | 0 | 0·5 | 1·0 | 1·5 | 2·0 | 2·5 | 3·0 | 3·5 | 4·0 |
|---|---|---|---|---|---|---|---|---|---|
| $v$ (cm/s) | 0 | 4·00 | 7·94 | 11·68 | 14·97 | 17·39 | 18·25 | 16·08 | 0 |

Apply Simpson's rule, using 8 intervals, to find the approximate total distance travelled by the pin between $t=0$ and $t=4$.

We must first interpret the problem, thus:

$$v=\frac{\mathrm{d}x}{\mathrm{d}t}\qquad \therefore x=\int_0^4 v\,\mathrm{d}t$$

and since we are given values of the function $v$ at regular intervals of $t$, and there is an even number of intervals, then we are all set to apply Simpson's rule.

Complete the problem then, entirely on your own.

*When you have finished it, check with Frame 34*

**34**

46·5 cm

| $t$ | $v$ | $F+L$ | $E$ | $R$ |
|---|---|---|---|---|
| 0 | 0·00 | 0·00 | | |
| 0·5 | 4·00 | | 4·00 | |
| 1·0 | 7·94 | | | 7·94 |
| 1·5 | 11·68 | | 11·68 | |
| 2·0 | 14·97 | | | 14·97 |
| 2·5 | 17·39 | | 17·39 | |
| 3·0 | 18·25 | | | 18·25 |
| 3·5 | 16·08 | | 16·08 | |
| 4·0 | 0·00 | 0·00 | | |
| | $F+L \longrightarrow$ | 0·00 | 49·15 | 41·16 |
| | $4E \longrightarrow$ | 196·60 | 4 | 2 |
| | $2R \longrightarrow$ | 82·32 | 196·60 | 82·32 |
| | $(F+L)+4E+2R \longrightarrow$ | 278·92 | | |

$$x = \frac{s}{3}\Big[(F+L)+4E+2R\Big] \text{ and } s = 0{\cdot}5$$

$$\therefore\ x = \frac{1}{6}\Big[278{\cdot}92\Big] = 46{\cdot}49 \quad \therefore \text{ Total distance} \approx 46{\cdot}5 \text{ cm}$$

## Proof of Simpson's rule

**35**

So far we have been using Simpson's rule, but we have not seen how it is established. You are not likely to be asked to prove it, but in case you are interested here is one proof.

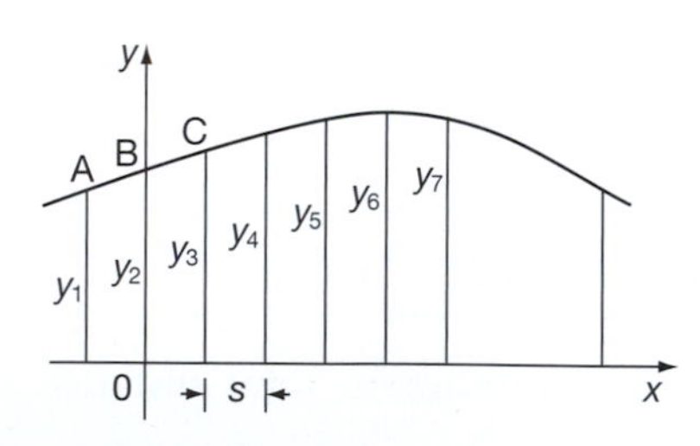


Divide into an even number of strips ($2n$) of equal widths ($s$). Let the ordinates be $y_1, y_2, y_3, y_{2n+1}$.

Take O$x$ and O$y$ as axes in the position shown.

Then A $= (-s,\ y_1)$
B $= (0,\ y_2)$
C $= (s,\ y_3)$

Let the curve through A, B, C be represented by $y = a + bx + cx^2$

$$y_1 = a + b(-s) + cs^2 \quad (1)$$
$$y_2 = a \quad (2)$$
$$y_3 = a + bs + cs^2 \quad (3)$$

(3) – (1) $\quad y_3 - y_1 = 2bs \quad \therefore\ b = \frac{1}{2s}(y_3 - y_1)$

(1) + (3) – 2 × (2) $\quad y_1 + y_3 - 2y_2 = 2cs^2 \quad \therefore\ c = \frac{1}{2s^2}(y_1 - 2y_2 + y_3)$

▶

Let $A_1$ = area of the first pair of strips, then:

$$A_1 = \int_{-s}^{s} y\,dx \approx \int_{-s}^{s} (a + bx + cx^2)\,dx \approx \left[ax + \frac{bx^2}{2} + \frac{cx^3}{3}\right]_{-s}^{s}$$

$$\approx 2as + \frac{2cs^3}{3} \approx 2sy_2 + \frac{2s^3}{3} \cdot \frac{1}{2s^2}(y_1 - 2y_2 + y_3)$$

$$\approx \frac{s}{3}(6y_2 + y_1 - 2y_2 + y_3) \approx \frac{s}{3}(y_1 + 4y_2 + y_3)$$

So $\quad A_1 \approx \frac{s}{3}(y_1 + 4y_2 + y_3)$

Similarly $\quad A_2 \approx \frac{s}{3}(y_3 + 4y_4 + y_5)$

$$A_3 \approx \frac{s}{3}(y_5 + 4y_6 + y_7)$$

............

$$A_n \approx \frac{s}{3}(y_{2n-1} + 4y_{2n} + y_{2n+1})$$

Total area $A = A_1 + A_2 + A_3 + \ldots + A_n$.

$$\therefore A \approx \frac{s}{3}\left[(y_1 + y_{2n+1}) + 4(y_2 + y_4 + \ldots + y_{2n}) + 2(y_3 + y_5 + \ldots + y_{2n-1})\right]$$

$$A \approx \frac{s}{3}\left[(F + L) + 4E + 2R\right]$$

*On to Frame 36*

**36**

We have almost reached the end of the Programme, except for the usual **Can You?** checklist and the **Test exercise** that await you. Before we turn to these, let us revise once again the requirements for applying Simpson's rule:

(a) The figure is divided into an *even* number of strips of equal width $s$. There will therefore be an *odd* number of ordinates or function values, including both boundary values.

(b) The value of the definite integral $\int_a^b f(x)\,dx$ is given by the numerical value of the area under the curve $y = f(x)$ between $x = a$ and $x = b$:

$$I = A \approx \frac{s}{3}\left[(F + L) + 4E + 2R\right]$$

where $s$ = width of strip (or interval)

$F + L$ = sum of the first and last ordinates

$4E = 4 \times$ sum of the even-numbered ordinates

$2R = 2 \times$ sum of remaining odd-numbered ordinates.

(c) A practical hint to finish with:

*Always* set your work out in the form of a table, as we have done in the examples. It prevents your making slips in method and calculation, and enables you to check without difficulty.

*On then to the* **Can You?** *checklist in Frame 37*

## Can You?

**37**

### Checklist 22

Check this list before and after you try the end of Programme test.

On a scale of 1 to 5 how confident are you that you can: Frames

- Recognize when an integral cannot be evaluated directly? 1 to 5
  *Yes* ☐ ☐ ☐ ☐ ☐ *No*
- Approximate integrals using series expansions? 6 to 15
  *Yes* ☐ ☐ ☐ ☐ ☐ *No*
- Use Simpson's rule to approximate the area beneath a curve? 16 to 36
  *Yes* ☐ ☐ ☐ ☐ ☐ *No*

## Test exercise 22

**38**

These problems are similar to those we have been considering in the Programme, so you will find them quite straightforward. Set the solutions out neatly. Take your time: it is very easy to make numerical slips in work of this kind.

1 Express $\sin x$ as a power series and hence evaluate $\displaystyle\int_0^1 \frac{\sin x}{x}\,\mathrm{d}x$ to 3 places of decimals.

2 Evaluate $\displaystyle\int_{0\cdot 1}^{0\cdot 2} x^{-1}e^{2x}\,\mathrm{d}x$ correct to 3 decimal places.

3 The values of a function $y = f(x)$ at stated values of $x$ are given below:

| $x$ | 2·0 | 2·5 | 3·0 | 3·5 | 4·0 | 4·5 | 5·0 | 5·5 | 6·0 |
|---|---|---|---|---|---|---|---|---|---|
| $y$ | 3·50 | 6·20 | 7·22 | 6·80 | 5·74 | 5·03 | 6·21 | 8·72 | 11·10 |

Using Simpson's rule, with 8 intervals, find an approximate value of $\displaystyle\int_2^6 y\,\mathrm{d}x$.

4 Evaluate $\displaystyle\int_0^{\pi/2} \sqrt{\cos\theta}\,\mathrm{d}\theta$, using 6 intervals.

5 Find an approximate value of $\displaystyle\int_0^{\pi/2} \sqrt{1 - 0.5\sin^2\theta}\,\mathrm{d}\theta$ using Simpson's rule with 6 intervals.

*Now you are ready for the next Programme*

## Further problems 22

39

1 Evaluate $\int_0^{\frac{1}{2}} \sqrt{1-x^2}\,dx$:

(a) by direct integration
(b) by expanding as a power series
(c) by Simpson's rule (8 intervals).

2 State the series for $\ln(1+x)$ and $\ln(1-x)$ and hence obtain a series for $\ln\left\{\frac{1+x}{1-x}\right\}$.

Evaluate $\int_0^{0\cdot3} \ln\left\{\frac{1+x}{1-x}\right\} dx$, correct to 3 decimal places.

3 In each of the following cases, apply Simpson's rule (6 intervals) to obtain an approximate value of the integral:

(a) $\int_0^{\pi/2} \frac{dx}{1+3\cos x}$

(b) $\int_0^{\pi} (5-4\cos\theta)^{\frac{1}{2}}\,d\theta$

(c) $\int_0^{\pi/2} \frac{d\theta}{\sqrt{1-\frac{1}{2}\sin^2\theta}}$

4 The coordinates of a point on a curve are given below:

| $x$ | 0 | 1 | 2 | 3 | 4 | 5 | 6 | 7 | 8 |
|---|---|---|---|---|---|---|---|---|---|
| $y$ | 4 | 5·9 | 7·0 | 6·4 | 4·8 | 3·4 | 2·5 | 1·7 | 1 |

The plane figure bounded by the curve, the $x$-axis and the ordinates at $x=0$ and $x=8$, rotates through a complete revolution about the $x$-axis. Use Simpson's rule (8 intervals) to obtain an approximate value of the volume generated.

5 The perimeter of an ellipse with parametric equations $x=3\cos\theta$, $y=2\sin\theta$, is $2\sqrt{2}\int_0^{\pi/2}(13-5\cos 2\theta)^{\frac{1}{2}}\,d\theta$. Evaluate this integral using Simpson's rule with 6 intervals.

6 Calculate the area bounded by the curve $y=e^{-x^2}$, the $x$-axis and the ordinates at $x=0$ and $x=1$. Use Simpson's rule with 6 intervals.

7 The voltage of a supply at regular intervals of 0·01 s, over a half-cycle, is found to be: 0, 19·5, 35, 45, 40·5, 25, 20·5, 29, 27, 12·5, 0. By Simpson's rule (10 intervals) find the rms value of the voltage over the half-cycle.

8 Show that the length of arc of the curve $x=3\theta-4\sin\theta$, $y=3-4\cos\theta$, between $\theta=0$ and $\theta=2\pi$, is given by the integral $\int_0^{2\pi}\sqrt{25-24\cos\theta}\,d\theta$. Evaluate the integral, using Simpson's rule with 8 intervals.

9 Obtain the first four terms of the expansion of $(1+x^3)^{\frac{1}{2}}$ and use them to determine the approximate value of $\int_0^{\frac{1}{2}}\sqrt{1+x^3}\,dx$, correct to three decimal places.

**10** Establish the integral in its simplest form representing the length of the curve $y = \frac{1}{2}\sin\theta$ between $\theta = 0$ and $\theta = \frac{\pi}{2}$. Apply Simpson's rule, using 6 intervals, to find an approximate value of this integral.

**11** Determine the first four non-zero terms of the series for $\tan^{-1} x$ and hence evaluate $\int_0^{\frac{1}{2}} \sqrt{x}.\tan^{-1} x\, dx$ correct to 3 decimal places.

**12** Evaluate, correct to three decimal places:

(a) $\int_0^1 \sqrt{x}.\cos x\, dx$ (b) $\int_0^1 \sqrt{x}.\sin x\, dx$.

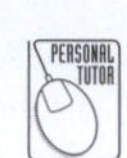

**13** Evaluate $\int_0^{\pi/2} \sqrt{2{\cdot}5 - 1{\cdot}5\cos 2\theta}\, d\theta$ by Simpson's rule, using 6 intervals.

**14** Determine the approximate value of $\int_0^1 (4 + x^4)^{\frac{1}{2}}\, dx$:

(a) by first expanding the expression in powers of $x$

(b) by applying Simpson's rule, using 4 intervals.

In each case, give the result to 2 places of decimals.

**40**

Now visit the companion website for this book at www.palgrave.com/stroud for more questions applying this mathematics to science and engineering.

# Polar coordinate systems

## Learning outcomes

When you have completed this Programme you will be able to:

- ☐ Convert expressions from Cartesian coordinates to polar coordinates and vice versa
- ☐ Plot the graphs of polar curves
- ☐ Recognize equations of standard polar curves
- ☐ Evaluate the areas enclosed by polar curves
- ☐ Evaluate the volumes of revolution generated by polar curves
- ☐ Evaluate the lengths of polar curves
- ☐ Evaluate the surface of revolution generated by polar curves

# Introduction to polar coordinates

**1**

We already know that there are two main ways in which the position of a point in a plane can be represented:

(a) by Cartesian coordinates, i.e. $(x, y)$

(b) by polar coordinates, i.e. $(r, \theta)$.

The relationship between the two systems can be seen from a diagram:

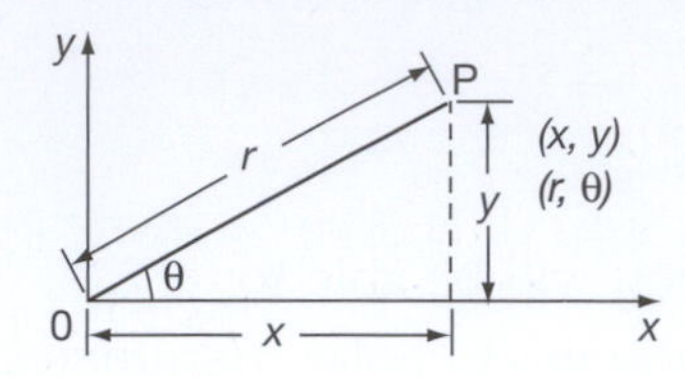


For instance, $x$ and $y$ can be expressed in terms of $r$ and $\theta$.

$x = \dots\dots\dots\dots;$
$y = \dots\dots\dots\dots$

**2**

$$x = r\cos\theta;\ y = r\sin\theta$$

Or, working in the reverse direction, the coordinates $r$ and $\theta$ can be found if we know the values of $x$ and $y$.

$r = \dots\dots\dots\dots;\quad \theta = \dots\dots\dots\dots$

**3**

$$r = \sqrt{x^2 + y^2};\quad \theta = \tan^{-1}\left(\frac{y}{x}\right)$$

This is just by way of review. We first met polar coordinates in an earlier Programme on complex numbers. In this Programme, we are going to direct a little more attention to the *polar coordinates system* and its applications.

First of all, some easy examples to warm up.

**Example 1**

Express in polar coordinates the position $(-5, 2)$.

Important hint: *always* draw a diagram; it will enable you to see which quadrant you are dealing with and prevent your making an initial slip.

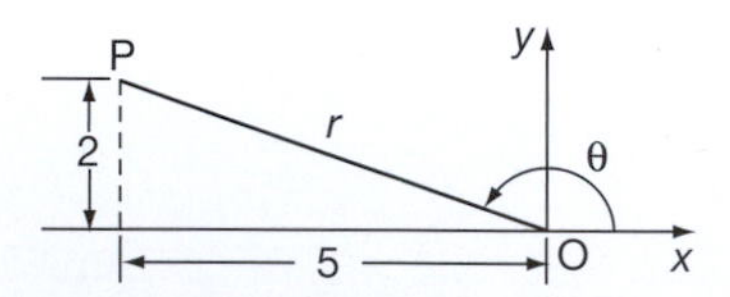


Remember that $\theta$ is measured from the positive O$x$ direction.

In this case, the polar coordinates of P are $\dots\dots\dots\dots$

**4**

$(5{\cdot}385, 158^\circ 12')$

Because

(a) $r^2 = 2^2 + 5^2 = 4 + 25 = 29$

$\therefore r = \sqrt{29} = 5{\cdot}385.$

(b) $\tan E = \dfrac{2}{5} = 0{\cdot}4$

$\therefore E = 21^\circ 48'$

$\therefore \theta = 158^\circ 12'$

Position of P is $(5{\cdot}385, 158^\circ 12')$

A sketch diagram will help you to check that $\theta$ is in the correct quadrant.

**Example 2**

Express $(4, -3)$ in polar coordinates. Draw a sketch and you cannot go wrong!

*When you are ready, move to Frame 5*

**5**

$5,\ 323^\circ 8'$

Here it is:

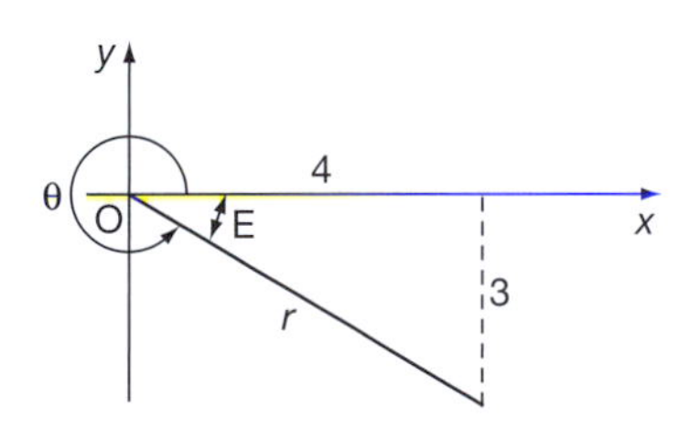


(a) $r^2 = 3^2 + 4^2 = 25 \quad \therefore r = 5$

(b) $\tan E = \dfrac{3}{4} = 0{\cdot}75 \quad \therefore E = 36^\circ 52'$

$\therefore \theta = 323^\circ 8'$

$(4, -3) = (5, 323^\circ 8')$

**Example 3**

Express in polar coordinates $(-2, -3)$.

*Finish it off and then move to Frame 6*

**6**

$3{\cdot}606,\ \ 236^\circ 19'$

Check your result:

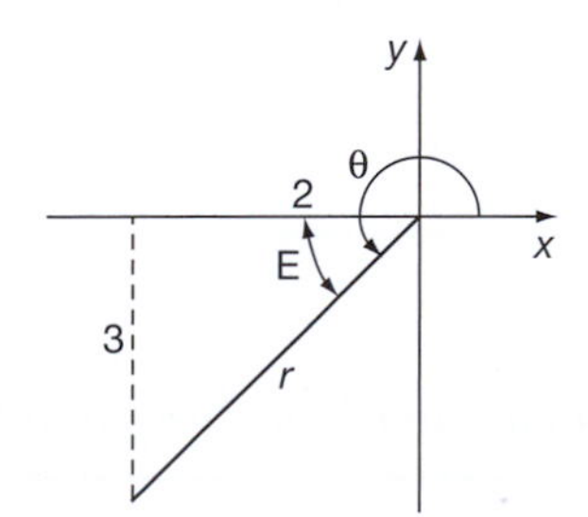


(a) $r^2 = 2^2 + 3^2 = 4 + 9 = 13$

$r = \sqrt{13} = 3{\cdot}606$

(b) $\tan E = \dfrac{3}{2} = 1{\cdot}5 \quad \therefore E = 56^\circ 19'$

$\therefore \theta = 236^\circ 19'$

$(-2, -3) = (3{\cdot}606, 236^\circ 19')$

Of course conversion in the opposite direction is just a matter of evaluating $x = r\cos\theta$ and $y = r\sin\theta$. Here is an example.

▶

**Example 4**

Express $(5, 124^\circ)$ in Cartesian coordinates.

*Do that, and then move on to Frame 7*

**7**

$(-2{\cdot}796,\ 4{\cdot}145)$

Working:

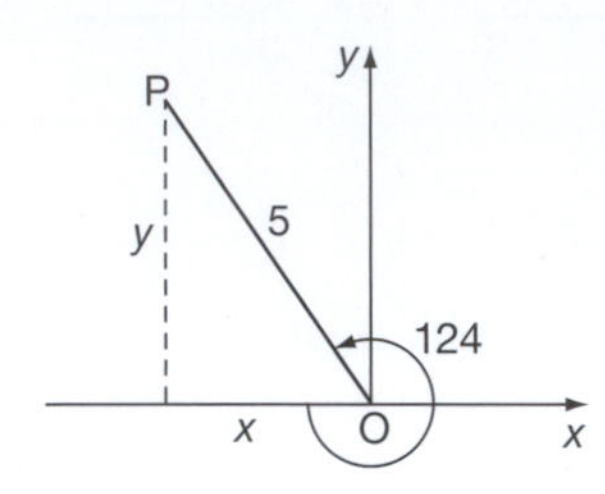


(a) $x = 5\cos 124^\circ = -5\cos 56^\circ$
$= -5(0{\cdot}5592) = -2{\cdot}7960$

(b) $y = 5\sin 124^\circ = 5\sin 56^\circ$
$= 5(0{\cdot}8290) = 4{\cdot}1450$

$\therefore\ (5, 124^\circ) = (-2{\cdot}796, 4{\cdot}145)$

That was all very easy.

*Now, on to the next frame*

## Polar curves

**8**

In Cartesian coordinates, the equation of a curve is given as the general relationship between $x$ and $y$, i.e. $y = f(x)$.

Similarly, in the polar coordinate system, the equation of a curve is given in the form $r = f(\theta)$. We can then take spot values for $\theta$, calculate the corresponding values of $r$, plot $r$ against $\theta$, and join the points up with a smooth curve to obtain the graph of $r = f(\theta)$.

**Example 1**

To plot the polar graph of $r = 2\sin\theta$ between $\theta = 0$ and $\theta = 2\pi$.

We take values of $\theta$ at convenient intervals and build up a table of values giving the corresponding values of $r$:

| $\theta$ | 0 | 30 | 60 | 90 | 120 | 150 | 180 |
|---|---|---|---|---|---|---|---|
| $\sin\theta$ | 0 | 0·5 | 0·866 | 1 | 0·866 | 0·5 | 0 |
| $r = 2\sin\theta$ | 0 | 1·0 | 1·732 | 2 | 1·732 | 1·0 | 0 |

| $\theta$ | 210 | 240 | 270 | 300 | 330 | 360 |
|---|---|---|---|---|---|---|
| $\sin\theta$ | | | | | | |
| $r = 2\sin\theta$ | | | | | | |

Complete the table, being careful of signs.

*When you have finished, move on to Frame 9*

**9**

Here is the complete table:

| $\theta$ | 0 | 30 | 60 | 90 | 120 | 150 | 180 |
|---|---|---|---|---|---|---|---|
| $\sin\theta$ | 0 | 0·5 | 0·866 | 1 | 0·866 | 0·5 | 0 |
| $r = 2\sin\theta$ | 0 | 1·0 | 1·732 | 2 | 1·732 | 1·0 | 0 |

| $\theta$ | 210 | 240 | 270 | 300 | 330 | 360 |
|---|---|---|---|---|---|---|
| $\sin\theta$ | −0·5 | −0·866 | −1 | −0·866 | −0·5 | 0 |
| $r = 2\sin\theta$ | −1·0 | −1·732 | −2 | −1·732 | −1·0 | 0 |

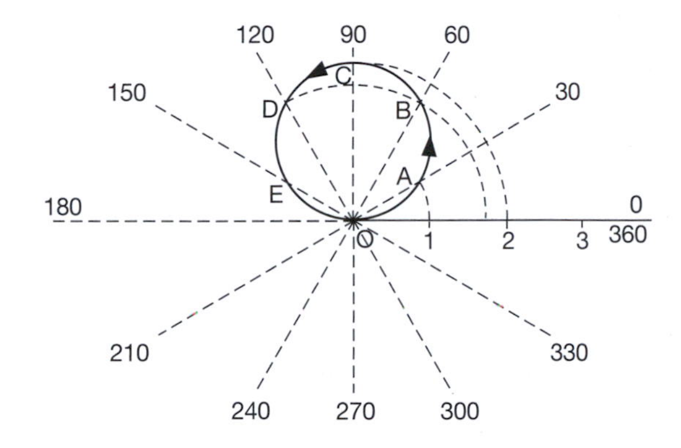


(a) We choose a linear scale for $r$ and indicate it along the initial line.

(b) The value of $r$ is then laid off along each direction in turn, points plotted, and finally joined up with a smooth curve. The resulting graph is as shown above.

*Note*: When we are dealing with the 210° direction, the value of $r$ is negative $(-1)$ and this distance is therefore laid off in the reverse direction which once again brings us to the point A. So for values of $\theta$ between $\theta = 180°$ and $\theta = 360°$, $r$ is negative and the first circle is retraced exactly. The graph, therefore, looks like one circle, but consists, in fact, of two circles, one on top of the other.

Now here is an example for you to do.

**Example 2**

In the same way, you can plot the graph of $r = 2\sin^2\theta$.

Compile a table of values at 30° intervals between $\theta = 0°$ and $\theta = 360°$ and proceed as we did above.

Take a little time over it.

*When you have finished, move on to Frame 10*

**10**

Here is the result in detail:

| $\theta$ | 0 | 30 | 60 | 90 | 120 | 150 | 180 |
|---|---|---|---|---|---|---|---|
| $\sin\theta$ | 0 | 0·5 | 0·866 | 1 | 0·866 | 0·5 | 0 |
| $\sin^2\theta$ | 0 | 0·25 | 0·75 | 1 | 0·75 | 0·25 | 0 |
| $r = 2\sin^2\theta$ | 0 | 0·5 | 1·5 | 2 | 1·5 | 0·5 | 0 |

| $\theta$ | 210 | 240 | 270 | 300 | 330 | 360 |
|---|---|---|---|---|---|---|
| $\sin\theta$ | −0·5 | −0·866 | −1 | −0·866 | −0·5 | 0 |
| $\sin^2\theta$ | 0·25 | 0·75 | 1 | 0·75 | 0·25 | 0 |
| $r = 2\sin^2\theta$ | 0·5 | 1·5 | 2 | 1·5 | 0·5 | 0 |

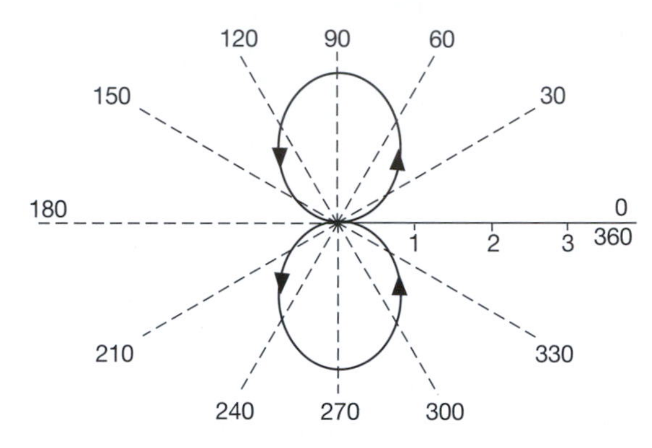


This time, $r$ is always positive and so there are, in fact, two distinct loops.

*Now on to the next frame*

## Standard polar curves

**11**

Polar curves can always be plotted from sample points as we have done above. However, it is often useful to know something of the shape of the curve without the rather tedious task of plotting points in detail.

In the next few frames, we will look at some of the more common polar curves.

*So on to Frame 12*

**12**

*Typical polar curves*

**1** $r = a\sin\theta$

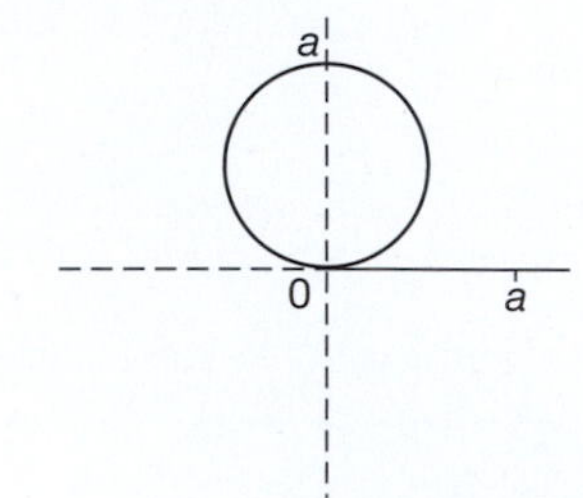


**2** $r = a\sin^2\theta$

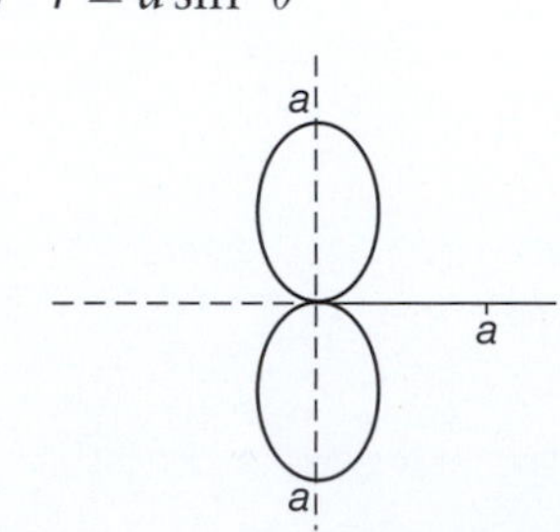


▶

**3** $r = a\cos\theta$

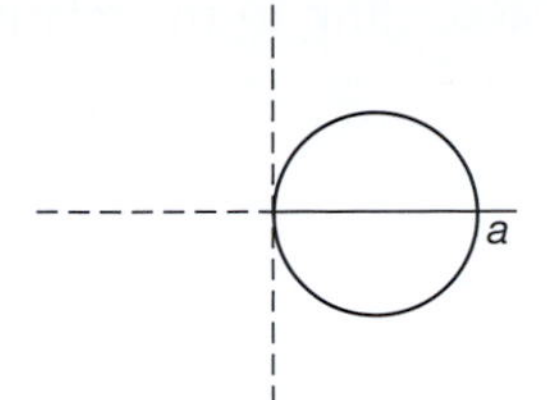


**4** $r = a\cos^2\theta$

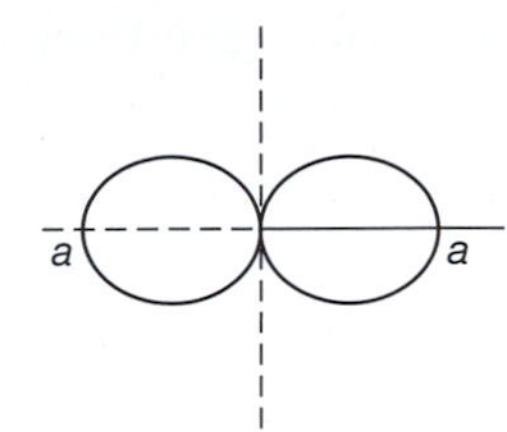


**5** $r = a\sin 2\theta$

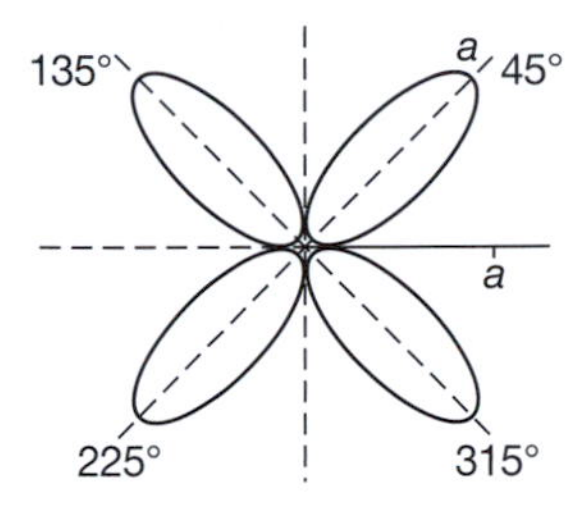


**6** $r = a\sin 3\theta$

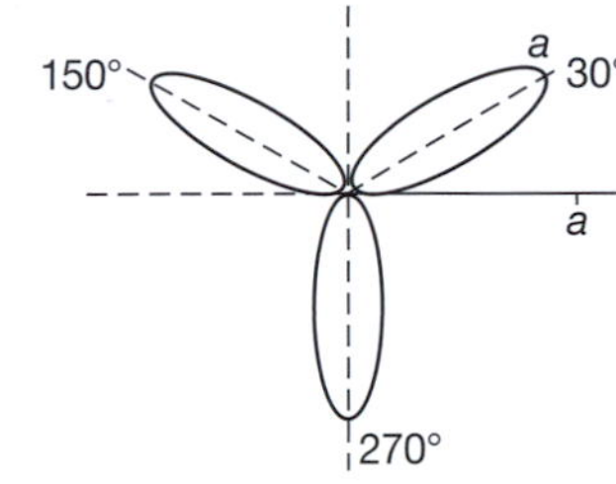


**7** $r = a\cos 2\theta$

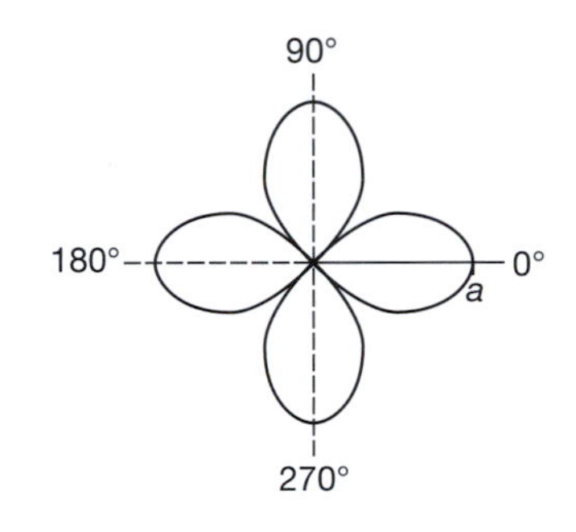


**8** $r = a\cos 3\theta$

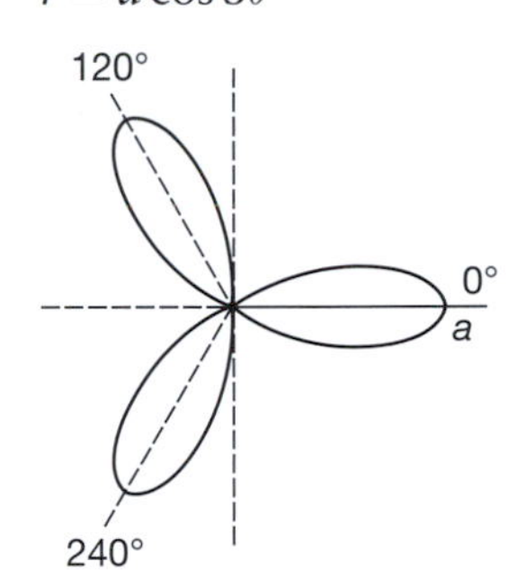


*There are some more interesting polar curves worth seeing, so move on to Frame 13*

---

**13**

**9** $r = a(1 + \cos\theta)$

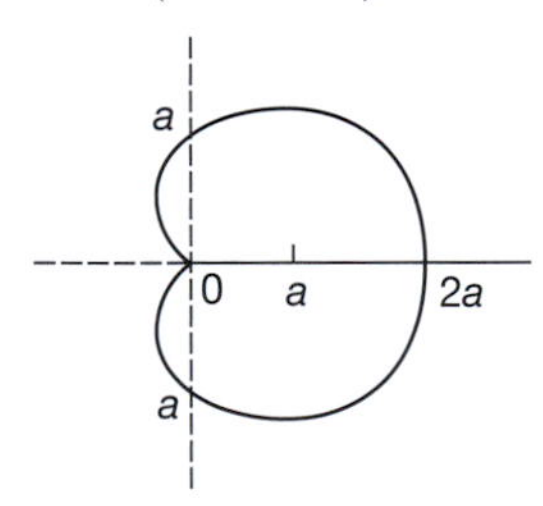


**10** $r = a(1 + 2\cos\theta)$

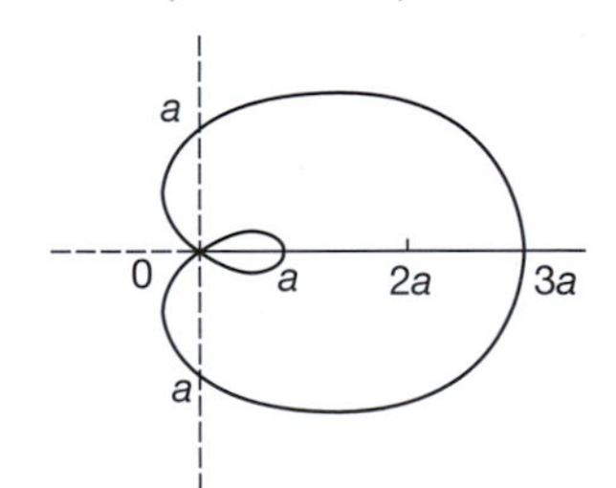


**11** $r^2 = a^2\cos 2\theta$

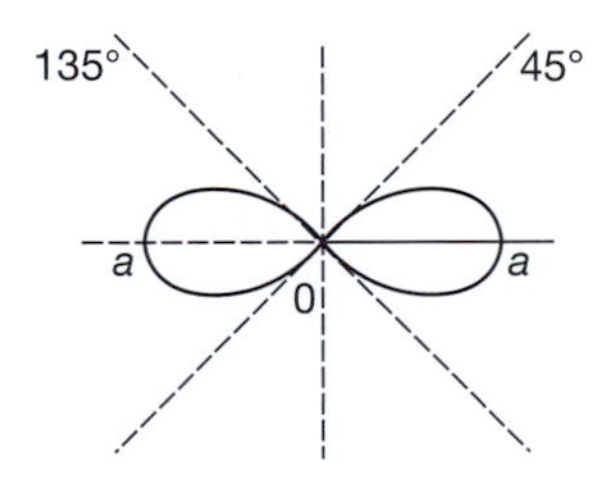


**12** $r = a\theta$

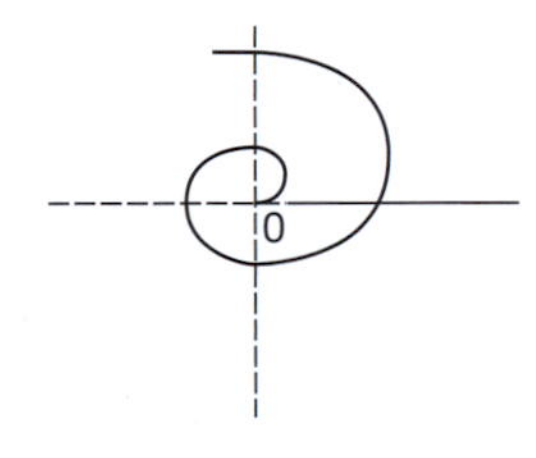


Sketch these 12 standard curves in your record book. They are quite common in use and worth remembering.

*Then on to the next frame*

**14**

The graphs of $r = a + b\cos\theta$ give three interesting results, according to the relative values of $a$ and $b$.

(a) If $a = b$, we get (cardioid)

(b) If $a < b$, we get (re-entrant loop)

(c) If $a > b$, we get (no cusp or re-entrant loop)

So sketch the graphs of the following. Do *not* compile tables of values.

(a) $r = 2 + 2\cos\theta$

(b) $r = 5 + 3\cos\theta$

(c) $r = 1 + 2\cos\theta$

(d) $r = 2 + \cos\theta$

**15**

Here they are. See how closely you agree.

(a) $r = 2 + 2\cos\theta$ $(a = b)$

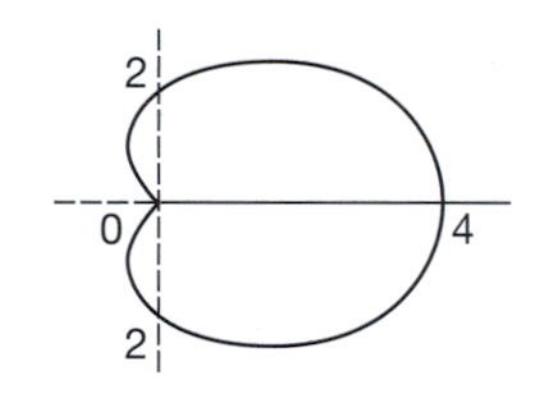


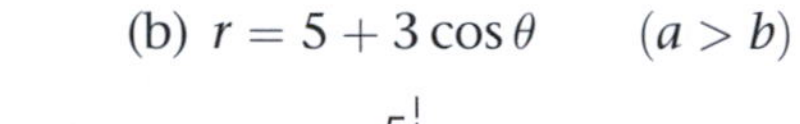

(b) $r = 5 + 3\cos\theta$ $(a > b)$

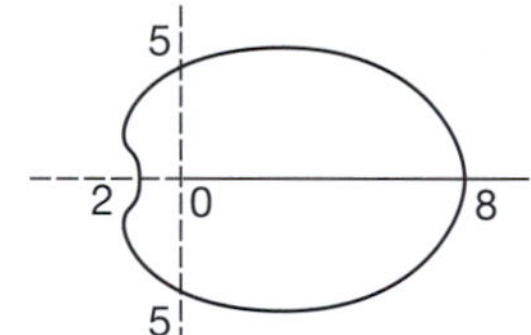


(c) $r = 1 + 2\cos\theta$ $(a < b)$

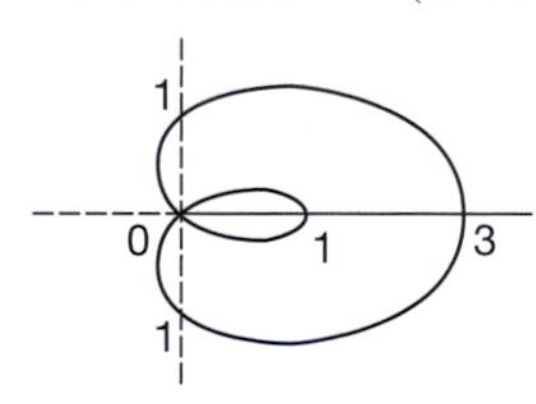


(d) $r = 2 + \cos\theta$ $(a > b)$

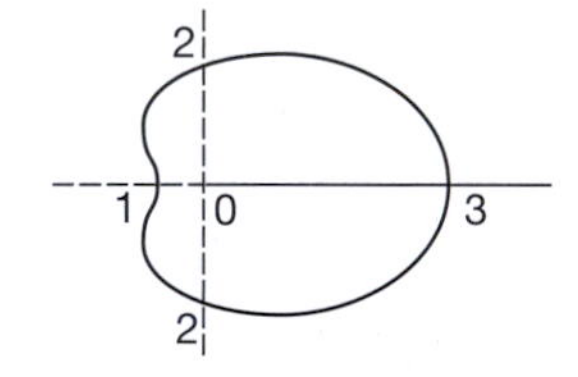


If you have slipped up with any of them, it would be worthwhile to plot a few points to confirm how the curve goes.

*On to Frame 16*

## Applications

16

**To find the area of the plane figure bounded by the polar curve $r = f(\theta)$ and the radius vectors at $\theta = \theta_1$ and $\theta = \theta_2$.**

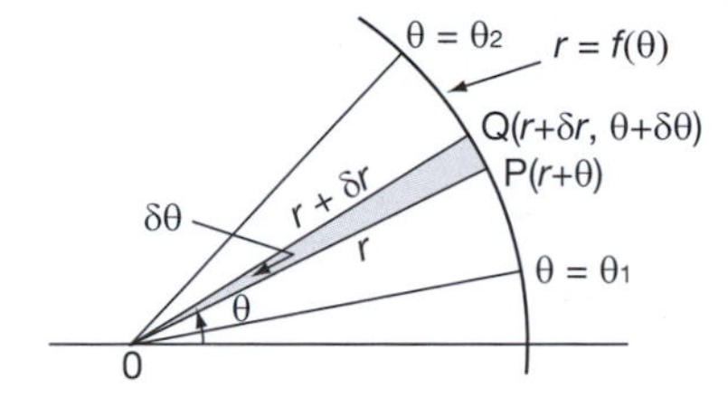


Area of sector OPQ $= \delta A \approx \frac{1}{2} r(r + \delta r)\sin\delta\theta$

$$\therefore \frac{\delta A}{\delta\theta} \approx \frac{1}{2} r(r + \delta r)\frac{\sin\delta\theta}{\delta\theta}$$

If $\delta\theta \to 0$, $\frac{\delta A}{\delta\theta} \to \frac{dA}{d\theta}$, $\delta r \to 0$, $\frac{\sin\delta\theta}{\delta\theta} \to$ ...........

*Next frame*

17

$$\boxed{\frac{\sin\delta\theta}{\delta\theta} \to 1}$$

$$\therefore \frac{dA}{d\theta} = \frac{1}{2} r(r + 0)1 = \frac{1}{2} r^2$$

$$\therefore A = \int_{\theta_1}^{\theta_2} \frac{1}{2} r^2 d\theta$$

**Example 1**

To find the area enclosed by the curve $r = 5\sin\theta$ and the radius vectors at $\theta = 0$ and $\theta = \pi/3$.

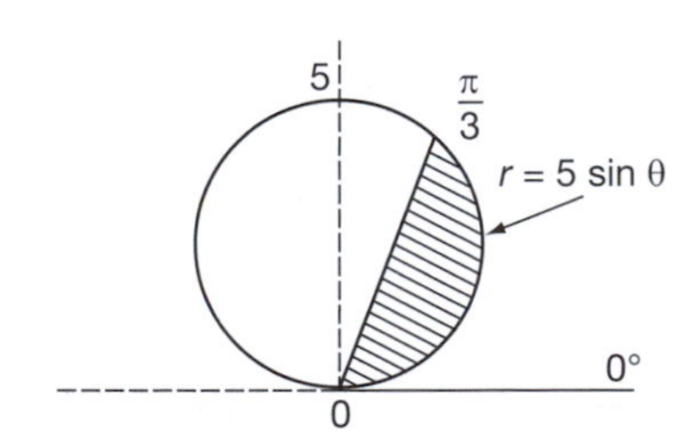


$$A = \int_0^{\pi/3} \frac{1}{2} r^2\, d\theta$$

$$A = \int_0^{\pi/3} \frac{25}{2}\sin^2\theta\, d\theta$$

$$\therefore A = \frac{25}{2}\int_0^{\pi/3} \frac{1}{2}(1 - \cos 2\theta)\, d\theta$$

$=$ ........... Finish it off.

## 18

$$A = \frac{25}{4}\left[\frac{\pi}{3} - \frac{\sqrt{3}}{4}\right] = 3{\cdot}84$$

Because

$$A = \frac{25}{4}\int_0^{\pi/3}(1 - \cos 2\theta)\mathrm{d}\theta = \frac{25}{4}\left[\theta - \frac{\sin 2\theta}{2}\right]_0^{\pi/3}$$

$$= \frac{25}{4}\left(\frac{\pi}{3} - \frac{\sin 2\pi/3}{2}\right)$$

$$= \frac{25}{4}\left(\frac{\pi}{3} - \frac{\sqrt{3}}{4}\right) = 3{\cdot}8386$$

$A = 3{\cdot}84$ to 2 decimal places

Now this one:

**Example 2**

Find the area enclosed by the curve $r = 1 + \cos\theta$ and the radius vectors at $\theta = 0$ and $\theta = \pi/2$.

First of all, what does the curve look like?

## 19

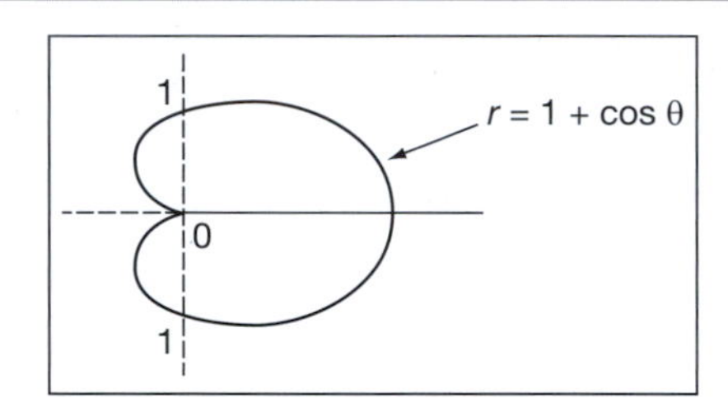


Right. So now calculate the value of $A$ between $\theta = 0$ and $\theta = \pi/2$.

*When you have finished, move on to Frame 20*

## 20

$$A = \frac{3\pi}{8} + 1 = 2{\cdot}178$$

Because

$$A = \frac{1}{2}\int_0^{\pi/2} r^2\mathrm{d}\theta = \frac{1}{2}\int_0^{\pi/2}(1 + 2\cos\theta + \cos^2\theta)\mathrm{d}\theta$$

$$= \frac{1}{2}\left[\theta + 2\sin\theta + \frac{\theta}{2} + \frac{\sin 2\theta}{4}\right]_0^{\pi/2}$$

$$= \frac{1}{2}\left\{\left(\frac{3\pi}{4} + 2 + 0\right) - \left(0\right)\right\}$$

$$\therefore A = \frac{3\pi}{8} + 1 = 2{\cdot}178$$

So the area of a polar sector is easy enough to obtain. It is simply

$$A = \int_{\theta_1}^{\theta_2}\frac{1}{2}r^2\mathrm{d}\theta$$

Make a note of this general result in your record book, if you have not already done so.

*Next frame*

**21**

**Example 3**

Find the total area enclosed by the curve $r = 2\cos 3\theta$. Notice that no limits are given, so we had better sketch the curve to see what is implied.

This was in fact one of the standard polar curves that we listed earlier in this Programme. Do you remember how it goes? If not, refer to your notes: it should be there.

*Then on to Frame 22*

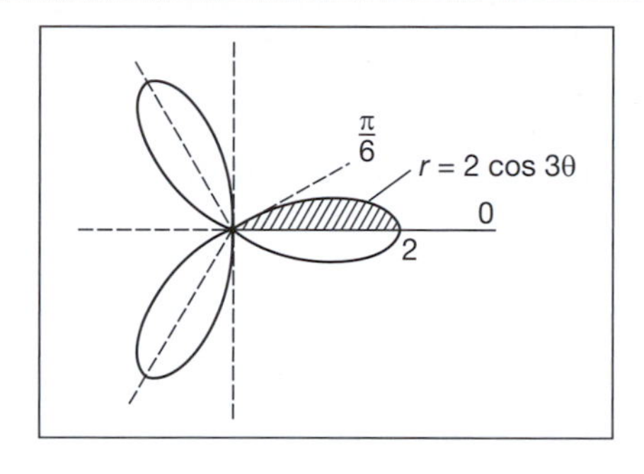


Since we are dealing with $r = 2\cos 3\theta$, $r$ will become zero when $\cos 3\theta = 0$, i.e. when $3\theta = \pi/2$, i.e. when $\theta = \pi/6$.

We see that the figure consists of 3 equal loops, so that the total area, $A$, is given by:

$$A = 3 \text{ (area of one loop)}$$

$$= 6 \text{ (area between } \theta = 0 \text{ and } \theta = \pi/6)$$

$$A = 6\int_0^{\pi/6} \frac{1}{2} r^2 \, d\theta$$

$$= 3\int_0^{\pi/6} 4\cos^2 3\theta \, d\theta$$

$$= \ldots\ldots\ldots\ldots$$

**23**

$$\boxed{\pi \text{ units}^2}$$

Because

$$A = 12\int_0^{\pi/6} \frac{1}{2}(1 + \cos 6\theta) \, d\theta$$

$$= 6\left[\theta + \frac{\sin 6\theta}{6}\right]_0^{\pi/6}$$

$$= \pi \text{ units}^2$$

Now here is one loop for you to do on your own.

**Example 4**

Find the area enclosed by one loop of the curve $r = a\sin 2\theta$.

First sketch the graph.

**24**

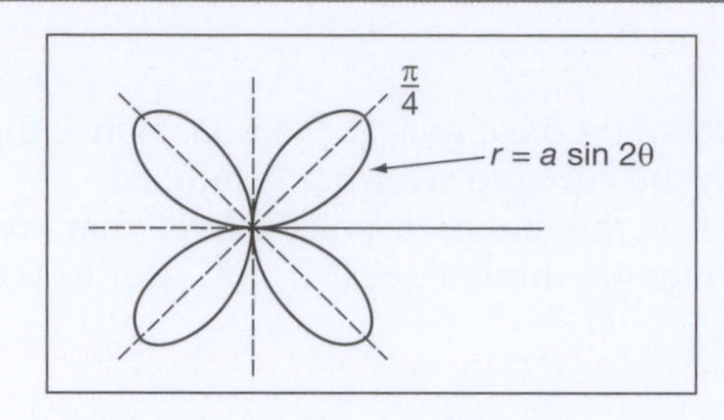


Arguing as before, $r = 0$ when $a \sin 2\theta = 0$, i.e. $\sin 2\theta = 0$, i.e. $2\theta = 0$, so that $2\theta = 0, \pi, 2\pi$, etc.

$\therefore \theta = 0, \pi/2, \pi$, etc.

So the integral denoting the area of the loop in the first quadrant will be

$A = \ldots\ldots\ldots\ldots$

**25**

$$A = \frac{1}{2}\int_0^{\pi/2} r^2 \, d\theta$$

Correct. Now go ahead and calculate the area.

**26**

$$A = \pi a^2/8 \text{ units}^2$$

Here is the working: check yours.

$$A = \frac{1}{2}\int_0^{\pi/2} r^2 \, d\theta = \frac{a^2}{2}\int_0^{\pi/2} \sin^2 2\theta \, d\theta$$

$$= \frac{a^2}{4}\int_0^{\pi/2} (1 - \cos 4\theta) \, d\theta$$

$$= \frac{a^2}{4}\left[\theta - \frac{\sin 4\theta}{4}\right]_0^{\pi/2}$$

$$= \frac{\pi a^2}{8} \text{ units}^2$$

*Now on to Frame 27*

**27**

**To find the volume generated when the plane figure bounded by $r = f(\theta)$ and the radius vectors at $\theta = \theta_1$ and $\theta = \theta_2$, rotates about the initial line.**

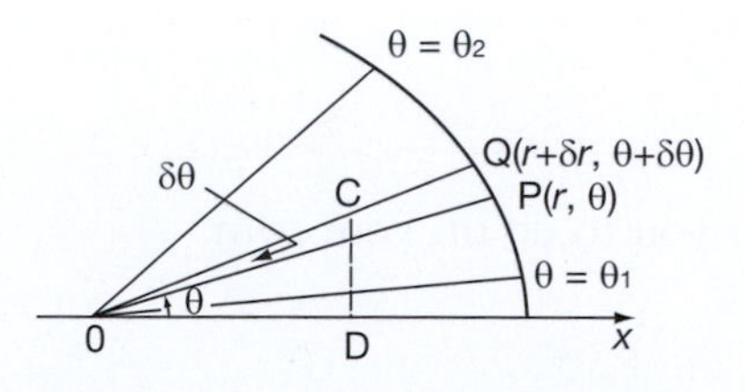


If we regard the elementary sector OPQ as approximately equal to the $\triangle$OPQ, then the centroid C is distance $\frac{2r}{3}$ from O.

▶

We have: Area OPQ $\approx \frac{1}{2}r(r+\delta r)\sin\delta\theta$

Volume generated when OPQ rotates about O$x = \delta V$

$\therefore\ \delta V =$ area OPQ $\times$ distance travelled by its centroid (Pappus)

$$= \frac{1}{2}r(r+\delta r)\sin\delta\theta.2\pi.\text{CD}$$

$$= \frac{1}{2}r(r+\delta r)\sin\delta\theta.2\pi.\frac{2}{3}r\sin\theta = \frac{2}{3}\pi r^2(r+\delta r)\sin\delta\theta.\sin\theta$$

$$\therefore\ \frac{\delta V}{\delta\theta} = \frac{2}{3}\pi r^2(r+\delta r)\frac{\sin\delta\theta}{\delta\theta}.\sin\theta$$

Then when $\delta\theta \to 0$, $\dfrac{dV}{d\theta} = \ldots\ldots\ldots\ldots$

---

**28**

$$\boxed{\frac{dV}{d\theta} = \frac{2}{3}\pi r^3\sin\theta}$$

and therefore $V = \ldots\ldots\ldots\ldots$

---

**29**

$$\boxed{V = \int_{\theta_1}^{\theta_2}\frac{2}{3}\pi r^3\sin\theta\,d\theta}$$

Correct. This is another standard result, so add it to your notes.

*Then move to the next frame for an example*

---

**30**

**Example 1**

Find the volume of the solid formed when the plane figure bounded by $r = 2\sin\theta$ and the radius vectors at $\theta = 0$ and $\theta = \pi/2$, rotates about the initial line.

Well now, $V = \displaystyle\int_0^{\pi/2}\frac{2}{3}\pi r^3\sin\theta\,d\theta$

$$= \int_0^{\pi/2}\frac{2}{3}.\pi.(2\sin\theta)^3.\sin\theta\,d\theta = \int_0^{\pi/2}\frac{16}{3}\pi\sin^4\theta\,d\theta$$

Since the limits are between 0 and $\pi/2$, we can use Wallis's formula for this. (Remember?)

So $V = \ldots\ldots\ldots\ldots$

---

**31**

$$\boxed{V = \pi^2\ \text{units}^3}$$

Because

$$V = \frac{16\pi}{3}\int_0^{\pi/2}\sin^4\theta\,d\theta$$

$$= \frac{16\pi}{3}.\frac{3.1}{4.2}.\frac{\pi}{2}$$

$$= \pi^2\ \text{units}^3$$

▶

**Example 2**

Find the volume of the solid formed when the plane figure bounded by $r = 2a\cos\theta$ and the radius vectors at $\theta = 0$ and $\theta = \pi/2$, rotates about the initial line.

Do that one entirely on your own.

*When you have finished it, move on to the next frame*

**32**

$$\boxed{V = \frac{4\pi a^3}{3} \text{ units}^3}$$

Because

$$V = \int_0^{\pi/2} \frac{2}{3}.\pi.r^3 \sin\theta \, d\theta \text{ and } r = 2a\cos\theta$$

$$= \int_0^{\pi/2} \frac{2}{3}.\pi.8a^3 \cos^3\theta . \sin\theta \, d\theta$$

$$= -\frac{16\pi a^3}{3} \int_0^{\pi/2} \cos^3\theta(-\sin\theta) \, d\theta$$

$$= -\frac{16\pi a^3}{3} \left[\frac{\cos^4\theta}{4}\right]_0^{\pi/2} = -\frac{16\pi a^3}{3}\left[-\frac{1}{4}\right]$$

$$V = \frac{4\pi a^3}{3} \text{ units}^3$$

So far then, we have had:

(a) $A = \int_{\theta_1}^{\theta_2} \frac{1}{2} r^2 \, d\theta$

(b) $V = \int_{\theta_1}^{\theta_2} \frac{2}{3} \pi r^3 \sin\theta \, d\theta$

Check that you have noted these results in your record book.

**33**

**To find the length of arc of the polar curve $r = f(\theta)$ between $\theta = \theta_1$ and $\theta = \theta_2$.**

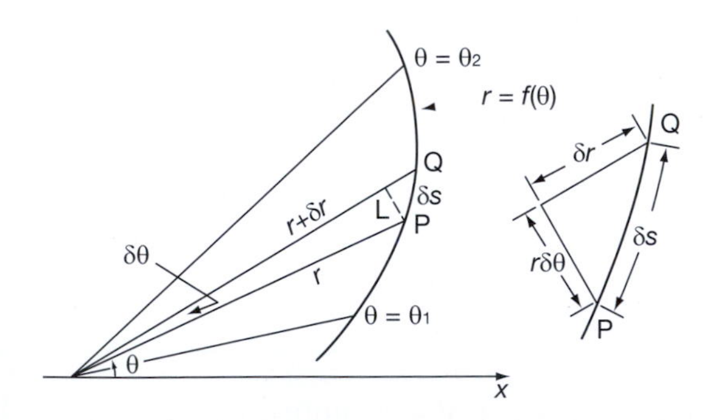


With the usual figure $\delta s^2 \approx r^2.\delta\theta^2 + \delta r^2 \quad \therefore \frac{\delta s^2}{\delta\theta^2} \approx r^2 + \frac{\delta r^2}{\delta\theta^2}$

If $\delta\theta \to 0$, $\left(\frac{ds}{d\theta}\right)^2 = r^2 + \left(\frac{dr}{d\theta}\right)^2 \quad \therefore \frac{ds}{d\theta} = \sqrt{r^2 + \left(\frac{dr}{d\theta}\right)^2}$

$\therefore s = \ldots\ldots\ldots\ldots$

34

$$s = \int_{\theta_1}^{\theta_2} \sqrt{r^2 + \left(\frac{dr}{d\theta}\right)^2}\, d\theta$$

**Example 1**

Find the length of arc of the spiral $r = ae^{3\theta}$ from $\theta = 0$ to $\theta = 2\pi$.

$$\text{Now, } r = ae^{3\theta} \quad \therefore \frac{dr}{d\theta} = 3ae^{3\theta}$$

$$\therefore r^2 + \left(\frac{dr}{d\theta}\right)^2 = a^2e^{6\theta} + 9a^2e^{6\theta} = 10a^2e^{6\theta}$$

$$\therefore s = \int_0^{2\pi} \sqrt{r^2 + \left(\frac{dr}{d\theta}\right)^2}\, d\theta$$

$$= \int_0^{2\pi} \sqrt{10}.ae^{3\theta}\, d\theta$$

$$= \ldots\ldots\ldots\ldots$$

35

$$s = \frac{a\sqrt{10}}{3}\{e^{6\pi} - 1\}$$

Because

$$\int_0^{2\pi} \sqrt{10}.a.e^{3\theta}\, d\theta = \frac{\sqrt{10}a}{3}\left[e^{3\theta}\right]_0^{2\pi}$$

$$= \frac{a\sqrt{10}}{3}\{e^{6\pi} - 1\}$$

As you can see, the method is very much the same every time. It is merely a question of substituting in the standard result, and, as usual, a knowledge of the shape of the polar curves is a very great help.

Here is our last result again:

$$s = \int_{\theta_1}^{\theta_2} \sqrt{r^2 + \left(\frac{dr}{d\theta}\right)^2}\, d\theta$$

Make a note of it: add it to the list.

36

Now here is an example for you to do.

**Example 2**

Find the length of the cardioid $r = a(1 + \cos\theta)$ between $\theta = 0$ and $\theta = \pi$.

*Finish it completely, and then check with the next frame*

**37**

$$\boxed{s = 4a \text{ units}}$$

Here is the working:

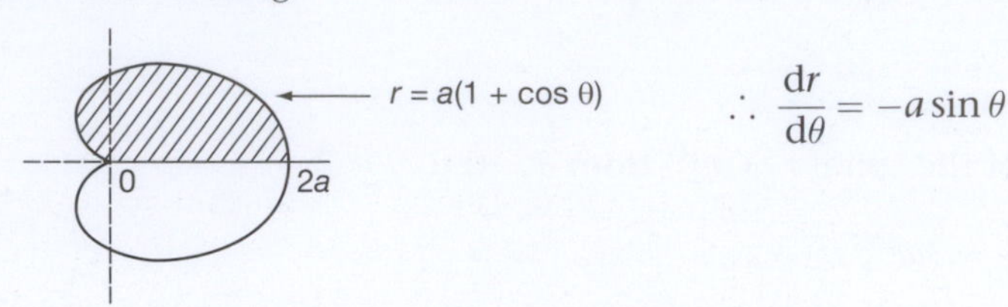


$$\therefore \frac{dr}{d\theta} = -a\sin\theta$$

$$\therefore r^2 + \left(\frac{dr}{d\theta}\right)^2 = a^2\{1 + 2\cos\theta + \cos^2\theta + \sin^2\theta\}$$

$$= a^2\{2 + 2\cos\theta\} = 2a^2(1 + \cos\theta)$$

Now $\cos\theta$ can be rewritten as $\left(2\cos^2\frac{\theta}{2} - 1\right)$

$$\therefore r^2 + \left(\frac{dr}{d\theta}\right)^2 = 2a^2.2\cos^2\frac{\theta}{2}$$

$$\therefore \sqrt{r^2 + \left(\frac{dr}{d\theta}\right)^2} = 2a\cos\frac{\theta}{2}$$

$$\therefore s = \int_0^{\pi} 2a\cos\frac{\theta}{2}\,d\theta = 2a\left[2\sin\frac{\theta}{2}\right]_0^{\pi}$$

$$= 4a(1 - 0) = 4a \text{ units}$$

*Next frame*

**38**

Let us pause for a moment and think back. So far we have established three useful results relating to polar curves. Without looking back in this Programme, or at your notes, complete the following.

If $r = f(\theta)$

(a) $A = \ldots\ldots\ldots\ldots$

(b) $V = \ldots\ldots\ldots\ldots$

(c) $s = \ldots\ldots\ldots\ldots$

*To see how well you have got on, move on to Frame 39*

**39**

(a) $A = \int_{\theta_1}^{\theta_2} \frac{1}{2} r^2\,d\theta$

(b) $V = \int_{\theta_1}^{\theta_2} \frac{2}{3}.\pi.r^3 \sin\theta\,d\theta$

(c) $s = \int_{\theta_1}^{\theta_2} \sqrt{r^2 + \left(\frac{dr}{d\theta}\right)^2}\,d\theta$

If you were uncertain of any of them, be sure to revise that particular result now. When you are ready, move on to the next section of the Programme.

40

Finally, we come to this topic.

**To find the area of the surface generated when the arc of the curve $r = f(\theta)$ between $\theta = \theta_1$ and $\theta = \theta_2$, rotates about the initial line.**

Once again, we refer to our usual figure:

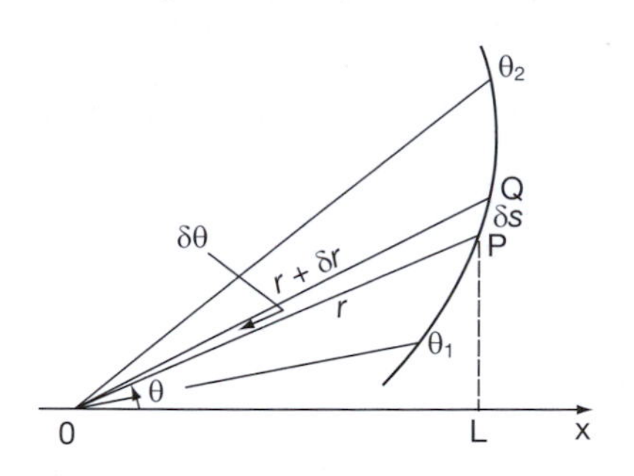


If the elementary arc PQ rotates about Ox, then, by the theorem of Pappus, the surface generated, $\delta S$, is given by (length of arc) $\times$ (distance travelled by its centroid).

$$\therefore \; \delta S \approx \delta s.2\pi.\text{PL} \approx \delta s.2\pi r \sin\theta$$

$$\therefore \; \frac{\delta S}{\delta\theta} \approx 2\pi r \sin\theta \frac{\delta s}{\delta\theta}$$

From our previous work, we know that $\dfrac{\delta s}{\delta\theta} \approx \sqrt{r^2 + \left(\dfrac{\delta r}{\delta\theta}\right)^2}$

$$\text{so that } \frac{\delta S}{\delta\theta} \approx 2\pi r \sin\theta \sqrt{r^2 + \left(\frac{\delta r}{\delta\theta}\right)^2}$$

$$\text{And now, if } \quad \delta\theta \to 0, \; \frac{\mathrm{d}S}{\mathrm{d}\theta} = 2\pi r \sin\theta \sqrt{r^2 + \left(\frac{\mathrm{d}r}{\mathrm{d}\theta}\right)^2}$$

$$\therefore \; S = \int_{\theta_1}^{\theta_2} 2\pi r \sin\theta \sqrt{r^2 + \left(\frac{\mathrm{d}r}{\mathrm{d}\theta}\right)^2} \, \mathrm{d}\theta$$

This is also an important result, so add it to your list.

41

$$\boxed{S = \int_{\theta_1}^{\theta_2} 2\pi r \sin\theta \sqrt{r^2 + \left(\frac{\mathrm{d}r}{\mathrm{d}\theta}\right)^2} \, \mathrm{d}\theta}$$

This looks a little more involved, but the method of attack is much the same. An example will show.

**Example 1**

Find the surface area generated when the arc of the curve $r = 5(1 + \cos\theta)$ between $\theta = 0$ and $\theta = \pi$, rotates completely about the initial line.

Now, $r = 5(1 + \cos\theta)$

$$\therefore \; \frac{\mathrm{d}r}{\mathrm{d}\theta} = -5\sin\theta$$

$$\therefore \; r^2 + \left(\frac{\mathrm{d}r}{\mathrm{d}\theta}\right)^2 = \dots\dots\dots\dots$$

**42**

$$\boxed{50(1+\cos\theta)}$$

Because

$$r^2+\left(\frac{dr}{d\theta}\right)^2 = 25(1+2\cos\theta+\cos^2\theta+\sin^2\theta)$$
$$= 25(2+2\cos\theta) = 50(1+\cos\theta)$$

We would like to express this as a square, since we have to take its root, so we now write $\cos\theta$ in terms of its half angle.

$$\therefore r^2+\left(\frac{dr}{d\theta}\right)^2 = 50\left(1+2\cos^2\frac{\theta}{2}-1\right)$$
$$= 100\cos^2\frac{\theta}{2}$$
$$\therefore \sqrt{r^2+\left(\frac{dr}{d\theta}\right)^2} = 10\cos\frac{\theta}{2}$$

So the formula in this case now becomes

$S = \ldots\ldots\ldots\ldots$

**43**

$$\boxed{S=\int_0^{\pi} 2\pi.5(1+\cos\theta)\sin\theta.10\cos\frac{\theta}{2}.\,d\theta}$$

$$\therefore S = 100\pi\int_0^{\pi}(1+\cos\theta)\sin\theta\cos\frac{\theta}{2}\,d\theta$$

We can make this more convenient if we express $(1+\cos\theta)$ and $\sin\theta$ also in terms of $\frac{\theta}{2}$.

What do we get?

**44**

$$\boxed{S=400\pi\int_0^{\pi}\cos^4\frac{\theta}{2}\sin\frac{\theta}{2}\,d\theta}$$

Because

$$S = 100\pi\int_0^{\pi}(1+\cos\theta)\sin\theta\cos\frac{\theta}{2}\,d\theta$$
$$= 100\pi\int_0^{\pi}2\cos^2\frac{\theta}{2}.2\sin\frac{\theta}{2}\cos\frac{\theta}{2}.\cos\frac{\theta}{2}\,d\theta$$
$$= 400\pi\int_0^{\pi}\cos^4\frac{\theta}{2}\sin\frac{\theta}{2}\,d\theta$$

Now the derivative of $\cos\frac{\theta}{2}$ is $\left\{-\frac{\sin\frac{\theta}{2}}{2}\right\}$

$$\therefore S = -800\pi\int_0^{\pi}\cos^4\frac{\theta}{2}\left\{-\frac{\sin\frac{\theta}{2}}{2}\right\}d\theta$$
$$= \ldots\ldots\ldots\ldots$$

Finish it off.

**45**

$$S = 160\pi \text{ units}^2$$

Because

$$S = -800\pi \int_0^{\pi} \cos^4 \frac{\theta}{2} \left\{ -\frac{\sin \frac{\theta}{2}}{2} \right\} d\theta$$

$$= -800\pi \left[ \frac{\cos^5 \frac{\theta}{2}}{5} \right]_0^{\pi} = \frac{-800\pi}{5}(0 - 1)$$

$$S = 160\pi \text{ units}^2$$

And finally, here is one for you to do.

**Example 2**

Find the area of the surface generated when the arc of the curve $r = ae^{\theta}$ between $\theta = 0$ and $\theta = \pi/2$, rotates about the initial line.

*Finish it completely and then check with the next frame*

**46**

$$S = \frac{2\sqrt{2}}{5}.\pi a^2(2e^{\pi} + 1)$$

Because we have:

$$S = \int_0^{\pi/2} 2\pi r \sin\theta \sqrt{r^2 + \left(\frac{dr}{d\theta}\right)^2}\, d\theta$$

And, in this case:

$$r = ae^{\theta} \quad \therefore \frac{dr}{d\theta} = ae^{\theta}$$

$$\therefore r^2 + \left(\frac{dr}{d\theta}\right)^2 = a^2e^{2\theta} + a^2e^{2\theta} = 2a^2e^{2\theta}$$

$$\therefore \sqrt{r^2 + \left(\frac{dr}{d\theta}\right)^2} = \sqrt{2}.a.e^{\theta}$$

$$\therefore S = \int_0^{\pi/2} 2\pi ae^{\theta} \sin\theta.\sqrt{2}ae^{\theta}\, d\theta$$

$$= 2\sqrt{2}\pi a^2 \int_0^{\pi/2} e^{2\theta} \sin\theta\, d\theta$$

Let $I = \int e^{2\theta} \sin\theta\, d\theta = e^{2\theta}(-\cos\theta) + 2\int \cos\theta\, e^{2\theta}\, d\theta$

$$= -e^{2\theta}\cos\theta + 2\left\{e^{2\theta}\sin\theta - 2\int \sin\theta\, e^{2\theta}\, d\theta\right\}$$

$$I = -e^{2\theta}\cos\theta + 2e^{2\theta}\sin\theta - 4I$$

$$\therefore 5I = e^{2\theta}\left\{2\sin\theta - \cos\theta\right\}$$

$$I = \frac{e^{2\theta}}{5}\left\{2\sin\theta - \cos\theta\right\}$$

▶

$$\therefore\ S = 2\sqrt{2}.\pi.a^2\left[\frac{e^{2\theta}}{5}\left\{2\sin\theta - \cos\theta\right\}\right]_0^{\pi/2}$$

$$= \frac{2\sqrt{2}.\pi.a^2}{5}\left\{e^{\pi}(2-0) - 1(0-1)\right\}$$

$$S = \frac{2\sqrt{2}.\pi.a^2}{5}(2e^{\pi}+1)\ \text{units}^2$$

We are almost at the end, but before we finish the Programme, let us collect our results together.

*So move to Frame 47*

## Review summary

**47**

*Polar curves* – applications.

1 *Area* $$A = \int_{\theta_1}^{\theta_2} \frac{1}{2} r^2\,d\theta$$

2 *Volume* $$V = \int_{\theta_1}^{\theta_2} \frac{2}{3}\pi r^3 \sin\theta\,d\theta$$

3 *Length of arc* $$s = \int_{\theta_1}^{\theta_2} \sqrt{r^2 + \left(\frac{dr}{d\theta}\right)^2}\,d\theta$$

4 *Surface of revolution* $$S = \int_{\theta_1}^{\theta_2} 2\pi r \sin\theta \sqrt{r^2 + \left(\frac{dr}{d\theta}\right)^2}\,d\theta$$

It is important to know these. The detailed working will depend on the particular form of the function $r = f(\theta)$, but as you have seen, the method of approach is mainly consistent.

The **Can You?** checklist and **Test exercise** now remain to be worked. Brush up any points on which you are not perfectly clear.

## Can You?

**Checklist 23** 48

Check this list before and after you try the end of Programme test.

On a scale of 1 to 5 how confident you are that you can: Frames

- Convert expressions from Cartesian coordinates to polar coordinates and vice versa? 1 to 7
  *Yes* ☐ ☐ ☐ ☐ ☐ *No*
- Plot the graphs of polar curves? 8 to 10
  *Yes* ☐ ☐ ☐ ☐ ☐ *No*
- Recognize equations of standard polar curves? 11 to 15
  *Yes* ☐ ☐ ☐ ☐ ☐ *No*
- Evaluate the areas enclosed by polar curves? 16 to 26
  *Yes* ☐ ☐ ☐ ☐ ☐ *No*
- Evaluate the volumes of revolution generated by polar curves? 27 to 32
  *Yes* ☐ ☐ ☐ ☐ ☐ *No*
- Evaluate the lengths of polar curves? 33 to 39
  *Yes* ☐ ☐ ☐ ☐ ☐ *No*
- Evaluate the surface of revolution generated by polar curves? 40 to 46
  *Yes* ☐ ☐ ☐ ☐ ☐ *No*

## Test exercise 23

49

All the questions are quite straightforward: there are no tricks. But take your time and work carefully.

1 Calculate the area enclosed by the curve $r\theta^2 = 4$ and the radius vectors at $\theta = \pi/2$ and $\theta = \pi$.

2 Sketch the polar curves:

(a) $r = 2\sin\theta$ (b) $r = 5\cos^2\theta$

(c) $r = \sin 2\theta$ (d) $r = 1 + \cos\theta$

(e) $r = 1 + 3\cos\theta$ (f) $r = 3 + \cos\theta$

3 The plane figure bounded by the curve $r = 2 + \cos\theta$ and the radius vectors at $\theta = 0$ and $\theta = \pi$, rotates about the initial line through a complete revolution. Determine the volume of the solid generated.

4 Find the length of the polar curve $r = 4\sin^2\dfrac{\theta}{2}$ between $\theta = 0$ and $\theta = \pi$.

5 Find the area of the surface generated when the arc of the curve $r = a(1 - \cos\theta)$ between $\theta = 0$ and $\theta = \pi$, rotates about the initial line.

*That completes the work on polar curves. You are now ready for the next Programme*

# Further problems 23

50

1 Sketch the curve $r = \cos^2\theta$. Find (a) the area of one loop and (b) the volume of the solid formed by rotating the curve about the initial line.

2 Show that $\sin^4\theta = \frac{3}{8} - \frac{1}{2}\cos 2\theta + \frac{1}{8}\cos 4\theta$. Hence find the area bounded by the curve $r = 4\sin^2\theta$ and the radius vectors at $\theta = 0$ and $\theta = \pi$.

3 Find the area of the plane figure enclosed by the curve $r = a\sec^2\left(\frac{\theta}{2}\right)$ and the radius vectors at $\theta = 0$ and $\theta = \pi/2$.

4 Determine the area bounded by the curve $r = 2\sin\theta + 3\cos\theta$ and the radius vectors at $\theta = 0$ and $\theta = \pi/2$.

5 Find the area enclosed by the curve $r = \dfrac{2}{1+\cos 2\theta}$ and the radius vectors at $\theta = 0$ and $\theta = \pi/4$.

6 Plot the graph of $r = 1 + 2\cos\theta$ at intervals of 30 and show that it consists of a small loop within a larger loop. The area between the two loops is rotated about the initial line through two right-angles. Find the volume generated.

7 Find the volume generated when the plane figure enclosed by the curve $r = 2a\sin^2\left(\frac{\theta}{2}\right)$ between $\theta = 0$ and $\theta = \pi$, rotates around the initial line.

8 The plane figure bounded by the cardioid $r = 2a(1+\cos\theta)$ and the parabola $r(1+\cos\theta) = 2a$, rotates around the initial line. Show that the volume generated is $18\pi a^3$.

9 Find the length of the arc of the curve $r = a\cos^3\left(\frac{\theta}{3}\right)$ between $\theta = 0$ and $\theta = 3\pi$.

10 Find the length of the arc of the curve $r = 3\sin\theta + 4\cos\theta$ between $\theta = 0$ and $\theta = \pi/2$.

11 Find the length of the spiral $r = a\theta$ between $\theta = 0$ and $\theta = 2\pi$.

12 Sketch the curve $r = a\sin^3\left(\frac{\theta}{3}\right)$ and calculate its total length.

13 Show that the length of arc of the curve $r = a\cos^2\theta$ between $\theta = 0$ and $\theta = \pi/2$ is $a[2\sqrt{3} + \ln(2+\sqrt{3})]/(2\sqrt{3})$.

14 Find the length of the spiral $r = ae^{b\theta}$ between $\theta = 0$ and $\theta = \theta_1$, and the area swept out by the radius vector between these two limits.

15 Find the area of the surface generated when the arc of the curve $r^2 = a^2\cos 2\theta$ between $\theta = 0$ and $\theta = \pi/4$, rotates about the initial line.

51

Now visit the companion website for this book at www.palgrave.com/stroud for more questions applying this mathematics to science and engineering.

# Multiple integrals

## Learning outcomes

When you have completed this Programme you will be able to:

- ☐ Determine the area of a rectangle using a double integral
- ☐ Evaluate double integrals over general areas
- ☐ Evaluate triple integrals over general volumes
- ☐ Apply double integrals to find areas and second moments of area
- ☐ Apply triple integrals to find volumes

## Summation in two directions

**1**

Let us consider the rectangle bounded by the straight lines, $x = r$, $x = s$, $y = k$ and $y = m$, as shown:

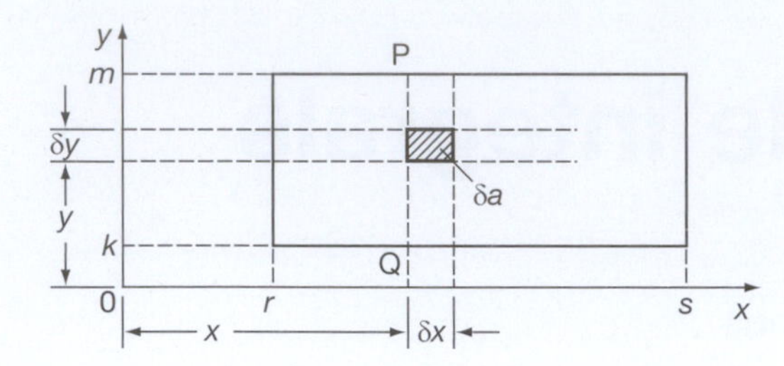


Then the area of the shaded element,

$\delta a = \ldots\ldots\ldots\ldots$

**2**

$$\boxed{\delta a = \delta y . \delta x}$$

If we add together all the elements of area, like $\delta a$, to form the vertical strip PQ, then $\delta A$, the area of the strip, can be expressed as

$\delta A = \ldots\ldots\ldots\ldots$

**3**

$$\boxed{\delta A = \sum_{y=k}^{y=m} \delta y . \delta x}$$

Did you remember to include the limits?

Note that during this summation in the $y$-direction, $\delta x$ is constant.

If we now sum all the strips across the figure from $x = r$ to $x = s$, we shall obtain the total area of the rectangle, $A$.

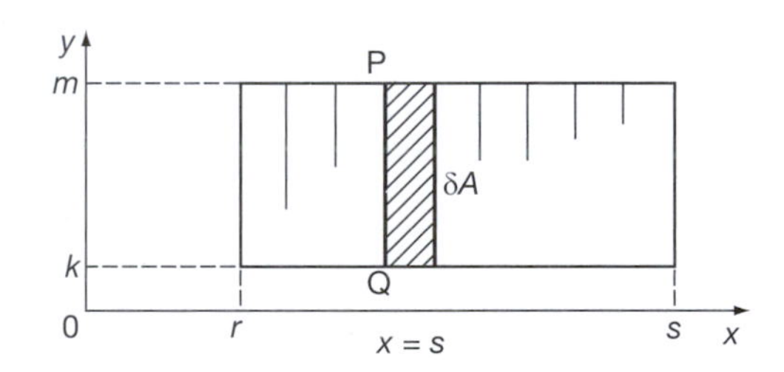


$$\therefore A = \sum_{x=r}^{x=s} (\text{all vertical strips like PQ})$$

$$= \sum_{x=r}^{x=s} \left\{ \sum_{y=k}^{y=m} \delta y . \delta x \right\}$$

Removing the brackets, this becomes:

$$A = \sum_{x=r}^{x=s} \sum_{y=k}^{y=m} \delta y . \delta x$$

If now $\delta y \to 0$ and $\delta x \to 0$, the finite summations become integrals, so the expression becomes

$A = \ldots\ldots\ldots\ldots$

4

$$A = \int_{x=r}^{x=s} \int_{y=k}^{y=m} \mathrm{d}y.\mathrm{d}x$$

To evaluate this expression, we start from the inside and work outwards.

$$A = \int_{x=r}^{x=s} \left[ \int_{y=k}^{y=m} \mathrm{d}y \right] \mathrm{d}x$$

$$= \int_{x=r}^{x=s} \Big[ y \Big]_{y=k}^{y=m} \mathrm{d}x$$

$$= \int_{x=r}^{x=s} (m-k)\mathrm{d}x$$

and since $m$ and $k$ are constants, this gives $A = \ldots\ldots\ldots\ldots$

5

$$A = (m-k).(s-r)$$

Because

$$A = \Big[ (m-k)x \Big]_{x=r}^{x=s}$$

$$= (m-k)\Big[ x \Big]_{x=r}^{x=s}$$

$$A = (m-k).(s-r)$$

which we know is correct, for it is merely $A =$ length $\times$ breadth.

That may seem a tedious way to find the area of a rectangle, but we have done it to introduce the method we are going to use.

First we define an element of area $\delta y.\delta x$.

Then we sum in the $y$-direction to obtain the area of a $\ldots\ldots\ldots\ldots$

Finally, we sum the result in the $x$-direction to obtain the area of the $\ldots\ldots\ldots\ldots$

6

vertical strip; whole figure

We could have worked slightly differently:

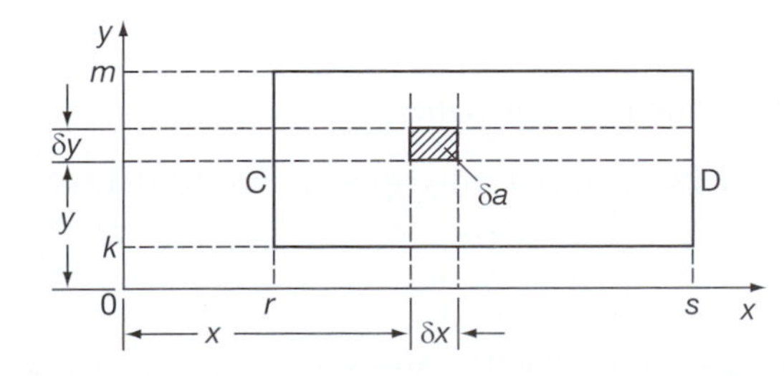


As before $\delta a = \delta x.\delta y$. If we sum the elements in the $x$-direction this time, we get the area $\delta A_1$ of the horizontal strip, CD.

$\therefore \delta A_1 = \ldots\ldots\ldots\ldots$

**7**

$$\delta A_1 = \sum_{x=r}^{x=s} \delta x.\delta y$$

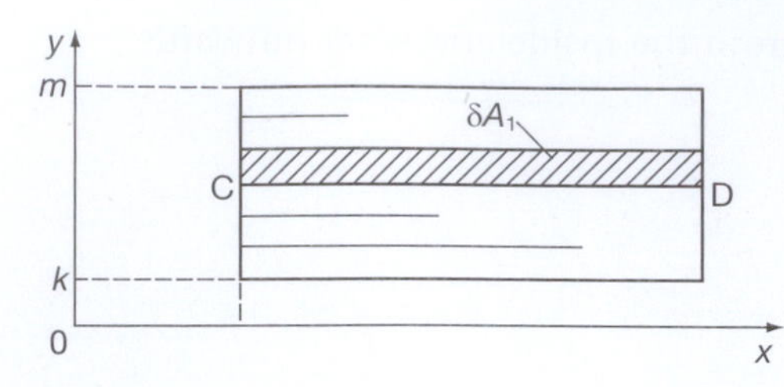


Now sum the strips vertically and we obtain once again the area of the whole rectangle.

$$A_1 = \sum_{y=k}^{y=m} \text{(all horizontal strips like CD)} = \sum_{y=k}^{y=m} \left\{ \sum_{x=r}^{x=s} \delta x.\delta y \right\}$$

As before, if we now remove the brackets and consider what this becomes when $\delta x \to 0$ and $\delta y \to 0$, we get:

$A_1 = \ldots\ldots\ldots\ldots$

**8**

$$A_1 = \int_{y=k}^{y=m} \int_{x=r}^{x=s} \mathrm{d}x\,\mathrm{d}y$$

To evaluate this we start from the centre:

$$A_1 = \int_{y=k}^{y=m} \left[ \int_{x=r}^{x=s} \mathrm{d}x \right] \mathrm{d}y$$

$= \ldots\ldots\ldots\ldots$

*Complete the working to find $A_1$ and then move on to Frame 9*

**9**

$$A_1 = (s - r).(m - k)$$

Because

$$A_1 = \int_{y=k}^{y=m} \left[ x \right]_r^s \mathrm{d}y$$

$$= \int_k^m (s - r)\mathrm{d}y$$

$$= (s - r)\left[ y \right]_k^m$$

$\therefore A_1 = (s - r).(m - k)$ which is the same result as before.

So the order in which we carry out our two summations appears not to matter.

*Remember:*

(a) We work from the inside integral.

(b) We integrate with respect to $x$ when the limits are values of $x$.

(c) We integrate with respect to $y$ when the limits are values of $y$.

*Move to the next frame*

## Double integrals

**10**

The expression $\int_{y_1}^{y_2}\int_{x_1}^{x_2} f(x,y)\,dx\,dy$ is called a *double integral* (for obvious reasons!) and indicates that:

(a) $f(x,y)$ is first integrated with respect to $x$ (regarding $y$ as being constant) between the limits $x = x_1$ and $x = x_2$

(b) the result is then integrated with respect to $y$ between the limits $y = y_1$ and $y = y_2$.

**Example 1**

Evaluate $\quad I = \int_1^2 \int_2^4 (x+2y)\,dx\,dy$

So $(x+2y)$ is first integrated with respect to $x$ between $x = 2$ and $x = 4$, with $y$ regarded as constant for the time being.

$$I = \int_1^2 \left[ \int_2^4 (x+2y)\,dx \right] dy$$

$$= \int_1^2 \left[ \frac{x^2}{2} + 2xy \right]_2^4 dy$$

$$= \int_1^2 \left\{ (8+8y) - (2+4y) \right\} dy$$

$$= \int_1^2 (6+4y)dy = \ldots\ldots\ldots\ldots$$

*Finish it off*

**11**

$$\boxed{I = 12}$$

Because

$$I = \int_1^2 (6+4y)dy$$

$$= \left[ 6y + 2y^2 \right]_1^2$$

$$= (12+8) - (6+2) = 20 - 8$$

$$= 12$$

Here is another.

**Example 2**

Evaluate $\quad I = \int_1^2 \int_0^3 x^2 y\,dx\,dy$

Do this one on your own. Remember to start with $\int_0^3 x^2 y\,dx$ with $y$ constant.

*Finish the double integral completely and then move on to Frame 12*

**12**

$$\boxed{I = 13{\cdot}5}$$

Check your working:

$$I = \int_1^2 \int_0^3 x^2 y \, dx \, dy$$

$$= \int_1^2 \left[ \int_0^3 x^2 y \, dx \right] dy$$

$$= \int_1^2 \left[ \frac{x^3}{3} . y \right]_{x=0}^{x=3} dy$$

$$= \int_1^2 (9y) dy = \left[ \frac{9y^2}{2} \right]_1^2$$

$$= 18 - 4{\cdot}5$$

$$= 13{\cdot}5$$

Now do this one in just the same way.

**Example 3**

Evaluate $\quad I = \int_1^2 \int_0^\pi (3 + \sin\theta) \, d\theta \, dr$

*When you have finished, check with the next frame*

**13**

$$\boxed{I = 3\pi + 2}$$

Here it is:

$$I = \int_1^2 \int_0^\pi (3 + \sin\theta) \, d\theta \, dr$$

$$= \int_1^2 \left[ 3\theta - \cos\theta \right]_0^\pi dr$$

$$= \int_1^2 \left\{ (3\pi + 1) - (-1) \right\} dr$$

$$= \int_1^2 (3\pi + 2) \, dr$$

$$= \left[ (3\pi + 2) r \right]_1^2$$

$$= (3\pi + 2)(2 - 1)$$

$$= 3\pi + 2$$

*On to the next frame*

## Triple integrals

14

Sometimes we have to deal with expressions such as

$$I = \int_a^b \int_c^d \int_e^f f(x, y, z)\,\mathrm{d}x\,\mathrm{d}y\,\mathrm{d}z$$

but the rules are as before. Start with the innermost integral and work outwards.

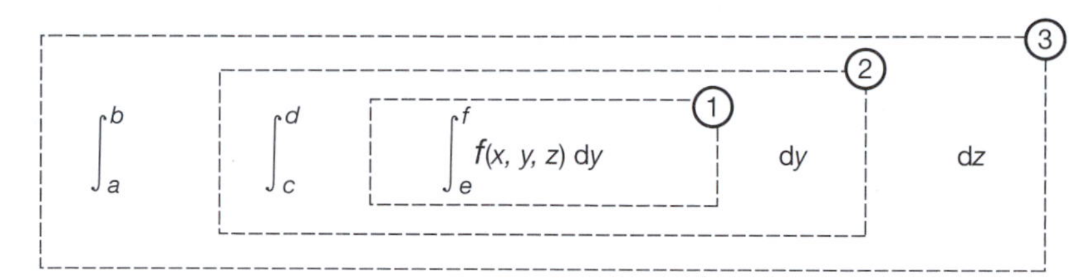


All symbols are regarded as constant for the time being, except the one variable with respect to which stage of integration is taking place. So try this one on your own straight away.

**Example 1**

Evaluate $I = \int_1^3 \int_{-1}^1 \int_0^2 (x + 2y - z)\,\mathrm{d}x\,\mathrm{d}y\,\mathrm{d}z$

15

$$\boxed{I = -8}$$

Did you manage it first time? Here is the working in detail.

$$I = \int_1^3 \int_{-1}^1 \int_0^2 (x + 2y - z)\,\mathrm{d}x\,\mathrm{d}y\,\mathrm{d}z$$

$$= \int_1^3 \int_{-1}^1 \left[\frac{x^2}{2} + 2xy - xz\right]_0^2 \mathrm{d}y\,\mathrm{d}z$$

$$= \int_1^3 \int_{-1}^1 (2 + 4y - 2z)\,\mathrm{d}y\,\mathrm{d}z$$

$$= \int_1^3 [2y + 2y^2 - 2yz]_{-1}^1\,\mathrm{d}z$$

$$= \int_1^3 \Big\{(2 + 2 - 2z) - (-2 + 2 + 2z)\Big\}\mathrm{d}z$$

$$= \int_1^3 (4 - 4z)\,\mathrm{d}z$$

$$= \Big[4z - 2z^2\Big]_1^3 = (12 - 18) - (4 - 2) = -8$$

And another.

**Example 2**

Evaluate $\int_1^2 \int_0^3 \int_0^1 (p^2 + q^2 - r^2)\,\mathrm{d}p\,\mathrm{d}q\,\mathrm{d}r$

*When you have finished it, move on to Frame 16*

**16**

$$I = 3$$

Because

$$
\begin{aligned}
I &= \int_1^2 \int_0^3 \int_0^1 (p^2 + q^2 - r^2)\, dp\, dq\, dr \\
&= \int_1^2 \int_0^3 \left[\frac{p^3}{3} + pq^2 - pr^2\right]_0^1 dq\, dr \\
&= \int_1^2 \int_0^3 \left\{\frac{1}{3} + q^2 - r^2\right\} dq\, dr \\
&= \int_1^2 \left[\frac{q}{3} + \frac{q^3}{3} - qr^2\right]_0^3 dr \\
&= \int_1^2 (1 + 9 - 3r^2) dr \\
&= \left[10r - r^3\right]_1^2 = (20 - 8) - (10 - 1) \\
&= 12 - 9 = 3
\end{aligned}
$$

It is all very easy if you take it steadily, step by step.

Now two quickies for revision.

Evaluate:

(a) $\int_1^2 \int_3^5 dy\, dx$

(b) $\int_0^4 \int_1^{3x} 2y\, dy\, dx$

*Finish them both and then move on to the next frame*

**17**

(a) $I = 2$ (b) $I = 188$

Here they are:

(a) $I = \int_1^2 \int_3^5 dy\, dx = \int_1^2 \left[y\right]_3^5 dx = \int_1^2 (5 - 3) dx = \int_1^2 2\, dx = \left[2x\right]_1^2$

$= 4 - 2 = 2$

(b) $I = \int_0^4 \int_1^{3x} 2y\, dy\, dx = \int_0^4 \left[y^2\right]_1^{3x} dx = \int_0^4 (9x^2 - 1) dx$

$= \left[3x^3 - x\right]_0^4 = 192 - 4 = 188$

And finally, do this one.

$I = \int_0^5 \int_1^2 (3x^2 - 4)\, dx\, dy = \dots\dots\dots\dots$

18

$$I = 15$$

Check this working.

$$I = \int_0^5 \int_1^2 (3x^2 - 4)\,dx\,dy$$

$$= \int_0^5 \Big[x^3 - 4x\Big]_1^2 dy$$

$$= \int_0^5 \Big\{(8 - 8) - (1 - 4)\Big\} dy$$

$$= \int_0^5 3\,dy = \Big[3y\Big]_0^5 = 15$$

Now let us see a few applications of multiple integrals.

*Move on then to the next frame*

## Applications

19

### Example 1

Find the area bounded by $y = \frac{4x}{5}$, the $x$-axis and the ordinate at $x = 5$.

Area of element $= \delta y.\delta x$

$\therefore$ Area of strip $\sum_{y=0}^{y=y_1} \delta y.\delta x$

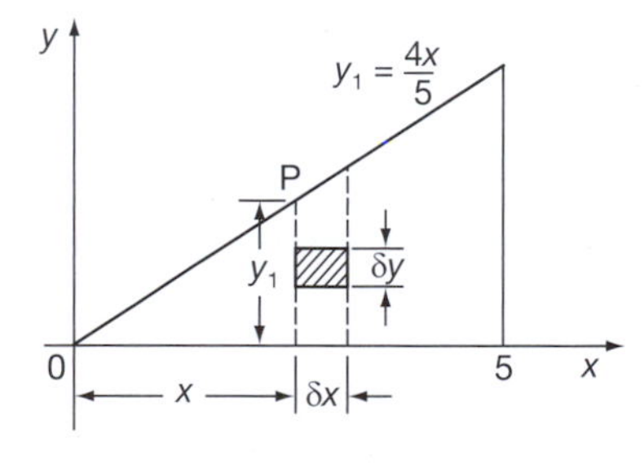


The sum of all such strips across the figure gives us:

$$A \approx \sum_{x=0}^{x=5} \left\{ \sum_{y=0}^{y=y_1} \delta y.\delta x \right\}$$

$$\approx \sum_{x=0}^{x=5} \sum_{y=0}^{y=y_1} \delta y.\delta x$$

Now, if $\delta y \to 0$ and $\delta x \to 0$, then:

$$A = \int_0^5 \int_0^{y_1} dy\,dx$$

$$= \int_0^5 \Big[y\Big]_0^{y_1} dx$$

$$= \int_0^5 y_1\,dx$$

But $y_1 = \frac{4x}{5}$

So $A = \ldots\ldots\ldots\ldots$

*Finish it off*

**20**

$$A = 10 \text{ unit}^2$$

Because

$$A = \int_0^5 \frac{4x}{5}\,dx = \left[\frac{2x^2}{5}\right]_0^5 = 10$$

Right. Now what about this one?

**Example 2**

Find the area under the curve $y = 4\sin\frac{x}{2}$ between $x = \frac{\pi}{3}$ and $x = \pi$, by the double integral method.

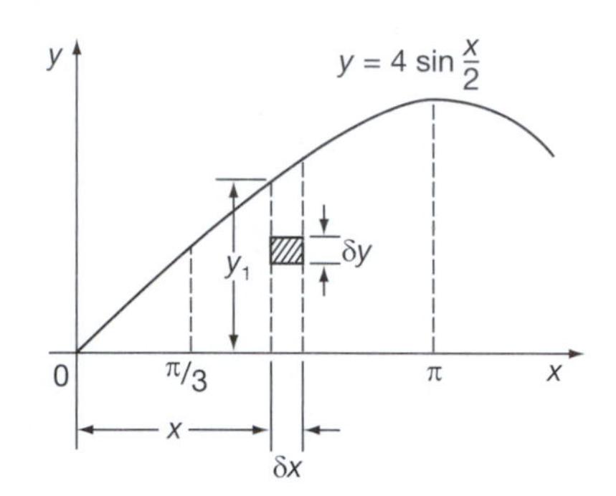


Steps as before:

Area of element $= \delta y.\delta x$

Area of vertical strip

$$\sum_{y=0}^{y=y_1} \delta y.\delta x$$

Total area of figure:

$$A \approx \sum_{x=\pi/3}^{x=\pi} \left\{ \sum_{y=0}^{y=y_1} \delta y.\delta x \right\}$$

If $\delta y \to 0$ and $\delta x \to 0$, then:

$$A = \int_{\pi/3}^{\pi} \int_0^{y_1} dy\,dx = \ldots\ldots\ldots\ldots$$

Complete it, remembering that $y_1 = 4\sin\frac{x}{2}$.

**21**

$$A = 4\sqrt{3} \text{ unit}^2$$

Because you get:

$$A = \int_{\pi/3}^{\pi} \int_0^{y_1} dy\,dx$$

$$= \int_{\pi/3}^{\pi} \Big[y\Big]_0^{y_1} dx = \int_{\pi/3}^{\pi} y_1\,dx$$

$$= \int_{\pi/3}^{\pi} 4\sin\frac{x}{2}\,dx = \left[-8\cos\frac{x}{2}\right]_{\pi/3}^{\pi}$$

$$= (-8\cos\pi/2) - (-8\cos\pi/6)$$

$$= 0 + 8.\frac{\sqrt{3}}{2} = 4\sqrt{3} \text{ unit}^2$$

*Now for a rather more worthwhile example – on to Frame 22*

22

**Example 3**

Find the area enclosed by the curves

$$y_1{}^2 = 9x \text{ and } y_2 = \frac{x^2}{9}$$

First we must find the points of intersection. For that, $y_1 = y_2$.

$$\therefore\ 9x = \frac{x^4}{81} \quad \therefore\ x = 0 \quad \text{or} \quad x^3 = 729, \quad \text{i.e. } x = 9.$$

So we have a diagram like this:

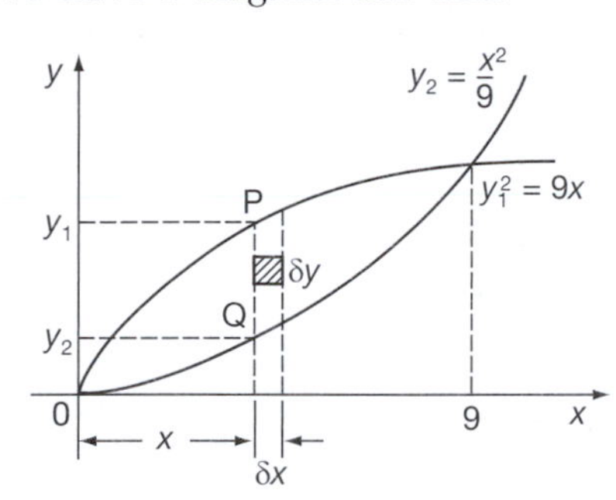


As usual:

Area of element $= \delta y.\delta x$

$\therefore$ Area of strip PQ

$$\sum_{y=y_2}^{y=y_1} \delta y.\delta x$$

Summing all strips between $x = 0$ and $x = 9$:

$$A \approx \sum_{x=0}^{x=9} \left\{ \sum_{y=y_2}^{y=y_1} \delta y.\delta x \right\} = \sum_{x=0}^{x=9} \sum_{y=y_2}^{y=y_1} \delta y.\delta x$$

If $\delta y \to 0$ and $\delta x \to 0$, $A = \int_0^9 \int_{y_2}^{y_1} \mathrm{d}y\,\mathrm{d}x$

Now finish it off, remembering that $y_1{}^2 = 9x$ and $y_2 = \dfrac{x^2}{9}$.

23

$$\boxed{A = 27 \text{ unit}^2}$$

Here it is.

$$A = \int_0^9 \int_{y_2}^{y_1} \mathrm{d}y\,\mathrm{d}x$$

$$= \int_0^9 \Big[ y \Big]_{y_2}^{y_1} \mathrm{d}x$$

$$= \int_0^9 (y_1 - y_2)\mathrm{d}x$$

$$= \int_0^9 \left\{ 3x^{\frac{1}{2}} - \frac{x^2}{9} \right\} \mathrm{d}x$$

$$= \left[ 2x^{3/2} - \frac{x^3}{27} \right]_0^9$$

$$= 54 - 27$$

$$= 27 \text{ unit}^2$$

*Now for a different one. So move on to the next frame*

**24**

Double integrals can conveniently be used for finding other values besides areas.

**Example 4**

Find the second moment of area of a rectangle 6 cm × 4 cm about an axis through one corner perpendicular to the plane of the figure.

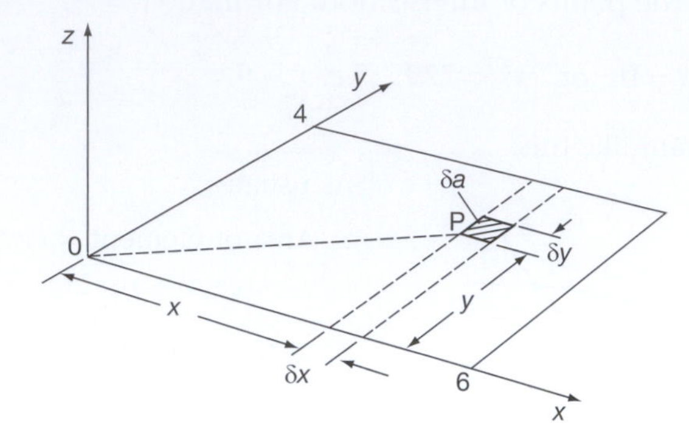


Second moment of element P about the $z$-axis $\approx \delta a(\text{OP})^2$

$$\approx \delta y.\delta x.(x^2 + y^2)$$

Total second moment about the $z$-axis

$$I \approx \sum_{x=0}^{x=6} \sum_{y=0}^{y=4} (x^2 + y^2)\,\text{d}y\,\text{d}x$$

If $\delta x \to 0$ and $\delta y \to 0$, this becomes:

$$I = \int_0^6 \int_0^4 (x^2 + y^2)\,\text{d}y\,\text{d}x$$

Now complete the working, $I = \dots\dots\dots\dots$

**25**

$$\boxed{I = 416 \text{ cm}^4}$$

Because

$$I = \int_0^6 \int_0^4 (x^2 + y^2)\,\text{d}y\,\text{d}x = \int_0^6 \left[x^2 y + \frac{y^3}{3}\right]_0^4 \text{d}xcr$$

$$= \left[\frac{4x^3}{3} + \frac{64x}{3}\right]_0^6$$

$$= 288 + 128$$

$$= 416 \text{ cm}^4$$

Now here is one for you to do on your own.

**Example 5**

Find the second moment of area of a rectangle 5 cm × 3 cm about one 5 cm side as axis.

*Complete it and then on to Frame 26*

**26**

$$I = 45\ \text{cm}^4$$

Here it is: check through the working.

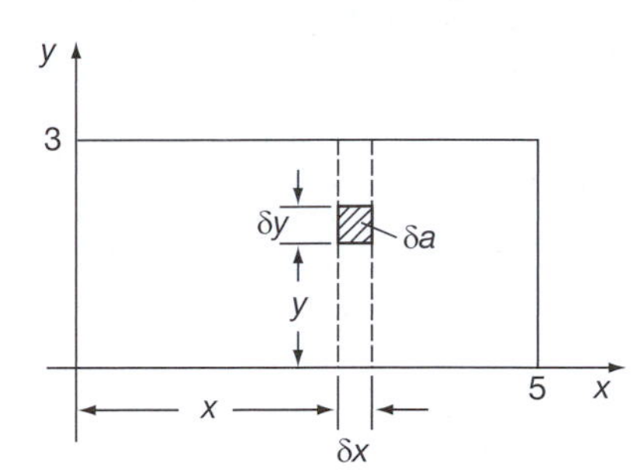


Area of element $= \delta a = \delta y.\delta x$

Second moment of area of $\delta a$ about the $x$-axis

$$= \delta a.y^2$$
$$= y^2.\delta y.\delta x$$

Second moment of strip $\approx \sum_{y=0}^{y=3} y^2.\delta y.\delta x$

Second moment of whole figure $\approx \sum_{x=0}^{x=5}\sum_{y=0}^{y=3} y^2.\delta y.\delta x$

If $\delta y \to 0$ and $\delta x \to 0$:

$$I = \int_0^5\int_0^3 y^2\,\mathrm{d}y\,\mathrm{d}x$$

$$\therefore I = \int_0^5 \left[\frac{y^3}{3}\right]_0^3 \mathrm{d}x$$

$$= \int_0^5 9\,\mathrm{d}x$$

$$= \Big[9x\Big]_0^5$$

$$I = 45\ \text{cm}^4$$

*On to Frame 27*

## Review exercise

**27**

Now a short review exercise. Finish both integrals, before moving on to the next frame. Here they are.

Evaluate the following:

(a) $\displaystyle\int_0^2\int_1^3 (y^2 - xy)\,\mathrm{d}y\,\mathrm{d}x$

(b) $\displaystyle\int_0^3\int_1^2 (x^2 + y^2)\,\mathrm{d}y\,\mathrm{d}x.$

*When you have finished both, move on*

**28**

$$\text{(a) } I = 9\frac{1}{3} \qquad \text{(b) } I = 16$$

Here they are in detail:

$$\text{(a) } I = \int_0^2 \int_1^3 (y^2 - xy)\,dy\,dx$$

$$= \int_0^2 \left[\frac{y^3}{3} - \frac{xy^2}{2}\right]_1^3 dx$$

$$= \int_0^2 \left\{\left(9 - \frac{9x}{2}\right) - \left(\frac{1}{3} - \frac{x}{2}\right)\right\} dx$$

$$= \int_0^2 \left(\frac{26}{3} - 4x\right) dx$$

$$= \left[\frac{26x}{3} - 2x^2\right]_0^2$$

$$= 17\frac{1}{3} - 8 = 9\frac{1}{3}$$

$$\text{(b) } I = \int_0^3 \int_1^2 (x^2 + y^2)\,dy\,dx$$

$$= \int_0^3 \left[x^2y + \frac{y^3}{3}\right]_1^2 dx$$

$$= \int_0^3 \left\{\left(2x^2 + \frac{8}{3}\right) - \left(x^2 + \frac{1}{3}\right)\right\} dx$$

$$= \int_0^3 \left(x^2 + \frac{7}{3}\right) dx$$

$$= \left[\frac{x^3}{3} + \frac{7x}{3}\right]_0^3$$

$$= 9 + 7 = 16$$

*Now on to Frame 29*

## Alternative notation

**29**

Sometimes double integrals are written in a slightly different way. For example, the last double integral $I = \int_0^3 \int_1^2 (x^2 + y^2)\,dy\,dx$ could have been written:

$$\int_0^3 dx \int_1^2 (x^2 + y^2)\,dy$$

The key now is that we start working from the *right-hand* side integral and gradually work back towards the front. Of course, we get the same result and the working is identical.

Let us have some examples, to get used to this notation.

*Move on then to Frame 30*

30

**Example 1**

$$I = \int_0^2 dx \int_0^{\pi/2} 5\cos\theta \, d\theta$$
$$= \int_0^2 dx \Big[5\sin\theta\Big]_0^{\pi/2}$$
$$= \int_0^2 dx \Big[5\Big]$$
$$= \int_0^2 5dx$$
$$= \Big[5x\Big]_0^2 = 10$$

It is all very easy, once you have seen the method.
You try this one.

**Example 2**

Evaluate $I = \int_3^6 dy \int_0^{\pi/2} 4\sin 3x \, dx$.

31

$$\boxed{I = 4}$$

Here it is:

$$I = \int_3^6 dy \int_0^{\pi/2} 4\sin 3x \, dx$$
$$= \int_3^6 dy \left[\frac{-4\cos 3x}{3}\right]_0^{\pi/2}$$
$$= \int_3^6 dy \left\{(0) - \left(-\frac{4}{3}\right)\right\}$$
$$= \int_3^6 dy \frac{4}{3}$$
$$= \left[\frac{4y}{3}\right]_3^6 = (8) - (4) = 4$$

Now do these two.

**Example 3**

$$\int_0^3 dx \int_0^1 (x - x^2) \, dy$$

**Example 4**

$$\int_1^2 dy \int_y^{2y} (x - y) \, dx$$

(Take care with the second one!)

*When you have finished them both, move on to the next frame*

**32**

*Example* 3: $I = -4{\cdot}5$ *Example* 4: $I = \dfrac{7}{6}$

Here is the working.

*Example 3*

$$I = \int_0^3 dx \int_0^1 (x - x^2)\, dy$$
$$= \int_0^3 dx \Big[xy - x^2 y\Big]_0^1$$
$$= \int_0^3 dx(x - x^2)$$
$$= \int_0^3 (x - x^2)\, dx$$
$$= \left[\frac{x^2}{2} - \frac{x^3}{3}\right]_0^3$$
$$= \frac{9}{2} - 9$$
$$= -4{\cdot}5$$

*Example 4*

$$I = \int_1^2 dy \int_y^{2y} (x - y)\, dx$$
$$= \int_1^2 dy \left[\frac{x^2}{2} - xy\right]_{x=y}^{x=2y}$$
$$= \int_1^2 dy \left\{(2y^2 - 2y^2) - \left(\frac{y^2}{2} - y^2\right)\right\}$$
$$= \int_1^2 dy \frac{y^2}{2} = \int_1^2 \frac{y^2}{2}\, dy$$
$$= \left[\frac{y^3}{6}\right]_1^2$$
$$= \frac{8}{6} - \frac{1}{6}$$
$$= \frac{7}{6}$$

*Next frame*

**33** Now, by way of revision, evaluate these:

(a) $\displaystyle\int_0^4 \int_y^{2y} (2x + 3y)\, dx\, dy$

(b) $\displaystyle\int_1^4 dx \int_0^{\sqrt{x}} (2y - 5x)\, dy$

*When you have completed both of them, move on to Frame 34*

**34**

(a) 128 (b) − 54·5

Working:

(a) $I = \int_0^4 \int_y^{2y} (2x + 3y)\,\mathrm{d}x\,\mathrm{d}y$

$= \int_0^4 \left[x^2 + 3xy\right]_{x=y}^{x=2y} \mathrm{d}y$

$= \int_0^4 \left\{(4y^2 + 6y^2) - (y^2 + 3y^2)\right\} \mathrm{d}y$

$= \int_0^4 \left\{10y^2 - 4y^2\right\} \mathrm{d}y$

$= \int_0^4 6y^2\,\mathrm{d}y$

$= \left[\dfrac{6y^3}{3}\right]_0^4$

$= \left[2y^3\right]_0^4$

$= 128$

(b) $I = \int_1^4 \mathrm{d}x \int_0^{\sqrt{x}} (2y - 5x)\,\mathrm{d}y$

$= \int_1^4 \mathrm{d}x \left[y^2 - 5xy\right]_{y=0}^{y=\sqrt{x}}$

$= \int_1^4 \mathrm{d}x \left\{x - 5x^{3/2}\right\}$

$= \int_1^4 (x - 5x^{3/2})\,\mathrm{d}x$

$= \left[\dfrac{x^2}{2} - 2x^{5/2}\right]_1^4$

$= (8 - 64) - \left(\dfrac{1}{2} - 2\right)$

$= -56 + 1{\cdot}5$

$= -54{\cdot}5$

So it is just a question of being able to recognize and to interpret the two notations.

Now let us look at one or two further examples of the use of multiple integrals.

*Move on then to Frame 35*

## Determination of areas by multiple integrals

**35**

To find the area of the plane figure bounded by the polar curve $r = f(\theta)$, and the radius vectors at $\theta = \theta_1$ and $\theta = \theta_2$.

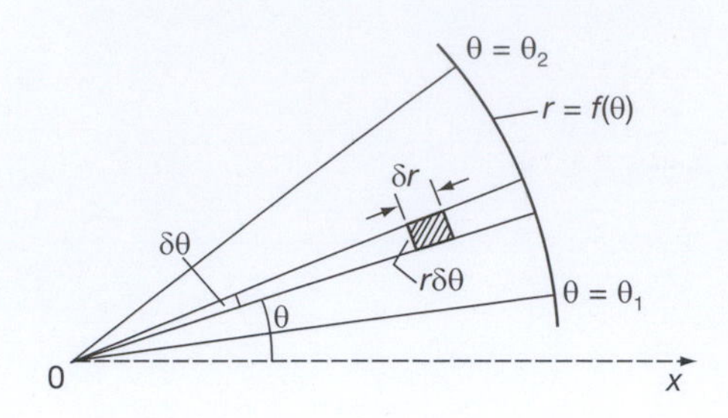


Small arc of a circle of radius $r$, subtending an angle $\delta\theta$ at the centre.

$\therefore$ arc $= r.\delta\theta$

We proceed very much as before:

Area of element $\approx \delta r.r\delta\theta$

$$\text{Area of thin sector} \approx \sum_{r=0}^{r=r_1} \delta r.r\delta\theta$$

$$\text{Total area} \approx \sum_{\theta=\theta_1}^{\theta=\theta_2} (\text{all such thin sectors})$$

$$\approx \sum_{\theta=\theta_1}^{\theta=\theta_2} \left\{ \sum_{r=0}^{r=r_1} r.\delta r.\delta\theta \right\}$$

$$\approx \sum_{\theta=\theta_1}^{\theta=\theta_2} \sum_{r=0}^{r=r_1} r.\delta r.\delta\theta$$

Then if $\delta\theta \to 0$ and $\delta r \to 0$:

$$A = \int_{\theta_1}^{\theta_2} \int_0^{r_1} r\,\mathrm{d}r\,\mathrm{d}\theta$$

$= \ldots\ldots\ldots\ldots$ Finish it off.

**36**

The working continues:

$$A = \int_{\theta_1}^{\theta_2} \left[\frac{r^2}{2}\right]_0^{r_1} \mathrm{d}\theta$$

$$= \int_{\theta_1}^{\theta_2} \left(\frac{{r_1}^2}{2}\right) \mathrm{d}\theta$$

$$\text{i.e. in general, } A = \int_{\theta_1}^{\theta_2} \frac{1}{2} r^2 \,\mathrm{d}\theta$$

$$= \int_{\theta_1}^{\theta_2} \frac{1}{2} f^2(\theta)\,\mathrm{d}\theta$$

which is the result we have met before.

*Let us work an actual example of this, so move on to Frame 37*

**37**

By the use of double integrals, find the area enclosed by the polar curve $r = 4(1 + \cos\theta)$ and the radius vectors at $\theta = 0$ and $\theta = \pi$.

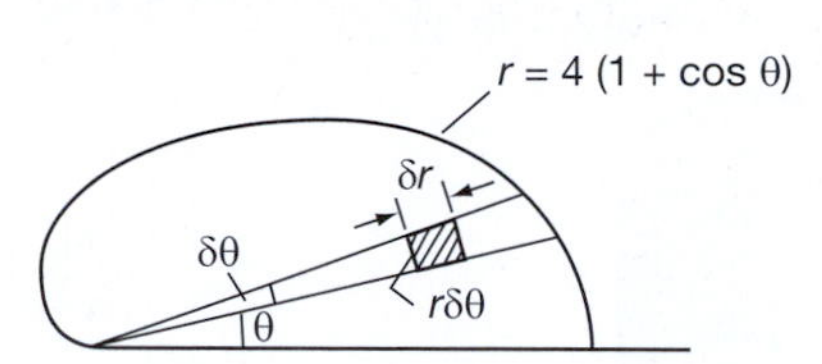


$$A \approx \sum_{\theta=0}^{\theta=\pi}\sum_{r=0}^{r=r_1} r\delta r.\delta\theta$$

$$A = \int_0^{\pi}\int_0^{r_1} r\,dr\,d\theta$$

$$= \int_0^{\pi}\left[\frac{r^2}{2}\right]_0^{r_1} d\theta$$

$$= \int_0^{\pi}\left[\frac{r_1{}^2}{2}\right] d\theta \qquad \text{But } r_1 = f(\theta) = 4(1+\cos\theta)$$

$$\therefore A = \int_0^{\pi} 8(1+\cos\theta)^2\,d\theta$$

$$= \int_0^{\pi} 8(1 + 2\cos\theta + \cos^2\theta)\,d\theta$$

$$= \dots\dots\dots\dots$$

**38**

$$\boxed{A = 12\pi \text{ unit}^2}$$

Because

$$A = 8\int_0^{\pi}(1 + 2\cos\theta + \cos^2\theta)\,d\theta$$

$$= 8\left[\theta + 2\sin\theta + \frac{\theta}{2} + \frac{\sin 2\theta}{4}\right]_0^{\pi}$$

$$= 8\left(\pi + \frac{\pi}{2}\right) - (0)$$

$$= 8\pi + 4\pi$$

$$= 12\pi \text{ unit}^2$$

*Now let us deal with volumes by the same method, so move on to the next frame*

# Determination of volumes by multiple integrals

**39**

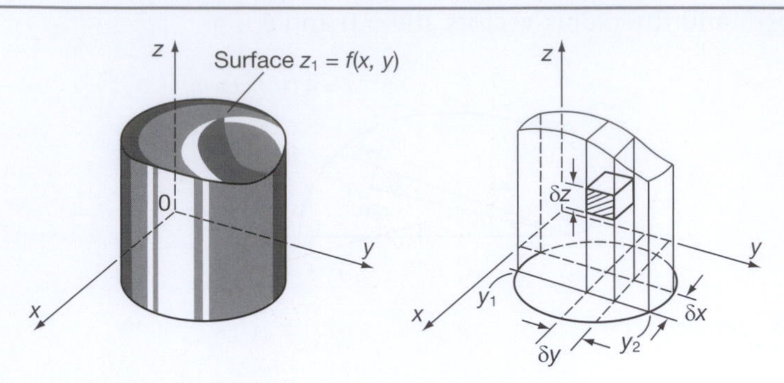


Element of volume $\delta v = \delta x.\delta y.\delta z$.
Summing the elements up the column, we have

$$\delta V_c = \sum_{z=0}^{z=z_1} \delta x.\delta y.\delta z$$

If we now sum the columns between $y = y_1$ and $y_1 = y_2$, we obtain the volume of the slice:

$$\delta V_s = \sum_{y=y_1}^{y=y_2} \sum_{z=0}^{z=z_1} \delta x.\delta y.\delta z$$

Then, summing all slices between $x = x_1$ and $x = x_2$, we have the total volume:

$$V = \sum_{x=x_1}^{x=x_2} \sum_{y=y_1}^{y=y_2} \sum_{z=z_0}^{z=z_1} \delta x.\delta y.\delta z$$

Then, as usual, if $\delta x \to 0$, $\delta y \to 0$ and $\delta z \to 0$:

$$V = \int_{x_1}^{x_2} \int_{y_1}^{y_2} \int_{0}^{z_1} \mathrm{d}x \, \mathrm{d}y \, \mathrm{d}z$$

The result this time is a triple integral, but the development is very much the same as in our previous examples.

Let us see this in operation in the following examples.

*Next frame*

**40**

**Example 1**

A solid is enclosed by the plane $z = 0$, the planes $x = 1$, $x = 4$, $y = 2$, $y = 5$ and the surface $z = x + y$. Find the volume of the solid.

First of all, what does the figure look like? The plane $z = 0$ is the $x$–$y$ plane and the plane $x = 1$ is positioned thus:

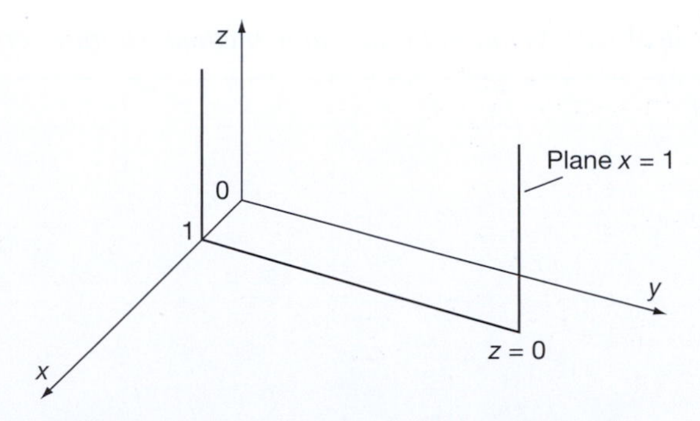


Working on the same lines, draw a sketch of the vertical sides.

**41**

The figure so far now looks like this:

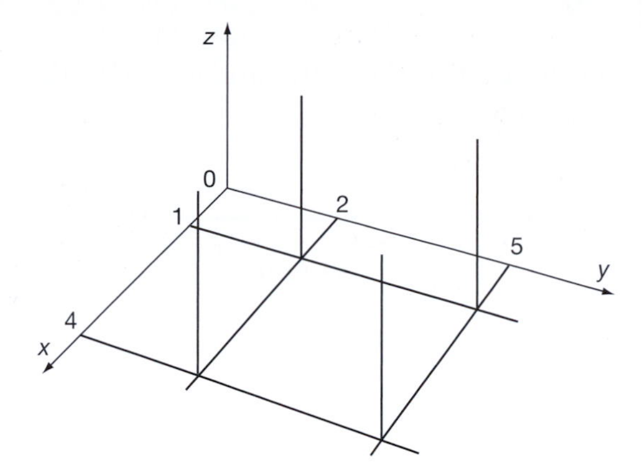


If we now mark in the calculated heights at each point of intersection $(z = x + y)$, we get:

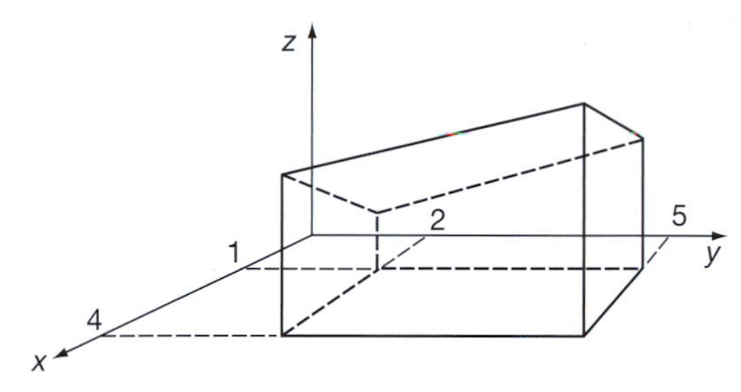


This is just preparing the problem, so that we can see how to develop the integral.

*For the calculation stage, move on to the next frame*

**42**

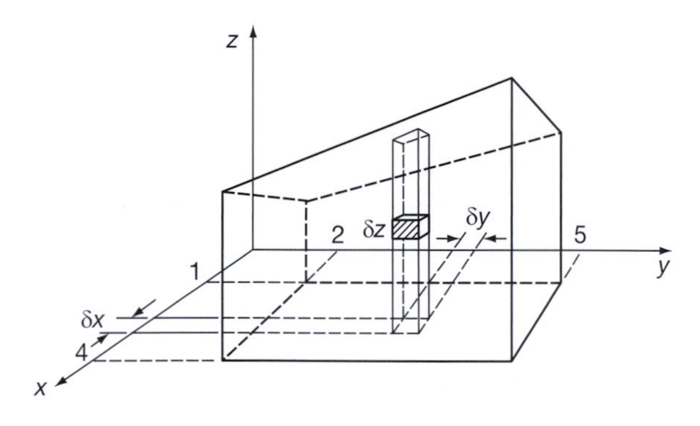


Volume element $\approx \delta x.\delta y.\delta z$

Volume of column $\approx \delta x.\delta y \displaystyle\sum_{z=0}^{z=(x+y)} \delta z$

Volume of slice $\approx \delta x \displaystyle\sum_{y=2}^{y=5} \delta y \sum_{z=0}^{z=x+y} \delta z$

Volume of total solid $\approx \displaystyle\sum_{x=1}^{x=4} \delta x \sum_{y=2}^{y=5} \delta y \sum_{z=0}^{z=x+y} \delta z$

▶

Then, as usual, if $\delta x \to 0$, $\delta y \to 0$, $\delta z \to 0$, this becomes:

$$V = \int_1^4 \mathrm{d}x \int_2^5 \mathrm{d}y \int_0^{x+y} \mathrm{d}z$$

And this you can now finish off without any trouble. (With this form of notation, start at the right-hand end. Remember?)

So $V = \ldots\ldots\ldots\ldots$

**43**

$$\boxed{V = 54 \text{ unit}^3}$$

$$V = \int_1^4 \mathrm{d}x \int_2^5 \mathrm{d}y \int_0^{x+y} \mathrm{d}z = \int_1^4 \mathrm{d}x \int_2^5 \mathrm{d}y(x+y)$$

$$= \int_1^4 \mathrm{d}x \int_2^5 (x+y)\,\mathrm{d}y = \int_1^4 \mathrm{d}x \left[xy + \frac{y^2}{2}\right]_2^5$$

$$= \int_1^4 \mathrm{d}x \left[5x + \frac{25}{2} - 2x - 2\right] = \int_1^4 \left(3x + \frac{21}{2}\right)\mathrm{d}x$$

$$= \left[\frac{3x^2}{2} + \frac{21x}{2}\right]_1^4 = \frac{1}{2}\left[3x^2 + 21x\right]_1^4$$

$$= \frac{1}{2}\left\{(48 + 84) - (3 + 21)\right\} = \frac{1}{2}\left\{132 - 24\right\} = 54 \text{ unit}^3$$

**44**

**Example 2**

Find the volume of the solid bounded by the planes $z = 0$, $x = 1$, $x = 2$, $y = -1$, $y = 1$ and the surface $z = x^2 + y^2$.

In the light of the previous example, can you conjure up a mental picture of what this solid looks like? As before it will give rise to a triple integral.

$$V = \int_1^2 \mathrm{d}x \int_{-1}^1 \mathrm{d}y \int_0^{x^2+y^2} \mathrm{d}z$$

Evaluate this and so find $V$. $V = \ldots\ldots\ldots\ldots$

**45**

$$\boxed{V = \frac{16}{3} \text{ unit}^3}$$

Because we have:

$$V = \int_1^2 \mathrm{d}x \int_{-1}^1 \mathrm{d}y \int_0^{x^2+y^2} \mathrm{d}z = \int_1^2 \mathrm{d}x \int_{-1}^1 \mathrm{d}y(x^2 + y^2)$$

$$= \int_1^2 \mathrm{d}x \left[x^2y + \frac{y^3}{3}\right]_{-1}^1 = \int_1^2 \left\{\left(x^2 + \frac{1}{3}\right) - \left(-x^2 - \frac{1}{3}\right)\right\}\mathrm{d}x$$

$$= \int_1^2 \left\{2x^2 + \frac{2}{3}\right\}\mathrm{d}x = \frac{2}{3}\left[x^3 + x\right]_1^2$$

$$= \frac{2}{3}\left\{(8 + 2) - (1 + 1)\right\} = \frac{16}{3} \text{ unit}^3$$

*Next frame*

46

That brings us almost to the end of this Programme.

In our work on multiple integrals, we have been developing a form of approach rather than compiling a catalogue of formulas. There is little therefore that we can list by way of revision on this occasion, except perhaps to remind you, once again, of the two forms of notation.

*Remember:*

(a) For integrals written $\int_c^d \int_a^b f(x,y)\,dx\,dy$, work from the centre outwards.

(b) For integrals written $\int_c^d dy \int_a^b f(x,y)\,dx$ work from the right-hand side.

Now there is the **Can You?** checklist and the **Test exercise** to follow, so work through them carefully at your own speed.

*On to Frames 47 and 48*

## Can You?

### Checklist 24

47

Check this list before and after you try the end of Programme test.

On a scale of 1 to 5 how confident are you that you can:

| | | | | | | | Frames |
|---|---|---|---|---|---|---|---|
| • Determine the area of a rectangle using a double integral? | | | | | | | 1 to 9 |
| *Yes* | ☐ | ☐ | ☐ | ☐ | ☐ | *No* | |
| • Evaluate double integrals over general areas? | | | | | | | 10 to 13 |
| *Yes* | ☐ | ☐ | ☐ | ☐ | ☐ | *No* | |
| • Evaluate triple integrals over general volumes? | | | | | | | 14 to 18 |
| *Yes* | ☐ | ☐ | ☐ | ☐ | ☐ | *No* | |
| • Apply double integrals to find areas and second moments of area? | | | | | | | 19 to 38 |
| *Yes* | ☐ | ☐ | ☐ | ☐ | ☐ | *No* | |
| • Apply triple integrals to find volumes? | | | | | | | 39 to 45 |
| *Yes* | ☐ | ☐ | ☐ | ☐ | ☐ | *No* | |

## Test exercise 24

48

The questions are just like those you have been doing quite successfully. They are all quite straightforward and should cause you no trouble.

**1** Evaluate:

(a) $\int_1^3 \int_0^2 (y^3 - xy)\,dy\,dx$

(b) $\int_0^a dx \int_0^{y_1} (x - y)\,dy$, where $y_1 = \sqrt{a^2 - x^2}$

▶

2 Determine:

(a) $\displaystyle\int_0^{\sqrt{3}+2}\int_0^{\pi/3}(2\cos\theta - 3\sin 3\theta)\,d\theta\,dr$

(b) $\displaystyle\int_2^4\int_1^2\int_0^4 xy(z+2)\,dx\,dy\,dz$

(c) $\displaystyle\int_0^1 dz\int_1^2 dx\int_0^x (x+y+z)\,dy$

3 The line $y = 2x$ and the parabola $y^2 = 16x$ intersect at $x = 4$. Find by a double integral, the area enclosed by $y = 2x$, $y^2 = 16x$ and the ordinate at $x = 1$, and the point of intersection at $x = 4$.

4 A triangle is bounded by the $x$-axis, the line $y = 2x$ and the ordinate at $x = 4$. Build up a double integral representing the second moment of area of this triangle about the $x$-axis and evaluate the integral.

5 Form a double integral to represent the area of the plane figure bounded by the polar curve $r = 3 + 2\cos\theta$ and the radius vectors at $\theta = 0$ and $\theta = \pi/2$, and evaluate it.

6 A solid is enclosed by the planes $z = 0$, $y = 1$, $y = 3$, $x = 0$, $x = 3$ and the surface $z = x^2 + xy$. Calculate the volume of the solid.

*That's it!*

## Further problems 24

49

1 Evaluate $\displaystyle\int_0^{\pi}\int_0^{\cos\theta} r\sin\theta\,dr\,d\theta$

2 Evaluate $\displaystyle\int_0^{2\pi}\int_0^3 r^3(9 - r^2)\,dr\,d\theta$

3 Evaluate $\displaystyle\int_{-2}^1\int_{x^2+4x}^{3x+2} dy\,dx$

4 Evaluate $\displaystyle\int_0^a\int_0^b\int_0^c (x^2 + y^2)\,dx\,dy\,dz$

5 Evaluate $\displaystyle\int_0^{\pi}\int_0^{\pi/2}\int_0^r x^2\sin\theta\,dx\,d\theta\,d\phi$

6 Find the area bounded by the curve $y = x^2$ and the line $y = x + 2$.

7 Find the area of the polar figure enclosed by the circle $r = 2$ and the cardioid $r = 2(1 + \cos\theta)$.

8 Evaluate $\displaystyle\int_0^2 dx\int_1^3 dy\int_1^2 xy^2z\,dz$

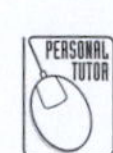

9 Evaluate $\displaystyle\int_0^2 dx\int_1^2 (x^2 + y^2)\,dy$

10 Evaluate $\displaystyle\int_0^1 dr\int_0^{\pi/4} r\cos^2\theta\,d\theta$

▶

**11** Determine the area bounded by the curves $x = y^2$ and $x = 2y - y^2$.

**12** Express as a double integral, the area contained by one loop of the curve $r = 2\cos 3\theta$, and evaluate the integral.

**13** Evaluate $\displaystyle\int_0^{\pi/2}\int_{\pi/4}^{\tan^{-1}(2)}\int_0^4 x\sin y\,dx\,dy\,dz$

**14** Evaluate $\displaystyle\int_0^{\pi}\int_0^{4\cos z}\int_0^{\sqrt{16-y^2}} y\,dx\,dy\,dz$

**15** A plane figure is bounded by the polar curve $r = a(1 + \cos\theta)$ between $\theta = 0$ and $\theta = \pi$, and the initial line OA. Express as a double integral the first moment of area of the figure about OA, and evaluate the integral. If the area of the figure is known to be $\dfrac{3\pi a^2}{4}$ unit$^2$, find the distance ($h$) of the centroid of the figure from OA.

**16** Using double integrals, find (a) the area and (b) the second moment about the $x$-axis of the plane figure bounded by the $x$-axis and that part of the ellipse $\dfrac{x^2}{a^2} + \dfrac{y^2}{b^2} = 1$ which lies above the $x$-axis. Find also the position of the centroid.

**17** The base of a solid is the plane figure in the $x$–$y$ plane bounded by $x = 0$, $x = 2$, $y = x$ and $y = x^2 + 1$. The sides are vertical and the top is the surface $z = x^2 + y^2$. Calculate the volume of the solid so formed.

**18** A solid consists of vertical sides standing on the plane figure enclosed by $x = 0$, $x = b$, $y = a$ and $y = c$. The top is the surface $z = xy$. Find the volume of the solid so defined.

**19** Show that the area outside the circle $r = a$ and inside the circle $r = 2a\cos\theta$ is given by

$$A = 2\int_0^{\pi/3}\int_a^{2a\cos\theta} r\,dr\,d\theta$$

Evaluate the integral.

**20** A rectangular block is bounded by the coordinate planes of reference and by the planes $x = 3$, $y = 4$, $z = 2$. Its density at any point is numerically equal to the square of its distance from the origin. Find the total mass of the solid.

---

Now visit the companion website for this book at www.palgrave.com/stroud for more questions applying this mathematics to science and engineering.

**50**

# First-order differential equations

## Learning outcomes

When you have completed this Programme you will be able to:

- ☐ Recognize the order of a differential equation
- ☐ Appreciate that a differential equation of order *n* can be derived from a function containing *n* arbitrary constants
- ☐ Solve certain first-order differential equations by direct integration
- ☐ Solve certain first-order differential equations by separating the variables
- ☐ Solve certain first-order homogeneous differential equations by an appropriate substitution
- ☐ Solve certain first-order differential equations by using an integrating factor
- ☐ Solve Bernoulli's equation

# Introduction

**1**

A *differential equation* is a relationship between an independent variable, $x$, a dependent variable $y$, and one or more derivatives of $y$ with respect to $x$.

e.g. $x^2\dfrac{dy}{dx} = y\sin x = 0$

$$xy\frac{d^2y}{dx^2} + y\frac{dy}{dx} + e^{3x} = 0$$

Differential equations represent dynamic relationships, i.e. quantities that change, and are thus frequently occurring in scientific and engineering problems.

The *order* of a differential equation is given by the highest derivative involved in the equation.

$x\dfrac{dy}{dx} - y^2 = 0$ is an equation of the 1st order

$xy\dfrac{d^2y}{dx^2} - y^2\sin x = 0$ is an equation of the 2nd order

$\dfrac{d^3y}{dx^3} - y\dfrac{dy}{dx} + e^{4x} = 0$ is an equation of the 3rd order

So that $\dfrac{d^2y}{dx^2} + 2\dfrac{dy}{dx} + 10y = \sin 2x$ is an equation of the ............ order.

**2**

second

Because

In the equation $\dfrac{d^2y}{dx^2} + 2\dfrac{dy}{dx} + 10y = \sin 2x$, the highest derivative involved is $\dfrac{d^2y}{dx^2}$.

Similarly:

(a) $x\dfrac{dy}{dx} = y^2 + 1$ is a ...........order equation

(b) $\cos^2 x\dfrac{dy}{dx} + y = 1$ is a ...........order equation

(c) $\dfrac{d^2y}{dx^2} - 3\dfrac{dy}{dx} + 2y = x^2$ is a ...........order equation

(d) $(y^3 + 1)\dfrac{dy}{dx} - xy^2 = x$ is a ...........order equation

*On to Frame 3*

**3**

(a) first (b) first (c) second (d) first

*Next frame*

# Formation of differential equations

**4**

Differential equations may be formed in practice from a consideration of the physical problems to which they refer. Mathematically, they can occur when arbitrary constants are eliminated from a given function. Here are a few examples.

**Example 1**

Consider $y = A\sin x + B\cos x$, where $A$ and $B$ are two arbitrary constants.

If we differentiate, we get:

$$\frac{dy}{dx} = A\cos x - B\sin x$$

$$\text{and} \quad \frac{d^2y}{dx^2} = -A\sin x - B\cos x$$

which is identical to the original equation, but with the sign changed.

$$\text{i.e.} \quad \frac{d^2y}{dx^2} = -y \qquad \therefore \quad \frac{d^2y}{dx^2} + y = 0$$

This is a differential equation of the ........... order.

**5**

second

**Example 2**

Form a differential equation from the function $y = x + \dfrac{A}{x}$.

We have $y = x + \dfrac{A}{x} = x + Ax^{-1}$

$$\therefore \frac{dy}{dx} = 1 - Ax^{-2} = 1 - \frac{A}{x^2}$$

From the given equation, $\dfrac{A}{x} = y - x \quad \therefore A = x(y - x)$

$$\therefore \frac{dy}{dx} = 1 - \frac{x(y-x)}{x^2}$$

$$= 1 - \frac{y-x}{x} = \frac{x - y + x}{x} = \frac{2x - y}{x}$$

$$\therefore x\frac{dy}{dx} = 2x - y$$

This is an equation of the ........... order.

**6**

first

Now one more.

**Example 3**

Form the differential equation for $y = Ax^2 + Bx$.

We have

$$y = Ax^2 + Bx \qquad (1)$$

$$\therefore \frac{dy}{dx} = 2Ax + B \qquad (2)$$

$$\therefore \frac{d^2y}{dx^2} = 2A \qquad (3) \quad A = \frac{1}{2}\frac{d^2y}{dx^2}$$

▶

Substitute for $2A$ in (2): $\dfrac{dy}{dx} = x\dfrac{d^2y}{dx^2} + B$

$\therefore B = \dfrac{dy}{dx} - x\dfrac{d^2y}{dx^2}$

Substituting for $A$ and $B$ in (1), we have:

$$y = x^2.\frac{1}{2}\frac{d^2y}{dx^2} + x\left(\frac{dy}{dx} - x\frac{d^2y}{dx^2}\right)$$

$$= \frac{x^2}{2}.\frac{d^2y}{dx^2} + x.\frac{dy}{dx} - x^2.\frac{d^2y}{dx^2}$$

$$\therefore y = x\frac{dy}{dx} - \frac{x^2}{2}.\frac{d^2y}{dx^2}$$

and this is an equation of the ........... order.

---

**7**

second

If we collect our last few results together, we have:

$y = A\sin x + B\cos x$ gives the equation $\dfrac{d^2y}{dx^2} + y = 0$ (2nd order)

$y = Ax^2 + Bx$ gives the equation $y = x\dfrac{dy}{dx} - \dfrac{x^2}{2}.\dfrac{d^2y}{dx^2}$ (2nd order)

$y = x + \dfrac{A}{x}$ gives the equation $x\dfrac{dy}{dx} = 2x - y$ (1st order)

If we were to investigate the following, we should also find that:

$y = Axe^x$ gives the differential equation $x\dfrac{dy}{dx} - y(1 + x) = 0$ (1st order)

$y = Ae^{-4x} + Be^{-6x}$ gives the differential equation $\dfrac{d^2y}{dx^2} + 10\dfrac{dy}{dx} + 24y = 0$ (2nd order)

Some of the functions give 1st-order equations: some give 2nd-order equations. Now look at the five results above and see if you can find any distinguishing features in the functions which decide whether we obtain a 1st-order equation or a 2nd-order equation in any particular case.

*When you have come to a conclusion, move on to Frame 8*

---

**8**

A function with 1 arbitrary constant gives a 1st-order equation.
A function with 2 arbitrary constants gives a 2nd-order equation.

Correct, and in the same way:

A function with 3 arbitrary constants would give a 3rd order equation.

So, without working each out in detail, we can say that:

(a) $y = e^{-2x}(A + Bx)$ would give a differential equation of ........... order.

(b) $y = A\dfrac{x-1}{x+1}$ would give a differential equation of ........... order.

(c) $y = e^{3x}(A\cos 3x + B\sin 3x)$ would give a differential equation of ........... order.

**9**

(a) 2nd (b) 1st (c) 2nd

Because

(a) and (c) each have 2 arbitrary constants,
while (b) has only 1 arbitrary constant.

Similarly:

(a) $x^2 \dfrac{dy}{dx} + y = 1$ is derived from a function having ........... arbitrary constants.

(b) $\cos^2 x \dfrac{dy}{dx} = 1 - y$ is derived from a function having ........... arbitrary constants.

(c) $\dfrac{d^2y}{dx^2} + 4\dfrac{dy}{dx} + y = e^{2x}$ is derived from a function having ........... arbitrary constants.

**10**

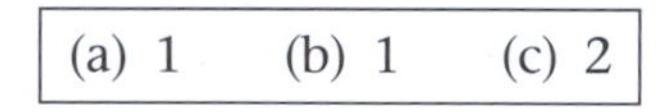

(a) 1 (b) 1 (c) 2

So, from all this, the following rule emerges:

A 1st-order differential equation is derived from a function having 1 arbitrary constant.

A 2nd-order differential equation is derived from a function having 2 arbitrary constants.

An $n$th-order differential equation is derived from a function having $n$ arbitrary constants.

Copy this last statement into your record book. It is important to remember this rule and we shall make use of it at various times in the future.

*Then on to Frame 11*

## Solution of differential equations

**11**

To solve a differential equation, we have to find the function for which the equation is true. This means that we have to manipulate the equation so as to eliminate all the derivatives and leave a relationship between $y$ and $x$. The rest of this particular Programme is devoted to the various methods of solving *first-order differential equations*. Second-order equations will be dealt with in the next Programme.

*So, for the first method, move on to Frame 12*

12

## Method 1: *By direct integration*

If the equation can be arranged in the form $\frac{dy}{dx} = f(x)$, then the equation can be solved by simple integration.

**Example 1**

$$\frac{dy}{dx} = 3x^2 - 6x + 5$$

Then $y = \int (3x^2 - 6x + 5)dx = x^3 - 3x^2 + 5x + C$

i.e. $y = x^3 - 3x^2 + 5x + C$

As always, of course, the constant of integration must be included. Here it provides the one arbitrary constant which we always get when solving a first-order differential equation.

**Example 2**

Solve $x\frac{dy}{dx} = 5x^3 + 4$

In this case, $\frac{dy}{dx} = 5x^2 + \frac{4}{x}$ So, $y = \ldots\ldots\ldots\ldots$

13

$$y = \frac{5x^3}{3} + 4\ln x + C$$

As you already know from your work on integration, the value of $C$ cannot be determined unless further information about the function is given. In this present form, the function is called the *general solution* (or *primitive*) of the given equation.

If we are told the value of $y$ for a given value of $x$, $C$ can be evaluated and the result is then a *particular solution* of the equation.

**Example 3**

Find the particular solution of the equation $e^x\frac{dy}{dx} = 4$, given that $y = 3$ when $x = 0$.

First rewrite the equation in the form $\frac{dy}{dx} = \frac{4}{e^x} = 4e^{-x}$.

Then $y = \int 4e^{-x}\,dx = -4e^{-x} + C$

Knowing that when $x = 0$, $y = 3$, we can evaluate $C$ in this case, so that the required particular solution is $y = \ldots\ldots\ldots\ldots$

14

$$y = -4e^{-x} + 7$$

## Method 2: *By separating the variables*

If the given equation is of the form $\frac{dy}{dx} = f(x, y)$, the variable $y$ on the right-hand side prevents solving by direct integration. We therefore have to devise some other method of solution.

▶

Let us consider equations of the form $\dfrac{dy}{dx} = f(x).F(y)$ and of the form $\dfrac{dy}{dx} = \dfrac{f(x)}{F(y)}$, i.e. equations in which the right-hand side can be expressed as products or quotients of functions of $x$ or of $y$.

A few examples will show how we proceed.

**Example 1**

Solve $\dfrac{dy}{dx} = \dfrac{2x}{y+1}$

We can rewrite this as $(y+1)\dfrac{dy}{dx} = 2x$

Now integrate both sides with respect to $x$:

$$\int (y+1)\frac{dy}{dx}\,dx = \int 2x\,dx \quad \text{i.e.}$$

$$\int (y+1)\,dy = \int 2x\,dx$$

and this gives $\dfrac{y^2}{2} + y = x^2 + C$

15

**Example 2**

Solve $\dfrac{dy}{dx} = (1+x)(1+y)$

$$\frac{1}{1+y}\frac{dy}{dx} = 1+x$$

Integrate both sides with respect to $x$:

$$\int \frac{1}{1+y}\frac{dy}{dx}\,dx = \int (1+x)\,dx \qquad \therefore \int \frac{1}{1+y}\,dy = \int (1+x)\,dx$$

$$\ln(1+y) = x + \frac{x^2}{2} + C$$

The method depends on our being able to express the given equation in the form $F(y).\dfrac{dy}{dx} = f(x)$. If this can be done, the rest is then easy, for we have

$$\int F(y).\frac{dy}{dx}\,dx = \int f(x)dx \qquad \therefore \int F(y)\,dy = \int f(x)\,dx$$

and we then continue as in the examples.

*Let us see another example, so move on to Frame 16*

16

**Example 3**

Solve $\dfrac{dy}{dx} = \dfrac{1+y}{2+x}$ (1)

This can be written as $\dfrac{1}{1+y}\dfrac{dy}{dx} = \dfrac{1}{2+x}$

Integrate both sides with respect to $x$:

$$\int \frac{1}{1+y}\frac{dy}{dx}\,dx = \int \frac{1}{2+x}\,dx$$

$$\therefore \int \frac{1}{1+y}\,dy = \int \frac{1}{2+x}\,dx \qquad (2)$$

$$\therefore \ln(1+y) = \ln(2+x) + C$$

▶

It is convenient to write the constant $C$ as the logarithm of some other constant $A$:

$$\ln(1+y) = \ln(2+x) + \ln A = \ln A(2+x)$$
$$\therefore\ 1+y = A(2+x)$$

*Note*: We can, in practice, get from the given equation (1) to the form of the equation in (2) by a simple routine, thus:

$$\frac{dy}{dx} = \frac{1+y}{2+x}$$

First multiply across by d$x$:

$$dy = \frac{1+y}{2+x}dx$$

Now collect the '$y$-factor' with the d$y$ on the left, i.e. divide by $(1+y)$:

$$\frac{1}{1+y}\,dy = \frac{1}{2+x}\,dx$$

Finally, add the integral signs:

$$\int \frac{1}{1+y}\,dy = \int \frac{1}{2+x}\,dx$$

and then continue as before.

This is purely a routine which enables us to sort out the equation algebraically, the whole of the work being done in one line. Notice, however, that the RHS of the given equation must be expressed as '$x$-factors' and '$y$-factors'.

Now for another example, using this routine.

**Example 4**

Solve $\quad \dfrac{dy}{dx} = \dfrac{y^2 + xy^2}{x^2y - x^2}$

First express the RHS in '$x$-factors' and '$y$-factors':

$$\frac{dy}{dx} = \frac{y^2(1+x)}{x^2(y-1)}$$

Now rearrange the equation so that we have the '$y$-factors' and d$y$ on the LHS and the '$x$-factors' and d$x$ on the RHS.

So we get ...........

---

**17**

$$\boxed{\frac{y-1}{y^2}\,dy = \frac{1+x}{x^2}\,dx}$$

We now add the integral signs:

$$\int \frac{y-1}{y^2}\,dy = \int \frac{1+x}{x^2}\,dx$$

and complete the solution:

$$\int\left\{\frac{1}{y} - y^{-2}\right\}dy = \int\left\{x^{-2} + \frac{1}{x}\right\}dx$$
$$\therefore\ \ln y + y^{-1} = \ln x - x^{-1} + C$$
$$\therefore\ \ln y + \frac{1}{y} = \ln x - \frac{1}{x} + C$$

Here is another.

▶

**Example 5**

Solve

$$\frac{dy}{dx} = \frac{y^2 - 1}{x}$$

$$dy = \frac{y^2 - 1}{x} dx$$

Rearranging, we have

$$\frac{1}{y^2 - 1} dy = \frac{1}{x} dx$$

$$\int \frac{1}{y^2 - 1} dy = \int \frac{1}{x} dx$$

which gives ...........

---

**18**

$$\boxed{\frac{1}{2} \ln \frac{y-1}{y+1} = \ln x + C}$$

$$\therefore \ln \frac{y-1}{y+1} = 2 \ln x + \ln A$$

$$\therefore \frac{y-1}{y+1} = Ax^2$$

$$y - 1 = Ax^2(y+1)$$

You see they are all done in the same way. Now here is one for you to do:

**Example 6**

Solve $xy\dfrac{dy}{dx} = \dfrac{x^2+1}{y+1}$

First of all, rearrange the equation into the form:

$$F(y)dy = f(x)dx$$

i.e. arrange the '$y$-factors' and $dy$ on the LHS and the '$x$-factors' and $dx$ on the RHS.

*What do you get?*

---

**19**

$$\boxed{y(y+1)dy = \frac{x^2+1}{x} dx}$$

Because

$$xy\frac{dy}{dx} = \frac{x^2+1}{y+1} \qquad \therefore xy\,dy = \frac{x^2+1}{y+1}\,dx \qquad \therefore y(y+1)\,dy = \frac{x^2+1}{x}\,dx$$

So we now have

$$\int (y^2 + y)\,dy = \int \left(x + \frac{1}{x}\right) dx$$

Now finish it off, then move on to the next frame.

20

$$\frac{y^3}{3}+\frac{y^2}{2}=\frac{x^2}{2}+\ln x+C$$

Provided that the RHS of the equation $\frac{dy}{dx}=f(x,y)$ can be separated into '$x$-factors' and '$y$-factors', the equation can be solved by the method of *separating the variables*. Now do this one entirely on your own.

**Example 7**

Solve $x\frac{dy}{dx}=y+xy$

*When you have finished it completely, move on to Frame 21 and check your solution*

21

Here is the result. Follow it through carefully, even if your own answer is correct.

$$x\frac{dy}{dx}=y+xy \qquad \therefore\ x\frac{dy}{dx}=y(1+x)$$

$$x\,dy=y(1+x)\,dx \qquad \therefore\ \frac{dy}{y}=\frac{1+x}{x}\,dx$$

$$\therefore\ \int\frac{1}{y}\,dy=\int\left(\frac{1}{x}+1\right)dx$$

$$\therefore\ \ln y=\ln x+x+C$$

At this stage we have eliminated the derivatives and so we have solved the equation. However, we can express the result in a neater form, thus:

$$\ln y-\ln x=x+C$$

$$\therefore\ \ln\left\{\frac{y}{x}\right\}=x+C$$

$$\therefore\ \frac{y}{x}=e^{x+C}=e^x.e^C \quad \text{Now } e^C \text{ is a constant; call it } A.$$

$$\therefore\ \frac{y}{x}=Ae^x \qquad \therefore\ y=Ax\,e^x$$

*Next frame*

22

This final example looks more complicated, but it is solved in just the same way. We go through the same steps as before. Here it is.

**Example 8**

Solve $y\tan x\frac{dy}{dx}=(4+y^2)\sec^2 x$

First separate the variables, i.e. arrange the '$y$-factors' and $dy$ on one side and the '$x$-factors' and $dx$ on the other.

So we get ...........

23

$$\frac{y}{4+y^2}\,dy=\frac{\sec^2 x}{\tan x}\,dx$$

Adding the integral signs, we get:

$$\int\frac{y}{4+y^2}\,dy=\int\frac{\sec^2 x}{\tan x}\,dx$$

Now determine the integrals, so that we have ...........

**24**

$$\frac{1}{2}\ln(4+y^2) = \ln\tan x + C$$

This result can now be simplified into:

$\ln(4+y^2) = 2\ln\tan x + \ln A$ (expressing the constant $2C$ as $\ln A$)

$\therefore\ 4+y^2 = A\tan^2 x$

$\therefore\ y^2 = A\tan^2 x - 4$

So there we are. Provided we can factorize the equation in the way we have indicated, solution by separating the variables is not at all difficult. So now for a short review exercise to wind up this part of the Programme.

*Move on now to Frame 25*

## Review exercise

**25** Work all the exercise before checking your results.

Find the general solutions of the following equations:

1 $\dfrac{dy}{dx} = \dfrac{y}{x}$

2 $\dfrac{dy}{dx} = (y+2)(x+1)$

3 $\cos^2 x\dfrac{dy}{dx} = y+3$

4 $\dfrac{dy}{dx} = xy - y$

5 $\dfrac{\sin x}{1+y}\cdot\dfrac{dy}{dx} = \cos x$

*When you have finished them all, move to Frame 26 and check your answers with the solutions given there*

**26**

1 $\dfrac{dy}{dx} = \dfrac{y}{x} \quad \therefore\ \displaystyle\int\frac{1}{y}dy = \int\frac{1}{x}dx$

$\therefore\ \ln y = \ln x + C$

$= \ln x + \ln A$

$\therefore\ y = Ax$

2 $\dfrac{dy}{dx} = (y+2)(x+1)$

$\therefore\ \displaystyle\int\frac{1}{y+2}dy = \int(x+1)dx$

$\therefore\ \ln(y+2) = \dfrac{x^2}{2} + x + C$

3 $\cos^2 x\dfrac{dy}{dx} = y+3$

$\therefore\ \displaystyle\int\frac{1}{y+3}dy = \int\frac{1}{\cos^2 x}dx$

$= \displaystyle\int\sec^2 x\,dx$

$\ln(y+3) = \tan x + C$

▶

**4** $\dfrac{dy}{dx} = xy - y \quad \therefore \dfrac{dy}{dx} = y(x-1)$

$\therefore \displaystyle\int \frac{1}{y}\,dy = \int (x-1)\,dx$

$\therefore \ln y = \dfrac{x^2}{2} - x + C$

**5** $\dfrac{\sin x}{1+y} \cdot \dfrac{dy}{dx} = \cos x$

$$\int \frac{1}{1+y}\,dy = \int \frac{\cos x}{\sin x}\,dx$$

$$\therefore \ln(1+y) = \ln \sin x + C$$

$$= \ln \sin x + \ln A$$

$$1 + y = A \sin x$$

$$\therefore y = A \sin x - 1$$

*If you are quite happy about those, we can start the next part of the Programme, so move on now to Frame 27*

## 27

## Method 3: *Homogeneous equations – by substituting $y = vx$*

Here is an equation:

$$\frac{dy}{dx} = \frac{x+3y}{2x}$$

This looks simple enough, but we find that we cannot express the RHS in the form of '$x$-factors' and '$y$-factors', so we cannot solve by the method of separating the variables.

In this case we make the substitution $y = vx$, where $v$ is a function of $x$. So $y = vx$. Differentiate with respect to $x$ (using the product rule):

$$\therefore \frac{dy}{dx} = v.1 + x\frac{dv}{dx} = v + x\frac{dv}{dx}$$

Also $\dfrac{x+3y}{2x} = \dfrac{x+3vx}{2x} = \dfrac{1+3v}{2}$

The equation now becomes $v + x\dfrac{dv}{dx} = \dfrac{1+3v}{2}$

$$\therefore x\frac{dv}{dx} = \frac{1+3v}{2} - v$$

$$= \frac{1+3v-2v}{2} = \frac{1+v}{2}$$

$$\therefore x\frac{dv}{dx} = \frac{1+v}{2}$$

The given equation is now expressed in terms of $v$ and $x$, and in this form we find that we can solve by separating the variables. Here goes:

$$\int \frac{2}{1+v}\,dv = \int \frac{1}{x}\,dx$$

$$\therefore 2\ln(1+v) = \ln x + C = \ln x + \ln A$$

$$(1+v)^2 = Ax$$

But $y = vx \quad \therefore v = \left\{\dfrac{y}{x}\right\} \quad \therefore \left(1 + \dfrac{y}{x}\right)^2 = Ax$

which gives $(x+y)^2 = Ax^3$

▶

*Note*: $\dfrac{dy}{dx} = \dfrac{x + 3y}{2x}$ is an example of a *homogeneous differential equation*.

This is determined by the fact that the total degree in $x$ and $y$ for each of the terms involved is the same (in this case, of degree 1). The key to solving every homogeneous equation is to substitute $y = vx$ where $v$ is a function of $x$. This converts the equation into a form which we can solve by separating the variables.

*Let us work some examples, so move on to Frame 28*

## 28

**Example 1**

Solve $\dfrac{dy}{dx} = \dfrac{x^2 + y^2}{xy}$

Here, all terms of the RHS are of degree 2, i.e. the equation is homogeneous. $\therefore$ We substitute $y = vx$ (where $v$ is a function of $x$)

$$\therefore \frac{dy}{dx} = v + x\frac{dv}{dx}$$

$$\text{and} \quad \frac{x^2 + y^2}{xy} = \frac{x^2 + v^2x^2}{vx^2} = \frac{1 + v^2}{v}$$

The equation now becomes:

$$v + x\frac{dv}{dx} = \frac{1 + v^2}{v}$$

$$\therefore x\frac{dv}{dx} = \frac{1 + v^2}{v} - v$$

$$= \frac{1 + v^2 - v^2}{v} = \frac{1}{v}$$

$$\therefore x\frac{dv}{dx} = \frac{1}{v}$$

Now you can separate the variables and get the result in terms of $v$ and $x$.

*Off you go: when you have finished, move on to Frame 29*

## 29

$$\boxed{\frac{v^2}{2} = \ln x + C}$$

Because

$$\int v\,dv = \int \frac{1}{x}\,dx$$

$$\therefore \frac{v^2}{2} = \ln x + C$$

All that remains now is to express $v$ back in terms of $x$ and $y$. The substitution we used was $y = vx \quad \therefore v = \dfrac{y}{x}$

$$\therefore \frac{1}{2}\left(\frac{y}{x}\right)^2 = \ln x + C$$

$$y^2 = 2x^2(\ln x + C)$$

Now, what about this one?

▶

**Example 2**

Solve $\frac{dy}{dx} = \frac{2xy + 3y^2}{x^2 + 2xy}$

Is this a homogeneous equation? If you think so, what are your reasons?

*When you have decided, move on to Frame 30*

---

**30**

Yes, because the degree of each term is the same

Correct. They are all, of course, of degree 2.

So we now make the substitution, $y = \dots\dots\dots$

---

**31**

$y = vx$, where $v$ is a function of $x$

Right. That is the key to the whole process.

$$\frac{dy}{dx} = \frac{2xy + 3y^2}{x^2 + 2xy}$$

So express each side of the equation in terms of $v$ and $x$:

$$\frac{dy}{dx} = \dots\dots\dots$$

and $$\frac{2xy + 3y^2}{x^2 + 2xy} = \dots\dots\dots$$

*When you have finished, move on to the next frame*

---

**32**

$$\frac{dy}{dx} = v + x\frac{dv}{dx}$$
$$\frac{2xy + 3y^2}{x^2 + 2xy} = \frac{2vx^2 + 3v^2x^2}{x^2 + 2vx^2} = \frac{2v + 3v^2}{1 + 2v}$$

So that $v + x\frac{dv}{dx} = \frac{2v + 3v^2}{1 + 2v}$

Now take the single $v$ over to the RHS and simplify, giving:

$$x\frac{dv}{dx} = \dots\dots\dots$$

---

**33**

$$x\frac{dv}{dx} = \frac{2v + 3v^2}{1 + 2v} - v$$
$$= \frac{2v + 3v^2 - v - 2v^2}{1 + 2v}$$
$$x\frac{dv}{dx} = \frac{v + v^2}{1 + 2v}$$

Now you can separate the variables, giving $\dots\dots\dots$

**34**

$$\int \frac{1+2v}{v+v^2}\,dv = \int \frac{1}{x}\,dx$$

Integrating both sides, we can now obtain the solution in terms of $v$ and $x$. What do you get?

**35**

$$\ln(v+v^2) = \ln x + C = \ln x + \ln A$$
$$\therefore v + v^2 = Ax$$

We have almost finished the solution. All that remains is to express $v$ back in terms of $x$ and $y$.

Remember the substitution was $y = vx$, so that $v = \dfrac{y}{x}$

So finish it off.

*Then move on*

**36**

$$xy + y^2 = Ax^3$$

Because

$$v + v^2 = Ax \text{ and } v = \frac{y}{x} \qquad \therefore \frac{y}{x} + \frac{y^2}{x^2} = Ax$$
$$xy + y^2 = Ax^3$$

And that is all there is to it.

*Move to Frame 37*

**37** Here is the solution of the previous equation, all in one piece. Follow it through again.

To solve $\dfrac{dy}{dx} = \dfrac{2xy + 3y^2}{x^2 + 2xy}$

This is homogeneous, all terms of degree 2. Put $y = vx$

$$\therefore \frac{dy}{dx} = v + x\frac{dv}{dx}$$

$$\frac{2xy + 3y^2}{x^2 + 2xy} = \frac{2vx^2 + 3v^2x^2}{x^2 + 2vx^2} = \frac{2v + 3v^2}{1 + 2v} \qquad \therefore v + x\frac{dv}{dx} = \frac{2v + 3v^2}{1 + 2v}$$

$$x\frac{dv}{dx} = \frac{2v + 3v^2}{1 + 2v} - v = \frac{2v + 3v^2 - v - 2v^2}{1 + 2v}$$

$$\therefore x\frac{dv}{dx} = \frac{v + v^2}{1 + 2v} \qquad \therefore \int \frac{1+2v}{v+v^2}\,dv = \int \frac{1}{x}\,dx$$

$$\therefore \ln(v+v^2) = \ln x + C = \ln x + \ln A$$

$$v + v^2 = Ax$$

$$\text{But } y = vx \qquad \therefore v = \frac{y}{x} \qquad \therefore \frac{y}{x} + \frac{y^2}{x^2} = Ax \qquad \therefore xy + y^2 = Ax^3$$

Now, in the same way, you do this one. Take your time and be sure that you understand each step.

▶

**Example 3**

Solve $(x^2+y^2)\dfrac{dy}{dx} = xy$

*When you have completely finished it, move to Frame 38 and check your solution*

---

**38**

Here is the solution in full.

$$(x^2+y^2)\frac{dy}{dx} = xy \qquad \therefore \frac{dy}{dx} = \frac{xy}{x^2+y^2}$$

$$\text{Put } y = vx \quad \therefore \frac{dy}{dx} = v + x\frac{dv}{dx}$$

$$\text{and } \frac{xy}{x^2+y^2} = \frac{vx^2}{x^2+v^2x^2} = \frac{v}{1+v^2}$$

$$\therefore v + x\frac{dv}{dx} = \frac{v}{1+v^2}$$

$$x\frac{dv}{dx} = \frac{v}{1+v^2} - v$$

$$x\frac{dv}{dx} = \frac{v-v-v^3}{1+v^2} = \frac{-v^3}{1+v^2}$$

$$\therefore \int \frac{1+v^2}{v^3}\,dv = -\int \frac{1}{x}\,dx$$

$$\therefore \int \left(v^{-3} + \frac{1}{v}\right) dv = -\ln x + C$$

$$\therefore \frac{-v^{-2}}{2} + \ln v = -\ln x + \ln A$$

$$\ln v + \ln x + \ln K = \frac{1}{2v^2} \qquad (\ln K = -\ln A)$$

$$\ln Kvx = \frac{1}{2v^2}$$

$$\text{But } v = \frac{y}{x} \qquad \therefore \ln Ky = \frac{x^2}{2y^2}$$

$$2y^2 \ln Ky = x^2$$

This is one form of the solution: there are of course other ways of expressing it.

*Now for a short review exercise on this part of the work, move on to Frame 39*

---

## Review exercise

**39**

Solve the following:

1 $(x-y)\dfrac{dy}{dx} = x+y$

2 $2x^2\dfrac{dy}{dx} = x^2+y^2$

3 $(x^2+xy)\dfrac{dy}{dx} = xy - y^2$

*When you have finished all three, move on and check your results*

**40**

The solution of equation **1** can be written as:

$$\tan^{-1}\left\{\frac{y}{x}\right\} = \ln A + \ln x + \frac{1}{2}\ln\left\{1 + \frac{y^2}{x^2}\right\}$$

Did you get that? If so, move straight on to Frame 41. If not, check your working with the following.

**1** $(x-y)\dfrac{dy}{dx} = x + y \quad \therefore \dfrac{dy}{dx} = \dfrac{x+y}{x-y}$

Put $y = vx \quad \therefore \dfrac{dy}{dx} = v + x\dfrac{dv}{dx} \qquad \dfrac{x+y}{x-y} = \dfrac{1+v}{1-v}$

$\therefore v + x\dfrac{dv}{dx} = \dfrac{1+v}{1-v} \qquad \therefore x\dfrac{dv}{dx} = \dfrac{1+v}{1-v} - v = \dfrac{1+v-v+v^2}{1-v} = \dfrac{1+v^2}{1-v}$

$\therefore \displaystyle\int \frac{1-v}{1+v^2}\,dv = \int \frac{1}{x}\,dx \quad \therefore \int\left\{\frac{1}{1+v^2} - \frac{v}{1+v^2}\right\}dv = \ln x + C$

$\therefore \tan^{-1} v - \dfrac{1}{2}\ln(1+v^2) = \ln x + \ln A$

But $v = \dfrac{y}{x} \quad \therefore \tan^{-1}\left\{\dfrac{y}{x}\right\} = \ln A + \ln x + \dfrac{1}{2}\ln\left(1 + \dfrac{y^2}{x^2}\right)$

This result can, in fact, be simplified further.

*Now on to Frame 41*

**41**

Equation **2** gives the solution: $\boxed{\dfrac{2x}{x-y} = \ln x + C}$

If you agree, move straight on to Frame 42. Otherwise, follow through the working. Here it is.

**2** $2x^2\dfrac{dy}{dx} = x^2 + y^2 \quad \therefore \dfrac{dy}{dx} = \dfrac{x^2+y^2}{2x^2}$

Put $y = vx \quad \therefore \dfrac{dy}{dx} = v + x\dfrac{dv}{dx}; \quad \dfrac{x^2+y^2}{2x^2} = \dfrac{x^2+v^2x^2}{2x^2} = \dfrac{1+v^2}{2}$

$\therefore v + x\dfrac{dv}{dx} = \dfrac{1+v^2}{2} \qquad \therefore x\dfrac{dv}{dx} = \dfrac{1+v^2}{2} - v = \dfrac{1-2v+v^2}{2} = \dfrac{(v-1)^2}{2}$

$\therefore \displaystyle\int \frac{2}{(v-1)^2}\,dv = \int \frac{1}{x}\,dx \quad \therefore -2\frac{1}{v-1} = \ln x + C$

But $v = \dfrac{y}{x}$ and $\dfrac{2}{1-v} = \ln x + C \quad \therefore \dfrac{2x}{x-y} = \ln x + C$

*On to Frame 42*

42

One form of the result for equation **3** is: $\boxed{xy = Ae^{x/y}}$. Follow through the working and check yours.

**3** $(x^2 + xy)\dfrac{dy}{dx} = xy - y^2 \quad \therefore \dfrac{dy}{dx} = \dfrac{xy - y^2}{x^2 + xy}$

Put $y = vx \quad \therefore \dfrac{dy}{dx} = v + x\dfrac{dv}{dx}; \quad \dfrac{xy - y^2}{x^2 + xy} = \dfrac{vx^2 - v^2x^2}{x^2 + vx^2} = \dfrac{v - v^2}{1 + v}$

$$\therefore v + x\frac{dv}{dx} = \frac{v - v^2}{1 + v}$$

$$x\frac{dv}{dx} = \frac{v - v^2}{1 + v} - v = \frac{v - v^2 - v - v^2}{1 + v} = \frac{-2v^2}{1 + v}$$

$$\therefore \int \frac{1 + v}{v^2}\,dv = \int \frac{-2}{x}\,dx$$

$$\int \left(v^{-2} + \frac{1}{v}\right) dv = -\int \frac{2}{x}\,dx$$

$$\therefore \ln v - \frac{1}{v} = -2\ln x + C \qquad \text{Let } C = \ln A$$

$$\ln v + 2\ln x = \ln A + \frac{1}{v}$$

$$\ln\left\{\frac{y}{x}.x^2\right\} = \ln A + \frac{x}{y} \quad \therefore xy = Ae^{x/y}$$

*Now move to the next frame*

## Method 4: *Linear equations – use of integrating factor*

43

Consider the equation $\dfrac{dy}{dx} + 5y = e^{2x}$

This is clearly an equation of the first order, but different from those we have dealt with so far. In fact, none of our previous methods could be used to solve this one, so we have to find a further method of attack.

In this case, we begin by multiplying both sides by $e^{5x}$. This gives

$$e^{5x}\frac{dy}{dx} + y5e^{5x} = e^{2x}.e^{5x} = e^{7x}$$

We now find that the LHS is, in fact, the derivative of $y.e^{5x}$.

$$\therefore \frac{d}{dx}\left\{y.e^{5x}\right\} = e^{7x}$$

Now, of course, the rest is easy. Integrate both sides with respect to $x$:

$$\therefore y.e^{5x} = \int e^{7x}\,dx = \frac{e^{7x}}{7} + C \quad \therefore y = \ldots\ldots\ldots\ldots$$

44

$$\boxed{y = \frac{e^{2x}}{7} + Ce^{-5x}}$$

Did you forget to divide the $C$ by the $e^{5x}$? It is a common error so watch out for it.

The equation we have just solved is an example of a set of equations of the form $\dfrac{dy}{dx} + Py = Q$, where $P$ and $Q$ are functions of $x$ (or constants). This equation is called a *linear equation of the first order* and to solve any such equation, we multiply both sides by an *integrating factor* which is always $e^{\int P dx}$. This converts the LHS into the derivative of a product.

▶

In our previous example, $\frac{dy}{dx} + 5y = e^{2x}$, $P = 5$

$\therefore \int P dx = 5x$ and the integrating factor was therefore $e^{5x}$.

*Note*: In determining $\int P\,dx$, we do not include a constant of integration. This omission is purely for convenience, for a constant of integration here would in practice give a constant factor on both sides of the equation, which would subsequently cancel. This is one of the rare occasions when we do not write down the constant of integration.

So: *To solve a differential equation of the form*

$$\frac{dy}{dx} + Py = Q$$

*where P and Q are constants or functions of x, multiply both sides by the integrating factor* $e^{\int P dx}$

This is important, so copy this rule down into your record book.

*Then move on to Frame 45*

**45**

**Example 1**

To solve $\frac{dy}{dx} - y = x$

If we compare this with $\frac{dy}{dx} + Py = Q$, we see that in this case

$P = -1$ and $Q = x$.

The integrating factor is always $e^{\int P dx}$ and here $P = -1$.

$\therefore \int P\,dx = -x$ and the integrating factor is therefore ...........

**46**

$e^{-x}$

We therefore multiply both sides by $e^{-x}$.

$$\therefore e^{-x}\frac{dy}{dx} - ye^{-x} = xe^{-x}$$

$$\frac{d}{dx}\left\{e^{-x}y\right\} = xe^{-x} \quad \therefore ye^{-x} = \int xe^{-x}\,dx$$

The RHS integral can now be determined by integrating by parts:

$$ye^{-x} = x(-e^{-x}) + \int e^{-x}\,dx = -xe^{-x} - e^{-x} + C$$

$$\therefore y = -x - 1 + Ce^{x}$$

$$\therefore y = Ce^{x} - x - 1$$

The whole method really depends on:

(a) being able to find the integrating factor

(b) being able to deal with the integral that emerges on the RHS.

Let us consider the general case.

47

Consider $\frac{dy}{dx} + Py = Q$ where $P$ and $Q$ are functions of $x$. Integrating factor, IF $= e^{\int P dx}$

$$\therefore \frac{dy}{dx}.e^{\int P dx} + Pye^{\int P dx} = Qe^{\int P dx}$$

You will now see that the LHS is the derivative of $ye^{\int P dx}$

$$\therefore \frac{d}{dx}\left\{ye^{\int P dx}\right\} = Qe^{\int P dx}$$

Integrate both sides with respect to $x$:

$$ye^{\int P dx} = \int Qe^{\int P dx}.dx$$

This result looks far more complicated than it really is. If we indicate the integrating factor by IF, this result becomes:

$$y.\text{IF} = \int Q.\text{IF}\, dx$$

and, in fact, we remember it in that way.

*So, the solution of an equation of the form*

$$\frac{dy}{dx} + Py = Q \text{ (where } P \text{ and } Q \text{ are functions of } x\text{)}$$

is given by $y.\text{IF} = \int Q.\text{IF}\, dx$ where IF $= e^{\int P dx}$

Copy this into your record book.

*Then move to Frame 48*

48

So if we have the equation:

$$\frac{dy}{dx} + 3y = \sin x \qquad \left[\frac{dy}{dx} + Py = Q\right]$$

then in this case:

(a) $P =$ ........... (b) $\int P\, dx =$ ........... (c) IF $=$ ...........

49

(a) $P = 3$ (b) $\int P\, dx = 3x$ (c) IF $= e^{3x}$

Before we work through any further examples, let us establish a very useful piece of simplification, which we can make good use of when we are finding integrating factors. We want to simplify $e^{\ln F}$, where $F$ is a function of $x$.

Let $y = e^{\ln F}$

Then, by the very definition of a logarithm, $\ln y = \ln F$

$\therefore y = F \quad \therefore F = e^{\ln F}$ i.e. $e^{\ln F} = F$

This means that $e^{\ln(\text{function})} =$ function. Always!

$$e^{\ln x} = x$$

$$e^{\ln \sin x} = \sin x$$

$$e^{\ln \tanh x} = \tanh x$$

$$e^{\ln(x^2)} = \ldots\ldots\ldots\ldots$$

**50**

$$\boxed{x^2}$$

Similarly, what about $e^{k \ln F}$? If the log in the index is multiplied by any external coefficient, this coefficient must be taken inside the log as a power.

e.g. $e^{2 \ln x} = e^{\ln(x^2)} = x^2$

$e^{3 \ln \sin x} = e^{\ln(\sin^3 x)} = \sin^3 x$

$e^{-\ln x} = e^{\ln(x^{-1})} = x^{-1} = \dfrac{1}{x}$

and $e^{-2 \ln x} = \dots\dots\dots\dots$

**51**

$$\boxed{\frac{1}{x^2}} \quad \text{because } e^{-2 \ln x} = e^{\ln(x^{-2})} = x^{-2} = \frac{1}{x^2}$$

So here is the rule once again: $e^{\ln F} = F$

Make a note of this rule in your record book.

*Then on to Frame 52*

**52** Now let us see how we can apply this result to our working.

**Example 2**

Solve $x\dfrac{dy}{dx} + y = x^3$

First we divide through by $x$ to reduce the first term to a single $\dfrac{dy}{dx}$

i.e. $\dfrac{dy}{dx} + \dfrac{1}{x}.y = x^2$

Compare with $\left[\dfrac{dy}{dx} + Py = Q\right]$

$\therefore P = \dfrac{1}{x}$ and $Q = x^2$

$\text{IF} = e^{\int P dx} \qquad \int P\,dx = \int \dfrac{1}{x}\,dx = \ln x$

$\therefore \text{IF} = e^{\ln x} = x$

$\therefore \text{IF} = x$

The solution is $y.\text{IF} = \int Q.\text{IF}\,dx$ so

$yx = \int x^2.x\,dx = \int x^3\,dx = \dfrac{x^4}{4} + C$

$\therefore xy = \dfrac{x^4}{4} + C$

*Move to Frame 53*

53

**Example 3**

Solve $\frac{dy}{dx} + y \cot x = \cos x$

Compare with $\left[\frac{dy}{dx} + Py = Q\right] \quad \therefore \begin{cases} P = \cot x \\ Q = \cos x \end{cases}$

$\text{IF} = e^{\int P dx} \qquad \int P\,dx = \int \cot x\,dx = \int \frac{\cos x}{\sin x}\,dx = \ln \sin x$

$\therefore \text{IF} = e^{\ln \sin x} = \sin x$

$y.\text{IF} = \int Q.\text{IF}\,dx \quad \therefore y \sin x = \int \sin x \cos x\,dx = \frac{\sin^2 x}{2} + C$

$\therefore y = \frac{\sin x}{2} + C\ x$

Now here is another.

**Example 4**

Solve $(x+1)\frac{dy}{dx} + y = (x+1)^2$

The first thing is to ...........

54

Divide through by $(x+1)$

Correct, since we must reduce the coefficient of $\frac{dy}{dx}$ to 1.

$\therefore \frac{dy}{dx} + \frac{1}{x+1}.y = x + 1$

Compare with $\frac{dy}{dx} + Py = Q$

In this case $P = \frac{1}{x+1}$ and $Q = x + 1$

Now determine the integrating factor, which simplifies to

$\text{IF} =$ ...........

55

$\text{IF} = x + 1$

Because

$\int P\,dx = \int \frac{1}{x+1}\,dx = \ln(x+1)$

$\therefore \text{IF} = e^{\ln(x+1)} = (x+1)$

The solution is always $y.\text{IF} = \int Q.\text{IF}\,dx$

and we know that, in this case, $\text{IF} = x + 1$ and $Q = x + 1$.

*So finish off the solution and then move on to Frame 56*

**56**

$$y = \frac{(x+1)^2}{3} + \frac{C}{x+1}$$

Here is the solution in detail:

$$y.(x+1) = \int (x+1)(x+1)\,dx$$

$$= \int (x+1)^2\,dx$$

$$= \frac{(x+1)^3}{3} + C$$

$$\therefore y = \frac{(x+1)^2}{3} + \frac{C}{x+1}$$

Now let us do another one.

**Example 5**

Solve $x\frac{dy}{dx} - 5y = x^7$

In this case, $P = \ldots\ldots\ldots\ldots$ $Q = \ldots\ldots\ldots\ldots$

**57**

$$P = -\frac{5}{x} \quad Q = x^6$$

Because if

$$x\frac{dy}{dx} - 5y = x^7$$

$$\therefore \frac{dy}{dx} - \frac{5}{x}.y = x^6$$

Compare with $\left[\frac{dy}{dx} + Py = Q\right]$ $\therefore P = -\frac{5}{x}$; $Q = x^6$

So integrating factor, IF $= \ldots\ldots\ldots\ldots$

**58**

$$\text{IF} = x^{-5} = \frac{1}{x^5}$$

Because

$$\text{IF} = e^{\int P dx} \qquad \int P\,dx = -\int \frac{5}{x}\,dx = -5\ln x$$

$$\therefore \text{IF} = e^{-5\ln x} = e^{\ln(x^{-5})} = x^{-5} = \frac{1}{x^5}$$

So the solution is:

$$y.\frac{1}{x^5} = \int x^6.\frac{1}{x^5}\,dx$$

$$\frac{y}{x^5} = \int x\,dx = \frac{x^2}{2} + C \qquad \therefore y = \ldots\ldots\ldots\ldots$$

59

$$y = \frac{x^7}{2} + Cx^5$$

Did you remember to multiply the $C$ by $x^5$?

Fine. Now you do this one entirely on your own.

**Example 6**

Solve $(1 - x^2)\frac{dy}{dx} - xy = 1$.

*When you have finished it, move to Frame 60*

60

$$y\sqrt{1 - x^2} = \sin^{-1} x + C$$

Here is the working in detail. Follow it through.

$$(1 - x^2)\frac{dy}{dx} - xy = 1$$

$$\therefore \frac{dy}{dx} - \frac{x}{1 - x^2}.y = \frac{1}{1 - x^2}$$

$$\text{IF} = e^{\int P dx} \qquad \int P\,dx = \int \frac{-x}{1 - x^2}\,dx = \frac{1}{2}\ln(1 - x^2)$$

$$\therefore \text{IF} = e^{\frac{1}{2}\ln(1 - x^2)} = e^{\ln\{(1 - x^2)^{\frac{1}{2}}\}} = (1 - x^2)^{\frac{1}{2}}$$

$$\text{Now } y.\text{IF} = \int Q.\text{IF}\,dx$$

$$\therefore y\sqrt{1 - x^2} = \int \frac{1}{1 - x^2}\sqrt{1 - x^2}.dx$$

$$= \int \frac{1}{\sqrt{1 - x^2}}\,dx = \sin^{-1} x + C$$

$$y\sqrt{1 - x^2} = \sin^{-1} x + C$$

*Now on to Frame 61*

61

In practically all the examples so far, we have been concerned with finding the general solutions. If further information is available, of course, particular solutions can be obtained. Here is one final example for you to do.

**Example 7**

Solve the equation

$$(x - 2)\frac{dy}{dx} - y = (x - 2)^3$$

given that $y = 10$ when $x = 4$.

Off you go then. It is quite straightforward.

*When you have finished, move on to Frame 62 and check your solution*

**62**

$$2y = (x-2)^3 + 6(x-2)$$

Here it is:

$$(x-2)\frac{dy}{dx} - y = (x-2)^3$$

$$\frac{dy}{dx} - \frac{1}{x-2}.y = (x-2)^2$$

$$\int P\,dx = \int \frac{-1}{x-2}\,dx = -\ln(x-2)$$

$$\therefore \text{IF} = e^{-\ln(x-2)} = e^{\ln\{(x-2)^{-1}\}} = (x-2)^{-1}$$

$$= \frac{1}{x-2}$$

$$\therefore y.\frac{1}{x-2} = \int (x-2)^2.\frac{1}{(x-2)}\,dx$$

$$= \int (x-2)\,dx$$

$$= \frac{(x-2)^2}{2} + C$$

$$\therefore y = \frac{(x-2)^3}{2} + C(x-2) \qquad \text{General solution}$$

When $x = 4$, $y = 10$:

$$10 = \frac{8}{2} + C.2 \quad \therefore 2C = 6 \quad \therefore C = 3$$

$$\therefore 2y = (x-2)^3 + 6(x-2)$$

## Review exercise

**63**

Finally, for this part of the Programme, here is a short revision exercise.

Solve the following:

1 $\dfrac{dy}{dx} + 3y = e^{4x}$

2 $x\dfrac{dy}{dx} + y = x\sin x$

3 $\tan x\dfrac{dy}{dx} + y = \sec x$

*Work through them all: then check your results with those given in Frame 64*

**64**

1 $y = \dfrac{e^{4x}}{7} + Ce^{-3x}$ $\quad(\text{IF} = e^{3x})$

2 $xy = \sin x - x\cos x + C$ $\quad(\text{IF} = x)$

3 $y\sin x = x + C$ $\quad(\text{IF} = \sin x)$

▶

There is just one other type of equation that we must consider. Here is an example: let us see how it differs from those we have already dealt with.

To solve $\dfrac{dy}{dx} + \dfrac{1}{x}.y = xy^2$

Note that if it were not for the factor $y^2$ on the right-hand side, this equation would be of the form $\dfrac{dy}{dx} + Py = Q$ that we know of old.

To see how we deal with this new kind of equation, we will consider the general form, so move on to Frame 65.

## Bernoulli's equation

**65**

The differential equation:

$$\frac{dy}{dx} + P(x)y = Q(x)y^n$$

is known as Bernoulli's equation and it is solved as follows:

(a) Divide both sides by $y^n$. This gives:

$$y^{-n}\frac{dy}{dx} + Py^{1-n} = Q$$

(b) Now put $z = y^{1-n}$

so that, differentiating, $\dfrac{dz}{dx} = \ldots\ldots\ldots\ldots$

**66**

$$\boxed{\frac{dz}{dx} = (1-n)y^{-n}\frac{dy}{dx}}$$

So we have:

$$\frac{dy}{dx} + Py = Qy^n \qquad (1)$$

$$\therefore\ y^{-n}\frac{dy}{dx} + Py^{1-n} = Q \qquad (2)$$

Put $z = y^{1-n}$ so that $\dfrac{dz}{dx} = (1-n)y^{-n}\dfrac{dy}{dx}$

If we now multiply (2) by $(1-n)$ we shall convert the first term into $\dfrac{dz}{dx}$.

$$(1-n)y^{-n}\frac{dy}{dx} + (1-n)Py^{1-n} = (1-n)Q$$

Remembering that $z = y^{1-n}$ and that $\dfrac{dz}{dx} = (1-n)y^{-n}\dfrac{dy}{dx}$, this last line can now be written $\dfrac{dz}{dx} + P_1z = Q_1$ with $P_1$ and $Q_1$ functions of $x$.

This we can solve by use of an integrating factor in the normal way.

Finally, having found $z$, we convert back to $y$ using $z = y^{1-n}$.

*Let us see this routine in operation – so on to Frame 67*

**67**

**Example 1**

Solve $\dfrac{dy}{dx}+\dfrac{1}{x}y=xy^2$

(a) Divide through by $y^2$, giving ...........

**68**

$$\boxed{y^{-2}\frac{dy}{dx}+\frac{1}{x}.y^{-1}=x}$$

(b) Now put $z=y^{1-n}$, i.e. in this case $z=y^{1-2}=y^{-1}$

$$z=y^{-1}\quad\therefore\ \frac{dz}{dx}=-y^{-2}\frac{dy}{dx}$$

(c) Multiply through the equation by $(-1)$, to make the first term $\dfrac{dz}{dx}$.

$$-y^{-2}\frac{dy}{dx}-\frac{1}{x}y^{-1}=-x$$

so that $\dfrac{dz}{dx}-\dfrac{1}{x}z=-x$ which is of the form $\dfrac{dz}{dx}+Pz=Q$ so that you can now solve the equation by the normal integrating factor method. What do you get?

*When you have done it, move on to the next frame*

**69**

$$\boxed{y=(Cx-x^2)^{-1}}$$

Check the working:

$$\frac{dz}{dx}-\frac{1}{x}z=-x$$

$$\text{IF}=e^{\int P dx}\qquad \int P\,dx=\int-\frac{1}{x}\,dx=-\ln x$$

$$\therefore\ \text{IF}=e^{-\ln x}=e^{\ln(x^{-1})}=x^{-1}=\frac{1}{x}$$

$$z.\text{IF}=\int Q.\text{IF}\,dx\quad\therefore\ z\frac{1}{x}=\int -x.\frac{1}{x}\,dx$$

$$\therefore\ \frac{z}{x}=\int -1\,dx=-x+C$$

$$\therefore\ z=Cx-x^2$$

But $z=y^{-1}\quad\therefore\ \dfrac{1}{y}=Cx-x^2\quad\therefore\ y=(Cx-x^2)^{-1}$

Right! Here is another.

**Example 2**

Solve $x^2y-x^3\dfrac{dy}{dx}=y^4\cos x$

First of all, we must rewrite this in the form $\dfrac{dy}{dx}+Py=Qy^n$

So, what do we do?

**70**

Divide both sides by $(-x^3)$

giving $\frac{dy}{dx} - \frac{1}{x}.y = -\frac{y^4 \cos x}{x^3}$

Now divide by the power of $y$ on the RHS giving . . . . . . . . . . .

**71**

$$y^{-4}\frac{dy}{dx} - \frac{1}{x}y^{-3} = -\frac{\cos x}{x^3}$$

Next we make the substitution $z = y^{1-n}$ which, in this example, is $z = y^{1-4} = y^{-3}$

$\therefore z = y^{-3}$ and $\therefore \frac{dz}{dx} =$ . . . . . . . . . . .

**72**

$$\frac{dz}{dx} = -3y^{-4}\frac{dy}{dx}$$

If we now multiply the equation by $(-3)$ to make the first term into $\frac{dz}{dx}$, we have:

$$-3y^{-4}\frac{dy}{dx} + 3\frac{1}{x}.y^{-3} = \frac{3\cos x}{x^3}$$

$$\text{i.e.} \quad \frac{dz}{dx} + \frac{3}{x}z = \frac{3\cos x}{x^3}$$

This you can now solve to find $z$ and so back to $y$.

*Finish it off and then check with the next frame*

**73**

$$y^3 = \frac{x^3}{3\sin x + C}$$

Because

$$\frac{dz}{dx} + \frac{3}{x}.z = \frac{3\cos x}{x^3}$$

$$\text{IF} = e^{\int P dx} \qquad \int P\,dx = \int \frac{3}{x}\,dx = 3\ln x$$

$$\therefore \text{IF} = e^{3\ln x} = e^{\ln(x^3)} = x^3$$

$$z.\text{IF} = \int Q.\text{IF}\,dx$$

$$\therefore zx^3 = \int \frac{3\cos x}{x^3}x^3\,dx = \int 3\cos x\,dx$$

$$\therefore zx^3 = 3\sin x + C$$

But, in this example, $z = y^{-3}$

$$\therefore \frac{x^3}{y^3} = 3\sin x + C$$

$$\therefore y^3 = \frac{x^3}{3\sin x + C}$$

*Let us look at the complete solution as a whole, so on to* *Frame 74*

**74**

Here it is:

To solve $x^2y - x^3\dfrac{dy}{dx} = y^4\cos x$

$$\frac{dy}{dx} - \frac{1}{x}y = -\frac{y^4\cos x}{x^3}$$

$$\therefore y^{-4}\frac{dy}{dx} - \frac{1}{x}y^{-3} = -\frac{\cos x}{x^3}$$

Put $z = y^{1-n} = y^{1-4} = y^{-3}$

$$\therefore \frac{dz}{dx} = -3y^{-4}\frac{dy}{dx}$$

Equation becomes

$$-3y^{-4}\frac{dy}{dx} + \frac{3}{x}.y^{-3} = \frac{3\cos x}{x^3}$$

$$\text{i.e.}\quad \frac{dz}{dx} + \frac{3}{x}.z = \frac{3\cos x}{x^3}$$

$$\text{IF} = e^{\int P dx} \qquad \int P\,dx = \int \frac{3}{x}\,dx = 3\ln x$$

$$\therefore \text{IF} = e^{3\ln x} = e^{\ln(x^3)} = x^3$$

$$\therefore zx^3 = \int \frac{3\cos x}{x^3}x^3\,dx$$

$$= \int 3\cos x\,dx$$

$$\therefore zx^3 = 3\sin x + C$$

But $\quad z = y^{-3}$

$$\therefore \frac{x^3}{y^3} = 3\sin x + C$$

$$\therefore y^3 = \frac{x^3}{3\sin x + C}$$

They are all done in the same way. Once you know the trick, the rest is very straightforward.

*On to the next frame*

**75**

Here is one for you to do entirely on your own.

**Example 3**

Solve $\quad 2y - 3\dfrac{dy}{dx} = y^4e^{3x}$

Work through the same steps as before. When you have finished, check your working with the solution in Frame 76.

76

$$y^3 = \frac{5e^{2x}}{e^{5x} + A}$$

Solution in detail:

$$2y - 3\frac{dy}{dx} = y^4 e^{3x}$$

$$\therefore \frac{dy}{dx} - \frac{2}{3}y = -\frac{y^4 e^{3x}}{3}$$

$$\therefore y^{-4}\frac{dy}{dx} - \frac{2}{3}y^{-3} = -\frac{e^{3x}}{3}$$

Put $z = y^{1-4} = y^{-3} \qquad \therefore \frac{dz}{dx} = -3y^{-4}\frac{dy}{dx}$

Multiplying through by $(-3)$, the equation becomes:

$$-3y^{-4}\frac{dy}{dx} + 2y^{-3} = e^{3x}$$

$$\text{i.e.} \quad \frac{dz}{dx} + 2z = e^{3x}$$

$$\text{IF} = e^{\int P\,dx} \qquad \int P\,dx = \int 2\,dx = 2x \quad \therefore \text{IF} = e^{2x}$$

$$\therefore ze^{2x} = \int e^{3x}e^{2x}dx = \int e^{5x}dx$$

$$= \frac{e^{5x}}{5} + C$$

But $z = y^{-3} \quad \therefore \frac{e^{2x}}{y^3} = \frac{e^{5x} + A}{5}$

$$\therefore y^3 = \frac{5e^{2x}}{e^{5x} + A}$$

*On to Frame 77*

77

Finally, one example for you, just to be sure.

**Example 4**

Solve $y - 2x\frac{dy}{dx} = x(x+1)y^3$

First rewrite the equation in standard form $\frac{dy}{dx} + Py = Qy^n$

This gives ............

78

$$\frac{dy}{dx} - \frac{1}{2x}.y = -\frac{(x+1)y^3}{2}$$

*Now off you go and complete the solution. When you have finished, check with the working in Frame 79.*

**79**

$$y^2 = \frac{6x}{2x^3 + 3x^2 + A}$$

Working:

$$\frac{dy}{dx} - \frac{1}{2x}.y = -\frac{(x+1)y^3}{2}$$

$$\therefore y^{-3}\frac{dy}{dx} - \frac{1}{2x}.y^{-2} = -\frac{(x+1)}{2}$$

Put $z = y^{1-3} = y^{-2}$ $\quad \therefore \frac{dz}{dx} = -2y^{-3}\frac{dy}{dx}$

Equation becomes:

$$-2y^{-3}\frac{dy}{dx} + \frac{1}{x}.y^{-2} = x + 1$$

$$\text{i.e. } \frac{dz}{dx} + \frac{1}{x}.z = x + 1$$

$$\text{IF} = e^{\int P dx} \qquad \int P\,dx = \int \frac{1}{x}\,dx = \ln x$$

$$\therefore \text{IF} = e^{\ln x} = x$$

$$z.\text{IF} = \int Q.\text{IF}\,dx \quad \therefore zx = \int (x+1)x\,dx$$

$$= \int (x^2 + x)\,dx$$

$$\therefore zx = \frac{x^3}{3} + \frac{x^2}{2} + C$$

$$\text{But } z = y^{-2} \quad \therefore \frac{x}{y^2} = \frac{2x^3 + 3x^2 + A}{6} \qquad (A = 6C)$$

$$\therefore y^2 = \frac{6x}{2x^3 + 3x^2 + A}$$

There we are. You have now reached the end of this Programme, except for the **Can You?** checklist and the **Test exercise** which follow. Before you tackle them, however, read down the **Review summary** presented in the next frame. It will remind you of the main points that we have covered in this Programme on first-order differential equations.

*Move on then to Frame 80*

## Review summary

**80**

1 The *order* of a differential equation is given by the highest derivative present. An equation of *order n* is derived from a function containing *n arbitrary constants.*

2 *Solution of first-order differential equations*

(a) By direct integration: $\frac{dy}{dx} = f(x)$

gives $y = \int f(x)\,dx$

(b) By separating the variables: $F(y).\dfrac{dy}{dx} = f(x)$

gives $\int F(y)\,dy = \int f(x)\,dx$

(c) Homogeneous equations: Substituting $y = vx$

gives $v + x\dfrac{dv}{dx} = F(v)$

(d) Linear equations: $\dfrac{dy}{dx} + Py = Q$

Integrating factor, $\text{IF} = e^{\int P dx}$

and remember that $e^{\ln F} = F$

gives $y\text{IF} = \int Q.\text{IF}\,dx$

(e) Bernoulli's equation: $\dfrac{dy}{dx} + Py = Qy^n$

Divide by $y^n$: then put $z = y^{1-n}$

Reduces to type (d) above.

---

## Can You?

### Checklist 25

81

Check this list before and after you try the end of Programme test.

On a scale of 1 to 5 how confident are you that you can:

| | Frames |
|---|---|
| • Recognize the order of a differential equation?<br>*Yes* ☐ ☐ ☐ ☐ ☐ *No* | 1 to 3 |
| • Appreciate that a differential equation of order $n$ can be derived from a function containing $n$ arbitrary constants?<br>*Yes* ☐ ☐ ☐ ☐ ☐ *No* | 4 to 10 |
| • Solve certain first-order differential equations by direct integration?<br>*Yes* ☐ ☐ ☐ ☐ ☐ *No* | 11 to 13 |
| • Solve certain first-order differential equations by separating the variables?<br>*Yes* ☐ ☐ ☐ ☐ ☐ *No* | 14 to 26 |
| • Solve certain first-order homogeneous differential equations by an appropriate substitution?<br>*Yes* ☐ ☐ ☐ ☐ ☐ *No* | 27 to 42 |
| • Solve certain first-order differential equations by using an integrating factor?<br>*Yes* ☐ ☐ ☐ ☐ ☐ *No* | 43 to 64 |
| • Solve Bernoulli's equation<br>*Yes* ☐ ☐ ☐ ☐ ☐ *No* | 65 to 79 |

## Test exercise 25

82

The questions are similar to the equations you have been solving in the Programme. They cover all the methods, but are quite straightforward. Do not hurry: take your time and work carefully and you will find no difficulty with them.

Solve the following differential equations:

1 $x\dfrac{dy}{dx} = x^2 + 2x - 3$

2 $(1+x)^2\dfrac{dy}{dx} = 1 + y^2$

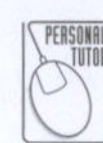


3 $\dfrac{dy}{dx} + 2y = e^{3x}$

4 $x\dfrac{dy}{dx} - y = x^2$

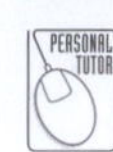


5 $x^2\dfrac{dy}{dx} = x^3 \sin 3x + 4$

6 $x\cos y\dfrac{dy}{dx} - \sin y = 0$

7 $(x^3 + xy^2)\dfrac{dy}{dx} = 2y^3$

8 $(x^2 - 1)\dfrac{dy}{dx} + 2xy = x$

9 $\dfrac{dy}{dx} + y\tanh x = 2\sinh x$

10 $x\dfrac{dy}{dx} - 2y = x^3 \cos x$

11 $\dfrac{dy}{dx} + \dfrac{y}{x} = y^3$

12 $x\dfrac{dy}{dx} + 3y = x^2y^2$

## Further problems 25

83

Solve the following equations.

I. *Separating the variables*

1 $x(y-3)\dfrac{dy}{dx} = 4y$

2 $(1+x^3)\dfrac{dy}{dx} = x^2y$, given that $x = 1$ when $y = 2$.

3 $x^3 + (y+1)^2\dfrac{dy}{dx} = 0$

4 $\cos y + (1 + e^{-x})\sin y\dfrac{dy}{dx} = 0$, given that $y = \pi/4$ when $x = 0$.

5 $x^2(y+1) + y^2(x-1)\dfrac{dy}{dx} = 0$

II. *Homogeneous equations*

6 $(2y - x)\dfrac{dy}{dx} = 2x + y$, given that $y = 3$ when $x = 2$.

7 $(xy + y^2) + (x^2 - xy)\dfrac{dy}{dx} = 0$

8 $(x^3 + y^3) = 3xy^2\dfrac{dy}{dx}$

9 $y - 3x + (4y + 3x)\dfrac{dy}{dx} = 0$

10 $(x^3 + 3xy^2)\dfrac{dy}{dx} = y^3 + 3x^2y$

III. *Integrating factor*

11 $x\frac{dy}{dx} - y = x^3 + 3x^2 - 2x$

12 $\frac{dy}{dx} + y\tan x = \sin x$

13 $x\frac{dy}{dx} - y = x^3 \cos x$, given that $y = 0$ when $x = \pi$.

14 $(1 + x^2)\frac{dy}{dx} + 3xy = 5x$, given that $y = 2$ when $x = 1$.

15 $\frac{dy}{dx} + y\cot x = 5e^{\cos x}$, given that $y = -4$ when $x = \pi/2$.

IV. *Transformations*. Make the given substitutions and work in much the same way as for first-order homogeneous equations.

16 $(3x + 3y - 4)\frac{dy}{dx} = -(x + y)$ Put $x + y = v$

17 $(y - xy^2) = (x + x^2y)\frac{dy}{dx}$ Put $y = \frac{v}{x}$

18 $(x - y - 1) + (4y + x - 1)\frac{dy}{dx} = 0$ Put $v = x - 1$

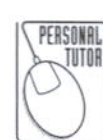

19 $(3y - 7x + 7) + (7y - 3x + 3)\frac{dy}{dx} = 0$ Put $v = x - 1$

20 $y(xy + 1) + x(1 + xy + x^2y^2)\frac{dy}{dx} = 0$ Put $y = \frac{v}{x}$

V. *Bernoulli's equation*

21 $\frac{dy}{dx} + y = xy^3$

22 $\frac{dy}{dx} + y = y^4 e^x$

23 $2\frac{dy}{dx} + y = y^3(x - 1)$

24 $\frac{dy}{dx} - 2y\tan x = y^2 \tan^2 x$

25 $\frac{dy}{dx} + y\tan x = y^3 \sec^4 x$

VI. *Miscellaneous*. Choose the appropriate method in each case.

26 $(1 - x^2)\frac{dy}{dx} = 1 + xy$

27 $xy\frac{dy}{dx} - (1 + x)\sqrt{y^2 - 1} = 0$

28 $(x^2 - 2xy + 5y^2) = (x^2 + 2xy + y^2)\frac{dy}{dx}$

29 $\frac{dy}{dx} - y\cot x = y^2 \sec^2 x$, given $y = -1$ when $x = \pi/4$.

30 $y + (x^2 - 4x)\frac{dy}{dx} = 0$

VII. *Further examples*

**31** Solve the equation $\frac{dy}{dx} - y\tan x = \cos x - 2x\sin x$, given that $y = 0$ when $x = \pi/6$.

**32** Find the general solution of the equation $\frac{dy}{dx} = \frac{2xy + y^2}{x^2 + 2xy}$.

**33** Find the general solution of $(1 + x^2)\frac{dy}{dx} = x(1 + y^2)$.

**34** Solve the equation $x\frac{dy}{dx} + 2y = 3x - 1$, given that $y = 1$ when $x = 2$.

**35** Solve $x^2\frac{dy}{dx} = y^2 - xy\frac{dy}{dx}$, given that $y = 1$ when $x = 1$.

**36** Solve $\frac{dy}{dx} = e^{3x-2y}$, given that $y = 0$ when $x = 0$.

**37** Find the particular solution of $\frac{dy}{dx} + \frac{1}{x}.y = \sin 2x$, such that $y = 2$ when $x = \pi/4$.

**38** Find the general solution of $y^2 + x^2\frac{dy}{dx} = xy\frac{dy}{dx}$.

**39** Obtain the general solution of the equation $2xy\frac{dy}{dx} = x^2 - y^2$.

**40** By substituting $z = x - 2y$, solve the equation $\frac{dy}{dx} = \frac{x - 2y + 1}{2x - 4y}$, given that $y = 1$ when $x = 1$.

**41** Find the general solution of $(1 - x^3)\frac{dy}{dx} + x^2y = x^2(1 - x^3)$.

**42** Solve $\frac{dy}{dx} + \frac{y}{x} = \sin x$, given $y = 0$ when $x = \pi/2$.

**43** Solve $\frac{dy}{dx} + x + xy^2 = 0$, given $y = 0$ when $x = 1$.

**44** Determine the general solution of the equation $\frac{dy}{dx} + \left\{\frac{1}{x} - \frac{2x}{1 - x^2}\right\}y = \frac{1}{1 - x^2}$

**45** Solve $(1 + x^2)\frac{dy}{dx} + xy = (1 + x^2)^{3/2}$.

**46** Solve $x(1 + y^2) - y(1 + x^2)\frac{dy}{dx} = 0$, given $y = 2$ when $x = 0$.

**47** Solve $\frac{r\tan\theta}{a^2 - r^2}\cdot\frac{dr}{d\theta} = 1$, given $r = 0$ when $\theta = \pi/4$.

**48** Solve $\frac{dy}{dx} + y\cot x = \cos x$, given $y = 0$ when $x = 0$.

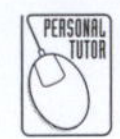
**49** Use the substitution $y = \frac{v}{x}$, where $v$ is a function of $x$ only, to transform the equation $\frac{dy}{dx} + \frac{y}{x} = xy^2$ into a differential equation in $v$ and $x$. Hence find $y$ in terms of $x$.

▶

**50** The rate of decay of a radioactive substance is proportional to the amount $A$ remaining at any instant. If $A = A_0$ at $t = 0$, prove that, if the time taken for the amount of the substance to become $\frac{1}{2}A_0$ is $T$, then $A = A_0 e^{-(t \ln 2)/T}$. Prove also that the time taken for the amount remaining to be reduced to $\frac{1}{20}A_0$ is $4{\cdot}32T$.

Now visit the companion website for this book at www.palgrave.com/stroud for more questions applying this mathematics to science and engineering.

**84**

# Second-order differential equations

## Learning outcomes

When you have completed this Programme you will be able to:

- Use the auxiliary equation to solve certain second-order homogeneous equations
- Use the complementary function and the particular integral to solve certain second-order inhomogeneous equations

# Homogeneous equations

**1**

Many practical problems in engineering give rise to second-order differential equations of the form

$$a\frac{d^2y}{dx^2}+b\frac{dy}{dx}+cy=f(x)$$

where $a$, $b$ and $c$ are constant coefficients and $f(x)$ is a given function of $x$. By the end of this Programme you will have no difficulty with equations of this type.

Let us first take the case where $f(x)=0$, so that the equation becomes

$$a\frac{d^2y}{dx^2}+b\frac{dy}{dx}+cy=0$$

This is called a linear, constant coefficient, second-order **homogeneous** differential equation. We shall now look at the solutions to this equation. Let $y=u$ and $y=v$ (where $u$ and $v$ are functions of $x$) be two solutions of the equation:

$$\therefore\ a\frac{d^2u}{dx^2}+b\frac{du}{dx}+cu=0 \quad \text{and} \quad a\frac{d^2v}{dx^2}+b\frac{dv}{dx}+cv=0$$

Adding these two lines together, we get:

$$a\left(\frac{d^2u}{dx^2}+\frac{d^2v}{dx^2}\right)+b\left(\frac{du}{dx}+\frac{dv}{dx}\right)+c(u+v)=0$$

Now $\frac{d}{dx}(u+v)=\frac{du}{dx}+\frac{dv}{dx}$ and $\frac{d^2}{dx^2}(u+v)=\frac{d^2u}{dx^2}+\frac{d^2v}{dx^2}$, therefore the equation can be written

$$a\frac{d^2}{dx^2}(u+v)+b\frac{d}{dx}(u+v)+c(u+v)=0$$

which is our original equation with $y$ replaced by $(u+v)$.

i.e. If $y=u$ and $y=v$ are solutions of the equation

$a\frac{d^2y}{dx^2}+b\frac{dy}{dx}+cy=0$, so also is $y=u+v$.

This is an important result and we shall be referring to it later, so make a note of it in your record book.

*Move on to Frame 2*

**2**

Our equation was $a\frac{d^2y}{dx^2}+b\frac{dy}{dx}+cy=0$. If $a=0$, we get the first-order equation of the same family:

$$b\frac{dy}{dx}+cy=0 \quad \text{i.e.} \quad \frac{dy}{dx}+ky=0 \quad \text{where } k=\frac{c}{b}$$

Solving this by the method of separating variables, we have

$$\frac{dy}{dx}=-ky \quad \therefore \quad \int\frac{dy}{y}=-\int k\,dx$$

which gives ...........

**3**

$$\ln y = -kx + c$$

$\therefore y = e^{-kx+c} = e^{-kx}.e^{c} = Ae^{-kx}$ (since $e^{c}$ is a constant)

i.e. $y = Ae^{-kx}$

If we write the symbol $m$ for $-k$, the solution is $y = Ae^{mx}$

In the same way, $y = Ae^{mx}$ will be a solution of the second-order equation $a\dfrac{d^2y}{dx^2} + b\dfrac{dy}{dx} + cy = 0$, if it satisfies this equation.

Now, if $y = Ae^{mx}$

$$\frac{dy}{dx} = Ame^{mx}$$

$$\frac{d^2y}{dx^2} = Am^2 e^{mx}$$

and substituting these expressions for the differential coefficients in the left-hand side of the equation, we get ...........

*On to Frame 4*

**4**

$$aAm^2 e^{mx} + bAme^{mx} + cAe^{mx} = 0$$

Right. So dividing both sides by $Ae^{mx}$ we obtain

$$am^2 + bm + c = 0$$

which is a quadratic equation giving two values for $m$. Let us call these

$$m = m_1 \text{ and } m = m_2$$

i.e. $y = Ae^{m_1x}$ and $y = Be^{m_2x}$ are two solutions of the given equation.

Now we have already seen that if $y = u$ and $y = v$ are two solutions so also is $y = u + v$.

$\therefore$ If $y = Ae^{m_1x}$ and $y = Be^{m_2x}$ are solutions so also is

$$y = Ae^{m_1x} + Be^{m_2x}$$

*Note*: This contains the necessary two arbitrary constants for a second-order differential equation, so there can be no further solution.

*Move to Frame 5*

**5**

The solution, then, of $a\dfrac{d^2y}{dx^2} + b\dfrac{dy}{dx} + cy = 0$ is seen to be

$$y = Ae^{m_1x} + Be^{m_2x}$$

where $A$ and $B$ are two arbitrary constants and $m_1$ and $m_2$ are the roots of the quadratic equation $am^2 + bm + c = 0$.

This quadratic equation is called the *auxiliary equation* and is obtained directly from the equation $a\dfrac{d^2y}{dx^2} + b\dfrac{dy}{dx} + cy = 0$, by writing $m^2$ for $\dfrac{d^2y}{dx^2}$, $m$ for $\dfrac{dy}{dx}$, 1 for $y$.

▶

**Example**

For the equation $2\dfrac{d^2y}{dx^2}+5\dfrac{dy}{dx}+6y=0$, the auxiliary equation is $2m^2+5m+6=0$.

In the same way, for the equation $\dfrac{d^2y}{dx^2}+3\dfrac{dy}{dx}+2y=0$, the auxiliary equation is

. . . . . . . . . . . .

*Then on to Frame 6*

---

**6**

$$\boxed{m^2+3m+2=0}$$

Since the auxiliary equation is always a quadratic equation, the values of $m$ can be determined in the usual way.

i.e. if $m^2+3m+2=0$

$(m+1)(m+2)=0 \quad \therefore\ m=-1$ and $m=-2$

$\therefore$ the solution of $\dfrac{d^2y}{dx^2}+3\dfrac{dy}{dx}+2y=0$ is $y=Ae^{-x}+Be^{-2x}$

In the same way, if the auxiliary equation were $m^2+4m-5=0$, this factorizes into $(m+5)(m-1)=0$ giving $m=1$ or $-5$, and in this case the solution would be

. . . . . . . . . . . .

---

**7**

$$\boxed{y=Ae^{x}+Be^{-5x}}$$

The type of solution we get depends on the roots of the auxiliary equation.

**1 *Real and different roots* to the auxiliary equation**

**Example 1**

$\dfrac{d^2y}{dx^2}+5\dfrac{dy}{dx}+6y=0$

Auxiliary equation: $m^2+5m+6=0$

$\therefore\ (m+2)(m+3)=0 \quad \therefore\ m=-2$ or $m=-3$

$\therefore$ Solution is $y=Ae^{-2x}+Be^{-3x}$

**Example 2**

$\dfrac{d^2y}{dx^2}-7\dfrac{dy}{dx}+12y=0$

Auxiliary equation: $m^2-7m+12=0$

$(m-3)(m-4)=0 \quad \therefore\ m=3$ or $m=4$

So the solution is . . . . . . . . . . .

*Move to Frame 8*

---

**8**

$$\boxed{y=Ae^{3x}+Be^{4x}}$$

Here you are. Do this one.

Solve the equation $\dfrac{d^2y}{dx^2}+3\dfrac{dy}{dx}-10y=0$

*When you have finished, move on to Frame 9*

**9**

$$\boxed{y = Ae^{2x} + Be^{-5x}}$$

Now consider the next case.

**2 *Real and equal roots* to the auxiliary equation**

Let us take $\dfrac{d^2y}{dx^2} + 6\dfrac{dy}{dx} + 9y = 0.$

The auxiliary equation is: $m^2 + 6m + 9 = 0$

$\therefore (m+3)(m+3) = 0 \quad \therefore m = -3$ (twice)

If $m_1 = -3$ and $m_2 = -3$ then these would give the solution $y = Ae^{-3x} + Be^{-3x}$ and their two terms would combine to give $y = Ce^{-3x}$. But every second-order differential equation has two arbitrary constants, so there must be another term containing a second constant. In fact, it can be shown that $y = Kxe^{-3x}$ also satisfies the equation, so that the complete general solution is of the form $y = Ae^{-3x} + Bxe^{-3x}$

i.e. $y = e^{-3x}(A + Bx)$

In general, if the auxiliary equation has real and equal roots, giving $m = m_1$ twice, the solution of the differential equation is

$$y = e^{m_1x}(A + Bx)$$

*Make a note of this general statement and then move on to Frame 10*

**10**

Here is an example:

**Example 1**

Solve $\dfrac{d^2y}{dx^2} + 4\dfrac{dy}{dx} + 4y = 0$

Auxiliary equation: $m^2 + 4m + 4 = 0$

$(m+2)(m+2) = 0 \quad \therefore m = -2$ (twice)

The solution is: $y = e^{-2x}(A + Bx)$

Here is another:

**Example 2**

Solve $\dfrac{d^2y}{dx^2} + 10\dfrac{dy}{dx} + 25y = 0$

Auxiliary equation: $m^2 + 10m + 25 = 0$

$(m+5)^2 = 0 \quad \therefore m = -5$ (twice)

$y = e^{-5x}(A + Bx)$

**Example 3**

Now here is one for you to do:

Solve $\dfrac{d^2y}{dx^2} + 8\dfrac{dy}{dx} + 16y = 0$

*When you have done it, move on to Frame 11*

11

$$y = e^{-4x}(A + Bx)$$

Because if $\frac{d^2y}{dx^2} + 8\frac{dy}{dx} + 16y = 0$

the auxiliary equation is

$$m^2 + 8m + 16 = 0$$

$$\therefore (m+4)^2 = 0 \quad \therefore m = -4 \text{ (twice)}$$

$$\therefore y = e^{-4x}(A + Bx)$$

So, for *real and different roots* $m = m_1$ and $m = m_2$ the solution is

$$y = Ae^{m_1x} + Be^{m_2x}$$

and for *real and equal roots* $m = m_1$ (twice) the solution is

$$y = e^{m_1x}(A + Bx)$$

Just find the values of $m$ from the auxiliary equation and then substitute these values in the appropriate form of the result.

*Move to Frame 12*

12

### 3 *Complex roots* to the auxiliary equation

Now let us see what we get when the roots of the auxiliary equation are complex.

Suppose $m = \alpha \pm j\beta$, i.e. $m_1 = \alpha + j\beta$ and $m_2 = \alpha - j\beta$. Then the solution would be of the form:

$$y = Ce^{(\alpha+j\beta)x} + De^{(\alpha-j\beta)x} = Ce^{\alpha x}.e^{j\beta x} + De^{\alpha x}.e^{-j\beta x}$$
$$= e^{\alpha x}\{Ce^{j\beta x} + De^{-j\beta x}\}$$

Now from our previous work on complex numbers, we know that:

$$e^{jx} = \cos x + j\sin x$$
$$e^{-jx} = \cos x - j\sin x$$

and that $\begin{cases} e^{j\beta x} = \cos\beta x + j\sin\beta x \\ e^{-j\beta x} = \cos\beta x - j\sin\beta x \end{cases}$

Our solution above can therefore be written:

$$y = e^{\alpha x}\{C(\cos\beta x + j\sin\beta x) + D(\cos\beta x - j\sin\beta x)\}$$
$$= e^{\alpha x}\{(C+D)\cos\beta x + j(C-D)\sin\beta x\}$$
$$y = e^{\alpha x}\{A\cos\beta x + B\sin\beta x\}$$

where $A = C + D$ and $B = j(C - D)$

$\therefore$ If $m = \alpha \pm j\beta$, the solution can be written in the form:

$$y = e^{\alpha x}\left\{A\cos\beta x + B\sin\beta x\right\}$$

Here is an example: If $m = -2 \pm j3$

then $y = e^{-2x}\left\{A\cos 3x + B\sin 3x\right\}$

Similarly, if $m = 5 \pm j2$ then $y = \ldots\ldots\ldots$

**13**

$$\boxed{y = e^{5x}[A\cos 2x + B\sin 2x]}$$

Here is one of the same kind:

Solve $\dfrac{d^2y}{dx^2} + 4\dfrac{dy}{dx} + 9y = 0$

Auxiliary equation: $m^2 + 4m + 9 = 0$

$$\therefore m = \frac{-4 \pm \sqrt{16 - 36}}{2} = \frac{-4 \pm \sqrt{-20}}{2} = \frac{-4 \pm 2j\sqrt{5}}{2} = -2 \pm j\sqrt{5}$$

In this case $\alpha = -2$ and $\beta = \sqrt{5}$

Solution is: $y = e^{-2x}(A\cos\sqrt{5}x + B\sin\sqrt{5}x)$

Now you can solve this one: $\dfrac{d^2y}{dx^2} - 2\dfrac{dy}{dx} + 10y = 0$

*When you have finished it, move on to Frame 14*

**14**

$$\boxed{y = e^{x}(A\cos 3x + B\sin 3x)}$$

Just check your working:

$$\frac{d^2y}{dx^2} - 2\frac{dy}{dx} + 10y = 0$$

Auxiliary equation: $m^2 - 2m + 10 = 0$

$$m = \frac{2 \pm \sqrt{4 - 40}}{2} = \frac{2 \pm \sqrt{-36}}{2} = 1 \pm j3$$

$$y = e^{x}(A\cos 3x + B\sin 3x)$$

*Move to Frame 15*

**15**

Here is a *summary* of the work so far.

Equations of the form $a\dfrac{d^2y}{dx^2} + b\dfrac{dy}{dx} + cy = 0$

Auxiliary equation: $am^2 + bm + c = 0$

**1** *Roots real and different* $m = m_1$ and $m = m_2$

Solution is $y = Ae^{m_1x} + Be^{m_2x}$

**2** *Real and equal roots* $m = m_1$ (twice)

Solution is $y = e^{m_1x}(A + Bx)$

**3** *Complex roots* $m = \alpha \pm j\beta$

Solution is $y = e^{\alpha x}(A\cos\beta x + B\sin\beta x)$

In each case, we simply solve the auxiliary equation to establish the values of $m$ and substitute in the appropriate form of the result.

*On to Frame 16*

16

We shall now consider equations of the form $\frac{d^2y}{dx^2} \pm n^2y = 0$

This is a special case of the equation

$$a\frac{d^2y}{dx^2} + b\frac{dy}{dx} + cy = 0 \text{ when } b = 0$$

$$\text{i.e.} \quad a\frac{d^2y}{dx^2} + cy = 0 \quad \text{i.e.} \quad \frac{d^2y}{dx^2} + \frac{c}{a}y = 0$$

which can be written as $\frac{d^2y}{dx^2} \pm n^2y = 0$ to cover the two cases when the coefficient of $y$ is positive or negative.

(a) If $\frac{d^2y}{dx^2} + n^2y = 0, \quad m^2 + n^2 = 0 \quad \therefore m^2 = -n^2 \quad \therefore m = \pm jn$

(This is like $m = \alpha \pm j\beta$, when $\alpha = 0$ and $\beta = n$)

$\therefore y = A\cos nx + B\sin nx$

(b) If $\frac{d^2y}{dx^2} - n^2y = 0, \quad m^2 - n^2 = 0 \quad \therefore m^2 = n^2 \quad \therefore m = \pm n$

$\therefore y = Ce^{nx} + De^{-nx}$

This last result can be written in another form which is sometimes more convenient, so move on to the next frame and we will see what it is.

17

You will remember from your work on hyperbolic functions that

$$\cosh nx = \frac{e^{nx} + e^{-nx}}{2} \qquad \therefore e^{nx} + e^{-nx} = 2\cosh nx$$

$$\sinh nx = \frac{e^{nx} - e^{-nx}}{2} \qquad \therefore e^{nx} - e^{-nx} = 2\sinh nx$$

Adding these two results: $2e^{nx} = 2\cosh nx + 2\sinh nx$

$\therefore e^{nx} = \cosh nx + \sinh nx$

Similarly, by subtracting: $e^{-nx} = \cosh nx - \sinh nx$

Therefore, the solution of our equation $y = Ce^{nx}$ can be written:

$$y = C(\cosh nx + \sinh nx) + D(\cosh nx - \sinh nx)$$
$$= (C + D)\cosh nx + (C - D)\sinh nx$$

i.e. $y = A\cosh nx + B\sinh nx$

*Note*: In this form the two results are very much alike:

(a) $\frac{d^2y}{dx^2} + n^2y = 0 \qquad y = A\cos nx + B\sin nx$

(b) $\frac{d^2y}{dx^2} - n^2y = 0 \qquad y = A\cosh nx + B\sinh nx$

Make a note of these results in your record book.

*Then – next frame*

18

Here are some examples:

**Example 1**

$\frac{d^2y}{dx^2} + 16y = 0 \quad \therefore m^2 = -16 \quad \therefore m = \pm j4$

$\therefore y = A\cos 4x + B\sin 4x$

▶

**Example 2**

$$\frac{d^2y}{dx^2} - 3y = 0 \quad \therefore m^2 = 3 \quad \therefore m = \pm\sqrt{3}$$

$$y = A\cosh\sqrt{3}x + B\sinh\sqrt{3}x$$

Similarly

**Example 3**

$$\frac{d^2y}{dx^2} + 5y = 0$$

$$y = \dots\dots\dots\dots$$

*Then move on to Frame 19*

**19**

$$\boxed{y = A\cos\sqrt{5}x + B\sin\sqrt{5}x}$$

And now this one:

**Example 4**

$$\frac{d^2y}{dx^2} - 4y = 0 \quad \therefore m^2 = 4 \quad \therefore m = \pm 2$$

$$y = \dots\dots\dots\dots$$

**20**

$$\boxed{y = A\cosh 2x + B\sinh 2x}$$

Now before we go on to the next section of the Programme, here is a Review exercise on what we have covered so far. The questions are set out in the next frame. Work them all before checking your results.

*So on you go to Frame 21*

##  Review exercise

**21**

Solve the following:

1 $\dfrac{d^2y}{dx^2} - 12\dfrac{dy}{dx} + 36y = 0$

2 $\dfrac{d^2y}{dx^2} + 7y = 0$

3 $\dfrac{d^2y}{dx^2} + 2\dfrac{dy}{dx} - 3y = 0$

4 $2\dfrac{d^2y}{dx^2} + 4\dfrac{dy}{dx} + 3y = 0$

5 $\dfrac{d^2y}{dx^2} - 9y = 0$

*For the answers, move to Frame 22*

22

Here are the answers:

1 $y = e^{6x}(A + Bx)$

2 $y = A\cos\sqrt{7}x + B\sin\sqrt{7}x$

3 $y = Ae^{x} + Be^{-3x}$

4 $y = e^{-x}\left(A\cos\dfrac{x}{\sqrt{2}} + B\sin\dfrac{x}{\sqrt{2}}\right)$

5 $y = A\cosh 3x + B\sinh 3x$

*By now we are ready for the next section of the Programme, so move on to Frame 23*

# Inhomogeneous equations

23

So far we have considered equations of the form:

$$a\frac{d^2y}{dx^2} + b\frac{dy}{dx} + cy = f(x) \text{ for the case where } f(x) = 0$$

If $f(x) = 0$, then $am^2 + bm + c = 0$ giving $m = m_1$ and $m = m_2$ and the solution is in general $y = Ae^{m_1x} + Be^{m_2x}$.

The equation $a\dfrac{d^2y}{dx^2} + b\dfrac{dy}{dx} + cy = f(x)$ where $f(x) \neq 0$ is called an **inhomogeneous** equation and the substitution $y = Ae^{m_1x} + Be^{m_2x}$ would make the left-hand side zero. Therefore, there must be a further term in the solution which will make the LHS equal to $f(x)$ and not zero. The complete solution will therefore be of the form

$y = Ae^{m_1x} + Be^{m_2x} + X$, where $X$ is the extra function yet to be found.

$y = Ae^{m_1x} + Be^{m_2x}$ is called the *complementary function* (CF)

$y = X$ (a function of $x$) is called the *particular integral* (PI)

*Note*: The complete general solution to the inhomogeneous equation is given by:

general solution = complementary function + particular integral

Our main problem at this stage is how are we to find the particular integral for any given equation? This is what we are now going to deal with.

*So on then to Frame 24*

24

To solve an equation $a\dfrac{d^2y}{dx^2} + b\dfrac{dy}{dx} + cy = f(x)$

(I) The *complementary function* is obtained by solving the equation with $f(x) = 0$, as in the previous part of this Programme. This will give one of the following types of solution:

(a) $y = Ae^{m_1x} + Be^{m_2x}$

(b) $y = e^{m_1x}(A + Bx)$

(c) $y = e^{\alpha x}(A\cos\beta x + B\sin\beta x)$

(d) $y = A\cos nx + B\sin nx$

(e) $y = A\cosh nx + B\sinh nx$

▶

(II) The *particular integral* is found by assuming the general form of the function on the right-hand side of the given equation, substituting this in the equation, and equating coefficients. An example will make this clear:

Solve $\dfrac{d^2y}{dx^2} - 5\dfrac{dy}{dx} + 6y = x^2$

(1) *To find the CF* solve LHS = 0, i.e. $m^2 - 5m + 6 = 0$

$\therefore (m-2)(m-3) = 0 \quad \therefore m = 2$ or $m = 3$

$\therefore$ Complementary function is $y = Ae^{2x} + Be^{3x}$ (1)

(2) *To find the PI* we assume the general form of the RHS which is a second-degree function. Let $y = Cx^2 + Dx + E$.

Then $\dfrac{dy}{dx} = 2Cx + D$ and $\dfrac{d^2y}{dx^2} = 2C$

Substituting these in the given equation, we get:

$$2C - 5(2Cx + D) + 6(Cx^2 + Dx + E) = x^2$$
$$2C - 10Cx - 5D + 6Cx^2 + 6Dx + 6E = x^2$$
$$6Cx^2 + (6D - 10C)x + (2C - 5D + 6E) = x^2$$

Equating coefficients of powers of $x$, we have:

$[x^2]$ $\quad 6C = 1$ $\qquad \therefore C = \dfrac{1}{6}$

$[x]$ $\quad 6D - 10C = 0$ $\qquad \therefore 6D = \dfrac{10}{6} = \dfrac{5}{3}$ $\qquad \therefore D = \dfrac{5}{18}$

$[CT]$ $\quad 2C - 5D + 6E = 0$ $\qquad \therefore 6E = \dfrac{25}{18} - \dfrac{2}{6} = \dfrac{19}{18}$ $\qquad \therefore E = \dfrac{19}{108}$

$\therefore$ Particular integral is $y = \dfrac{x^2}{6} + \dfrac{5x}{18} + \dfrac{19}{108}$ (2)

Complete general solution = CF + PI

General solution is $y = Ae^{2x} + Be^{3x} + \dfrac{x^2}{6} + \dfrac{5x}{18} + \dfrac{19}{108}$

This frame is quite important, since all equations of this type are solved in this way.

*On to Frame 25*

## 25

We have seen that to find the particular integral, we assume the general form of the function on the RHS of the equation and determine the values of the constants by substitution in the whole equation and equating coefficients. These will be useful:

| If | Assume |
|---|---|
| $f(x) = k$ ... ... | $y = C$ |
| $f(x) = kx$ ... ... | $y = Cx + D$ |
| $f(x) = kx^2$ ... ... | $y = Cx^2 + Dx + E$ |
| $f(x) = k\sin x$ or $k\cos x$ | $y = C\cos x + D\sin x$ |
| $f(x) = k\sinh x$ or $k\cosh x$ | $y = C\cosh x + D\sinh x$ |
| $f(x) = e^{kx}$ ... ... | $y = Ce^{kx}$ |

This list covers all the cases you are likely to meet at this stage.

So if the function on the RHS of the equation is $f(x) = 2x^2 + 5$, you would take as the assumed PI:

$y = \dots\dots\dots\dots$

**26**

$$y = Cx^2 + Dx + E$$

Correct, since the assumed PI will be the general form of the second-degree function.

What would you take as the assumed PI in each of the following cases:

1 $f(x) = 2x - 3$

2 $f(x) = e^{5x}$

3 $f(x) = \sin 4x$

4 $f(x) = 3 - 5x^2$

5 $f(x) = 27$

6 $f(x) = 5 \cosh 4x$

*When you have decided all six, check you answers with those in Frame 27*

**27**

Here are the answers:

1 $f(x) = 2x - 3$ PI is of the form $y = Cx + D$

2 $f(x) = e^{5x}$ $y = Ce^{5x}$

3 $f(x) = \sin 4x$ $y = C\cos 4x + D\sin 4x$

4 $f(x) = 3 - 5x^2$ $y = Cx^2 + Dx + E$

5 $f(x) = 27$ $y = C$

6 $f(x) = 5 \cosh 4x$ $y = C\cosh 4x + D\sinh 4x$

All correct? If you have made a slip with any one of them, be sure that you understand where and why your result was incorrect before moving on.

*Next frame*

**28**

Let us work through a few examples. Here is the first:

**Example 1**

Solve $\dfrac{d^2y}{dx^2} - 5\dfrac{dy}{dx} + 6y = 24$

(1) *CF* Solve LHS $= 0$ $\therefore m^2 - 5m + 6 = 0$

$\therefore (m-2)(m-3) = 0$ $\therefore m = 2$ and $m = 3$

$\therefore y = Ae^{2x} + Be^{3x}$ (1)

(2) *PI* $f(x) = 24$, i.e. a constant. Assume $y = C$

Then $\dfrac{dy}{dx} = 0$ and $\dfrac{d^2y}{dx^2} = 0$

Substituting in the given equation:

$0 - 5(0) + 6C = 24$ $C = 24/6 = 4$

$\therefore$ PI is $y = 4$ (2)

General solution is $y = \text{CF} + \text{PI}$, i.e. $y = \underbrace{Ae^{2x} + Be^{3x}}_{\text{CF}} \underbrace{+4}_{\text{PI}}$

▶

Now another:

**Example 2**

Solve $\dfrac{d^2y}{dx^2} - 5\dfrac{dy}{dx} + 6y = 2\sin 4x$

(1) *CF* This will be the same as in the previous example, since the LHS of this equation is the same

i.e. $y = Ae^{2x} + Be^{3x}$

(2) *PI* The general form of the PI in this case will be ...........

**29**

$$y = C\cos 4x + D\sin 4x$$

*Note*: Although the RHS is $f(x) = 2\sin 4x$, it is necessary to include the full general function $y = C\cos 4x + D\sin 4x$ since, in finding the derivatives, the cosine term will also give rise to $\sin 4x$.

So we have:

$$y = C\cos 4x + D\sin 4x$$

$$\frac{dy}{dx} = -4C\sin 4x + 4D\cos 4x$$

$$\frac{d^2y}{dx^2} = -16C\cos 4x - 16D\sin 4x$$

We now substitute these expressions in the LHS of the equation and by equating coefficients, find the values of $C$ and $D$.

Away you go then.

*Complete the job and then move on to Frame 30*

**30**

$$C = \frac{2}{25} \quad D = -\frac{1}{25} \quad y = \frac{1}{25}(2\cos 4x - \sin 4x)$$

Here is the working:

$$-16C\cos 4x - 16D\sin 4x + 20C\sin 4x - 20D\cos 4x$$
$$+ 6C\cos 4x + 6D\sin 4x = 2\sin 4x$$

$$(20C - 10D)\sin 4x - (10C + 20D)\cos 4x = 2\sin 4x$$

$$\begin{array}{ll} 20C - 10D = 2 & 40C - 20D = 4 \\ 10C + 20D = 0 & 10C + 20D = 0 \end{array} \Bigg\} \; 50C = 4 \qquad \therefore C = \frac{2}{25}$$

$$D = -\frac{1}{25}$$

In each case the PI is $y = \dfrac{1}{25}(2\cos 4x - \sin 4x)$

The CF was $y = Ae^{2x} + Be^{3x}$

The general solution is: $y = Ae^{2x} + Be^{3x} + \dfrac{1}{25}(2\cos 4x - \sin 4x)$

31

Here is an example we can work through together:

**Example 3**

Solve $\dfrac{d^2y}{dx^2}+14\dfrac{dy}{dx}+49y=4e^{5x}$

First we have to find the CF. To do this we solve the equation ...........

32

$$\frac{d^2y}{dx^2}+14\frac{dy}{dx}+49y=0$$

Correct. So start off by writing down the auxiliary equation, which is ...........

33

$$m^2+14m+49=0$$

This gives $(m+7)(m+7)=0$, i.e. $m=-7$ (twice)

$\therefore$ The CF is $y=e^{-7x}(A+Bx)$ (1)

Now for the PI. To find this we take the general form of the RHS of the given equation, i.e. we assume $y=$ ...........

34

$$y=Ce^{5x}$$

Right. So we now differentiate twice which gives us:

$\dfrac{dy}{dx}=$ ........... and $\dfrac{d^2y}{dx^2}=$ ...........

35

$$\frac{dy}{dx}=5Ce^{5x},\quad \frac{d^2y}{dx^2}=25\,Ce^{5x}$$

The equation now becomes:

$25Ce^{5x}+14.5Ce^{5x}+49Ce^{5x}=4e^{5x}$

Dividing through by $e^{5x}$: $25C+70C+49C=4$

$$144C=4 \quad \therefore C=\frac{1}{36}$$

$$\text{The PI is } y=\frac{e^{5x}}{36} \qquad (2)$$

So there we are. The CF is $y=e^{-7x}(A+Bx)$ and the PI is $y=\dfrac{e^{5x}}{36}$

and the complete general solution is therefore ...........

36

$$y=e^{-7x}(A+Bx)+\frac{e^{5x}}{36}$$

Correct, because in every case, the general solution is the sum of the complementary function and the particular integral.

Here is another example.

**Example 4**

Solve $\dfrac{d^2y}{dx^2}+6\dfrac{dy}{dx}+10y=2\sin 2x$

(1) *To find CF* Solve LHS = 0 $\therefore m^2+6m+10=0$

$$\therefore m=\frac{-6\pm\sqrt{36-40}}{2}=\frac{-6\pm\sqrt{-4}}{2}=-3\pm j$$

$$y=e^{-3x}(A\cos x+B\sin x) \qquad (1)$$

(2) *To find PI* Assume the general form of the RHS

i.e. $y=\ldots\ldots\ldots\ldots$

*On to Frame 37*

## 37

$$\boxed{y=C\cos 2x+D\sin 2x}$$

Do not forget that we have to include the cosine term as well as the sine term, since that will also give $\sin 2x$ when the derivatives are found.

As usual, we now differentiate twice and substitute in the given equation $\dfrac{d^2y}{dx^2}+6\dfrac{dy}{dx}+10y=2\sin 2x$ and equate coefficients of $\sin 2x$ and of $\cos 2x$.

Off you go then. Find the PI on your own.

*When you have finished, check your result with that in Frame 38*

## 38

$$\boxed{y=\frac{1}{15}(\sin 2x-2\cos 2x)}$$

Because if:

$$y=C\cos 2x+D\sin 2x$$

$$\therefore \frac{dy}{dx}=-2C\sin 2x+2D\cos 2x$$

$$\therefore \frac{d^2y}{dx^2}=-4C\cos 2x-4D\sin 2x$$

Substituting in the equation gives:

$$-4C\cos 2x-4D\sin 2x-12C\sin 2x+12D\cos 2x$$
$$+10C\cos 2x+10D\sin 2x=2\sin 2x$$

$$(6C+12D)\cos 2x+(6D-12C)\sin 2x=2\sin 2x$$

$$6C+12D=0 \quad \therefore C=-2D$$

$$6D-12C=2 \quad \therefore 6D+24D=2 \quad \therefore 30D=2 \quad \therefore D=\frac{1}{15}$$

$$\therefore C=-\frac{2}{15}$$

PI is $y=\dfrac{1}{15}(\sin 2x-2\cos 2x)$ (2)

So the CF is $y=e^{-3x}(A\cos x+B\sin x)$

and the PI is $y=\dfrac{1}{15}(\sin 2x-2\cos 2x)$

The complete general solution is therefore $y=\ldots\ldots\ldots\ldots$

**39**

$$y = e^{-3x}(A\cos x + B\sin x) + \frac{1}{15}(\sin 2x - 2\cos 2x)$$

Before we do another example, list what you would assume for the PI in an equation when the RHS function was:

1 $f(x) = 3\cos 4x$
2 $f(x) = 2e^{7x}$
3 $f(x) = 3\sinh x$
4 $f(x) = 2x^2 - 7$
5 $f(x) = x + 2e^x$

*Jot down all five results before turning to Frame 40 to check your answers*

**40**

1 $y = C\cos 4x + D\sin 4x$
2 $y = Ce^{7x}$
3 $y = C\cosh x + D\sinh x$
4 $y = Cx^2 + Dx + E$
5 $y = Cx + D + Ee^x$

Note that in **5** we use the general form of both the terms.

General form for $x$ is $Cx + D$
and for $e^x$ is $Ee^x$

$\therefore$ The general form of $x + e^x$ is $y = Cx + D + Ee^x$

Now do this one all on your own:

**Example 5**

Solve $\dfrac{d^2y}{dx^2} - 3\dfrac{dy}{dx} + 2y = x^2$

Do not forget: find (1) the CF and (2) the PI. Then the general solution is $y = \text{CF} + \text{PI}$.

Off you go.

*When you have finished completely, move to Frame 41*

**41**

$$y = Ae^x + Be^{2x} + \frac{1}{4}(2x^2 + 6x + 7)$$

Here is the solution in detail:

$$\frac{d^2y}{dx^2} - 3\frac{dy}{dx} + 2y = x^2$$

(1) CF $m^2 - 3m + 2 = 0$

$\therefore (m - 1)(m - 2) = 0 \quad \therefore m = 1$ or $2$

$\therefore y = Ae^x + Be^{2x}$ (1)

▶

(2) $PI \quad y = Cx^2 + Dx + E \quad \therefore \ \frac{dy}{dx} = 2Cx + D \quad \therefore \ \frac{d^2y}{dx^2} = 2C$

$$2C - 3(2Cx + D) + 2(Cx^2 + Dx + E) = x^2$$

$$2Cx^2 + (2D - 6C)x + (2C - 3D + 2E) = x^2$$

$$2C = 1 \quad \therefore \ C = \frac{1}{2}$$

$$2D - 6C = 0 \quad \therefore \ D = 3C \quad \therefore \ D = \frac{3}{2}$$

$$2C - 3D + 2E = 0 \quad \therefore \ 2E = 3D - 2C = \frac{9}{2} - 1 = \frac{7}{2} \quad \therefore \ E = \frac{7}{4}$$

$$\therefore \ \text{PI is } y = \frac{x^2}{2} + \frac{3x}{2} + \frac{7}{4} = \frac{1}{4}(2x^2 + 6x + 7) \qquad (2)$$

General solution: $y = Ae^x + Be^{2x} + \frac{1}{4}(2x^2 + 6x + 7)$

*Next frame*

---

# Particular solution

**42**

All our solutions to the equation

$$a\frac{d^2y}{dx^2} + b\frac{dy}{dx} + cy = f(x) \text{ where } f(x) \neq 0$$

have contained two unknown arbitrary constants. For instance, in the previous example the general solution to the differential equation

$$\frac{d^2y}{dx^2} - 3\frac{dy}{dx} + 2y = x^2 \text{ was seen to be } y = Ae^x + Be^{2x} + \frac{1}{4}(2x^2 + 6x + 7)$$

which contains the two arbitrary constants $A$ and $B$. These two constants are arbitrary because whatever values are chosen for them, when inserted into the above equation for $y$ it will still be a solution to the differential equation. This means, of course, that there is an infinite number of solutions to the differential equation, each one having specific values for $A$ and $B$. To select just one solution requires additional information and this information is provided by what are called **boundary conditions** that take the form of a given specific value of $y$ and its derivative for a particular value of $x$. When these boundary conditions are then imposed on the general solution we obtain a **particular solution**. For example, in the problem of Frame 40 we might have been told that at $x = 0$, $y = \frac{3}{4}$ and $\frac{dy}{dx} = \frac{5}{2}$.

*It is important* to note that the values of $A$ and $B$ can be found only from the complete general solution and not from the CF as soon as you obtain it. This is a common error so do not be caught out by it. Get the complete general solution before substituting to find $A$ and $B$.

In this case, we are told that when $x = 0$, $y = \frac{3}{4}$, so inserting these values gives

............

*Move on to Frame 43*

43

$$\boxed{A + B = -1}$$

Because

$$\frac{3}{4} = A + B + \frac{7}{4} \quad \therefore A + B = -1$$

We are also told that when $x = 0$, $\frac{dy}{dx} = \frac{5}{2}$, so we must first differentiate the general solution

$$y = Ae^x + Be^{2x} + \frac{1}{4}(2x^2 + 6x + 7)$$

to obtain an expression for $\frac{dy}{dx}$. So, $\frac{dy}{dx} = \ldots\ldots\ldots\ldots$

44

$$\boxed{\frac{dy}{dx} = Ae^x + 2Be^{2x} + \frac{1}{2}(2x + 3)}$$

Now we are given that when $x = 0$, $\frac{dy}{dx} = \frac{5}{2}$

$$\therefore \frac{5}{2} = A + 2B + \frac{3}{2} \quad \therefore A + 2B = 1$$

So we have $\left.\begin{aligned} A + B &= -1 \\ \text{and } A + 2B &= 1 \end{aligned}\right\}$

and these simultaneous equations give:

$A = \ldots\ldots\ldots\ldots \quad B = \ldots\ldots\ldots\ldots$

*Then on to Frame 45*

45

$$\boxed{A = -3 \qquad B = 2}$$

Substituting these values in the general solution

$$y = Ae^x + Be^{2x} + \frac{1}{4}(2x^2 + 6x + 7)$$

gives the *particular solution*:

$$y = 2e^{2x} - 3e^x + \frac{1}{4}(2x^2 + 6x + 7)$$

And here is one for you, all on your own:

Solve the equation $\frac{d^2y}{dx^2} + 4\frac{dy}{dx} + 5y = 13e^{3x}$, given that when $x = 0$, $y = \frac{5}{2}$ and $\frac{dy}{dx} = \frac{1}{2}$. Remember:

(1) Find the CF.
(2) Find the PI.
(3) The general solution is $y = \text{CF} + \text{PI}$.
(4) Finally insert the given conditions to obtain the particular solution.

*When you have finished, check with the solution in Frame 46*

**46**

$$y = e^{-2x}(2\cos x + 3\sin x) + \frac{e^{3x}}{2}$$

Because

$$\frac{d^2y}{dx^2} + 4\frac{dy}{dx} + 5y = 13e^{3x}$$

(1) *CF* $m^2 + 4m + 5 = 0 \quad \therefore m = \frac{-4 \pm \sqrt{16-20}}{2} = \frac{-4 \pm j2}{2}$

$\therefore m = -2 \pm j \quad \therefore y = e^{-2x}(A\cos x + B\sin x)$ (1)

(2) *PI* $y = Ce^{3x} \quad \therefore \frac{dy}{dx} = 3Ce^{3x}, \quad \frac{d^2y}{dx^2} = 9Ce^{3x}$

$\therefore 9Ce^{3x} + 12Ce^{3x} + 5Ce^{3x} = 13e^{3x}$

$26C = 13 \quad \therefore C = \frac{1}{2} \quad \therefore \text{PI is } y = \frac{e^{3x}}{2}$ (2)

General solution $y = e^{-2x}(A\cos x + B\sin x) + \frac{e^{3x}}{2}; \quad x = 0, \quad y = \frac{5}{2}$

$\therefore \frac{5}{2} = A + \frac{1}{2} \quad \therefore A = 2 \qquad y = e^{-2x}(2\cos x + B\sin x) + \frac{e^{3x}}{2}$

$\frac{dy}{dx} = e^{-2x}(-2\sin x + B\cos x) - 2e^{-2x}(2\cos x + B\sin x) + \frac{3e^{3x}}{2}$

$x = 0, \quad \frac{dy}{dx} = \frac{1}{2} \quad \therefore \frac{1}{2} = B - 4 + \frac{3}{2} \quad \therefore B = 3$

$\therefore$ Particular solution is $y = e^{-2x}(2\cos x + 3\sin x) + \frac{e^{3x}}{2}$

**47**

Since the CF makes the LHS = 0, it is pointless to use as a PI a term already contained in the CF. If this occurs, multiply the assumed PI by $x$ and proceed as before. If this too is already included in the CF, multiply by a further $x$ and proceed as usual. Here is an example:

Solve $\frac{d^2y}{dx^2} - 2\frac{dy}{dx} - 8y = 3e^{-2x}$

(1) *CF* $m^2 - 2m - 8 = 0 \quad \therefore (m+2)(m-4) = 0 \quad \therefore m = -2$ or $4$

$y = Ae^{4x} + Be^{-2x}$ (1)

(2) *PI* The general form of the RHS is $Ce^{-2x}$, but this term in $e^{-2x}$ is already contained in the CF. Assume $y = Cxe^{-2x}$, and continue as usual:

$y = Cxe^{-2x}$

$\frac{dy}{dx} = Cx(-2e^{-2x}) + Ce^{-2x} = Ce^{-2x}(1 - 2x)$

$\frac{d^2y}{dx^2} = Ce^{-2x}(-2) - 2Ce^{-2x}(1 - 2x) = Ce^{-2x}(4x - 4)$

Substituting in the given equation, we get:

$$Ce^{-2x}(4x-4)-2.Ce^{-2x}(1-2x)-8Cxe^{-2x}=3e^{-2x}$$

$$(4C+4C-8C)x-4C-2C=3$$

$$-6C=3 \quad \therefore C=-\frac{1}{2}$$

$$\text{PI is } y=-\frac{1}{2}xe^{-2x} \qquad (2)$$

General solution: $y=Ae^{4x}+Be^{-2x}-\dfrac{xe^{-2x}}{2}$

So remember, if the general form of the RHS is already included in the CF, multiply the assumed general form of the PI by $x$ and continue as before.

Here is one final example for you to work:

$$\text{Solve} \quad \frac{d^2y}{dx^2}+\frac{dy}{dx}-2y=e^x$$

*Finish it off and then move to Frame 48*

48

$$\boxed{y=Ae^x+Be^{-2x}+\frac{xe^x}{3}}$$

Here is the working:

$$\text{To solve} \quad \frac{d^2y}{dx^2}+\frac{dy}{dx}-2y=e^x$$

(1) *CF* $m^2+m-2=0$

$(m-1)(m+2)=0 \quad \therefore m=1 \text{ or } -2$

$$\therefore y=Ae^x+Be^{-2x} \qquad (1)$$

(2) *PI* Take $y=Ce^x$. But this is already included in the CF. Therefore, assume $y=Cxe^x$.

$$\text{Then } \frac{dy}{dx}=Cxe^x+Ce^x=Ce^x(x+1)$$

$$\frac{d^2y}{dx^2}=Ce^x+Cxe^x+Ce^x=Ce^x(x+2)$$

$$\therefore Ce^x(x+2)+Ce^x(x+1)-2Cxe^x=e^x$$

$$C(x+2)+C(x+1)-2Cx=1$$

$$3C=1 \quad \therefore C=\frac{1}{3}$$

$$\text{PI is } y=\frac{xe^x}{3} \qquad (2)$$

and so the general solution is

$$y=Ae^x+Be^{-2x}+\frac{xe^x}{3}$$

You are now almost at the end of the Programme. Before you work through the **Can You?** checklist and the **Test exercise**, however, look down the **Review summary** given in Frame 49. It lists the main points that we have established during this Programme, and you may find it very useful.

*So on now to Frame 49*

## Review summary

**49**

1 Solution of equations of the form $a\dfrac{d^2y}{dx^2}+b\dfrac{dy}{dx}+cy=f(x)$

(1) Auxiliary equation: $am^2+bm+c=0$

(2) Types of solutions:

(a) Real and different roots $\quad m=m_1$ and $m=m_2$

$y=Ae^{m_1x}+Be^{m_2x}$

(b) Real and equal roots $\quad m=m_1$ (twice)

$y=e^{m_1x}(A+Bx)$

(c) Complex roots $\quad m=\alpha\pm j\beta$

$y=e^{\alpha x}(A\cos\beta x+B\sin\beta x)$

2 Equations of the form $\dfrac{d^2y}{dx^2}+n^2y=0$

$y=A\cos nx+B\sin nx$

3 Equations of the form $\dfrac{d^2y}{dx^2}-n^2y=0$

$y=A\cosh nx+B\sinh nx$

4 General solution

$y=$ complementary function $+$ particular integral

5 (1) To find CF solve $a\dfrac{d^2y}{dx^2}+b\dfrac{dy}{dx}+cy=0$

(2) To find PI assume the general form of the RHS.
*Note*: If the general form of the RHS is already included in the CF, multiply by $x$ and proceed as before, etc. Determine the complete general solution before substituting to find the values of the arbitrary constants $A$ and $B$.

*Now all that remains is the* **Can You?** *checklist and the* **Test exercise**, *so on to Frame 50*

## Can You?

**50**

### Checklist 26

Check this list before and after you try the end of Programme test.

On a scale of 1 to 5 how confident are you that you can: Frames

- Use the auxiliary equation to solve certain second-order homogeneous equations? 1 to 22
  *Yes* ☐ ☐ ☐ ☐ ☐ *No*
- Use the complementary function and the particular integral to solve certain second-order inhomogeneous equations? 23 to 48
  *Yes* ☐ ☐ ☐ ☐ ☐ *No*

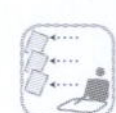

## Test exercise 26

51

Here are eight differential equations for you to solve, similar to those we have dealt with in the Programme. They are quite straightforward, so you should have no difficulty with them. Set your work out neatly and take your time: this will help you to avoid making unnecessary slips.

Solve the following:

1 $\dfrac{d^2y}{dx^2}-\dfrac{dy}{dx}-2y=8$

2 $\dfrac{d^2y}{dx^2}-4y=10e^{3x}$

3 $\dfrac{d^2y}{dx^2}+2\dfrac{dy}{dx}+y=e^{-2x}$

4 $\dfrac{d^2y}{dx^2}+25y=5x^2+x$

5 $\dfrac{d^2y}{dx^2}-2\dfrac{dy}{dx}+y=4\sin x$

6 $\dfrac{d^2y}{dx^2}+4\dfrac{dy}{dx}+5y=2e^{-2x}$, given that $x=0$, $y=1$ and $\dfrac{dy}{dx}=-2$.

7 $3\dfrac{d^2y}{dx^2}-2\dfrac{dy}{dx}-y=2x-3$

8 $\dfrac{d^2y}{dx^2}-6\dfrac{dy}{dx}+8y=8e^{4x}$

## Further problems 26

52

Solve the following equations:

1 $2\dfrac{d^2y}{dx^2}-7\dfrac{dy}{dx}-4y=e^{3x}$

2 $\dfrac{d^2y}{dx^2}-6\dfrac{dy}{dx}+9y=54x+18$

3 $\dfrac{d^2y}{dx^2}-5\dfrac{dy}{dx}+6y=100\sin 4x$

4 $\dfrac{d^2y}{dx^2}+2\dfrac{dy}{dx}+y=4\sinh x$

5 $\dfrac{d^2y}{dx^2}+\dfrac{dy}{dx}-2y=2\cosh 2x$

6 $\dfrac{d^2y}{dx^2}-6\dfrac{dy}{dx}+10y=20-e^{2x}$

7 $\dfrac{d^2y}{dx^2}+4\dfrac{dy}{dx}+4y=2\cos^2 x$

8 $\dfrac{d^2y}{dx^2}-4\dfrac{dy}{dx}+3y=x+e^{2x}$

9 $\dfrac{d^2y}{dx^2}-2\dfrac{dy}{dx}+3y=x^2-1$

10 $\dfrac{d^2y}{dx^2}-9y=e^{3x}+\sin 3x$

11 For a horizontal cantilever of length $l$, with load $w$ per unit length, the equation of bending is

$$EI\frac{d^2y}{dx^2}=\frac{w}{2}(l-x)^2$$

where $E$, $I$, $w$ and $l$ are constants. If $y=0$ and $\dfrac{dy}{dx}=0$ at $x=0$, find $y$ in terms of $x$. Hence find the value of $y$ when $x=l$.

▶

12 Solve the equation

$$\frac{d^2x}{dt^2} + 4\frac{dx}{dt} + 3x = e^{-3t}$$

given that at $t = 0$, $x = \frac{1}{2}$ and $\frac{dx}{dt} = -2$.

13 Obtain the general solution of the equation

$$\frac{d^2y}{dt^2} + 4\frac{dy}{dt} + 5y = 6\sin t$$

and determine the amplitude and frequency of the steady-state function. [*Note*: The steady state function describes the behaviour of the solution as $t \to \infty$]

14 Solve the equation

$$\frac{d^2x}{dt^2} - 3\frac{dx}{dt} + 2x = \sin t$$

given that at $t = 0$, $x = 0$ and $\frac{dx}{dt} = 0$.

15 Solve $\frac{d^2y}{dx^2} + 3\frac{dy}{dx} + 2y = 3\sin x$, given that when $x = 0$, $y = -0{\cdot}9$ and $\frac{dy}{dx} = -0{\cdot}7$.

16 Obtain the general solution of the equation

$$\frac{d^2y}{dx^2} + 6\frac{dy}{dx} + 10y = 50x$$

17 Solve the equation

$$\frac{d^2x}{dt^2} + 2\frac{dx}{dt} + 2x = 85\sin 3t$$

given that when $t = 0$, $x = 0$ and $\frac{dx}{dt} = -20$. Show that the values of $t$ for stationary values of the steady-state solution are the roots of $6\tan 3t = 7$ [*see* 13].

18 Solve the equation $\frac{d^2y}{dx^2} = 3\sin x - 4y$, given that $y = 0$, at $x = 0$ and that $\frac{dy}{dx} = 1$ at $x = \pi/2$. Find the maximum value of $y$ in the interval $0 < x < \pi$.

19 A mass suspended from a spring performs vertical oscillations and the displacement $x$ (cm) of the mass at time $t$ (s) is given by $\frac{1}{2}\frac{d^2x}{dt^2} = -48x$. If $x = \frac{1}{6}$ and $\frac{dx}{dt} = 0$ when $t = 0$, determine the period and amplitude of the oscillations.

20 The equation of motion of a body performing damped forced vibrations is $\frac{d^2x}{dt^2} + 5\frac{dx}{dt} + 6x = \cos t$. Solve this equation, given that $x = 0{\cdot}1$ and $\frac{dx}{dt} = 0$ when $t = 0$. Write the steady-state solution in the form $K\sin(t + a)$ [*see* 13].

**53**

Now visit the companion website for this book at www.palgrave.com/stroud for more questions applying this mathematics to science and engineering.

# Introduction to Laplace transforms

## Learning outcomes

When you have completed this Programme you will be able to:

- Derive the Laplace transform of an expression by using the integral definition
- Obtain inverse Laplace transforms with the help of a Table of Laplace transforms
- Derive the Laplace transform of the derivative of an expression
- Solve first-order, constant-coefficient, inhomogeneous differential equations using the Laplace transform
- Derive further Laplace transforms from known transforms
- Use the Laplace transform to obtain the solution to linear, constant-coefficient, inhomogeneous differential equations of second and higher order

# The Laplace transform

**1**

All the differential equations you have looked at so far have had solutions containing a number of unknown integration constants $A$, $B$, $C$ etc. The values of these constants have then been found by applying boundary conditions to the solution, a procedure that can often prove to be tedious. Fortunately, for a certain type of differential equation there is a method of obtaining the solution where these unknown integration constants are evaluated *during the process of solution*. Furthermore, rather than employing integration as the way of unravelling the differential equation, you use straightforward algebra.

The method hinges on what is called the *Laplace transform*. If $f(t)$ represents some expression in $t$ defined for $t \geq 0$, the *Laplace transform* of $f(t)$, denoted by $L\{f(t)\}$, is defined to be:

$$L\{f(t)\} = \int_{t=0}^{\infty} e^{-st} f(t)\,dt$$

where $s$ is a variable whose values are chosen so as to ensure that the semi-infinite integral converges. More will be said about the variable $s$ in Frame 3. For now, what would you say is the Laplace transform $f(t) = 2$ for $t \geq 0$?

*Substitute for $f(t)$ in the integral above and then perform the integration.*
*The answer is in the next frame*

**2**

$$\boxed{L\{2\} = \frac{2}{s} \text{ provided } s > 0}$$

Because:

$$L\{f(t)\} = \int_{t=0}^{\infty} e^{-st} f(t)\,dt$$

so

$$\begin{aligned} L\{2\} &= \int_{t=0}^{\infty} e^{-st}2\,dt \\ &= 2\left[\frac{e^{-st}}{-s}\right]_{t=0}^{\infty} \\ &= 2(0 - (-1/s)) \\ &= \frac{2}{s} \end{aligned}$$

Notice that $s > 0$ is demanded because if $s < 0$ then $e^{-st} \to \infty$ as $t \to \infty$ and if $s = 0$ then $L\{2\}$ is not defined (in both of these two cases the integral diverges), so that

$$L\{2\} = \frac{2}{s} \text{ provided } s > 0$$

By the same reasoning, if $k$ is some constant then

$$L\{k\} = \frac{k}{s} \text{ provided } s > 0$$

Now, how about the Laplace transform of $f(t) = e^{-kt}$, $t \geq 0$ where $k$ is a constant?

*Go back to the integral definition and work it out.*
*Again, the answer is in the next frame*

3

$$L\{e^{-kt}\} = \frac{1}{s+k} \text{ provided } s > -k$$

Because

$$\begin{aligned} L\{e^{-kt}\} &= \int_{t=0}^{\infty} e^{-st} e^{-kt}\, dt \\ &= \int_{t=0}^{\infty} e^{-(s+k)t}\, dt \\ &= \left[\frac{e^{-(s+k)t}}{-(s+k)}\right]_{t=0}^{\infty} \\ &= \left(0 - \left(-\frac{1}{(s+k)}\right)\right) \\ &= \frac{1}{(s+k)} \quad \text{provided } s+k > 0, \text{ that is provided } s > -k \end{aligned}$$

These two examples have demonstrated that you need to be careful about the finite existence of the Laplace transform and not just take the integral definition without some thought. For the Laplace transform to exist the integrand

$$e^{-st} f(t)$$

must converge to zero as $t \to \infty$ and this will impose some conditions on the values of $s$ for which the integral does converge and, hence, the Laplace transform exists. In this Programme you can be assured that there are no problems concerning the existence of any of the Laplace transforms that you will meet.

*Move on to the next frame*

## The inverse Laplace transform

4

The Laplace transform is an expression in the variable $s$ which is denoted by $F(s)$. It is said that $f(t)$ and $F(s) = L\{f(t)\}$ form a *transform pair*. This means that if $F(s)$ is the *Laplace transform* of $f(t)$ then $f(t)$ is the *inverse Laplace transform* of $F(s)$. We write:

$$f(t) = L^{-1}\{F(s)\}$$

There is no simple integral definition of the inverse transform so you have to find it by working backwards. For example:

if $f(t) = 4$ then the Laplace transform $L\{f(t)\} = F(s) = \dfrac{4}{s}$

so

if $F(s) = \dfrac{4}{s}$ then the inverse Laplace transform $L^{-1}\{F(s)\} = f(t) = 4$

It is this ability to find the Laplace transform of an expression and then reverse it that makes the Laplace transform so useful in the solution of differential equations, as you will soon see.

For now, what is the inverse Laplace transform of $F(s) = \dfrac{1}{s-1}$?

*To answer this, look at the Laplace transforms you now know.*
*The answer is in the next frame*

5

$$L^{-1}\{F(s)\} = f(t) = e^t$$

Because you know that:

$L\{e^{-kt}\} = \dfrac{1}{s+k}$ you can say that $L^{-1}\left\{\dfrac{1}{s+k}\right\} = e^{-kt}$

so when $k = -1$, $L^{-1}\left\{\dfrac{1}{s-1}\right\} = e^{-(-1)t} = e^t$

To assist in the process of finding Laplace transforms and their inverses a table is used. In the next frame is a short table containing what you know to date.

6

## Table of Laplace transforms

| $f(t) = L^{-1}\{F(s)\}$ | $F(s) = L\{f(t)\}$ | |
|---|---|---|
| $k$ | $\dfrac{k}{s}$ | $s > 0$ |
| $e^{-kt}$ | $\dfrac{1}{s+k}$ | $s > -k$ |

Reading the table from left to right gives the Laplace transform and reading the table from right to left gives the inverse Laplace transform.

*Use these, where possible, to answer the questions in the* **Review exercise** *that follows. Otherwise use the basic definition in Frame 1.*

# Review summary

7

1 The *Laplace transform* of $f(t)$, denoted by $L\{f(t)\}$, is defined to be:

$$L\{f(t)\} = \int_{t=0}^{\infty} e^{-st} f(t)\,dt$$

where $s$ is a variable whose values are chosen so as to ensure that the semi-infinite integral converges.

2 If $F(s)$ is the *Laplace transform* of $f(t)$ then $f(t)$ is the *inverse Laplace transform* of $F(s)$. We write:

$f(t) = L^{-1}\{F(s)\}$

There is no simple integral definition of the inverse transform so you have to find it by working backwards using a *Table of Laplace transforms.*

# Review exercise

8

1 Find the Laplace transform of each of the following. In each case $f(t)$ is defined for $t \geq 0$:

(a) $f(t) = -3$ (b) $f(t) = e$ (c) $f(t) = e^{2t}$

(d) $f(t) = -5e^{-3t}$ (e) $f(t) = 2e^{7t-2}$

▶

2 Find the inverse Laplace transform of each of the following:

(a) $F(s) = -\frac{1}{s}$ (b) $F(s) = \frac{1}{s-5}$ (c) $F(s) = \frac{3}{s+2}$

(d) $F(s) = -\frac{3}{4s}$ (e) $F(s) = \frac{1}{2s-3}$

*Solutions in next frame*

---

**9**

1 (a) $f(t) = -3$

Because $L\{k\} = \frac{k}{s}$ provided $s > 0$, $L\{-3\} = -\frac{3}{s}$ provided $s > 0$

(b) $f(t) = e$

Because $L\{k\} = \frac{k}{s}$ provided $s > 0$, $L\{e\} = \frac{e}{s}$ provided $s > 0$

(c) $f(t) = e^{2t}$

Because $L\{e^{-kt}\} = \frac{1}{s+k}$ provided $s > -k$, $L\{e^{2t}\} = \frac{1}{s-2}$ provided $s > 2$

(d) $f(t) = -5e^{-3t}$

$$L\{-5e^{-3t}\} = \int_{t=0}^{\infty} e^{-st}(-5e^{-3t})\,dt = -5\int_{t=0}^{\infty} e^{-st}e^{-3t}\,dt = -5L\{e^{-3t}\}$$

$$L\{-5e^{-3t}\} = -\frac{5}{s+3} \text{ provided } s > -3$$

(e) $f(t) = 2e^{7t-2}$

$$L\{2e^{7t-2}\} = \int_{t=0}^{\infty} e^{-st}(2e^{7t-2})\,dt = 2e^{-2}\int_{t=0}^{\infty} e^{-st}e^{7t}\,dt = 2e^{-2}L\{e^{7t}\}$$

$$L\{2e^{7t-2}\} = \frac{2e^{-2}}{s-7} \text{ provided } s > 7$$

2 (a) $F(s) = -\frac{1}{s}$

Because $L^{-1}\left\{\frac{k}{s}\right\} = k$, $L^{-1}\left\{-\frac{1}{s}\right\} = L^{-1}\left\{\frac{-1}{s}\right\} = -1$

(b) $F(s) = \frac{1}{s-5}$

Because $L^{-1}\left\{\frac{1}{s+k}\right\} = e^{-kt}$, $L^{-1}\left\{\frac{1}{s-5}\right\} = e^{-(-5)t} = e^{5t}$

(c) $F(s) = \frac{3}{s+2}$

Because $L^{-1}\left\{\frac{1}{s+2}\right\} = e^{-2t}$ and $L\{3e^{-2t}\} = 3L\{e^{-2t}\} = \frac{3}{s+2}$ so

$$L^{-1}\left\{\frac{3}{s+2}\right\} = 3e^{-2t}$$

(d) $F(s) = -\frac{3}{4s}$

$$F(s) = -\frac{3}{4s} = \frac{(-3/4)}{s} \text{ so that } L^{-1}\left\{-\frac{3}{4s}\right\} = L^{-1}\left\{\frac{-3/4}{s}\right\} = -3/4$$

(e) $F(s) = \frac{1}{2s-3}$

$$F(s) = \frac{1}{2s-3} = \frac{\frac{1}{2}}{s-\frac{3}{2}} \text{ so that } f(t) = L^{-1}\left\{\frac{1}{2s-3}\right\} = L^{-1}\left\{\frac{\frac{1}{2}}{s-\frac{3}{2}}\right\} = \frac{1}{2}e^{\frac{3}{2}t}$$

*Next frame*

## 10 Laplace transform of a derivative

Before you can use the Laplace transform to solve a differential equation you need to know the Laplace transform of a derivative. Given some expression $f(t)$ with Laplace transform $L\{f(t)\} = F(s)$, the Laplace transform of the derivative $f'(t)$ is:

$$L\{f'(t)\} = \int_{t=0}^{\infty} e^{-st} f'(t)\,dt$$

This can be integrated by parts as follows:

$$\begin{aligned} L\{f'(t)\} &= \int_{t=0}^{\infty} e^{-st} f'(t)\,dt \\ &= \int_{t=0}^{\infty} u(t)dv(t) \\ &= \Big[u(t)v(t)\Big]_{t=0}^{\infty} - \int_{t=0}^{\infty} v(t)du(t) \quad \text{(the Parts formula – see Programme 16, Frame 21)} \end{aligned}$$

where $u(t) = e^{-st}$ so $du(t) = -se^{-st}dt$ and where $dv(t) = f'(t)dt$ so $v(t) = f(t)$.

Therefore, substitution in the Parts formula gives:

$$\begin{aligned} L\{f'(t)\} &= \Big[e^{-st}f(t)\Big]_{t=0}^{\infty} + s\int_{t=0}^{\infty} e^{-st}f(t)dt \\ &= (0 - f(0)) + sF(s) \text{ assuming } e^{-st}f(t) \to 0 \text{ as } t \to \infty \end{aligned}$$

That is:

$$L\{f'(t)\} = sF(s) - f(0)$$

So the Laplace transform of the derivative of $f(t)$ is given in terms of the Laplace transform of $f(t)$ itself and the value of $f(t)$ when $t = 0$. Before you use this fact just consider two properties of the Laplace transform in the next frame.

## 11 Two properties of Laplace transforms

Both the Laplace transform and its inverse are *linear transforms*, by which is meant that:

(1) *The transform of a sum (or difference) of expressions is the sum (or difference) of the individual transforms. That is:*

$$L\{f(t) \pm g(t)\} = L\{f(t)\} \pm L\{g(t)\}$$

and $L^{-1}\{F(s) \pm G(s)\} = L^{-1}\{F(s)\} \pm L^{-1}\{G(s)\}$

(2) *The transform of an expression that is multiplied by a constant is the constant multiplied by the transform of the expression. That is:*

$L\{kf(t)\} = kL\{f(t)\}$ and $L^{-1}\{kF(s)\} = kL^{-1}\{F(s)\}$ where $k$ is a constant

These are easily proved using the basic definition of the Laplace transform in Frame 1.

Armed with this information let's try a simple differential equation. By using

$$L\{f'(t)\} = sF(s) - f(0)$$

take the Laplace transform of both sides of the equation

$$f'(t) + f(t) = 1 \text{ where } f(0) = 0$$

and find an expression for the Laplace transform $F(s)$.

*Work through this steadily using what you know;*
*you will find the answer in Frame 12*

**12**

$$F(s) = \frac{1}{s(s+1)}$$

Because, taking Laplace transforms of both sides of the equation you have that:

$L\{f'(t) + f(t)\} = L\{1\}$ The Laplace transform of the left-hand side equals the Laplace transform of the right-hand side

That is:

$L\{f'(t)\} + L\{f(t)\} = L\{1\}$ The transform of a sum is the sum of the transforms.

From what you know about the Laplace transform of $f(t)$ and its derivative $f'(t)$ this gives:

$$[sF(s) - f(0)] + F(s) = \frac{1}{s}$$

That is:

$(s+1)F(s) - f(0) = \frac{1}{s}$ and you are given that $f(0) = 0$ so

$(s+1)F(s) = \frac{1}{s}$, that is $F(s) = \frac{1}{s(s+1)}$

Well done. Now, separate the right-hand side into partial fractions.

*You have done plenty of this before in Programme F.8; the answer is in Frame 13*

**13**

$$F(s) = \frac{1}{s} - \frac{1}{s+1}$$

Because

Assume that $\frac{1}{s(s+1)} = \frac{A}{s} + \frac{B}{s+1}$ then, $1 = A(s+1) + Bs$ from which you find that $A = 1$ and $B = -1$ so that $F(s) = \frac{1}{s} - \frac{1}{s+1}$

That was straightforward enough. Now take the inverse Laplace transform and find the solution to the differential equation.

*The answer is in Frame 14*

**14**

$$f(t) = 1 - e^{-t}$$

Because

$f(t) = L^{-1}\{F(s)\}$

$= L^{-1}\left\{\frac{1}{s} - \frac{1}{s+1}\right\}$

$= L^{-1}\left\{\frac{1}{s}\right\} - L^{-1}\left\{\frac{1}{s+1}\right\}$ The inverse Laplace transform of a difference is the difference of the inverse transforms

$= 1 - e^{-t}$ Using the Table of Laplace transforms in Frame 6

▶

You now have a method for solving a differential equation of the form:

$af'(t) + bf(t) = g(t)$ given that $f(0) = k$

where $a$, $b$ and $k$ are known constants and $g(t)$ is a known expression in $t$:

(a) Take the Laplace transform of both sides of the differential equation
(b) Find the expression $F(s) = L\{f(t)\}$ in the form of an algebraic fraction
(c) Separate $F(s)$ into its partial fractions
(d) Find the inverse Laplace transform $L^{-1}\{F(s)\}$ to find the solution $f(t)$ to the differential equation.

*Now you try some but before you do just look at the Table of Laplace transforms in the next frame. You will need them to solve the equations in the* **Review exercise** *that follows.*

**15**

## Table of Laplace transforms

| $f(t) = L^{-1}\{F(s)\}$ | $F(s) = L\{f(t)\}$ | |
|---|---|---|
| $k$ | $\frac{k}{s}$ | $s > 0$ |
| $e^{-kt}$ | $\frac{1}{s+k}$ | $s > -k$ |
| $te^{-kt}$ | $\frac{1}{(s+k)^2}$ | $s > -k$ |

*We will derive this third transform later in the Programme. For now, use these to answer the questions that follow the* **Review summary** *in the next frame*

## Review summary

**16**

1 If $F(s)$ is the Laplace transform of $f(t)$ then the Laplace transform of $f'(t)$ is:

$$L\{f'(t)\} = sF(s) - f(0)$$

2 (a) The Laplace transform of a sum (or difference) of expressions is the sum (or difference) of the individual transforms. That is:

$$L\{f(t) \pm g(t)\} = L\{f(t)\} \pm L\{(g(t)\}$$

and $L^{-1}\{F(s) \pm G(s)\} = L^{-1}\{F(s)\} \pm L^{-1}\{G(s)\}$

(b) The transform of an expression multiplied by a constant is the constant multiplied by the transform of the expression. That is:

$$L\{kf(t)\} = kL\{f(t)\} \text{ and } L^{-1}\{kF(s)\} = kL^{-1}\{F(s)\}$$

where $k$ is a constant.

3 To solve a differential equation of the form:

$af'(t) + bf(t) = g(t)$ given that $f(0) = k$

where $a$, $b$ and $k$ are known constants and $g(t)$ is a known expression in $t$:

(a) Take the Laplace transform of both sides of the differential equation

(b) Find the expression $F(s) = L\{f(t)\}$ in the form of an algebraic fraction

(c) Separate $F(s)$ into its partial fractions

(d) Find the inverse Laplace transform $L^{-1}\{F(s)\}$ to find the solution $f(t)$ to the differential equation.

---

##  Review exercise

**17**

Solve each of the following differential equations:

(a) $f'(t) - f(t) = 2$ where $f(0) = 0$

(b) $f'(t) + f(t) = e^{-t}$ where $f(0) = 0$

(c) $f'(t) + f(t) = 3$ where $f(0) = -2$

(d) $f'(t) - f(t) = e^{2t}$ where $f(0) = 1$

(e) $3f'(t) - 2f(t) = 4e^{-t} + 2$ where $f(0) = 0$

*Solutions in next frame*

---

**18**

(a) $f'(t) - f(t) = 2$ where $f(0) = 0$

Taking Laplace transforms of both sides of this equation gives:

$$sF(s) - f(0) - F(s) = \frac{2}{s} \text{ so that } F(s) = \frac{2}{s(s-1)} = -\frac{2}{s} + \frac{2}{s-1}$$

The inverse transform then gives the solution as

$f(t) = -2 + 2e^t = 2(e^t - 1)$

(b) $f'(t) + f(t) = e^{-t}$ where $f(0) = 0$

Taking Laplace transforms of both sides of this equation gives:

$$sF(s) - f(0) + F(s) = \frac{1}{s+1} \text{ so that } F(s) = \frac{1}{(s+1)^2}$$

The Table of inverse transforms then gives the solution as $f(t) = te^{-t}$

(c) $f'(t) + f(t) = 3$ where $f(0) = -2$

Taking Laplace transforms of both sides of this equation gives:

$$sF(s) - f(0) + F(s) = \frac{3}{s} \text{ so that}$$

$$F(s) = -\frac{2}{s+1} + \frac{3}{s(s+1)} = \frac{3-2s}{s(s+1)} = \frac{3}{s} - \frac{5}{s+1}$$

The inverse transform then gives the solution as $f(t) = 3 - 5e^{-t}$

(d) $f'(t) - f(t) = e^{2t}$ where $f(0) = 1$

Taking Laplace transforms of both sides of this equation gives:

$$sF(s) - f(0) - F(s) = \frac{1}{s-2} \text{ giving } (s-1)F(s) - 1 = \frac{1}{s-2}$$

$$\text{so that } F(s) = \frac{1}{s-1} + \frac{1}{(s-1)(s-2)} = \frac{1}{s-2}$$

The inverse transform then gives the solution as $f(t) = e^{2t}$

▶

(e) $3f'(t) - 2f(t) = 4e^{-t} + 2$ where $f(0) = 0$

Taking Laplace transforms of both sides of this equation gives:

$$3[sF(s) - f(0)] - 2F(s) = \frac{4}{s+1} + \frac{2}{s} = \frac{6s+2}{s(s+1)} \text{ so that}$$

$$F(s) = \frac{6s+2}{s(s+1)(3s-2)} = \frac{27}{5}\left(\frac{1}{3s-2}\right) - \frac{1}{s} - \frac{4}{5}\left(\frac{1}{s+1}\right)$$

$$= \frac{27}{15}\left(\frac{1}{s-\frac{2}{3}}\right) - \frac{1}{s} - \frac{4}{5}\left(\frac{1}{s+1}\right)$$

The inverse transform then gives the solution as:

$$f(t) = \frac{9}{5}e^{2t/3} - \frac{4}{5}e^{-t} - 1$$

*On now to Frame 19*

19

## Generating new transforms

Deriving the Laplace transform of $f(t)$ often requires you to integrate by parts, sometimes repeatedly. However, because $L\{f'(t)\} = sL\{f(t)\} - f(0)$ you can sometimes avoid this involved process when you know the transform of the derivative $f'(t)$. Take as an example the problem of finding the Laplace transform of the expression $f(t) = t$. Now $f'(t) = 1$ and $f(0) = 0$ so that substituting in the equation:

$$L\{f'(t)\} = sL\{f(t)\} - f(0)$$

gives

$$L\{1\} = sL\{t\} - 0$$

that is

$$\frac{1}{s} = sL\{t\}$$

therefore

$$L\{t\} = \frac{1}{s^2}$$

That was easy enough, so what is the Laplace transform of $f(t) = t^2$?

*The answer is in the next frame*

20

$$\boxed{\frac{2}{s^3}}$$

Because

$$f(t) = t^2,\ f'(t) = 2t \text{ and } f(0) = 0$$

Substituting in

$$L\{f'(t)\} = sL\{f(t)\} - f(0)$$

gives

$$L\{2t\} = sL\{t^2\} - 0$$

that is

$$2L\{t\} = sL\{t^2\} \text{ so } \frac{2}{s^2} = sL\{t^2\}$$

▶

therefore

$$L\{t^2\} = \frac{2}{s^3}$$

Just try another one. Verify the third entry in the Table of Laplace transforms in Frame 15 for $k = 1$, that is:

$$L\{te^{-t}\} = \frac{1}{(s+1)^2}$$

*This is a littler harder but just follow the procedure laid out in the previous two frames and try it. The explanation is in the next frame*

---

**21**

Because

$$f(t) = te^{-t},\ f'(t) = e^{-t} - te^{-t} \text{ and } f(0) = 0$$

Substituting in

$$L\{f'(t)\} = sL\{f(t)\} - f(0)$$

gives

$$L\{e^{-t} - te^{-t}\} = sL\{te^{-t}\} - 0$$

that is

$$L\{e^{-t}\} - L\{te^{-t}\} = sL\{te^{-t}\}$$

therefore

$$L\{e^{-t}\} = (s+1)L\{te^{-t}\}$$

giving

$$\frac{1}{s+1} = (s+1)L\{te^{-t}\} \text{ and so } L\{te^{-t}\} = \frac{1}{(s+1)^2}$$

*On now to Frame 22*

---

## Laplace transforms of higher derivatives

**22**

The Laplace transforms of derivatives higher than the first are readily derived. Let $F(s)$ and $G(s)$ be the respective Laplace transforms of $f(t)$ and $g(t)$. That is

$$L\{f(t)\} = F(s) \text{ so that } L\{f'(t)\} = sF(s) - f(0)$$

and

$$L\{g(t)\} = G(s) \text{ and } L\{g'(t)\} = sG(s) - g(0)$$

Now let $g(t) = f'(t)$ so that $L\{g(t)\} = L\{f'(t)\}$ where

$$g(0) = f'(0) \text{ and } G(s) = sF(s) - f(0)$$

Now, because $g(t) = f'(t)$

$$g'(t) = f''(t)$$

This means that

$$L\{g'(t)\} = L\{f''(t)\} = sG(s) - g(0) = s[sF(s) - f(0)] - f'(0)$$

so

$$L\{f''(t)\} = s^2F(s) - sf(0) - f'(0)$$

▶

By a similar argument it can be shown that

$$L\{f'''(t)\} = s^3F(s) - s^2f(0) - sf'(0) - f''(0)$$

and so on. Can you see the pattern developing here?

The Laplace transform of $f^{iv}(t)$ is ...........

*Next frame*

**23**

$$\boxed{L\{f^{iv}(t)\} = s^4F(s) - s^3f(0) - s^2f'(0) - sf''(0) - f'''(0)}$$

Now, using $L\{f''(t)\} = s^2F(s) - sf(0) - f'(0)$ the Laplace transform of $f(t) = \sin kt$ where $k$ is a constant is ...........

*Differentiate $f(t)$ twice and follow the procedure that you used in Frames 19 to 21. Take it carefully, the answer and working are in the following frame*

**24**

$$\boxed{L\{\sin kt\} = \frac{k}{s^2 + k^2}}$$

Because

$f(t) = \sin kt$, $f'(t) = k\cos kt$ and $f''(t) = -k^2 \sin kt$.
Also $f(0) = 0$ and $f'(0) = k$.

Substituting in

$$L\{f''(t)\} = s^2F(s) - sf(0) - f'(0) \text{ where } F(s) = L\{f(t)\}$$

gives

$$L\{-k^2 \sin kt\} = s^2L\{\sin kt\} - s.0 - k$$

that is

$$-k^2L\{\sin kt\} = s^2L\{\sin kt\} - k$$

so

$$(s^2 + k^2)L\{\sin kt\} = k \text{ and } L\{\sin kt\} = \frac{k}{s^2 + k^2}$$

And $L\{\cos kt\} =$ ...........

*It's just the same method*

**25**

$$\boxed{L\{\cos kt\} = \frac{s}{s^2 + k^2}}$$

Because

$f(t) = \cos kt$, $f'(t) = -k\sin kt$ and $f''(t) = -k^2 \cos kt$.
Also $f(0) = 1$ and $f'(0) = 0$.

Substituting in

$$L\{f''(t)\} = s^2F(s) - sf(0) - f'(0) \text{ where } F(s) = L\{f(t)\}$$

gives

$$L\{-k^2 \cos kt\} = s^2L\{\cos kt\} - s.1 - 0$$

▶

that is

$$-k^2L\{\cos kt\} = s^2L\{\cos kt\} - s$$

so

$$(s^2 + k^2)L\{\cos kt\} = s \text{ and } L\{\cos kt\} = \frac{s}{s^2 + k^2}$$

The Table of transforms is now extended in the next frame.

## Table of Laplace transforms

26

| $f(t) = L^{-1}\{F(s)\}$ | $F(s) = L\{f(t)\}$ | |
|---|---|---|
| $k$ | $\frac{k}{s}$ | $s > 0$ |
| $e^{-kt}$ | $\frac{1}{s+k}$ | $s > -k$ |
| $te^{-kt}$ | $\frac{1}{(s+k)^2}$ | $s > -k$ |
| $t$ | $\frac{1}{s^2}$ | $s > 0$ |
| $t^2$ | $\frac{2}{s^3}$ | $s > 0$ |
| $\sin kt$ | $\frac{k}{s^2 + k^2}$ | $s^2 + k^2 > 0$ |
| $\cos kt$ | $\frac{s}{s^2 + k^2}$ | $s^2 + k^2 > 0$ |

## Linear, constant-coefficient, inhomogeneous differential equations

27

The Laplace transform can be used to solve equations of the form:

$$a_nf^{(n)}(t) + a_{n-1}f^{(n-1)}(t) + \cdots + a_2f''(t) + a_1f'(t) + a_0f(t) = g(t)$$

where $a_n, a_{n-1}, \ldots, a_2, a_1, a_0$ are known constants, $g(t)$ is a known expression in $t$ and the values of $f(t)$ and its derivatives are known at $t = 0$. This type of equation is called a *linear, constant-coefficient, inhomogeneous differential equation* and the values of $f(t)$ and its derivatives at $t = 0$ are called *boundary conditions*. The method of obtaining the solution follows the procedure laid down in Frame 14. For example:

To find the solution of:

$$f''(t) + 3f'(t) + 2f(t) = 4t \text{ where } f(0) = f'(0) = 0$$

(a) *Take the Laplace transform of both sides of the equation*

$$L\{f''(t)\} + 3L\{f'(t)\} + 2L\{f(t)\} = 4L\{t\}$$

to give $\left[s^2F(s) - sf(0) - f'(0)\right] + 3[sF(s) - f(0)] + 2F(s) = \frac{4}{s^2}$

▶

(b) *Find the expression $F(s) = L\{f(t)\}$ in the form of an algebraic fraction*
Substituting the values for $f(0)$ and $f'(0)$ and then rearranging gives

$$(s^2 + 3s + 2)F(s) = \frac{4}{s^2}$$

so that

$$F(s) = \frac{4}{s^2(s+1)(s+2)}$$

(c) *Separate $F(s)$ into its partial fractions*

$$\frac{4}{s^2(s+1)(s+2)} = \frac{A}{s} + \frac{B}{s^2} + \frac{C}{s+1} + \frac{D}{s+2}$$

Adding the right-hand side partial fractions together and then equating the left-hand side numerator with the right-hand side numerator gives

$$4 = As(s+1)(s+2) + B(s+1)(s+2) + Cs^2(s+2) + Ds^2(s+1)$$

Let $s = 0$ $\quad 4 = 2B$ therefore $B = 2$

$s = -1$ $\quad 4 = C(-1)^2(-1+2) = C$

$s = -2$ $\quad 4 = D(-2)^2(-2+1) = -4D$ therefore $D = -1$

Equate the coefficients of $s$:

$0 = 2A + 3B = 2A + 6$ therefore $A = -3$

Consequently:

$$F(s) = -\frac{3}{s} + \frac{2}{s^2} + \frac{4}{s+1} - \frac{1}{s+2}$$

(d) *Use the Tables to find the inverse Laplace transform $L^{-1}\{F(s)\}$ and so find the solution $f(t)$ to the differential equation*

$$f(t) = -3 + 2t + 4e^{-t} - e^{-2t}$$

*So that was all very straightforward even if it was involved.*
*Now try your hand at the differential equations in Frame 29*

---

## Review summary

**28**

1 If $F(s)$ is the Laplace transform of $f(t)$ then:

$$L\{f''(t)\} = s^2F(s) - sf(0) - f'(0)$$
and $$L\{f'''(t)\} = s^3F(s) - s^2f(0) - sf'(0) - f''(0)$$

2 Equations of the form:

$$a_nf^{(n)}(t) + a_{n-1}f^{(n-1)}(t) + \cdots + a_2f''(t) + a_1f'(t) + a_0f(t) = g(t)$$

where $a_n, a_{n-1}, \ldots, a_2, a_1, a_0$ are constants are called linear, constant-coefficient, inhomogeneous differential equations.

3 The Laplace transform can be used to solve constant-coefficient, inhomogeneous differential equations provided $a_n, a_{n-1}, \ldots, a_2, a_1, a_0$ are known constants, $g(t)$ is a known expression in $t$, and the values of $f(t)$ and its derivatives are known at $t = 0$.

▶

4 The procedure for solving these equations of second and higher order is the same as that for solving the equations of first order. Namely:

(a) Take the Laplace transform of both sides of the differential equation
(b) Find the expression $F(s) = L\{f(t)\}$ in the form of an algebraic fraction
(c) Separate $F(s)$ into its partial fractions
(d) Find the inverse Laplace transform $L^{-1}\{F(s)\}$ to find the solution $f(t)$ to the differential equation.

---

## Review exercise

**29**

Use the Laplace transform to solve each of the following equations:

(a) $f'(t) + f(t) = 3$ where $f(0) = 0$

(b) $3f'(t) + 2f(t) = t$ where $f(0) = -2$

(c) $f''(t) + 5f'(t) + 6f(t) = 2e^{-t}$ where $f(0) = 0$ and $f'(0) = 0$

(d) $f''(t) - 4f(t) = \sin 2t$ where $f(0) = 1$ and $f'(0) = -2$

*Answers in next frame*

**30**

(a) $f'(t) + f(t) = 3$ where $f(0) = 0$

Taking Laplace transforms of both sides of the equation gives

$L\{f'(t)\} + L\{f(t)\} = L\{3\}$ so that $[sF(s) - f(0)] + F(s) = \dfrac{3}{s}$

That is $(s+1)F(s) = \dfrac{3}{s}$ so $F(s) = \dfrac{3}{s(s+1)} = \dfrac{3}{s} - \dfrac{3}{s+1}$

giving the solution as $f(t) = 3 - 3e^{-t} = 3(1 - e^{-t})$

(b) $3f'(t) + 2f(t) = t$ where $f(0) = -2$

Taking Laplace transforms of both sides of the equation gives

$L\{3f'(t)\} + L\{2f(t)\} = L\{t\}$ so that $3[sF(s) - f(0)] + 2F(s) = \dfrac{1}{s^2}$

That is $(3s+2)F(s) - (-6) = \dfrac{1}{s^2}$ so $F(s) = \dfrac{1 - 6s^2}{s^2(3s+2)}$

The partial fraction breakdown gives

$$F(s) = -\frac{3}{4}\cdot\frac{1}{s} + \frac{1}{2}\cdot\frac{1}{s^2} - \frac{15}{4}\cdot\frac{1}{(3s+2)} = -\frac{3}{4}\cdot\frac{1}{s} + \frac{1}{2}\cdot\frac{1}{s^2} - \frac{5}{4}\cdot\frac{1}{(s+\frac{2}{3})}$$

giving the solution as

$$f(t) = -\frac{3}{4} + \frac{t}{2} - \frac{5e^{-2t/3}}{4}$$

▶

(c) $f''(t) + 5f'(t) + 6f(t) = 2e^{-t}$ where $f(0) = 0$ and $f'(0) = 0$

Taking Laplace transforms of both sides of the equation gives

$L\{f''(t)\} + L\{5f'(t)\} + L\{6f(t)\} = L\{2e^{-t}\}$

so that $[s^2F(s) - sf(0) - f'(0)] + 5[sF(s) - f(0)] + 6F(s) = \dfrac{2}{s+1}$

That is $(s^2 + 5s + 6)F(s) = \dfrac{2}{s+1}$

so $F(s) = \dfrac{2}{(s+1)(s+2)(s+3)} = \dfrac{1}{s+1} - \dfrac{2}{s+2} + \dfrac{1}{s+3}$

giving the solution as

$f(t) = e^{-t} - 2e^{-2t} + e^{-3t}$

(d) $f''(t) - 4f(t) = \sin 2t$ where $f(0) = 1$ and $f'(0) = -2$

Taking Laplace transforms of both sides of the equation gives

$L\{f''(t)\} - L\{4f(t)\} = L\{\sin 2t\}$

so that $[s^2F(s) - sf(0) - f'(0)] - 4F(s) = \dfrac{2}{s^2 + 2^2}$

That is $(s^2 - 4)F(s) - s.1 - (-2) = \dfrac{2}{s^2 + 2^2}$

$$\text{so } F(s) = \frac{2}{(s^2-4)(s^2+2^2)} + \frac{s-2}{s^2-4}$$

$$= \frac{15}{16}\cdot\frac{1}{s+2} + \frac{1}{16}\cdot\frac{1}{s-2} - \frac{1}{8}\cdot\frac{2}{s^2+2^2}$$

giving the solution as

$$f(t) = \frac{15}{16}e^{-2t} + \frac{1}{16}e^{2t} - \frac{\sin 2t}{8}$$

*So, finally, the* **Can You?** *checklist followed by the* **Test exercise** *and* **Further problems**

## Can You?

### Checklist 27

31

Check this list before and after you try the end of Programme test.

On a scale of 1 to 5 how confident are you that you can: Frames

- Derive the Laplace transform of an expression by using the integral definition? 1 to 3
  *Yes* ☐ ☐ ☐ ☐ ☐ *No*
- Obtain inverse Laplace transforms with the help of a Table of Laplace transforms? 4 to 9
  *Yes* ☐ ☐ ☐ ☐ ☐ *No*
- Derive the Laplace transform of the derivative of an expression? 10
  *Yes* ☐ ☐ ☐ ☐ ☐ *No*
- Solve first-order, constant-coefficient, inhomogeneous differential equations using the Laplace transform? 11 to 18
  *Yes* ☐ ☐ ☐ ☐ ☐ *No*
- Derive further Laplace transforms from known transforms? 19 to 26
  *Yes* ☐ ☐ ☐ ☐ ☐ *No*
- Use the Laplace transform to obtain the solution to linear, constant-coefficient, inhomogeneous differential equations of higher order than the first? 27 to 30
  *Yes* ☐ ☐ ☐ ☐ ☐ *No*

## Test exercise 27

32

1 Using the integral definition, find the Laplace transforms for each of the following:

(a) $f(t) = 8$ (b) $f(t) = e^{5t}$ (c) $f(t) = -4e^{2t+3}$

2 Using the Table of Laplace transforms, find the inverse Laplace transforms of each of the following:

(a) $F(s) = -\dfrac{5}{(s-2)^2}$ (b) $F(s) = \dfrac{2e^3}{s^3}$

(c) $F(s) = \dfrac{3}{s^2+9}$ (d) $F(s) = -\dfrac{2s-5}{s^2+3}$

3 Given that the Laplace transform of $te^{-kt}$ is $F(s) = \dfrac{1}{(s+k)^2}$ derive the Laplace transform of $t^2e^{3t}$ without using the integral definition.

4 Use the Laplace transform to solve each of the following equations:

(a) $f'(t) + 2f(t) = t$ where $f(0) = 0$

(b) $f'(t) - f(t) = e^{-t}$ where $f(0) = -1$

(c) $f''(t) + 4f'(t) + 4f(t) = e^{-2t}$ where $f(0) = 0$ and $f'(0) = 0$

(d) $4f''(t) - 9f(t) = -18$ where $f(0) = 0$ and $f'(0) = 0$

## Further problems 27

**33**

1 Find the Laplace transform of each of the following expressions (each being defined for $t \geq 0$):

(a) $f(t) = a^{kt}, a > 0$ (b) $f(t) = \sinh kt$ (c) $f(t) = \cosh kt$

(d) $f(t) = \begin{cases} k & \text{for } 0 \leq t \leq a \\ 0 & \text{for } t > a \end{cases}$

2 Find the inverse Laplace transform of each of the following:

(a) $F(s) = -\dfrac{2}{3s - 4}$ (b) $F(s) = \dfrac{1}{s^2 - 8}$ (c) $F(s) = \dfrac{3s - 4}{s^2 + 16}$

(d) $F(s) = \dfrac{7s^2 + 27}{s^3 + 9s}$ (e) $F(s) = \dfrac{4s}{(s^2 - 1)^2}$ (f) $F(s) = -\dfrac{s^2 - 6s + 14}{s^3 - s^2 + 4s - 4}$

3 Show that if $F(s) = L\{f(t)\} = \displaystyle\int_{t=0}^{\infty} e^{-st} f(t)\,dt$ then:

(a) (i) $F'(s) = -L\{tf(t)\}$ (ii) $F''(s) = L\{t^2 f(t)\}$

Use part (a) to find

(b) (i) $L\{t \sin 2t\}$ (ii) $L\{t^2 \cos 3t\}$

(c) What would you say the $n$th derivative of $F(s)$ is equal to?

4 Show that if $L\{f(t)\} = F(s)$ then $L\{e^{kt} f(t)\} = F(s - k)$ where $k$ is a constant. Hence find:

(a) $L\{e^{at} \sin bt\}$

(b) $L\{e^{at} \cos bt\}$ where $a$ and $b$ are constants in both cases.

5 Solve each of the following differential equations:

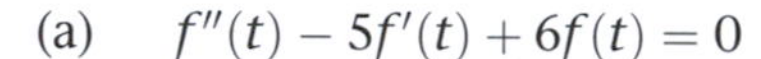

(a) $f''(t) - 5f'(t) + 6f(t) = 0$ where $f(0) = 0$ and $f'(0) = 1$

(b) $f''(t) - 5f'(t) + 6f(t) = 1$ where $f(0) = 0$ and $f'(0) = 0$

(c) $f''(t) - 5f'(t) + 6f(t) = e^{2t}$ where $f(0) = 0$ and $f'(0) = 0$

(d) $2f''(t) - f'(t) - f(t) = e^{-3t}$ where $f(0) = 2$ and $f'(0) = 1$

(e) $f(t) + f'(t) - 2f''(t) = te^{-t}$ where $f(0) = 0$ and $f'(0) = 1$

(f) $f''(t) + 16f(t) = 0$ where $f(0) = 1$ and $f'(0) = 4$

(g) $2f''(t) - f'(t) - f(t) = \sin t - \cos t$ where $f(0) = 0$ and $f'(0) = 0$

**34**

Now visit the companion website for this book at www.palgrave.com/stroud for more questions applying this mathematics to science and engineering.

# Data handling and statistics

## Learning outcomes

When you have completed this Programme you will be able to:

- ☐ Distinguish between discrete and continuous data
- ☐ Construct frequency and relative frequency tables for grouped and ungrouped discrete data
- ☐ Determine class boundaries, class intervals and central values for discrete and continuous data
- ☐ Construct a histogram and a frequency polygon
- ☐ Determine the mean, median and mode of grouped and ungrouped data
- ☐ Determine the range, variance and standard deviation of discrete data
- ☐ Measure the dispersion of data using the normal and standard normal curves

## Introduction

**1**

*Statistics* is concerned with the collection, ordering and analysis of data. *Data* consist of sets of recorded observations or values. Any quantity that can have a number of values is a *variable*. A variable may be one of two kinds:

(a) *Discrete* – a variable that can be counted, or for which there is a fixed set of values.

(b) *Continuous* – a variable that can be measured on a continuous scale, the result depending on the precision of the measuring instrument, or the accuracy of the observer.

A statistical exercise normally consists of four stages:

(a) collection of data by counting or measuring
(b) ordering and presentation of the data in a convenient form
(c) analysis of the collected data
(d) interpretation of the results and conclusions formulated.

State whether each of the following is a discrete or continuous variable:

(a) the number of components in a machine
(b) the capacity of a container
(c) the size of workforce in a factory
(d) the speed of rotation of a shaft
(e) the temperature of a coolant.

**2**

(a) and (c) discrete; (b), (d) and (e) continuous

*Next frame*

## Arrangement of data

**3**

The contents of each of 30 packets of washers are recorded:

| | | | | | | | | | |
|---|---|---|---|---|---|---|---|---|---|
| 28 | 31 | 29 | 27 | 30 | 29 | 29 | 26 | 30 | 28 |
| 28 | 29 | 27 | 26 | 32 | 28 | 32 | 31 | 25 | 30 |
| 27 | 30 | 29 | 30 | 28 | 29 | 31 | 27 | 28 | 28 |

We can appreciate this set of numbers better if we now arrange the values in ascending order, writing them still in 3 lines of 10. If we do this, we get ...........

4

| 25 | 26 | 26 | 27 | 27 | 27 | 27 | 28 | 28 | 28 |
|---|---|---|---|---|---|---|---|---|---|
| 28 | 28 | 28 | 28 | 29 | 29 | 29 | 29 | 29 | 29 |
| 30 | 30 | 30 | 30 | 30 | 31 | 31 | 31 | 32 | 32 |

Some values occur more than once. Therefore, we can form a table showing how many times each value occurs:

| Value | Number of times |
|---|---|
| 25 | 1 |
| 26 | 2 |
| 27 | 4 |
| 28 | 7 |
| 29 | 6 |
| 30 | 5 |
| 31 | 3 |
| 32 | 2 |

The number of occasions on which any particular value occurs is called its *frequency*, denoted by the symbol $f$.

The total frequency is therefore . . . . . . . . . . .

5

30, the total number of readings

## Tally diagram

When dealing with large numbers of readings, instead of writing all the values in ascending order, it is more convenient to compile a *tally diagram*, recording the range of values of the variable and adding a stroke for each occurrence of that reading, thus:

| Variable ($x$) | Tally marks | Frequency ($f$) |
|---|---|---|
| 25 | / | 1 |
| 26 | // | 2 |
| 27 | //// | 4 |
| 28 | ~~////~~ // | 7 |
| 29 | ~~////~~ / | 6 |
| 30 | ~~////~~ | 5 |
| 31 | /// | 3 |
| 32 | // | 2 |

It is usual to denote the variable by $x$ and the frequency by $f$. The right-hand column gives the *frequency distribution* of the values of the variable.

### Exercise

The number of components per hour turned out on a lathe was measured on 40 occasions:

| 18 | 17 | 21 | 18 | 19 | 17 | 18 | 20 | 16 | 17 |
|---|---|---|---|---|---|---|---|---|---|
| 19 | 19 | 16 | 17 | 15 | 19 | 17 | 17 | 20 | 18 |
| 17 | 18 | 19 | 19 | 18 | 19 | 18 | 18 | 19 | 20 |
| 18 | 15 | 18 | 17 | 20 | 18 | 16 | 17 | 18 | 17 |

Compile a tally diagram and so determine the frequency distribution of the values. This gives . . . . . . . . . . .

6

| Variable ($x$) | Tally marks | Frequency ($f$) |
|---|---|---|
| 15 | / | 2 |
| 16 | /// | 3 |
| 17 | ~~////~~ ~~////~~ | 10 |
| 18 | ~~////~~ ~~////~~ // | 12 |
| 19 | ~~////~~ /// | 8 |
| 20 | //// | 4 |
| 21 | / | 1 |
| | $n = \sum f = 40$ | |

## Grouped data

If the range of values of the variable is large, it is often helpful to consider these values arranged in regular groups, or *classes*.

For example, the numbers of the overtime hours per week worked by employees at a factory are as follows:

| | | | | | | | | | |
|---|---|---|---|---|---|---|---|---|---|
| 45 | 31 | 46 | 25 | 57 | 39 | 42 | 55 | 20 | 37 |
| 40 | 59 | 11 | 38 | 34 | 22 | 62 | 33 | 48 | 43 |
| 57 | 37 | 43 | 51 | 29 | 41 | 35 | 66 | 45 | 32 |
| 44 | 47 | 42 | 46 | 54 | 65 | 17 | 35 | 53 | 27 |
| 38 | 22 | 33 | 39 | 45 | 32 | 43 | 41 | 57 | 45 |

$$\left.\begin{array}{r}\text{Lowest value of the variable} = 11\\ \text{Highest value of the variable} = 66\end{array}\right\} \quad \therefore \text{ Arrange 6 classes of 10 h each.}$$

To determine the frequency distribution, we set up a table as follows:

| Overtime hours ($x$) | Tally marks | Frequency ($f$) |
|---|---|---|
| 10–19 | | |
| 20–29 | | |
| 30–39 | | |
| etc. | | |

Complete the frequency distribution.

7

| Overtime hours ($x$) | Tally marks | Frequency ($f$) |
|---|---|---|
| 10–19 | // | 2 |
| 20–29 | ~~////~~ / | 6 |
| 30–39 | ~~////~~ ~~////~~ //// | 14 |
| 40–49 | ~~////~~ ~~////~~ ~~////~~ // | 17 |
| 50–59 | ~~////~~ /// | 8 |
| 60–70 | /// | 3 |
| | $n = \sum f = 50$ | |

▶

## Grouping with continuous data

In the previous example using discrete data, there is no difficulty in allocating any given value to its appropriate group, since, for example, there is no value between, say, 29 and 30. However, with continuous data, the variable is measured on ........... and may well have values lying between 29 and 30, e.g. 29·7, 29·8 etc.

**8**

a continuous scale

In practice, where the values of the variable are all given to the same number of significant figures or decimal places, there is no trouble and we form the groups accordingly.

For example, the lengths (in mm) of 40 spindles were measured with the following results:

| | | | | | | | | | |
|---|---|---|---|---|---|---|---|---|---|
| 20·90 | 20·57 | 20·86 | 20·74 | 20·82 | 20·63 | 20·53 | 20·89 | 20·75 | 20·65 |
| 20·71 | 21·03 | 20·72 | 20·41 | 20·94 | 20·75 | 20·79 | 20·65 | 21·08 | 20·89 |
| 20·50 | 20·88 | 20·97 | 20·78 | 20·61 | 20·92 | 21·07 | 21·16 | 20·80 | 20·77 |
| 20·82 | 20·72 | 20·60 | 20·90 | 20·86 | 20·68 | 20·75 | 20·88 | 20·56 | 20·94 |

Lowest value = 20·41
Highest value = 21·16 } ∴ Form classes from 20·40 to 21·20 at 0·10 intervals.

| Length (mm) ($x$) | Tally marks | Frequency ($f$) |
|---|---|---|
| 20·40–20·49 | | |
| 20·50–20·59 | | |
| 20·60–20·69 | | |
| etc. | | |

Complete the table and so determine the frequency distribution ...........

**9**

| Length (mm) ($x$) | Tally marks | Frequency ($f$) |
|---|---|---|
| 20·40–20·49 | / | 1 |
| 20·50–20·59 | //// | 4 |
| 20·60–20·69 | ~~////~~ / | 6 |
| 20·70–20·79 | ~~////~~ ~~////~~ | 10 |
| 20·80–20·89 | ~~////~~ //// | 9 |
| 20·90–20·99 | ~~////~~ / | 6 |
| 21·00–21·09 | /// | 3 |
| 21·10–21·20 | / | 1 |
| | $n = \sum f = 40$ | |

Note that the last class is slightly larger than the others, but this has negligible effect, since there are very few entries in the end classes.

▶

## Relative frequency

In the frequency distribution just determined, if the frequency of any one class is compared with the sum of the frequencies of all classes (i.e. the total frequency), the ratio is the *relative frequency* of that class. The result is generally expressed as a percentage.

Add a fourth column to the table above showing the relative frequency of each class expressed as a percentage.

*Check with the next frame*

**10**

| Length (mm) ($x$) | Frequency ($f$) | Relative frequency (%) |
|---|---|---|
| 20·40–20·49 | 1 | 2·5 |
| 20·50–20·59 | 4 | 10·0 |
| 20·60–20·69 | 6 | 15·0 |
| 20·70–20·79 | 10 | 25·0 |
| 20·80–20·89 | 9 | 22·5 |
| 20·90–20·99 | 6 | 15·0 |
| 21·00–21·09 | 3 | 7·5 |
| 21·10–21·20 | 1 | 2·5 |
| | $n = \sum f = 40$ | 100·0% |

The sum of the relative frequencies of all classes must add up to the whole and therefore has a value 1 or 100 per cent.

## Rounding off data

If the value 21·7 is expressed to two significant figures, the result is, of course, 22. Similarly, 21·4 is rounded off to 21. In order to maintain consistency of class boundaries, 'middle values' will, in what follows, always be rounded up. For example, 21·5 is rounded up to 22 and 42·5 is rounded up to 43.

Therefore, when a result is quoted to two significant figures as 37 on a continuous scale, this includes all possible values between 36·500000… and 37·499999… (37·5 itself would be rounded off to 38). This could be expressed by saying that 37 included all values between 36·5 and $37{\cdot}5^-$, the small negative sign in the index position indicating that the value is just under 37·5 without actually reaching it.

So 42 includes all values between ………… and …………
31·4 includes all values between ………… and …………
17·63 includes all values between ………… and …………

**11**

| | | | |
|---|---|---|---|
| 42: | 41·5 | and | $42{\cdot}5^-$ |
| 31·4: | 31·35 | and | $31{\cdot}45^-$ |
| 17·63: | 17·625 | and | $17{\cdot}635^-$ |

**Exercise**

The thicknesses of 20 samples of steel plate are measured and the results (in millimetres) to two significant figures are as follows:

| | | | | | | | | | |
|---|---|---|---|---|---|---|---|---|---|
| 7·3 | 7·1 | 6·6 | 7·0 | 7·8 | 7·3 | 7·5 | 6·2 | 6·9 | 6·7 |
| 6·5 | 6·8 | 7·2 | 7·4 | 6·5 | 6·9 | 7·2 | 7·6 | 7·0 | 6·8 |

Compile a table showing the frequency distribution and the relative frequency distribution for regular classes of 0·2 mm from 6·2 mm to 7·9 mm.



| Length (mm) ($x$) | Frequency ($f$) | Relative frequency (%) |
|---|---|---|
| 6·2–6·4 | 1 | 5 |
| 6·5–6·7 | 4 | 20 |
| 6·8–7·0 | 6 | 30 |
| 7·1–7·3 | 5 | 25 |
| 7·4–7·6 | 3 | 15 |
| 7·7–7·9 | 1 | 5 |
| | $n = \sum f = 20$ | 100% |

## Class boundaries

In the example above, the values of the variable are given to 2 significant figures. With the usual rounding-off procedure, each class in effect extends from 0·05 below the first stated value of the class to just under 0·05 above the second stated value of the class. So, the class

7·1–7·3 includes all values between 7·05 and $7{\cdot}35^-$
7·4–7·6 includes all values between 7·35 and $7{\cdot}65^-$ etc.

We can use this example to define a number of terms we shall need. Let us consider in particular the class labelled 7·1–7·3.

(a) The class values stated in the table are the *lower* and *upper limits* of the class and their difference gives the *class width*.

(b) The *class boundaries* are 0·05 below the lower class limit and 0·05 above the upper class limit, that is:

the lower class boundary is $7{\cdot}1 - 0{\cdot}05 = 7{\cdot}05$

the upper class boundary is $7{\cdot}3 + 0{\cdot}05 = 7{\cdot}35$.

(c) The *class interval* is the difference between the upper and lower class boundaries.

$$\text{class interval} = \text{upper class boundary} - \text{lower class boundary}$$
$$= 7{\cdot}35 - 7{\cdot}05 = 0{\cdot}30$$

Where the classes are regular, the class interval can also be found by subtracting any lower class limit from the lower class limit of the following class.

(d) The *central value* (or mid-value) of the class is the average of the upper and lower class boundaries. So, in the particular class we are considering, the central value is

. . . . . . . . . . . .

13

7·20

For the central value $= \frac{1}{2}(7{\cdot}05 + 7{\cdot}35) = 7{\cdot}20$.

We can summarize these terms in the following diagram, using the class 7·1–7·3 (inclusive) as our example.

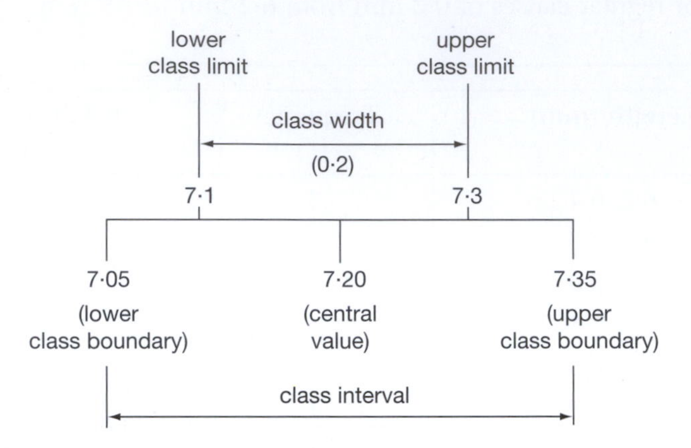


So now you can complete the following table:

| Thickness (mm) ($x$) | Central value | Lower class boundary | Upper class boundary |
|---|---|---|---|
| 6·2–6·4 | | | |
| 6·5–6·7 | | | |
| 6·8–7·0 | | | |
| 7·1–7·3 | | | |
| 7·4–7·6 | | | |
| 7·7–7·9 | | | |

14

| Thickness (mm) ($x$) | Central value | Lower class boundary | Upper class boundary |
|---|---|---|---|
| 6·2–6·4 | 6·3 | 6·15 | 6·45 |
| 6·5–6·7 | 6·6 | 6·45 | 6·75 |
| 6·8–7·0 | 6·9 | 6·75 | 7·05 |
| 7·1–7·3 | 7·2 | 7·05 | 7·35 |
| 7·4–7·6 | 7·5 | 7·35 | 7·65 |
| 7·7–7·9 | 7·8 | 7·65 | 7·95 |

▶

**Exercise**

A machine is set to produce metal washers of nominal diameter 20·0 mm. The diameters of 34 samples are measured and the following results in millimetres obtained:

| | | | | | | | | |
|---|---|---|---|---|---|---|---|---|
| 19·63 | 19·82 | 19·96 | 19·75 | 19·86 | 19·82 | 19·61 | 19·97 | 20·07 |
| 19·89 | 20·16 | 19·56 | 20·05 | 19·72 | 19·96 | 19·68 | 19·87 | 19·90 |
| 19·73 | 19·93 | 20·03 | 19·86 | 19·81 | 19·77 | 19·78 | 19·75 | 19·87 |
| 19·66 | 19·77 | 19·99 | 20·00 | 20·11 | 20·01 | 19·84 | | |

Arrange the values into 7 equal classes of width 0·09 mm for the range 19·50 mm to 20·19 mm and determine the frequency distribution.

*Check your result with the next frame*

**15**

| Diameter (mm) ($x$) | Frequency ($f$) |
|---|---|
| 19·50–19·59 | 1 |
| 19·60–19·69 | 4 |
| 19·70–19·79 | 7 |
| 19·80–19·89 | 9 |
| 19·90–19·99 | 6 |
| 20·00–20·09 | 5 |
| 20·10–20·19 | 2 |
| | $n = \sum f = 34$ |

So (a) the class having the highest frequency is ...........

(b) the lower class boundary of the third class is ...........

(c) the upper class boundary of the seventh class is ...........

(d) the central value of the fifth class is ...........

**16**

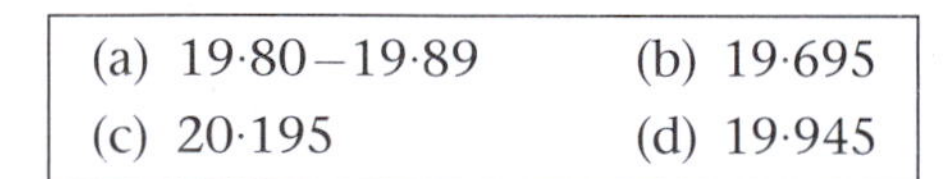

(a) 19·80 – 19·89 (b) 19·695
(c) 20·195 (d) 19·945

*Next frame*

# Histograms

**17**

## Frequency histogram

A histogram is a graphical representation of a frequency distribution, in which vertical rectangular blocks are drawn so that:

(a) the centre of the base indicates the central value of the class and

(b) the area of the rectangle represents the class frequency.

If the class intervals are regular, the frequency is then denoted by the height of the rectangle.

▶

For example, measurement of the lengths of 50 brass rods gave the following frequency distribution:

| Length (mm) ($x$) | Lower class boundary | Upper class boundary | Central value | Frequency ($f$) |
|---|---|---|---|---|
| 3·45–3·47 | 3·445 | 3·475 | 3·460 | 2 |
| 3·48–3·50 | 3·475 | 3·505 | 3·490 | 6 |
| 3·51–3·53 | 3·505 | 3·535 | 3·520 | 12 |
| 3·54–3·56 | 3·535 | 3·565 | 3·550 | 14 |
| 3·57–3·59 | 3·565 | 3·595 | 3·580 | 10 |
| 3·60–3·62 | 3·595 | 3·625 | 3·610 | 5 |
| 3·63–3·65 | 3·625 | 3·655 | 3·640 | 1 |

First, we draw a base line and on it mark a scale of $x$ on which we can indicate the central values of the classes. Do that for a start.

**18**

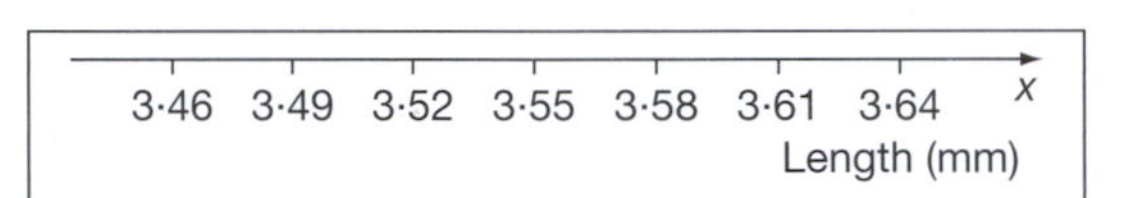


Since the classes are of regular class interval, the class boundaries will coincide with the points mid-way between the central values, thus:

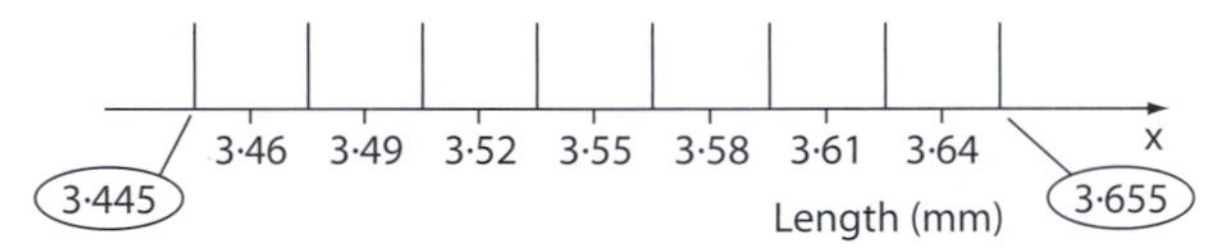


Note that the lower boundary of the first class extends to 3·445 and that the upper boundary of the seventh class extends to 3·655. Because the class intervals are regular, we can now erect a vertical scale to represent class frequencies and rectangles can be drawn to the appropriate height. Complete the work and give it a title.

**19**

## Relative frequency histogram

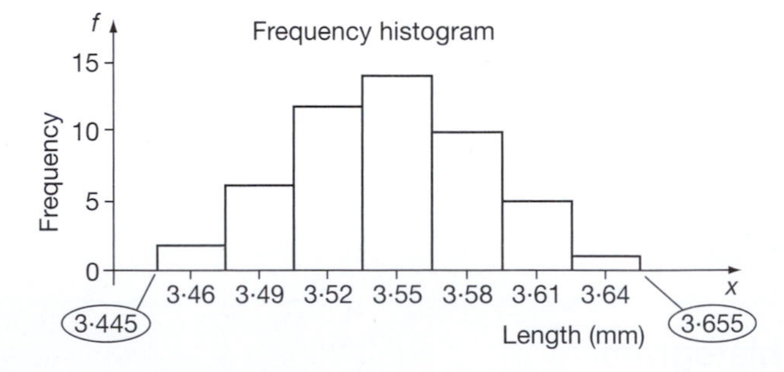


The same diagram can be made to represent the relative frequency distribution by replacing the scale of frequency along the $y$-axis with a scale of relative frequencies (percentages).

Considerable information can be gleaned from a histogram on sight. For instance, we see that the class having the highest frequency is the ........... class.

20

fourth

and the class with the lowest frequency is the ........... class.

21

seventh

Most of the whole range of values is clustered within the middle classes and knowledge of the centre region of the histogram is important. We can put a numerical value on this by determining a *measure of central tendency* which we shall now deal with.

*So, on to the next frame*

## Measures of central tendency

22

A **statistic** is a number or a fact that summarizes a collection of data and is derived from the data by various arithmetic methods. The simplest single statistics that are used to summarize data are the averages also known as the *measures of central tendency*. We shall consider three, namely the **arithmetic mean**, the **mode** and the **median**.

### Mean

The arithmetic mean $\bar{x}$ of a set of $n$ observations $x$ is simply their average,

i.e. $\text{mean} = \dfrac{\text{sum of the observations}}{\text{number of observations}} \quad \therefore \bar{x} = \dfrac{\sum x}{n}$

When calculating the mean from a frequency distribution, this becomes

$$\text{mean} = \bar{x} = \frac{\sum xf}{n} = \frac{\sum xf}{\sum f}$$

For example, for the following frequency distribution, we need to add a third column showing the values of the product $x \times f$, after which the mean can be found:

| Variable ($x$) | Frequency ($f$) | Product ($xf$) |
|---|---|---|
| 15 | 1 | 15 |
| 16 | 4 | 64 |
| 17 | 9 | 153 |
| 18 | 10 | 180 |
| 19 | 6 | 114 |
| 20 | 2 | 40 |

So, $\bar{x} =$ ...........

## 23

17·69

Because

$$n = \sum f = 32 \text{ and } \sum xf = 566 \qquad \therefore \bar{x} = \frac{\sum xf}{n} = \frac{566}{32} = 17{\cdot}69$$

When calculating the mean from a frequency distribution with grouped data, the central value, $x_m$, of the class is taken as the $x$-value in forming the product $xf$. So, for the frequency distribution

| Variable ($x$) | 12–14 | 15–17 | 18–20 | 21–23 | 24–26 | 27–29 |
|---|---|---|---|---|---|---|
| Frequency ($f$) | 2 | 6 | 9 | 8 | 4 | 1 |

$\bar{x} =$ ...........

## 24

$\bar{x} = 19{\cdot}9$

Because

$$n = \sum f = 30 \text{ and } \sum x_m f = 597 \qquad \therefore \bar{x} = \frac{\sum x_m f}{n} = \frac{597}{30} = 19{\cdot}9$$

Here is one more.

Measurement in millimetres of 60 bolts gave the following frequency distribution:

| Length $x$ (mm) | 30·2 | 30·4 | 30·6 | 30·8 | 31·0 | 31·2 | 31·4 |
|---|---|---|---|---|---|---|---|
| Frequency $f$ | 3 | 7 | 12 | 17 | 11 | 8 | 2 |

The mean $\bar{x} =$ ...........

## 25

$\bar{x} = 30{\cdot}79$

| Length (mm) ($x$) | Frequency ($f$) | Product ($xf$) |
|---|---|---|
| 30·2 | 3 | 90·6 |
| 30·4 | 7 | 212·8 |
| 30·6 | 12 | 367·2 |
| 30·8 | 17 | 523·6 |
| 31·0 | 11 | 341·0 |
| 31·2 | 8 | 249·6 |
| 31·4 | 2 | 62·8 |

$n = \sum f = 60$ ↓ $\sum xf = 1847{\cdot}6$

$$\therefore \bar{x} = \frac{\sum xf}{n} = \frac{1847{\cdot}6}{60} = 30{\cdot}79$$

$$\therefore \bar{x} = 30{\cdot}79$$

## Coding for calculating the mean

26

We can save ourselves some of the tedious work by using a system of *coding* which involves converting the $x$-values into simpler values for the calculation and then converting back again for the final result. An example will show the method in detail: we will use the same frequency distribution as above.

First, we choose a convenient value of $x$ (near the middle of the range) and subtract this from every other value of $x$ to give the second column.

| Length (mm) ($x$) | Deviation from chosen value ($x - 30{\cdot}8$) | In units of 0·2 mm $\left(x_c = \dfrac{x - 30{\cdot}8}{0{\cdot}2}\right)$ |
|---|---|---|
| 30·2 | −0·6 | |
| 30·4 | −0·4 | |
| 30·6 | −0·2 | |
| 30·8 | 0 | |
| 31·0 | 0·2 | |
| 31·2 | 0·4 | |
| 31·4 | 0·6 | |

Then we change the values into an even simpler form by dividing the values in column 2 by 0·2 mm. These are entered in column 3 and give the coded values of $x$, i.e. $x_c$

All we have to do then is to add column 4 showing the class frequencies (from the table above) and then column 5 containing values of the product $x_c f$. Using the last two columns, the mean value of $x_c$ can be found as before.

$\bar{x}_c = \dots\dots\dots\dots$

27

$\bar{x}_c = -0{\cdot}0333$

| Length (mm) ($x$) | Deviation from chosen value ($x - 30{\cdot}8$) | In units of 0·2 mm $\left(x_c = \dfrac{x - 30{\cdot}8}{0{\cdot}2}\right)$ | Frequency ($f$) | Product ($x_c f$) |
|---|---|---|---|---|
| 30·2 | −0·6 | −3 | 3 | −9 |
| 30·4 | −0·4 | −2 | 7 | −14 |
| 30·6 | −0·2 | −1 | 12 | −12 |
| 30·8 | 0 | 0 | 17 | 0 |
| 31·0 | 0·2 | 1 | 11 | 11 |
| 31·2 | 0·4 | 2 | 8 | 16 |
| 31·4 | 0·6 | 3 | 2 | 6 |

$n = \sum f = 60; \qquad \sum x_c f = -2{\cdot}0$

$$\therefore \bar{x}_c = \frac{-2}{60} = -0{\cdot}0333$$

**28**

So we have $\bar{x}_c = -0{\cdot}0333$. Now we have to retrace our steps back to the original units of $x$.

### Decoding

In the coding procedure, our last step was to divide by 0·2. We therefore now multiply by 0·2 to return to the correct units of $(x - 30{\cdot}8)$:

$$\therefore \bar{x} - 30{\cdot}8 = -0{\cdot}0333 \times 0{\cdot}2 = -0{\cdot}00667$$

We now add the 30·8 to both sides:

$$\therefore \bar{x} = 30{\cdot}8 - 0{\cdot}00667 = 30{\cdot}79333 \qquad \therefore \bar{x} = 30{\cdot}79$$

Note that in decoding, we reverse the operations used in the original coding process. The value (30·8) which was subtracted from all values of $x$ is near the centre of the range of $x$-values and is therefore sometimes referred to as a *false mean*.

**29**

### Coding with a grouped frequency distribution

The method is precisely the same, except that we work with the centre values of the classes of $x$, i.e. $x_m$, for the calculation purposes.

Here is an exercise:

The thicknesses of 50 spacing pieces were measured:

| Thickness $x$ (mm) | 2·20–2·22 | 2·23–2·25 | 2·26–2·28 | 2·29–2·31 | 2·32–2·34 |
|---|---|---|---|---|---|
| Frequency $f$ | 1 | 5 | 8 | 15 | 12 |

| Thickness $x$ (mm) | 2·35–2·37 | 2·38–2·40 |
|---|---|---|
| Frequency $f$ | 7 | 2 |

Using coding, determine the mean value of the thickness. $\bar{x} = \ldots\ldots\ldots\ldots$

**30**

$\bar{x} = 2{\cdot}307$ mm

| Thickness (mm) $(x)$ | Central value $(x_m)$ | Deviation from chosen value $(x_m - 2{\cdot}30)$ | In units of 0·03 mm $\left(x_c = \dfrac{x_m - 2{\cdot}30}{0{\cdot}03}\right)$ | Frequency $(f)$ | Product $(x_c f)$ |
|---|---|---|---|---|---|
| 2·20–2·22 | 2·21 | −0·09 | −3 | 1 | −3 |
| 2·23–2·25 | 2·24 | −0·06 | −2 | 5 | −10 |
| 2·26–2·28 | 2·27 | −0·03 | −1 | 8 | −8 |
| 2·29–2·31 | 2·30 | 0 | 0 | 15 | 0 |
| 2·32–2·34 | 2·33 | 0·03 | 1 | 12 | 12 |
| 2·35–2·37 | 2·36 | 0·06 | 2 | 7 | 14 |
| 2·38–2·40 | 2·39 | 0·09 | 3 | 2 | 6 |

$n = \sum f = 50$ $\qquad \sum x_c f = 11$

▶

$$\therefore\ \bar{x}_c = \frac{\sum x_c f}{n} = \frac{11}{50} = 0{\cdot}22$$

$$\therefore\ \bar{x}_m - 2{\cdot}30 = \bar{x}_c \times 0{\cdot}03 = 0{\cdot}22 \times 0{\cdot}03 = 0{\cdot}0066$$

$$\therefore\ \bar{x}_m = 2{\cdot}30 + 0{\cdot}0066 = 2{\cdot}3067 \qquad \therefore\ \bar{x} = 2{\cdot}307$$

*Now let us deal with the mode, so move on to the next frame*

## Mode

**31**

The *mode* of a set of data is that value of the variable that occurs most often. For instance, in the set of values 2, 2, 6, 7, 7, 7, 10, 13, the mode is clearly 7. There could, of course, be more than one mode in a set of observations, e.g. 23, 25, 25, 25, 27, 27, 28, 28, 28, has two modes, 25 and 28, each of which appears three times.

## Mode with grouped data

**32**

The masses of 50 castings gave the following frequency distribution:

| Mass $x$ (kg) | 10–12 | 13–15 | 16–18 | 19–21 | 22–24 | 25–27 | 28–30 |
|---|---|---|---|---|---|---|---|
| Frequency $f$ | 3 | 7 | 16 | 10 | 8 | 5 | 1 |

If we draw the histogram, using the central values as the mid-points of the bases of the rectangles, we obtain . . . . . . . . . . .

**33**

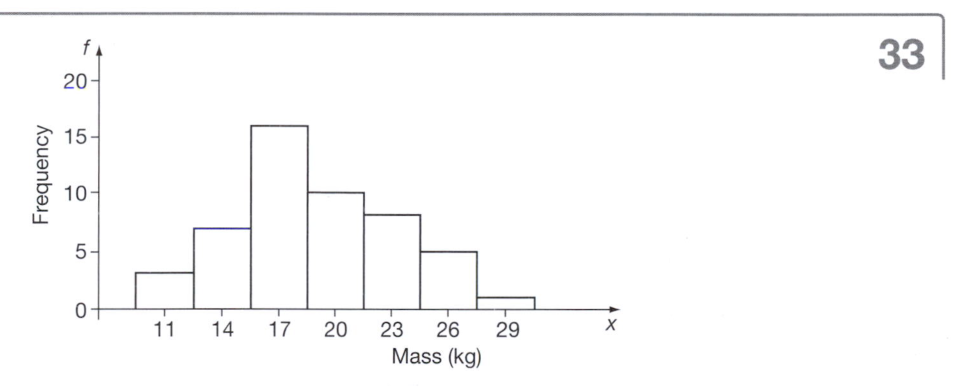


The class having the highest frequency is the . . . . . . . . . . . class

**34**

third

The modal class is therefore the third class, with boundaries 15·5 and 18·5 kg. The value of the mode itself lies somewhere within that range and its value can be found by a simple construction.

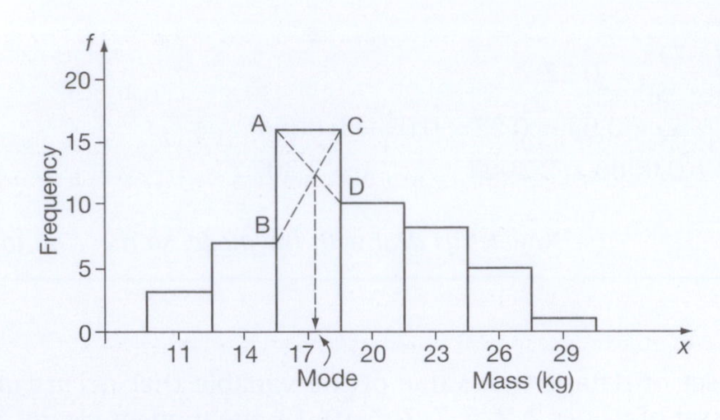


The two diagonal lines AD and BC are drawn as shown. The $x$-value of their point of intersection is taken as the mode of the set of observations.
Carry out the construction on your histogram and you will find that, for this set of observations, the mode = ...........

## 35

mode = 17·3

The value of the mode can also be calculated.

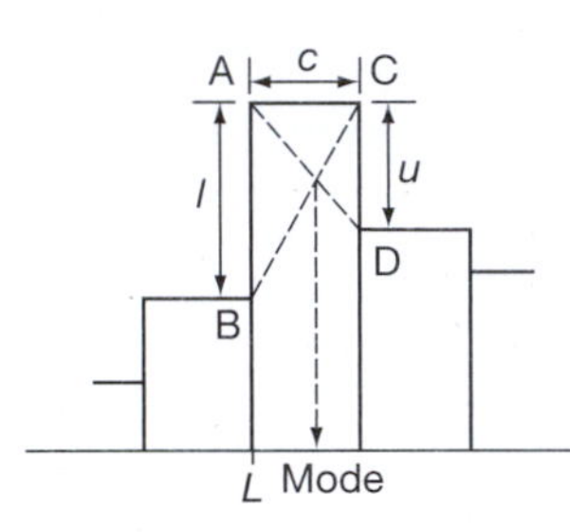


If $L$ = lower boundary value

$l$ = AB = difference in frequency on the lower boundary

$u$ = CD = difference in frequency on the upper boundary

$c$ = class interval

then the mode $= L + \left(\dfrac{l}{l+u}\right)c$

In the example we have just considered:

$L =$ ...........

$l =$ ...........

$u =$ ...........

$c =$ ...........

## 36

$L = 15{\cdot}5$; $l = 16 - 7 = 9$; $u = 16 - 10 = 6$; $c = 3$

$$\text{Then the mode} = 15{\cdot}5 + \left(\frac{9}{9+6}\right) \times 3$$

$$= 15{\cdot}5 + 1{\cdot}8 = 17{\cdot}3$$

$\therefore$ Mode = 17·3 kg

This result agrees with the graphical method we used before.

## Median

37

The third measure of central tendency is the *median*, which is the value of the middle term when all the observations are arranged in ascending or descending order.

For example, with 4, 7, 8, 9, 12, 15, 26, the median = 9.

↑
median

Where there is an even number of values, the median is then the average of the two middle terms, e.g. 5, 6, 10, $\underbrace{12, 14,}$ 17, 23, 30 has a median of

$$\frac{12+14}{2} = 13$$

Similarly, for the set of values 13, 4, 18, 23, 9, 16, 18, 10, 20, 6, the median is . . . . . . . . . . . .

38

14·5

Because arranging the terms in order, we have:

$$4, 6, 9, 10, 13, 16, 18, 18, 20, 23 \text{ and median} = \frac{13+16}{2} = 14{\cdot}5$$

*Now we must see how we can get the median with grouped data, so move on to Frame 39*

## Median with grouped data

39

Since the median is the value of the middle term, it divides the frequency histogram into two equal areas. This fact gives us a method for determining the median.

For example, the temperature of a component was monitored at regular intervals on 80 occasions. The frequency distribution was as follows:

| Temperature $x$ (C) | 30·0–30·2 | 30·3–30·5 | 30·6–30·8 | 30·9–31·1 |
|---|---|---|---|---|
| Frequency $f$ | 6 | 12 | 15 | 20 |

| Temperature $x$ (C) | 31·2–31·4 | 31·5–31·7 | 31·8–32·0 |
|---|---|---|---|
| Frequency $f$ | 13 | 9 | 5 |

First we draw the frequency histogram.

**40**

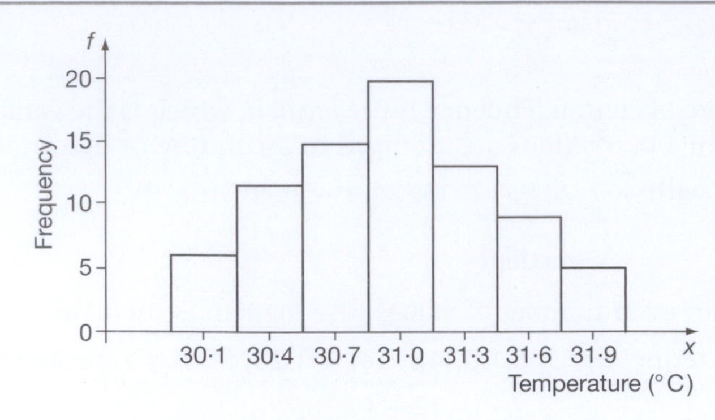


The median is the average of the 40th and 41st terms and, if we count up the frequencies of the rectangles, these terms are included in the ........... class.

**41**

fourth

If we insert a dashed line to represent the value of the median, it will divide the area of the histogram into two equal parts.

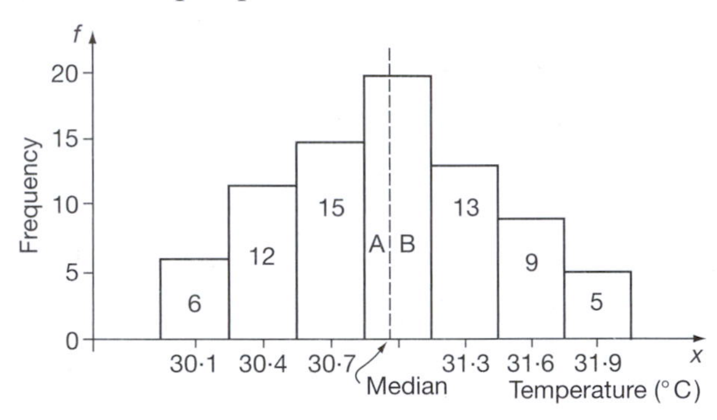


$6 + 12 + 15 + A = B + 13 + 9 + 5$

$\therefore\ 33 + A = B + 37$ But $A + B = 20$ $\therefore\ B = 20 - A$

$\therefore\ 33 + A = 20 - A + 27$ $\therefore\ 2A = 14$ $\therefore\ A = 7$

$\therefore$ Width of $A = \dfrac{7}{20} \times$ class interval

$= 0{\cdot}35 \times 0{\cdot}3 = 0{\cdot}105$ $\therefore$ Median $= 30{\cdot}85 + 0{\cdot}105$

$= 30{\cdot}96^\circ\text{C}$

Now, by way of revision, here is a short exercise.

**Exercise**

Determine (a) the mean, (b) the mode and (c) the median of the following:

| $x$ | 5–9 | 10–14 | 15–19 | 20–24 | 25–29 | 30–34 |
|---|---|---|---|---|---|---|
| $f$ | 4 | 9 | 16 | 12 | 6 | 3 |

**42**

(a) mean = 18·6 (b) mode = 17·7 (c) median = 18·25

# Measures of dispersion



The mean, mode and median give important information regarding the general mass of the observations recorded. They do not, however, tell us anything about the dispersion of the values around the central value.

The set 26, 27, 28, 29, 30 has a mean of 28
and 5, 19, 20, 36, 60 also has a mean of 28

These two sets have the same mean, but clearly the first is more tightly arranged around the mean than is the second. We therefore need a measure to indicate the spread of the values about the mean.

## Mean deviation

The deviation of a datum from the mean is obtained by subtracting the value of the datum from the mean. However, adding up all the mean deviations for each datum to find the average always results in zero. For example, the numbers

10 20 30 40 50

have an average of 30 and deviations

$$30 - 10 = 20$$
$$30 - 20 = 10$$
$$30 - 30 = 0$$
$$30 - 40 = -10$$
$$30 - 50 = -20$$

which have an average of 0 because the negative numbers cancel out the positive numbers.

## Range

The simplest value to indicate the spread or dispersion is the **range**. This is the difference between the highest and lowest values in the set of data; here it is $50 - 10 = 40$. However, this gives no indication of how intermediate values are spread.

## Standard deviation

To eliminate the negative numbers encountered in the deviations we can square the deviations thus:

$$(30 - 10)^2 = 400$$
$$(30 - 20)^2 = 100$$
$$(30 - 30)^2 = 0$$
$$(30 - 40)^2 = 100$$
$$(30 - 50)^2 = 400$$

These squared deviations have an average of $\frac{1000}{5} = 200$. This is called the **variance** of the data. The variance is a perfectly respectable measure of spread but it suffers from the fact that it is measured in the square of the units of the data. To overcome this we take the square root of the variance to produce what is called the **standard deviation**.

standard deviation $= \sqrt{\text{variance}}$, which, in this case, is $\sqrt{200} = 14{\cdot}14$ to 2 dp

▶

In summary, the *standard deviation from the mean* is used widely in statistics to indicate the degree of dispersion. It takes into account the deviation of every value from the mean and it is found as follows:

(a) The mean $\bar{x}$ of the set of $n$ values is first calculated

(b) The deviation of each of these $n$ values, $x_1, x_2, x_3, \ldots, x_n$ from the mean is calculated and the results squared, i.e. $(x_1 - \bar{x})^2; (x_2 - \bar{x})^2; (x_3 - \bar{x})^2; \ldots; (x_n - \bar{x})^2$

(c) The average of these results is then found and the result is called the *variance* of the set of observations,

$$\text{i.e. variance} = \frac{(x_1 - \bar{x})^2 + (x_2 - \bar{x})^2 + \ldots + (x_n - \bar{x})^2}{n}$$

(d) The square root of the variance gives the standard deviation, denoted by the Greek letter 'sigma'

$$\text{standard deviation} = \sigma = \sqrt{\frac{\sum(x - \bar{x})^2}{n}}$$

So, for the set of values 3, 6, 7, 8, 11, 13, the mean $\bar{x} = \ldots\ldots\ldots\ldots$, the variance $= \ldots\ldots\ldots\ldots$ and the standard deviation $= \ldots\ldots\ldots\ldots$

**44**

$\bar{x} = 8{\cdot}0$; variance $= 10{\cdot}67$; $\sigma = 3{\cdot}27$

Because

| $x$ | $f$ |
|---|---|
| 3 | 1 |
| 6 | 1 |
| 7 | 1 |
| 8 | 1 |
| 11 | 1 |
| 13 | 1 |

$$\sum x = 48$$
$$n = 6$$
$$\therefore \bar{x} = 8$$

| $x - 8$ | $(x - 8)^2$ |
|---|---|
| −5 | 25 |
| −2 | 4 |
| −1 | 1 |
| 0 | 0 |
| 3 | 9 |
| 5 | 25 |

$$\therefore \sum(x - \bar{x})^2 = 64$$

$$\text{variance} = \frac{\sum(x - \bar{x})^2}{n} = \frac{64}{6} = 10.67$$

$$\therefore \text{standard deviation} = \sigma = \sqrt{10{\cdot}67} = 3{\cdot}266 \quad \therefore \sigma = 3{\cdot}27$$

## Alternative formula for the standard deviation

The formula $\sigma = \sqrt{\frac{\sum(x - \bar{x})^2}{n}}$ can also be written in another and more convenient form, $\sigma = \sqrt{\frac{\sum x^2}{n} - (\bar{x})^2}$ which requires only the mean and the squares of the values of $x$. For grouped data, this then becomes $\sigma = \sqrt{\frac{\sum x^2 f}{n} - (\bar{x})^2}$ and the working is even more simplified if we use the coding procedure as we did earlier in the Programme.

*Move on to the next frame for a typical example*

**45**

Here is an example. The lengths of 70 bars were measured and the following frequency distribution obtained:

| Length $x$ (mm) | 21·2–21·4 | 21·5–21·7 | 21·8–22·0 | 22·1–22·3 |
|---|---|---|---|---|
| Frequency $f$ | 3 | 5 | 10 | 16 |

| Length $x$ (mm) | 22·4–22·6 | 22·7–22·9 | 23·0–23·2 |
|---|---|---|---|
| Frequency $f$ | 18 | 12 | 6 |

First we prepare a table with the following headings:

| Length (mm) $(x)$ | Central value $(x_m)$ | Deviation from chosen value $(x_m - 22{\cdot}2)$ | Units of 0·3 $\left(x_c = \dfrac{x_m - 22{\cdot}2}{0{\cdot}3}\right)$ | Freq. $f$ | $x_c f$ | | |
|---|---|---|---|---|---|---|---|

Leave room on the right-hand side for two more columns yet to come. Now you can complete the six columns shown and from the values entered, determine the coded value of the mean and also the actual mean.

*Do that and then move on*

**46**

So far, the work looks like this:

| Length (mm) $(x)$ | Central value $(x_m)$ | Deviation from chosen value $(x_m - 22{\cdot}2)$ | Units of 0·3 $\left(x_c = \dfrac{x_m - 22{\cdot}2}{0{\cdot}3}\right)$ | Freq. $f$ | $x_c f$ | | |
|---|---|---|---|---|---|---|---|
| 21·2–21·4 | 21·3 | −0·9 | −3 | 3 | −9 | | |
| 21·5–21·7 | 21·6 | −0·6 | −2 | 5 | −10 | | |
| 21·8–22·0 | 21·9 | −0·3 | −1 | 10 | −10 | | |
| 22·1–22·3 | 22·2 | 0 | 0 | 16 | 0 | | |
| 22·4–22·6 | 22·5 | 0·3 | 1 | 18 | 18 | | |
| 22·7–22·9 | 22·8 | 0·6 | 2 | 12 | 24 | | |
| 23·0–23·2 | 23·1 | 0·9 | 3 | 6 | 18 | | |

$n = \sum f = 70$ ↓

$\sum x_c f = 31$

$$\text{Coded mean} = \bar{x}_c = \frac{\sum x_c f}{n} = \frac{31}{70} = 0{\cdot}4429 \text{ (in units of } 0{\cdot}\,3)$$

$$\therefore\ \bar{x}_m - 22{\cdot}2 = 0{\cdot}4429 \times 0{\cdot}3 = 0{\cdot}1329$$

$$\bar{x}_m = 22{\cdot}2 + 0{\cdot}1329 = 22{\cdot}333 \qquad \therefore\ \bar{x} = 22{\cdot}33$$

Now we can complete the remaining two columns. Head these $x_c{}^2$ and $x_c{}^2 f$. Fill in the appropriate values and using $\sigma_c = \sqrt{\dfrac{\sum x_c{}^2 f}{n} - (\bar{x}_c)^2}$ we can find the coded value of the standard deviation $(\sigma_c)$.

*Complete the table then and determine the coded standard deviation on your own*

**47** Finally the table now becomes:

| Length (mm) $(x)$ | Central value $(x_m)$ | Deviation from chosen value $(x_m - 22{\cdot}2)$ | Units of 0·3 $\left(x_c = \frac{x_m - 22{\cdot}2}{0{\cdot}3}\right)$ | Freq. $f$ | $x_c f$ | $x_c^2$ | $x_c^2 f$ |
|---|---|---|---|---|---|---|---|
| 21·2–21·4 | 21·3 | −0·9 | −3 | 3 | −9 | 9 | 27 |
| 21·5–21·7 | 21·6 | −0·6 | −2 | 5 | −10 | 4 | 20 |
| 21·8–22·0 | 21·9 | −0·3 | −1 | 10 | −10 | 1 | 10 |
| 22·1–22·3 | 22·2 | 0 | 0 | 16 | 0 | 0 | 0 |
| 22·4–22·6 | 22·5 | 0·3 | 1 | 18 | 18 | 1 | 18 |
| 22·7–22·9 | 22·8 | 0·6 | 2 | 12 | 24 | 4 | 48 |
| 23·0–23·2 | 23·1 | 0·9 | 3 | 6 | 18 | 9 | 54 |
| | | | | 70 | 31 | | 177 |

Now $\sigma_c = \sqrt{\frac{\sum x_c^2 f}{n} - (\bar{x}_c)^2}$ and from the table, $n = 70$ and $\sum x_c^2 f = 177$.

Also, from the previous work, $\bar{x}_c = 0{\cdot}443$. $\therefore\ \sigma_c = \ldots\ldots\ldots\ldots$

**48**

$$\therefore\ \sigma_c = 1{\cdot}527 \text{ (in units of } 0{\cdot}3)$$

$\therefore\ \sigma = 1{\cdot}527 \times 0{\cdot}3 = 0{\cdot}4581 \quad \therefore\ \sigma = 0{\cdot}458$

*Note*: In calculating the standard deviation, we do not restore the 'false mean' subtracted from the original values, since the standard deviation is relative to the mean and not to the zero or origin of the set of observations.

*Now on to something different*

# Distribution curves

**49**

## Frequency polygons

If the centre points of the tops of the rectangular blocks of a frequency histogram are joined, the resulting figure is a *frequency polygon*. If the polygon is extended to include the mid-points of the zero frequency classes at each end of the histogram, then the area of the complete polygon is equal to the area of the histogram and therefore represents the total frequency of the variable.

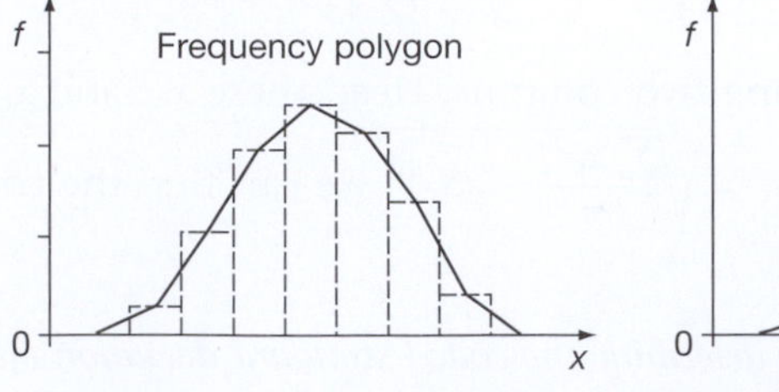


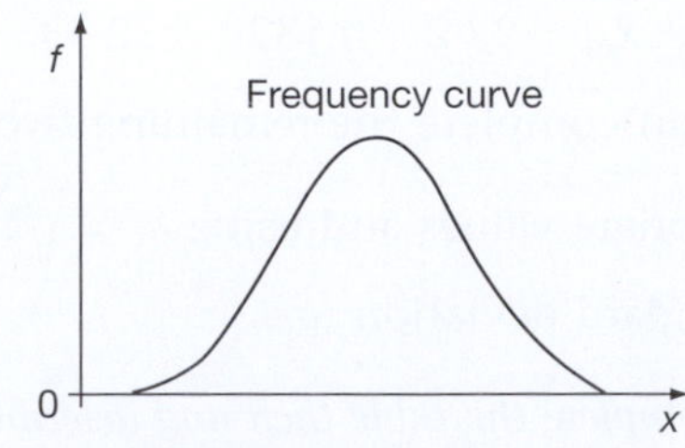

## Frequency curves

50

If the frequency polygon is 'smoothed out', or if we plot the frequency against the central value of each class and draw a smooth curve, the result is a frequency curve.

## Normal distribution curve

51

When very large numbers of observations are made and the range is divided into a very large number of 'narrow' classes, the resulting frequency curve, in many cases, approximates closely to a standard curve known as the *normal distribution curve*, which has a characteristic bell-shaped formation.

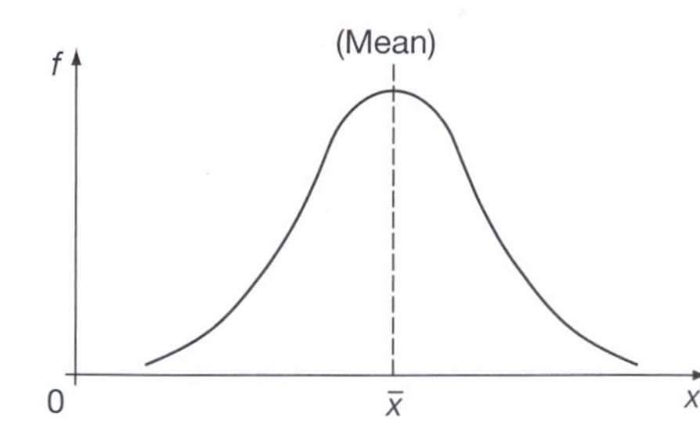


The normal distribution curve (or normal curve) is symmetrical about its centre line which coincides with the mean $\bar{x}$ of the observations.

There is, in fact, a close connection between the standard deviation (sd) from the mean of a set of values and the normal curve.

### 1 Values within 1 sd of the mean

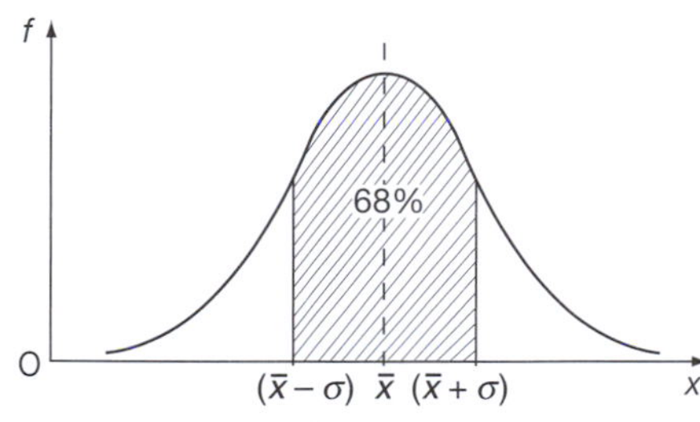


The normal curve has a complicated equation, but it can be shown that the shaded area is 68 per cent of the total area under the normal curve, i.e. 68 per cent of the observations occur within the range $(\bar{x}-\sigma)$ to $(\bar{x}+\sigma)$.

For example, on a manufacturing run to produce 1000 bolts of nominal length 32·5 mm, sampling gave a mean of 32·58 and a standard deviation of 0·06 mm. From this information, $\bar{x} = 32{\cdot}58$ mm and $\sigma = 0{\cdot}06$ mm.

$$\left.\begin{aligned}\bar{x}-\sigma &= 32{\cdot}58 - 0{\cdot}06 = 32{\cdot}52\\ \bar{x}+\sigma &= 32{\cdot}58 + 0{\cdot}06 = 32{\cdot}64\end{aligned}\right\}$$

$\therefore$ 68 per cent of the bolts, i.e. 680, are likely to have lengths between 32.52 mm and 32.64 mm.

### 2 Values within 2 sd of the mean

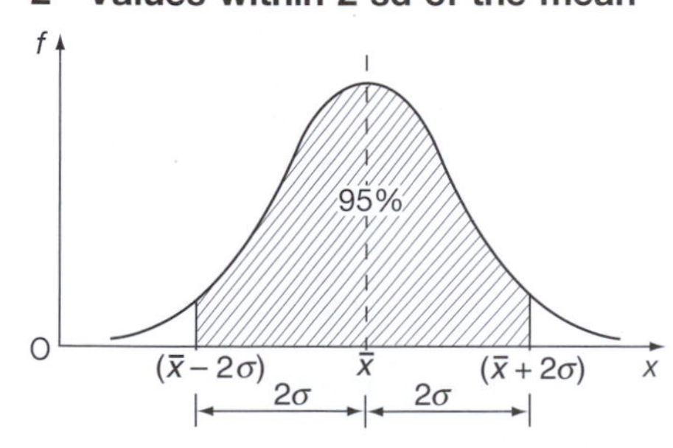


Between $(\bar{x}-2\sigma)$ and $(\bar{x}+2\sigma)$ the shaded area accounts for 95 per cent of the area of the whole figure, i.e. 95 per cent of the observations occur between these two values.

▶

**3 Values within 3 sd of the mean**

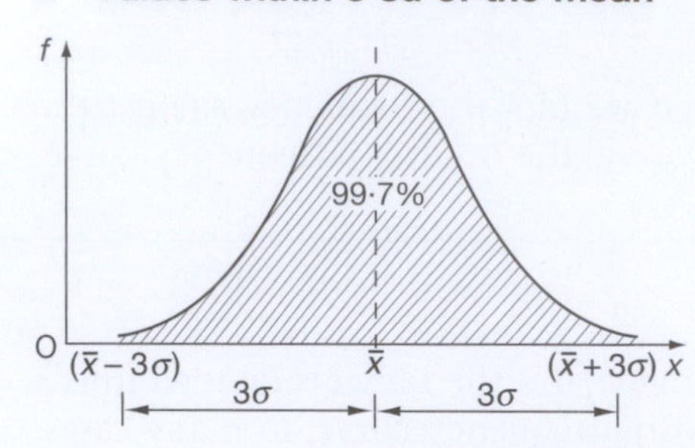


Between $(\bar{x}-3\sigma)$ and $(\bar{x}+3\sigma)$ the shaded area is 99·7 per cent of the total area under the normal curve and therefore 99·7 per cent of the observations occur within this range, i.e. almost all the values occur within $3\sigma$ or 3 sd of the mean.

Therefore, in our previous example where $\bar{x} = 32{\cdot}58$ and $\sigma = 0{\cdot}06$ mm:

68 per cent of the bolts are likely to have lengths between 32·52 and 32·64 mm,
95 per cent of the bolts are likely to have lengths between $32{\cdot}5 - 0{\cdot}12$ and $32{\cdot}5 + 0{\cdot}12$, i.e. between 32·38 and 32·62 mm,
99·7 per cent, i.e. almost all, are likely to have lengths between $32{\cdot}5 - 0{\cdot}18$ and $32{\cdot}5 + 0{\cdot}18$, i.e. between 32·32 and 32·68 mm.

We can enter the same information in a slightly different manner, dividing the figure into columns of $1\sigma$ width on each side of the mean.

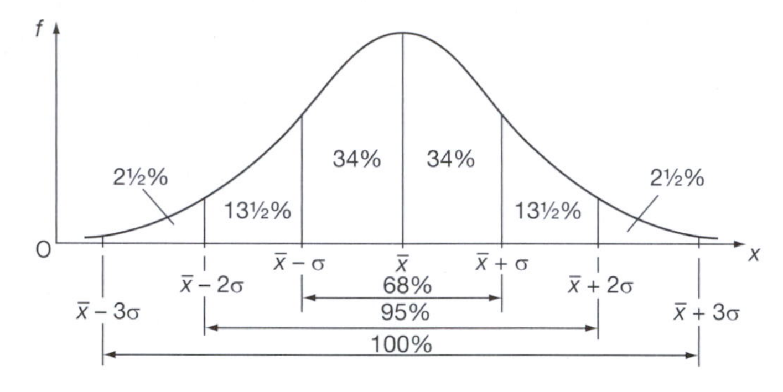


**Exercise**

Measurement of the diameters of 1600 plugs gave a mean of 74·82 mm and a standard deviation of 0·14 mm. Calculate:

(a) the number of plugs likely to have diameters less than 74·54 mm

(b) the number of plugs likely to have diameters between 74·68 mm and 75·10 mm.

**52**

(a) 40 (b) 1304

Because

(a) $74{\cdot}54 = 74{\cdot}82 - 2(0{\cdot}14) = \bar{x} - 2\sigma$ $\therefore$ number of plugs with diameters less than 74·54 mm $= 2\frac{1}{2}$ per cent $\times 1600 = 40$

(b) $74{\cdot}68 = 74{\cdot}82 - 0{\cdot}14 = \bar{x} - \sigma$

$75{\cdot}10 = 74{\cdot}82 + 2(0{\cdot}14) = \bar{x} + 2\sigma$

$\therefore$ Numbers of plugs between these values $= (34 + 34 + 13\frac{1}{2})$ per cent

$= 81\frac{1}{2}$ per cent

$81\frac{1}{2}$ per cent of 1600 $= 1304$

## Standardized normal curve

**53**

This is the same shape as the normal curve, but the axis of symmetry now becomes the vertical axis with a scale of relative frequency. The horizontal axis now carries a scale of $z$-values indicated as multiples of the standard deviation $\sigma$. The area under the standardized normal curve has a total value of 1. The standard normal curve has a mean of 0 and a standard deviation of 1.

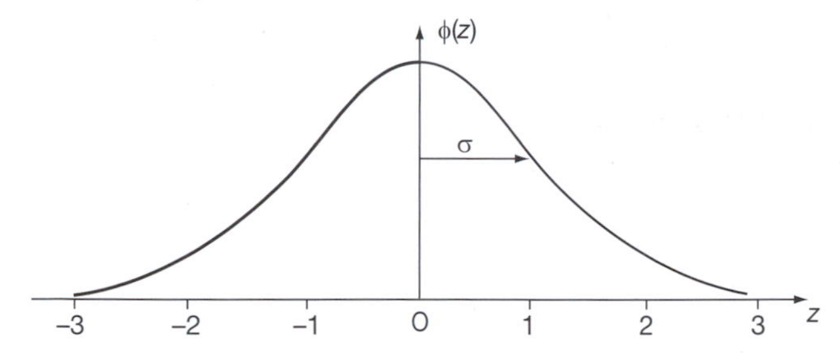


The standardized normal curve is obtained from the normal curve by the substitution $z = \dfrac{x - \bar{x}}{\sigma}$ and it converts the original distribution into one with zero mean and unit standard deviation. The equation of the standardized normal curve is somewhat complicated:

$$y = \phi(z) = \frac{1}{\sqrt{2\pi}} e^{-(1/2)z^2}$$

but, if required, the curve can be drawn from the following table:

| $z$ | 0 | 0·2 | 0·4 | 0·6 | 0·8 | 1·0 | 1·2 | 1·4 |
|---|---|---|---|---|---|---|---|---|
| Relative $f$ | 0·399 | 0·391 | 0·368 | 0·333 | 0·290 | 0·242 | 0·194 | 0·150 |

| $z$ | 1·6 | 1·8 | 2·0 | 2·2 | 2·4 | 2·6 | 2·8 | 3·0 |
|---|---|---|---|---|---|---|---|---|
| Relative $f$ | 0·111 | 0·080 | 0·054 | 0·035 | 0·022 | 0·014 | 0·008 | 0·004 |

Since the curve is symmetrical about $z = 0$, the curve for negative values of $z$ is the mirror image of that for positive values of $z$.

**One further problem**: Components are machined to a nominal diameter of 32·65 mm. A sample of 400 components gave a mean diameter of 32·66 mm with a standard deviation of 0·02 mm. If a total of 3500 components was produced, calculate:

(a) the limits between which all the diameters are likely to lie

(b) the total number of components with diameters between 32·64 and 32·72 mm.

**54**

(a) 32·60 mm to 32·72 mm (b) 2940 components

(a) $\bar{x} = 32{\cdot}66$ mm; $\sigma = 0{\cdot}02$ mm

Limits $(\bar{x} - 3\sigma)$ to $(\bar{x} + 3\sigma) = (32{\cdot}66 - 0{\cdot}06)$ to $(32{\cdot}66 + 0{\cdot}06)$

$= 32{\cdot}60$ to $32{\cdot}72$ mm

(b) $32{\cdot}64 = 32{\cdot}66 - 0{\cdot}02 = (\bar{x} - \sigma)$

$32{\cdot}72 = 32{\cdot}66 + 0{\cdot}06 = (\bar{x} + 3\sigma)$

Between $(\bar{x} - \sigma)$ and $(\bar{x} + 3\sigma)$ there are $(34 + 34 + 13\frac{1}{2} + 2\frac{1}{2})$ per cent, i.e. 84 per cent of the values. 84 per cent of 3500 = 2940 components

▶

We have now come to the end of this Programme on Data handling and statistics with this brief introduction to the standard normal curve. We will return to the standard normal curve and deal with it in much more detail at the end of the next Programme on Probability. In the meantime, check down the **Review summary** that follows and refer back to any points that may need clarification. All that remains after that is to work through the **Can You?** checklist and the **Test exercise**. You should have no trouble.

---

## Review summary

**55**

1 *Data* (a) Discrete data – values that can be precisely counted.
(b) Continuous data – values measured on a continuous scale.

2 *Grouped data*

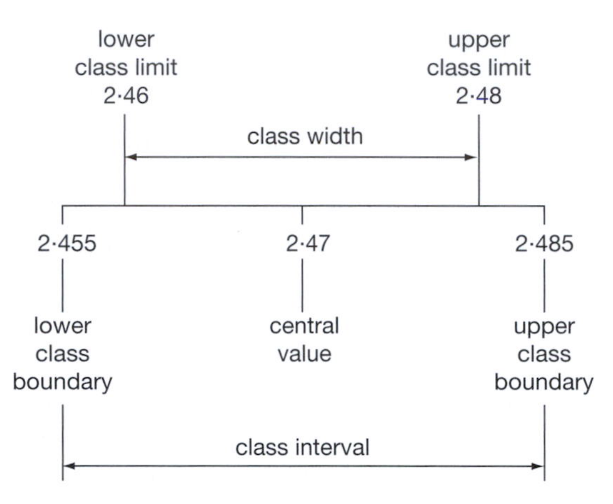


3 *Frequency* ($f$) – the number of occasions on which each value, or class, occurs.
*Relative frequency* – each frequency expressed as a percentage or fraction of the total frequency.

4 *Histogram* – graphical representation of a frequency or relative frequency distribution.

The frequency of any one class is given by the *area* of its column. If the class intervals are constant, the height of the rectangle indicates the frequency on the vertical scale.

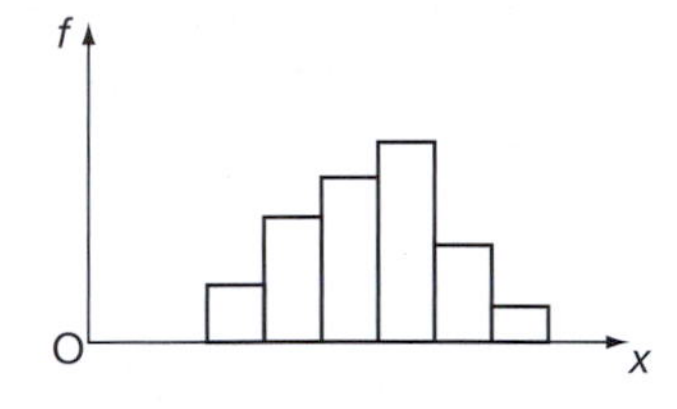


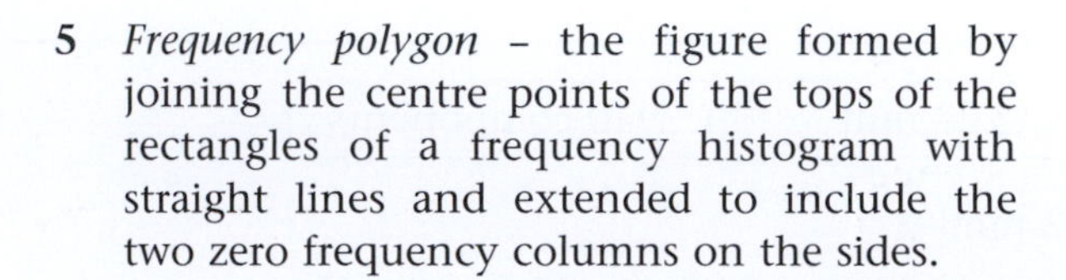
5 *Frequency polygon* – the figure formed by joining the centre points of the tops of the rectangles of a frequency histogram with straight lines and extended to include the two zero frequency columns on the sides.

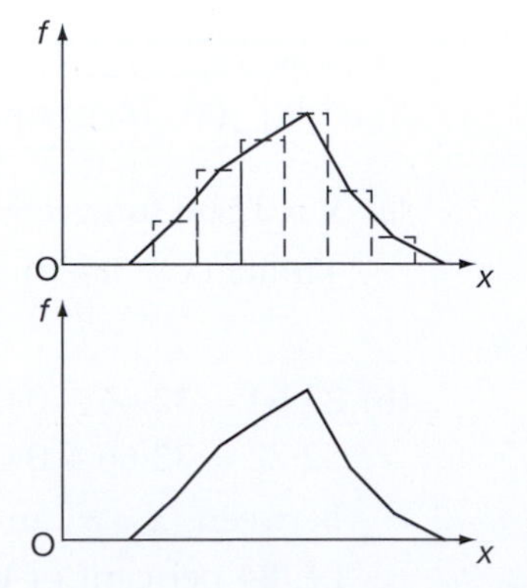


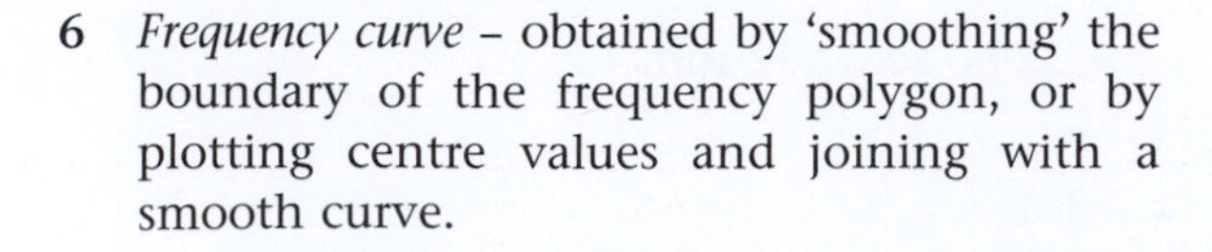
6 *Frequency curve* – obtained by 'smoothing' the boundary of the frequency polygon, or by plotting centre values and joining with a smooth curve.

▶

7 *Mean* (arithmetic mean) $\bar{x} = \dfrac{\sum xf}{n} = \dfrac{\sum xf}{\sum f}$

8 *Mode* – the value of the variable that occurs most often. For grouped distribution:

$$\text{mode} = L + \left(\frac{l}{l+u}\right)c$$

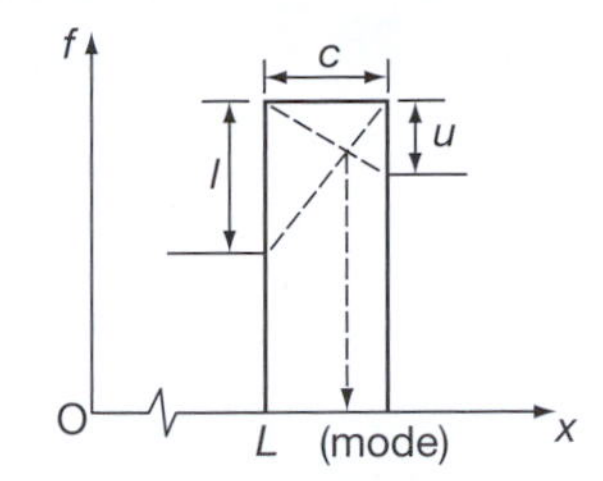


9 *Median* – the value of the middle term when all values are put in ascending or descending order. With an even number of terms, the median is the average of the two middle terms.

10 *Standard deviation*

$$\sigma = \sqrt{\frac{\sum x^2 f}{n} - (\bar{x})^2}$$

$n =$ number of observations
$\bar{x} =$ mean
$\sigma =$ standard deviation from the mean.

With coding, $\bar{x}_c =$ coded mean

$$\sigma_c = \sqrt{\frac{\sum {x_c}^2 f}{n} - (\bar{x}_c)^2}$$

11 *Normal distribution curve* – large numbers of observations.

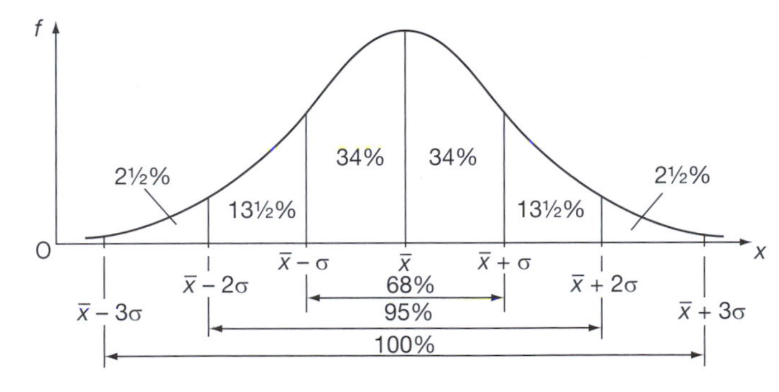


Symmetrical about the mean.
68 per cent of observations lie within ±1 sd of the mean.
95 per cent of observations lie within ±2 sd of the mean.
99·7 per cent of observations lie within ±3 sd of the mean.

12 *Standardized normal curve* – the axis of symmetry of the normal curve becomes the vertical axis with a scale of relative frequency. The horizontal axis carries a scale of $z$-values indicated as multiples of the standard deviation. The curve therefore represents a distribution with zero mean and unit standard deviation.

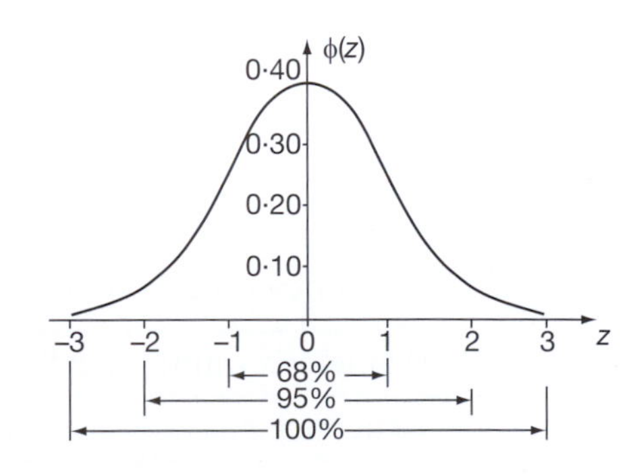

## Can You?

**56**

### Checklist 28

Check this list before and after you try the end of Programme test.

On a scale of 1 to 5 how confident are you that you can:

Frames

- Distinguish between discrete and continuous data? — Frames 1 to 2
  *Yes* ☐ ☐ ☐ ☐ ☐ *No*
- Construct frequency and relative frequency tables for grouped and ungrouped discrete data? — Frames 3 to 11
  *Yes* ☐ ☐ ☐ ☐ ☐ *No*
- Determine class boundaries, class intervals and central values for discrete and continuous data? — Frames 12 to 16
  *Yes* ☐ ☐ ☐ ☐ ☐ *No*
- Construct a histogram and a frequency polygon? — Frames 17 to 21
  *Yes* ☐ ☐ ☐ ☐ ☐ *No*
- Determine the mean, median and mode of grouped and ungrouped data? — Frames 22 to 42
  *Yes* ☐ ☐ ☐ ☐ ☐ *No*
- Determine the range, variance and standard deviation of discrete data? — Frames 43 to 48
  *Yes* ☐ ☐ ☐ ☐ ☐ *No*
- Measure the dispersion of data using the normal and standard normal curves? — Frames 49 to 54
  *Yes* ☐ ☐ ☐ ☐ ☐ *No*

## Test exercise 28

**57**

**1** The masses of 50 castings were measured. The results in kilograms were as follows:

| | | | | | | | | | |
|---|---|---|---|---|---|---|---|---|---|
| 4·6 | 4·7 | 4·5 | 4·6 | 4·7 | 4·4 | 4·8 | 4·3 | 4·2 | 4·8 |
| 4·7 | 4·5 | 4·7 | 4·4 | 4·5 | 4·5 | 4·6 | 4·4 | 4·6 | 4·6 |
| 4·8 | 4·3 | 4·8 | 4·5 | 4·5 | 4·6 | 4·6 | 4·7 | 4·6 | 4·7 |
| 4·4 | 4·6 | 4·5 | 4·4 | 4·3 | 4·7 | 4·7 | 4·6 | 4·6 | 4·8 |
| 4·9 | 4·4 | 4·5 | 4·7 | 4·4 | 4·5 | 4·9 | 4·7 | 4·5 | 4·6 |

(a) Arrange the data in 8 equal classes between 4·2 and 4·9 mm.

(b) Determine the frequency distribution.

(c) Draw the frequency histogram.

**2** The diameters of 75 rollers gave the following frequency distribution:

| Diameter $x$ (mm) | 8·82–8·86 | 8·87–8·91 | 8·92–8·96 | 8·97–9·01 |
|---|---|---|---|---|
| Frequency $f$ | 1 | 8 | 16 | 18 |

| Diameter $x$ (mm) | 9·02–9·06 | 9·07–9·11 | 9·12–9·16 | 9·17–9·21 |
|---|---|---|---|---|
| Frequency $f$ | 15 | 10 | 5 | 2 |

▶

(a) For each class, calculate (i) the central value, (ii) the relative frequency.

(b) Draw the relative frequency histogram.

(c) State (i) the lower boundary of the third class, (ii) the upper boundary of the sixth class, (iii) the class interval.

**3** The thicknesses of 40 samples of steel plate were measured:

| Thickness $x$ (mm) | 9·60–9·80 | 9·90–10·1 | 10·2–10·4 | 10·5–10·7 |
|---|---|---|---|---|
| Frequency $f$ | 1 | 4 | 10 | 11 |

| Thickness $x$ (mm) | 10·8–11·0 | 11·1–11·3 | 11·4–11·6 |
|---|---|---|---|
| Frequency $f$ | 7 | 4 | 3 |

Using coding procedure, calculate:

(a) the mean

(b) the standard deviation

(c) the mode

(d) the median of the set of values given.

**4** The lengths of 50 copper plugs gave the following frequency distribution:

| Length $x$ (mm) | 14·0–14·2 | 14·3–14·5 | 14·6–14·8 | 14·9–15·1 |
|---|---|---|---|---|
| Frequency $f$ | 2 | 4 | 9 | 15 |

| Length $x$ (mm) | 15·2–15·4 | 15·5–15·7 | 15·8–16·00 |
|---|---|---|---|
| Frequency $f$ | 11 | 6 | 3 |

(a) Calculate the mean and the standard deviation.

(b) For a full batch of 2400 plugs, calculate (i) the limits between which all the lengths are likely to occur, (ii) the number of plugs with lengths greater than 15·09 mm.

## Further problems 28

58

**1** The number of components processed in one hour on a new machine was recorded on 40 occasions:

| | | | | | | | | | |
|---|---|---|---|---|---|---|---|---|---|
| 66 | 87 | 79 | 74 | 84 | 72 | 81 | 78 | 68 | 74 |
| 80 | 71 | 91 | 62 | 77 | 86 | 87 | 72 | 80 | 77 |
| 76 | 83 | 75 | 71 | 83 | 67 | 94 | 64 | 82 | 78 |
| 77 | 67 | 76 | 82 | 78 | 88 | 66 | 79 | 74 | 64 |

(a) Divide the set of values into seven equal width classes from 60 to 94.

(b) Calculate (i) the frequency distribution, (ii) the mean, (iii) the standard deviation.

2 The lengths, in millimetres, of 40 bearings were determined with the following results:

| | | | | | | | | | |
|---|---|---|---|---|---|---|---|---|---|
| 16·6 | 15·3 | 16·3 | 14·2 | 16·7 | 17·3 | 18·2 | 15·6 | 14·9 | 17·2 |
| 18·7 | 16·4 | 19·0 | 15·8 | 18·4 | 15·1 | 17·0 | 18·9 | 18·3 | 15·9 |
| 13·6 | 18·3 | 17·2 | 18·0 | 15·8 | 19·3 | 16·8 | 17·7 | 16·8 | 17·9 |
| 17·3 | 16·6 | 15·3 | 16·4 | 17·3 | 16·9 | 14·7 | 16·2 | 17·4 | 15·6 |

(a) Group the data into six equal width classes between 13·5 and 19·4 mm.

(b) Obtain the frequency distribution.

(c) Calculate (i) the mean, (ii) the standard deviation.

3 Masses of 80 brass junctions gave the following frequency distribution:

| Mass $x$ (kg) | 4·12–4·16 | 4·17–4·21 | 4·22–4·26 | 4·27–4·31 |
|---|---|---|---|---|
| Frequency $f$ | 5 | 12 | 16 | 20 |

| Mass $x$ (kg) | 4·32–4·36 | 4·37–4·41 | 4·42–4·46 |
|---|---|---|---|
| Frequency $f$ | 14 | 9 | 4 |

(a) Calculate (i) the mean, (ii) the standard deviation.

(b) For a batch of 1800 such components, calculate (i) the limits between which all the masses are likely to lie, (ii) the number of junctions with masses greater than 4·36 kg.

4 The values of the resistance of 90 carbon resistors were determined:

| Resistance $x$ (MΩ) | 2·35 | 2·36 | 2·37 | 2·38 | 2·39 | 2·40 | 2·41 |
|---|---|---|---|---|---|---|---|
| Frequency $f$ | 3 | 10 | 19 | 20 | 18 | 13 | 7 |

Calculate (a) the mean, (b) the standard deviation, (c) the mode and (d) the median of the set of values.

5 Forty concrete cubes were subjected to failure tests in a crushing machine. Failure loads were as follows:

| Load $x$ (kN) | 30·6 | 30·8 | 31·0 | 31·2 | 31·4 | 31·6 |
|---|---|---|---|---|---|---|
| Frequency $f$ | 2 | 8 | 14 | 10 | 4 | 2 |

Calculate (a) the mean, (b) the standard deviation, (c) the mode and (d) the median of the set of results.

6 The time taken by employees to complete an operation was recorded on 80 occasions:

| Time (min) | 10·0 | 10·5 | 11·0 | 11·5 | 12·0 | 12·5 | 13·0 |
|---|---|---|---|---|---|---|---|
| Frequency $f$ | 4 | 8 | 14 | 22 | 19 | 10 | 3 |

(a) Determine (i) the mean, (ii) the standard deviation, (iii) the mode and (iv) the median of the set of observations.

(b) State (i) the class interval, (ii) the lower boundary of the third class, (iii) the upper boundary of the seventh class.

▶

**7** Components are machined to a nominal diameter of 32·65 mm. A sample batch of 400 components gave a mean diameter of 32·66 mm with a standard deviation of 0·02 mm. For a production total of 2400 components, calculate:

(a) the limits between which all the diameters are likely to lie

(b) the number of acceptable components if those with diameters less than 32·62 mm or greater than 32·68 mm are rejected.

**8** The masses of 80 castings were determined with the following results:

| Mass $x$ (kg) | 7·3 | 7·4 | 7·5 | 7·6 | 7·7 | 7·8 |
|---|---|---|---|---|---|---|
| Frequency $f$ | 4 | 13 | 21 | 23 | 14 | 5 |

(a) Calculate (i) the mean mass, (ii) the standard deviation from the mean.

(b) For a batch of 2000 such castings, determine (i) the likely limits of all the masses and (ii) the number of castings likely to have a mass greater than 7·43 kg.

**9** The heights of 120 pivot blocks were measured:

| Height $x$ (mm) | 29·4 | 29·5 | 29·6 | 29·7 | 29·8 | 29·9 |
|---|---|---|---|---|---|---|
| Frequency $f$ | 6 | 25 | 34 | 32 | 18 | 5 |

(a) Calculate (i) the mean height and (ii) the standard deviation.

(b) For a batch of 2500 such blocks, calculate (i) the limits between which all the heights are likely to lie and (ii) the number of blocks with height greater than 29·52 mm.

**10** A machine is set to produce bolts of nominal diameter 25·0 mm. Measurement of the diameters of 60 bolts gave the following frequency distribution:

| Diameter $x$ (mm) | 23·3–23·7 | 23·8–24·2 | 24·3–24·7 | 24·8–25·2 |
|---|---|---|---|---|
| Frequency $f$ | 2 | 4 | 10 | 17 |

| Diameter $x$ (mm) | 25·3–25·7 | 25·8–26·2 | 26·3–26·7 |
|---|---|---|---|
| Frequency $f$ | 16 | 8 | 3 |

(a) Calculate (i) the mean diameter and (ii) the standard deviation from the mean.

(b) For a full run of 3000, calculate (i) the limits between which all the diameters are likely to lie, (ii) the number of bolts with diameters less than 24·45 mm.

Now visit the companion website for this book at www.palgrave.com/stroud for more questions applying this mathematics to science and engineering.

59

# Probability

## Learning outcomes

When you have completed this Programme you will be able to:

- ☐ Understand what is meant by a random experiment
- ☐ Distinguish between the result and an outcome of a random experiment
- ☐ Recognize that, whilst outcomes are mutually exclusive, events may not be
- ☐ Combine events and construct an outcome tree for a sequence of random experiments
- ☐ Assign probabilities to events and distinguish between *a priori* and statistical regularity
- ☐ Distinguish between mutually exclusive and non-mutually exclusive events and compute their probabilities
- ☐ Distinguish between dependent and independent events and apply the multiplication law of probabilities
- ☐ Compute conditional probabilities
- ☐ Use the binomial and Poisson probability distributions to calculate probabilities
- ☐ Use the standard normal probability distribution to calculate probabilities

# Probability

## Random experiments

**1**

An experiment is any action or sequence of actions performed to find the value or values of some unknown quantity or quantities. For example, an experiment could be conducted to find the average height of a group of people. The experimental method would involve measuring the heights of everyone in the group and, from these measurements, calculating the average height. The unknown quantity whose value we are trying to find by performing the experiment is called the **result** of the experiment and its value, found from an experiment, is called an **outcome** of the result. By the very nature of an experiment the outcome of an experiment is unknown before the experiment is performed; a result can be anticipated but its actual value – the outcome – is unknown until the experiment is completed.

Take, for example, the simple experiment of tossing a coin. Before the coin is tossed it is not known whether it will show a head or a tail but it is known that it will show one or the other – it will produce a result. We can, therefore, list the possible outcomes of the result as

Head, Tail.

An experiment that has a result with more than one possible outcome is referred to as a **random experiment**. The only requirement that is made of the outcomes of a random experiment is that they be **mutually exclusive**, that is, only one of the possible outcomes can actually occur when the experiment is performed.

So, measuring lengths up to 1 cm using a measuring scale graded to 1 mm gives the list of possible outcomes as:

. . . . . . . . . . .

*The answer is in the next frame*

---

**2**

1 mm, 2 mm, 3 mm, 4 mm, 5 mm, 6 mm, 7 mm, 8 mm, 9 mm, 10 mm

Because

These are the only measurements possible using that particular graded scale.

## Events

Whilst the completion of a random experiment will be a single outcome we may not be interested in the specific outcome but whether the outcome lies within a range of possible outcomes. To cater for ranges of possible outcomes we define an **event**. An event consists of one or more outcomes selected from a list of all possible outcomes. An event consisting of a single outcome is called a **simple event**. For example, in the random experiment of throwing a six-sided die two possible events could be:

$A$ : An even number

$B$ : A number less than 4

In these cases $A$ consists of the outcomes 2, 4 and 6, and $B$ consists of the outcomes 1, 2 and 3.

From the rolling of a die the outcome that is common to the two events:

$A$ : An odd number $\geq 2$

$B$ : An odd number $< 4$

is . . . . . . . . . . .

*Next frame*

**3**

3

Because

$A$ consists of the outcomes 3 and 5 and $B$ consists of 1 and 3 so 3 is the only outcome common to both events.

## Sequences of random experiments

When two or more random experiments are performed one after the other, the final outcome of the sequence of experiments will consist of combinations of the outcomes of the individual experiments. For example, consider the two random experiments of tossing a silver coin followed by tossing a copper coin. We shall list the possible outcomes of the first experiment as

SH, ST

where S stands for silver and of the second as

CH, CT

where C stands for copper. We can describe this sequence of experiments using an **outcome tree**.

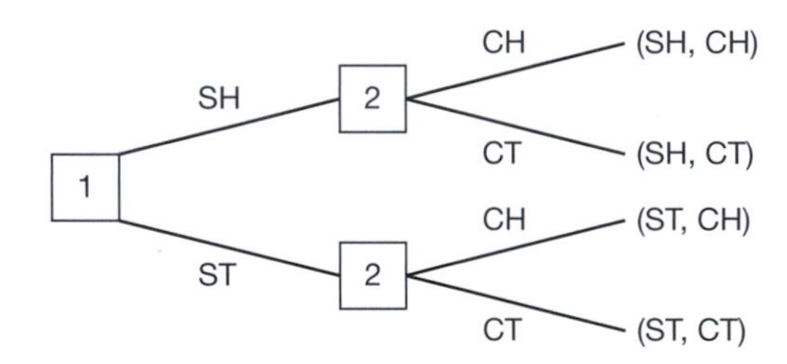


Experiment 1 (represented by the square containing the number 1) is performed first. The two branches labelled SH and ST represent the two possible outcomes of the result. Experiment 2 (represented by the square containing the number 2) is then performed and the two branches labelled CH and CT represent the two possible outcomes of the result. Each experiment has two possible outcomes so the sequence of two experiments has four possible outcomes:

(SH, CH), (SH, CT), (ST, CH) and (ST, CT)

Defining the events:

$A$ : One silver head
$B$ : Two heads
$C$ : At least one tail

Then $A$ consists of outcomes (SH, CH) and (SH, CT) and $B$ and $C$ consist of

............

*Next frame*

**4**

$B$ : (SH, CH)
$C$ : (SH, CT), (ST, CH) and (ST, CT)

Because

From the outcome tree these are the only branches that satisfy the criteria.

## Combining events

Events can be combined using *or* and *and*. For example, in the previous sequence of experiments of tossing a silver coin and then tossing a copper coin we could define the events:

$D$ : One silver head *or* At least one tail

$E$ : One silver head *and* A copper tail

in which case:

$D$ consists of (SH, CH), (SH, CT), (ST, CH) and (ST, CT)

These are all the outcomes of $A$ and $C$ combined *with no repetition*. And

$E$ consists of (SH, CT)

This is just the single outcome that is common to $A$ and $C$.

Therefore, if

$F$ : No silver head

then $F$ consists of . . . . . . . . . . .

*Next frame*

5

$F$ : (ST, CH) and (ST, CT)

Because

$F$ consists of those outcomes that are **not** in $A$

*Next frame*

# Events and probabilities

6

## Probability

If you tossed a fair coin then there would be 1 chance in 2 of it landing head face uppermost. If you select six numbers between 1 and 49 for a lottery ticket then there are nearly 14 million possible different selections of six numbers so your selection has a very small chance of winning – in fact it is 1 chance in 13 983 816. The chance of something happening can be high and can be low but we really want to be more precise than that and quantify chance in a way that makes predicting the future more accurate and more consistent. To do this we use the idea of **probability**.

Consider the random experiment of tossing a coin that has a head on both sides – a double-headed coin. If you toss this coin then it is **certain** to fall down heads. The event $H$ is a certain event and we define the *probability of a certain event as unity*:

$$P(H) = P(\text{Certainty}) = 1$$

Similarly it is impossible for it to fall down tails – the event $T$ is impossible. We define the *probability of an impossible event as zero*:

$$P(T) = P(\text{Impossibility}) = 0$$

If the double-headed coin is replaced by a normal coin possessing a tail as well as a head then the event $H$ is no longer certain and the event T is no longer impossible. In both cases the events lie somewhere between certainty and impossibility so their probabilities lie somewhere between zero and unity:

$$0 < P(H) < 1 \quad \text{and} \quad 0 < P(T) < 1$$

▶

Assigning numbers to these probabilities can be problematic but what we do say is that when a normal coin is tossed it is certain to show either a head or a tail and so the two probabilities must add up to the probability of certainty, that is unity:

$P(H) + P(T) = 1$

When we have assigned probabilities to every possible simple event of a random experiment we have what is called a **probability distribution**. From this we see that if a random experiment has four possible simple events of $E1$, $E2$, $E3$ and $E4$ then:

$$P(E1) + P(E2) + P(E3) + P(E4) = \ldots\ldots\ldots\ldots$$

*Next frame*

**7**

1

Because

One of the four events is certain to occur so the four probabilities must add up to the probability of certainty, namely unity.

## Assigning probabilities

Probabilities can be assigned to the events of a random experiment either beforehand – we call it **a priori** – or afterwards by **statistical regularity**. For example, if we make the assumption that a coin is fair we have tacitly assigned probabilities beforehand because the notion of fairness means that there is an equal chance of the tossed coin falling down head or tails. That is:

$P(H) + P(T) = 1$ and $P(H) = P(T)$

so that

$P(H) = 1/2 = 0{\cdot}5$ and $P(T) = 1/2 = 0{\cdot}5$

If, instead of assuming the coin to be fair, we tossed it 1000 times and found that we ended up with 510 heads and 490 tails then we could say that, based on statistical reularity, the probabilities were defined as:

$$\frac{\text{number of times an outcome occurs in a sequence of trials}}{\text{total number of trials}}$$

Consequently,

probability of a head, $P(H) = 510/1000 = 0{\cdot}51$
probability of a tail, $P(T) = 490/1000 = 0{\cdot}49$

Remember that the probabilities must add up to 1.

Try this. Consider the random experiment of picking a ball out of a bag that contains 1 white, 4 red and 5 blue balls and noting its colour (simple events $W$, $R$ and $B$. In order to assign probabilities to this experiment a ball was picked out and its colour noted. This process was repeated 1000 times and the colours occurred with the following frequencies:

| Colour | Frequency | Relative frequency |
|---|---|---|
| $W$ | 110 | ........... |
| $R$ | 380 | ........... |
| $B$ | 510 | ........... |
| Total | 1000 | ........... |

▶

Complete the table and define the probabilities as:

$P(W) = \ldots\ldots\ldots\ldots$
$P(R) = \ldots\ldots\ldots\ldots$
$P(B) = \ldots\ldots\ldots\ldots$

*Next frame*

---

**8**

| Colour | Frequency | Relative frequency |
|---|---|---|
| $W$ | 110 | 0·11 |
| $R$ | 380 | 0·38 |
| $B$ | 510 | 0·51 |
| Total | 1000 | 1·00 |

$P(W) = 0{\cdot}11 \quad P(R) = 0{\cdot}38 \quad P(B) = 0{\cdot}51$

Because

| Colour | Frequency | Relative frequency |
|---|---|---|
| $W$ | 110 | 110/1000 = 0·11 |
| $R$ | 380 | 380/1000 = 0·38 |
| $B$ | 510 | 510/1000 = 0·51 |
| Total | 1000 | 1·00 |

and so by statistical regularity

$P(W) = 0{\cdot}11$
$P(R) = 0{\cdot}38$
$P(B) = 0{\cdot}51$

Had every ball the same chance of being drawn then the probabilities would have been defined as:

$P(W) = \ldots\ldots\ldots\ldots$
$P(R) = \ldots\ldots\ldots\ldots$
$P(B) = \ldots\ldots\ldots\ldots$

*Next frame*

---

**9**

$P(W) = 0{\cdot}1$
$P(R) = 0{\cdot}4$
$P(B) = 0{\cdot}5$

Because

By assuming the chance of drawing any particular ball had been the same an *a priori* assignment had been made so the probabilities are simply linked to the number of balls of each colour present. So:

$P(W) = 1/10 = 0{\cdot}1$
$P(R) = 4/10 = 0{\cdot}4$
$P(B) = 5/10 = 0{\cdot}5$

## Review summary

10

1 *Random experiment*

An experiment is any action or sequence of actions performed to find the value or values of some unknown quantity or quantities. The unknown quantity whose value we are trying to find is called the **result** of the experiment and its value, found from an experiment, is called an **outcome** of the result. Any experiment that has a result with more than one possible outcome is referred to as a **random experiment**. The only requirement that is made of outcomes of a random experiment is that they be **mutually exclusive**.

2 *Event*

An event is a collection of outcomes taken from a list of possible outcomes. Events can be combined using *or* and *and*.

3 *Probability*

Probability is a measure of the chance of something happening. Probabilities can be assigned to the events of a random experiment beforehand – we call it **a priori** – or afterwards by **statistical regularity**. Probabilities lie between zero (impossibility) and unity (certainty) and all probabilities associated with the simple events of a random experiment add up to unity.

## Probabilities of combined events

11

### Or

Let $A$ and $B$ be two events associated with a random experiment. These two events can be connected using *or* to form the event $C$:

$$C = A \text{ or } B$$

In other words *either* event $A$ occurs *or* event $B$ occurs or *both*. This is an **inclusive** *or* because it permits both events to occur simultaneously. If $A$ and $B$ are mutually exclusive they contain no outcomes in common, in which case

$$P(A \text{ or } B) = P(A) + P(B)$$

For example, a bead is drawn from a bag containing 6 red beads, 4 green beads, 2 yellow beads and 3 blue beads, and each bead has an equal chance of being drawn from the bag. Let events $A$, $B$, $C$ and $D$ be:

$A$ : drawing a red bead with probability $P(A) = 6/15$

$B$ : drawing a blue bead with probability $P(B) = 3/15$

$C$ : drawing a yellow bead with probability $P(C) = 2/15$

$D$ : drawing a green bead with probability $P(D) = 4/15$

Because events $A$ and $B$ are mutually exclusive the probability of drawing either a red bead or a blue bead is then

$$P(A \text{ or } B) = P(A) + P(B) = 6/15 + 3/15 = 9/15$$

Nine of the 15 beads are either red or blue.

Also, the probability of drawing a yellow bead or drawing a green bead is:

. . . . . . . . . . . .

*Next frame*

12

6/15

Because

The events are mutually exclusive and so:

$P(C \text{ or } D) = P(C) + P(D) = 2/15 + 4/15 = 6/15$

Six of the 15 beads are either yellow or green.

## Non-mutually exclusive events

Now, how about the random experiment of rolling a six-sided die and calculating the probability of rolling a number that is either divisible by 2 or by 3. We define the two events:

$A$ : rolling a number divisible by 2

$B$ : rolling a number divisible by 3

The first aspect of this experiment to note is that both events can be achieved simultaneously by rolling a 6, so they are not mutually exclusive events. How do we cater for this?

Start with the assumption that the die is a 'fair die' so that there is an equal probability of any one of the numbers 1 to 6 being rolled. There are 6 numbers so the probability of any one of the numbers being rolled is 1/6. This means that:

$A$ consists of the numbers 2, 4, and 6 so $P(A) = 3/6$

$B$ consists of the numbers 3 and 6 so $P(B) = 2/6$

Now $P(A) + P(B) = 5/6$ but there are only 4 numbers that satisfy either criterion of being divisible by 2 or by 3, namely 2, 3, 4 and 6 so

$P(A \text{ or } B) = 4/6$

How do we cater for the discrepancy?

We recognize the fact that the *or* is *inclusive* so that when we add together outcomes that are in either $A$ or $B$ we add in *twice* those outcomes that are in both. Therefore we must subtract *once* those that are in both:

$$\begin{aligned} P(A \text{ or } B) &= P(A) + P(B) - P(\text{both } A \text{ and } B) \\ &= 5/6 - 1/6 \\ &= 4/6 \end{aligned}$$

Since

$P(\text{both } A \text{ and } B) = 1/6$

the number 6 is the only number that satisfies both criteria.

A car park contains 10 red Ford cars, 8 white Ford cars, 12 red Honda cars and 3 white Honda cars. A car is selected at random. The probability that it is either a white car or a Ford is

. . . . . . . . . . . .

*Next frame*

13

21/33

Because

Defining the events:

$A$ : a white car is selected with probability $P(A) = 11/33$
$B$ : a Ford is selected with probability $P(B) = 18/33$

The events are not mutually exclusive since there are 8 white Fords and the probability of selecting a white Ford is 8/33. Therefore:

$$P(A \text{ or } B) = P(A) + P(B) - P(\text{both } A \text{ and } B)$$
$$= 11/33 + 18/33 - 8/33$$
$$= 21/33$$

Let's move on.

*Next frame*

14

## And

Two random experiments are performed in sequence. Let $A$ be an event associated with the first experiment and event $B$ an event associated with the second experiment. These two events can be connected via the word *and* to form the event $C$ where:

$C = A \text{ and } B$

That is, both events $A$ and $B$ occur. Furthermore:

$P(A \text{ and } B) = P(A)P(B)$

This is the probability *multiplication rule* for a sequence of two random experiments.

## Dependent events

If two random experiments are performed in sequence, one after the other, then it may be possible for the outcome of the first experiment to affect the outcome of the second experiment. If this is the case then the outcomes are **dependent** upon each other and the probabilities change after the first experiment has been performed. For example, a drawer contains one blue sock, one red sock and one white sock. Assuming each sock has an equal chance of being selected from the drawer then:

$P(\text{selection of a blue sock}) = 1/3$
$P(\text{selection of a red sock}) = 1/3$
$P(\text{selection of a white sock}) = 1/3$

Let the result of the first experiment be the *selection of a blue sock*. This now affects the probabilities of the second experiment because there are only two socks left in the drawer with the consequence that:

$P(\text{selection of a blue sock}) = 0$
$P(\text{selection of a red sock}) = 1/2$
$P(\text{selection of a white sock}) = 1/2$

The selection of a second blue sock is **impossible** so it has a probability of zero.

So, the possibility of selecting a blue sock and then selecting a red sock is

...........

*Next frame*

15

1/6

Because

$P(\text{blue sock and red sock}) = P(\text{blue sock})P(\text{red sock}) = (1/3)(1/2) = 1/6$

## Independent events

If the outcome of the second experiment is unaffected by the outcome of the first experiment then the events are **independent** of each other and the probabilities will not change after the first experiment has been performed. For example, if we had replaced the sock selected by the first experiment back in the drawer before the second experiment was performed then we should have had independent events. There would have been three socks in the drawer ready for experiment 2 with probabilities:

$P(\text{selection of a blue sock}) = 1/3$
$P(\text{selection of a red sock}) = 1/3$
$P(\text{selection of a white sock}) = 1/3$

So, the probability of selecting a blue sock, replacing it and then selecting a red sock is

. . . . . . . . . . . .

*Next frame*

16

1/9

Because

$P(\text{blue sock and red sock}) = P(\text{blue sock})P(\text{red sock}) = (1/3)(1/3) = 1/9$

*Next frame*

17

## Probability trees

We are already familiar with the idea of a sequence of random experiments and the outcome tree that results from it. If we now list the probabilities against each outcome of the tree we construct what is called a *probability tree*. For example, in a factory items pass through two processes, namely cleaning and painting. The probability that an item has a cleaning fault is 0·2 and the probability that an item has a painting fault is 0·3. Cleaning and painting faults occur independently of each other so that:

Probability of a cleaning fault $P(C) = 0{\cdot}2$
Probability of no cleaning fault $P(NC) = 0{\cdot}8$

and

Probability of a painting fault $P(P) = 0{\cdot}3$
Probability of no painting fault $P(NP) = 0{\cdot}7$

The probability tree for this is:

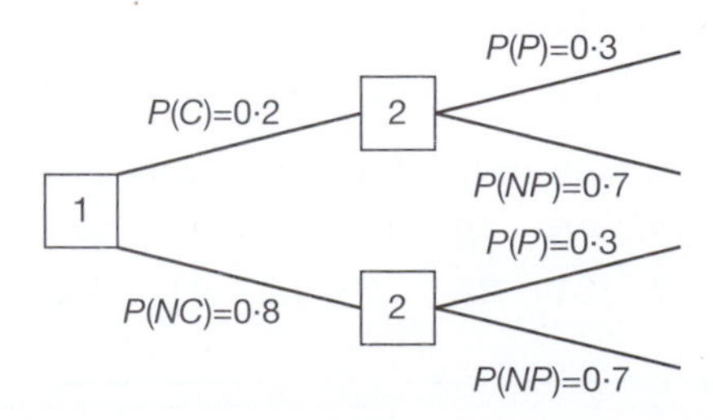

From the diagram we can see that:

(a) Probability of a cleaning fault *and* a painting fault:

$P(C)P(P) = (0{\cdot}2)(0{\cdot}3) = 0{\cdot}06$

(b) Probability of no cleaning fault *and* no painting fault:

$P(NC)P(NP) = (0{\cdot}8)(0{\cdot}7) = 0{\cdot}56$

(c) Probability of a cleaning fault but no painting fault:

$P(C)P(NP) = (0{\cdot}2)(0{\cdot}7) = 0{\cdot}14$

A car park contains 10 Ford cars and 12 Honda cars. Two cars are stolen at different times, being selected at random. The probability that the stolen cars are:

(a) A Honda and a Ford is ............

(b) Both Fords is ............

*Next frame*

## 18

(a) 120/462
(b) 90/462

Because

Letting $P(H)$ be the probability that a Honda is stolen and $P(F)$ be the probability that a Ford is stolen, the probability tree is:

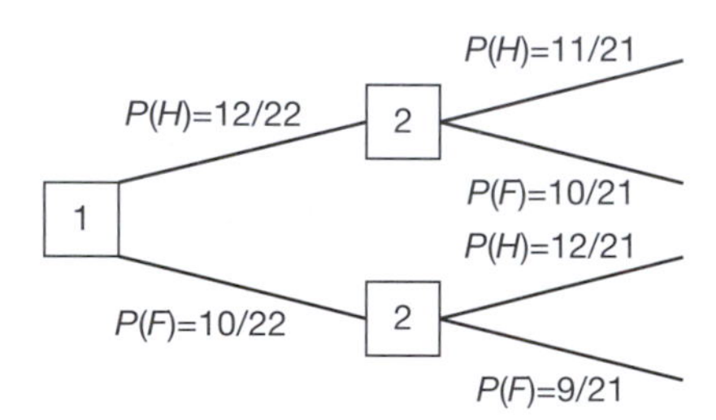


From the diagram we can see that:

(a) Probability a Honda is stolen: $P(H) = 12/22$ leaving 21 cars.

Probability a Ford is stolen: $P(F) = 10/21$

Therefore

$P$(Honda is stolen *and* a Ford is stolen):

$P(H)P(F) = (12/22)(10/21) = 120/462$

(b) Probability a Ford is stolen: $P(F1) = 10/22$ leaving 21 cars.

Probability a second Ford is stolen: $P(F2) = 9/21$

Therefore

Probability two Fords are stolen:

$P(F1)P(F2) = (10/22)(9/21) = 90/462$

Notice that the sum of the probabilities is unity as is required:

$$\frac{12}{22}\times\frac{11}{21}+\frac{12}{22}\times\frac{10}{21}+\frac{10}{22}\times\frac{12}{21}+\frac{10}{22}\times\frac{9}{21}=\frac{462}{462}=1$$

## Review summary

19

1 *Combining events using 'or'*
Events $A$ and $B$ can be combined using the connective *or* to form event $C$. If $A$ and $B$ have no outcomes in common they are said to be mutually exclusive and $P(C) = P(A) + P(B)$. If they do have outcomes in common they are not mutually exclusive and

$$P(C) = P(A) + P(B) - P(\text{both } A \text{ and } B)$$

2 *Combining events using 'and'*
A sequence of two random experiments has event $A$ associated with the first experiment and event $B$ associated with the second experiment. Event $C$ can be constructed as the combination $A$ *and* $B$. Furthermore, $P(C) = P(A)P(B)$. If the outcomes of experiment are affected by the outcome of the first experiment the probabilities associated with the second experiment will change after the performance of the first experiment. Such events are said to be dependent events. If the probabilities do not change then the events are independent of each other.

3 *Probability trees*
A probability tree can aid in the calculation of probabilities arising from sequences of random experiments.

## Conditional probability

20

We are concerned here with the probability of an event $B$ occurring, given that an event $A$ has already taken place. This is denoted by the symbol $P(B|A)$. If $A$ and $B$ are *independent* events, the fact that event $A$ has already occurred will not affect the probability of event $B$. In that case:

$P(B|A) = \ldots\ldots\ldots\ldots$

21

$$\boxed{P(B|A) = P(B)}$$

If $A$ and $B$ are *dependent* events, then event $A$ having occurred will affect the probability of the occurrence of event $B$. Let us see an example.

**Example**

A box contains five 10 ohm resistors and twelve 30 ohm resistors. The resistors are all unmarked and of the same physical size.

(a) If one resistor is picked out at random, determine the probability of its resistance being 10 ohms.

(b) If this first resistor is found to be 10 ohms and it is retained on one side, find the probability that a second selected resistor will be of resistance 30 ohms.

Let events $A$ and $B$ be:

$A$ : selection of a 10 ohm resistor
$B$ : selection of a 30 ohm resistor

(a) The number of resistors $N = 5 + 12 = 17$ Then $P(A) = \ldots\ldots\ldots\ldots$

**22**

$$P(A) = \frac{5}{17}$$

(b) The box now contains four 10 ohm resistors and twelve 30 ohm resistors. Then the probability of $B$, $A$ having occurred

$= P(B|A) = \dots\dots\dots\dots$

**23**

$$P(B|A) = \frac{12}{16} = \frac{3}{4}$$

So the probability of getting a 10 ohm resistor at the first selection, retaining it, and getting a 30 ohm resistor at the second selection is

$P(A \text{ and } B|A) = \dots\dots\dots\dots$

**24**

$$\frac{15}{68}$$

Because $P(A \text{ and } B|A) = P(A) \times P(B|A) = \frac{5}{17} \times \frac{3}{4} = \frac{15}{68}$

So, if $A$ and $B$ are independent events:

$P(A \text{ and } B) = P(A) \times P(B)$

and if $A$ and $B$ are dependent events:

$P(A \text{ and } B) = P(A) \times P(B|A)$

*Make a note of these results. Then on to the example in the next frame*

**25**

A box contains 100 copper plugs, 27 of which are oversize and 16 undersize. A plug is taken from the box, tested and replaced: a second plug is then similarly treated. Determine the probability that (a) both plugs are acceptable, (b) the first is oversize and the second undersize, (c) one is oversize and the other undersize.

Let events $A$ and $B$ be:

$A$ : selection of an oversize plug

$B$ : selection of an undersize plug

$N = 100$; 27 oversize; 16 undersize; $\therefore$ 57 acceptable

(a) $P_1(\text{first plug acceptable}) = \frac{57}{100}$

$P_2(\text{second acceptable}) = \frac{57}{100}$

$\therefore P_{12}(\text{first acceptable and second acceptable}) = \dots\dots\dots\dots$

26

$$P_1 \times P_2 = \frac{57}{100} \times \frac{57}{100} = \frac{3249}{10\,000} = 0{\cdot}3249$$

(b) $P_1(\text{first oversize}) = \dfrac{27}{100}$

$P_2(\text{second undersize}) = \dfrac{16}{100}$

$\therefore\ P_{12}(\text{first oversize and second undersize}) = \ldots\ldots\ldots\ldots$

27

$$P_1 \times P_2 = \frac{27}{100} \times \frac{16}{100} = \frac{432}{10\,000} = 0{\cdot}0432$$

(c) This section, of course, includes part (b) of the problem, but also covers the case when the first is undersize and the second oversize.

$P_3(\text{first undersize}) = \dfrac{16}{100}$; $P_4(\text{second oversize}) = \dfrac{27}{100}$

$\therefore\ P_{34}(\text{first undersize and second oversize})$

$$= \frac{16}{100} \times \frac{27}{100} = \frac{432}{10\,000} = 0{\cdot}0432$$

$\therefore\ P(\text{one oversize and one undersize})$

$= P\{(\text{first oversize and second undersize}) \text{ or } (\text{first undersize and second oversize})\} = \ldots\ldots\ldots\ldots$

28

$$P_{12} + P_{34} = \frac{432}{10\,000} + \frac{432}{10\,000} = \frac{864}{10\,000} = 0{\cdot}0864$$

One must be careful to read the precise requirements of the problem. Then the solution is straightforward, the main tools being:

*Independent events*

$P(A \text{ or } B) = P(A) + P(B)$ i.e. 'or' is associated with $+$

$P(A \text{ and } B) = P(A) \times P(B)$ i.e. 'and' is associated with $\times$

*Dependent events*

$P(A \text{ or } B) = P(A) + P(B)$; $P(A \text{ and } B) = P(A) \times P(B|A)$

Now one more on your own. It is the previous example repeated with one important variation.

A box contains 100 copper plugs, 27 oversize and 16 undersize. A plug is taken, tested but *not* replaced: a second plug is then treated similarly. Determine the probability that (a) both plugs are acceptable, (b) the first is oversize and the second undersize, (c) one is oversize and the other undersize.

*Complete all three sections and then check with the next frame*

**29**

See if you agree.

Let events $A$ and $B$ be:

$A$ : selection of an oversize plug

$B$ : selection of an undersize plug

$N = 100$; 27 oversize; 16 undersize; $\therefore$ 57 acceptable

(a) $P_1(\text{first plug acceptable}) = \dfrac{57}{100}$ Plug *not* replaced

$P_2(\text{second acceptable}) = \dfrac{56}{99}$

$\therefore P_{12}(\text{first acceptable and second acceptable})$

$= \dfrac{57}{100} \times \dfrac{56}{99} = \dfrac{3192}{9900} = 0{\cdot}3224$

(b) $P_1(\text{first oversize}) = \dfrac{27}{100}$ Plug *not* replaced

$P_2(\text{second undersize}) = \dfrac{16}{99}$

$\therefore P_{12}(\text{first oversize and second undersize})$

$= \dfrac{27}{100} \times \dfrac{16}{99} = \dfrac{432}{9900} = 0{\cdot}0436$

(c) Here again, we must include the two cases of either (first oversize and second undersize) or (first undersize and second oversize). The first of these we have already covered above:

$P_{12} = \dfrac{432}{9900} = 0{\cdot}0436$

Now we have:

$P_3(\text{first undersize}) = \dfrac{16}{100}$ Plug *not* replaced

$P_4(\text{second oversize}) = \dfrac{27}{99}$

$\therefore P_{34}(\text{first undersize and second oversize})$

$= \dfrac{16}{100} \times \dfrac{27}{99} = \dfrac{432}{9900} = 0{\cdot}0436$

$\therefore P(\text{one oversize and one undersize}) =$

$P\{(\text{first oversize and second undersize})$ or $(\text{first undersize and second oversize})\} = P_{12} + P_{34}$

$= 0{\cdot}0436 + 0{\cdot}0436 = 0{\cdot}0872$

*Now let us move on to something rather different*

# Probability distributions

**30**

## Random variables

Every random experiment gives rise to a collection of mutually exclusive outcomes, each with an associated probability of that outcome occurring. This collection of probabilities is called the **probability distribution** of the random experiment and we have seen how a probability distribution can be divined either from a relative frequency distribution using the notion of statistical regularity or from *a priori* considerations.

▶

Whichever method we choose to create the probability distribution the process can be greatly assisted with the notion of a **random variable** that is created by coding each outcome with a number. For example, the random experiment of tossing a coin with simple events $H$ and $T$ where:

$H$ is coded as 1 $\quad$ $T$ is coded as 2

Next, the result of the random experiment that *a face will show* will be represented by the variable $x$ where 1 and 2 are the only allowed values of the variable. If the coin is fair then:

$P(H) = P(x = 1) = P(1) = 0{\cdot}5$ and $P(T) = P(x = 2) = P(2) = 0{\cdot}5$

By coding in this manner a relationship is formed between the values of the random variable and their associated probabilities – this is *the probability distribution*. Because the variable $x$ is associated with a random experiment it is called a *random variable*.

A sample of 30 packets of cereal were taken from a supermarket shelf and weighed. They were all supposed to contain 200 g of cereal but the weights were as follows:

2 weighed 190 g
4 weighed 195 g
15 weighed 200 g
5 weighed 205 g
3 weighed 210 g
1 weighed 215 g

The associated probability distribution for this random experiment was:

| Weight ($x$) | Number | Probability $P(x)$ |
|---|---|---|
| 190 | 2 | ........... |
| 195 | 4 | ........... |
| 200 | 15 | ........... |
| 205 | 5 | ........... |
| 210 | 3 | ........... |
| 215 | 1 | ........... |
| Total | 30 | 1.00 |

*Next frame*

**31**

| Weight ($x$) | Number | Probability $P(x)$ |
|---|---|---|
| 190 | 2 | $2/30 = 0{\cdot}07$ |
| 195 | 4 | $4/30 = 0{\cdot}13$ |
| 200 | 15 | $15/30 = 0{\cdot}50$ |
| 205 | 5 | $5/30 = 0{\cdot}10$ |
| 210 | 3 | $3/30 = 0{\cdot}10$ |
| 215 | 1 | $1/30 = 0{\cdot}03$ |
| Total | 30 | 1.00 |

Because

In this experiment the weight ($x$) is a random variable.

*Next frame*

32

## Expectation

Every permitted value of a random variable $x$ associated with a random experiment has a probability $P(x)$ of being realized. In complete analogy with defining the average value of a collection of data as the sum of the product of each datum value with its relative frequency, we can define the average value of a random variable as the sum of the product of each of its values with its respective probability:

$$\mu = \sum_{r=1}^{n} x_r P(x_r) \qquad \text{(the symbol } \mu \text{ is a lower-case Greek m, pronounced 'myou')}$$

Here, the average value of the random variable is called the **expectation** of $x$, denoted by $E(x)$. That is:

$\mu = E(x)$, the expectation of $x$.

For example, the random experiment of tossing the fair coin of Frame 30 has an expectation

$$\begin{aligned}\mu &= 1P(1) + 2P(2)\\ &= 1(0{\cdot}5) + 2(0{\cdot}5)\\ &= 1{\cdot}5\end{aligned}$$

The expectation of the weight of the box of cereals in the question at the end of Frame 30 is

. . . . . . . . . . . to 1 dp

*Next frame*

33

201·0

Because

| Weight ($x$) | Number | Probability $P(x)$ | $xP(x)$ |
|---|---|---|---|
| 190 | 2 | 2/30 | $190(2/30) = 12{\cdot}67$ |
| 195 | 4 | 4/30 | $195(4/30) = 26{\cdot}00$ |
| 200 | 15 | 15/30 | $200(15/30) = 100{\cdot}00$ |
| 205 | 5 | 5/30 | $205(5/30) = 34{\cdot}17$ |
| 210 | 3 | 3/30 | $210(3/30) = 21{\cdot}00$ |
| 215 | 1 | 1/30 | $215(1/30) = 7{\cdot}17$ |
| Total | 30 | 1·00 | $\mu = 201{\cdot}01$ |

Note that the use of the word 'expectation' can be misleading. It does not mean the most likely value of $x$ to be expected is $\mu$, because the most likely value to be expected is that with the highest probability; the expectation is the *average value* of the random variable $x$ and as such need not have the same value as one of the values of the random variable.

▶

## Variance and standard deviation

The spread of the values of the random variable about the mean (the expectation) is given as the variance ($\sigma^2$) which is the expectation of the square of the deviations from the mean. That is

$$\sigma^2 = E\left([x-\mu]^2\right)$$
$$= \sum_{r=1}^{n}(x_r-\mu)^2 P(x_r)$$

where $\sigma$ is the standard deviation (see Programme 28, Frame 43). So, for example, the variance of the random variable associated with the tossing of the fair coin of Frame 30 is

$$\sigma^2 = \sum_{r=1}^{2}(x_r-\mu)^2 P(x_r)$$
$$= (1-1{\cdot}5)^2(0{\cdot}5)+(2-1{\cdot}5)^2(0{\cdot}5)$$
$$= (0{\cdot}25)(0{\cdot}5)+(0{\cdot}25)(0{\cdot}5)$$
$$= 0{\cdot}25$$

and standard deviation $\sigma = \sqrt{0{\cdot}25} = 0{\cdot}5$.

The variance and standard deviation of the weight of the boxes of cereal in the question at the end of Frame 30 are:

............

*Next frame*

---

**34**

$\sigma^2 = 30{\cdot}67$
$\sigma = 5{\cdot}54$

Because

| Weight $(x)$ | No. | Prob. $P(x)$ | $xP(x)$ | $x-\mu$ | $(x-\mu)^2$ | $(x-\mu)^2P(x)$ |
|---|---|---|---|---|---|---|
| 190 | 2 | 2/30 | 190(2/30) = 12·67 | −11·00 | 121·00 | 8·07 |
| 195 | 4 | 4/30 | 195(4/30) = 26·00 | −6·00 | 36·00 | 4·80 |
| 200 | 15 | 15/30 | 200(15/30) = 100·00 | −1·00 | 1·00 | 0·50 |
| 205 | 5 | 5/30 | 205(5/30) = 34·17 | 4·00 | 16·00 | 2·67 |
| 210 | 3 | 3/30 | 210(3/30) = 21·00 | 9·00 | 81·00 | 8·10 |
| 215 | 1 | 1/30 | 215(1/30) = 7·17 | 14·00 | 196·00 | 6·53 |
| Total | 30 | 1·00 | $\mu = 201{\cdot}0$ | | | $\sigma^2 = 30{\cdot}67$ $\sigma = 5{\cdot}54$ |

*Next frame*

35

## Bernoulli trials

A Bernoulli trial is any random experiment whose result has only two outcomes which we shall call *success* with probability $p$ and *failure* with probability $q$.

$P(\text{success}) = p$ and $P(\text{failure}) = q$ where, naturally, $p + q = 1$

A typical Bernoulli trial is the tossing of a coin where a head could be considered as success and a tail as failure. The actual connotations of success and failure are unimportant. What is important is that the random experiment has *only two possible outcomes*.

We shall not be concerned with just a single Bernoulli trial but rather with a succession of such trials where each trial is independent of the previous trials. For example, consider the five successive tossings of a single coin – the fivefold repetition of a single random experiment where the result of each repetition is independent of its predecessor. One such sequence of outcomes could be the event:

H, H, T, T, T

Here there are 2 heads and 3 tails. If we assume the coin could possibly be unfair with a probability of a head being $p$ and of a tail being $q = 1 - p$ then the probability of the event is:

$$p \times p \times q \times q \times q = p^2q^3$$

This combination of two heads and three tails is not unique. For example, the following is another such sequence

H, T, T, H, T

and this has the same probability as the other sequence, namely:

$$p \times q \times q \times p \times q = p^2q^3$$

The total number of such combinations of 2 heads and 3 tails is ...........

*Next frame*

36

| 10 |
|---|

Because

We are dealing here with combinations and considering the number of combinations of 2 identical items [heads] among 5 places (or alternatively 3 identical items [tails] among 5 places). This is given by the combinatorial coefficient ${}^5C_2$ which is:

$$\begin{aligned} {}^5C_2 &= \frac{5!}{(5-2)!2!} \\ &= \frac{5!}{3!2!} \\ &= \frac{5 \times 4}{2} \\ &= 10 \end{aligned}$$

(See Programme F.7, Frames 15 and 16.)

▶

We shall denote the probability of 2 successes out of 5 Bernoulli trials as $P(2:5)$ where

$$P(2:5) = {}^5C_2 p^2 q^3$$

Therefore, if the probability of being allocated your preferred day for a weekday late shift is 0·2, the probability of being allocated your only preferred day twice in five successive weeks is:

............

*Next frame*

**37**

0·2048

Because

We are considering the distribution of 2 successes in 5 trials where the probability of success is $p = 0{\cdot}2$. Then, since $p + q = 1$, we have the probability of failure as $q = 0{\cdot}8$ so

$$\begin{aligned} P(2:5) &= {}^5C_2 p^2 q^3 \\ &= 10(0{\cdot}2)^2(0{\cdot}8)^3 \\ &= 0{\cdot}2048 \end{aligned}$$

*Let's take this further.*
*Next frame*

**38**

## Binomial probability distribution

The probability of two successes in five Bernoulli trials is given by:

$$P(2:5) = {}^5C_2 p^2 q^3$$

You will recognize this as the third term in the binomial expansion

$$\begin{aligned} (p+q)^5 &= \sum_{r=0}^{5} 5C_r p^r q^{5-r} \\ &= {}^5C_0 p^0 q^5 + {}^5C_1 p^1 q^4 + {}^5C_2 p^2 q^3 + {}^5C_3 p^3 q^2 + {}^5C_4 p^4 q^1 + {}^5C_5 p^5 q^0 \\ &= 1 \end{aligned}$$

This can be generalized to the probability of $r$ successes in $n$ Bernoulli trials to

$$P(r:n) = {}^nC_r p^r q^{n-r}$$

where

$$\begin{aligned} \sum_{r=0}^{n} {}^nC_r p^r q^{n-r} &= {}^nC_0 p^0 q^n + {}^nC_1 p^1 q^{n-1} + \ldots + {}^nC_n p^n q^0 \\ &= (p+q)^n \\ &= 1 \end{aligned}$$

The terms of this expansion add up to 1 and so form a probability distribution. It is called, unsurprisingly, the **binomial probability distribution.**

The binomial probability distribution is concerned with the probability of $r$ successes in $n$ Bernoulli trials and is given by:

$$P(r:n) = {}^nC_r p^r q^{n-r} \text{ where } {}^nC_r = \frac{n!}{(n-r)!r!}$$

$p$ is the probability of success, $q$ is the probability of failure and $p + q = 1$.

▶

**Example 1**

A production line produced items of which 1% are defective so the probability of selecting a defective item is $p = 1/100 = 0{\cdot}01$. If a sample of 10 is taken from the line then the probability of the sample containing 1 defective is

$$\begin{aligned} P(1:10) &= {}^{10}C_1 p^1 q^{10-1} \\ &= \frac{10!}{9!1!} pq^9 \\ &= 10pq^9 \quad \text{where } p = 1\% = 0{\cdot}01 \text{ so } q = 0{\cdot}99 \end{aligned}$$

That is, the probability of finding 1 defective out of the sample of 10 is

$$10(0{\cdot}01)(0{\cdot}99)^9 = 0{\cdot}091 \text{ to 3 dp}$$

So, the probability of finding 2 defectives is ...........

*Next frame*

**39**

0·004 to 3 dp

Because

$$\begin{aligned} P(2:10) &= {}^{10}C_2 p^2 q^8 \\ &= \frac{10!}{8!2!}(0{\cdot}01)^2(0{\cdot}99)^8 \\ &= 45(0{\cdot}0001)(0{\cdot}9227) \\ &= 0{\cdot}004 \text{ to 3 dp} \end{aligned}$$

The probability of finding no defectives is ...........

*Next frame*

**40**

0·904 to 3 dp

Because

$$\begin{aligned} P(0:10) &= {}^{10}C_0 p^0 q^{10} \\ &= (0{\cdot}99)^{10} \\ &= 0{\cdot}904 \text{ to 3 dp} \end{aligned}$$

And just to make sure, the probability of finding no more than 2 defectives is

...........

*Next frame*

**41**

0·999 to 3 dp

Because

The probability of finding no more than 2 defective items is the sum of the probabilities of finding no defective items, finding one defective item and finding two defective items. That is

$$0{\cdot}904 + 0{\cdot}091 + 0{\cdot}004 = 0{\cdot}999 \text{ to 3 dp}$$

*Next frame*

42

**Example 2**

Twenty per cent of items produced on a machine are outside stated tolerances. Determine the probability distribution of the number of defectives in a pack of five items.

The probability distribution is the set of probabilities for $x = 0, 1, 2, 3, 4, 5$ successes, i.e. defectives in this case. We have $p = 0{\cdot}2$; $q = 0{\cdot}8$ and we compile the table shown:

| $x$ | 0 | 1 | 2 | 3 | 4 | 5 |
|---|---|---|---|---|---|---|
| $P$ | $q^5$<br>$\left(\frac{4}{5}\right)^5$<br>$\frac{4^5}{5^5}$<br>0·3277 | $5q^4p$<br>$5\left(\frac{4}{5}\right)^4\left(\frac{1}{5}\right)$<br>$\frac{5 \times 4^4}{5^5}$<br>... | $10q^3p^2$<br>$10\left(\frac{4}{5}\right)^3\left(\frac{1}{5}\right)^2$<br>...<br>... | $10q^2p^3$<br>...<br>...<br>... | $5qp^4$<br>...<br>...<br>... | $p^5$<br>...<br>...<br>... |

Complete the table and check with the next frame.

43

| $x$ | 0 | 1 | 2 | 3 | 4 | 5 |
|---|---|---|---|---|---|---|
| $P$ | $q^5$<br>$\left(\frac{4}{5}\right)^5$<br>$\frac{4^5}{5^5}$<br>0·3277 | $5q^4p$<br>$5\left(\frac{4}{5}\right)^4\left(\frac{1}{5}\right)$<br>$\frac{5 \times 4^4}{5^5}$<br>0·4096 | $10q^3p^2$<br>$10\left(\frac{4}{5}\right)^3\left(\frac{1}{5}\right)^2$<br>$\frac{10 \times 4^3}{5^5}$<br>0·2048 | $10q^2p^3$<br>$10\left(\frac{4}{5}\right)^2\left(\frac{1}{5}\right)^3$<br>$\frac{10 \times 4^2}{5^5}$<br>0·0512 | $5qp^4$<br>$5\left(\frac{4}{5}\right)\left(\frac{1}{5}\right)^4$<br>$\frac{5 \times 4}{5^5}$<br>0·0064 | $p^5$<br>$\left(\frac{1}{5}\right)^5$<br>$\frac{1}{5^5}$<br>0·0003 |

*Move on to Frame 44*

44

In this case, then, we have the results:

Probability distribution $n = 5$; $p = 0{\cdot}2$; $q = 0{\cdot}8$

| Successes $x$ | 0 | 1 | 2 | 3 | 4 | 5 |
|---|---|---|---|---|---|---|
| Probability $P$ | 0·3277 | 0·4096 | 0·2048 | 0·0512 | 0·0064 | 0·0003 |

and these can be displayed as a probability histogram:

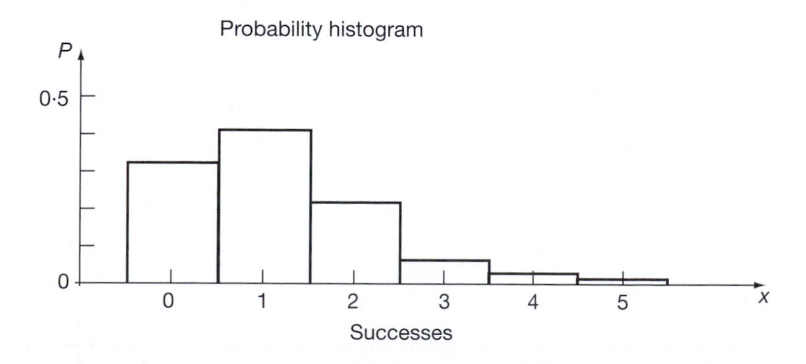


▶

(a) The probability of any particular outcome is given by the height of each column, but since the columns are 1 unit wide, the probability is also represented by the area of each column.

(b) The total probability is 1, i.e. the total area of the probability histogram is also 1.

Now for another example to work on your own.

**Example 3**

A run of 600 components was found to contain 75 defectives. Determine the probability distribution of the number of defectives in a random sample set of 6 components.

First we find $p = \dots\dots\dots$; $q = \dots\dots\dots$

**45**

$$p = \frac{1}{8}; \quad q = \frac{7}{8}$$

So we have $p = \frac{75}{600} = \frac{1}{8}$; $q = \frac{7}{8}$; $n = 6$

and now we have to find the probabilities of 0, 1, 2, 3, 4, 5, 6 successes (defectives) in any sample set of 6 components.

Proceed exactly as before. Complete the table on your own and then check the results with the next frame.

**46**

| $x$ | 0 | 1 | 2 | 3 | 4 | 5 | 6 |
|---|---|---|---|---|---|---|---|
| $P$ | $q^6$ | $6q^5p$ | $15q^4p^2$ | $20q^3p^3$ | $15q^2p^4$ | $6qp^5$ | $p^6$ |
| | Substituting $q = \frac{7}{8}$ and $p = \frac{1}{8}$ gives | | | | | | |
| | $\frac{7^6}{8^6}$ | $\frac{6 \times 7^5}{8^6}$ | $\frac{15 \times 7^4}{8^6}$ | $\frac{20 \times 7^3}{8^6}$ | $\frac{15 \times 7^2}{8^6}$ | $\frac{6 \times 7}{8^6}$ | $\frac{1}{8^6}$ |
| | 0·4488 | 0·3847 | 0·1374 | 0·0262 | 0·0028 | 0·0002 | 0·0000 |

From these results we can now draw the probability histogram which is ...........

**47**

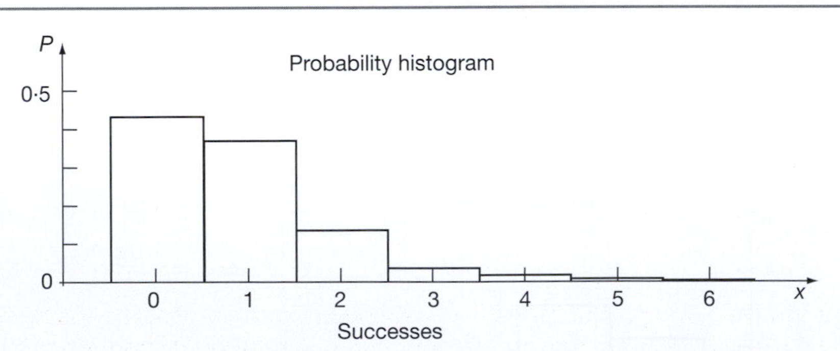


We can use the probability distribution obtained above to determine the theoretical frequencies of sets of 6 components having 0, 1, 2, 3, 4, 5, 6 defectives per set in a total of $N$ sets.

Frequency $f = N \times P$ where $N$ = number of trials (sets).

$\therefore$ If $N = 5000$, the theoretical frequencies are:

| $x$ | 0 | 1 | 2 | 3 | 4 | 5 | 6 |
|---|---|---|---|---|---|---|---|
| $P$ | 0·4488 | 0·3847 | 0·1374 | 0·0262 | 0·0028 | 0·0002 | 0·0000 |
| $f$ | | | | | | | |

The bottom line becomes . . . . . . . . . . .

48

| $x$ | 0 | 1 | 2 | 3 | 4 | 5 | 6 |
|---|---|---|---|---|---|---|---|
| $f$ | 2244 | 1924 | 687 | 131 | 14 | 1 | 0 |

The total value of $f$ in this case is, of course, 5000 (to 3 sig fig) and the total value of $P$ is . . . . . . . . . . .

49

1

*Move on*

50

## Expectation and standard deviation

The permitted values of the random variable $x_r$ associated with a binomial distribution are the number of successes in $n$ trials. That is $x_r = r$, so the expectation (mean) of a binomial probability distribution is

$$
\begin{aligned}
\mu &= \sum_{r=0}^{n} x_r P(x_r) \\
&= 0 + \sum_{r=1}^{n} r \frac{n!}{(n-r)!r!} p^r q^{n-r} \\
&= np \sum_{r=1}^{n} \frac{(n-1)!}{(n-r)!(r-1)!} p^{r-1} q^{n-r} \quad \text{since } n! = n(n-1)! \text{ and } p^r = p.p^{r-1} \\
&= np \sum_{r=1}^{n} \frac{(n-1)!}{([n-1]-[r-1])!(r-1)!} p^{r-1} q^{[n-1]-[r-1]} \quad \text{since } n-r = [n-1]-[r-1] \\
&= np \sum_{s=0}^{n-1} \frac{(n-1)!}{([n-1]-s)!(s)!} p^s q^{[n-1]-s} \quad \text{where } r-1 = s \\
&= np(p+q)^{n-1} \\
&= np \quad \text{since } p+q = 1
\end{aligned}
$$

**Example 1**

In our production line producing 1% defectives giving the probability of successfully selecting a defective as $p = 0{\cdot}01$, in a production run of 4000 items we would expect to find

$$
\begin{aligned}
\mu &= np \\
&= 4000 \times 0{\cdot}01 \\
&= 40 \text{ defectives}
\end{aligned}
$$

▶

It can be shown that the variance of a binomial probability distribution is given as:

$\sigma^2 = npq$ and so the standard deviation is $\sigma = \sqrt{npq}$

Therefore the standard deviation of the binomial distribution governing our production run of 4000 items is:

. . . . . . . . . . . to 3 dp

*Next frame*

**51**

6·293 to 3 dp

Because

$$\sigma = \sqrt{npq}$$
$$= \sqrt{4000(0\cdot01)(0\cdot99)}$$
$$= 6\cdot293 \text{ to 3 dp}$$

*Next frame*

**52**

**Example 2**

In the example we have just completed, $n = 6$ and $p = \frac{1}{8}$

Then $\mu = np = 6 \times \frac{1}{8} = 0\cdot7500 \quad \therefore \mu = 0\cdot750.$

Using the equation in Frame 50 for the standard deviation with $n = 6, p = \frac{1}{8}, q = \frac{7}{8}$

$$\sigma = \sqrt{npq} = \sqrt{6 \times \frac{1}{8} \times \frac{7}{8}} = \sqrt{\frac{42}{64}} = 0\cdot810 \quad \therefore \sigma = 0\cdot810$$

It is just as easy as that!

Now another example by way of revision.

**Example 3**

Twelve per cent of a batch of transistors are defective. Determine the binomial distribution that a packet of five transistors will contain up to 5 defectives. Calculate the mean and standard deviation of the distribution.

When you have completed it, check with the next frame.

**53**

Here is the working:

$P(\text{defective}) = p = 0\cdot12; \quad q = 0\cdot88; \quad n = 5;$

$x$ = number of defectives.

| $x$ | 0 | 1 | 2 | 3 | 4 | 5 |
|---|---|---|---|---|---|---|
| $P$ | $q^5$ | $5q^4p$ | $10q^3p^2$ | $10q^2p^3$ | $5qp^4$ | $p^5$ |
| | 0·5277 | 0·3598 | 0·0981 | 0·0134 | 0·0009 | 0·0000 |

(a) Mean $\mu = np = 5 \times 0\cdot12 = 0\cdot6 \quad \mu = 0\cdot60$

(b) Standard deviation $\sigma = \sqrt{npq} = \sqrt{5 \times 0\cdot12 \times 0\cdot88}$
$= \sqrt{0\cdot528} \quad \sigma = 0\cdot727$

To wind up this section of the work, complete the following formulas – without reference to your notes!

(a) ${}^nP_r = \ldots\ldots\ldots\ldots = \ldots\ldots\ldots\ldots$

(b) ${}^nP_n = \ldots\ldots\ldots\ldots$

(c) ${}^nC_r = \ldots\ldots\ldots\ldots$

(d) $P(r$ successes in $n$ trials$) = \ldots\ldots\ldots\ldots$

(e) $\mu = \ldots\ldots\ldots\ldots = \ldots\ldots\ldots\ldots$

(f) $\sigma = \ldots\ldots\ldots\ldots$

*Answers in next frame*

---

**54**

Here they are:

(a) ${}^nP_r = n(n-1)(n-2)\ldots(n-r+1) = \dfrac{n!}{(n-r)!}$

(b) ${}^nP_n = n!$

(c) ${}^nC_r = \dfrac{n!}{(n-r)!r!}$

(d) $P(r \text{ successes in } n \text{ trials}) = \dfrac{n!}{(n-r)!r!}q^{n-r}p^r$

(e) $\mu = np$

(f) $\sigma = \sqrt{npq}$

*Now on to the next stage*

---

## The Poisson probability distribution

**55**

If we know the average number of occurrences of an event during a fixed period of time then the Poisson probability distribution will enable us to compute the probabilities of 0, 1, 2, 3, ... occurrences during that same interval of time. The probabilities are given by:

$$P(r) = \frac{\lambda^r e^{-\lambda}}{r!}$$

with mean and variance both equal to $\lambda$ (so standard deviation is $\sqrt{\lambda}$).

Here $\lambda$ is the average number of occurrences during the fixed period of time and $r$ is a positive integer. For example, if it is known that at a particular busy traffic intersection there were 104 collisions last year this averages out at 2 per week. We can then use the Poisson probability distribution to calculate the probability of 0, 1, 2, 3, ... collisions in any particular week. They are (each to 3 dp):

$$0 \text{ collisions} = \frac{2^0e^{-2}}{0!} = 0{\cdot}135$$

$$1 \text{ collision} = \frac{2^1e^{-2}}{1!} = 0{\cdot}271$$

$$2 \text{ collisions} = \frac{2^2e^{-2}}{2!} = 0{\cdot}271$$

$$3 \text{ collisions} = \frac{2^3e^{-2}}{3!} = 0{\cdot}180$$

$$4 \text{ collisions} = \frac{2^4e^{-2}}{4!} = 0{\cdot}090$$

5 or more collisions $= 0{\cdot}053$ (since the sum of all probabilities is unity)

So, if a particular website is accessed on average 45 times each minute the probability that it will be accessed 5 times in the next 4 seconds is:

$\ldots\ldots\ldots\ldots$

*Next frame*

**56**

$\boxed{0{\cdot}101 \text{ to } 3 \text{ dp}}$

Because

If the website is accessed on average 45 times per minute then in 4 seconds its average access rate will be $\frac{45}{60}\times 4 = 3$ times. The probability of 5 accesses is then:

$$\text{5 accesses} \quad \frac{3^5 e^{-3}}{5!} = 0{\cdot}101 \text{ to } 3 \text{ dp}$$

We have considered events occurring in a given period of time but we need not restrict ourselves just to time. We could consider the number of events in a given area or volume, for example. What is important is that:

- $\lambda$ is known to begin with.
- The size of the intervals under consideration are all the same size.
- Events occurring in one interval do so independently of events occurring in a different interval.

Notice that the sum of all probabilities is unity as expected:

$$\begin{aligned}\sum_{r=0}^{\infty} P(r) &= \sum_{r=0}^{\infty} \frac{\lambda^r e^{-\lambda}}{r!}\\ &= e^{-\lambda}\left\{1 + \lambda + \frac{\lambda^2}{2!} + \frac{\lambda^3}{3!} + \frac{\lambda^4}{4!} + \dots\right\}\\ &= e^{-\lambda}e^{\lambda}\\ &= 1\end{aligned}$$

*Next frame*

**57**

## Binomial and Poisson compared

If the mean of the Poisson probability distribution $\lambda$ is less than 5, the probabilities obtained from the Poisson distribution are a good approximation to those obtained using the binomial probability distribution, particularly if the number of trials $n$ is large ($n \geq 50$) and the probability of success $p$ is small ($p \leq 0{\cdot}1$); what are called *rare events*. For this reason, it can be more convenient to calculate probabilities using the Poisson distribution because the calculations involved are more easily performed. In such a case we take $\lambda = \mu$ where $\mu = np$, the mean of the binomial probability distribution.

### Example 1

A machine produces on average 2 per cent defectives. In a random sample of 60 items, determine the probability of there being three defectives.

$$n = 60; \; p = \frac{2}{100} = 0{\cdot}02 \qquad \mu = \dots\dots\dots\dots$$

**58**

$\boxed{\mu = np = 60 \times 0{\cdot}02 = 1{\cdot}2}$

Then $P(x = 3) = \dfrac{e^{-\mu}\mu^3}{3!} = \dots\dots\dots\dots$

59

$$P = 0{\cdot}0867$$

Because $P = \dfrac{e^{-1{\cdot}2}1{\cdot}2^3}{3!} = 0{\cdot}0867$

If we use the binomial expression ${}^nC_r q^{n-r} p^r$:

$${}^{60}C_3 q^{57} p^3 = \frac{60 \times 59 \times 58}{1 \times 2 \times 3} \times 0{\cdot}98^{57} \times 0{\cdot}02^3$$

and we get 0·0865, so the agreement is close.

*Move on for another example*

60

**Example 2**

Items processed on a certain machine are found to be 1 per cent defective. Determine the probabilities of obtaining 0, 1, 2, 3, 4 defectives in a random sample batch of 80 such items.

$$P(x = r) = \frac{e^{-\mu}\mu^r}{r!}$$

$\mu = np$; $n = 80$; $p = 0{\cdot}01$; $\therefore$ $\mu = 80 \times 0{\cdot}01 = 0{\cdot}8$

Now we can calculate the Poisson probability distribution for $r = 0$ to 4, which is

...........

61

| $x$ | 0 | 1 | 2 | 3 | 4 |
|---|---|---|---|---|---|
| $P$ | 0·4493 | 0·3595 | 0·1438 | 0·0383 | 0·0077 |

For example, $P(x = 2) = \dfrac{e^{-0{\cdot}8}0{\cdot}8^2}{2!} = \dfrac{(0{\cdot}4493)(0{\cdot}64)}{2} = 0{\cdot}1438$, and similarly for the others.

If we compare these results with the binomial distribution for the same data, we get:

| $x = r$ | Probabilities | | Difference |
|---|---|---|---|
| | Binomial | Poisson | |
| 0 | 0·4475 | 0·4493 | 0·0018 |
| 1 | 0·3616 | 0·3595 | −0·0021 |
| 2 | 0·1443 | 0·1438 | −0·0005 |
| 3 | 0·0379 | 0·0383 | 0·0004 |
| 4 | 0·0074 | 0·0077 | 0·0003 |

As we see, the two sets of results agree closely.

*Move on*

62

**Example 3**

Fifteen per cent of carbon resistors drawn from stock are, on average, outside acceptable tolerances. Determine the probabilities of obtaining 0, 1, 2, 3, 4 defectives in a random batch of 20 such resistors, expressed as (a) a binomial distribution, (b) a Poisson distribution.

There is nothing new here, so you will have no trouble.

Dealing with part (a), the binomial distribution is ...........

63

*Binomial distribution*

| $x$ | 0 | 1 | 2 | 3 | 4 |
|---|---|---|---|---|---|
| $P$ | 0·0388 | 0·1368 | 0·2293 | 0·2428 | 0·1821 |

Because $p = 0{\cdot}15$; $q = 0{\cdot}85$; $n = 20$; $x =$ defectives

$$P(x = r) = \frac{n!}{(n-r)!r!} q^{n-r} p^r$$

Now for the second part (b), the Poisson distribution is ...........

64

*Poisson distribution*

| $x$ | 0 | 1 | 2 | 3 | 4 |
|---|---|---|---|---|---|
| $P$ | 0·0498 | 0·1494 | 0·2240 | 0·2240 | 0·1680 |

Because $n = 20$; $p = 0{\cdot}15$; $\mu = np = 20 \times 0{\cdot}15 = 3{\cdot}0$

$$P(x = r) = \frac{e^{-\mu}\mu^r}{r!}$$

Comparing the two sets of results as before, we now have:

| | Probabilities | | |
|---|---|---|---|
| $x = r$ | Binomial | Poisson | Difference |
| 0 | 0·0388 | 0·0498 | 0·0110 |
| 1 | 0·1368 | 0·1494 | 0·0126 |
| 2 | 0·2293 | 0·2240 | −0·0053 |
| 3 | 0·2428 | 0·2240 | −0·0188 |
| 4 | 0·1821 | 0·1680 | −0·0141 |

65

Finally, let us recall the results of the previous two examples

| | $x$ | Binomial | Poisson | Difference | Percentage difference |
|---|---|---|---|---|---|
| *Example 2* | 0 | 0·4475 | 0·4493 | 0·0018 | 0·4022 |
| $n = 80$ | 1 | 0·3616 | 0·3595 | −0·0021 | −0·5808 |
| $p = 0{\cdot}01$ | 2 | 0·1443 | 0·1438 | −0·0005 | −0·3465 |
| | 3 | 0·0379 | 0·0383 | 0·0004 | 1·0554 |
| | 4 | 0·0074 | 0·0077 | 0·0003 | 4·0541 |
| *Example 3* | 0 | 0·0388 | 0·0498 | 0·0110 | 28·3505 |
| $n = 20$ | 1 | 0·1368 | 0·1494 | 0·0126 | 9·2105 |
| $p = 0{\cdot}15$ | 2 | 0·2293 | 0·2240 | −0·0053 | −2·3114 |
| | 3 | 0·2428 | 0·2240 | −0·0188 | −7·7430 |
| | 4 | 0·1821 | 0·1680 | −0·0141 | −7·7430 |

In Example 3, the differences between the binomial and Poisson probabilities are considerably greater than in Example 2 – and this is even more apparent if we compare percentage differences.

The reason for this is ........... (Suggestions?)

66

For the Poisson distribution to represent closely the binomial distribution, the size of the sample ($n$) should be large and the probability ($p$) of obtaining success in any one trial very small

The basic data of these two problems differ in these respects. In practice, for the satisfactory use of the Poisson distribution, we should have

$p =$ ........... and $n =$ ...........

67

$p \leq 0{\cdot}1$ and $n \geq 50$

*Now on to the next topic*

# Continuous probability distributions

68

The binomial and Poisson distributions refer to discrete events, e.g. the number of successes probable in a trial. Where continuous variables are involved, e.g. measurements of length, mass, time, etc., we are concerned with the probability that a particular dimension lies between certain limiting values of that variable and for this we refer to the *normal distribution curve*.

69

## Normal distribution curve (or normal curve)

We introduced the normal curve in Programme 28 (Frame 51) as the limiting curve to which a relative frequency polygon approaches as the number of classes is greatly increased.

From a more theoretical approach, the equation of the normal curve is, in fact,

$$y = \frac{1}{\sigma\sqrt{2\pi}} e^{-\frac{1}{2}(x-\mu)^2/\sigma^2}$$

where $\mu =$ mean and $\sigma =$ standard deviation of the distribution

This equation is not at all easy to deal with! In practice, it is convenient to convert a normal distribution into a standardized normal distribution having a mean of 0 and a standard deviation of 1.

# Standard normal curve

70

The conversion from normal distribution to standard normal distribution is achieved by the substitution $z = \frac{x-\mu}{\sigma}$ which effectively moves the distribution curve along the $x$-axis and reduces the scale of the horizontal units by dividing by $\sigma$. To keep the total area under the curve at unity, we multiply the $y$-values by $\sigma$. The equation of the standardized normal curve then becomes

$$y = \phi(z) = \frac{1}{\sqrt{2\pi}} e^{-z^2/2}$$

$z = \frac{x-\mu}{\sigma}$ is called the *standard normal variable*,

$\phi(z)$ is the *probability density function*.

*Make a note of these*

**71**

Standard normal curve:

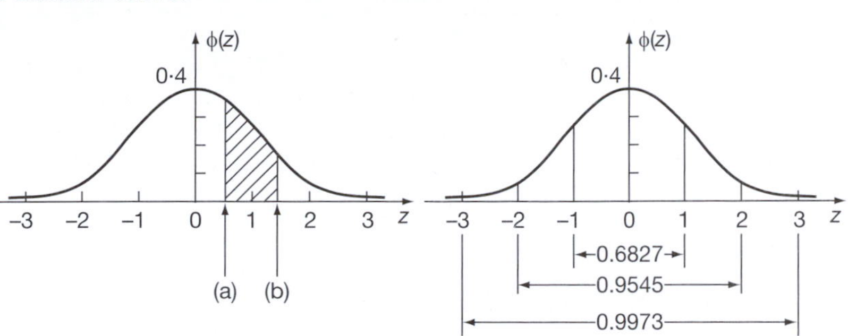


Note the following:

(a) Mean $\mu = 0$.

(b) $z$-values are in standard deviation units.

(c) Total area under the curve from $z = -\infty$ to $z = +\infty = 1$.

(d) Area between $z = a$ and $z = b$ represents the probability that $z$ lies between the values $z = a$ and $z = b$, i.e. $P(a \leq z \leq b) =$ area shaded.

(e) The probability of a value of $z$ being

between $z = -1$ and $z = 1$ is 68·27 per cent $= 0{\cdot}6827$

between $z = -2$ and $z = 2$ is 95·45 per cent $= 0{\cdot}9545$

between $z = -3$ and $z = 3$ is 99·73 per cent $= 0{\cdot}9973$

In each case the probability is given by ...........

**72**

the area under the curve between the stated limits.

Similarly, the probability of a randomly selected value of $z$ lying between $z = 0{\cdot}5$ and $z = 1{\cdot}5$ is given by the area shaded.

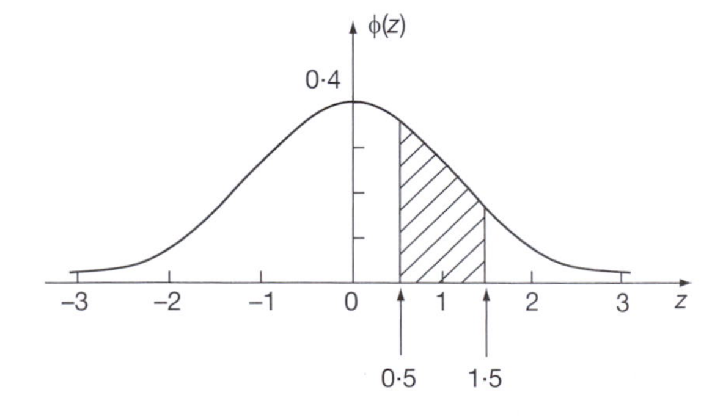


That is $P(0{\cdot}5 \leq z \leq 1{\cdot}5) = \int_{0\cdot5}^{1\cdot5} \frac{1}{\sqrt{2\pi}} e^{-z^2/2} dz$

This integral cannot be evaluated by ordinary means, so we use a table giving the area under the standard normal curve from $z = 0$ to $z = z_1$.

*Move to Frame 73*

**73**

Area under the standard normal curve

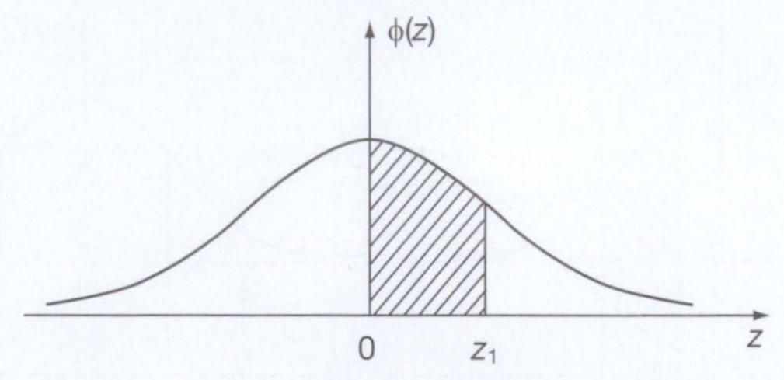


| $z_1$ | 0·00 | 0·01 | 0·02 | 0·03 | 0·04 | 0·05 | 0·06 | 0·07 | 0·08 | 0·09 |
|---|---|---|---|---|---|---|---|---|---|---|
| 0·0 | 0·0000 | 0·0040 | 0·0080 | 0·0120 | 0·0160 | 0·0199 | 0·0239 | 0·0279 | 0·0319 | 0·0359 |
| 0·1 | 0·0398 | 0·0438 | 0·0478 | 0·0517 | 0·0557 | 0·0596 | 0·0636 | 0·0675 | 0·0714 | 0·0753 |
| 0·2 | 0·0793 | 0·0832 | 0·0871 | 0·0910 | 0·0948 | 0·0987 | 0·1026 | 0·1064 | 0·1103 | 0·1141 |
| 0·3 | 0·1179 | 0·1217 | 0·1255 | 0·1293 | 0·1331 | 0·1368 | 0·1406 | 0·1443 | 0·1480 | 0·1517 |
| 0·4 | 0·1554 | 0·1591 | 0·1628 | 0·1664 | 0·1700 | 0·1736 | 0·1772 | 0·1808 | 0·1844 | 0·1879 |
| 0·5 | 0·1915 | 0·1950 | 0·1985 | 0·2019 | 0·2054 | 0·2088 | 0·2123 | 0·2157 | 0·2190 | 0·2224 |
| 0·6 | 0·2257 | 0·2291 | 0·2324 | 0·2357 | 0·2389 | 0·2422 | 0·2454 | 0·2486 | 0·2517 | 0·2549 |
| 0·7 | 0·2580 | 0·2611 | 0·2642 | 0·2673 | 0·2704 | 0·2734 | 0·2764 | 0·2794 | 0·2823 | 0·2852 |
| 0·8 | 0·2881 | 0·2910 | 0·2939 | 0·2967 | 0·2995 | 0·3023 | 0·3051 | 0·3078 | 0·3106 | 0·3133 |
| 0·9 | 0·3159 | 0·3186 | 0·3212 | 0·3238 | 0·3264 | 0·3289 | 0·3315 | 0·3340 | 0·3365 | 0·3389 |
| 1·0 | 0·3413 | 0·3438 | 0·3461 | 0·3485 | 0·3508 | 0·3531 | 0·3554 | 0·3577 | 0·3599 | 0·3621 |
| 1·1 | 0·3643 | 0·3665 | 0·3686 | 0·3708 | 0·3729 | 0·3749 | 0·3770 | 0·3790 | 0·3810 | 0·3830 |
| 1·2 | 0·3849 | 0·3869 | 0·3888 | 0·3907 | 0·3925 | 0·3944 | 0·3962 | 0·3980 | 0·3997 | 0·4015 |
| 1·3 | 0·4032 | 0·4049 | 0·4066 | 0·4082 | 0·4099 | 0·4115 | 0·4131 | 0·4147 | 0·4162 | 0·4177 |
| 1·4 | 0·4192 | 0·4207 | 0·4222 | 0·4236 | 0·4251 | 0·4265 | 0·4279 | 0·4292 | 0·4306 | 0·4319 |
| 1·5 | 0·4332 | 0·4345 | 0·4357 | 0·4370 | 0·4382 | 0·4394 | 0·4406 | 0·4418 | 0·4429 | 0·4441 |
| 1·6 | 0·4452 | 0·4463 | 0·4474 | 0·4484 | 0·4495 | 0·4505 | 0·4515 | 0·4525 | 0·4535 | 0·4545 |
| 1·7 | 0·4554 | 0·4564 | 0·4573 | 0·4582 | 0·4591 | 0·4599 | 0·4608 | 0·4616 | 0·4625 | 0·4633 |
| 1·8 | 0·4641 | 0·4649 | 0·4656 | 0·4664 | 0·4671 | 0·4678 | 0·4686 | 0·4693 | 0·4699 | 0·4706 |
| 1·9 | 0·4713 | 0·4719 | 0·4726 | 0·4732 | 0·4738 | 0·4744 | 0·4750 | 0·4756 | 0·4761 | 0·4767 |
| 2·0 | 0·4773 | 0·4778 | 0·4783 | 0·4788 | 0·4793 | 0·4798 | 0·4803 | 0·4808 | 0·4812 | 0·4817 |
| 2·1 | 0·4821 | 0·4826 | 0·4830 | 0·4834 | 0·4838 | 0·4842 | 0·4846 | 0·4850 | 0·4854 | 0·4857 |
| 2·2 | 0·4861 | 0·4864 | 0·4868 | 0·4871 | 0·4875 | 0·4878 | 0·4881 | 0·4884 | 0·4887 | 0·4890 |
| 2·3 | 0·4893 | 0·4896 | 0·4898 | 0·4901 | 0·4904 | 0·4906 | 0·4909 | 0·4911 | 0·4913 | 0·4916 |
| 2·4 | 0·4918 | 0·4920 | 0·4922 | 0·4925 | 0·4927 | 0·4929 | 0·4931 | 0·4932 | 0·4934 | 0·4936 |
| 2·5 | 0·4938 | 0·4940 | 0·4941 | 0·4943 | 0·4945 | 0·4946 | 0·4948 | 0·4949 | 0·4951 | 0·4952 |
| 2·6 | 0·4953 | 0·4955 | 0·4956 | 0·4957 | 0·4959 | 0·4960 | 0·4961 | 0·4962 | 0·4963 | 0·4964 |
| 2·7 | 0·4965 | 0·4966 | 0·4967 | 0·4968 | 0·4969 | 0·4970 | 0·4971 | 0·4972 | 0·4973 | 0·4974 |
| 2·8 | 0·4974 | 0·4975 | 0·4976 | 0·4977 | 0·4977 | 0·4978 | 0·4979 | 0·4979 | 0·4980 | 0·4981 |
| 2·9 | 0·4981 | 0·4982 | 0·4983 | 0·4983 | 0·4984 | 0·4984 | 0·4985 | 0·4985 | 0·4986 | 0·4986 |
| 3·0 | 0·4987 | 0·4987 | 0·4987 | 0·4988 | 0·4988 | 0·4989 | 0·4989 | 0·4989 | 0·4989 | 0·4990 |
| 3·1 | 0·4990 | 0·4991 | 0·4991 | 0·4991 | 0·4992 | 0·4992 | 0·4992 | 0·4992 | 0·4993 | 0·4993 |
| 3·2 | 0·4993 | 0·4993 | 0·4994 | 0·4994 | 0·4994 | 0·4994 | 0·4994 | 0·4995 | 0·4995 | 0·4995 |

*Now move on to Frame 74*

We need to find the area between $z = 0{\cdot}5$ and $z = 1{\cdot}5$:

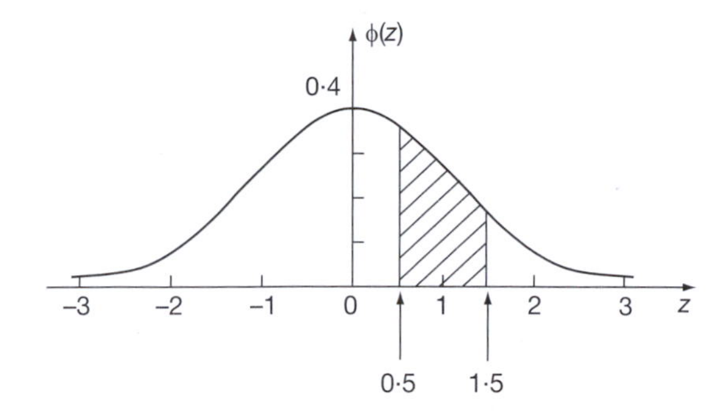


From the table:

$$\begin{aligned} &\text{area from } z = 0 \text{ to } z = 1{\cdot}5 = 0{\cdot}4332 \\ &\text{area from } z = 0 \text{ to } z = 0{\cdot}5 = 0{\cdot}1915 \\ \therefore\ &\text{area from } z = 0{\cdot}5 \text{ to } z = 1{\cdot}5 = 0{\cdot}2417 \\ \therefore\ &P(0{\cdot}5 \le z \le 1{\cdot}5) = 0{\cdot}2417 = 24{\cdot}17 \text{ per cent} \end{aligned}$$

Although the table gives areas for only positive values of $z$, the symmetry of the curve enables us to deal equally well with negative values. Now for some examples.

**Example 1**

Determine the probability that a random value of $z$ lies between $z = -1{\cdot}4$ and $z = 0{\cdot}7$.

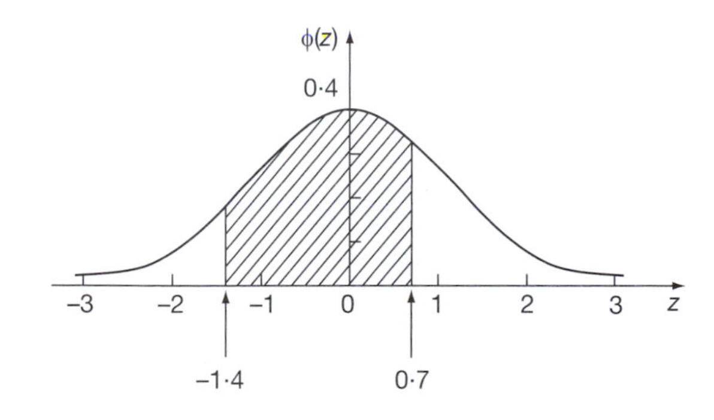


$$\begin{aligned} &\text{area from } z = -1{\cdot}4 \text{ to } z = 0 = \text{ area from } z = 0 \text{ to } z = 1{\cdot}4 \\ &\qquad = 0{\cdot}4192 \\ &\text{area from } z = 0 \text{ to } z = 0{\cdot}7 = 0{\cdot}2580 \quad \text{(from the table)} \\ \therefore\ &\text{area from } z = -1{\cdot}4 \text{ to } z = 0{\cdot}7 = \dots\dots\dots\dots \end{aligned}$$

---

**75**

$$\boxed{0{\cdot}4192 + 0{\cdot}2580 = 0{\cdot}6772}$$

$\therefore\ P(-1{\cdot}4 \le z \le 0{\cdot}7) = 0{\cdot}6772 = 67{\cdot}72$ per cent

**Example 2**

Determine the probability that a value of $z$ is greater than $2{\cdot}5$.

Draw a diagram: then it is easy enough. Finish it off.

**76**

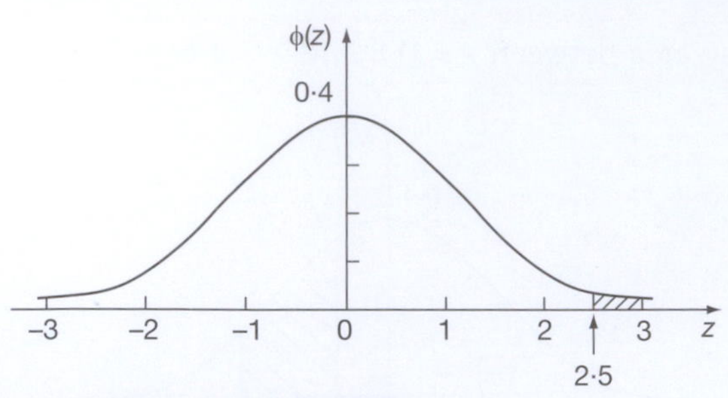


The total area from $z = 0$ to $z = \infty$ is 0·5000.

The total area from $z = 0$ to $z = 2{\cdot}5 = 0{\cdot}4938$ (from the table).

$\therefore$ The area for $z \geq 2{\cdot}5 = 0{\cdot}5000 - 0{\cdot}4938 = 0{\cdot}0062$

$\therefore$ $P(z \geq 2{\cdot}5) = 0{\cdot}0062 = 0{\cdot}62$ per cent

**Example 3**

The mean diameter of a sample of 400 rollers is 22·50 mm and the standard deviation 0·50 mm. Rollers are acceptable with diameters $22{\cdot}36 \pm 0{\cdot}53$ mm. Determine the probability of any one roller being within the acceptable limits.

We have $\mu = 22{\cdot}50$ mm; $\sigma = 0{\cdot}50$ mm

Limits of $x_1 = 22{\cdot}36 - 0{\cdot}53 = 21{\cdot}83$ mm

$x_2 = 22{\cdot}36 + 0{\cdot}53 = 22{\cdot}89$ mm

Using $z = \dfrac{x - \mu}{\sigma}$ we convert $x_1$ and $x_2$ into $z_1$ and $z_2$

$z_1 = \ldots\ldots\ldots\ldots$; $z_2 = \ldots\ldots\ldots\ldots$

**77**

$z_1 = -1{\cdot}34$; $z_2 = 0{\cdot}78$

Now, using the table, we can find the area under the normal curve between $z = -1{\cdot}34$ and $z = 0{\cdot}78$ which gives us the required result.

$P(21{\cdot}83 \leq x \leq 22{\cdot}89) = P(-1{\cdot}34 \leq z \leq 0{\cdot}78) = \ldots\ldots\ldots\ldots$

**78**

0·6922

So that is it. Convert the given values into $z$-values and apply the table as required. They are all done in much the same way.

**79**

Finally one more entirely on your own.

**Example 4**

A thermostat set to switch at 20°C operates at a range of temperatures having a mean of 20·4°C and a standard deviation of 1·3°C. Determine the probability of its opening at temperatures between 19·5°C and 20·5°C.

Complete it and then check your working with the next frame.

80

$$P(19{\cdot}5 \le x \le 20{\cdot}5) = 0{\cdot}2868$$

Here it is:

$$x_1 = 19{\cdot}5 \quad \therefore \quad z_1 = \frac{19{\cdot}5 - 20{\cdot}4}{1{\cdot}3} = -\frac{0{\cdot}9}{1{\cdot}3} = -0{\cdot}692$$

$$x_2 = 20{\cdot}5 \quad \therefore \quad z_2 = \frac{20{\cdot}5 - 20{\cdot}4}{1{\cdot}3} = \frac{0{\cdot}1}{1{\cdot}3} = 0{\cdot}077$$

$$\begin{aligned} \text{Area } (z = -0{\cdot}69 \text{ to } z = 0) &= \text{ area } (z = 0 \text{ to } z = 0{\cdot}69) \\ &= 0{\cdot}2549 \\ \text{Area } (z = 0 \text{ to } z = 0{\cdot}08) &= 0{\cdot}0319 \\ \therefore \quad \text{Area } (z = -0{\cdot}69 \text{ to } z = 0{\cdot}08) &= 0{\cdot}2868 \end{aligned}$$

$$\therefore \quad P(19{\cdot}5 \le x \le 20{\cdot}5) = P(-0{\cdot}692 \le z \le 0{\cdot}077) = 0{\cdot}2868$$

That brings us to the end of this introduction to probability. Check down the **Review summary** before dealing with the **Can You?** checklist and **Test exercise**.

There are no tricks, so you will have no difficulty.

## Review summary

81

1 *Types of probability*: empirical (experimental); classical (theoretical).

2 *Expectation*: $E = N \times P(A)$ $\quad N =$ number of trials

3 Complement of $A =$ not $A = \bar{A}$

4 $P(A) + P(\bar{A}) = 1$

5 *Classical*: $P(A) = \dfrac{\text{number of ways in which event } A \text{ can occur}}{\text{total number of all possible outcomes}}$

6 $P(\text{certainty}) = 1$; $\quad P(\text{impossibility}) = 0$

7 *Addition law*

(a) Mutually exclusive events, $P(A \text{ or } B) = P(A) + P(B)$

(b) Non-exclusive events, $P(A \text{ or } B) = P(A) + P(B) - P(A \text{ and } B)$

8 *Multiplication law*

(a) Independent events: $P(A \text{ and } B) = P(A) \times P(B)$

(b) Dependent events: $P(A \text{ and } B) = P(A) \times P(B|A)$

9 *Permutations*

(a) ${}^nP_r = n(n-1)(n-2)\ldots(n-r+1) = \dfrac{n!}{(n-r)!}$

(b) ${}^nP_n = n!$

*Combinations*

$${}^nC_r = \frac{n!}{(n-r)!r!} = \binom{n}{r}$$

10 *Binomial distribution*

$$P(r \text{ successes in } n \text{ trials}) = \frac{n!}{(n-r)!r!}q^{n-r}p^r$$

11 *Probability distribution*:

$$\mu = \sum xP(x) = np$$

$$\sigma = \sqrt{npq}$$

12 *Poisson distribution*: $P(x = r) = \dfrac{e^{-\mu}\mu^r}{r!}$

13 *Standard normal curve*: $y = \phi(z) = \dfrac{1}{\sqrt{2\pi}} e^{-z^2/2}$

$z = \dfrac{x - \mu}{\sigma} =$ *standard normal variable*

$\phi(z) =$ *probability density function*

$P(a \le z \le b) =$ area under the standard normal curve between $z = a$ and $z = b$.

## Can You?

**82**

### Checklist 29

Check this list before and after you try the end of Programme test.

On a scale of 1 to 5 how confident are you that you can: **Frames**

- Understand what is meant by a random experiment? 1 to 2
  *Yes* ☐ ☐ ☐ ☐ ☐ *No*
- Distinguish between the result and an outcome of a random experiment? 1 to 2
  *Yes* ☐ ☐ ☐ ☐ ☐ *No*
- Recognize that, whilst outcomes are mutually exclusive, events may not be? 1 to 2
  *Yes* ☐ ☐ ☐ ☐ ☐ *No*
- Combine events and construct an outcome tree for a sequence of random experiments? 3 to 5
  *Yes* ☐ ☐ ☐ ☐ ☐ *No*
- Assign probabilities to events and distinguish between *a priori* and statistical regularity? 6 to 9
  *Yes* ☐ ☐ ☐ ☐ ☐ *No*
- Distinguish between mutually exclusive and non-mutually exclusive events and compute their probabilities? 11 to 13
  *Yes* ☐ ☐ ☐ ☐ ☐ *No*
- Distinguish between dependent and independent events and apply the multiplication law of probabilities? 14 to 18
  *Yes* ☐ ☐ ☐ ☐ ☐ *No*
- Compute conditional probabilities? 20 to 29
  *Yes* ☐ ☐ ☐ ☐ ☐ *No*
- Use the binomial and Poisson probability distributions to calculate probabilities? 30 to 67
  *Yes* ☐ ☐ ☐ ☐ ☐ *No*
- Use the standard normal probability distribution to calculate probabilities? 68 to 80
  *Yes* ☐ ☐ ☐ ☐ ☐ *No*

## Test exercise 29

83

There are no tricks here, so you will have no difficulty.

1 Twelve per cent of a type of plastic bushes are rejects. Determine:

(a) the probability that any one item drawn at random is

(i) defective (ii) acceptable

(b) the number of acceptable bushes likely to be found in a sample batch of 4000.

2 A box contains 12 transistors of type *A* and 18 of type *B*, all identical in appearance. If one transistor is taken at random, tested and returned to the box, and a second transistor then treated in the same manner, determine the probability that:

(a) the first is type *A* and the second type *B*

(b) both are type *A*

(c) neither is type *A*.

3 A packet contains 100 washers, 24 of which are brass, 36 copper and the remainder steel. One washer is taken at random, retained, and a second washer similarly drawn. Determine the probability that:

(a) both washers are steel

(b) the first is brass and the second copper

(c) one is brass and one is steel.

4 A large stock of resistors is known to have 20 per cent defectives. If 5 resistors are drawn at random, determine:

(a) the probabilities that

(i) none is defective (ii) at least two are defective

(b) the mean and standard deviation of the distribution of defects.

5 A firm, on average, receives 4 enquiries per week relating to its new product. Determine the probability that the number of enquiries in any one week will be:

(a) none (b) two (c) 3 or more.

6 A machine delivers rods having a mean length of 18·0 mm and a standard deviation of 1·33 mm. If the lengths are normally distributed, determine the number of rods between 16·0 mm and 19·0 mm long likely to occur in a run of 300.

## Further problems 29

84

*Binomial distribution*

1 A box contains a large number of transistors, 30 per cent of which are type *A* and the rest type *B*. A random sample of 4 transistors is taken. Determine the probabilities that they are:

(a) all of type *A*

(b) all of type *B*

(c) two of type *A* and two of type *B*

(d) three of type *A* and one of type *B*.

▶

2 A milling machine produces products with an average of 4 per cent rejects. If a random sample of 5 components is taken, determine the probability that it contains:

(a) no reject

(b) fewer than 2 rejects.

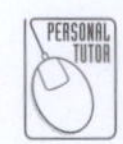

3 If 12 per cent of resistors produced in a run are defective, determine the probability distribution of defectives in a random sample of 5 resistors.

4 Production of steel rollers includes, on average, 8 per cent defectives. Determine the probability that a random sample of 6 rollers contains:

(a) 2 defectives

(b) fewer than 3 defectives.

5 A machine produces, on average, 95 per cent of mouldings within tolerance values. Determine the probability that a random sample of 5 mouldings shall contain:

(a) no defective

(b) more than one defective.

6 A large stock of resistors has 80 per cent within tolerance values. If 7 resistors are drawn at random, determine the probability that:

(a) at least 5 are acceptable

(b) all 7 are acceptable.

7 Twenty per cent of the output from a production run are rejects. In a random sample of 5 items, determine the probability of there being:

(a) 0, 1, 2, 3, 4, 5 rejects

(b) more than 1 reject

(c) fewer than 4 rejects.

8 A production line produces 6 per cent defectives. For a random sample of 10 components, determine the probability of obtaining:

(a) no defective

(b) 2 defectives

(c) more than 3 defectives.

*Poisson distribution*

9 Small metal springs are packed in boxes of 100 and 0·5 per cent of the total output of springs are defective. Determine the probability that any one box chosen at random shall have:

(a) no defective

(b) 2 or more defectives.

10 In a long production run, 1 per cent of the components are normally found to be defective. In a random sample of 10 components, determine the probability that there will be fewer than 2 defectives in the sample.

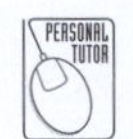

11 Three per cent of stampings are rejects. For a production run of 400, determine:

(a) the mean and standard deviation of the probability distribution

(b) the probability of obtaining 10 rejects.

12 If 2 per cent of a certain brand of light bulbs are defective, find the probability that, in a random sample of 80 bulbs, a total of 0, 1, 2, 3 and more than 5 will be defective.

**13** Brass terminals are packed in boxes of 200. If the production process is known to produce 1·5 per cent defectives on average, determine the probability that a box chosen at random will contain:

(a) no defective (b) 1 defective

(c) 2 defectives (d) 3 defectives

(e) more than 3 defectives.

**14** A product output is known to be 1 per cent defective. In a random sample of 400 components, determine the probability of including:

(a) 2 or fewer defectives

(b) 7 or more defectives.

**15** If 4 per cent of the total output of panels are substandard, find the probability of obtaining 0, 1, 2, 3, 4-or-more defectives in a random sample batch of 60 panels.

**16** Six per cent, on average, of manufactured spindles are outside stated tolerances. The spindles are randomly packed in boxes of 50 and dispatched in cartons of 100 boxes. For any one carton, determine the number of boxes likely to contain 0, 1, 2, 3, 4, 5-or-more defectives.

*Normal distribution*

**17** Boxes of screws, nominally containing 250, have a mean content of 248 screws with a standard deviation of 7. If the contents are normally distributed, determine the probability that a randomly chosen box will contain fewer than 240 screws.

**18** Samples of 10 A fuses have a mean fusing current of 9·9 A and a standard deviation of 1·2 A. Determine the probability of a fuse blowing with a current:

(a) less than 7·0 A

(b) between 8·0 A and 12·0 A.

**19** Resistors of a certain type have a mean resistance of 420 ohms with a standard deviation of 12 ohms. Determine the percentage of resistors having resistance values:

(a) between 400 ohms and 430 ohms

(b) equal to 450 ohms

**20** Washers formed on a machine have a mean diameter of 12·60 mm with a standard deviation of 0·52 mm. Determine the number of washers in a random sample of 400 likely to have diameters between 12·00 mm and 13·50 mm.

**21** The life of a drill bit has a mean of 16 hours and a standard deviation of 2·6 hours. Assuming a normal distribution, determine the probability of a sample bit lasting for:

(a) more than 20 hours

(b) fewer than 14 hours

**22** A type of bearing has an average life of 1500 hours and a standard deviation of 40 hours. Assuming a normal distribution, determine the number of bearings in a batch of 1200 likely:

(a) to fail before 1400 hours

(b) to last for more than 1550 hours.

**23** Telephone calls from an office are monitored and found to have a mean duration of 452 s and a standard deviation of 123 s. Determine:

(a) the probability of the length of a call being between 300 s and 480 s

(b) the proportion of calls likely to last for more than 720 s.

**24** Light bulbs, having a mean life of 2400 hours and standard deviation of 62 hours, are used for a consignment of 4000 bulbs. Determine:

(a) the number of bulbs likely to have a life in excess of 2500 hours

(b) the percentage of bulbs with a life length between 2300 hours and 2500 hours

(c) the probability of any one bulb having a life of 2500 hours (to the nearest hour).

**85**

Now visit the companion website for this book at www.palgrave.com/stroud for more questions applying this mathematics to science and engineering.

# Answers

## Part I

### Quiz F.1 (page 4)

1 (a) $-3 < -2$ (b) $8 > -13$ (c) $-25 < 0$

2 (a) 6 (b) 110

3 (a) 1350, 1400, 1000
(b) 2500, 2500, 3000
(c) −2450, −2500, −2000
(d) −23 630, −23 600, −24 000

4 (a) $2 \times 5 \times 17$ (b) $5 \times 7 \times 13$
(c) $3 \times 5 \times 5 \times 11 \times 11$
(d) $2 \times 2 \times 3 \times 5 \times 19$

5 (a) HCF = 4, LCM = 1848
(b) HCF = 3, LCM = 2310

6 (a) $\frac{2}{3}$ (b) $\frac{48}{7}$ (c) $-\frac{7}{2}$ (d) 16

7 (a) $\frac{2}{7}$ (b) $\frac{11}{25}$ (c) $\frac{67}{91}$ (d) $-\frac{49}{48}$

8 (a) 5 : 2 : 3 (b) 15 : 20 : 12 : 13

9 (a) 80 per cent (b) 24
(c) 64·3 per cent (to 1 dp) (d) 3·75

10 (a) 21·4, 21·36 (b) 0·0246, 0·02
(c) 0·311, 0·31 (d) 5130, 5134·56

11 (a) 0·267 (b) −0·538 (c) 1·800
(d) −2·154

12 (a) $\frac{4}{5}$ (b) $\frac{14}{5}$ (c) $\frac{329}{99}$ (d) $-\frac{11}{2}$

13 (a) $1{\cdot}\dot{0}\dot{1}$ (b) $9{\cdot}2\dot{4}5\dot{6}$

14 (a) $3^9$ (b) $2^1 = 2$ (c) $9^6$ (d) 1

15 (a) 2·466 (b) 1·380 (c) −3
(d) not possible

16 (a) 3204·4 (b) 1611·05

17 (a) $1{\cdot}3465 \times 10^2$ (b) $2{\cdot}401 \times 10^{-3}$

18 (a) $1{\cdot}61105 \times 10^3$ (b) 9·3304

19 (a) 0·024 (b) 5·21

20 (a) $11{\cdot}25_{10}$ (b) 302·908...
(c) 171·9357... (d) 3233·6958...

21 $17{\cdot}465_8\ldots$, $1111{\cdot}100_2\ldots$,
$13{\cdot}731_{12}\ldots$, $\mathrm{F{\cdot}9AE}_{16}\ldots$

### Test exercise F.1 (page 59)

1 (a) $-12 > -15$ (b) $9 > -17$
(c) $-11 < 10$

2 (a) 14 (b) 48

3 (a) $2 \times 2 \times 3 \times 13$ (b) $2 \times 3 \times 7 \times 13$
(c) $5 \times 17 \times 17$ (d) $3 \times 3 \times 3 \times 5 \times 11$

4 (a) 5050, 5000, 5000
(b) 1100, 1100, 1000
(c) −1550, −1600, −2000
(d) −5000, −5000, −5000

5 (a) HCF = 13, LCM = 19 110
(b) HCF = 2, LCM = 576

6 (a) $\frac{4}{7}$ (b) $\frac{9}{2}$ (c) $-\frac{31}{3}$ (d) −27

7 (a) $\frac{14}{15}$ (b) $\frac{11}{63}$ (c) $\frac{16}{5}$ (d) $\frac{8}{75}$
(e) 3 (f) $\frac{9}{4}$

8 (a) 3 : 1 (b) 10 : 1 : 4 (c) 6 : 18 : 5 : 1

9 (a) 60 per cent (b) $\frac{4}{25}$ (c) £2·19

10 (a) 83·54, 83·543 (b) 83·54, 83·543
(c) −2692, −2692·228
(d) −550·3, −550·341

11 (a) 0·176 (b) −0·133 (c) 5·667
(d) −2·182

12 (a) $6{\cdot}\dot{7}$ (b) $0{\cdot}\dot{0}1\dot{0}$

13 (a) $\frac{2}{5}$ (b) $\frac{92}{25}$ (c) $\frac{13}{9}$ (d) $-\frac{61}{10}$

14 (a) $2^{11}$ (b) $\left(\frac{6}{5}\right)^2 = (1{\cdot}2)^2$ (c) $(-4)^{-16}$
(d) 1

15 (a) 1·821 (b) 1·170 (c) −2·408
(d) not possible

16 (a) $5{\cdot}376 \times 10^2$ (b) $3{\cdot}64 \times 10^{-1}$
(c) $4{\cdot}902 \times 10^3$ (d) $1{\cdot}25 \times 10^{-4}$

17 (a) $61{\cdot}47 \times 10^6$ (b) $243{\cdot}9 \times 10^{-6}$
(c) $528{\cdot}6 \times 10^3$ (d) $437{\cdot}1 \times 10^{-9}$

18 $4{\cdot}72605 \times 10^8$, $472{\cdot}605 \times 10^6$

19 (a) 1 (b) 0·55

20 (a) 15·75 (b) 511·876... (c) 567·826...
(d) 3586·9792...

21 $23{\cdot}676_8\ldots$, $10011{\cdot}110_2\ldots$, $17{\cdot}\mathrm{X}56_{12}\ldots$,
$13{\cdot}\mathrm{DF}3_{16}\ldots$

### Further problems F.1 (page 60)

1 (a) $-4 > -11$ (b) $7 > -13$ (c) $-15 < 13$

2 (a) 2 (b) 12

3 (a) 3510, 3500, 4000 (b) 500, 500, 1000
(c) −2470, −2500, −2000
(d) −9010, −9000, −9000

4 (a) (i) 4·3460 (ii) 4·3460 (iii) 4·35
(b) (i) 94·541 (ii) 94·5414 (iii) 94·5

(c) (i) 0·48058 (ii) 0·4806 (iii) 0·481
(d) (i) 1·5692 (ii) 1·5692 (iii) 1·57
(e) (i) 0·86276 (ii) 0·8628 (iii) 0·86
(f) (i) 0·87927 (ii) 0·8793 (iii) 0·879

**5** (a) $2 \times 2 \times 3 \times 7 \times 11$
(b) $3 \times 5 \times 5 \times 11$
(c) $2 \times 3 \times 5 \times 7 \times 11$
(d) $2 \times 5 \times 11 \times 17 \times 19$

**6** (a) HCF = 3, LCM = 63
(b) HCF = 5, LCM = 255
(c) HCF = 6, LCM = 462
(d) HCF = 8, LCM = 2880

**7** (a) $\frac{1}{4}$ (b) $\frac{13}{6}$ (c) $-8$ (d) $-\frac{51}{7}$

**8** (a) $\frac{441}{110}$ (b) $\frac{513}{104}$ (c) $\frac{16\,641}{64}$ (d) $\frac{32\,725}{8168}$

**9** (a) $\frac{9}{25}$ (b) $\frac{7}{40}$ (c) $\frac{87}{1000}$ (d) $\frac{18}{25}$

**10** (a) 80% (b) 27·3%
(c) 22·2% (d) 14·3%
(e) 47·4% (f) 48·1%
(g) 6·9% (h) 99·5%

**11** (a) 20 (b) 0·54 to 2 sig fig (c) −1·8225
(d) 0·12 to 2 sig fig

**12** (a) 3 : 10 : 2 (b) 9 : 5 : 10 (c) 4 : 10 : 11
(d) 13 : 91 : 63

**13** (a) $8{\cdot}\dot{7}\dot{6}$ (b) $212{\cdot}\dot{2}1\dot{1}$

**14** (a) $\frac{3}{25}$ (b) $\frac{21}{4}$ (c) $\frac{589}{111}$ (d) $-\frac{93}{10}$

**15** (a) $8^7$ (b) $2^3$ (c) $5^{15}$ (d) 1

**16** (a) 3·106 (b) 1·899 (c) 2·924 (d) 25

**17** (a) 0·238 (b) −0·118 (c) 2·667
(d) −1·684

**18** (a) $5{\cdot}2876 \times 10^1$ (b) $1{\cdot}5243 \times 10^4$
(c) $8{\cdot}765 \times 10^{-2}$ (d) $4{\cdot}92 \times 10^{-5}$
(e) $4{\cdot}362 \times 10^2$ (f) $5{\cdot}728 \times 10^{-1}$

**19** (a) $4{\cdot}285 \times 10^3$ (b) $16{\cdot}9 \times 10^{-3}$
(c) $852{\cdot}6 \times 10^{-6}$ (d) $362{\cdot}9 \times 10^3$
(e) $10{\cdot}073 \times 10^6$ (f) $569{\cdot}4 \times 10^6$

**20** $1{\cdot}257 \times 10^8$, $125{\cdot}7 \times 10^6$

**21** (a) $1{\cdot}110_2\ldots$, $1{\cdot}650_8\ldots$, $1{\cdot}9\Lambda6_{12}\ldots$,
$1{\cdot}\text{D}47_{16}\ldots$
(b) $101010110{\cdot}1_2$, $526{\cdot}4_8$, $246{\cdot}6_{12}$, $156{\cdot}8_{16}$

**22** (a) 0·101 . . ., 0·748 . . ., 0·8Λ8 . . ., 0·BF8 . . .
(b) 101110011, 371, 26Λ, 173

**23** (a) 0·111 . . ., 0·770 . . ., 0·986 . . ., 0·FC7 . . .
(b) 10110010101, 2625, 1429, 595

**24** (a) 0·59375, 0·46, 0·716, 0·98
(b) 460, 714, 324, 1CC

**25** (a) 0·111 . . ., 0·751 . . ., 0·Λ57 . . ., 0·955 . . .
(b) 1110100101, 1645, 659, 933

## Quiz F.2 (page 64)

**1** (a) $4pq - pr - 2qr$ (b) $mn(8l^2 + ln - m)$
(c) $w^{4+a-b}$ (d) $s^{\frac{1}{2}}t^{\frac{7}{4}}$

**2** (a) $8xy^2 - 4x^2y - 24x^3$
(b) $6a^3 - 29a^2b + 46ab^2 - 24b^3$
(c) $2x - 6y + 5z + 54$

**3** (a) −1·996 (b) 1·244 (c) 1·660

**4** $F = \frac{GmM}{r^2}$

**5** $\log T = \log 2 + \log \pi + \frac{1}{2}(\log l - \log g)$

**6** (a) $4n^3 + 13n^2 - 2n - 15$
(b) $2v^5 - 5v^4 + 4v^3 - 5v^2 + 6v - 2$

**7** (a) $2y - 5$ (b) $q^2 + 2q + 4$ (c) $2r + 1$

**8** (a) $\left(\frac{p}{q}\right)^4$ (b) $\frac{a^4}{2b^2}$

**9** (a) $6xy(3x - 2y)$ (b) $(x^2 - 3y^2)(x + 4y)$
(c) $(3x - 5y)(x + y)$ (d) $(4x - 3)(3x - 4)$

## Test exercise F.2 (page 98)

**1** (a) $2(3ab - 2ac - cb)$ (b) $yz(7xz - 3y)$
(c) $c^{p-q+2}$ (d) $x^{-\frac{13}{12}}y^{-\frac{7}{2}}$

**2** (a) $6fg^2 + 4fgh - 16fh^2$
(b) $50x^3 + 80x^2y - 198xy^2 + 36y^3$
(c) $12p - 24q - 12r + 96$

**3** (a) −1·5686 (b) 3·8689 (c) 2·4870

**4** (a) $\log V = \log \pi + \log h + \log(D - h) + \log(D + h) - \log 4$
(b) $\log P = 2\log(2d - 1) + \log N + \frac{1}{2}\log S - \log 16$

**5** (a) $x = \frac{PQ^2}{1000K}$ (b) $R = 10S^3.\sqrt[3]{M}$
(c) $P = \frac{e^2\sqrt{Q + 1}}{R^3}$

**6** (a) $3x^3 - 4x^2 - 2x + 1$
(b) $3a^4 + 10a^3 + 18a^2 + 16a + 8$

**7** (a) $x^2y$ (b) $\frac{b^3}{ac}$

**8** (a) $x + 3$ (b) $n^2 - 3n + 9$ (c) $3a^2 - a + 1$

**9** (a) $4x^2y(9xy - 2)$ (b) $(x + 3y)(x^2 + 2y^2)$
(c) $(2x + 3)^2$ (d) $(x + 5y)(5x + 3y)$
(e) $(x + 6)(x + 4)$ (f) $(x - 2)(x - 8)$
(g) $(x - 9)(x + 4)$ (h) $(2x + 3)(3x - 2)$

## Further problems F.2 (page 99)

**1** (a) $8x^3 + 6x^2 - 47x + 21$
(b) $20x^3 - 11x^2 - 27x + 18$
(c) $15x^3 - 34x^2 + 3x + 20$ (d) $3x^2 + 2x - 4$
(e) $3x^2 + 5x - 7$ (f) $6x^2 - 4x + 7$

**2** $(x + 1)^2\sqrt{x^2 - 1}$

3 (a) $ab^{\frac{3}{2}}c^{\frac{2}{3}}$ (b) 16 (c) $\dfrac{12x^3y^3}{z}$ (d) $(x-y)^2$

4 (a) $-2{\cdot}0720$ (b) $3{\cdot}2375$ (c) $2{\cdot}8660$

5 (a) $\log f = -\left(\log\pi + \log d + \dfrac{1}{2}\log L + \dfrac{1}{2}\log C\right)$

(b) $\log K = 3\log a + \dfrac{1}{2}\log b - \dfrac{1}{6}\log c - \dfrac{2}{5}\log d$

6 (a) $W = \dfrac{A^2w^2}{32\pi^2r^2c}$ (b) $S = \dfrac{K}{2}\cdot\dfrac{\pi^2n^2yrL^2}{h^2g}$

(c) $I = \dfrac{2VKe^{-KL}}{KR+r}$

7 (a) $5xy^2(3x+4y)$ (b) $2a^2b(7a-6b)$

(c) $(2x+3y)(x-5)$ (d) $(4x-7y)(y-3)$

(e) $(5x+2)(3x+4y)$ (f) $(3x-4)(2y+5)$

(g) $(3x+4y)^2$ (h) $(4x-5y)^2$

(i) $xy^2(5xy-4)(5xy+4)$

(j) $(x+8y)(3x+2y)$

8 (a) $(x+2)(5x+3)$ (b) $(2x-3)(x-4)$

(c) $(2x-3)(3x+2)$ (d) no factors

(e) $(5x-4)(x-3)$ (f) no factors

(g) no factors (h) $(3x-2)(3x-4)$

(i) $(5x-2)(2x+3)$ (j) $(5x-3)(3x-2)$

(k) $(2x+3)(4x-5)$

## Quiz F.3 (page 102)

1 $16\,739{\cdot}41$

2 (a) $l = \sqrt{\dfrac{T^2 3g(r-t)}{4\pi^2} - 4t^2}$

(b) $r = t + \dfrac{4\pi^2}{3gT^2}(l^2 + 4t^2)$

3 $f(x) = ((7x-6)x+4)x+1$ so $f(-2) = ((7(-2)-6)(-2)+4)(-2)+1 = -87$

4 remainder 235

5 $(2x+1)(x-2)(3x-2)(x+3)$

## Test exercise F.3 (page 124)

1 (a) $1{\cdot}74$ (b) $0{\cdot}401$

2 $p = \sqrt{\dfrac{\sqrt{q^2+2}+1}{5}}$

3 (a) 38 (b) 24 (c) 136 (d) 250

4 81

5 $(x-2)(2x^2+6x-5)$

6 $(x+2)(x+4)(2x-5)$

7 $(x+1)(x-2)(x-3)(2x+1)$

## Further problems F.3 (page 125)

1 (a) $K = 8{\cdot}755$ (b) $P = 130{\cdot}6$

(c) $Q = 0{\cdot}8628$

2 (a) $I = \dfrac{V}{R}$ (b) $u = v - at$ (c) $u = \dfrac{2s - at^2}{2t}$

(d) $L = \dfrac{1}{4\pi^2f^2C}$ (e) $C = \dfrac{SF}{S-P}$

(f) $L = \dfrac{8S^2 + 3D^2}{3D}$

(g) $M = \dfrac{T+m}{1-Tm}$ (h) $h = \dfrac{\sqrt{A^2 - \pi^2r^4}}{\pi r}$

(i) $R = \sqrt{\dfrac{6V - \pi h^3}{3\pi h}}$

3 (a) 18 (b) 143 (c) $-79$ (d) 69 (e) 226

4 (a) 363 (b) 261 (c) $-76$ (d) $-59$

(e) 595

5 (a) $(x-1)(x+3)(x+4)$

(b) $(x-2)(x+5)(2x+3)$

(c) $(x+2)(x-4)(3x+2)$

(d) $(x+1)(3x^2-4x+5)$

(e) $(x-2)(2x+5)(3x-4)$

(f) $(x-1)(2x-5)(2x+7)$

(g) $(x-2)(2x^2+3x-4)$

(h) $(x+3)(3x+4)(5x-4)$

(i) $(x-2)(x^2+3x+4)$

(j) $(x+3)(2x+5)(3x+2)$

6 (a) $(x-1)(x+2)(x-4)(2x+1)$

(b) $(x-2)(x-3)(x+3)(3x-1)$

(c) $(x-2)(x+3)(4x^2-8x-3)$

(d) $(x-1)(x+1)(x^2+2x-5)$

(e) $(x+2)(x-3)(2x+1)(3x-4)$

(f) $(x-2)(x+2)(x+4)(2x-3)$

## Quiz F.4 (page 128)

1 (a) $y = \pm\sqrt{1+x^2}$

(b) $(-10, \pm 10{\cdot}0)$, $(-8, \pm 8{\cdot}1)$, $(-6, \pm 6{\cdot}1)$, $(-4, \pm 4{\cdot}1)$, $(-2, \pm 2{\cdot}2)$, $(0, \pm 1{\cdot}0)$, $(2, \pm 2{\cdot}2)$, $(4, \pm 4{\cdot}1)$, $(6, \pm 6{\cdot}1)$, $(8, \pm 8{\cdot}1)$, $(10, \pm 10{\cdot}0)$

(c)

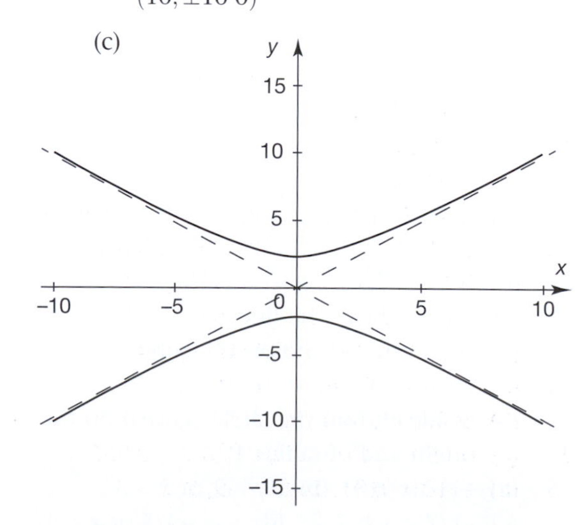

2 (a)

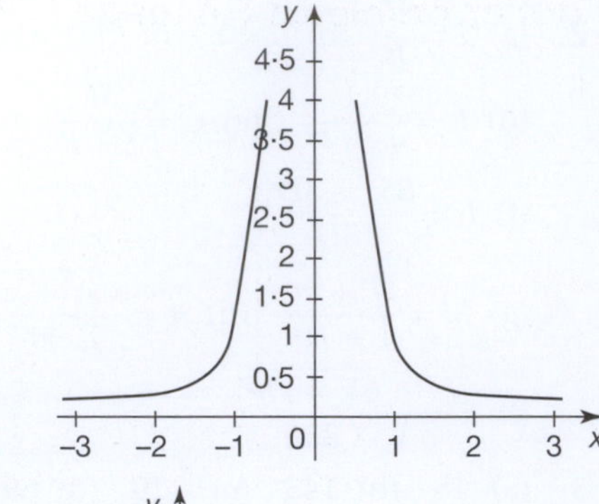


(b)

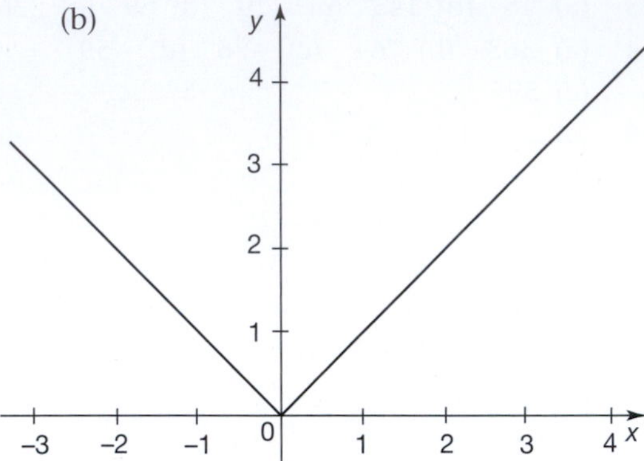


3 (a)

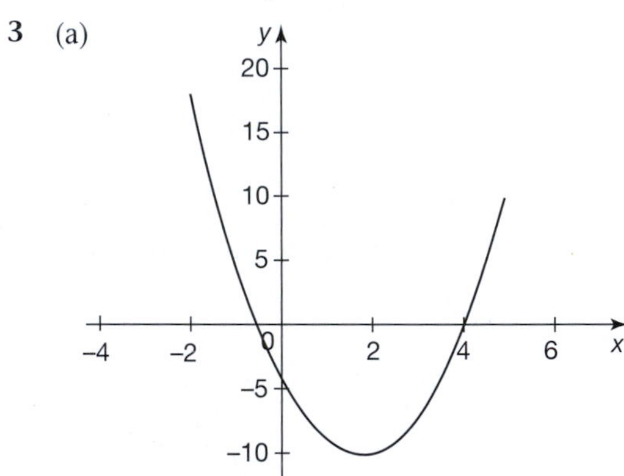


(b)

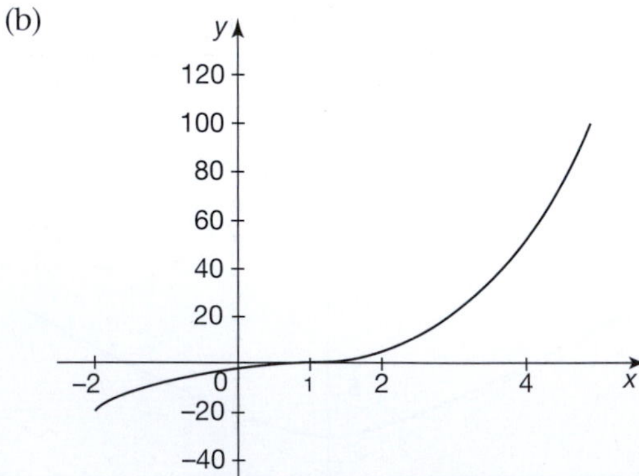


4 (a) region above the line $y = -x$
(b) region below and on the curve $y = x - 3x^3$
(c) inside and on the circle centred on the origin and of radius 1

5 (a) $-7 < x < 3$ (b) $x < -7$ or $x > 3$
(c) $-1/2 < x < 7/2$ (d) $x < -4/5$ or $x > 4$

## Test exercise F.4 (page 158)

1 (a) $y = \pm\sqrt{1 - x^2}$ (b) $(-1{\cdot}0, 0)$, $(-0{\cdot}8, \pm0{\cdot}6)$, $(-0{\cdot}6, \pm0{\cdot}8)$, $(-0{\cdot}4, \pm0{\cdot}9)$, $(-0{\cdot}2, \pm1{\cdot}0)$, $(0{\cdot}0, \pm1{\cdot}0)$, $(0{\cdot}2, \pm1{\cdot}0)$, $(0{\cdot}4, \pm0{\cdot}9)$, $(0{\cdot}6, \pm0{\cdot}8)$, $(0{\cdot}8, \pm0{\cdot}6)$, $(1{\cdot}0, 0)$

(c)

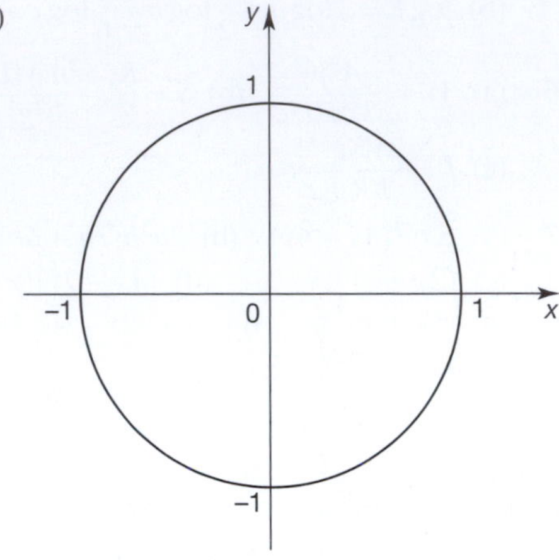


2 (a)

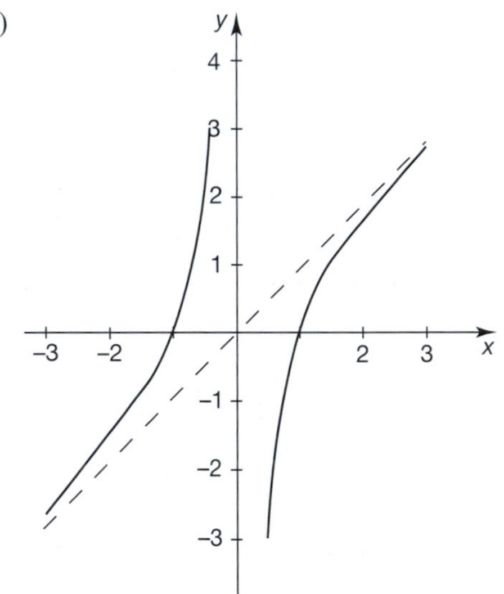


(b)

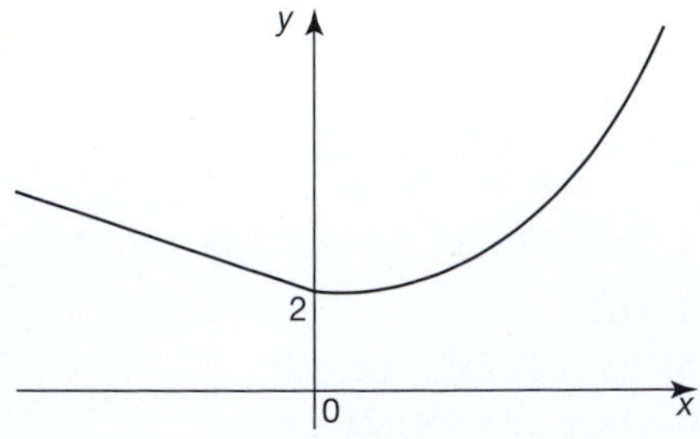

**3** (a)

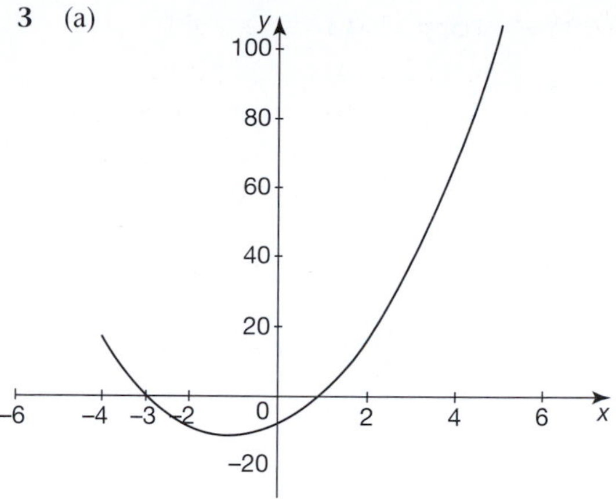


(b)

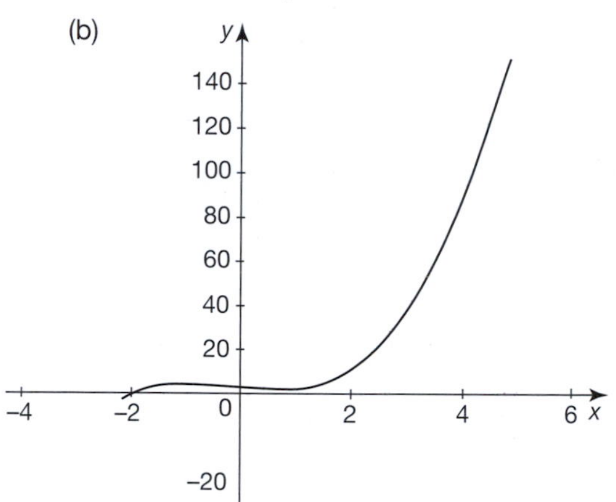


**4** (a) region below the line $y = 2 - 3x$

(b) region above the line $y = x - \dfrac{2}{x}$

(c) region on and below the line $y = 1 - x$

**5** (a) $-21 < x < 5$ (b) $x < -21$ or $x > 5$

(c) $-3/4 < x < 13/4$

(d) $x < -8/3$ or $x > 20/3$

## Further problems F.4 (page 159)

**1**

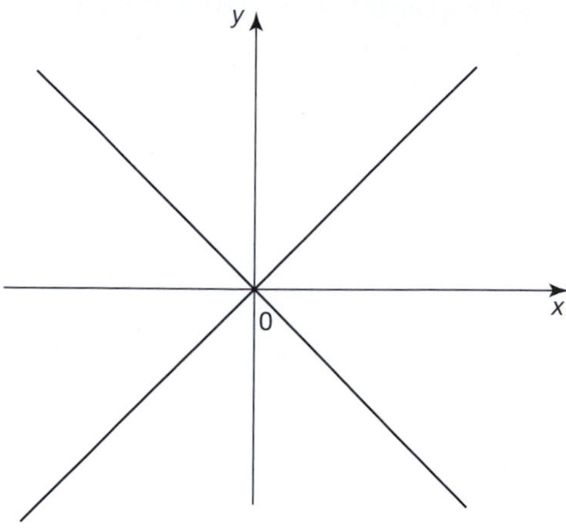


**2**

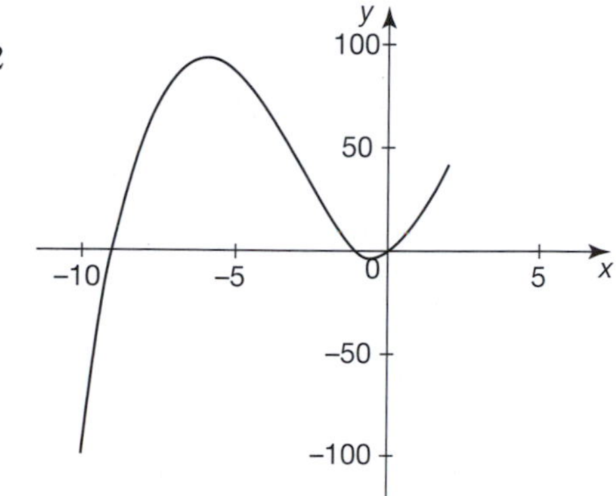


**3** See below

**4** $y = \pm\sqrt{4 - x^2}$ circle centre $(0, 0)$ radius 2

**5** $y = \pm 4\sqrt{1 - \dfrac{x^2}{4}}$ ellipse

**6** $(x + 1)^2 + (y + 1)^2 = 1$ so that $y = \pm\sqrt{1 - (1 + x)^2} - 1$; circle centre $(-1, -1)$ radius 1

---

**Further problems F.4**

**3**

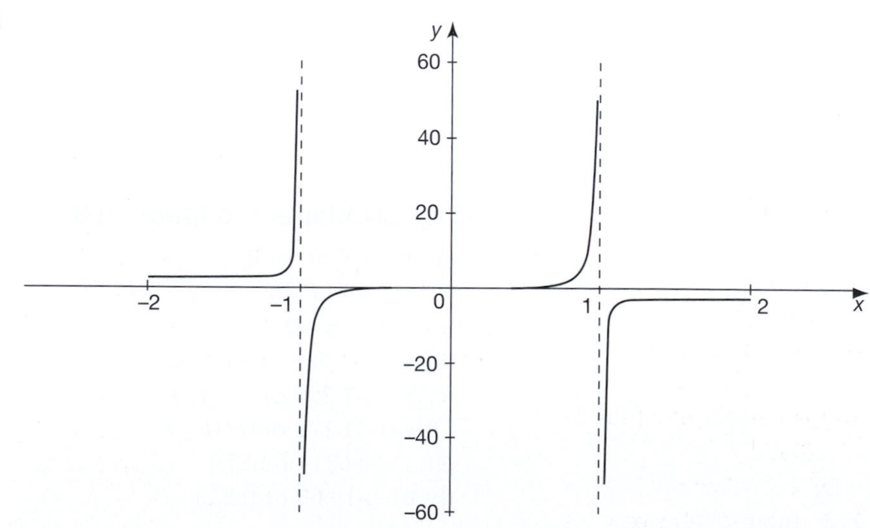

7 (a) region on and above the line $y = 2x + 4$
(b) region below the line $y = 3 - x$
(c) region on and below the line $y = \dfrac{3x}{4} - \dfrac{1}{4}$
(d) region outside the circle centred on the origin and radius $\sqrt{2}$

8 

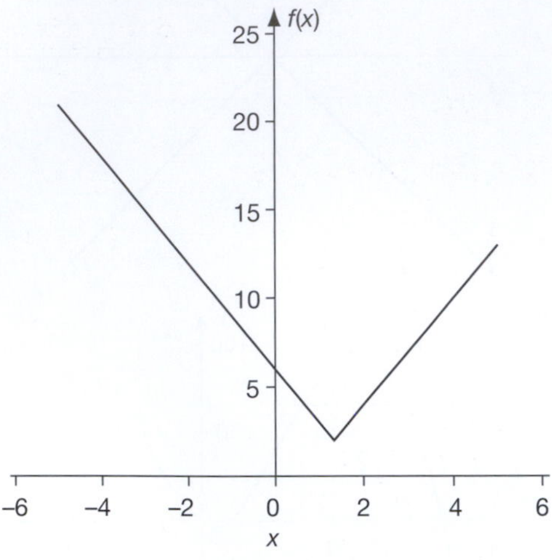


No difference since $|-x| = |x|$

9 (a) $0 \le x < \pi/6$, $5\pi/6 < x < 7\pi/6$ and $11\pi/6 < x \le 2\pi$
(b) $0 \le x < \pi/3$, $2\pi/3 < x < 4\pi/3$ and $5\pi/3 < x \le 2\pi$

## Quiz F.5 (page 162)

1 (a) $x = -2$ (b) $x = 2$ (c) $x = -3$
2 $x = 1, y = -1$
3 $x = 1, y = 3, z = 5$

## Test exercise F.5 (page 174)

1 (a) $x = -3$ (b) $x = 25$ (c) $x = \dfrac{3}{2}$
(d) $x = 1$
2 (a) $x = 2, y = 1$ (b) $x = 6{\cdot}5$ $y = -10{\cdot}5$
3 $x = 3, y = -2, z = 5$
4 $x = 1, y = 3, z = -5$

## Further problems F.5 (page 174)

1 $x = -\dfrac{3}{4}$
2 $x = \dfrac{10}{3}$
3 $x = -3$
4 $x = \dfrac{3}{4}$
5 $x = \dfrac{2}{3}$
6 $x = \dfrac{11}{4}$
7 $x = -7{\cdot}5, y = -14$
8 $x = -\dfrac{3}{2}$
9 $x = 2$
10 $x = 1, y = 2, z = 3$
11 $x = 6, y = 7, z = 5$
12 $x = 2, y = -2, z = 0$
13 $x = 3, y = 5, z = 2$
14 $x = -\dfrac{3}{2}, y = \dfrac{1}{2}$
15 $x = -\dfrac{35}{6}$
16 $x = 4{\cdot}5$
17 $x = 5, y = 4$
18 $x = \dfrac{1}{15}, y = \dfrac{1}{30}$
19 $x = -14, y = -1$
20 $x = -0{\cdot}5, y = 0{\cdot}25$

## Quiz F.6 (page 178)

1 $x = \dfrac{2}{3}$ or $x = \dfrac{3}{2}$
2 $x = 0{\cdot}215$ or $x = -1{\cdot}549$
3 $x = 2{\cdot}618$ or $x = 0{\cdot}382$
4 $x = 3$ or $x = 0{\cdot}697$ or $x = 4{\cdot}303$
5 $x = 1$ or $x = 2$ or $x = -0{\cdot}281$ or $x = 1{\cdot}781$

## Test exercise F.6 (page 193)

1 (a) $x = -2$ or $x = -9$ (b) $x = 6$ or $x = 7$
(c) $x = 3$ or $x = -7$ (d) $x = -4$ or $x = -2{\cdot}5$
(e) $x = -3$ or $x = \dfrac{4}{3}$ (f) $x = 4$ or $x = 1{\cdot}2$
2 (a) $x = -0{\cdot}581$ or $2{\cdot}581$
(b) $x = -1{\cdot}775$ or $-0{\cdot}225$
(c) $x = -5{\cdot}162$ or $1{\cdot}162$
3 (a) $x = -2{\cdot}554$ or $-0{\cdot}196$
(b) $x = -0{\cdot}696$ or $1{\cdot}196$
(c) $x = -0{\cdot}369$ or $2{\cdot}169$
4 $x = -1, -4{\cdot}812$ or $0{\cdot}312$
5 $x = 1, x = 2, x = -0{\cdot}396$ or $1{\cdot}896$

## Further problems F.6 (page 193)

1 (a) $x = -8$ or $5$ (b) $x = 4$ or $7$
(c) $x = -4$ or $-6$ (d) $x = 9$ or $-5$
2 (a) $x = 2$ or $-0{\cdot}75$
(b) $x = -1{\cdot}768$ or $-0{\cdot}566$
(c) $x = -1{\cdot}290$ or $-0{\cdot}310$
(d) $x = -1{\cdot}178$ or $0{\cdot}606$
(e) $x = 0{\cdot}621$ or $1{\cdot}879$
(f) $x = -1{\cdot}608$ or $0{\cdot}233$

3 (a) $x = -1, -2, 0{\cdot}2$
(b) $x = 1, -2{\cdot}175, -0{\cdot}575$
(c) $x = -2, -0{\cdot}523, 3{\cdot}189$
(d) $x = 1, -2{\cdot}366, -0{\cdot}634$
(e) $x = 2, -2{\cdot}151, -0{\cdot}349$
(f) $x = 3, 0{\cdot}319, -1{\cdot}569$

4 (a) $x = -1, -2, 0{\cdot}321$ or $4{\cdot}679$
(b) $x = 1, -1, -1{\cdot}914$ or $0{\cdot}314$
(c) $x = 1, -3, 0{\cdot}431$ or $2{\cdot}319$
(d) $x = -1, -2, -3{\cdot}225$ or $-0{\cdot}775$
(e) $x = 2, 3, -2{\cdot}758$ or $-0{\cdot}242$
(f) $x = -1, 2, -0{\cdot}344$ or $1{\cdot}744$
(g) $x = 1, 2, 0{\cdot}321$ or $4{\cdot}679$

## Quiz F.7 (page 196)

1 $^{12}C_7 = \frac{12!}{(12-7)!7!} = 792$

2 (a) 720 (b) 39 916 800 (c) 2730 (d) 720 (e) 1/6

3 (a) 6 (b) 210 (c) 1 (d) 36

4 $81a^4 + 432a^3b + 864a^2b^2 + 768ab^3 + 256b^4$

5 (a) $-\frac{21}{2}x^3$ (b) $-\frac{63}{16}$

6 (a) 210 (b) $n(n+4)$

7 12, 110

8 0·638

## Test exercise F.7 (page 220)

1 13 983 816

2 (a) 40 320 (b) 3 628 800 (c) 4080 (d) 24 (e) 24

3 (a) 56 (b) 455 (c) 159 (d) 1

4 $128a^7 - 2240a^6b + 16\,800a^5b^2 - 70\,000a^4b^3 + 175\,000a^3b^4 - 262\,500a^2b^5 + 218\,750ab^6 - 78\,125b^7$

5 (a) $\frac{1{\cdot}2 \times 10^9}{x^7}$ (b) $4{\cdot}5 \times 10^9$

6 (a) 1035 (b) $\frac{3n(5-n)}{2}$

7 9, 400

8 0·135

## Further problems F.7 (page 221)

1 (a) $^{12}C_5 = 792$ (b) $^{12}C_5 \times {}^7C_4 = 27\,720$
(c) $^{12}C_5 \times {}^7C_4 \times {}^3C_2 = 83\,160$

3 (a) $1 - 12x + 54x^2 - 108x^3 + 81x^4$
(b) $32 + 40x + 20x^2 + 5x^3 + \frac{5}{8}x^4 + \frac{1}{32}x^5$
(c) $1 - 5x^{-1} + 10x^{-2} - 10x^{-3} + 5x^{-4} - x^{-5}$
(d) $x^6 + 6x^4 + 15x^2 + 20 + 15x^{-2} + 6x^{-4} + x^{-6}$

4 (a) 568 (b) $\frac{n(3-5n)}{2}$

5 $e^2 \approx 7{\cdot}267$, $\sqrt{e} \approx 1{\cdot}649$

6 $e^{3x} = 1 + 3x + \frac{9x^2}{2} + \frac{9x^3}{2} + \frac{27x^4}{8} + \frac{81x^5}{40} + \dots$

$e^{-2x} = 1 - 2x + 2x^2 - \frac{4x^3}{3} + \frac{2x^4}{3} - \frac{4x^5}{15} + \dots$

$e^{3x} - e^{-2x} = 5x + \frac{5x^2}{2} + \frac{35x^3}{6} + \frac{65x^4}{24} + \frac{55x^5}{24} + \dots$

7 0·095

## Quiz F.8 (page 224)

1 $\frac{1}{4}\left(\frac{3}{x+2} + \frac{9}{x+6}\right)$

2 $1 - \frac{3}{x+2} + \frac{1}{x+1}$

3 $\frac{6}{x+1} + \frac{x-1}{x^2+x+1}$

4 $\frac{5}{x-1} + \frac{11}{(x-1)^2}$

5 $2x - 16 + \frac{91}{x+4} - \frac{95}{(x+4)^2}$

## Test exercise F.8 (page 240)

1 $\frac{5}{x-4} - \frac{4}{x-6}$

2 $\frac{4}{5x-3} + \frac{1}{2x-1}$

3 $4 + \frac{3}{x-4} + \frac{2}{x+5}$

4 $\frac{4}{x+1} + \frac{2x-3}{x^2+5x-2}$

5 $\frac{5}{2x-3} + \frac{2}{(2x-3)^2}$

6 $\frac{3}{x-5} - \frac{4}{(x-5)^2} + \frac{2}{(x-5)^3}$

7 $\frac{2}{x+4} + \frac{1}{2x-3} - \frac{3}{3x+1}$

8 $\frac{4}{x-1} + \frac{3}{x+2} + \frac{1}{x+3}$

## Further problems F.8 (page 240)

1 $\frac{2}{x+4} + \frac{5}{x+8}$

2 $\frac{3}{x-7} + \frac{2}{x+4}$

3 $\frac{4}{x-5} - \frac{3}{x-2}$

4 $\frac{1}{x-6} + \frac{2}{x+3}$

5 $\frac{2}{x-5} + \frac{3}{2x+3}$

6 $\frac{4}{3x-2} + \frac{2}{2x+1}$

7 $\frac{4}{5x-3}+\frac{1}{2x-1}$

8 $\frac{5}{2x+3}-\frac{4}{3x+1}$

9 $\frac{6}{3x+4}-\frac{4}{(3x+4)^2}$

10 $\frac{7}{5x+2}+\frac{3}{(5x+2)^2}$

11 $\frac{3}{4x-5}-\frac{1}{(4x-5)^2}$

12 $\frac{5}{x-2}+\frac{7}{(x-2)^2}-\frac{1}{(x-2)^3}$

13 $\frac{3}{5x+2}-\frac{5}{(5x+2)^2}-\frac{6}{(5x+2)^3}$

14 $\frac{4}{4x-5}+\frac{3}{(4x-5)^2}-\frac{7}{(4x-5)^3}$

15 $\frac{3}{x+2}+\frac{5}{x-1}+\frac{2}{(x-1)^2}$

16 $\frac{3}{x+2}+\frac{1}{x-3}-\frac{5}{(x-3)^2}$

17 $\frac{2}{x-1}+\frac{3}{2x+3}+\frac{4}{(2x+3)^2}$

18 $4-\frac{2}{x-5}+\frac{7}{x-8}$

19 $5+\frac{4}{x+3}+\frac{6}{x-5}$

20 $\frac{6}{x-2}+\frac{2x-3}{x^2-2x-5}$

21 $\frac{1}{2x+3}+\frac{2x-1}{x^2+5x+2}$

22 $\frac{2}{3x-1}+\frac{x-3}{2x^2-4x-5}$

23 $\frac{6}{x+1}-\frac{1}{x+2}-\frac{5}{x+3}$

24 $\frac{4}{x-1}-\frac{2}{x+3}+\frac{3}{x+4}$

25 $\frac{4}{4x+1}+\frac{2}{2x+3}-\frac{5}{3x-2}$

26 $\frac{2}{2x-1}+\frac{4}{3x+2}+\frac{3}{4x-3}$

27 $\frac{3}{x-1}+\frac{2}{2x+1}-\frac{4}{3x-1}$

28 $\frac{4}{x+1}-\frac{5}{2x+3}+\frac{1}{3x-5}$

29 $\frac{3}{x-2}-\frac{2}{2x-3}-\frac{4}{3x-4}$

30 $\frac{3}{5x+3}+\frac{6}{4x-1}-\frac{2}{x+2}$

## Quiz F.9 (page 244)

1 $253{\cdot}311\dot{6}^\circ$

2 $73^\circ 24'54''$

3 (a) 0·82 rad (b) 0·22 rad (c) $3\pi/4$ rad

4 (a) 264·76° (b) 405° (c) 468°

5 (a) 0·9135 (b) 0·9659 (c) 0·5774 (d) 3·2535 (e) 1·0045 (f) 0·0987

6 11·0 cm

8 18·4 km, 13·5 km

## Test exercise F.9 (page 264)

1 39·951° to 3 dp

2 $52^\circ 30'18''$

3 (a) 1·47 rad (b) 1·21 rad (c) $4\pi/3$ rad

4 (a) 122·56° (b) 300·00° (c) 162·00°

5 (a) 0·9511 (b) 0·2817 (c) 0·6235 (d) 0·8785 (e) 1·1595 (f) 1·5152

6 10·3 cm to 1 dp

8 yes, it is a 3, 4, 5 triangle

## Further problems F.9 (page 265)

1 81·306° to 3 dp

2 $63^\circ 12'58''$ to the nearest second

3 (a) 0·54 rad (b) 0·84 rad (c) $5\pi/4$ rad

4 (a) 102·22° (b) 135·00° (c) 144·00°

5 (a) 0·5095 (b) 0·5878 (c) 5·6713 (d) 71·6221 (e) 1432·3946 (f) 0·4120

6 6·5 cm to 1 dp

8 2

## Quiz F.10 (page 268)

1 (a) and (b) do, (c) does not

2 (a) $h(x)=\frac{1}{4-x}+3x-9$;
domain $0<x<4$,
range $-35/4<h(x)<\infty$

(b) $k(x)=\frac{1}{2(4-x)(x-3)}$;
domain $0<x<4$, $x\neq 3$,
range $-\infty<k(x)<\infty$

3 (a) inverse is a function (b) inverse is not a function (c) inverse is not a function

4 (a) $f(x)=-2(\sqrt{x}-1)^3$ (b) $f(x)=4\sqrt{x}$
(c) $f(x)=((x-1)^3-1)^3$

5 $f(x)=b[a(b[c(x)])]$ where $a(x)=x^2$, $b(x)=x+4$ and $c(x)=3x$

$a^{-1}(x)=x^{1/2}$, $b^{-1}(x)=x-4$, $c^{-1}(x)=\frac{x}{3}$

$f^{-1}(x)=\frac{(x-4)^{1/2}-4}{3}$: not a function

## Test exercise F.10 (page 285)

1 (a) does not, (b) and (c) do

2 (a) $h(x)=\frac{2}{x-2}-3x+3$;
domain $2<x<3$, range $-4<h(x)<\infty$

(b) $k(x) = -\dfrac{3}{5(x-2)(x-1)}$;
domain $2 < x < 3$,
range $-\infty < k(x) < -0{\cdot}3$

4 (a) $f(x) = 5(\sqrt{x}+3)^4$ (b) $f(x) = 25\sqrt{x}$
(c) $f(x) = ((x+3)^4+3)^4$

5 $f(x) = a[c(a[b(x)])]$ where $a(x) = x-4$, $b(x) = 5x$ and $c(x) = x^3$
$f^{-1}(x) = b^{-1}[a^{-1}(c^{-1}[a^{-1}(x)])]$:
$a^{-1}(x) = x+4$, $b^{-1}(x) = x/5$, $c^{-1}(x) = x^{\frac{1}{3}}$
$f^{-1}(x) = \dfrac{(x+4)^{\frac{1}{3}}+4}{5}$: a function

## Further problems F.10 (page 285)

1 (a) Yes, Yes (b) Yes, No (c) Yes, Yes
(d) No, Yes (e) Yes, Yes (f) No, Yes

2 (a) Function (b) Not a function
(c) Function (d) Not a function

3 (a) Domain $0 < x < \infty$,
Range $-\infty < h(x) < \infty$
(b) Domain $0 < x < \infty$,
Range $0 < h(x) < \infty$
(c) Domain $0 < x < \infty$,
Range $0 < h(x) < \infty$
(d) Domain $0 < x < \infty$,
Range $0 < h(x) < \infty$

4 (a)

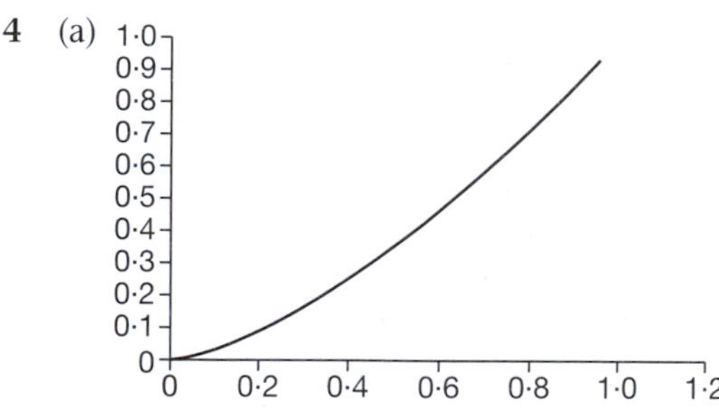


(b)

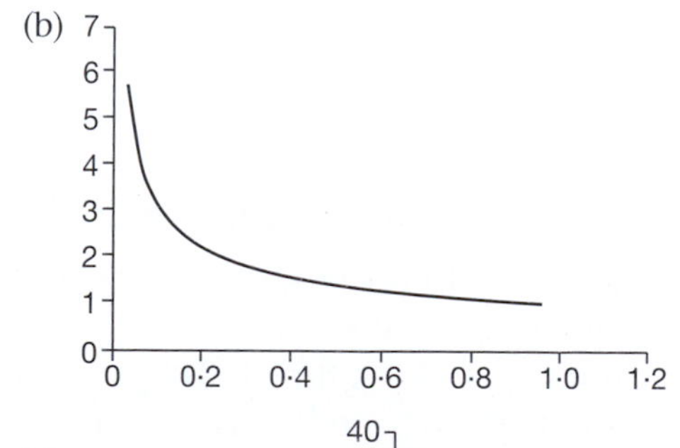


(c)

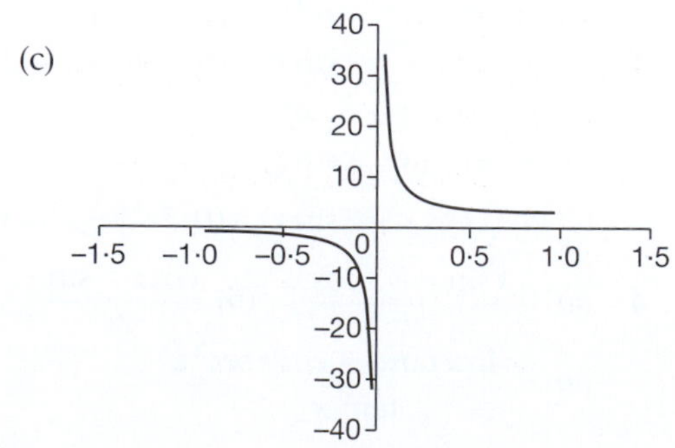


(d)

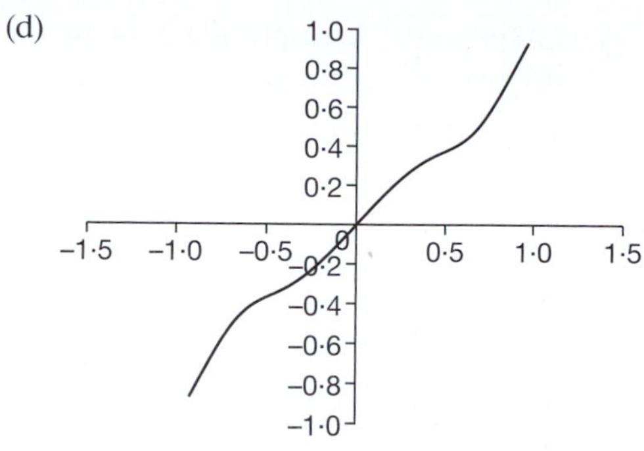


(e)

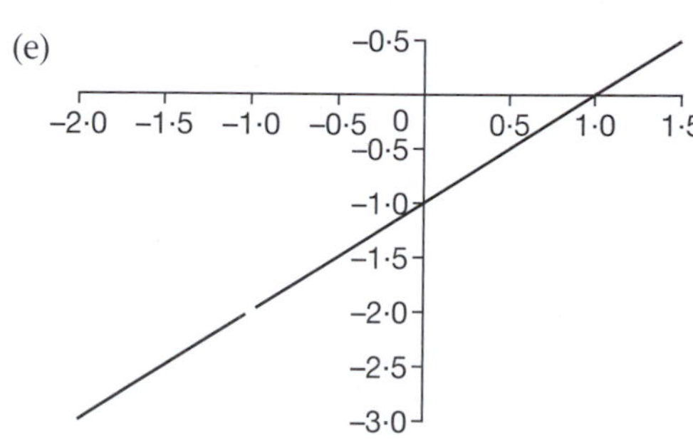


(f)

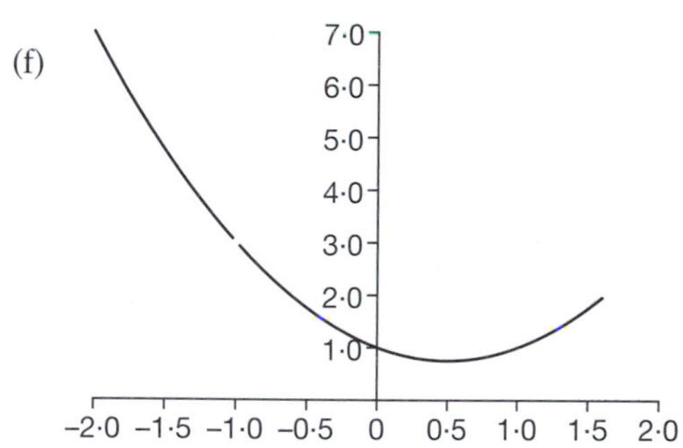


5 (a) $-\dfrac{3\sqrt{x}}{2} - 5$ (b) $3(x-5)-5 = 3x-20$
(c) $x-20$ (d) $\sqrt[8]{x}$

6 (a) $d(c[b(a[x])])$ where $a(x) = 3x$, $b(x) = x-1$, $c(x) = x^3$, $d(x) = 5x$
$a^{-1}(x) = \dfrac{x}{3}$, $b^{-1}(x) = x+1$, $c^{-1}(x) = x^{1/3}$,
$d^{-1}(x) = \dfrac{x}{5}$, $f^{-1}(x) = \dfrac{\left(\dfrac{x}{5}\right)^{1/3}+1}{3}$
(b) $c(d[b(a[x])])$, $f^{-1}(x) = \dfrac{\dfrac{x^{1/3}}{5}+1}{3}$
(c) $b(d[c(x)])$, $f^{-1}(x) = \left(\dfrac{x+1}{5}\right)^{1/3}$
(d) $c(b[x])$, $f^{-1}(x) = x^{1/3}+1$

7 $f(x) = x$

8 No

## Quiz F.11 (page 288)

1 (a) 0 (b) 0 (c) undefined

2 (a) 0·5 (b) −2·3662 (c) −0·2309

3 (a) period $\pi/3$, amplitude 2, phase 0
(b) period $\pi/2$, amplitude $\infty$, phase $-1$
(c) period $2\pi$, amplitude 1, phase $-\pi$
4 (a) 2 (b) 2 (c) 0
5 (a) $1{\cdot}440 \pm n\pi$ (b) $\pi/4 \pm n\pi$, $1{\cdot}249 \pm n\pi$
(c) 0·86
6 (a) 0 (b) −0·75 (c) 1·1763
(d) 1·3195 (e) 7·3891
7 (a) 2·6021 (b) 1, 3
(c) 0·5848 (d) ±1·732
8 $f_e(x) = x^2$ and $f_o(x) = x(x^2+1)$
9 (a) 6 (b) 2/3 (c) $-1/2e$

## Test exercise F.11 (page 313)

1 (a) 1 (b) 1 (c) 0
2 (a) 0·6428 (b) 3·5495 (c) 0
3 (a) period $2\pi/7$, amplitude 4, phase 0
(b) period $\pi$, amplitude 2, phase 0 relative to $\cos 2\theta$
(c) period $2\pi/3$, amplitude $\infty$, phase 4/3
4 (a) 3 (b) 4·5 (c) 0
5 (a) $\pm n\pi$
(b) $\pm 35{\cdot}26^\circ \pm n \times 360^\circ$, $\pm 30^\circ \pm n \times 360^\circ$
(c) $1{\cdot}380 \pm 2n\pi$ radians
6 (a) 0 (b)1/3 (c) 3·9949 (d) 625
(e) 4·2618 (f) 4·6052
7 (a) 2 (b) 0·6931, 1·0986 (c) 6
(d) ±32/81 (e) 1/9
8 $f_e(x) = \dfrac{a^x + a^{-x}}{2}$ and $f_o(x) = \dfrac{a^x - a^{-x}}{2}$
9 (a) 0 (b) 6 (c) −5/28 (d) −1/9 (e) $\pi/4$

## Further problems F.11 (page 314)

1 $x = \sqrt{3}$
2 (a) domain $-1 < x < 1$, range $-\infty < f(x) < \infty$
3 the two straight lines $y = x/3$ and $y = -x/3$
6 yes; consider $a = \sqrt{2}$ and $x = 2$ as an example
7 (a) $f^{-1}(x) = \left(\dfrac{x}{6}+2\right)^{\frac{1}{3}}$
(b) $f^{-1}(x) = \left(x^{\frac{1}{3}}+2\right)^{\frac{1}{3}}$
(c) $f^{-1}(x) = \left(\dfrac{(x+2)^{\frac{1}{3}}}{6}+2\right)^{\frac{1}{3}}$
10 (b) $f'(x) = \begin{cases} -1 & \text{if } x < 0 \\ 1 & \text{if } x > 0 \end{cases}$
(c) $f'(0)$ does not exist
13 (a) −5·702 (b) 0·8594 (c) −2·792
14 (a) 0 (b) −4 (c) 1/2

## Quiz F.12 (page 316)

1 (a) 2 (b) −11/9 (c) 2
2 14·4
3 $12x^3 - 15x^2 + 2x + 2$, 538
4 (a) (i) $12x^2 + 2x - 6$, $24x + 2$
(ii) 0·629 or −0·795 to 3 dp
(b) $-18\sin(9x-1) - 3\cos(3x+7)$, $-162\cos(9x-1) + 9\sin(3x+7)$
5 (a) $x^3(4\cos x - x\sin x)$ (b) $x(2\ln x + 1)$
(c) $\dfrac{x\cos x - \sin x}{x^2}$ (d) $\dfrac{3}{x^5}(1 - 4\ln x)$
6 (a) $15(3x-2)^4$ (b) $6\cos(3x-1)$
(c) $-2e^{(2x+3)}$ (d) $\dfrac{2}{2x-3}$
(e) $e^{2x}(2\cos 3x - 3\sin 3x)$
(f) $\dfrac{\sin(2-5x)}{x} - 5(\ln x)\cos(2-5x)$
7 1·442250

## Test exercise F.12 (page 351)

1 (a) $\dfrac{4}{3}$ (b) $-\dfrac{2}{5}$ (c) $-\dfrac{6}{7}$ (d) $-\dfrac{13}{5}$
2 7·2
3 $\dfrac{dy}{dx} = 4x^3 + 15x^2 - 12x + 7$, $\dfrac{dy}{dx} = 59$
4 (a) (i) $6x^2 - 22x + 12$, $12x - 22$ (ii) $\dfrac{2}{3}$, 3
(b) $\dfrac{dy}{dx} = 6\cos(2x+1) - 12\sin(3x-1)$,
$\dfrac{d^2y}{dx^2} = -12\sin(2x+1) - 36\cos(3x-1)$
5 (a) $x(x\cos x + 2\sin x)$ (b) $x^2e^x(x+3)$
(c) $\dfrac{-x\sin x - 2\cos x}{x^3}$ (d) $\dfrac{2e^x(\tan x - \sec^2 x)}{\tan^2 x}$
6 (a) $20(5x+2)^3$ (b) $3\cos(3x+2)$
(c) $4e^{4x-1}$ (d) $-10\sin(2x+3)$
(e) $-3\cos^2 x.\sin x$ (f) $\dfrac{4}{4x-5}$
7 1·179509

## Further problems F.12 (page 352)

1 (a) $m = -4$ (b) $m = 25$ (c) $m = 14$
2 (a) $6x^2 + 8x - 2$, 6
(b) $12x^3 - 15x^2 + 8x - 1$, 212
(c) $20x^4 + 8x^3 - 9x^2 + 14x - 2$, 31
3 (a) $x^4(x\cos x + 5\sin x)$ (b) $e^x(\cos x - \sin x)$
(c) $x^2(x\sec^2 x + 3\tan x)$
(d) $x^3(4\cos x - x\sin x)$
(e) $5x(x\cos x + 2\sin x)$ (f) $2e^x\left(\dfrac{1}{x} + \ln x\right)$
4 (a) $\dfrac{-x\sin x - 3\cos x}{x^4}$ (b) $\dfrac{\cos x - \sin x}{2e^x}$
(c) $\dfrac{-\sin x\tan x - \cos x\sec^2 x}{\tan^2 x}$

(d) $\frac{4x^2(3\cos x + x\sin x)}{\cos^2 x}$ (e) $\frac{\sec^2 x - \tan x}{e^x}$
(f) $\frac{1 - 3\ln x}{x^4}$

**5** (a) $6(2x-3)^2$ (b) $3e^{3x+2}$
(c) $-12\sin(3x+1)$ (d) $\frac{2x}{x^2+4}$
(e) $2\cos(2x-3)$ (f) $2x\sec^2(x^2-3)$
(g) $40(4x+5)$ (h) $12x.e^{x^2+2}$ (i) $\frac{2}{x}$
(j) $-15\cos(4-5x)$

**6** (a) $-2{\cdot}456$ (b) $1{\cdot}765$ (c) $0{\cdot}739$
(d) $1{\cdot}812$ (e) $1{\cdot}8175$ (f) $0{\cdot}5170$
(g) $0{\cdot}8449$ (h) $0{\cdot}8806$

## Quiz F.13 (page 354)

**1** (a) $\frac{x^8}{8} + C$ (b) $4\sin x + C$ (c) $2e^x + C$
(d) $12x + C$ (e) $-\frac{x^{-3}}{3} + C$ (f) $\frac{6^x}{\ln 6} + C$
(g) $\frac{27}{4}x^{\frac{4}{3}} + C$ (h) $4\tan x + C$
(i) $-9\cos x + C$ (j) $8\ln x + C$

**2** (a) $I = \frac{x^4}{4} - \frac{x^3}{3} + \frac{x^2}{2} - x + C$
(b) $I = x^4 - 3x^3 + 4x^2 - 2x + 1$

**3** (a) $\frac{(5x-1)^5}{25} + C$ (b) $-\frac{\cos(6x-1)}{18} + C$
(c) $-\frac{(4-2x)^{\frac{3}{2}}}{3} + C$ (d) $\frac{2e^{3x+2}}{3} + C$
(e) $-\frac{5^{1-x}}{\ln 5} + C$ (f) $\frac{3\ln(2x-3)}{2} + C$
(g) $-\frac{\tan(2-5x)}{25} + C$

**4** (a) $\frac{\ln(3x+1)}{3} + \frac{\ln(2x-1)}{2} + C$
(b) $-\frac{\ln(1-4x)}{2} - 3\ln(x+2) + C$
(c) $2\frac{\ln(1+3x)}{3} + \frac{\ln(1-3x)}{3} + C$

**5** $e(e-1)$

**6** $5{\cdot}171$ units$^2$

## Test exercise F.13 (page 380)

**1** (a) $\frac{x^5}{5} + C$ (b) $-3\cos x + C$ (c) $4e^x + C$
(d) $6x + C$ (e) $2x^{\frac{3}{2}} + C$ (f) $5^x/\ln 5 + C$
(g) $-\frac{1}{3x^3} + C$ (h) $4\ln x + C$ (i) $3\sin x + C$
(j) $2\tan x + C$

**2** (a) $2x^4 + 2x^3 - \frac{5}{2}x^2 + 4x + C$ (b) $I = 236$

**3** (a) $\frac{-2\cos(3x+1)}{3} + C$ (b) $\frac{-(5-2x)^{\frac{3}{2}}}{3} + C$
(c) $-2e^{1-3x} + C$ (d) $\frac{(4x+1)^4}{16} + C$
(e) $\frac{4^{2x-3}}{2\ln 4} + C$ (f) $-3\sin(1-2x) + C$
(g) $\frac{5\ln(3x-2)}{3} + C$ (h) $\frac{3\tan(1+4x)}{4} + C$

**4** $x + \frac{1}{2}\ln(2x+1) + 2\ln(x+4) + C$

**5** (a) $95{\cdot}43$ unit$^2$ (b) $36$ unit$^2$

## Further problems F.13 (page 381)

**1** (a) $-\frac{(5-6x)^3}{18} + C$ (b) $-\frac{4\cos(3x+2)}{3} + C$
(c) $\frac{5\ln(2x+3)}{2} + C$ (d) $\frac{1}{2}3^{2x-1}/\ln 3 + C$
(e) $-\frac{2}{9}(5-3x)^{\frac{3}{2}} + C$ (f) $2e^{3x+1} + C$
(g) $\frac{1}{3}(4-3x)^{-1} + C$ (h) $-2\sin(1-2x) + C$
(i) $-\tan(4-3x) + C$ (j) $-2e^{3-4x} + C$

**2** $2e^{3x-5} + \frac{4}{3}\ln(3x-2) - \frac{1}{2}5^{2x+1}/\ln 5 + C$

**3** $I = 95$

**4** (a) $\frac{1}{2}\ln(2x-1) + \ln(2x+3) + C$
(b) $-2\ln(x+1) + 3\ln(x-4) + C$
(c) $-\ln(3x-4) + \frac{5}{4}\ln(4x-1) + C$
(d) $2\ln(2x-5) + \frac{1}{2}\ln(4x+3) + C$
(e) $x + \frac{3}{2}\ln(2x-1) - \frac{2}{3}\ln(3x-2) + C$
(f) $2x + \ln(x-4) + \frac{3}{2}\ln(2x+1) + C$
(g) $3x - 3\ln(x+2) + \frac{5}{2}\ln(2x-3) + C$
(h) $2x + 5\ln(x-4) + \ln(2x+1) + C$

**5** (a) $13\frac{1}{3}$ unit$^2$ (b) $60$ unit$^2$ (c) $14$ unit$^2$
(d) $57$ unit$^2$ (e) $33{\cdot}75$ unit$^2$

**6** (a) $21\frac{1}{3}$ unit$^2$ (b) $45\frac{1}{3}$ unit$^2$ (c) $791$ unit$^2$
(d) $196$ unit$^2$ (e) $30{\cdot}6$ unit$^2$

# Part II

## Test exercise 1 (page 409)

**1** (a) $-j$ (b) $j$ (c) $1$ (d) $-1$

**2** (a) $29 - j2$ (b) $-j2$ (c) $111 + j56$
(d) $1 + j2$

**3** $x = 10{\cdot}5,\ y = 4{\cdot}3$

**4** (a) $5{\cdot}831\underline{|59^\circ 2'}$ (b) $6{\cdot}708\underline{|153^\circ 26'}$
(c) $6{\cdot}403\underline{|231^\circ 20'}$

**5** (a) $-3{\cdot}5355(1+j)$ (b) $3{\cdot}464 - j2$

6 (a) $10e^{j0{\cdot}650}$
(b) $10e^{-j0{\cdot}650}$, $2{\cdot}303 + j0{\cdot}650$, $2{\cdot}303 - j0{\cdot}650$

7 $je$

## Further problems 1 (page 410)

1 (a) $115 + j133$ (b) $2{\cdot}52 + j0{\cdot}64$ (c) $\cos 2x + j\sin 2x$

2 $(22 - j75)/41$

3 $0{\cdot}35 + j0{\cdot}17$

4 $0{\cdot}7$, $0{\cdot}9$

5 $-24{\cdot}4 + j22{\cdot}8$

6 $1{\cdot}2 + j1{\cdot}6$

8 $x = 18$, $y = 1$

9 $a = 2$, $b = -20$

10 $x = \pm 2$, $y = \pm 3/2$

12 $a = 1{\cdot}5$, $b = -2{\cdot}5$

13 $\sqrt{2}e^{j2{\cdot}3562}$

14 $2{\cdot}6$

16 $R = (R_2C_3 - R_1C_4)/C_4$, $L = R_2R_4C_3$

18 $E = (1811 + j1124)/34$

20 $2 + j3$, $-2 + j3$

## Test exercise 2 (page 434)

1 $5{\cdot}831\lfloor\underline{210^\circ 58'}$

2 (a) $-1{\cdot}827 + j0{\cdot}813$ (b) $3{\cdot}993 - j3{\cdot}009$

3 (a) $36\lfloor\underline{197^\circ}$ (b) $4\lfloor\underline{53^\circ}$

4 $8\lfloor\underline{75^\circ}$

5 $2\lfloor\underline{88^\circ}$, $2\lfloor\underline{208^\circ}$, $2\lfloor\underline{328^\circ}$;
principal root $= 2\lfloor\underline{328^\circ}$

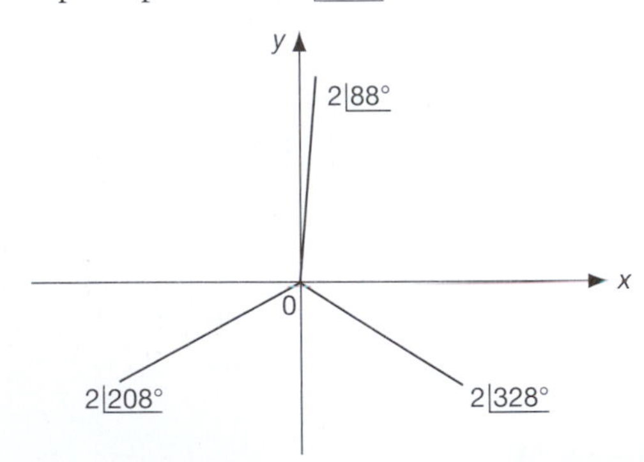


6 $\sin 4\theta = 4\sin\theta\cos^3\theta - 4\sin^3\theta\cos\theta$

7 $\cos^4\theta = \dfrac{1}{8}[\cos 4\theta + 4\cos 2\theta + 3]$

8 (a) $x^2 + y^2 - 8x + 7 = 0$ (b) $y = \dfrac{x + 2}{\sqrt{3}}$

## Further problems 2 (page 435)

1 $x = 0{\cdot}27$, $y = 0{\cdot}53$

2 $-3 + j\sqrt{3}$; $-j2\sqrt{3}$

3 $3{\cdot}606\lfloor\underline{56^\circ 19'}$, $2{\cdot}236\lfloor\underline{296^\circ 34'}$; $24{\cdot}2 - j71{\cdot}6$; $75{\cdot}6e^{-j1{\cdot}245}$

4 $1{\cdot}335(\lfloor\underline{27^\circ}, \lfloor\underline{99^\circ}, \lfloor\underline{171^\circ}, \lfloor\underline{243^\circ}, \lfloor\underline{315^\circ})$;
$1{\cdot}335(e^{j0{\cdot}4712}, e^{j1{\cdot}7279}, e^{j2{\cdot}9845}, e^{-j2{\cdot}0420}, e^{-j0{\cdot}7854})$

5 $2{\cdot}173 + j0{\cdot}898$, $2{\cdot}351e^{j0{\cdot}392}$

6 $\sqrt{2}(1 + j)$, $\sqrt{2}(-1 + j)$, $\sqrt{2}(-1 - j)$, $\sqrt{2}(1 - j)$

7 $1\lfloor\underline{36^\circ}$, $1\lfloor\underline{108^\circ}$, $1\lfloor\underline{180^\circ}$, $1\lfloor\underline{252^\circ}$, $1\lfloor\underline{324^\circ}$;
$e^{j0{\cdot}6283}$

8 $x = -4$ and $x = 2 \pm j3{\cdot}464$

9 $1\lfloor\underline{102^\circ 17'}$, $1\lfloor\underline{222^\circ 17'}$, $1\lfloor\underline{342^\circ 17'}$;
$0{\cdot}953 - j0{\cdot}304$

11 $1{\cdot}4(\lfloor\underline{58^\circ 22'}, \lfloor\underline{130^\circ 22'}, \lfloor\underline{202^\circ 22'}, \lfloor\underline{274^\circ 22'}, \lfloor\underline{346^\circ 22'})$
principal root $= 1{\cdot}36 - j0{\cdot}33 = 1{\cdot}4e^{-j0{\cdot}2379}$

12 $-0{\cdot}36 + j0{\cdot}55$, $-1{\cdot}64 - j2{\cdot}55$

13 $-je$, i.e. $-j2{\cdot}718$

14 $\sin 7\theta = 7s - 56s^3 + 112s^5 - 64s^7$ $(s \equiv \sin\theta)$

15 $\dfrac{1}{32}[10 - 15\cos 2x + 6\cos 4x - \cos 6x]$

16 $x^2 + y^2 + \dfrac{20}{3}x + 4 = 0$; centre $\left(-\dfrac{10}{3}, 0\right)$, radius $8/3$

17 $x^2 + y^2 - (1 + \sqrt{3})x - (1 + \sqrt{3})y + \sqrt{3} = 0$,
centre $\left(\dfrac{1 + \sqrt{3}}{2}, \dfrac{1 + \sqrt{3}}{2}\right)$, radius $\sqrt{2}$

18 $x^2 + y^2 = 16$

19 (a) $2x^2 + 2y^2 - x - 1 = 0$
(b) $x^2 + y^2 + 2x + 2y = 0$

20 (a) $x^2 + y^2 - 4x = 0$
(b) $x^2 + y^2 + x - 2 = 0$

22 (a) $y = 3$ (b) $x^2 + y^2 = 4k^2$

## Test exercise 3 (page 457)

1

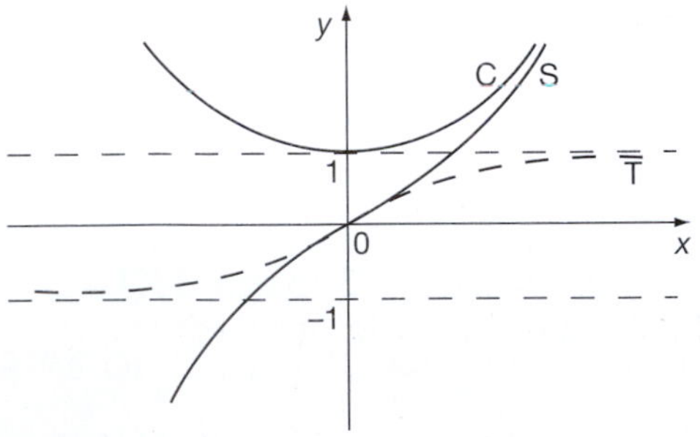

2 $67{\cdot}25$

3 $\pm 19{\cdot}40$

4 (a) $1{\cdot}2124$ (b) $\pm 0{\cdot}6931$

5 $x = 0{\cdot}3466$

6 (a) $y = 224$ (b) $x = \pm 48{\cdot}12$

7 $-\coth A$

8 $\sin x\cosh y - j\cos x\sinh y$

## Further problems 3 (page 457)

2 $x = 0,\ x = 0{\cdot}549$
5 (a) $0{\cdot}9731$ (b) $\pm 1{\cdot}317$
7 (a) $0{\cdot}9895 + j0{\cdot}2498$ (b) $0{\cdot}4573 + j0{\cdot}5406$
10 $x = 0,\ x = \dfrac{1}{2}\ln 2$
12 $x = 0{\cdot}3677$ or $-1{\cdot}0986$
14 $1{\cdot}528 + j0{\cdot}427$
17 $1{\cdot}007$

## Test exercise 4 (page 485)

1 (a) 4 (b) 18
2 $x = 2,\ y = -1,\ z = 4$
3 $x = 3,\ y = -2,\ z = -1$
4 $k = 3$ or $-25$
5 $x = 3,\ 1{\cdot}653,\ -6{\cdot}653$

## Further problems 4 (page 486)

1 (a) 144 (b) 0
2 (a) 0 (b) 666
3 $x = 5,\ y = 4,\ z = -2$
4 $x = 2{\cdot}5,\ y = 3,\ z = -4$
5 $x = 2{\cdot}286,\ y = 1{\cdot}095,\ z = -3{\cdot}476$
6 4 or $-14$
7 5 or $-2{\cdot}7$
8 (a) 0 or $\pm\sqrt{2}$
(b) $(a-b)(b-c)(c-a)(a+b+c)$
9 $x = 1$ or $x = -5 \pm \sqrt{34}$
10 $x = -1{\cdot}5$
11 $2(a-b)(b-c)(c-a)(a+b+c)$
12 $i_2 = 5{\cdot}2$
13 $(a+b+c)^2(a-b)(b-c)(c-a)$
14 2 or $-16/3$
15 $(x-y)(y-z)(z-x)(x+y+z)$
16 $x = -3$ or $\pm\sqrt{3}$
17 $x = \dfrac{(2M_1 + M_2)W}{M_1(M_1 + 2M_2)}$
18 $i_1 = 0,\ i_2 = 2,\ i_3 = 3$
20 $\theta = \dfrac{7\pi}{12}$ or $\dfrac{11\pi}{12}$

## Test exercise 5 (page 515)

1 (a) $\begin{pmatrix} 5 & 9 & 8 & 10 \\ 10 & 8 & 6 & 7 \end{pmatrix}$
(b) $\begin{pmatrix} -1 & -1 & 4 & -4 \\ -8 & 6 & -6 & 1 \end{pmatrix}$
2 (a) $\begin{pmatrix} 18 & 0 & 12 \\ 3 & 15 & -9 \end{pmatrix}$ (b) $\begin{pmatrix} -4 & 82 \\ 54 & -12 \end{pmatrix}$
(c) $\begin{pmatrix} 21 & 45 & -19 \\ 48 & 0 & 32 \\ -17 & 35 & -37 \end{pmatrix}$
3 $\mathbf{A}^{\mathrm{T}} = \begin{pmatrix} 2 & 1 & 8 \\ 3 & 7 & 0 \\ 5 & 4 & 6 \end{pmatrix}$ $\mathbf{A}^{\mathrm{T}}.\mathbf{I} = \begin{pmatrix} 2 & 1 & 8 \\ 3 & 7 & 0 \\ 5 & 4 & 6 \end{pmatrix}$
i.e. $\mathbf{A}^{\mathrm{T}}$
4 $|a| = 0$
5 (a) 105 (b) $\begin{pmatrix} -35 & 21 & 14 \\ 5 & -3 & 13 \\ 40 & -3 & -22 \end{pmatrix}$
6 $\begin{pmatrix} 7{\cdot}5 & -1{\cdot}5 & -4{\cdot}75 \\ -4 & 1 & 2{\cdot}5 \\ -2{\cdot}5 & 0{\cdot}5 & 1{\cdot}75 \end{pmatrix}$
7 $\begin{pmatrix} 2 & 4 & -5 \\ 1 & -3 & 1 \\ 3 & 5 & 3 \end{pmatrix} \cdot \begin{pmatrix} x_1 \\ x_2 \\ x_3 \end{pmatrix} = \begin{pmatrix} -7 \\ 10 \\ 2 \end{pmatrix}$
8 $x_1 = 2;\ x_2 = -3;\ x_3 = 5$
9 (b) $x_1 = -1;\ x_2 = -3;\ x_3 = 4$
10 $\lambda_1 = 1,\ \mathbf{x}_1 = \{-2 \quad 1 \quad 0\},\ \lambda_2 = 2,$
$\mathbf{x}_2 = \{-2 \quad 1 \quad 1\},\ \lambda_3 = 4,\ \mathbf{x}_3 = \{0 \quad 1 \quad 1\}$

## Further problems 5 (page 516)

1 (a) $\begin{pmatrix} 11 & 8 \\ 8 & 9 \end{pmatrix}$ (b) $\begin{pmatrix} 3 & -4 \\ -2 & -7 \end{pmatrix}$
(c) $\begin{pmatrix} 38 & 58 \\ 17 & 26 \end{pmatrix}$ (d) $\begin{pmatrix} 46 & 14 \\ 59 & 18 \end{pmatrix}$
2 (a) **C** (b) **A** (c) **B** (d) $-\mathbf{I}$
3 (a) $\begin{pmatrix} -3 & 2 \\ 2 & -1 \end{pmatrix}$ (b) $\begin{pmatrix} 2 & 3{\cdot}5 \\ 0{\cdot}7 & 1{\cdot}3 \end{pmatrix}$
(c) $\begin{pmatrix} -2 & -1{\cdot}3 \\ 1{\cdot}5 & 0{\cdot}9 \end{pmatrix}$
4 $k = -2$
5 (a) $\begin{pmatrix} 2 & -3 & -1 \\ 1 & 4 & 2 \\ 1 & -1 & 1 \end{pmatrix} \cdot \begin{pmatrix} x_1 \\ x_2 \\ x_3 \end{pmatrix} = \begin{pmatrix} 2 \\ 3 \\ 5 \end{pmatrix}$
(b) $\begin{pmatrix} 1 & -2 & -1 & 3 \\ 2 & 3 & 0 & 1 \\ 1 & 0 & -4 & -2 \\ 0 & -1 & 3 & 1 \end{pmatrix} \cdot \begin{pmatrix} x_1 \\ x_2 \\ x_3 \\ x_4 \end{pmatrix} = \begin{pmatrix} 10 \\ 8 \\ 3 \\ -7 \end{pmatrix}$
6 $i_1 = 2;\ i_2 = 3;\ i_3 = 1$
7 $x = 1;\ y = 2;\ z = -1$
8 $i_1 = 2;\ i_2 = 1;\ i_3 = 0$
9 no unique solution
10 $x_1 = 2;\ x_2 = 1{\cdot}5;\ x_3 = -3{\cdot}5$
11 $i_1 = 0{\cdot}5;\ i_2 = 1{\cdot}5;\ i_3 = 1{\cdot}0$
12 $i_1 = 3{\cdot}0;\ i_2 = 2{\cdot}5;\ i_3 = -4{\cdot}0$
13 $i_1 = 2{\cdot}26;\ i_2 = 0{\cdot}96;\ i_3 = 0{\cdot}41$
14 $i_1 = 12{\cdot}5;\ i_2 = 7{\cdot}5;\ i_3 = -20$
15 $i_1 = \dfrac{(Z_2 + Z_3)V}{Z};\ i_2 = \dfrac{Z_3 V}{Z};\ i_3 = \dfrac{Z_2 V}{Z};$
where $Z = Z_1Z_2 + Z_2Z_3 + Z_3Z_1$

16 $\lambda_1 = 1\ \mathbf{x}_1 = \{0\ \ 1\ \ -1\}$,
$\lambda_2 = 2\ \mathbf{x}_2 = \{1\ \ 1\ \ -1\}$,
$\lambda_3 = 4\ \mathbf{x}_3 = \{3\ \ 5\ \ 1\}$

17 $\lambda_1 = -1\ \mathbf{x}_1 = \{1\ \ 0\ \ -1\}$,
$\lambda_2 = 1\ \mathbf{x}_2 = \{1\ \ -1\ \ 1\}$,
$\lambda_3 = 5\mathbf{x}_3 = \{1\ \ 1\ \ 1\}$

18 $\lambda_1 = 1\ \ \mathbf{x}_1 = \{1\ \ 0\ \ -1\}$,
$\lambda_2 = 2\ \mathbf{x}_2 = \{2\ \ 1\ \ 0\}$,
$\lambda_3 = 3\ \mathbf{x}_3 = \{1\ \ 2\ \ 1\}$

19 $\lambda_1 = 1\ \mathbf{x}_1 = \{4\ \ 1\ \ -2\}$,
$\lambda_2 = 3\ \mathbf{x}_2 = \{-2\ \ 1\ \ 0\}$,
$\lambda_3 = 4\ \mathbf{x}_3 = \{-2\ \ 1\ \ 1\}$

20 $\lambda_1 = -2\ \mathbf{x}_1 = \{3\ \ 3\ \ -5\}$,
$\lambda_2 = 3\ \mathbf{x}_2 = \{3\ \ -2\ \ 0\}$,
$\lambda_3 = 6\ \mathbf{x}_3 = \{1\ \ 1\ \ 1\}$

## Test exercise 6 (page 542)

1 $\overline{AB} = 2\mathbf{i} - 5\mathbf{j}$, $\overline{BC} = -4\mathbf{i} + \mathbf{j}$, $\overline{CA} = 2\mathbf{i} + 4\mathbf{j}$,
AB $= \sqrt{29}$, BC $= \sqrt{17}$, CA $= \sqrt{20}$

2 $l = 3/13$, $m = 4/13$, $n = 12/13$

3 (a) $-8$ (b) $-2\mathbf{i} - 7\mathbf{j} - 18\mathbf{k}$

4 (a) 6, $\theta = 82°44'$ (b) 47·05, $\theta = 19°31'$

## Further problems 6 (page 542)

1 $\overline{OG} = \frac{1}{3}(10\mathbf{i} + 2\mathbf{j})$

2 $\frac{1}{\sqrt{50}}(3, 4, 5)$; $\frac{1}{\sqrt{14}}(1, 2, -3)$; $\theta = 98°42'$

3 moduli: $\sqrt{74}$, $3\sqrt{10}$, $2\sqrt{46}$;
direction cosines: $\frac{1}{\sqrt{74}}(3, 7, -4)$,
$\frac{1}{3\sqrt{10}}(1, -5, -8)$, $\frac{1}{\sqrt{46}}(3, -1, 6)$;
sum = $10\mathbf{i}$, direction cosines (1, 0, 0)

4 8, $17\mathbf{i} - 7\mathbf{j} + 2\mathbf{k}$, $\theta = 66°36'$

5 (a) $-7$ (b) $7(\mathbf{i} - \mathbf{j} - \mathbf{k})$ (c) $\cos\theta = -0{\cdot}5$

6 $\cos\theta = -0{\cdot}4768$

7 (a) 7, $5\mathbf{i} - 3\mathbf{j} - \mathbf{k}$ (b) 8, $11\mathbf{i} + 18\mathbf{j} - 19\mathbf{k}$

8 $-\frac{3}{\sqrt{155}}\mathbf{i} + \frac{5}{\sqrt{155}}\mathbf{j} + \frac{11}{\sqrt{155}}\mathbf{k}$; $\sin\theta = 0{\cdot}997$

9 $\frac{2}{\sqrt{13}}, \frac{-3}{\sqrt{13}}, 0; \frac{5}{\sqrt{30}}, \frac{1}{\sqrt{30}}, \frac{-2}{\sqrt{30}}$

10 $6\sqrt{5}; \frac{-2}{3\sqrt{5}}, \frac{4}{3\sqrt{5}}, \frac{5}{3\sqrt{5}}$

11 (a) 0, $\theta = 90°$ (b) 68·53,
$(-0{\cdot}1459, -0{\cdot}5982, -0{\cdot}7879)$

12 $4\mathbf{i} - 5\mathbf{j} + 11\mathbf{k}$; $\frac{1}{9\sqrt{2}}(4, -5, 11)$

13 (a) $\mathbf{i} + 3\mathbf{j} - 7\mathbf{k}$ (b) $-4\mathbf{i} + \mathbf{j} + 2\mathbf{k}$
(c) $13(\mathbf{i} + 2\mathbf{j} + \mathbf{k})$ (d) $\frac{\sqrt{6}}{6}(\mathbf{i} + 2\mathbf{j} + \mathbf{k})$

14 (a) $-2\mathbf{i} + 2\mathbf{j} - 6\mathbf{k}$ (b) $-40$
(c) $-30\mathbf{i} - 10\mathbf{j} + 30\mathbf{k}$

15 $2\mathbf{i} + 12\mathbf{j} + 34\mathbf{k}$

20 $2x - 3y + z = 0$

## Test exercise 7 (page 560)

1 (a) $2\sec^2 2x$ (b) $30(5x + 3)^5$ (c) $\sinh 2x$
(d) $\frac{2x - 3}{(x^2 - 3x - 1)\ln 10}$ (e) $-3\tan 3x$
(f) $12\sin^2 4x\cos 4x$
(g) $e^{2x}(3\cos 3x + 2\sin 3x)$ (h) $\frac{2x^3(x + 2)}{(x + 1)^3}$
(i) $\frac{e^{4x}\sin x}{x\cos 2x}\left[4 + \cot x - \frac{1}{x} + 2\tan 2x\right]$

2 $\frac{3}{4}, -\frac{25}{64}$

3 $-\frac{3x^2 + 4y^2}{3y^2 + 8xy}$

4 $\frac{1 - \cos\theta}{\sin\theta}, \frac{1 - \cos\theta}{3\sin^3\theta}$

## Further problems 7 (page 560)

1 (a) $\frac{2}{\cos 2x}$ (b) $\sec x$
(c) $4\cos^4 x\sin^3 x - 3\cos^2 x\sin^5 x$

2 (a) $\frac{x\sin x}{1 + \cos x}\left[\frac{1}{x} + \cot x + \frac{\sin x}{1 + \cos x}\right]$
(b) $\frac{-4x}{1 - x^4}$

4 $\frac{y^2 - x^2}{y^2 - 2xy}$

5 (a) $5\sin 10x e^{\sin^2 5x}$ (b) $\frac{2}{\sinh x}$ (c) $\frac{x^2 - 1}{x^2 - 4}$

6 (a) $2x\cos^2 x - 2x^2\sin x\cos x$
(b) $\frac{2}{x} - \frac{x}{1 - x^2}$
(c) $\frac{e^{2x}\ln x}{(x - 1)^3}\left[2 + \frac{1}{x\ln x} - \frac{3}{x - 1}\right]$

8 $-4, -42$

12 $-\frac{1}{\sqrt{3}}, -\frac{8\sqrt{3}}{9}; x^2 + y^2 - 2y = 0$

14 $-\tan\theta; \frac{1}{3a\sin\theta\cos^4\theta}$

15 $-\cot^3\theta; -\cot^2\theta\,\mathrm{cosec}^5\theta$

## Test exercise 8 (page 580)

1 (a) 130° (b) $-37°$

2 (a) $\frac{3}{\sqrt{-9x^2 - 12x - 3}}$

(b) $\dfrac{-1}{x\sqrt{1-x^2}} - \dfrac{\cos^{-1} x}{x^2}$

(c) $\dfrac{2x^2}{4+x^2} + 2x\tan^{-1}\left(\dfrac{x}{2}\right)$ (d) $\dfrac{-3}{\sqrt{9x^2-6x}}$

(e) $\dfrac{-\sin x}{\sqrt{\cos^2 x+1}}$ (f) $\dfrac{5}{1-25x^2}$

3 (a) $y_{max} = 10$ at $x = 1$; $y_{min} = 6$ at $x = 3$; P of I at $(2, 8)$
(b) $y_{max} = -2$ at $x = -1$; $y_{min} = 2$ at $x = 1$
(c) $y_{max} = e^{-1} = 0{\cdot}3679$ at $x = 1$; P of I at $(2, 0{\cdot}271)$

## Further problems 8 (page 580)

1 (a) 1 (b) $2\sqrt{1-x^2}$

3 (a) $\dfrac{2}{\sqrt{x}(1+4x)}$ (b) $\dfrac{2}{1-x^2}$

4 (a) $\left(\dfrac{11}{3}, -\dfrac{250}{27}\right)$ (b) $(-0{\cdot}25, -4{\cdot}375)$

5 $y_{max} = 0$ at $x = \dfrac{1}{3}$; $y_{min} = 4$ at $x = 1$

6 $y_{max}$ at $x = 2$; $y_{min}$ at $x = 3$; P of I at $x = \sqrt{6}$

7 $y_{max} = \dfrac{16}{5}$ at $x = -\dfrac{11}{5}$; $y_{min} = 0$ at $x = 1$

8 $x = 1{\cdot}5$

10 $\dfrac{dy}{dx} = \sqrt{2}.e^{-x}\cos\left(x + \dfrac{\pi}{4}\right)$

11 (a) $y_{max}$ at $\left(\dfrac{2}{3}, \dfrac{1}{27}\right)$, $y_{min}$ at $(1, 0)$;
P of I at $\left(\dfrac{5}{6}, \dfrac{1}{54}\right)$
(b) $y_{max}$ at $(2-\sqrt{2}, 3-2\sqrt{2})$;
$y_{min}$ at $(2+\sqrt{2}, 3+2\sqrt{2})$
(c) P of I at $(n\pi, n\pi)$

12 (a) $\pm 0{\cdot}7071$ (b) $0, 1$ (c) $\pm 1{\cdot}29$

13 $0{\cdot}606$

14 $v = \sqrt{\dfrac{gT}{2w}}$

16 $y_{max} = 0{\cdot}514$

17 $17{\cdot}46$ cm

18 $\theta = 77^\circ$

20 $A = C$, $B = 0$

## Test exercise 9 (page 602)

1 $\theta = 37^\circ 46'$

2 $16y + 5x = 94$, $5y = 16x - 76$

3 $y = x$

4 $y = 2{\cdot}598x - 3{\cdot}849$

5 $R = 477$; $C$: $(-470, 50{\cdot}2)$

6 $R = 5{\cdot}59$; $C$: $(-3{\cdot}5, 2{\cdot}75)$

## Further problems 9 (page 602)

1 $20y = 125x - 363$; $y = 2x$

2 $y + 2x = 2$; $2y = x + 4$; $x = 1$, $y = 0$

3 $\dfrac{x\cos\theta}{13} + \dfrac{y\sin\theta}{5} = 1$;
$5y = 13\tan\theta.x - 144\sin\theta$; ON.OT $= 144$

4 $\dfrac{3x+y}{x+3y}$; $3y + 5x = 14$

5 $R = {y_1}^2/c$

6 $5y + 8x = 43$

7 $a^2\cos^3 t\sin t$

8 $\dfrac{2a^2 - b^2}{a}$; $b$

9 (a) $y = x$; $y = -x$ (b) $R = -\sqrt{2}$ (c) $(1, -1)$

10 (a) $R = -6{\cdot}25$; $C$: $(0, -2{\cdot}25)$
(b) $R = 1$; $C$: $(2, 0)$
(c) $R = -11{\cdot}68$; $C$: $(12{\cdot}26, -6{\cdot}5)$

11 $R = -0{\cdot}177$

14 $R = 2{\cdot}744$

17 $\rho = t$; $(h, k) = (\cos t, \sin t)$

18 $R = -10{\cdot}54$, $C$: $(11, -3{\cdot}33)$

20 (a) $y = \pm\dfrac{x}{\sqrt{2}}$ (c) $R = 0{\cdot}5$

## Test exercise 10 (page 636)

1 (a) 9, 11, $2n+1$ (b) 8, 11, $-4+3n$
(c) 1, 1/2, $16(1/2)^n : n \geq 0$
(d) $-48$, 96, $-3(-2)^n$
(e) 17, 31: $f(n+3) = f(n+2) + f(n+1) + f(n)$ where $f(1) = 1$, $f(2) = 1$, $f(3) = 1$
(f) 208, 568: $f(n+2) = 2[f(n+1) + f(n)]$ where $f(1) = 1$, $f(2) = 1$

2 (a)

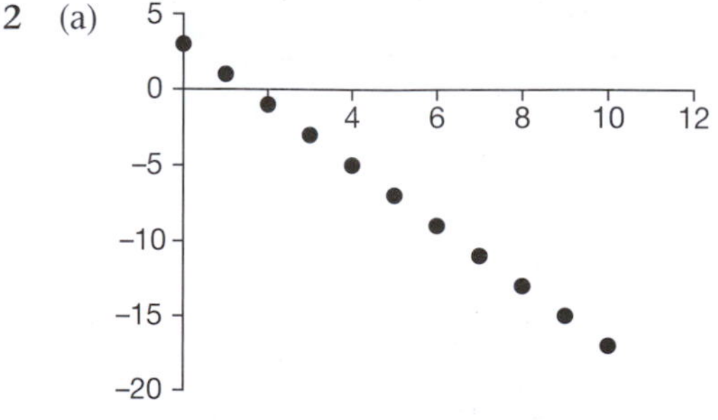


(b)

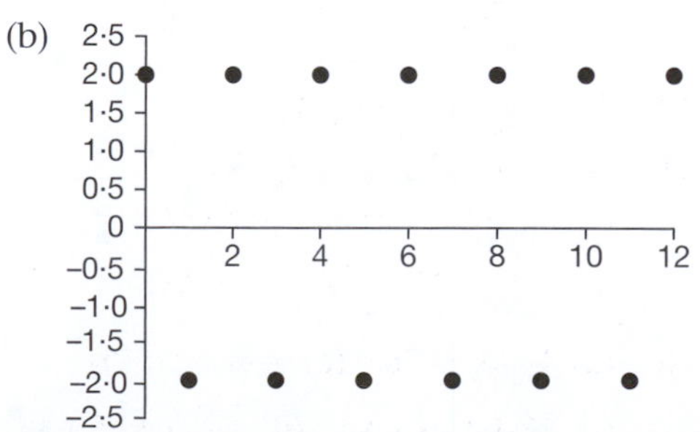

(c)

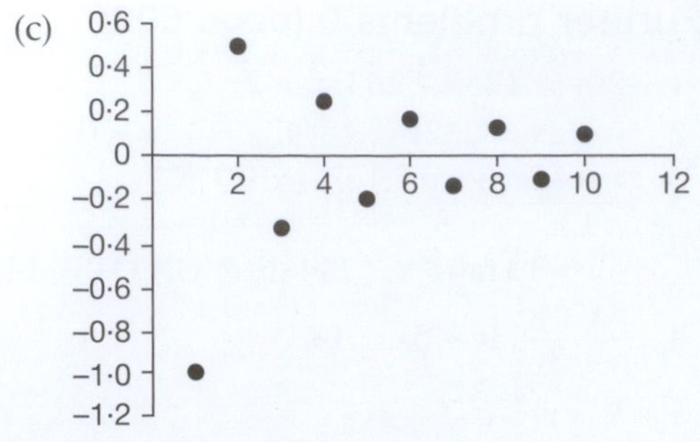

3 (a) $f(n+1) = f(n) + 5 : f(1) = -4$
(b) $f(n+1) = f(n) - 4 : f(0) = 23$
(c) $f(n+1) = \dfrac{f(n)}{3} : f(-2) = 9$

4 (a) $f(n) = 5^n$ (b) $f(n) = \frac{1}{7}\{3^{n+1} + 4(-4)^n\}$
(c) $f(n) = n8^{n-1}$

5 (a) $-\infty$ (b) $\infty$ (c) 0 (d) undefined
(e) 1 (f) 1

6 (a) diverges (b) converges to 0
(c) converges to 0·5

## Further problems 10 (page 637)

1 (a) 21, 26, $5n+1$
(b) $-93$, $-91$, $-101+2n$
(c) 0·001, 0·0001, $10^{-n+1}$
(d) 1·2345, 0·12345, $1234{\cdot}5(10)^{-n}$
(e) 14, $-22$: $f(n+2) = f(n) - f(n+1)$ where $f(1) = 2$, $f(2) = 4$
(f) 34, 55: $f(n+2) = f(n+1) + f(n)$ where $f(1) = 5$, $f(2) = 8$

2 (a)

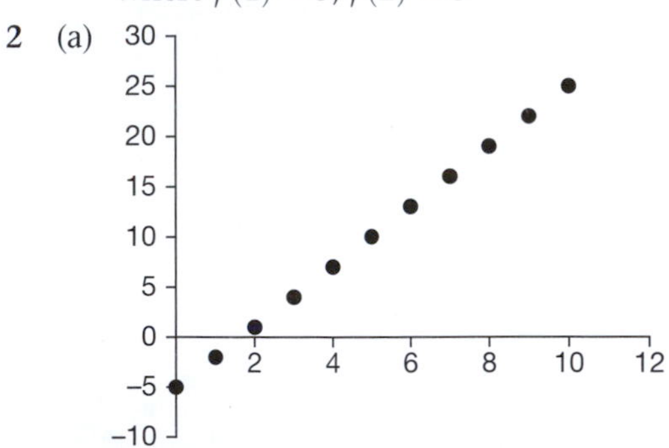

(b)

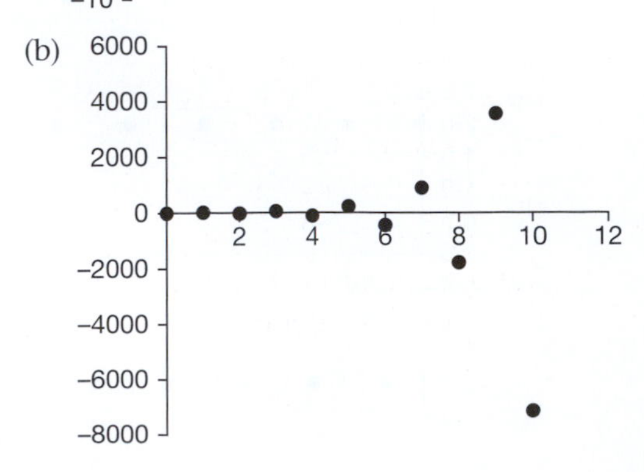

(c)

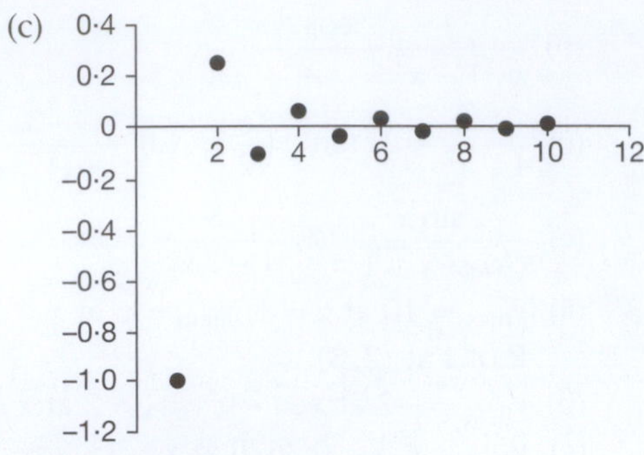

3 (a) $f(n) = (-2)^n - (-3)^n$
(b) $f(n) = \frac{1}{5}\{6 \times 3^{-n} - 2^n\}$
(c) $f(n) = \left\{3(-7)^{n-1} + 4(7)^{n-1}\right\}$

4 (a) $\infty$ (b) $\infty$ (c) undefined (d) 0
(e) 1 (f) $-1/5$

5 (a) diverges (b) diverges
(c) converges to $-0{\cdot}125$ (d) converges to 0
(e) diverges

6 26

7 12, 18, 27: $8(3/2)^n$

8 20, 40, 80

9 $n = 196$

10 (a) 1 (b) $-1$ (c) 0 (d) undefined

12 First term $b-c$, common difference $b+c$

13 39, 51, 63, 75

14 The reciprocals form an arithmetic sequence

15 3/7

19 $g(n) = \frac{1}{5}\{3^n - (-2)^n\}$: $f(n) = g(n-1)$

20 $f(n) = \dfrac{1}{\sqrt{5}}\left\{\left(\dfrac{1+\sqrt{5}}{2}\right)^n - \left(\dfrac{1-\sqrt{5}}{2}\right)^n\right\}$

## Test exercise 11 (page 660)

1 230

2 2·488, 25·958

3 1812

4 (a) convergent (b) divergent
(c) divergent (d) convergent

5 (a) convergent for all values of $x$
(b) convergent for $-1 \le x \le 1$
(c) convergent for $-1 \le x \le 1$

## Further problems 11 (page 661)

1 $\dfrac{n}{3}(4n^2 - 1)$

2 $\dfrac{n(3n+1)}{4(n+1)(n+2)}$

3 $\dfrac{n}{4}(n+1)(n+4)(n+5)$

4 (a) $\frac{n}{3}(n+1)(n+5)$
(b) $\frac{1}{4}(n^2+3n)(n^2+3n+4)$

5 2

6 $S_n = \frac{10}{3}\left\{1+\frac{(-1)^{n+1}}{2^n}\right\};\ S_\infty = \frac{10}{3}$

7 (a) 0·6 (b) 0·5

8 (a) diverges (b) diverges (c) converges (d) converges

9 $-1 \leq x \leq 1$

11 $-1 \leq x \leq 1$

12 All values of $x$

13 $0 < x \leq 1$

16 (a) convergent (b) divergent (c) divergent (d) divergent

18 (a) convergent (b) convergent

19 $1 \leq x \leq 3$

20 $\frac{n}{6}(n+1)(4n+5)+2^{n+2}-4$

## Test exercise 12 (page 685)

1 $f(x) = f(0) + xf'(0) + \frac{x^2}{2!}f''(0) + \dots$

2 $1 - x^2 + \frac{x^4}{3} - \frac{2x^6}{45} + \dots$

3 $1 + \frac{x^2}{2} + \frac{5x^4}{24} + \dots$

5 $x + x^2 + \frac{5x^3}{6} + \frac{x^4}{2} + \dots$

6 1·0247

7 (a) $-\frac{1}{10}$ (b) $\frac{2}{9}$ (c) $-\frac{1}{2}$

8 0·85719

## Further problems 12 (page 685)

3 (a) $-\frac{1}{10}$ (b) $\frac{1}{3}$ (c) $\frac{1}{2}$ (d) $-\frac{1}{6}$ (e) 2

6 $-\frac{1}{4}$

7 $-\frac{3}{2} - \frac{5x}{2} - \frac{11x^2}{4} - \frac{13x^3}{4}$

9 $\frac{2}{3}$

10 (a) $-\frac{1}{6}$ (b) $\frac{1}{2}$ (c) 2

11 $\frac{(n-r+2)x}{r-1}$; 1·426

13 $\ln\cos x = -\frac{x^2}{2} - \frac{x^4}{12} - \dots$

16 (a) $-\frac{1}{6}$ (b) $\frac{1}{2}$

17 $1 - \frac{7x}{2} + 8x^2$

19 $x^2 - x^3 + \frac{11x^4}{12}$; max. at $x = 0$

## Test exercise 13 (page 727)

1 (a)

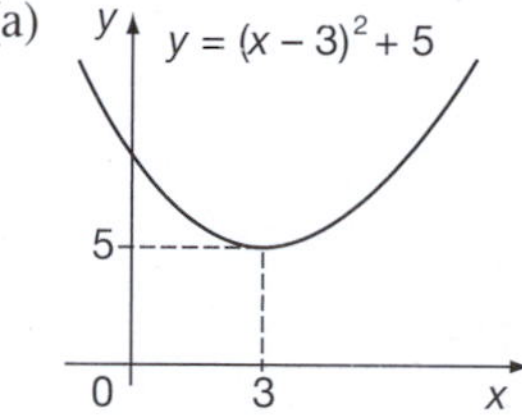


(b)

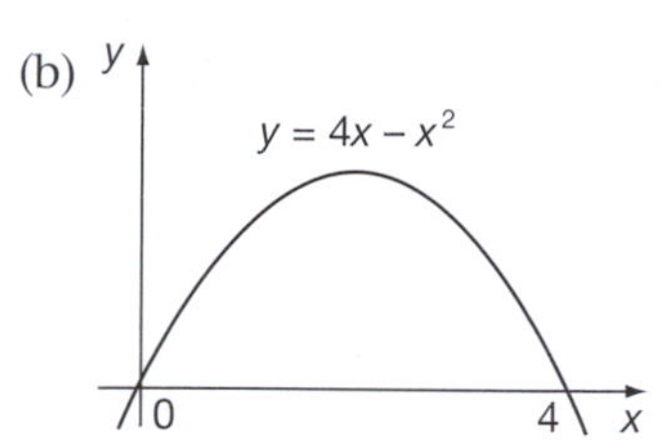


(c)

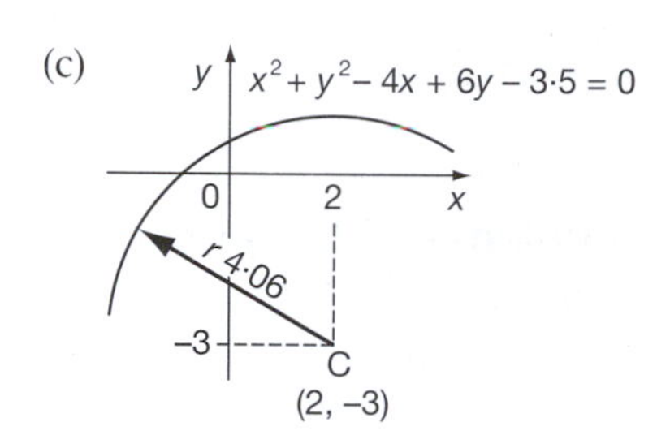


(d)

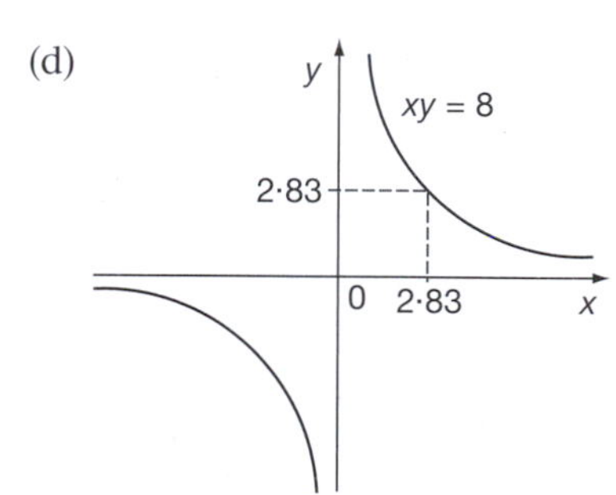


(e)

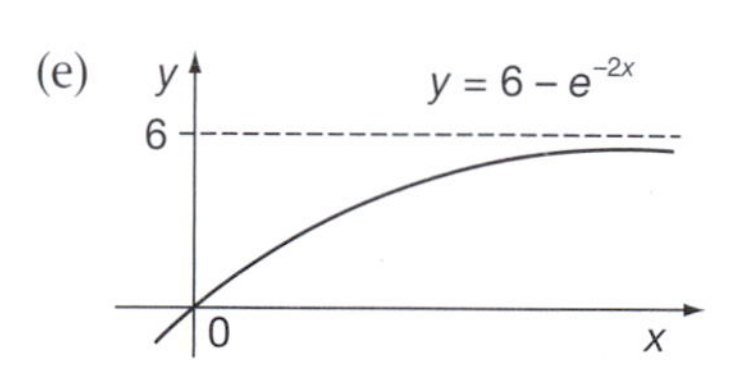


2 (a) $x = -3$; $x = 3$; $y = -x$
(b) $x = 4$; $y = x + 3$; $y = -x - 3$

3 (i) no symmetry
(ii) $\begin{cases} x = 0,\ y = -7{\cdot}5 \\ y = 0,\ x = 3 \text{ or } -5 \end{cases}$
(iii) asymptotes: $x = -2$ and $y = x$
(iv) no turning points

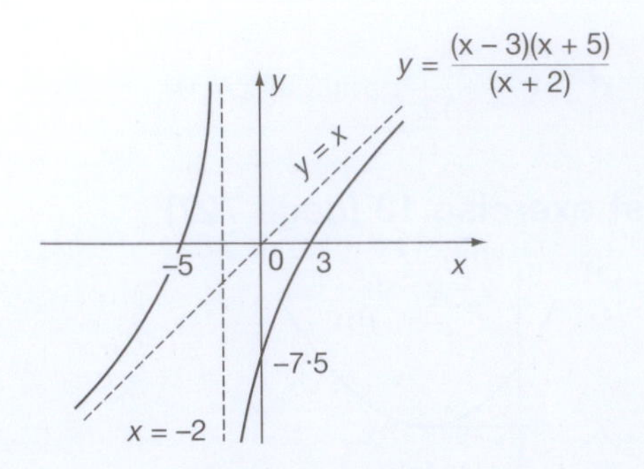


| | | x-axis | y-axis |
|---|---|---|---|
| 4 | (a) $\frac{y}{x} = A + Bx$ | $x$ | $\frac{y}{x}$ |
| | (b) $\ln(y - x) = kx + \ln A$ | $x$ | $\ln(y - x)$ |
| | (c) $xy = A - By$ | $y$ | $xy$ |
| | (d) $(xy)^2 = x^2 + k$ | $x^2$ | $(xy)^2$ |

5 $V = 0{\cdot}14p^{0{\cdot}54}$

6 $r = 0{\cdot}99999984$

7 $r_s = 0$

## Further problems 13 (page 728)

1 (a) $x = 0$ (b) $x = \pm 2, y = \pm 2$ (c) $x = 3$
(d) $x = -2; x = -3; y = 1$
(e) $x = \pm 1; y = \pm 1$ (f) $x = 2; y = \pm 1$

2 (a) $x = 0; y = x; y = -x$
(b) $x = 4; y = 0$ (c) $y = 0; y = x - 1;$
$y = -x - 1$

3 (a)

(b)

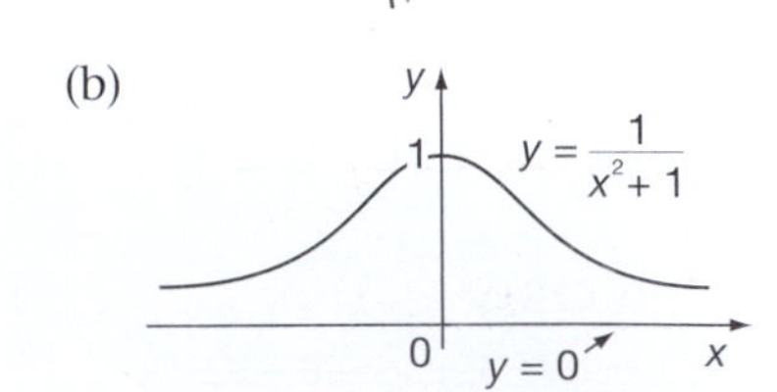


(c)

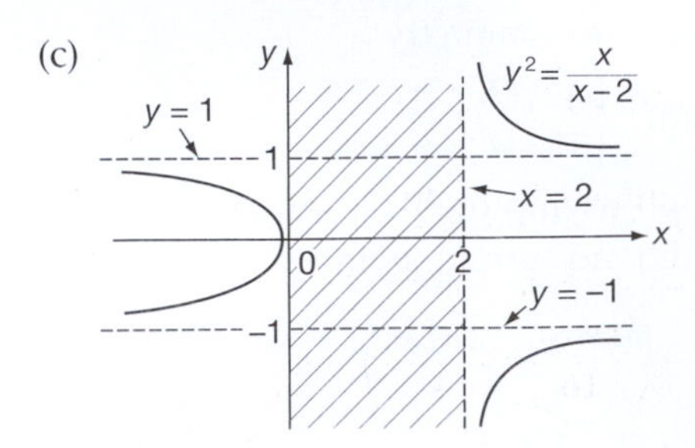


(d)

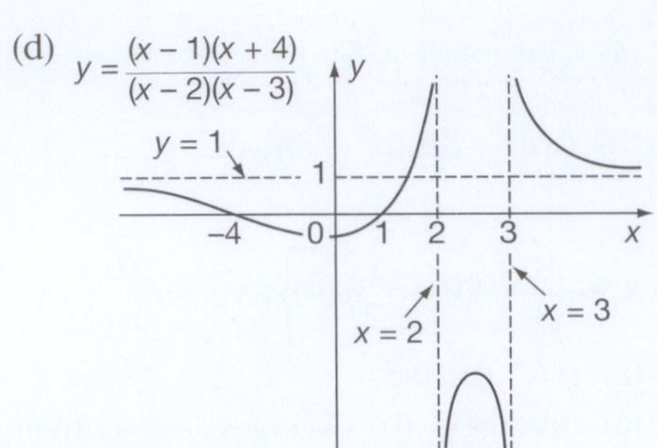


(e)

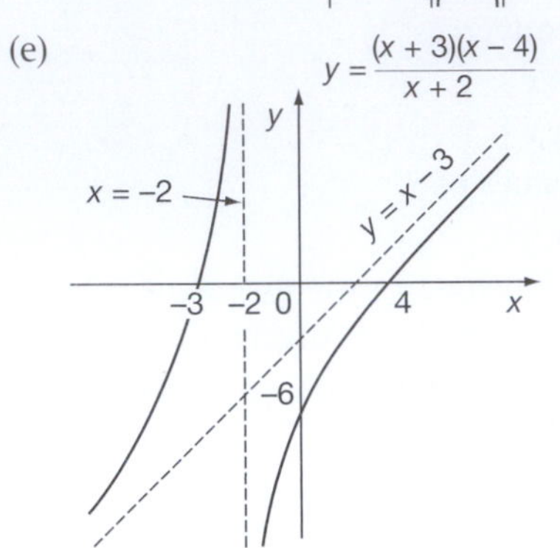


(f)

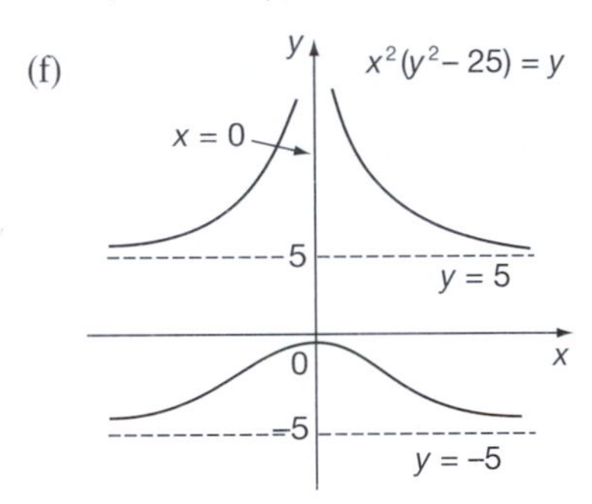


(g)

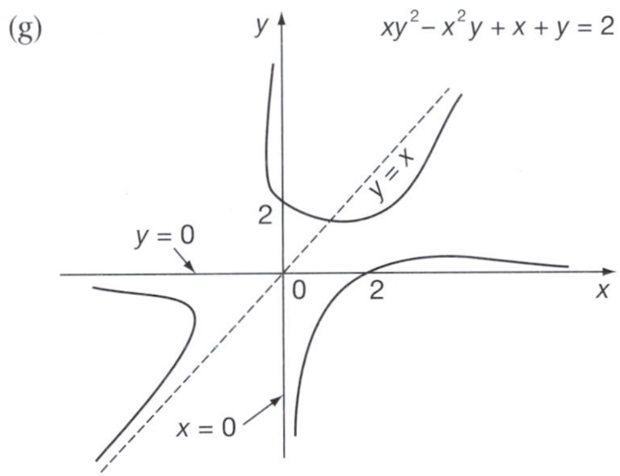


4 $y = 1{\cdot}08 + 0{\cdot}48x^2$

5 $P = 2{\cdot}48\sqrt{W} + 3{\cdot}20$

6 $R = 1{\cdot}08 + \dfrac{0{\cdot}047}{d^2}$

7 $y = \dfrac{4{\cdot}5}{1 - 0{\cdot}005x^2}$

8 $pv^{1{\cdot}39} = 118$

9 $I = 8{\cdot}63V^{0{\cdot}75}$

10 $y = 0{\cdot}044x + 3{\cdot}3\ln x$

11 $y = 1{\cdot}6x^2 + \dfrac{3{\cdot}7}{x}$

12 $V = 96t^{0.67}$
13 $T = 0.045D^{1.58}$
14 $r = 0.92$
15 $r_s = 0.71$ Strong correlation

## Test exercise 14 (page 749)

1 (a) $\frac{\partial z}{\partial x} = 12x^2 - 5y^2 \quad \frac{\partial z}{\partial y} = -10xy + 9y^2$

$\frac{\partial^2 z}{\partial x^2} = 24x \quad \frac{\partial^2 z}{\partial y^2} = -10x + 18y$

$\frac{\partial^2 z}{\partial y.\partial x} = -10y \quad \frac{\partial^2 z}{\partial x.\partial y} = -10y$

(b) $\frac{\partial z}{\partial x} = -2\sin(2x + 3y)$

$\frac{\partial z}{\partial y} = -3\sin(2x + 3y)$

$\frac{\partial^2 z}{\partial x^2} = -4\cos(2x + 3y)$

$\frac{\partial^2 z}{\partial y^2} = -9\cos(2x + 3y)$

$\frac{\partial^2 z}{\partial y.\partial x} = -6\cos(2x + 3y)$

$\frac{\partial^2 z}{\partial x.\partial y} = -6\cos(2x + 3y)$

(c) $\frac{\partial z}{\partial x} = 2xe^{x^2-y^2} \quad \frac{\partial z}{\partial y} = -2ye^{x^2-y^2}$

$\frac{\partial^2 z}{\partial x^2} = 2e^{x^2-y^2}(2x^2 + 1)$

$\frac{\partial^2 z}{\partial y^2} = 2e^{x^2-y^2}(2y^2 - 1)$

$\frac{\partial^2 z}{\partial y.\partial x} = -4xye^{x^2-y^2} \quad \frac{\partial^2 z}{\partial x.\partial y} = -4xye^{x^2-y^2}$

(d) $\frac{\partial z}{\partial x} = 2x^2\cos(2x + 3y) + 2x\sin(2x + 3y)$

$\frac{\partial^2 z}{\partial x^2} = (2 - 4x^2)\sin(2x + 3y) + 8x\cos(2x + 3y)$

$\frac{\partial^2 z}{\partial y.\partial x} = -6x^2\sin(2x + 3y) + 6x\cos(2x + 3y)$

$\frac{\partial z}{\partial y} = 3x^2\cos(2x + 3y)$

$\frac{\partial^2 z}{\partial y^2} = -9x^2\sin(2x + 3y)$

$\frac{\partial^2 z}{\partial x.\partial y} = -6x^2\sin(2x + 3y) + 6x\cos(2x + 3y)$

2 (a) $2V$
3 $P$ decreases 375 W
4 $\pm 2.5\%$

## Further problems 14 (page 750)

10 $\pm 0.67E \times 10^{-5}$ approx.
12 $\pm(x + y + z)\%$
13 $y$ decreases by 19% approx.
14 $\pm 4.25\%$
16 19%
18 $\delta y = y\{\delta x.p\cot(px + a) - \delta t.q\tan(qt + b)\}$

## Test exercise 15 (page 765)

1 (a) $\frac{4xy - 3x^2}{3y^2 - 2x^2}$ (b) $\frac{e^x\cos y - e^y\cos x}{e^x\sin y + e^y\sin x}$

(c) $\frac{5\cos x\cos y - 2\sin x\cos x}{5\sin x\sin y + \sec^2 y}$

2 $V$ decreases at 0.419 $cm^3/s$
3 $y$ decreases at 1.524 cm/s
4 $\frac{\partial x}{\partial r} = (4x^3 + 4xy)\cos\theta + (2x^2 + 3y^2)\sin\theta$

$\frac{\partial z}{\partial \theta} = r\{(2x^2 + 3y^2)\cos\theta - (4x^3 + 4xy)\sin\theta\}$

## Further problems 15 (page 766)

2 $3x^2 - 3xy$
3 $\tan\theta = 17/6 = 2.8333$
9 (a) $\frac{1 - y}{x + 2}$ (b) $\frac{8y - 3y^2 + 4xy - 3x^2y^2}{2x^3y - 2x^2 + 6xy - 8x}$

(c) $\frac{y}{x}$

14 $a = -\frac{5}{2}, b = -\frac{3}{2}$
16 $\frac{\cos x(5\cos y - 2\sin x)}{5\sin x\sin y + \sec^2 y}$
17 $\frac{y\cos x - \tan y}{x\sec^2 y - \sin x}$
20 (a) $-\left\{\frac{2xy + y\cos xy}{x^2 + x\cos xy}\right\}$ (b) $-\left\{\frac{xy + \tan xy}{x^2}\right\}$

## Test exercise 16 (page 792)

1 $-e^{\cos x} + C$
2 $2\sqrt{x}(\ln x - 2) + C$
3 $\tan x - x + C$
4 $\frac{x\sin 2x}{2} - \frac{x^2\cos 2x}{2} + \frac{\cos 2x}{4} + C$
5 $\frac{2e^{-3x}}{13}\left\{\sin 2x - \frac{3}{2}\cos 2x\right\} + C$
6 $-\cos x + \frac{2\cos^3 x}{3} - \frac{\cos^5 x}{5} + C$
7 $\frac{3x}{8} + \frac{\sin 2x}{4} + \frac{\sin 4x}{32} + C$
8 $2\ln(x^2 + x + 5) + C$
9 $\frac{1}{3}(1 + x^2)^{3/2} + C$
10 $\frac{9}{2}\ln(x - 5) - \frac{5}{2}\ln(x - 3) + C$
11 $2\ln(x - 1) + \tan^{-1} x + C$
12 $-\left(\frac{\cos 8x}{16} + \frac{\cos 2x}{4}\right) + C$

## Further problems 16 (page 792)

1 $\ln\{(x-1)(x^2+x+1)\}+C$

2 $\frac{1}{2}$

3 $-\ln(1+\cos^2 x)+C$

4 $\frac{1}{\sqrt{3}}-\frac{\pi}{6}$

5 $\frac{\pi^2}{4}$

6 $C-\frac{2}{(x^2+x+1)^{1/2}}$

7 $\frac{2}{3}\ln(x-1)-\frac{1}{3}\ln(x^2+x+1)+C$

8 $\frac{x^2}{2}-x+\ln(x+1)+C$

9 $\ln((x+1)\sqrt{x^2+1})+C$

10 2

11 $\frac{2n^{p+3}}{(p+1)(p+2)(p+3)}$

12 $3\ln(x-2)+\frac{1}{2}\ln(x^2+1)-5\tan^{-1}x+C$

13 $\frac{1}{2}$

14 $\frac{(\sin^{-1}x)^2}{2}+C$

15 $\frac{1}{4}(2\ln 3-\pi)$

16 $\pi^2-4$

17 $\frac{\pi^3}{6}-\frac{\pi}{4}$

18 $\frac{\pi}{4}-\frac{1}{2}$

19 $-\frac{1}{x}-\tan^{-1}x+C$

20 $\frac{1}{3}(1+x^2)^{3/2}+C$

21 $\ln(x+1)-\ln(x-2)-\frac{2}{x-2}+C$

22 $\frac{1}{10}(e^{2\pi}-1)=53{\cdot}45$

23 $\frac{1}{24}$

24 $\frac{1}{13}\left\{3e^{\pi/3}-2\right\}$

25 $-\frac{2}{3\omega}$

26 $\frac{\tan^3 x}{3}+C$

27 $\frac{1}{2}\ln(x-4)-\frac{1}{10}\ln(5x+2)+C$

28 $\ln(x+2)+C$

29 $2\ln(x+5)+\frac{3}{2}\ln(x^2+9)-\frac{4}{3}\tan^{-1}\left(\frac{x}{3}\right)+C$

30 $\ln(9x^2-18x+17)^{1/18}+C$

31 $2x^2+\ln\{(x^2-1)/(x^2+1)\}+C$

32 $\frac{1}{9}\{3x^3\ln(1+x^2)-2x^3+6x-6\tan^{-1}x+C\}$

33 $\ln(\cos\theta+\sin\theta)+C$

34 $\tan\theta-\sec\theta+C$

35 $\frac{1}{4}\ln(x-1)+\frac{1}{5}\ln(x-2)-\frac{9}{20}\ln(x+3)+C$

36 $\frac{1}{6}$

37 $\frac{2}{3}\ln 2-\frac{5}{18}$

38 $3\ln x+\frac{1}{2}\ln(x^2+4)-\frac{1}{2}\tan^{-1}\left(\frac{x}{2}\right)+C$

39 $\ln x-\tan^{-1}x-\frac{1}{x}+C$

## Test exercise 17 (page 819)

1 $\sin^{-1}\left(\frac{x}{7}\right)+C$

2 $\frac{1}{\sqrt{29}}\ln\left\{\frac{2x+3-\sqrt{29}}{2x+3+\sqrt{29}}\right\}+C$

3 $\frac{1}{\sqrt{2}}\tan^{-1}\left\{(x+2)\sqrt{2}\right\}+C$

4 $\frac{1}{\sqrt{3}}\sinh^{-1}\left(\frac{x\sqrt{3}}{4}\right)+C$

5 $\frac{1}{10}\ln\left\{\frac{x+9}{1-x}\right\}+C$

6 $\frac{5}{8}\left\{\sin^{-1}\left(\frac{2x+1}{\sqrt{5}}\right)+\frac{2(2x+1)}{5}\sqrt{1-x-x^2}\right\}+C$

7 $\frac{1}{\sqrt{5}}\cosh^{-1}\left(\frac{x+1}{\sqrt{21/5}}\right)+C$

8 $\frac{1}{\sqrt{3}}\tan^{-1}(\sqrt{3}\tan x)+C$

9 $\frac{1}{\sqrt{13}}\ln\left\{\frac{\sqrt{13}-3+2\tan x/2}{\sqrt{13}+3-2\tan x/2}\right\}+C$

10 $\ln\left\{\frac{1+\tan x/2}{1-\tan x/2}\right\}+C$

## Further problems 17 (page 820)

1 $\frac{1}{2\sqrt{21}}\ln\left\{\frac{x+6-\sqrt{21}}{x+6+\sqrt{21}}\right\}+C$

2 $\frac{1}{4\sqrt{11}}\ln\left\{\frac{2\sqrt{11}+x+6}{2\sqrt{11}-x-6}\right\}+C$

3 $\frac{1}{\sqrt{11}}\tan^{-1}\left(\frac{x+7}{\sqrt{11}}\right)+C$

4 $\frac{1}{2}\ln(x^2+4x+16)-\frac{5}{\sqrt{3}}\tan^{-1}\left(\frac{x+2}{2\sqrt{3}}\right)+C$

5 $\sinh^{-1}\left(\frac{x+6}{2\sqrt{3}}\right) = \ln\left\{\frac{x+6+\sqrt{x^2+12x+48}}{2\sqrt{3}}\right\} + C$

6 $\sin^{-1}\left(\frac{x+7}{\sqrt{66}}\right) + C$

7 $\cosh^{-1}\left(\frac{x+8}{2\sqrt{7}}\right) + C$

8 $6\sqrt{x^2-12x+52} + 31\sinh^{-1}\left(\frac{x-6}{4}\right) + C$

9 $\frac{2}{\sqrt{3}}\tan^{-1}\left\{\frac{1}{\sqrt{3}}\tan\left(\frac{x}{2}\right)\right\} + C$

10 $\frac{\sqrt{5}.\pi}{20} = 0{\cdot}3511$

11 $\frac{1}{\sqrt{5}}\ln\left\{\frac{2x+5-\sqrt{5}}{2x+5+\sqrt{5}}\right\} + C$

12 $\frac{3x^2}{2} - 4x + 4\tan^{-1}x + C$

13 $\frac{x+1}{2}\sqrt{3-2x-x^2} + 2\sin^{-1}\left(\frac{x+1}{2}\right) + C$

14 $\pi$

15 $\cosh^{-1}\left(\frac{x+2}{5}\right) + C$

16 $\frac{1}{6}\tan^{-1}\left\{\frac{2}{3}\tan x\right\} + C$

17 $\frac{1}{5}\ln\left\{\frac{2\tan\left(\frac{x}{2}\right)-1}{\tan\left(\frac{x}{2}\right)+4}\right\} + C$

18 $\frac{\pi}{2} - 1$

19 $3\sin^{-1}x - \sqrt{1-x^2} + C$

20 $\frac{4}{\sqrt{3}}\tan^{-1}\left\{\sqrt{3}\tan\frac{x}{2}\right\} - x + C$

21 $\frac{5}{2}\ln(x+2) - \frac{3}{4}\ln(x^2+4) + \tan^{-1}\left(\frac{x}{2}\right) + C$

22 $\frac{1}{3\sqrt{5}}\tan^{-1}\left(\frac{\sqrt{5}\tan x}{3}\right) + C$

23 $\sqrt{x^2+9} + 2\ln\left\{\frac{x}{3} + \frac{\sqrt{x^2+9}}{3}\right\} + C$

24 $\frac{1}{\sqrt{2}}\cosh^{-1}\left(\frac{4x-7}{3}\right) + C$

25 $\frac{\pi}{2}$

26 $\frac{1}{2\sqrt{2}}\ln\left\{\frac{\sqrt{2}\tan\theta-1}{\sqrt{2}\tan\theta+1}\right\} + C$

27 $\sqrt{x^2+2x+10} + 2\sinh^{-1}\left(\frac{x+1}{3}\right) + C$

28 $8\sin^{-1}\left(\frac{x+1}{4}\right) + \frac{x+1}{2}\sqrt{15-2x-x^2} + C$

29 $\frac{1}{8a^3}(\pi+2)$

30 $\frac{1}{3\sqrt{2}}\tan^{-1}\left(\frac{x}{a\sqrt{2}}\right) + \frac{1}{6}\ln\left\{\frac{(x+a)^2}{x^2+2a^2}\right\} + C$

## Test exercise 18 (page 831)

1 $e^{2x}\left\{\frac{x^3}{2} - \frac{3x^2}{4} + \frac{3x}{4} - \frac{3}{8}\right\} + C$

2 (a) $\frac{5\pi}{256}$ (b) $\frac{8}{315}$

3 $\frac{2a^7}{35}$

4 $I_n = \frac{1}{n-1}\tan^{n-1}x - I_{n-2}$

5 $\frac{3\pi}{256}$

## Further problems 18 (page 832)

2 $-\frac{1}{7}s^6c - \frac{6}{35}s^4c - \frac{8}{35}s^2c - \frac{16}{35}c + C_1$

where $\left\{\begin{array}{l} s \equiv \sin x \\ c \equiv \cos x \end{array}\right\}$

3 $\frac{2835}{8}$

5 $I_3 = \frac{3\pi^2}{4} - 6;\ I_4 = \frac{\pi^3}{2} - 12\pi + 24$

6 $I_n = x^n e^x - nI_{n-1}$;

$I_4 = e^x(x^4 - 4x^3 + 12x^2 - 24x + 24)$

7 $\frac{1328\sqrt{3}}{2835}$

10 $I_6 = -\frac{\cot^5 x}{5} + \frac{\cot^3 x}{3} - \cot x - x + C$

11 $I_3 = x\left\{(\ln x)^3 - 3(\ln x)^2 + 6\ln x - 6\right\} + C$

12 $\frac{4}{3}$

## Test exercise 19 (page 849)

1 70·12

2 $\frac{80}{\pi} + 2\pi = 31{\cdot}75$

3 $\frac{3}{2}\ln 6 = 2{\cdot}688$

4 73·485

5 $\frac{1}{2}RI^2$

6 ±132·3

## Further problems 19 (page 849)

1 2·5

2 1·101

3 $3\pi$

4 $\frac{1}{4}$

5 0

7 2

8 $\frac{1}{2}v_0 i_0 \cos a$

9 $\sqrt{\frac{E^2}{R^2}+\frac{1}{2}I^2}$

11 $\ln(2^{11}\times 3^{-6})-1$

12 $a^2\left(\ln 2-\frac{2}{3}\right)$

15 $a(1-2e^{-1})$

16 2·83

17 39·01

18 $\sqrt{\frac{1}{2}(I_1{}^2+I_2{}^2)}$

20 1·361

## Test exercise 20 (page 870)

1 $(0{\cdot}75, 1{\cdot}6)$

2 $(0{\cdot}4, 0)$

3 $5\pi^2 a^3$

4 $\frac{e^2+7}{8}$

5 $70{\cdot}35\pi$

6 $\frac{5.\pi^2}{8}$

7 $2\sqrt{2}.\pi\left(\frac{e^{\pi}-2}{5}\right)$

## Further problems 20 (page 870)

1 $\frac{3}{16}+\frac{1}{2}\ln 2$

2 (a) 2·054 (b) 66·28

3 $\frac{64\pi a^3}{15}$

4 (a) $(0{\cdot}4, 1)$ (b) $(0{\cdot}5, 0)$

6 24

7 $\frac{17}{12}$

8 $-\frac{19}{20}$

9 $\frac{11}{5}$

10 $A = 2{\cdot}457$, $V = 4\pi\sqrt{3}$, $\bar{y} = 1{\cdot}409$

12 (a) 8 (b) $\frac{64\pi}{3}$ (c) $\frac{4}{3}$

13 1·175

16 $V = 25{\cdot}4\ \text{cm}^3$, $A = 46{\cdot}65\ \text{cm}^2$

17 $S = 15{\cdot}32a^2$, $y = 1{\cdot}062a$

## Test exercise 21 (page 899)

1 (a) $\text{I}_z = \frac{ab\rho}{12}(b^2+a^2)$

(b) $\text{I}_{\text{AB}} = \frac{ab\rho}{3}(a^2+b^2)$ (c) $k = \sqrt{\frac{a^2+b^2}{3}}$

2 $k = \frac{l}{\sqrt{2}}$

3 (a) $\frac{1}{\ln 4}$ (b) $\frac{6}{\ln 2}$

5 $\frac{wa^3}{8}$, $0{\cdot}433a$

## Further problems 21 (page 899)

2 $\frac{1}{2}Ma^2$

6 (a) $\sqrt{\frac{4ac}{5}}$ (b) $\sqrt{\frac{3c^2}{7}}$

9 $\frac{a^4}{12}$

10 $\text{I} = M\left\{\frac{h^2}{10}+\frac{3r^2}{20}\right\}$; $k = \sqrt{\frac{h^2}{10}+\frac{3r^2}{20}}$

12 $\frac{\pi ab^3}{4}$

14 $\frac{2wa^3}{3}, \frac{3\pi a}{16}$

15 (a) $\frac{1}{3}\sqrt{e^2+e+1}$ (b) $\sqrt{\frac{e-2}{e-1}}$

16 $51{\cdot}2w$

17 9·46 cm

19 $\frac{(15\pi-32)a}{4(3\pi-4)}$

## Test exercise 22 (page 918)

1 0·946

2 0·926

3 26·7

4 1·188

5 1·351

## Further problems 22 (page 919)

1 0·478

2 0·091

3 (a) 0·625 (b) 6·682 (c) 1·854

4 560

5 15·86

6 0·747

7 28·4

8 28·92

9 0·508

10 $\frac{\sqrt{2}}{4}\int_0^{\pi/2}\sqrt{9+\cos 2\theta}.\text{d}\theta$; 1·66

11 $\tan^{-1} x = x - \frac{x^3}{3} + \frac{x^5}{5} - \frac{x^7}{7}$; 0·076

12 (a) 0·5314 (b) 0·364

13 2·422

14 2·05

## Test exercise 23 (page 941)

1 $\frac{56}{3\pi^3}$

2 (a) $r = 2\sin\theta$

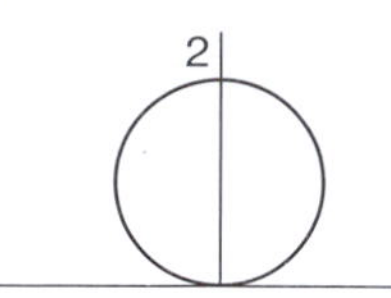


(b) $r = 5\cos^2\theta$

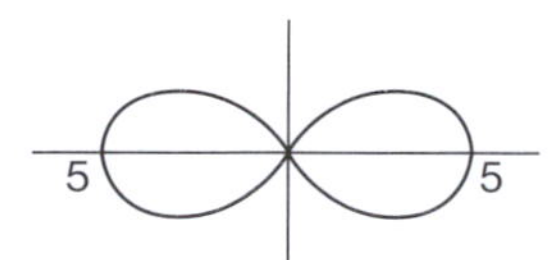


(c) $r = \sin 2\theta$

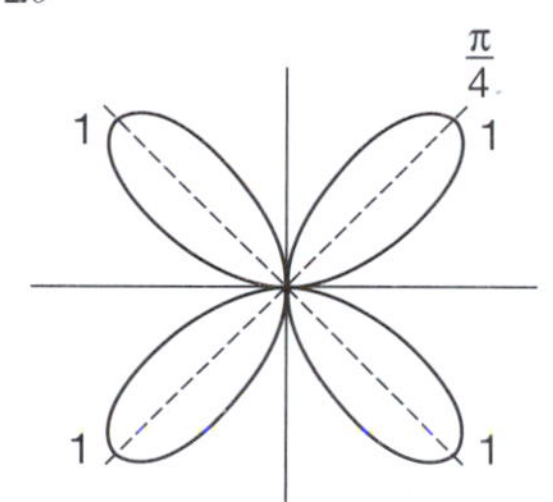


(d) $r = 1 + \cos\theta$

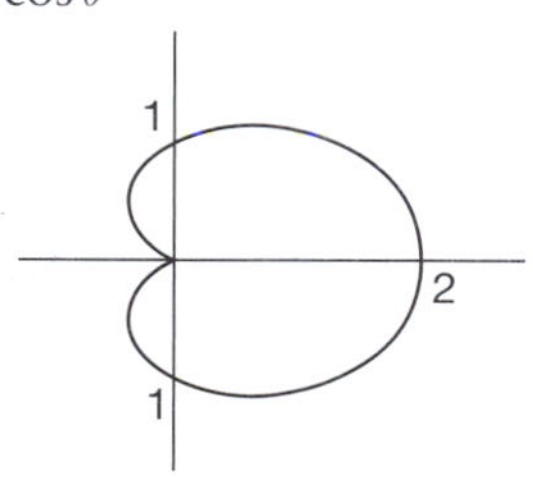


(e) $r = 1 + 3\cos\theta$

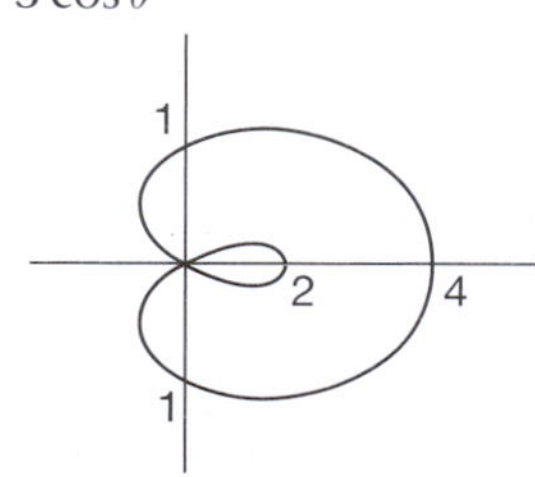


(f) $r = 3 + \cos\theta$

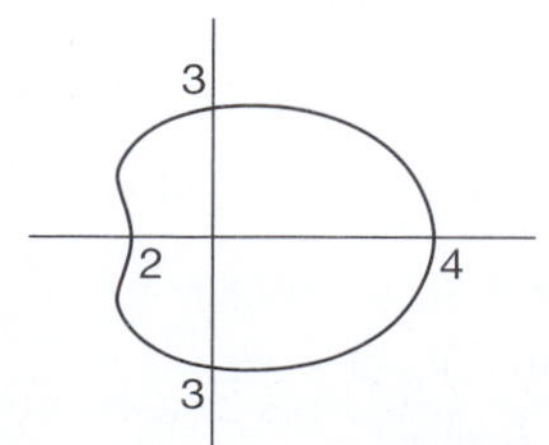


3 $\frac{40\pi}{3}$

4 8

5 $\frac{32\pi a^2}{5}$

## Further problems 23 (page 941)

1 (a) $A = \frac{3\pi}{16}$ (b) $V = \frac{2\pi}{21}$

2 $3\pi$

3 $\frac{4a^2}{3}$

4 $\frac{13\pi}{8} + 3$

5 $\frac{2}{3}$

6 $\frac{20\pi}{3}$

7 $\frac{8\pi a^3}{3}$

9 $\frac{3\pi a}{2}$

10 $\frac{5\pi}{2}$

11 $21{\cdot}25a$

12 $\frac{3\pi a}{2}$

14 $\frac{a}{b}\left\{\sqrt{b^2+1}\right\}(e^{b\theta_1} - 1)$; $\frac{a^2}{4b}(e^{2b\theta_1} - 1)$

15 $\pi a^2(2 - \sqrt{2})$

## Test exercise 24 (page 965)

1 (a) 0 (b) 0

2 (a) $-1$ (b) 120 (c) $\frac{17}{4}$

3 3·67

4 170·67

5 $\frac{11\pi}{4} + 6$

6 36

## Further problems 24 (page 966)

1 $\frac{1}{3}$

2 $\frac{243\pi}{2}$

3 4·5

4 $\frac{abc}{3}(b^2 + c^2)$

5 $\frac{\pi r^3}{3}$

6 4·5

7 $\pi + 8$

8 26

9 $\dfrac{22}{3}$

10 $\dfrac{1}{8}\left(\dfrac{\pi}{2}+1\right)$

11 $\dfrac{1}{3}$

12 $A = 2\displaystyle\int_0^{\pi/6}\int_0^{2\cos 3\theta} r\,\mathrm{d}r\,\mathrm{d}\theta = \dfrac{\pi}{3}$

13 $4\pi\left\{\dfrac{1}{\sqrt{2}} - \dfrac{1}{\sqrt{5}}\right\}$

14 $\dfrac{64}{9}(3\pi - 4)$

15 $M = \displaystyle\int_0^{\pi}\int_0^{a(1+\cos\theta)} r^2 \sin\theta\,\mathrm{d}r\,\mathrm{d}\theta = \dfrac{4a^3}{3}$; $h = \dfrac{16a}{9\pi}$

16 (a) $\dfrac{1}{2}\pi ab$

(b) $\dfrac{1}{8}\pi ab^3$; centroid $\left(0, \dfrac{4b}{3\pi}\right)$

17 19·56

18 $\dfrac{b^2}{4}(c^2 - a^2)$

19 $\dfrac{a^2}{6}(2\pi + 3\sqrt{3})$

20 232

## Test exercise 25 (page 1000)

1 $y = \dfrac{x^2}{2} + 2x - 3\ln x + C$

2 $\tan^{-1} y = C - \dfrac{1}{1+x}$

3 $y = \dfrac{e^{3x}}{5} + Ce^{-2x}$

4 $y = x^2 + Cx$

5 $y = -\dfrac{x\cos 3x}{3} + \dfrac{\sin 3x}{9} - \dfrac{4}{x} + C$

6 $\sin y = Ax$

7 $y^2 - x^2 = Ax^2y$

8 $y(x^2 - 1) = \dfrac{x^2}{2} + C$

9 $y = \cosh x + \dfrac{C}{\cosh x}$

10 $y = x^2(\sin x + C)$

11 $xy^2(Cx + 2) = 1$

12 $y = 1/(Cx^3 + x^2)$

## Further problems 25 (page 1000)

1 $x^4y^3 = Ae^y$

2 $y^3 = 4(1 + x^3)$

3 $3x^4 + 4(y+1)^3 = A$

4 $(1 + e^x)\sec y = 2\sqrt{2}$

5 $x^2+y^2+2x-2y+2\ln(x-1) + 2\ln(y+1) = A$

6 $y^2 - xy - x^2 + 1 = 0$

7 $xy = Ae^{y/x}$

8 $x^3 - 2y^3 = Ax$

9 $A(x - 2y)^5(3x + 2y)^3 = 1$

10 $(x^2 - y^2)^2 = Axy$

11 $2y = x^3 + 6x^2 - 4x\ln x + Ax$

12 $y = \cos x(A + \ln\sec x)$

13 $y = x(1 + x\sin x + \cos x)$

14 $(3y - 5)(1 + x^2)^{3/2} = 2\sqrt{2}$

15 $y\sin x + 5e^{\cos x} = 1$

16 $x + 3y + 2\ln(x + y - 2) = A$

17 $x = Aye^{xy}$

18 $\ln\left\{4y^2 + (x-1)^2\right\} + \tan^{-1}\left\{\dfrac{2y}{x-1}\right\} = A$

19 $(y - x + 1)^2(y + x - 1)^5 = A$

20 $2x^2y^2\ln y - 2xy - 1 = Ax^2y^2$

21 $\dfrac{2}{y^2} = 2x + 1 + Ce^{2x}$

22 $\dfrac{1}{y^3} = \dfrac{3e^x}{2} + Ce^{3x}$

23 $y^2(x + Ce^x) = 1$

24 $\dfrac{\sec^2 x}{y} = C - \dfrac{\tan^3 x}{3}$

25 $\cos^2 x = y^2(C - 2\tan x)$

26 $y\sqrt{1 - x^2} = A + \sin^{-1} x$

27 $x + \ln Ax = \sqrt{y^2 - 1}$

28 $\ln(x - y) = A + \dfrac{2x^2}{(x-y)^2} - \dfrac{4x}{(x-y)}$

29 $y = \dfrac{\sqrt{2}\sin 2x}{2(\cos x - \sqrt{2})}$

30 $(x - 4)y^4 = Ax$

31 $y = x\cos x - \dfrac{\pi}{8}\sec x$

32 $(x - y)^3 - Axy = 0$

33 $2\tan^{-1} y = \ln(1 + x^2) + A$

34 $2x^2y = 2x^3 - x^2 - 4$

35 $y = e^{\frac{x-y}{x}}$

36 $3e^{2y} = 2e^{3x} + 1$

37 $4xy = \sin 2x - 2x\cos 2x + 2\pi - 1$

38 $y = Ae^{y/x}$

39 $x^3 - 3xy^2 = A$

40 $x^2 - 4xy + 4y^2 + 2x - 3 = 0$

41 $y(1 - x^3)^{-1/3} = -\dfrac{1}{2}(1 - x^3)^{2/3} + C$

42 $xy + x\cos x - \sin x + 1 = 0$

43 $2\tan^{-1} y = 1 - x^2$

44 $y = \dfrac{x^2 + C}{2x(1 - x^2)}$

45 $y\sqrt{1+x^2} = x + \frac{x^3}{3} + C$

46 $1 + y^2 = 5(1 + x^2)$

47 $\sin^2\theta(a^2 - r^2) = \frac{a^2}{2}$

48 $y = \frac{1}{2}\sin x$

49 $y = \frac{1}{x(A-x)}$

## Test exercise 26 (page 1025)

1 $y = Ae^{-x} + Be^{2x} - 4$

2 $y = Ae^{2x} + Be^{-2x} + 2e^{3x}$

3 $y = e^{-x}(A + Bx) + e^{-2x}$

4 $y = A\cos 5x + B\sin 5x + \frac{1}{125}(25x^2 + 5x - 2)$

5 $y = e^x(A + Bx) + 2\cos x$

6 $y = e^{-2x}(2 - \cos x)$

7 $y = Ae^x + Be^{-x/3} - 2x + 7$

8 $y = Ae^{2x} + Be^{4x} + 4xe^{4x}$

## Further problems 26 (page 1025)

1 $y = Ae^{4x} + Be^{-x/2} - \frac{e^{3x}}{7}$

2 $y = e^{3x}(A + Bx) + 6x + 6$

3 $y = 4\cos 4x - 2\sin 4x + Ae^{2x} + Be^{3x}$

4 $y = e^{-x}(Ax + B) + \frac{e^x}{2} - x^2e^{-x}$

5 $y = Ae^x + Be^{-2x} + \frac{e^{2x}}{4} - \frac{xe^{-2x}}{3}$

6 $y = e^{3x}(A\cos x + B\sin x) + 2 - \frac{e^{2x}}{2}$

7 $y = e^{-2x}(A + Bx) + \frac{1}{4} + \frac{1}{8}\sin 2x$

8 $y = Ae^x + Be^{3x} + \frac{1}{9}(3x + 4) - e^{2x}$

9 $y = e^x(A\cos 2x + B\sin 2x) + \frac{x^2}{3} + \frac{4x}{9} - \frac{7}{27}$

10 $y = Ae^{3x} + Be^{-3x} - \frac{1}{18}\sin 3x + \frac{1}{6}xe^{3x}$

11 $y = \frac{wx^2}{24EI}\{x^2 - 4lx + 6l^2\}$; $y = \frac{wl^4}{8EI}$

12 $x = \frac{1}{2}(1 - t)e^{-3t}$

13 $y = e^{-2t}(A\cos t + B\sin t) - \frac{3}{4}(\cos t - \sin t)$;
amplitude $\frac{3\sqrt{2}}{4}$, frequency $\frac{1}{2\pi}$

14 $x = -\frac{1}{2}e^t + \frac{1}{5}e^{2t} + \frac{1}{10}(\sin t + 3\cos t)$

15 $y = e^{-2x} - e^{-x} + \frac{3}{10}(\sin x - 3\cos x)$

16 $y = e^{-3x}(A\cos x + B\sin x) + 5x - 3$

17 $x = e^{-t}(6\cos t + 7\sin t) - 6\cos 3t - 7\sin 3t$

18 $y = \sin x - \frac{1}{2}\sin 2x$; $y_{\text{max}} = 1{\cdot}299$ at $x = \frac{2\pi}{3}$

19 $T = \frac{\pi}{2\sqrt{6}} = 0{\cdot}641\text{s}; A = \frac{1}{6}$

20 $x = \frac{1}{10}\{e^{-3t} - e^{-2t} + \cos t + \sin t\}$;
steady state: $x = \frac{\sqrt{2}}{10}\sin\left(t + \frac{\pi}{4}\right)$

## Test exercise 27 (page 1043)

1 (a) $F(s) = \frac{8}{s}$ (b) $F(s) = \frac{1}{s-5}$
(c) $F(s) = -\frac{4e^3}{s-2}$

2 (a) $f(t) = -5te^{2t}$ (b) $f(t) = e^3t^2$
(c) $f(t) = \sin 3t$
(d) $f(t) = \frac{5}{\sqrt{3}}\sin\sqrt{3}t - 2\cos\sqrt{3}t$

3 $F(s) = \frac{2}{(s-3)^3}$

4 (a) $f(t) = \frac{1}{4}(e^{-2t} + 2t - 1)$
(b) $f(t) = -\frac{1}{2}(e^t + e^{-t})$
(c) $f(t) = \frac{1}{2}t^2e^{-2t}$
(d) $f(t) = 2 - e^{-3t/2} - e^{3t/2}$

## Further problems 27 (page 1043)

1 (a) $\frac{1}{s - k\ln a}$ (b) $\frac{k}{s^2 - k^2}$ (c) $\frac{s}{s^2 - k^2}$
(d) $\frac{k(1 - e^{-sa})}{s}$

2 (a) $-\frac{2}{3}e^{4t/3}$ (b) $\frac{\sinh 2\sqrt{2}t}{2\sqrt{2}}$
(c) $3\cos 4t - \sin 4t$ (d) $f(t) = 3 + 4\cos 3t$
(e) $f(t) = t(e^t - e^{-t})$
(f) $f(t) = \frac{4}{5}\cos 2t + \frac{17}{5}\sin 2t - \frac{9}{5}e^t$

3 (b) (i) $\frac{4s}{(s^2+4)^2}$ (ii) $\frac{2s(s^2 - 27)}{(s^2+9)^3}$
(c) $F^{(n)}(s) = (-1)^nL\{t^nf(t)\}$

4 (a) $\frac{b}{(s-a)^2 + b^2}$ (b) $\frac{s-a}{(s-a)^2 + b^2}$

5 (a) $e^{3t} - e^{2t}$ (b) $\frac{1}{6} + \frac{1}{3}e^{3t} - \frac{1}{2}e^{2t}$
(c) $e^{3t} - (1 + t)e^{2t}$
(d) $\frac{1}{20}e^{-3t} + \frac{8}{15}e^{-t/2} + \frac{17}{12}e^t$

(e) $\frac{1}{12}\left\{8e^{-t/2}+7e^{t}-15e^{-t}-6te^{-t}\right\}$

(f) $\sin 4t+\cos 4t$

(g) $-\frac{2}{5}e^{-t/2}-\frac{1}{5}\sin t+\frac{2}{5}\cos t$

## Test exercise 28 (page 1072)

**1** (a), (b)

| Mass (kg) $x$ | Frequency $f$ |
|---|---|
| 4·2 | 1 |
| 4·3 | 3 |
| 4·4 | 7 |
| 4·5 | 10 |
| 4·6 | 12 |
| 4·7 | 10 |
| 4·8 | 5 |
| 4·9 | 2 |
| | $n=\sum f=50$ |

(c)

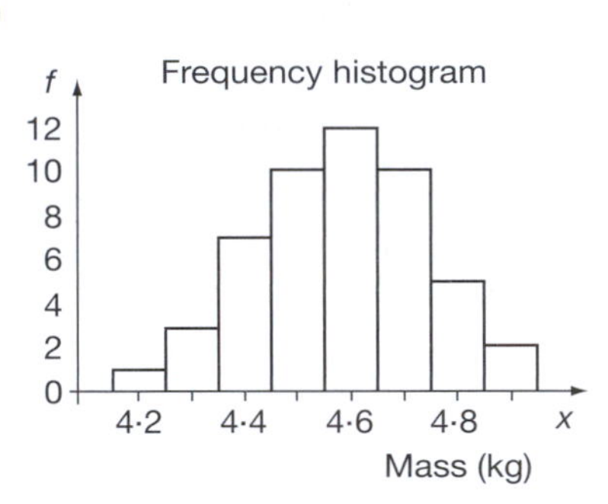


**2** (a)

| Diam (mm) $x$ | $x_m$ | Rel $f$ (%) |
|---|---|---|
| 8·82–8·86 | 8·84 | 1·33 |
| 8·87–8·91 | 8·89 | 10·67 |
| 8·92–8·96 | 8·94 | 21·33 |
| 8·97–9·01 | 8·99 | 24·00 |
| 9·02–9·06 | 9·04 | 20·00 |
| 9·07–9·11 | 9·09 | 13·33 |
| 9·12–9·16 | 9·14 | 6·67 |
| 9·17–9·21 | 9·19 | 2·67 |
| | | 100·00% |

(b)

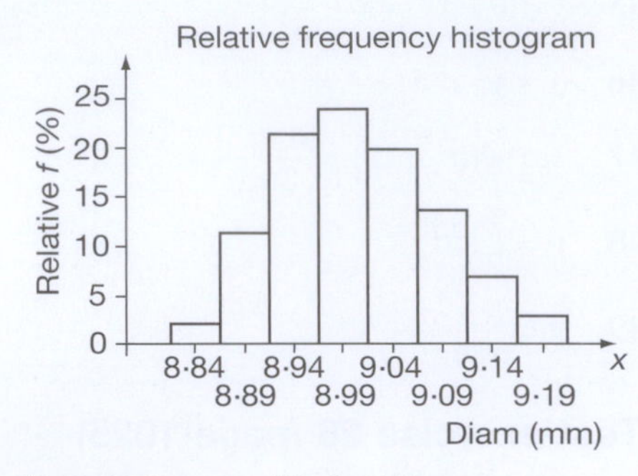


(c) (i) 8·915 mm (ii) 9·115 mm (iii) 0·05 mm

**3** (a) $\bar{x}=10{\cdot}62$ mm (b) $\sigma=0{\cdot}437$ mm
(c) mode = 10·51 mm
(d) median = 10·59 mm

**4** (a) $\bar{x}=15{\cdot}05$ mm; $\sigma=0{\cdot}4314$ mm
(b) (i) 13·76 mm to 16·34 mm (ii) 1200

## Further problems 28 (page 1073)

**1** (a), (b) (i)

| $x$ | $f$ |
|---|---|
| 60–64 | 3 |
| 65–69 | 5 |
| 70–74 | 7 |
| 75–79 | 11 |
| 80–84 | 8 |
| 85–89 | 4 |
| 90–94 | 2 |
| | $n=\sum f=40$ |

(b) (ii) $\bar{x}=76{\cdot}5$ (iii) $\sigma=7{\cdot}73$

**2** (a), (b)

| Length (mm) $x$ | Frequency $f$ |
|---|---|
| 13·5–14·4 | 2 |
| 14·5–15·4 | 5 |
| 15·5–16·4 | 9 |
| 16·5–17·4 | 13 |
| 17·5–18·4 | 7 |
| 18·5–19·4 | 4 |
| | $n=\sum f=40$ |

(c) (i) $\bar{x}=16{\cdot}77$ mm (ii) $\sigma=1{\cdot}343$ mm

**3** (a) (i) $\bar{x}=4{\cdot}283$ kg (ii) $\sigma=0{\cdot}078$ kg
(b) (i) 4·050 kg to 4·516 kg (ii) 288

**4** (a) $\bar{x}=2{\cdot}382$ MΩ (b) $\sigma=0{\cdot}0155$ MΩ
(c) mode = 2·378 MΩ
(d) median = 2·382 MΩ

5 (a) $\bar{x} = 31{\cdot}06$ kN
(b) $\sigma = 0{\cdot}237$ kN (c) mode $= 31{\cdot}02$ kN
(d) median $= 31{\cdot}04$ kN
6 (a) (i) $\bar{x} = 11{\cdot}538$ min (ii) $\sigma = 0{\cdot}724$ min
(iii) mode $= 11{\cdot}6$ min
(iv) median $= 11{\cdot}57$ min
(b) (i) 0·5 min (ii) 10·75 min
(iii) 13·25 min
7 (a) 32·60 to 32·72 mm (b) 1956
8 (a) (i) $\bar{x} = 7{\cdot}556$ kg (ii) $\sigma = 0{\cdot}126$ kg
(b) (i) 7·18 kg to 7·94 kg (ii) 1680
9 (a) (i) $\bar{x} = 29{\cdot}64$ mm (ii) $\sigma = 0{\cdot}123$ mm
(b) (i) 29·27 mm to 30·01 mm (ii) 2100
10 (a) (i) $\bar{x} = 25{\cdot}14$ mm
(ii) $\sigma = 0{\cdot}690$ mm
(b) (i) 23·07 mm to 27·21 mm (ii) 480

## Test exercise 29 (page 1113)

1 (a) (i) 0·12 (ii) 0·88 (b) 3520
2 (a) 0·2400 (b) 0·1600 (c) 0·3600
3 (a) 0·1576 (b) 0·0873 (c) 0·1939
4 (a) (i) 0·3277 (ii) 0·2627
(b) $\mu = 1{\cdot}0$, $\sigma = 0{\cdot}8944$
5 (a) 0·0183 (b) 0·1465 (c) 0·7619
6 212

## Further problems 29 (page 1113)

1 (a) 0·0081 (b) 0·2401 (c) 0·2646
(d) 0·0756
2 (a) 0·8154 (b) 0·9852
3 0·5277, 0·3598, 0·0981, 0·0134, 0·0009, 0·0000
4 (a) 0·0688 (b) 0·9915
5 (a) 0·7738 (b) 0·0226
6 (a) 0·8520 (b) 0·2097
7 (a) 0·3277, 0·4096, 0·2048, 0·0512, 0·0064, 0·0003
(b) 0·2627; (c) 0·9933
8 (a) 0·5386 (b) 0·0988 (c) 0·0020
9 (a) 0·6065 (b) 0·0902
10 0·9953
11 (a) $\mu = 12, \sigma = 3{\cdot}412$ (b) 0·1048
12 0·202, 0·323, 0·258, 0·138, 0·006
13 (a) 0·0498 (b) 0·1494 (c) 0·2240
(d) 0·2240 (e) 0·3528
14 (a) 0·2381
(b) 0·1107
15 0·0907, 0·2177, 0·2613, 0·2090, 0·2213
16 5, 15, 22, 22, 17, 18
17 0·1271
18 (a) 0·0078 (b) 0·9028
19 (a) 74·9% (b) 0·14%
20 333
21 (a) 0·0618 (b) 0·2206
22 (a) 7 (b) 127
23 (a) 0·4835 (b) 1·46%
24 (a) 215 (b) 89·3% (c) 0·0022

# Index